Population	Growth Rate, k	Doubling Time, T
c) India	1.4% per year	
d) Finland	0.2% per year	
e) Egypt		38.5 yr
f) Philippines		36.5 yr
g) United States	0.9% per year	
h) Japan		69.3 yr
i) Ireland	1.2% per year	
j) Kenya	2.6% per year	

4. *DVD Videos.* The total number of DVD videos produced and shipped in 1998 was 0.5 million. In 2004, the total number of units reached 29.01 million.

DVD Videos

Source: Recording Industry Association of America

Assuming the exponential growth model applies:
a) Find the value of k and write the function.
b) Estimate the number of DVD videos produced and shipped in 2005, in 2008, and in 2011.

5. *Population Growth of Israel.* The population of Israel has a growth rate of 1.3% per year. In 2 the population was 6,276,883. The land area of Israel is 24,839,654,400 square yards. (*Source: Statistical Abstract of the United States*) Assum this growth rate continues and is exponential,

after how long will there be one person for every square yard of land?

6. *Value of Manhattan Island.* In 1626, Peter Minuit of the Dutch West India Company purchased Manhattan Island from Native Americans for $24. Assuming an exponential rate of inflation of 6% per year, find the value of Manhattan Island in 2010.

7. *Interest Compounded Continuously.* Suppose that $10,000 is invested at an interest rate of 5.4% per year, compounded continuously.
a) Find the exponential function that describes the amount in the account after time t, in years.
b) What is the balance after 1 yr? 2 yr? 5 yr? 10 yr?
c) What is the doubling time?

8. *Interest Compounded Continuously.* Complete the following table.

Initial Investment at $t = 0$, P_0	Interest Rate, k	Doubling Time, T	Amount After 5 yr
a) $35,000	6.2%		

Collaborative Discussion and Writing

103. Explain why an even function f does not have an inverse f^{-1}.

104. The following formulas for the conversion between Fahrenheit temperature and Celsius temperature have been considered several times in this text:
$$C = \tfrac{5}{9}(F - 32)$$
and
$$F = \tfrac{9}{5}C + 32.$$
Discuss these formulas from the standpoint of inverses.

Skill Maintenance

Consider the following quadratic functions. Without graphing them, answer the questions below.
a) $f(x) = 2x^2$
b) $f(x) = -x^2$
c) $f(x) = \tfrac{1}{4}x^2$
d) $f(x) = -5x^2 + 3$
e) $f(x) = \tfrac{2}{3}(x - 1)^2 - 3$
f) $f(x) = -2(x + 3)^2 + 1$
g) $f(x) = (x - 3)^2 + 1$
h) $f(x) = -4(x + 1)^2 - 3$
105. Which functions have a maximum value?
106. Which graphs open up?
107. Consider (a) and (c). Which graph is narrower?
108. Consider (d) and (e). Which graph is narrower?
109. Which graph has vertex $(-3, 1)$?
110. For which is the line of symmetry $x = 0$?

Synthesis

Using only a graphing calculator, determine whether the functions are inverses of each other.

111. $f(x) = \sqrt[3]{\dfrac{x - 3.2}{1.4}}, \ g(x) = 1.4x^3 + 3.2$

112. $f(x) = \dfrac{2x - 5}{4x + 7}, \ g(x) = \dfrac{7x - 4}{5x + 2}$

113. $f(x) = \dfrac{2}{3}, \ g(x) = \dfrac{3}{2}$

114. $f(x) = x^4, x \geq 0; \ g(x) = \sqrt[4]{x}$

115. The function $f(x) = x^2 - 3$ is not one-to-one. Restrict the domain of f so that its inverse is a function. Find the inverse and state the restriction on the domain of the inverse.

116. Consider the function f given by
$$f(x) = \begin{cases} x^3 + 2, & x \leq -1, \\ x^2, & -1 < x < 1, \\ x + 1, & x \geq 1. \end{cases}$$
Does f have an inverse that is a function? Why or why not?

117. Find three examples of functions that are their own inverses; that is, $f = f^{-1}$.

118. Given the function $f(x) = ax + b, a \neq 0$, find the values of a and b for which $f^{-1}(x) = f(x)$.

REAL-DATA APPLICATIONS

Through a focus on real-data applications, the authors provide a variety of examples and exercises that connect the mathematical content with everyday life.

VARIETY OF EXERCISES

The exercise sets are designed to provide opportunities to practice the concepts presented in each section. Exercise types include *Collaborative Discussion and Writing, Skill Maintenance, Classifying the Function, Vocabulary Review,* and *Synthesis.*

Algebra and Trigonometry

EDITION **4**

GRAPHS & MODELS

MARVIN L. BITTINGER
Indiana University Purdue University Indianapolis

JUDITH A. BEECHER
Indiana University Purdue University Indianapolis

DAVID J. ELLENBOGEN
Community College of Vermont

JUDITH A. PENNA
Indiana University Purdue University Indianapolis

PEARSON

Addison
Wesley

Boston   San Francisco   New York
London   Toronto   Sydney   Tokyo   Singapore   Madrid
Mexico City   Munich   Paris   Cape Town   Hong Kong   Montreal

Executive Editor	Anne Kelly
Senior Project Editor	Rachel S. Reeve
Editorial Assistant	Leah Goldberg
Senior Managing Editor	Karen Wernholm
Senior Production Supervisor	Kathleen A. Manley
Senior Designer	Barbara T. Atkinson
Supplements Production Coordinator	Kayla Smith-Tarbox
Media Associate Producer	Jennifer Thomas
Software Development	Bob Carroll, MathXL; Eric Gregg, TestGen
Executive Marketing Manager	Becky Anderson
Marketing Assistant	Bonnie Gill
Senior Author Support/Technology Specialist	Joe Vetere
Senior Prepress Supervisor	Caroline Fell
Rights and Permissions Advisor	Shannon Barbe
Manufacturing Manager	Evelyn Beaton
Cover Design	Night & Day Design
Design, Art Editing, and Photo Research	The Davis Group, Inc.
Editorial and Production Coordination	Martha Morong/Quadrata, Inc.
Composition	Pre-Press PMG
Illustrations	Network Graphics and William Melvin
Cover Photo	Kenneth Libbrecht, SnowCrystals.com

Photo credits can be found on p. 993.

Many of the designations used by manufacturers and sellers to distinguish their products are claimed as trademarks. Where those designations appear in this book, and Addison-Wesley was aware of a trademark claim, the designations have been printed in initial caps or all caps.

Library of Congress Cataloging-in-Publication Data
Algebra and trigonometry : graphs & models / Marvin L. Bittinger ... [et al.]. -- 4th ed.
    p. cm.
  Includes index.
  ISBN 0-321-50112-8 --
  1. Algebra--Textbooks. 2. Algebra--Graphic methods--Textbooks. 3. Trigonometry--Textbooks. 4. Functional analysis--Textbooks. I. Bittinger, Marvin L.
  QA152.3.A49 2009
  512'.13--dc22   2007060114

ISBN-13: 978-0-321-50112-7   ISBN-10: 0-321-50112-8

5 6 7 8 9 10—DOW—12 11 10

Contents

Preface   xiii

BASIC CONCEPTS OF ALGEBRA   1

R.1 The Real-Number System   2
Real Numbers • Interval Notation • Properties of the Real Numbers •
Absolute Value

R.2 Integer Exponents, Scientific Notation, and
Order of Operations   9
Integers as Exponents • Scientific Notation • Order of Operations

R.3 Addition, Subtraction, and Multiplication of
Polynomials   18
Polynomials • Addition and Subtraction • Multiplication

R.4 Factoring   23
Terms with Common Factors • Factoring by Grouping • Trinomials of the
Type $x^2 + bx + c$ • Trinomials of the Type $ax^2 + bx + c, a \neq 1$ • Special
Factorizations

R.5 The Basics of Equation Solving   32
Linear and Quadratic Equations

R.6 Rational Expressions   36
The Domain of a Rational Expression • Simplifying, Multiplying, and Dividing
Rational Expressions • Adding and Subtracting Rational Expressions •
Complex Rational Expressions

R.7 Radical Notation and Rational Exponents   45
Simplifying Radical Expressions • An Application • Rationalizing Denominators
or Numerators • Rational Exponents
SUMMARY AND REVIEW   55
TEST   59

1 **GRAPHS, FUNCTIONS, AND MODELS**   **61**

1.1 Introduction to Graphing   62
Graphs • Solutions of Equations • Graphs of Equations • The Distance Formula • Midpoints of Segments • Circles

VISUALIZING THE GRAPH   74

1.2 Functions and Graphs   80
Functions • Notation for Functions • Graphs of Functions • Finding Domains of Functions • Visualizing Domain and Range • Applications of Functions

1.3 Linear Functions, Slope, and Applications   97
Linear Functions • The Linear Function $f(x) = mx + b$ and Slope • Applications of Slope • Slope–Intercept Equations of Lines • Graphing $f(x) = mx + b$ Using m and b • Applications of Linear Functions

VISUALIZING THE GRAPH   109

1.4 Equations of Lines and Modeling   115
Slope–Intercept Equations of Lines • Point–Slope Equations of Lines • Parallel Lines • Perpendicular Lines • Mathematical Models • Curve Fitting • Linear Regression

1.5 Linear Equations, Functions, Zeros, and Applications   129
Linear Equations • Applications Using Linear Models • Zeros of Linear Functions • Formulas

1.6 Solving Linear Inequalities   150
Linear Inequalities • Compound Inequalities • An Application

SUMMARY AND REVIEW   156

TEST   161

2 **MORE ON FUNCTIONS**   **165**

2.1 Increasing, Decreasing, and Piecewise Functions; Applications   166
Increasing, Decreasing, and Constant Functions • Relative Maximum and Minimum Values • Applications of Functions • Functions Defined Piecewise

2.2 The Algebra of Functions   181
The Algebra of Functions: Sums, Differences, Products, and Quotients • Difference Quotients

2.3 The Composition of Functions   189
The Composition of Functions • Decomposing a Function as a Composition

2.4 Symmetry and Transformations   198

Symmetry • Even and Odd Functions • Transformations of Functions •
Vertical and Horizontal Translations • Reflections • Vertical and Horizontal
Stretchings and Shrinkings

VISUALIZING THE GRAPH   213

2.5 Variation and Applications   219

Direct Variation • Inverse Variation • Combined Variation

SUMMARY AND REVIEW   228

TEST   232

QUADRATIC FUNCTIONS AND EQUATIONS; INEQUALITIES   235

3.1 The Complex Numbers   236

The Complex-Number System • Addition and Subtraction •
Multiplication • Conjugates and Division

3.2 Quadratic Equations, Functions, Zeros, and Models   244

Quadratic Equations and Quadratic Functions • Completing the Square •
Using the Quadratic Formula • The Discriminant • Equations Reducible to
Quadratic • Applications

3.3 Analyzing Graphs of Quadratic Functions   261

Graphing Quadratic Functions of the Type $f(x) = a(x - h)^2 + k$ • Graphing
Quadratic Functions of the Type $f(x) = ax^2 + bx + c, a \neq 0$ • Applications

VISUALIZING THE GRAPH   271

3.4 Solving Rational Equations and Radical Equations   276

Rational Equations • Radical Equations

3.5 Solving Equations and Inequalities with Absolute Value   284

Equations with Absolute Value • Inequalities with Absolute Value

SUMMARY AND REVIEW   289

TEST   293

POLYNOMIAL AND RATIONAL FUNCTIONS   295

4.1 Polynomial Functions and Modeling   296

The Leading-Term Test • Finding Zeros of Factored Polynomial
Functions • Finding Real Zeros on a Calculator • Polynomial Models

4.2 Graphing Polynomial Functions   313

Graphing Polynomial Functions • The Intermediate Value Theorem

VISUALIZING THE GRAPH   320

4.3 Polynomial Division; The Remainder and Factor
Theorems   323
Division and Factors • The Remainder Theorem and Synthetic
Division • Finding Factors of Polynomials

4.4 Theorems about Zeros of Polynomial Functions   332
The Fundamental Theorem of Algebra • Finding Polynomials with Given
Zeros • Zeros of Polynomial Functions with Real Coefficients •
Rational Coefficients • Integer Coefficients and the Rational Zeros
Theorem • Descartes' Rule of Signs

4.5 Rational Functions   342
The Domain of a Rational Function • Asymptotes • Applications

VISUALIZING THE GRAPH   356

4.6 Polynomial and Rational Inequalities   360
Polynomial Inequalities • Rational Inequalities

SUMMARY AND REVIEW   372
TEST   377

EXPONENTIAL AND LOGARITHMIC FUNCTIONS   379

5.1 Inverse Functions   380
Inverses • Inverses and One-to-One Functions • Finding Formulas for
Inverses • Inverse Functions and Composition • Restricting a Domain

5.2 Exponential Functions and Graphs   394
Graphing Exponential Functions • Applications • The Number e • Graphs
of Exponential Functions, Base e

5.3 Logarithmic Functions and Graphs   408
Logarithmic Functions • Finding Certain Logarithms • Converting Between
Exponential Equations and Logarithmic Equations • Finding Logarithms on a
Calculator • Natural Logarithms • Changing Logarithmic Bases • Graphs of
Logarithmic Functions • Applications

VISUALIZING THE GRAPH   422

5.4 Properties of Logarithmic Functions   426
Logarithms of Products • Logarithms of Powers • Logarithms of
Quotients • Applying the Properties • Simplifying Expressions of the Type
$\log_a a^x$ and $a^{\log_a x}$

5.5 Solving Exponential and Logarithmic Equations   435
Solving Exponential Equations • Solving Logarithmic Equations

5.6 Applications and Models: Growth and Decay; Compound Interest 446

Population Growth • Interest Compounded Continuously • Models of Limited Growth • Exponential Decay • Exponential and Logarithmic Curve Fitting

SUMMARY AND REVIEW 464
TEST 469

THE TRIGONOMETRIC FUNCTIONS 471

6.1 Trigonometric Functions of Acute Angles 472

The Trigonometric Ratios • The Six Functions Related • Function Values of 30°, 45°, and 60° • Function Values of Any Acute Angle • Cofunctions and Complements

6.2 Applications of Right Triangles 485

Solving Right Triangles • Applications

6.3 Trigonometric Functions of Any Angle 498

Angles, Rotations, and Degree Measure • Trigonometric Functions of Angles or Rotations • The Six Functions Related • Terminal Side on an Axis • Reference Angles: 30°, 45°, and 60° • Function Values for Any Angle

6.4 Radians, Arc Length, and Angular Speed 514

Distances on the Unit Circle • Radian Measure • Arc Length and Central Angles • Linear Speed and Angular Speed

6.5 Circular Functions: Graphs and Properties 530

Reflections on the Unit Circle • Finding Function Values • Graphs of the Sine and Cosine Functions • Graphs of the Tangent, Cotangent, Cosecant, and Secant Functions

6.6 Graphs of Transformed Sine and Cosine Functions 547

Variations of Basic Graphs • Graphs of Sums: Addition of Ordinates • Damped Oscillation: Multiplication of Ordinates

VISUALIZING THE GRAPH 561

SUMMARY AND REVIEW 566
TEST 571

TRIGONOMETRIC IDENTITIES, INVERSE FUNCTIONS, AND EQUATIONS   573

7.1 Identities: Pythagorean and Sum and Difference   574
Pythagorean Identities • Simplifying Trigonometric Expressions • Sum and Difference Identities

7.2 Identities: Cofunction, Double-Angle, and Half-Angle   588
Cofunction Identities • Double-Angle Identities • Half-Angle Identities

7.3 Proving Trigonometric Identities   598
The Logic of Proving Identities • Proving Identities • Product-to-Sum and Sum-to-Product Identities

7.4 Inverses of the Trigonometric Functions   606
Restricting Ranges to Define Inverse Functions • Composition of Trigonometric Functions and Their Inverses

7.5 Solving Trigonometric Equations   618

VISUALIZING THE GRAPH   629

SUMMARY AND REVIEW   633
TEST   638

APPLICATIONS OF TRIGONOMETRY   639

8.1 The Law of Sines   640
Solving Oblique Triangles • The Law of Sines • Solving Triangles (AAS and ASA) • Solving Triangles (SSA) • The Area of a Triangle

8.2 The Law of Cosines   654
The Law of Cosines • Solving Triangles (SAS) • Solving Triangles (SSS)

8.3 Complex Numbers: Trigonometric Form   664
Graphical Representation • Trigonometric Notation for Complex Numbers • Multiplication and Division with Trigonometric Notation • Powers of Complex Numbers • Roots of Complex Numbers

8.4 Polar Coordinates and Graphs   676
Polar Coordinates • Polar and Rectangular Equations • Graphing Polar Equations

VISUALIZING THE GRAPH   685

8.5 Vectors and Applications   688
Vectors • Vector Addition • Applications • Components

8.6 Vector Operations 697

Position Vectors • Operations on Vectors • Unit Vectors •
Direction Angles • Angle Between Vectors • Forces in Equilibrium

SUMMARY AND REVIEW 712

TEST 717

9

SYSTEMS OF EQUATIONS AND MATRICES 719

9.1 Systems of Equations in Two Variables 720

Solving Systems of Equations Graphically • The Substitution Method •
The Elimination Method • Applications

VISUALIZING THE GRAPH 730

9.2 Systems of Equations in Three Variables 736

Solving Systems of Equations in Three Variables • Applications •
Mathematical Models and Applications

9.3 Matrices and Systems of Equations 748

Matrices and Row-Equivalent Operations • Gaussian Elimination with
Matrices • Gauss–Jordan Elimination

9.4 Matrix Operations 756

Matrix Addition and Subtraction • Scalar Multiplication • Products of
Matrices • Matrix Equations

9.5 Inverses of Matrices 768

The Identity Matrix • The Inverse of a Matrix • Solving Systems of Equations

9.6 Determinants and Cramer's Rule 776

Determinants of Square Matrices • Evaluating Determinants Using
Cofactors • Cramer's Rule

9.7 Systems of Inequalities and Linear Programming 784

Graphs of Linear Inequalities • Systems of Linear Inequalities •
Applications: Linear Programming

9.8 Partial Fractions 798

Partial Fraction Decompositions

SUMMARY AND REVIEW 805

TEST 811

ANALYTIC GEOMETRY TOPICS   813

10.1 The Parabola   814
Parabolas • Finding Standard Form by Completing the Square • Applications

10.2 The Circle and the Ellipse   822
Circles • Ellipses • Applications

10.3 The Hyperbola   833
Standard Equations of Hyperbolas • Applications

10.4 Nonlinear Systems of Equations and Inequalities   843
Nonlinear Systems of Equations • Modeling and Problem Solving • Nonlinear Systems of Inequalities

VISUALIZING THE GRAPH   850

10.5 Rotation of Axes   855
Rotation of Axes • The Discriminant

10.6 Polar Equations of Conics   865
Polar Equations of Conics • Converting from Polar Equations to Rectangular Equations • Finding Polar Equations of Conics

10.7 Parametric Equations   872
Graphing Parametric Equations • Determining a Rectangular Equation for Given Parametric Equations • Determining Parametric Equations for a Given Rectangular Equation • Applications

SUMMARY AND REVIEW   881

TEST   886

SEQUENCES, SERIES, AND COMBINATORICS   889

11.1 Sequences and Series   890
Sequences • Finding the General Term • Sums and Series • Sigma Notation • Recursive Definitions

11.2 Arithmetic Sequences and Series   900
Arithmetic Sequences • Sum of the First n Terms of an Arithmetic Sequence • Applications

11.3 Geometric Sequences and Series   910
Geometric Sequences • Sum of the First n Terms of a Geometric Sequence • Infinite Geometric Series • Applications

VISUALIZING THE GRAPH   918

11.4 Mathematical Induction   922
Sequences of Statements • Proving Infinite Sequences of Statements

11.5 Combinatorics: Permutations   928
Permutations · Factorial Notation · Permutations of n Objects Taken k at a Time · Permutations of Sets with Nondistinguishable Objects

11.6 Combinatorics: Combinations   938
Combinations

11.7 The Binomial Theorem   946
Binomial Expansions Using Pascal's Triangle · Binomial Expansion Using Factorial Notation · Finding a Specific Term · Total Number of Subsets

11.8 Probability   954
Experimental and Theoretical Probability · Computing Experimental Probabilities · Theoretical Probability

SUMMARY AND REVIEW   965
TEST   969

Appendix: Basic Concepts from Geometry   971
Photo Credits   993
Answers   A-1
Index   I-1
Index of Applications   I-15

Preface

Our challenge, and our goal, when writing this textbook was to do everything possible to help you learn the concepts and skills contained between its covers. Every feature we have included was put here with this in mind. We realize that your time is both valuable and limited, so we communicate in a highly visual way that allows you to focus easily and learn quickly and efficiently.

Take advantage of the side-by-side algebraic and graphical solutions, the Visualizing the Graph and the Connecting the Concepts features, the Vocabulary Reviews, the Classify the Function exercises, and the Study Tips. We included them to enable you to make the most of your study time and to be successful in this course.

In an effort to encourage you to observe and interpret the mathematics that appears daily in the world around you, we conducted a vigorous search for real-data applications during our writing process. These applied problems connect the mathematical concepts in this course with your everyday life.

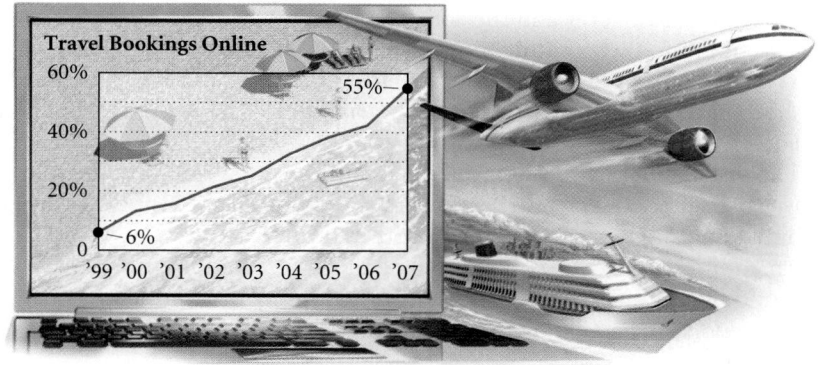

Source: PhoCusWright

Best wishes for a positive learning experience,

Marv Bittinger
Judy Beecher
David Ellenbogen
Judy Penna

NEW IN THIS EDITION

In this Fourth Edition of the text, we have changed the table of contents to make the material at the beginning of the text more easily taught and learned. We have balanced and evened out the lengths of Chapters 1–3 in the Third Edition. By presenting this material in four chapters rather than in three and also by rearranging the content, we have balanced the level of difficulty of the material covered in these early chapters as well.

We think students and instructors alike will be pleased with the changes that have been made. By spreading this material out over four chapters and presenting it in a different order, we provide the student with a more even-handed and consistent introduction to Algebra and Trigonometry. The new arrangement is better suited to a one-section-per-lecture class format. In addition, chapter tests cover more manageable amounts of material.

An overview of the changes made in the Fourth Edition follows.

❖ The material on increasing, decreasing, and piecewise functions, along with the discussion of the algebra of functions, composition of functions, and symmetry and transformations that was in Chapter 1 in the Third Edition has been moved to a separate chapter (Chapter 2) in the Fourth Edition.

❖ The algebra of functions and composition of functions are now presented in two sections rather than one.

❖ The material on linear equations, functions, and inequalities has been moved from Chapter 2 in the Third Edition to Chapter 1 in the Fourth Edition where functions are introduced.

❖ The chapter on polynomial functions and rational functions has been shortened by moving the section on variation to Chapter 2 in the Fourth Edition.

❖ The discussion of equations and inequalities with absolute value that was presented in different sections in the Third Edition is now combined in a single section in the Fourth Edition.

❖ An additional objective on graphing damped oscillations, functions found by multiplying trigonometric functions by other functions, has been added to Section 6.6, *Graphs of Transformed Sine and Cosine Functions.*

❖ Section 7.3, *Proving Trigonometric Identities*, now includes using the product-to-sum identities and the sum-to-product identities to derive other identities.

❖ The section that reviews equation solving in Chapter R, the review chapter, has been moved to appear now before the sections on rational expressions, radical notation, and rational exponents.

In addition to the changes described above, we have increased our coverage of certain topics by responding to instructors' requests for more examples and exercises. These topics include difference quotients, finding the domain of a composed function, finding the asymptotes of rational functions, and solving polynomial and rational inequalities.

Also new to this edition are many new and updated applied problems that contain real and relevant data. These applications help answer the question, "What is all this math good for?"

To encourage students to put pencil to paper and practice skills at the time they are presented in the text, we have added a "Now Try" feature. At the end of nearly every example, the student is given the number of an exercise that corresponds to that example and is prompted to "Now Try" doing the exercise.

To help build test-taking skills, we have added true/false and multiple-choice exercises in each chapter review.

In a continuing effort to emphasize graphing and visualization, we have added exercises in which the student is given an equation and four graphs and is asked to match the equation with the correct graph. This feature appears in both the review exercises and the chapter test in each chapter with the exception of Chapter R, in which graphing has not yet been introduced.

OUR APPROACH

Algebra and Trigonometry: Graphs & Models, Fourth Edition, covers college-level algebra and trigonometry and is appropriate for a one- or two-term course in precalculus mathematics. Our goal is to enhance the learning process through the use of technology. Our approach is more interactive and more visual than most algebra and trigonometry texts. Although a course in intermediate algebra is a prerequisite for using this text, Chapter R, "Basic Concepts of Algebra," provides sufficient review to unify the diverse mathematical backgrounds of most students.

✧ **Function Emphasis** Functions are the core of this course and should be presented as a thread that runs throughout the course rather than as an isolated topic. We introduce functions in Chapter 1, whereas most traditional algebra and trigonometry textbooks cover equation-solving in Chapter 1. (See pp. 80, 88–90, 139, 244, and 297.) Our approach introduces students to a relatively new concept at the beginning of the course rather than requiring them to begin with a review of what was previously covered in intermediate algebra. The concept of a function can be challenging for students. By repeatedly exposing them to the language, notation, and use of functions, demonstrating visually how functions relate to equations and graphs, and also showing how functions can be used to model real data, we hope that students will not only become comfortable with functions but will also come to understand and appreciate them.

✧ **Visual Emphasis** Our early introduction of functions allows graphs to be used to provide a visual aspect to solving equations and inequalities. For example, we are able to show the students both algebraically and visually that the solutions of a quadratic equation $ax^2 + bx + c = 0$ are the zeros of the quadratic function $f(x) = ax^2 + bx + c$ as well as the

x-intercepts of the graph of that function. This makes it possible for students, particularly visual learners, to gain a quick understanding of these concepts. (See pp. 252, 256, 297, 361, and 422.)

❖   **Emphasis on Graphing**   To further enhance the visual aspect of the material, there is an increased emphasis on graphs and graphing in this edition, again reminding students how important and informative visualization can be in the discipline of mathematics. (See pp. 66, 85, 261–267, 315–317, 395, and 417.)

The feature titled Visualizing the Graph also contributes to this emphasis, as do many of the exercises. This feature is described in more detail under the "Features" heading.

❖   **Zeros, Solutions, and *x*-Intercepts Theme**   We find that when students understand the connections among the real zeros of a function, the solutions of its associated equation, and the first coordinates of the x-intercepts of its graph, a door opens to a new level of mathematical comprehension that increases the probability of success in this course. We emphasize zeros, solutions, and x-intercepts throughout the text by using consistent, precise terminology and including exceptional graphics. Seeing this theme repeated in different contexts leads to a better understanding and retention of these concepts. (See pp. 245–246, 256, and 301.)

❖   **Real-Data Applications and Regression**   We encourage students to see and interpret the mathematics that appears every day in the world around them. Throughout the writing process, we conducted an energetic search for real-data applications, and the result is a variety of examples and exercises that connect the mathematical content with everyday life. Most of these applications feature source lines and frequently include charts and graphs. Many are drawn from the fields of health, business and economics, life and physical sciences, social science, and areas of general interest such as sports and travel. (See pp. 104, 111, 126–127, 134, 144–145, 305–306, and 457–458.)

Credit Card Debt

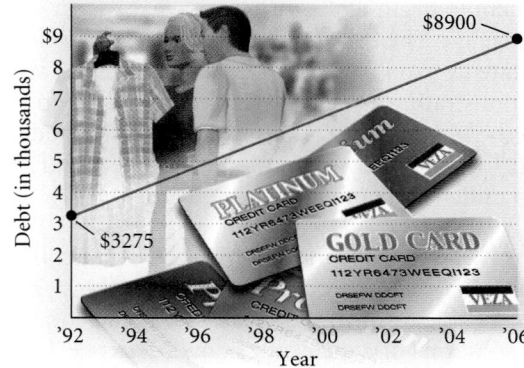

Prescription Drug Sales

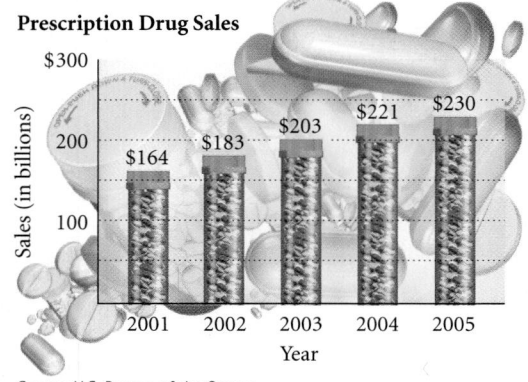

Source: U.S. Bureau of the Census

69. *Filing Tax Returns Electronically.* The number of tax returns filed electronically in 2005 reflects an increase of approximately 480% over the number of electronic returns in 1995.

Year, x	Number of Tax Returns Filed Electronically, y (in millions)
1995, 0	11.807
1997, 2	19.136
1999, 4	29.349
2001, 6	40.245
2003, 8	52.945
2005, 10	68.476

Source: 2005 IRS Data Book

a) Use a graphing calculator to model the data with a linear function.
b) Predict the number of tax returns that will be filed electronically in 2010 and in 2014.
c) Find the correlation coefficient for the regression line and determine whether the line fits the data closely.

In Chapter 1, we introduce the use of regression or curve fitting to model data with linear functions, and continue this visual theme with quadratic, cubic, quartic, exponential, logarithmic, and logistic functions. (See pp. 122–123, 305–307, and 404.) Although the theoretical aspects of curve fitting cannot be developed in this course, the power of the graphing calculator is very apparent in this area as the technique is applied to real data. Students can quickly make the "Aha!" connection between real data and the interpolated and extrapolated results of the curve fitting, giving them a better conceptual understanding of the material.

❖ **Integrated Technology** In order to increase students' understanding of the course content through a visual means, we integrate graphing calculator technology throughout. The use of the graphing calculator is woven throughout the text's exposition, exercise sets, and testing program without sacrificing algebraic skills. Graphing calculator technology is included in order to enhance—not replace—students' mathematical skills, and to alleviate the tedium associated with certain procedures. (See pp. 250, 353–354, and 435.) In addition, each copy of the student text is bundled with a complimentary graphing calculator manual created by author Judy Penna. A GCM icon $\boxed{\text{GCM}}$ indicates the correlation of the manual to the main text content.

❖ **Five-Step Problem-Solving Process** The basis for problem solving is a distinctive five-step process established early in the text (Section 1.5) to help students learn strategic ways to approach and solve applied problems. This process is then used throughout the text to give students a consistent framework for problem solving. (See pp. 133–134 and 254–255.)

FEATURES

❖ **Chapter Openers** Each chapter opens with an application relevant to the content of the chapter. Also included is a table of contents for the chapter, listing section titles. (See pp. 61 and 379.)

❖ **Section Objectives** Content objectives are listed at the beginning of each section. Together with subheadings throughout the section, these objectives provide a useful outline of the section for both instructors and students. (See pp. 80 and 244.)

❖ **Annotated Examples** We have included over 875 examples designed to prepare the student fully to do the exercises. Learning is carefully guided with the use of numerous color-coded art pieces and step-by-step annotations. Substitutions and annotations are highlighted in red for emphasis. (See pp. 248 and 429.)

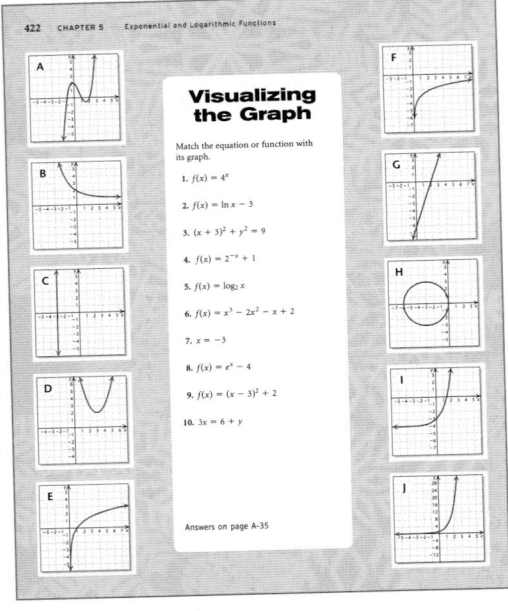

Now Try Feature   Now Try exercises are found after nearly every example. This feature encourages active learning by asking students to do an exercise in the exercise set that is similar to the example the student has just read. (See pp. 245 and 395.)

Visualizing the Graph   This feature provides students with an opportunity to match an equation with its graph by focusing on the characteristics of the equation and the corresponding attributes of the graph. This feature appears at least once in every chapter, beginning with Chapter 1. (See pp. 213, 271, 356, and 422.)

In addition to the full-page feature shown at left, many of the exercise sets include exercises in which the student is asked to match an equation with its graph or to find an equation of a function from its graph. (See pp. 216, 217, 308, and 403.)

Side-by-Side Feature   Many examples are presented in a side-by-side, two-column format in which the algebraic solution of an equation appears in the left column and a graphical solution appears in the right column. (See pp. 245, 364–365, and 438.) This enables students to visualize and comprehend the connections among the solutions of an equation, the zeros of a function, and the x-intercepts of the graph of a function.

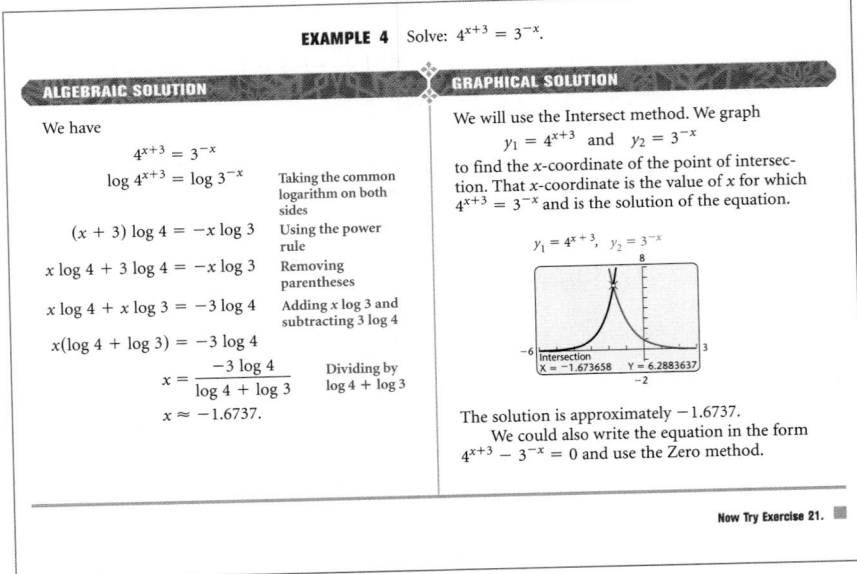

EXAMPLE 4   Solve: $4^{x+3} = 3^{-x}$.

ALGEBRAIC SOLUTION

We have

$$4^{x+3} = 3^{-x}$$

$$\log 4^{x+3} = \log 3^{-x} \qquad \text{Taking the common logarithm on both sides}$$

$$(x + 3)\log 4 = -x\log 3 \qquad \text{Using the power rule}$$

$$x\log 4 + 3\log 4 = -x\log 3 \qquad \text{Removing parentheses}$$

$$x\log 4 + x\log 3 = -3\log 4 \qquad \text{Adding } x\log 3 \text{ and subtracting } 3\log 4$$

$$x(\log 4 + \log 3) = -3\log 4$$

$$x = \frac{-3\log 4}{\log 4 + \log 3} \qquad \text{Dividing by } \log 4 + \log 3$$

$$x \approx -1.6737.$$

GRAPHICAL SOLUTION

We will use the Intersect method. We graph

$$y_1 = 4^{x+3} \quad \text{and} \quad y_2 = 3^{-x}$$

to find the x-coordinate of the point of intersection. That x-coordinate is the value of x for which $4^{x+3} = 3^{-x}$ and is the solution of the equation.

$y_1 = 4^{x+3}, \quad y_2 = 3^{-x}$

Intersection
X = -1.673658   Y = 6.2883637

The solution is approximately -1.6737.
We could also write the equation in the form $4^{x+3} - 3^{-x} = 0$ and use the Zero method.

Now Try Exercise 21. ■

❖ **Connecting the Concepts**   This feature highlights the importance of connecting concepts. When students are presented with concepts in a visual form—using graphs, an outline, or a chart—rather than merely in paragraphs of text, comprehension is streamlined and retention is maximized. The visual aspect of this feature invites students to stop and check their understanding of how concepts work together in one section or in several sections. This concept check in turn enhances student performance on homework assignments and exams. (See pp. 143, 256, and 364.)

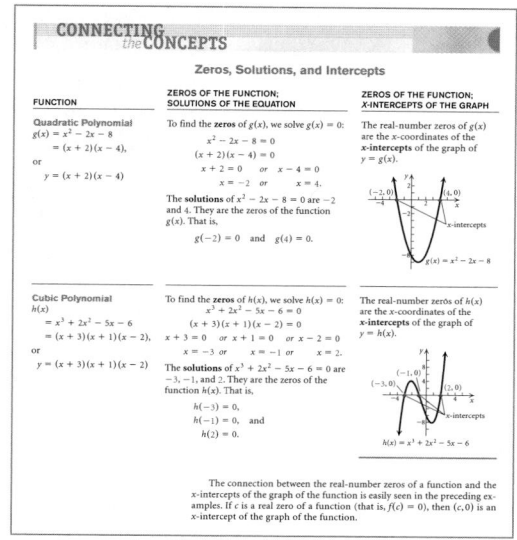

❖ **Study Tips**   Appearing in the text margin, the Study Tips provide helpful study hints throughout the text and also briefly remind students to use the electronic and print supplements that accompany the text. (See pp. 168, 199, and 326.)

❖ **Review Icons**   Placed next to the concept that a student is currently studying, a review icon references a section of the text in which the student can find and review the topics on which the current concept is built. (See pp. 343 and 380.)

❖ **Graphing Calculator Manual Icons**   Placed next to relevant examples and exercises, a GCM icon directs students to the Graphing Calculator Manual for just-in-time, keystroke level instruction. (See pp. 66, 238, 343, and 414.)

❖ **Variety of Exercises**   There are over 7950 exercises in this text. The exercise sets are enhanced with real-data applications and source lines, detailed art pieces, tables, graphs, and photographs. In addition to the exercises that provide students practice with the concepts presented in the section, the exercise sets feature the following elements.

Collaborative Discussion and Writing Exercises   These exercises can be used in small groups or by the class as a whole to encourage students to talk and write about the key mathematical concepts in each section. (See pp. 260, 322, and 425.)

Skill Maintenance Exercises   These exercises provide an ongoing review of concepts previously presented in the course, enhancing students' retention of these concepts. These exercises include Vocabulary Review (see pp. 288 and 462) and Classifying the Function (see pp. 341 and 434). Answers to *all* Skill Maintenance exercises appear in the answer section at the back of the book along with a section reference that directs students quickly and efficiently to the appropriate section of the text if they need help with an exercise. (See pp. 217, 284, 371, and 425.)

Classifying the Function Exercises   With a focus on conceptual understanding, students are asked periodically to identify a number of functions by their type (for example, linear, quadratic, rational, and so on). As students progress through the text, the variety of functions they know increases and these exercises become more challenging. The "classifying the function" exercises appear with the review exercises in the Skill Maintenance portion of an exercise set. (See pp. 341 and 434.)

Vocabulary Review Exercises   This feature checks and reviews students' understanding of the vocabulary introduced throughout the text. It appears once in every chapter, in the Skill Maintenance portion of an exercise set, and is intended to provide a continuing review of the terms that students must know in order to be able to communicate effectively in the language of mathematics. (See pp. 180–181, 288, and 462.)

Synthesis Exercises   These exercises appear at the end of each exercise set and encourage critical thinking by requiring students to synthesize concepts from several sections or to take a concept a step further than in the general exercises. (See pp. 331, 408, and 462–463.)

❖ **Highlighted Information**   Important definitions, properties, and rules are displayed in screened boxes, and summaries and procedures are listed in boxes outlined in black. This organization and presentation provides for efficient learning and review. (See pp. 166, 335, and 383.)

❖ **Summary and Review**   The Summary and Review at the end of each chapter contains a list of important properties and formulas covered in that chapter followed by an extensive set of review exercises. A section reference is provided for each exercise. This directs the student to the appropriate place in the chapter to find help, if needed, in doing the exercise. To help students build and/or polish test-taking skills, we have added true/false and multiple-choice exercises to the review exercises. This includes multiple-choice exercises that involve identifying the graph of an equation. The chapter summary and the review exercises provide excellent preparation for chapter tests and also for the final examination. Answers to *all* review exercises appear at the back of the book. (See pp. 289–292 and 464–468.)

❖ **Chapter Tests** The test at the end of each chapter allows students to test themselves and target areas that need further study before taking the in-class test. Each chapter test includes a multiple-choice exercise involving graphing. Answers to *all* Chapter Test questions appear in the answer section at the back of the book, along with corresponding section references. (See pp. 232–233 and 377–378.)

❖ **Use of Color** The text uses full color in an extremely functional manner, as seen in the design elements and numerous pieces of art. The use of color has been carefully thought out so that it carries a consistent meaning that enhances students' ability to read and comprehend the exposition. (See pp. 211, 380–381, and 383.)

❖ **Art Package** The text contains over 2130 art pieces including photographs, situational art, and statistical graphs that not only highlight the abundance of real-world applications but also help students visualize the mathematics being discussed. (See pp. 75, 177, 353, 394, and 456.)

❖ **Optional Review Chapter** Chapter R, "Basic Concepts of Algebra," provides an optional review of intermediate algebra. Some or all of the topics in this chapter can be taught at the beginning of the course, or it can be used as a convenient source of information throughout the term for students who need a quick review of particular topics.

❖ **Appendix Covering Basic Concepts from Geometry** This appendix covers basic concepts from geometry that students need to know in order to work with topics pertaining to angles and triangles. Formulas from geometry are also listed near the back of the book.

SUPPLEMENTS

STUDENT SUPPLEMENTS

Graphing Calculator Manual
ISBN: 0-321-53198-1, 978-0-321-53198-8
- By Judith A. Penna
- Contains keystroke level instruction for the Texas Instruments TI-83 Plus, TI-84 Plus, and TI-89
- Teaches students how to use a graphing calculator using actual examples and exercises from the main text
- Mirrors the topic order to the main text to provide a just-in-time mode of instruction
- Automatically ships with each new copy of the text

Student's Solutions Manual
ISBN: 0-321-53197-3, 978-0-321-53197-1
- By Judith A. Penna
- Contains completely worked-out solutions with step-by-step annotations for all the odd-numbered exercises in the exercise sets, with the exception of the Collaborative Discussion and Writing exercises, and for all the odd-numbered review exercises and all chapter test exercises

Video Lectures on CD with Optional Captioning
ISBN: 0-321-54314-9, 978-0-321-54314-1
- Complete set of digitized videos on CD-ROMs for student use at home or on campus
- Ideal for distance learning or supplemental instruction
- Features authors Judy Beecher and Judy Penna working through and explaining key examples in the text

INSTRUCTOR SUPPLEMENTS

Annotated Instructor's Edition
ISBN: 0-321-53194-9, 978-0-321-53194-0
- Includes all the answers to the exercise sets, usually right on the page where the exercises appear
- Readily accessible answers help both new and experienced instructors prepare for class efficiently

Insider's Guide
ISBN: 0-321-53195-7, 978-0-321-53195-7
- Includes resources to help faculty with course preparation and classroom management
- Provides helpful teaching tips, conversion guide, and sample syllabi
- Includes black-line masters of grids and number lines for transparency masters or test preparation

Pearson Adjunct Support Center
- Offers consultation on suggested syllabi, helpful tips for using the textbook support package, assistance with content, and advice on classroom strategies
- Available Sunday through Thursday evenings from 5 P.M. to midnight EST
- e-mail: AdjunctSupport@aw.com; Telephone: 1-800-435-4084; fax: 1-877-262-9774

Instructor's Solutions Manual
ISBN: 0-321-53203-1, 978-0-321-53203-9
- By Judith A. Penna
- Contains worked-out solutions to all exercises in the exercise sets, including the Collaborative Discussion and Writing exercises, and solutions for all end-of-chapter material

Printed Test Bank
ISBN: 0-321-54316-5, 978-0-321-54316-5
- By Laurie Hurley
- Contains four free-response test forms for each chapter following the same format and having the same level of difficulty as the tests in the main text, plus two multiple-choice test forms for each chapter
- Provides six forms of the final examination, four with free-response questions and two with multiple-choice questions

MEDIA SUPPLEMENTS

❖ **MyMathLab®** MyMathLab® is a series of text-specific, easily customizable online courses for Pearson Education's textbooks in mathematics and statistics. Powered by CourseCompass™ (our online teaching and learning environment) and MathXL® (our online homework, tutorial, and assessment system), MyMathLab gives you the tools you need to deliver all or a portion of your course online, whether your students are in a lab setting or working from home. MyMathLab provides a rich and flexible set of course materials, featuring free-response exercises that are algorithmically generated for unlimited practice and mastery. Students can also use online tools, such as video lectures, animations, and a multimedia textbook, to independently improve their understanding and performance. Instructors can use MyMathLab's homework and test managers to select and assign online exercises correlated directly to the textbook, and they can also create and assign their own online exercises and import TestGen tests for added flexibility. MyMathLab's online gradebook—designed specifically for mathematics and statistics—automatically tracks students' homework and test results and gives the instructor control over how to calculate final grades. Instructors can also add offline (paper-and-pencil) grades to the gradebook. MyMathLab also includes access to the **Pearson Tutor Center** (www.pearsontutorservices.com), which provides students with tutoring via toll-free phone, fax, email, and interactive Web sessions. MyMathLab is available to qualified adopters. For more information, visit our Web site at www.mymathlab.com or contact your sales representative.

❖ **MathXL®** MathXL® is a powerful online homework, tutorial, and assessment system that accompanies Pearson Education's textbooks in mathematics and statistics. With MathXL, instructors can create, edit, and assign online homework and tests using algorithmically generated exercises correlated to the textbook at the objective level. They can also create and assign their own online exercises and import TestGen tests for added flexibility. All student work is tracked in MathXL's online gradebook. Students can take chapter tests in MathXL and receive personalized study plans based on their test results. The study plan diagnoses weaknesses and links students directly to tutorial exercises for the objectives they need to study and retest. Students can also access supplemental animations and video clips directly from selected exercises. MathXL is available to qualified adopters. For more information, visit our Web site at www.mathxl.com, or contact your sales representative.

❖ **MathXL® Tutorials on CD**
ISBN: 0-321-53196-5, 978-0-321-53196-4
This interactive tutorial CD-ROM provides algorithmically generated practice exercises that are correlated at the objective level to the exercises in the textbook. Every practice exercise is accompanied by an example and a guided solution designed to involve students in the solution process. Selected exercises may also include a video clip to help students visualize concepts. The software provides helpful feedback for incorrect answers and can generate printed summaries of students' progress.

❖ **InterAct Math® Tutorial Web site   www.interactmath.com**
Get practice and tutorial help online! This interactive tutorial Web site provides algorithmically generated practice exercises that correlate directly to the exercises in the textbook. Students can retry an exercise as many times as they like with new values each time for unlimited practice and mastery. Every exercise is accompanied by an interactive guided solution that provides helpful feedback for incorrect answers, and students can also view a worked-out sample problem that steps them through an exercise similar to the one they're working on.

❖ **TestGen®**
TestGen enables instructors to build, edit, print, and administer tests using a computerized bank of questions developed to cover all the objectives of the text. TestGen is algorithmically based, allowing instructors to create multiple but equivalent versions of the same question or test with the click of a button. Instructors can also modify test bank questions or add new questions. Tests can be printed or administered online. The software and testbank are available for download from www.aw-bc.com/irc.

❖ **PowerPoint Lecture Presentation with Active Learning Questions**   The PowerPoint Lecture slides feature presentations written and designed specifically for this text, including figures and examples from the text. The Active Learning Questions, prepared in PowerPoint, are intended for use with classroom response systems and include multiple-choice questions to review lecture material. The PowerPoint Lecture Presentation with Active Learning Questions are available for download from within MyMathLab and from www.aw-bc.com/irc.

ACKNOWLEDGMENTS

We wish to express our heartfelt thanks to a number of people who have contributed in special ways to the development of this textbook. Our editor, Anne Kelly, encouraged and supported our vision. We are very appreciative of the marketing insight provided by Becky Anderson, our marketing manager, and of the support that we received from the entire Addison-Wesley Higher Education Group, including Kathy Manley, senior production supervisor, Rachel S. Reeve, senior project editor, Leah Goldberg, editorial assistant, and Bonnie Gill, marketing assistant. We also thank Jennifer Thomas, media producer, for her creative work on the media products that accompany this text. And we are immensely grateful to Martha Morong, for her production services, and to Geri Davis, for her text design and art editing, and for the endless hours of hard work they have done to make this a book of which we are proud. We also thank Laurie Hurley, Barbara Johnson, Patty LaGree, and Jennifer Rosenberg for their meticulous accuracy checking and proofreading of the text.

The following reviewers made invaluable contributions to the development of the recent revisions and we thank them for that:

Gerald Allen, *Howard College*
Robin Ayers, *Western Kentucky University*
Heidi Barrett, *Arapahoe Community College*
George Behr, *Coastline Community College*
Kimberly Bennekin, *Georgia Perimeter College*
Marc Campbell, *Daytona Beach Community College*
Brad Feldser, *Kennesaw State University*
Homa Ghaussi-Mujtaba, *Lansing Community College*
Bob Gravelle, *Colorado Technical University, Colorado Springs*
Judy Hayes, *Lake-Sumter Community College*
Michelle Hollis, *Bowling Green Community College*
Bridgette Jacob, *Onondaga Community College*
Deanna Kindhart, *Illinois Central College*
Daniel Olson, *Purdue University, North Central*
Randy K. Ross, *Morehead State University*
Daniel Russow, *Arizona Western College*
Brian Schworm, *Morehead State University*
Judith Staver, *Florida Community College at Jacksonville, South Campus*
Douglas Windham, *Tallahassee Community College*

M.L.B.
J.A.B.
D.J.E.
J.A.P.

Basic Concepts of Algebra

APPLICATION The area of Hong Kong is 412 square miles. It is estimated that the population of Hong Kong will be 9,600,000 in 2050. Find the number of square miles of land per person in 2050.

This problem appears as Exercise 82 in Section R.2.

R.1 The Real-Number System

R.2 Integer Exponents, Scientific Notation, and Order of Operations

R.3 Addition, Subtraction, and Multiplication of Polynomials

R.4 Factoring

R.5 The Basics of Equation Solving

R.6 Rational Expressions

R.7 Radical Notation and Rational Exponents

R.1 The Real-Number System

❖ Identify various kinds of real numbers.
❖ Use interval notation to write a set of numbers.
❖ Identify the properties of real numbers.
❖ Find the absolute value of a real number.

❖ Real Numbers

In applications of algebraic concepts, we use real numbers to represent quantities such as distance, time, speed, area, profit, loss, and temperature. Some frequently used sets of real numbers and the relationships among them are shown below.

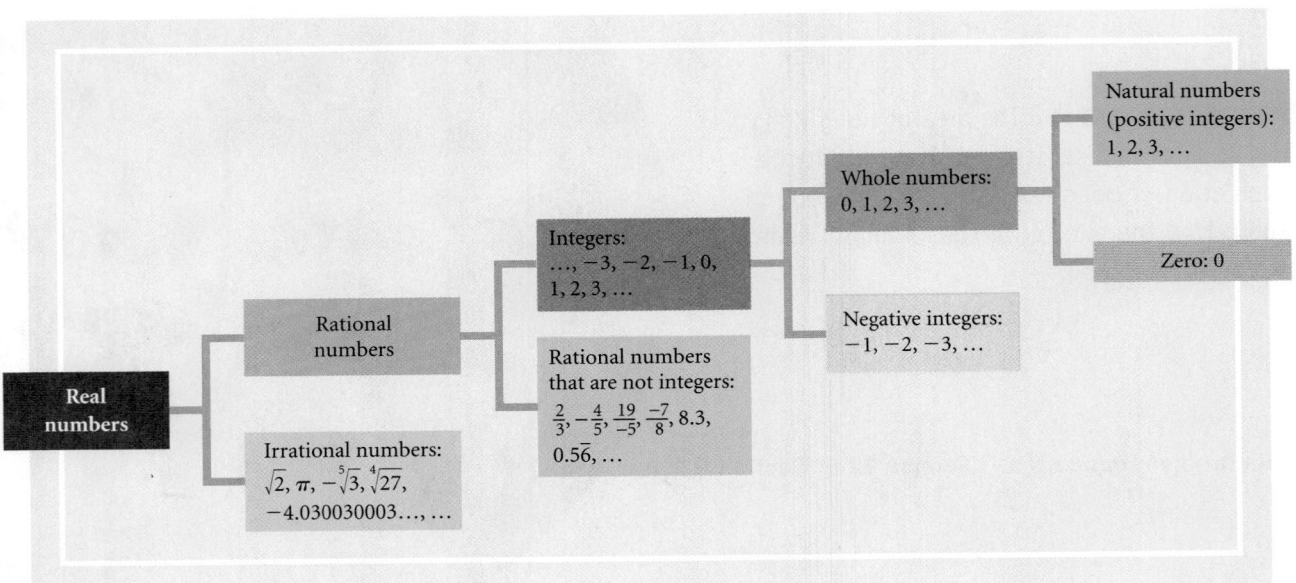

Numbers that can be expressed in the form p/q, where p and q are integers and $q \neq 0$, are **rational numbers**. Decimal notation for rational numbers either *terminates* (ends) or *repeats*. Each of the following is a rational number.

a) 0 $0 = \dfrac{0}{a}$ **for any nonzero integer** a

b) -7 $-7 = \dfrac{-7}{1}$, **or** $\dfrac{7}{-1}$

c) $\dfrac{1}{4} = 0.25$ **Terminating decimal**

d) $-\dfrac{5}{11} = -0.454545\ldots = -0.\overline{45}$ **Repeating decimal**

The real numbers that are not rational are **irrational numbers**. Decimal notation for irrational numbers neither terminates nor repeats. Each of the following is an irrational number.

a) $\pi = 3.1415926535\ldots$ There is no repeating block of digits.
    $\left(\frac{22}{7}\right.$ and 3.14 are rational *approximations* of the irrational number π.$\left.\right)$

b) $\sqrt{2} = 1.414213562\ldots$ There is no repeating block of digits.

c) $-6.12122122212222\ldots$ Although there is a pattern, there is no repeating block of digits.

The set of all rational numbers combined with the set of all irrational numbers gives us the set of **real numbers**. The real numbers are *modeled* using a **number line**, as shown below.

Each point on the line represents a real number, and every real number is represented by a point on the line.

The order of the real numbers can be determined from the number line. If a number a is to the left of a number b, then a **is less than** b $(a < b)$. Similarly, a **is greater than** b $(a > b)$ if a is to the right of b on the number line. For example, we see from the number line above that $-2.9 < -\frac{3}{5}$, because -2.9 is to the left of $-\frac{3}{5}$. Also, $\frac{17}{4} > \sqrt{3}$, because $\frac{17}{4}$ is to the right of $\sqrt{3}$.

The statement $a \le b$, read "a is less than or equal to b," is true if either $a < b$ is true or $a = b$ is true.

The symbol $\in$ is used to indicate that a member, or **element**, belongs to a set. Thus if we let $\mathbb{Q}$ represent the set of rational numbers, we can see from the diagram on p. 2 that $0.5\overline{6} \in \mathbb{Q}$. We can also write $\sqrt{2} \notin \mathbb{Q}$ to indicate that $\sqrt{2}$ is *not* an element of the set of rational numbers.

When *all* the elements of one set are elements of a second set, we say that the first set is a **subset** of the second set. The symbol $\subseteq$ is used to denote this. For instance, if we let $\mathbb{R}$ represent the set of real numbers, we can see from the diagram that $\mathbb{Q} \subseteq \mathbb{R}$ (read "$\mathbb{Q}$ is a subset of $\mathbb{R}$").

❈ Interval Notation

Sets of real numbers can be expressed using **interval notation**. For example, for real numbers a and b such that $a < b$, the **open interval** (a, b) is the set of real numbers between, but not including, a and b. That is,

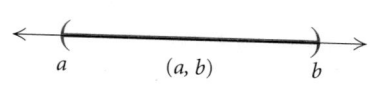

$$(a, b) = \{x \mid a < x < b\}.$$

The points a and b are **endpoints** of the interval. The parentheses indicate that the endpoints are not included in the interval.

Some intervals extend without bound in one or both directions. The interval $[a, \infty)$, for example, begins at a and extends to the right without bound; that is,

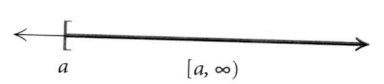

$$[a, \infty) = \{x \mid x \ge a\}.$$

The bracket indicates that a is included in the interval.

The various types of intervals are listed below.

Intervals: Types, Notation, and Graphs

TYPE	INTERVAL NOTATION	SET NOTATION	GRAPH
Open	(a, b)	$\{x \mid a < x < b\}$	
Closed	$[a, b]$	$\{x \mid a \leq x \leq b\}$	
Half-open	$[a, b)$	$\{x \mid a \leq x < b\}$	
Half-open	$(a, b]$	$\{x \mid a < x \leq b\}$	
Open	(a, ∞)	$\{x \mid x > a\}$	
Half-open	$[a, \infty)$	$\{x \mid x \geq a\}$	
Open	$(-\infty, b)$	$\{x \mid x < b\}$	
Half-open	$(-\infty, b]$	$\{x \mid x \leq b\}$	

The interval $(-\infty, \infty)$, graphed below, names the set of all real numbers, $\mathbb{R}$.

EXAMPLE 1 Write interval notation for each set and graph the set.

a) $\{x \mid -4 < x < 5\}$

b) $\{x \mid x \geq 1.7\}$

c) $\{x \mid -5 < x \leq -2\}$

d) $\{x \mid x < \sqrt{5}\}$

Solution

a) $\{x \mid -4 < x < 5\} = (-4, 5)$;

b) $\{x \mid x \geq 1.7\} = [1.7, \infty)$;

c) $\{x \mid -5 < x \leq -2\} = (-5, -2]$;

d) $\{x \mid x < \sqrt{5}\} = (-\infty, \sqrt{5})$;

Now Try Exercises 13 and 15. ■

❄ Properties of the Real Numbers

The following properties can be used to manipulate algebraic expressions as well as real numbers.

Properties of the Real Numbers

For any real numbers a, b, and c:

$a + b = b + a$ and $ab = ba$	Commutative properties of addition and multiplication
$a + (b + c) = (a + b) + c$ and $a(bc) = (ab)c$	Associative properties of addition and multiplication
$a + 0 = 0 + a = a$	Additive identity property
$-a + a = a + (-a) = 0$	Additive inverse property
$a \cdot 1 = 1 \cdot a = a$	Multiplicative identity property
$a \cdot \dfrac{1}{a} = \dfrac{1}{a} \cdot a = 1 \ (a \neq 0)$	Multiplicative inverse property
$a(b + c) = ab + ac$	Distributive property

Note that the distributive property is also true for subtraction since $a(b - c) = a[b + (-c)] = ab + a(-c) = ab - ac$.

EXAMPLE 2   State the property being illustrated in each sentence.

a) $8 \cdot 5 = 5 \cdot 8$

b) $5 + (m + n) = (5 + m) + n$

c) $14 + (-14) = 0$

d) $6 \cdot 1 = 1 \cdot 6 = 6$

e) $2(a - b) = 2a - 2b$

Solution

SENTENCE	PROPERTY
a) $8 \cdot 5 = 5 \cdot 8$	Commutative property of multiplication: $ab = ba$
b) $5 + (m + n) = (5 + m) + n$	Associative property of addition: $a + (b + c) = (a + b) + c$
c) $14 + (-14) = 0$	Additive inverse property: $a + (-a) = 0$
d) $6 \cdot 1 = 1 \cdot 6 = 6$	Multiplicative identity property: $a \cdot 1 = 1 \cdot a = a$
e) $2(a - b) = 2a - 2b$	Distributive property: $a(b + c) = ab + ac$

Now Try Exercises 49 and 55. ■

❖ Absolute Value

The number line can be used to provide a geometric interpretation of *absolute value*. The **absolute value** of a number a, denoted $|a|$, is its distance from 0 on the number line. For example, $|-5| = 5$, because the distance of -5 from 0 is 5. Similarly, $\left|\frac{3}{4}\right| = \frac{3}{4}$, because the distance of $\frac{3}{4}$ from 0 is $\frac{3}{4}$.

Absolute Value

For any real number a,

$$|a| = \begin{cases} a, & \text{if } a \geq 0, \\ -a, & \text{if } a < 0. \end{cases}$$

When a is nonnegative, the absolute value of a is a. When a is negative, the absolute value of a is the opposite, or additive inverse, of a. Thus, $|a|$ is never negative; that is, for any real number a, $|a| \geq 0$.

Absolute value can be used to find the distance between two points on the number line.

Distance Between Two Points on the Number Line

For any real numbers a and b, the **distance between a and b** is $|a - b|$, or equivalently, $|b - a|$.

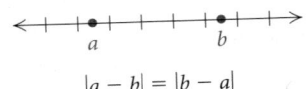

$$|a - b| = |b - a|$$

GCM   **EXAMPLE 3**   Find the distance between -2 and 3.

Solution   The distance is

$$|-2 - 3| = |-5| = 5, \quad \text{or equivalently,}$$
$$|3 - (-2)| = |3 + 2| = |5| = 5.$$

We can also use the absolute-value operation on a graphing calculator to find the distance between two points. On many graphing calculators, absolute value is denoted "abs" and is found in the MATH NUM menu and also in the CATALOG.

```
abs (−2−3)
                        5
abs (3−(−2))
                        5
```

Now Try Exercise 69. ◼

R.1 Exercise Set

In Exercises 1–10, consider the numbers

$$\tfrac{2}{3}, \; 6, \; \sqrt{3}, \; -2.45, \; \sqrt[6]{26}, \; 18.\overline{4}, \; -11, \; \sqrt[3]{27},$$
$$5\tfrac{1}{6}, \; 7.151551555\ldots, \; -\sqrt{35}, \; \sqrt[5]{3}, \; -\tfrac{8}{7}, \; 0, \; \sqrt{16}.$$

1. Which are rational numbers?

2. Which are natural numbers?

3. Which are irrational numbers?

4. Which are integers?

5. Which are whole numbers?

6. Which are real numbers?

7. Which are integers but not natural numbers?

8. Which are integers but not whole numbers?

9. Which are rational numbers but not integers?

10. Which are real numbers but not integers?

Write interval notation. Then graph the interval.

11. $\{x \mid -5 \le x \le 5\}$ **12.** $\{x \mid -2 < x < 2\}$

13. $\{x \mid -3 < x \le -1\}$ **14.** $\{x \mid 4 \le x < 6\}$

15. $\{x \mid x \le -2\}$ **16.** $\{x \mid x > -5\}$

17. $\{x \mid x > 3.8\}$ **18.** $\{x \mid x \ge \sqrt{3}\}$

19. $\{x \mid 7 < x\}$ **20.** $\{x \mid -3 > x\}$

Write interval notation for the graph.

21.

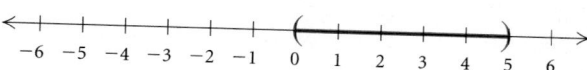

22.

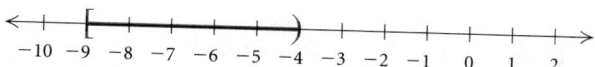

23.

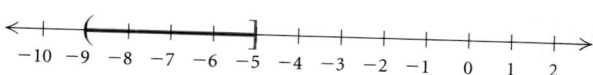

24.

25.

26.

27.

28.

In Exercises 29–46, the following notation is used:
$\mathbb{N}$ = the set of natural numbers, $\mathbb{W}$ = the set of whole numbers, $\mathbb{Z}$ = the set of integers, $\mathbb{Q}$ = the set of rational numbers, $\mathbb{I}$ = the set of irrational numbers, and $\mathbb{R}$ = the set of real numbers. Classify the statement as true or false.

29. $6 \in \mathbb{N}$ **30.** $0 \notin \mathbb{N}$

31. $3.2 \in \mathbb{Z}$ **32.** $-10.\overline{1} \in \mathbb{R}$

33. $-\dfrac{11}{5} \in \mathbb{Q}$ **34.** $-\sqrt{6} \in \mathbb{Q}$

35. $\sqrt{11} \notin \mathbb{R}$ **36.** $-1 \in \mathbb{W}$

37. $24 \notin \mathbb{W}$ **38.** $1 \in \mathbb{Z}$

39. $1.089 \notin \mathbb{I}$ **40.** $\mathbb{N} \subseteq \mathbb{W}$

41. $\mathbb{W} \subseteq \mathbb{Z}$ **42.** $\mathbb{Z} \subseteq \mathbb{N}$

43. $\mathbb{Q} \subseteq \mathbb{R}$ **44.** $\mathbb{Z} \subseteq \mathbb{Q}$

45. $\mathbb{R} \subseteq \mathbb{Z}$ **46.** $\mathbb{Q} \subseteq \mathbb{I}$

Name the property illustrated by the sentence.

47. $3 + y = y + 3$

48. $6(xz) = (6x)z$

49. $-3 \cdot 1 = -3$

50. $4(y - z) = 4y - 4z$

51. $5 \cdot x = x \cdot 5$

52. $7 + (x + y) = (7 + x) + y$

53. $2(a + b) = (a + b)2$

54. $-11 + 11 = 0$

55. $-6(m + n) = -6(n + m)$

56. $t + 0 = t$

57. $8 \cdot \dfrac{1}{8} = 1$

58. $9x + 9y = 9(x + y)$

Simplify.

59. $|-8.15|$

60. $|-14.7|$

61. $|295|$

62. $|-93|$

63. $\left|-\sqrt{97}\right|$

64. $\left|\dfrac{12}{19}\right|$

65. $|0|$

66. $|15|$

67. $\left|\dfrac{5}{4}\right|$

68. $\left|-\sqrt{3}\right|$

Find the distance between the given pair of points on the number line.

69. $-8,\ 14$

70. $-5.2,\ 0$

71. $-9,\ -3$

72. $\dfrac{15}{8},\ \dfrac{23}{12}$

73. $6.7,\ 12.1$

74. $-15,\ -6$

75. $-\dfrac{3}{4},\ \dfrac{15}{8}$

76. $-3.4,\ 10.2$

77. $-7,\ 0$

78. $3,\ 19$

Collaborative Discussion and Writing

To the student and the instructor: The Collaborative Discussion and Writing exercises are meant to be answered with one or more sentences. These exercises can also be discussed and answered collaboratively by the entire class or by small groups. Because of their open-ended nature, the answers to these exercises do not appear at the back of the book. They are denoted by the words "Discussion and Writing."

79. How would you convince a classmate that division is not associative?

80. Under what circumstances is $\sqrt{a}$ a rational number?

Synthesis

To the student and the instructor: The Synthesis exercises found at the end of every exercise set challenge students to combine concepts or skills studied in that section or in preceding parts of the text.

Between any two (different) real numbers there are many other real numbers. Find each of the following. Answers may vary.

81. An irrational number between 0.124 and 0.125

82. A rational number between $-\sqrt{2.01}$ and $-\sqrt{2}$

83. A rational number between $-\dfrac{1}{101}$ and $-\dfrac{1}{100}$

84. An irrational number between $\sqrt{5.99}$ and $\sqrt{6}$

85. The hypotenuse of an isosceles right triangle with legs of length 1 unit can be used to "measure" a value for $\sqrt{2}$ by using the Pythagorean theorem, as shown.

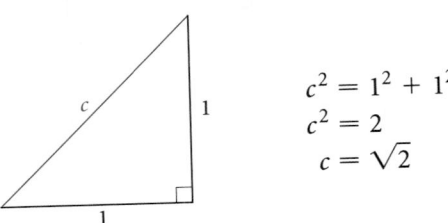

$$c^2 = 1^2 + 1^2$$
$$c^2 = 2$$
$$c = \sqrt{2}$$

Draw a right triangle that could be used to "measure" $\sqrt{10}$ units.

R.2

Integer Exponents, Scientific Notation, and Order of Operations

❖ Simplify expressions with integer exponents.
❖ Solve problems using scientific notation.
❖ Use the rules for order of operations.

❖ Integers as Exponents

When a positive integer is used as an *exponent*, it indicates the number of times that a factor appears in a product. For example, 7^3 means $7 \cdot 7 \cdot 7$ and 5^1 means 5.

> For any positive integer n,
> $$a^n = \underbrace{a \cdot a \cdot a \cdots a}_{n \text{ factors}},$$
> where a is the **base** and n is the **exponent**.

Zero and negative-integer exponents are defined as follows.

> For any nonzero real number a and any integer m,
> $$a^0 = 1 \quad \text{and} \quad a^{-m} = \frac{1}{a^m}.$$

EXAMPLE 1 Simplify each of the following.

a) 6^0

b) $(-3.4)^0$

Solution

a) $6^0 = 1$

b) $(-3.4)^0 = 1$ **Now Try Exercise 7.** ■

EXAMPLE 2 Write each of the following with positive exponents.

a) 4^{-5}

b) $\dfrac{1}{(0.82)^{-7}}$

c) $\dfrac{x^{-3}}{y^{-8}}$

Solution

a) $4^{-5} = \dfrac{1}{4^5}$

b) $\dfrac{1}{(0.82)^{-7}} = (0.82)^{-(-7)} = (0.82)^7$

c) $\dfrac{x^{-3}}{y^{-8}} = x^{-3} \cdot \dfrac{1}{y^{-8}} = \dfrac{1}{x^3} \cdot y^8 = \dfrac{y^8}{x^3}$

Now Try Exercise 1. ■

The results in Example 2 can be generalized as follows.

For any nonzero numbers a and b and any integers m and n,

$$\frac{a^{-m}}{b^{-n}} = \frac{b^n}{a^m}.$$

(A factor can be moved to the other side of the fraction bar if the sign of the exponent is changed.)

EXAMPLE 3 Write an equivalent expression without negative exponents:

$$\frac{x^{-3}y^{-8}}{z^{-10}}.$$

Solution Since each exponent is negative, we move each factor to the other side of the fraction bar and change the sign of each exponent:

$$\frac{x^{-3}y^{-8}}{z^{-10}} = \frac{z^{10}}{x^3y^8}.$$

Now Try Exercise 5. ◼

The following properties of exponents can be used to simplify expressions.

Properties of Exponents

For any real numbers a and b and any integers m and n, assuming 0 is not raised to a nonpositive power:

$a^m \cdot a^n = a^{m+n}$ Product rule

$\dfrac{a^m}{a^n} = a^{m-n}$ $(a \neq 0)$ Quotient rule

$(a^m)^n = a^{mn}$ Power rule

$(ab)^m = a^m b^m$ Raising a product to a power

$\left(\dfrac{a}{b}\right)^m = \dfrac{a^m}{b^m}$ $(b \neq 0)$ Raising a quotient to a power

EXAMPLE 4 Simplify each of the following.

a) $y^{-5} \cdot y^3$

b) $\dfrac{48x^{12}}{16x^4}$

c) $(t^{-3})^5$

d) $(2s^{-2})^5$

e) $\left(\dfrac{45x^{-4}y^2}{9z^{-8}}\right)^{-3}$

Solution

a) $y^{-5} \cdot y^3 = y^{-5+3} = y^{-2}$, or $\dfrac{1}{y^2}$

b) $\dfrac{48x^{12}}{16x^4} = \dfrac{48}{16}x^{12-4} = 3x^8$

c) $(t^{-3})^5 = t^{-3 \cdot 5} = t^{-15}$, or $\dfrac{1}{t^{15}}$

d) $(2s^{-2})^5 = 2^5(s^{-2})^5 = 32s^{-10}$, or $\dfrac{32}{s^{10}}$

e) $\left(\dfrac{45x^{-4}y^2}{9z^{-8}}\right)^{-3} = \left(\dfrac{5x^{-4}y^2}{z^{-8}}\right)^{-3}$

$$= \dfrac{5^{-3}x^{12}y^{-6}}{z^{24}} = \dfrac{x^{12}}{5^3y^6z^{24}}, \text{ or } \dfrac{x^{12}}{125y^6z^{24}}$$

Now Try Exercises 15 and 39. ▪

❋ Scientific Notation

We can use scientific notation to name very large and very small positive numbers and to perform computations.

> **Scientific Notation**
>
> **Scientific notation** for a number is an expression of the type
>
> $$N \times 10^m,$$
>
> where $1 \le N < 10$, N is in decimal notation, and m is an integer.

Keep in mind that in scientific notation positive exponents are used for numbers greater than or equal to 10 and negative exponents for numbers between 0 and 1.

EXAMPLE 5 *Business Travel.* In a recent year, Americans made about 151,700,000 business trips that took them at least 50 mi from home (*Source*: Travel Industry Association of America). Convert the number 151,700,000 to scientific notation.

Solution We want the decimal point to be positioned between the first 1 and the 5, so we move it 8 places to the left. Since the number to be converted is greater than 10, the exponent must be positive. Thus we have

$$151,700,000 = 1.517 \times 10^8.$$

Now Try Exercise 51. ▪

EXAMPLE 6 *Mass of a Neutron.* The mass of a neutron is about 0.0000000000000000000000000167 kg. Convert this number to scientific notation.

Solution We want the decimal point to be positioned between the 1 and the 6, so we move it 27 places to the right. Since the number to be converted is between 0 and 1, the exponent must be negative.

$$0.0000000000000000000000000167 = 1.67 \times 10^{-27}$$

Now Try Exercise 59. ■

EXAMPLE 7 Convert each of the following to decimal notation.

a) 7.632×10^{-4} **b)** 9.4×10^{5}

Solution

a) The exponent is negative, so the number is between 0 and 1. We move the decimal point 4 places to the left.

$$7.632 \times 10^{-4} = 0.0007632$$

b) The exponent is positive, so the number is greater than 10. We move the decimal point 5 places to the right.

$$9.4 \times 10^{5} = 940,000$$

Now Try Exercises 61 and 63. ■

GCM   **EXAMPLE 8** *Distance to a Star.* The nearest star, Alpha Centauri C, is about 4.22 light-years from Earth. One **light-year** is the distance that light travels in one year and is about 5.88×10^{12} miles. How many miles is it from Earth to Alpha Centauri C? Express your answer in scientific notation.

Solution

$$\begin{aligned}
4.22 \times (5.88 \times 10^{12}) &= (4.22 \times 5.88) \times 10^{12} \\
&= 24.8136 \times 10^{12} \quad \text{This is not scientific notation} \\
&\qquad\qquad\qquad\qquad\quad \text{because } 24.8136 + 10. \\
&= (2.48136 \times 10^{1}) \times 10^{12} \\
&= 2.48136 \times (10^{1} \times 10^{12}) \\
&= 2.48136 \times 10^{13} \text{ miles} \quad \text{Writing scientific} \\
&\qquad\qquad\qquad\qquad\qquad \text{notation}
\end{aligned}$$

Now Try Exercise 79. ■

Most calculators make use of scientific notation. For example, the number 48,000,000,000,000 might be expressed as shown on the left below. The computation in Example 8 can be performed on a calculator as shown on the right below.

✺ Order of Operations

Recall that to simplify the expression $3 + 4 \cdot 5$, first we multiply 4 and 5 to get 20 and then add 3 to get 23. Mathematicians have agreed on the following procedure, or rules for order of operations.

Rules for Order of Operations

1. Do all calculations within grouping symbols before operations outside. When nested grouping symbols are present, work from the inside out.
2. Evaluate all exponential expressions.
3. Do all multiplications and divisions in order from left to right.
4. Do all additions and subtractions in order from left to right.

GCM **EXAMPLE 9** Calculate each of the following.

a) $8(5 - 3)^3 - 20$

b) $\dfrac{10 \div (8 - 6) + 9 \cdot 4}{2^5 + 3^2}$

Solution

a)
$$
\begin{aligned}
8(5 - 3)^3 - 20 &= 8 \cdot 2^3 - 20 \qquad &&\text{Doing the calculation within} \\
&&&\text{parentheses} \\
&= 8 \cdot 8 - 20 \qquad &&\text{Evaluating the exponential expression} \\
&= 64 - 20 \qquad &&\text{Multiplying} \\
&= 44 \qquad &&\text{Subtracting}
\end{aligned}
$$

b)
$$
\frac{10 \div (8 - 6) + 9 \cdot 4}{2^5 + 3^2} = \frac{10 \div 2 + 9 \cdot 4}{32 + 9}
$$
$$
= \frac{5 + 36}{41} = \frac{41}{41} = 1
$$

Note that fraction bars act as grouping symbols. That is, the given expression is equivalent to $[10 \div (8 - 6) + 9 \cdot 4] \div (2^5 + 3^2)$.

We can also enter these computations on a graphing calculator as shown below.

```
8(5−3)^3−20
                        44
(10/(8−6)+9∗4)/(2^5+3²)
                         1
```

Now Try Exercises 87 and 91. ■

To confirm that it is essential to include parentheses around the numerator and around the denominator when the computation in Example 9(b) is entered in a calculator, enter the computation without using these parentheses. Observe that the result is 15.125 rather than 1.

GCM **EXAMPLE 10** *Compound Interest.* If a principal P is invested at an interest rate r, compounded n times per year, in t years it will grow to an amount A given by

$$A = P\left(1 + \frac{r}{n}\right)^{nt}.$$

Suppose that $1250 is invested at 4.6% interest, compounded quarterly. How much is in the account at the end of 8 years?

Solution We have $P = 1250$, $r = 4.6\%$, or 0.046, $n = 4$, and $t = 8$. Substituting, we find that the amount in the account at the end of 8 years is given by

$$A = 1250\left(1 + \frac{0.046}{4}\right)^{4 \cdot 8}.$$

Next, we evaluate this expression:

$$A = 1250(1 + 0.0115)^{4 \cdot 8} \quad \text{Dividing}$$
$$= 1250(1.0115)^{4 \cdot 8} \quad \text{Adding}$$
$$= 1250(1.0115)^{32} \quad \text{Multiplying in the exponent}$$
$$\approx 1250(1.441811175) \quad \text{Evaluating the exponential expression}$$
$$\approx 1802.263969 \quad \text{Multiplying}$$
$$\approx 1802.26 \quad \text{Rounding to the nearest cent}$$

The amount in the account at the end of 8 years is $1802.26. The TVM SOLVER option in the FINANCE APP on a graphing calculator can be used to do this computation.

```
N=32
I%=4.6
PV=-1250
PMT=0
FV=1802.263969
P/Y=4
C/Y=4
PMT: END BEGIN
```

Now Try Exercise 93. ■

R.2 Exercise Set

Write an equivalent expression without negative exponents.

1. 3^{-7}

2. $\dfrac{1}{(5.9)^{-4}}$

3. $\dfrac{x^{-5}}{y^{-4}}$

4. $\dfrac{a^{-2}}{b^{-8}}$

5. $\dfrac{m^{-1}n^{-12}}{t^{-6}}$

6. $\dfrac{x^{-9}y^{-17}}{z^{-11}}$

Simplify.

7. 23^0

8. $\left(-\dfrac{2}{5}\right)^0$

9. $z^0 \cdot z^7$

10. $x^{10} \cdot x^0$

11. $5^8 \cdot 5^{-6}$

12. $6^2 \cdot 6^{-7}$

13. $m^{-5} \cdot m^5$

14. $n^9 \cdot n^{-9}$

15. $y^3 \cdot y^{-7}$

16. $b^{-4} \cdot b^{12}$

17. $3^{-3} \cdot 3^8 \cdot 3$

18. $6^7 \cdot 6^{-10} \cdot 6^2$

19. $2x^3 \cdot 3x^2$

20. $3y^4 \cdot 4y^3$

21. $(-3a^{-5})(5a^{-7})$

22. $(-6b^{-4})(2b^{-7})$

23. $(5a^2b)(3a^{-3}b^4)$

24. $(4xy^2)(3x^{-4}y^5)$

25. $(6x^{-3}y^5)(-7x^2y^{-9})$

26. $(8ab^7)(-7a^{-5}b^2)$

27. $(2x)^4(3x)^3$

28. $(4y)^2(3y)^3$

29. $(-2n)^3(5n)^2$

30. $(2x)^5(3x)^2$

31. $\dfrac{y^{35}}{y^{31}}$

32. $\dfrac{x^{26}}{x^{13}}$

33. $\dfrac{b^{-7}}{b^{12}}$

34. $\dfrac{a^{-18}}{a^{-13}}$

35. $\dfrac{x^2y^{-2}}{x^{-1}y}$

36. $\dfrac{x^3y^{-3}}{x^{-1}y^2}$

37. $\dfrac{32x^{-4}y^3}{4x^{-5}y^8}$

38. $\dfrac{20a^5b^{-2}}{5a^7b^{-3}}$

39. $(2x^2y)^4$

40. $(3ab^5)^3$

41. $(-2x^3)^5$

42. $(-3x^2)^4$

43. $(-5c^{-1}d^{-2})^{-2}$

44. $(-4x^{-5}z^{-2})^{-3}$

45. $(3m^4)^3(2m^{-5})^4$

46. $(4n^{-1})^2(2n^3)^3$

47. $\left(\dfrac{2x^{-3}y^7}{z^{-1}}\right)^3$

48. $\left(\dfrac{3x^5y^{-8}}{z^{-2}}\right)^4$

49. $\left(\dfrac{24a^{10}b^{-8}c^7}{12a^6b^{-3}c^5}\right)^{-5}$

50. $\left(\dfrac{125p^{12}q^{-14}r^{22}}{25p^8q^6r^{-15}}\right)^{-4}$

Convert to scientific notation.

51. 16,500,000

52. 359,000

53. 0.000000437

54. 0.0056

55. 234,600,000,000

56. 8,904,000,000

57. 0.00104

58. 0.00000000514

59. One cubic inch is approximately equal to 0.000016 m³.

60. The United States government collected $1,137,000,000,000 in individual income taxes in a recent year (*Source*: U.S. Internal Revenue Service).

Convert to decimal notation.

61. 7.6×10^5

62. 3.4×10^{-6}

63. 1.09×10^{-7}

64. 5.87×10^8

65. 3.496×10^{10}

66. 8.409×10^{11}

67. 5.41×10^{-8}

68. 6.27×10^{-10}

69. The amount of solid waste generated in the United States in a recent year was 2.319×10^8 tons (*Source*: Franklin Associates, Ltd.).

70. The mass of a proton is about 1.67×10^{-24} g.

Compute. Write the answer using scientific notation.

71. $(4.2 \times 10^7)(3.2 \times 10^{-2})$

72. $(8.3 \times 10^{-15})(7.7 \times 10^4)$

73. $(2.6 \times 10^{-18})(8.5 \times 10^7)$

74. $(6.4 \times 10^{12})(3.7 \times 10^{-5})$

75. $\dfrac{6.4 \times 10^{-7}}{8.0 \times 10^6}$ **76.** $\dfrac{1.1 \times 10^{-40}}{2.0 \times 10^{-71}}$

77. $\dfrac{1.8 \times 10^{-3}}{7.2 \times 10^{-9}}$ **78.** $\dfrac{1.3 \times 10^4}{5.2 \times 10^{10}}$

Solve. Write the answer using scientific notation.

79. *Nanowires.* A **nanometer** is 0.000000001 m. Scientists have developed optical nanowires to transmit light waves short distances. A nanowire with a diameter of 360 nanometers has been used in experiments on the transmission of light (*Source*: *The New York Times*, January 29, 2004). Find the diameter of such a wire in meters.

80. *Lost Luggage.* In 2005, airlines worldwide lost a record 30 million pieces of luggage (*Source*: SITA). On average, how many pieces of luggage were lost each day of the year? (Use 1 year = 365 days.)

81. *Chesapeake Bay Bridge-Tunnel.* The 17.6-mile-long Chesapeake Bay Bridge-Tunnel was completed in 1964. Construction costs were $210 million. Find the average cost per mile.

82. *Personal Space in Hong Kong.* The area of Hong Kong is 412 square miles. It is estimated that the population of Hong Kong will be 9,600,000 in 2050. Find the number of square miles of land per person in 2050.

83. *Distance to Pluto.* The distance from Earth to the sun is defined as 1 **astronomical unit**, or AU. It is about 93 million miles. The average distance from Earth to Pluto is 39 AUs. Find this distance in miles.

84. *Parsecs.* One **parsec** is about 3.26 light-years and 1 light-year is about 5.88×10^{12} miles. Find the number of miles in 1 parsec.

85. *Nuclear Disintegration.* One gram of radium produces 37 billion disintegrations per second. How many disintegrations are produced in 1 hour?

86. *Length of Earth's Orbit.* The average distance from Earth to the sun is 93 million miles. About how far does Earth travel in a yearly orbit? (Assume a circular orbit.)

Calculate.

87. $5 \cdot 3 + 8 \cdot 3^2 + 4(6 - 2)$

88. $5[3 - 8 \cdot 3^2 + 4 \cdot 6 - 2]$

89. $16 \div 4 \cdot 4 \div 2 \cdot 256$

90. $2^6 \cdot 2^{-3} \div 2^{10} \div 2^{-8}$

91. $\dfrac{4(8 - 6)^2 - 4 \cdot 3 + 2 \cdot 8}{3^1 + 19^0}$

92. $\dfrac{[4(8-6)^2+4](3-2\cdot 8)}{2^2(2^3+5)}$

Compound Interest. *Use the compound interest formula from Example 10 in Exercises 93–96. Round to the nearest cent.*

93. Suppose that $3225 is invested at 5.1%, compounded semiannually. How much is in the account at the end of 4 years?

94. Suppose that $7550 is invested at 4.8%, compounded semiannually. How much is in the account at the end of 5 years?

95. Suppose that $4100 is invested at 4.3%, compounded quarterly. How much is in the account at the end of 6 years?

96. Suppose that $4875 is invested at 5.8%, compounded quarterly. How much is in the account at the end of 9 years?

Collaborative Discussion and Writing

97. Are the parentheses necessary in the expression $4 \cdot 25 \div (10 - 5)$? Why or why not?

98. Is $x^{-2} < x^{-1}$ for any negative value(s) of x? Why or why not?

Synthesis

Savings Plan. *The formula*

$$S = P\left[\dfrac{\left(1+\dfrac{r}{12}\right)^{12\cdot t}-1}{\dfrac{r}{12}}\right]$$

gives the amount S accumulated in a savings plan when a deposit of P dollars is made each month for t years in an account with interest rate r, compounded monthly. Use this formula for Exercises 99–102.

99. James deposits $250 in a retirement account each month beginning at age 40. If the investment earns 5% interest, compounded monthly, how much will have accumulated in the account when he retires 27 years later?

100. Kayla deposits $100 in a retirement account each month beginning at age 25. If the investment earns 4% interest, compounded monthly, how much will have accumulated in the account when she retires at age 65?

101. Sue wants to establish a college fund for her newborn daughter that will have accumulated $120,000 at the end of 18 years. If she can count on an interest rate of 6%, compounded monthly, how much should she deposit each month to accomplish this?

102. Lamont wants to have $200,000 accumulated in a retirement account by age 70. If he starts making monthly deposits to the plan at age 30 and can count on an interest rate of 4.5%, compounded monthly, how much should he deposit each month in order to accomplish this?

Simplify. Assume that all exponents are integers, all denominators are nonzero, and zero is not raised to a nonpositive power.

103. $(x^t \cdot x^{3t})^2$

104. $(x^y \cdot x^{-y})^3$

105. $(t^{a+x} \cdot t^{x-a})^4$

106. $(m^{x-b} \cdot n^{x+b})^x (m^b n^{-b})^x$

107. $\left[\dfrac{(3x^a y^b)^3}{(-3x^a y^b)^2}\right]^2$

108. $\left[\left(\dfrac{x^r}{y^t}\right)^2 \left(\dfrac{x^{2r}}{y^{4t}}\right)^{-2}\right]^{-3}$

R.3

Addition, Subtraction, and Multiplication of Polynomials

❖ Identify the terms, the coefficients, and the degree of a polynomial.

❖ Add, subtract, and multiply polynomials.

❖ Polynomials

Polynomials are a type of algebraic expression that you will often encounter in your study of algebra. Some examples of polynomials are

$$3x - 4y, \quad 5y^3 - \tfrac{7}{3}y^2 + 3y - 2, \quad -2.3a^4, \quad \text{and} \quad z^6 - \sqrt{5}.$$

All but the first are polynomials in one variable.

Polynomials in One Variable

A **polynomial in one variable** is any expression of the type

$$a_n x^n + a_{n-1}x^{n-1} + \cdots + a_2 x^2 + a_1 x + a_0,$$

where n is a nonnegative integer and $a_n, \ldots, a_0$ are real numbers, called **coefficients**. The parts of a polynomial separated by plus signs are called **terms**. The **leading coefficient** is a_n, and the **constant term** is a_0. If $a_n \neq 0$, the **degree** of the polynomial is n. The polynomial is said to be written in **descending order**, because the exponents decrease from left to right.

EXAMPLE 1 Identify the terms of the polynomial

$$2x^4 - 7.5x^3 + x - 12.$$

Solution Writing plus signs between the terms, we have

$$2x^4 - 7.5x^3 + x - 12 = 2x^4 + (-7.5x^3) + x + (-12),$$

so the terms are

$$2x^4, \quad -7.5x^3, \quad x, \quad \text{and} \quad -12.$$

A polynomial consisting of only a nonzero constant term, like 23, has degree 0. It is agreed that the polynomial consisting only of 0 has *no* degree.

EXAMPLE 2   Find the degree of each polynomial.

a) $2x^3 - 9$ **b)** $y^2 - \frac{3}{2} + 5y^4$ **c)** 7

Solution

POLYNOMIAL	DEGREE
a) $2x^3 - 9$	3
b) $y^2 - \frac{3}{2} + 5y^4 = 5y^4 + y^2 - \frac{3}{2}$	4
c) $7 = 7x^0$	0

Now Try Exercise 1. ▨

Algebraic expressions like $3ab^3 - 8$ and $5x^4y^2 - 3x^3y^8 + 7xy^2 + 6$ are **polynomials in several variables**. The **degree of a term** is the sum of the exponents of the variables in that term. The **degree of a polynomial** is the degree of the term of highest degree.

EXAMPLE 3   Find the degree of the polynomial

$$7ab^3 - 11a^2b^4 + 8.$$

Solution   The degrees of the terms of $7ab^3 - 11a^2b^4 + 8$ are 4, 6, and 0, respectively, so the degree of the polynomial is 6.   **Now Try Exercise 3.** ▨

A polynomial with just one term, like $-9y^6$, is a **monomial**. If a polynomial has two terms, like $x^2 + 4$, it is a **binomial**. A polynomial with three terms, like $4x^2 - 4xy + 1$, is a **trinomial**.

Expressions like

$$2x^2 - 5x + \frac{3}{x}, \qquad 9 - \sqrt{x}, \quad \text{and} \quad \frac{x + 1}{x^4 + 5}$$

are not polynomials, because they cannot be written in the form $a_nx^n + a_{n-1}x^{n-1} + \cdots + a_1x + a_0$, where the exponents are all nonnegative integers and the coefficients are all real numbers.

❋ Addition and Subtraction

If two terms of an expression have the same variables raised to the same powers, they are called **like terms**, or **similar terms**. We can **combine**, or **collect**, **like terms** using the distributive property. For example, $3y^2$ and $5y^2$ are like terms and

$$3y^2 + 5y^2 = (3 + 5)y^2$$
$$= 8y^2.$$

We add or subtract polynomials by combining like terms.

EXAMPLE 4 Add or subtract each of the following.

a) $(-5x^3 + 3x^2 - x) + (12x^3 - 7x^2 + 3)$

b) $(6x^2y^3 - 9xy) - (5x^2y^3 - 4xy)$

Solution

a) $(-5x^3 + 3x^2 - x) + (12x^3 - 7x^2 + 3)$

$\quad = (-5x^3 + 12x^3) + (3x^2 - 7x^2) - x + 3$ Rearranging using the commutative and associative properties

$\quad = (-5 + 12)x^3 + (3 - 7)x^2 - x + 3$ Using the distributive property

$\quad = 7x^3 - 4x^2 - x + 3$

b) We can subtract by adding an opposite:

$\quad (6x^2y^3 - 9xy) - (5x^2y^3 - 4xy)$

$\quad = (6x^2y^3 - 9xy) + (-5x^2y^3 + 4xy)$ Adding the opposite of $5x^2y^3 - 4xy$

$\quad = 6x^2y^3 - 9xy - 5x^2y^3 + 4xy$

$\quad = x^2y^3 - 5xy.$ Combining like terms

Now Try Exercises 5 and 9. ◼

❖ Multiplication

Multiplication of polynomials is based on the distributive property—for example,

$(x + 4)(x + 3) = x(x + 3) + 4(x + 3)$ Using the distributive property

$\quad = x^2 + 3x + 4x + 12$ Using the distributive property two more times

$\quad = x^2 + 7x + 12.$ Combining like terms

In general, to multiply two polynomials, we multiply each term of one by each term of the other and add the products.

EXAMPLE 5 Multiply: $(4x^4y - 7x^2y + 3y)(2y - 3x^2y)$.

Solution We have

$(4x^4y - 7x^2y + 3y)(2y - 3x^2y)$

$\quad = 4x^4y(2y - 3x^2y) - 7x^2y(2y - 3x^2y) + 3y(2y - 3x^2y)$ Using the distributive property

$\quad = 8x^4y^2 - 12x^6y^2 - 14x^2y^2 + 21x^4y^2 + 6y^2 - 9x^2y^2$ Using the distributive property three more times

$\quad = 29x^4y^2 - 12x^6y^2 - 23x^2y^2 + 6y^2.$ Combining like terms

We can also use columns to organize our work, aligning like terms under each other in the products.

$$4x^4y - 7x^2y + 3y$$
$$2y - 3x^2y$$
$$\overline{\hspace{3cm}}$$
$$-12x^6y^2 + 21x^4y^2 - 9x^2y^2 \qquad \text{Multiplying by } -3x^2y$$
$$\underline{\qquad\quad 8x^4y^2 - 14x^2y^2 + 6y^2} \quad \text{Multiplying by } 2y$$
$$-12x^6y^2 + 29x^4y^2 - 23x^2y^2 + 6y^2 \quad \text{Adding}$$

Now Try Exercise 13. ▪

We can find the product of two binomials by multiplying the **F**irst terms, then the **O**uter terms, then the **I**nner terms, then the **L**ast terms. Then we combine like terms, if possible. This procedure is sometimes called **FOIL**.

EXAMPLE 6   Multiply: $(2x - 7)(3x + 4)$.

Solution   We have

$$(2x - 7)(3x + 4) = 6x^2 + 8x - 21x - 28$$
$$= 6x^2 - 13x - 28$$

Now Try Exercise 15. ▪

We can use FOIL to find some special products.

Special Products of Binomials

$(A + B)^2 = A^2 + 2AB + B^2$ Square of a sum
$(A - B)^2 = A^2 - 2AB + B^2$ Square of a difference
$(A + B)(A - B) = A^2 - B^2$ Product of a sum and a difference

EXAMPLE 7   Multiply each of the following.

a) $(4x + 1)^2$ **b)** $(3y^2 - 2)^2$ **c)** $(x^2 + 3y)(x^2 - 3y)$

Solution

a) $(4x + 1)^2 = (4x)^2 + 2 \cdot 4x \cdot 1 + 1^2 = 16x^2 + 8x + 1$
b) $(3y^2 - 2)^2 = (3y^2)^2 - 2 \cdot 3y^2 \cdot 2 + 2^2 = 9y^4 - 12y^2 + 4$
c) $(x^2 + 3y)(x^2 - 3y) = (x^2)^2 - (3y)^2 = x^4 - 9y^2$

Now Try Exercises 23 and 33. ▪

Division of polynomials is discussed in Section 4.3.

R.3   Exercise Set

Determine the terms and the degree of the polynomial.

1. $7x^3 - 4x^2 + 8x + 5$

2. $-3n^4 - 6n^3 + n^2 + 2n - 1$

3. $3a^4b - 7a^3b^3 + 5ab - 2$

4. $6p^3q^2 - p^2q^4 - 3pq^2 + 5$

Perform the indicated operations.

5. $(3ab^2 - 4a^2b - 2ab + 6) + (-ab^2 - 5a^2b + 8ab + 4)$

6. $(-6m^2n + 3mn^2 - 5mn + 2) + (4m^2n + 2mn^2 - 6mn - 9)$

7. $(2x + 3y + z - 7) + (4x - 2y - z + 8) + (-3x + y - 2z - 4)$

8. $(2x^2 + 12xy - 11) + (6x^2 - 2x + 4) + (-x^2 - y - 2)$

9. $(3x^2 - 2x - x^3 + 2) - (5x^2 - 8x - x^3 + 4)$

10. $(5x^2 + 4xy - 3y^2 + 2) - (9x^2 - 4xy + 2y^2 - 1)$

11. $(x^4 - 3x^2 + 4x) - (3x^3 + x^2 - 5x + 3)$

12. $(2x^4 - 3x^2 + 7x) - (5x^3 + 2x^2 - 3x + 5)$

13. $(a - b)(2a^3 - ab + 3b^2)$

14. $(n + 1)(n^2 - 6n - 4)$

15. $(y - 3)(y + 5)$

16. $(z + 4)(z - 2)$

17. $(x + 6)(x + 3)$

18. $(a - 8)(a - 1)$

19. $(2a + 3)(a + 5)$

20. $(3b + 1)(b - 2)$

21. $(2x + 3y)(2x + y)$

22. $(2a - 3b)(2a - b)$

23. $(x + 3)^2$

24. $(z + 6)^2$

25. $(y - 5)^2$

26. $(x - 4)^2$

27. $(5x - 3)^2$

28. $(3x - 2)^2$

29. $(2x + 3y)^2$

30. $(5x + 2y)^2$

31. $(2x^2 - 3y)^2$

32. $(4x^2 - 5y)^2$

33. $(n + 6)(n - 6)$

34. $(m + 1)(m - 1)$

35. $(3y + 4)(3y - 4)$

36. $(2x - 7)(2x + 7)$

37. $(3x - 2y)(3x + 2y)$

38. $(3x + 5y)(3x - 5y)$

39. $(2x + 3y + 4)(2x + 3y - 4)$

40. $(5x + 2y + 3)(5x + 2y - 3)$

41. $(x + 1)(x - 1)(x^2 + 1)$

42. $(y - 2)(y + 2)(y^2 + 4)$

Collaborative Discussion and Writing

43. Is the sum of two polynomials of degree n always a polynomial of degree n? Why or why not?

44. Explain how you would convince a classmate that $(A + B)^2 \neq A^2 + B^2$.

Synthesis

Multiply. Assume that all exponents are natural numbers.

45. $(a^n + b^n)(a^n - b^n)$

46. $(t^a + 4)(t^a - 7)$

47. $(a^n + b^n)^2$

48. $(x^{3m} - t^{5n})^2$

49. $(x - 1)(x^2 + x + 1)(x^3 + 1)$

50. $[(2x - 1)^2 - 1]^2$

51. $(x^{a-b})^{a+b}$

52. $(t^{m+n})^{m+n} \cdot (t^{m-n})^{m-n}$

53. $(a + b + c)^2$

Factoring

❖ Factor polynomials by removing a common factor.
❖ Factor polynomials by grouping.
❖ Factor trinomials of the type $x^2 + bx + c$.
❖ Factor trinomials of the type $ax^2 + bx + c$, $a \neq 1$, using the FOIL method and the grouping method.
❖ Factor special products of polynomials.

To factor a polynomial, we do the reverse of multiplying; that is, we find an equivalent expression that is written as a product.

❖ Terms with Common Factors

When a polynomial is to be factored, we should always look first to factor out a factor that is common to all the terms using the distributive property. We generally look for the constant common factor with the largest absolute value and for variables with the largest exponent common to all the terms. In this sense, we factor out the "largest" common factor.

EXAMPLE 1 Factor each of the following.

a) $15 + 10x - 5x^2$ **b)** $12x^2y^2 - 20x^3y$

Solution

a) $15 + 10x - 5x^2 = 5 \cdot 3 + 5 \cdot 2x - 5 \cdot x^2 = 5(3 + 2x - x^2)$

We can always check a factorization by multiplying:

$$5(3 + 2x - x^2) = 15 + 10x - 5x^2.$$

b) There are several factors common to the terms of $12x^2y^2 - 20x^3y$, but $4x^2y$ is the "largest" of these.

$$12x^2y^2 - 20x^3y = 4x^2y \cdot 3y - 4x^2y \cdot 5x$$
$$= 4x^2y(3y - 5x)$$

Now Try Exercise 3. ◼

❖ Factoring by Grouping

In some polynomials, pairs of terms have a common binomial factor that can be removed in a process called **factoring by grouping**.

EXAMPLE 2 Factor: $x^3 + 3x^2 - 5x - 15$.

Solution We have

$$x^3 + 3x^2 - 5x - 15 = (x^3 + 3x^2) + (-5x - 15)$$ Grouping. Each group of terms has a common factor.

$$= x^2(x + 3) - 5(x + 3)$$ Factoring a common factor out of each group

$$= (x + 3)(x^2 - 5).$$ Factoring out the common binomial factor

Now Try Exercise 9. ■

❋ Trinomials of the Type $x^2 + bx + c$

Some trinomials can be factored into the product of two binomials. To factor a trinomial of the form $x^2 + bx + c$, we look for binomial factors of the form

$$(x + p)(x + q),$$

where $p \cdot q = c$ and $p + q = b$. That is, we look for two numbers p and q whose sum is the coefficient of the middle term of the polynomial, b, and whose product is the constant term, c.

When we factor any polynomial, we should always check first to determine whether there is a factor common to all the terms. If there is, we factor it out first.

EXAMPLE 3 Factor: $x^2 + 5x + 6$.

Solution First, we look for a common factor. There is none. Next, we look for two numbers whose product is 6 and whose sum is 5. Since the constant term, 6, and the coefficient of the middle term, 5, are both positive, we look for a factorization of 6 in which both factors are positive.

Pairs of Factors	Sums of Factors
1, 6	7
2, 3	5 ←

The numbers we need are 2 and 3.

The factorization is $(x + 2)(x + 3)$. We have

$$x^2 + 5x + 6 = (x + 2)(x + 3).$$

We can check this by multiplying:

$$(x + 2)(x + 3) = x^2 + 3x + 2x + 6 = x^2 + 5x + 6.$$

Now Try Exercise 19. ■

EXAMPLE 4   Factor: $x^4 - 6x^3 + 8x^2$.

Solution   First, we look for a common factor. Each term has a factor of x^2, so we factor it out first:

$$x^4 - 6x^3 + 8x^2 = x^2(x^2 - 6x + 8).$$

Now we consider the trinomial $x^2 - 6x + 8$. We look for two numbers whose product is 8 and whose sum is -6. Since the constant term, 8, is positive and the coefficient of the middle term, -6, is negative, we look for a factorization of 8 in which both factors are negative.

Pairs of Factors	Sums of Factors
$-1, -8$	-9
$-2, -4$	-6 ←

The numbers we need are -2 and -4.

The factorization of $x^2 - 6x + 8$ is $(x - 2)(x - 4)$. We must also include the common factor that we factored out earlier. Thus we have

$$x^4 - 6x^3 + 8x^2 = x^2(x - 2)(x - 4).$$

Now Try Exercise 31. ■

EXAMPLE 5   Factor: $x^2 + xy - 12y^2$.

Solution   Our thinking is much the same as if we were factoring $x^2 + x - 12$. We look for factors of -12 whose sum is the coefficient of the xy-term, 1. Since the last term, $-12y^2$, is negative, one factor will be positive and the other will be negative.

Pairs of Factors	Sums of Factors
$-1, \quad 12$	11
$1, -12$	-11
$-2, \quad 6$	4
$2, -6$	-4
$-3, \quad 4$	1 ←
$3, -4$	-1

The numbers we need are -3 and 4.

We might have observed at the outset that since the sum of the factors must be 1, a positive number, we need consider only pairs of factors for which the positive factor has the greater absolute value. Thus only the pairs -1 and 12, -2 and 6, -3 and 4 need have been considered.

The factorization is

$$(x - 3y)(x + 4y).$$

Now Try Exercise 25. ■

❖ Trinomials of the Type $ax^2 + bx + c$, $a \neq 1$

We consider two methods for factoring trinomials of the type $ax^2 + bx + c$, $a \neq 1$.

The FOIL Method

We first consider the **FOIL method** for factoring trinomials of the type $ax^2 + bx + c$, $a \neq 1$. Consider the following multiplication.

$$(3x + 2)(4x + 5) = 12x^2 + \underline{15x + 8x} + 10$$

$$= 12x^2 + 23x + 10$$

To factor $12x^2 + 23x + 10$, we must reverse what we just did. We look for two binomials whose product is this trinomial. The product of the First terms must be $12x^2$. The product of the Outer terms plus the product of the Inner terms must be $23x$. The product of the Last terms must be 10. We know from the preceding discussion that the answer is $(3x + 2)(4x + 5)$. In general, however, finding such an answer involves trial and error. We use the following method.

To factor trinomials of the type $ax^2 + bx + c$, $a \neq 1$, using the **FOIL method**:

1. Factor out the largest common factor.
2. Find two First terms whose product is ax^2:

$$(\blacksquare x + \quad)(\blacksquare x + \quad) = ax^2 + bx + c.$$
$$\text{FOIL}$$

3. Find two Last terms whose product is c:

$$(\ x + \blacksquare)(\ x + \blacksquare) = ax^2 + bx + c.$$
$$\text{FOIL}$$

4. Repeat steps (2) and (3) until a combination is found for which the sum of the Outer product and the Inner product is bx:

$$(\blacksquare x + \blacksquare)(\blacksquare x + \blacksquare) = ax^2 + bx + c.$$
$$\text{FOIL}$$

EXAMPLE 6   Factor: $3x^2 - 10x - 8$.

Solution

1. There is no common factor (other than 1 or -1).
2. Factor the first term, $3x^2$. The only possibility (with positive integer coefficients) is $3x \cdot x$. The factorization, if it exists, must be of the form $(3x + \blacksquare)(x + \blacksquare)$.
3. Next, factor the constant term, -8. The possibilities are $-8(1)$, $8(-1)$, $-2(4)$, and $2(-4)$. The factors can be written in the opposite order as well: $1(-8)$, $-1(8)$, $4(-2)$, and $-4(2)$.
4. Find a pair of factors for which the sum of the outer products and the inner products is the middle term, $-10x$. Each possibility should be checked by multiplying. Some trials show that the desired factorization is $(3x + 2)(x - 4)$.

Now Try Exercise 35. ■

The Grouping Method

The second method for factoring trinomials of the type $ax^2 + bx + c$, $a \neq 1$, is known as the **grouping method**, or the **ac-method**.

To factor $ax^2 + bx + c$, $a \neq 1$, using the **grouping method**:

1. Factor out the largest common factor.
2. Multiply the leading coefficient a and the constant c.
3. Try to factor the product ac so that the sum of the factors is b. That is, find integers p and q such that $pq = ac$ and $p + q = b$.
4. Split the middle term. That is, write it as a sum using the factors found in step (3).
5. Factor by grouping.

EXAMPLE 7   Factor: $12x^3 + 10x^2 - 8x$.

Solution

1. Factor out the largest common factor, $2x$:
$$12x^3 + 10x^2 - 8x = 2x(6x^2 + 5x - 4).$$

2. Now consider $6x^2 + 5x - 4$. Multiply the leading coefficient, 6, and the constant, -4: $6(-4) = -24$.

3. Try to factor -24 so that the sum of the factors is the coefficient of the middle term, 5.

Pairs of Factors	Sums of Factors
1, -24	-23
-1, 24	23
2, -12	-10
-2, 12	10
3, -8	-5
-3, 8	5 ← $\quad -3 \cdot 8 = -24; -3 + 8 = 5$
4, -6	-2
-4, 6	2

4. Split the middle term using the numbers found in step (3):

$$5x = -3x + 8x.$$

5. Finally, factor by grouping:

$$\begin{aligned} 6x^2 + 5x - 4 &= 6x^2 - 3x + 8x - 4 \\ &= 3x(2x - 1) + 4(2x - 1) \\ &= (2x - 1)(3x + 4). \end{aligned}$$

Be sure to include the common factor to get the complete factorization of the original trinomial:

$$12x^3 + 10x^2 - 8x = 2x(2x - 1)(3x + 4). \qquad \textbf{Now Try Exercise 45.} \ ■$$

❖ Special Factorizations

We reverse the equation $(A + B)(A - B) = A^2 - B^2$ to factor a **difference of squares**.

$$A^2 - B^2 = (A + B)(A - B)$$

EXAMPLE 8 Factor each of the following.

a) $x^2 - 16$ b) $9a^2 - 25$ c) $6x^4 - 6y^4$

Solution

a) $x^2 - 16 = x^2 - 4^2 = (x + 4)(x - 4)$

b) $9a^2 - 25 = (3a)^2 - 5^2 = (3a + 5)(3a - 5)$

c) $6x^4 - 6y^4 = 6(x^4 - y^4)$
$$= 6[(x^2)^2 - (y^2)^2]$$
$$= 6(x^2 + y^2)(x^2 - y^2) \qquad \text{$x^2 - y^2$ can be factored further.}$$
$$= 6(x^2 + y^2)(x + y)(x - y) \qquad \text{Because none of these factors can be factored further, we have } \textit{factored completely}.$$

Now Try Exercise 49. ■

The rules for squaring binomials can be reversed to factor trinomials that are squares of binomials:

$$A^2 + 2AB + B^2 = (A + B)^2;$$
$$A^2 - 2AB + B^2 = (A - B)^2.$$

EXAMPLE 9 Factor each of the following.

a) $x^2 + 8x + 16$ **b)** $25y^2 - 30y + 9$

Solution

$$A^2 + 2 \cdot A \cdot B + B^2 = (A + B)^2$$
a) $x^2 + 8x + 16 = x^2 + 2 \cdot x \cdot 4 + 4^2 = (x + 4)^2$

$$A^2 - 2 \cdot A \cdot B + B^2 = (A - B)^2$$
b) $25y^2 - 30y + 9 = (5y)^2 - 2 \cdot 5y \cdot 3 + 3^2 = (5y - 3)^2$

Now Try Exercise 57. ■

We can use the following rules to factor a **sum** or a **difference of cubes**:

$$A^3 + B^3 = (A + B)(A^2 - AB + B^2);$$
$$A^3 - B^3 = (A - B)(A^2 + AB + B^2).$$

These rules can be verified by multiplying.

EXAMPLE 10 Factor each of the following.

a) $x^3 + 27$ **b)** $16y^3 - 250$

Solution

a) $x^3 + 27 = x^3 + 3^3$
$$= (x + 3)(x^2 - 3x + 9)$$
b) $16y^3 - 250 = 2(8y^3 - 125)$
$$= 2[(2y)^3 - 5^3]$$
$$= 2(2y - 5)(4y^2 + 10y + 25) \qquad \text{**Now Try Exercise 67.** ■}$$

Not all polynomials can be factored into polynomials with integer coefficients. An example is $x^2 - x + 7$. There are no real factors of 7 whose sum is -1. In such a case, we say that the polynomial is "not factorable," or **prime**.

CONNECTING the CONCEPTS

A Strategy for Factoring

A. Always factor out the largest common factor first.

B. Look at the number of terms.

Two terms: Try factoring as a difference of squares first. Next, try factoring as a sum or a difference of cubes. Do *not* try to factor a *sum* of squares.

Three terms: Try factoring as the square of a binomial. Next, try using the FOIL method or the grouping method for factoring a trinomial.

Four or more terms: Try factoring by grouping and factoring out a common binomial factor.

C. Always *factor completely*. If a factor with more than one term can itself be factored further, do so.

R.4   Exercise Set

Factor out the largest common factor.

1. $3x + 18$

2. $5y - 20$

3. $2z^3 - 8z^2$

4. $12m^2 + 3m^6$

5. $4a^2 - 12a + 16$

6. $6n^2 + 24n - 18$

7. $a(b - 2) + c(b - 2)$

8. $a(x^2 - 3) - 2(x^2 - 3)$

Factor by grouping.

9. $3x^3 - x^2 + 18x - 6$

10. $x^3 + 3x^2 + 6x + 18$

11. $y^3 - y^2 + 2y - 2$

12. $y^3 - y^2 + 3y - 3$

13. $24x^3 - 36x^2 + 72x - 108$

14. $5a^3 - 10a^2 + 25a - 50$

15. $x^3 - x^2 - 5x + 5$

16. $t^3 + 6t^2 - 2t - 12$

17. $a^3 - 3a^2 - 2a + 6$

18. $x^3 - x^2 - 6x + 6$

Factor the trinomial.

19. $w^2 - 7w + 10$

20. $p^2 + 6p + 8$

21. $x^2 + 6x + 5$

22. $x^2 - 8x + 12$

23. $t^2 + 8t + 15$

24. $y^2 + 12y + 27$

25. $x^2 - 6xy - 27y^2$

26. $t^2 - 2t - 15$

27. $2n^2 - 20n - 48$

28. $2a^2 - 2ab - 24b^2$

29. $y^2 - 4y - 21$

30. $m^2 - m - 90$

31. $y^4 - 9y^3 + 14y^2$

32. $3z^3 - 21z^2 + 18z$

33. $2x^3 - 2x^2y - 24xy^2$

34. $a^3b - 9a^2b^2 + 20ab^3$

35. $2n^2 + 9n - 56$

36. $3y^2 + 7y - 20$

37. $12x^2 + 11x + 2$

38. $6x^2 - 7x - 20$

39. $4x^2 + 15x + 9$

40. $2y^2 + 7y + 6$

41. $2y^2 + y - 6$

42. $20p^2 - 23p + 6$

43. $6a^2 - 29ab + 28b^2$

44. $10m^2 + 7mn - 12n^2$

45. $12a^2 - 4a - 16$

46. $12a^2 - 14a - 20$

Factor the difference of squares.

47. $z^2 - 81$

48. $m^2 - 4$

49. $16x^2 - 9$

50. $4z^2 - 81$

51. $6x^2 - 6y^2$

52. $8a^2 - 8b^2$

53. $4xy^4 - 4xz^2$

54. $5x^2y - 5yz^4$

55. $7pq^4 - 7py^4$

56. $25ab^4 - 25az^4$

Factor the square of a binomial.

57. $x^2 + 12x + 36$

58. $y^2 - 6y + 9$

59. $9z^2 - 12z + 4$

60. $4z^2 + 12z + 9$

61. $1 - 8x + 16x^2$

62. $1 + 10x + 25x^2$

63. $a^3 + 24a^2 + 144a$

64. $y^3 - 18y^2 + 81y$

65. $4p^2 - 8pq + 4q^2$

66. $5a^2 - 10ab + 5b^2$

Factor the sum or the difference of cubes.

67. $x^3 + 64$

68. $y^3 - 8$

69. $m^3 - 216$

70. $n^3 + 1$

71. $8t^3 + 8$

72. $2y^3 - 128$

73. $3a^5 - 24a^2$

74. $250z^4 - 2z$

75. $t^6 + 1$

76. $27x^6 - 8$

Factor completely.

77. $18a^2b - 15ab^2$

78. $4x^2y + 12xy^2$

79. $x^3 - 4x^2 + 5x - 20$

80. $z^3 + 3z^2 - 3z - 9$

81. $8x^2 - 32$

82. $6y^2 - 6$

83. $4y^2 - 5$

84. $16x^2 - 7$

85. $m^2 - 9n^2$

86. $25t^2 - 16$

87. $x^2 + 9x + 20$

88. $y^2 + y - 6$

89. $y^2 - 6y + 5$

90. $x^2 - 4x - 21$

91. $2a^2 + 9a + 4$

92. $3b^2 - b - 2$

93. $6x^2 + 7x - 3$

94. $8x^2 + 2x - 15$

95. $y^2 - 18y + 81$

96. $n^2 + 2n + 1$

97. $9z^2 - 24z + 16$

98. $4z^2 + 20z + 25$

99. $x^2y^2 - 14xy + 49$

100. $x^2y^2 - 16xy + 64$

101. $4ax^2 + 20ax - 56a$

102. $21x^2y + 2xy - 8y$

103. $3z^3 - 24$

104. $4t^3 + 108$

105. $16a^7b + 54ab^7$

106. $24a^2x^4 - 375a^8x$

107. $y^3 - 3y^2 - 4y + 12$

108. $p^3 - 2p^2 - 9p + 18$

109. $x^3 - x^2 + x - 1$

110. $x^3 - x^2 - x + 1$

111. $5m^4 - 20$

112. $2x^2 - 288$

113. $2x^3 + 6x^2 - 8x - 24$

114. $3x^3 + 6x^2 - 27x - 54$

115. $4c^2 - 4cd + d^2$

116. $9a^2 - 6ab + b^2$

117. $m^6 + 8m^3 - 20$

118. $x^4 - 37x^2 + 36$

119. $p - 64p^4$

120. $125a - 8a^4$

Collaborative Discussion and Writing

121. Under what circumstances can $A^2 + B^2$ be factored?

122. Explain how the rule for factoring a sum of cubes can be used to factor a difference of cubes.

Synthesis

Factor.

123. $y^4 - 84 + 5y^2$

124. $11x^2 + x^4 - 80$

125. $y^2 - \frac{8}{49} + \frac{2}{7}y$

126. $t^2 - \frac{27}{100} + \frac{3}{5}t$

127. $x^2 + 3x + \frac{9}{4}$

128. $x^2 - 5x + \frac{25}{4}$

129. $x^2 - x + \frac{1}{4}$

130. $x^2 - \frac{2}{3}x + \frac{1}{9}$

131. $(x + h)^3 - x^3$

132. $(x + 0.01)^2 - x^2$

133. $(y - 4)^2 + 5(y - 4) - 24$

134. $6(2p + q)^2 - 5(2p + q) - 25$

Factor. Assume that variables in exponents represent natural numbers.

135. $x^{2n} + 5x^n - 24$

136. $4x^{2n} - 4x^n - 3$

137. $x^2 + ax + bx + ab$

138. $bdy^2 + ady + bcy + ac$

139. $25y^{2m} - (x^{2n} - 2x^n + 1)$

140. $x^{6a} - t^{3b}$

141. $(y - 1)^4 - (y - 1)^2$

142. $x^6 - 2x^5 + x^4 - x^2 + 2x - 1$

R.5   The Basics of Equation Solving

✿ Solve linear equations.
✿ Solve quadratic equations.

An **equation** is a statement that two expressions are equal. To **solve** an equation in one variable is to find all the values of the variable that make the equation true. Each of these numbers is a **solution** of the equation. The set of all solutions of an equation is its **solution set**. Equations that have the same solution set are called **equivalent equations**.

✿ Linear and Quadratic Equations

> A **linear equation in one variable** is an equation that is equivalent to one of the form $ax + b = 0$, where a and b are real numbers and $a \neq 0$.
>
> A **quadratic equation** is an equation that is equivalent to one of the form $ax^2 + bx + c = 0$, where a, b, and c are real numbers and $a \neq 0$.

The following principles allow us to solve many linear and quadratic equations.

Equation-Solving Principles

For any real numbers a, b, and c,

The Addition Principle: If $a = b$ is true, then $a + c = b + c$ is true.

The Multiplication Principle: If $a = b$ is true, then $ac = bc$ is true.

The Principle of Zero Products: If $ab = 0$ is true, then $a = 0$ or $b = 0$, and if $a = 0$ or $b = 0$, then $ab = 0$.

The Principle of Square Roots: If $x^2 = k$, then $x = \sqrt{k}$ or $x = -\sqrt{k}$.

First we consider a linear equation. We will use the addition and multiplication principles to solve it.

EXAMPLE 1   Solve: $2x + 3 = 1 - 6(x - 1)$.

Solution   We begin by using the distributive property to remove the parentheses.

$2x + 3 = 1 - 6(x - 1)$	
$2x + 3 = 1 - 6x + 6$	Using the distributive property
$2x + 3 = 7 - 6x$	Combining like terms
$8x + 3 = 7$	Using the addition principle to add $6x$ on both sides
$8x = 4$	Using the addition principle to add -3, or subtract 3, on both sides
$x = \dfrac{4}{8}$	Using the multiplication principle to multiply by $\frac{1}{8}$, or divide by 8, on both sides
$x = \dfrac{1}{2}$	Simplifying

We check the result in the original equation.

Check:

$$2x + 3 = 1 - 6(x - 1)$$

$$\begin{array}{c|c} 2 \cdot \tfrac{1}{2} + 3 \;\;?\;\; 1 - 6\left(\tfrac{1}{2} - 1\right) & \text{Substituting } \tfrac{1}{2} \text{ for } x \\ 1 + 3 \;\bigm|\; 1 - 6\left(-\tfrac{1}{2}\right) & \\ 4 \;\bigm|\; 1 + 3 & \\ 4 \;\bigm|\; 4 & \text{TRUE} \end{array}$$

The solution is $\frac{1}{2}$.

Now Try Exercise 13. ◼

Now we consider a quadratic equation that can be solved using the principle of zero products.

EXAMPLE 2 Solve: $x^2 - 3x = 4$.

Solution First we write the equation with 0 on one side.

$$x^2 - 3x = 4$$
$$x^2 - 3x - 4 = 0 \qquad \text{Subtracting 4 on both sides}$$
$$(x + 1)(x - 4) = 0 \qquad \text{Factoring}$$
$$x + 1 = 0 \quad \text{or} \quad x - 4 = 0 \qquad \begin{array}{l}\text{Using the principle of}\\ \text{zero products}\end{array}$$
$$x = -1 \quad \text{or} \quad x = 4$$

Check: For -1:

$$\begin{array}{c} x^2 - 3x = 4 \\ \hline (-1)^2 - 3(-1) \;?\; 4 \\ 1 + 3 \quad | \\ 4 \;\bigm|\; 4 \quad \text{TRUE} \end{array}$$

For 4:

$$\begin{array}{c} x^2 - 3x = 4 \\ \hline 4^2 - 3 \cdot 4 \;?\; 4 \\ 16 - 12 \quad | \\ 4 \;\bigm|\; 4 \quad \text{TRUE} \end{array}$$

The solutions are -1 and 4.   **Now Try Exercise 35.** ■

The principle of square roots can be used to solve some quadratic equations, as we see in the next example.

EXAMPLE 3 Solve: $3x^2 - 6 = 0$.

Solution We will use the principle of square roots:

$$3x^2 - 6 = 0$$
$$3x^2 = 6 \qquad \text{Adding 6 on both sides}$$
$$x^2 = 2 \qquad \text{Dividing by 3 on both sides to isolate } x^2$$
$$x = \sqrt{2} \quad \text{or} \quad x = -\sqrt{2}. \qquad \text{Using the principle of square roots}$$

Both numbers check. The solutions are $\sqrt{2}$ and $-\sqrt{2}$, or $\pm\sqrt{2}$ (read "plus or minus $\sqrt{2}$").   **Now Try Exercise 53.** ■

(R.5) Exercise Set

Solve.

1. $x - 5 = 7$

2. $y + 3 = 4$

3. $3x + 4 = -8$

4. $5x - 7 = 23$

5. $5y - 12 = 3$

6. $6x + 23 = 5$

7. $6x - 15 = 45$

8. $4x - 7 = 81$

9. $5x - 10 = 45$

10. $6x - 7 = 11$

11. $9t + 4 = -5$

12. $5x + 7 = -13$

13. $8x + 48 = 3x - 12$

14. $15x + 40 = 8x - 9$

15. $7y - 1 = 23 - 5y$

16. $3x - 15 = 15 - 3x$

17. $3x - 4 = 5 + 12x$

18. $9t - 4 = 14 + 15t$

19. $5 - 4a = a - 13$

20. $6 - 7x = x - 14$

21. $3m - 7 = -13 + m$

22. $5x - 8 = 2x - 8$

23. $11 - 3x = 5x + 3$

24. $20 - 4y = 10 - 6y$

25. $2(x + 7) = 5x + 14$

26. $3(y + 4) = 8y$

27. $24 = 5(2t + 5)$

28. $9 = 4(3y - 2)$

29. $5y - (2y - 10) = 25$

30. $8x - (3x - 5) = 40$

31. $7(3x + 6) = 11 - (x + 2)$

32. $9(2x + 8) = 20 - (x + 5)$

33. $4(3y - 1) - 6 = 5(y + 2)$

34. $3(2n - 5) - 7 = 4(n - 9)$

35. $x^2 + 3x - 28 = 0$

36. $y^2 - 4y - 45 = 0$

37. $x^2 + 5x = 0$

38. $t^2 + 6t = 0$

39. $y^2 + 6y + 9 = 0$

40. $n^2 + 4n + 4 = 0$

41. $x^2 + 100 = 20x$

42. $y^2 + 25 = 10y$

43. $x^2 - 4x - 32 = 0$

44. $t^2 + 12t + 27 = 0$

45. $3y^2 + 8y + 4 = 0$

46. $9y^2 + 15y + 4 = 0$

47. $12z^2 + z = 6$

48. $6x^2 - 7x = 10$

49. $12a^2 - 28 = 5a$

50. $21n^2 - 10 = n$

51. $14 = x(x - 5)$

52. $24 = x(x - 2)$

53. $x^2 - 36 = 0$

54. $y^2 - 81 = 0$

55. $z^2 = 144$

56. $t^2 = 25$

57. $2x^2 - 20 = 0$

58. $3y^2 - 15 = 0$

59. $6z^2 - 18 = 0$

60. $5x^2 - 75 = 0$

Collaborative Discussion and Writing

61. When using the addition and multiplication principles to solve an equation, how do you determine which number to add or multiply by on both sides of the equation?

62. Explain how to write a quadratic equation with solutions -3 and 4.

Synthesis

Solve.

63. $3[5 - 3(4 - t)] - 2 = 5[3(5t - 4) + 8] - 26$

64. $6[4(8 - y) - 5(9 + 3y)] - 21 = -7[3(7 + 4y) - 4]$

65. $x - \{3x - [2x - (5x - (7x - 1))]\} = x + 7$

66. $23 - 2[4 + 3(x - 1)] + 5[x - 2(x + 3)] = 7\{x - 2[5 - (2x + 3)]\}$

67. $(5x^2 + 6x)(12x^2 - 5x - 2) = 0$

68. $(3x^2 + 7x - 20)(x^2 - 4x) = 0$

69. $3x^3 + 6x^2 - 27x - 54 = 0$

70. $2x^3 + 6x^2 = 8x + 24$

Rational Expressions

❖ Determine the domain of a rational expression.
❖ Simplify rational expressions.
❖ Multiply, divide, add, and subtract rational expressions.
❖ Simplify complex rational expressions.

A **rational expression** is the quotient of two polynomials. For example,

$$\frac{3}{5}, \quad \frac{2}{x-3}, \quad \text{and} \quad \frac{x^2-4}{x^2-4x-5}$$

are rational expressions.

❖ The Domain of a Rational Expression

The **domain** of an algebraic expression is the set of all real numbers for which the expression is defined. Since division by zero is not defined, any number that makes the denominator zero is not in the domain of a rational expression.

EXAMPLE 1   Find the domain of each of the following.

a) $\dfrac{2}{x-3}$

b) $\dfrac{x^2-4}{x^2-4x-5}$

Solution

a) We solve the equation $x - 3 = 0$ to determine the numbers that are *not* in the domain:

$$x - 3 = 0$$
$$x = 3. \qquad \text{Adding 3 on both sides}$$

Since the denominator is 0 when $x = 3$, the domain of $2/(x-3)$ is the set of all real numbers except 3.

b) We solve the equation $x^2 - 4x - 5 = 0$ to find the numbers that are not in the domain:

$$x^2 - 4x - 5 = 0$$
$$(x + 1)(x - 5) = 0 \qquad\qquad \text{Factoring}$$
$$x + 1 = 0 \quad or \quad x - 5 = 0 \qquad \text{Using the principle of zero products}$$
$$x = -1 \quad or \qquad\quad x = 5.$$

Since the denominator is 0 when $x = -1$ or $x = 5$, the domain is the set of all real numbers except -1 and 5.   **Now Try Exercise 3.** ■

We can describe the domains found in Example 1 using *set-builder notation*. For example, we write "The set of all real numbers x such that x is not equal to 3" as

$$\{x \mid x \text{ is a real number } and \; x \neq 3\}.$$

Similarly, we write "The set of all real numbers x such that x is not equal to -1 and x is not equal to 5" as

$$\{x \mid x \text{ is a real number } and \; x \neq -1 \; and \; x \neq 5\}.$$

❖ Simplifying, Multiplying, and Dividing Rational Expressions

To simplify rational expressions, we use the fact that

$$\frac{a \cdot c}{b \cdot c} = \frac{a}{b} \cdot \frac{c}{c} = \frac{a}{b} \cdot 1 = \frac{a}{b}.$$

EXAMPLE 2 Simplify: $\dfrac{9x^2 + 6x - 3}{12x^2 - 12}$.

Solution

$$\frac{9x^2 + 6x - 3}{12x^2 - 12} = \frac{3(3x^2 + 2x - 1)}{12(x^2 - 1)}$$

$$= \frac{3(x + 1)(3x - 1)}{3 \cdot 4(x + 1)(x - 1)}$$

Factoring the numerator and the denominator

$$= \frac{3(x + 1)}{3(x + 1)} \cdot \frac{3x - 1}{4(x - 1)}$$

Factoring the rational expression

$$= 1 \cdot \frac{3x - 1}{4(x - 1)}$$

$\dfrac{3(x + 1)}{3(x + 1)} = 1$

$$= \frac{3x - 1}{4(x - 1)}$$

Removing a factor of 1

Now Try Exercise 9. ■

Canceling is a shortcut that is often used to remove a factor of 1.

EXAMPLE 3 Simplify each of the following.

a) $\dfrac{4x^3 + 16x^2}{2x^3 + 6x^2 - 8x}$ b) $\dfrac{2 - x}{x^2 + x - 6}$

Solution

a) $\dfrac{4x^3 + 16x^2}{2x^3 + 6x^2 - 8x} = \dfrac{2 \cdot 2 \cdot x \cdot x(x + 4)}{2 \cdot x(x + 4)(x - 1)}$

Factoring the numerator and the denominator

$$= \frac{2 \cdot 2 \cdot \cancel{x} \cdot x\cancel{(x + 4)}}{2 \cdot \cancel{x}\cancel{(x + 4)}(x - 1)}$$

Removing a factor of 1: $\dfrac{2x(x + 4)}{2x(x + 4)} = 1$

$$= \frac{2x}{x - 1}$$

b) $\dfrac{2 - x}{x^2 + x - 6} = \dfrac{2 - x}{(x + 3)(x - 2)}$ Factoring the denominator

$= \dfrac{-1(x - 2)}{(x + 3)(x - 2)}$ $2 - x = -1(x - 2)$

$= \dfrac{-1(x - 2)}{(x + 3)(x - 2)}$ Removing a factor of 1: $\dfrac{x - 2}{x - 2} = 1$

$= \dfrac{-1}{x + 3}, \text{ or } -\dfrac{1}{x + 3}$ **Now Try Exercise 11.** ▣

In Example 3(b), we saw that

$$\dfrac{2 - x}{x^2 + x - 6} \quad \text{and} \quad -\dfrac{1}{x + 3}$$

are **equivalent expressions**. This means that they have the same value for all numbers that are in *both* domains. Note that -3 is not in the domain of *either* expression, whereas 2 is in the domain of $-1/(x + 3)$ but not in the domain of $(2 - x)/(x^2 + x - 6)$ and thus is not in the domain of *both* expressions.

To multiply rational expressions, we multiply numerators and multiply denominators and, if possible, simplify the result. To divide rational expressions, we multiply the dividend by the reciprocal of the divisor and, if possible, simplify the result; that is,

$$\dfrac{a}{b} \cdot \dfrac{c}{d} = \dfrac{ac}{bd} \quad \text{and} \quad \dfrac{a}{b} \div \dfrac{c}{d} = \dfrac{a}{b} \cdot \dfrac{d}{c} = \dfrac{ad}{bc}.$$

EXAMPLE 4 Multiply or divide and simplify each of the following.

a) $\dfrac{a^2 - 4}{16a} \cdot \dfrac{20a^2}{a + 2}$

b) $\dfrac{x + 4}{2x^2 - 6x} \cdot \dfrac{x^4 - 9x^2}{x^2 + 2x - 8}$

c) $\dfrac{x - 2}{12} \div \dfrac{x^2 - 4x + 4}{3x^3 + 15x^2}$

d) $\dfrac{y^3 - 1}{y^2 - 1} \div \dfrac{y^2 + y + 1}{y^2 + 2y + 1}$

Solution

a) $\dfrac{a^2 - 4}{16a} \cdot \dfrac{20a^2}{a + 2} = \dfrac{(a^2 - 4)(20a^2)}{16a(a + 2)}$ Multiplying the numerators and the denominators

$= \dfrac{(a + 2)(a - 2) \cdot 4 \cdot 5 \cdot a \cdot a}{4 \cdot 4 \cdot a \cdot (a + 2)}$ Factoring and removing a factor of 1

$= \dfrac{5a(a - 2)}{4}$

b) $\dfrac{x+4}{2x^2-6x} \cdot \dfrac{x^4-9x^2}{x^2+2x-8}$

$= \dfrac{(x+4)\,(x^4-9x^2)}{(2x^2-6x)(x^2+2x-8)}$ Multiplying the numerators and the denominators

$= \dfrac{(x+4)(x^2)(x^2-9)}{2x\,(x-3)(x+4)(x-2)}$ Factoring

$= \dfrac{(x+4)\cdot x \cdot x \cdot (x+3)\,(x-3)}{2\cdot x \cdot (x-3)(x+4)(x-2)}$ Factoring further and removing a factor of 1

$= \dfrac{x(x+3)}{2(x-2)}$

c) $\dfrac{x-2}{12} \div \dfrac{x^2-4x+4}{3x^3+15x^2}$

$= \dfrac{x-2}{12} \cdot \dfrac{3x^3+15x^2}{x^2-4x+4}$ Multiplying by the reciprocal of the divisor

$= \dfrac{(x-2)(3x^3+15x^2)}{12(x^2-4x+4)}$

$= \dfrac{(x-2)(3)(x^2)(x+5)}{3 \cdot 4(x-2)(x-2)}$ Factoring and removing a factor of 1

$= \dfrac{x^2(x+5)}{4(x-2)}$

d) $\dfrac{y^3-1}{y^2-1} \div \dfrac{y^2+y+1}{y^2+2y+1}$

$= \dfrac{y^3-1}{y^2-1} \cdot \dfrac{y^2+2y+1}{y^2+y+1}$ Multiplying by the reciprocal of the divisor

$= \dfrac{(y^3-1)(y^2+2y+1)}{(y^2-1)(y^2+y+1)}$

$= \dfrac{(y-1)(y^2+y+1)(y+1)(y+1)}{(y+1)(y-1)(y^2+y+1)}$ Factoring and removing a factor of 1

$= y+1$

Now Try Exercises 19 and 25. ■

❈ Adding and Subtracting Rational Expressions

When rational expressions have the same denominator, we can add or subtract by adding or subtracting the numerators and retaining the common denominator. If the denominators differ, we must find equivalent rational expressions that have a common denominator. In general, it is most efficient to find the **least common denominator (LCD)** of the expressions.

> To find the least common denominator of rational expressions, factor each denominator and form the product that uses each factor the greatest number of times it occurs in any factorization.

EXAMPLE 5 Add or subtract and simplify each of the following.

a) $\dfrac{x^2 - 4x + 4}{2x^2 - 3x + 1} + \dfrac{x + 4}{2x - 2}$
b) $\dfrac{x - 2}{x^2 - 25} - \dfrac{x + 5}{x^2 - 5x}$

c) $\dfrac{x}{x^2 + 11x + 30} - \dfrac{5}{x^2 + 9x + 20}$

Solution

a) $\dfrac{x^2 - 4x + 4}{2x^2 - 3x + 1} + \dfrac{x + 4}{2x - 2}$

$= \dfrac{x^2 - 4x + 4}{(2x - 1)(x - 1)} + \dfrac{x + 4}{2(x - 1)}$ Factoring the denominators

The LCD is $(2x - 1)(x - 1)(2)$, or $2(2x - 1)(x - 1)$.

$= \dfrac{x^2 - 4x + 4}{(2x - 1)(x - 1)} \cdot \dfrac{2}{2} + \dfrac{x + 4}{2(x - 1)} \cdot \dfrac{2x - 1}{2x - 1}$ Multiplying each term by 1 to get the LCD

$= \dfrac{2x^2 - 8x + 8}{(2x - 1)(x - 1)(2)} + \dfrac{2x^2 + 7x - 4}{2(x - 1)(2x - 1)}$

$= \dfrac{4x^2 - x + 4}{2(2x - 1)(x - 1)}$ Adding the numerators

b) $\dfrac{x - 2}{x^2 - 25} - \dfrac{x + 5}{x^2 - 5x}$

$= \dfrac{x - 2}{(x + 5)(x - 5)} - \dfrac{x + 5}{x(x - 5)}$ Factoring the denominators

The LCD is $(x + 5)(x - 5)(x)$, or $x(x + 5)(x - 5)$.

$= \dfrac{x - 2}{(x + 5)(x - 5)} \cdot \dfrac{x}{x} - \dfrac{x + 5}{x(x - 5)} \cdot \dfrac{x + 5}{x + 5}$ Multiplying each term by 1 to get the LCD

$= \dfrac{x^2 - 2x}{x(x + 5)(x - 5)} - \dfrac{x^2 + 10x + 25}{x(x + 5)(x - 5)}$

$= \dfrac{x^2 - 2x - (x^2 + 10x + 25)}{x(x + 5)(x - 5)}$ Subtracting numerators and keeping the common denominator

$= \dfrac{x^2 - 2x - x^2 - 10x - 25}{x(x + 5)(x - 5)}$ Removing parentheses

$= \dfrac{-12x - 25}{x(x + 5)(x - 5)}$

c) $\dfrac{x}{x^2 + 11x + 30} - \dfrac{5}{x^2 + 9x + 20}$

$= \dfrac{x}{(x + 5)(x + 6)} - \dfrac{5}{(x + 5)(x + 4)}$ Factoring the denominators

The LCD is $(x + 5)(x + 6)(x + 4)$.

$= \dfrac{x}{(x + 5)(x + 6)} \cdot \dfrac{x + 4}{x + 4} - \dfrac{5}{(x + 5)(x + 4)} \cdot \dfrac{x + 6}{x + 6}$

Multiplying each term by 1 to get the LCD

$= \dfrac{x^2 + 4x}{(x + 5)(x + 6)(x + 4)} - \dfrac{5x + 30}{(x + 5)(x + 4)(x + 6)}$

$= \dfrac{x^2 + 4x - 5x - 30}{(x + 5)(x + 6)(x + 4)}$

Be sure to change the sign of *every* term in the numerator of the expression being subtracted:
$-(5x + 30) = -5x - 30$

$= \dfrac{x^2 - x - 30}{(x + 5)(x + 6)(x + 4)}$

$= \dfrac{(x + 5)(x - 6)}{(x + 5)(x + 6)(x + 4)}$

Factoring and removing a factor of 1:
$\dfrac{x + 5}{x + 5} = 1$

$= \dfrac{x - 6}{(x + 6)(x + 4)}$

Now Try Exercises 37 and 39. ■

❈ Complex Rational Expressions

A **complex rational expression** has rational expressions in its numerator or its denominator or both.

To simplify a complex rational expression:

Method 1. Find the LCD of all the denominators within the complex rational expression. Then multiply by 1 using the LCD as the numerator and the denominator of the expression for 1.

Method 2. First add or subtract, if necessary, to get a single rational expression in the numerator and in the denominator. Then divide by multiplying by the reciprocal of the denominator.

EXAMPLE 6   Simplify: $\dfrac{\dfrac{1}{a} + \dfrac{1}{b}}{\dfrac{1}{a^3} + \dfrac{1}{b^3}}.$

Solution

Method 1.   The LCD of the four rational expressions in the numerator and the denominator is a^3b^3.

$$\dfrac{\dfrac{1}{a} + \dfrac{1}{b}}{\dfrac{1}{a^3} + \dfrac{1}{b^3}} = \dfrac{\dfrac{1}{a} + \dfrac{1}{b}}{\dfrac{1}{a^3} + \dfrac{1}{b^3}} \cdot \dfrac{a^3b^3}{a^3b^3} \qquad \text{Multiplying by 1 using } \dfrac{a^3b^3}{a^3b^3}$$

$$= \dfrac{\left(\dfrac{1}{a} + \dfrac{1}{b}\right)(a^3b^3)}{\left(\dfrac{1}{a^3} + \dfrac{1}{b^3}\right)(a^3b^3)}$$

$$= \dfrac{a^2b^3 + a^3b^2}{b^3 + a^3}$$

$$= \dfrac{a^2b^2(b + a)}{(b + a)(b^2 - ba + a^2)} \qquad \begin{array}{l} \text{Factoring and removing a factor} \\ \text{of 1:} \dfrac{b + a}{b + a} = 1 \end{array}$$

$$= \dfrac{a^2b^2}{b^2 - ba + a^2}$$

Method 2.   We add in the numerator and in the denominator.

$$\dfrac{\dfrac{1}{a} + \dfrac{1}{b}}{\dfrac{1}{a^3} + \dfrac{1}{b^3}} = \dfrac{\dfrac{1}{a} \cdot \dfrac{b}{b} + \dfrac{1}{b} \cdot \dfrac{a}{a}}{\dfrac{1}{a^3} \cdot \dfrac{b^3}{b^3} + \dfrac{1}{b^3} \cdot \dfrac{a^3}{a^3}} \begin{array}{l} \longleftarrow \text{The LCD is } ab. \\ \\ \longleftarrow \text{The LCD is } a^3b^3. \end{array}$$

$$= \dfrac{\dfrac{b}{ab} + \dfrac{a}{ab}}{\dfrac{b^3}{a^3b^3} + \dfrac{a^3}{a^3b^3}}$$

$$= \dfrac{\dfrac{b + a}{ab}}{\dfrac{b^3 + a^3}{a^3b^3}} \qquad \begin{array}{l} \text{We have a single rational} \\ \text{expression in both the} \\ \text{numerator and the} \\ \text{denominator.} \end{array}$$

$$= \dfrac{b + a}{ab} \cdot \dfrac{a^3b^3}{b^3 + a^3} \qquad \begin{array}{l} \text{Multiplying by the reciprocal} \\ \text{of the denominator} \end{array}$$

$$= \dfrac{(b + a)(a)(b)(a^2b^2)}{(a)(b)(b + a)(b^2 - ba + a^2)}$$

$$= \dfrac{a^2b^2}{b^2 - ba + a^2} \qquad\qquad\qquad \textbf{Now Try Exercise 57.} \ \blacksquare$$

(R.6) Exercise Set

Find the domain of the rational expression.

1. $-\dfrac{5}{3}$

2. $\dfrac{4}{7-x}$

3. $\dfrac{3x-3}{x(x-1)}$

4. $\dfrac{15x-10}{2x(3x-2)}$

5. $\dfrac{x+5}{x^2+4x-5}$

6. $\dfrac{(x^2-4)(x+1)}{(x+2)(x^2-1)}$

7. $\dfrac{7x^2-28x+28}{(x^2-4)(x^2+3x-10)}$

8. $\dfrac{7x^2+11x-6}{x(x^2-x-6)}$

Simplify.

9. $\dfrac{x^2-4}{x^2-4x+4}$

10. $\dfrac{x^2+2x-3}{x^2-9}$

11. $\dfrac{x^3-6x^2+9x}{x^3-3x^2}$

12. $\dfrac{y^5-5y^4+4y^3}{y^3-6y^2+8y}$

13. $\dfrac{6y^2+12y-48}{3y^2-9y+6}$

14. $\dfrac{2x^2-20x+50}{10x^2-30x-100}$

15. $\dfrac{4-x}{x^2+4x-32}$

16. $\dfrac{6-x}{x^2-36}$

Multiply or divide and, if possible, simplify.

17. $\dfrac{r-s}{r+s}\cdot\dfrac{r^2-s^2}{(r-s)^2}$

18. $\dfrac{x^2-y^2}{(x-y)^2}\cdot\dfrac{1}{x+y}$

19. $\dfrac{x^2+2x-35}{3x^3-2x^2}\cdot\dfrac{9x^3-4x}{7x+49}$

20. $\dfrac{x^2-2x-35}{2x^3-3x^2}\cdot\dfrac{4x^3-9x}{7x-49}$

21. $\dfrac{a^2-a-6}{a^2-7a+12}\cdot\dfrac{a^2-2a-8}{a^2-3a-10}$

22. $\dfrac{a^2-a-12}{a^2-6a+8}\cdot\dfrac{a^2+a-6}{a^2-2a-24}$

23. $\dfrac{m^2-n^2}{r+s}\div\dfrac{m-n}{r+s}$

24. $\dfrac{a^2-b^2}{x-y}\div\dfrac{a+b}{x-y}$

25. $\dfrac{3x+12}{2x-8}\div\dfrac{(x+4)^2}{(x-4)^2}$

26. $\dfrac{a^2-a-2}{a^2-a-6}\div\dfrac{a^2-2a}{2a+a^2}$

27. $\dfrac{x^2-y^2}{x^3-y^3}\cdot\dfrac{x^2+xy+y^2}{x^2+2xy+y^2}$

28. $\dfrac{c^3+8}{c^2-4}\div\dfrac{c^2-2c+4}{c^2-4c+4}$

29. $\dfrac{(x-y)^2-z^2}{(x+y)^2-z^2}\div\dfrac{x-y+z}{x+y-z}$

30. $\dfrac{(a+b)^2-9}{(a-b)^2-9}\cdot\dfrac{a-b-3}{a+b+3}$

Add or subtract and, if possible, simplify.

31. $\dfrac{7}{5x}+\dfrac{3}{5x}$

32. $\dfrac{7}{12y}-\dfrac{1}{12y}$

33. $\dfrac{4}{3a+4}+\dfrac{3a}{3a+4}$

34. $\dfrac{a-3b}{a+b}+\dfrac{a+5b}{a+b}$

35. $\dfrac{5}{4z}-\dfrac{3}{8z}$

36. $\dfrac{12}{x^2y}+\dfrac{5}{xy^2}$

37. $\dfrac{3}{x+2}+\dfrac{2}{x^2-4}$

38. $\dfrac{5}{a-3}-\dfrac{2}{a^2-9}$

39. $\dfrac{y}{y^2-y-20}-\dfrac{2}{y+4}$

40. $\dfrac{6}{y^2+6y+9}-\dfrac{5}{y+3}$

41. $\dfrac{3}{x+y}+\dfrac{x-5y}{x^2-y^2}$

42. $\dfrac{a^2 + 1}{a^2 - 1} - \dfrac{a - 1}{a + 1}$

43. $\dfrac{y}{y - 1} + \dfrac{2}{1 - y}$
(*Note*: $1 - y = -1(y - 1)$.)

44. $\dfrac{a}{a - b} + \dfrac{b}{b - a}$
(*Note*: $b - a = -1(a - b)$.)

45. $\dfrac{x}{2x - 3y} - \dfrac{y}{3y - 2x}$

46. $\dfrac{3a}{3a - 2b} - \dfrac{2a}{2b - 3a}$

47. $\dfrac{9x + 2}{3x^2 - 2x - 8} + \dfrac{7}{3x^2 + x - 4}$

48. $\dfrac{3y}{y^2 - 7y + 10} - \dfrac{2y}{y^2 - 8y + 15}$

49. $\dfrac{5a}{a - b} + \dfrac{ab}{a^2 - b^2} + \dfrac{4b}{a + b}$

50. $\dfrac{6a}{a - b} - \dfrac{3b}{b - a} + \dfrac{5}{a^2 - b^2}$

51. $\dfrac{7}{x + 2} - \dfrac{x + 8}{4 - x^2} + \dfrac{3x - 2}{4 - 4x + x^2}$

52. $\dfrac{6}{x + 3} - \dfrac{x + 4}{9 - x^2} + \dfrac{2x - 3}{9 - 6x + x^2}$

53. $\dfrac{1}{x + 1} + \dfrac{x}{2 - x} + \dfrac{x^2 + 2}{x^2 - x - 2}$

54. $\dfrac{x - 1}{x - 2} - \dfrac{x + 1}{x + 2} - \dfrac{x - 6}{4 - x^2}$

Simplify.

55. $\dfrac{\dfrac{a - b}{b}}{\dfrac{a^2 - b^2}{ab}}$

56. $\dfrac{\dfrac{x^2 - y^2}{xy}}{\dfrac{x - y}{y}}$

57. $\dfrac{\dfrac{x}{y} - \dfrac{y}{x}}{\dfrac{1}{y} + \dfrac{1}{x}}$

58. $\dfrac{\dfrac{a}{b} - \dfrac{b}{a}}{\dfrac{1}{a} - \dfrac{1}{b}}$

59. $\dfrac{c + \dfrac{8}{c^2}}{1 + \dfrac{2}{c}}$

60. $\dfrac{a - \dfrac{a}{b}}{b - \dfrac{b}{a}}$

61. $\dfrac{x^2 + xy + y^2}{\dfrac{x^2}{y} - \dfrac{y^2}{x}}$

62. $\dfrac{\dfrac{a^2}{b} + \dfrac{b^2}{a}}{a^2 - ab + b^2}$

63. $\dfrac{a - a^{-1}}{a + a^{-1}}$

64. $\dfrac{x^{-1} + y^{-1}}{x^{-3} + y^{-3}}$

65. $\dfrac{\dfrac{1}{x - 3} + \dfrac{2}{x + 3}}{\dfrac{3}{x - 1} - \dfrac{4}{x + 2}}$

66. $\dfrac{\dfrac{5}{x + 1} - \dfrac{3}{x - 2}}{\dfrac{1}{x - 5} + \dfrac{2}{x + 2}}$

67. $\dfrac{\dfrac{a}{1 - a} + \dfrac{1 + a}{a}}{\dfrac{1 - a}{a} + \dfrac{a}{1 + a}}$

68. $\dfrac{\dfrac{1 - x}{x} + \dfrac{x}{1 + x}}{\dfrac{1 + x}{x} + \dfrac{x}{1 - x}}$

69. $\dfrac{\dfrac{1}{a^2} + \dfrac{2}{ab} + \dfrac{1}{b^2}}{\dfrac{1}{a^2} - \dfrac{1}{b^2}}$

70. $\dfrac{\dfrac{1}{x^2} - \dfrac{1}{y^2}}{\dfrac{1}{x^2} - \dfrac{2}{xy} + \dfrac{1}{y^2}}$

Collaborative Discussion and Writing

71. When adding or subtracting rational expressions, we can always find a common denominator by forming the product of all the denominators. Explain why it is usually preferable to find the least common denominator.

72. How would you determine which method to use to simplify a particular complex rational expression?

Synthesis

Simplify.

73. $\dfrac{(x + h)^2 - x^2}{h}$

74. $\dfrac{\dfrac{1}{x + h} - \dfrac{1}{x}}{h}$

75. $\dfrac{(x + h)^3 - x^3}{h}$

76. $\dfrac{\dfrac{1}{(x + h)^2} - \dfrac{1}{x^2}}{h}$

77. $\left[\dfrac{\dfrac{x + 1}{x - 1} + 1}{\dfrac{x + 1}{x - 1} - 1}\right]^5$

78. $1 + \dfrac{1}{1 + \dfrac{1}{1 + \dfrac{1}{1 + \dfrac{1}{x}}}}$

Perform the indicated operations and, if possible, simplify.

79. $\dfrac{n(n + 1)(n + 2)}{2 \cdot 3} + \dfrac{(n + 1)(n + 2)}{2}$

80. $\dfrac{n(n + 1)(n + 2)(n + 3)}{2 \cdot 3 \cdot 4} + \dfrac{(n + 1)(n + 2)(n + 3)}{2 \cdot 3}$

81. $\dfrac{x^2 - 9}{x^3 + 27} \cdot \dfrac{5x^2 - 15x + 45}{x^2 - 2x - 3} + \dfrac{x^2 + x}{4 + 2x}$

82. $\dfrac{x^2 + 2x - 3}{x^2 - x - 12} \div \dfrac{x^2 - 1}{x^2 - 16} - \dfrac{2x + 1}{x^2 + 2x + 1}$

R.7   Radical Notation and Rational Exponents

❋ Simplify radical expressions.
❋ Rationalize denominators or numerators in rational expressions.
❋ Convert between exponential notation and radical notation.
❋ Simplify expressions with rational exponents.

A number c is said to be a **square root** of a if $c^2 = a$. Thus, 3 is a square root of 9, because $3^2 = 9$, and -3 is also a square root of 9, because $(-3)^2 = 9$. Similarly, 5 is a third root (called a **cube root**) of 125, because $5^3 = 125$. The number 125 has no other real-number cube root.

> **nth Root**
>
> A number c is said to be an **nth root** of a if $c^n = a$.

The symbol $\sqrt{a}$ denotes the nonnegative square root of a, and the symbol $\sqrt[3]{a}$ denotes the real-number cube root of a. The symbol $\sqrt[n]{a}$ denotes the nth root of a; that is, a number whose nth power is a. The symbol $\sqrt[n]{}$ is called a **radical**, and the expression under the radical is called the **radicand**. The number n (which is omitted when it is 2) is called the **index**. Examples of roots for $n = 3$, 4, and 2, respectively, are

$$\sqrt[3]{125}, \quad \sqrt[4]{16}, \quad \text{and} \quad \sqrt{3600}.$$

Any real number has only one real-number odd root. Any positive number has two square roots, one positive and one negative. Similarly, for any even index, a positive number has two real-number roots. The positive root is called the **principal root**. When an expression such as $\sqrt{4}$ or $\sqrt[6]{23}$ is used, it is understood to represent the principal (nonnegative) root. To denote a negative root, we use $-\sqrt{4}$, $-\sqrt[6]{23}$, and so on.

GCM **EXAMPLE 1** Simplify each of the following.

a) $\sqrt{36}$ **b)** $-\sqrt{36}$ **c)** $\sqrt[3]{-8}$

d) $\sqrt[5]{\dfrac{32}{243}}$ **e)** $\sqrt[4]{-16}$

Solution

a) $\sqrt{36} = 6$, because $6^2 = 36$.

b) $-\sqrt{36} = -6$, because $6^2 = 36$ and $-\left(\sqrt{36}\right) = -(6) = -6$.

c) $\sqrt[3]{-8} = -2$, because $(-2)^3 = -8$.

d) $\sqrt[5]{\dfrac{32}{243}} = \dfrac{2}{3}$, because $\left(\dfrac{2}{3}\right)^5 = \dfrac{2^5}{3^5} = \dfrac{32}{243}$.

e) $\sqrt[4]{-16}$ is not a real number, because we cannot find a real number that can be raised to the fourth power to get -16.   **Now Try Exercise 11.** ■

We can generalize Example 1(e) and say that when a is negative and n is even, $\sqrt[n]{a}$ is not a real number. For example, $\sqrt{-4}$ and $\sqrt[4]{-81}$ are not real numbers.

We can find $\sqrt{36}$ and $-\sqrt{36}$ in Example 1 using the square-root feature on the keypad of a graphing calculator, and we can use the cube-root feature to find $\sqrt[3]{-8}$. We can use the xth-root feature to find higher roots.

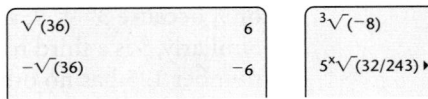

STUDY TIP

The keystrokes for entering the radical expressions in Example 1 on a graphing calculator are found in the *Graphing Calculator Manual* that accompanies this text.

When we try to find $\sqrt[4]{-16}$ on a graphing calculator set in REAL mode, we get an error message indicating that the answer is nonreal.

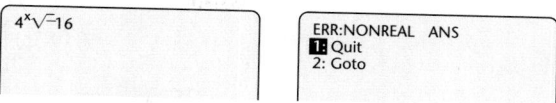

❋ Simplifying Radical Expressions

Consider the expression $\sqrt{(-3)^2}$. This is equivalent to $\sqrt{9}$, or 3. Similarly, $\sqrt{3^2} = \sqrt{9} = 3$. This illustrates the first of several properties of radicals, listed below.

Properties of Radicals

Let a and b be any real numbers or expressions for which the given roots exist. For any natural numbers m and n ($n \neq 1$):

1. If n is even, $\sqrt[n]{a^n} = |a|$.

2. If n is odd, $\sqrt[n]{a^n} = a$.

3. $\sqrt[n]{a} \cdot \sqrt[n]{b} = \sqrt[n]{ab}$.

4. $\sqrt[n]{\dfrac{a}{b}} = \dfrac{\sqrt[n]{a}}{\sqrt[n]{b}}$ ($b \neq 0$).

5. $\sqrt[n]{a^m} = \left(\sqrt[n]{a}\right)^m$.

EXAMPLE 2   Simplify each of the following.

a) $\sqrt{(-5)^2}$ b) $\sqrt[3]{(-5)^3}$ c) $\sqrt[4]{4} \cdot \sqrt[4]{5}$ d) $\sqrt{50}$

e) $\dfrac{\sqrt{72}}{\sqrt{6}}$ f) $\sqrt[3]{8^5}$ g) $\sqrt{216x^5y^3}$ h) $\sqrt{\dfrac{x^2}{16}}$

Solution

a) $\sqrt{(-5)^2} = |-5| = 5$ Using Property 1

b) $\sqrt[3]{(-5)^3} = -5$ Using Property 2

c) $\sqrt[4]{4} \cdot \sqrt[4]{5} = \sqrt[4]{4 \cdot 5} = \sqrt[4]{20}$ Using Property 3

d) $\sqrt{50} = \sqrt{25 \cdot 2} = \sqrt{25} \cdot \sqrt{2} = 5\sqrt{2}$ Using Property 3

e) $\dfrac{\sqrt{72}}{\sqrt{6}} = \sqrt{\dfrac{72}{6}}$ Using Property 4

$= \sqrt{12} = \sqrt{4 \cdot 3} = \sqrt{4} \cdot \sqrt{3}$ Using Property 3

$= 2\sqrt{3}$

f) $\sqrt[3]{8^5} = \left(\sqrt[3]{8}\right)^5$ Using Property 5

$= 2^5 = 32$

g) $\sqrt{216x^5y^3} = \sqrt{36 \cdot 6 \cdot x^4 \cdot x \cdot y^2 \cdot y}$

$\qquad\qquad = \sqrt{36x^4y^2}\,\sqrt{6xy}$ Using Property 3

$\qquad\qquad = |6x^2y|\sqrt{6xy}$ Using Property 1

$\qquad\qquad = 6x^2|y|\sqrt{6xy}$ $6x^2$ cannot be negative, so absolute-value signs are not needed for it.

h) $\sqrt{\dfrac{x^2}{16}} = \dfrac{\sqrt{x^2}}{\sqrt{16}}$ Using Property 4

$\qquad\quad = \dfrac{|x|}{4}$ Using Property 1   **Now Try Exercise 19.** ■

In many situations, radicands are never formed by raising negative quantities to even powers. In such cases, absolute-value notation is not required. For this reason, **we will henceforth assume that no radicands are formed by raising negative quantities to even powers** and, consequently, we will not use absolute-value notation when we simplify radical expressions. For example, we will write $\sqrt{x^2} = x$ and $\sqrt[4]{a^5b} = a\sqrt[4]{ab}$.

Radical expressions with the same index and the same radicand can be combined (added or subtracted) in much the same way that we combine like terms.

EXAMPLE 3 Perform the operations indicated.

a) $3\sqrt{8x^2} - 5\sqrt{2x^2}$

b) $\left(4\sqrt{3} + \sqrt{2}\right)\left(\sqrt{3} - 5\sqrt{2}\right)$

Solution

a) $3\sqrt{8x^2} - 5\sqrt{2x^2} = 3\sqrt{4x^2 \cdot 2} - 5\sqrt{x^2 \cdot 2}$

$\qquad\qquad\qquad\qquad = 3 \cdot 2x\sqrt{2} - 5x\sqrt{2}$

$\qquad\qquad\qquad\qquad = 6x\sqrt{2} - 5x\sqrt{2}$

$\qquad\qquad\qquad\qquad = (6x - 5x)\sqrt{2}$ Using the distributive property

$\qquad\qquad\qquad\qquad = x\sqrt{2}$

b) $\left(4\sqrt{3} + \sqrt{2}\right)\left(\sqrt{3} - 5\sqrt{2}\right) = 4\left(\sqrt{3}\right)^2 - 20\sqrt{6} + \sqrt{6} - 5\left(\sqrt{2}\right)^2$

$\qquad\qquad\qquad\qquad\qquad\qquad\qquad$ Multiplying

$\qquad\qquad\qquad\qquad\qquad = 4 \cdot 3 + (-20 + 1)\sqrt{6} - 5 \cdot 2$

$\qquad\qquad\qquad\qquad\qquad = 12 - 19\sqrt{6} - 10$

$\qquad\qquad\qquad\qquad\qquad = 2 - 19\sqrt{6}$ **Now Try Exercise 45.** ■

❉ An Application

The Pythagorean theorem relates the lengths of the sides of a right triangle. The side opposite the triangle's right angle is called the **hypotenuse**. The other sides are the **legs**.

> **The Pythagorean Theorem**
>
> The sum of the squares of the lengths of the legs of a right triangle is equal to the square of the length of the hypotenuse:
>
> $$a^2 + b^2 = c^2.$$

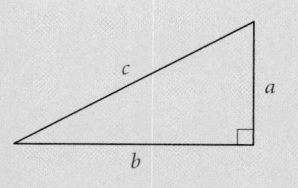

EXAMPLE 4 *Surveying.* A surveyor places poles at points A, B, and C in order to measure the distance across a pond. The distances AC and BC are measured as shown. Find the distance AB across the pond.

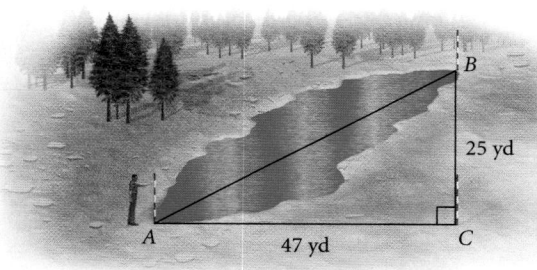

Solution We see that the lengths of the legs of a right triangle are given. Thus we use the Pythagorean theorem to find the length of the hypotenuse:

$$c^2 = a^2 + b^2$$
$$c = \sqrt{a^2 + b^2} \qquad \text{Solving for } c$$
$$= \sqrt{25^2 + 47^2}$$
$$= \sqrt{625 + 2209}$$
$$= \sqrt{2834}$$
$$\approx 53.2.$$

√(25²+47²)

53.23532662

The distance across the pond is about 53.2 yd. **Now Try Exercise 59.** ■

❖ Rationalizing Denominators or Numerators

There are times when we need to remove the radicals in a denominator or a numerator. This is called **rationalizing the denominator** or **rationalizing the numerator**. It is done by multiplying by 1 in such a way as to obtain a perfect nth power.

EXAMPLE 5 Rationalize the denominator of each of the following.

a) $\sqrt{\dfrac{3}{2}}$

b) $\dfrac{\sqrt[3]{7}}{\sqrt[3]{9}}$

Solution

a) $\sqrt{\dfrac{3}{2}} = \sqrt{\dfrac{3}{2} \cdot \dfrac{2}{2}} = \sqrt{\dfrac{6}{4}} = \dfrac{\sqrt{6}}{\sqrt{4}} = \dfrac{\sqrt{6}}{2}$

b) $\dfrac{\sqrt[3]{7}}{\sqrt[3]{9}} = \dfrac{\sqrt[3]{7}}{\sqrt[3]{9}} \cdot \dfrac{\sqrt[3]{3}}{\sqrt[3]{3}} = \dfrac{\sqrt[3]{21}}{\sqrt[3]{27}} = \dfrac{\sqrt[3]{21}}{3}$ **Now Try Exercise 65.** ■

Pairs of expressions of the form $a\sqrt{b} + c\sqrt{d}$ and $a\sqrt{b} - c\sqrt{d}$ are called **conjugates**. The product of such a pair contains no radicals and can be used to rationalize a denominator or a numerator.

EXAMPLE 6   Rationalize the numerator: $\dfrac{\sqrt{x} - \sqrt{y}}{5}$.

Solution

$$\dfrac{\sqrt{x} - \sqrt{y}}{5} = \dfrac{\sqrt{x} - \sqrt{y}}{5} \cdot \dfrac{\sqrt{x} + \sqrt{y}}{\sqrt{x} + \sqrt{y}} \qquad \text{The conjugate of } \sqrt{x} - \sqrt{y} \text{ is } \sqrt{x} + \sqrt{y}.$$

$$= \dfrac{\left(\sqrt{x}\right)^2 - \left(\sqrt{y}\right)^2}{5\sqrt{x} + 5\sqrt{y}} \qquad (A + B)(A - B) = A^2 - B^2$$

$$= \dfrac{x - y}{5\sqrt{x} + 5\sqrt{y}} \qquad \text{**Now Try Exercise 85.** ■}$$

❖ Rational Exponents

We are motivated to define *rational exponents* so that the properties for integer exponents hold for them as well. For example, we must have

$$a^{1/2} \cdot a^{1/2} = a^{1/2+1/2} = a^1 = a.$$

Thus we are led to define $a^{1/2}$ to mean $\sqrt{a}$. Similarly, $a^{1/n}$ would mean $\sqrt[n]{a}$. Again, if the laws of exponents are to hold, we must have

$$(a^{1/n})^m = (a^m)^{1/n} = a^{m/n}.$$

Thus we are led to define $a^{m/n}$ to mean $\sqrt[n]{a^m}$, or, equivalently, $\left(\sqrt[n]{a}\right)^m$.

Rational Exponents

For any real number a and any natural numbers m and n, $n \neq 1$, for which $\sqrt[n]{a}$ exists,

$$a^{1/n} = \sqrt[n]{a},$$
$$a^{m/n} = \sqrt[n]{a^m} = \left(\sqrt[n]{a}\right)^m, \quad \text{and}$$

$$a^{-m/n} = \dfrac{1}{a^{m/n}}.$$

We can use the definition of rational exponents to convert between radical notation and exponential notation.

EXAMPLE 7 Convert to radical notation and, if possible, simplify each of the following.

a) $7^{3/4}$ **b)** $8^{-5/3}$

c) $m^{1/6}$ **d)** $(-32)^{2/5}$

Solution

a) $7^{3/4} = \sqrt[4]{7^3}$, or $\left(\sqrt[4]{7}\right)^3$

b) $8^{-5/3} = \dfrac{1}{8^{5/3}} = \dfrac{1}{\left(\sqrt[3]{8}\right)^5} = \dfrac{1}{2^5} = \dfrac{1}{32}$

c) $m^{1/6} = \sqrt[6]{m}$

d) $(-32)^{2/5} = \sqrt[5]{(-32)^2} = \sqrt[5]{1024} = 4$,   or

       $(-32)^{2/5} = \left(\sqrt[5]{-32}\right)^2 = (-2)^2 = 4$ **Now Try Exercise 87.** ■

EXAMPLE 8 Convert each of the following to exponential notation.

a) $\left(\sqrt[4]{7xy}\right)^5$ **b)** $\sqrt[6]{x^3}$

Solution

a) $\left(\sqrt[4]{7xy}\right)^5 = (7xy)^{5/4}$

b) $\sqrt[6]{x^3} = x^{3/6} = x^{1/2}$ **Now Try Exercise 97.** ■

We can use the laws of exponents to simplify exponential expressions and radical expressions.

`GCM` **EXAMPLE 9** Simplify and then, if appropriate, write radical notation for each of the following.

a) $x^{5/6} \cdot x^{2/3}$ **b)** $(x + 3)^{5/2}(x + 3)^{-1/2}$

c) $\sqrt[3]{\sqrt{7}}$

Solution

a) $x^{5/6} \cdot x^{2/3} = x^{5/6+2/3} = x^{9/6} = x^{3/2} = \sqrt{x^3} = \sqrt{x^2}\sqrt{x} = x\sqrt{x}$

b) $(x + 3)^{5/2}(x + 3)^{-1/2} = (x + 3)^{5/2-1/2} = (x + 3)^2$

c) $\sqrt[3]{\sqrt{7}} = \sqrt[3]{7^{1/2}} = (7^{1/2})^{1/3} = 7^{1/6} = \sqrt[6]{7}$ **Now Try Exercise 107.** ■

We can add and subtract rational exponents on a graphing calculator. The FRAC feature from the MATH menu allows us to express the result as a fraction. The addition of the exponents in Example 9(a) is shown here.

```
5/6+2/3▸Frac
                      3/2
```

EXAMPLE 10   Write an expression containing a single radical: $\sqrt{a}\,\sqrt[6]{b^5}$.

Solution $\sqrt{a}\,\sqrt[6]{b^5} = a^{1/2}b^{5/6} = a^{3/6}b^{5/6} = (a^3b^5)^{1/6} = \sqrt[6]{a^3b^5}$

Now Try Exercise 117. ▪

(R.7) Exercise Set

Simplify. Assume that variables can represent any real number.

1. $\sqrt{(-21)^2}$

1. $\sqrt{(-7)^2}$

3. $\sqrt{9y^2}$

4. $\sqrt{64t^2}$

5. $\sqrt{(a-2)^2}$

6. $\sqrt{(2b+5)^2}$

7. $\sqrt[3]{-27x^3}$

8. $\sqrt[3]{-8y^3}$

9. $\sqrt[4]{81x^8}$

10. $\sqrt[4]{16z^{12}}$

11. $\sqrt[5]{32}$

12. $\sqrt[5]{-32}$

13. $\sqrt{180}$

14. $\sqrt{48}$

15. $\sqrt{72}$

16. $\sqrt{250}$

17. $\sqrt[3]{54}$

18. $\sqrt[3]{135}$

19. $\sqrt{128c^2d^4}$

20. $\sqrt{162c^4d^6}$

21. $\sqrt[4]{48x^6y^4}$

22. $\sqrt[4]{243m^5n^{10}}$

23. $\sqrt{x^2 - 4x + 4}$

24. $\sqrt{x^2 + 16x + 64}$

Simplify. Assume that no radicands were formed by raising negative quantities to even powers.

25. $\sqrt{15}\,\sqrt{35}$

26. $\sqrt{21}\,\sqrt{6}$

27. $\sqrt{8}\,\sqrt{10}$

28. $\sqrt{12}\,\sqrt{15}$

29. $\sqrt{2x^3y}\,\sqrt{12xy}$

30. $\sqrt{3y^4z}\,\sqrt{20z}$

31. $\sqrt[3]{3x^2y}\,\sqrt[3]{36x}$

32. $\sqrt[5]{8x^3y^4}\,\sqrt[5]{4x^4y}$

33. $\sqrt[3]{2(x+4)}\,\sqrt[3]{4(x+4)^4}$

34. $\sqrt[3]{4(x+1)^2}\,\sqrt[3]{18(x+1)^2}$

35. $\sqrt[8]{\dfrac{m^{16}n^{24}}{2^8}}$

36. $\sqrt[6]{\dfrac{m^{12}n^{24}}{64}}$

37. $\dfrac{\sqrt{40xy}}{\sqrt{8x}}$

38. $\dfrac{\sqrt[3]{40m}}{\sqrt[3]{5m}}$

39. $\dfrac{\sqrt[3]{3x^2}}{\sqrt[3]{24x^5}}$

40. $\dfrac{\sqrt{128a^2b^4}}{\sqrt{16ab}}$

41. $\sqrt[3]{\dfrac{64a^4}{27b^3}}$

42. $\sqrt{\dfrac{9x^7}{16y^8}}$

43. $\sqrt{\dfrac{7x^3}{36y^6}}$

44. $\sqrt[3]{\dfrac{2yz}{250z^4}}$

45. $5\sqrt{2} + 3\sqrt{32}$

46. $7\sqrt{12} - 2\sqrt{3}$

47. $6\sqrt{20} - 4\sqrt{45} + \sqrt{80}$

48. $2\sqrt{32} + 3\sqrt{8} - 4\sqrt{18}$

49. $8\sqrt{2x^2} - 6\sqrt{20x} - 5\sqrt{8x^2}$

50. $2\sqrt[3]{8x^2} + 5\sqrt[3]{27x^2} - 3\sqrt{x^3}$

51. $\left(\sqrt{8} + 2\sqrt{5}\right)\left(\sqrt{8} - 2\sqrt{5}\right)$

52. $\left(\sqrt{3} - \sqrt{2}\right)\left(\sqrt{3} + \sqrt{2}\right)$

53. $\left(2\sqrt{3} + \sqrt{5}\right)\left(\sqrt{3} - 3\sqrt{5}\right)$

54. $\left(\sqrt{6} - 4\sqrt{7}\right)\left(3\sqrt{6} + 2\sqrt{7}\right)$

55. $\left(\sqrt{2} - 5\right)^2$

56. $\left(1 + \sqrt{3}\right)^2$

57. $\left(\sqrt{5} - \sqrt{6}\right)^2$

58. $\left(\sqrt{3} + \sqrt{2}\right)^2$

59. *Distance from Airport.* An airplane is flying at an altitude of 3700 ft. The slanted distance directly to the airport is 14,200 ft. How far horizontally is the airplane from the airport?

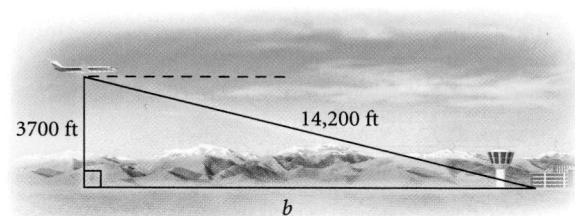

3700 ft   14,200 ft

b

60. *Bridge Expansion.* During a summer heat wave, a 2-mi bridge expands 2 ft in length. Assuming that the bulge occurs straight up the middle, estimate the height of the bulge. (In reality, bridges are built with expansion joints to control such buckling.)

61. An *equilateral triangle* is shown below.

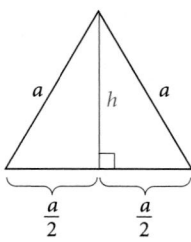

a h a

$\dfrac{a}{2}$ $\dfrac{a}{2}$

a) Find an expression for its height h in terms of a.
b) Find an expression for its area A in terms of a.

62. An isosceles right triangle has legs of length s. Find an expression for the length of the hypotenuse in terms of s.

63. The diagonal of a square has length $8\sqrt{2}$. Find the length of a side of the square.

64. The area of square $PQRS$ is 100 ft^2, and A, B, C, and D are the midpoints of the sides. Find the area of square $ABCD$.

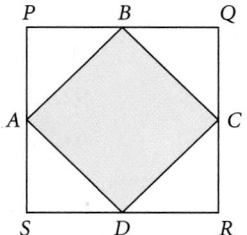

Rationalize the denominator.

65. $\sqrt{\dfrac{3}{7}}$ **66.** $\sqrt{\dfrac{2}{3}}$

67. $\dfrac{\sqrt[3]{7}}{\sqrt[3]{25}}$ **68.** $\dfrac{\sqrt[3]{5}}{\sqrt[3]{4}}$

69. $\sqrt[3]{\dfrac{16}{9}}$ **70.** $\sqrt[3]{\dfrac{3}{5}}$

71. $\dfrac{2}{\sqrt{3}-1}$ **72.** $\dfrac{6}{3+\sqrt{5}}$

73. $\dfrac{1-\sqrt{2}}{2\sqrt{3}-\sqrt{6}}$ **74.** $\dfrac{\sqrt{5}+4}{\sqrt{2}+3\sqrt{7}}$

75. $\dfrac{6}{\sqrt{m}-\sqrt{n}}$ **76.** $\dfrac{3}{\sqrt{v}+\sqrt{w}}$

Rationalize the numerator.

77. $\dfrac{\sqrt{50}}{3}$ **78.** $\dfrac{\sqrt{12}}{5}$

79. $\sqrt[3]{\dfrac{2}{5}}$ **80.** $\sqrt[3]{\dfrac{7}{2}}$

81. $\dfrac{\sqrt{11}}{\sqrt{3}}$

82. $\dfrac{\sqrt{5}}{\sqrt{2}}$

83. $\dfrac{9 - \sqrt{5}}{3 - \sqrt{3}}$

84. $\dfrac{8 - \sqrt{6}}{5 - \sqrt{2}}$

85. $\dfrac{\sqrt{a} + \sqrt{b}}{3a}$

86. $\dfrac{\sqrt{p} - \sqrt{q}}{1 + \sqrt{q}}$

Convert to radical notation and, if possible, simplify.

87. $y^{5/6}$

88. $x^{2/3}$

89. $16^{3/4}$

90. $4^{7/2}$

91. $125^{-1/3}$

92. $32^{-4/5}$

93. $a^{5/4}b^{-3/4}$

94. $x^{2/5}y^{-1/5}$

95. $m^{5/3}n^{7/3}$

96. $p^{7/6}q^{11/6}$

Convert to exponential notation.

97. $\sqrt[5]{17^3}$

98. $\left(\sqrt[4]{13}\right)^5$

99. $\left(\sqrt[5]{12}\right)^4$

100. $\sqrt[3]{20^2}$

101. $\sqrt[3]{\sqrt{11}}$

102. $\sqrt[3]{\sqrt[4]{7}}$

103. $\sqrt{5}\,\sqrt[3]{5}$

104. $\sqrt[3]{2}\,\sqrt{2}$

105. $\sqrt[5]{32^2}$

106. $\sqrt[3]{64^2}$

Simplify and then, if appropriate, write radical notation.

107. $(2a^{3/2})(4a^{1/2})$

108. $(3a^{5/6})(8a^{2/3})$

109. $\left(\dfrac{x^6}{9b^{-4}}\right)^{1/2}$

110. $\left(\dfrac{x^{2/3}}{4y^{-2}}\right)^{1/2}$

111. $\dfrac{x^{2/3}y^{5/6}}{x^{-1/3}y^{1/2}}$

112. $\dfrac{a^{1/2}b^{5/8}}{a^{1/4}b^{3/8}}$

113. $(m^{1/2}n^{5/2})^{2/3}$

114. $(x^{5/3}y^{1/3}z^{2/3})^{3/5}$

115. $a^{3/4}(a^{2/3} + a^{4/3})$

116. $m^{2/3}(m^{7/4} - m^{5/4})$

Write an expression containing a single radical and simplify.

117. $\sqrt[3]{6}\,\sqrt{2}$

118. $\sqrt{2}\,\sqrt[4]{8}$

119. $\sqrt[4]{xy}\,\sqrt[3]{x^2y}$

120. $\sqrt[3]{ab^2}\,\sqrt{ab}$

121. $\sqrt[3]{a^4\sqrt{a^3}}$

122. $\sqrt{a^3\sqrt[3]{a^2}}$

123. $\dfrac{\sqrt{(a + x)^3}\,\sqrt[3]{(a + x)^2}}{\sqrt[4]{a + x}}$

124. $\dfrac{\sqrt[4]{(x + y)^2}\,\sqrt[3]{x + y}}{\sqrt{(x + y)^3}}$

Collaborative Discussion and Writing

125. Explain how you would convince a classmate that $\sqrt{a + b}$ is not equivalent to $\sqrt{a} + \sqrt{b}$, for positive real numbers a and b.

126. Explain how you would determine whether $10\sqrt{26} - 50$ is positive or negative without carrying out the actual computation.

Synthesis

Simplify.

127. $\sqrt{1 + x^2} + \dfrac{1}{\sqrt{1 + x^2}}$

128. $\sqrt{1 - x^2} - \dfrac{x^2}{2\sqrt{1 - x^2}}$

129. $\left(\sqrt{a^{\sqrt{a}}}\right)^{\sqrt{a}}$

130. $(2a^3b^{5/4}c^{1/7})^4 \div (54a^{-2}b^{2/3}c^{6/5})^{-1/3}$

CHAPTER R   Summary and Review

Important Properties and Formulas

Properties of the Real Numbers

Commutative: $a + b = b + a;$
$ab = ba$

Associative: $a + (b + c) = (a + b) + c;$
$a(bc) = (ab)c$

Additive Identity: $a + 0 = 0 + a = a$

Additive Inverse: $-a + a = a + (-a) = 0$

Multiplicative Identity: $a \cdot 1 = 1 \cdot a = a$

Multiplicative Inverse: $a \cdot \dfrac{1}{a} = \dfrac{1}{a} \cdot a = 1$
$(a \neq 0)$

Distributive: $a(b + c) = ab + ac$

Absolute Value

For any real number a,

$$|a| = \begin{cases} a, & \text{if } a \geq 0, \\ -a, & \text{if } a < 0. \end{cases}$$

Properties of Exponents

For any real numbers a and b and any integers m and n, assuming 0 is not raised to a nonpositive power:

The Product Rule: $a^m \cdot a^n = a^{m+n}$

The Quotient Rule: $\dfrac{a^m}{a^n} = a^{m-n} \quad (a \neq 0)$

The Power Rule: $(a^m)^n = a^{mn}$

Raising a Product to a Power: $(ab)^m = a^m b^m$

Raising a Quotient to a Power:

$$\left(\dfrac{a}{b}\right)^m = \dfrac{a^m}{b^m} \quad (b \neq 0)$$

Compound Interest Formula

$$A = P\left(1 + \dfrac{r}{n}\right)^{nt}$$

Special Products of Binomials

$(A + B)^2 = A^2 + 2AB + B^2$
$(A - B)^2 = A^2 - 2AB + B^2$
$(A + B)(A - B) = A^2 - B^2$

Sum or Difference of Cubes

$A^3 + B^3 = (A + B)(A^2 - AB + B^2)$
$A^3 - B^3 = (A - B)(A^2 + AB + B^2)$

Equation-Solving Principles

The Addition Principle: If $a = b$ is true, then $a + c = b + c$ is true.

The Multiplication Principle: If $a = b$ is true, then $ac = bc$ is true.

The Principle of Zero Products: If $ab = 0$ is true, then $a = 0$ or $b = 0$, and if $a = 0$ or $b = 0$, then $ab = 0$.

The Principle of Square Roots: If $x^2 = k$, then $x = \sqrt{k}$ or $x = -\sqrt{k}$.

Properties of Radicals

Let a and b be any real numbers or expressions for which the given roots exist. For any natural numbers m and n $(n \neq 1)$:

If n is even, $\sqrt[n]{a^n} = |a|$.
If n is odd, $\sqrt[n]{a^n} = a$.
$\sqrt[n]{a} \cdot \sqrt[n]{b} = \sqrt[n]{ab}.$
$\sqrt[n]{\dfrac{a}{b}} = \dfrac{\sqrt[n]{a}}{\sqrt[n]{b}} \quad (b \neq 0).$
$\sqrt[n]{a^m} = \left(\sqrt[n]{a}\right)^m.$

(continued)

Rational Exponents

For any real number a and any natural numbers m and n, $n \neq 1$, for which $\sqrt[n]{a}$ exists,

$$a^{1/n} = \sqrt[n]{a},$$
$$a^{m/n} = \sqrt[n]{a^m} = \left(\sqrt[n]{a}\right)^m, \quad \text{and}$$
$$a^{-m/n} = \frac{1}{a^{m/n}}.$$

Pythagorean Theorem

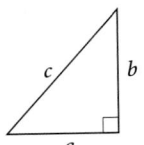

$$a^2 + b^2 = c^2$$

Review Exercises

Answers for all the review exercises appear in the answer section at the back of the book. If you get an incorrect answer, restudy the section indicated in red next to the exercise or the direction line that precedes it.

Determine whether the statement is true or false.

1. If $a < 0$, then $|a| = -a$.   [R.1]

2. For any real number a, $a \neq 0$, and any integers m and n, $a^m \cdot a^n = a^{mn}$.   [R.2]

3. If $a = b$ is true, then $a + c = b + c$ is true. [R.5]

4. The domain of an algebraic expression is the set of all real numbers for which the expression is defined.   [R.6]

In Exercises 5–10, consider the numbers

$$-7, \ 43, \ -\frac{4}{9}, \ \sqrt{17}, \ 0, \ 2.191191119\ldots, \ \sqrt[3]{64},$$

$$-\sqrt{2}, \ 4\frac{3}{4}, \ \frac{12}{7}, \ 102, \ \sqrt[5]{5}.$$

5. Which are rational numbers?   [R.1]

6. Which are whole numbers?   [R.1]

7. Which are integers?   [R.1]

8. Which are real numbers?   [R.1]

9. Which are natural numbers?   [R.1]

10. Which are irrational numbers?   [R.1]

11. Write interval notation for $\{x \mid -4 < x \leq 7\}$. [R.1]

Simplify.   [R.1]

12. $|24|$

13. $\left| -\frac{7}{8} \right|$

14. Find the distance between -5 and 5 on the number line.   [R.1]

Calculate.   [R.2]

15. $3 \cdot 2 - 4 \cdot 2^2 + 6(3 - 1)$

16. $\dfrac{3^4 - (6 - 7)^4}{2^3 - 2^4}$

Convert to decimal notation.   [R.2]

17. 8.3×10^{-5}

18. 2.07×10^7

Convert to scientific notation.   [R.2]

19. $405{,}000$

20. 0.00000039

Compute. Write the answer using scientific notation.   [R.2]

21. $(3.1 \times 10^5)(4.5 \times 10^{-3})$

22. $\dfrac{2.5 \times 10^{-8}}{3.2 \times 10^{13}}$

Simplify.

23. $(-3x^4y^{-5})(4x^{-2}y)$ [R.2]

24. $\dfrac{48a^{-3}b^2c^5}{6a^3b^{-1}c^4}$ [R.2]

25. $\sqrt[4]{81}$ [R.7]   **26.** $\sqrt[5]{-32}$ [R.7]

27. $\dfrac{b - a^{-1}}{a - b^{-1}}$ [R.6]

28. $\dfrac{\dfrac{x^2}{y} + \dfrac{y^2}{x}}{y^2 - xy + x^2}$ [R.6]

29. $\left(\sqrt{3} - \sqrt{7}\right)\left(\sqrt{3} + \sqrt{7}\right)$ [R.7]

30. $\left(5 - \sqrt{2}\right)^2$ [R.7]

31. $8\sqrt{5} + \dfrac{25}{\sqrt{5}}$ [R.7]

32. $(x + t)(x^2 - xt + t^2)$ [R.3]

33. $(5a + 4b)(2a - 3b)$ [R.3]

34. $(6x^2y - 3xy^2 + 5xy - 3) -$
$(-4x^2y - 4xy^2 + 3xy + 8)$ [R.3]

Factor.   [R.4]

35. $32x^4 - 40xy^3$

36. $y^3 + 3y^2 - 2y - 6$

37. $24x + 144 + x^2$

38. $9x^3 + 35x^2 - 4x$

39. $9x^2 - 30x + 25$

40. $8x^3 - 1$

41. $18x^2 - 3x + 6$

42. $4x^3 - 4x^2 - 9x + 9$

43. $6x^3 + 48$

44. $a^2b^2 - ab - 6$

45. $2x^2 + 5x - 3$

Solve.   [R.5]

46. $2x - 7 = 7$

47. $5x - 7 = 3x - 9$

48. $8 - 3x = -7 + 2x$

49. $6(2x - 1) = 3 - (x + 10)$

50. $y^2 + 16y + 64 = 0$

51. $x^2 - x = 20$

52. $2x^2 + 11x - 6 = 0$

53. $x(x - 2) = 3$

54. $y^2 - 16 = 0$

55. $n^2 - 7 = 0$

56. Divide and simplify:

$$\dfrac{3x^2 - 12}{x^2 + 4x + 4} \div \dfrac{x - 2}{x + 2}.$$ [R.6]

57. Subtract and simplify:

$$\dfrac{x}{x^2 + 9x + 20} - \dfrac{4}{x^2 + 7x + 12}.$$ [R.6]

Write an expression containing a single radical.   [R.7]

58. $\sqrt{y^5}\,\sqrt[3]{y^2}$

59. $\dfrac{\sqrt{(a + b)^3}\,\sqrt[3]{a + b}}{\sqrt[6]{(a + b)^7}}$

60. Convert to radical notation: $b^{7/5}$.   [R.7]

61. Convert to exponential notation:

$$\sqrt[8]{\dfrac{m^{32}n^{16}}{3^8}}.$$ [R.7]

62. Rationalize the denominator:

$$\dfrac{4 - \sqrt{3}}{5 + \sqrt{3}}.$$ [R.7]

63. How long is a guy wire that reaches from the top of a 17-ft pole to a point on the ground 8 ft from the bottom of the pole?   [R.7]

64. Calculate: $128 \div (-2)^3 \div (-2) \cdot 3$.   [R.2]

A. $\dfrac{8}{3}$ **B.** 24   **C.** 96   **D.** $\dfrac{512}{3}$

65. Factor completely: $9x^2 - 36y^2$.   [R.4]

A. $(3x + 6y)(3x - 6y)$
B. $3(x + 2y)(x - 2y)$
C. $9(x + 2y)(x - 2y)$
D. $9(x - 2y)^2$

Collaborative Discussion and Writing

66. Anya says that $15 - 6 \div 3 \cdot 4$ is 12. What mistake is she probably making?   [R.2]

67. A calculator indicates that
$$4^{21} = 4.398046511 \times 10^{12}.$$
How can you tell that this is an approximation? [R.2]

Synthesis

Mortgage Payments. *The formula*

$$M = P\left[\dfrac{\dfrac{r}{12}\left(1 + \dfrac{r}{12}\right)^{n}}{\left(1 + \dfrac{r}{12}\right)^{n} - 1}\right]$$

gives the monthly mortgage payment M on a home loan of P dollars at interest rate r, where n is the total number of payments (12 times the number of years). Use this formula in Exercises 68–71.   [R.2]

68. The cost of a house is $98,000. The down payment is $16,000, the interest rate is $6\frac{1}{2}$%, and the loan period is 25 yr. What is the monthly mortgage payment?

69. The cost of a house is $124,000. The down payment is $20,000, the interest rate is $5\frac{3}{4}$%, and the loan period is 30 yr. What is the monthly mortgage payment?

70. The cost of a house is $135,000. The down payment is $18,000, the interest rate is $7\frac{1}{2}$%, and the loan period is 20 yr. What is the monthly mortgage payment?

71. The cost of a house is $151,000. The down payment is $21,000, the interest rate is $6\frac{1}{4}$%, and the loan period is 25 yr. What is the monthly mortgage payment?

Multiply. Assume that all exponents are integers.   [R.3]

72. $(x^n + 10)(x^n - 4)$

73. $(t^a + t^{-a})^2$

74. $(y^b - z^c)(y^b + z^c)$

75. $(a^n - b^n)^3$

Factor.   [R.4]

76. $y^{2n} + 16y^n + 64$

77. $x^{2t} - 3x^t - 28$

78. $m^{6n} - m^{3n}$

CHAPTER R   Test

1. Consider the numbers
$$6\tfrac{6}{7},\ \sqrt{12},\ 0,\ -\frac{13}{4},\ \sqrt[3]{8},\ -1.2,\ 29,\ -5.$$
   a) Which are whole numbers?
   b) Which are irrational numbers?
   c) Which are integers but not natural numbers?
   d) Which are rational numbers but not integers?

Simplify.

2. $|-17.6|$ 3. $\left|\dfrac{15}{11}\right|$ 4. $|-5xy|$

5. Write interval notation for $\{x \mid -3 < x \le 6\}$. Then graph the interval.

6. Find the distance between -9 and 6 on the number line.

7. Calculate: $32 \div 2^3 - 12 \div 4 \cdot 3$.

8. Convert to scientific notation: $4{,}509{,}000$.

9. Convert to decimal notation: 8.6×10^{-5}.

10. Compute and write scientific notation for the answer:
$$\frac{2.7 \times 10^4}{3.6 \times 10^{-3}}.$$

Simplify.

11. $x^{-8} \cdot x^5$

12. $(2y^2)^3(3y^4)^2$

13. $(-3a^5b^{-4})(5a^{-1}b^3)$

14. $(5xy^4 - 7xy^2 + 4x^2 - 3) -$
    $(-3xy^4 + 2xy^2 - 2y + 4)$

15. $(y - 2)(3y + 4)$

16. $(4x - 3)^2$

17. $\dfrac{\dfrac{x}{y} - \dfrac{y}{x}}{x + y}$

18. $\sqrt{45}$

19. $\sqrt[3]{56}$

20. $3\sqrt{75} + 2\sqrt{27}$

21. $\sqrt{18}\ \sqrt{10}$

22. $(2 + \sqrt{3})(5 - 2\sqrt{3})$

Factor.

23. $8x^2 - 18$

24. $y^2 - 3y - 18$

25. $2n^2 + 5n - 12$

26. $x^3 + 10x^2 + 25x$

27. $m^3 - 8$

Solve.

28. $7x - 4 = 24$

29. $3(y - 5) + 6 = 8 - (y + 2)$

30. $2x^2 + 5x + 3 = 0$

31. $z^2 - 11 = 0$

32. Multiply and simplify:
$$\frac{x^2 + x - 6}{x^2 + 8x + 15} \cdot \frac{x^2 - 25}{x^2 - 4x + 4}.$$

33. Subtract and simplify: $\dfrac{x}{x^2 - 1} - \dfrac{3}{x^2 + 4x - 5}.$

34. Rationalize the denominator: $\dfrac{5}{7 - \sqrt{3}}.$

35. Convert to radical notation: $m^{3/8}$.

36. Convert to exponential notation: $\sqrt[6]{3^5}$.

37. How long is a guy wire that reaches from the top of a 12-ft pole to a point on the ground 5 ft from the bottom of the pole?

Synthesis

38. Multiply: $(x - y - 1)^2$.

Graphs, Functions, and Models

APPLICATION U.S. honey producers turned out 22% less honey in 2006 than in 1997. In 1997, approximately 198.2 million pounds were produced. This amount decreased to 154.8 million pounds in 2006. (*Source:* National Agricultural Statistics Services) Find the average rate of change in the number of pounds of honey produced from 1997 to 2006.

This problem appears as Exercise 39 in Section 1.3.

1.1 Introduction to Graphing

1.2 Functions and Graphs

1.3 Linear Functions, Slope, and Applications

1.4 Equations of Lines and Modeling

1.5 Linear Equations, Functions, Zeros, and Applications

1.6 Solving Linear Inequalities

1.1

Introduction to Graphing

❖ Plot points.
❖ Determine whether an ordered pair is a solution of an equation.
❖ Find the *x*- and *y*-intercepts of an equation of the form *Ax* + *By* = *C*.
❖ Graph equations.
❖ Find the distance between two points in the plane and find the midpoint of a segment.
❖ Find an equation of a circle with a given center and radius, and given an equation of a circle in standard form, find the center and the radius.
❖ Graph equations of circles.

❖ Graphs

Graphs provide a means of displaying, interpreting, and analyzing data in a visual format. It is not uncommon to open a newspaper or a magazine and encounter graphs. Examples of bar, circle, and line graphs are shown below.

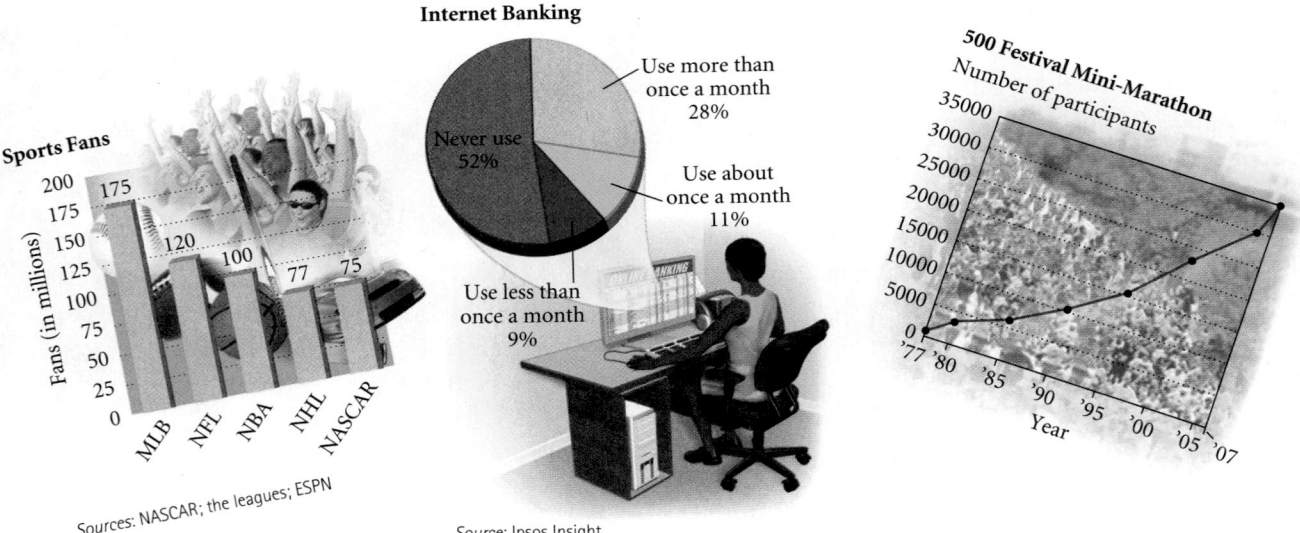

Sources: NASCAR; the leagues; ESPN

Source: Ipsos Insight

Many real-world situations can be modeled, or described mathematically, using equations in which two variables appear. We use a plane to graph a pair of numbers. To locate points on a plane, we use two perpendicular number lines, called **axes,** which intersect at (0,0). We call this point the **origin.** The horizontal axis is called the **x-axis,** and the vertical axis is called the **y-axis.** (Other variables, such as *a* and *b*, can also be used.) The axes divide the plane into four regions, called **quadrants,** denoted by Roman numerals and numbered counterclockwise from the upper right. Arrows show the positive direction of each axis.

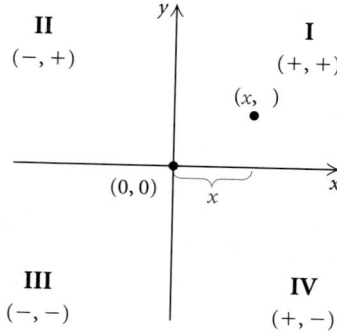

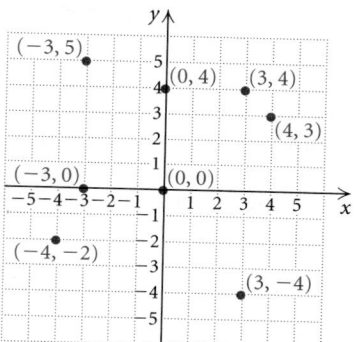

Each point (x, y) in the plane is described by an **ordered pair.** The first number, x, indicates the point's horizontal location with respect to the y-axis, and the second number, y, indicates the point's vertical location with respect to the x-axis. We call x the **first coordinate, x-coordinate,** or **abscissa.** We call y the **second coordinate, y-coordinate,** or **ordinate.** Such a representation is called the **Cartesian coordinate system** in honor of the French mathematician and philosopher René Descartes (1596–1650).

In the first quadrant, both coordinates of a point are positive. In the second quadrant, the first coordinate is negative and the second is positive. In the third quadrant, both coordinates are negative, and in the fourth quadrant, the first coordinate is positive and the second is negative.

EXAMPLE 1   Graph and label the points $(-3, 5)$, $(4, 3)$, $(3, 4)$, $(-4, -2)$, $(3, -4)$, $(0, 4)$, $(-3, 0)$, and $(0, 0)$.

Solution   To graph or **plot** $(-3, 5)$, we note that the x-coordinate, -3, tells us to move from the origin 3 units to the left of the y-axis. Then we move 5 units up from the x-axis.* To graph the other points, we proceed in a similar manner. (See the graph at left.) Note that the point $(4, 3)$ is different from the point $(3, 4)$.

Now Try Exercise 3. ■

❄ Solutions of Equations

Equations in two variables, like $2x + 3y = 18$, have solutions (x, y) that are ordered pairs such that when the first coordinate is substituted for x and the second coordinate is substituted for y, the result is a true equation. The first coordinate in an ordered pair generally represents the variable that occurs first alphabetically.

EXAMPLE 2   Determine whether each ordered pair is a solution of $2x + 3y = 18$.

a) $(-5, 7)$ **b)** $(3, 4)$

Solution   We substitute the ordered pair into the equation and determine whether the resulting equation is true.

a)

$$2x + 3y = 18$$

$$
\begin{array}{c|c}
2(-5) + 3(7) \ ? \ 18 & \text{We substitute } -5 \text{ for } x \text{ and} \\
-10 + 21 & 7 \text{ for } y \text{ (alphabetical order).} \\
11 \ \big| \ 18 & \text{FALSE}
\end{array}
$$

The equation $11 = 18$ is false, so $(-5, 7)$ is not a solution.

---

*We first saw notation such as $(-3, 5)$ in Section R.1. There the notation represented an open interval. Here the notation represents an ordered pair. The context in which the notation appears usually makes the meaning clear.

b)

$$2x + 3y = 18$$

$2(3) + 3(4)$? 18	We substitute 3 for x
$6 + 12$	and 4 for y.
18 $\mid$ 18 TRUE	

The equation $18 = 18$ is true, so $(3, 4)$ is a solution.

We can also perform these substitutions on a graphing calculator. When we substitute -5 for x and 7 for y, we get 11. Since $11 \neq 18$, $(-5, 7)$ is not a solution of the equation. When we substitute 3 for x and 4 for y, we get 18, so $(3, 4)$ is a solution.

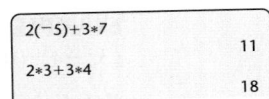

$2(-5)+3*7$	
	11
$2*3+3*4$	
	18

Now Try Exercise 11. ■

❋ Graphs of Equations

The equation considered in Example 2 actually has an infinite number of solutions. Since we cannot list all the solutions, we will make a drawing, called a **graph**, that represents them. On the following page are some suggestions for drawing graphs.

> **To Graph an Equation**
>
> To **graph an equation** is to make a drawing that represents the solutions of that equation.

Graphs of equations of the type $Ax + By = C$ are straight lines. Many such equations can be graphed conveniently using intercepts. The **x-intercept** of the graph of an equation is the point at which the graph crosses the x-axis. The **y-intercept** is the point at which the graph crosses the y-axis. We know from geometry that only one line can be drawn through two given points. Thus, if we know the intercepts, we can graph the line. To ensure that a computational error has not been made, it is a good idea to calculate and plot a third point as a check.

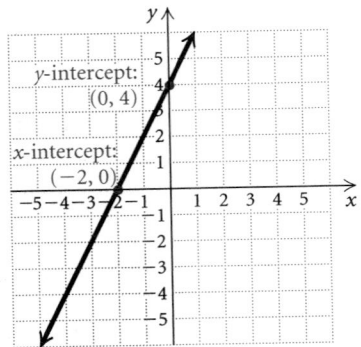

> **x- and y-Intercepts**
>
> An **x-intercept** is a point $(a, 0)$. To find a, let $y = 0$ and solve for x.
> A **y-intercept** is a point $(0, b)$. To find b, let $x = 0$ and solve for y.

EXAMPLE 3   Graph: $2x + 3y = 18$.

Solution   The graph is a line. To find ordered pairs that are solutions of this equation, we can replace either x or y with any number and then solve for the other variable. In this case, it is convenient to find the intercepts of the graph. For instance, if x is replaced with 0, then

$$2 \cdot 0 + 3y = 18$$
$$3y = 18$$
$$y = 6. \qquad \text{Dividing by 3}$$

Thus, $(0, 6)$ is a solution. It is the *y-intercept* of the graph. If y is replaced with 0, then

$$2x + 3 \cdot 0 = 18$$
$$2x = 18$$
$$x = 9. \qquad \text{Dividing by 2}$$

Thus, $(9, 0)$ is a solution. It is the *x-intercept* of the graph. We find a third solution as a check. If x is replaced with 5, then

$$2 \cdot 5 + 3y = 18$$
$$10 + 3y = 18$$
$$3y = 8 \qquad \text{Subtracting 10}$$
$$y = \tfrac{8}{3}. \qquad \text{Dividing by 3}$$

Thus, $\left(5, \tfrac{8}{3}\right)$ is a solution.

We list the solutions in a table and then plot the points. Note that the points appear to lie on a straight line.

Suggestions for Drawing Graphs

1. Calculate solutions and list the ordered pairs in a table.
2. Use graph paper.
3. Draw axes and label them with the variables.
4. Use arrows on the axes to indicate positive directions.
5. Scale the axes; that is, label the tick marks on the axes. Consider the ordered pairs found in part (1) above when choosing the scale.
6. Plot the ordered pairs, look for patterns, and complete the graph. Label the graph with the equation being graphed.

x	y	(x, y)
0	6	$(0, 6)$
9	0	$(9, 0)$
5	$\tfrac{8}{3}$	$\left(5, \tfrac{8}{3}\right)$

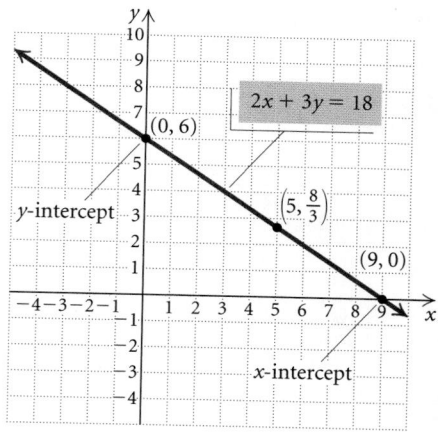

Were we to graph additional solutions of $2x + 3y = 18$, they would be on the same straight line. Thus, to complete the graph, we use a straightedge to draw a line, as shown in the figure. This line represents all solutions of the equation. Every point on the line represents a solution; every solution is represented by a point on the line.

Now Try Exercise 17. ■

When graphing some equations, it is convenient to first solve for y and then find ordered pairs. We can use the addition and multiplication principles to solve for y.

EQUATION SOLVING

REVIEW SECTION **R.5.**

GCM | **EXAMPLE 4**   Graph: $3x - 5y = -10$.

Solution   We first solve for y:

$$3x - 5y = -10$$
$$-5y = -3x - 10 \qquad \text{Subtracting } 3x \text{ on both sides}$$
$$y = \tfrac{3}{5}x + 2. \qquad \text{Multiplying by } -\tfrac{1}{5} \text{ on both sides}$$

By choosing multiples of 5 for x, we can avoid fraction values when calculating y. For example, if we choose -5 for x, we get

$$y = \tfrac{3}{5}x + 2 = \tfrac{3}{5}(-5) + 2 = -3 + 2 = -1.$$

The following table lists a few more points. We plot the points and draw the graph.

x	y	(x, y)
-5	-1	$(-5, -1)$
0	2	$(0, 2)$
5	5	$(5, 5)$

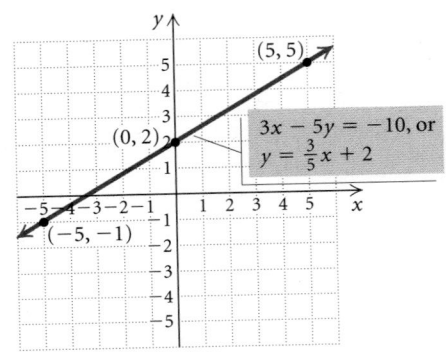

$3x - 5y = -10$, or $y = \tfrac{3}{5}x + 2$

Now Try Exercise 29. ■

In the equation $y = \tfrac{3}{5}x + 2$ in Example 4, the value of y *depends* on the value chosen for x, so x is said to be the **independent variable** and y the **dependent variable**.

We can graph an equation on a graphing calculator. Many calculators require an equation to be entered in the form "$y =$." In such a case, if the equation is not initially given in this form, it must be solved for y before it is entered in the calculator. For the equation $3x - 5y = -10$ in Example 4, we enter $y = \tfrac{3}{5}x + 2$ on the equation-editor, or "$y =$", screen in the form $y = (3/5)x + 2$, as shown in the window at left.

Next, we determine the portion of the xy-plane that will appear on the calculator's screen. That portion of the plane is called the **viewing window**.

The notation used in this text to denote a window setting consists of four numbers [L, R, B, T], which represent the **L**eft and **R**ight endpoints of the x-axis and the **B**ottom and **T**op endpoints of the y-axis, respectively. The window with the settings $[-10, 10, -10, 10]$ is the **standard viewing**

window. On some graphing calculators, the standard window can be selected quickly using the ZSTANDARD feature from the ZOOM menu.

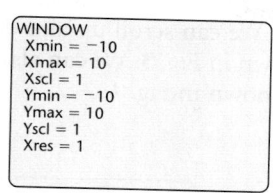

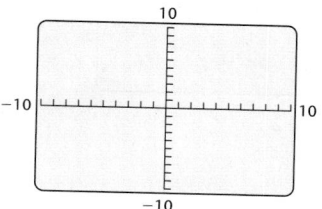

Xmin and Xmax are used to set the left and right endpoints of the x-axis, respectively; Ymin and Ymax are used to set the bottom and top endpoints of the y-axis, respectively. The settings Xscl and Yscl give the scales for the axes. For example, Xscl = 1 and Yscl = 1 means that there is 1 unit between tick marks on each of the axes. In this text, scaling factors other than 1 will be listed by the window unless they are readily apparent.

$y = \frac{3}{5}x + 2$

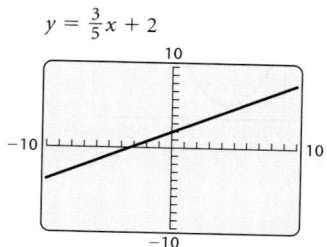

After entering the equation $y = (3/5)x + 2$ and choosing a viewing window, we can then draw the graph shown at left.

GCM **EXAMPLE 5**   Graph: $y = x^2 - 9x - 12$.

Solution   Note that since this equation is not of the form $Ax + By = C$, its graph is not a straight line. We make a table of values, plot enough points to obtain an idea of the shape of the curve, and connect them with a smooth curve. It is important to scale the axes to include most of the ordered pairs listed in the table. Here it is appropriate to use a larger scale on the y-axis than on the x-axis.

x	y	(x, y)
-3	24	$(-3, 24)$
-1	-2	$(-1, -2)$
0	-12	$(0, -12)$
2	-26	$(2, -26)$
4	-32	$(4, -32)$
5	-32	$(5, -32)$
10	-2	$(10, -2)$
12	24	$(12, 24)$

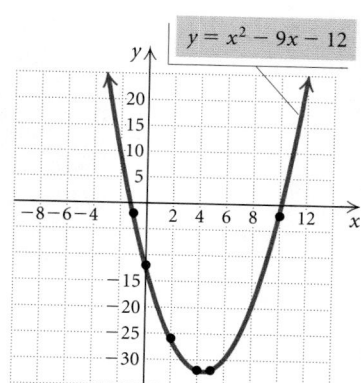

① Select values for x.

② Compute values for y.

Now Try Exercise 39. ■

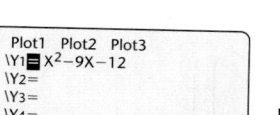

FIGURE 1

GCM A graphing calculator can be used to create a table of ordered pairs that are solutions of an equation. For the equation in Example 5, $y = x^2 - 9x - 12$, we first enter the equation on the equation-editor screen. (See Fig. 1.) Then we set up a table in AUTO mode by designating a value for TBLSTART and a value for ΔTBL. The calculator will produce a table

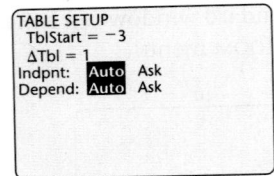

FIGURE 2

starting with the value of TBLSTART and continuing by adding ΔTBL to supply succeeding x-values. For the equation $y = x^2 - 9x - 12$, we let TBLSTART $= -3$ and ΔTBL $= 1$. (See Fig. 2.)

We can scroll up and down in the table to find values other than those shown in Fig. 3. We can also graph this equation on the graphing calculator, as shown in Fig. 4.

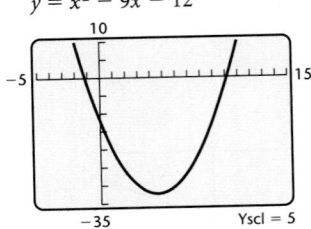

FIGURE 3

FIGURE 4

❋ The Distance Formula

Suppose that a photographer assigned to a story on the Panama Canal needs to determine the distance between two points, A and B, on opposite sides of the canal. One way in which he or she might proceed is to measure two legs of a right triangle that is situated as shown below. The Pythagorean equation, $a^2 + b^2 = c^2$, where c is the length of the hypotenuse and a and b are the lengths of the legs, can then be used to find the length of the hypotenuse, which is the distance from A to B.

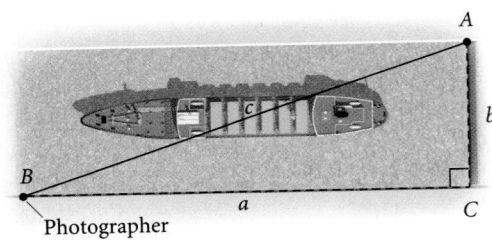

A similar strategy is used to find the distance between two points in a plane. For two points (x_1, y_1) and (x_2, y_2), we can draw a right triangle in which the legs have lengths $|x_2 - x_1|$ and $|y_2 - y_1|$.

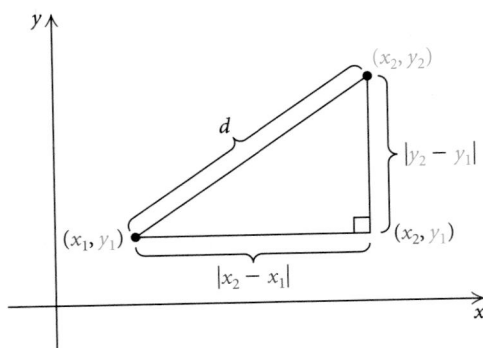

Using the Pythagorean equation, we have

$$d^2 = |x_2 - x_1|^2 + |y_2 - y_1|^2.$$

Substituting d for c, $|x_2 - x_1|$ for a, and $|y_2 - y_1|$ for b in the Pythagorean equation, $c^2 = a^2 + b^2$

Because we are squaring, parentheses can replace the absolute-value symbols:

$$d^2 = (x_2 - x_1)^2 + (y_2 - y_1)^2.$$

Taking the principal square root, we obtain the distance formula.

The Distance Formula

The **distance d** between any two points (x_1, y_1) and (x_2, y_2) is given by

$$d = \sqrt{(x_2 - x_1)^2 + (y_2 - y_1)^2}.$$

The subtraction of the x-coordinates can be done in any order, as can the subtraction of the y-coordinates. Although we derived the distance formula by considering two points not on a horizontal line or a vertical line, the distance formula holds for *any* two points.

EXAMPLE 6 Find the distance between each pair of points.

a) $(-2, 2)$ and $(3, -6)$ **b)** $(-1, -5)$ and $(-1, 2)$

Solution We substitute into the distance formula.

a) $d = \sqrt{[3 - (-2)]^2 + (-6 - 2)^2}$
$= \sqrt{5^2 + (-8)^2} = \sqrt{25 + 64}$
$= \sqrt{89} \approx 9.4$

b) $d = \sqrt{[-1 - (-1)]^2 + (-5 - 2)^2}$
$= \sqrt{0^2 + (-7)^2} = \sqrt{0 + 49}$
$= \sqrt{49} = 7$

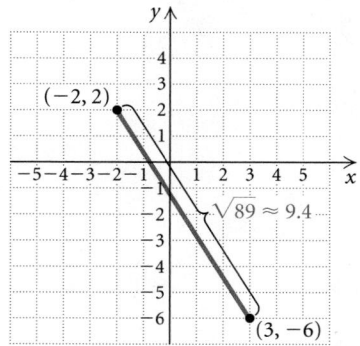

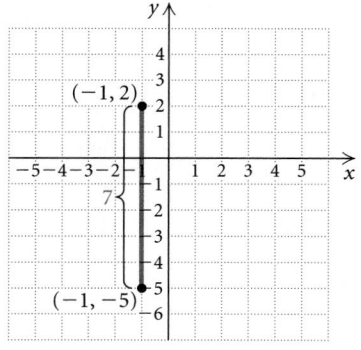

Now Try Exercises 63 and 71. ■

EXAMPLE 7 The point $(-2, 5)$ is on a circle that has $(3, -1)$ as its center. Find the length of the radius of the circle.

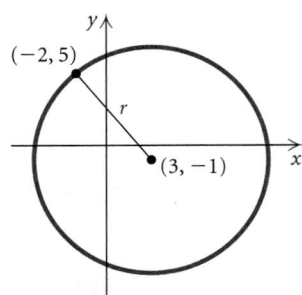

Solution Since the length of the radius is the distance from the center to a point on the circle, we substitute into the distance formula:

$$d = \sqrt{(x_2 - x_1)^2 + (y_2 - y_1)^2} \,;$$
$$r = \sqrt{[3 - (-2)]^2 + (-1 - 5)^2}$$

Substituting r for d, $(3, -1)$ for (x_2, y_2), and $(-2, 5)$ for (x_1, y_1). Either point can serve as (x_1, y_1).

$$= \sqrt{5^2 + (-6)^2} = \sqrt{61} \approx 7.8.$$

Rounded to the nearest tenth

The radius of the circle is approximately 7.8.

Now Try Exercise 77. ▩

✹ Midpoints of Segments

The distance formula can be used to develop a way of determining the *midpoint* of a segment when the endpoints are known. We state the formula and leave its proof to the exercises.

The Midpoint Formula

If the endpoints of a segment are (x_1, y_1) and (x_2, y_2), then the coordinates of the **midpoint** of the segment are

$$\left(\frac{x_1 + x_2}{2}, \frac{y_1 + y_2}{2} \right).$$

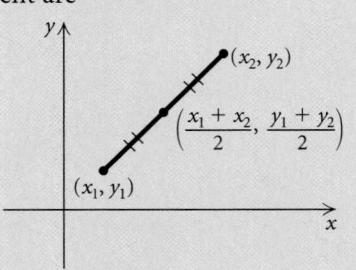

Note that we obtain the coordinates of the midpoint by averaging the coordinates of the endpoints. This is a good way to remember the midpoint formula.

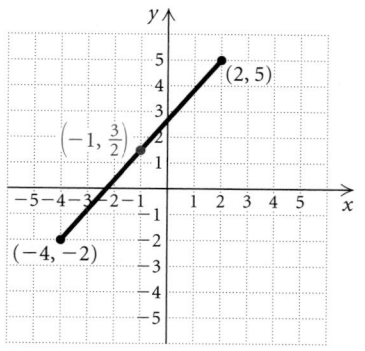

EXAMPLE 8 Find the midpoint of the segment whose endpoints are $(-4, -2)$ and $(2, 5)$.

Solution Using the midpoint formula, we obtain

$$\left(\frac{-4 + 2}{2}, \frac{-2 + 5}{2}\right) = \left(\frac{-2}{2}, \frac{3}{2}\right) = \left(-1, \frac{3}{2}\right).$$

Now Try Exercise 83. ■

EXAMPLE 9 The diameter of a circle connects the points $(2, -3)$ and $(6, 4)$ on the circle. Find the coordinates of the center of the circle.

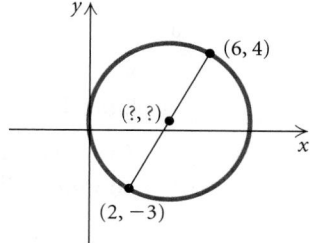

Solution Since the center of the circle is the midpoint of the diameter, we use the midpoint formula:

$$\left(\frac{2 + 6}{2}, \frac{-3 + 4}{2}\right), \quad \text{or} \quad \left(\frac{8}{2}, \frac{1}{2}\right), \quad \text{or} \quad \left(4, \frac{1}{2}\right).$$

The coordinates of the center are $\left(4, \frac{1}{2}\right)$.

Now Try Exercise 95. ■

❋ Circles

A **circle** is the set of all points in a plane that are a fixed distance r from a *center* (h, k). Thus if a point (x, y) is to be r units from the center, we must have

$$r = \sqrt{(x - h)^2 + (y - k)^2}. \qquad \text{Using the distance formula,} \\ d = \sqrt{(x_2 - x_1)^2 + (y_2 - y_1)^2}$$

Squaring both sides gives an equation of a circle. The distance r is the length of a *radius* of the circle.

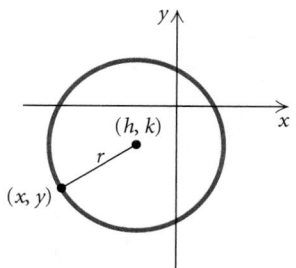

The Equation of a Circle

The standard form of the equation of a circle with center (h, k) and radius r is

$$(x - h)^2 + (y - k)^2 = r^2.$$

EXAMPLE 10   Find an equation of the circle having radius 5 and center $(3, -7)$.

Solution   Using the standard form, we have

$$[x - 3]^2 + [y - (-7)]^2 = 5^2 \qquad \text{Substituting}$$
$$(x - 3)^2 + (y + 7)^2 = 25. \qquad \qquad \text{\textbf{Now Try Exercise 99.}} \blacksquare$$

EXAMPLE 11   Graph the circle $(x + 5)^2 + (y - 2)^2 = 16$.

Solution   We write the equation in standard form to determine the center and the radius:

$$[x - (-5)]^2 + [y - 2]^2 = 4^2.$$

The center is $(-5, 2)$ and the radius is 4. We locate the center and draw the circle using a compass.

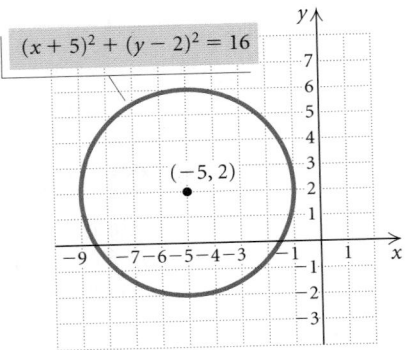

Now Try Exercise 111. ◼

Circles can also be graphed using a graphing calculator. We show one method of doing so here. Another method is discussed in the conics chapter.

When we graph a circle, we select a viewing window in which the distance between units is visually the same on both axes. This procedure is called **squaring the viewing window.** We do this so that the graph will not be distorted. A graph of the circle $x^2 + y^2 = 36$ in a nonsquared window is shown in Fig. 1.

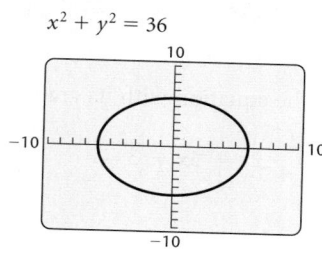

$x^2 + y^2 = 36$

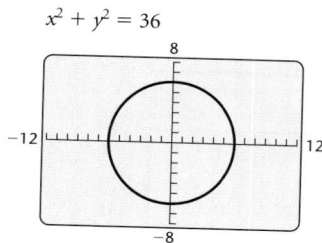

$x^2 + y^2 = 36$

FIGURE 1   FIGURE 2

On many graphing calculators, the ratio of the height to the width of the viewing screen is $\frac{2}{3}$. When we choose a window in which Xscl = Yscl and the length of the y-axis is $\frac{2}{3}$ the length of the x-axis, the window will be squared. The windows with dimensions $[-6, 6, -4, 4]$, $[-9, 9, -6, 6]$, and $[-12, 12, -8, 8]$ are examples of squared windows. A graph of the circle $x^2 + y^2 = 36$ in a squared window is shown in Fig. 2. Many graphing calculators have an option on the ZOOM menu that squares the window automatically.

GCM **EXAMPLE 12**   Graph the circle $(x - 2)^2 + (y + 1)^2 = 16$.

Solution   The circle $(x - 2)^2 + (y + 1)^2 = 16$ has center $(2, -1)$ and radius 4, so the viewing window $[-9, 9, -6, 6]$ is a good choice for the graph.

To graph a circle, we select the CIRCLE feature from the DRAW menu and enter the coordinates of the center and the length of the radius. The graph of the circle $(x - 2)^2 + (y + 1)^2 = 16$ is shown here. For more on graphing circles with a graphing calculator, see Section 7.2.

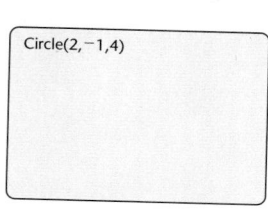

Circle(2,−1,4)

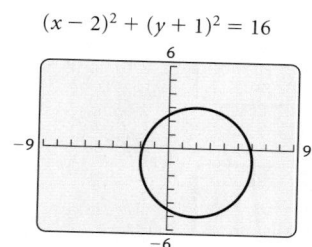

$(x - 2)^2 + (y + 1)^2 = 16$

Now Try Exercise 113. ■

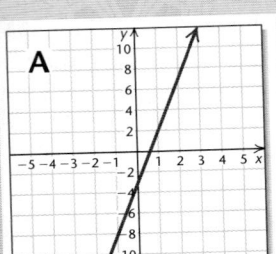

A

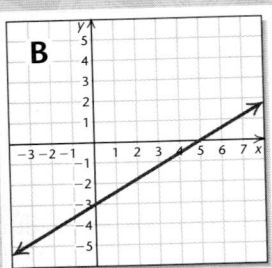

B

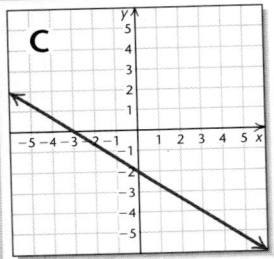

C

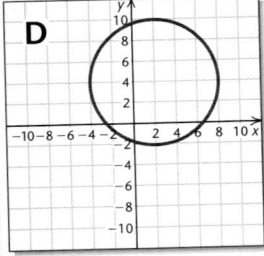

D

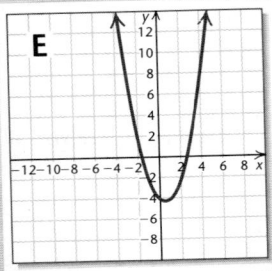

E

Visualizing the Graph

Match the equation with its graph.

1. $y = -x^2 + 5x - 3$

2. $3x - 5y = 15$

3. $(x - 2)^2 + (y - 4)^2 = 36$

4. $y - 5x = -3$

5. $x^2 + y^2 = \dfrac{25}{4}$

6. $15y - 6x = 90$

7. $y = -\dfrac{2}{3}x - 2$

8. $(x + 3)^2 + (y - 1)^2 = 16$

9. $3x + 5y = 15$

10. $y = x^2 - x - 4$

Answers on page A-4

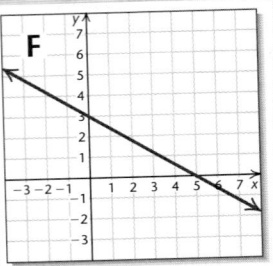

F

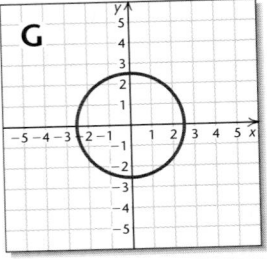

G

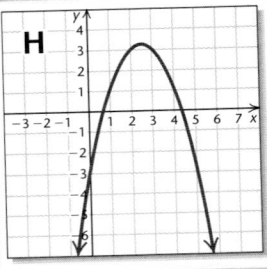

H

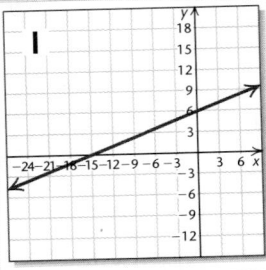

I

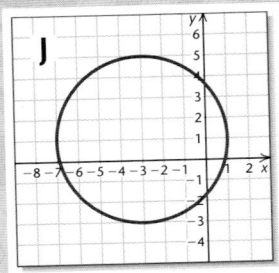

J

Exercise Set

Use this graph for Exercises 1 and 2.

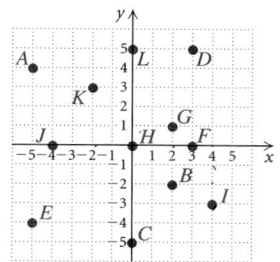

1. Find the coordinates of points *A, B, C, D, E,* and *F.*

2. Find the coordinates of points *G, H, I, J, K,* and *L.*

Graph and label the given points by hand.

3. $(4, 0), (-3, -5), (-1, 4), (0, 2), (2, -2)$

4. $(1, 4), (-4, -2), (-5, 0), (2, -4), (4, 0)$

5. $(-5, 1), (5, 1), (2, 3), (2, -1), (0, 1)$

6. $(4, 0), (4, -3), (-5, 2), (-5, 0), (-1, -5)$

Express the data pictured in the graph as ordered pairs, letting the first coordinate represent the year and the second coordinate the amount.

7. **Spring Break Vacation:**
The length of the average spring vacation is decreasing.

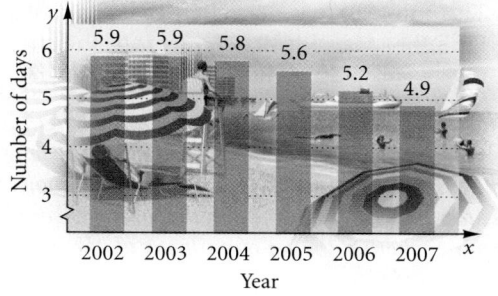

Source: Travelocity

8. **National Collegiate Athletic Association (NCAA):**
Total Advertisement Spending for Basketball Tournament

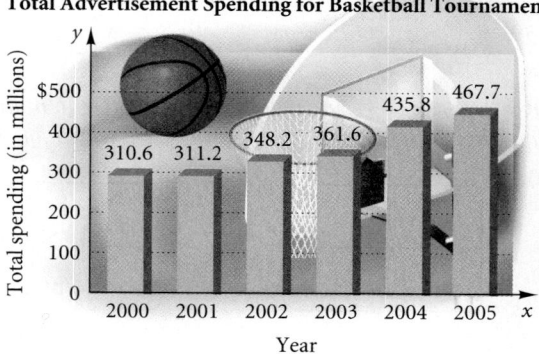

Source: NCAA

Use substitution to determine whether the given ordered pairs are solutions of the given equation.

9. $(-1, -9), (0, 2); \ y = 7x - 2$

10. $\left(\frac{1}{2}, 8\right), (-1, 6); \ y = -4x + 10$

11. $\left(\frac{2}{3}, \frac{3}{4}\right), \left(1, \frac{3}{2}\right); \ 6x - 4y = 1$

12. $(1.5, 2.6), (-3, 0); \ x^2 + y^2 = 9$

13. $\left(-\frac{1}{2}, -\frac{4}{5}\right), \left(0, \frac{3}{5}\right); \ 2a + 5b = 3$

14. $\left(0, \frac{3}{2}\right), \left(\frac{2}{3}, 1\right); \ 3m + 4n = 6$

15. $(-0.75, 2.75), (2, -1); \ x^2 - y^2 = 3$

16. $(2, -4), (4, -5); \ 5x + 2y^2 = 70$

Find the intercepts and then graph the line.

17. $5x - 3y = -15$ 18. $2x - 4y = 8$

19. $2x + y = 4$ 20. $3x + y = 6$

21. $4y - 3x = 12$ 22. $3y + 2x = -6$

Graph the equation.

23. $y = 3x + 5$ 24. $y = -2x - 1$

25. $x - y = 3$ 26. $x + y = 4$

27. $y = -\frac{3}{4}x + 3$ 28. $3y - 2x = 3$

29. $5x - 2y = 8$

30. $y = 2 - \frac{4}{3}x$

31. $x - 4y = 5$

32. $6x - y = 4$

33. $2x + 5y = -10$

34. $4x - 3y = 12$

35. $y = -x^2$

36. $y = x^2$

37. $y = x^2 - 3$

38. $y = 4 - x^2$

39. $y = -x^2 + 2x + 3$

40. $y = x^2 + 2x - 1$

In Exercises 41–44, use a graphing calculator to match the equation with one of the graphs (a)–(d), which follow.

a)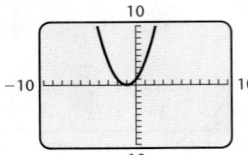

b)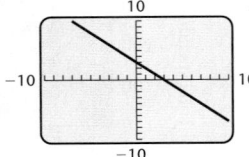

c)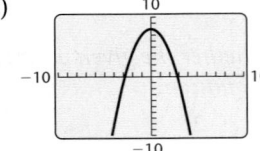

d)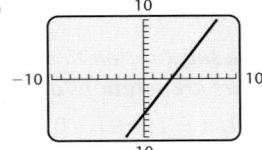

41. $y = 3 - x$

42. $2x - y = 6$

43. $y = x^2 + 2x + 1$

44. $y = 8 - x^2$

Use a graphing calculator to graph the equation in the standard window.

45. $y = 2x + 1$

46. $y = 3x - 4$

47. $4x + y = 7$

48. $5x + y = -8$

49. $y = \frac{1}{3}x + 2$

50. $y = \frac{3}{2}x - 4$

51. $2x + 3y = -5$

52. $3x + 4y = 1$

53. $y = x^2 + 6$

54. $y = x^2 - 8$

55. $y = 2 - x^2$

56. $y = 5 - x^2$

57. $y = x^2 + 4x - 2$

58. $y = x^2 - 5x + 3$

Graph the equation in the standard window and in the given window. Determine which window better shows the shape of the graph and the x- and y-intercepts.

59. $y = 3x^2 - 6$
$[-4, 4, -4, 4]$

60. $y = -2x + 24$
$[-15, 15, -10, 30]$, with Xscl $= 3$ and Yscl $= 5$

61. $y = -\frac{1}{6}x^2 + \frac{1}{12}$
$[-1, 1, -0.3, 0.3]$, with Xscl $= 0.1$ and Yscl $= 0.1$

62. $y = 6 - x^2$
$[-3, 3, -3, 3]$

Find the distance between the pair of points. Give an exact answer and, where appropriate, an approximation to three decimal places.

63. $(4, 6)$ and $(5, 9)$

64. $(-3, 7)$ and $(2, 11)$

65. $(-11, -8)$ and $(1, -13)$

66. $(-60, 5)$ and $(-20, 35)$

67. $(6, -1)$ and $(9, 5)$

68. $(-4, -7)$ and $(-1, 3)$

69. $\left(-8, \frac{7}{11}\right)$ and $\left(8, \frac{7}{11}\right)$

70. $\left(\frac{1}{2}, -\frac{4}{25}\right)$ and $\left(\frac{1}{2}, -\frac{13}{25}\right)$

71. $\left(-\frac{3}{5}, -4\right)$ and $\left(-\frac{3}{5}, \frac{2}{3}\right)$

72. $\left(-\frac{11}{3}, -\frac{1}{2}\right)$ and $\left(\frac{1}{3}, \frac{5}{2}\right)$

73. $(-4.2, 3)$ and $(2.1, -6.4)$

74. $(0.6, -1.5)$ and $(-8.1, -1.5)$

75. $(0, 0)$ and (a, b)

76. (r, s) and $(-r, -s)$

77. The points $(-3, -1)$ and $(9, 4)$ are the endpoints of the diameter of a circle. Find the length of the radius of the circle.

78. The point $(0, 1)$ is on a circle that has center $(-3, 5)$. Find the length of the diameter of the circle.

The converse of the Pythagorean theorem is also a true statement: If the sum of the squares of the lengths of two sides of a triangle is equal to the square of the length of the third side, then the triangle is a right triangle. Use the distance formula and the Pythagorean theorem to determine whether the set of points could be vertices of a right triangle.

79. $(-4, 5), (6, 1),$ and $(-8, -5)$

80. $(-3, 1), (2, -1),$ and $(6, 9)$

81. $(-4, 3)$, $(0, 5)$, and $(3, -4)$

82. The points $(-3, 4)$, $(2, -1)$, $(5, 2)$, and $(0, 7)$ are vertices of a quadrilateral. Show that the quadrilateral is a rectangle. (*Hint*: Show that the quadrilateral's opposite sides are the same length and that the two diagonals are the same length.)

Find the midpoint of the segment having the given endpoints.

83. $(4, -9)$ and $(-12, -3)$

84. $(7, -2)$ and $(9, 5)$

85. $\left(0, \frac{1}{2}\right)$ and $\left(-\frac{2}{5}, 0\right)$

86. $(0, 0)$ and $\left(-\frac{7}{13}, \frac{2}{7}\right)$

87. $(6.1, -3.8)$ and $(3.8, -6.1)$

88. $(-0.5, -2.7)$ and $(4.8, -0.3)$

89. $(-6, 5)$ and $(-6, 8)$

90. $(1, -2)$ and $(-1, 2)$

91. $\left(-\frac{1}{6}, -\frac{3}{5}\right)$ and $\left(-\frac{2}{3}, \frac{5}{4}\right)$

92. $\left(\frac{2}{9}, \frac{1}{3}\right)$ and $\left(-\frac{2}{5}, \frac{4}{5}\right)$

93. Graph the rectangle described in Exercise 82. Then determine the coordinates of the midpoint of each of the four sides. Are the midpoints vertices of a rectangle?

94. Graph the square with vertices $(-5, -1)$, $(7, -6)$, $(12, 6)$, and $(0, 11)$. Then determine the midpoint of each of the four sides. Are the midpoints vertices of a square?

95. The points $\left(\sqrt{7}, -4\right)$ and $\left(\sqrt{2}, 3\right)$ are endpoints of the diameter of a circle. Determine the center of the circle.

96. The points $\left(-3, \sqrt{5}\right)$ and $\left(1, \sqrt{2}\right)$ are endpoints of the diagonal of a square. Determine the center of the square.

In Exercises 97 and 98, how would you change the window so that the circle is not distorted? Answers may vary.

97. $(x + 3)^2 + (y - 2)^2 = 36$

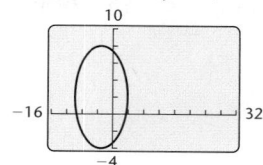

98. $(x - 4)^2 + (y + 5)^2 = 49$

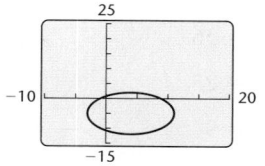

Find an equation for a circle satisfying the given conditions.

99. Center $(2, 3)$, radius of length $\frac{5}{3}$

100. Center $(4, 5)$, diameter of length 8.2

101. Center $(-1, 4)$, passes through $(3, 7)$

102. Center $(6, -5)$, passes through $(1, 7)$

103. The points $(7, 13)$ and $(-3, -11)$ are at the ends of a diameter.

104. The points $(-9, 4)$, $(-2, 5)$, $(-8, -3)$, and $(-1, -2)$ are vertices of an inscribed square.

105. Center $(-2, 3)$, tangent (touching at one point) to the y-axis

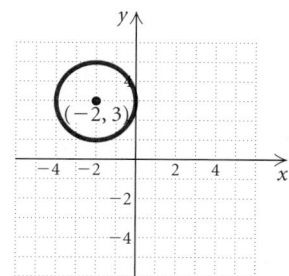

106. Center $(4, -5)$, tangent to the x-axis

Find the center and the radius of the circle. Then graph the circle by hand. Check your graph with a graphing calculator.

107. $x^2 + y^2 = 4$

108. $x^2 + y^2 = 81$

109. $x^2 + (y - 3)^2 = 16$

110. $(x + 2)^2 + y^2 = 100$

111. $(x - 1)^2 + (y - 5)^2 = 36$

112. $(x - 7)^2 + (y + 2)^2 = 25$

113. $(x + 4)^2 + (y + 5)^2 = 9$

114. $(x + 1)^2 + (y - 2)^2 = 64$

Find the equation of the circle. Express the equation in standard form.

115.

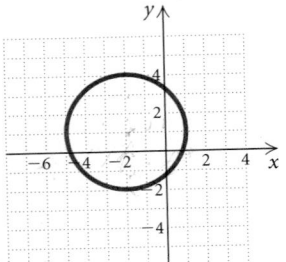

116.

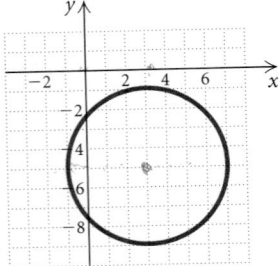

117.

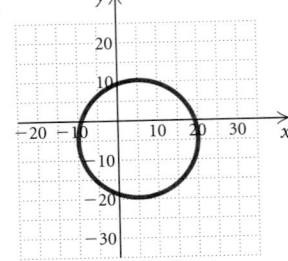

118.

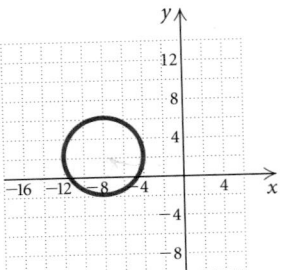

Collaborative Discussion and Writing

To the student and the instructor: The Collaborative Discussion and Writing exercises are meant to be answered with one or more sentences. They can be discussed and answered collaboratively by the entire class or by small groups. Because of their open-ended nature, the answers to these exercises do not appear at the back of the book. They are denoted by the words "Discussion and Writing."

119. Explain how the Pythagorean theorem is used to develop the equation of a circle in standard form.

120. Explain how you could find the coordinates of a point $\frac{7}{8}$ of the way from point A to point B.

Synthesis

To the student and the instructor: The Synthesis exercises found at the end of every exercise set challenge students to combine concepts or skills studied in that section or in preceding parts of the text.

121. If the point (p, q) is in the fourth quadrant, in which quadrant is the point $(q, -p)$?

Find the distance between the pair of points and find the midpoint of the segment having the given points as endpoints.

122. $\left(a, \dfrac{1}{a}\right)$ and $\left(a + h, \dfrac{1}{a + h}\right)$

123. $\left(a, \sqrt{a}\right)$ and $\left(a + h, \sqrt{a + h}\right)$

Find an equation of a circle satisfying the given conditions.

124. Center $(-5, 8)$ with a circumference of 10π units

125. Center $(2, -7)$ with an area of 36π square units

126. Find the point on the x-axis that is equidistant from the points $(-4, -3)$ and $(-1, 5)$.

127. Find the point on the y-axis that is equidistant from the points $(-2, 0)$ and $(4, 6)$.

128. Determine whether the points $(-1, -3)$, $(-4, -9)$, and $(2, 3)$ are collinear.

129. *Swimming Pool.*   A swimming pool is being constructed in the corner of a yard, as shown. Before installation, the contractor needs to know measurements a_1 and a_2. Find them.

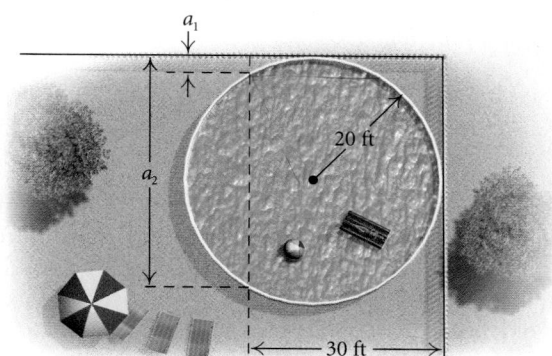

130. *An Arch of a Circle in Carpentry.*   Ace Carpentry needs to cut an arch for the top of an entranceway. The arch needs to be 8 ft wide and 2 ft high. To draw the arch, the carpenters will use a stretched string with chalk attached at an end as a compass.

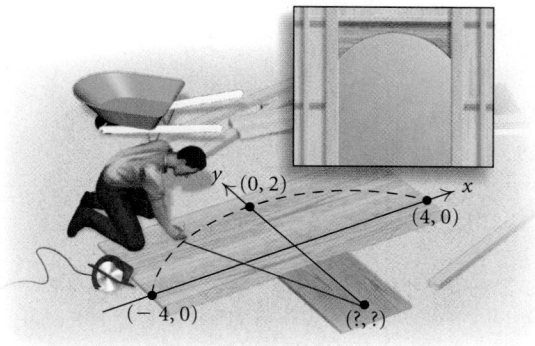

a) Using a coordinate system, locate the center of the circle.
b) What radius should the carpenters use to draw the arch?

*Determine whether each of the following points lies on the **unit circle,** $x^2 + y^2 = 1$.*

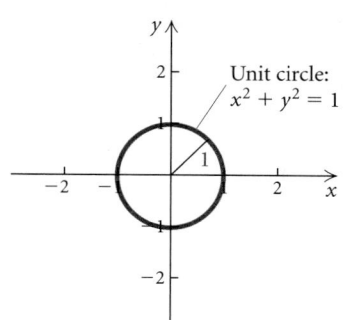

131. $\left(\dfrac{\sqrt{3}}{2}, -\dfrac{1}{2}\right)$ **132.** $(0, -1)$

133. $\left(-\dfrac{\sqrt{2}}{2}, \dfrac{\sqrt{2}}{2}\right)$ **134.** $\left(\dfrac{1}{2}, -\dfrac{\sqrt{3}}{2}\right)$

135. Prove the midpoint formula by showing that:

a) $\left(\dfrac{x_1 + x_2}{2}, \dfrac{y_1 + y_2}{2}\right)$ is equidistant from the points (x_1, y_1) and (x_2, y_2); and
b) the distance from (x_1, y_1) to the midpoint plus the distance from (x_2, y_2) to the midpoint equals the distance from (x_1, y_1) to (x_2, y_2).

136. Consider any right triangle with base b and height h, situated as shown. Show that the midpoint of the hypotenuse P is equidistant from the three vertices of the triangle.

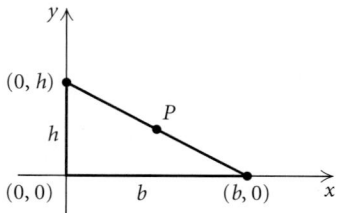

1.2 Functions and Graphs

❀ Determine whether a correspondence or a relation is a function.

❀ Find function values, or outputs, using a formula or a graph.

❀ Graph functions.

❀ Determine whether a graph is that of a function.

❀ Find the domain and the range of a function.

❀ Solve applied problems using functions.

We now focus our attention on a concept that is fundamental to many areas of mathematics—the idea of a *function*.

❀ Functions

We first consider an application.

Child's Age Related to Recommended Daily Amount of Fiber.   The recommended minimum amount of dietary fiber needed each day for children age 3 and older is the child's age, in years, plus 5 grams of fiber. If a child is 7 years old, the minimum recommendation per day is $7 + 5$, or 12, grams of fiber. Similarly, a 3-year-old needs $3 + 5$, or 8, grams of fiber. (*Sources: American Family Physician,* April 1996; American Health Foundation) We can express this relationship with a set of ordered pairs, a graph, and an equation.

x	y	Ordered Pairs: (x, y)	Correspondence
3	8	$(3, 8)$	$3 \longrightarrow 8$
$4\frac{1}{2}$	$9\frac{1}{2}$	$\left(4\frac{1}{2}, 9\frac{1}{2}\right)$	$4\frac{1}{2} \longrightarrow 9\frac{1}{2}$
7	12	$(7, 12)$	$7 \longrightarrow 12$
$10\frac{2}{3}$	$15\frac{2}{3}$	$\left(10\frac{2}{3}, 15\frac{2}{3}\right)$	$10\frac{2}{3} \longrightarrow 15\frac{2}{3}$
14	19	$(14, 19)$	$14 \longrightarrow 19$

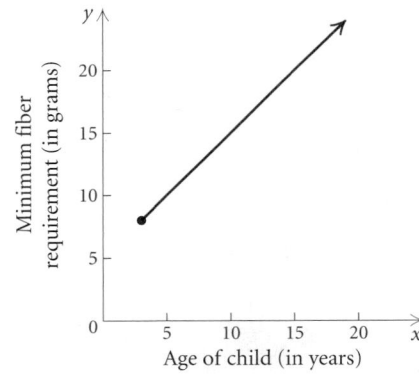

The ordered pairs express a relationship, or correspondence, between the first and second coordinates. We can see this relationship in the graph as well. The equation that describes the correspondence is

$$y = x + 5, \ x \geq 3.$$

This is an example of a *function*. In this case, grams of fiber y is a function of age x; that is, y is a function of x, where x is the independent variable and y is the dependent variable.

Let's consider some other correspondences before giving the definition of a function.

First Set	Correspondence	Second Set
To each registered student	there corresponds	an I. D. number.
To each laptop sold	there corresponds	its price.
To each real number	there corresponds	the square of that number.

In each correspondence, the first set is called the **domain** and the second set is called the **range.** For each member, or **element,** in the domain, there is *exactly one* member in the range to which it corresponds. Thus each registered student has exactly *one* I. D. number, each laptop has exactly *one* price, and each real number has exactly *one* square. Each correspondence is a *function.*

Function

A **function** is a correspondence between a first set, called the **domain,** and a second set, called the **range,** such that each member of the domain corresponds to *exactly one* member of the range.

It is important to note that not every correspondence between two sets is a function.

EXAMPLE 1 Determine whether each of the following correspondences is a function.

a) $-6 \to 36$, $6 \to 36$
$-3 \to 9$, $3 \to 9$
$0 \to 0$

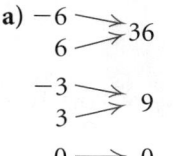

b) Helen Mirren → *The Queen*
Jennifer Hudson → *Dreamgirls*
Leonardo DiCaprio → *Blood Diamond*, *The Departed*
Jamie Foxx → *Dreamgirls*

Solution

a) This correspondence *is* a function because each member of the domain corresponds to exactly one member of the range. Note that the definition of a function allows more than one member of the domain to correspond to the same member of the range.

b) This correspondence *is not* a function because there is a member of the domain (Leonardo DiCaprio) that is paired with more than one member of the range (*Blood Diamond* and *The Departed*). **Now Try Exercise 7.**

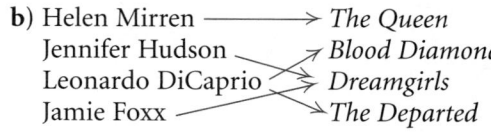

EXAMPLE 2   Determine whether each of the following correspondences is a function.

DOMAIN	CORRESPONDENCE	RANGE
a) Years in which a presidential election occurs	The person elected	A set of presidents
b) The integers	Each integer's cube root	A subset of the real numbers
c) All states in the United States	A senator from that state	The set of all U.S. senators
d) The set of all U.S. senators	The state a senator represents	All states in the United States

Solution

a) This correspondence *is* a function because in each presidential election *exactly one* president is elected.

b) This correspondence *is* a function because each integer has *exactly one* cube root.

c) This correspondence *is not* a function because each state can be paired with *two* different senators.

Wayne Allard

Ken Salazar

d) This correspondence *is* a function because each senator represents only one state.

Now Try Exercise 11. ■

When a correspondence between two sets is not a function, it may still be an example of a **relation**.

> **Relation**
> A **relation** is a correspondence between a first set, called the **domain**, and a second set, called the **range**, such that each member of the domain corresponds to *at least one* member of the range.

All the correspondences in Examples 1 and 2 are relations, but, as we have seen, not all are functions. Relations are sometimes written as sets of ordered pairs (as we saw earlier in the example on dietary fiber) in which elements of the domain are the first coordinates of the ordered pairs and elements of the range are the second coordinates. For example, instead of writing $-3 \longrightarrow 9$, as we did in Example 1(a), we could write the ordered pair $(-3, 9)$.

EXAMPLE 3 Determine whether each of the following relations is a function. Identify the domain and the range.

a) $\{(9, -5), (9, 5), (2, 4)\}$

b) $\{(-2, 5), (5, 7), (0, 1), (4, -2)\}$

c) $\{(-5, 3), (0, 3), (6, 3)\}$

Solution

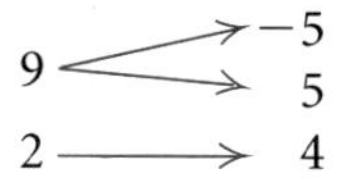

a) The relation *is not* a function because the ordered pairs $(9, -5)$ and $(9, 5)$ have the same first coordinate and different second coordinates. (See Fig. 1.)

The domain is the set of all first coordinates: $\{9, 2\}$.

The range is the set of all second coordinates: $\{-5, 5, 4\}$.

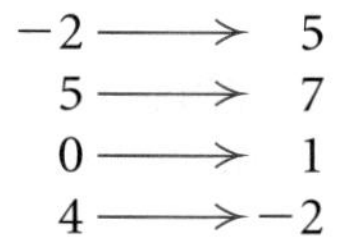

b) The relation *is* a function because *no* two ordered pairs have the same first coordinate and different second coordinates. (See Fig. 2.)

The domain is the set of all first coordinates: $\{-2, 5, 0, 4\}$.

The range is the set of all second coordinates: $\{5, 7, 1, -2\}$.

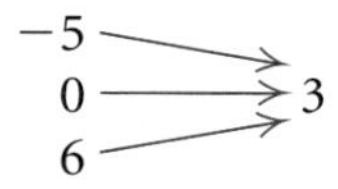

c) The relation *is* a function because *no* two ordered pairs have the same first coordinate and different second coordinates. (See Fig. 3.)

The domain is $\{-5, 0, 6\}$.

The range is $\{3\}$.   **Now Try Exercise 17.** ■

9 → −5
9 → 5
2 → 4

FIGURE 1

−2 → 5
5 → 7
0 → 1
4 → −2

FIGURE 2

−5 →
0 → 3
6 →

FIGURE 3

❇ Notation for Functions

Functions used in mathematics are often given by equations. They generally require that certain calculations be performed in order to determine which member of the range is paired with each member of the domain. For example, in Section 1.1 we graphed the function $y = x^2 - 9x - 12$ by doing calculations like the following:

for $x = -2, y = (-2)^2 - 9(-2) - 12 = 10$,

for $x = 0, y = 0^2 - 9 \cdot 0 - 12 = -12$,   and

for $x = 1, y = 1^2 - 9 \cdot 1 - 12 = -20$.

A more concise notation is often used. For $y = x^2 - 9x - 12$, the **inputs** (members of the domain) are values of x substituted into the equation. The **outputs** (members of the range) are the resulting values of y. If we call the

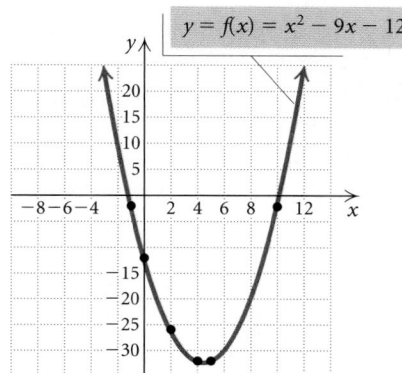

$y = f(x) = x^2 - 9x - 12$

function f, we can use x to represent an arbitrary *input* and $f(x)$—read "f of x," or "f at x," or "the value of f at x"—to represent the corresponding *output*. In this notation, the function given by $y = x^2 - 9x - 12$ is written as $f(x) = x^2 - 9x - 12$ and the above calculations would be

$$f(-2) = (-2)^2 - 9(-2) - 12 = 10,$$
$$f(0) = 0^2 - 9 \cdot 0 - 12 = -12,$$
$$f(1) = 1^2 - 9 \cdot 1 - 12 = -20. \qquad \text{Keep in mind that } f(x)$$
$$\text{does not mean } f \cdot x.$$

Thus, instead of writing "when $x = -2$, the value of y is 10," we can simply write "$f(-2) = 10$," which can be read as "f of -2 is 10" or "for the input -2, the output of f is 10." The letters g and h are also often used to name functions.

GCM   **EXAMPLE 4**   A function f is given by $f(x) = 2x^2 - x + 3$. Find each of the following.

a) $f(0)$ **b)** $f(-7)$
c) $f(5a)$ **d)** $f(a - 4)$

Solution   We can think of this formula as follows:

$$f(\blacksquare) = 2(\blacksquare)^2 - (\blacksquare) + 3.$$

Then to find an output for a given input we think: "Whatever goes in the blank on the left goes in the blank(s) on the right." This gives us a "recipe" for finding outputs.

a) $f(0) = 2(0)^2 - 0 + 3 = 0 - 0 + 3 = 3$
b) $f(-7) = 2(-7)^2 - (-7) + 3 = 2 \cdot 49 + 7 + 3 = 108$
c) $f(5a) = 2(5a)^2 - 5a + 3 = 2 \cdot 25a^2 - 5a + 3 = 50a^2 - 5a + 3$
d) $f(a - 4) = 2(a - 4)^2 - (a - 4) + 3$
$$= 2(a^2 - 8a + 16) - a + 4 + 3$$
$$= 2a^2 - 16a + 32 - a + 4 + 3$$
$$= 2a^2 - 17a + 39$$

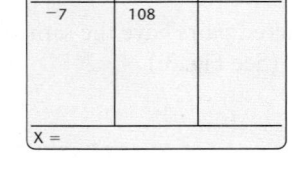

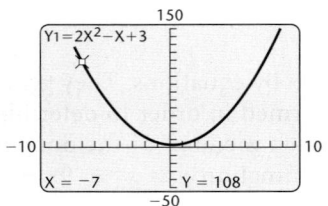

We can find function values with a graphing calculator. Most calculators do not use function notation "$f(x) = \ldots$" to enter a function formula. Instead, we must enter the function using "$y = \ldots$." At left, we illustrate finding $f(-7)$ from part (b), first with the TABLE feature set in ASK mode and then with the VALUE feature from the CALC menu. In both screens, we see that $f(-7) = 108$. **Now Try Exercise 21.** ◼

✤ Graphs of Functions

We graph functions the same way we graph equations. We find ordered pairs (x, y), or $(x, f(x))$, plot points, and complete the graph.

EXAMPLE 5 Graph each of the following functions.

a) $f(x) = x^2 - 5$ b) $f(x) = x^3 - x$ c) $f(x) = \sqrt{x + 4}$

Solution We select values for x and find the corresponding values of $f(x)$. Then we plot the points and connect them with a smooth curve.

a) $f(x) = x^2 - 5$

x	$f(x)$	$(x, f(x))$
-3	4	$(-3, 4)$
-2	-1	$(-2, -1)$
-1	-4	$(-1, -4)$
0	-5	$(0, -5)$
1	-4	$(1, -4)$
2	-1	$(2, -1)$
3	4	$(3, 4)$

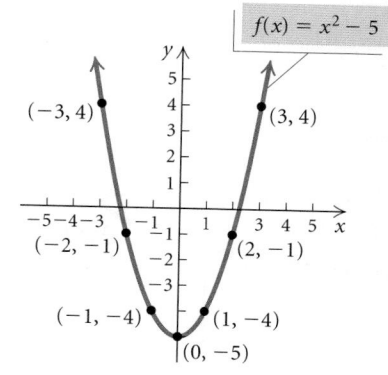

b) $f(x) = x^3 - x$

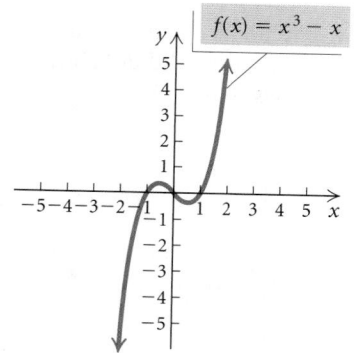

c) $f(x) = \sqrt{x + 4}$

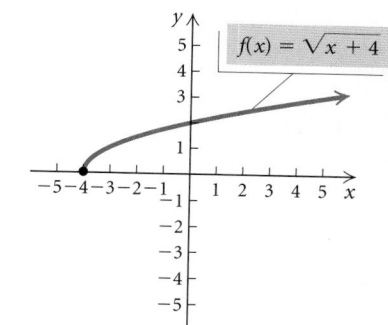

We can check the graphs with a graphing calculator. The checks for parts (b) and (c) are shown at left.

Now Try Exercise 33. ■

Function values can be also determined from a graph.

EXAMPLE 6 For the function $f(x) = x^2 - 5$, use the graph to find each of the following function values.

a) $f(3)$ b) $f(-2)$

Solution

a) To find the function value $f(3)$ from the graph at left, we locate the input 3 on the horizontal axis, move vertically to the graph of the function, and then move horizontally to find the output on the vertical axis. We see that $f(3) = 4$.

$y = x^3 - x$

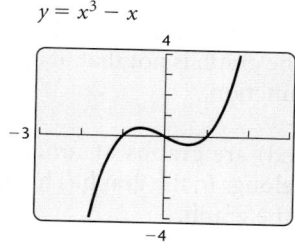

$y = \sqrt{x + 4}$

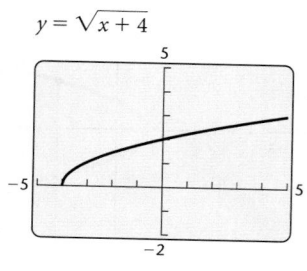

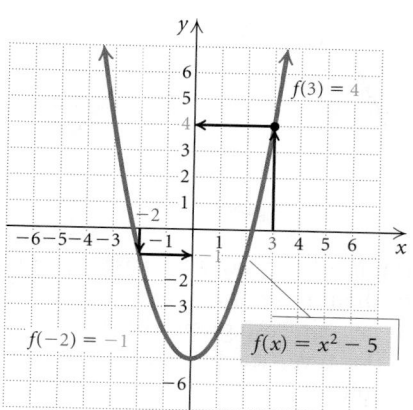

b) To find the function value $f(-2)$, we locate the input -2 on the horizontal axis, move vertically to the graph, and then move horizontally to find the output on the vertical axis. We see that $f(-2) = -1$.

Now Try Exercise 37. ■

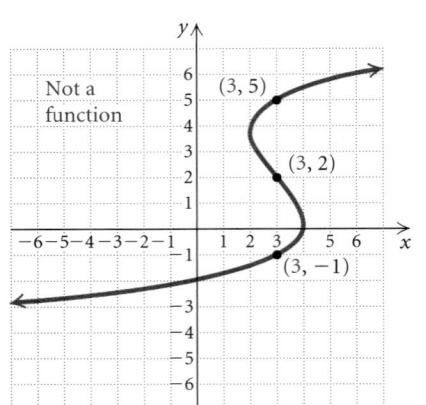

Since 3 is paired with more than one member of the range, the graph does not represent a function.

We know that when one member of the domain is paired with two or more different members of the range, the correspondence *is not* a function. Thus, when a graph contains two or more different points with the same first coordinate, the graph cannot represent a function. (See the graph at left; note that 3 is paired with $-1, 2$, and 5.) Points sharing a common first coordinate are vertically above or below each other. This leads us to the *vertical-line test*.

The Vertical-Line Test

If it is possible for a vertical line to cross a graph more than once, then the graph *is not* the graph of a function.

To apply the vertical-line test, we try to find a vertical line that crosses the graph more than once. If we succeed, then the graph is not that of a function. If we do not, then the graph is that of a function.

EXAMPLE 7 Which of graphs (a)–(f) (in red) are graphs of functions? In graph (f), the solid dot shows that $(-1, 1)$ belongs to the graph. The open circle shows that $(-1, -2)$ does *not* belong to the graph.

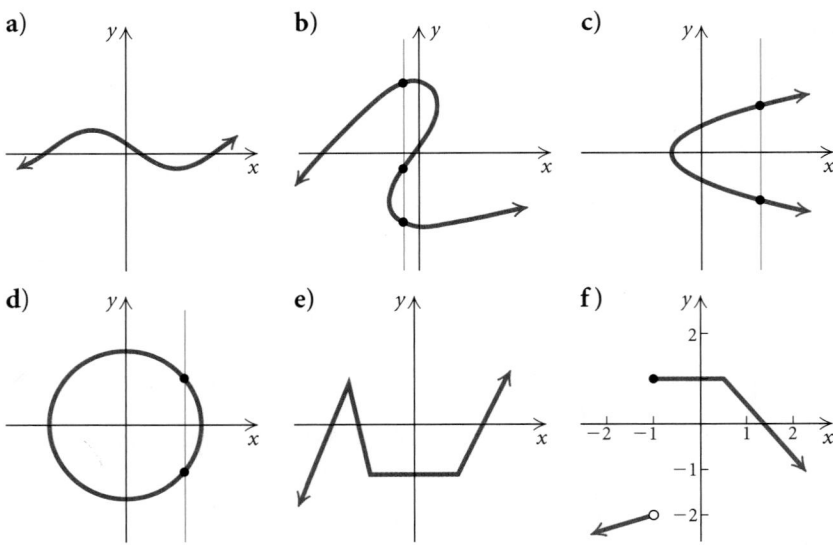

Solution Graphs (a), (e), and (f) are graphs of functions because we cannot find a vertical line that crosses any of them more than once. In (b), the vertical line drawn crosses the graph at three points, so graph (b) is not that of a function. Also, in (c) and (d), we can find a vertical line that crosses the graph more than once, so these are not graphs of functions.

Now Try Exercise 49. ■

❄ Finding Domains of Functions

When a function f whose inputs and outputs are real numbers is given by a formula, the *domain* is understood to be the set of all inputs for which the expression is defined as a real number. When an input results in an expression that is not defined as a real number, we say that the function value *does not exist* and that the number being substituted *is not* in the domain of the function.

REAL NUMBERS

REVIEW SECTION **R.1.**

EXAMPLE 8 Find the indicated function values, if possible, and determine whether the given values are in the domain of the function.

a) $f(1)$ and $f(3)$, for $f(x) = \dfrac{1}{x - 3}$

b) $g(16)$ and $g(-7)$, for $g(x) = \sqrt{x} + 5$

Solution

a) $f(1) = \dfrac{1}{1 - 3} = \dfrac{1}{-2} = -\dfrac{1}{2}$

Since $f(1)$ is defined, 1 is in the domain of f.

$$f(3) = \dfrac{1}{3 - 3} = \dfrac{1}{0}$$

Since division by 0 is not defined, $f(3)$ does not exist and the number 3 is not in the domain of f. In a table from a graphing calculator, this is indicated with an ERROR message.

b) $g(16) = \sqrt{16} + 5 = 4 + 5 = 9$

Since $g(16)$ is defined, 16 is in the domain of g.

$$g(-7) = \sqrt{-7} + 5$$

Since $\sqrt{-7}$ is not defined as a real number, $g(-7)$ does not exist and the number -7 is not in the domain of g. Note the ERROR message in the table at left. ■

$y = 1/(x - 3)$

X	Y₁	
1	-.5	
3	ERROR	

X =

$y = \sqrt{x} + 5$

X	Y₁	
16	9	
-7	ERROR	

X =

As we see in Example 8, inputs that make a denominator 0 or that yield a negative radicand in an even root are not in the domain of a function.

EXAMPLE 9 Find the domain of each of the following functions.

a) $f(x) = \dfrac{1}{x - 3}$ b) $h(x) = \dfrac{3x^2 - x + 7}{x^2 + 2x - 3}$ c) $f(x) = x^3 + |x|$

Solution

INTERVAL NOTATION

REVIEW SECTION **R.1.**

a) Because $x - 3 = 0$ when $x = 3$, the only input that results in a denominator of 0 is 3. The domain is $\{x \mid x \neq 3\}$. We can also write the solution using interval notation and the symbol $\cup$ for the **union,** or inclusion, of both sets: $(-\infty, 3) \cup (3, \infty)$.

b) We can substitute any real number in the numerator, but we must avoid inputs that make the denominator 0. To find those inputs, we solve $x^2 + 2x - 3 = 0$, or $(x + 3)(x - 1) = 0$. Since $x^2 + 2x - 3$ is 0 for -3 and 1, the domain consists of the set of all real numbers except -3 and 1, or $\{x \mid x \neq -3 \text{ and } x \neq 1\}$, or $(-\infty, -3) \cup (-3, 1) \cup (1, \infty)$.

c) We can substitute any real number for x. The domain is the set of all real numbers, $\mathbb{R}$, or $(-\infty, \infty)$.

Now Try Exercises 53, 59, and 63. ■

❖ Visualizing Domain and Range

Keep the following in mind regarding the *graph* of a function:

Domain = the set of a function's inputs, found on the horizontal x-axis;

Range = the set of a function's outputs, found on the vertical y-axis.

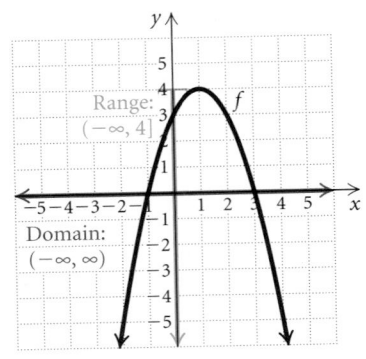

Consider the graph of function f, shown at left. To determine the domain of f, we look for the inputs on the x-axis that correspond to a point on the graph. We see that they include the entire set of real numbers, illustrated in red on the x-axis. Thus the domain is $(-\infty, \infty)$. To find the range, we look for the outputs on the y-axis that correspond to a point on the graph. We see that they include 4 and all real numbers less than 4, illustrated in blue on the y-axis. The bracket at 4 indicates that 4 is included in the interval. The range is $\{x \mid x \leq 4\}$, or $(-\infty, 4]$.

Let's now consider the graph of function g, shown at left. The solid dot shows that $(-4, 5)$ belongs to the graph. The open circle shows that $(3, 2)$ does *not* belong to the graph.

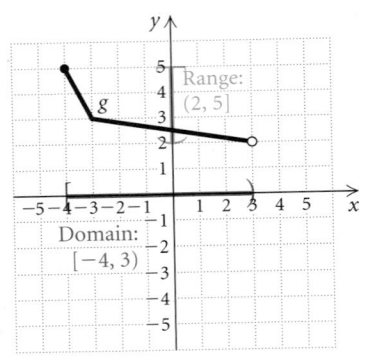

We see that the inputs of the function include -4 and all real numbers between -4 and 3, illustrated in red on the x-axis. The bracket at -4 indicates that -4 is included in the interval. The parenthesis at 3 indicates that 3 is not included in the interval. The domain is $\{x \mid -4 \leq x < 3\}$, or $[-4, 3)$. The outputs of the function include 5 and all real numbers between 2 and 5, illustrated in blue on the y-axis. The parenthesis at 2 indicates that 2 is not included in the interval. The bracket at 5 indicates that 5 is included in the interval. The range is $\{x \mid 2 < x \leq 5\}$, or $(2, 5]$.

EXAMPLE 10 Graph each of the following functions. Then estimate the domain and the range of each.

a) $f(x) = \dfrac{1}{2}x + 1$

b) $f(x) = \sqrt{x + 4}$

c) $f(x) = x^3 - x$

d) $f(x) = \dfrac{1}{x - 2}$

e) $f(x) = x^4 - 2x^2 - 3$

f) $f(x) = \sqrt{4 - (x - 3)^2}$

Solution

a)

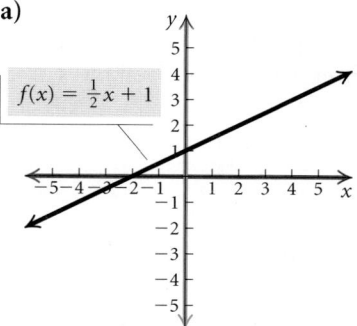

Domain = all real numbers, $(-\infty, \infty)$; range = all real numbers, $(-\infty, \infty)$

b)

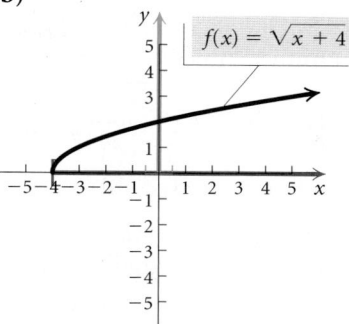

Domain = $[-4, \infty)$; range = $[0, \infty)$

c)

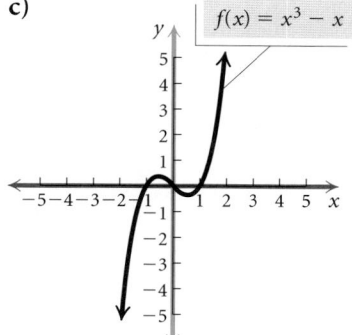

Domain = all real numbers, $(-\infty, \infty)$; range = all real numbers, $(-\infty, \infty)$

d)

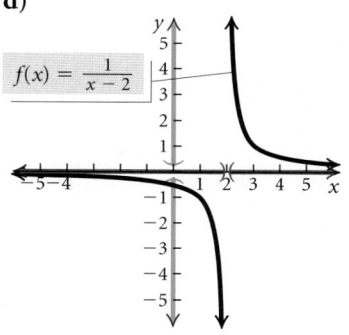

Since the graph does not touch or cross either the vertical line $x = 2$ or the x-axis $y = 0$, 2 is excluded from the domain and 0 is excluded from the range. Domain = $(-\infty, 2) \cup (2, \infty)$; range = $(-\infty, 0) \cup (0, \infty)$

e)

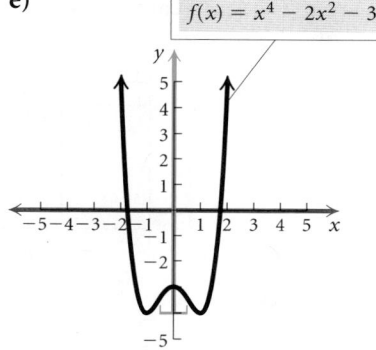

Domain = all real numbers, $(-\infty, \infty)$; range = $[-4, \infty)$

f)

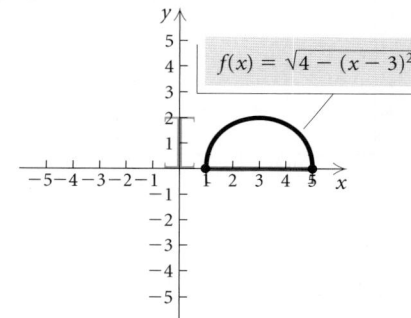

Domain = $[1, 5]$; range = $[0, 2]$

Now Try Exercises 73 and 79. ■

Always consider adding the reasoning of Example 9 to a graphical analysis. Think, "What can I input?" to find the domain. Think, "What do I get out?" to find the range. Thus, in Examples 10(c) and 10(e), it might not appear as though the domain is all real numbers because the graph rises steeply, but by examining the equation we see that we can indeed substitute any real number for x.

❖ Applications of Functions

EXAMPLE 11   *Speed of Sound in Air.*   The speed S of sound in air is a function of the temperature t, in degrees Fahrenheit, and is given by

$$S(t) = 1087.7 \sqrt{\frac{5t + 2457}{2457}},$$

where S is in feet per second.

a) Using the viewing window $[-1000, 2000, -1000, 3000]$, with Xscl = 500 and Yscl = 500, graph the function.

b) Find the speed of sound in air when the temperature is 0°, 32°, 70°, and $-10°$ Fahrenheit.

Solution

a) The graph is shown at left. Note that $S(t)$ must be changed to y and t must be changed to x when the function is entered in a graphing calculator.

b) We use a graphing calculator with the TABLE feature set in ASK mode to compute the function values.

$$y = 1087.7 \sqrt{\frac{5x + 2457}{2457}}$$

X	Y1
0	1087.7
32	1122.6
70	1162.6
−10	1076.6

X =

We find that

$$S(0) = 1087.7 \text{ ft/sec}, \qquad S(32) \approx 1122.6 \text{ ft/sec},$$
$$S(70) \approx 1162.6 \text{ ft/sec}, \quad \text{and} \quad S(-10) \approx 1076.6 \text{ ft/sec}.$$

Now Try Exercise 83. ■

CONNECTING *the* CONCEPTS

FUNCTION CONCEPTS

Formula for f: $f(x) = 5 + 2x^2 - x^4$.

For every input, there is exactly one output.

$(1, 6)$ is on the graph.

For the input 1, the output is 6.

$f(1) = 6$

Domain: set of all inputs $= (-\infty, \infty)$

Range: set of all outputs $= (-\infty, 6]$

GRAPH

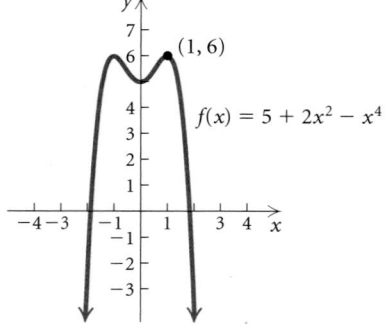

$f(x) = 5 + 2x^2 - x^4$

1.2 Exercise Set

In Exercises 1–14, determine whether the correspondence is a function.

1. $a \longrightarrow w$
$b \longrightarrow y$
$c \longrightarrow z$

2. $m \longrightarrow q$
$n \searrow\nearrow r$
$o \nearrow\searrow s$

3. $-6 \longrightarrow 36$
$-2 \nearrow 4$
$2 \nearrow$

4. $-3 \longrightarrow 2$
$1 \searrow\nearrow 4$
$5 \longrightarrow 6$
$9 \longrightarrow 8$

5. $m \searrow A$
$n \searrow B$
$r \times C$
$s \nearrow D$

6. $a \longrightarrow r$
$b \searrow\nearrow s$
$c \times t$
$d \nearrow$

7. World's Ten Largest Earthquakes (1900–2006)

Location and Date	Magnitude
Chile (May 22, 1960)	9.5
Prince William Sound, Alaska (March 28, 1964)	9.2
Andreanof Islands, Aleutian Islands (March 9, 1957)	9.1
Kamchatka (November 4, 1952)	9.0
Off the coast of Sumatra, Indonesia (December 26, 2004)	
Off the coast of Ecuador (January 31, 1906)	8.8
Rat Islands, Aleutian Islands (February 4, 1965)	8.7
Northern Sumatra, Indonesia (March 28, 2005)	
India–China border (August 15, 1950)	8.6
Kamchatka (February 3, 1923)	8.5

(*Source*: National Earthquake Information Center, U.S. Geological Survey)

8.

Animal	Maximum Speed on the Ground (in miles per hour)
Cheetah	70
Lion	50
Ostrich	40
Reindeer	32
Giraffe	
Grizzly bear	30
Elephant	25
Squirrel	12
Giant tortoise	0.17

(*Source*: *Time Almanac*, 2007, p. 554)

Domain	Correspondence	Range
9. A set of cars in a parking lot	Each car's license number	A set of letters and numbers
10. A set of people in a town	A doctor a person uses	A set of doctors
11. A set of members of a family	Each person's eye color	A set of colors

Domain	Correspondence	Range
12. A set of members of a rock band	An instrument each person plays	A set of instruments
13. A set of students in a class	A student sitting in a neighboring seat	A set of students
14. A set of bags of chips on a shelf	Each bag's weight	A set of weights

Determine whether the relation is a function. Identify the domain and the range.

15. $\{(2, 10), (3, 15), (4, 20)\}$

16. $\{(3, 1), (5, 1), (7, 1)\}$

17. $\{(-7, 3), (-2, 1), (-2, 4), (0, 7)\}$

18. $\{(1, 3), (1, 5), (1, 7), (1, 9)\}$

19. $\{(-2, 1), (0, 1), (2, 1), (4, 1), (-3, 1)\}$

20. $\{(5, 0), (3, -1), (0, 0), (5, -1), (3, -2)\}$

21. Given that $g(x) = 3x^2 - 2x + 1$, find each of the following.
a) $g(0)$
b) $g(-1)$
c) $g(3)$
d) $g(-x)$
e) $g(1 - t)$

22. Given that $f(x) = 5x^2 + 4x$, find each of the following.
a) $f(0)$
b) $f(-1)$
c) $f(3)$
d) $f(t)$
e) $f(t - 1)$

23. Given that $g(x) = x^3$, find each of the following.
a) $g(2)$
b) $g(-2)$
c) $g(-x)$
d) $g(3y)$
e) $g(2 + h)$

24. Given that $f(x) = 2|x| + 3x$, find each of the following.
a) $f(1)$
b) $f(-2)$
c) $f(-x)$
d) $f(2y)$
e) $f(2 - h)$

25. Given that
$$g(x) = \frac{x - 4}{x + 3},$$
find each of the following.
a) $g(5)$
b) $g(4)$
c) $g(-3)$
d) $g(-16.25)$
e) $g(x + h)$

26. Given that
$$f(x) = \frac{x}{2 - x},$$
find each of the following.
a) $f(2)$
b) $f(1)$
c) $f(-16)$
d) $f(-x)$
e) $f\left(-\frac{2}{3}\right)$

27. Find $g(0), g(-1), g(5)$, and $g\left(\frac{1}{2}\right)$ for
$$g(x) = \frac{x}{\sqrt{1 - x^2}}.$$

28. Find $h(0), h(2)$, and $h(-x)$ for
$$h(x) = x + \sqrt{x^2 - 1}.$$

In Exercises 29 and 30, use a graphing calculator and the TABLE *feature set in* ASK *mode.*

29. Given that
$$g(x) = 0.06x^3 - 5.2x^2 - 0.8x,$$
find $g(-2.1), g(5.08)$, and $g(10.003)$. Round answers to the nearest tenth.

30. Given that
$$h(x) = 3x^4 - 10x^3 + 5x^2 - x + 6,$$
find $h(-11), h(7)$, and $h(15)$.

Graph the function.

31. $f(x) = \frac{1}{2}x + 3$

32. $f(x) = \sqrt{x} - 1$

33. $f(x) = -x^2 + 4$

34. $f(x) = x^2 + 1$

35. $f(x) = \sqrt{x-1}$

36. $f(x) = x - \dfrac{1}{2}x^3$

A graph of a function is shown. Using the graph, find the indicated function values; that is, given the inputs, find the outputs.

37. $h(1), h(3),$ and $h(4)$

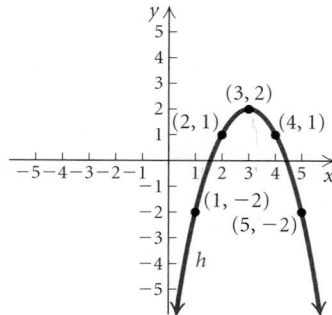

38. $t(-4), t(0),$ and $t(3)$

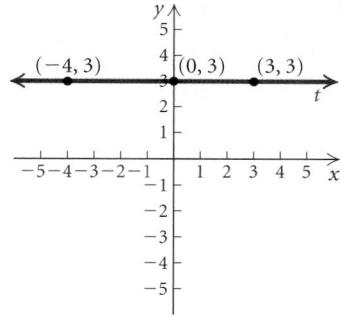

39. $s(-4), s(-2),$ and $s(0)$

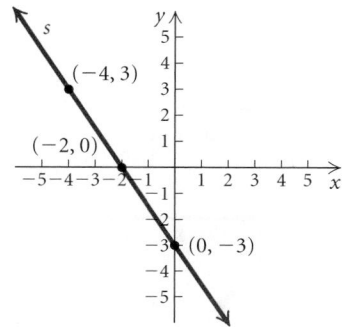

40. $g(-4), g(-1),$ and $g(0)$

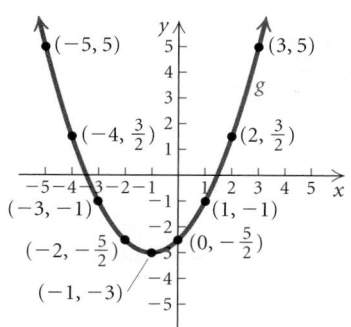

41. $f(-1), f(0),$ and $f(1)$

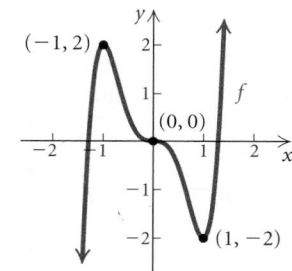

42. $g(-2), g(0),$ and $g(2.4)$

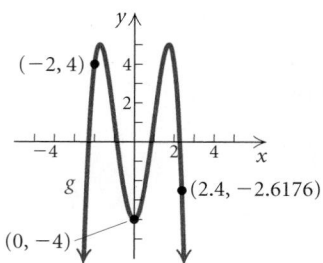

In Exercises 43–50, determine whether the graph is that of a function. An open circle indicates that the point does not belong to the graph.

43.

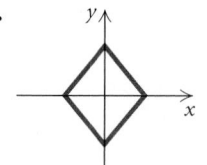

44.

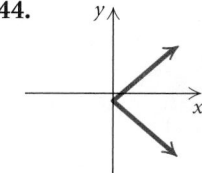

45.

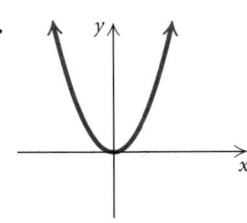

46.

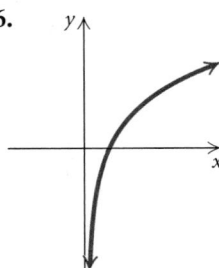

47.

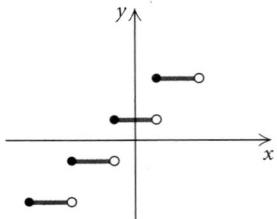

48.

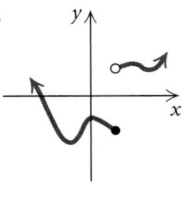

49.

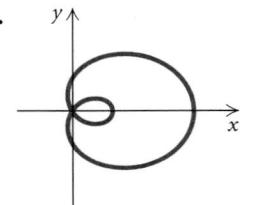

50.

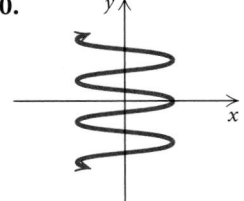

Find the domain of the function. Do not use a graphing calculator.

51. $f(x) = 7x + 4$

52. $f(x) = |3x - 2|$

53. $f(x) = |6 - x|$

54. $f(x) = \dfrac{1}{x^4}$

55. $f(x) = 4 - \dfrac{2}{x}$

56. $f(x) = \dfrac{1}{5}x^2 - 5$

57. $f(x) = \dfrac{x + 5}{2 - x}$

58. $f(x) = \dfrac{8}{x + 4}$

59. $f(x) = \dfrac{1}{x^2 - 4x - 5}$

60. $f(x) = \dfrac{(x - 2)(x + 9)}{x^3}$

61. $f(x) = \dfrac{8 - x}{x^2 - 7x}$

62. $f(x) = \dfrac{x^4 - 2x^3 + 7}{3x^2 - 10x - 8}$

63. $f(x) = \frac{1}{10}|x|$

64. $f(x) = x^2 - 2x$

In Exercises 65–72, determine the domain and the range of the function.

65.

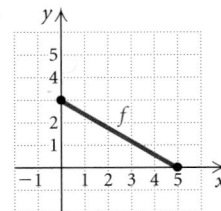

66.

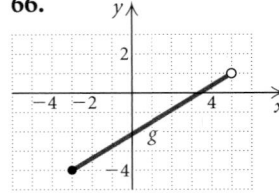

67.

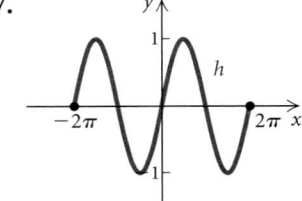

68.

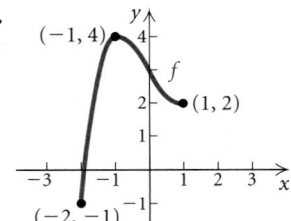

69.

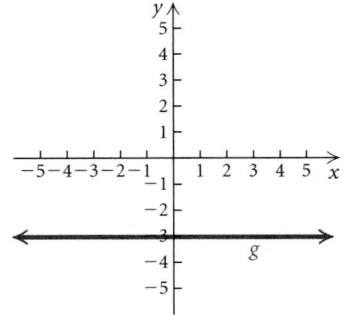

70.

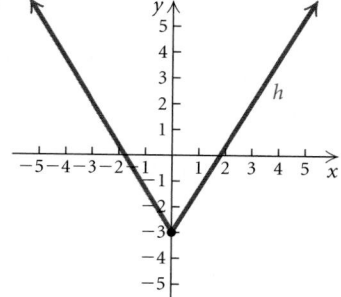

71.

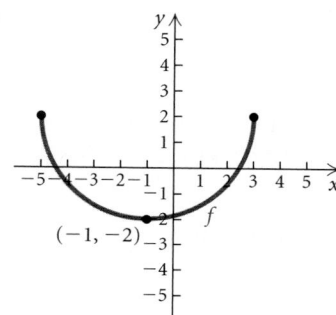

72.

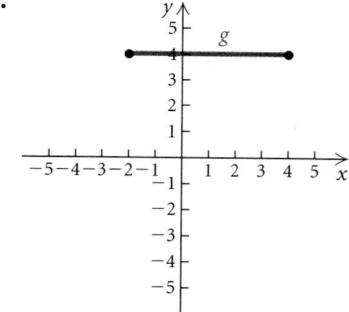

Graph the function with a graphing calculator. Then visually estimate the domain and the range.

73. $f(x) = |x|$

74. $f(x) = |x| - 2$

75. $f(x) = \sqrt{9 - x^2}$

76. $f(x) = -\sqrt{25 - x^2}$

77. $f(x) = (x - 1)^3 + 2$

78. $f(x) = (x - 2)^4 + 1$

79. $f(x) = \sqrt{7 - x}$

80. $f(x) = \sqrt{x + 8}$

81. $f(x) = -x^2 + 4x - 1$

82. $f(x) = 2x^2 - x^4 + 5$

83. *Boiling Point and Elevation.* The elevation E, in meters, above sea level at which the boiling point of water is t degrees Celsius is given by the function

$$E(t) = 1000(100 - t) + 580(100 - t)^2.$$

At what elevation is the boiling point 99.5°? 100°?

84. *Territorial Area of an Animal.* The territorial area of an animal is defined to be its defended, or exclusive, region. For example, a lion has a certain region over which it is considered ruler. It has been shown that the territorial area T, in acres, of predatory animals is a function of body weight w, in pounds, and is given by the function

$$T(w) = w^{1.31}.$$

Find the territorial area of animals whose body weights are 0.5 lb, 10 lb, 20 lb, 100 lb, and 200 lb.

85. *Decreasing Value of the Dollar.* In 2005, it took $19.37 to equal the value of $1 in 1913. In 1990, it took only $13.20 to equal the value of $1 in 1913. The amount it takes to equal the value of $1 in 1913 can be estimated by the linear function V given by

$$V(x) = 0.4123x + 13.2617,$$

where x is the number of years since 1990. Thus, $V(11)$ gives the amount it took in 2001 to equal the value of $1 in 1913.

Source: U.S. Bureau of Labor Statistics

a) Use this function to predict the amount it will take in 2008 and in 2015 to equal the value of $1 in 1913.

b) When will it take approximately $30 to equal the value of $1 in 1913?

Collaborative Discussion and Writing

86. Explain in your own words what a function is.

87. Explain in your own words the difference between the domain of a function and the range of a function.

Skill Maintenance

To the student and the instructor: The Skill Maintenance exercises review skills covered previously in the text. You can expect such exercises in every exercise set. They provide excellent review for a final examination. Answers to all skill maintenance exercises, along with section references, appear in the answer section at the back of the book.

Use substitution to determine whether the given ordered pairs are solutions of the given equation.

88. $(-3, -2), (2, -3); \ y^2 - x^2 = -5$

89. $(0, -7), (8, 11); \ y = 0.5x + 7$

90. $\left(\frac{4}{5}, -2\right), \left(\frac{11}{5}, \frac{1}{10}\right); \ 15x - 10y = 32$

Graph the equation.

91. $y = (x - 1)^2$

92. $y = \frac{1}{3}x - 6$

93. $-2x - 5y = 10$

94. $(x - 3)^2 + y^2 = 4$

Synthesis

Find the domain of the function. Do not use a graphing calculator.

95. $f(x) = \sqrt[3]{x - 1}$

96. $f(x) = \sqrt[4]{2x + 5} + 3$

97. $f(x) = \sqrt{8 - x}$

98. $f(x) = \dfrac{\sqrt{x + 1}}{x}$

99. $f(x) = \dfrac{\sqrt{x + 6}}{(x + 2)(x - 3)}$ } overlap

100. $f(x) = \dfrac{\sqrt{x - 1}}{x^2 + x - 6}$ ← AND

101. $f(x) = \sqrt{3 - x} + \sqrt{x + 5}$

102. $f(x) = \sqrt{x} - \sqrt{4 - x}$

103. Give an example of two different functions that have the same domain and the same range, but have no pairs in common. Answers may vary.

104. Draw a graph of a function for which the domain is $[-4, 4]$ and the range is $[1, 2] \cup [3, 5]$. Answers may vary.

105. Draw a graph of a function for which the domain is $[-3, -1] \cup [1, 5]$ and the range is $\{1, 2, 3, 4\}$. Answers may vary.

106. Suppose that for some function f, $f(x - 1) = 5x$. Find $f(6)$.

107. Suppose that for some function g, $g(x + 3) = 2x + 1$. Find $g(-1)$.

108. Suppose $f(x) = |x + 3| - |x - 4|$. Write $f(x)$ without using absolute-value notation if x is in each of the following intervals.
a) $(-\infty, -3)$
b) $[-3, 4)$
c) $[4, \infty)$

109. Suppose $g(x) = |x| + |x - 1|$. Write $g(x)$ without using absolute-value notation if x is in each of the following intervals.
a) $(-\infty, 0)$
b) $[0, 1)$
c) $[1, \infty)$

Linear Functions, Slope, and Applications

❖ Determine the slope of a line given two points on the line.
❖ Solve applied problems involving slope.
❖ Find the slope and the y-intercept of a line given the equation $y = mx + b$, or $f(x) = mx + b$.
❖ Graph a linear equation using the slope and the y-intercept.
❖ Solve applied problems involving linear functions.

In real-life situations, we often need to make decisions on the basis of limited information. When the given information is used to formulate an equation or inequality that at least approximates the situation mathematically, we have created a **model**. One of the most frequently used mathematical models is *linear*. The graph of a linear model is a straight line.

❖ Linear Functions

Let's begin to examine the connections among equations, functions, and graphs that are straight lines. Compare the graphs of linear and nonlinear functions shown here.

Linear Functions

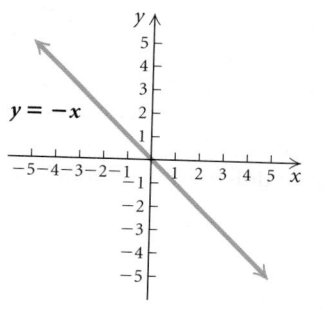

Nonlinear Functions

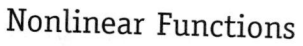

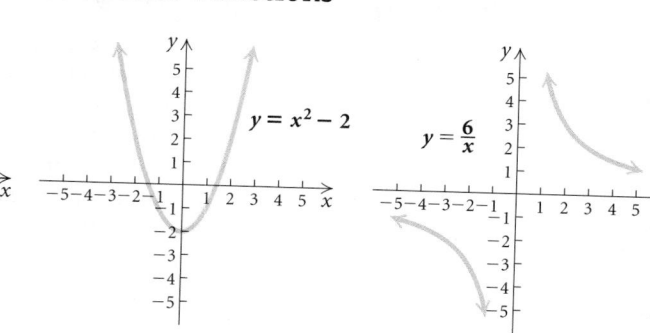

We have the following conclusions and related terminology.

Linear Functions

A function f is a **linear function** if it can be written as

$$f(x) = mx + b,$$

where m and b are constants.

If $m = 0$, the function is a **constant function** $f(x) = b$. If $m = 1$ and $b = 0$, the function is the **identity function** $f(x) = x$.

Linear function:
$y = mx + b$

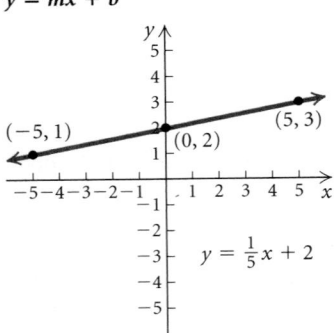

$y = \frac{1}{5}x + 2$

Identity function:
$y = 1 \cdot x + 0$, or $y = x$

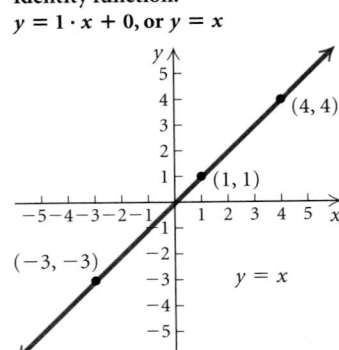

$y = x$

Constant function:
$y = 0 \cdot x + b$, or $y = b$ (Horizontal line)

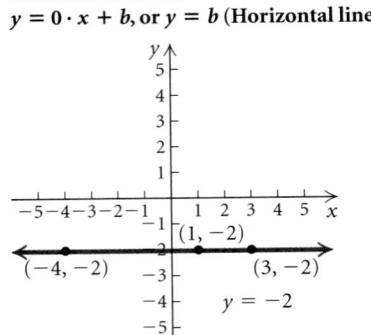

$y = -2$

Vertical line: $x = a$
(*not* a function)

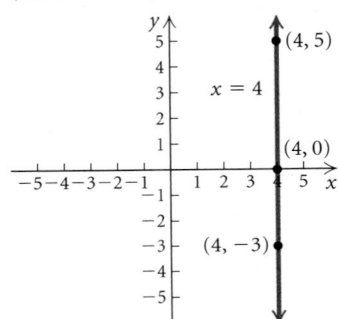

$x = 4$

Horizontal and Vertical Lines

Horizontal lines are given by equations of the type $y = b$ or $f(x) = b$. (They are functions.)

Vertical lines are given by equations of the type $x = a$. (They are *not* functions.)

❋ The Linear Function $f(x) = mx + b$ and Slope

To attach meaning to the constant m in the equation $f(x) = mx + b$, we first consider an application. Suppose FaxMax is an office machine business that currently has stores in locations A and B in the same city. Their total operating costs for the same time period are given by the two functions shown in the tables and graphs that follow. The variable x represents time, in months. The variable y represents total costs, in thousands of dollars, over that period of time. Look for a pattern.

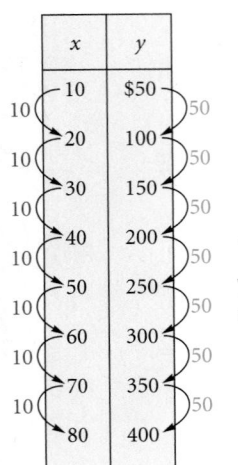

x	y
10	$50
20	100
30	150
40	200
50	250
60	300
70	350
80	400

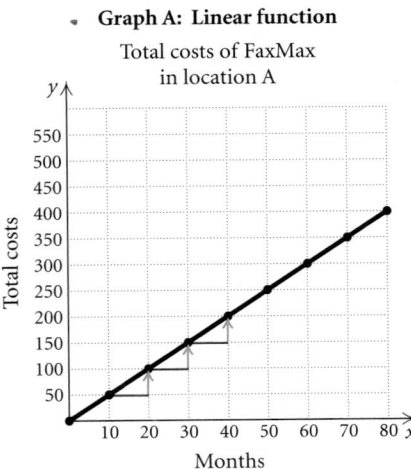

Graph A: Linear function

Total costs of FaxMax in location A

Months

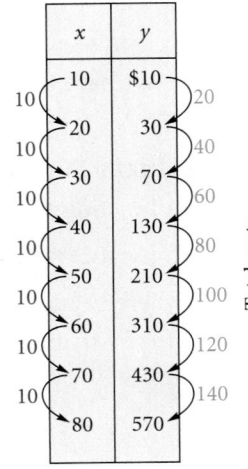

x	y
10	$10
20	30
30	70
40	130
50	210
60	310
70	430
80	570

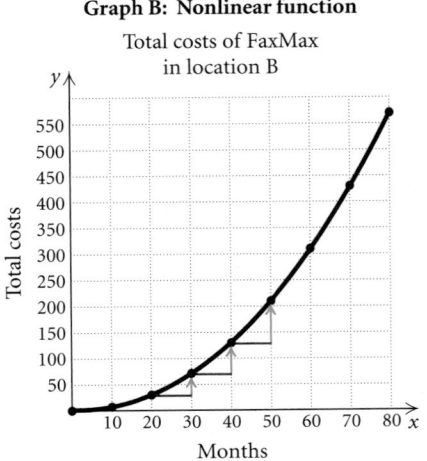

Graph B: Nonlinear function

Total costs of FaxMax in location B

Months

We see in graph A that *every* change of 10 months results in a $50 thousand change in total costs. But in graph B, changes of 10 months do *not* result in constant changes in total costs. This is a way to distinguish linear functions from nonlinear functions. The rate at which a linear function changes, or the steepness of its graph, is constant.

Mathematically, we define the steepness, or **slope**, of a line as the ratio of its vertical change (*rise*) to the corresponding horizontal change (*run*). Slope represents the **rate of change** of y with respect to x.

Slope

The **slope m** of a line containing points (x_1, y_1) and (x_2, y_2) is given by

$$m = \frac{\text{rise}}{\text{run}}$$

$$= \frac{\text{the change in } y}{\text{the change in } x}$$

$$= \frac{y_2 - y_1}{x_2 - x_1} = \frac{y_1 - y_2}{x_1 - x_2}.$$

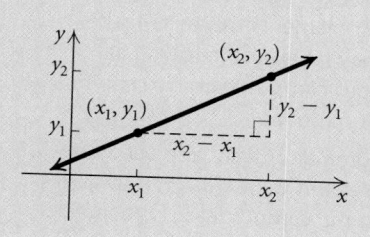

EXAMPLE 1   Graph the function $f(x) = -\frac{2}{3}x + 1$ and determine its slope.

Solution   Since the equation for f is in the form $f(x) = mx + b$, we know that it is a linear function. We can graph it by connecting two points on the graph with a straight line. We calculate two ordered pairs, plot the points, graph the function, and determine the slope:

$$f(3) = -\frac{2}{3} \cdot 3 + 1 = -2 + 1 = -1;$$

$$f(9) = -\frac{2}{3} \cdot 9 + 1 = -6 + 1 = -5;$$

Pairs: $(3, -1), (9, -5)$;

$$\text{Slope} = m = \frac{y_2 - y_1}{x_2 - x_1}$$

$$= \frac{-5 - (-1)}{9 - 3} = \frac{-4}{6} = -\frac{2}{3}.$$

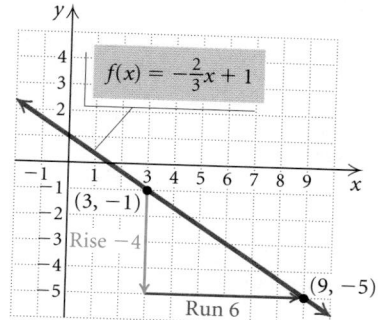

The slope is the same for any two points on a line. Thus, to check our work, note that $f(6) = -\frac{2}{3} \cdot 6 + 1 = -4 + 1 = -3$. Using the points $(6, -3)$ and $(3, -1)$, we have

$$m = \frac{-1 - (-3)}{3 - 6} = \frac{2}{-3} = -\frac{2}{3}.$$

STUDY TIP

Are you aware of all the learning resources that exist for this textbook? Many details are given in the preface.

- The *Student's Solutions Manual* contains worked-out solutions to the odd-numbered exercises in the exercise sets and to all the exercises in the end-of-chapter material.
- An extensive set of *videos* supplements this text. These may be available to you at your campus learning center or math lab.
- *Tutorial software* also accompanies this text. If it is not available in the campus learning center, you can order it by calling 1-800-824-7799.
- You can call the Addison Wesley *Tutor Center* for "live" help with the odd-numbered exercises. This service is available with a student access number.

We can also use the points in the opposite order when computing slope:

$$m = \frac{-3 - (-1)}{6 - 3} = \frac{-2}{3} = -\frac{2}{3}.$$

Note too that the slope of the line is the number m in the equation for the function $f(x) = -\frac{2}{3}x + 1$. **Now Try Exercise 11.**

The *slope* of the line given by $f(x) = mx + b$ is m.

Exploring with Technology

We can animate the effect of the slope m in linear functions of the type $f(x) = mx$ with the graphing calculator set in SEQUENTIAL mode. Graph the following equations:

$$y_1 = x, \quad y_2 = 2x, \quad y_3 = 5x, \quad \text{and} \quad y_4 = 10x.$$

Enter these equations as $y_1 = \{1, 2, 5, 10\}x$. What do you think the graph of $y = 128x$ will look like?

Clear the screen and graph the following equations:

$$y_1 = x, \quad y_2 = 0.75x, \quad y_3 = 0.48x, \quad \text{and} \quad y_4 = 0.12x.$$

What do you think the graph of $y = 0.000029x$ will look like?

Again clear the screen and graph each set of equations:

$$y_1 = -x, \quad y_2 = -2x, \quad y_3 = -4x, \quad \text{and} \quad y_4 = -10x$$

and then

$$y_1 = -x, \quad y_2 = -\tfrac{2}{3}x, \quad y_3 = -\tfrac{7}{20}x, \quad \text{and} \quad y_4 = -\tfrac{1}{10}x.$$

From your observations, what do you think the graphs of $y = -200x$ and $y = -\frac{17}{100{,}000}x$ will look like?

If a line slants up from left to right, the change in x and the change in y have the same sign, so the line has a positive slope. The larger the slope, the steeper the line, as shown in Fig. 1. If a line slants down from left to right, the change in x and the change in y are of opposite signs, so the line has a negative slope. The larger the absolute value of the slope, the steeper the line, as shown in Fig. 2. Considering $y = mx$ when $m = 0$, we have $y = 0x$, or $y = 0$. Note that this horizontal line is the x-axis, as shown in Fig. 3.

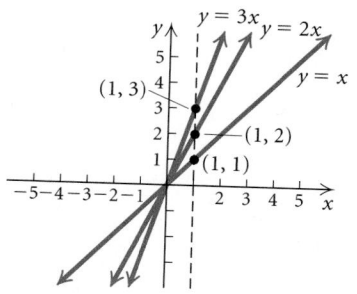

FIGURE 1

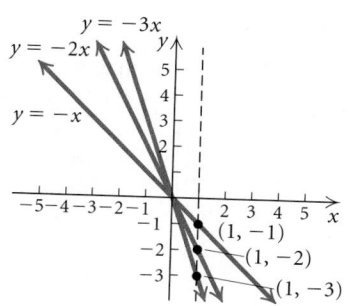

FIGURE 2

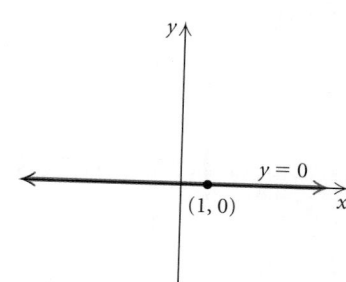

FIGURE 3

Horizontal and Vertical Lines

If a line is horizontal, the change in y for any two points is 0 and the change in x is nonzero. Thus a horizontal line has slope 0. (See Fig. 4.)

If a line is vertical, the change in y for any two points is nonzero and the change in x is 0. Thus the slope is *not defined* because we cannot divide by 0. (See Fig. 5.)

Horizontal lines

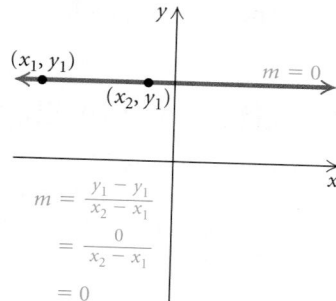

FIGURE 4

Vertical lines

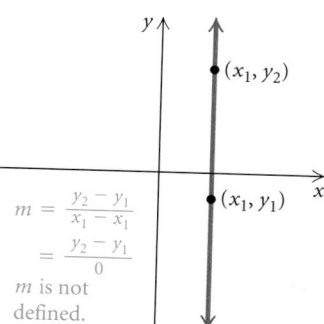

FIGURE 5

Note that zero slope and an undefined slope are two very different concepts.

EXAMPLE 2 Graph each linear equation and determine its slope.

a) $x = -2$

b) $y = \dfrac{5}{2}$

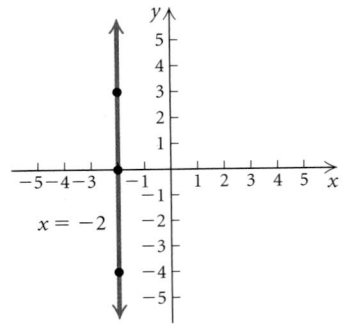

FIGURE 6

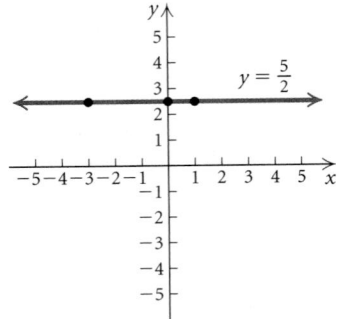

FIGURE 7

Solution

a) Since y is missing in $x = -2$, any value for y will do.

x	y
-2	0
-2	3
-2	-4

Choose any number for y; x must be -2.

The graph is a *vertical line* 2 units to the left of the y-axis. (See Fig. 6.) The slope is not defined. The graph is *not* the graph of a function.

b) Since x is missing in $y = \frac{5}{2}$, any value for x will do.

x	y
0	$\frac{5}{2}$
-3	$\frac{5}{2}$
1	$\frac{5}{2}$

Choose any number for x; y must be $\frac{5}{2}$.

The graph is a *horizontal line* $\frac{5}{2}$, or $2\frac{1}{2}$, units above the x-axis. (See Fig. 7.) The slope is 0. The graph is the graph of a constant function.

Now Try Exercises 17 and 23. ■

❈ Applications of Slope

Slope has many real-world applications. Numbers like 2%, 4%, and 7% are often used to represent the **grade** of a road. Such a number is meant to tell how steep a road is on a hill or mountain. For example, a 4% grade means that the road rises 4 ft for every horizontal distance of 100 ft.

Road grade = $\frac{a}{b}$
(expressed as a percent)

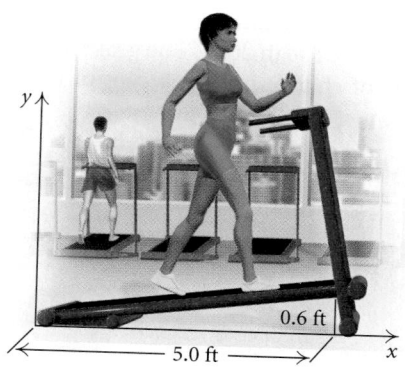

The concept of grade is also used with a treadmill. During a treadmill test, a cardiologist might change the slope, or grade, of the treadmill to measure its effect on heart rate. Another example occurs in hydrology. The strength or force of a river depends on how far the river falls vertically compared to how far it flows horizontally.

EXAMPLE 3   *Ramp for the Disabled.*   Construction laws regarding access ramps for the disabled state that every vertical rise of 1 ft requires a horizontal run of 12 ft. What is the grade, or slope, of such a ramp?

Solution   The grade, or slope, is given by $m = \frac{1}{12} \approx 0.083 \approx 8.3\%$.

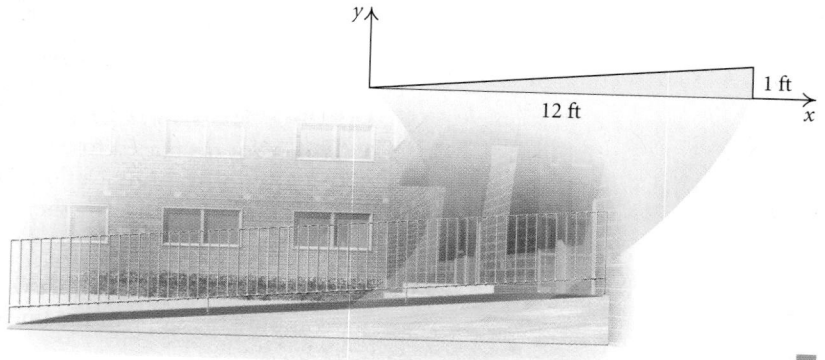

Average Rate of Change

Slope can also be considered as an **average rate of change**. To find the average rate of change between any two data points on a graph, we determine the slope of the line that passes through the two points.

EXAMPLE 4   *Travel Bookings Online.*   The percent of travel bookings made online has increased from 6% in 1999 to 55% in 2007. The graph below illustrates this trend. Find the average rate of change in the percent of travel bookings made online from 1999 to 2007.

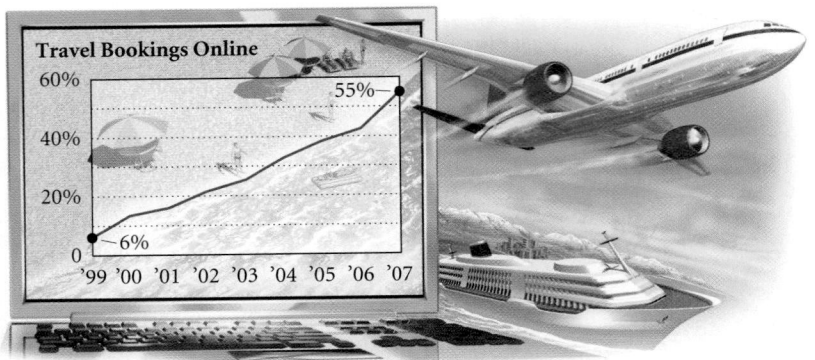

Source: PhoCusWright

Solution   We determine the coordinates of two points on the graph. In this case, we use (1999, 6%) and (2007, 55%). Then we compute the slope, or average rate of change, as follows:

$$\text{Slope} = \text{Average rate of change} = \frac{\text{Change in } y}{\text{Change in } x}$$

$$= \frac{55 - 6}{2007 - 1999} = \frac{49}{8}, \text{ or } 6\frac{1}{8}.$$

This result tells us that each year from 1999 to 2007, online travel bookings increased on average by $6\frac{1}{8}$%. The average rate of change over the 8-yr period was an increase of $6\frac{1}{8}$% per year.   **Now Try Exercise 39.** ■

EXAMPLE 5   *Helmet Use.*   In 1994, 63% of motorcyclists wore helmets. That percent decreased to 51% in 2006. (*Source*: National Highway Traffic Safety Administration's National Center for Statistics and Analysis) Find the average rate of change in the percent of motorcyclists who wore helmets from 1994 to 2006.

Solution   Using the coordinates of the points (1994, 63%) and (2006, 51%), we compute the slope of the line containing these two points.

$$\text{Slope} = \text{Average rate of change} = \frac{\text{Change in } y}{\text{Change in } x}$$

$$= \frac{51\% - 63\%}{2006 - 1994} = \frac{-12\%}{12} = -1\%.$$

The result tells us that each year from 1994 to 2006, the percent of motorcyclists who wore helmets decreased on average by 1%. The average rate of change over the 12-yr period was a decrease of 1% per year.

Now Try Exercise 43. ▪

❊ Slope–Intercept Equations of Lines

Exploring with Technology

We can use a graphing calculator to explore the effect of the constant b in linear equations of the type $f(x) = mx + b$. Begin with the graph of $y = x$. Now graph the lines $y = x + 3$ and $y = x - 4$ in the same viewing window. Try entering these equations as $y = x + \{0, 3, -4\}$ and compare the graphs. How do the last two lines differ from $y = x$? What do you think the line $y = x - 6$ will look like?

Clear the first set of equations and graph $y = -0.5x$, $y = -0.5x - 4$, and $y = -0.5x + 3$ in the same viewing window. Describe what happens to the graph of $y = -0.5x$ when a number b is added.

> **_y_-INTERCEPT**
>
> REVIEW SECTION **1.1.**

Compare the graphs of the equations

$$y = 3x \quad \text{and} \quad y = 3x - 2.$$

Note that the graph of $y = 3x - 2$ is a shift of the graph of $y = 3x$ down 2 units and that $y = 3x - 2$ has y-intercept $(0, -2)$. That is, the graph is parallel to $y = 3x$ and it crosses the y-axis at $(0, -2)$. The point $(0, -2)$ is the **_y_-intercept** of the graph.

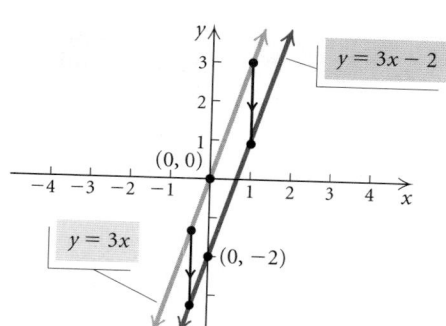

The Slope–Intercept Equation

The linear function f given by

$$f(x) = mx + b$$

is written in slope–intercept form. The graph of an equation in this form is a straight line parallel to $f(x) = mx$. The constant m is called the slope, and the y-intercept is $(0, b)$.

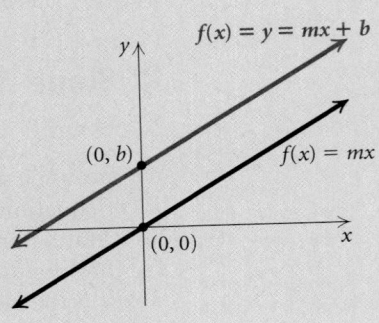

We can read the slope m and the y-intercept $(0, b)$ directly from the equation of a line written in slope–intercept form $y = mx + b$.

EXAMPLE 6   Find the slope and the y-intercept of the line with equation $y = -0.25x - 3.8$.

Solution

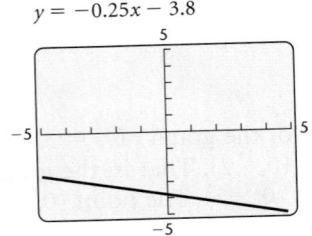

$y = -0.25x - 3.8$

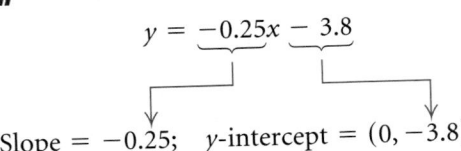

Slope $= -0.25$; y-intercept $= (0, -3.8)$ **Now Try Exercise 47.** ▪

Any equation whose graph is a straight line is a **linear equation.** To find the slope and the y-intercept of the graph of a linear equation, we can solve for y, and then read the information from the equation.

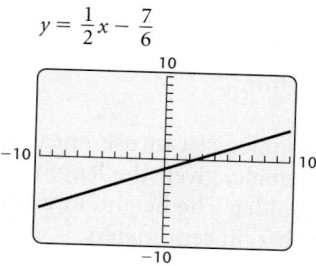

$y = \frac{1}{2}x - \frac{7}{6}$

EXAMPLE 7 Find the slope and the y-intercept of the line with equation $3x - 6y - 7 = 0$.

Solution We solve for y:

$$3x - 6y - 7 = 0$$
$$-6y = -3x + 7 \qquad \text{Adding } -3x \text{ and 7 on both sides}$$
$$-\frac{1}{6}(-6y) = -\frac{1}{6}(-3x + 7) \qquad \text{Multiplying by } -\frac{1}{6}$$
$$y = \frac{1}{2}x - \frac{7}{6}.$$

Thus the slope is $\frac{1}{2}$, and the y-intercept is $\left(0, -\frac{7}{6}\right)$. **Now Try Exercise 53.** ■

❖ Graphing $f(x) = mx + b$ Using m and b

We can also graph a linear equation using its slope and y-intercept.

EXAMPLE 8 Graph: $y = -\frac{2}{3}x + 4$.

Solution This equation is in slope–intercept form, $y = mx + b$. The y-intercept is $(0, 4)$. We plot this point. We can think of the slope $\left(m = -\frac{2}{3}\right)$ as $\frac{-2}{3}$.

$$m = \frac{\text{rise}}{\text{run}} = \frac{\text{change in } y}{\text{change in } x} = \frac{-2}{3} \begin{array}{l} \leftarrow \text{Move 2 units down.} \\ \leftarrow \text{Move 3 units to the right.} \end{array}$$

Starting at the y-intercept and using the slope, we find another point by moving 2 units down and 3 units to the right. We get a new point $(3, 2)$. In a similar manner, we can move from $(3, 2)$ to find another point $(6, 0)$.

We could also think of the slope $\left(m = -\frac{2}{3}\right)$ as $\frac{2}{-3}$. Then we can start at $(0, 4)$ and move 2 units up and 3 units to the left. We get to another point on the graph, $(-3, 6)$. We now plot the points and draw the line. Note that we need only the y-intercept and one other point in order to graph the line, but it's a good idea to find a third point as a check that the first two points are correct.

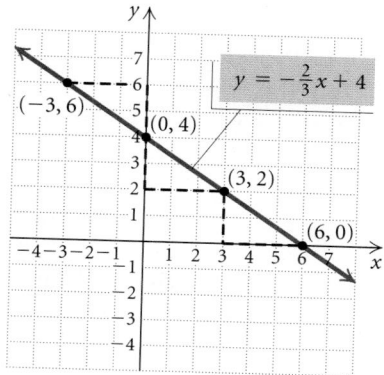

Now Try Exercise 61. ■

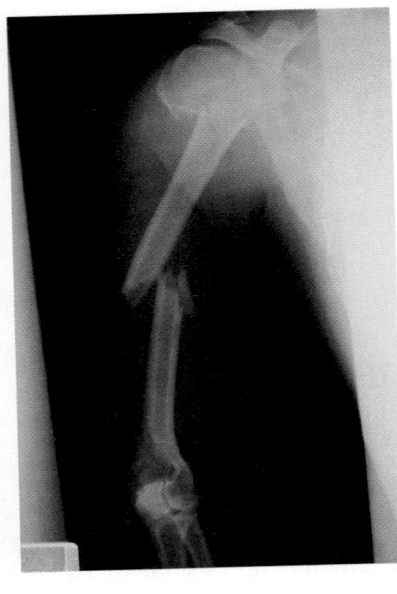

✳ Applications of Linear Functions

We now consider an application of linear functions.

EXAMPLE 9 *Height Estimates.*   An anthropologist can use linear functions to estimate the height of a male or a female, given the length of the humerus, the bone from the elbow to the shoulder. The height, in centimeters, of an adult male with a humerus of length x, in centimeters, is given by the function

$$M(x) = 2.89x + 70.64.$$

The height, in centimeters, of an adult female with a humerus of length x is given by the function

$$F(x) = 2.75x + 71.48.$$

A 26-cm humerus was uncovered in a ruins. Assuming it was from a female, how tall was she? What is the domain of this function?

ALGEBRAIC SOLUTION

We substitute into the function:

$$F(26) = 2.75(26) + 71.48$$
$$= 142.98.$$

Thus the female was 142.98 cm tall.

GRAPHICAL SOLUTION

We graph $F(x) = 2.75x + 71.48$ and use the VALUE feature to find $F(26)$.

$$y = 2.75x + 71.48$$

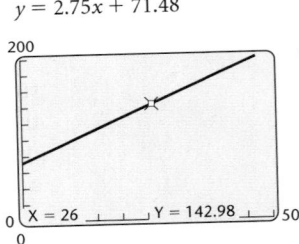

The female was 142.98 cm tall.

Theoretically, the domain of the function is the set of all real numbers. However, the context of the problem dictates a different domain. One could not find a bone with a length of 0 or less. Thus the domain consists of positive real numbers, that is, the interval $(0, \infty)$. A more realistic domain might be 20 cm to 50 cm, the interval $[20, 50]$.   **Now Try Exercise 71.** ■

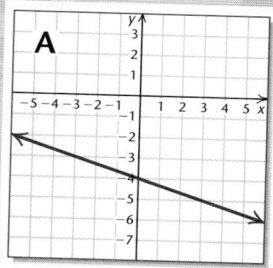

A

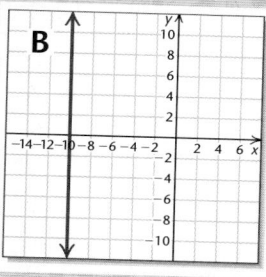

B

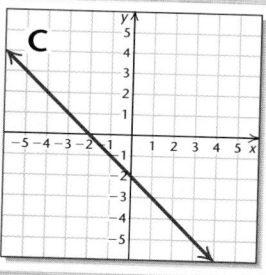

C

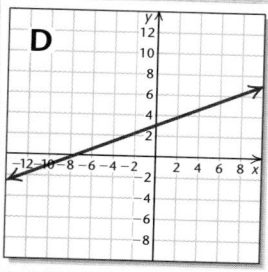

D

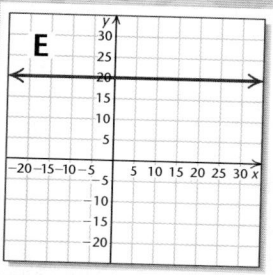

E

Visualizing the Graph

Match the equation with its graph.

1. $y = 20$

2. $5y = 2x + 15$

3. $y = -\dfrac{1}{3}x - 4$

4. $x = \dfrac{5}{3}$

5. $y = -x - 2$

6. $y = 2x$

7. $y = -3$

8. $3y = -4x$

9. $x = -10$

10. $y = x + \dfrac{7}{2}$

Answers on page A-7

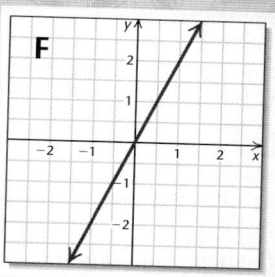

F

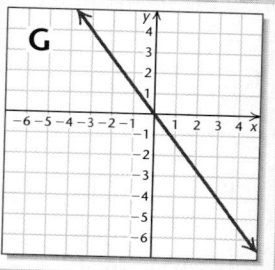

G

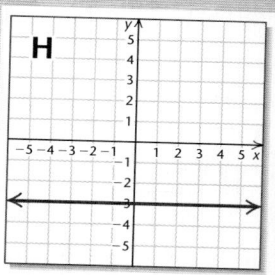

H

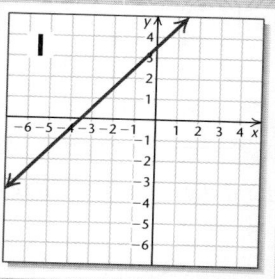

I

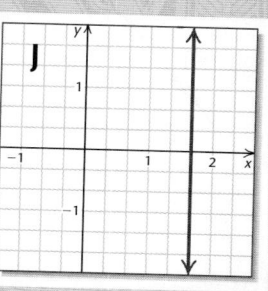

J

1.3   Exercise Set

In Exercises 1–4, the table of data contains input–output values for a function. Answer the following questions for each table.

a) *Is the change in the inputs x the same?*
b) *Is the change in the outputs y the same?*
c) *Is the function linear?*

1.

x	y
−3	7
−2	10
−1	13
0	16
1	19
2	22
3	25

2.

x	y
20	12.4
30	24.8
40	49.6
50	99.2
60	198.4
70	396.8
80	793.6

3.

x	y
11	3.2
26	5.7
41	8.2
56	9.3
71	11.3
86	13.7
101	19.1

4.

x	y
2	−8
4	−12
6	−16
8	−20
10	−24
12	−28
14	−32

Find the slope of the line containing the given points.

5.

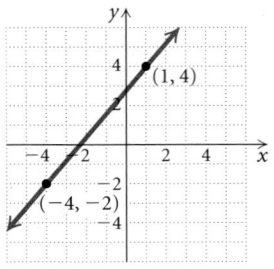

6.

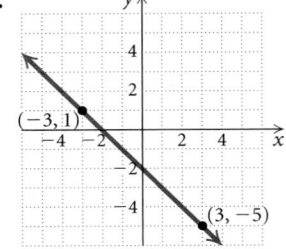

7.

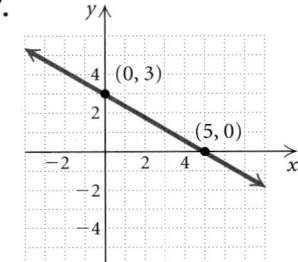

8.

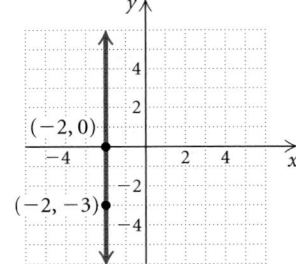

9.

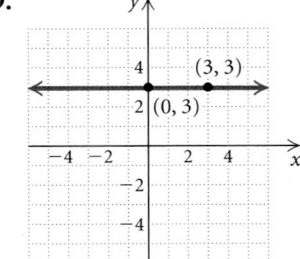

10.

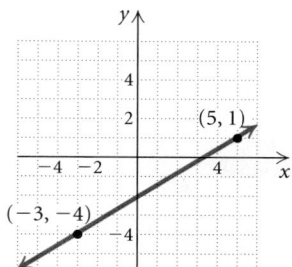

11. $(9, 4)$ and $(-1, 2)$

12. $(-3, 7)$ and $(5, -1)$

13. $(4, -9)$ and $(4, 6)$

14. $(-6, -1)$ and $(2, -13)$

15. $(0.7, -0.1)$ and $(-0.3, -0.4)$

16. $\left(-\frac{3}{4}, -\frac{1}{4}\right)$ and $\left(\frac{2}{7}, -\frac{5}{7}\right)$

17. $(2, -2)$ and $(4, -2)$

18. $(-9, 8)$ and $(7, -6)$

19. $\left(\frac{1}{2}, -\frac{3}{5}\right)$ and $\left(-\frac{1}{2}, \frac{3}{5}\right)$

20. $(-8.26, 4.04)$ and $(3.14, -2.16)$

21. $(16, -13)$ and $(-8, -5)$

22. $(\pi, -3)$ and $(\pi, 2)$

23. $(-10, -7)$ and $(-10, 7)$

24. $\left(\sqrt{2}, -4\right)$ and $(0.56, -4)$

25. $f(4) = 3$ and $f(-2) = 15$

26. $f(-4) = -5$ and $f(4) = 1$

27. $f\left(\frac{1}{5}\right) = \frac{1}{2}$ and $f(-1) = -\frac{11}{2}$

28. $f(8) = -1$ and $f\left(-\frac{2}{3}\right) = \frac{10}{3}$

Determine the slope, if it exists, of the graph of the given linear equation.

29. $y = 1.3x - 5$

30. $y = -\frac{2}{5}x + 7$

31. $x = -2$

32. $f(x) = 4x - \frac{1}{4}$

33. $f(x) = -\frac{1}{2}x + 3$

34. $y = \frac{3}{4}$

35. $y = 9 - x$

36. $x = 8$

37. $y = 0.7$

38. $y = \frac{4}{5} - 2x$

39. *Honey Production.* U.S. honey producers turned out 22% less honey in 2006 than in 1997. In 1997, approximately 198.2 million lb were produced. This amount decreased to 154.8 million lb in 2006. (*Source:* National Agricultural Statistics Services) Find the average rate of change in the number of pounds of honey produced from 1997 to 2006.

40. *HIV Cases.* HIV, human immunodeficiency virus, spreads to 10 people every minute. It is estimated that there were about 40.3 million cases of HIV worldwide in 2005. The estimated number of cases in 1985 was about 2 million. (*Source:* UNAIDS) Find the average rate of change in the number of adults and children worldwide with HIV from 1985 to 2005.

41. *Richmond International Raceway.* The 29 NASCAR races from March 1992 to May 2006 at Richmond International Raceway, Richmond, Virginia, were sellouts. Ticket sales increased over 80%. Use the data in the graph below to find the average rate of change in attendance for the consecutive races.

NASCAR Races at Richmond International Raceway

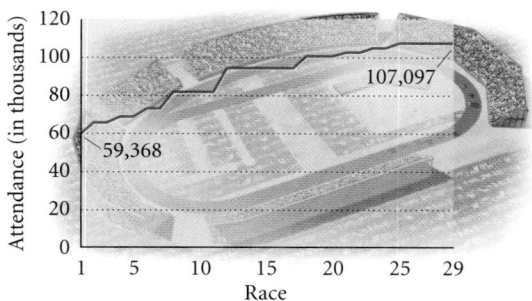

Source: NASCAR, Richmond International Raceway

42. *Decline in Teen Smoking.*   The percent of 10th-grade students who have smoked daily in the last 30 days has greatly decreased, from 16.3% in 1995 to 8.3% in 2004 (*Source: Monitoring the Future*, University of Michigan Institute for Social Research and National Institute on Drug Abuse). Find the average rate of change over the 9-yr period in the percent of 10th-grade students who have smoked daily in the last 30 days.

43. *Decline in Land-Line Phones.*   The number of land-line phones in the United States has decreased from 191 million in 2001 to 172 million in 2006 (*Source*: Federal Communications Commission). Find the average rate of change in the number of land-line phones over the 5-yr period.

44. *Credit-Card Debt.*   From 1992 to 2006, the average household credit-card balance has risen 172%. Use the data in the graph below to find the average rate of change in the average credit-card balance from 1992 to 2006.

Credit Card Debt

Source: CardWeb.com, Paul Bannister, Bankrate.com, and "High Credit Card Debt Shouldn't Be Taken Lightly," Jeff Reish, *Canon Daily Record*, 4/22/2006

45. *Running Rate.*   Lucie Mays-Sulewski, from Westfield, Indiana, won the women's division of the 2006 Indianapolis 500 Mini-Marathon race. The first American woman to win this race since 1993, she reached the 5-mi point after 31 min 10 sec and arrived at the 10-mi point after 1 hr 1 min 38 sec.

(*Source*: www.500festival.com) Find Lucie's speed in miles per minute (average rate of change) from the 5-mi point to the 10-mi point.

46. *Work Rate.*   As a typist resumes work on a research paper, $\frac{1}{6}$ of the paper has already been keyboarded. Six hours later, the paper is $\frac{3}{4}$ done. Calculate the worker's typing rate.

Find the slope and the y-intercept of the line with the given equation.

47. $y = \frac{3}{5}x - 7$

48. $f(x) = -2x + 3$

49. $x = -\frac{2}{5}$

50. $y = \frac{4}{7}$

51. $f(x) = 5 - \frac{1}{2}x$

52. $y = 2 + \frac{3}{7}x$

53. $3x + 2y = 10$

54. $2x - 3y = 12$

55. $y = -6$

56. $x = 10$

57. $5y - 4x = 8$

58. $5x - 2y + 9 = 0$

59. $4y - x + 2 = 0$

60. $f(x) = 0.3 + x$

Graph the equation using the slope and the y-intercept.

61. $y = -\frac{1}{2}x - 3$

62. $y = \frac{3}{2}x + 1$

63. $f(x) = 3x - 1$

64. $f(x) = -2x + 5$

65. $3x - 4y = 20$

66. $2x + 3y = 15$

67. $x + 3y = 18$

68. $5y - 2x = -20$

69. *Minimum Ideal Weight.* One way to estimate the minimum ideal weight of a woman in pounds is to multiply her height, in inches, by 4 and subtract 130. Let W = the minimum ideal weight and h = height.

a) Express W as a linear function of h.
b) Graph W.
c) Find the minimum ideal weight of a woman whose height is 62 in.

70. *Pressure at Sea Depth.* The function P, given by
$$P(d) = \frac{1}{33}d + 1,$$
gives the pressure, in atmospheres (atm), at a depth d, in feet, under the sea.

a) Graph P.
b) Find $P(0)$, $P(5)$, $P(10)$, $P(33)$, and $P(200)$.

71. *Stopping Distance on Glare Ice.* The stopping distance (at some fixed speed) of regular tires on glare ice is a function of the air temperature F, in degrees Fahrenheit. This function is estimated by
$$D(F) = 2F + 115,$$
where $D(F)$ is the stopping distance, in feet, when the air temperature is F, in degrees Fahrenheit.

a) Graph D.
b) Find $D(0°)$, $D(-20°)$, $D(10°)$, and $D(32°)$.
c) Explain why the domain should be restricted to $[-57.5°, 32°]$.

72. *Anthropology Estimates.* Consider Example 9 and the function
$$M(x) = 2.89x + 70.64$$
for estimating the height of a male.

a) If a 26-cm humerus from a male is found in an archeological dig, estimate the height of the male.
b) What is the domain of M?

73. *Reaction Time.* Suppose that while driving a car, you suddenly see a school crossing guard standing in the road. Your brain registers the information and sends a signal to your foot to hit the brake. The car travels a distance D, in feet, during this time, where D is a function of the speed r, in miles per hour, of the car when you see the crossing guard. That reaction distance is a linear function given by
$$D(r) = \frac{11}{10}r + \frac{1}{2}.$$

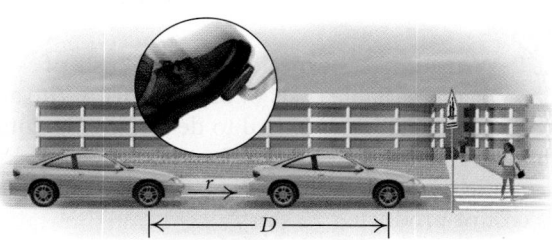

a) Find the slope of this line and interpret its meaning in this application.
b) Graph D.
c) Find $D(5)$, $D(10)$, $D(20)$, $D(50)$, and $D(65)$.
d) What is the domain of this function? Explain.

74. *Straight-Line Depreciation.* A marketing firm buys a new color printer for $5200 to print banners for a sales campaign. The printer is purchased on January 1 and is expected to last 8 yr, at the end of which time its *trade-in*, or *salvage, value* will be $1100. If the company figures the decline or depreciation in value to be the same each year, then the salvage value V, after t years, is given by the linear function
$$V(t) = \$5200 - \$512.50t, \quad \text{for } 0 \le t \le 8.$$

a) Graph V.
b) Find $V(0)$, $V(1)$, $V(2)$, $V(3)$, and $V(8)$.
c) Find the domain and the range of this function.

75. *Total Cost.*   The Cellular Connection charges $60 for a phone and $29 per month under its economy plan. Write an equation that can be used to determine the total cost, $C(t)$, of operating a Cellular Connection phone for t months. Then find the total cost for 6 months.

76. *Total Cost.*   Superior Cable Television charges a $65 installation fee and $105 per month for "deluxe" service. Write an equation that can be used to determine the total cost, $C(t)$, for t months of deluxe cable television service. Then find the total cost for 8 months of service.

*In Exercises 77 and 78, the term **fixed costs** refers to the start-up costs of operating a business. This includes machinery and building costs. The term **variable costs** refers to what it costs a business to produce or service one item.*

77. Ty's Custom Tees experienced fixed costs of $800 and variable costs of $3 per shirt. Write an equation that can be used to determine the total costs encountered by Ty's Custom Tees when x shirts are produced. Then determine the total cost of producing 75 shirts.

78. It's My Racquet experienced fixed costs of $950 and variable costs of $24 for each tennis racquet that is restrung. Write an equation that can be used to determine the total costs encountered by It's My Racquet when x racquets are restrung. Then determine the total cost of restringing 150 tennis racquets.

Collaborative Discussion and Writing

79. Discuss why the graph of a vertical line $x = a$ cannot represent a function.

80. Explain as you would to a fellow student how the numerical value of slope can be used to describe the slant and the steepness of a line.

Skill Maintenance

If $f(x) = x^2 - 3x$, find each of the following.

81. $f\left(\frac{1}{2}\right)$

82. $f(5)$

83. $f(-5)$

84. $f(-a)$

85. $f(a + h)$

Synthesis

86. *Grade of Treadmills.*   A treadmill is 5 ft long and is set at an 8% grade. How high is the end of the treadmill?

Find the slope of the line containing the given points.

87. $(-c, -d)$ and $(9c, -2d)$

88. $(r, s + t)$ and (r, s)

89. $(z + q, z)$ and $(z - q, z)$

90. $(-a - b, p + q)$ and $(a + b, p - q)$

91. (a, a^2) and $(a + h, (a + h)^2)$

92. $(a, 3a + 1)$ and $(a + h, 3(a + h) + 1)$

Suppose that f is a linear function. Determine whether the statement is true or false.

93. $f(cd) = f(c)f(d)$

94. $f(c + d) = f(c) + f(d)$

95. $f(c - d) = f(c) - f(d)$

96. $f(kx) = kf(x)$

Let $f(x) = mx + b$. Find a formula for $f(x)$ given each of the following.

97. $f(x + 2) = f(x) + 2$

98. $f(3x) = 3f(x)$

Equations of Lines and Modeling

❖ Determine equations of lines.

❖ Given the equations of two lines, determine whether their graphs are parallel or perpendicular.

❖ Model a set of data with a linear function.

❖ Fit a regression line to a set of data; then use the linear model to make predictions.

❖ Slope–Intercept Equations of Lines

In Section 1.3, we developed the slope–intercept equation $y = mx + b$, or $f(x) = mx + b$. If we know the slope and the y-intercept of a line, we can find an equation of the line using the slope–intercept equation.

EXAMPLE 1   A line has slope $-\frac{7}{9}$ and y-intercept $(0, 16)$. Find an equation of the line.

Solution   We use the slope–intercept equation and substitute $-\frac{7}{9}$ for m and 16 for b:

$$y = mx + b$$
$$y = -\frac{7}{9}x + 16, \text{ or } f(x) = -\frac{7}{9}x + 16.$$ **Now Try Exercise 7.** ◼

EXAMPLE 2   A line has slope $-\frac{2}{3}$ and contains the point $(-3, 6)$. Find an equation of the line.

Solution   We use the slope–intercept equation, $y = mx + b$, and substitute $-\frac{2}{3}$ for m: $y = -\frac{2}{3}x + b$. Using the point $(-3, 6)$, we substitute -3 for x and 6 for y in $y = -\frac{2}{3}x + b$. Then we solve for b.

$$
\begin{aligned}
y &= mx + b \\
y &= -\frac{2}{3}x + b & \text{Substituting } -\frac{2}{3} \text{ for } m \\
6 &= -\frac{2}{3}(-3) + b & \text{Substituting } -3 \text{ for } x \text{ and } 6 \text{ for } y \\
6 &= 2 + b \\
4 &= b & \text{Solving for } b. \text{ The } y\text{-intercept is } (0, b).
\end{aligned}
$$

The equation of the line is $y = -\frac{2}{3}x + 4$, or $f(x) = -\frac{2}{3}x + 4$.

Now Try Exercise 13. ◼

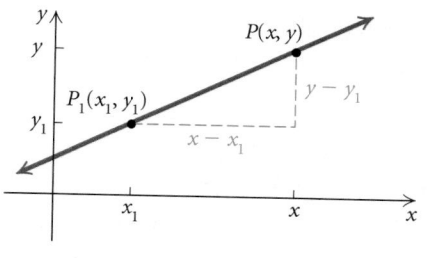

❖ Point–Slope Equations of Lines

Another formula that can be used to determine an equation of a line is the *point–slope equation*. Suppose that we have a nonvertical line and that the coordinates of point P_1 on the line are (x_1, y_1). We can think of P_1 as fixed and imagine another point P on the line with coordinates (x, y). Thus the slope is given by

$$\frac{y - y_1}{x - x_1} = m.$$

Multiplying by $x - x_1$ on both sides, we get the *point–slope equation* of the line:

$$(x - x_1) \cdot \frac{y - y_1}{x - x_1} = m \cdot (x - x_1)$$

$$y - y_1 = m(x - x_1).$$

Point–Slope Equation

The **point–slope equation** of the line with slope m passing through (x_1, y_1) is

$$y - y_1 = m(x - x_1).$$

If we know the slope of a line and the coordinates of one point on the line, we can find an equation of the line using either the point–slope equation,

$$y - y_1 = m(x - x_1),$$

or the slope–intercept equation,

$$y = mx + b.$$

EXAMPLE 3   Find an equation of the line containing the points $(2, 3)$ and $(1, -4)$.

Solution   We first determine the slope:

$$m = \frac{-4 - 3}{1 - 2} = \frac{-7}{-1} = 7.$$

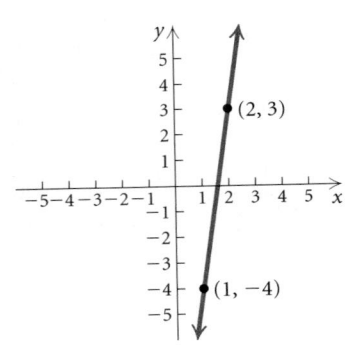

Using the Point–Slope Equation: We substitute 7 for m and either of the points $(2, 3)$ or $(1, -4)$ for (x_1, y_1) in the point–slope equation. In this case, we use $(2, 3)$.

$$y - y_1 = m(x - x_1) \qquad \text{Point–slope equation}$$
$$y - 3 = 7(x - 2) \qquad \text{Substituting}$$
$$y - 3 = 7x - 14$$
$$y = 7x - 11, \text{ or } f(x) = 7x - 11$$

Using the Slope–Intercept Equation: We substitute 7 for m and either of the points $(2, 3)$ or $(1, -4)$ for (x, y) in the slope–intercept equation and solve for b. Here we use $(1, -4)$.

$$y = mx + b \qquad \text{Slope–intercept equation}$$
$$-4 = 7 \cdot 1 + b \qquad \text{Substituting}$$
$$-4 = 7 + b$$
$$-11 = b \qquad \text{Solving for } b$$

We substitute 7 for m and -11 for b in $y = mx + b$ to get

$$y = 7x - 11, \text{ or } f(x) = 7x - 11.$$

Now Try Exercise 19. ◼

❖ Parallel Lines

Can we determine whether the graphs of two linear equations are parallel without graphing them? Let's observe three pairs of equations and their graphs.

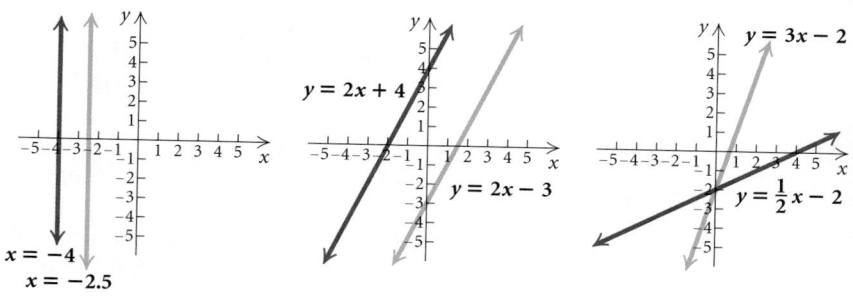

If two different lines, such as $x = -4$ and $x = -2.5$, are vertical, then they are parallel. Thus two equations such as $x = a_1$ and $x = a_2$, where $a_1 \neq a_2$, have graphs that are *parallel lines*. Two nonvertical lines, such as $y = 2x + 4$ and $y = 2x - 3$, or, in general, $y = mx + b_1$ and $y = mx + b_2$, where the slopes are the *same* and $b_1 \neq b_2$, also have graphs that are *parallel lines*.

Parallel Lines

Vertical lines are **parallel**. Nonvertical lines are **parallel** if and only if they have the same slope and different y-intercepts.

❖ Perpendicular Lines

Can we examine a pair of equations to determine whether their graphs are perpendicular without graphing the equations? Let's look at the following pairs of equations and their graphs.

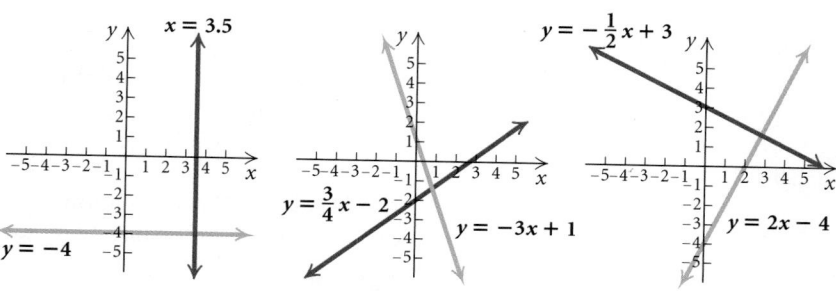

> **Perpendicular Lines**
>
> Two lines with slopes m_1 and m_2 are **perpendicular** if and only if the product of their slopes is -1:
>
> $$m_1 m_2 = -1.$$
>
> Lines are also **perpendicular** if one is vertical $(x = a)$ and the other is horizontal $(y = b)$.

$y_1 = 5x - 2, \quad y_2 = -\frac{1}{5}x - 3$

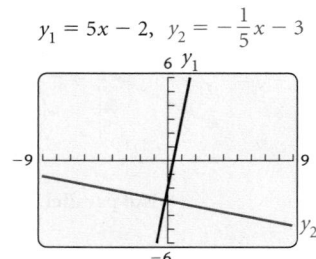

FIGURE 1

If a line has slope m_1, the slope m_2 of a line perpendicular to it is $-1/m_1$. (The slope of one line is the *opposite of the reciprocal* of the other.)

EXAMPLE 4   Determine whether each of the following pairs of lines is parallel, perpendicular, or neither.

a) $y + 2 = 5x, \quad 5y + x = -15$
b) $2y + 4x = 8, \quad 5 + 2x = -y$
c) $2x + 1 = y, \quad y + 3x = 4$

Solution   We use the slopes of the lines to determine whether the lines are parallel or perpendicular.

a) We solve each equation for y:

$$y = 5x - 2, \qquad y = -\tfrac{1}{5}x - 3.$$

The slopes are 5 and $-\frac{1}{5}$. Their product is -1, so the lines are perpendicular. (See Fig. 1.) Note in the graphs at left that the graphing calculator windows have been squared to avoid distortion. (Review squaring windows in Section 1.1.)

$y_1 = -2x + 4, \quad y_2 = -2x - 5$

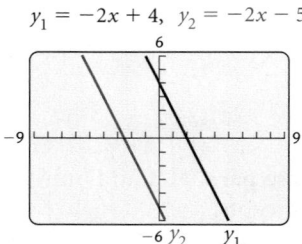

FIGURE 2

b) Solving each equation for y, we get

$$y = -2x + 4, \qquad y = -2x - 5.$$

We see that $m_1 = -2$ and $m_2 = -2$. Since the slopes are the same and the y-intercepts, $(0, 4)$ and $(0, -5)$, are different, the lines are parallel. (See Fig. 2.)

c) Solving each equation for y, we get

$$y = 2x + 1, \qquad y = -3x + 4.$$

We see that $m_1 = 2$ and $m_2 = -3$. Since the slopes are not the same and their product is not -1, it follows that the lines are neither parallel nor perpendicular. (See Fig. 3.)   **Now Try Exercise 35.** ■

$y_1 = 2x + 1, \quad y_2 = -3x + 4$

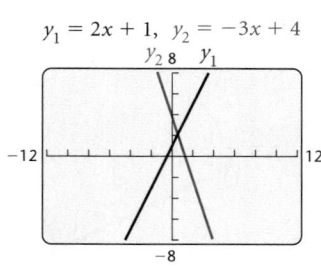

FIGURE 3

EXAMPLE 5   Write equations of the lines **(a)** parallel and **(b)** perpendicular to the graph of the line $4y - x = 20$ and containing the point $(2, -3)$.

Solution   We first solve $4y - x = 20$ for y to get $y = \frac{1}{4}x + 5$. Thus the slope of the given line is $\frac{1}{4}$.

a) The line parallel to the given line will have slope $\frac{1}{4}$. We use either the slope–intercept equation or the point–slope equation for a line with slope $\frac{1}{4}$ and containing the point $(2, -3)$. Here we use the point–slope equation:

$$y - y_1 = m(x - x_1)$$
$$y - (-3) = \tfrac{1}{4}(x - 2)$$
$$y + 3 = \tfrac{1}{4}x - \tfrac{1}{2}$$
$$y = \tfrac{1}{4}x - \tfrac{7}{2}.$$

b) The slope of the perpendicular line is the opposite of the reciprocal of $\frac{1}{4}$, or -4. Again we use the point–slope equation to write an equation for a line with slope -4 and containing the point $(2, -3)$:

$$y - y_1 = m(x - x_1)$$
$$y - (-3) = -4(x - 2)$$
$$y + 3 = -4x + 8$$
$$y = -4x + 5.$$

Now Try Exercise 43. ▨

$y_1 = \frac{1}{4}x + 5, \ y_2 = \frac{1}{4}x - \frac{7}{2},$

$y_3 = -4x + 5$

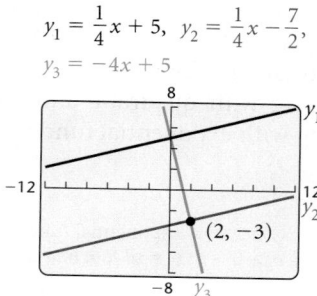

Summary of Terminology about Lines

Terminology	Mathematical Interpretation
Slope	$m = \dfrac{y_2 - y_1}{x_2 - x_1}$, or $\dfrac{y_1 - y_2}{x_1 - x_2}$
Slope–intercept equation	$y = mx + b$
Point–slope equation	$y - y_1 = m(x - x_1)$
Horizontal line	$y = b$
Vertical line	$x = a$
Parallel lines	$m_1 = m_2, b_1 \neq b_2$; or $x = a_1, x = a_2, a_1 \neq a_2$
Perpendicular lines	$m_1 m_2 = -1$

❖ Mathematical Models

When a real-world problem can be described in mathematical language, we have a **mathematical model**. For example, the natural numbers constitute a mathematical model for situations in which counting is essential. Situations in which algebra can be brought to bear often require the use of functions as models.

Mathematical models are abstracted from real-world situations. The mathematical model gives results that allow one to predict what will happen

Creating a Mathematical Model

1. Recognize real-world problem.

2. Collect data.

3. Analyze data.

4. Construct model.

5. Test and refine model

6. Explain and predict.

in that real-world situation. If the predictions are inaccurate or the results of experimentation do not conform to the model, the model must be changed or discarded.

Mathematical modeling can be an ongoing process. For example, finding a mathematical model that will provide an accurate prediction of population growth is not a simple problem. Any population model that one might devise would need to be reshaped as further information is acquired.

❖ Curve Fitting

We will develop and use many kinds of mathematical models in this text. In this chapter, we have used *linear* functions as models. Other types of functions, such as quadratic, cubic, and exponential functions, can also model data. These functions are *nonlinear*. Modeling with quadratic and cubic functions is discussed in Chapter 4. Modeling with exponential functions is discussed in Chapter 5.

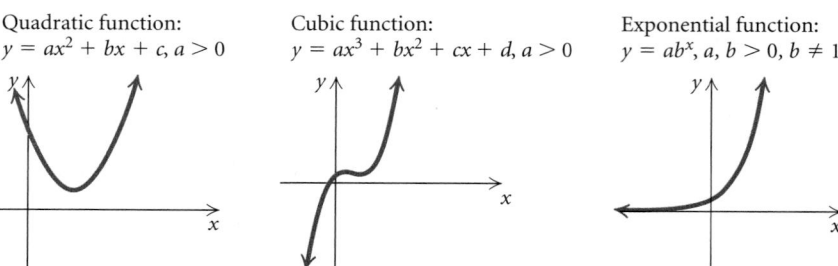

Quadratic function:
$y = ax^2 + bx + c, a > 0$

Cubic function:
$y = ax^3 + bx^2 + cx + d, a > 0$

Exponential function:
$y = ab^x, a, b > 0, b \neq 1$

In general, we try to find a function that fits, as well as possible, observations (data), theoretical reasoning, and common sense. We call this **curve fitting**; it is one aspect of mathematical modeling.

Let's look at some data and related graphs or **scatterplots** and determine whether a linear function seems to fit the set of data.

Year, x	Number of Cable Television Subscribers (in millions)	Scatterplot
1999, 0	76.4	
2000, 1	78.6	
2001, 2	81.5	
2002, 3	87.8	
2003, 4	88.4	
2004, 5	92.4	
2005, 6	94.0	
2006, 7	95.0	

It appears that the data points can be represented or modeled by a linear function.

The graph is **linear**.

Source: Nielsen Media Research

Year, x	Number of Gallons of Fuel Ethanol Produced in the United States (in billions)	Scatterplot
1998, 0	1.4	
1999, 1	1.5	
2000, 2	1.6	
2001, 3	1.8	
2002, 4	2.1	
2003, 5	2.8	
2004, 6	3.4	
2005, 7	3.9	
2006, 8	4.9	

It appears that the data points cannot be modeled accurately by a linear function.

The graph is **nonlinear**.

Source: Renewable Fuels Association

Looking at the scatterplots, we see that the data on cable subscribers seem to be rising in a manner to suggest that a *linear function* might fit, although a "perfect" straight line cannot be drawn through the data points. A linear function does not seem to fit the data on fuel ethanol production.

EXAMPLE 6 *Cable Subscribers.* Model the data in the table on the number of U.S. households with cable television on the preceding page with a linear function. Then predict the number of cable subscribers in 2010.

Solution We can choose any two of the data points to determine an equation. Let's use (2, 81.5) and (6, 94.0).

We first determine the slope of the line:

$$m = \frac{94.0 - 81.5}{6 - 2} = \frac{12.5}{4} = 3.125.$$

Then we substitute 3.125 for m and either of the points (2, 81.5) or (6, 94.0) for (x_1, y_1) in the point–slope equation. In this case, we use (6, 94.0). We get

$$y - 94.0 = 3.125(x - 6),$$

which simplifies to

$$y = 3.125x + 75.25,$$

where x is the number of years after 1999 and y is in millions.

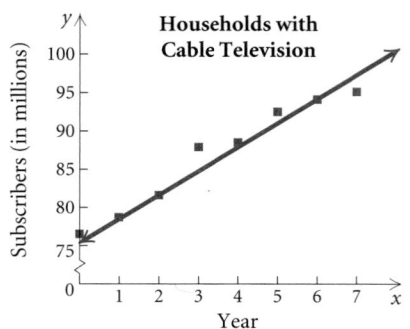

Now we can estimate the number of cable subscribers in 2010 by substituting 11 for x in the model ($2010 - 1999 = 11$):

$$y = 3.125x + 75.25 \qquad \text{Model}$$
$$ = 3.125(11) + 75.25 \qquad \text{Substituting}$$
$$ \approx 109.6. \qquad \text{Rounding to the nearest tenth}$$

We predict that there will be about 109.6 million U.S. households with cable television in 2010.   **Now Try Exercise 61.** ▨

In Example 6, if we use the data points $(1, 78.6)$ and $(7, 95.0)$, our model would be

$$y = 2.733x + 75.867$$

and our estimate for the number of households with cable television in 2010 would be about 105.9 million, about 3.7 million fewer than the estimate provided by the first model. This illustrates that a model and the estimates it produces are dependent on the pair of data points used.

Models that consider all the data points, not just two, are generally better models. The model that best fits the data can be found using a graphing calculator and a procedure called **linear regression**.

❈ Linear Regression

We now consider **linear regression**, a procedure that can be used to model a set of data using a linear function. Although discussion leading to a complete understanding of this method belongs in a statistics course, we present the procedure here because we can carry it out easily using technology. The graphing calculator gives us the powerful capability to find linear models and to make predictions using them.

Consider the data presented before Example 6 on the number of U.S. households with cable television. We can fit a regression line of the form $y = mx + b$ to the data using the LINEAR REGRESSION feature on a graphing calculator.

GCM   **EXAMPLE 7**   *Cable Subscribers.*   Fit a regression line to the data given in the table on U.S. households with cable television on p. 120. Then use the function to predict the number of cable subscribers in 2010.

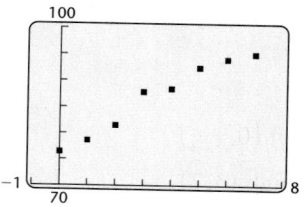

FIGURE 2

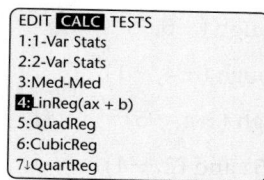

FIGURE 3

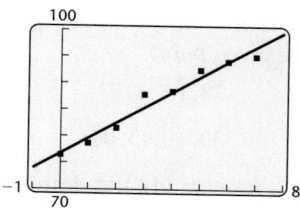

FIGURE 4

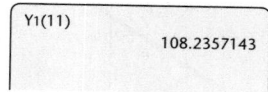

100

70

−1   8

FIGURE 5

Solution   First, we enter the data in lists on the calculator. We enter the values of the independent variable x in list L1 and the corresponding values of the dependent variable y in L2. (See Fig. 1.) The graphing calculator can then create a scatterplot of the data, as shown at left in Fig. 2.

L1	L2	L3
0	76.4	-----
1	78.6	
2	81.5	
3	87.8	
4	88.4	
5	92.4	
6	94	
L2(7) = 94		

L1	L2	L3
1	78.6	-----
2	81.5	
3	87.8	
4	88.4	
5	92.4	
6	94	
7	95	
L2(8) = 95		

FIGURE 1

When we select the LINEAR REGRESSION feature from the STAT CALC menu, we find the linear equation that best models the data. It is

$$y = 2.863095238x + 76.74166667. \qquad \textbf{Regression line}$$

(See Figs. 3 and 4.) We can then graph the regression line on the same graph as the scatterplot, as shown in Fig. 5.

To predict the number of U.S. households with cable television in 2010, we substitute 11 for x in the regression equation. Using this model, we see that the number of cable subscribers in 2010 is predicted to be about 108.2 million.

Note that 108.2 million is closer to the value 109.6 million found with data points $(2, 81.5)$ and $(6, 94.0)$ than to the value 105.9 million found with the data points $(1, 78.6)$ and $(7, 95.0)$ after Example 6.

Now Try Exercise 71. ■

The Correlation Coefficient

On some graphing calculators with the DIAGNOSTIC feature turned on, a constant r between -1 and 1, called the **coefficient of linear correlation**, appears with the equation of the regression line. Though we cannot develop a formula for calculating r in this text, keep in mind that it is used to describe the strength of the linear relationship between x and y. The closer $|r|$ is to 1, the better the correlation. A positive value of r also indicates that the regression line has a positive slope, and a negative value of r indicates that the regression line has a negative slope. For the cable television data just discussed,

$$r = 0.9813757551,$$

which indicates a good linear correlation.

1.4 Exercise Set

Find the slope and the y-intercept of the graph of the linear equation. Then write the equation of the line in slope–intercept form.

1.

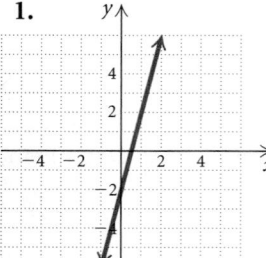

2.

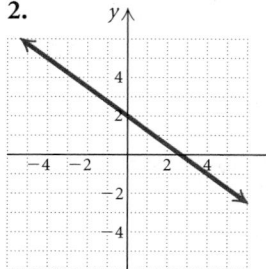

3.

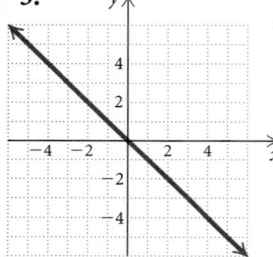

4.

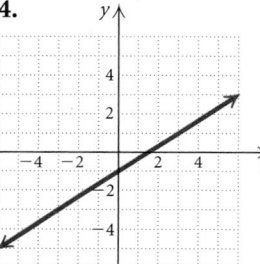

5.

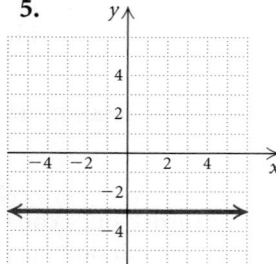

6.

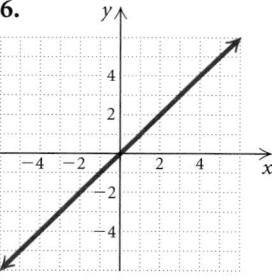

Write a slope–intercept equation for a line with the given characteristics.

7. $m = \frac{2}{9}$, y-intercept $(0, 4)$

8. $m = -\frac{3}{8}$, y-intercept $(0, 5)$

9. $m = -4$, y-intercept $(0, -7)$

10. $m = \frac{2}{7}$, y-intercept $(0, -6)$

11. $m = -4.2$, y-intercept $\left(0, \frac{3}{4}\right)$

12. $m = -4$, y-intercept $\left(0, -\frac{3}{2}\right)$

13. $m = \frac{2}{9}$, passes through $(3, 7)$

14. $m = -\frac{3}{8}$, passes through $(5, 6)$

15. $m = 0$, passes through $(-2, 8)$

16. $m = -2$, passes through $(-5, 1)$

17. $m = -\frac{3}{5}$, passes through $(-4, -1)$

18. $m = \frac{2}{3}$, passes through $(-4, -5)$

19. Passes through $(-1, 5)$ and $(2, -4)$

20. Passes through $\left(-3, \frac{1}{2}\right)$ and $\left(1, \frac{1}{2}\right)$

21. Passes through $(7, 0)$ and $(-1, 4)$

22. Passes through $(-3, 7)$ and $(-1, -5)$

23. Passes through $(0, -6)$ and $(3, -4)$

24. Passes through $(-5, 0)$ and $\left(0, \frac{4}{5}\right)$

25. Passes through $(-4, 7.3)$ and $(0, 7.3)$

26. Passes through $(-13, -5)$ and $(0, 0)$

Write equations of the horizontal lines and the vertical lines that pass through the given point.

27. $(0, -3)$ **28.** $\left(-\frac{1}{4}, 7\right)$

29. $\left(\frac{2}{11}, -1\right)$ **30.** $(0.03, 0)$

31. Find a linear function h given $h(1) = 4$ and $h(-2) = 13$. Then find $h(2)$.

32. Find a linear function g given $g\left(-\frac{1}{4}\right) = -6$ and $g(2) = 3$. Then find $g(-3)$.

33. Find a linear function f given $f(5) = 1$ and $f(-5) = -3$. Then find $f(0)$.

34. Find a linear function h given $h(-3) = 3$ and $h(0) = 2$. Then find $h(-6)$.

Determine whether the pair of lines is parallel, perpendicular, or neither.

35. $y = \frac{26}{3}x - 11$,   **36.** $y = -3x + 1$,
$\quad y = -\frac{3}{26}x - 11$ $\quad\quad y = -\frac{1}{3}x + 1$

37. $y = \frac{2}{5}x - 4,$
$y = -\frac{2}{5}x + 4$

38. $y = \frac{3}{2}x - 8,$
$y = 8 + 1.5x$

39. $x + 2y = 5,$
$2x + 4y = 8$

40. $2x - 5y = -3,$
$2x + 5y = 4$

41. $y = 4x - 5,$
$4y = 8 - x$

42. $y = 7 - x,$
$y = x + 3$

Write a slope–intercept equation for a line passing through the given point that is parallel to the given line. Then write a second equation for a line passing through the given point that is perpendicular to the given line.

43. $(3, 5), \; y = \frac{2}{7}x + 1$

44. $(-1, 6), \; f(x) = 2x + 9$

45. $(-7, 0), \; y = -0.3x + 4.3$

46. $(-4, -5), \; 2x + y = -4$

47. $(3, -2), \; 3x + 4y = 5$

48. $(8, -2), \; y = 4.2(x - 3) + 1$

49. $(3, -3), \; x = -1$

50. $(4, -5), \; y = -1$

In Exercises 51–56, determine whether the statement is true or false.

51. The lines $x = -3$ and $y = 5$ are perpendicular.

52. The lines $y = 2x - 3$ and $y = -2x - 3$ are perpendicular.

53. The lines $y = \frac{2}{5}x + 4$ and $y = \frac{2}{5}x - 4$ are parallel.

54. The intersection of the lines $y = 2$ and $x = -\frac{3}{4}$ is $\left(-\frac{3}{4}, 2\right).$

55. The lines $x = -1$ and $x = 1$ are perpendicular.

56. The lines $2x + 3y = 4$ and $3x - 2y = 4$ are perpendicular.

In Exercises 57–60, determine whether a linear model might fit the data.

57.

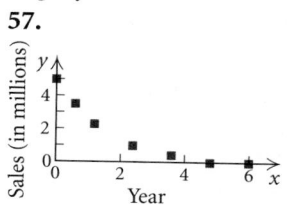

58.

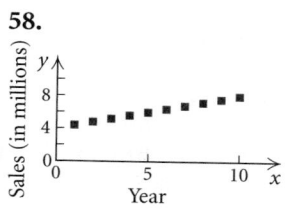

59.

60.

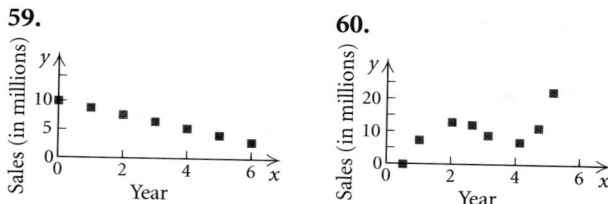

61. *Twin Births.* The table below illustrates the upward trend in twin births.

a) Model the data with a linear function. Let the independent variable represent the number of years after 1990; that is, the data points are $(0, 93.8)$, $(5, 96.7)$, and so on. Answers may vary depending on the data points used.

b) With the function found in part (a), estimate the number of twin births in 2009 and in 2012.

Year, x	Number of Twin Births, y (in thousands)
1990, 0	93.8
1995, 5	96.7
2000, 10	118.9
2001, 11	121.2
2002, 12	125.1
2003, 13	128.7
2004, 14	139.2

Source: National Center for Health Statistics, *National Vital Statistics Reports*, vol 54, no. 2, Sept. 8, 2005

62. *Triplet Births.* The table on the following page illustrates the trend in triplet births in recent years.

a) Model the data given in the table below with a linear function. Let the independent variable represent the number of years after 1990. Answers may vary depending on the data points used.

b) With the function found in part (a), estimate the number of triplet births in 2009 and in 2012.

Year, x	Number of Triplet Births, y
1990, 0	2830
1995, 5	4551
2000, 10	6742
2001, 11	6885
2002, 12	6898
2003, 13	7110
2004, 14	6750

Source: National Center for Health Statistics, *National Vital Statistics Reports*, vol 54, no. 2, Sept. 8, 2005

63. *Theme Park Attendance.* Model the data given in the table below with a linear function. Then estimate the attendance at U.S. amusement/theme parks in 2006 and predict the attendance in 2010. Answers may vary depending on the data points used.

Year, x	Amusement/Theme Park Attendance (in millions)
1995, 0	280
1997, 2	300
2000, 5	317
2001, 6	319
2002, 7	324
2003, 8	322
2004, 9	328
2005, 10	335

Source: International Association of Amusement Parks and Attractions

64. *Prescription Drug Sales.* The following graph illustrates the retail sales of prescription drugs for certain years. Model the data with a linear function, estimate the total sales for 2007, and predict the total sales for 2011. Let $x =$ the number of years after 2001. Answers may vary depending on the data points used.

Prescription Drug Sales

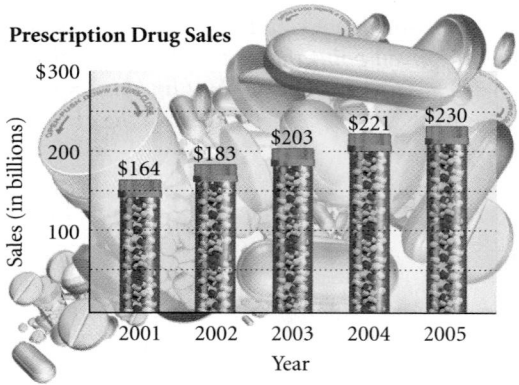

Source: U.S. Bureau of the Census

65. *Social-Security Benefits.* Model the data given in the table below with a linear function, estimate the average monthly social-security benefits for retired workers in 2009, and predict the average benefits in 2012 and in 2020. Answers may vary depending on the data points used.

Year, x	Average Monthly Social-Security Benefits for Retired Workers, y
1970, 0	$124
1980, 10	321
1990, 20	551
2000, 30	845
2003, 33	922
2004, 34	930
2005, 35	955
2006, 36	1011
2007, 37	1044

Source: Social Security Administration, *Social Security Bulletin: Annual Statistical Supplement*, 2006

66. *Sheep and Lambs.* The number of sheep and lambs on farms in the United States has declined in recent years. Model the data given in the table on the following page with a linear function and estimate the number of sheep and lambs on farms in 2008 and in 2013. Answers may vary depending on the data points used.

Year, x	Sheep and Lambs on Farms in the United States, y (in millions)
1970, 0	20.423
1975, 5	14.515
1980, 10	12.699
1985, 15	10.716
1990, 20	11.358
1995, 25	8.886
2000, 30	7.032
2005, 35	6.135

Source: National Agricultural Statistics Service, U.S. Department of Agriculture

67. *Maximum Heart Rate.*   A person who is exercising should not exceed his or her maximum heart rate, which is determined on the basis of that person's sex, age, and resting heart rate. The following table relates resting heart rate and maximum heart rate for a 20-year-old man.

Resting Heart Rate, H (in beats per minute)	Maximum Heart Rate, M (in beats per minute)
50	166
60	168
70	170
80	172

Source: American Heart Association

a) Use a graphing calculator to model the data with a linear function.
b) Estimate the maximum heart rate if the resting heart rate is 40, 65, 76, and 84.

c) What is the correlation coefficient? How confident are you about using the regression line to estimate function values?

68. *Study Time versus Grades.*   A math instructor asked her students to keep track of how much time each spent studying a chapter on functions in her algebra–trigonometry course. She collected the information together with test scores from that chapter's test. The data are listed in the following table.

Study Time, x (in hours)	Test Grade, y (in percent)
23	81%
15	85
17	80
9	75
21	86
13	80
16	85
11	93

a) Use a graphing calculator to model the data with a linear function.
b) Predict a student's score if he or she studies 24 hr, 6 hr, and 18 hr.
c) What is the correlation coefficient? How confident are you about using the regression line to predict function values?

69. *Filing Tax Returns Electronically.*   The number of tax returns filed electronically in 2005 reflects an increase of approximately 480% over the number of electronic returns in 1995.

Year, x	Number of Tax Returns Filed Electronically, y (in millions)
1995, 0	11.807
1997, 2	19.136
1999, 4	29.349
2001, 6	40.245
2003, 8	52.945
2005, 10	68.476

Source: 2005 IRS Data Book

a) Use a graphing calculator to model the data with a linear function.
b) Predict the number of tax returns that will be filed electronically in 2010 and in 2014.
c) Find the correlation coefficient for the regression line and determine whether the line fits the data closely.

70. a) Use a graphing calculator to fit a regression line to the data in Exercise 62.
b) Estimate the number of triplet births in 2012 and compare the value with the result found in Exercise 62.
c) Find the correlation coefficient for the regression line and determine whether the line fits the data closely.

71. a) Use a graphing calculator to fit a regression line to the data in Exercise 61.
b) Estimate the number of twin births in 2012 and compare the result with the estimate found with the model in Exercise 61.
c) Find the correlation coefficient for the regression line and determine whether the line fits the data closely.

Collaborative Discussion and Writing

72. Answers to Exercise 62 vary depending on which two data points are used for the model. Visualize a line that seems to best model the data and discuss which points in the data set would not be good choices. Explain why.

73. Answers to Exercise 61 vary depending on which two data points are used for the model. Visualize a line that seems to best model the data and discuss which points in the data set would not be good choices. Explain why.

Skill Maintenance

Find the slope of the line containing the given points.
74. $(5, 7)$ and $(5, -7)$

75. $(2, -8)$ and $(-5, -1)$

Find an equation for a circle satisfying the given conditions.
76. Center $(0, 3)$, diameter of length 5

77. Center $(-7, -1)$, radius of length $\frac{9}{5}$

Synthesis

78. *Road Grade.*   Using the following figure, find the road grade and an equation giving the height y as a function of the horizontal distance x.

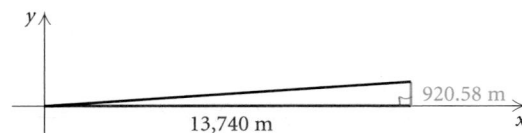

79. Find k so that the line containing the points $(-3, k)$ and $(4, 8)$ is parallel to the line containing the points $(5, 3)$ and $(1, -6)$.

80. Find an equation of the line passing through the point $(4, 5)$ and perpendicular to the line passing through the points $(-1, 3)$ and $(2, 9)$.

Linear Equations, Functions, Zeros, and Applications

❖ Solve linear equations.

❖ Solve applied problems using linear models.

❖ Find zeros of linear functions.

❖ Solve a formula for a given variable.

An **equation** is a statement that two expressions are equal. To **solve** an equation in one variable is to find all the values of the variable that make the equation true. Each of these numbers is a **solution** of the equation. The set of all solutions of an equation is its **solution set**. Some examples of **equations in one variable** are

$$2x + 3 = 5, \quad 3(x - 1) = 4x + 5, \quad x^2 - 3x + 2 = 0, \quad \text{and} \quad \frac{x - 3}{x + 4} = 1.$$

❖ Linear Equations

The first two equations above are *linear equations* in one variable. We define such equations as follows.

> A **linear equation in one variable** is an equation that can be expressed in the form $mx + b = 0$, where m and b are real numbers and $m \neq 0$.

Equations that have the same solution set are **equivalent equations**. For example, $2x + 3 = 5$ and $x = 1$ are equivalent equations because 1 is the solution of each equation. On the other hand, $x^2 - 3x + 2 = 0$ and $x = 1$ are not equivalent equations because 1 and 2 are both solutions of $x^2 - 3x + 2 = 0$ but 2 is not a solution of $x = 1$.

To solve an equation, we find an equivalent equation in which the variable is isolated. The following principles allow us to solve linear equations.

Equation-Solving Principles

For any real numbers a, b, and c:

The Addition Principle: If $a = b$ is true, then $a + c = b + c$ is true.

The Multiplication Principle: If $a = b$ is true, then $ac = bc$ is true.

Note to the student and the instructor: We assume that students come to a College Algebra course with some equation-solving skills from their study of Intermediate Algebra. Thus a portion of the material in this section might be considered by some to be review in nature. We present this material here in order to use linear functions, with which students are familiar, to lay the groundwork for zeros of higher-order polynomial functions and their connection to solutions of equations and x-intercepts of graphs.

GCM   **EXAMPLE 1**   Solve: $\dfrac{3}{4}x - 1 = \dfrac{7}{5}$.

Solution   When we have an equation that contains fractions, it is often convenient to multiply both sides of the equation by the least common denominator (LCD) of the fractions in order to clear the equation of fractions. We have

$$\frac{3}{4}x - 1 = \frac{7}{5} \qquad \text{The LCD is } 4 \cdot 5, \text{ or } 20.$$

$$20\left(\frac{3}{4}x - 1\right) = 20 \cdot \frac{7}{5} \qquad \begin{array}{l}\text{Multiplying by the LCD on both} \\ \text{sides to clear fractions}\end{array}$$

$$20 \cdot \frac{3}{4}x - 20 \cdot 1 = 28$$

$$15x - 20 = 28$$

$$15x - 20 + 20 = 28 + 20 \qquad \begin{array}{l}\text{Using the addition principle to add} \\ 20 \text{ on both sides}\end{array}$$

$$15x = 48$$

$$\frac{15x}{15} = \frac{48}{15} \qquad \begin{array}{l}\text{Using the multiplication principle to} \\ \text{multiply by } \frac{1}{15}, \text{ or divide by 15, on} \\ \text{both sides}\end{array}$$

$$x = \frac{48}{15}$$

$$x = \frac{16}{5}. \qquad \begin{array}{l}\text{Simplifying. Note that } \frac{3}{4}x - 1 = \frac{7}{5} \\ \text{and } x = \frac{16}{5} \text{ are equivalent equations.}\end{array}$$

Check:
$$\frac{3}{4}x - 1 = \frac{7}{5}$$

$$\frac{3}{4} \cdot \frac{16}{5} - 1 \; ? \; \frac{7}{5} \qquad \text{Substituting } \frac{16}{5} \text{ for } x$$

$$\frac{12}{5} - \frac{5}{5}$$

$$\frac{7}{5} \; \bigg| \; \frac{7}{5} \quad \text{TRUE}$$

The solution is $\dfrac{16}{5}$.

Now Try Exercise 15. ■

We can use the INTERSECT feature on a graphing calculator to solve equations. We call this the **Intersect method**. To use the Intersect method to solve the equation in Example 1, for instance, we graph $y_1 = \frac{3}{4}x - 1$ and $y_2 = \frac{7}{5}$. The value of x for which $y_1 = y_2$ is the solution of the equation $\frac{3}{4}x - 1 = \frac{7}{5}$. This value of x is the first coordinate of the point of intersection of the graphs of y_1 and y_2. Using the INTERSECT feature, we find that the first coordinate of this point is 3.2. We can find fraction notation for the solution by using the ▶FRAC feature. The solution is 3.2, or $\frac{16}{5}$.

$$y_1 = \frac{3}{4}x - 1, \quad y_2 = \frac{7}{5}$$

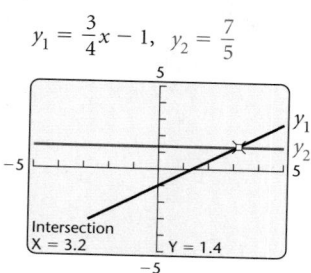

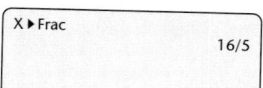

EXAMPLE 2 Solve: $2(5 - 3x) = 8 - 3(x + 2)$.

ALGEBRAIC SOLUTION

We have

$$2(5 - 3x) = 8 - 3(x + 2)$$

$10 - 6x = 8 - 3x - 6$ **Using the distributive property**

$10 - 6x = 2 - 3x$ **Collecting like terms**

$10 - 6x + 6x = 2 - 3x + 6x$ **Using the addition principle to add $6x$ on both sides**

$$10 = 2 + 3x$$

$10 - 2 = 2 + 3x - 2$ **Using the addition principle to add -2, or subtract 2, on both sides**

$$8 = 3x$$

$\dfrac{8}{3} = \dfrac{3x}{3}$ **Using the multiplication principle to multiply by $\frac{1}{3}$, or divide by 3, on both sides**

$$\frac{8}{3} = x.$$

Check:

$$2(5 - 3x) = 8 - 3(x + 2)$$

$2\left(5 - 3 \cdot \frac{8}{3}\right) \; ? \; 8 - 3\left(\frac{8}{3} + 2\right)$ **Substituting $\frac{8}{3}$ for x**

$$2(5 - 8) \;\bigg|\; 8 - 3\left(\frac{14}{3}\right)$$

$$2(-3) \;\bigg|\; 8 - 14$$

$$-6 \;\bigg|\; -6 \qquad \text{TRUE}$$

The solution is $\dfrac{8}{3}$.

GRAPHICAL SOLUTION

We graph $y_1 = 2(5 - 3x)$ and $y_2 = 8 - 3(x + 2)$. The first coordinate of the point of intersection of the graphs is the value of x for which $2(5 - 3x) = 8 - 3(x + 2)$ and is thus the solution of the equation.

$$y_1 = 2(5 - 3x), \quad y_2 = 8 - 3(x + 2)$$

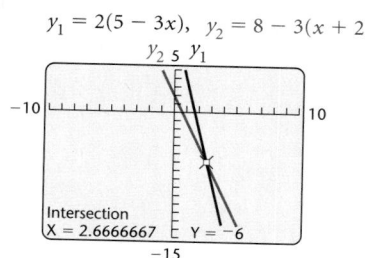

The solution is approximately 2.6666667.

We can find fraction notation for the exact solution by using the ▶FRAC feature. The solution is $\frac{8}{3}$.

Now Try Exercise 27. ■

We can use the TABLE feature on a graphing calculator, set in ASK mode, to check the solutions of equations. In Example 2, for instance, let $y_1 = 2(5 - 3x)$ and $y_2 = 8 - 3(x + 2)$. When $\frac{8}{3}$ is entered for x, we see that $y_1 = y_2$, or $2(5 - 3x) = 8 - 3(x + 2)$. Thus, $\frac{8}{3}$ is the solution of the equation. (Note that the calculator converts $\frac{8}{3}$ to decimal notation in the table.)

X	Y₁	Y₂
2.6667	⁻6	⁻6

X =

Special Cases

Some equations have *no* solution.

EXAMPLE 3 Solve: $-24x + 7 = 17 - 24x$.

Solution We have

$$-24x + 7 = 17 - 24x$$
$$24x - 24x + 7 = 24x + 17 - 24x \qquad \text{Adding } 24x$$
$$7 = 17. \qquad \text{We get a false equation.}$$

No matter what number we substitute for x, we get a false sentence. Thus the equation has *no* solution.

Now Try Exercise 11. ■

There are some equations for which *any* real number is a solution.

EXAMPLE 4 Solve: $3 - \dfrac{1}{3}x = -\dfrac{1}{3}x + 3$.

Solution We have

$$3 - \frac{1}{3}x = -\frac{1}{3}x + 3$$

$$\frac{1}{3}x + 3 - \frac{1}{3}x = \frac{1}{3}x - \frac{1}{3}x + 3 \qquad \text{Adding } \frac{1}{3}x$$

$$3 = 3. \qquad \text{We get a true equation.}$$

Replacing x with any real number gives a true sentence. Thus *any* real number is a solution. This equation has *infinitely* many solutions. The solution set is the set of real numbers, $\{x \mid x \text{ is a real number}\}$, or $(-\infty, \infty)$.

Now Try Exercise 3. ■

❖ Applications Using Linear Models

Mathematical techniques are used to answer questions arising from real-world situations. Linear equations and functions *model* many of these situations.

The following strategy is of great assistance in problem solving.

Five Steps for Problem Solving

1. **Familiarize** yourself with the problem situation. If the problem is presented in words, this means to read carefully. Some or all of the following can also be helpful.
   a) Make a drawing, if it makes sense to do so.
   b) Make a written list of the known facts and a list of what you wish to find out.
   c) Assign variables to represent unknown quantities.
   d) Organize the information in a chart or a table.
   e) Find further information. Look up a formula, consult a reference book or an expert in the field, or do research on the Internet.
   f) Guess or estimate the answer and check your guess or estimate.

2. **Translate** the problem situation to mathematical language or symbolism. For most of the problems you will encounter in algebra, this means to write one or more equations, but sometimes an inequality or some other mathematical symbolism may be appropriate.

3. **Carry out** some type of mathematical manipulation. Use your mathematical skills to find a possible solution. In algebra, this usually means to solve an equation, an inequality, or a system of equations or inequalities.

4. **Check** to see whether your possible solution actually fits the problem situation and is thus really a solution of the problem. Although you may have solved an equation, the solution(s) of the equation might not be solution(s) of the original problem.

5. **State** the answer clearly using a complete sentence.

EXAMPLE 5 *Dining Out.* Americans spent about $511 billion dining out in 2006. This was a 5.1% increase over the amount spent in 2005. (*Source*: National Restaurant Association) How much was spent dining out in 2005?

Solution

1. **Familiarize.** Let's estimate that $500 billion was spent dining out in 2005. Then the amount spent in 2006, in billions of dollars, would be

$$500 + 5.1\% \cdot 500 = 1(500) + 0.051(500)$$
$$= 1.051(500)$$
$$= 525.5.$$

Since we know that $511 billion was actually spent dining out in 2006, the estimate is too high. Nevertheless, the calculations performed in checking the estimate indicate how we can translate the problem to an equation. We let $x =$ the amount spent dining out in 2005, in billions of dollars. Then $x + 5.1\%x$, or $1 \cdot x + 0.051 \cdot x$, or $1.051x$, is the amount spent dining out in 2006.

2. **Translate.**   We translate to an equation:

$$\underbrace{\text{Amount spent dining out}}_{1.051x} \quad \underset{=}{\text{was}} \quad \underbrace{\$511 \text{ billion.}}_{511}$$

3. **Carry out.**   We solve the equation, as follows:

$$1.051x = 511$$

$$x = \frac{511}{\cdot 1.051} \qquad \text{Dividing by 1.051 on both sides}$$

$$x \approx 486.$$

4. **Check.**   Since 5.1% of $486 billion is about $25 billion and $486 billion + $25 billion = $511 billion, the answer checks.

5. **State.**   Americans spent about $486 billion dining out in 2005.

Now Try Exercise 33. ▪

EXAMPLE 6   *Veterinary Expenses.*   Together, a dog owner and a cat owner spend an average of $376 annually for veterinary-related expenses. A dog owner spends $150 more per year than a cat owner. (*Source:* The Humane Society of the United States) Find the average annual veterinary-related expenses of a dog owner and of a cat owner.

Solution

1. **Familiarize.**   A dog owner's spending is described in terms of a cat owner's spending, so we will let $x =$ the amount spent annually for veterinary-related expenses by a cat owner. Then $x + 150 =$ the amount spent annually by a dog owner.

2. **Translate.**   We translate to an equation:

$$\underbrace{\text{Dog owner's expenses}}_{x + 150} \quad \underset{+}{\text{plus}} \quad \underbrace{\text{cat owner's expenses}}_{x} \quad \underset{=}{\text{is}} \quad \underset{376}{\$376.}$$

3. **Carry out.**   We solve the equation, as follows:

$$x + 150 + x = 376$$

$$2x + 150 = 376 \qquad \text{Collecting like terms}$$

$$2x = 226 \qquad \text{Subtracting 150 on both sides}$$

$$x = 113. \qquad \text{Dividing by 2 on both sides}$$

If $x = 113$, then $x + 150 = 113 + 150 = 263$.

4. **Check.** If a dog owner spends $263 annually for veterinary-related expenses and a cat owner spends $113 annually, then together they spend $263 + $113, or $376. Also, $263 is $150 more than $113. The answer checks.

5. **State.** A dog owner spends an average of $263 annually for veterinary-related expenses and a cat owner spends an average of $113 annually.

Now Try Exercise 35. ■

In some applications, we need to use a formula that describes the relationships among variables. When a situation involves distance, rate (also called speed or velocity), and time, for example, we use the following formula.

The Motion Formula

The distance d traveled by an object moving at rate r in time t is given by

$$d = r \cdot t.$$

EXAMPLE 7 *Airplane Speed.* America West Airlines' fleet includes Boeing 737-200's, each with a cruising speed of 500 mph, and Bombardier deHavilland Dash 8-200's, each with a cruising speed of 302 mph (*Source:* America West Airlines). Suppose that a Dash 8-200 takes off and travels at its cruising speed. One hour later, a 737-200 takes off and follows the same route, traveling at its cruising speed. How long will it take the 737-200 to overtake the Dash 8-200?

Solution

1. **Familiarize.** We make a drawing showing both the known and the unknown information. We let $t =$ the time, in hours, that the 737-200 travels before it overtakes the Dash 8-200. Since the Dash 8-200 takes off 1 hr before the 737, it will travel for $t + 1$ hr before being overtaken. The planes will have traveled the same distance, d, when one overtakes the other.

We can also organize the information in a table, as follows.

$$d \quad = \quad r \quad \cdot \quad t$$

	Distance	Rate	Time	
737-200	d	500	t	$\longrightarrow d = 500t$
Dash 8-200	d	302	$t + 1$	$\longrightarrow d = 302(t + 1)$

2. **Translate.**   Using the formula $d = rt$ in each row of the table, we get two expressions for d:

$$d = 500t \quad \text{and} \quad d = 302(t + 1).$$

Since the distances are the same, we have the following equation:

$$500t = 302(t + 1).$$

3. **Carry out.**   We solve the equation, as follows:

$$500t = 302(t + 1)$$
$$500t = 302t + 302 \qquad \text{Using the distributive property}$$
$$198t = 302 \qquad \text{Subtracting } 302t \text{ on both sides}$$
$$t \approx 1.53. \qquad \text{Dividing by 198 on both sides and rounding to the nearest hundredth}$$

4. **Check.**   If the 737-200 travels for about 1.53 hr, then the Dash 8-200 travels for about $1.53 + 1$, or 2.53 hr. In 2.53 hr, the Dash 8-200 travels $302(2.53)$, or 764.06 mi, and in 1.53 hr, the 737-200 travels $500(1.53)$, or 765 mi. Since 764.06 mi $\approx$ 765 mi, the answer checks. (Remember that we rounded the value of t.)

5. **State.**   About 1.53 hr after the 737-200 has taken off, it will overtake the Dash 8-200.   **Now Try Exercise 57.** ■

For some applications, we need to use a formula to find the amount of interest earned by an investment or the amount of interest due on a loan.

> **The Simple-Interest Formula**
>
> The **simple interest** I on a principal of P dollars at interest rate r for t years is given by
>
> $$I = Prt.$$

EXAMPLE 8   *Student Loans.*   Jared's two student loans total $12,000. One loan is at 5% simple interest and the other is at 8% simple interest. After 1 yr, Jared owes $750 in interest. What is the amount of each loan?

Solution

1. **Familiarize.** We let x = the amount borrowed at 5% interest. Then the remainder of the $12,000, or $12,000 - x$, is borrowed at 8%. We organize the information in a table, keeping in mind the formula $I = Prt$.

	Amount Borrowed	Interest Rate	Time	Amount of Interest
5% Loan	x	5%, or 0.05	1 yr	$x(0.05)(1)$, or $0.05x$
8% Loan	$12,000 - x$	8%, or 0.08	1 yr	$(12,000 - x)(0.08)(1)$, or $0.08(12,000 - x)$
Total	12,000			750

2. **Translate.** The total amount of interest on the two loans is $750. Thus we write the following equation.

Interest on 5% loan ⟶ plus ⟶ interest on 8% loan ⟶ is ⟶ $750.

$$0.05x \quad + \quad 0.08(12,000 - x) \quad = \quad 750$$

3. **Carry out.** We solve the equation, as follows:

$$0.05x + 0.08(12,000 - x) = 750$$
$$0.05x + 960 - 0.08x = 750 \qquad \text{Using the distributive property}$$
$$-0.03x + 960 = 750 \qquad \text{Collecting like terms}$$
$$-0.03x = -210 \qquad \text{Subtracting 960 on both sides}$$
$$x = 7000. \qquad \text{Dividing by } -0.03 \text{ on both sides}$$

If $x = 7000$, then $12,000 - x = 12,000 - 7000 = 5000$.

4. **Check.** The interest on $7000 at 5% for 1 yr is $7000(0.05)(1)$, or $350. The interest on $5000 at 8% for 1 yr is $5000(0.08)(1)$, or $400. Since $350 + $400 = $750, the answer checks.

5. **State.** Jared borrowed $7000 at 5% interest and $5000 at 8% interest.

Now Try Exercise 63. ■

Sometimes we use formulas from geometry in solving applied problems. In the following example, we use the formula for the perimeter P of a rectangle with length l and width w: $P = 2l + 2w$. There is a summary of geometric formulas at the back of the book.

EXAMPLE 9 *Soccer Fields.* The length of the largest regulation soccer field is 30 yd greater than the width and the perimeter is 460 yd. Find the length and the width.

Solution

1. **Familiarize.** We first make a drawing. Since the length of the field is described in terms of the width, we let w = the width, in yards. Then $w + 30$ = the length, in yards.

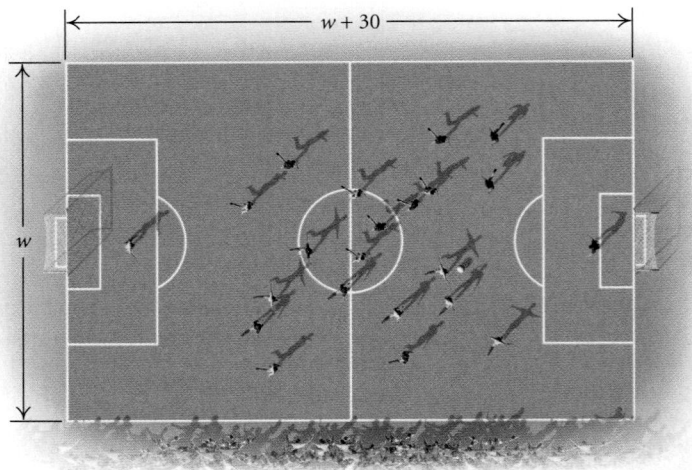

2. **Translate.** We use the formula for the perimeter of a rectangle:

$$P = 2l + 2w$$
$$460 = 2(w + 30) + 2w. \quad \textbf{Substituting 460 for } P \textbf{ and } w + 30 \textbf{ for } l$$

3. **Carry out.** We solve the equation:

$$460 = 2(w + 30) + 2w$$
$$460 = 2w + 60 + 2w \quad \textbf{Using the distributive property}$$
$$460 = 4w + 60 \quad \textbf{Collecting like terms}$$
$$400 = 4w \quad \textbf{Subtracting 60 on both sides}$$
$$100 = w. \quad \textbf{Dividing by 4 on both sides}$$

If $w = 100$, then $w + 30 = 100 + 30 = 130$.

4. **Check.** The length, 130 yd, is 30 yd more than the width, 100 yd. Also,

$$2 \cdot 130 \text{ yd} + 2 \cdot 100 \text{ yd} = 260 \text{ yd} + 200 \text{ yd} = 460 \text{ yd}.$$

The answer checks.

5. **State.** The length of the largest regulation soccer field is 130 yd and the width is 100 yd. **Now Try Exercise 53.** ■

EXAMPLE 10 *Cab Fare.* Metro Taxi charges a $1.25 pickup fee and $2 per mile traveled. Cecilia's cab fare from the airport to her hotel is $31.25. How many miles did she travel in the cab?

Solution

1. **Familiarize.** Let's guess that Cecilia traveled 12 mi in the cab. Then her fare would be

$$\$1.25 + \$2 \cdot 12 = \$1.25 + \$24 = \$25.25.$$

We see that our guess is low, but the calculation we did shows us how to translate the problem to an equation. We let $m =$ the number of miles that Cecilia traveled in the cab.

2. **Translate.** We translate to an equation.

$\underbrace{\text{Pickup fee}}$	plus	$\underbrace{\text{cost per mile}}$	times	$\underbrace{\text{number of miles traveled}}$	is	$\underbrace{\text{total charge.}}$
↓	↓	↓	↓	↓	↓	↓
1.25	$+$	2	$\cdot$	m	$=$	31.25

3. **Carry out.** We solve the equation:

$$1.25 + 2 \cdot m = 31.25$$

$$2m = 30 \qquad \text{Subtracting 1.25 on both sides}$$

$$m = 15. \qquad \text{Dividing by 2 on both sides}$$

4. **Check.** If Cecilia travels 15 mi in the cab, the mileage charge is $\$2 \cdot 15$, or $\$30$. Then, with the $\$1.25$ pickup fee included, her total charge is $\$1.25 + \$30$, or $\$31.25$. The answer checks.

5. **State.** Cecilia traveled 15 mi in the cab.

Now Try Exercise 47. ■

❖ Zeros of Linear Functions

An input for which a function's output is 0 is called a **zero** of the function. We will restrict our attention in this section to zeros of linear functions. This allows us to become familiar with the concept of a zero, and it lays the groundwork for working with zeros of other types of functions in succeeding chapters.

Zeros of Functions

An input c of a function f is called a **zero** of the function if the output for the function is 0 when the input is c. That is, c is a zero of f if $f(c) = 0$.

> **LINEAR FUNCTIONS**
>
> REVIEW SECTION **1.3.**

Recall that a linear function is given by $f(x) = mx + b$, where m and b are constants. For the linear function $f(x) = 2x - 4$, we have $f(2) = 2 \cdot 2 - 4 = 0$, so 2 is a **zero** of the function. In fact, 2 is the *only* zero of this function. In general, a **linear function $f(x) = mx + b$, with $m \neq 0$, has exactly one zero.**

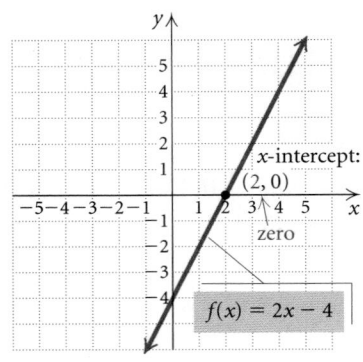

Consider the graph of $f(x) = 2x - 4$, shown at left. We see from the graph that the zero, 2, is the first coordinate of the point at which the graph crosses the x-axis. This point, $(2, 0)$, is the *x-intercept* of the graph. Thus when we find the zero of a linear function, we are also finding the first coordinate of the x-intercept of the graph of the function.

For every linear function $f(x) = mx + b$, there is an associated linear equation $mx + b = 0$. When we find the zero of a function $f(x) = mx + b$, we are also finding the solution of the equation $mx + b = 0$.

GCM | **EXAMPLE 11**   Find the zero of $f(x) = 5x - 9$.

ALGEBRAIC SOLUTION

We find the value of x for which $f(x) = 0$:

$$5x - 9 = 0 \qquad \text{Setting } f(x) = 0$$

$$5x = 9 \qquad \begin{array}{l}\text{Adding 9 on both}\\ \text{sides}\end{array}$$

$$x = \frac{9}{5}, \text{ or } 1.8. \qquad \begin{array}{l}\text{Dividing by 5}\\ \text{on both sides}\end{array}$$

Using a table, set in ASK mode, we can check the solution. We enter $y = 5x - 9$ on the equation-editor screen and then enter the value $x = \frac{9}{5}$, or 1.8, in the table.

X	Y₁	
1.8	0	
X =		

We see that $y = 0$ when $x = 1.8$, so the number 1.8 checks. The zero is $\frac{9}{5}$, or 1.8. This means that $f\left(\frac{9}{5}\right) = 0$, or $f(1.8) = 0$. Note that the *zero* of the function $f(x) = 5x - 9$ is the *solution* of the equation $5x - 9 = 0$.

GRAPHICAL SOLUTION

The solution of $5x - 9 = 0$ is also the zero of $f(x) = 5x - 9$. Thus we can solve an equation by finding the zeros of the function associated with it. We call this the **zero method**.

We graph $y = 5x - 9$ in the standard window and use the ZERO feature from the CALC menu to find the zero of $f(x) = 5x - 9$. Note that the x-intercept must appear in the window when the ZERO feature is used.

$y = 5x - 9$

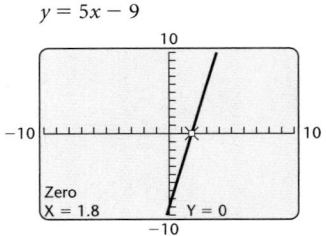

We can check algebraically by substituting 1.8 for x:

$$f(1.8) = 5(1.8) - 9 = 9 - 9 = 0.$$

The zero of $f(x) = 5x - 9$ is 1.8, or $\frac{9}{5}$.

Now Try Exercise 73. ▪

CONNECTING *the* CONCEPTS

The Intersect and Zero Methods

An equation such as $x - 1 = 2x - 6$ can be solved using the Intersect method by graphing $y_1 = x - 1$ and $y_2 = 2x - 6$ and using the INTERSECT feature to find the first coordinate of the point of intersection of the graphs. The equation can also be solved using the Zero method by writing it with 0 on one side of the equals sign and then using the ZERO feature.

Solve: $x - 1 = 2x - 6$.

The Intersect Method
Graph

$$y_1 = x - 1$$

and

$$y_2 = 2x - 6.$$

Point of intersection: $(5, 4)$
Solution: 5

The Zero Method
First add $-2x$ and 6 on both sides of the equation to get 0 on one side.

$$x - 1 = 2x - 6$$
$$x - 1 - 2x + 6 = 0$$

Graph

$$y_3 = x - 1 - 2x + 6.$$

Zero: 5
Solution: 5

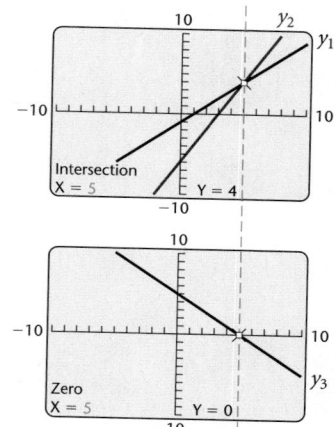

❋ Formulas

A **formula** is an equation that can be used to *model* a situation. For example, the formula $P = 2l + 2w$ in Example 9 gives the perimeter of a rectangle with length l and width w. We also used the motion formula, $d = r \cdot t$, in Example 7 and the simple-interest formula, $I = Prt$, in Example 8.

The equation-solving principles presented earlier can be used to solve a formula for a given variable.

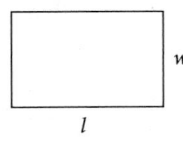

EXAMPLE 12   Solve $P = 2l + 2w$ for l.

Solution   We have

$$P = 2l + 2w \qquad \text{We want to isolate } l.$$

$$P - 2w = 2l \qquad \text{Subtracting } 2w \text{ on both sides}$$

$$\frac{P - 2w}{2} = l. \qquad \text{Dividing by 2 on both sides}$$

Now Try Exercise 95. ■

The formula $l = \dfrac{P - 2w}{2}$ can be used to determine a rectangle's length if we are given its perimeter and its width.

EXAMPLE 13   The formula $A = P + Prt$ gives the amount A to which a principal of P dollars will grow when invested at simple interest rate r for t years. Solve the formula for P.

Solution   We have

$$A = P + Prt \qquad \text{We want to isolate } P.$$

$$A = P(1 + rt) \qquad \text{Factoring}$$

$$\frac{A}{1 + rt} = \frac{P(1 + rt)}{1 + rt} \qquad \text{Dividing by } 1 + rt \text{ on both sides}$$

$$\frac{A}{1 + rt} = P.$$

Now Try Exercise 109. ■

The formula $P = \dfrac{A}{1 + rt}$ can be used to determine how much should be invested at simple interest rate r in order to have A dollars t years later.

EXAMPLE 14   Solve $A = \frac{1}{2}h(b_1 + b_2)$ for b_1.

Solution   We have

$$A = \frac{1}{2}h(b_1 + b_2) \qquad \text{Formula for the area of a trapezoid}$$

$$2A = h(b_1 + b_2) \qquad \text{Multiplying by 2}$$

$$2A = hb_1 + hb_2 \qquad \text{Removing parentheses}$$

$$2A - hb_2 = hb_1 \qquad \text{Subtracting } hb_2$$

$$\frac{2A - hb_2}{h} = b_1.$$

Dividing by h *Now Try Exercise 97.* ■

CONNECTING *the* CONCEPTS

Zeros, Solutions, and Intercepts

The zero of a linear function $f(x) = mx + b$, with $m \neq 0$, is the solution of the linear equation $mx + b = 0$ and is the first coordinate of the x-intercept of the graph of $f(x) = mx + b$. To find the zero of $f(x) = mx + b$, we solve $f(x) = 0$, or $mx + b = 0$.

FUNCTION	ZERO OF THE FUNCTION; SOLUTION OF THE EQUATION	ZERO OF THE FUNCTION; X-INTERCEPT OF THE GRAPH

Linear Function

$$f(x) = 2x - 4, \text{ or}$$
$$y = 2x - 4$$

To find the **zero** of $f(x)$, we solve $f(x) = 0$:

$$2x - 4 = 0$$
$$2x = 4$$
$$x = 2.$$

The **solution** of $2x - 4 = 0$ is 2. This is the zero of the function $f(x) = 2x - 4$. That is, $f(2) = 0$.

The zero of $f(x)$ is the first coordinate of the **x-intercept** of the graph of $y = f(x)$.

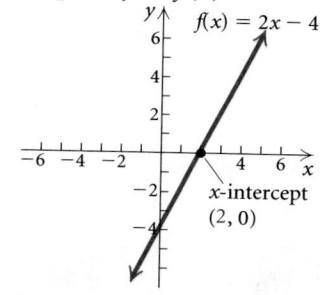

x-intercept
$(2, 0)$

1.5 Exercise Set

Solve.

1. $4x + 5 = 21$

2. $2y - 1 = 3$

3. $23 - \frac{2}{5}x = -\frac{2}{5}x + 23$

4. $\frac{6}{5}y + 3 = \frac{3}{10}$

5. $4x + 3 = 0$

6. $3x - 16 = 0$

7. $3 - x = 12$

8. $4 - x = -5$

9. $3 - \frac{1}{4}x = \frac{3}{2}$

10. $10x - 3 = 8 + 10x$

11. $\frac{2}{11} - 4x = -4x + \frac{9}{11}$

12. $8 - \frac{2}{9}x = \frac{5}{6}$

13. $8 = 5x - 3$

14. $9 = 4x - 8$

15. $\frac{2}{5}y - 2 = \frac{1}{3}$

16. $-x + 1 = 1 - x$

17. $y + 1 = 2y - 7$

18. $5 - 4x = x - 13$

19. $2x + 7 = x + 3$

20. $5x - 4 = 2x + 5$

21. $3x - 5 = 2x + 1$

22. $4x + 3 = 2x - 7$

23. $4x - 5 = 7x - 2$

24. $5x + 1 = 9x - 7$

25. $5x - 2 + 3x = 2x + 6 - 4x$

26. $5x - 17 - 2x = 6x - 1 - x$

27. $7(3x + 6) = 11 - (x + 2)$

28. $4(5y + 3) = 3(2y - 5)$

29. $3(x + 1) = 5 - 2(3x + 4)$

30. $4(3x + 2) - 7 = 3(x - 2)$

31. $2(x - 4) = 3 - 5(2x + 1)$

32. $3(2x - 5) + 4 = 2(4x + 3)$

33. *Bottled-Water Sales.*   Wholesale sales of bottled water in the United States in 2006 were $10.6 billion. This was an increase of about 54% over sales in 2001. (*Source*: Beverage Marketing Corporation) Find the wholesale sales of bottled water in the United States in 2001.

34. *Oil Consumption.*   Projections indicate that the global demand for oil will be 103 million barrels per day in 2015. This is about a 23% increase over the daily demand in 2005. (*Source*: U.S. Energy Information Administration) What was the daily global demand for oil in 2005?

35. *Credit-Card Debt.*   For households with at least one credit card, the average U.S. credit-card debt per household was $9312 in 2004. This was $6346 more than the average credit-card debt in 1990. (*Source*: CardWeb.com) What was the average credit-card debt per household in 1990?

36. *Bald-Eagle Population.*   There are 7066 known nesting pairs of bald eagles in the lower 48 states today. This is 6649 more pairs than in 1963. (*Source*: U.S. Fish and Wildlife Service) How many nesting pairs of bald eagles were there in the lower 48 states in 1963?

37. *Data Storage.*   As consumers fill electronic devices with digital photos and music, storage needs for digital data are soaring. It is estimated that the amount of digital data stored on various devices in a typical U.S. household in 2010 will be 4430 gigabytes (GB). This is 4024 GB more than the amount of data stored in 2004. (*Source*: Coughlin Associates) Find the amount of digital data stored on various devices in a typical household in 2004.

38. *Where the Textbook Dollar Goes.*   Of each dollar spent on textbooks at college bookstores, 23.2 cents goes to the college store for profit, store operations, and personnel. On average, a college student at a four-year college spends $940 per year for textbooks. (*Source*: College Board) How much of this expenditure goes to the college store?

39. *Visitors to National Parks.* The two most visited national parks are the Great Smoky Mountains and the Grand Canyon. In 2005, there was a total of 13.6 million visitors to these two parks. Great Smoky Mountains Park had 4.8 million more visitors than the Grand Canyon. (*Source*: National Park Service) How many visitors did each park have?

40. *Nutrition.* A slice of carrot cake from the popular restaurant The Cheesecake Factory contains 1560 calories. This is three-fourths of the average daily calorie requirement for many adults. (*Source*: The Center for Science in the Public Interest) Find the average daily calorie requirement for these adults.

41. *Television Viewers.* In a recent week, the television networks CBS, ABC, and NBC together averaged a total of 29.1 million viewers. CBS had 1.7 million more viewers than ABC and NBC had 1.7 million fewer viewers than ABC. (*Source*: Nielsen Media Research) How many viewers did each network have?

42. *Nielsen Ratings.* Nielsen Media Research surveys TV-watching habits and provides a list of the 20 most-watched TV programs each week. Each rating point in the survey represents 1,102,000 households. One week "60 Minutes" had a rating of 11.0. How many households did this represent?

43. *Amount Borrowed.* Tamisha borrowed money from her father at 5% simple interest to help pay her tuition at Wellington Community College. At the end of 1 yr, she owed a total of $1365 in principal and interest. How much did she borrow?

44. *Amount of an Investment.* Khalid makes an investment at 4% simple interest. At the end of 1 yr, the total value of the investment is $1560. How much was originally invested?

45. *Sales Commission.* Ryan, a consumer electronics salesperson, earns a base salary of $1500 per month and a commission of 8% on the amount of sales he makes. One month Ryan received a paycheck for $2284. Find the amount of his sales for the month.

46. *Commission vs. Salary.* Juliet has a choice between receiving an $1800 monthly salary from Furniture by Design or a base salary of $1600 and a 4% commission on the amount of furniture she sells during the month. For what amount of sales will the two choices be equal?

47. *Cab Fare.* City Cabs charges a $1.75 pickup fee and $1.50 per mile traveled. Diego's fare for a cross-town cab ride is $19.75. How far did he travel in the cab?

48. *Hourly Wage.* Soledad worked 48 hr one week and earned a $442 paycheck. She earns time and a half (1.5 times her regular hourly wage) for the hours she works in excess of 40. What is Soledad's regular hourly wage?

49. *Angle Measure.* In triangle *ABC*, angle *B* is five times as large as angle *A*. The measure of angle *C* is 2° less than that of angle *A*. Find the measures of the angles. (*Hint*: The sum of the angle measures is 180°.)

50. *Angle Measure.*   In triangle *ABC*, angle *B* is twice as large as angle *A*. Angle *C* measures 20° more than angle *A*. Find the measures of the angles.

51. *Test-Plot Dimensions.*   Morgan's Seeds has a rectangular test plot with a perimeter of 322 m. The length is 25 m more than the width. Find the dimensions of the plot.

52. *Garden Dimensions.*   The children at Tiny Tots Day Care plant a rectangular vegetable garden with a perimeter of 39 m. The length is twice the width. Find the dimensions of the garden.

53. *Soccer-Field Dimensions.*   The width of the soccer field recommended for players under the age of 12 is 35 yd less than the length. The perimeter of the field is 330 yd. (*Source*: U.S. Youth Soccer) Find the dimensions of the field.

54. *Poster Dimensions.*   Marissa is designing a poster to promote the Talbot Street Art Fair. The width of the poster will be two-thirds of its height and its perimeter will be 100 in. Find the dimensions of the poster.

55. *Water Weight.*   Water accounts for 50% of a woman's weight (*Source*: National Institute for Fitness and Sport). Kimiko weighs 135 lb. How much of her body weight is water?

56. *Water Weight.*   Water accounts for 60% of a man's weight (*Source*: National Institute for Fitness and Sport). Emilio weighs 186 lb. How much of his body weight is water?

57. *Train Speeds.*   A Central Railway freight train leaves a station and travels due north at a speed of 60 mph. One hour later, an Amtrak passenger train leaves the same station and travels due

north on a parallel track at a speed of 80 mph. How long will it take the passenger train to overtake the freight train?

58. *Distance Traveled.*   A private airplane leaves Midway Airport and flies due east at a speed of 180 km/h. Two hours later, a jet leaves Midway and flies due east at a speed of 900 km/h. How far from the airport will the jet overtake the private plane?

59. *Traveling Upstream.*   A kayak moves at a rate of 12 mph in still water. If the river's current flows at a rate of 4 mph, how long does it take the boat to travel 36 mi upstream?

60. *Traveling Downstream.*   Angelo's kayak travels 14 km/h in still water. If the river's current flows at a rate of 2 km/h, how long will it take him to travel 20 km downstream?

61. *Flying into a Headwind.*   An airplane that travels 450 mph in still air encounters a 30-mph headwind. How long will it take the plane to travel 1050 mi into the wind?

62. *Flying with a Tailwind.*   An airplane that can travel 375 mph in still air is flying with a 25-mph tailwind. How long will it take the plane to travel 700 mi with the wind?

63. *Investment Income.*   Erica invested a total of $5000, part at 3% simple interest and part at 4% simple interest. At the end of 1 yr, the investments had earned $176 interest. How much was invested at each rate?

64. *Student Loans.*   Dimitri's two student loans total $9000. One loan is at 5% simple interest and the other is at 6% simple interest. At the end of 1 yr, Dimitri owes $492 in interest. What is the amount of each loan?

65. *Calcium Content of Foods.* Together, one 8-oz serving of plain nonfat yogurt and one 1-oz serving of Swiss cheese contain 676 mg of calcium. The yogurt contains 4 mg more than twice the calcium in the cheese. (*Source*: U.S. Department of Agriculture) Find the calcium content of each food.

66. *Uninsured.* There were 55.8 million people in the United States without health insurance in 2005. This was 17.2 million less than twice the number of uninsured in 1987. (*Source*: U.S. Census Bureau) How many people were without health insurance in 1987?

67. *NFL Stadium Elevation.* The elevations of the 31 NFL stadiums range from 3 ft at Giants Stadium in East Rutherford, New Jersey, to 5210 ft at Invesco Field at Mile High in Denver, Colorado. The elevation of Invesco Field at Mile High is 247 ft higher than seven times the elevation of Lucas Oil Stadium in Indianapolis, Indiana. What is the elevation of Lucas Oil Stadium?

68. *Prison Population.* It is estimated that there will be 1.7 million adults in prison in the United States in 2011. This is about 5.4 times the number of adults in prison in 1980. (*Sources*: Pew Charitable Trusts; Bureau of Justice Statistics) Find the number of adults in prison in the United States in 1980.

69. *Volcanic Activity.* A volcano that is currently about one-half mile below the surface of the Pacific Ocean near the Big Island of Hawaii will eventually become a new Hawaiian island, Loihi. The volcano will break the surface of the ocean in about 50,000 yr. (*Source*: U.S. Geological Survey) On average, how many inches does the volcano rise in a year?

70. *Erosion.* Because of erosion, Horseshoe Falls, one of the two falls that make up Niagara Falls, is migrating upstream at a rate of 2 ft per year (*Source*: *Indianapolis Star*, February 14, 1999). At this rate, how long will it take the falls to move one-fourth mile?

Find the zero of the linear function.

71. $f(x) = x + 5$

72. $f(x) = 5x + 20$

73. $f(x) = -2x + 11$

74. $f(x) = 8 + x$

75. $f(x) = 16 - x$

76. $f(x) = -2x + 7$

77. $f(x) = x + 12$

78. $f(x) = 8x + 2$

79. $f(x) = -x + 6$

80. $f(x) = 4 + x$

81. $f(x) = 20 - x$

82. $f(x) = -3x + 13$

83. $f(x) = \frac{2}{5}x - 10$

84. $f(x) = 3x - 9$

85. $f(x) = -x + 15$

86. $f(x) = 4 - x$

In Exercises 87–92, use the given graph to find each of the following: **(a)** *the x-intercept and* **(b)** *the zero of the function.*

87.

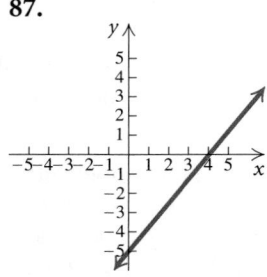

88.

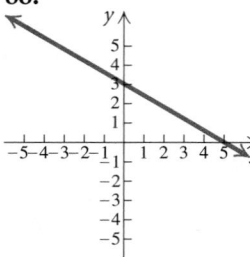

89.

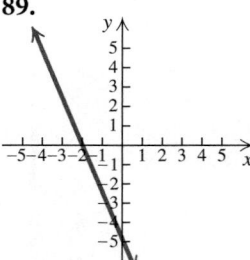

90.

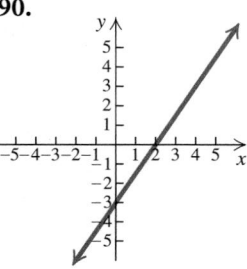

91.

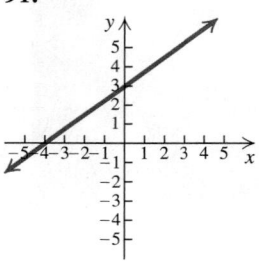

92.

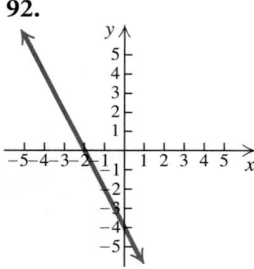

Solve.

93. $A = \frac{1}{2}bh$, for b
(Area of a triangle)

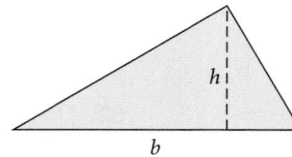

94. $A = \pi r^2$, for π
(Area of a circle)

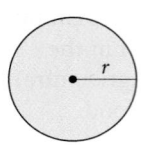

95. $P = 2l + 2w$, for w
(Perimeter of a rectangle)

96. $A = P + Prt$, for r
(Simple interest)

97. $A = \frac{1}{2}h(b_1 + b_2)$, for b_2

98. $A = \frac{1}{2}h(b_1 + b_2)$, for h
(Area of a trapezoid)

99. $V = \frac{4}{3}\pi r^3$, for π
(Volume of a sphere)

100. $V = \frac{4}{3}\pi r^3$, for r^3

101. $F = \frac{9}{5}C + 32$, for C
(Temperature conversion)

102. $Ax + By = C$, for y
(Standard linear equation)

103. $Ax + By = C$, for A

104. $2w + 2h + l = p$, for w

105. $2w + 2h + l = p$, for h

106. $3x + 4y = 12$, for y

107. $2x - 3y = 6$, for y

108. $T = \frac{3}{10}(I - 12{,}000)$, for I

109. $a = b + bcd$, for b

110. $q = p - np$, for p

111. $z = xy - xy^2$, for x

112. $st = t - 4$, for t

Collaborative Discussion and Writing

113. Explain in your own words why a linear function $f(x) = mx + b$, with $m \neq 0$, has exactly one zero.

114. The formula in Exercise 101, $F = \frac{9}{5}C + 32$, can be used to convert Celsius temperature to Fahrenheit temperature. Under what circumstances would it be useful to solve this formula for C?

Skill Maintenance

115. Write a slope–intercept equation for the line containing the point $(-1, 4)$ and parallel to the line $3x + 4y = 7$.

116. Write an equation of the line containing the points $(-5, 4)$ and $(3, -2)$.

117. Find the distance between $(2, 2)$ and $(-3, -10)$.

118. Find the midpoint of the segment with endpoints $\left(-\frac{1}{2}, \frac{2}{5}\right)$ and $\left(-\frac{3}{2}, \frac{3}{5}\right)$.

119. Given that $f(x) = \dfrac{x}{x - 3}$, find $f(-3), f(0)$, and $f(3)$.

120. Find the slope and the y-intercept of the line with the equation $7x - y = \frac{1}{2}$.

Synthesis

State whether each of the following is a linear function.

121. $f(x) = 7 - \dfrac{3}{2}x$

122. $f(x) = \dfrac{3}{2x} + 5$

123. $f(x) = x^2 + 1$

124. $f(x) = \dfrac{3}{4}x - (2.4)^2$

Solve.

125. $2x - \{x - [3x - (6x + 5)]\} = 4x - 1$

126. $14 - 2[3 + 5(x - 1)] = 3\{x - 4[1 + 6(2 - x)]\}$

127. *Packaging and Price.* Dannon recently replaced its 8-oz cup of yogurt with a 6-oz cup and reduced the suggested retail price from 89 cents to 71 cents (*Source*: IRI). Was the price per ounce reduced by the same percent as the size of the cup? If not, find the price difference per ounce in terms of a percent.

128. *Packaging and Price.* Wisk laundry detergent recently replaced its 100-oz container with an 80-oz container and reduced the suggested retail price from \$6.99 to \$5.75 (*Source*: IRI). Was the price per ounce reduced by the same percent as the size of the container? If not, find the price difference per ounce in terms of a percent.

129. *Running vs. Walking.* A 150-lb person who runs at 6 mph for 1 hr burns about 720 calories. The same person, walking at 4 mph for 90 min, burns about 480 calories. (*Source*: FitSmart, *USA Weekend*, July 19–21, 2002) Suppose a 150-lb person runs at 6 mph for 75 min. How far would the person have to walk at 4 mph in order to burn the same number of calories used running?

130. *Bestsellers.* One week 10 copies of the novel *The Secret* by Rhonda Byrne were sold for every 3.9 copies of Sidney Poitier's *The Measure of a Man* that were sold (*Source*: USA Today Best-Selling Books). If a total of 7367 copies of the two books were sold, how many copies of each were sold?

1.6 Solving Linear Inequalities

✿ Solve linear inequalities.

✿ Solve compound inequalities.

✿ Solve applied problems using inequalities.

An **inequality** is a sentence with $<$, $>$, $\leq$, or $\geq$ as its verb. An example is $3x - 5 < 6 - 2x$. To **solve** an inequality is to find all values of the variable that make the inequality true. Each of these numbers is a **solution** of the inequality, and the set of all such solutions is its **solution set.** Inequalities that have the same solution set are called **equivalent inequalities.**

✿ Linear Inequalities

The principles for solving inequalities are similar to those for solving equations.

Principles for Solving Inequalities

For any real numbers a, b, and c:

The Addition Principle for Inequalities: If $a < b$ is true, then $a + c < b + c$ is true.

The Multiplication Principle for Inequalities: If $a < b$ and $c > 0$ are true, then $ac < bc$ is true. If $a < b$ and $c < 0$ are true, then $ac > bc$ is true.

Similar statements hold for $a \leq b$.

> When both sides of an inequality are multiplied by a negative number, we must reverse the inequality sign.

First-degree inequalities with one variable, like those in Example 1 below, are **linear inequalities**.

EXAMPLE 1   Solve each of the following. Then graph the solution set.

a) $3x - 5 < 6 - 2x$

b) $13 - 7x \geq 10x - 4$

Solution

a) $3x - 5 < 6 - 2x$

$\qquad 5x - 5 < 6$ Using the addition principle for inequalities; adding $2x$

$\qquad\qquad 5x < 11$ Using the addition principle for inequalities; adding 5

$\qquad\qquad\quad x < \frac{11}{5}$ Using the multiplication principle for inequalities; multiplying by $\frac{1}{5}$, or dividing by 5

INTERVAL NOTATION

REVIEW SECTION **R.1.**

Any number less than $\frac{11}{5}$ is a solution. The solution set is $\left\{x \mid x < \frac{11}{5}\right\}$, or $\left(-\infty, \frac{11}{5}\right)$. The graph of the solution set is shown below.

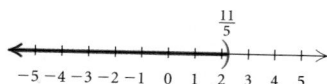

To check, we can graph $y_1 = 3x - 5$ and $y_2 = 6 - 2x$. The graph shows that for $x < 2.2$, or $x < \frac{11}{5}$, the graph of y_1 lies below the graph of y_2, or $y_1 < y_2$.

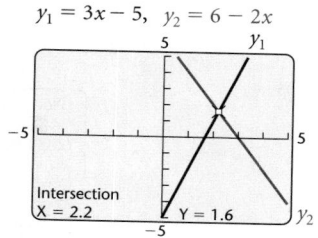

$$y_1 = 3x - 5, \quad y_2 = 6 - 2x$$

b) $13 - 7x \geq 10x - 4$

$$13 - 17x \geq -4 \qquad \text{Subtracting } 10x$$
$$-17x \geq -17 \qquad \text{Subtracting } 13$$
$$x \leq 1 \qquad \text{Dividing by } -17 \text{ and reversing the inequality sign}$$

The solution set is $\{x \mid x \leq 1\}$, or $(-\infty, 1]$. The graph of the solution set is shown below.

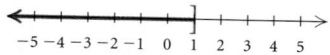

Now Try Exercise 1. ■

✿ Compound Inequalities

When two inequalities are joined by the word *and* or the word *or*, a **compound inequality** is formed. A compound inequality like

$$-3 < 2x + 5 \quad and \quad 2x + 5 \leq 7$$

is called a **conjunction,** because it uses the word *and*. The sentence $-3 < 2x + 5 \leq 7$ is an abbreviation for the preceding conjunction.

Compound inequalities can be solved using the addition and multiplication principles for inequalities.

EXAMPLE 2 Solve $-3 < 2x + 5 \leq 7$. Then graph the solution set.

Solution We have

$$-3 < 2x + 5 \leq 7$$
$$-8 < 2x \leq 2 \qquad \text{Subtracting } 5$$
$$-4 < x \leq 1. \qquad \text{Dividing by } 2$$

The solution set is $\{x \mid -4 < x \leq 1\}$, or $(-4, 1]$. The graph of the solution set is shown below.

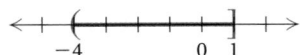

GCM We can perform a partial check of the solution graphically using operations from the TEST menu and its LOGIC submenu on a graphing calculator. We graph $y_1 = (-3 < 2x + 5)$ *and* $(2x + 5 \leq 7)$ in DOT mode. The calculator graphs a segment 1 unit above the x-axis for the values of x for which this expression for y is true. Here the number 1 corresponds to "true."

$$y = (-3 < 2x + 5) \quad \text{and} \quad (2x + 5 \leq 7)$$

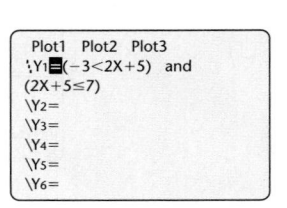

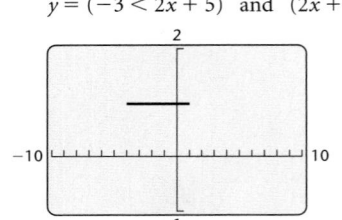

The segment extends from -4 to 1, confirming that all x-values from -4 to 1 are in the solution set. The algebraic solution indicates that the endpoint 1 is also in the solution set. **Now Try Exercise 17.** ◼

A compound inequality like $2x - 5 \leq -7$ *or* $2x - 5 > 1$ is called a **disjunction,** because it contains the word *or*. Unlike some conjunctions, it cannot be abbreviated; that is, it cannot be written without the word *or*.

EXAMPLE 3 Solve $2x - 5 \leq -7$ *or* $2x - 5 > 1$. Then graph the solution set.

Solution We have

$$\begin{array}{llll} 2x - 5 \leq -7 & or & 2x - 5 > 1 \\ 2x \leq -2 & or & 2x > 6 & \text{Adding 5} \\ x \leq -1 & or & x > 3. & \text{Dividing by 2} \end{array}$$

The solution set is $\{x \mid x \leq -1 \text{ or } x > 3\}$. We can also write the solution set using interval notation and the symbol $\cup$ for the **union** or inclusion of both sets: $(-\infty, -1] \cup (3, \infty)$. The graph of the solution set is shown below.

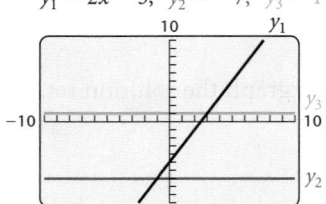

$y_1 = 2x - 5$, $y_2 = -7$, $y_3 = 1$

To check, we graph $y_1 = 2x - 5$, $y_2 = -7$, and $y_3 = 1$. Note that for $\{x \mid x \leq -1 \text{ or } x > 3\}$, $y_1 \leq y_2$ or $y_1 > y_3$. **Now Try Exercise 29.** ◼

❋ An Application

EXAMPLE 4 *Income Plans.*   For her house-painting job, Erica can be paid in one of two ways:

   Plan A: $250 plus $10 per hour;

   Plan B: $20 per hour.

Suppose that a job takes n hours. For what values of n is plan B better for Erica?

Solution

1. **Familiarize.**   Suppose that a job takes 20 hr. Then $n = 20$, and under plan A, Erica would earn $250 + $10 \cdot 20$, or $250 + 200, or 450. Her earnings under plan B would be $20 \cdot 20$, or 400. This shows that plan A is better for Erica if a job takes 20 hr. If a job takes 30 hr, then $n = 30$, and under plan A, Erica would earn $250 + $10 \cdot 30$, or $250 + 300, or 550. Under plan B, she would earn $20 \cdot 30$, or 600, so plan B is better in this case. To determine *all* values of n for which plan B is better for Erica, we solve an inequality. Our work in this step helps us write the inequality.

2. **Translate.**   We translate to an inequality:

$$\underbrace{\text{Income from plan B}}_{20n} \quad \underbrace{\text{is greater than}}_{>} \quad \underbrace{\text{income from plan A.}}_{250 + 10n}$$

3. **Carry out.**   We solve the inequality:

   $20n > 250 + 10n$

   $10n > 250$ Subtracting $10n$ on both sides

   $n > 25.$ Dividing by 10 on both sides

4. **Check.**   For $n = 25$, the income from plan A is $250 + $10 \cdot 25$, or $250 + 250, or 500, and the income from plan B is $20 \cdot 25$, or 500. This shows that for a job that takes 25 hr to complete, the income is the same under either plan. In the *Familiarize* step, we saw that plan B pays more for a 30-hr job. Since $30 > 25$, this provides a partial check of the result. We cannot check all values of n.

5. **State.**   For values of n greater than 25 hr, plan B is better for Erica.

Now Try Exercise 43. ■

1.6 Exercise Set

Solve and graph the solution set.

1. $4x - 3 > 2x + 7$
2. $8x + 1 \geq 5x - 5$
3. $x + 6 < 5x - 6$
4. $3 - x < 4x + 7$
5. $4 - 2x \leq 2x + 16$
6. $3x - 1 > 6x + 5$
7. $14 - 5y \leq 8y - 8$
8. $8x - 7 < 6x + 3$
9. $7x - 7 > 5x + 5$
10. $12 - 8y \geq 10y - 6$
11. $3x - 3 + 2x \geq 1 - 7x - 9$
12. $5y - 5 + y \leq 2 - 6y - 8$
13. $-\frac{3}{4}x \geq -\frac{5}{8} + \frac{2}{3}x$
14. $-\frac{5}{6}x \leq \frac{3}{4} + \frac{8}{3}x$
15. $4x(x - 2) < 2(2x - 1)(x - 3)$
16. $(x + 1)(x + 2) > x(x + 1)$

Solve and write interval notation for the solution set. Then graph the solution set.

17. $-2 \leq x + 1 < 4$
18. $-3 < x + 2 \leq 5$
19. $5 \leq x - 3 \leq 7$
20. $-1 < x - 4 < 7$
21. $-3 \leq x + 4 \leq 3$
22. $-5 < x + 2 < 15$
23. $-2 < 2x + 1 < 5$
24. $-3 \leq 5x + 1 \leq 3$
25. $-4 \leq 6 - 2x < 4$
26. $-3 < 1 - 2x \leq 3$
27. $-5 < \frac{1}{2}(3x + 1) < 7$
28. $\frac{2}{3} \leq -\frac{4}{5}(x - 3) < 1$
29. $3x \leq -6 \ or \ x - 1 > 0$
30. $2x < 8 \ or \ x + 3 \geq 10$
31. $2x + 3 \leq -4 \ or \ 2x + 3 \geq 4$
32. $3x - 1 < -5 \ or \ 3x - 1 > 5$
33. $2x - 20 < -0.8 \ or \ 2x - 20 > 0.8$
34. $5x + 11 \leq -4 \ or \ 5x + 11 \geq 4$
35. $x + 14 \leq -\frac{1}{4} \ or \ x + 14 \geq \frac{1}{4}$
36. $x - 9 < -\frac{1}{2} \ or \ x - 9 > \frac{1}{2}$

37. *Cost of Business on the Internet.* The equation $y = 12.7x + 15.2$ estimates the amount that businesses will spend, in billions of dollars, on Internet software to conduct transactions via the Web, where x is the number of years after 2002 (*Source*: IDC). For what years will the spending be more than $66 billion?

38. *Digital Hubs.* The equation $y = 5x + 5$ estimates the number of U.S. households, in millions, expected to install devices that receive and manage broadband TV and Internet content to the home, where x is the number of years after 2002 (*Source*: Forrester Research). For what years will there be at least 20 million homes with these devices?

39. *Moving Costs.* Acme Movers charges $100 plus $30 per hour to move a household across town. Hank's Movers charges $55 per hour. For what lengths of time does it cost less to hire Hank's Movers?

40. *Investment Income.* Gina plans to invest $12,000, part at 4% simple interest and the rest at 6% simple interest. What is the most she can invest at 4% and still be guaranteed at least $650 in interest per year?

41. *Investment Income.* Kyle plans to invest $7500, part at 4% simple interest and the rest at 5% simple interest. What is the most that he can invest at 4% and still be guaranteed at least $325 in interest per year?

42. *Checking-account Plans.* The Addison Bank offers two checking-account plans. The Smart Checking plan charges 20¢ per check whereas the Consumer Checking plan costs $6 per month plus 5¢ per check. For what number of checks per month will the Smart Checking plan cost less?

43. *Checking-account Plans.* Parson's Bank offers two checking-account plans. The No Frills plan charges 35¢ per check whereas the Simple Checking plan costs $5 per month plus 10¢ per check. For what number of checks per month will the Simple Checking plan cost less?

44. *Income Plans.* Karen can be paid in one of two ways for selling insurance policies:

Plan A: A salary of $750 per month, plus a commission of 10% of sales;
Plan B: A salary of $1000 per month, plus a commission of 8% of sales in excess of $2000.

For what amount of monthly sales is plan A better than plan B if we can assume that sales are always more than $2000?

45. *Income Plans.* Curt can be paid in one of two ways for the furniture he sells:

Plan A: A salary of $900 per month, plus a commission of 10% of sales;
Plan B: A salary of $1200 per month, plus a commission of 15% of sales in excess of $8000.

For what amount of monthly sales is plan B better than plan A if we can assume that Curt's sales are always more than $8000?

46. *Income Plans.* Jeanette can be paid in one of two ways for painting a house:

Plan A: $200 plus $12 per hour;
Plan B: $20 per hour.

Suppose a job takes n hours to complete. For what values of n is plan A better for Jeanette?

Collaborative Discussion and Writing

47. Is it possible for a disjunction to have no solution? Why or why not?

48. Why can the conjunction $3 < x$ and $x < 4$ be written as $3 < x < 4$, but the disjunction $x < 3$ or $x > 4$ cannot be written $3 > x > 4$?

Skill Maintenance

Solve.

49. $5x - 7 = 8$

50. $2 - 3x = x - 8$

51. $3(x + 1) = 4 - 2(x - 3)$

52. $4(2x - 5) + 12 = 3(2x - 6)$

53. *U.S. Transportation Systems.* Together, 15.8 million Americans travel on mass transit systems and airlines each day. Users of mass transit systems outnumber airline passengers by 12.2 million. How many passengers use each type of transportation system each day?

54. *Organic Pet Food.* U.S. sales of organic pet food are rising. Sales totaled $51 million in 2007. This is an increase of $37 million over sales in 2003. What were the total sales of organic pet food in 2003?

Synthesis

Solve.

55. $2x \leq 5 - 7x < 7 + x$

56. $x \leq 3x - 2 \leq 2 - x$

57. $3y < 4 - 5y < 5 + 3y$

58. $y - 10 < 5y + 6 \leq y + 10$

CHAPTER 1   Summary and Review

Important Properties and Formulas

The Distance Formula

$$d = \sqrt{(x_2 - x_1)^2 + (y_2 - y_1)^2}$$

The Midpoint Formula

$$\left(\frac{x_1 + x_2}{2}, \frac{y_1 + y_2}{2}\right)$$

Equation of a Circle

$$(x - h)^2 + (y - k)^2 = r^2$$

Terminology about Lines

Slope: $\quad m = \dfrac{y_2 - y_1}{x_2 - x_1}$

The Slope–Intercept Equation: $\quad y = mx + b$

The Point–Slope Equation: $\quad y - y_1 = m(x - x_1)$

Horizontal Lines: $\quad y = b$

Vertical Lines: $\quad x = a$

Parallel Lines: $\quad m_1 = m_2, b_1 \neq b_2;$ or $x = a_1, x = a_2, a_1 \neq a_2$

Perpendicular Lines: $\quad m_1 m_2 = -1$, or $x = a, y = b$

Equation-Solving Principles

The Addition Principle:
If $a = b$ is true, then $a + c = b + c$ is true.

The Multiplication Principle:
If $a = b$ is true, then $ac = bc$ is true.

Five Steps for Problem Solving

1. Familiarize.
2. Translate.
3. Carry out.
4. Check.
5. State.

The Motion Formula

$$d = rt$$

The Simple–Interest Formula

$$I = Prt$$

Zero of a Function

An input c of a function f is a zero of f if $f(c) = 0$.

Principles for Solving Inequalities

The Addition Principle for Inequalities:
If $a < b$ is true, then $a + c < b + c$ is true.

The Multiplication Principle for Inequalities:
If $a < b$ and $c > 0$ are true, then $ac < bc$ is true.

If $a < b$ and $c < 0$ are true, then $ac > bc$ is true.

Similar statements hold for $a \leq b$.

Review Exercises

Answers to all of the review exercises appear in the answer section at the back of the book. If you get an incorrect answer, restudy the objective indicated in red next to the exercise or the direction line that precedes it.

Determine whether the statement is true or false.

1. The x-intercept of the line that passes through $\left(-\frac{2}{3}, \frac{3}{2}\right)$ and the origin is $\left(-\frac{2}{3}, 0\right)$.   [1.1]

2. All functions are relations, but not all relations are functions.   [1.2]

3. If the line $ax + y = c$ is perpendicular to the line $x - by = d$, then $\dfrac{a}{b} = 1$.   [1.4]

4. The line parallel to the y-axis that passes through $(-5, 25)$ is $y = -5$.   [1.3]

5. The intersection of the lines $y = \frac{1}{2}$ and $x = -5$ is $\left(-5, \frac{1}{2}\right)$.   [1.3]

6. The domain of the function $f(x) = \dfrac{\sqrt{3 - x}}{x}$ does not contain -3 and 0.   [1.2]

Use substitution to determine whether the given ordered pairs are solutions of the given equation.   [1.1]

7. $\left(3, \frac{24}{9}\right), (0, -9)$; $2x - 9y = -18$

8. $(0, 7), (7, 1)$; $y = 7$

Find the intercepts and then graph the line.   [1.1]

9. $2x - 3y = 6$

10. $10 - 5x = 2y$

Graph the equation.   [1.1]

11. $y = -\frac{2}{3}x + 1$

12. $2x - 4y = 8$

13. $y = 2 - x^2$

14. Find the distance between $(3, 7)$ and $(-2, 4)$.   [1.1]

15. Find the midpoint of the segment with endpoints $(3, 7)$ and $(-2, 4)$.   [1.1]

16. Find the center and the radius of the circle with equation $(x + 1)^2 + (y - 3)^2 = 9$. Then graph the circle.   [1.1]

Find an equation for a circle satisfying the given conditions.   [1.1]

17. Center: $(0, -4)$, radius of length $\frac{3}{2}$

18. Center: $(-2, 6)$, radius of length $\sqrt{13}$

19. Diameter with endpoints $(-3, 5)$ and $(7, 3)$

Determine whether the correspondence is a function. [1.2]

20. $-6 \longrightarrow 1$
    $-1 \longrightarrow 3$
    $2 \longleftrightarrow 10$
    $7 \longrightarrow 12$

21. $h \longrightarrow r$
    $i \longrightarrow s$
    $j \longrightarrow t$
    $k \nearrow$

Determine whether the relation is a function. Identify the domain and the range.   [1.2]

22. $\{(3, 1), (5, 3), (7, 7), (3, 5)\}$

23. $\{(2, 7), (-2, -7), (7, -2), (0, 2), (1, -4)\}$

24. Given that $f(x) = x^2 - x - 3$, find each of the following.   [1.2]

    a) $f(0)$ b) $f(-3)$
    c) $f(a - 1)$ d) $f(-x)$

25. Given that $f(x) = \dfrac{x - 7}{x + 5}$, find each of the following.   [1.2]

    a) $f(7)$ b) $f(x + 1)$
    c) $f(-5)$ d) $f\left(-\frac{1}{2}\right)$

26. A graph of a function is shown. Find $f(2), f(-4)$, and $f(0)$. [1.2]

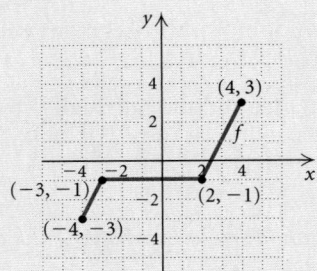

Determine whether the graph is that of a function. [1.2]

27.

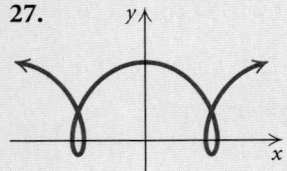

28.

29.

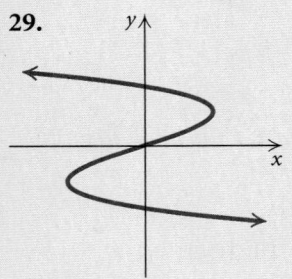

30.

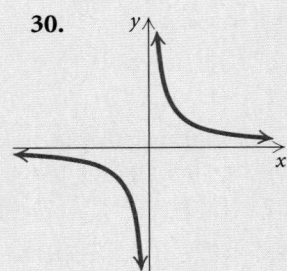

Find the domain of the function. [1.2]

31. $f(x) = 4 - 5x + x^2$

32. $f(x) = \dfrac{3}{x} + 2$

33. $f(x) = \dfrac{1}{x^2 - 6x + 5}$

34. $f(x) = \dfrac{-5x}{|16 - x^2|}$

Graph the function. Then visually estimate the domain and the range. [1.2]

35. $f(x) = \sqrt{16 - x^2}$

36. $g(x) = |x - 5|$

37. $f(x) = x^3 - 7$

38. $h(x) = x^4 + x^2$

In Exercises 39 and 40, the table of data contains input–output values for a function. Answer the following questions. [1.3]

a) Is the change in the inputs, x, the same?
b) Is the change in the outputs, y, the same?
c) Is the function linear?

39.

x	y
-3	8
-2	11
-1	14
0	17
1	20
2	22
3	26

40.

x	y
20	11.8
30	24.2
40	36.6
50	49.0
60	61.4
70	73.8
80	86.2

Find the slope of the line containing the given points. [1.3]

41. $(2, -11), (5, -6)$

42. $(5, 4), (-3, 4)$

43. $\left(\frac{1}{2}, 3\right), \left(\frac{1}{2}, 0\right)$

44. *Minimum Wage.* The minimum wage was $3.10 in 1980 and $5.15 in 2006 (*Source*: U.S. Department of Labor). Find the average rate of change in the minimum wage from 1980 to 2006. [1.3]

Find the slope and the y-intercept of the line with the given equation. [1.3]

45. $y = -\frac{7}{11}x - 6$

46. $-2x - y = 7$

47. Graph $y = -\frac{1}{4}x + 3$ using the slope and the y-intercept. [1.3]

48. *Total Cost.* Clear County Cable Television charges a $60 installation fee and $44 per month for basic service. Write an equation that can be used to determine the total cost, $C(t)$, of t months of basic cable television service. Find the total cost of 1 year of service. [1.3]

49. *Temperature and Depth of the Earth.*   The function T given by $T(d) = 10d + 20$ can be used to determine the temperature T, in degrees Celsius, at a depth d, in kilometers, inside the earth.

a) Find $T(5)$, $T(20)$, and $T(1000)$.   [1.3]
b) The radius of the earth is about 5600 km. Use this fact to determine the domain of the function.   [1.3]

Write a slope–intercept equation for a line with the following characteristics.   [1.4]

50. $m = -\frac{2}{3}$, y-intercept $(0, -4)$

51. $m = 3$, passes through $(-2, -1)$

52. Passes through $(4, 1)$ and $(-2, -1)$

53. Write equations of the horizontal line and the vertical line that pass through $\left(-4, \frac{2}{5}\right)$.   [1.4]

54. Find a linear function h given $h(-2) = -9$ and $h(4) = 3$. Then find $h(0)$.   [1.4]

Determine whether the lines are parallel, perpendicular, or neither.   [1.4]

55. $3x - 2y = 8$,
$6x - 4y = 2$

56. $y - 2x = 4$,
$2y - 3x = -7$

57. $y = \frac{3}{2}x + 7$,
$y = -\frac{2}{3}x - 4$

Given the point $(1, -1)$ and the line $2x + 3y = 4$:

58. Find an equation of the line containing the given point and parallel to the given line.   [1.4]

59. Find an equation of the line containing the given point and perpendicular to the given line.   [1.4]

60. *Drive-in Movie Sites.*   The data in the following table show the decrease in the number of drive-in movie sites since 1990.

Year, x	Drive-in Movie Sites in the United States, y
1990, 0	910
1993, 3	837
1996, 6	826
1999, 9	683
2002, 12	666
2005, 15	648

Source: National Association of Theater Owners

a) Without using a graphing calculator, model the data with a linear function and, using this function, estimate the number of drive-in movie sites in 2009. Answers may vary depending on the data points used.   [1.4]

b) Using a graphing calculator, fit a regression line to the data and use it to estimate the number of drive-in movie sites in 2009. What is the correlation coefficient for the regression line? How close a fit is the regression line?   [1.4]

Solve.   [1.5]

61. $4y - 5 = 1$

62. $3x - 4 = 5x + 8$

63. $5(3x + 1) = 2(x - 4)$

64. $2(n - 3) = 3(n + 5)$

65. $\frac{3}{5}y - 2 = \frac{3}{8}$

66. $5 - 2x = -2x + 3$

67. $x - 13 = -13 + x$

68. *International Adoptions.*   In 2006, U.S. international adoptions from China and Russia totaled 10,199. The number of orphans adopted from China exceeded those from Russia by 2787. (*Source*: U.S. State Department) Find the number of adoptions from China and from Russia.   [1.5]

69. *Amount of Investment.*   Kaleb makes an investment at 5.2% simple interest. At the end of 1 yr, the total value of the investment is $2419.60. How much was originally invested?   [1.5]

70. *Flying into a Headwind.*   An airplane that can travel 550 mph in still air encounters a 20-mph headwind. How long will it take the plane to travel 1802 mi?   [1.5]

In Exercises 71–74, find the zero(s) of the function.   [1.5]

71. $f(x) = 6x - 18$

72. $f(x) = x - 4$

73. $f(x) = 2 - 10x$

74. $f(x) = 8 - 2x$

75. Solve $V = lwh$ for h.   [1.5]

76. Solve $M = n + 0.3s$ for s.   [1.5]

77. Solve $A = P + Prt$ for t.   [1.5]

Solve and write interval notation for the solution set. Then graph the solution set.   [1.6]

78. $2x - 5 < x + 7$

79. $3x + 1 \geq 5x + 9$

80. $-3 \leq 3x + 1 \leq 5$

81. $-2 < 5x - 4 \leq 6$

82. $2x < -1 \ or \ x - 3 > 0$

83. $3x + 7 \leq 2 \ or \ 2x + 3 \geq 5$

84. *Faculty at Two-Year Colleges.*   The equation $y = 6x + 121$ estimates the number of faculty members at two-year colleges, in thousands, where x is the number of years after 1970 (*Source*: U.S. National Center for Education Statistics). For what years will there be more than 325 thousand faculty members?   [1.6]

85. *Temperature Conversion.*   The formula $C = \frac{5}{9}(F - 32)$ can be used to convert Fahrenheit temperatures F to Celsius temperatures C. For what Fahrenheit temperatures is the Celsius temperature lower than 45°C?   [1.6]

86. The domain of the function

$$f(x) = \frac{x + 3}{8 - 4x}$$

is which of the following?   [1.2]

A. $(-3, 2)$
B. $(-\infty, 2) \cup (2, \infty)$
C. $(-\infty, -3) \cup (-3, 2) \cup (2, \infty)$
D. $(-\infty, -3) \cup (-3, \infty)$

87. The center of the circle described by the equation $(x - 1)^2 + y^2 = 9$ is which of the following?   [1.1]

A. $(-1, 0)$
B. $(1, 0)$
C. $(0, -3)$
D. $(-1, 3)$

88. The graph of $f(x) = -\frac{1}{2}x - 2$ is which of the following? [1.3]

A.
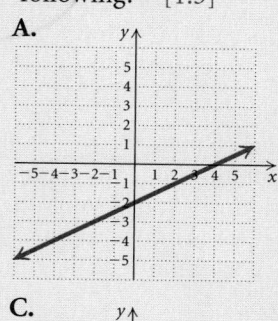

B.
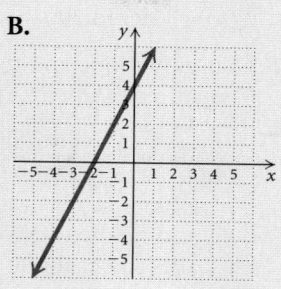

C.
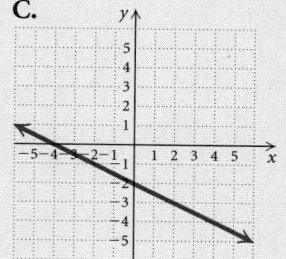

D.
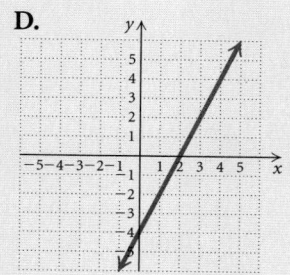

Collaborative Discussion and Writing

89. As the first step in solving
$$3x - 1 = 8,$$
Stella multiplies by $\frac{1}{3}$ on both sides. What advice would you give her about the procedure for solving equations? [1.5]

90. Discuss why the graph of $f(x) = -\frac{3}{5}x + 4$ is steeper than the graph of $g(x) = \frac{1}{2}x - 6$. [1.3]

Synthesis

91. Find the point on the x-axis that is equidistant from the points $(1, 3)$ and $(4, -3)$ [1.1]

Find the domain. [1.2]

92. $f(x) = \dfrac{\sqrt{1 - x}}{x - |x|}$

93. $f(x) = (x - 9x^{-1})^{-1}$

CHAPTER 1   Test

1. Determine whether the ordered pair $\left(\frac{1}{2}, \frac{9}{10}\right)$ is a solution of the equation $5y - 4 = x$.

2. Find the intercepts of $5x - 2y = -10$ and graph the line.

3. Find the distance between $(5, 8)$ and $(-1, 5)$.

4. Find the midpoint of the segment with endpoints $(-2, 6)$ and $(-4, 3)$.

5. Find the center and the radius of the circle
$$(x + 4)^2 + (y - 5)^2 = 36.$$

6. Find an equation of the circle with center $(-1, 2)$ and radius $\sqrt{5}$.

7. a) Determine whether the relation
$$\{(-4, 7), (3, 0), (1, 5), (0, 7)\}$$
is a function. Answer yes or no.
b) Find the domain of the relation.
c) Find the range of the relation.

8. Given that $f(x) = 2x^2 - x + 5$, find each of the following.
a) $f(-1)$ **b)** $f(a + 2)$

9. Given that $f(x) = \dfrac{1 - x}{x}$, find each of the following.
a) $f(0)$ **b)** $f(1)$

10. Using the graph below, find $f(-3)$.

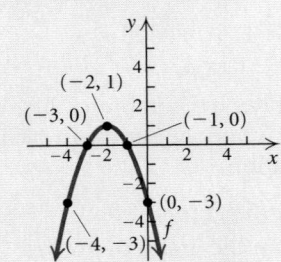

11. Determine whether each graph is that of a function. Answer yes or no.

a)

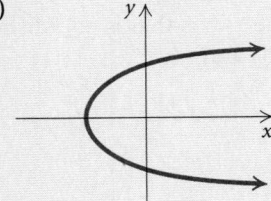

b)

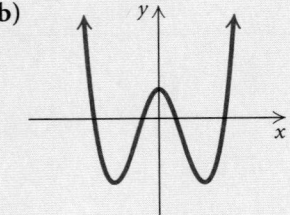

Find the domain of the function.

12. $f(x) = \dfrac{1}{x - 4}$

13. $g(x) = x^3 + 2$

14. $h(x) = \sqrt{25 - x^2}$

15. a) Graph: $f(x) = |x - 2| + 3$.
　　b) Visually estimate the domain of $f(x)$.
　　c) Visually estimate the range of $f(x)$.

Find the slope of the line containing the given points.

16. $\left(-2, \frac{2}{3}\right), (-2, 5)$

17. $(4, -10), (-8, 12)$

18. $(-5, 6), \left(\frac{3}{4}, 6\right)$

19. *NASCAR Attendance.* The weekend attendance at NASCAR events has increased from 5.3 million in 1995 to 6.8 million in 2004 (*Sources*: Goodyear Tire & Rubber Co. (1995–1999); NASCAR). Find the average rate of change in weekend attendance from 1995 to 2004.

20. Find the slope and the *y*-intercept of the line with equation $-3x + 2y = 5$.

21. *Total Cost.* Clear Signal charges \$80 for a cell phone and \$39.95 per month under its standard plan. Write an equation that can be used to determine the total cost, $C(t)$, of operating a Clear Signal cell phone for t months. Then find the total cost for 2 yr.

22. Write an equation for the line with $m = -\frac{5}{8}$ and *y*-intercept $(0, -5)$.

23. Write an equation for the line that passes through $(-5, 4)$ and $(3, -2)$.

24. Write the equation of the vertical line that passes through $\left(-\frac{3}{8}, 11\right)$.

25. Determine whether the lines are parallel, perpendicular, or neither.
$$2x + 3y = -12,$$
$$2y - 3x = 8$$

26. Find an equation of the line containing the point $(-1, 3)$ and parallel to the line $x + 2y = -6$.

27. Find an equation of the line containing the point $(-1, 3)$ and perpendicular to the line $x + 2y = -6$.

28. *U.S. Mail.* The data in the following table show an increase in the number of pieces of mail delivered by the U.S. Postal Service from 2002 to 2006.

Year, x	Number of Pieces of Mail (in billions)
2002, 0	203
2003, 1	202
2004, 2	206
2005, 3	212
2006, 4	213

Source: U.S. Postal Service

a) Without using a graphing calculator, model the data with a linear function and, using this function, predict the number of pieces of mail delivered in 2010. Answers may vary depending on the data points used.

b) Using a graphing calculator, fit a regression line to the data and use it to predict the number of pieces of mail delivered in 2010. What is the correlation coefficient for the regression line?

Solve.

29. $6x + 7 = 1$

30. $2.5 - x = -x + 2.5$

31. $\frac{3}{2}y - 4 = \frac{5}{3}y + 6$

32. $2(4x + 1) = 8 - 3(x - 5)$

33. *Parking-Lot Dimensions.*   The parking lot behind Kai's Kafé has a perimeter of 210 m. The width is three-fourths of the length. What are the dimensions of the parking lot?

34. *Pricing.*   Jessie's Juice Bar prices its bottled juices by raising the wholesale price 50% and then adding 25¢. What is the wholesale price of a bottle of juice that sells for $2.95?

35. Find the zero(s) of the function $f(x) = 3x + 9$.

36. Solve $V = \frac{2}{3}\pi r^2 h$ for h.

37. Solve $r = s - ts$ for s.

Solve and write interval notation for the solution set. Then graph the solution set.

38. $5 - x \geq 4x + 20$

39. $-7 < 2x + 3 < 9$

40. $2x - 1 \leq 3 \text{ or } 5x + 6 \geq 26$

41. *Moving Costs.*   Morgan Movers charges $90 plus $25 per hour to move households across town. McKinley Movers charges $40 per hour for crosstown moves. For what lengths of time does it cost less to hire Morgan Movers?

42. The graph of $g(x) = 1 - \frac{1}{2}x$ is which of the following?

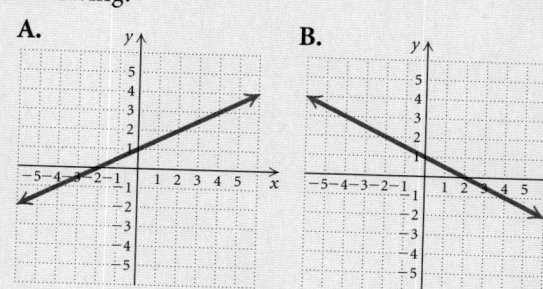

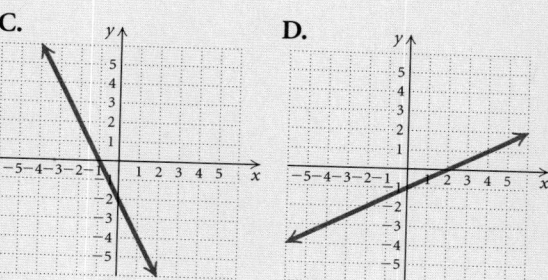

Synthesis

43. Suppose that for some function h, $h(x + 2) = \frac{1}{2}x$. Find $h(-2)$.

More on Functions

APPLICATION A hot-air balloon rises straight up from the ground at a rate of 120 ft/min. The balloon is tracked from a rangefinder on the ground that is located 400 ft from the release point of the balloon. Let d = the distance from the balloon to the rangefinder and t = the time, in minutes, since the balloon was released. Express d as a function of t.

This problem appears as Exercise 35 in Section 2.1.

2.1 Increasing, Decreasing, and Piecewise Functions; Applications

2.2 The Algebra of Functions

2.3 The Composition of Functions

2.4 Symmetry and Transformations

2.5 Variation and Applications

2.1

Increasing, Decreasing, and Piecewise Functions; Applications

❖ Graph functions, looking for intervals on which the function is increasing, decreasing, or constant, and estimate relative maxima and minima.

❖ Given an application, find a function that models the application. Find the domain of the function and function values, and then graph the function.

❖ Graph functions defined piecewise.

Because functions occur in so many real-world situations, it is important to be able to analyze them carefully.

❖ Increasing, Decreasing, and Constant Functions

On a given interval, if the graph of a function rises from left to right, it is said to be **increasing** on that interval. If the graph drops from left to right, it is said to be **decreasing**. If the function values stay the same from left to right, the function is said to be **constant**.

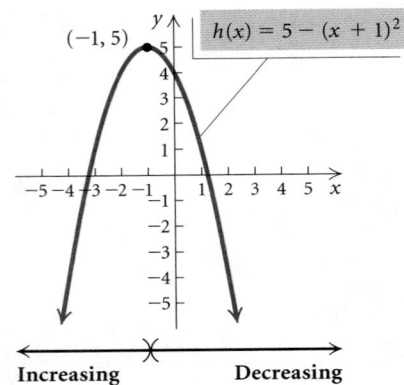

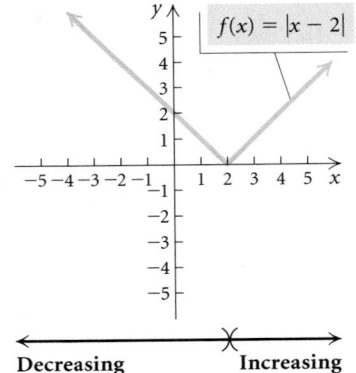

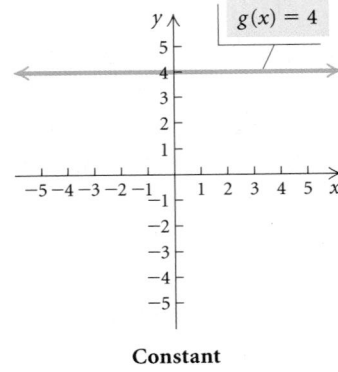

We are led to the following definitions.

Increasing, Decreasing, and Constant Functions

A function f is said to be **increasing** on an *open* interval I, if for all a and b in that interval, $a < b$ implies $f(a) < f(b)$. (See Fig. 1 on the following page.)

A function f is said to be **decreasing** on an *open* interval I, if for all a and b in that interval, $a < b$ implies $f(a) > f(b)$. (See Fig. 2.)

A function f is said to be **constant** on an *open* interval I, if for all a and b in that interval, $f(a) = f(b)$. (See Fig. 3.)

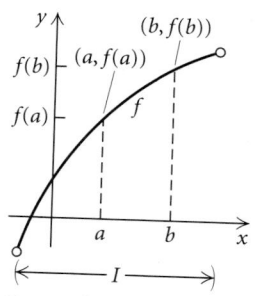

For $a < b$ in $I, f(a) < f(b)$;
f is **increasing** on I.

FIGURE 1

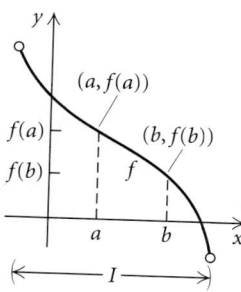

For $a < b$ in $I, f(a) > f(b)$;
f is **decreasing** on I.

FIGURE 2

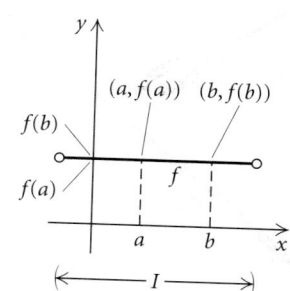

For all a and b in $I, f(a) = f(b)$;
f is **constant** on I.

FIGURE 3

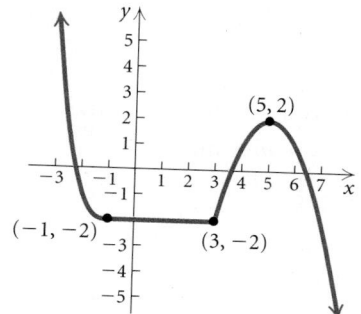

EXAMPLE 1 Determine the intervals on which the function in the figure at left is **(a)** increasing; **(b)** decreasing; **(c)** constant.

Solution When expressing interval(s) on which a function is increasing, decreasing, or constant, we consider only values in the *domain* of the function. Since the domain of this function is $(-\infty, \infty)$, we consider all real values of x.

a) As x-values (that is, values in the domain) increase from $x = 3$ to $x = 5$, the y-values (that is, values in the range) increase from -2 to 2. Thus the function is increasing on the interval $(3, 5)$.

b) As x-values increase from negative infinity to -1, y-values decrease; y-values also decrease as x-values increase from 5 to positive infinity. Thus the function is decreasing on the intervals $(-\infty, -1)$ and $(5, \infty)$.

c) As x-values increase from -1 to 3, y remains -2. The function is constant on the interval $(-1, 3)$.

Now Try Exercise 5. ▧

In calculus, the slope of a line tangent to the graph of a function at a particular point is used to determine whether the function is increasing, decreasing, or constant at that point. If the slope is positive, the function is increasing; if the slope is negative, the function is decreasing; if the slope is 0, the function is constant. Since slope cannot be both positive and negative at the same time, a function cannot be both increasing and decreasing at a specific point. For this reason, increasing, decreasing, and constant intervals are expressed in *open interval* notation. In Example 1, if $[3, 5]$ had been used for the increasing interval and $[5, \infty)$ for a decreasing interval, the function would be both increasing and decreasing at $x = 5$. This is not possible.

❊ Relative Maximum and Minimum Values

Consider the graph shown at the top of the next page. Note the "peaks" and "valleys" at the x-values c_1, c_2, and c_3. The function value $f(c_2)$ is called a **relative maximum** (plural, **maxima**). Each of the function values $f(c_1)$ and $f(c_3)$ is called a **relative minimum** (plural, **minima**).

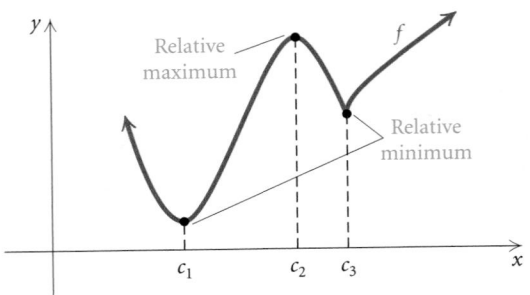

Relative Maxima and Minima

Suppose that f is a function for which $f(c)$ exists for some c in the domain of f. Then:

$f(c)$ is a **relative maximum** if there exists an *open* interval I containing c such that $f(c) > f(x)$, for all x in I where $x \neq c$; and

$f(c)$ is a **relative minimum** if there exists an *open* interval I containing c such that $f(c) < f(x)$, for all x in I where $x \neq c$.

Simply stated, $f(c)$ is a *relative maximum* if $f(c)$ is the highest point in some *open* interval, and $f(c)$ is a *relative minimum* if $f(c)$ is the lowest point in some *open* interval.

If you take a calculus course, you will learn a method for determining exact values of relative maxima and minima. In Section 3.3, we will find exact maximum and minimum values of quadratic functions algebraically. The MAXIMUM and MINIMUM features on a graphing calculator can be used to approximate relative maxima and minima.

GCM | **EXAMPLE 2** Use a graphing calculator to determine any relative maxima or minima of the function $f(x) = 0.1x^3 - 0.6x^2 - 0.1x + 2$ and to determine intervals on which the function is increasing or decreasing.

Solution We first graph the function, experimenting with the window dimensions as needed. The curvature is seen fairly well with window settings of $[-4, 6, -3, 3]$. Using the MAXIMUM and MINIMUM features, we determine the relative maximum value and the relative minimum value of the function.

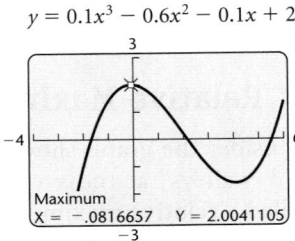

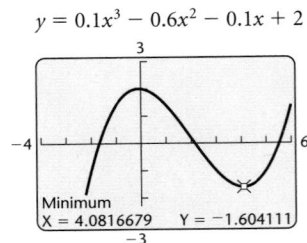

We see that the relative maximum value of the function is about 2.004. It occurs when $x \approx -0.082$. We also approximate the relative minimum, -1.604 at $x \approx 4.082$.

We note that the graph rises, or increases, from the left and stops increasing at the relative maximum. From this point, the graph decreases to the relative minimum and then begins to rise again. Thus the function is *increasing* on the intervals

$$(-\infty, -0.082) \quad \text{and} \quad (4.082, \infty)$$

and *decreasing* on the interval

$$(-0.082, 4.082).$$

Now Try Exercise 15. ■

❈ Applications of Functions

Many real-world situations can be modeled by functions.

EXAMPLE 3 *Car Distance.* Emma and Brandt drive away from a campground at right angles to each other. Emma's speed is 65 mph and Brandt's is 55 mph.

a) Express the distance between the cars as a function of time, $d(t)$.
b) Find the domain of the function.

Solution

a) Suppose 1 hr goes by. At that time, Emma has traveled 65 mi and Brandt has traveled 55 mi. We can use the Pythagorean theorem to find the distance between them. This distance would be the length of the hypotenuse of a triangle with legs measuring 65 mi and 55 mi. After 2 hr, the triangle's legs would measure 130 mi and 110 mi. Observing that the distances will always be changing, we make a drawing and let t = the time, in hours, that Emma and Brandt have been driving since leaving the campground.

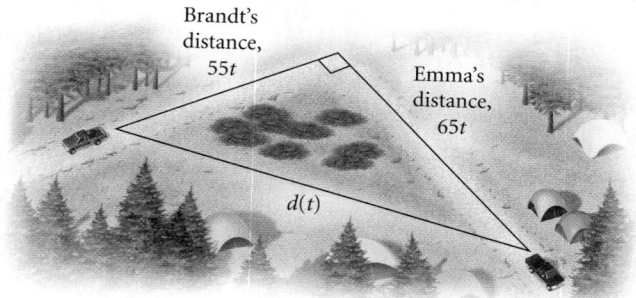

After t hours, Emma has traveled $65t$ miles and Brandt $55t$ miles. We now use the Pythagorean theorem:

$$[d(t)]^2 = (65t)^2 + (55t)^2.$$

Because distance must be nonnegative, we need consider only the positive square root when solving for $d(t)$:

$$\begin{aligned} d(t) &= \sqrt{(65t)^2 + (55t)^2} \\ &= \sqrt{4225t^2 + 3025t^2} \\ &= \sqrt{7250t^2} \\ &= \sqrt{7250}\,\sqrt{t^2} \\ &\approx 85.15|t| \qquad \text{Approximating the root to two decimal places} \\ &\approx 85.15t. \qquad \text{Since } t \geq 0, |t| = t. \end{aligned}$$

Thus, $d(t) = 85.15t$, $t \geq 0$.

b) Since the time traveled, t, must be nonnegative, the domain is the set of nonnegative real numbers $[0, \infty)$.   **Now Try Exercise 35.** ■

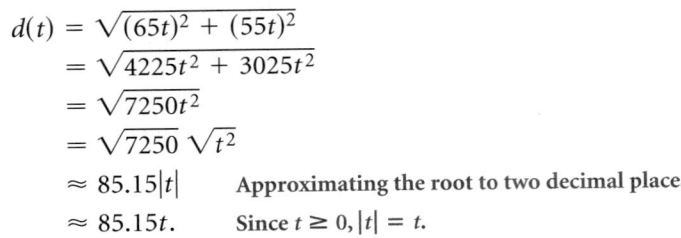
$y = 85.15x$

EXAMPLE 4 *Storage Area.* The Sound Shop has 20 ft of dividers with which to set off a rectangular area for the storage of overstock. If a corner of the store is used for the storage area, the partition need only form two sides of a rectangle.

a) Express the floor area of the storage space as a function of the length of the partition.

b) Find the domain of the function.

c) Graph the function.

d) Find the dimensions that maximize the floor area.

Solution

a) Note that the dividers will form two sides of a rectangle. If, for example, 14 ft of dividers is used for the length of the rectangle, that would leave $20 - 14$, or 6 ft of dividers for the width. Thus if $x =$ the length of the rectangle, in feet, then $20 - x =$ the width. We represent this information in a sketch, as shown at left.

The area, $A(x)$, is given by

$$\begin{aligned} A(x) &= x(20 - x) \qquad \text{Area} = \text{length} \cdot \text{width.} \\ &= 20x - x^2. \end{aligned}$$

The function $A(x) = 20x - x^2$ can be used to express the rectangle's area as a function of the length of the partition.

b) Because the rectangle's length must be positive and only 20 ft of dividers is available, we restrict the domain of A to $\{x \mid 0 < x < 20\}$, that is, the interval $(0, 20)$.

c) The graph is shown at left.

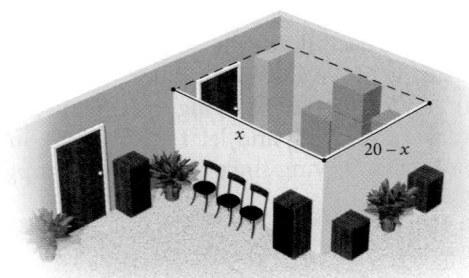

$y = 20x - x^2$

Maximum
X = 10.000007   Y = 100

d) We use the MAXIMUM feature as shown on the graph on the previous page. The maximum value of the area function on the interval $(0, 20)$ is 100 when $x = 10$. Thus the dimensions that maximize the area are

$$\text{Length} = x = 10 \text{ ft} \quad \text{and}$$
$$\text{Width} = 20 - x = 20 - 10 = 10 \text{ ft}.$$

Now Try Exercise 41. ◼

❈ Functions Defined Piecewise

Sometimes functions are defined **piecewise** using different output formulas for different pieces, or parts, of the domain.

EXAMPLE 5 For the function defined as

$$f(x) = \begin{cases} x + 1, & \text{for } x < -2, \\ 5, & \text{for } -2 \le x \le 3, \\ x^2, & \text{for } x > 3, \end{cases}$$

find $f(-5), f(-3), f(0), f(3), f(4),$ and $f(10)$.

Solution First, we determine which part of the domain contains the given input. Then we use the corresponding formula to find the output.

Since $-5 < -2$, we use the formula $f(x) = x + 1$:

$$f(-5) = -5 + 1 = -4.$$

Since $-3 < -2$, we use the formula $f(x) = x + 1$ again:

$$f(-3) = -3 + 1 = -2.$$

Since $-2 \le 0 \le 3$, we use the formula $f(x) = 5$:

$$f(0) = 5.$$

Since $-2 \le 3 \le 3$, we use the formula $f(x) = 5$ a second time:

$$f(3) = 5.$$

Since $4 > 3$, we use the formula $f(x) = x^2$:

$$f(4) = 4^2 = 16.$$

Since $10 > 3$, we once again use the formula $f(x) = x^2$:

$$f(10) = 10^2 = 100.$$

Now Try Exercise 47. ◼

EXAMPLE 6 Graph the function defined as

$$g(x) = \begin{cases} \dfrac{1}{3}x + 3, & \text{for } x < 3, \\ -x, & \text{for } x \ge 3. \end{cases}$$

Solution Since the function is defined in two pieces, or parts, we create the graph in two parts.

a) We graph $g(x) = \frac{1}{3}x + 3$ *only* for inputs x less than 3. That is, we use $g(x) = \frac{1}{3}x + 3$ only for x-values in the interval $(-\infty, 3)$. Some ordered pairs that are solutions of this piece of the function are shown in Table 1.

TABLE 1

x $(x < 3)$	$g(x) = \frac{1}{3}x + 3$
-3	2
0	3
2	$3\frac{2}{3}$

TABLE 2

x (x ≥ 3)	g(x) = −x
3	−3
4	−4
6	−6

b) We graph $g(x) = -x$ *only* for inputs x greater than or equal to 3. That is, we use $g(x) = -x$ only for x-values in the interval $[3, \infty)$. Some ordered pairs that are solutions of this piece of the function are shown in Table 2.

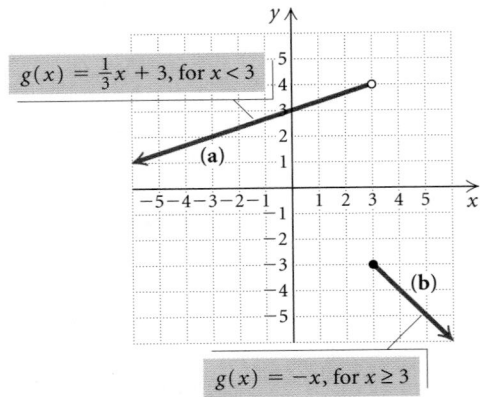

Now Try Exercise 51. ■

TABLE 3

x (x ≤ 0)	f(x) = 4
−5	4
−2	4
0	4

TABLE 4

x (0 < x ≤ 2)	f(x) = 4 − x²
$\frac{1}{2}$	$3\frac{3}{4}$
1	3
2	0

TABLE 5

x (x > 2)	f(x) = 2x − 6
$2\frac{1}{2}$	−1
3	0
5	4

GCM **EXAMPLE 7**   Graph the function defined as

$$f(x) = \begin{cases} 4, & \text{for } x \le 0, \\ 4 - x^2, & \text{for } 0 < x \le 2, \\ 2x - 6, & \text{for } x > 2. \end{cases}$$

Solution   We create the graph in three pieces, or parts.

a) We graph $f(x) = 4$ *only* for inputs x less than or equal to 0. That is, we use $f(x) = 4$ only for x-values in the interval $(-\infty, 0]$. Some ordered pairs that are solutions of this piece of the function are shown in Table 3.

b) We graph $f(x) = 4 - x^2$ *only* for inputs x greater than 0 and less than or equal to 2. That is, we use $f(x) = 4 - x^2$ only for x-values in the interval $(0, 2]$. Some ordered pairs that are solutions of this piece of the function are shown in Table 4.

c) We graph $f(x) = 2x - 6$ *only* for inputs x greater than 2. That is, we use $f(x) = 2x - 6$ only for x-values in the interval $(2, \infty)$. Some ordered pairs that are solutions of this piece of the function are shown in Table 5.

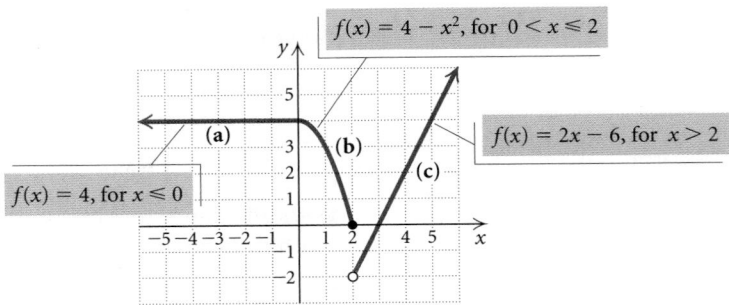

Now Try Exercise 55. ■

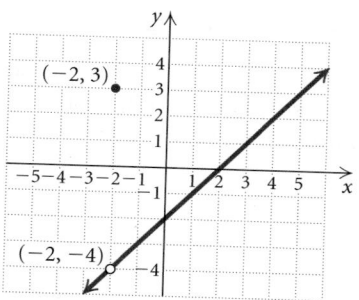

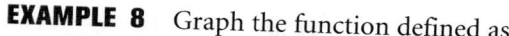

EXAMPLE 8 Graph the function defined as

$$f(x) = \begin{cases} \dfrac{x^2 - 4}{x + 2}, & \text{for } x \neq -2, \\ 3, & \text{for } x = -2. \end{cases}$$

Solution When $x \neq -2$, the denominator of $(x^2 - 4)/(x + 2)$ is nonzero, so we can simplify:

$$\frac{x^2 - 4}{x + 2} = \frac{(x + 2)(x - 2)}{x + 2} = x - 2.$$

Thus,

$$f(x) = x - 2, \quad \text{for } x \neq -2.$$

The graph of this part of the function consists of a line with a "hole" at the point $(-2, -4)$, indicated by the open circle. The hole occurs because the piece of the function represented by $(x^2 - 4)/(x + 2)$ is not defined for $x = -2$. By the definition of the function, we see that $f(-2) = 3$, so we plot the point $(-2, 3)$ above the open circle.

Now Try Exercise 59. ■

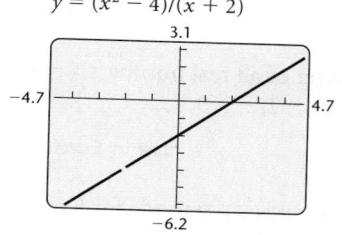

$y = (x^2 - 4)/(x + 2)$

The hand-drawn graph of the piece of the function in Example 8 represented by $y = (x^2 - 4)/(x + 2)$, $x \neq 2$, can be checked using a graphing calculator. When $y = (x^2 - 4)/(x + 2)$ is graphed, the hole may or may not be visible, depending on the window dimensions chosen. When we use the ZDECIMAL feature from the ZOOM menu, note the hole will appear, as shown in the graph at left. (See the *Graphing Calculator Manual* that accompanies this text for further details on selecting window dimensions.) If we examine a table of values for this function, we see that an ERROR message corresponds to the *x*-value -2. This indicates that -2 is *not* in the domain of the function $y = (x^2 - 4)/(x + 2)$. However, -2 *is* in the domain of the function f in Example 8 because $f(-2)$ is defined to be 3.

A piecewise function with importance in calculus and computer programming is the **greatest integer function**, f, denoted $f(x) = [\![x]\!]$, or int(x).

> ### Greatest Integer Function
>
> $f(x) = [\![x]\!] = $ the greatest integer *less than or equal to x*.

The greatest integer function pairs each input with the greatest integer *less than or equal to* that input. Thus, *x*-values 1, $1\frac{1}{2}$, and 1.8 are all paired with the *y*-value 1. Other pairings are shown below.

$$\begin{array}{cccc} -4 & -1 & 0 & 2 \\ -3.6 \searrow -4 & -\frac{7}{8} \searrow -1 & \frac{1}{10} \searrow 0 & 2.1 \searrow 2 \\ -3\frac{1}{4} \nearrow & -0.25 \nearrow & 0.99 \nearrow & 2\frac{3}{4} \nearrow \end{array}$$

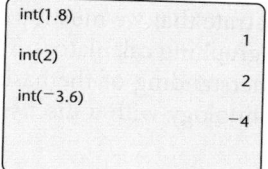

These values can be checked with a graphing calculator using the INT(feature from the NUM submenu in the MATH menu.

GCM **EXAMPLE 9**   Graph $f(x) = [\![x]\!]$ and determine its domain and range.

Solution   The greatest integer function can also be defined as a piecewise function with an infinite number of statements. When plotting points by hand, it can be helpful to use the TABLE feature on a graphing calculator to find ordered pairs.

X	Y1
−2.5	−3
−2	−2
−1.45	−2
−1	−1
−.75	−1
0	0
1.8	1

X = 1.8

$$
f(x) = [\![x]\!] =
\begin{cases}
\;\vdots & \\
-3, & \text{for } -3 \le x < -2, \\
-2, & \text{for } -2 \le x < -1, \\
-1, & \text{for } -1 \le x < 0, \\
0, & \text{for } 0 \le x < 1, \\
1, & \text{for } 1 \le x < 2, \\
2, & \text{for } 2 \le x < 3, \\
3, & \text{for } 3 \le x < 4, \\
\;\vdots &
\end{cases}
$$

We see that the domain of this function is the set of all real numbers, $(-\infty, \infty)$, and the range is the set of all integers, $\{\ldots, -3, -2, -1, 0, 1, 2, 3, \ldots\}$.

Now Try Exercise 63. ◼

If we had used a graphing calculator set in CONNECTED mode for Example 9, we would see the graph on the left below. The nearly vertical line segments are not part of the graph. CONNECTED mode connects points (or rectangles called **pixels** on a graphing calculator) with line segments. We would see the graph on the right if we used DOT mode. The DOT mode is preferable, though even it does not show the open circles at the righthand endpoints of the segments.

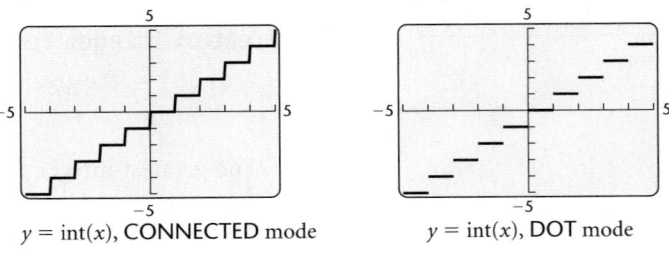

$y = \text{int}(x)$, CONNECTED mode $y = \text{int}(x)$, DOT mode

The graphs in the two windows above illustrate that we must be careful when analyzing graphs of functions drawn on graphing calculators. The use of technology must be combined with an understanding of the basic concepts of functions. This is why we balance technology with a discussion of hand-drawn graphs.

2.1 Exercise Set

Determine the intervals on which the function is
(a) *increasing,* **(b)** *decreasing, and* **(c)** *constant.*

1.

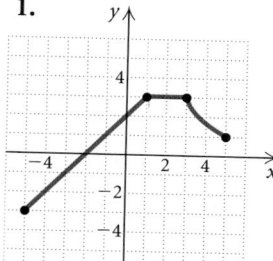

2.

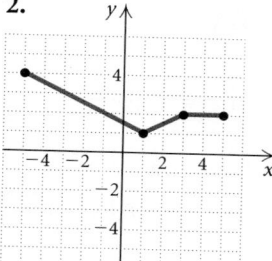

3.

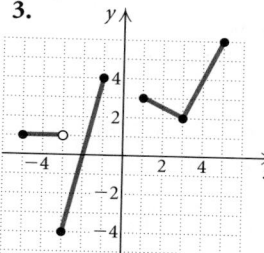

4.

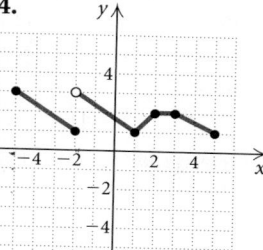

5.

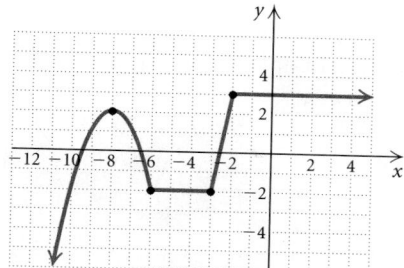

6.

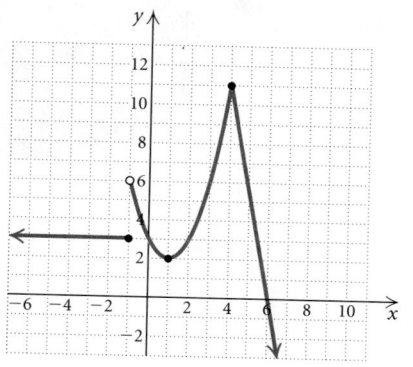

7.–12. Determine the domain and the range of each of the functions graphed in Exercises 1–6.

Using the graph, determine any relative maxima or minima of the function and the intervals on which the function is increasing or decreasing.

13. $f(x) = -x^2 + 5x - 3$

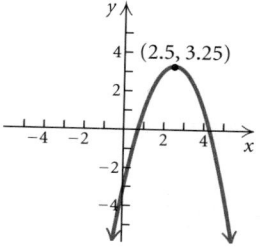

14. $f(x) = x^2 - 2x + 3$

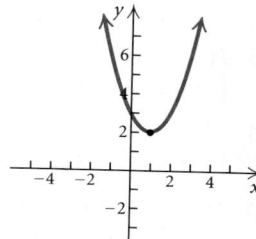

15. $f(x) = \frac{1}{4}x^3 - \frac{1}{2}x^2 - x + 2$

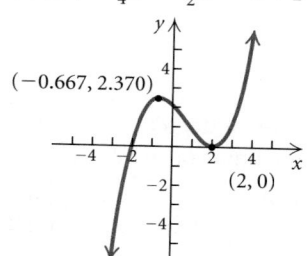

16. $f(x) = -0.09x^3 + 0.5x^2 - 0.1x + 1$

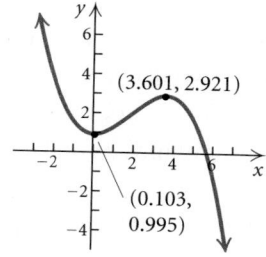

Graph the function. Estimate the intervals on which the function is increasing or decreasing and any relative maxima or minima.

17. $f(x) = x^2$

18. $f(x) = 4 - x^2$

19. $f(x) = 5 - |x|$

20. $f(x) = |x + 3| - 5$

21. $f(x) = x^2 - 6x + 10$

22. $f(x) = -x^2 - 8x - 9$

Graph the function using the given viewing window. Find the intervals on which the function is increasing or decreasing and find any relative maxima or minima. Change the viewing window if it seems appropriate for further analysis.

23. $f(x) = -x^3 + 6x^2 - 9x - 4$,
    $[-3, 7, -20, 15]$

24. $f(x) = 0.2x^3 - 0.2x^2 - 5x - 4$,
    $[-10, 10, -30, 20]$

25. $f(x) = 1.1x^4 - 5.3x^2 + 4.07$,
    $[-4, 4, -4, 8]$

26. $f(x) = 1.2(x + 3)^4 + 10.3(x + 3)^2 + 9.78$,
    $[-9, 3, -40, 100]$

27. *Temperature During an Illness.* The temperature of a patient during an illness is given by the function

    $T(t) = -0.1t^2 + 1.2t + 98.6, \quad 0 \le t \le 12,$

    where T is the temperature, in degrees Fahrenheit, at time t, in days, after the onset of the illness.

    a) Graph the function using a graphing calculator.
    b) Use the MAXIMUM feature to determine at what time the patient's temperature was the highest. What was the highest temperature?

28. *Advertising Effect.* A software firm estimates that it will sell N units of a new video game after spending a dollars on advertising, where

    $N(a) = -a^2 + 300a + 6, \quad 0 \le a \le 300,$

    and a is measured in thousands of dollars.

    a) Graph the function using a graphing calculator.
    b) Use the MAXIMUM feature to find the relative maximum.

    c) For what advertising expenditure will the greatest number of games be sold? How many games will be sold for that amount?

Use a graphing calculator to find the intervals on which the function is increasing or decreasing. Consider the entire set of real numbers if no domain is given.

29. $f(x) = \dfrac{8x}{x^2 + 1}$

30. $f(x) = \dfrac{-4}{x^2 + 1}$

31. $f(x) = x\sqrt{4 - x^2}, \quad \text{for } -2 \le x \le 2$

32. $f(x) = -0.8x\sqrt{9 - x^2}, \quad \text{for } -3 \le x \le 3$

33. *Garden Area.* Creative Landscaping has 60 yd of fencing with which to enclose a rectangular flower garden. If the garden is x yards long, express the garden's area as a function of the length.

34. *Triangular Scarf.* A seamstress is designing a triangular scarf so that the length of the base of the triangle, in inches, is 7 less than twice the height h. Express the area of the scarf as a function of the height.

35. *Rising Balloon.* A hot-air balloon rises straight up from the ground at a rate of 120 ft/min. The balloon is tracked from a rangefinder on the ground at point P, which is 400 ft from the release point Q of the balloon. Let $d =$ the distance from the balloon to the rangefinder and $t =$

the time, in minutes, since the balloon was released. Express d as a function of t.

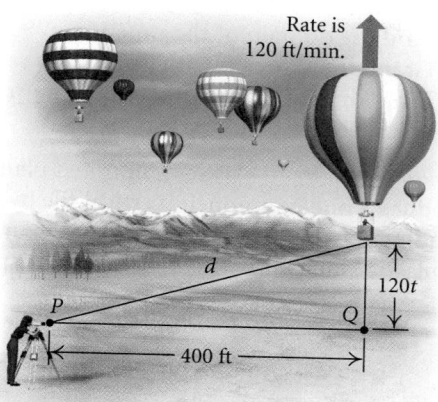

Rate is 120 ft/min.

36. *Airplane Distance.* An airplane is flying at an altitude of 3700 ft. The slanted distance directly to the airport is d feet. Express the horizontal distance h as a function of d.

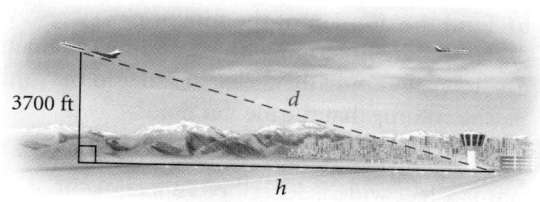

3700 ft

d

h

37. *Inscribed Rhombus.* A rhombus is inscribed in a rectangle that is w meters wide with a perimeter of 40 m. Each vertex of the rhombus is a midpoint of a side of the rectangle. Express the area of the rhombus as a function of the rectangle's width.

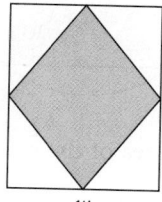

w

38. *Carpet Area.* A carpet installer uses 46 ft of linen tape to bind the edges of a rectangular hall runner. If the runner is w feet wide, express its area as a function of the width.

39. *Golf Distance Finder.* A device used in golf to estimate the distance d, in yards, to a hole

measures the size s, in inches, that the 7-ft pin appears to be in a viewfinder. Express the distance d as a function of s.

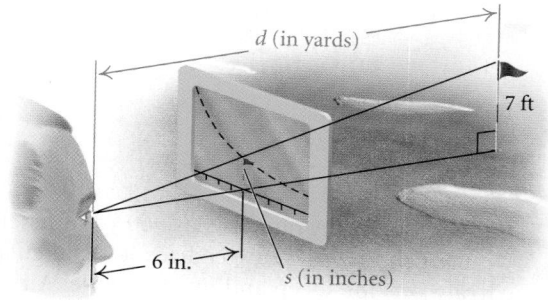

d (in yards)

7 ft

6 in.

s (in inches)

40. *Gas Tank Volume.* A gas tank has ends that are hemispheres of radius r ft. The cylindrical midsection is 6 ft long. Express the volume of the tank as a function of r.

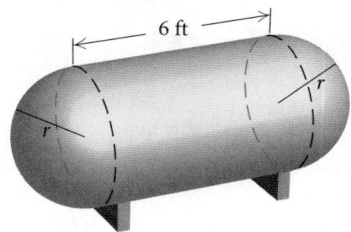

6 ft

r

r

41. *Play Space.* A daycare center has 30 ft of dividers with which to enclose a rectangular play space in a corner of a large room. The sides against the wall require no partition. Suppose the play space is x feet long.

a) Express the area of the play space as a function of x.

b) Find the domain of the function.
c) Using the graph shown below, determine the dimensions that yield the maximum area.

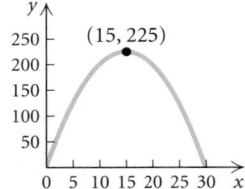

42. *Corral Design.*   A rancher has 360 yd of fencing with which to enclose two adjacent rectangular corrals, one for sheep and one for cattle. A river forms one side of the corrals. Suppose the width of each corral is x yards.

a) Express the total area of the two corrals as a function of x.
b) Find the domain of the function.
c) Using the graph shown below, determine the dimensions that yield the maximum area.

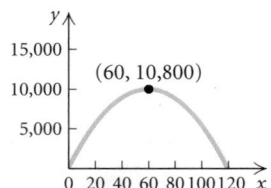

43. *Volume of a Box.*   From a 12-cm by 12-cm piece of cardboard, square corners are cut out so that the sides can be folded up to make a box.

a) Express the volume of the box as a function of the side x, in centimeters, of a cut-out square.
b) Find the domain of the function.
c) Graph the function with a graphing calculator.
d) What dimensions yield the maximum volume?

44. *Molding Plastics.*   Plastics Unlimited plans to produce a one-component vertical file by bending the long side of an 8-in. by 14-in. sheet of plastic along two lines to form a ⊔ shape.

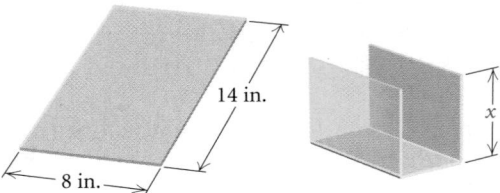

a) Express the volume of the file as a function of the height x, in inches, of the file.
b) Find the domain of the function.
c) Graph the function with a graphing calculator.
d) How tall should the file be in order to maximize the volume that the file can hold?

45. *Area of an Inscribed Rectangle.*   A rectangle that is x feet wide is inscribed in a circle of radius 8 ft.

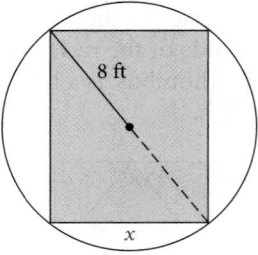

a) Express the area of the rectangle as a function of x.
b) Find the domain of the function.
c) Graph the function with a graphing calculator.
d) What dimensions maximize the area of the rectangle?

46. *Cost of Material.*   A rectangular box with volume 320 ft³ is built with a square base and top. The cost is $1.50/ft² for the bottom, $2.50/ft² for the sides, and $1/ft² for the top. Let $x =$ the length of the base, in feet.

a) Express the cost of the box as a function of x.
b) Find the domain of the function.
c) Graph the function with a graphing calculator.
d) What dimensions minimize the cost of the box?

For each piecewise function, find the specified function values.

47. $g(x) = \begin{cases} x + 4, & \text{for } x \le 1, \\ 8 - x, & \text{for } x > 1 \end{cases}$
$g(-4), g(0), g(1),$ and $g(3)$

48. $f(x) = \begin{cases} 3, & \text{for } x \le -2, \\ \frac{1}{2}x + 6, & \text{for } x > -2 \end{cases}$
$f(-5), f(-2), f(0),$ and $f(2)$

49. $h(x) = \begin{cases} -3x - 18, & \text{for } x < -5, \\ 1, & \text{for } -5 \le x < 1, \\ x + 2, & \text{for } x \ge 1 \end{cases}$
$h(-5), h(0), h(1),$ and $h(4)$

50. $f(x) = \begin{cases} -5x - 8, & \text{for } x < -2, \\ \frac{1}{2}x + 5, & \text{for } -2 \le x \le 4, \\ 10 - 2x, & \text{for } x > 4 \end{cases}$
$f(-4), f(-2), f(4),$ and $f(6)$

Make a hand-drawn graph of each of the following. Check your results using a graphing calculator.

51. $f(x) = \begin{cases} \frac{1}{2}x, & \text{for } x < 0, \\ x + 3, & \text{for } x \ge 0 \end{cases}$

52. $f(x) = \begin{cases} -\frac{1}{3}x + 2, & \text{for } x \le 0, \\ x - 5, & \text{for } x > 0 \end{cases}$

53. $f(x) = \begin{cases} -\frac{3}{4}x + 2, & \text{for } x < 4, \\ -1, & \text{for } x \ge 4 \end{cases}$

54. $h(x) = \begin{cases} 2x - 1, & \text{for } x < 2, \\ 2 - x, & \text{for } x \ge 2 \end{cases}$

55. $f(x) = \begin{cases} x + 1, & \text{for } x \le -3, \\ -1, & \text{for } -3 < x < 4, \\ \frac{1}{2}x, & \text{for } x \ge 4 \end{cases}$

56. $f(x) = \begin{cases} 4, & \text{for } x \le -2, \\ x + 1, & \text{for } -2 < x < 3, \\ -x, & \text{for } x \ge 3 \end{cases}$

57. $g(x) = \begin{cases} \frac{1}{2}x - 1, & \text{for } x < 0, \\ 3, & \text{for } 0 \le x \le 1, \\ -2x, & \text{for } x > 1 \end{cases}$

58. $f(x) = \begin{cases} \dfrac{x^2 - 9}{x + 3}, & \text{for } x \ne -3, \\ 5, & \text{for } x = -3 \end{cases}$

59. $f(x) = \begin{cases} 2, & \text{for } x = 5, \\ \dfrac{x^2 - 25}{x - 5}, & \text{for } x \ne 5 \end{cases}$

60. $f(x) = \begin{cases} \dfrac{x^2 + 3x + 2}{x + 1}, & \text{for } x \ne -1, \\ 7, & \text{for } x = -1 \end{cases}$

61. $f(x) = [\![x]\!]$

62. $f(x) = 2[\![x]\!]$

63. $g(x) = 1 + [\![x]\!]$

64. $h(x) = \frac{1}{2}[\![x]\!] - 2$

65.–70. Find the domain and the range of each of the functions defined in Exercises 51–56.

Determine the domain and the range of the piecewise function. Then write an equation for the function.

71.

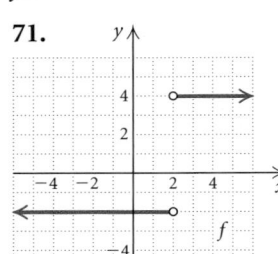

72.

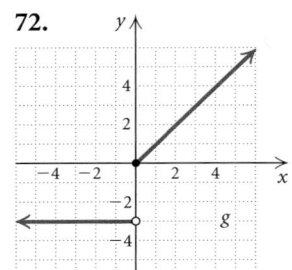

73.

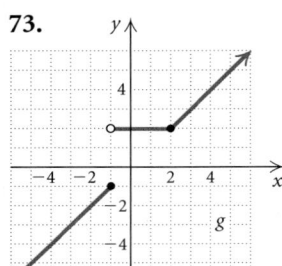

74.

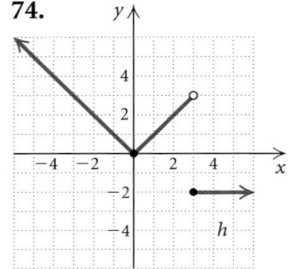

75.

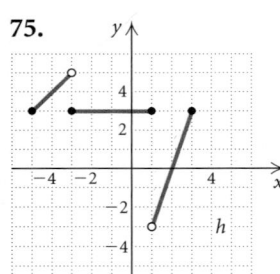

76.

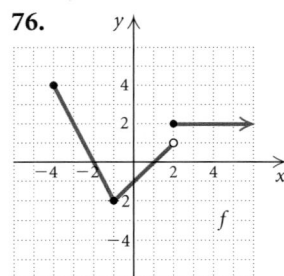

Collaborative Discussion and Writing

77. Describe a real-world situation that could be modeled by a function that is, in turn, increasing, then constant, and finally decreasing.

78. Simply stated, a *continuous function* is a function whose graph can be drawn without lifting the pencil from the paper. Examine several functions in this exercise set to see if they are continuous. Then explore the continuous functions to estimate the relative maxima and minima. For continuous functions, how can you connect the ideas of increasing and decreasing on an interval to relative maxima and minima?

Skill Maintenance

In each of Exercises 79–82, fill in the blank(s) with the correct term(s). Some of the given choices will not be used; others will be used more than once.

constant
function
any
midpoint formula
y-intercept
range
domain
distance formula
exactly one
identity
x-intercept

79. A(n) _____ is a correspondence between a first set, called the _____ , and a second set called the _____ , such that each member of the _____ corresponds to _____ member of the _____ .

80. The _____ is $\left(\dfrac{x_1 + x_2}{2}, \dfrac{y_1 + y_2}{2} \right)$.

81. A(n) _____ is a point $(a, 0)$.

82. A function f is a linear function if it can be written as $f(x) = mx + b$, where m and b are constants. If $m = 0$, the function is a(n) _____ function $f(x) = b$. If $m = 1$ and $b = 0$, the function is the _____ function $f(x) = x$.

Synthesis

Using a graphing calculator, estimate the interval on which the function is increasing or decreasing and any relative maxima or minima.

83. $f(x) = x^4 + 4x^3 - 36x^2 - 160x + 400$

84. $f(x) = 3.22x^5 - 5.208x^3 - 11$

85. *Parking Costs.* A parking garage charges $2 for up to (but not including) 1 hr of parking, $4 for up to 2 hr of parking, $6 for up to 3 hr of parking, and so on. Let $C(t)$ = the cost of parking for t hours.

a) Graph the function.

b) Write an equation for $C(t)$ using the greatest integer notation $[\![t]\!]$.

86. If $[\![x + 2]\!] = -3$, what are the possible inputs for x?

87. If $([\![x]\!])^2 = 25$, what are the possible inputs for x?

88. *Minimizing Power Line Costs.* A power line is constructed from a power station at point A to an island at point I, which is 1 mi directly out in the water from a point B on the shore. Point B is 4 mi downshore from the power station at A. It costs $5000 per mile to lay the power line under water and $3000 per mile to lay the power line under ground. The line comes to the shore at point S downshore from A. Let x = the distance from B to S.

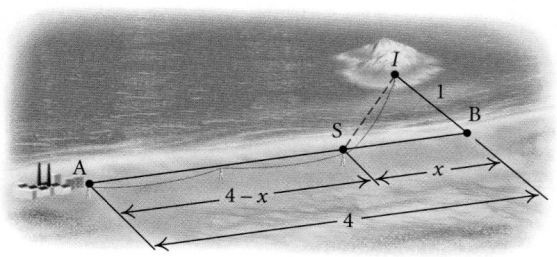

a) Express the cost C of laying the line as a function of x.

b) At what distance x from point B should the line come to shore in order to minimize cost?

89. *Volume of an Inscribed Cylinder.* A right circular cylinder of height h and radius r is inscribed in a right circular cone with a height of 10 ft and a base with radius 6 ft.

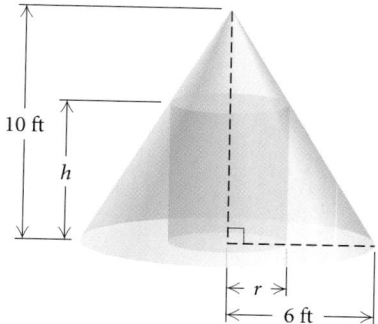

a) Express the height h of the cylinder as a function of r.

b) Express the volume V of the cylinder as a function of r.

c) Express the volume V of the cylinder as a function of h.

2.2 The Algebra of Functions

❖ Find the sum, the difference, the product, and the quotient of two functions, and determine the domains of the resulting functions.

❖ Find the difference quotient for a function.

❖ The Algebra of Functions: Sums, Differences, Products, and Quotients

We now use addition, subtraction, multiplication, and division to combine functions and obtain new functions.

Consider the following two functions f and g:

$$f(x) = x + 2 \quad \text{and} \quad g(x) = x^2 + 1.$$

Since $f(3) = 3 + 2 = 5$ and $g(3) = 3^2 + 1 = 10$, we have

$$f(3) + g(3) = 5 + 10 = 15,$$
$$f(3) - g(3) = 5 - 10 = -5,$$
$$f(3) \cdot g(3) = 5 \cdot 10 = 50,$$

and

$$\frac{f(3)}{g(3)} = \frac{5}{10} = \frac{1}{2}.$$

In fact, so long as x is in the domain of *both* f and g, we can easily compute $f(x) + g(x)$, $f(x) - g(x)$, $f(x) \cdot g(x)$, and, assuming $g(x) \neq 0$, $f(x)/g(x)$. We use the notation shown below.

> ### Sums, Differences, Products, and Quotients of Functions
>
> If f and g are functions and x is in the domain of each function, then
>
> $$(f + g)(x) = f(x) + g(x),$$
> $$(f - g)(x) = f(x) - g(x),$$
> $$(fg)(x) = f(x) \cdot g(x),$$
> $$(f/g)(x) = f(x)/g(x), \text{ provided } g(x) \neq 0.$$

EXAMPLE 1 Given that $f(x) = x + 1$ and $g(x) = \sqrt{x + 3}$, find each of the following.

a) $(f + g)(x)$ **b)** $(f + g)(6)$ **c)** $(f + g)(-4)$

Solution

a) $(f + g)(x) = f(x) + g(x)$
$$= x + 1 + \sqrt{x + 3} \qquad \text{This cannot be simplified.}$$

b) We can find $(f + g)(6)$ provided 6 is in the domain of *each* function. The domain of f is all real numbers. The domain of g is all real numbers x for which $x + 3 \geq 0$, or $x \geq -3$. This is the interval $[-3, \infty)$. We see that 6 is in both domains, so we have

$$f(6) = 6 + 1 = 7, \qquad g(6) = \sqrt{6 + 3} = \sqrt{9} = 3,$$
$$(f + g)(6) = f(6) + g(6) = 7 + 3 = 10.$$

Another method is to use the formula found in part (a):

$$(f + g)(6) = 6 + 1 + \sqrt{6 + 3} = 7 + \sqrt{9} = 7 + 3 = 10.$$

We can check our work using a graphing calculator by entering $y_1 = x + 1$, $y_2 = \sqrt{x + 3}$, and $y_3 = y_1 + y_2$ on the $y =$ screen. Then on the home screen, we find Y3(6).

```
Plot1  Plot2  Plot3
\Y1 ■ X+1
\Y2 ■ √(X+3)
\Y3 ■ Y1+Y2
\Y4=
\Y5=
\Y6=
\Y7=
```

```
Y3(6)
                10
```

c) To find $(f + g)(-4)$, we must first determine whether -4 is in the domain of each function. We note that -4 is not in the domain of g, $[-3, \infty)$. That is, $\sqrt{-4 + 3}$ is not a real number. Thus, $(f + g)(-4)$ does not exist.

Now Try Exercise 15. ▦

It is useful to view the concept of the sum of two functions graphically. In the graph below, we see the graphs of two functions f and g and their sum, $f + g$. Consider finding $(f + g)(4)$, or $f(4) + g(4)$. We can locate $g(4)$ on the graph of g and measure it. Then we add that length on top of $f(4)$ on the graph of f. The sum gives us $(f + g)(4)$.

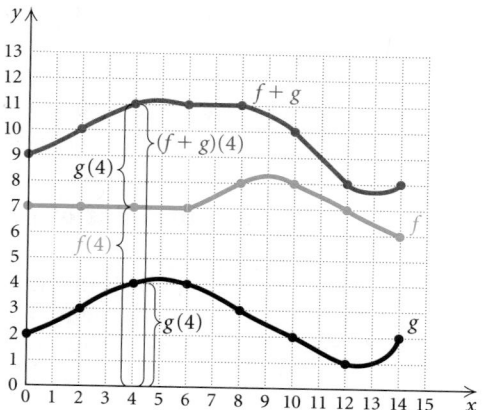

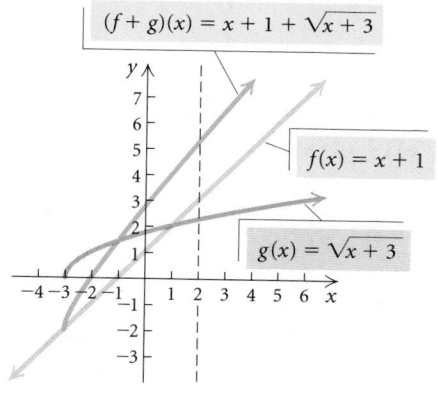

$(f + g)(x) = x + 1 + \sqrt{x + 3}$

$f(x) = x + 1$

$g(x) = \sqrt{x + 3}$

With this in mind, let's view Example 1 from a graphical perspective. Let's look at the graphs of

$$f(x) = x + 1, \qquad g(x) = \sqrt{x + 3}, \quad \text{and}$$
$$(f + g)(x) = x + 1 + \sqrt{x + 3}.$$

See the graph at left. Note that the domain of f is the set of all real numbers. The domain of g is $[-3, \infty)$. The domain of $f + g$ is the set of numbers in the intersection of the domains. This is the set of numbers in both domains.

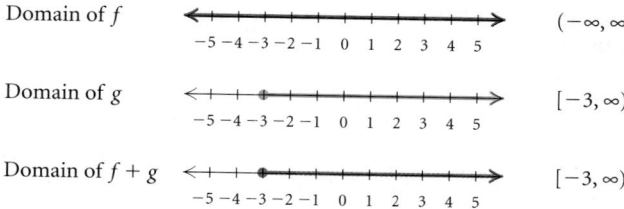

Domain of f $(-\infty, \infty)$

Domain of g $[-3, \infty)$

Domain of $f + g$ $[-3, \infty)$

Thus the domain of $f + g$ is $[-3, \infty)$.

We can confirm that the y-coordinates of the graph of $(f + g)(x)$ are the sums of the corresponding y-coordinates of the graphs of $f(x)$ and $g(x)$. Here we confirm it for $x = 2$.

$$f(x) = x + 1 \qquad\qquad g(x) = \sqrt{x + 3}$$
$$f(2) = 2 + 1 = 3; \qquad g(2) = \sqrt{2 + 3} = \sqrt{5};$$
$$(f + g)(x) = x + 1 + \sqrt{x + 3}$$
$$(f + g)(2) = 2 + 1 + \sqrt{2 + 3}$$
$$= 3 + \sqrt{5} = f(2) + g(2).$$

Let's also examine the domains of $f - g$, fg, and f/g for the functions $f(x) = x + 1$ and $g(x) = \sqrt{x + 3}$ of Example 1. The domains of $f - g$ and fg are the same as the domain of $f + g$, $[-3, \infty)$, because numbers in this interval are in the domains of *both* functions. For f/g, $g(x)$ cannot be 0. Since $\sqrt{x + 3} = 0$ when $x = -3$, we must exclude -3 and the domain of f/g is $(-3, \infty)$.

Domains of $f + g$, $f - g$, fg, and f/g

If f and g are functions, then the domain of the functions $f + g$, $f - g$, and fg is the intersection of the domain of f and the domain of g. The domain of f/g is also the intersection of the domains of f and g with the exclusion of any x-values for which $g(x) = 0$.

EXAMPLE 2   Given that $f(x) = x^2 - 4$ and $g(x) = x + 2$, find each of the following.

a) The domain of $f + g$, $f - g$, fg, and f/g

b) $(f + g)(x)$ c) $(f - g)(x)$

d) $(fg)(x)$ e) $(f/g)(x)$

f) $(gg)(x)$

Solution

a) The domain of f is the set of all real numbers. The domain of g is also the set of all real numbers. The domains of $f + g$, $f - g$, and fg are the set of numbers in the intersection of the domains—that is, the set of numbers in both domains, which is again the set of real numbers. For f/g, we must exclude -2, since $g(-2) = 0$. Thus the domain of f/g is the set of real numbers excluding -2, or $(-\infty, -2) \cup (-2, \infty)$.

b) $(f + g)(x) = f(x) + g(x) = (x^2 - 4) + (x + 2) = x^2 + x - 2$

c) $(f - g)(x) = f(x) - g(x) = (x^2 - 4) - (x + 2) = x^2 - x - 6$

d) $(fg)(x) = f(x) \cdot g(x) = (x^2 - 4)(x + 2) = x^3 + 2x^2 - 4x - 8$

e) $(f/g)(x) = \dfrac{f(x)}{g(x)}$

$= \dfrac{x^2 - 4}{x + 2}$ Note that $g(x) = 0$ when $x = -2$, so $(f/g)(x)$ is not defined when $x = -2$.

$= \dfrac{(x + 2)(x - 2)}{x + 2}$ Factoring

$= x - 2$ Removing a factor of 1: $\dfrac{x + 2}{x + 2} = 1$

Thus, $(f/g)(x) = x - 2$ with the added stipulation that $x \neq -2$ since -2 is not in the domain of $(f/g)(x)$.

f) $(gg)(x) = g(x) \cdot g(x) = [g(x)]^2 = (x + 2)^2 = x^2 + 4x + 4$

Now Try Exercise 21. ■

❋ Difference Quotients

In Section 1.3, we learned that the slope of a line can be considered as an average *rate of change*. Here let's consider a nonlinear function f and draw a line through two points $(x, f(x))$ and $(x + h, f(x + h))$ as shown at left.

The slope of the line, called a **secant line**, is

$$\frac{f(x + h) - f(x)}{x + h - x},$$

which simplifies to

$$\frac{f(x + h) - f(x)}{h}. \qquad \text{Difference quotient}$$

This ratio is called the **difference quotient**, or the **average rate of change**. In calculus, it is important to be able to find and simplify difference quotients.

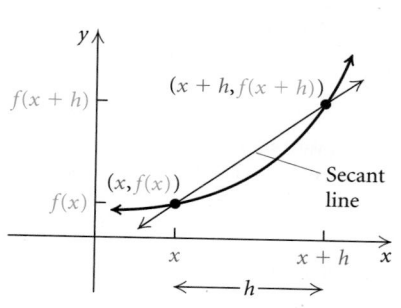

EXAMPLE 3 For the function f given by $f(x) = 2x - 3$, find and simplify the difference quotient

$$\frac{f(x + h) - f(x)}{h}.$$

Solution

$$\frac{f(x + h) - f(x)}{h} = \frac{2(x + h) - 3 - (2x - 3)}{h} \qquad \text{Substituting}$$

$$= \frac{2x + 2h - 3 - 2x + 3}{h} \qquad \text{Removing parentheses}$$

$$= \frac{2h}{h}$$

$$= 2 \qquad \text{Simplifying}$$

Now Try Exercise 47. ■

EXAMPLE 4   For the function f given by $f(x) = \dfrac{1}{x}$, find and simplify the difference quotient

$$\frac{f(x + h) - f(x)}{h}.$$

Solution

$$\frac{f(x + h) - f(x)}{h} = \frac{\dfrac{1}{x + h} - \dfrac{1}{x}}{h} \qquad \text{Substituting}$$

$$= \frac{\dfrac{1}{x + h} \cdot \dfrac{x}{x} - \dfrac{1}{x} \cdot \dfrac{x + h}{x + h}}{h} \qquad \begin{array}{l}\text{The LCD of } \dfrac{1}{x + h} \\[4pt] \text{and } \dfrac{1}{x} \text{ is } x(x + h).\end{array}$$

$$= \frac{\dfrac{x}{x(x + h)} - \dfrac{x + h}{x(x + h)}}{h}$$

$$= \frac{\dfrac{x - (x + h)}{x(x + h)}}{h} \qquad \begin{array}{l}\text{Subtracting in} \\ \text{the numerator}\end{array}$$

$$= \frac{\dfrac{x - x - h}{x(x + h)}}{h} \qquad \text{Removing parentheses}$$

$$= \frac{\dfrac{-h}{x(x + h)}}{h} \qquad \begin{array}{l}\text{Simplifying the} \\ \text{numerator}\end{array}$$

$$= \frac{-h}{x(x + h)} \cdot \frac{1}{h} \qquad \begin{array}{l}\text{Multiplying by the} \\ \text{reciprocal of the} \\ \text{divisor}\end{array}$$

$$= \frac{-h \cdot 1}{x \cdot (x + h) \cdot h}$$

$$= \frac{-1 \cdot \cancel{h}}{x \cdot (x + h) \cdot \cancel{h}} \qquad \text{Simplifying}$$

$$= \frac{-1}{x(x + h)}, \text{ or } -\frac{1}{x(x + h)} \qquad \text{Now Try Exercise 53.} \ \blacksquare$$

EXAMPLE 5   For the function f given by $f(x) = 2x^2 - x - 3$, find and simplify the difference quotient

$$\frac{f(x + h) - f(x)}{h}.$$

Solution   We first find $f(x + h)$:

$$f(x + h) = 2(x + h)^2 - (x + h) - 3 \qquad \begin{array}{l}\text{Substituting } x + h \text{ for } x \text{ in} \\ f(x) = 2x^2 - x - 3\end{array}$$

$$= 2[x^2 + 2xh + h^2] - (x + h) - 3$$

$$= 2x^2 + 4xh + 2h^2 - x - h - 3.$$

Then

$$\frac{f(x+h)-f(x)}{h}=\frac{[2x^2+4xh+2h^2-x-h-3]-[2x^2-x-3]}{h}$$

$$=\frac{2x^2+4xh+2h^2-x-h-3-2x^2+x+3}{h}$$

$$=\frac{4xh+2h^2-h}{h}$$

$$=\frac{h(4x+2h-1)}{h\cdot 1}=\frac{h}{h}\cdot\frac{4x+2h-1}{1}=4x+2h-1.$$

Now Try Exercise 61. ■

2.2 Exercise Set

Given that $f(x)=x^2-3$ and $g(x)=2x+1$, find each of the following, if it exists.

1. $(f+g)(5)$

2. $(fg)(0)$

3. $(f-g)(-1)$

4. $(fg)(2)$

5. $(f/g)\left(-\frac{1}{2}\right)$

6. $(f-g)(0)$

7. $(fg)\left(-\frac{1}{2}\right)$

8. $(f/g)\left(-\sqrt{3}\right)$

9. $(g-f)(-1)$

10. $(g/f)\left(-\frac{1}{2}\right)$

Given that $h(x)=x+4$ and $g(x)=\sqrt{x-1}$, find each of the following, if it exists.

11. $(h-g)(-4)$

12. $(gh)(10)$

13. $(g/h)(1)$

14. $(h/g)(1)$

15. $(g+h)(1)$

16. $(hg)(3)$

For each pair of functions in Exercises 17–32:

a) *Find the domain of f, g, $f+g$, $f-g$, fg, ff, f/g, and g/f.*

b) *Find $(f+g)(x)$, $(f-g)(x)$, $(fg)(x)$, $(ff)(x)$, $(f/g)(x)$, and $(g/f)(x)$.*

17. $f(x)=2x+3$, $g(x)=3-5x$

18. $f(x)=-x+1$, $g(x)=4x-2$

19. $f(x)=x-3$, $g(x)=\sqrt{x+4}$

20. $f(x)=x+2$, $g(x)=\sqrt{x-1}$

21. $f(x)=2x-1$, $g(x)=-2x^2$

22. $f(x)=x^2-1$, $g(x)=2x+5$

23. $f(x)=\sqrt{x-3}$, $g(x)=\sqrt{x+3}$

24. $f(x)=\sqrt{x}$, $g(x)=\sqrt{2-x}$

25. $f(x)=x+1$, $g(x)=|x|$

26. $f(x)=4|x|$, $g(x)=1-x$

27. $f(x)=x^3$, $g(x)=2x^2+5x-3$

28. $f(x)=x^2-4$, $g(x)=x^3$

29. $f(x)=\dfrac{4}{x+1}$, $g(x)=\dfrac{1}{6-x}$

30. $f(x)=2x^2$, $g(x)=\dfrac{2}{x-5}$

31. $f(x)=\dfrac{1}{x}$, $g(x)=x-3$

32. $f(x)=\sqrt{x+6}$, $g(x)=\dfrac{1}{x}$

In Exercises 33–38, consider the functions F and G as shown in the following graph.

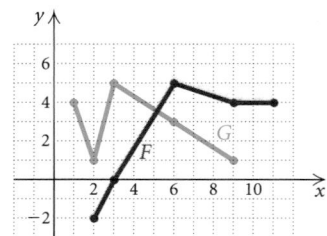

33. Find the domain of F, the domain of G, and the domain of $F + G$.

34. Find the domain of $F - G$, FG, and F/G.

35. Find the domain of G/F.

36. Graph $F + G$.

37. Graph $G - F$.

38. Graph $F - G$.

In Exercises 39–44, consider the functions F and G as shown in the following graph.

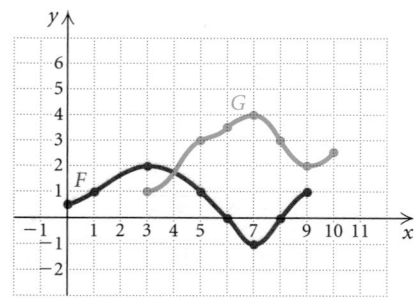

39. Find the domain of F, the domain of G, and the domain of $F + G$.

40. Find the domain of $F - G$, FG, and F/G.

41. Find the domain of G/F.

42. Graph $F + G$.

43. Graph $G - F$.

44. Graph $F - G$.

45. *Total Cost, Revenue, and Profit.* In economics, functions that involve revenue, cost, and profit are used. For example, suppose that $R(x)$ and $C(x)$ denote the total revenue and the total cost, respectively, of producing a new kind of tool for King Hardware Wholesalers. Then the difference

$$P(x) = R(x) - C(x)$$

represents the total profit for producing x tools. Given

$$R(x) = 60x - 0.4x^2 \quad \text{and} \quad C(x) = 3x + 13,$$

find each of the following.

a) $P(x)$
b) $R(100)$, $C(100)$, and $P(100)$
c) Using a graphing calculator, graph the three functions in the viewing window $[0, 160, 0, 3000]$.

46. *Total Cost, Revenue, and Profit.* Given that

$$R(x) = 200x - x^2 \quad \text{and} \quad C(x) = 5000 + 8x,$$

for a new radio produced by Clear Communication, find each of the following. (See Exercise 45.)

a) $P(x)$
b) $R(175)$, $C(175)$, and $P(175)$
c) Using a graphing calculator, graph the three functions in the viewing window $[0, 200, 0, 10,000]$.

For each function f, construct and simplify the difference quotient

$$\frac{f(x + h) - f(x)}{h}.$$

47. $f(x) = 3x - 5$ **48.** $f(x) = 4x - 1$

49. $f(x) = 6x + 2$ **50.** $f(x) = 5x + 3$

51. $f(x) = \dfrac{1}{3}x + 1$ **52.** $f(x) = -\dfrac{1}{2}x + 7$

53. $f(x) = \dfrac{1}{3x}$ **54.** $f(x) = \dfrac{1}{2x}$

55. $f(x) = -\dfrac{1}{4x}$ **56.** $f(x) = -\dfrac{1}{x}$

57. $f(x) = x^2 + 1$ **58.** $f(x) = x^2 - 3$

59. $f(x) = 4 - x^2$ **60.** $f(x) = 2 - x^2$

61. $f(x) = 3x^2 - 2x + 1$ **62.** $f(x) = 5x^2 + 4x$

63. $f(x) = 4 + 5|x|$ **64.** $f(x) = 2|x| + 3x$

65. $f(x) = x^3$

66. $f(x) = x^3 - 2x$

67. $f(x) = \dfrac{x - 4}{x + 3}$

68. $f(x) = \dfrac{x}{2 - x}$

Collaborative Discussion and Writing

69. If $g(x) = b$, where b is a positive constant, describe how the graphs of $y = h(x)$ and $y = (h - g)(x)$ will differ.

70. If the domain of a function f is the set of real numbers and the domain of a function g is also the set of real numbers, under what circumstances do $(f + g)(x)$ and $(f/g)(x)$ have different domains?

Skill Maintenance

Consider the following linear equations. Without graphing them, answer the questions below.

a) $y = x$
b) $y = -5x + 4$
c) $y = \frac{2}{3}x + 1$
d) $y = -0.1x + 6$
e) $y = 3x - 5$
f) $y = -x - 1$
g) $2x - 3y = 6$
h) $6x + 3y = 9$

71. Which, if any, have y-intercept $(0, 1)$?

72. Which, if any, have the same y-intercept?

73. Which slope down from left to right?

74. Which has the steepest slope?

75. Which pass(es) through the origin?

76. Which, if any, have the same slope?

77. Which, if any, are parallel?

78. Which, if any, are perpendicular?

Synthesis

79. Write equations for two functions f and g such that the domain of $f - g$ is
$$\{x \,|\, x \neq -7 \text{ and } x \neq 3\}.$$

80. For functions h and f, find the domain of $h + f$, $h - f$, hf, and h/f if
$$h = \{(-4, 13), (-1, 7), (0, 5), (\tfrac{5}{2}, 0), (3, -5)\}, \quad \text{and}$$
$$f = \{(-4, -7), (-2, -5), (0, -3), (3, 0), (5, 2), (9, 6)\}.$$

81. Find the domain of $(h/g)(x)$ given that
$$h(x) = \frac{5x}{3x - 7} \quad \text{and} \quad g(x) = \frac{x^4 - 1}{5x - 15}.$$

2.3 ## The Composition of Functions

❖ Find the composition of two functions and the domain of the composition.

❖ Decompose a function as a composition of two functions.

❖ The Composition of Functions

In real-world situations, it is not uncommon for the output of a function to depend on some input that is itself an output of another function. For instance, the amount a person pays as state income tax usually depends on the amount of adjusted gross income on the person's federal tax return, which, in turn, depends on his or her annual earnings. Such functions are called **composite functions**.

To illustrate how composite functions work, suppose a chemistry student needs a formula to convert Fahrenheit temperatures to Kelvin units. The formula

$$c(t) = \frac{5}{9}(t - 32)$$

gives the Celsius temperature $c(t)$ that corresponds to the Fahrenheit temperature t. The formula

$$k(c(t)) = c(t) + 273$$

gives the Kelvin temperature $k(c(t))$ that corresponds to the Celsius temperature $c(t)$. Thus, 50° Fahrenheit corresponds to

$$c(50) = \frac{5}{9}(50 - 32) = \frac{5}{9}(18) = 10° \text{ Celsius}$$

and 10° Celsius corresponds to

$$k(c(50)) = k(10) = 10 + 273 = 283 \text{ Kelvin units,}$$

which is usually written 283K. We see that 50° Fahrenheit is the same as 283K. This two-step procedure can be used to convert any Fahrenheit temperature to Kelvin units.

	°F Fahrenheit	°C Celsius	K Kelvin
Boiling point of water	212°	100°	373K
	50° ➡	10° ➡	283K
Freezing point of water	32°	0°	273K
Absolute zero	−460°	−273°	0K

In the table shown at left, we use a graphing calculator to convert Fahrenheit temperatures x to Celsius temperatures y_1, using $y_1 = \frac{5}{9}(x - 32)$. We also convert Celsius temperatures to Kelvin units y_2, using $y_2 = y_1 + 273$.

A student making numerous conversions might look for a formula that converts directly from Fahrenheit to Kelvin. Such a formula can be found by substitution:

$$k(c(t)) = c(t) + 273$$

$$= \frac{5}{9}(t - 32) + 273 \qquad \text{Substituting } \frac{5}{9}(t - 32) \text{ for } c(t)$$

$$= \frac{5}{9}t - \frac{160}{9} + 273$$

$$= \frac{5}{9}t - \frac{160}{9} + \frac{2457}{9}$$

$$= \frac{5t + 2297}{9}. \qquad \text{Simplifying}$$

$y_1 = \frac{5}{9}(x - 32), \ y_2 = y_1 + 273$

X	Y₁	Y₂
50	10	283
59	15	288
68	20	293
77	25	298
86	30	303
95	35	308
104	40	313

X = 50

We can show on a graphing calculator that the same values that appear in the table for y_2 will appear when y_2 is entered as

$$y_2 = \frac{5x + 2297}{9}.$$

Since the formula found above expresses the Kelvin temperature as a new function K of the Fahrenheit temperature t, we can write

$$K(t) = \frac{5t + 2297}{9},$$

where $K(t)$ is the Kelvin temperature corresponding to the Fahrenheit temperature, t. Here we have $K(t) = k(c(t))$. The new function K is called the **composition** of k and c and can be denoted $k \circ c$ (read "k composed with c," "the composition of k and c," or "k circle c").

Composition of Functions

The **composite function** $f \circ g$, the **composition** of f and g, is defined as

$$(f \circ g)(x) = f(g(x)),$$

where x is in the domain of g and $g(x)$ is in the domain of f.

GCM | **EXAMPLE 1** Given that $f(x) = 2x - 5$ and $g(x) = x^2 - 3x + 8$, find each of the following.

a) $(f \circ g)(x)$ and $(g \circ f)(x)$ **b)** $(f \circ g)(7)$ and $(g \circ f)(7)$

Solution Consider each function separately:

$$f(x) = 2x - 5 \qquad \text{This function multiplies each input by 2 and then subtracts 5.}$$

and

$$g(x) = x^2 - 3x + 8. \qquad \text{This function squares an input, subtracts three times the input from the result, and then adds 8.}$$

a) To find $(f \circ g)(x)$, we substitute $g(x)$ for x in the equation for $f(x)$:

$$(f \circ g)(x) = f(g(x)) = f(x^2 - 3x + 8) \qquad \text{$x^2 - 3x + 8$ is the input for f.}$$

$$= 2(x^2 - 3x + 8) - 5 \qquad \text{f multiplies the input by 2 and then subtracts 5.}$$

$$= 2x^2 - 6x + 16 - 5$$

$$= 2x^2 - 6x + 11.$$

To find $(g \circ f)(x)$, we substitute $f(x)$ for x in the equation for $g(x)$:

$(g \circ f)(x) = g(f(x)) = g(2x - 5)$ 2x − 5 is the input for g.

$= (2x - 5)^2 - 3(2x - 5) + 8$ g squares the input, subtracts three times the input, and then adds 8.

$= 4x^2 - 20x + 25 - 6x + 15 + 8$

$= 4x^2 - 26x + 48.$

b) To find $(f \circ g)(7)$, we first find $g(7)$. Then we use $g(7)$ as an input for f:

$(f \circ g)(7) = f(g(7)) = f(7^2 - 3 \cdot 7 + 8)$

$= f(36) = 2 \cdot 36 - 5$

$= 67.$

To find $(g \circ f)(7)$, we first find $f(7)$. Then we use $f(7)$ as an input for g:

$(g \circ f)(7) = g(f(7)) = g(2 \cdot 7 - 5)$

$= g(9) = 9^2 - 3 \cdot 9 + 8$

$= 62.$

We could also find $(f \circ g)(7)$ and $(g \circ f)(7)$ by substituting 7 for x in the equations that we found in part (a):

$(f \circ g)(x) = 2x^2 - 6x + 11$

$(f \circ g)(7) = 2 \cdot 7^2 - 6 \cdot 7 + 11 = 67;$

$(g \circ f)(x) = 4x^2 - 26x + 48$

$(g \circ f)(7) = 4 \cdot 7^2 - 26 \cdot 7 + 48 = 62.$

We can check our work using a graphing calculator. On the equation-editor screen, we enter $f(x)$ as

$y_1 = 2x - 5$

and $g(x)$ as

$y_2 = x^2 - 3x + 8.$

Then, on the home screen, we find $(f \circ g)(7)$ and $(g \circ f)(7)$ using the function notations Y1(Y2(7)) and Y2(Y1(7)), respectively.

$y_1 = 2x - 5, \quad y_2 = x^2 - 3x + 8$

Y1(Y2(7))	
	67
Y2(Y1(7))	
	62

Now Try Exercise 1. ■

$y_1 = (f \circ g)(x) = 2x^2 - 6x + 11,$
$y_2 = (g \circ f)(x) = 4x^2 - 26x + 48$

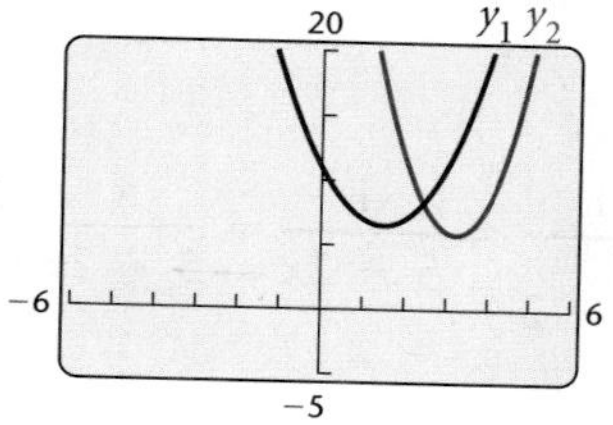

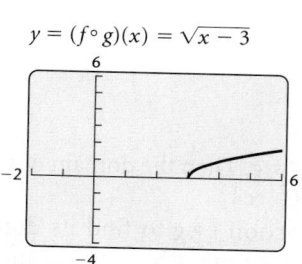

Example 1 illustrates that, as a rule, $(f \circ g)(x) \neq (g \circ f)(x)$. We can see this graphically, as shown in the graphs at left.

EXAMPLE 2   Given that $f(x) = \sqrt{x}$ and $g(x) = x - 3$:

a) Find $f \circ g$ and $g \circ f$.

b) Find the domain of $f \circ g$ and the domain of $g \circ f$.

Solution

a) $(f \circ g)(x) = f(g(x)) = f(x - 3) = \sqrt{x - 3}$

$(g \circ f)(x) = g(f(x)) = g(\sqrt{x}) = \sqrt{x} - 3$

b) Since $f(x)$ is not defined for negative radicands, the domain of $f(x)$ is $\{x \mid x \geq 0\}$, or $[0, \infty)$. Any real number can be an input for $g(x)$, so the domain of $g(x)$ is $(-\infty, \infty)$.

Since the inputs of $f \circ g$ are outputs of g, the domain of $f \circ g$ consists of the values of x in the domain of g, $(-\infty, \infty)$, for which $g(x)$ is nonnegative. (Recall that the inputs of $f(x)$ must be nonnegative.) Thus we have

$g(x) \geq 0$

$x - 3 \geq 0$ **Substituting $x - 3$ for $g(x)$**

$x \geq 3.$

We see that the domain of $f \circ g$ is $\{x \mid x \geq 3\}$, or $[3, \infty)$.

$y = (f \circ g)(x) = \sqrt{x - 3}$

FIGURE 1

We can also find the domain of $f \circ g$ by examining the composite function itself, $(f \circ g)(x) = \sqrt{x - 3}$. Since any real number can be an input for g, the only restriction on $f \circ g$ is that the radicand must be nonnegative. We have

$x - 3 \geq 0$

$x \geq 3.$

Again, we see that the domain of $f \circ g$ is $\{x \mid x \geq 3\}$, or $[3, \infty)$. The graph in Fig. 1 confirms this.

$y = (g \circ f)(x) = \sqrt{x} - 3$

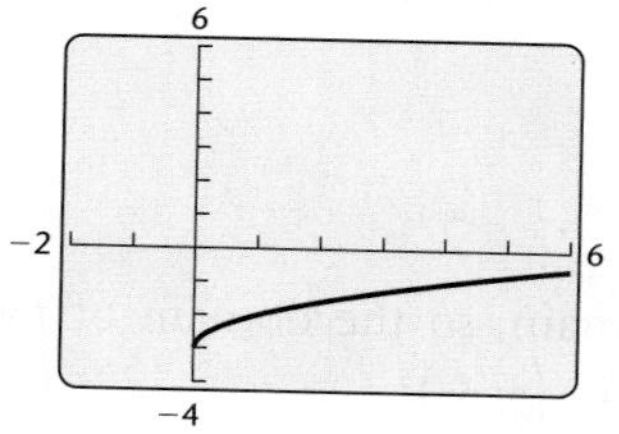

FIGURE 2

The inputs of $g \circ f$ are outputs of f, so the domain of $g \circ f$ consists of the values of x in the domain of f, $[0, \infty)$, for which $g(x)$ is defined. Since g can accept *any* real number as an input, any output from f is acceptable, so the entire domain of f is the domain of $g \circ f$. That is, the domain of $g \circ f$ is $\{x \mid x \geq 0\}$, or $[0, \infty)$.

We can also examine the composite function itself to find its domain. First recall that the domain of f is $\{x \mid x \geq 0\}$, or $[0, \infty)$. Then consider $(g \circ f)(x) = \sqrt{x} - 3$. The radicand cannot be negative, so we have $x \geq 0$. As above, we see that the domain of $g \circ f$ is the domain of f, $\{x \mid x \geq 0\}$, or $[0, \infty)$. The graph in Fig. 2 confirms this.   **Now Try Exercise 19.** ■

EXAMPLE 3   Given that $f(x) = \dfrac{1}{x-2}$ and $g(x) = \dfrac{5}{x}$, find $f \circ g$ and $g \circ f$ and the domain of each.

Solution   We have

$$(f \circ g)(x) = f(g(x)) = f\left(\frac{5}{x}\right) = \frac{1}{\dfrac{5}{x} - 2} = \frac{1}{\dfrac{5-2x}{x}} = \frac{x}{5-2x};$$

$$(g \circ f)(x) = g(f(x)) = g\left(\frac{1}{x-2}\right) = \frac{5}{\dfrac{1}{x-2}} = 5(x-2).$$

Values of x that make the denominator 0 are not in the domains of these functions. Since $x - 2 = 0$ when $x = 2$, the domain of f is $\{x \mid x \neq 2\}$. The denominator of g is x, so the domain of g is $\{x \mid x \neq 0\}$.

The inputs of $f \circ g$ are outputs of g, so the domain of $f \circ g$ consists of the values of x in the domain of g for which $g(x) \neq 2$. (Recall that 2 cannot be an input of f.) Since the domain of g is $\{x \mid x \neq 0\}$, 0 is not in the domain of $f \circ g$. In addition, we must find the value(s) of x for which $g(x) = 2$. We have

$$g(x) = 2$$

$$\frac{5}{x} = 2 \qquad \text{Substituting } \frac{5}{x} \text{ for } g(x)$$

$$5 = 2x$$

$$\frac{5}{2} = x.$$

This tells us that $\frac{5}{2}$ is also *not* in the domain of $f \circ g$. Then the domain of $f \circ g$ is $\left\{x \mid x \neq 0 \text{ and } x \neq \frac{5}{2}\right\}$, or $(-\infty, 0) \cup \left(0, \frac{5}{2}\right) \cup \left(\frac{5}{2}, \infty\right)$.

We can also examine the composite function $f \circ g$ to find its domain. First recall that 0 is not in the domain of g, so it cannot be in the domain of $(f \circ g)(x) = x/(5 - 2x)$. We must also exclude the value(s) of x for which the denominator of $f \circ g$ is 0. We have

$$5 - 2x = 0$$

$$5 = 2x$$

$$\frac{5}{2} = x.$$

Again, we see that $\frac{5}{2}$ is also not in the domain, so the domain of $f \circ g$ is $\left\{x \mid x \neq 0 \text{ and } x \neq \frac{5}{2}\right\}$, or $(-\infty, 0) \cup \left(0, \frac{5}{2}\right) \cup \left(\frac{5}{2}, \infty\right)$.

Since the inputs of $g \circ f$ are outputs of f, the domain of $g \circ f$ consists of the values of x in the domain of f for which $f(x) \neq 0$. (Recall that 0 cannot be an input of g.) The domain of f is $\{x \mid x \neq 2\}$, so 2 is not in the domain of $g \circ f$. Next, we determine whether there are values of x for which $f(x) = 0$:

$$f(x) = 0$$

$$\frac{1}{x - 2} = 0 \qquad\qquad \text{Substituting } \frac{1}{x - 2} \text{ for } f(x)$$

$$(x - 2) \cdot \frac{1}{x - 2} = (x - 2) \cdot 0 \qquad \text{Multiplying by } x - 2$$

$$1 = 0. \qquad\qquad \text{False equation}$$

We see that there are no values of x for which $f(x) = 0$, so there are no additional restrictions on the domain of $g \circ f$. Thus the domain of $g \circ f$ is $\{x \mid x \neq 2\}$, or $(-\infty, 2) \cup (2, \infty)$.

We can also examine $g \circ f$ to find its domain. First recall that 2 is not in the domain of f, so it cannot be in the domain of $(g \circ f)(x) = 5(x - 2)$. Since $5(x - 2)$ is defined for all real numbers, there are no additional restrictions on the domain of $g \circ f$. The domain is $\{x \mid x \neq 2\}$, or $(-\infty, 2) \cup (2, \infty)$.

Now Try Exercise 15. ■

❈ Decomposing a Function as a Composition

In calculus, one often needs to recognize how a function can be expressed as the composition of two functions. In this way, we are "decomposing" the function.

EXAMPLE 4   If $h(x) = (2x - 3)^5$, find $f(x)$ and $g(x)$ such that $h(x) = (f \circ g)(x)$.

Solution   The function $h(x)$ raises $(2x - 3)$ to the 5th power. Two functions that can be used for the composition are

$$f(x) = x^5 \quad \text{and} \quad g(x) = 2x - 3.$$

We can check by forming the composition:

$$h(x) = (f \circ g)(x) = f(g(x)) = f(2x - 3) = (2x - 3)^5.$$

This is the most "obvious" solution. There can be other less obvious solutions. For example, if

$$f(x) = (x + 7)^5 \quad \text{and} \quad g(x) = 2x - 10,$$

then

$$h(x) = (f \circ g)(x) = f(g(x))$$
$$= f(2x - 10)$$
$$= [2x - 10 + 7]^5 = (2x - 3)^5.$$

Now Try Exercise 31. ■

EXAMPLE 5   If $h(x) = \dfrac{1}{(x + 3)^3}$, find $f(x)$ and $g(x)$ such that $h(x) = (f \circ g)(x)$.

Solution   Two functions that can be used are

$$f(x) = \frac{1}{x} \quad \text{and} \quad g(x) = (x + 3)^3.$$

We check by forming the composition:

$$h(x) = (f \circ g)(x) = f(g(x)) = f((x + 3)^3) = \frac{1}{(x + 3)^3}.$$

There are other functions that can be used as well. For example, if

$$f(x) = \frac{1}{x^3} \quad \text{and} \quad g(x) = x + 3,$$

then

$$h(x) = (f \circ g)(x) = f(g(x)) = f(x + 3) = \frac{1}{(x + 3)^3}.$$

Now Try Exercise 33. ■

2.3   Exercise Set

Given that $f(x) = 3x + 1$, $g(x) = x^2 - 2x - 6$, *and* $h(x) = x^3$, *find each of the following.*

1. $(f \circ g)(-1)$

2. $(g \circ f)(-2)$

3. $(h \circ f)(1)$

4. $(g \circ h)\left(\frac{1}{2}\right)$

5. $(g \circ f)(5)$

6. $(f \circ g)\left(\frac{1}{3}\right)$

7. $(f \circ h)(-3)$

8. $(h \circ g)(3)$

Find $(f \circ g)(x)$ *and* $(g \circ f)(x)$ *and the domain of each.*

9. $f(x) = x + 3$, $g(x) = x - 3$

10. $f(x) = \frac{4}{5}x$, $g(x) = \frac{5}{4}x$

11. $f(x) = x + 1$, $g(x) = 3x^2 - 2x - 1$

12. $f(x) = 3x - 2$, $g(x) = x^2 + 5$

13. $f(x) = x^2 - 3$, $g(x) = 4x - 3$

14. $f(x) = 4x^2 - x + 10$, $g(x) = 2x - 7$

15. $f(x) = \dfrac{4}{1 - 5x}$, $g(x) = \dfrac{1}{x}$

16. $f(x) = \dfrac{6}{x}$, $g(x) = \dfrac{1}{2x + 1}$

17. $f(x) = 3x - 7$, $g(x) = \dfrac{x + 7}{3}$

18. $f(x) = \frac{2}{3}x - \frac{4}{5}$, $g(x) = 1.5x + 1.2$

19. $f(x) = 2x + 1$, $g(x) = \sqrt{x}$

20. $f(x) = \sqrt{x}$, $g(x) = 2 - 3x$

21. $f(x) = 20$, $g(x) = 0.05$

22. $f(x) = x^4$, $g(x) = \sqrt[4]{x}$

23. $f(x) = \sqrt{x + 5}$, $g(x) = x^2 - 5$

24. $f(x) = x^5 - 2$, $g(x) = \sqrt[5]{x + 2}$

25. $f(x) = x^2 + 2$, $g(x) = \sqrt{3 - x}$

26. $f(x) = 1 - x^2$, $g(x) = \sqrt{x^2 - 25}$

27. $f(x) = \dfrac{1 - x}{x}$, $g(x) = \dfrac{1}{1 + x}$

28. $f(x) = \dfrac{1}{x-2}$, $g(x) = \dfrac{x+2}{x}$

29. $f(x) = x^3 - 5x^2 + 3x + 7$, $g(x) = x + 1$

30. $f(x) = x - 1$, $g(x) = x^3 + 2x^2 - 3x - 9$

Find $f(x)$ and $g(x)$ such that $h(x) = (f \circ g)(x)$.
Answers may vary.

31. $h(x) = (4 + 3x)^5$

32. $h(x) = \sqrt[3]{x^2 - 8}$

33. $h(x) = \dfrac{1}{(x-2)^4}$

34. $h(x) = \dfrac{1}{\sqrt{3x+7}}$

35. $h(x) = \dfrac{x^3 - 1}{x^3 + 1}$

36. $h(x) = |9x^2 - 4|$

37. $h(x) = \left(\dfrac{2 + x^3}{2 - x^3}\right)^6$

38. $h(x) = \left(\sqrt{x} - 3\right)^4$

39. $h(x) = \sqrt{\dfrac{x-5}{x+2}}$

40. $h(x) = \sqrt{1 + \sqrt{1 + x}}$

41. $h(x) = (x + 2)^3 - 5(x + 2)^2 + 3(x + 2) - 1$

42. $h(x) = 2(x - 1)^{5/3} + 5(x - 1)^{2/3}$

43. *Blouse Sizes.* A blouse that is size x in Japan is size $s(x)$ in the United States, where $s(x) = x - 3$. A blouse that is size x in the United States is size $t(x)$ in Australia, where $t(x) = x + 4$. (*Source:* www.onlineconversion.com) Find a function that will convert Japanese blouse sizes to Australian blouse sizes.

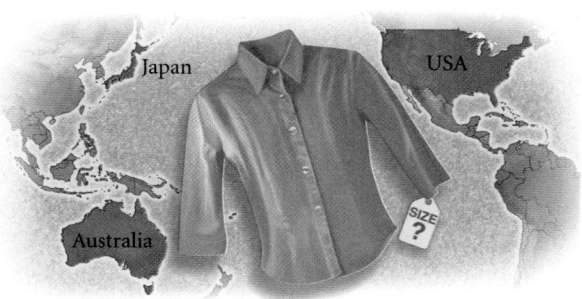

44. *Ripple Spread.* A stone is thrown into a pond, creating a circular ripple that spreads over the pond in such a way that the radius is increasing at the rate of 3 ft/sec.

a) Find a function $r(t)$ for the radius in terms of t.
b) Find a function $A(r)$ for the area of the ripple in terms of the radius r.
c) Find $(A \circ r)(t)$. Explain the meaning of this function.

Collaborative Discussion and Writing

45. If f and g are linear functions, what can you say about the domain of $f \circ g$ and the domain of $g \circ f$?

46. Nora determines the domain of $f \circ g$ by examining only the formula for $(f \circ g)(x)$. Is her approach valid? Why or why not?

Skill Maintenance

Graph the equation.

47. $y = 3x - 1$

48. $2x + y = 4$

49. $x - 3y = 3$

50. $y = x^2 + 1$

Synthesis

51. Let $p(a)$ represent the number of pounds of grass seed required to seed a lawn with area a. Let $c(s)$ represent the cost of s pounds of grass seed. Which composition makes sense: $(c \circ p)(a)$ or $(p \circ c)(s)$? What does it represent?

52. Write equations of two functions f and g such that $f \circ g = g \circ f = x$. (In Section 5.1, we will study inverse functions. If $f \circ g = g \circ f = x$, functions f and g are *inverses* of each other.)

2.4   Symmetry and Transformations

❀ Determine whether a graph is symmetric with respect to the *x*-axis, the *y*-axis, and the origin.

❀ Determine whether a function is even, odd, or neither even nor odd.

❀ Given the graph of a function, graph its transformation under translations, reflections, stretchings, and shrinkings.

❀ Symmetry

Symmetry occurs often in nature and in art. For example, when viewed from the front, the bodies of most animals are at least approximately symmetric. This means that each eye is the same distance from the center of the bridge of the nose, each shoulder is the same distance from the center of the chest, and so on. Architects have used symmetry for thousands of years to enhance the beauty of buildings.

A knowledge of symmetry in mathematics helps us graph and analyze equations and functions.

STUDY TIP

Try to keep one section ahead of your syllabus. If you study ahead of your lectures, you can concentrate on what is being explained in them, rather than trying to write everything down. You can then take notes only on special points or of questions related to what is happening in class.

Consider the points $(4, 2)$ and $(4, -2)$ that appear on the graph of $x = y^2$. (See Fig. 1.) Points like these have the same x-value but opposite y-values and are **reflections** of each other across the x-axis. If, for any point (x, y) on a graph, the point $(x, -y)$ is also on the graph, then the graph is said to be **symmetric with respect to the x-axis**. If we fold the graph on the x-axis, the parts above and below the x-axis will coincide.

Consider the points $(3, 4)$ and $(-3, 4)$ that appear on the graph of $y = x^2 - 5$. (See Fig. 2.) Points like these have the same y-value but opposite x-values and are **reflections** of each other across the y-axis. If, for any point (x, y) on a graph, the point $(-x, y)$ is also on the graph, then the graph is said to be **symmetric with respect to the y-axis**. If we fold the graph on the y-axis, the parts to the left and right of the y-axis will coincide.

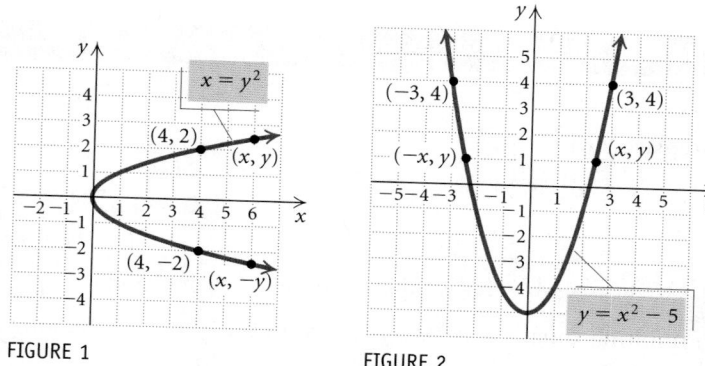

FIGURE 1

FIGURE 2

Consider the points $\left(-3, \sqrt{7}\right)$ and $\left(3, -\sqrt{7}\right)$ that appear on the graph of $x^2 = y^2 + 2$. (See Fig. 3.) Note that if we take the opposites of the coordinates of one pair, we get the other pair. If, for any point (x, y) on a graph, the point $(-x, -y)$ is also on the graph, then the graph is said to be **symmetric with respect to the origin**. Visually, if we rotate the graph $180°$ about the origin, the resulting figure coincides with the original.

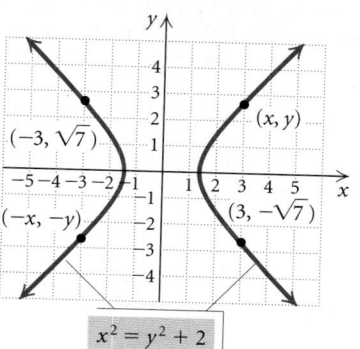

FIGURE 3

> ## Algebraic Tests of Symmetry
>
> *x-axis*: If replacing y with $-y$ produces an equivalent equation, then the graph is *symmetric with respect to the x-axis*.
>
> *y-axis*: If replacing x with $-x$ produces an equivalent equation, then the graph is *symmetric with respect to the y-axis*.
>
> *Origin*: If replacing x with $-x$ and y with $-y$ produces an equivalent equation, then the graph is *symmetric with respect to the origin*.

EXAMPLE 1   Test $y = x^2 + 2$ for symmetry with respect to the x-axis, the y-axis, and the origin.

ALGEBRAIC SOLUTION

***x*-AXIS:**
We replace y with $-y$:

$$y = x^2 + 2$$
$$\downarrow$$
$$-y = x^2 + 2$$
$$y = -x^2 - 2. \qquad \text{Multiplying by } -1 \text{ on both sides}$$

The resulting equation *is not* equivalent to the original equation, so the graph *is not* symmetric with respect to the x-axis.

***y*-AXIS:**
We replace x with $-x$:

$$y = x^2 + 2$$
$$y = (-x)^2 + 2$$
$$y = x^2 + 2. \qquad \text{Simplifying}$$

The resulting equation *is* equivalent to the original equation, so the graph *is* symmetric with respect to the y-axis.

ORIGIN:
We replace x with $-x$ and y with $-y$:

$$y = x^2 + 2$$
$$\downarrow \qquad \downarrow$$
$$-y = (-x)^2 + 2$$
$$-y = x^2 + 2 \qquad \text{Simplifying}$$
$$y = -x^2 - 2.$$

The resulting equation *is not* equivalent to the original equation, so the graph *is not* symmetric with respect to the origin.

GRAPHICAL SOLUTION

We use a graphing calculator to graph the equation.

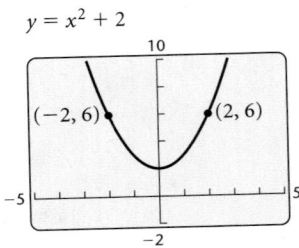

$y = x^2 + 2$

Note that if the graph were folded on the x-axis, the parts above and below the x-axis would not co-incide so the graph *is not* symmetric with respect to the x-axis. If it were folded on the y-axis, the parts to the left and right of the y-axis would coincide so the graph *is* symmetric with respect to the y-axis. If we rotated it 180° around the origin, the resulting graph would not coin-cide with the original graph so the graph *is not* symmetric with respect to the origin.

Now Try Exercise 11. ■

The algebraic method is often easier to apply than the graphical method, especially with equations that we may not be able to graph easily. It is also often more precise.

EXAMPLE 2 Test $x^2 + y^4 = 5$ for symmetry with respect to the x-axis, the y-axis, and the origin.

ALGEBRAIC SOLUTION

x-AXIS:
We replace y with $-y$:

$$x^2 + y^4 = 5$$
$$x^2 + (-y)^4 = 5$$
$$x^2 + y^4 = 5.$$

The resulting equation *is* equivalent to the original equation. Thus the graph *is* symmetric with respect to the x-axis.

y-AXIS:
We replace x with $-x$:

$$x^2 + y^4 = 5$$
$$(-x)^2 + y^4 = 5$$
$$x^2 + y^4 = 5.$$

The resulting equation *is* equivalent to the original equation, so the graph *is* symmetric with respect to the y-axis.

ORIGIN:
We replace x with $-x$ and y with $-y$:

$$x^2 + y^4 = 5$$
$$(-x)^2 + (-y)^4 = 5$$
$$x^2 + y^4 = 5.$$

The resulting equation *is* equivalent to the original equation, so the graph *is* symmetric with respect to the origin.

GRAPHICAL SOLUTION

To graph $x^2 + y^4 = 5$ using a graphing calculator, we first solve the equation for y:

$$y = \pm\sqrt[4]{5 - x^2}.$$

Then on the $y =$ screen we enter the equations

$$y_1 = \sqrt[4]{5 - x^2} \quad \text{and}$$
$$y_2 = -\sqrt[4]{5 - x^2}.$$

From the graph of the equation, we see symmetry with respect to both axes and with respect to the origin.

Now Try Exercise 21. ▪

❊ Even and Odd Functions

Now we relate symmetry to graphs of functions.

> ### Algebraic Procedure for Determining Even and Odd Functions
>
> Given the function $f(x)$:
>
> 1. Find $f(-x)$ and simplify. If $f(x) = f(-x)$, then f is even.
> 2. Find $-f(x)$, simplify, and compare with $f(-x)$ from step (1). If $f(-x) = -f(x)$, then f is odd.
>
> Except for the function $f(x) = 0$, a function cannot be *both* even and odd. Thus if $f(x) \neq 0$ and we see in step (1) that $f(x) = f(-x)$ (that is, f is even), we need not continue.

> ### Even and Odd Functions
>
> If the graph of a function f is symmetric with respect to the y-axis, we say that it is an **even function**. That is, for each x in the domain of f, $f(x) = f(-x)$.
>
> If the graph of a function f is symmetric with respect to the origin, we say that it is an **odd function**. That is, for each x in the domain of f, $f(-x) = -f(x)$.

An algebraic procedure for determining even and odd functions is shown at left. Below we show an even function and an odd function. Many functions are neither even nor odd.

EXAMPLE 3   Determine whether each of the following functions is even, odd, or neither.

a) $f(x) = 5x^7 - 6x^3 - 2x$

b) $h(x) = 5x^6 - 3x^2 - 7$

ALGEBRAIC SOLUTION

a) $f(x) = 5x^7 - 6x^3 - 2x$

1. $f(-x) = 5(-x)^7 - 6(-x)^3 - 2(-x)$
   $= 5(-x^7) - 6(-x^3) + 2x$
   $(-x)^7 = (-1 \cdot x)^7 = (-1)^7 x^7 = -x^7;$
   $(-x)^3 = (-1 \cdot x)^3 = (-1)^3 x^3 = -x^3$
   $= -5x^7 + 6x^3 + 2x$

   We see that $f(x) \neq f(-x)$. Thus, f is *not* even.

2. $-f(x) = -(5x^7 - 6x^3 - 2x)$
   $= -5x^7 + 6x^3 + 2x$

   We see that $f(-x) = -f(x)$. Thus, f is odd.

b) $h(x) = 5x^6 - 3x^2 - 7$

1. $h(-x) = 5(-x)^6 - 3(-x)^2 - 7$
   $= 5x^6 - 3x^2 - 7$

   We see that $h(x) = h(-x)$. Thus the function is even.

GRAPHICAL SOLUTION

a) We see that the graph appears to be symmetric with respect to the origin. The function is odd.

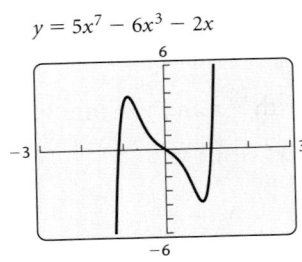

$y = 5x^7 - 6x^3 - 2x$

b) We see that the graph appears to be symmetric with respect to the y-axis. The function is even.

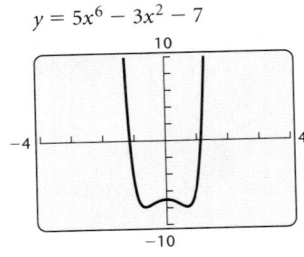

$y = 5x^6 - 3x^2 - 7$

Now Try Exercises 39 and 41. ■

❋ Transformations of Functions

The graphs of some basic functions are shown below. Others can be seen on the inside back cover.

Identity function:
$y = x$

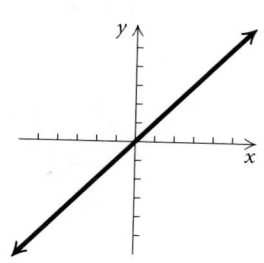

Squaring function:
$y = x^2$

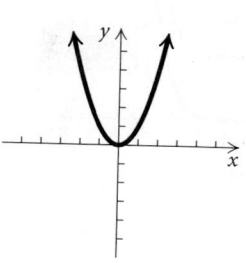

Square root function:
$y = \sqrt{x}$

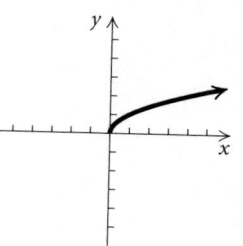

Cubing function:
$y = x^3$

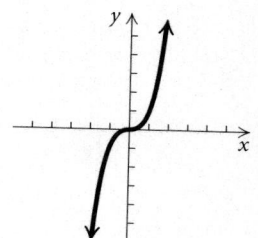

Cube root function:
$y = \sqrt[3]{x}$

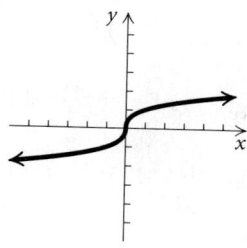

Reciprocal function:
$y = \dfrac{1}{x}$

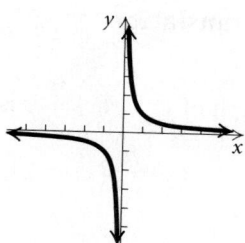

Absolute-value function:
$y = |x|$

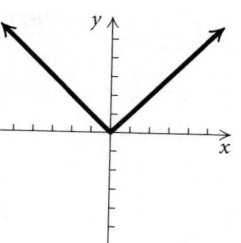

These functions can be considered building blocks for many other functions. We can create graphs of new functions by shifting them horizontally or vertically, stretching or shrinking them, and reflecting them across an axis. We now consider these **transformations**.

❋ Vertical and Horizontal Translations

Suppose that we have a function given by $y = f(x)$. Let's explore the graphs of the new functions $y = f(x) + b$ and $y = f(x) - b$, for $b > 0$.

Consider the functions $y = \frac{1}{5}x^4$, $y = \frac{1}{5}x^4 + 5$, and $y = \frac{1}{5}x^4 - 3$ and compare their graphs. What pattern do you see? Test it with some other graphs.

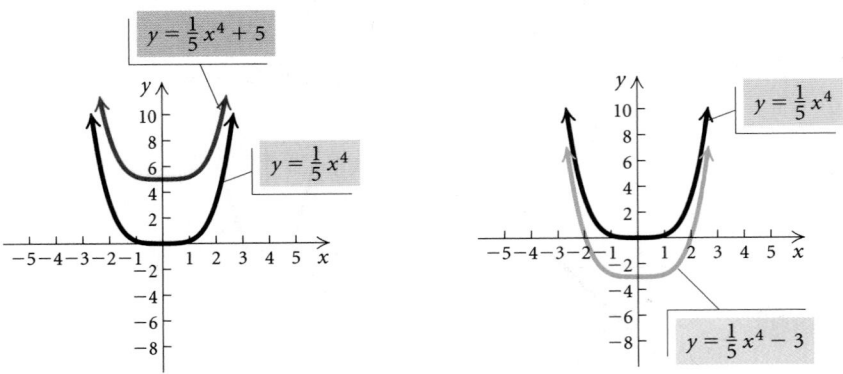

The effect of adding a constant to or subtracting a constant from $f(x)$ in $y = f(x)$ is a shift of the graph of $f(x)$ up or down. Such a shift is called a **vertical translation.**

> **Vertical Translation**
>
> For $b > 0$:
>
> the graph of $y = f(x) + b$ is the graph of $y = f(x)$ shifted *up* b units;
>
> the graph of $y = f(x) - b$ is the graph of $y = f(x)$ shifted *down* b units.

Suppose that we have a function given by $y = f(x)$. Let's explore the graphs of the new functions $y = f(x - d)$ and $y = f(x + d)$, for $d > 0$.

Consider the functions $y = \frac{1}{5}x^4$, $y = \frac{1}{5}(x - 3)^4$, and $y = \frac{1}{5}(x + 7)^4$ and compare their graphs. What pattern do you observe? Test it with some other graphs.

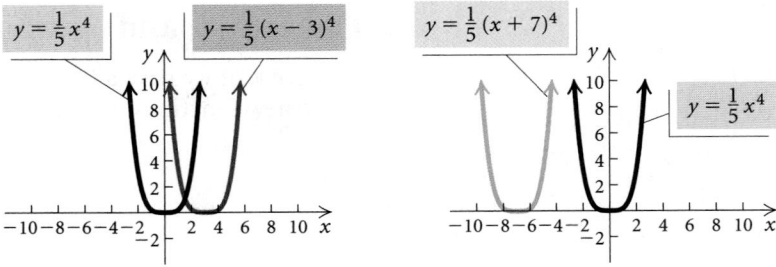

The effect of subtracting a constant from the x-value or adding a constant to the x-value in $y = f(x)$ is a shift of the graph of $f(x)$ to the right or left. Such a shift is called a **horizontal translation.**

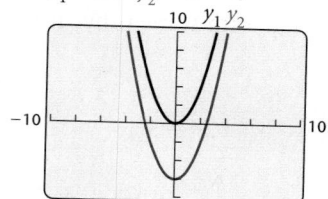

$y_1 = x^2, \quad y_2 = x^2 - 6$

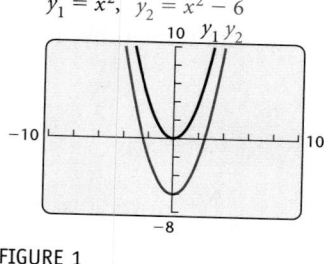

FIGURE 1

$y_1 = |x|, \quad y_2 = |x - 4|$

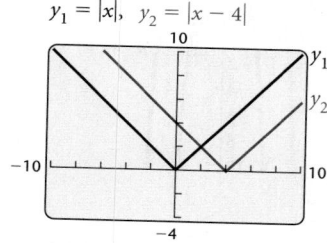

FIGURE 2

$y_1 = \sqrt{x}, \quad y_2 = \sqrt{x + 2}$

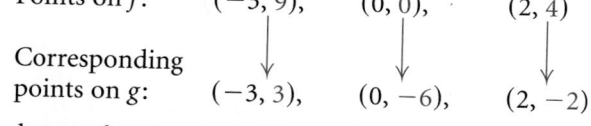

FIGURE 3

$y_1 = \sqrt{x}, \quad y_2 = \sqrt{x + 2},$
$y_3 = \sqrt{x + 2} - 3$

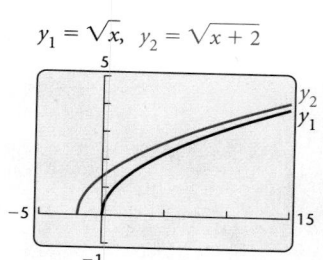

FIGURE 4

> ### Horizontal Translation
> For $d > 0$:
>
> the graph of $y = f(x - d)$ is the graph of $y = f(x)$ shifted *right* d units;
>
> the graph of $y = f(x + d)$ is the graph of $y = f(x)$ shifted *left* d units.

EXAMPLE 4 Graph each of the following. Before doing so, describe how each graph can be obtained from one of the basic graphs shown on the preceding pages.

a) $g(x) = x^2 - 6$ **b)** $g(x) = |x - 4|$
c) $g(x) = \sqrt{x + 2}$ **d)** $h(x) = \sqrt{x + 2} - 3$

Solution

a) To graph $g(x) = x^2 - 6$, think of the graph of $f(x) = x^2$. Since $g(x) = f(x) - 6$, the graph of $g(x) = x^2 - 6$ is the graph of $f(x) = x^2$, shifted, or translated, *down* 6 units. (See Fig. 1.)
Let's compare some points on the graphs of f and g.

Points on f: $(-3, 9),$ $(0, 0),$ $(2, 4)$

Corresponding
points on g: $(-3, 3),$ $(0, -6),$ $(2, -2)$

We observe that the y-coordinate of a point on the graph of g is 6 less than the corresponding y-coordinate on the graph of f.

b) To graph $g(x) = |x - 4|$, think of the graph of $f(x) = |x|$. Since $g(x) = f(x - 4)$, the graph of $g(x) = |x - 4|$ is the graph of $f(x) = |x|$ shifted *right* 4 units. (See Fig. 2.)
Let's again compare points on the two graphs.

Points on f: $(-4, 4),$ $(-2, 2),$ $(0, 0)$

Corresponding
points on g: $(0, 4),$ $(2, 2),$ $(4, 0)$

We note that the x-coordinate of a point on the graph of g is 4 more than the x-coordinate of the corresponding point on f.

c) To graph $g(x) = \sqrt{x + 2}$, think of the graph of $f(x) = \sqrt{x}$. Since $g(x) = f(x + 2)$, the graph of $g(x) = \sqrt{x + 2}$ is the graph of $f(x) = \sqrt{x}$, shifted *left* 2 units. (See Fig. 3.)

d) To graph $h(x) = \sqrt{x + 2} - 3$, think of the graph of $f(x) = \sqrt{x}$. In part (c), we found that the graph of $g(x) = \sqrt{x + 2}$ is the graph of $f(x) = \sqrt{x}$ shifted left 2 units. Since $h(x) = g(x) - 3$, we shift the graph of $g(x) = \sqrt{x + 2}$ *down* 3 units. Thus the graph of h is obtained by shifting the graph of $f(x) = \sqrt{x}$ *left* 2 units and *down* 3 units. (See Fig. 4.)

Now Try Exercises 49 and 59. ■

❋ Reflections

Suppose that we have a function given by $y = f(x)$. Let's explore the graphs of the new functions $y = -f(x)$ and $y = f(-x)$.

Compare the functions $y = f(x)$ and $y = -f(x)$ by observing the graphs of $y = \frac{1}{5}x^4$ and $y = -\frac{1}{5}x^4$ shown on the left below. What do you see? Test your observation with some other functions y_1 and y_2 where $y_2 = -y_1$.

Compare the functions $y = f(x)$ and $y = f(-x)$ by observing the graphs of $y = 2x^3 - x^4 + 5$ and $y = 2(-x)^3 - (-x)^4 + 5$ shown on the right below. What do you see? Test your observation with some other functions in which x is replaced with $-x$.

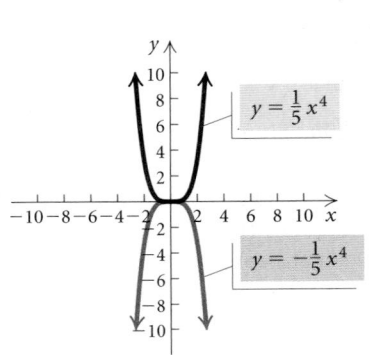

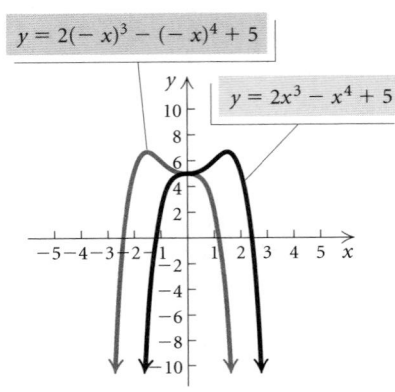

Given the graph of $y = f(x)$, we can reflect each point *across the x-axis* to obtain the graph of $y = -f(x)$. We can reflect each point of $y = f(x)$ *across the y-axis* to obtain the graph of $y = f(-x)$. The new graphs are called **reflections** of $y = f(x)$.

$y_1 = x^3 - 4x^2,$
$y_2 = (-x)^3 - 4(-x)^2$

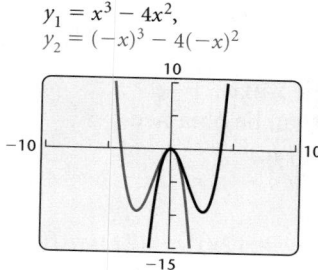

FIGURE 1

$y_1 = x^3 - 4x^2, \quad y_2 = 4x^2 - x^3$

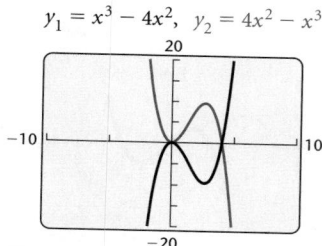

FIGURE 2

Reflections

The graph of $y = -f(x)$ is the **reflection** of the graph of $y = f(x)$ across the x-axis.

The graph of $y = f(-x)$ is the **reflection** of the graph of $y = f(x)$ across the y-axis.

If a point (x, y) is on the graph of $y = f(x)$, then $(x, -y)$ is on the graph of $y = -f(x)$, and $(-x, y)$ is on the graph of $y = f(-x)$.

EXAMPLE 5 Graph each of the following. Before doing so, describe how each graph can be obtained from the graph of $f(x) = x^3 - 4x^2$.

a) $g(x) = (-x)^3 - 4(-x)^2$ **b)** $h(x) = 4x^2 - x^3$

Solution

a) We first note that

$$f(-x) = (-x)^3 - 4(-x)^2 = g(x).$$

Thus the graph of g is a *reflection* of the graph of f across the y-axis. (See Fig. 1.) If (x, y) is on the graph of f, then $(-x, y)$ is on the graph of g. For example, $(2, -8)$ is on f and $(-2, -8)$ is on g.

b) We first note that

$$\begin{aligned} -f(x) &= -(x^3 - 4x^2) \\ &= -x^3 + 4x^2 \\ &= 4x^2 - x^3 \\ &= h(x). \end{aligned}$$

Thus the graph of h is a reflection of the graph of f across the x-axis. (See Fig. 2.) If (x, y) is on the graph of f, then $(x, -y)$ is on the graph of h. For example, $(2, -8)$ is on f and $(2, 8)$ is on h. ∎

❋ Vertical and Horizontal Stretchings and Shrinkings

Suppose that we have a function given by $y = f(x)$. Let's explore the graphs of the new functions $y = af(x)$ and $y = f(cx)$.

Consider the functions $y = f(x) = x^3 - x$, $y = \frac{1}{10}(x^3 - x) = \frac{1}{10}f(x)$, $y = 2(x^3 - x) = 2f(x)$, and $y = -2(x^3 - x) = -2f(x)$ and compare their graphs. What pattern do you observe? Test it with some other graphs.

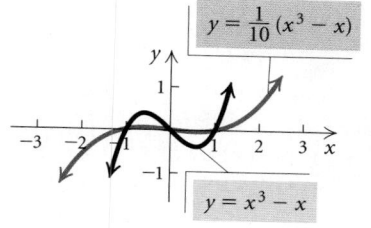

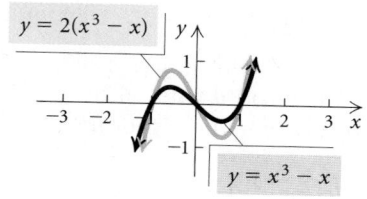

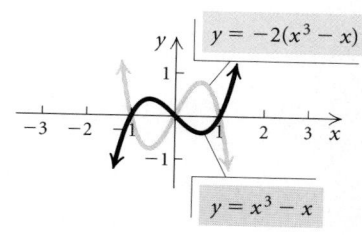

Consider any function f given by $y = f(x)$. Multiplying $f(x)$ by any constant a, where $|a| > 1$, to obtain $g(x) = af(x)$ will *stretch* the graph vertically away from the x-axis. If $0 < |a| < 1$, then the graph will be flattened or *shrunk* vertically toward the x-axis. If $a < 0$, the graph is also reflected across the x-axis.

Vertical Stretching and Shrinking

The graph of $y = af(x)$ can be obtained from the graph of $y = f(x)$ by

stretching vertically for $|a| > 1$, or

shrinking vertically for $0 < |a| < 1$.

For $a < 0$, the graph is also reflected across the x-axis. (The y-coordinates of the graph of $y = af(x)$ can be obtained by multiplying the y-coordinates of $y = f(x)$ by a.)

Consider the functions $y = f(x) = x^3 - x$, $y = (2x)^3 - (2x) = f(2x)$, $y = \left(\frac{1}{2}x\right)^3 - \left(\frac{1}{2}x\right) = f\left(\frac{1}{2}x\right)$, and $y = \left(-\frac{1}{2}x\right)^3 - \left(-\frac{1}{2}x\right) = f\left(-\frac{1}{2}x\right)$ and compare their graphs. What pattern do you observe? Test it with some other graphs.

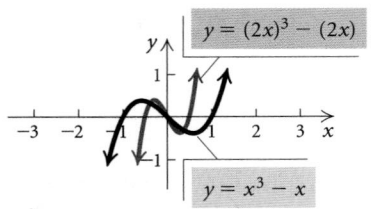

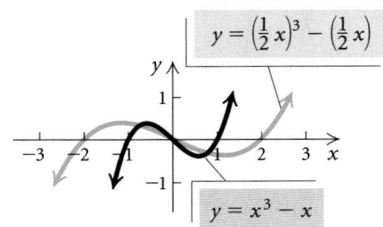

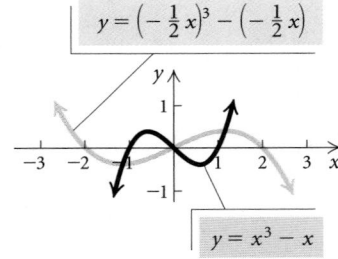

The constant c in the equation $g(x) = f(cx)$ will *shrink* the graph of $y = f(x)$ horizontally toward the y-axis if $|c| > 1$. If $0 < |c| < 1$, the graph will be *stretched* horizontally away from the y-axis. If $c < 0$, the graph is also reflected across the y-axis.

Horizontal Stretching and Shrinking

The graph of $y = f(cx)$ can be obtained from the graph of $y = f(x)$ by

shrinking horizontally for $|c| > 1$, or

stretching horizontally for $0 < |c| < 1$.

For $c < 0$, the graph is also reflected across the y-axis. (The x-coordinates of the graph of $y = f(cx)$ can be obtained by dividing the x-coordinates of the graph of $y = f(x)$ by c.)

It is instructive to use these concepts to create transformations of a given graph.

EXAMPLE 6 Shown at left is a graph of $y = f(x)$ for some function f. No formula for f is given. Graph each of the following.

a) $g(x) = 2f(x)$ **b)** $h(x) = \frac{1}{2}f(x)$

c) $r(x) = f(2x)$ **d)** $s(x) = f\left(\frac{1}{2}x\right)$

e) $t(x) = f\left(-\frac{1}{2}x\right)$

Solution

a) Since $|2| > 1$, the graph of $g(x) = 2f(x)$ is a vertical stretching of the graph of $y = f(x)$ by a factor of 2. We can consider the key points $(-5, 0)$, $(-2, 2)$, $(0, 0)$, $(2, -4)$, and $(4, 0)$ on the graph of $y = f(x)$. The transformation multiplies each y-coordinate by 2 to obtain the key points $(-5, 0)$, $(-2, 4)$, $(0, 0)$, $(2, -8)$, and $(4, 0)$ on the graph of $g(x) = 2f(x)$. The graph is shown below.

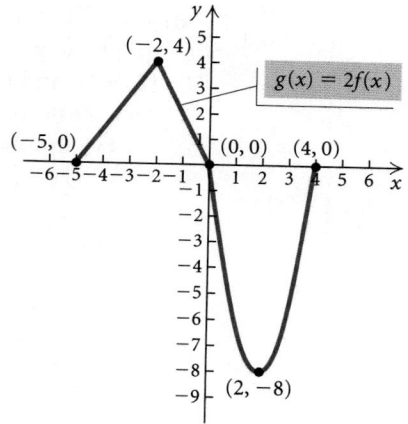

b) Since $\left|\frac{1}{2}\right| < 1$, the graph of $h(x) = \frac{1}{2}f(x)$ is a vertical shrinking of the graph of $y = f(x)$ by a factor of $\frac{1}{2}$. We again consider the key points $(-5, 0)$, $(-2, 2)$, $(0, 0)$, $(2, -4)$, and $(4, 0)$ on the graph of $y = f(x)$. The transformation multiplies each y-coordinate by $\frac{1}{2}$ to obtain the key points $(-5, 0)$, $(-2, 1)$, $(0, 0)$, $(2, -2)$, and $(4, 0)$ on the graph of $h(x) = \frac{1}{2}f(x)$. The graph is shown below.

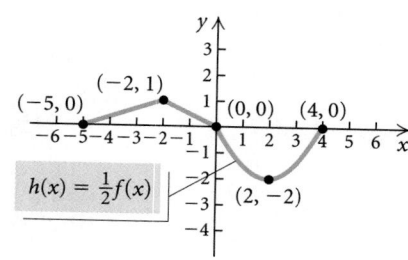

c) Since $|2| > 1$, the graph of $r(x) = f(2x)$ is a horizontal shrinking of the graph of $y = f(x)$. We consider the key points $(-5, 0)$, $(-2, 2)$, $(0, 0)$, $(2, -4)$, and $(4, 0)$ on the graph of $y = f(x)$. The transformation divides each x-coordinate by 2 to obtain the key points $(-2.5, 0)$, $(-1, 2)$, $(0, 0)$, $(1, -4)$, and $(2, 0)$ on the graph of $r(x) = f(2x)$. The graph is shown below.

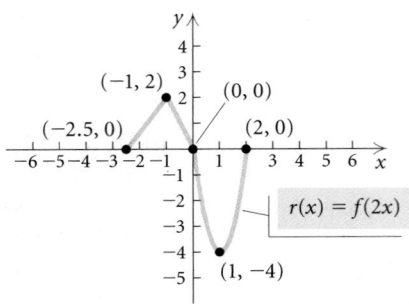

d) Since $\left|\frac{1}{2}\right| < 1$, the graph of $s(x) = f\left(\frac{1}{2}x\right)$ is a horizontal stretching of the graph of $y = f(x)$. We consider the key points $(-5, 0)$, $(-2, 2)$, $(0, 0)$, $(2, -4)$, and $(4, 0)$ on the graph of $y = f(x)$. The transformation divides each x-coordinate by $\frac{1}{2}$ (which is the same as multiplying by 2) to obtain the key points $(-10, 0)$, $(-4, 2)$, $(0, 0)$, $(4, -4)$, and $(8, 0)$ on the graph of $s(x) = f\left(\frac{1}{2}x\right)$. The graph is shown below.

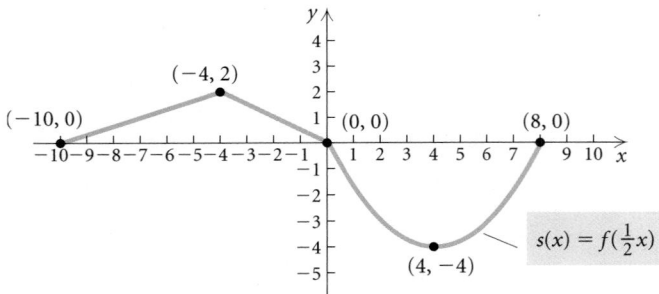

e) The graph of $t(x) = f\left(-\frac{1}{2}x\right)$ can be obtained by reflecting the graph in part (d) across the y-axis.

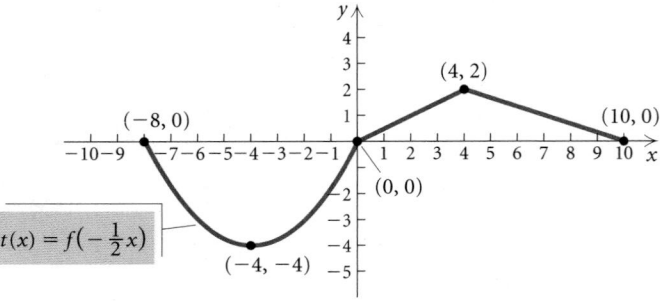

Now Try Exercises 107 and 109. ■

EXAMPLE 7 Use the graph of $y = f(x)$ shown at left to graph
$y = -2f(x - 3) + 1$.

Solution

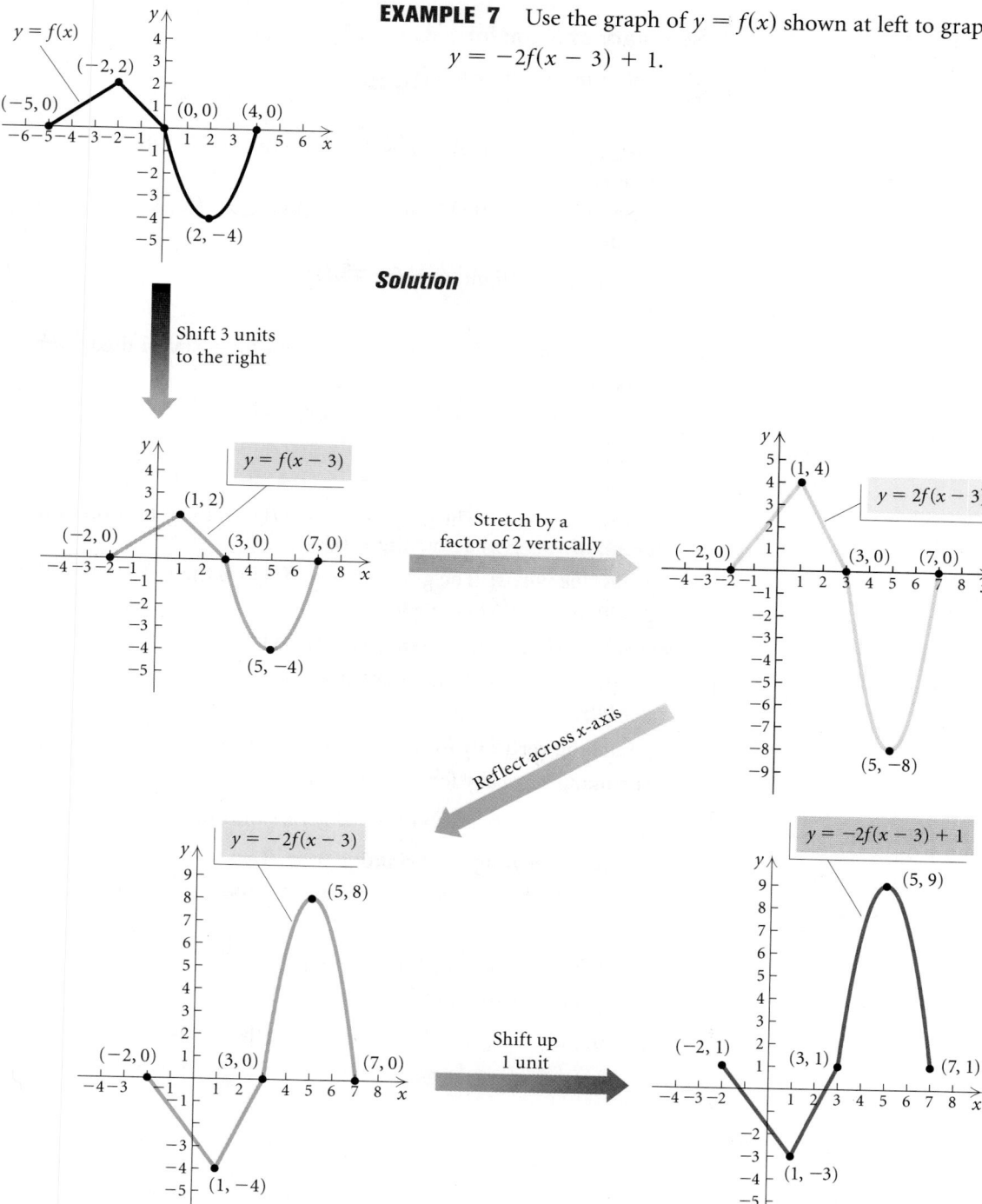

> ## Summary of Transformations of $y = f(x)$
>
> ### Vertical Translation: $y = f(x) \pm b$
> For $b > 0$:
>
> the graph of $y = f(x) + b$ is the graph of $y = f(x)$ shifted *up* b units;
>
> the graph of $y = f(x) - b$ is the graph of $y = f(x)$ shifted *down* b units.
>
> ### Horizontal Translation: $y = f(x \mp d)$
> For $d > 0$:
>
> the graph of $y = f(x - d)$ is the graph of $y = f(x)$ shifted *right* d units;
>
> the graph of $y = f(x + d)$ is the graph of $y = f(x)$ shifted *left* d units.
>
> ### Reflections
>
> *Across the x-axis*: The graph of $y = -f(x)$ is the reflection of the graph of $y = f(x)$ across the x-axis.
>
> *Across the y-axis*: The graph of $y = f(-x)$ is the reflection of the graph of $y = f(x)$ across the y-axis.
>
> ### Vertical Stretching or Shrinking: $y = af(x)$
> The graph of $y = af(x)$ can be obtained from the graph of $y = f(x)$ by
>
> stretching vertically for $|a| > 1$, or
>
> shrinking vertically for $0 < |a| < 1$.
>
> For $a < 0$, the graph is also reflected across the x-axis.
>
> ### Horizontal Stretching or Shrinking: $y = f(cx)$
> The graph of $y = f(cx)$ can be obtained from the graph of $y = f(x)$ by
>
> shrinking horizontally for $|c| > 1$, or
>
> stretching horizontally for $0 < |c| < 1$.
>
> For $c < 0$, the graph is also reflected across the y-axis.

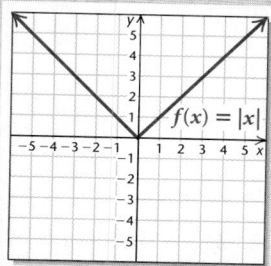

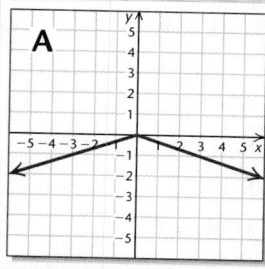

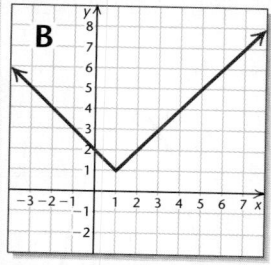

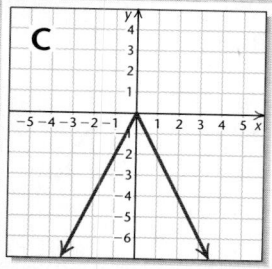

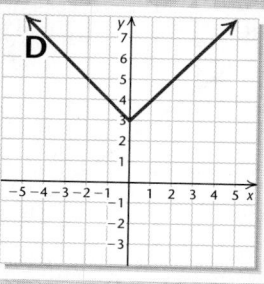

Visualizing the Graph

Match the function with its graph. Use transformation graphing techniques to obtain the graph of g from the basic function $f(x) = |x|$ shown at top left.

1. $g(x) = -2|x|$

2. $g(x) = |x - 1| + 1$

3. $g(x) = -\left|\frac{1}{3}x\right|$

4. $g(x) = |2x|$

5. $g(x) = |x + 2|$

6. $g(x) = |x| + 3$

7. $g(x) = -\frac{1}{2}|x - 4|$

8. $g(x) = \frac{1}{2}|x| - 3$

9. $g(x) = -|x| - 2$

Answers on page A-14

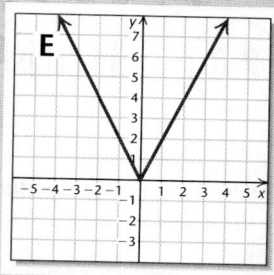

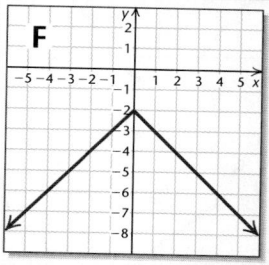

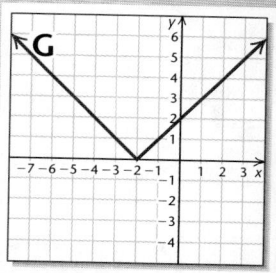

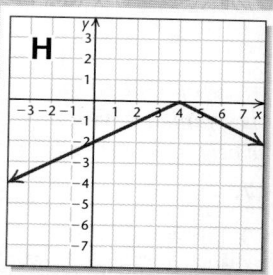

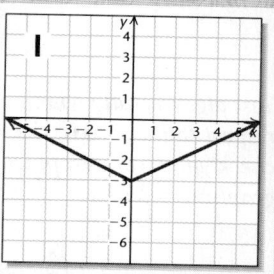

2.4   Exercise Set

Determine visually whether the graph is symmetric with respect to the x-axis, the y-axis, and the origin.

1.

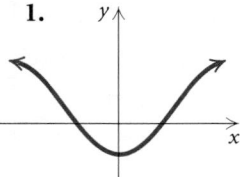

2.

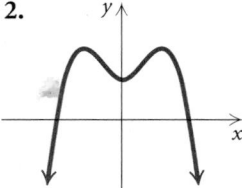

3.

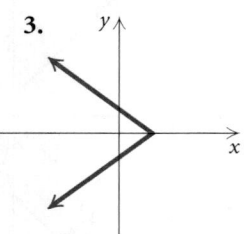

4.

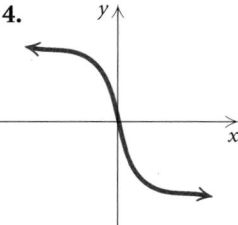

5.

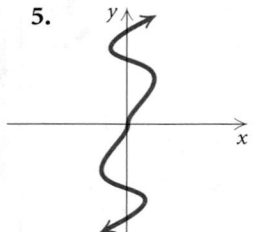

6.

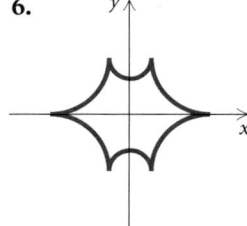

First, graph the equation and determine visually whether it is symmetric with respect to the x-axis, the y-axis, and the origin. Then verify your assertion algebraically.

7. $y = |x| - 2$

8. $y = |x + 5|$

9. $5y = 4x + 5$

10. $2x - 5 = 3y$

11. $5y = 2x^2 - 3$

12. $x^2 + 4 = 3y$

13. $y = \dfrac{1}{x}$

14. $y = -\dfrac{4}{x}$

Test algebraically whether the graph is symmetric with respect to the x-axis, the y-axis, and the origin. Then check your work graphically, if possible, using a graphing calculator.

15. $5x - 5y = 0$

16. $6x + 7y = 0$

17. $3x^2 - 2y^2 = 3$

18. $5y = 7x^2 - 2x$

19. $y = |2x|$

20. $y^3 = 2x^2$

21. $2x^4 + 3 = y^2$

22. $2y^2 = 5x^2 + 12$

23. $3y^3 = 4x^3 + 2$

24. $3x = |y|$

25. $xy = 12$

26. $xy - x^2 = 3$

Find the point that is symmetric to the given point with respect to the x-axis, the y-axis, and the origin.

27. $(-5, 6)$

28. $\left(\frac{7}{2}, 0\right)$

29. $(-10, -7)$

30. $\left(1, \frac{3}{8}\right)$

31. $(0, -4)$

32. $(8, -3)$

Determine visually whether the function is even, odd, or neither even nor odd.

33.

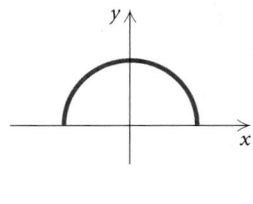

34.

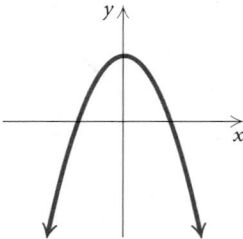

35.

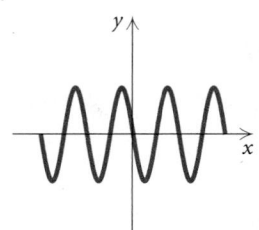

36.

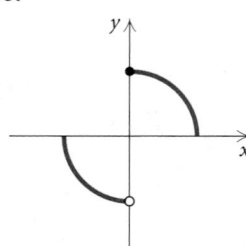

37.

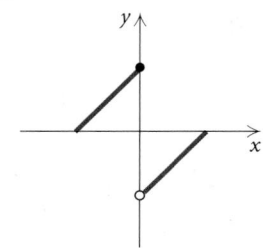

38.

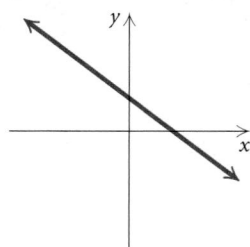

Test algebraically whether the function is even, odd, or neither even nor odd. Then check your work graphically, where possible, using a graphing calculator.

39. $f(x) = -3x^3 + 2x$

40. $f(x) = 7x^3 + 4x - 2$

41. $f(x) = 5x^2 + 2x^4 - 1$

42. $f(x) = x + \dfrac{1}{x}$

43. $f(x) = x^{17}$

44. $f(x) = \sqrt[3]{x}$

45. $f(x) = x - |x|$

46. $f(x) = \dfrac{1}{x^2}$

47. $f(x) = 8$

48. $f(x) = \sqrt{x^2 + 1}$

Describe how the graph of the function can be obtained from one of the basic graphs on p. 203. Then graph the function by hand or with a graphing calculator.

49. $f(x) = (x - 3)^2$

50. $g(x) = x^2 + \frac{1}{2}$

51. $g(x) = x - 3$

52. $g(x) = -x - 2$

53. $h(x) = -\sqrt{x}$

54. $g(x) = \sqrt{x - 1}$

55. $h(x) = \dfrac{1}{x} + 4$

56. $g(x) = \dfrac{1}{x - 2}$

57. $h(x) = -3x + 3$

58. $f(x) = 2x + 1$

59. $h(x) = \frac{1}{2}|x| - 2$

60. $g(x) = -|x| + 2$

61. $g(x) = -(x - 2)^3$

62. $f(x) = (x + 1)^3$

63. $g(x) = (x + 1)^2 - 1$

64. $h(x) = -x^2 - 4$

65. $g(x) = \frac{1}{3}x^3 + 2$

66. $h(x) = (-x)^3$

67. $f(x) = \sqrt{x + 2}$

68. $f(x) = -\frac{1}{2}\sqrt{x - 1}$

69. $f(x) = \sqrt[3]{x} - 2$

70. $h(x) = \sqrt[3]{x + 1}$

Describe how the graph of the function can be obtained from one of the basic graphs on p. 203.

71. $g(x) = |3x|$

72. $f(x) = \frac{1}{2}\sqrt[3]{x}$

73. $h(x) = \dfrac{2}{x}$

74. $f(x) = |x - 3| - 4$

75. $f(x) = 3\sqrt{x} - 5$

76. $f(x) = 5 - \dfrac{1}{x}$

77. $g(x) = \left|\frac{1}{3}x\right| - 4$

78. $f(x) = \frac{2}{3}x^3 - 4$

79. $f(x) = -\frac{1}{4}(x - 5)^2$

80. $f(x) = (-x)^3 - 5$

81. $f(x) = \dfrac{1}{x + 3} + 2$

82. $g(x) = \sqrt{-x} + 5$

83. $h(x) = -(x - 3)^2 + 5$

84. $f(x) = 3(x + 4)^2 - 3$

The point $(-12, 4)$ is on the graph of $y = f(x)$. Find the corresponding point on the graph of $y = g(x)$.

85. $g(x) = \frac{1}{2}f(x)$

86. $g(x) = f(x - 2)$

87. $g(x) = f(-x)$

88. $g(x) = f(4x)$

89. $g(x) = f(x) - 2$

90. $g(x) = f(\frac{1}{2}x)$

91. $g(x) = 4f(x)$

92. $g(x) = -f(x)$

Given that $f(x) = x^2 + 3$, match the function g with a transformation of f from one of A–D.

93. $g(x) = x^2 + 4$ **A.** $f(x - 2)$

94. $g(x) = 9x^2 + 3$ **B.** $f(x) + 1$

95. $g(x) = (x - 2)^2 + 3$ **C.** $2f(x)$

96. $g(x) = 2x^2 + 6$ **D.** $f(3x)$

Write an equation for a function that has a graph with the given characteristics.

97. The shape of $y = x^2$, but upside-down and shifted right 8 units

98. The shape of $y = \sqrt{x}$, but shifted left 6 units and down 5 units

99. The shape of $y = |x|$, but shifted left 7 units and up 2 units

100. The shape of $y = x^3$, but upside-down and shifted right 5 units

101. The shape of $y = 1/x$, but shrunk horizontally by a factor of 2 and shifted down 3 units

102. The shape of $y = x^2$, but shifted right 6 units and up 2 units

103. The shape of $y = x^2$, but upside-down and shifted right 3 units and up 4 units

104. The shape of $y = |x|$, but stretched horizontally by a factor of 2 and shifted down 5 units

105. The shape of $y = \sqrt{x}$, but reflected across the y-axis and shifted left 2 units and down 1 unit

106. The shape of $y = 1/x$, but reflected across the x-axis and shifted up 1 unit

A graph of $y = f(x)$ follows. No formula for f is given. In Exercises 107–114, graph the given equation.

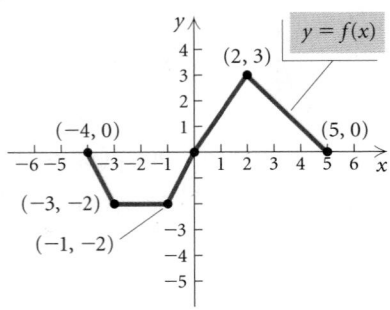

107. $g(x) = -2f(x)$ **108.** $g(x) = \frac{1}{2}f(x)$

109. $g(x) = f\left(-\frac{1}{2}x\right)$ **110.** $g(x) = f(2x)$

111. $g(x) = -\frac{1}{2}f(x - 1) + 3$

112. $g(x) = -3f(x + 1) - 4$

113. $g(x) = f(-x)$

114. $g(x) = -f(x)$

A graph of $y = g(x)$ follows. No formula for g is given. In Exercises 115–118, graph the given equation.

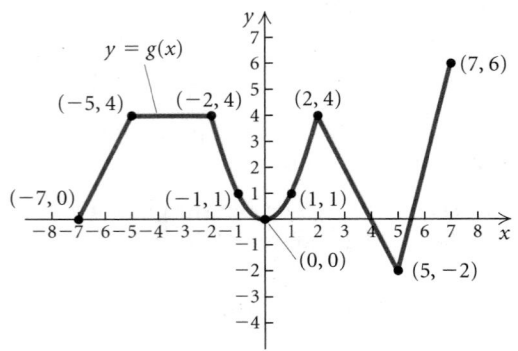

115. $h(x) = -g(x + 2) + 1$

116. $h(x) = \frac{1}{2}g(-x)$

117. $h(x) = g(2x)$

118. $h(x) = 2g(x - 1) - 3$

The graph of the function f is shown in figure (a). In Exercises 119–126, match the function g with one of the graphs (a)–(h), which follow. Some graphs may be used more than once and some may not be used at all.

a)
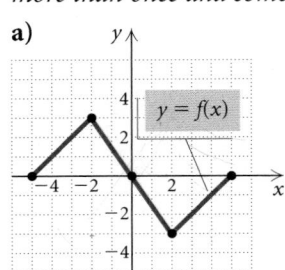

b)
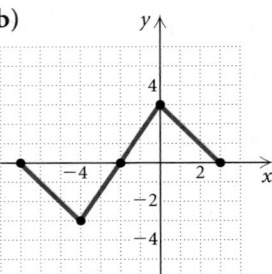

c)
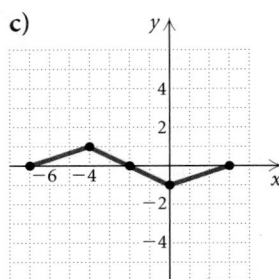

d)
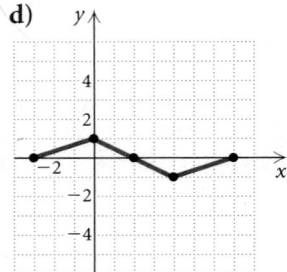

e)
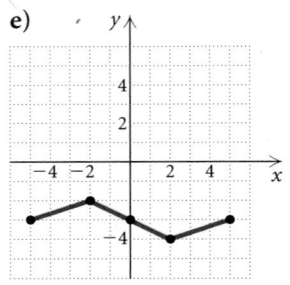

f)
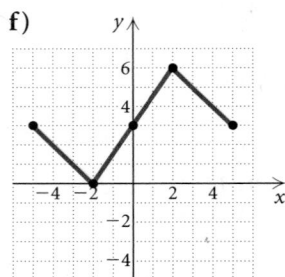

g)
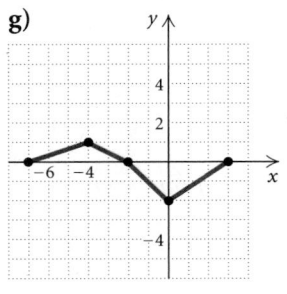

h)
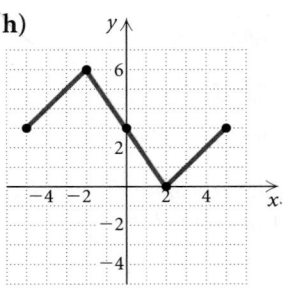

119. $g(x) = f(-x) + 3$ **120.** $g(x) = f(x) + 3$

121. $g(x) = -f(x) + 3$ **122.** $g(x) = -f(-x)$

123. $g(x) = \frac{1}{3}f(x - 2)$ **124.** $g(x) = \frac{1}{3}f(x) - 3$

125. $g(x) = \frac{1}{3}f(x + 2)$ **126.** $g(x) = -f(x + 2)$

For each pair of functions, determine if $g(x) = f(-x)$.

127. $f(x) = 2x^4 - 35x^3 + 3x - 5$,
$g(x) = 2x^4 + 35x^3 - 3x - 5$

128. $f(x) = \frac{1}{4}x^4 + \frac{1}{5}x^3 - 81x^2 - 17$,
$g(x) = \frac{1}{4}x^4 + \frac{1}{5}x^3 + 81x^2 - 17$

A graph of the function $f(x) = x^3 - 3x^2$ is shown below. Exercises 129–132 show graphs of functions transformed from this one. Find a formula for each function.

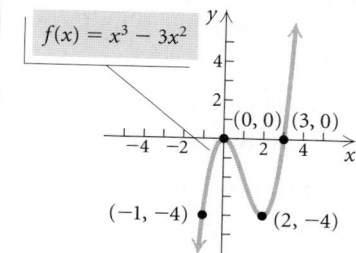

129.

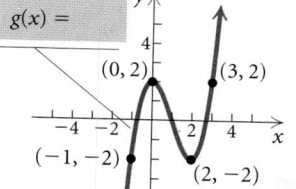

130.

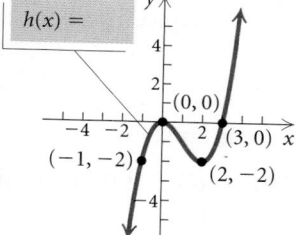

131.

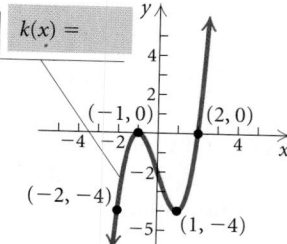

132.

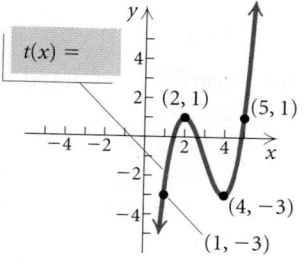

Collaborative Discussion and Writing

133. Consider the constant function $f(x) = 0$. Determine whether the graph of this function is symmetric with respect to the x-axis, the y-axis, and/or the origin. Determine whether this function is even or odd.

134. Describe conditions under which you would know whether a polynomial function $f(x) = a_n x^n + a_{n-1} x^{n-1} + \cdots + a_2 x^2 + a_1 x + a_0$ is even or odd without using an algebraic procedure. Explain.

135. Explain in your own words why the graph of $y = f(-x)$ is a reflection of the graph of $y = f(x)$ across the y-axis.

136. Without drawing the graph, describe what the graph of $f(x) = |x^2 - 9|$ looks like.

Skill Maintenance

137. Given $f(x) = 5x^2 - 7$, find each of the following.
a) $f(-3)$
b) $f(3)$
c) $f(a)$
d) $f(-a)$

138. Given $f(x) = 4x^3 - 5x$, find each of the following.
a) $f(2)$
b) $f(-2)$
c) $f(a)$
d) $f(-a)$

139. Write an equation of the line perpendicular to the graph of the line $8x - y = 10$ and containing the point $(-1, 1)$.

140. Find the slope and the y-intercept of the line with equation $2x - 9y + 1 = 0$.

Synthesis

Use the graph of the function f shown below in Exercises 141 and 142.

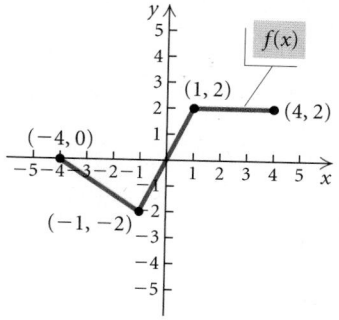

141. Graph: $y = |f(x)|$. **142.** Graph: $y = f(|x|)$.

Use the graph of the function g shown below in Exercises 143 and 144.

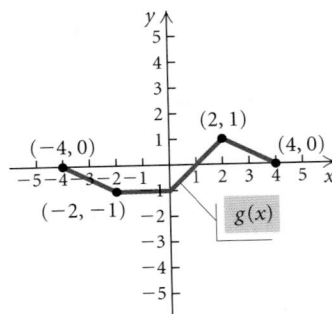

143. Graph: $y = g(|x|)$.   **144.** Graph: $y = |g(x)|$.

Graph each of the following using a graphing calculator. Before doing so, describe how the graph can be obtained from a more basic graph. Give the domain and the range of the function.

145. $f(x) = \left[\!\left[x - \tfrac{1}{2} \right]\!\right]$ **146.** $f(x) = \left| \sqrt{x} - 1 \right|$

Determine whether the function is even, odd, or neither even nor odd.

147. $f(x) = x\sqrt{10 - x^2}$ **148.** $f(x) = \dfrac{x^2 + 1}{x^3 - 1}$

Determine whether the graph is symmetric with respect to the x-axis, the y-axis, and the origin.

149. $x^3 = y^2(2 - x)$ **150.** $(x^2 + y^2)^2 = 2xy$

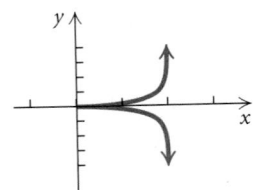 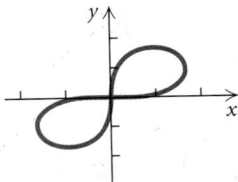

151. $y^2 + 4xy^2 - y^4 =$
$\quad x^4 - 4x^3 + 3x^2 + 2x^2y^2$

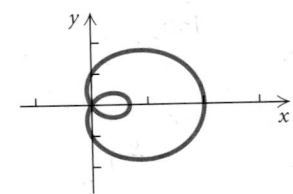

152. The graph of $f(x) = |x|$ passes through the points $(-3, 3)$, $(0, 0)$, and $(3, 3)$. Transform this function to one whose graph passes through the points $(5, 1)$, $(8, 4)$, and $(11, 1)$.

153. If $(-1, 5)$ is a point on the graph of $y = f(x)$, find b such that $(2, b)$ is on the graph of $y = f(x - 3)$.

154. Find the zeros of $f(x) = 3x^5 - 20x^3$. Then, without using a graphing calculator, state the zeros of $f(x - 3)$ and $f(x + 8)$.

155. If $(3, 4)$ is a point on the graph of $y = f(x)$, what point do you know is on the graph of $y = 2f(x)$? of $y = 2 + f(x)$? of $y = f(2x)$?

State whether each of the following is true or false.

156. The product of two odd functions is odd.

157. The sum of two even functions is even.

158. The product of an even function and an odd function is odd.

159. Show that if f is *any* function, then the function E defined by

$$E(x) = \frac{f(x) + f(-x)}{2}$$

is even.

160. Show that if f is *any* function, then the function O defined by

$$O(x) = \frac{f(x) - f(-x)}{2}$$

is odd.

161. Consider the functions E and O of Exercises 159 and 160.
   a) Show that $f(x) = E(x) + O(x)$. This means that every function can be expressed as the sum of an even function and an odd function.
   b) Let $f(x) = 4x^3 - 11x^2 + \sqrt{x} - 10$. Express f as a sum of an even function and an odd function.

2.5 Variation and Applications

❖ Find equations of direct variation, inverse variation, and combined variation given values of the variables.

❖ Solve applied problems involving variation.

We now extend our study of formulas and functions by considering applications involving variation.

❖ Direct Variation

Suppose an experienced auto mechanic earns $15 per hour. In 1 hr, $15 is earned; in 2 hr, $30 is earned; in 3 hr, $45 is earned; and so on. This gives rise to a set of ordered pairs:

$$(1, 15), \quad (2, 30), \quad (3, 45), \quad (4, 60), \quad \text{and so on.}$$

Auto Mechanic's Earnings

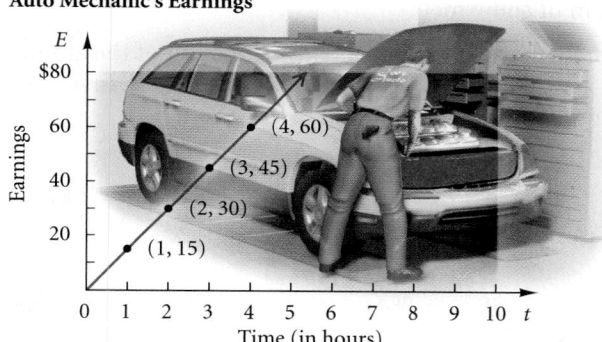

Note that the ratio of the second coordinate to the first coordinate is the same number for each pair:

$$\frac{15}{1} = 15, \quad \frac{30}{2} = 15, \quad \frac{45}{3} = 15,$$

$$\frac{60}{4} = 15, \quad \text{and so on.}$$

Whenever a situation produces pairs of numbers in which the *ratio is constant*, we say that there is **direct variation**. Here the amount earned E varies directly as the time worked t:

$$\frac{E}{t} = 15 \text{ (a constant)}, \quad \text{or} \quad E = 15t,$$

or, using function notation, $E(t) = 15t$. This equation is an equation of **direct variation**. The coefficient, 15, is called the **variation constant**. In this case, it is the rate of change of earnings with respect to time.

The graph of $y = kx$, $k > 0$, always goes through the origin and rises from left to right. Note that as x increases, y increases; that is, the function is increasing on the interval $(0, \infty)$. The constant k is also the slope of the line.

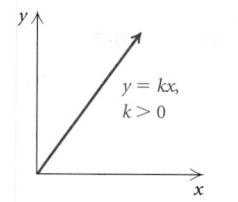

Direct Variation

If a situation gives rise to a linear function $f(x) = kx$, or $y = kx$, where k is a positive constant, we say that we have **direct variation**, or that y **varies directly as** x, or that y **is directly proportional to** x. The number k is called the **variation constant**, or **constant of proportionality**.

EXAMPLE 1 Find the variation constant and an equation of variation in which y varies directly as x, and $y = 32$ when $x = 2$.

Solution We know that $(2, 32)$ is a solution of $y = kx$. Thus,

$$y = kx$$
$$32 = k \cdot 2 \qquad \text{Substituting}$$
$$\frac{32}{2} = k \qquad \text{Solving for } k$$
$$16 = k. \qquad \text{Simplifying}$$

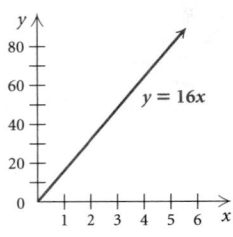

The variation constant, 16, is the rate of change of y with respect to x. The equation of variation is $y = 16x$.

Now Try Exercise 1. ◼

EXAMPLE 2 *Water from Melting Snow.* The number of centimeters W of water produced from melting snow varies directly as S, the number of centimeters of snow. Meteorologists have found that under certain conditions 150 cm of snow will melt to 16.8 cm of water. To how many centimeters of water will 200 cm of snow melt?

Solution We can express the amount of water as a function of the amount of snow. Thus, $W(S) = kS$, where k is the variation constant. We first find k using the given data and then find an equation of variation:

$$W(S) = kS \qquad W \text{ varies directly as } S.$$
$$W(150) = k \cdot 150 \qquad \text{Substituting 150 for } S$$
$$16.8 = k \cdot 150 \qquad \text{Replacing } W(150) \text{ with 16.8}$$
$$\frac{16.8}{150} = k \qquad \text{Solving for } k$$
$$0.112 = k. \qquad \text{This is the variation constant.}$$

The equation of variation is $W(S) = 0.112S$.

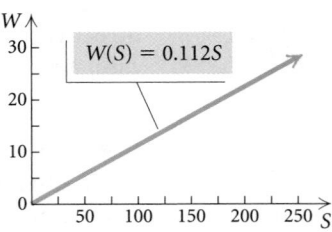

Next, we use the equation to find how many centimeters of water will result from melting 200 cm of snow:

$$W(S) = 0.112S$$
$$W(200) = 0.112(200) \qquad \text{Substituting}$$
$$= 22.4.$$

Thus, 200 cm of snow will melt to 22.4 cm of water.

Now Try Exercise 15. ◼

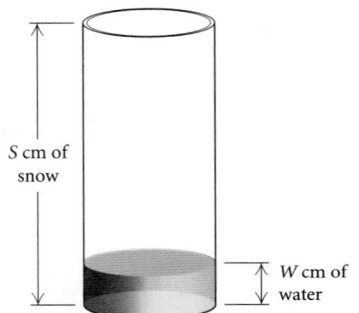

S cm of snow

W cm of water

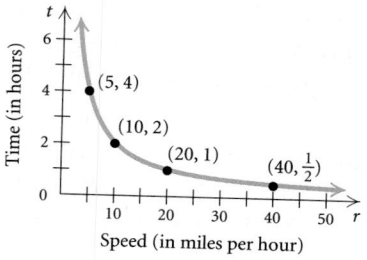

Time (in hours)

Speed (in miles per hour)

(5, 4)
(10, 2)
(20, 1)
$(40, \frac{1}{2})$

❊ Inverse Variation

Suppose a bus is traveling a distance of 20 mi. At a speed of 5 mph, the trip will take 4 hr; at 10 mph, it will take 2 hr; at 20 mph, it will take 1 hr; at 40 mph, it will take $\frac{1}{2}$ hr; and so on. We plot this information on a graph, using speed as the first coordinate and time as the second coordinate to determine a set of ordered pairs:

$$(5, 4), \quad (10, 2), \quad (20, 1), \quad \left(40, \tfrac{1}{2}\right), \quad \text{and so on.}$$

Note that the products of the coordinates are all the same number:

$$5 \cdot 4 = 20, \quad 10 \cdot 2 = 20, \quad 20 \cdot 1 = 20, \quad 40 \cdot \tfrac{1}{2} = 20, \quad \text{and so on.}$$

Whenever a situation produces pairs of numbers in which the *product is constant*, we say that there is **inverse variation**. Here the time varies inversely as the speed:

$$rt = 20 \text{ (a constant),} \quad \text{or} \quad t = \frac{20}{r},$$

or, using function notation, $t(r) = 20/r$. This equation is an equation of **inverse variation**. The coefficient, 20, is called the **variation constant**. Note that as the first number increases, the second number decreases.

The graph of $y = k/x$, $k > 0$, is like the one shown below. Note that as x increases, y decreases; that is, the function is decreasing on the interval $(0, \infty)$.

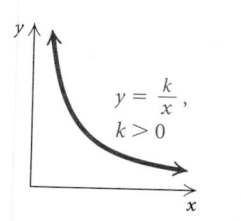

$y = \dfrac{k}{x}$, $k > 0$

Inverse Variation

If a situation gives rise to a function $f(x) = k/x$, or $y = k/x$, where k is a positive constant, we say that we have **inverse variation**, or that **y varies inversely as x**, or that **y is inversely proportional to x**. The number k is called the **variation constant**, or **constant of proportionality**.

EXAMPLE 3 Find the variation constant and an equation of variation in which y varies inversely as x, and $y = 16$ when $x = 0.3$.

Solution We know that $(0.3, 16)$ is a solution of $y = k/x$. We substitute:

$$y = \frac{k}{x}$$

$$16 = \frac{k}{0.3} \qquad \text{Substituting}$$

$$(0.3)16 = k \qquad \text{Solving for } k$$

$$4.8 = k.$$

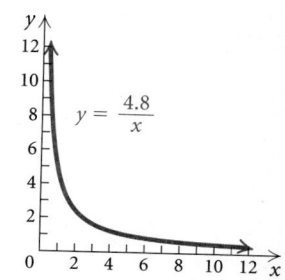

$y = \dfrac{4.8}{x}$

The variation constant is 4.8. The equation of variation is $y = 4.8/x$.

Now Try Exercise 3. ■

There are many problems that translate to an equation of inverse variation.

EXAMPLE 4 *Framing a House.* The time t required to do a job varies inversely as the number of people P who work on the job (assuming that all work at the same rate). If it takes 72 hr for 9 people to frame a house, how long will it take 12 people to complete the same job?

Solution   We can express the amount of time required, in hours, as a function of the number of people working. Thus we have $t(P) = k/P$. We first find k using the given information and then find an equation of variation:

$$t(P) = \frac{k}{P} \qquad \text{t varies inversely as P.}$$

$$t(9) = \frac{k}{9} \qquad \text{Substituting 9 for P}$$

$$72 = \frac{k}{9} \qquad \text{Replacing $t(9)$ with 72}$$

$$9 \cdot 72 = k \qquad \text{Solving for k}$$

$$648 = k. \qquad \text{This is the variation constant.}$$

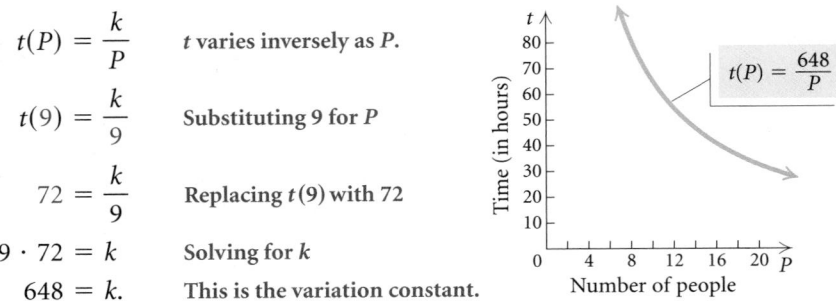

The equation of variation is $t(P) = 648/P$.

Next, we use the equation to find the time that it would take 12 people to do the job. We compute $t(12)$:

$$t(P) = \frac{648}{P}$$

$$t(12) = \frac{648}{12} \qquad \text{Substituting}$$

$$t = 54.$$

Thus it would take 54 hr for 12 people to complete the job.

Now Try Exercise 13. ■

❖ Combined Variation

We now look at other kinds of variation.

> y varies **directly as the nth power of x** if there is some positive constant k such that
>
> $$y = kx^n.$$
>
> y varies **inversely as the nth power of x** if there is some positive constant k such that
>
> $$y = \frac{k}{x^n}.$$
>
> y varies **jointly as x and z** if there is some positive constant k such that
>
> $$y = kxz.$$

There are other types of combined variation as well. Consider the formula for the volume of a right circular cylinder, $V = \pi r^2 h$, in which V, r, and h are variables and π is a constant. We say that V varies jointly as h and the square of r. In this formula, π is the variation constant.

EXAMPLE 5 Find an equation of variation in which y varies directly as the square of x, and $y = 12$ when $x = 2$.

Solution We write an equation of variation and find k:

$$y = kx^2$$
$$12 = k \cdot 2^2 \quad \text{Substituting}$$
$$12 = k \cdot 4$$
$$3 = k.$$

Thus, $y = 3x^2$.

Now Try Exercise 25. ■

EXAMPLE 6 Find an equation of variation in which y varies jointly as x and z, and $y = 42$ when $x = 2$ and $z = 3$.

Solution We have

$$y = kxz$$
$$42 = k \cdot 2 \cdot 3 \quad \text{Substituting}$$
$$42 = k \cdot 6$$
$$7 = k.$$

Thus, $y = 7xz$.

Now Try Exercise 29. ■

EXAMPLE 7 Find an equation of variation in which y varies jointly as x and z and inversely as the square of w, and $y = 105$ when $x = 3$, $z = 20$, and $w = 2$.

Solution We have

$$y = k \cdot \frac{xz}{w^2}$$
$$105 = k \cdot \frac{3 \cdot 20}{2^2} \quad \text{Substituting}$$
$$105 = k \cdot 15$$
$$7 = k.$$

Thus, $y = 7\dfrac{xz}{w^2}$.

Now Try Exercise 33. ■

Many applied problems can be modeled using equations of combined variation.

EXAMPLE 8 *Volume of a Tree.* The volume of wood V in a tree varies jointly as the height h and the square of the girth g. (Girth is distance around.) If the volume of a redwood tree is 216 m^3 when the height is 30 m and the girth is 1.5 m, what is the height of a tree whose volume is 960 m^3 and whose girth is 2 m?

Solution We first find k using the first set of data. Then we solve for h using the second set of data.

$$V = khg^2$$
$$216 = k \cdot 30 \cdot 1.5^2$$
$$216 = k \cdot 30 \cdot 2.25$$
$$216 = k \cdot 67.5$$
$$3.2 = k$$

Then the equation of variation is $V = 3.2hg^2$. We substitute the second set of data into the equation:

$$960 = 3.2 \cdot h \cdot 2^2$$
$$960 = 12.8 \cdot h$$
$$75 = h.$$

The height of the tree is 75 m.

Now Try Exercise 35.

Exercise Set

2.5

Find the variation constant and an equation of variation for the given situation.

1. y varies directly as x, and $y = 54$ when $x = 12$

2. y varies directly as x, and $y = 0.1$ when $x = 0.2$

3. y varies inversely as x, and $y = 3$ when $x = 12$

4. y varies inversely as x, and $y = 12$ when $x = 5$

5. y varies directly as x, and $y = 1$ when $x = \frac{1}{4}$

6. y varies inversely as x, and $y = 0.1$ when $x = 0.5$

7. y varies inversely as x, and $y = 32$ when $x = \frac{1}{8}$

8. y varies directly as x, and $y = 3$ when $x = 33$

9. y varies directly as x, and $y = \frac{3}{4}$ when $x = 2$

10. y varies inversely as x, and $y = \frac{1}{5}$ when $x = 35$

11. y varies inversely as x, and $y = 1.8$ when $x = 0.3$

12. y varies directly as x, and $y = 0.9$ when $x = 0.4$

13. *Work Rate.* The time T required to do a job varies inversely as the number of people P working. It takes 5 hr for 7 bricklayers to build a park wall. (See the graph below.) How long will it take 10 bricklayers to complete the job?

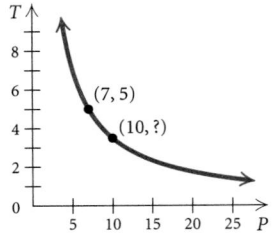

14. *Weekly Allowance.* According to Fidelity Investments *Investment Vision Magazine*, the average weekly allowance A of children varies directly as their grade level G. It is known that the average allowance of a 5th-grade student is $5.88 per week. What then is the average allowance of a 9th-grade student?

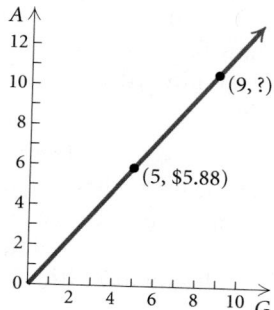

15. *Fat Intake.* The maximum number of grams of fat that should be in a diet varies directly as a person's weight. A person weighing 120 lb should have no more than 60 g of fat per day. What is the maximum daily fat intake for a person weighing 180 lb?

16. *Rate of Travel.* The time t required to drive a fixed distance varies inversely as the speed r. It takes 5 hr at a speed of 80 km/h to drive a fixed distance. How long will it take to drive the same distance at a speed of 70 km/h?

17. *Beam Weight.* The weight W that a horizontal beam can support varies inversely as the length L of the beam. Suppose an 8-m beam can support 1200 kg. How many kilograms can a 14-m beam support?

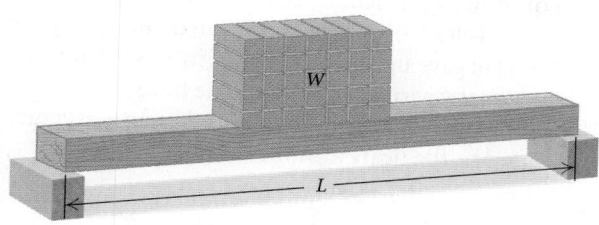

18. *House of Representatives.* The number of representatives N that each state has varies directly as the number of people P living in the state. If New York, with 19,254,630 residents, has 29 representatives, how many representatives does Colorado, with a population of 4,665,177, have?

19. *Weight on Mars.* The weight M of an object on Mars varies directly as its weight E on Earth. A person who weighs 95 lb on Earth weighs 38 lb on Mars. How much would a 100-lb person weigh on Mars?

20. *Pumping Rate.* The time t required to empty a tank varies inversely as the rate r of pumping. If a pump can empty a tank in 45 min at the rate of 600 kL/min, how long will it take the pump to empty the same tank at the rate of 1000 kL/min?

21. *Hooke's Law.* Hooke's law states that the distance d that a spring will stretch varies directly as the mass m of an object hanging from the spring. If a 3-kg mass stretches a spring 40 cm, how far will a 5-kg mass stretch the spring?

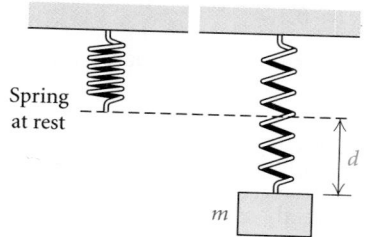

22. *Relative Aperture.* The relative aperture, or f-stop, of a 23.5-mm diameter lens is directly proportional to the focal length F of the lens. If a 150-mm focal length has an f-stop of 6.3, find the f-stop of a 23.5-mm diameter lens with a focal length of 80 mm.

23. *Musical Pitch.* The pitch P of a musical tone varies inversely as its wavelength W. One tone has a pitch of 330 vibrations per second and a wavelength of 3.2 ft. Find the wavelength of another tone that has a pitch of 550 vibrations per second.

24. *Recycling Rechargeable Batteries.*   The Indiana Household Hazardous Waste Task Force was awarded the 2002 National Community Recycling Leadership Award, with special recognition given to Monroe County Solid Waste Management District. Monroe County, Indiana, with a population of 9880, collected 4445 lb of rechargeable batteries in a recent recycling effort. (*Source*: Rechargeable Battery Recycling Corporation) If the number of pounds collected varies directly as the population, how many pounds of rechargeable batteries could a county with a population of 74,650 collect for recycling?

Find an equation of variation for the given situation.

25. y varies inversely as the square of x, and $y = 0.15$ when $x = 0.1$

26. y varies inversely as the square of x, and $y = 6$ when $x = 3$

27. y varies directly as the square of x, and $y = 0.15$ when $x = 0.1$

28. y varies directly as the square of x, and $y = 6$ when $x = 3$

29. y varies jointly as x and z, and $y = 56$ when $x = 7$ and $z = 8$

30. y varies directly as x and inversely as z, and $y = 4$ when $x = 12$ and $z = 15$

31. y varies jointly as x and the square of z, and $y = 105$ when $x = 14$ and $z = 5$

32. y varies jointly as x and z and inversely as w, and $y = \frac{3}{2}$ when $x = 2$, $z = 3$, and $w = 4$

33. y varies jointly as x and z and inversely as the product of w and p, and $y = \frac{3}{28}$ when $x = 3$, $z = 10$, $w = 7$, and $p = 8$

34. y varies jointly as x and z and inversely as the square of w, and $y = \frac{12}{5}$ when $x = 16$, $z = 3$, and $w = 5$

35. *Intensity of Light.*   The intensity I of light from a light bulb varies inversely as the square of the distance d from the bulb. Suppose that I is 90 W/m^2 (watts per square meter) when the distance is 5 m. How much *farther* would it be to a point where the intensity is 40 W/m^2?

36. *Atmospheric Drag.*   Wind resistance, or atmospheric drag, tends to slow down moving objects. Atmospheric drag varies jointly as an object's surface area A and velocity v. If a car traveling at a speed of 40 mph with a surface area of 37.8 ft^2 experiences a drag of 222 N (Newtons), how fast must a car with 51 ft^2 of surface area travel in order to experience a drag force of 430 N?

37. *Stopping Distance of a Car.*   The stopping distance d of a car after the brakes have been applied varies directly as the square of the speed r. If a car traveling 60 mph can stop in 200 ft, how fast can a car travel and still stop in 72 ft?

38. *Weight of an Astronaut.*   The weight W of an object varies inversely as the square of the distance d from the center of the earth. At sea level (3978 mi from the center of the earth), an astronaut weighs 220 lb. Find his weight when he is 200 mi above the surface of the earth.

39. *Earned-Run Average.*   A pitcher's earned-run average E varies directly as the number R of earned runs allowed and inversely as the number I of innings pitched. In 2006, Brad Penny of the Los Angeles Dodgers had an earned-run average of 4.33. He gave up 91 earned runs in 189.0 innings. How many earned runs would he have given up had he pitched 225 innings with the same average? Round to the nearest whole number.

40. *Boyle's Law.*   The volume V of a given mass of a gas varies directly as the temperature T and inversely as the pressure P. If $V = 231$ cm^3 when $T = 42°$ and $P = 20$ kg/cm^2, what is the volume when $T = 30°$ and $P = 15$ kg/cm^2?

Collaborative Discussion and Writing

41. If y varies directly as x^2, explain why doubling x would not cause y to be doubled as well.

42. If y varies directly as x and x varies inversely as z, how does y vary with regard to z? Why?

Skill Maintenance

43. Graph: $f(x) = \begin{cases} x - 2, & \text{for } x \le -1, \\ 3, & \text{for } -1 < x \le 2, \\ x, & \text{for } x > 2. \end{cases}$

Determine algebraically whether the graph is symmetric with respect to the x-axis, the y-axis, and the origin.

44. $y = 3x^4 - 3$

45. $y^2 = x$

46. $2x - 5y = 0$

Synthesis

47. *Volume and Cost.* An 18-oz jar of peanut butter in the shape of a right circular cylinder is 5 in. high and 3 in. in diameter and sells for $1.80. In the same store, a 28-oz jar of the same brand is $5\frac{1}{2}$ in. high and $3\frac{1}{4}$ in. in diameter. If the cost is directly proportional to volume, what should the price of the larger jar be? If the cost is directly proportional to weight, what should the price of the larger jar be?

48. In each of the following equations, state whether y varies directly as x, inversely as x, or neither directly nor inversely as x.

a) $7xy = 14$
b) $x - 2y = 12$
c) $-2x + 3y = 0$
d) $x = \frac{3}{4}y$
e) $\frac{x}{y} = 2$

49. *Area of a Circle.* The area of a circle varies directly as the square of the length of a diameter. What is the variation constant?

50. Describe in words the variation given by the equation

$$Q = \frac{kp^2}{q^3}.$$

CHAPTER 2   Summary and Review

Important Properties and Formulas

The Algebra of Functions

The Sum of Two Functions:
$(f + g)(x) = f(x) + g(x)$

The Difference of Two Functions:
$(f - g)(x) = f(x) - g(x)$

The Product of Two Functions:
$(fg)(x) = f(x) \cdot g(x)$

The Quotient of Two Functions:
$(f/g)(x) = f(x)/g(x), \ g(x) \neq 0$

The Composition of Two Functions:
$(f \circ g)(x) = f(g(x))$

Tests for Symmetry

x-axis: If replacing y with $-y$ produces an equivalent equation, then the graph is symmetric with respect to the x-axis.

y-axis: If replacing x with $-x$ produces an equivalent equation, then the graph is symmetric with respect to the y-axis.

Origin: If replacing x with $-x$ and y with $-y$ produces an equivalent equation, then the graph is symmetric with respect to the origin.

Even Function: $f(-x) = f(x)$
Odd Function: $f(-x) = -f(x)$

Transformations

Vertical Translation:	$y = f(x) \pm b$
Horizontal Translation:	$y = f(x \mp d)$
Reflection across the x-axis:	$y = -f(x)$
Reflection across the y-axis:	$y = f(-x)$
Vertical Stretching or Shrinking:	$y = af(x)$
Horizontal Stretching or Shrinking:	$y = f(cx)$

Variation

Direct: $y = kx$

Inverse: $y = \dfrac{k}{x}$

Joint: $y = kxz$

Review Exercises

Determine whether the statement is true or false.

1. The greatest integer function pairs each input with the greatest integer less than or equal to that input. [2.1]

2. For functions f and g, the domains of $f + g$, $f - g$, fg, and f/g are always the intersection of the domains of f and g. [2.2]

3. The graph of $y = (x - 2)^2$ is the graph of $y = x^2$ shifted right 2 units. [2.4]

4. The graph of $y = -x^2$ is the reflection of the graph of $y = x^2$ across the x-axis. [2.4]

Determine the intervals on which the function is **(a)** *increasing,* **(b)** *decreasing, and* **(c)** *constant.* [2.1]

5.
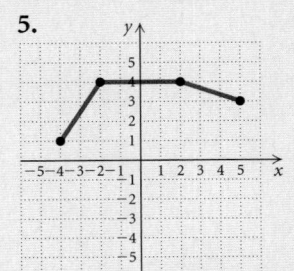

6.
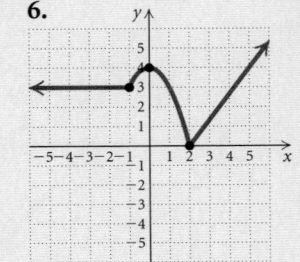

Graph the function. Estimate the intervals on which the function is increasing or decreasing and estimate any relative maxima or minima. [2.1]

7. $f(x) = x^2 - 1$
8. $f(x) = 2 - |x|$

Use a graphing calculator to find the intervals on which the function is increasing or decreasing and find any relative maxima or minima. [2.1]

9. $f(x) = x^2 - 4x + 3$
10. $f(x) = -x^2 + x + 6$

11. $f(x) = x^3 - 4x$
12. $f(x) = 2x - 0.5x^3$

13. *Tablecloth Area.* A seamstress uses 20 ft of lace to trim the edges of a rectangular tablecloth. If the tablecloth is l feet long, express its area as a function of the length. [2.1]

14. *Inscribed Rectangle.* A rectangle is inscribed in a semicircle of radius 2, as shown. The variable $x =$ half the length of the rectangle. Express the area of the rectangle as a function of x. [2.1]

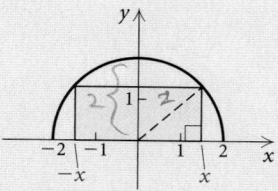

15. *Dog Pen.* Mamie has 66 ft of fencing with which to enclose a rectangular dog pen. The side of her garage forms one side of the pen. Suppose the side of the pen parallel to the garage is x feet long.

a) Express the area of the dog pen as a function of x. [2.1]
b) Find the domain of the function. [2.1]
c) Graph the function using a graphing calculator. [2.1]
d) Determine the dimensions that yield the maximum area. [2.1]

16. *Minimizing Surface Area.* A container firm is designing an open-top rectangular box, with a square base, that will hold 108 in³. Let $x =$ the length of a side of the base.

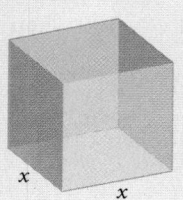
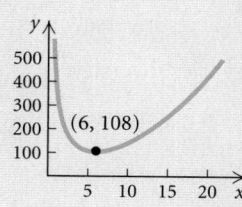

a) Express the surface area as a function of x. [2.1]
b) Find the domain of the function. [2.1]
c) Using the accompanying graph, determine the dimensions that will minimize the surface area of the box. [2.1]

Graph each of the following. [2.1]

17. $f(x) = \begin{cases} -x, & \text{for } x \le -4, \\ \frac{1}{2}x + 1, & \text{for } x > -4 \end{cases}$

18. $f(x) = \begin{cases} x^3, & \text{for } x < -2, \\ |x|, & \text{for } -2 \le x \le 2, \\ \sqrt{x - 1}, & \text{for } x > 2 \end{cases}$

19. $f(x) = \begin{cases} \dfrac{x^2 - 1}{x + 1}, & \text{for } x \ne -1, \\ 3, & \text{for } x = -1 \end{cases}$

20. $f(x) = [\![x]\!]$
21. $f(x) = [\![x - 3]\!]$

22. For the function in Exercise 18, find $f(-1)$, $f(5)$, $f(-2)$, and $f(-3)$. [2.1]

23. For the function in Exercise 19, find $f(-2)$, $f(-1)$, $f(0)$, and $f(4)$. [2.1]

Given that $f(x) = \sqrt{x - 2}$ and $g(x) = x^2 - 1$, find each of the following if it exists. [2.2]

24. $(f - g)(6)$

25. $(fg)(2)$

26. $(f + g)(-1)$

For each pair of functions in Exercises 27 and 28:

a) *Find the domain of f, g, $f + g$, $f - g$, fg, and f/g.* [2.2]
b) *Find $(f + g)(x)$, $(f - g)(x)$, $fg(x)$, and $(f/g)(x)$.* [2.2]

27. $f(x) = \dfrac{4}{x^2}$; $g(x) = 3 - 2x$

28. $f(x) = 3x^2 + 4x$; $g(x) = 2x - 1$

29. Given the total-revenue and total-cost functions $R(x) = 120x - 0.5x^2$ and $C(x) = 15x + 6$, find the total-profit function $P(x)$. [2.2]

For each function f, construct and simplify the difference quotient. [2.2]

30. $f(x) = 2x + 7$
31. $f(x) = 3 - x^2$

32. $f(x) = \dfrac{4}{x}$

Given that $f(x) = 2x - 1$, $g(x) = x^2 + 4$, and $h(x) = 3 - x^3$, find each of the following.   [2.3]

33. $(f \circ g)(1)$

34. $(g \circ f)(1)$

35. $(h \circ f)(-2)$

36. $(g \circ h)(3)$

37. $(f \circ h)(-1)$

38. $(h \circ g)(2)$

In Exercises 39 and 40, for the pair of functions:

a) *Find $(f \circ g)(x)$ and $(g \circ f)(x)$.*   [2.3]

b) *Find the domain of $f \circ g$ and $g \circ f$.*   [2.3]

39. $f(x) = \dfrac{4}{x^2}$; $g(x) = 3 - 2x$

40. $f(x) = 3x^2 + 4x$; $g(x) = 2x - 1$

Find $f(x)$ and $g(x)$ such that $h(x) = (f \circ g)(x)$.   [2.3]

41. $h(x) = \sqrt{5x + 2}$

42. $h(x) = 4(5x - 1)^2 + 9$

Graph the given equation and determine visually whether it is symmetric with respect to the x-axis, the y-axis, and the origin. Then verify your assertion algebraically.   [2.4]

43. $x^2 + y^2 = 4$

44. $y^2 = x^2 + 3$

45. $x + y = 3$

46. $y = x^2$

47. $y = x^3$

48. $y = x^4 - x^2$

Determine visually whether the function is even, odd, or neither even nor odd.   [2.4]

49.
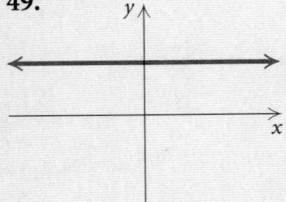

50.
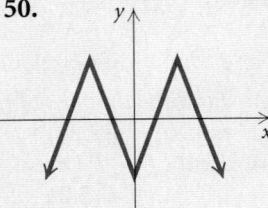

51.
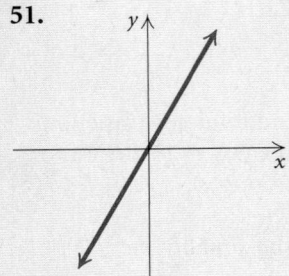

52.
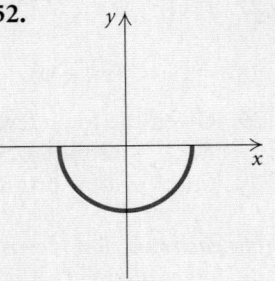

Test whether the function is even, odd, or neither even nor odd.   [2.4]

53. $f(x) = 9 - x^2$

54. $f(x) = x^3 - 2x + 4$

55. $f(x) = x^7 - x^5$

56. $f(x) = |x|$

57. $f(x) = \sqrt{16 - x^2}$

58. $f(x) = \dfrac{10x}{x^2 + 1}$

Write an equation for a function that has a graph with the given characteristics.   [2.4]

59. The shape of $y = x^2$, but shifted left 3 units

60. The shape of $y = \sqrt{x}$, but upside down and shifted right 3 units and up 4 units

61. The shape of $y = |x|$, but stretched vertically by a factor of 2 and shifted right 3 units

A graph of $y = f(x)$ is shown below. No formula for f is given. Graph each of the following.   [2.4]

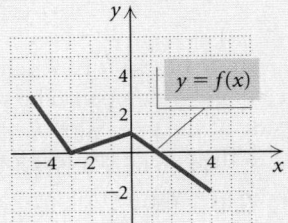

62. $y = f(x - 1)$

63. $y = f(2x)$

64. $y = -2f(x)$

65. $y = 3 + f(x)$

Find an equation of variation for the given situation. [2.5]

66. y varies directly as x, and $y = 100$ when $x = 25$.

67. y varies directly as x, and $y = 6$ when $x = 9$.

68. y varies inversely as x, and $y = 100$ when $x = 25$.

69. y varies inversely as x, and $y = 6$ when $x = 9$.

70. y varies inversely as the square of x, and $y = 12$ when $x = 2$.

71. y varies jointly as x and the square of z and inversely as w, and $y = 2$ when $x = 16$, $w = 0.2$, and $z = \frac{1}{2}$.

72. *Pumping Time.* The time t required to empty a tank varies inversely as the rate r of pumping. If a pump can empty a tank in 35 min at the rate of 800 kL/min, how long will it take the pump to empty the same tank at the rate of 1400 kL/min? [2.5]

73. *Test Score.* The score N on a test varies directly as the number of correct responses a. Ellen answers 29 questions correctly and earns a score of 87. What would Ellen's score have been if she had answered 25 questions correctly? [2.5]

74. *Power of Electric Current.* The power P expended by heat in an electric circuit of fixed resistance varies directly as the square of the current C in the circuit. A circuit expends 180 watts when a current of 6 amperes is flowing. What is the amount of heat expended when the current is 10 amperes? [2.5]

75. For $f(x) = x + 1$ and $g(x) = \sqrt{x}$, the domain of $(g \circ f)(x)$ is which of the following? [2.3]

A. $[-1, \infty)$
B. $[-1, 0)$
C. $[0, \infty)$
D. $(-\infty, \infty)$

76. For $b > 0$, the graph of $y = f(x) + b$ is the graph of $y = f(x)$ shifted in which of the following ways? [2.4]

A. Right b units
B. Left b units
C. Up b units
D. Down b units

77. The graph of the function f is shown below.

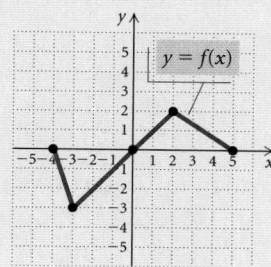

The graph of $g(x) = -\frac{1}{2}f(x) + 1$ is which of the following? [2.4]

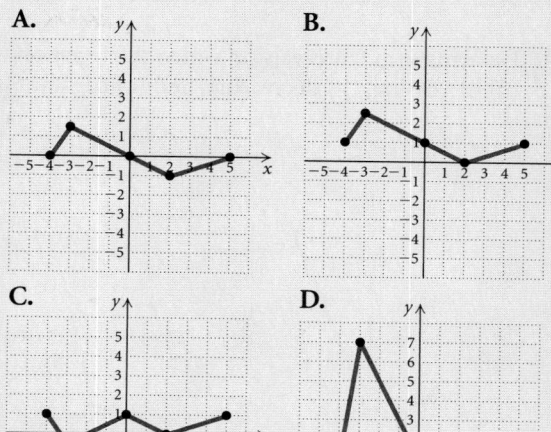

Collaborative Discussion and Writing

78. Given that $f(x) = 4x^3 - 2x + 7$, find each of the following. Then discuss how each expression differs from the other. [1.2], [2.4]

a) $f(x) + 2$
b) $f(x + 2)$
c) $f(x) + f(2)$

79. Given the graph of $y = f(x)$, explain and contrast the effect of the constant c on the graphs of $y = f(cx)$ and $y = cf(x)$. [2.4]

80. a) Graph several functions of the type $y_1 = f(x)$ and $y_2 = |f(x)|$. Describe a procedure, involving transformations, for creating the graph of y_2 from y_1. [2.4]
b) Describe a procedure, involving transformations, for creating the graph of $y_2 = f(|x|)$ from $y_1 = f(x)$. [2.4]

Synthesis

81. Prove that the sum of two odd functions is odd. [2.2], [2.4]

82. Describe how the graph of $y = -f(-x)$ is obtained from the graph of $y = f(x)$. [2.4]

CHAPTER 2   Test

1. Determine the intervals on which the function is (a) increasing, (b) decreasing, and (c) constant.

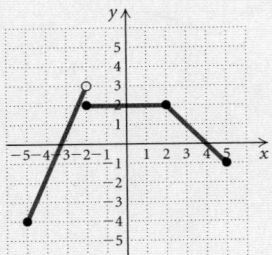

2. Graph the function $f(x) = 2 - x^2$. Estimate the intervals on which the function is increasing or decreasing and estimate any relative maxima or minima.

3. Use a graphing calculator to find the intervals on which the function $f(x) = x^3 + 4x^2$ is increasing or decreasing and find any relative maxima or minima.

4. *Triangular Pennant.* A softball team is designing a triangular pennant such that the height is 6 in. less than four times the length of the base b. Express the area of the pennant as a function of b.

5. Graph:
$$f(x) = \begin{cases} x^2, & \text{for } x < -1, \\ |x|, & \text{for } -1 \le x \le 1, \\ \sqrt{x - 1}, & \text{for } x > 1. \end{cases}$$

6. For the function in Exercise 5, find $f\left(-\frac{7}{8}\right)$, $f(5)$, and $f(-4)$.

Given that $f(x) = x^2 - 4x + 3$ and $g(x) = \sqrt{3 - x}$, find each of the following, if it exists.

7. $(f + g)(-6)$

8. $(f - g)(-1)$

9. $(fg)(2)$

10. $(f/g)(1)$

For $f(x) = x^2$ and $g(x) = \sqrt{x - 3}$, find each of the following.

11. The domain of f

12. The domain of g

13. The domain of $f + g$

14. The domain of $f - g$

15. The domain of fg

16. The domain of f/g

17. $(f + g)(x)$

18. $(f - g)(x)$

19. $(fg)(x)$

20. $(f/g)(x)$

For each function, construct and simplify the difference quotient.

21. $f(x) = \frac{1}{2}x + 4$

22. $f(x) = 2x^2 - x + 3$

Given that $f(x) = x^2 - 1$, $g(x) = 4x + 3$, and $h(x) = 3x^2 + 2x + 4$, find each of the following.

23. $(g \circ h)(2)$

24. $(f \circ g)(-1)$

25. $(h \circ f)(1)$

For $f(x) = \sqrt{x - 5}$ and $g(x) = x^2 + 1$:

26. Find $(f \circ g)(x)$ and $(g \circ f)(x)$.

27. Find the domain of $(f \circ g)(x)$ and $(g \circ f)(x)$.

28. Find $f(x)$ and $g(x)$ such that $h(x) = (f \circ g)(x) = (2x - 7)^4$.

29. Determine whether the graph of $y = x^4 - 2x^2$ is symmetric with respect to the x-axis, the y-axis, and the origin.

30. Test whether the function
$$f(x) = \frac{2x}{x^2 + 1}$$
is even, odd, or neither even nor odd. Show your work.

31. Write an equation for a function that has the shape of $y = x^2$, but shifted right 2 units and down 1 unit.

32. Write an equation for a function that has the shape of $y = x^2$, but shifted left 2 units and down 3 units.

33. The graph of a function $y = f(x)$ is shown below. No formula for f is given. Graph $y = -\frac{1}{2}f(x)$.

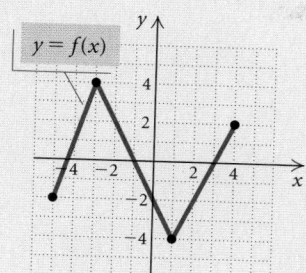

34. Find an equation of variation in which y varies inversely as x, and $y = 5$ when $x = 6$.

35. Find an equation of variation in which y varies directly as x, and $y = 60$ when $x = 12$.

36. Find an equation of variation where y varies jointly as x and the square of z and inversely as w, and $y = 100$ when $x = 0.1$, $z = 10$, and $w = 5$.

37. The stopping distance d of a car after the brakes have been applied varies directly as the square of the speed r. If a car traveling 60 mph can stop in 200 ft, how long will it take a car traveling 30 mph to stop?

38. The graph of the function f is shown below.

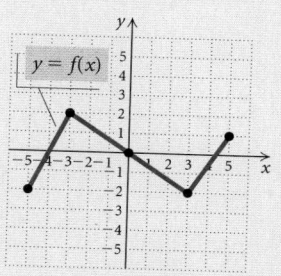

The graph of $g(x) = 2f(x) - 1$ is which of the following?

A.   **B.**

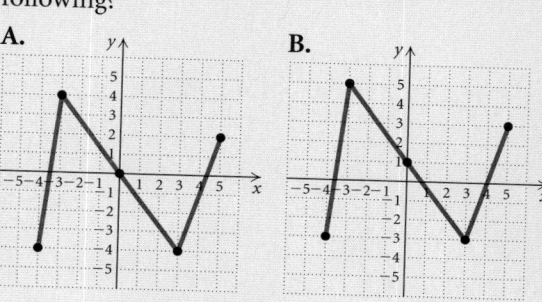

C.   **D.**

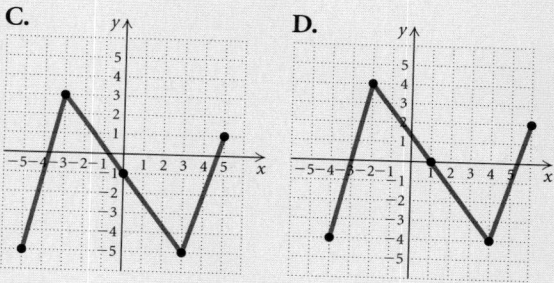

Synthesis

39. If $(-3, 1)$ is a point on the graph of $y = f(x)$, what point do you know is on the graph of $y = f(3x)$?

Quadratic Functions and Equations; Inequalities

APPLICATION The function $w(x) = 0.04x^2 - 0.12x + 10.16$ can be used to estimate the number of self-employed workers in the United States, in millions, x years after 2000 (*Source*: U.S. Bureau of Labor Statistics). In what year will there be 13 million self-employed workers in the United States?

This problem appears as Exercise 107 in Section 3.2.

3.1 The Complex Numbers

3.2 Quadratic Equations, Functions, Zeros, and Models

3.3 Analyzing Graphs of Quadratic Functions

3.4 Solving Rational Equations and Radical Equations

3.5 Solving Equations and Inequalities with Absolute Value

3.1 | The Complex Numbers

❖ Perform computations involving complex numbers.

Some functions have zeros that are not real numbers. In order to find the zeros of such functions, we must consider the **complex-number system**.

❖ The Complex-Number System

We know that the square root of a negative number is not a real number. For example, $\sqrt{-1}$ is not a real number because there is no real number x such that $x^2 = -1$. This means that certain equations, like $x^2 = -1$, or $x^2 + 1 = 0$, do not have real-number solutions, and certain functions, like $f(x) = x^2 + 1$, do not have real-number zeros. Consider the graph of $f(x) = x^2 + 1$.

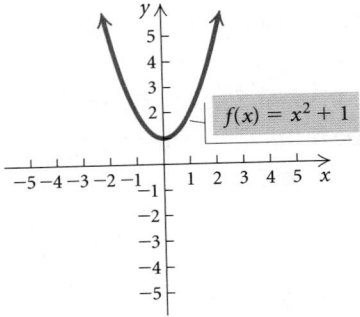

RADICAL EXPRESSIONS
REVIEW SECTION **R.7.**

We see that the graph does not cross the x-axis and thus has no x-intercepts. This illustrates that the function $f(x) = x^2 + 1$ has no real-number zeros. Thus there are no real-number solutions of the corresponding equation $x^2 + 1 = 0$.

We can define a non-real number that is a solution of the equation $x^2 + 1 = 0$.

The Number i

The number i is defined such that

$$i = \sqrt{-1} \quad \text{and} \quad i^2 = -1.$$

To express roots of negative numbers in terms of i, we can use the fact that

$$\sqrt{-p} = \sqrt{-1 \cdot p} = \sqrt{-1} \cdot \sqrt{p} = i\sqrt{p}$$

when p is a positive real number.

STUDY TIP

Don't hesitate to ask questions in class at appropriate times. Most instructors welcome questions and encourage students to ask them. Other students in your class probably have the same questions you do.

EXAMPLE 1 Express each number in terms of i.

a) $\sqrt{-7}$ b) $\sqrt{-16}$ c) $-\sqrt{-13}$

d) $-\sqrt{-64}$ e) $\sqrt{-48}$

Solution

a) $\sqrt{-7} = \sqrt{-1 \cdot 7} = \sqrt{-1} \cdot \sqrt{7}$
$$= i\sqrt{7}, \text{ or } \sqrt{7}i \leftarrow$$

*i is **not** under the radical.*

b) $\sqrt{-16} = \sqrt{-1 \cdot 16} = \sqrt{-1} \cdot \sqrt{16}$
$$= i \cdot 4 = 4i$$

c) $-\sqrt{-13} = -\sqrt{-1 \cdot 13} = -\sqrt{-1} \cdot \sqrt{13}$
$$= -i\sqrt{13}, \text{ or } -\sqrt{13}i \leftarrow$$

d) $-\sqrt{-64} = -\sqrt{-1 \cdot 64} = -\sqrt{-1} \cdot \sqrt{64}$
$$= -i \cdot 8 = -8i$$

e) $\sqrt{-48} = \sqrt{-1 \cdot 48} = \sqrt{-1} \cdot \sqrt{48}$
$$= i\sqrt{16 \cdot 3}$$
$$= i \cdot 4\sqrt{3}$$
$$= 4i\sqrt{3}, \text{ or } 4\sqrt{3}i \leftarrow$$

Now Try Exercise 1. ■

The complex numbers are formed by adding real numbers and multiples of i.

Complex Numbers

A **complex number** is a number of the form $a + bi$, where a and b are real numbers. The number a is said to be the **real part** of $a + bi$ and the number b is said to be the **imaginary part** of $a + bi$.*

Note that either a or b or both can be 0. When $b = 0$, $a + bi = a + 0i = a$, so every real number is a complex number. A complex number like $3 + 4i$ or $17i$, in which $b \neq 0$, is called an **imaginary number**. A complex number like $17i$ or $-4i$, in which $a = 0$ and $b \neq 0$, is sometimes called a **pure imaginary number**. The relationships among various types of complex numbers are shown in the figure on the following page.

---

*Sometimes bi is considered to be the imaginary part.

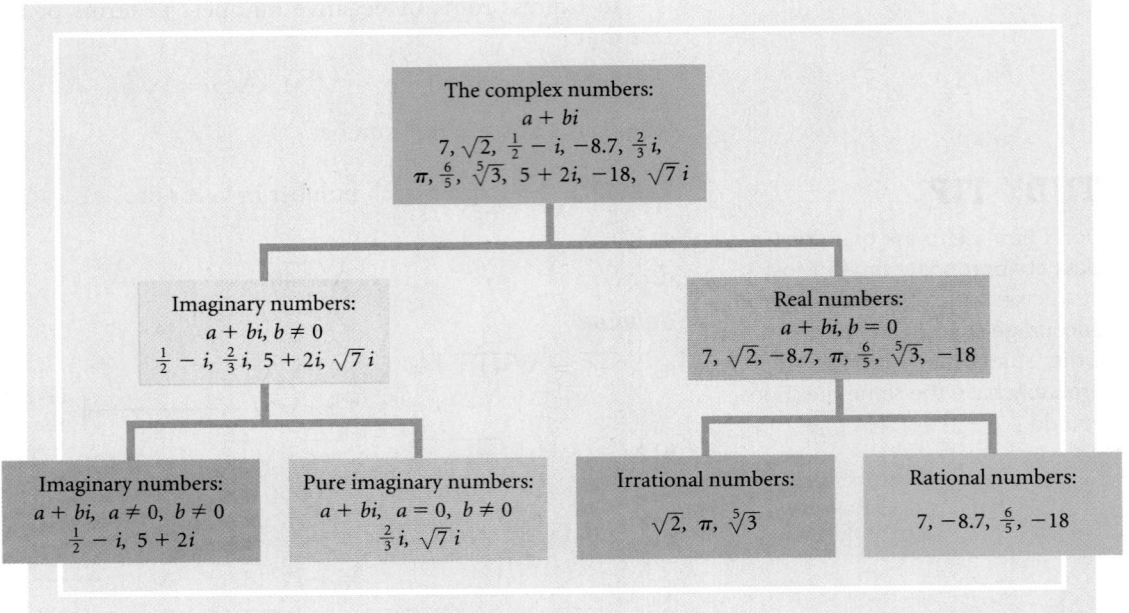

✿ Addition and Subtraction

The complex numbers obey the commutative, associative, and distributive laws. Thus we can add and subtract them as we do binomials. We collect the real parts and the imaginary parts of complex numbers just as we collect like terms in binomials.

GCM | **EXAMPLE 2** Add or subtract and simplify each of the following.

a) $(8 + 6i) + (3 + 2i)$ **b)** $(4 + 5i) - (6 - 3i)$

Solution

a) $(8 + 6i) + (3 + 2i) = (8 + 3) + (6i + 2i)$

                       Collecting the real parts and the imaginary parts

                 $= 11 + (6 + 2)i = 11 + 8i$

b) $(4 + 5i) - (6 - 3i) = (4 - 6) + [5i - (-3i)]$

                     Note that 6 and $-3i$ are both being subtracted.

                 $= -2 + 8i$ **Now Try Exercise 11.** ■

When set in $a + bi$ mode, most graphing calculators can perform operations on complex numbers. The operations in Example 2 are shown in the window below. Some calculators will express a complex number in the form (a, b) rather than $a + bi$.

```
(8+6i)+(3+2i)
                    11+8i
(4+5i)−(6−3i)
                    −2+8i
```

❈ Multiplication

When $\sqrt{a}$ and $\sqrt{b}$ are real numbers, $\sqrt{a} \cdot \sqrt{b} = \sqrt{ab}$, but this is not true when $\sqrt{a}$ and $\sqrt{b}$ are not real numbers. Thus,

$$\sqrt{-2} \cdot \sqrt{-5} = \sqrt{-1} \cdot \sqrt{2} \cdot \sqrt{-1} \cdot \sqrt{5}$$
$$= i\sqrt{2} \cdot i\sqrt{5}$$
$$= i^2\sqrt{10} = -1\sqrt{10} = -\sqrt{10} \quad \text{is correct!}$$

But

$$\sqrt{-2} \cdot \sqrt{-5} = \sqrt{(-2)(-5)} = \sqrt{10} \quad \text{is wrong!}$$

Keeping this and the fact that $i^2 = -1$ in mind, we multiply with imaginary numbers in much the same way that we do with real numbers.

GCM **EXAMPLE 3** Multiply and simplify each of the following.

a) $\sqrt{-16} \cdot \sqrt{-25}$ b) $(1 + 2i)(1 + 3i)$ c) $(3 - 7i)^2$

Solution

a) $\sqrt{-16} \cdot \sqrt{-25} = \sqrt{-1} \cdot \sqrt{16} \cdot \sqrt{-1} \cdot \sqrt{25}$
$$= i \cdot 4 \cdot i \cdot 5$$
$$= i^2 \cdot 20$$
$$= -1 \cdot 20 \qquad i^2 = -1$$
$$= -20$$

b) $(1 + 2i)(1 + 3i) = 1 + 3i + 2i + 6i^2$ **Multiplying each term of one number by every term of the other (FOIL)**

$$= 1 + 3i + 2i - 6 \qquad i^2 = -1$$
$$= -5 + 5i \qquad\qquad \textbf{Collecting like terms}$$

c) $(3 - 7i)^2 = 3^2 - 2 \cdot 3 \cdot 7i + (7i)^2$ **Recall that $(A - B)^2 = A^2 - 2AB + B^2$.**

$$= 9 - 42i + 49i^2$$
$$= 9 - 42i - 49 \qquad i^2 = -1$$
$$= -40 - 42i$$

Now Try Exercise 35. ■

We can multiply complex numbers on a graphing calculator set in $a + bi$ mode. The products found in Example 3 are shown below.

```
√(-16)√(-25)
                    -20
(1+2i)(1+3i)
                  -5+5i
(3-7i)²
                -40-42i
```

Recall that -1 raised to an *even* power is 1, and -1 raised to an *odd* power is -1. Simplifying powers of i can then be done by using the fact that $i^2 = -1$ and expressing the given power of i in terms of i^2. Consider the following:

$$i = \sqrt{-1},$$
$$i^2 = -1,$$
$$i^3 = i^2 \cdot i = (-1)i = -i,$$
$$i^4 = (i^2)^2 = (-1)^2 = 1,$$
$$i^5 = i^4 \cdot i = (i^2)^2 \cdot i = (-1)^2 \cdot i = 1 \cdot i = i,$$
$$i^6 = (i^2)^3 = (-1)^3 = -1,$$
$$i^7 = i^6 \cdot i = (i^2)^3 \cdot i = (-1)^3 \cdot i = -1 \cdot i = -i,$$
$$i^8 = (i^2)^4 = (-1)^4 = 1.$$

Note that the powers of i cycle through the values i, -1, $-i$, and 1.

EXAMPLE 4   Simplify each of the following.

a) i^{37} b) i^{58}

c) i^{75} d) i^{80}

Solution

a) $i^{37} = i^{36} \cdot i = (i^2)^{18} \cdot i = (-1)^{18} \cdot i = 1 \cdot i = i$

b) $i^{58} = (i^2)^{29} = (-1)^{29} = -1$

c) $i^{75} = i^{74} \cdot i = (i^2)^{37} \cdot i = (-1)^{37} \cdot i = -1 \cdot i = -i$

d) $i^{80} = (i^2)^{40} = (-1)^{40} = 1$ **Now Try Exercise 75.** ■

These powers of i can also be simplified in terms of i^4 rather than i^2. Consider i^{37} in Example 4(a), for instance. When we divide 37 by 4, we get 9 with a remainder of 1. Then $37 = 4 \cdot 9 + 1$, so

$$i^{37} = (i^4)^9 \cdot i = 1^9 \cdot i = 1 \cdot i = i.$$

The other examples shown above can be done in a similar manner.

✿ Conjugates and Division

Conjugates of complex numbers are defined as follows.

> ### Conjugate of a Complex Number
> The **conjugate** of a complex number $a + bi$ is $a - bi$. The numbers $a + bi$ and $a - bi$ are **complex conjugates.**

Each of the following pairs of numbers are complex conjugates:

$$-3 + 7i \text{ and } -3 - 7i; \quad 14 - 5i \text{ and } 14 + 5i; \quad \text{and} \quad 8i \text{ and } -8i.$$

The product of a complex number and its conjugate is a real number.

EXAMPLE 5 Multiply each of the following.

a) $(5 + 7i)(5 - 7i)$ b) $(8i)(-8i)$

Solution

```
(5+7i)(5-7i)
                    74
(8i)(-8i)
                    64
```

a) $(5 + 7i)(5 - 7i) = 5^2 - (7i)^2$ Using $(A + B)(A - B) = A^2 - B^2$

$$= 25 - 49i^2$$

$$= 25 - 49(-1)$$

$$= 25 + 49$$

$$= 74$$

b) $(8i)(-8i) = -64i^2$

$$= -64(-1)$$

$$= 64$$

Now Try Exercise 45. ▪

Conjugates are used when we divide complex numbers.

GCM **EXAMPLE 6**   Divide $2 - 5i$ by $1 - 6i$.

Solution   We write fraction notation and then multiply by 1, using the conjugate of the denominator to form the symbol for 1.

$$\frac{2 - 5i}{1 - 6i} = \frac{2 - 5i}{1 - 6i} \cdot \frac{1 + 6i}{1 + 6i}$$ Note that $1 + 6i$ is the conjugate of the divisor, $1 - 6i$.

$$= \frac{(2 - 5i)(1 + 6i)}{(1 - 6i)(1 + 6i)}$$

$$= \frac{2 + 7i - 30i^2}{1 - 36i^2}$$

$$= \frac{2 + 7i + 30}{1 + 36}$$ $i^2 = -1$

$$= \frac{32 + 7i}{37}$$

$$= \frac{32}{37} + \frac{7}{37}i.$$ Writing the quotient in the form $a + bi$

Now Try Exercise 65. ▪

With a graphing calculator set in $a + bi$ mode, we can divide complex numbers and express the real and imaginary parts in fraction form, just as we did in Example 6.

```
(2-5i)/(1-6i) ▶Frac
              32/37+7/37i
```

3.1 Exercise Set

Express the number in terms of i.

1. $\sqrt{-3}$ **2.** $\sqrt{-21}$

3. $\sqrt{-25}$ **4.** $\sqrt{-100}$

5. $-\sqrt{-33}$ **6.** $-\sqrt{-59}$

7. $-\sqrt{-81}$ **8.** $-\sqrt{-9}$

9. $\sqrt{-98}$ **10.** $\sqrt{-28}$

Simplify. Write answers in the form a + bi, where a and b are real numbers.

11. $(-5 + 3i) + (7 + 8i)$

12. $(-6 - 5i) + (9 + 2i)$

13. $(4 - 9i) + (1 - 3i)$

14. $(7 - 2i) + (4 - 5i)$

15. $(12 + 3i) + (-8 + 5i)$

16. $(-11 + 4i) + (6 + 8i)$

17. $(-1 - i) + (-3 - i)$

18. $(-5 - i) + (6 + 2i)$

19. $\left(3 + \sqrt{-16}\right) + \left(2 + \sqrt{-25}\right)$

20. $\left(7 - \sqrt{-36}\right) + \left(2 + \sqrt{-9}\right)$

21. $(10 + 7i) - (5 + 3i)$

22. $(-3 - 4i) - (8 - i)$

23. $(13 + 9i) - (8 + 2i)$

24. $(-7 + 12i) - (3 - 6i)$

25. $(6 - 4i) - (-5 + i)$

26. $(8 - 3i) - (9 - i)$

27. $(-5 + 2i) - (-4 - 3i)$

28. $(-6 + 7i) - (-5 - 2i)$

29. $(4 - 9i) - (2 + 3i)$

30. $(10 - 4i) - (8 + 2i)$

31. $7i(2 - 5i)$

32. $3i(6 + 4i)$

33. $-2i(-8 + 3i)$

34. $-6i(-5 + i)$

35. $(1 + 3i)(1 - 4i)$

36. $(1 - 2i)(1 + 3i)$

37. $(2 + 3i)(2 + 5i)$

38. $(3 - 5i)(8 - 2i)$

39. $(-4 + i)(3 - 2i)$

40. $(5 - 2i)(-1 + i)$

41. $(8 - 3i)(-2 - 5i)$

42. $(7 - 4i)(-3 - 3i)$

43. $\left(3 + \sqrt{-16}\right)\left(2 + \sqrt{-25}\right)$

44. $\left(7 - \sqrt{-16}\right)\left(2 + \sqrt{-9}\right)$

45. $(5 - 4i)(5 + 4i)$

46. $(5 + 9i)(5 - 9i)$

47. $(3 + 2i)(3 - 2i)$

48. $(8 + i)(8 - i)$

49. $(7 - 5i)(7 + 5i)$

50. $(6 - 8i)(6 + 8i)$

51. $(4 + 2i)^2$ **52.** $(5 - 4i)^2$

53. $(-2 + 7i)^2$ **54.** $(-3 + 2i)^2$

55. $(1 - 3i)^2$ **56.** $(2 - 5i)^2$

57. $(-1 - i)^2$ **58.** $(-4 - 2i)^2$

59. $(3 + 4i)^2$ **60.** $(6 + 5i)^2$

61. $\dfrac{3}{5 - 11i}$ **62.** $\dfrac{i}{2 + i}$

63. $\dfrac{5}{2 + 3i}$ **64.** $\dfrac{-3}{4 - 5i}$

65. $\dfrac{4 + i}{-3 - 2i}$

66. $\dfrac{5 - i}{-7 + 2i}$

67. $\dfrac{5 - 3i}{4 + 3i}$

68. $\dfrac{6 + 5i}{3 - 4i}$

69. $\dfrac{2 + \sqrt{3}i}{5 - 4i}$

70. $\dfrac{\sqrt{5} + 3i}{1 - i}$

71. $\dfrac{1 + i}{(1 - i)^2}$

72. $\dfrac{1 - i}{(1 + i)^2}$

73. $\dfrac{4 - 2i}{1 + i} + \dfrac{2 - 5i}{1 + i}$

74. $\dfrac{3 + 2i}{1 - i} + \dfrac{6 + 2i}{1 - i}$

Simplify.

75. i^{11}

76. i^7

77. i^{35}

78. i^{24}

79. i^{64}

80. i^{42}

81. $(-i)^{71}$

82. $(-i)^6$

83. $(5i)^4$

84. $(2i)^5$

Collaborative Discussion and Writing

85. Is the sum of two imaginary numbers always an imaginary number? Explain your answer.

86. Is the product of two imaginary numbers always an imaginary number? Explain your answer.

Skill Maintenance

87. Write a slope–intercept equation for the line containing the point $(3, -5)$ and perpendicular to the line $3x - 6y = 7$.

Given that $f(x) = x^2 + 4$ and $g(x) = 3x + 5$, find each of the following.

88. The domain of $f - g$

89. The domain of f/g

90. $(f - g)(x)$

91. $(f/g)(2)$

92. For the function $f(x) = x^2 - 3x + 4$, construct and simplify the difference quotient

$$\dfrac{f(x + h) - f(x)}{h}.$$

Synthesis

Determine whether each of the following is true or false.

93. The sum of two numbers that are conjugates of each other is always a real number.

94. The conjugate of a sum is the sum of the conjugates of the individual complex numbers.

95. The conjugate of a product is the product of the conjugates of the individual complex numbers.

Let $z = a + bi$ and $\bar{z} = a - bi$.

96. Find a general expression for $1/z$.

97. Find a general expression for $z\bar{z}$.

98. Solve $z + 6\bar{z} = 7$ for z.

99. Multiply and simplify:
$$[x - (3 + 4i)][x - (3 - 4i)].$$

3.2 Quadratic Equations, Functions, Zeros, and Models

❖ Find zeros of quadratic functions and solve quadratic equations by using the principle of zero products, by using the principle of square roots, by completing the square, and by using the quadratic formula.

❖ Solve equations that are reducible to quadratic.

❖ Solve applied problems using quadratic equations.

❖ Quadratic Equations and Quadratic Functions

In this section, we will explore the relationship between the solutions of quadratic equations and the zeros of quadratic functions. We define quadratic equations and functions as follows.

> **Quadratic Equations**
>
> A **quadratic equation** is an equation that can be written in the form
> $$ax^2 + bx + c = 0, \quad a \neq 0,$$
> where a, b, and c are real numbers.
>
> **Quadratic Functions**
>
> A **quadratic function** f is a function that can be written in the form
> $$f(x) = ax^2 + bx + c, \quad a \neq 0,$$
> where a, b, and c are real numbers.

A quadratic equation written in the form $ax^2 + bx + c = 0$ is said to be in **standard form.**

The *zeros* of a quadratic function $f(x) = ax^2 + bx + c$ are the *solutions* of the associated quadratic equation $ax^2 + bx + c = 0$. (These solutions are sometimes called *roots* of the equation.) Quadratic functions can have real-number or imaginary-number zeros and quadratic equations can have real-number or imaginary-number solutions. If the zeros or solutions are real numbers, they are also the first coordinates of the x-intercepts of the graph of the quadratic function.

The following principles allow us to solve many quadratic equations.

```
ZEROS OF A FUNCTION
REVIEW SECTION 1.5.
```

> **Equation-Solving Principles**
>
> ***The Principle of Zero Products*****:** If $ab = 0$ is true, then $a = 0$ or $b = 0$, and if $a = 0$ or $b = 0$, then $ab = 0$.
>
> ***The Principle of Square Roots*****:** If $x^2 = k$, then $x = \sqrt{k}$ or $x = -\sqrt{k}$.

FACTORING TRINOMIALS

REVIEW SECTION **R.4.**

EXAMPLE 1 Solve: $2x^2 - x = 3$.

ALGEBRAIC SOLUTION

We have

$$2x^2 - x = 3$$
$$2x^2 - x - 3 = 0 \qquad \text{Subtracting 3 on both sides}$$
$$(x + 1)(2x - 3) = 0 \qquad \text{Factoring}$$
$$x + 1 = 0 \quad \text{or } 2x - 3 = 0 \qquad \begin{array}{l}\text{Using the principle of}\\ \text{zero products}\end{array}$$
$$x = -1 \quad \text{or} \qquad 2x = 3$$
$$x = -1 \quad \text{or} \qquad x = \tfrac{3}{2}.$$

Check: For $x = -1$:

$$\begin{array}{c|c} \dfrac{2x^2 - x = 3}{2(-1)^2 - (-1) \;\; ? \;\; 3} \\ 2 \cdot 1 + 1 \\ 2 + 1 \\ \hphantom{xxx} 3 & 3 \quad \text{TRUE} \end{array}$$

For $x = \tfrac{3}{2}$:

$$\begin{array}{c|c} \dfrac{2x^2 - x = 3}{2\left(\tfrac{3}{2}\right)^2 - \tfrac{3}{2} \;\; ? \;\; 3} \\ 2 \cdot \tfrac{9}{4} - \tfrac{3}{2} \\ \tfrac{9}{2} - \tfrac{3}{2} \\ \tfrac{6}{2} \\ 3 & 3 \quad \text{TRUE} \end{array}$$

The solutions are -1 and $\tfrac{3}{2}$.

GRAPHICAL SOLUTION

The solutions of the equation $2x^2 - x = 3$, or the equivalent equation $2x^2 - x - 3 = 0$, are the zeros of the function $f(x) = 2x^2 - x - 3$. They are also the first coordinates of the x-intercepts of the graph of $f(x) = 2x^2 - x - 3$.

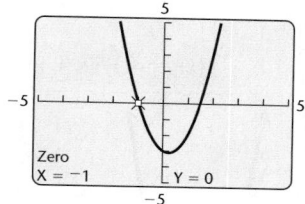

$y = 2x^2 - x - 3$

Zero
X = -1   Y = 0

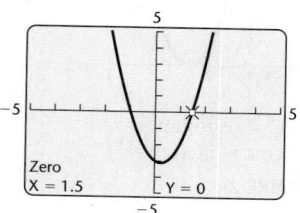

$y = 2x^2 - x - 3$

Zero
X = 1.5   Y = 0

The solutions are -1 and 1.5, or -1 and $\tfrac{3}{2}$.

Now Try Exercise 3. ■

EXAMPLE 2 Solve: $2x^2 - 10 = 0$.

Solution We have

$$2x^2 - 10 = 0$$
$$2x^2 = 10 \qquad \text{Adding 10 on both sides}$$
$$x^2 = 5 \qquad \text{Dividing by 2 on both sides}$$
$$x = \sqrt{5} \;\; \text{or} \;\; x = -\sqrt{5}. \qquad \text{Using the principle of square roots}$$

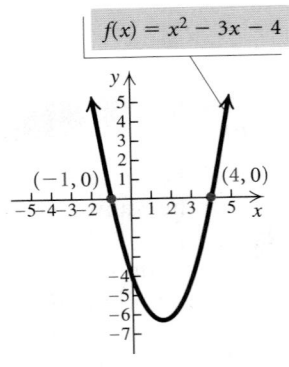

$f(x) = x^2 - 3x - 4$

$(-1, 0)$ $(4, 0)$

Two real-number zeros
Two x-intercepts

FIGURE 1

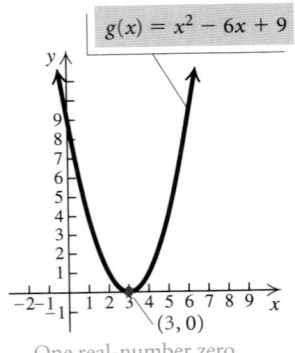

$g(x) = x^2 - 6x + 9$

$(3, 0)$

One real-number zero
One x-intercept

FIGURE 2

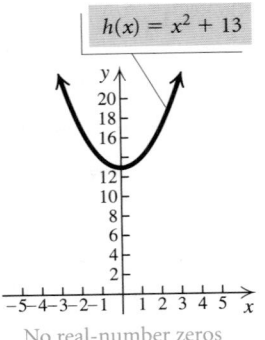

$h(x) = x^2 + 13$

No real-number zeros
No x-intercepts

FIGURE 3

Check:
$$2x^2 - 10 = 0$$
$$\overline{2(\pm\sqrt{5})^2 - 10 \; ? \; 0} \qquad \text{We can check both solutions at once.}$$
$$2 \cdot 5 - 10$$
$$10 - 10$$
$$0 \;\big|\; 0 \quad \text{TRUE}$$

The solutions are $\sqrt{5}$ and $-\sqrt{5}$, or $\pm\sqrt{5}$.   **Now Try Exercise 7.** ■

We have seen that some quadratic equations can be solved by factoring and using the principle of zero products. For example, consider the equation $x^2 - 3x - 4 = 0$:

$$x^2 - 3x - 4 = 0$$
$$(x + 1)(x - 4) = 0 \qquad \text{Factoring}$$
$$x + 1 = 0 \quad \text{or} \; x - 4 = 0 \qquad \begin{array}{l}\text{Using the principle of}\\ \text{zero products}\end{array}$$
$$x = -1 \; \text{or} \qquad x = 4.$$

The equation $x^2 - 3x - 4 = 0$ has *two real-number* solutions, -1 and 4. These are the zeros of the associated quadratic function $f(x) = x^2 - 3x - 4$ and the first coordinates of the x-intercepts of the graph of this function. (See Fig. 1.)

Next, consider the equation $x^2 - 6x + 9 = 0$. Again, we factor and use the principle of zero products:

$$x^2 - 6x + 9 = 0$$
$$(x - 3)(x - 3) = 0 \qquad \text{Factoring}$$
$$x - 3 = 0 \; \text{or} \; x - 3 = 0 \qquad \begin{array}{l}\text{Using the principle of}\\ \text{zero products}\end{array}$$
$$x = 3 \; \text{or} \qquad x = 3.$$

The equation $x^2 - 6x + 9 = 0$ has *one real-number* solution, 3. It is the zero of the quadratic function $g(x) = x^2 - 6x + 9$ and the first coordinate of the x-intercept of the graph of this function. (See Fig. 2.)

The principle of square roots can be used to solve quadratic equations like $x^2 + 13 = 0$:

$$x^2 + 13 = 0$$
$$x^2 = -13$$
$$x = \pm\sqrt{-13} \qquad \text{Using the principle of square roots}$$
$$x = \pm\sqrt{13}i. \qquad \sqrt{-13} = \sqrt{-1} \cdot \sqrt{13} = i \cdot \sqrt{13} = \sqrt{13}i$$

The equation has *two imaginary-number* solutions, $-\sqrt{13}i$ and $\sqrt{13}i$. These are the zeros of the associated quadratic function $h(x) = x^2 + 13$. Since the zeros are not real numbers, the graph of the function has no x-intercepts. (See Fig. 3.)

❋ Completing the Square

Neither the principle of zero products nor the principle of square roots would yield the *exact* zeros of a function like $f(x) = x^2 - 6x - 10$ or the *exact* solutions of the associated equation $x^2 - 6x - 10 = 0$. If we wish to find exact zeros or solutions, we can use a procedure called **completing the square** and then use the principle of square roots.

EXAMPLE 3   Find the zeros of $f(x) = x^2 - 6x - 10$ by completing the square.

Solution   We find the values of x for which $f(x) = 0$; that is, we solve the associated equation $x^2 - 6x - 10 = 0$. Our goal is to find an equivalent equation of the form $x^2 + bx + c = d$ in which $x^2 + bx + c$ is a perfect square. Since

$$x^2 + bx + \left(\frac{b}{2}\right)^2 = \left(x + \frac{b}{2}\right)^2,$$

the number c is found by taking half the coefficient of the x-term and squaring it. Thus for the equation $x^2 - 6x - 10 = 0$, we have

$$x^2 - 6x - 10 = 0$$
$$x^2 - 6x \qquad = 10 \qquad \text{Adding 10}$$
$$x^2 - 6x + 9 = 10 + 9 \qquad \text{Adding 9 to complete the square:}$$
$$\left(\frac{b}{2}\right)^2 = \left(\frac{-6}{2}\right)^2 = (-3)^2 = 9$$
$$x^2 - 6x + 9 = 19.$$

Because $x^2 - 6x + 9$ is a perfect square, we are able to write it as $(x - 3)^2$, the square of a binomial. We can then use the principle of square roots to finish the solution:

$$(x - 3)^2 = 19 \qquad \text{Factoring}$$
$$x - 3 = \pm\sqrt{19} \qquad \text{Using the principle of square roots}$$
$$x = 3 \pm \sqrt{19}. \qquad \text{Adding 3}$$

Therefore, the solutions of the equation are $3 + \sqrt{19}$ and $3 - \sqrt{19}$, or simply $3 \pm \sqrt{19}$. The zeros of $f(x) = x^2 - 6x - 10$ are also $3 + \sqrt{19}$ and $3 - \sqrt{19}$, or $3 \pm \sqrt{19}$.

Decimal approximations for $3 \pm \sqrt{19}$ can be found using a calculator:

$$3 + \sqrt{19} \approx 7.359 \quad \text{and} \quad 3 - \sqrt{19} \approx -1.359.$$

The zeros are approximately 7.359 and -1.359.   **Now Try Exercise 29.** ■

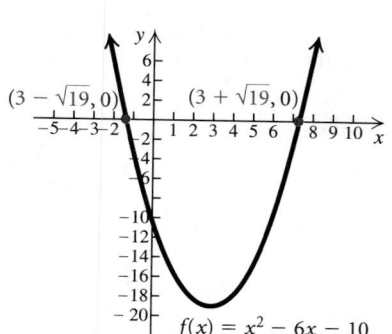

$(3 - \sqrt{19}, 0)$ $(3 + \sqrt{19}, 0)$

$f(x) = x^2 - 6x - 10$

Approximations for the zeros of the quadratic function $f(x) = x^2 - 6x - 10$ in Example 3 can be found using the Zero method.

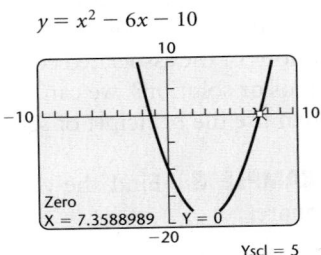

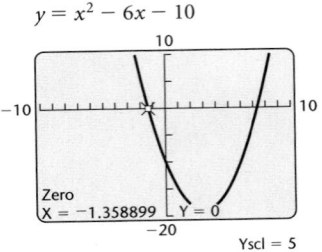

Before we can complete the square, the coefficient of the x^2-term must be 1. When it is not, we divide both sides of the equation by the x^2-coefficient.

EXAMPLE 4   Solve: $2x^2 - 1 = 3x$.

Solution   We have

$$2x^2 - 1 = 3x$$

$$2x^2 - 3x - 1 = 0 \qquad \text{Subtracting } 3x. \text{ We are unable to factor the result.}$$

$$2x^2 - 3x = 1 \qquad \text{Adding 1}$$

$$x^2 - \frac{3}{2}x = \frac{1}{2} \qquad \text{Dividing by 2 to make the } x^2\text{-coefficient 1}$$

$$x^2 - \frac{3}{2}x + \frac{9}{16} = \frac{1}{2} + \frac{9}{16} \qquad \text{Completing the square: } \frac{1}{2}\left(-\frac{3}{2}\right) = -\frac{3}{4} \text{ and } \left(-\frac{3}{4}\right)^2 = \frac{9}{16}; \text{ adding } \frac{9}{16}$$

$$\left(x - \frac{3}{4}\right)^2 = \frac{17}{16} \qquad \text{Factoring and simplifying}$$

$$x - \frac{3}{4} = \pm\frac{\sqrt{17}}{4} \qquad \text{Using the principle of square roots and the quotient rule for radicals}$$

$$x = \frac{3}{4} \pm \frac{\sqrt{17}}{4} \qquad \text{Adding } \frac{3}{4}$$

$$x = \frac{3 \pm \sqrt{17}}{4}.$$

The solutions are

$$\frac{3 + \sqrt{17}}{4} \quad \text{and} \quad \frac{3 - \sqrt{17}}{4}, \quad \text{or} \quad \frac{3 \pm \sqrt{17}}{4}.$$

Now Try Exercise 33. ■

> To solve a quadratic equation by completing the square:
>
> **1.** Isolate the terms with variables on one side of the equation and arrange them in descending order.
> **2.** Divide by the coefficient of the squared term if that coefficient is not 1.
> **3.** Complete the square by taking half the coefficient of the first-degree term and adding its square on both sides of the equation.
> **4.** Express one side of the equation as the square of a binomial.
> **5.** Use the principle of square roots.
> **6.** Solve for the variable.

❖ Using the Quadratic Formula

Because completing the square works for *any* quadratic equation, it can be used to solve the general quadratic equation $ax^2 + bx + c = 0$ for x. The result will be a formula that can be used to solve any quadratic equation quickly.

Consider any quadratic equation in standard form:

$$ax^2 + bx + c = 0, \quad a \neq 0.$$

For now, we assume that $a > 0$ and solve by completing the square. As the steps are carried out, compare them with those of Example 4.

$$ax^2 + bx + c = 0 \qquad \text{Standard form}$$

$$ax^2 + bx = -c \qquad \text{Adding } -c$$

$$x^2 + \frac{b}{a}x = -\frac{c}{a} \qquad \text{Dividing by } a$$

Half of $\dfrac{b}{a}$ is $\dfrac{b}{2a}$, and $\left(\dfrac{b}{2a}\right)^2 = \dfrac{b^2}{4a^2}$. Thus we add $\dfrac{b^2}{4a^2}$:

$$x^2 + \frac{b}{a}x + \frac{b^2}{4a^2} = -\frac{c}{a} + \frac{b^2}{4a^2} \qquad \text{Adding } \frac{b^2}{4a^2} \text{ to complete the square}$$

$$\left(x + \frac{b}{2a}\right)^2 = -\frac{4ac}{4a^2} + \frac{b^2}{4a^2} \qquad \begin{array}{l} \text{Factoring on the left; finding} \\ \text{a common denominator on the} \\ \text{right: } -\frac{c}{a} = \frac{c}{a} \cdot -\frac{4a}{4a} = -\frac{4ac}{4a^2} \end{array}$$

$$\left(x + \frac{b}{2a}\right)^2 = \frac{b^2 - 4ac}{4a^2}$$

$$x + \frac{b}{2a} = \pm\frac{\sqrt{b^2 - 4ac}}{2a} \qquad \begin{array}{l} \text{Using the principle of square roots} \\ \text{and the quotient rule for radicals.} \\ \text{Since } a > 0, \sqrt{4a^2} = 2a. \end{array}$$

$$x = -\frac{b}{2a} \pm \frac{\sqrt{b^2 - 4ac}}{2a} \qquad \text{Adding } -\frac{b}{2a}$$

$$x = \frac{-b \pm \sqrt{b^2 - 4ac}}{2a}.$$

It can also be shown that this result holds if $a < 0$.

> ### The Quadratic Formula
> The solutions of $ax^2 + bx + c = 0$, $a \neq 0$, are given by
> $$x = \frac{-b \pm \sqrt{b^2 - 4ac}}{2a}.$$

EXAMPLE 5   Solve $3x^2 + 2x = 7$. Find exact solutions and approximate solutions rounded to the nearest thousandth.

ALGEBRAIC SOLUTION

After writing the equation in standard form, we are unable to factor, so we identify a, b, and c in order to use the quadratic formula:

$$3x^2 + 2x = 7$$
$$3x^2 + 2x - 7 = 0;$$
$$a = 3, \quad b = 2, \quad c = -7.$$

We then use the quadratic formula:

$$x = \frac{-b \pm \sqrt{b^2 - 4ac}}{2a}$$

$$= \frac{-2 \pm \sqrt{2^2 - 4(3)(-7)}}{2(3)} \quad \text{Substituting}$$

$$= \frac{-2 \pm \sqrt{4 + 84}}{6} = \frac{-2 \pm \sqrt{88}}{6}$$

$$= \frac{-2 \pm \sqrt{4 \cdot 22}}{6} = \frac{-2 \pm 2\sqrt{22}}{6} = \frac{2(-1 \pm \sqrt{22})}{2 \cdot 3}$$

$$= \frac{2}{2} \cdot \frac{-1 \pm \sqrt{22}}{3} = \frac{-1 \pm \sqrt{22}}{3}.$$

The exact solutions are

$$\frac{-1 - \sqrt{22}}{3} \quad \text{and} \quad \frac{-1 + \sqrt{22}}{3}.$$

Using a calculator, we approximate the solutions to be -1.897 and 1.230.

GRAPHICAL SOLUTION

Using the Intersect method, we graph $y_1 = 3x^2 + 2x$ and $y_2 = 7$ and use the INTERSECT feature to find the coordinates of the points of intersection. The first coordinates of these points are the solutions of the equation $y_1 = y_2$, or $3x^2 + 2x = 7$.

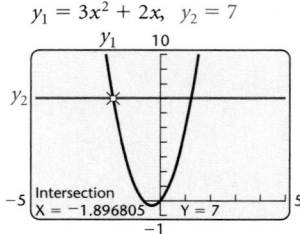

$y_1 = 3x^2 + 2x, \ y_2 = 7$

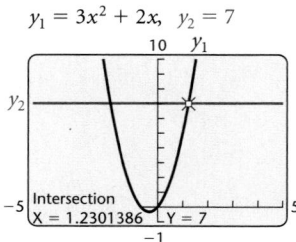

$y_1 = 3x^2 + 2x, \ y_2 = 7$

The solutions of $3x^2 + 2x = 7$ are approximately -1.897 and 1.230. We could also write the equation in standard form, $3x^2 + 2x - 7 = 0$, and use the Zero method.

Now Try Exercise 39. ■

Not all quadratic equations can be solved graphically.

EXAMPLE 6 Solve: $x^2 + 5x + 8 = 0$.

ALGEBRAIC SOLUTION

To find the solutions, we use the quadratic formula. For $x^2 + 5x + 8 = 0$, we have

$$a = 1, \quad b = 5, \quad c = 8;$$

$$x = \frac{-b \pm \sqrt{b^2 - 4ac}}{2a}$$

$$= \frac{-5 \pm \sqrt{5^2 - 4(1)(8)}}{2 \cdot 1} \quad \text{Substituting}$$

$$= \frac{-5 \pm \sqrt{25 - 32}}{2}$$

$$= \frac{-5 \pm \sqrt{-7}}{2}$$

$$= \frac{-5 \pm \sqrt{7}i}{2}.$$

The solutions are $-\dfrac{5}{2} - \dfrac{\sqrt{7}}{2}i$ and $-\dfrac{5}{2} + \dfrac{\sqrt{7}}{2}i$.

GRAPHICAL SOLUTION

The graph of the function $f(x) = x^2 + 5x + 8$ shows no x-intercepts.

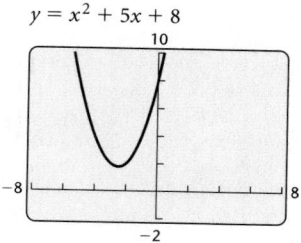

Thus the function has no real-number zeros and there are no real-number solutions of the associated equation $x^2 + 5x + 8 = 0$. This is a quadratic equation that cannot be solved graphically.

Now Try Exercise 45. ■

❋ The Discriminant

From the quadratic formula, we know that the solutions x_1 and x_2 of a quadratic equation are given by

$$x_1 = \frac{-b + \sqrt{b^2 - 4ac}}{2a} \quad \text{and} \quad x_2 = \frac{-b - \sqrt{b^2 - 4ac}}{2a}.$$

The expression $b^2 - 4ac$ shows the nature of the solutions. This expression is called the **discriminant**. If it is 0, then it makes no difference whether we choose the plus sign or the minus sign in the formula. That is, $x_1 = -\dfrac{b}{2a} = x_2$, so there is just one solution. In this case, we sometimes say that there is one repeated real solution. If the discriminant is positive, there will be two real solutions. If it is negative, we will be taking the square root of a negative number; hence there will be two imaginary-number solutions, and they will be complex conjugates.

> **Discriminant**
>
> For $ax^2 + bx + c = 0$, where a, b, and c are real numbers:
>
> $$b^2 - 4ac = 0 \longrightarrow \text{One real-number solution;}$$
> $$b^2 - 4ac > 0 \longrightarrow \text{Two different real-number solutions;}$$
> $$b^2 - 4ac < 0 \longrightarrow \text{Two different imaginary-number solutions,}$$
> $$\text{complex conjugates.}$$

In Example 5, the discriminant, 88, is positive, indicating that there are two different real-number solutions. The negative discriminant, -7, in Example 6 indicates that there are two different imaginary-number solutions.

❈ Equations Reducible to Quadratic

Some equations can be treated as quadratic, provided that we make a suitable substitution. For example, consider the following:

$$x^4 - 5x^2 + 4 = 0$$
$$(x^2)^2 - 5x^2 + 4 = 0 \qquad x^4 = (x^2)^2$$
$$u^2 - 5u + 4 = 0. \qquad \text{Substituting } u \text{ for } x^2$$

The equation $u^2 - 5u + 4 = 0$ can be solved for u by factoring or using the quadratic formula. Then we can reverse the substitution, replacing u with x^2, and solve for x. Equations like the one above are said to be **reducible to quadratic**, or **quadratic in form**.

EXAMPLE 7   Solve: $x^4 - 5x^2 + 4 = 0$.

ALGEBRAIC SOLUTION

We let $u = x^2$ and substitute:

$$u^2 - 5u + 4 = 0 \qquad \text{Substituting } u \text{ for } x^2$$
$$(u - 1)(u - 4) = 0 \qquad \text{Factoring}$$
$$u - 1 = 0 \ \ or \ \ u - 4 = 0 \qquad \text{Using the principle of zero products}$$
$$u = 1 \ \ or \qquad u = 4.$$

Don't stop here! We must solve for the original variable. We substitute x^2 for u and solve for x:

$$x^2 = 1 \quad or \ \ x^2 = 4$$
$$x = \pm 1 \ \ or \ \ x = \pm 2. \qquad \text{Using the principle of square roots}$$

The solutions are -1, 1, -2, and 2.

GRAPHICAL SOLUTION

Using the Zero method, we graph the function $y = x^4 - 5x^2 + 4$ and use the ZERO feature to find the zeros.

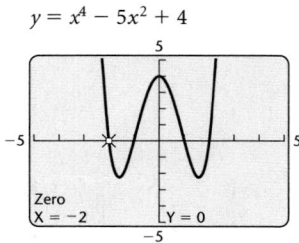

$y = x^4 - 5x^2 + 4$

The leftmost zero is -2. Using the ZERO feature three more times, we find that the other zeros are -1, 1, and 2. Thus the solutions of $x^4 - 5x^2 + 4 = 0$ are -2, -1, 1, and 2.

Now Try Exercise 89. ■

❖ Applications

Some applied problems can be translated to quadratic equations.

EXAMPLE 8 *Time of a Free Fall.* The Petronas Towers in Kuala Lumpur, Malaysia, are 1482 ft tall. How long would it take an object dropped from the top to reach the ground?

Solution

1. **Familiarize.** The formula $s = 16t^2$ is used to approximate the distance s, in feet, that an object falls freely from rest in t seconds. In this case, the distance is 1482 ft.

2. **Translate.** We substitute 1482 for s in the formula:

$$1482 = 16t^2.$$

3. **Carry out.**

ALGEBRAIC SOLUTION	**GRAPHICAL SOLUTION**

ALGEBRAIC SOLUTION

We use the principle of square roots:

$$1482 = 16t^2$$

$$\frac{1482}{16} = t^2 \qquad \text{Dividing by 16}$$

$$\sqrt{\frac{1482}{16}} = t \qquad \begin{array}{l}\text{Taking the positive}\\\text{square root. Time}\\\text{cannot be negative in}\\\text{this application.}\end{array}$$

$$9.624 \approx t.$$

GRAPHICAL SOLUTION

Using the Intersect method, we replace t with x, graph $y_1 = 1482$ and $y_2 = 16x^2$, and find the first coordinates of the points of intersection. Time cannot be negative in this application, so we need find only the point of intersection with a positive first coordinate. Since $y_1 = 1482$, we must choose a viewing window with Ymax greater than this value. Trial and error shows that a good choice is $[-15, 15, 0, 1650]$, with Xscl = 3 and Yscl = 100.

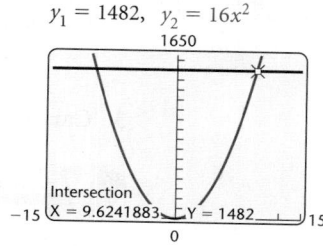

$$y_1 = 1482, \quad y_2 = 16x^2$$

We see that $x \approx 9.624$.

4. **Check.** In 9.624 sec, a dropped object would travel a distance of $16(9.624)^2$, or about 1482 ft. The answer checks.

5. **State.** It would take about 9.624 sec for an object dropped from the top of the Petronas Towers to reach the ground. **Now Try Exercise 105.** ■

EXAMPLE 9 *Bicycling Speed.* Logan and Cassidy leave a campsite, Logan biking due north and Cassidy biking due east. Logan bikes 7 km/h slower than Cassidy. After 4 hr, they are 68 km apart. Find the speed of each bicyclist.

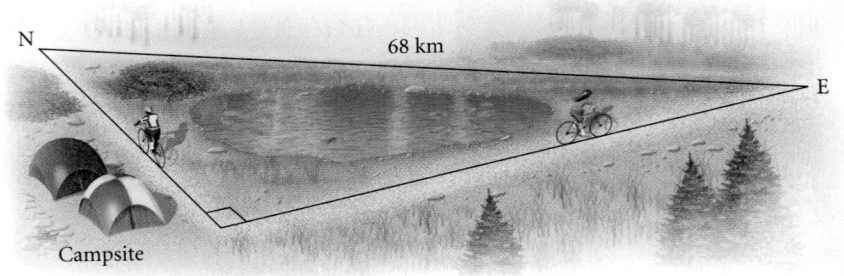

Solution

1. **Familiarize.** We let r = Cassidy's speed, in kilometers per hour. Then $r - 7$ = Logan's speed, in kilometers per hour. We will use the motion formula $d = rt$, where d is the distance, r is the rate (or speed), and t is the time. After 4 hr, Cassidy has traveled $4r$ kilometers and Logan has traveled $4(r - 7)$ kilometers. We add these distances to the drawing, as shown below.

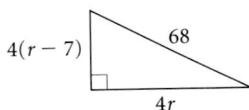

THE PYTHAGOREAN THEOREM

REVIEW SECTION **R.7.**

2. **Translate.** We use the Pythagorean theorem, $a^2 + b^2 = c^2$, where a and b are the lengths of the legs of a right triangle and c is the length of the hypotenuse:

$$(4r)^2 + [4(r - 7)]^2 = 68^2.$$

3. **Carry out.**

ALGEBRAIC SOLUTION

We have

$$(4r)^2 + [4(r - 7)]^2 = 68^2$$

$$16r^2 + 16(r^2 - 14r + 49) = 4624$$

$$16r^2 + 16r^2 - 224r + 784 = 4624$$

$$32r^2 - 224r - 3840 = 0 \qquad \text{Subtracting 4624}$$

$$r^2 - 7r - 120 = 0 \qquad \text{Dividing by 32}$$

$$(r + 8)(r - 15) = 0 \qquad \text{Factoring}$$

$$r + 8 = 0 \quad or \quad r - 15 = 0 \qquad \text{Principle of zero products}$$

$$r = -8 \quad or \qquad r = 15.$$

GRAPHICAL SOLUTION

Using the Intersect method, we enter the functions

$$y_1 = (4x)^2 + (4(x - 7))^2 \quad \text{and} \quad y_2 = 68^2$$

and find the points of intersection. It will probably be necessary to experiment with several viewing windows before an appropriate one is found. Note that $y_2 = 68^2$, or 4624, so Ymax must be greater than 4624. We try $[-10, 10, 0, 6000]$, with Yscl = 500.

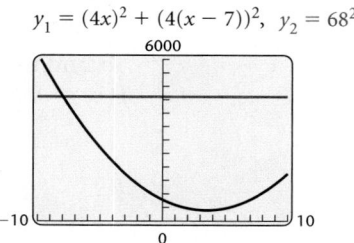

$y_1 = (4x)^2 + (4(x - 7))^2, \; y_2 = 68^2$

We note that we must increase Xmax in order to see the second point of intersection. The window $[-10, 20, 0, 6000]$, with Xscl = 5 and Yscl = 500, works well. The first coordinates of the points of intersection are the solutions of the equation $y_1 = y_2$, or $(4x)^2 + [4(x - 7)]^2 = 68^2$. Since the speed cannot be negative, we need find only the point of intersection with a positive first coordinate. We find that $x = 15$ at this point.

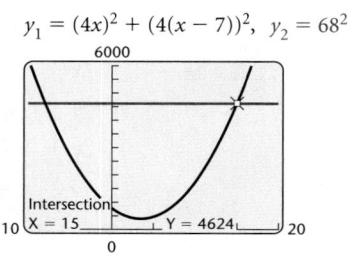

$y_1 = (4x)^2 + (4(x - 7))^2, \; y_2 = 68^2$

4. **Check.** Since speed cannot be negative, we need check only 15. If Cassidy's speed is 15 km/h, then Logan's speed is $15 - 7$, or 8 km/h. In 4 hr, Cassidy travels $4 \cdot 15$, or 60 km, and Logan travels $4 \cdot 8$, or 32 km. Then they are $\sqrt{60^2 + 32^2}$, or 68 km apart. The answer checks.

5. **State.** Cassidy's speed is 15 km/h, and Logan's speed is 8 km/h.

Now Try Exercise 109. ▨

CONNECTING *the* CONCEPTS

Zeros, Solutions, and Intercepts

The zeros of a function $y = f(x)$ are also the solutions of the equation $f(x) = 0$, and the real-number zeros are the first coordinates of the x-intercepts of the graph of the function.

FUNCTION	ZEROS OF THE FUNCTION; SOLUTIONS OF THE EQUATION	X-INTERCEPTS OF THE GRAPH
Linear Function $f(x) = 2x - 4$, or $y = 2x - 4$	To find the **zero** of $f(x)$, we solve $f(x) = 0$: $$2x - 4 = 0$$ $$2x = 4$$ $$x = 2.$$ The **solution** of the equation $2x - 4 = 0$ is 2. This is the zero of the function $f(x) = 2x - 4$; that is, $f(2) = 0$.	The zero of $f(x)$ is the first coordinate of the **x-intercept** of the graph of $y = f(x)$. 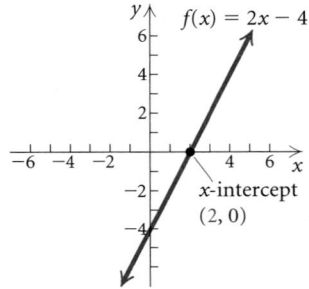
Quadratic Function $g(x) = x^2 - 3x - 4$, or $y = x^2 - 3x - 4$	To find the **zeros** of $g(x)$, we solve $g(x) = 0$: $$x^2 - 3x - 4 = 0$$ $$(x + 1)(x - 4) = 0$$ $$x + 1 = 0 \quad or \quad x - 4 = 0$$ $$x = -1 \quad or \quad x = 4.$$ The **solutions** of the equation $x^2 - 3x - 4 = 0$ are -1 and 4. They are the zeros of the function $g(x)$; that is, $g(-1) = 0$ and $g(4) = 0$.	The real-number zeros of $g(x)$ are the first coordinates of the **x-intercepts** of the graph of $y = g(x)$. 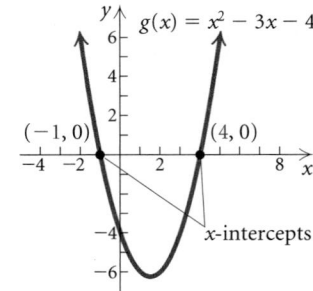

3.2 Exercise Set

Solve.

1. $(2x - 3)(3x - 2) = 0$

2. $(5x - 2)(2x + 3) = 0$

3. $x^2 - 8x - 20 = 0$

4. $x^2 + 6x + 8 = 0$

5. $3x^2 + x - 2 = 0$

6. $10x^2 - 16x + 6 = 0$

7. $4x^2 - 12 = 0$

8. $6x^2 = 36$

9. $3x^2 = 21$

10. $2x^2 - 20 = 0$

11. $5x^2 + 10 = 0$

12. $4x^2 + 12 = 0$

13. $x^2 + 16 = 0$

14. $x^2 + 25 = 0$

15. $2x^2 = 6x$

16. $18x + 9x^2 = 0$

17. $3y^3 - 5y^2 - 2y = 0$

18. $3t^3 + 2t = 5t^2$

19. $7x^3 + x^2 - 7x - 1 = 0$
(*Hint*: Factor by grouping.)

20. $3x^3 + x^2 - 12x - 4 = 0$
(*Hint*: Factor by grouping.)

In Exercises 21–26, use the given graph to find each of the following: (**a**) *the x-intercepts and* (**b**) *the zeros of the function.*

21.

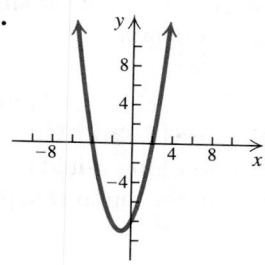

22.

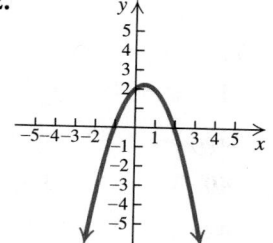

23.

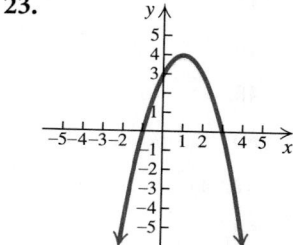

24.

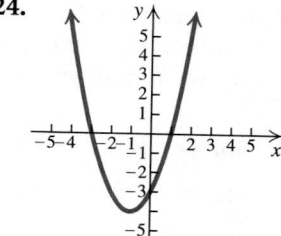

25.

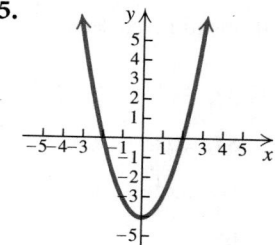

26.

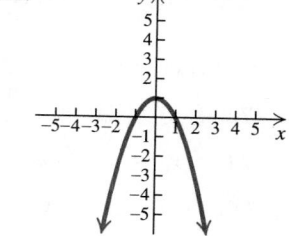

Solve by completing the square to obtain exact solutions.

27. $x^2 + 6x = 7$ **28.** $x^2 + 8x = -15$

29. $x^2 = 8x - 9$ **30.** $x^2 = 22 + 10x$

31. $x^2 + 8x + 25 = 0$ **32.** $x^2 + 6x + 13 = 0$

33. $3x^2 + 5x - 2 = 0$ **34.** $2x^2 - 5x - 3 = 0$

Use the quadratic formula to find exact solutions.

35. $x^2 - 2x = 15$ **36.** $x^2 + 4x = 5$

37. $5m^2 + 3m = 2$ **38.** $2y^2 - 3y - 2 = 0$

39. $3x^2 + 6 = 10x$ **40.** $3t^2 + 8t + 3 = 0$

41. $x^2 + x + 2 = 0$ **42.** $x^2 + 1 = x$

43. $5t^2 - 8t = 3$ **44.** $5x^2 + 2 = x$

45. $3x^2 + 4 = 5x$ **46.** $2t^2 - 5t = 1$

47. $x^2 - 8x + 5 = 0$ **48.** $x^2 - 6x + 3 = 0$

49. $3x^2 + x = 5$ **50.** $5x^2 + 3x = 1$

51. $2x^2 + 1 = 5x$ **52.** $4x^2 + 3 = x$

53. $5x^2 + 2x = -2$ **54.** $3x^2 + 3x = -4$

For each of the following, find the discriminant, $b^2 - 4ac$, and then determine whether one real-number solution, two different real-number solutions, or two different imaginary-number solutions exist.

55. $4x^2 = 8x + 5$ **56.** $4x^2 - 12x + 9 = 0$

57. $x^2 + 3x + 4 = 0$ **58.** $x^2 - 2x + 4 = 0$

59. $5t^2 - 7t = 0$ **60.** $5t^2 - 4t = 11$

Solve graphically. Round solutions to three decimal places, where appropriate.

61. $x^2 - 8x + 12 = 0$ **62.** $5x^2 + 42x + 16 = 0$

63. $7x^2 - 43x + 6 = 0$ **64.** $10x^2 - 23x + 12 = 0$

65. $6x + 1 = 4x^2$ **66.** $3x^2 + 5x = 3$

67. $2x^2 - 4 = 5x$ **68.** $4x^2 - 2 = 3x$

Find the zeros of the function algebraically. Give exact answers.

69. $f(x) = x^2 + 6x + 5$ **70.** $f(x) = x^2 - x - 2$

71. $f(x) = x^2 - 3x - 3$ **72.** $f(x) = 3x^2 + 8x + 2$

73. $f(x) = x^2 - 5x + 1$ **74.** $f(x) = x^2 - 3x - 7$

75. $f(x) = x^2 + 2x - 5$ **76.** $f(x) = x^2 - x - 4$

77. $f(x) = 2x^2 - x + 4$ **78.** $f(x) = 2x^2 + 3x + 2$

79. $f(x) = 3x^2 - x - 1$ **80.** $f(x) = 3x^2 + 5x + 1$

81. $f(x) = 5x^2 - 2x - 1$ **82.** $f(x) = 4x^2 - 4x - 5$

83. $f(x) = 4x^2 + 3x - 3$ **84.** $f(x) = x^2 + 6x - 3$

Use a graphing calculator to find the zeros of the function. Round to three decimal places.

85. $f(x) = 3x^2 + 2x - 4$ **86.** $f(x) = 9x^2 - 8x - 7$

87. $f(x) = 5.02x^2 - 4.19x - 2.057$

88. $f(x) = 1.21x^2 - 2.34x - 5.63$

Solve.

89. $x^4 - 3x^2 + 2 = 0$

90. $x^4 + 3 = 4x^2$

91. $x^4 + 3x^2 = 10$

92. $x^4 - 8x^2 = 9$

93. $y^4 + 4y^2 - 5 = 0$

94. $y^4 - 15y^2 - 16 = 0$

95. $x - 3\sqrt{x} - 4 = 0$
 (*Hint*: Let $u = \sqrt{x}$.)

96. $2x - 9\sqrt{x} + 4 = 0$

97. $m^{2/3} - 2m^{1/3} - 8 = 0$
 (*Hint*: Let $u = m^{1/3}$.)

98. $t^{2/3} + t^{1/3} - 6 = 0$

99. $x^{1/2} - 3x^{1/4} + 2 = 0$

100. $x^{1/2} - 4x^{1/4} = -3$

101. $(2x - 3)^2 - 5(2x - 3) + 6 = 0$
 (*Hint*: Let $u = 2x - 3$.)

102. $(3x + 2)^2 + 7(3x + 2) - 8 = 0$

103. $(2t^2 + t)^2 - 4(2t^2 + t) + 3 = 0$

104. $12 = (m^2 - 5m)^2 + (m^2 - 5m)$

Time of a Free Fall. *The formula $s = 16t^2$ is used to approximate the distance s, in feet, that an object falls freely from rest in t seconds. Use this formula for Exercises 105 and 106.*

105. The Warszawa Radio Mast in Poland, at 2120 ft, is the world's tallest structure (*Source: The Cambridge Fact Finder*). How long would it take an object falling freely from the top to reach the ground?

106. The tallest structure in the United States, at 2063 ft, is the KTHI-TV tower in North Dakota (*Source: The Cambridge Fact Finder*). How long would it take an object falling freely from the top to reach the ground?

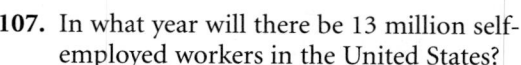

Self-Employed Workers.   *The function* $w(x) = 0.04x^2 - 0.12x + 10.16$ *can be used to estimate the number of self-employed workers in the United States, in millions, x years after 2000 (Source: U.S. Bureau of Labor Statistics). Use this function for Exercises 107 and 108.*

107. In what year will there be 13 million self-employed workers in the United States?

108. In what year will there be 21 million self-employed workers in the United States?

109. The length of a rectangular poster is 1 ft more than the width and a diagonal of the poster is 5 ft. Find the length and the width.

110. One leg of a right triangle is 7 cm less than the length of the other leg. The length of the hypotenuse is 13 cm. Find the lengths of the legs.

111. One number is 5 greater than another. The product of the numbers is 36. Find the numbers.

112. One number is 6 less than another. The product of the numbers is 72. Find the numbers.

113. *Box Construction.*   An open box is made from a 10-cm by 20-cm piece of tin by cutting a square from each corner and folding up the edges. The

area of the resulting base is 96 cm². What is the length of the sides of the squares?

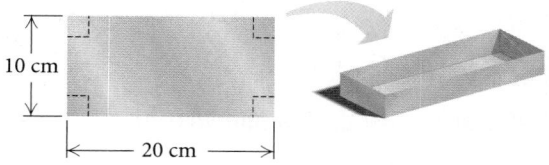

10 cm

20 cm

114. *Petting Zoo Dimensions.*   At the Glen Island Zoo, 170 m of fencing was used to enclose a petting area of 1750 m². Find the dimensions of the petting area.

115. *Dimensions of a Rug.*   Find the dimensions of a Persian rug whose perimeter is 28 ft and whose area is 48 ft².

116. *Picture Frame Dimensions.*   The frame on a picture is 28 cm by 32 cm outside and is of uniform width. What is the width of the frame if 192 cm² of the picture shows?

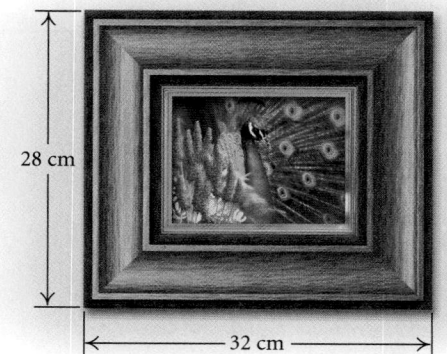

28 cm

32 cm

State whether the function is linear or quadratic.

117. $f(x) = 4 - 5x$

118. $f(x) = 4 - 5x^2$

119. $f(x) = 7x^2$

120. $f(x) = 23x + 6$

121. $f(x) = 1.2x - (3.6)^2$

122. $f(x) = 2 - x - x^2$

Collaborative Discussion and Writing

123. Is it possible for a quadratic function to have one real zero and one imaginary zero? Why or why not?

124. The graph of a quadratic function can have 0, 1, or 2 x-intercepts. How can you predict the number of x-intercepts without drawing the graph or (completely) solving an equation?

Skill Maintenance

Associate's Degrees Conferred.   The function $a(x) = 9096x + 387,725$ can be used to estimate the number of associate's degrees conferred x years after 1980 (Source: U.S. National Center for Education Statistics). Use this function for Exercises 125 and 126.

125. Estimate the number of associate's degrees conferred in 1998.

126. Estimate the number of associate's degrees conferred in 2010.

Determine whether the graph is symmetric with respect to the x-axis, the y-axis, and the origin.

127. $3x^2 + 4y^2 = 5$

128. $y^3 = 6x^2$

Determine whether the function is even, odd, or neither even nor odd.

129. $f(x) = 2x^3 - x$

130. $f(x) = 4x^2 + 2x - 3$

Synthesis

*For each equation in Exercises 131–134, under the given condition: **(a)** Find k and **(b)** find a second solution.*

131. $kx^2 - 17x + 33 = 0$; one solution is 3

132. $kx^2 - 2x + k = 0$; one solution is -3

133. $x^2 - kx + 2 = 0$; one solution is $1 + i$

134. $x^2 - (6 + 3i)x + k = 0$; one solution is 3

Solve.

135. $(x - 2)^3 = x^3 - 2$

136. $(x + 1)^3 = (x - 1)^3 + 26$

137. $(6x^3 + 7x^2 - 3x)(x^2 - 7) = 0$

138. $\left(x - \frac{1}{5}\right)\left(x^2 - \frac{1}{4}\right) + \left(x - \frac{1}{5}\right)\left(x^2 + \frac{1}{8}\right) = 0$

139. $x^2 + x - \sqrt{2} = 0$

140. $x^2 + \sqrt{5}x - \sqrt{3} = 0$

141. $2t^2 + (t - 4)^2 = 5t(t - 4) + 24$

142. $9t(t + 2) - 3t(t - 2) = 2(t + 4)(t + 6)$

143. $\sqrt{x - 3} - \sqrt[4]{x - 3} = 2$

144. $x^6 - 28x^3 + 27 = 0$

145. $\left(y + \frac{2}{y}\right)^2 + 3y + \frac{6}{y} = 4$

146. $x^2 + 3x + 1 - \sqrt{x^2 + 3x + 1} = 8$

147. Solve $\frac{1}{2}at^2 + v_0t + x_0 = 0$ for t.

3.3

Analyzing Graphs of Quadratic Functions

❖ Find the vertex, the axis of symmetry, and the maximum or minimum value of a quadratic function using the method of completing the square.

❖ Graph quadratic functions.

❖ Solve applied problems involving maximum and minimum function values.

❖ Graphing Quadratic Functions of the Type $f(x) = a(x - h)^2 + k$

The graph of a quadratic function is called a **parabola**. The graph of every parabola evolves from the graph of the squaring function $f(x) = x^2$ using transformations.

Exploring with Technology

Think of transformations and look for patterns. Consider the following functions:

$$y_1 = x^2, \quad y_2 = -0.4x^2,$$
$$y_3 = -0.4(x - 2)^2, \quad y_4 = -0.4(x - 2)^2 + 3.$$

Graph y_1 and y_2. How do you get from the graph of y_1 to y_2?
Graph y_2 and y_3. How do you get from the graph of y_2 to y_3?
Graph y_3 and y_4. How do you get from the graph of y_3 to y_4?

Consider the following functions:

$$y_1 = x^2, \quad y_2 = 2x^2,$$
$$y_3 = 2(x + 3)^2, \quad y_4 = 2(x + 3)^2 - 5.$$

Graph y_1 and y_2. How do you get from the graph of y_1 to y_2?
Graph y_2 and y_3. How do you get from the graph of y_2 to y_3?
Graph y_3 and y_4. How do you get from the graph of y_3 to y_4?

> **TRANSFORMATIONS**
>
> REVIEW SECTION **2.4.**

We get the graph of $f(x) = a(x - h)^2 + k$ from the graph of $f(x) = x^2$ as follows:

$$f(x) = x^2$$
$$\downarrow$$
$$f(x) = ax^2 \qquad \text{Vertical stretching or shrinking with a reflection across the } x\text{-axis if } a < 0$$
$$\downarrow$$
$$f(x) = a(x - h)^2 \qquad \text{Horizontal translation}$$
$$\downarrow$$
$$f(x) = a(x - h)^2 + k. \qquad \text{Vertical translation}$$

Consider the following graphs of the form $f(x) = a(x - h)^2 + k$. The point (h, k) at which the graph turns is called the **vertex**. The maximum or minimum value of $f(x)$ occurs at the vertex. Each graph has a line $x = h$ that is called the **axis of symmetry**.

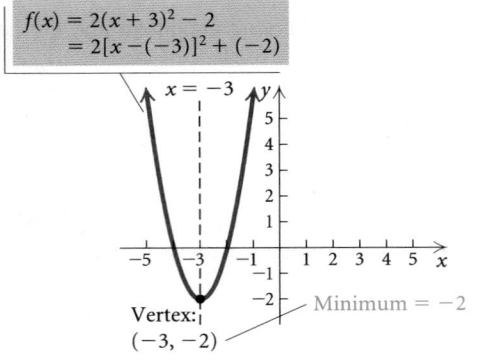

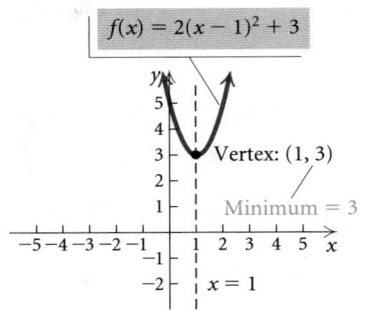

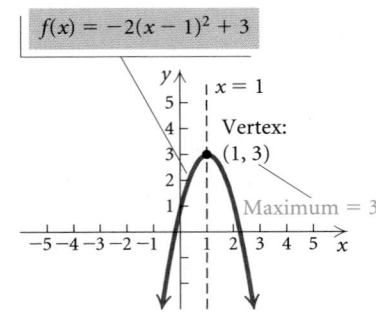

CONNECTING the CONCEPTS

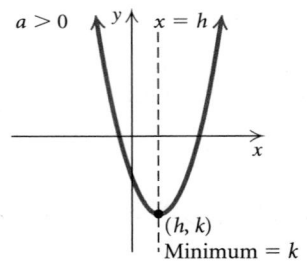

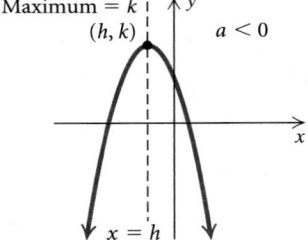

Graphing Quadratic Functions

The graph of the function $f(x) = a(x - h)^2 + k$ is a parabola that

- opens up if $a > 0$ and down if $a < 0$;
- has (h, k) as the vertex;
- has $x = h$ as the axis of symmetry;
- has k as a minimum value (output) if $a > 0$;
- has k as a maximum value if $a < 0$.

As we saw in Section 2.4, the constant a serves to stretch or shrink the graph vertically. As a parabola is stretched vertically, it becomes narrower, and as it is shrunk vertically, it becomes wider. That is, as $|a|$ increases, the graph becomes narrower, and as $|a|$ gets close to 0, the graph becomes wider.

If the equation is in the form $f(x) = a(x - h)^2 + k$, we can learn a great deal about the graph without actually graphing the function.

Function	$f(x) = 3\left(x - \frac{1}{4}\right)^2 - 2$ $= 3\left(x - \frac{1}{4}\right)^2 + (-2)$	$g(x) = -3(x + 5)^2 + 7$ $= -3[x - (-5)]^2 + 7$
Vertex	$\left(\frac{1}{4}, -2\right)$	$(-5, 7)$
Axis of Symmetry	$x = \frac{1}{4}$	$x = -5$
Maximum	None (3 > 0, so the graph opens up.)	7 (−3 < 0, so the graph opens down.)
Minimum	−2 (3 > 0, so the graph opens up.)	None (−3 < 0, so the graph opens down.)

Note that the vertex (h, k) is used to find the maximum or the minimum value of the function. The maximum or minimum value is the number k, *not* the ordered pair (h, k).

❖ Graphing Quadratic Functions of the Type $f(x) = ax^2 + bx + c, a \neq 0$

We now use a modification of the method of completing the square as an aid in graphing and analyzing quadratic functions of the form $f(x) = ax^2 + bx + c, a \neq 0$.

EXAMPLE 1   Find the vertex, the axis of symmetry, and the maximum or minimum value of $f(x) = x^2 + 10x + 23$. Then graph the function.

Solution   To express $f(x) = x^2 + 10x + 23$ in the form $f(x) = a(x - h)^2 + k$, we complete the square on the terms involving x. To do so, we take half the coefficient of x and square it, obtaining $(10/2)^2$, or 25. We now add and subtract that number on the *right side*:

$$f(x) = x^2 + 10x + 23 = x^2 + 10x + 25 - 25 + 23.$$

Since $25 - 25 = 0$, the new expression for the function is equivalent to the original expression. Note that this process differs from the one we used to complete the square in order to solve a quadratic equation, where we added the same number on both sides of the equation to obtain an equivalent equation. Instead, when we complete the square to write a function in the form $f(x) = a(x - h)^2 + k$, we add and subtract the same number on the one side. The entire process is shown below:

$f(x) = x^2 + 10x + 23$ Note that 25 completes the square for $x^2 + 10x$.

$= x^2 + 10x + 25 - 25 + 23$ Adding 25 − 25, or 0, to the right side

$= (x^2 + 10x + 25) - 25 + 23$ Regrouping

$= (x + 5)^2 - 2$ Factoring and simplifying

$= [x - (-5)]^2 + (-2).$ Writing in the form $f(x) = a(x - h)^2 + k$

Keeping in mind that this function will have a minimum value since $a > 0$ ($a = 1$), from this form of the function we know the following:

Vertex: $(-5, -2)$;

Axis of symmetry: $x = -5$;

Minimum value of the function: -2.

To graph the function by hand, we first plot the vertex and find several points on either side of it. Then we plot these points and connect them with a smooth curve.

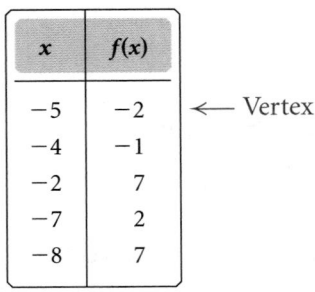

x	$f(x)$	
-5	-2	← Vertex
-4	-1	
-2	7	
-7	2	
-8	7	

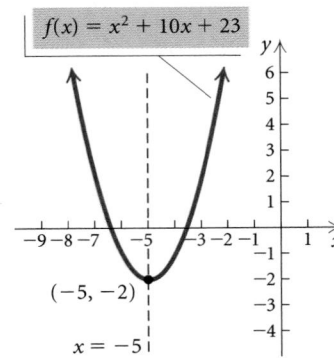

The graph of $f(x) = x^2 + 10x + 23$, or $f(x) = [x - (-5)]^2 + (-2)$, shown above, is a shift of the graph of $y = x^2$ left 5 units and down 2 units.

Now Try Exercise 5. ■

Keep in mind that the axis of symmetry is not part of the graph; it is a characteristic of the graph. If you fold the graph on its axis of symmetry, the two halves of the graph will coincide.

EXAMPLE 2 Find the vertex, the axis of symmetry, and the maximum or minimum value of $g(x) = x^2/2 - 4x + 8$. Then graph the function.

Solution We complete the square in order to write the function in the form $g(x) = a(x - h)^2 + k$. First, we factor $\frac{1}{2}$ out of the first two terms. This makes the coefficient of x^2 within the parentheses 1:

$$g(x) = \frac{x^2}{2} - 4x + 8$$

$$= \frac{1}{2}(x^2 - 8x) + 8. \qquad \text{Factoring } \tfrac{1}{2} \text{ out of the first two terms:}$$
$$\frac{x^2}{2} - 4x = \frac{1}{2} \cdot x^2 - \frac{1}{2} \cdot 8x$$

Next, we complete the square inside the parentheses: Half of -8 is -4, and $(-4)^2 = 16$. We add and subtract 16 inside the parentheses:

$$g(x) = \tfrac{1}{2}(x^2 - 8x + 16 - 16) + 8$$
$$= \tfrac{1}{2}(x^2 - 8x + 16) - \tfrac{1}{2} \cdot 16 + 8 \qquad \text{Using the distributive law to remove } -16 \text{ from within the parentheses}$$
$$= \tfrac{1}{2}(x^2 - 8x + 16) - 8 + 8$$
$$= \tfrac{1}{2}(x - 4)^2 + 0, \text{ or } \tfrac{1}{2}(x - 4)^2. \qquad \text{Factoring and simplifying}$$

We know the following:

Vertex: $(4, 0)$;

Axis of symmetry: $x = 4$;

Minimum value of the function: 0.

Finally, we plot the vertex and several points on either side of it and draw the graph of the function. The graph of g is a vertical shrinking of the graph of $y = x^2$ along with a shift 4 units to the right.

Now Try Exercise 9. ■

EXAMPLE 3 Find the vertex, the axis of symmetry, and the maximum or minimum value of $f(x) = -2x^2 + 10x - \tfrac{23}{2}$. Then graph the function.

Solution We have

$$f(x) = -2x^2 + 10x - \tfrac{23}{2}$$
$$= -2(x^2 - 5x) - \tfrac{23}{2} \qquad \text{Factoring } -2 \text{ out of the first two terms}$$
$$= -2\left(x^2 - 5x + \tfrac{25}{4} - \tfrac{25}{4}\right) - \tfrac{23}{2} \qquad \text{Completing the square inside the parentheses}$$
$$= -2\left(x^2 - 5x + \tfrac{25}{4}\right) - 2\left(-\tfrac{25}{4}\right) - \tfrac{23}{2} \qquad \text{Using the distributive law to remove } -\tfrac{25}{4} \text{ from within the parentheses}$$
$$= -2\left(x^2 - 5x + \tfrac{25}{4}\right) + \tfrac{25}{2} - \tfrac{23}{2}$$
$$= -2\left(x - \tfrac{5}{2}\right)^2 + 1.$$

This form of the function yields the following:

Vertex: $\left(\tfrac{5}{2}, 1\right)$;

Axis of symmetry: $x = \tfrac{5}{2}$;

Maximum value of the function: 1.

The graph is found by shifting the graph of $f(x) = x^2$ right $\tfrac{5}{2}$ units, reflecting it across the x-axis, stretching it vertically, and shifting it up 1 unit.

Now Try Exercise 13. ■

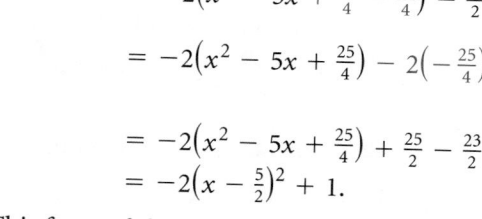

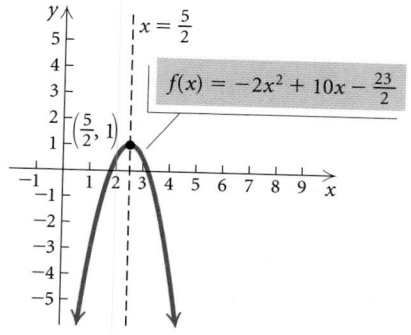

In many situations, we want to use a formula to find the coordinates of the vertex directly from the equation $f(x) = ax^2 + bx + c$. One way to develop such a formula is to observe that the x-coordinate of the vertex is centered between the x-intercepts, or zeros, of the function. By averaging the two solutions of $ax^2 + bx + c = 0$, we find a formula for the x-coordinate of the vertex:

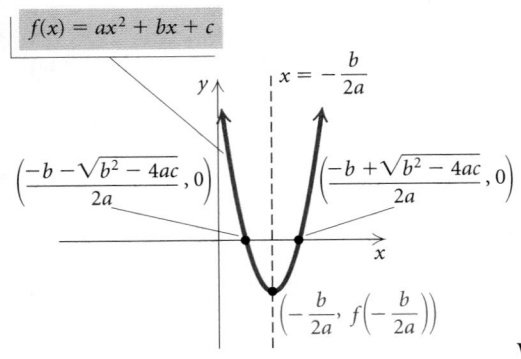

$$x\text{-coordinate of vertex } = \dfrac{\dfrac{-b - \sqrt{b^2 - 4ac}}{2a} + \dfrac{-b + \sqrt{b^2 - 4ac}}{2a}}{2}$$

$$= \dfrac{\dfrac{-2b}{2a}}{2} = \dfrac{-\dfrac{b}{a}}{2}$$

$$= -\dfrac{b}{a} \cdot \dfrac{1}{2} = -\dfrac{b}{2a}.$$

We use this value of x to find the y-coordinate of the vertex, $f\left(-\dfrac{b}{2a}\right)$.

The Vertex of a Parabola

The **vertex** of the graph of $f(x) = ax^2 + bx + c$ is

$$\left(-\frac{b}{2a}, f\left(-\frac{b}{2a}\right)\right).$$

    ↑   ↑

We calculate the   We substitute to
x-coordinate.   find the y-coordinate.

EXAMPLE 4   For the function $f(x) = -x^2 + 14x - 47$:

a) Find the vertex.

b) Determine whether there is a maximum or minimum value and find that value.

c) Find the range.

d) On what intervals is the function increasing? decreasing?

Solution   There is no need to graph the function.

a) The x-coordinate of the vertex is

$$-\frac{b}{2a} = -\frac{14}{2(-1)}, \text{ or } 7.$$

Since

$$f(7) = -7^2 + 14 \cdot 7 - 47 = 2,$$

the vertex is $(7, 2)$.

b) Since a is negative ($a = -1$), the graph opens down so the second coordinate of the vertex, 2, is the maximum value of the function.

c) The range is $(-\infty, 2]$.

d) Since the graph opens down, function values increase as we approach the vertex from the left and decrease as we move to the right from the vertex. Thus the function is increasing on the interval $(-\infty, 7)$ and decreasing on $(7, \infty)$.

Now Try Exercise 31. ▮

We can use a graphing calculator to do Example 4. Once we have graphed $y = -x^2 + 14x - 47$, we see that the graph opens down and thus has a maximum value. We can use the MAXIMUM feature to find the coordinates of the vertex. Using these coordinates, we can then find the maximum value and the range of the function along with the intervals on which the function is increasing or decreasing.

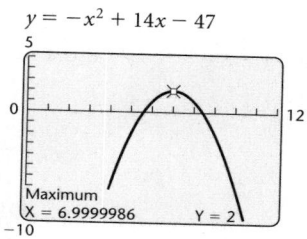

$y = -x^2 + 14x - 47$

Maximum
X = 6.9999986   Y = 2

❀ Applications

Many real-world situations involve finding the maximum or minimum value of a quadratic function.

EXAMPLE 5 *Maximizing Area.* A stonemason has enough stones to enclose a rectangular patio with 60 ft of stone wall. If the house forms one side of the rectangle, what is the maximum area that the mason can enclose? What should the dimensions of the patio be in order to yield this area?

Solution We will use the five-step problem-solving strategy.

PROBLEM-SOLVING STRATEGY

REVIEW SECTION **1.5.**

1. **Familiarize.** We make a drawing of the situation, using w to represent the width of the patio, in feet. Then $(60 - 2w)$ feet of stone is available for the length. Suppose the patio were 10 ft wide. It would then be $60 - 2 \cdot 10 = 40$ ft long. The area would be $(10 \text{ ft})(40 \text{ ft}) = 400 \text{ ft}^2$. If the patio were 12 ft wide, it would be $60 - 2 \cdot 12 = 36$ ft long. The area would be $(12 \text{ ft})(36 \text{ ft}) = 432 \text{ ft}^2$. If it were 16 ft wide, it would be $60 - 2 \cdot 16 = 28$ ft long and the area would be $(16 \text{ ft})(28 \text{ ft}) = 448 \text{ ft}^2$. There are more combinations of length and width than we could possibly try. Instead we will find a function that represents the area and then determine the maximum value of the function.

2. **Translate.** Since the area of a rectangle is given by length times width, we have

$$A(w) = (60 - 2w)w \qquad A = lw; l = 60 - 2w$$
$$= -2w^2 + 60w,$$

where $A(w)$ is the area of the patio, in square feet, as a function of the width w.

3. **Carry out.** To solve this problem, we need to determine the maximum value of $A(w)$ and find the dimensions for which that maximum occurs. Since A is a quadratic function and w^2 has a negative coefficient, we know that the function has a maximum value that occurs at the vertex of the graph of the function. The first coordinate of the vertex, $(w, A(w))$, is

$$w = -\frac{b}{2a} = -\frac{60}{2(-2)} = 15.$$

Thus, if $w = 15$ ft, then the length $l = 60 - 2 \cdot 15 = 30$ ft, and the area is $15 \cdot 30$, or 450 ft^2.

4. **Check.** As a partial check, we note that 450 ft$^2 > 448$ ft^2, which is the largest area we found in a guess in the *Familiarize* step. As a more complete check, assuming that the function $A(w)$ is correct, we could examine a table of values for $A(w) = (60 - 2w)w$ and/or examine its graph.

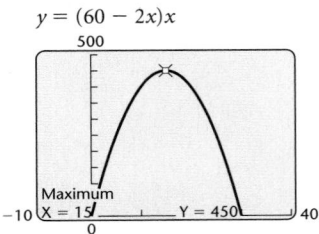

X	Y₁
14.7	449.82
14.8	449.92
14.9	449.98
15	450
15.1	449.98
15.2	449.92
15.3	449.82

X = 15

5. **State.** The maximum possible area is 450 ft^2 when the patio is 15 ft wide and 30 ft long.

Now Try Exercise 45. ■

EXAMPLE 6 *Height of a Rocket.* A model rocket is launched with an initial velocity of 100 ft/sec from the top of a hill that is 20 ft high. Its height t seconds after it has been launched is given by the function $s(t) = -16t^2 + 100t + 20$. Determine the time at which the rocket reaches its maximum height and find the maximum height.

Solution

1., 2. Familiarize and Translate. We are given the function in the statement of the problem: $s(t) = -16t^2 + 100t + 20$.

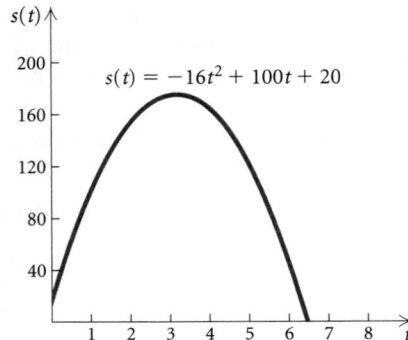

3. **Carry out.** We need to find the maximum value of the function and the value of t for which it occurs. Since $s(t)$ is a quadratic function and t^2 has a negative coefficient, we know that the maximum value of the function occurs at the vertex of the graph of the function. The first coordinate of the vertex gives the time t at which the rocket reaches its maximum height. It is

$$t = -\frac{b}{2a} = -\frac{100}{2(-16)} = 3.125.$$

The second coordinate of the vertex gives the maximum height of the rocket. We substitute in the function to find it:

$$s(3.125) = -16(3.125)^2 + 100(3.125) + 20 = 176.25.$$

4. Check. As a check, we can complete the square to write the function in the form $s(t) = a(t - h)^2 + k$ and determine the coordinates of the vertex from this form of the function. We get

$$s(t) = -16(t - 3.125)^2 + 176.25.$$

This confirms that the vertex is $(3.125, 176.25)$, so the answer checks.

5. State. The rocket reaches a maximum height of 176.25 ft 3.125 sec after it has been launched.

Now Try Exercise 41. ■

EXAMPLE 7 *Finding the Depth of a Well.* Two seconds after a chlorine tablet has been dropped into a well, a splash is heard. The speed of sound is 1100 ft/sec. How far is the top of the well from the water?

Solution

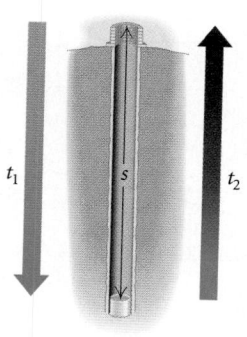

1. Familiarize. We first make a drawing and label it with known and unknown information. We let s = the depth of the well, in feet, t_1 = the time, in seconds, that it takes for the tablet to hit the water, and t_2 = the time, in seconds, that it takes for the sound to reach the top of the well. This gives us the equation

$$t_1 + t_2 = 2. \tag{1}$$

2. Translate. Can we find any relationship between the two times and the distance s? Often in problem solving you may need to look up related formulas in a physics book, another mathematics book, or on the Internet. We find that the formula

$$s = 16t^2$$

gives the distance, in feet, that a dropped object falls in t seconds. The time t_1 that it takes the tablet to hit the water can be found as follows:

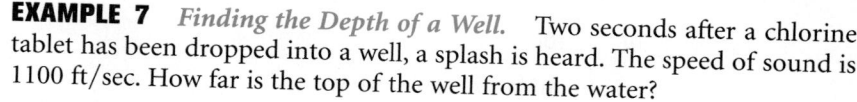

$$s = 16t_1^2, \quad \text{or} \quad \frac{s}{16} = t_1^2, \quad \text{so} \quad t_1 = \frac{\sqrt{s}}{4}. \qquad \text{Taking the positive square root} \tag{2}$$

To find an expression for t_2, the time it takes the sound to travel to the top of the well, recall that *Distance* = *Rate · Time*. Thus,

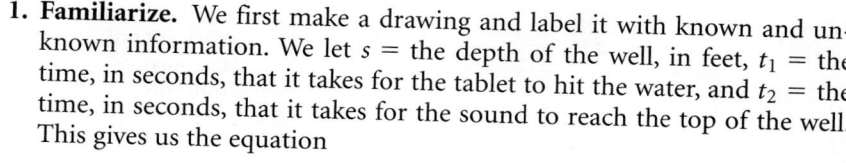

$$s = 1100t_2, \quad \text{or} \quad t_2 = \frac{s}{1100}. \tag{3}$$

We now have expressions for t_1 and t_2, both in terms of s. Substituting into equation (1), we obtain

$$t_1 + t_2 = 2, \quad \text{or} \quad \frac{\sqrt{s}}{4} + \frac{s}{1100} = 2. \tag{4}$$

3. Carry out.

ALGEBRAIC SOLUTION

We solve equation (4) for s. Multiplying by 1100, we get

$$275\sqrt{s} + s = 2200, \quad \text{or} \quad s + 275\sqrt{s} - 2200 = 0.$$

This equation is reducible to quadratic with $u = \sqrt{s}$. Substituting, we get

$$u^2 + 275u - 2200 = 0.$$

Using the quadratic formula, we can solve for u:

$$u = \frac{-b \pm \sqrt{b^2 - 4ac}}{2a}$$

$$= \frac{-275 + \sqrt{275^2 - 4 \cdot 1 \cdot (-2200)}}{2 \cdot 1} \qquad \begin{array}{l} \text{We want only} \\ \text{the positive} \\ \text{solution.} \end{array}$$

$$= \frac{-275 + \sqrt{84,425}}{2}$$

$$\approx 7.78.$$

Since $u \approx 7.78$, we have

$$\sqrt{s} = 7.78$$

$$s \approx 60.5. \qquad \text{Squaring both sides}$$

GRAPHICAL SOLUTION

We use the Intersect method. It will probably require some trial and error to determine an appropriate window.

$$y_1 = \frac{\sqrt{x}}{4} + \frac{x}{1100}, \quad y_2 = 2$$

4. **Check.** To check, we can substitute 60.5 for s in equation (4) and see that $t_1 + t_2 \approx 2$. We leave the mathematics to the student.

5. **State.** The top of the well is about 60.5 ft above the water.

Now Try Exercise 55. ■

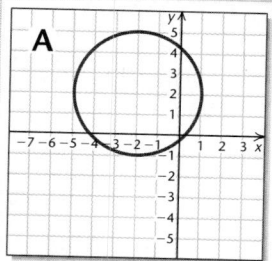

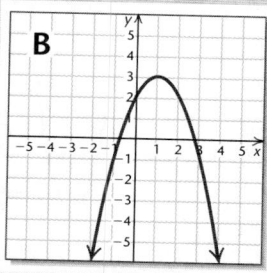

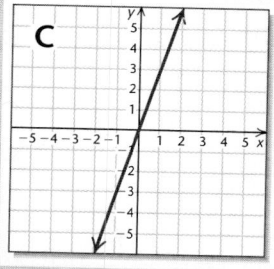

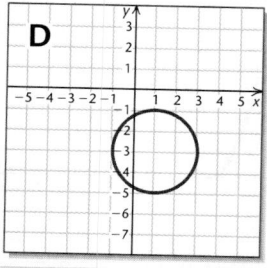

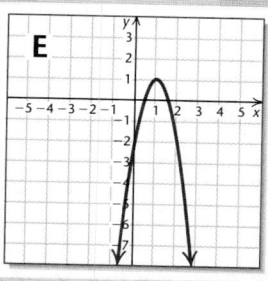

Visualizing the Graph

Match the equation with its graph.

1. $y = 3x$

2. $y = -(x - 1)^2 + 3$

3. $(x + 2)^2 + (y - 2)^2 = 9$

4. $y = 3$

5. $2x - 3y = 6$

6. $(x - 1)^2 + (y + 3)^2 = 4$

7. $y = -2x + 1$

8. $y = 2x^2 - x - 4$

9. $x = -2$

10. $y = -3x^2 + 6x - 2$

Answers on page A-19

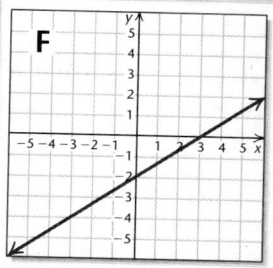

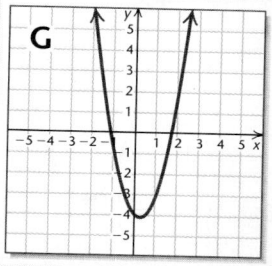

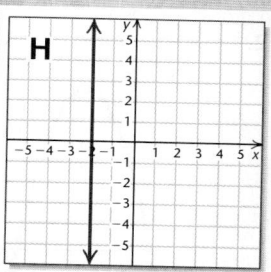

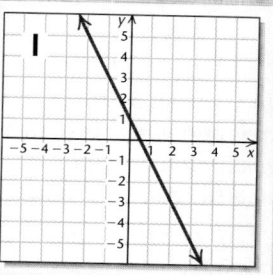

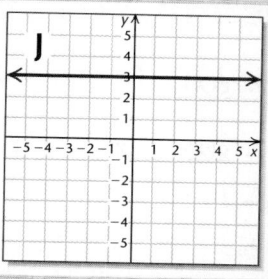

(3.3)   Exercise Set

In Exercises 1 and 2, use the given graph to find each of the following: (a) *the vertex;* (b) *the axis of symmetry; and* (c) *the maximum or minimum value of the function.*

1.

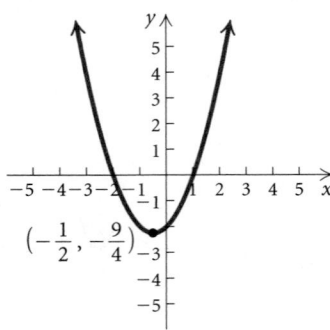

$\left(-\dfrac{1}{2}, -\dfrac{9}{4}\right)$

2.

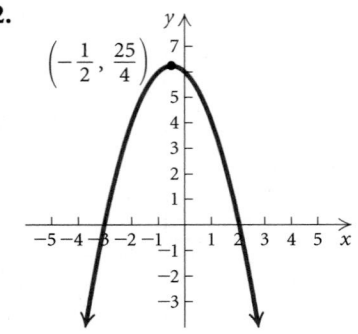

$\left(-\dfrac{1}{2}, \dfrac{25}{4}\right)$

practice

In Exercises 3–16, (a) *find the vertex;* (b) *find the axis of symmetry;* (c) *determine whether there is a maximum or minimum value and find that value; and* (d) *graph the function.*

3. $f(x) = x^2 - 8x + 12$ **4.** $g(x) = x^2 + 7x - 8$

5. $f(x) = x^2 - 7x + 12$ **6.** $g(x) = x^2 - 5x + 6$

7. $f(x) = x^2 + 4x + 5$ **8.** $f(x) = x^2 + 2x + 6$

9. $g(x) = \dfrac{x^2}{2} + 4x + 6$ **10.** $g(x) = \dfrac{x^2}{3} - 2x + 1$

11. $g(x) = 2x^2 + 6x + 8$

12. $f(x) = 2x^2 - 10x + 14$

13. $f(x) = -x^2 - 6x + 3$

14. $f(x) = -x^2 - 8x + 5$

15. $g(x) = -2x^2 + 2x + 1$

16. $f(x) = -3x^2 - 3x + 1$

In Exercises 17–24, match the equation with one of the graphs (a)–(h), which follow.

a)

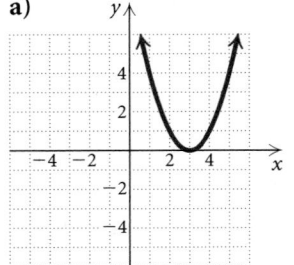

b)

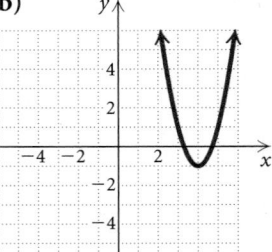

c)

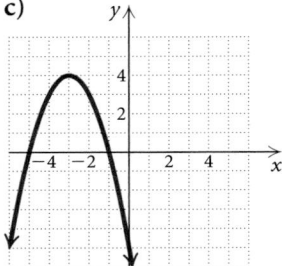

d)

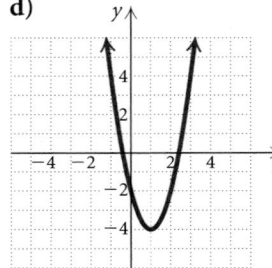

e)

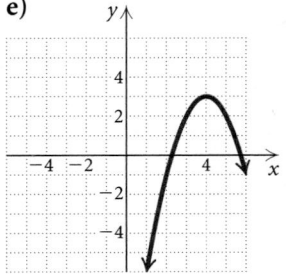

f)

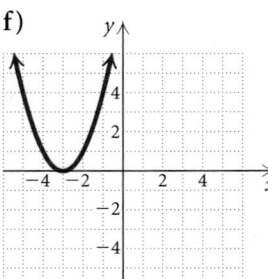

g)

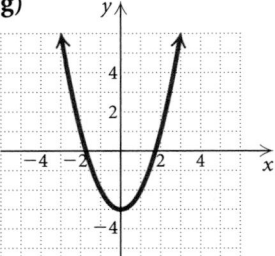

h)

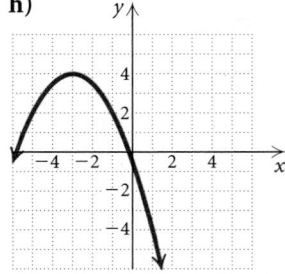

17. $y = (x + 3)^2$

18. $y = -(x - 4)^2 + 3$

19. $y = 2(x - 4)^2 - 1$

20. $y = x^2 - 3$

21. $y = -\frac{1}{2}(x + 3)^2 + 4$

22. $y = (x - 3)^2$

23. $y = -(x + 3)^2 + 4$

24. $y = 2(x - 1)^2 - 4$

Determine whether the statement is true or false.

25. The function $f(x) = -3x^2 + 2x + 5$ has a maximum value.

26. The vertex of the graph of $f(x) = ax^2 + bx + c$ is $-\dfrac{b}{2a}$.

27. The graph of $h(x) = (x + 2)^2$ can be obtained by translating the graph of $h(x) = x^2$ right 2 units.

28. The vertex of the graph of the function $g(x) = 2(x - 4)^2 - 1$ is $(-4, -1)$.

29. The axis of symmetry of the function $f(x) = -(x + 2)^2 - 4$ is $x = -2$.

30. The minimum value of the function $f(x) = 3(x - 1)^2 + 5$ is 5.

In Exercises 31–40:

a) *Find the vertex.*
b) *Determine whether there is a maximum or minimum value and find that value.*
c) *Find the range.*
d) *Find the intervals on which the function is increasing and the intervals on which the function is decreasing.*

31. $f(x) = x^2 - 6x + 5$

32. $f(x) = x^2 + 4x - 5$

33. $f(x) = 2x^2 + 4x - 16$

34. $f(x) = \frac{1}{2}x^2 - 3x + \frac{5}{2}$

35. $f(x) = -\frac{1}{2}x^2 + 5x - 8$

36. $f(x) = -2x^2 - 24x - 64$

37. $f(x) = 3x^2 + 6x + 5$

38. $f(x) = -3x^2 + 24x - 49$

39. $g(x) = -4x^2 - 12x + 9$

40. $g(x) = 2x^2 - 6x + 5$

41. *Height of a Ball.* A ball is thrown directly upward from a height of 6 ft with an initial velocity of 20 ft/sec. The function $s(t) = -16t^2 + 20t + 6$ gives the height of the ball t seconds after it has been thrown. Determine the time at which the ball reaches its maximum height and find the maximum height.

42. *Height of a Projectile.* A stone is thrown directly upward from a height of 30 ft with an initial velocity of 60 ft/sec. The height of the stone t seconds after it has been thrown is given by the function $s(t) = -16t^2 + 60t + 30$. Determine the time at which the stone reaches its maximum height and find the maximum height.

43. *Height of a Rocket.* A model rocket is launched with an initial velocity of 120 ft/sec from a height of 80 ft. The height of the rocket t seconds after it has been launched is given by the function $s(t) = -16t^2 + 120t + 80$. Determine the time at which the rocket reaches its maximum height and find the maximum height.

44. *Height of a Rocket.* A model rocket is launched with an initial velocity of 150 ft/sec from a height of 40 ft. The function $s(t) = -16t^2 + 150t + 40$ gives the height of the rocket t seconds after it has been launched. Determine the time at which the rocket reaches its maximum height and find the maximum height.

45. *Maximizing Volume.* Mendoza Manufacturing plans to produce a one-compartment vertical file by bending the long side of a 10-in. by 18-in. sheet of plastic along two lines to form a ⊔-shape. How tall should the file be in order to maximize the volume that it can hold?

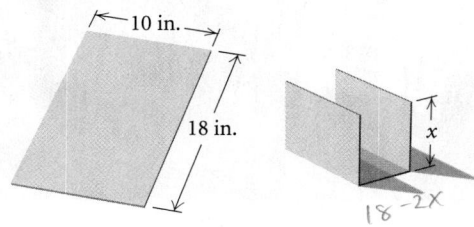

10 in.

18 in.

x

$18 - 2x$

46. *Maximizing Area.* A fourth-grade class decides to enclose a rectangular garden, using the side of the school as one side of the rectangle. What is the maximum area that the class can enclose with 32 ft of fence? What should the dimensions of the garden be in order to yield this area?

47. *Maximizing Area.* The sum of the base and the height of a triangle is 20 cm. Find the dimensions for which the area is a maximum.

48. *Maximizing Area.* The sum of the base and the height of a parallelogram is 69 cm. Find the dimensions for which the area is a maximum.

49. *Minimizing Cost.* Classic Furniture Concepts has determined that when x hundred wooden chairs are built, the average cost per chair is given by

$$C(x) = 0.1x^2 - 0.7x + 1.625,$$

where $C(x)$ is in hundreds of dollars. How many chairs should be built in order to minimize the average cost per chair?

Maximizing Profit. *In business, profit is the difference between revenue and cost; that is,*

$$\text{Total profit} = \text{Total revenue} - \text{Total cost},$$
$$P(x) = R(x) - C(x),$$

where x is the number of units sold. Find the maximum profit and the number of units that must be sold in order to yield the maximum profit for each of the following.

50. $R(x) = 5x$, $C(x) = 0.001x^2 + 1.2x + 60$

51. $R(x) = 50x - 0.5x^2$, $C(x) = 10x + 3$

52. $R(x) = 20x - 0.1x^2$, $C(x) = 4x + 2$

53. *Maximizing Area.* A rancher needs to enclose two adjacent rectangular corrals, one for cattle and one for sheep. If a river forms one side of the corrals and 240 yd of fencing is available, what is the largest total area that can be enclosed?

54. *Norman Window.* A Norman window is a rectangle with a semicircle on top. Sky Blue Windows is designing a Norman window that will require 24 ft of trim on the outer edges. What dimensions will allow the maximum amount of light to enter a house?

A Norman window

55. *Finding the Height of an Elevator Shaft.*   Jenelle drops a screwdriver from the top of an elevator shaft. Exactly 5 sec later, she hears the sound of the screwdriver hitting the bottom of the shaft. How tall is the elevator shaft? (*Hint*: See Example 7.)

56. *Finding the Height of a Cliff.*   A water balloon is dropped from a cliff. Exactly 3 sec later, the sound of the balloon hitting the ground reaches the top of the cliff. How high is the cliff? (*Hint*: See Example 7.)

Collaborative Discussion and Writing

57. Write a problem for a classmate to solve. Design it so that it is a maximum or minimum problem using a quadratic function.

58. Discuss two ways in which we used completing the square in this chapter.

59. Suppose that the graph of $f(x) = ax^2 + bx + c$ has x-intercepts $(x_1, 0)$ and $(x_2, 0)$. What are the x-intercepts of $g(x) = -ax^2 - bx - c$? Explain.

Skill Maintenance

For each function f, construct and simplify the difference quotient

$$\frac{f(x + h) - f(x)}{h}.$$

60. $f(x) = 3x - 7$

61. $f(x) = 2x^2 - x + 4$

A graph of $y = f(x)$ follows. No formula is given for f. Make a hand-drawn graph of each of the following.

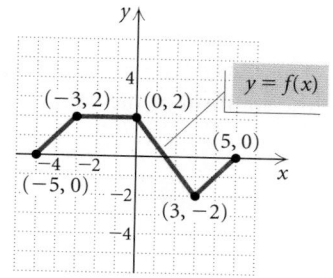

62. $g(x) = f(2x)$

63. $g(x) = -2f(x)$

Synthesis

64. Find b such that
$$f(x) = -4x^2 + bx + 3$$
has a maximum value of 50.

65. Find c such that
$$f(x) = -0.2x^2 - 3x + c$$
has a maximum value of -225.

66. Find a quadratic function with vertex $(4, -5)$ and containing the point $(-3, 1)$.

67. Graph: $f(x) = (|x| - 5)^2 - 3$.

68. *Minimizing Area.*   A 24-in. piece of string is cut into two pieces. One piece is used to form a circle while the other is used to form a square. How should the string be cut so that the sum of the areas is a minimum?

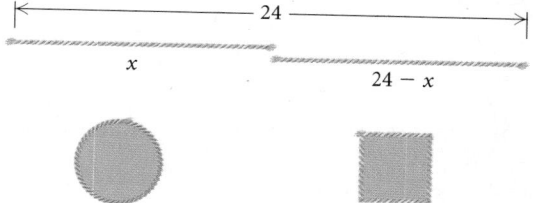

3.4 Solving Rational Equations and Radical Equations

❖ Solve rational equations.
❖ Solve radical equations.

❖ Rational Equations

Equations containing rational expressions are called **rational equations**. Solving such equations involves multiplying both sides by the least common denominator (LCD) to *clear the equation of fractions*.

EXAMPLE 1 Solve: $\dfrac{x-8}{3} + \dfrac{x-3}{2} = 0.$

ALGEBRAIC SOLUTION

We have

$$\dfrac{x-8}{3} + \dfrac{x-3}{2} = 0$$ The LCD is $3 \cdot 2$, or 6.

$$6\left(\dfrac{x-8}{3} + \dfrac{x-3}{2}\right) = 6 \cdot 0$$ Multiplying by the LCD on both sides to clear fractions

$$6 \cdot \dfrac{x-8}{3} + 6 \cdot \dfrac{x-3}{2} = 0$$

$$2(x-8) + 3(x-3) = 0$$

$$2x - 16 + 3x - 9 = 0$$

$$5x - 25 = 0$$

$$5x = 25$$

$$x = 5.$$

The possible solution is 5. We check using a table in ASK mode.

$$y = \dfrac{x-8}{3} + \dfrac{x-3}{2}$$

X	Y₁
5	0

X =

Since the value of $\dfrac{x-8}{3} + \dfrac{x-3}{2}$ is 0 when $x = 5$, the number 5 is the solution.

GRAPHICAL SOLUTION

We use the Zero method. The solution of the equation

$$\dfrac{x-8}{3} + \dfrac{x-3}{2} = 0$$

is the zero of the function

$$f(x) = \dfrac{x-8}{3} + \dfrac{x-3}{2}.$$

$$y = \dfrac{x-8}{3} + \dfrac{x-3}{2}$$

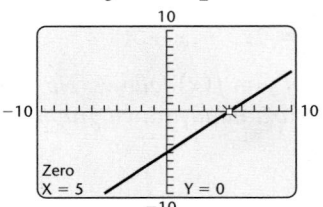

The zero of the function is 5. Thus the solution of the equation is 5.

Now Try Exercise 3.

> **CAUTION!** Clearing fractions is a valid procedure when solving rational equations but not when adding, subtracting, multiplying, or dividing rational expressions. A rational expression may have operation signs but it will have no equals sign. A rational equation *always* has an equals sign. For example, $\dfrac{x-8}{3} + \dfrac{x-3}{2}$ is a rational expression but $\dfrac{x-8}{3} + \dfrac{x-3}{2} = 0$ is a rational equation.
>
> To *simplify* the rational *expression* $\dfrac{x-8}{3} + \dfrac{x-3}{2}$, we first find the LCD and write each fraction with that denominator. The final result is usually a rational expression.
>
> To *solve* the rational *equation* $\dfrac{x-8}{3} + \dfrac{x-3}{2} = 0$, we first multiply both sides by the LCD to clear fractions. The final result is one or more numbers. As we will see in Example 2, these numbers must be checked in the original equation.

When we use the multiplication principle to multiply (or divide) on both sides of an equation by an expression with a variable, we might not obtain an equivalent equation. We must check the possible solutions obtained in this manner by substituting them in the original equation. The next example illustrates this.

EXAMPLE 2 Solve: $\dfrac{x^2}{x-3} = \dfrac{9}{x-3}$.

Solution The LCD is $x - 3$.

$$(x-3) \cdot \frac{x^2}{x-3} = (x-3) \cdot \frac{9}{x-3}$$

$$x^2 = 9$$

$$x = -3 \quad or \quad x = 3 \qquad \text{Using the principle of square roots}$$

The possible solutions are -3 and 3. We check.

Check: For -3:

$$\frac{x^2}{x-3} = \frac{9}{x-3}$$

$$\frac{(-3)^2}{-3-3} \;\overset{?}{\underset{|}{\,}}\; \frac{9}{-3-3}$$

$$\frac{9}{-6} \;\bigg|\; \frac{9}{-6} \qquad \text{TRUE}$$

For 3:

$$\frac{x^2}{x-3} = \frac{9}{x-3}$$

$$\frac{3^2}{3-3} \;\overset{?}{\underset{|}{\,}}\; \frac{9}{3-3}$$

$$\frac{9}{0} \;\bigg|\; \frac{9}{0} \qquad \text{NOT DEFINED}$$

The number -3 checks, so it is a solution. Since division by 0 is not defined, 3 is not a solution.

We can also use a table on a graphing calculator to check the possible solutions.

$$\text{Enter } y_1 = \frac{x^2}{x-3} \quad \text{and} \quad y_2 = \frac{9}{x-3}.$$

$$y_1 = \frac{x^2}{x-3}, \ y_2 = \frac{9}{x-3}$$

X	Y₁	Y₂
−3	−1.5	−1.5
3	ERROR	ERROR

X =

When $x = -3$, we see that $y_1 = -1.5 = y_2$, so -3 is a solution. When $x = 3$, we get ERROR messages. This indicates that 3 is not in the domain of y_1 or y_2 and thus is not a solution. **Now Try Exercise 23.** ◼

EXAMPLE 3 Solve: $\dfrac{2}{3x+6} + \dfrac{1}{x^2-4} = \dfrac{4}{x-2}$.

Solution We first factor the denominators in order to determine the LCD:

$$\frac{2}{3(x+2)} + \frac{1}{(x+2)(x-2)} = \frac{4}{x-2} \qquad \begin{matrix}\text{The LCD is}\\ 3(x+2)(x-2).\end{matrix}$$

$$3(x+2)(x-2)\left(\frac{2}{3(x+2)} + \frac{1}{(x+2)(x-2)}\right) = 3(x+2)(x-2)\cdot\frac{4}{x-2}$$

$$\qquad\qquad \text{Multiplying by the LCD to clear fractions}$$

$$2(x-2) + 3 = 3\cdot 4(x+2)$$
$$2x - 4 + 3 = 12x + 24$$
$$2x - 1 = 12x + 24$$
$$-10x = 25$$
$$x = -\frac{5}{2}.$$

The possible solution is $-\dfrac{5}{2}$. We check this on a graphing calculator.

$$\text{Enter } y_1 = \frac{2}{3x+6} + \frac{1}{x^2-4} \quad \text{and} \quad y_2 = \frac{4}{x-2}.$$

$$y_1 = \frac{2}{3x+6} + \frac{1}{x^2-4}, \ y_2 = \frac{4}{x-2}$$

X	Y₁	Y₂
−2.5	−.8889	−.8889

X =

We see that $y_1 = y_2$ when $x = -\dfrac{5}{2}$, or -2.5, so $-\dfrac{5}{2}$ is the solution.

Now Try Exercise 19. ▉

✿ Radical Equations

A **radical equation** is an equation in which variables appear in one or more radicands. For example,

$$\sqrt{2x - 5} - \sqrt{x - 3} = 1$$

is a radical equation. The following principle is used to solve such equations.

> ### The Principle of Powers
> For any positive integer n:
>
> If $a = b$ is true, then $a^n = b^n$ is true.

EXAMPLE 4 Solve: $\sqrt{3x + 1} = 4$.

ALGEBRAIC SOLUTION

We have

$$\sqrt{3x + 1} = 4$$
$$\left(\sqrt{3x + 1}\right)^2 = 4^2$$

Using the principle of powers; squaring both sides

$$3x + 1 = 16$$
$$3x = 15$$
$$x = 5.$$

Check:

$$\begin{array}{c|c} \sqrt{3x + 1} = 4 \\ \hline \sqrt{3 \cdot 5 + 1} \; ? \; 4 \\ \sqrt{15 + 1} \\ \sqrt{16} \\ 4 \;\big|\; 4 \quad \text{TRUE} \end{array}$$

The solution is 5.

GRAPHICAL SOLUTION

We graph $y_1 = \sqrt{3x + 1}$ and $y_2 = 4$ and then use the INTERSECT feature. We see that the solution is 5.

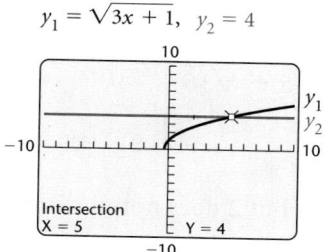

$y_1 = \sqrt{3x + 1}, \quad y_2 = 4$

Check:

X	Y₁	Y₂
5	4	4

X =

The check confirms that the solution is 5.

Now Try Exercise 29. ▉

In Example 4, the radical was isolated on one side of the equation. If this had not been the case, our first step would have been to isolate the radical. We do so in the next example.

EXAMPLE 5   Solve: $5 + \sqrt{x + 7} = x$.

ALGEBRAIC SOLUTION

We have

$$5 + \sqrt{x + 7} = x$$

$$\sqrt{x + 7} = x - 5 \qquad \text{Subtracting 5 on both sides to isolate the radical}$$

$$\left(\sqrt{x + 7}\right)^2 = (x - 5)^2 \qquad \text{Using the principle of powers; squaring both sides}$$

$$x + 7 = x^2 - 10x + 25$$

$$0 = x^2 - 11x + 18 \qquad \text{Subtracting } x \text{ and 7}$$

$$0 = (x - 9)(x - 2) \qquad \text{Factoring}$$

$$x - 9 = 0 \quad \text{or} \quad x - 2 = 0$$

$$x = 9 \quad \text{or} \qquad x = 2.$$

The possible solutions are 9 and 2.

Check:

For 9:

$$\begin{array}{c|c} 5 + \sqrt{x + 7} = x \\ \hline 5 + \sqrt{9 + 7} \ ? \ 9 \\ 5 + \sqrt{16} \\ 5 + 4 \\ 9 \ \bigm| \ 9 \quad \text{TRUE} \end{array}$$

For 2:

$$\begin{array}{c|c} 5 + \sqrt{x + 7} = x \\ \hline 5 + \sqrt{2 + 7} \ ? \ 2 \\ 5 + \sqrt{9} \\ 5 + 3 \\ 8 \ \bigm| \ 2 \quad \text{FALSE} \end{array}$$

Since 9 checks but 2 does not, the only solution is 9.

GRAPHICAL SOLUTION

We graph $y_1 = 5 + \sqrt{x + 7}$ and $y_2 = x$. Using the INTERSECT feature, we see that the solution is 9.

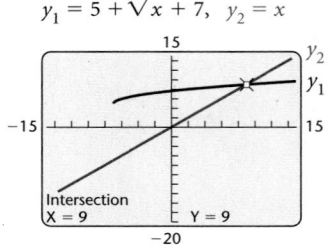

$y_1 = 5 + \sqrt{x + 7}, \quad y_2 = x$

We can also use the ZERO feature to get this result. To do so, we first write the equivalent equation $5 + \sqrt{x + 7} - x = 0$. The zero of the function $f(x) = 5 + \sqrt{x + 7} - x$ is 9, so the solution of the original equation is 9.

$y = 5 + \sqrt{x + 7} - x$

Note that the graphs show that the equation has only one solution.

Now Try Exercise 53. ■

When we raise both sides of an equation to an even power, the resulting equation can have solutions that the original equation does not. This is because the converse of the principle of powers is not necessarily true. That is, if $a^n = b^n$ is true, we do not know that $a = b$ is true. For example, $(-2)^2 = 2^2$, but $-2 \neq 2$. Thus, as we see in Example 5, it is necessary to check the possible solutions in the original equation when the principle of powers is used to raise both sides of an equation to an even power.

When a radical equation has two radical terms on one side, we isolate one of them and then use the principle of powers. If, after doing so, a radical term remains, we repeat these steps.

STUDY TIP

Consider forming a study group with some of your fellow students. Exchange e-mail addresses, telephone numbers, and schedules so you can coordinate study time for homework and tests.

EXAMPLE 6   Solve: $\sqrt{x - 3} + \sqrt{x + 5} = 4$.

Solution   We have

$$\sqrt{x - 3} = 4 - \sqrt{x + 5} \qquad \text{Isolating one radical}$$

$$\left(\sqrt{x - 3}\right)^2 = \left(4 - \sqrt{x + 5}\right)^2 \qquad \begin{array}{l}\text{Using the principle of powers;}\\ \text{squaring both sides}\end{array}$$

$$x - 3 = 16 - 8\sqrt{x + 5} + (x + 5)$$

$$x - 3 = 21 - 8\sqrt{x + 5} + x \qquad \text{Combining like terms}$$

$$-24 = -8\sqrt{x + 5} \qquad \begin{array}{l}\text{Isolating the remaining}\\ \text{radical; subtracting } x \text{ and } 21\\ \text{on both sides}\end{array}$$

$$3 = \sqrt{x + 5} \qquad \text{Dividing by } -8 \text{ on both sides}$$

$$3^2 = \left(\sqrt{x + 5}\right)^2 \qquad \begin{array}{l}\text{Using the principle of powers;}\\ \text{squaring both sides}\end{array}$$

$$9 = x + 5$$

$$4 = x. \qquad \text{Subtracting 5 on both sides}$$

We check the possible solution, 4, on a graphing calculator.

$$y_1 = \sqrt{x - 3} + \sqrt{x + 5}, \ y_2 = 4$$

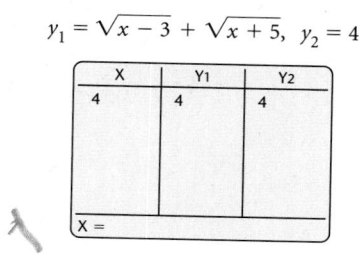

Since $y_1 = y_2$ when $x = 4$, the number 4 checks. It is the solution.

Now Try Exercise 63. ■

3.4 Exercise Set

Solve.

1. $\dfrac{1}{4} + \dfrac{1}{5} = \dfrac{1}{t}$

2. $\dfrac{1}{3} - \dfrac{5}{6} = \dfrac{1}{x}$

3. $\dfrac{x+2}{4} - \dfrac{x-1}{5} = 15$

4. $\dfrac{t+1}{3} - \dfrac{t-1}{2} = 1$

5. $\dfrac{1}{2} + \dfrac{2}{x} = \dfrac{1}{3} + \dfrac{3}{x}$

6. $\dfrac{1}{t} + \dfrac{1}{2t} + \dfrac{1}{3t} = 5$

7. $\dfrac{5}{3x+2} = \dfrac{3}{2x}$

8. $\dfrac{2}{x-1} = \dfrac{3}{x+2}$

9. $x + \dfrac{6}{x} = 5$

10. $x - \dfrac{12}{x} = 1$

11. $\dfrac{6}{y+3} + \dfrac{2}{y} = \dfrac{5y-3}{y^2-9}$

12. $\dfrac{3}{m+2} + \dfrac{2}{m} = \dfrac{4m-4}{m^2-4}$

13. $\dfrac{2x}{x-1} = \dfrac{5}{x-3}$

14. $\dfrac{2x}{x+7} = \dfrac{5}{x+1}$

15. $\dfrac{2}{x+5} + \dfrac{1}{x-5} = \dfrac{16}{x^2-25}$

16. $\dfrac{2}{x^2-9} + \dfrac{5}{x-3} = \dfrac{3}{x+3}$

17. $\dfrac{3x}{x+2} + \dfrac{6}{x} = \dfrac{12}{x^2+2x}$

18. $\dfrac{3y+5}{y^2+5y} + \dfrac{y+4}{y+5} = \dfrac{y+1}{y}$

19. $\dfrac{1}{5x+20} - \dfrac{1}{x^2-16} = \dfrac{3}{x-4}$

20. $\dfrac{1}{4x+12} - \dfrac{1}{x^2-9} = \dfrac{5}{x-3}$

21. $\dfrac{2}{5x+5} - \dfrac{3}{x^2-1} = \dfrac{4}{x-1}$

22. $\dfrac{1}{3x+6} - \dfrac{1}{x^2-4} = \dfrac{3}{x-2}$

23. $\dfrac{8}{x^2-2x+4} = \dfrac{x}{x+2} + \dfrac{24}{x^3+8}$

24. $\dfrac{18}{x^2-3x+9} - \dfrac{x}{x+3} = \dfrac{81}{x^3+27}$

25. $\dfrac{x}{x-4} - \dfrac{4}{x+4} = \dfrac{32}{x^2-16}$

26. $\dfrac{x}{x-1} - \dfrac{1}{x+1} = \dfrac{2}{x^2-1}$

27. $\dfrac{1}{x-6} - \dfrac{1}{x} = \dfrac{6}{x^2-6x}$

28. $\dfrac{1}{x-15} - \dfrac{1}{x} = \dfrac{15}{x^2-15x}$

29. $\sqrt{3x-4} = 1$

30. $\sqrt{4x+1} = 3$

31. $\sqrt{2x-5} = 2$

32. $\sqrt{3x+2} = 6$

33. $\sqrt{7-x} = 2$

34. $\sqrt{5-x} = 1$

35. $\sqrt{1-2x} = 3$

36. $\sqrt{2-7x} = 2$

37. $\sqrt[3]{5x - 2} = -3$

38. $\sqrt[3]{2x + 1} = -5$

39. $\sqrt[4]{x^2 - 1} = 1$

40. $\sqrt[5]{3x + 4} = 2$

41. $\sqrt{y - 1} + 4 = 0$

42. $\sqrt{m + 1} - 5 = 8$

43. $\sqrt{b + 3} - 2 = 1$

44. $\sqrt{x - 4} + 1 = 5$

45. $\sqrt{z + 2} + 3 = 4$

46. $\sqrt{y - 5} - 2 = 3$

47. $\sqrt{2x + 1} - 3 = 3$

48. $\sqrt{3x - 1} + 2 = 7$

49. $\sqrt{2 - x} - 4 = 6$

50. $\sqrt{5 - x} + 2 = 8$

51. $\sqrt[3]{6x + 9} + 8 = 5$

52. $\sqrt[5]{2x - 3} - 1 = 1$

53. $\sqrt{x + 4} + 2 = x$

54. $\sqrt{x + 1} + 1 = x$

55. $\sqrt{x - 3} + 5 = x$

56. $\sqrt{x + 3} - 1 = x$

57. $\sqrt{x + 7} = x + 1$

58. $\sqrt{6x + 7} = x + 2$

59. $\sqrt{3x + 3} = x + 1$

60. $\sqrt{2x + 5} = x - 5$

61. $\sqrt{5x + 1} = x - 1$

62. $\sqrt{7x + 4} = x + 2$

63. $\sqrt{x - 3} + \sqrt{x + 2} = 5$

64. $\sqrt{x} - \sqrt{x - 5} = 1$

65. $\sqrt{3x - 5} + \sqrt{2x + 3} + 1 = 0$

66. $\sqrt{2m - 3} = \sqrt{m + 7} - 2$

67. $\sqrt{x} - \sqrt{3x - 3} = 1$

68. $\sqrt{2x + 1} - \sqrt{x} = 1$

69. $\sqrt{2y - 5} - \sqrt{y - 3} = 1$

70. $\sqrt{4p + 5} + \sqrt{p + 5} = 3$

71. $\sqrt{y + 4} - \sqrt{y - 1} = 1$

72. $\sqrt{y + 7} + \sqrt{y + 16} = 9$

73. $\sqrt{x + 5} + \sqrt{x + 2} = 3$

74. $\sqrt{6x + 6} = 5 + \sqrt{21 - 4x}$

75. $x^{1/3} = -2$

76. $t^{1/5} = 2$

77. $t^{1/4} = 3$

78. $m^{1/2} = -7$

Solve.

79. $\dfrac{P_1 V_1}{T_1} = \dfrac{P_2 V_2}{T_2}$, for T_1
(A chemistry formula for gases)

80. $\dfrac{1}{F} = \dfrac{1}{m} + \dfrac{1}{p}$, for F
(A formula from optics)

81. $W = \sqrt{\dfrac{1}{LC}}$, for C
(An electricity formula)

82. $s = \sqrt{\dfrac{A}{6}}$, for A
(A geometry formula)

83. $\dfrac{1}{R} = \dfrac{1}{R_1} + \dfrac{1}{R_2}$, for R_2
(A formula for resistance)

84. $\dfrac{1}{t} = \dfrac{1}{a} + \dfrac{1}{b}$, for t
(A formula for work rate)

85. $I = \sqrt{\dfrac{A}{P}} - 1$, for P
(A compound-interest formula)

86. $T = 2\pi \sqrt{\dfrac{1}{g}}$, for g
(A pendulum formula)

87. $\dfrac{1}{F} = \dfrac{1}{m} + \dfrac{1}{p}$, for p
(A formula from optics)

88. $\dfrac{V^2}{R^2} = \dfrac{2g}{R + h}$, for h
(A formula for escape velocity)

Collaborative Discussion and Writing

89. Explain why it is necessary to check the possible solutions of a rational equation.

90. Explain in your own words why it is necessary to check the possible solutions when the principle of powers is used to solve an equation.

Skill Maintenance

Find the zero of the function.

91. $f(x) = 15 - 2x$

92. $f(x) = -3x + 9$

93. *Sleep-starved Americans.* Nearly half of Americans say they do not get enough sleep. Many turn to sleeping pills, filling 42 million prescriptions for these medications in 2005. This amount was 60% more than the number of prescriptions filled for sleeping pills in 2000. (*Sources*: "NBC Today Show"/Zogby International poll; IMS Health) How many prescriptions for sleeping pills were filled in 2000?

94. *Big Sites.* Together, the Mall of America in Minnesota and the Disneyland theme park in California occupy 181 acres of land. The Mall of America occupies 11 acres more than Disneyland. (*Sources*: Mall of America; Disneyland) How much land does each occupy?

Synthesis

Solve.

95. $(x - 3)^{2/3} = 2$

96. $\dfrac{x + 3}{x + 2} - \dfrac{x + 4}{x + 3} = \dfrac{x + 5}{x + 4} - \dfrac{x + 6}{x + 5}$

97. $\sqrt{x + 5} + 1 = \dfrac{6}{\sqrt{x + 5}}$

98. $\sqrt{15 + \sqrt{2x + 80}} = 5$

99. $x^{2/3} = x$

3.5

Solving Equations and Inequalities with Absolute Value

❖ Solve equations with absolute value.
❖ Solve inequalities with absolute value.

ABSOLUTE VALUE

REVIEW SECTION **R.1.**

❖ Equations with Absolute Value

Recall that the absolute value of a number is its distance from 0 on the number line. We use this concept to solve equations with absolute value.

> For $a > 0$ and an algebraic expression X:
>
> $|X| = a$ is equivalent to $X = -a$ or $X = a$.

EXAMPLE 1 Solve: $|x| = 5$.

ALGEBRAIC SOLUTION

We have

$$|x| = 5$$
$$x = -5 \quad or \quad x = 5.$$
Writing an equivalent statement

The solutions are -5 and 5.

To check, note that -5 and 5 are both 5 units from 0 on the number line.

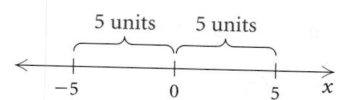

GRAPHICAL SOLUTION

Using the Intersect method, we graph $y_1 = |x|$ and $y_2 = 5$ and find the first coordinates of the points of intersection.

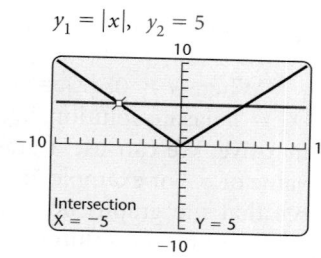 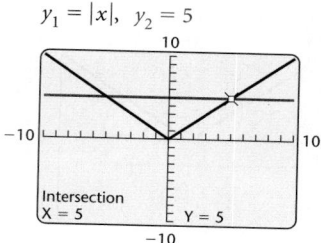

The solutions are -5 and 5.

We could also have used the Zero method to get this result, graphing $y = |x| - 5$ and using the ZERO feature twice.

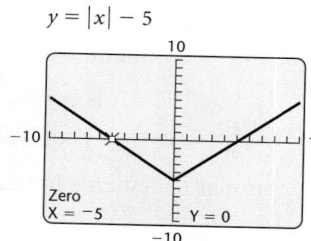 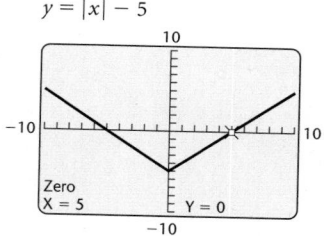

The zeros of $f(x) = |x| - 5$ are -5 and 5, so the solutions of the original equation are -5 and 5.

Now Try Exercise 1. ◼

EXAMPLE 2 Solve: $|x - 3| - 1 = 4$.

Solution First we add 1 on both sides to get an expression of the form $|X| = a$:

$$|x - 3| - 1 = 4$$
$$|x - 3| = 5$$
$$x - 3 = -5 \quad or \quad x - 3 = 5 \qquad |X| = a \text{ is equivalent to } X = -a \text{ or } X = a.$$
$$x = -2 \quad or \qquad x = 8. \qquad \text{Adding 3}$$

Check: For -2:

$$\frac{|x - 3| - 1 = 4}{}$$

$|-2 - 3| - 1 \; ? \; 4$
$|-5| - 1$
$5 - 1$
$4 \quad | \quad 4 \quad$ TRUE

For 8:

$$\frac{|x - 3| - 1 = 4}{}$$

$|8 - 3| - 1 \; ? \; 4$
$|5| - 1$
$5 - 1$
$4 \quad | \quad 4 \quad$ TRUE

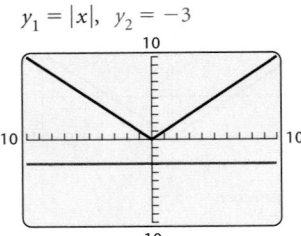

$y_1 = |x|, \; y_2 = -3$

The solutions are -2 and 8.

Now Try Exercise 21. ■

When $a = 0$, $|X| = a$ is equivalent to $X = 0$. Note that for $a < 0$, $|X| = a$ has no solution, because the absolute value of an expression is never negative. We can use a graph to illustrate the last statement for a specific value of a. For example, if we let $a = -3$ and graph $y = |x|$ and $y = -3$, we see that the graphs do not intersect, as shown at left. Thus the equation $|x| = -3$ has no solution. The solution set is the **empty set**, denoted $\varnothing$.

❈ Inequalities with Absolute Value

Inequalities sometimes contain absolute-value notation. The following properties are used to solve them.

> For $a > 0$ and an algebraic expression X:
>
> $\quad |X| < a \quad$ is equivalent to $\quad -a < X < a.$
> $\quad |X| > a \quad$ is equivalent to $\quad X < -a \; or \; X > a.$
>
> Similar statements hold for $|X| \le a$ and $|X| \ge a$.

For example,

$\quad |x| < 3$ is equivalent to $-3 < x < 3;$
$\quad |y| \ge 1$ is equivalent to $y \le -1 \; or \; y \ge 1;$ and
$\quad |2x + 3| \le 4$ is equivalent to $-4 \le 2x + 3 \le 4.$

EXAMPLE 3   Solve and graph the solution set: $|3x + 2| < 5.$

Solution   We have

$\quad |3x + 2| < 5$
$\qquad -5 < 3x + 2 < 5 \qquad$ Writing an equivalent inequality
$\qquad -7 < 3x < 3 \qquad$ Subtracting 2
$\qquad -\frac{7}{3} < x < 1. \qquad$ Dividing by 3

The solution set is $\left\{x \middle| -\frac{7}{3} < x < 1\right\}$, or $\left(-\frac{7}{3}, 1\right)$. The graph of the solution set is shown below.

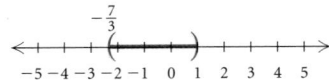

To perform a partial check with a graphing calculator, we graph $y = |3x + 2| < 5$ in DOT mode. The calculator graphs a segment 1 unit above the x-axis for the values of x for which this expression for y is true. The graph shows that the solution is probably correct.

$$y = |3x + 2| < 5$$

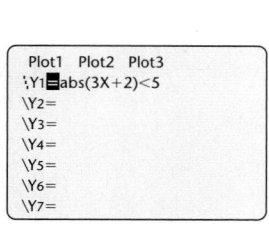

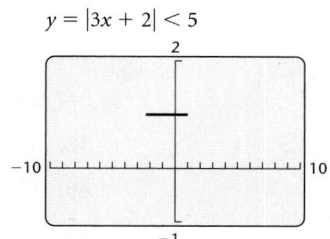

Now Try Exercise 45. ◼

EXAMPLE 4   Solve and graph the solution set: $|5 - 2x| \geq 1$.

Solution   We have

$$|5 - 2x| \geq 1$$

$5 - 2x \leq -1$	*or*	$5 - 2x \geq 1$	Writing an equivalent inequality
$-2x \leq -6$	*or*	$-2x \geq -4$	Subtracting 5
$x \geq 3$	*or*	$x \leq 2.$	Dividing by -2 and reversing the inequality signs

The solution set is $\{x \mid x \leq 2 \text{ or } x \geq 3\}$, or $(-\infty, 2] \cup [3, \infty)$. The graph of the solution set is shown below.

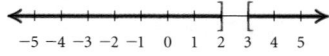

Now Try Exercise 47. ◼

3.5 Exercise Set

Solve.

1. $|x| = 7$

2. $|x| = 4.5$

3. $|x| = 0$

4. $|x| = \frac{3}{2}$

5. $|x| = \frac{5}{6}$

6. $|x| = -\frac{3}{5}$

7. $|x| = -10.7$

8. $|x| = 12$

9. $|3x| = 1$

10. $|5x| = 4$

11. $|8x| = 24$

12. $|6x| = 0$

13. $|x - 1| = 4$

14. $|x - 7| = 5$

15. $|x + 2| = 6$

16. $|x + 5| = 1$

17. $|3x + 2| = 1$

18. $|7x - 4| = 8$

19. $\left|\frac{1}{2}x - 5\right| = 17$

20. $\left|\frac{1}{3}x - 4\right| = 13$

21. $|x - 1| + 3 = 6$

22. $|x + 2| - 5 = 9$

23. $|x + 3| - 2 = 8$

24. $|x - 4| + 3 = 9$

25. $|3x + 1| - 4 = -1$

26. $|2x - 1| - 5 = -3$

27. $|4x - 3| + 1 = 7$

28. $|5x + 4| + 2 = 5$

29. $12 - |x + 6| = 5$

30. $9 - |x - 2| = 7$

31. $7 - |2x - 1| = 6$

32. $5 - |4x + 3| = 2$

Solve and write interval notation for the solution set. Then graph the solution set.

33. $|x| < 7$

34. $|x| \leq 4.5$

35. $|x| \leq 2$

36. $|x| < 3$

37. $|x| \geq 4.5$

38. $|x| > 7$

39. $|x| > 3$

40. $|x| \geq 2$

41. $|3x| < 1$

42. $|5x| \leq 4$

43. $|2x| \geq 6$

44. $|4x| > 20$

45. $|x + 8| < 9$

46. $|x + 6| \leq 10$

47. $|x + 8| \geq 9$

48. $|x + 6| > 10$

49. $\left|x - \frac{1}{4}\right| < \frac{1}{2}$

50. $|x - 0.5| \leq 0.2$

51. $|2x + 3| \leq 9$

52. $|3x + 4| < 13$

53. $|x - 5| > 0.1$

54. $|x - 7| \geq 0.4$

55. $|6 - 4x| \geq 8$

56. $|5 - 2x| > 10$

57. $\left|x + \frac{2}{3}\right| \leq \frac{5}{3}$

58. $\left|x + \frac{3}{4}\right| < \frac{1}{4}$

59. $\left|\dfrac{2x + 1}{3}\right| > 5$

60. $\left|\dfrac{2x - 1}{3}\right| \geq \dfrac{5}{6}$

61. $|2x - 4| < -5$

62. $|3x + 5| < 0$

Collaborative Discussion and Writing

63. Explain why $|x| < p$ has no solution for $p \leq 0$.

64. Explain why all real numbers are solutions of $|x| > p$, for $p < 0$.

Skill Maintenance

In each of Exercises 65–72, fill in the blank with the correct term. Some of the given choices will not be used.

distance formula	symmetric with respect
midpoint formula	to the x-axis
function	symmetric with respect
relation	to the y-axis
x-intercept	symmetric with respect
y-intercept	to the origin
perpendicular	increasing
parallel	decreasing
horizontal lines	constant
vertical lines	

65. A(n) _____ is a point $(0, b)$.

66. The _____ is $d = \sqrt{(x_2 - x_1)^2 + (y_2 - y_1)^2}$.

67. A(n) _____ is a correspondence such that each member of the domain corresponds to at least one member of the range.

68. A(n) _____ is a correspondence such that each member of the domain corresponds to exactly one member of the range.

69. _____ are given by equations of the type $y = b$, or $f(x) = b$.

70. Nonvertical lines are _____ if and only if they have the same slope and different y-intercepts.

71. A function f is said to be _____ on an open interval I if, for all a and b in that interval, $a < b$ implies $f(a) > f(b)$.

72. For an equation $y = f(x)$, if replacing x with $-x$ produces an equivalent equation, then the graph is _____ .

Synthesis

Solve.

73. $|3x - 1| > 5x - 2$

74. $|x + 2| \leq |x - 5|$

75. $|p - 4| + |p + 4| < 8$

76. $|x| + |x + 1| < 10$

77. $|x - 3| + |2x + 5| > 6$

CHAPTER 3   Summary and Review

Important Properties and Formulas

Complex Number: $a + bi$, a, b real, $i^2 = -1$
Imaginary Number: $a + bi$, $b \neq 0$
Complex Conjugates: $a + bi$, $a - bi$

Quadratic Equation:

$$ax^2 + bx + c = 0, \ a \neq 0, \ a, b, c \text{ real}$$

Quadratic Function:

$$f(x) = ax^2 + bx + c, \ a \neq 0, \ a, b, c \text{ real}$$

Quadratic Formula:

For $ax^2 + bx + c = 0, a \neq 0$,

$$x = \frac{-b \pm \sqrt{b^2 - 4ac}}{2a}.$$

Graphing Quadratic Functions

The graph of the function $f(x) = a(x - h)^2 + k$ is a parabola that

- opens up if $a > 0$ and down if $a < 0$;
- has (h, k) as the vertex;
- has $x = h$ as the axis of symmetry;
- has k as a minimum value (output) if $a > 0$;
- has k as a maximum value if $a < 0$.

The Vertex of a Parabola

The vertex of the graph of $f(x) = ax^2 + bx + c$ is

$$\left(-\frac{b}{2a}, f\left(-\frac{b}{2a}\right)\right).$$

The Principle of Powers

For any positive integer n:

If $a = b$ is true, then $a^n = b^n$ is true.

Principles for Solving Inequalities

The Addition Principle for Inequalities:
If $a < b$ is true, then $a + c < b + c$ is true.

The Multiplication Principle for Inequalities:
If $a < b$ and $c > 0$ are true, then $ac < bc$ is true.
If $a < b$ and $c < 0$ are true, then $ac > bc$ is true.

Similar statements hold for $\leq$.

Equations and Inequalities with Absolute Value

For $a > 0$,

$$|X| = a \longrightarrow X = -a \text{ or } X = a,$$
$$|X| < a \longrightarrow -a < X < a,$$
$$|X| > a \longrightarrow X < -a \text{ or } X > a.$$

Review Exercises

Determine whether the statement is true or false.

1. We can use the quadratic formula to solve any quadratic equation.   [3.2]

2. The function $f(x) = -3(x + 4)^2 - 1$ has a maximum value.   [3.3]

3. For any positive integer n, if $a^n = b^n$ is true, then $a = b$ is true.   [3.4]

4. An equation with absolute value cannot have two negative-number solutions.   [3.5]

Solve.   [3.2]

5. $(2y + 5)(3y - 1) = 0$

6. $x^2 + 4x - 5 = 0$

7. $3x^2 + 2x = 8$

8. $5x^2 = 15$

9. $x^2 + 10 = 0$

Find the zero(s) of the function.   [3.2]

10. $f(x) = x^2 - 2x + 1$

11. $f(x) = x^2 + 2x - 15$

12. $f(x) = 2x^2 - x - 5$

13. $f(x) = 3x^2 + 2x + 3$

Solve.

14. $\dfrac{5}{2x + 3} + \dfrac{1}{x - 6} = 0$ [3.4]

15. $\dfrac{3}{8x + 1} + \dfrac{8}{2x + 5} = 1$ [3.4]

16. $\sqrt{5x + 1} - 1 = \sqrt{3x}$ [3.4]

17. $\sqrt{x - 1} - \sqrt{x - 4} = 1$ [3.4]

18. $|x - 4| = 3$ [3.5]

19. $|2y + 7| = 9$ [3.5]

Solve and write interval notation for the solution set. Then graph the solution set.   [3.5]

20. $|5x| \geq 15$ 21. $|3x + 4| < 10$

22. $|6x - 1| < 5$ 23. $|x + 4| \geq 2$

24. Solve $\dfrac{1}{M} + \dfrac{1}{N} = \dfrac{1}{P}$ for P.   [3.4]

Express in terms of i.   [3.1]

25. $-\sqrt{-40}$

26. $\sqrt{-12} \cdot \sqrt{-20}$

27. $\dfrac{\sqrt{-49}}{-\sqrt{-64}}$

Simplify each of the following. Write the answer in the form $a + bi$, where a and b are real numbers.   [3.1]

28. $(6 + 2i) + (-4 - 3i)$

29. $(3 - 5i) - (2 - i)$

30. $(6 + 2i)(-4 - 3i)$

31. $\dfrac{2 - 3i}{1 - 3i}$

32. i^{23}

Solve by completing the square to obtain exact solutions. Show your work.   [3.2]

33. $x^2 - 3x = 18$

34. $3x^2 - 12x - 6 = 0$

Solve. Give exact solutions.   [3.2]

35. $3x^2 + 10x = 8$

36. $r^2 - 2r + 10 = 0$

37. $x^2 = 10 + 3x$

38. $x = 2\sqrt{x} - 1$

39. $y^4 - 3y^2 + 1 = 0$

40. $(x^2 - 1)^2 - (x^2 - 1) - 2 = 0$

41. $(p - 3)(3p + 2)(p + 2) = 0$

42. $x^3 + 5x^2 - 4x - 20 = 0$

In Exercises 43 and 44, complete the square to:

a) *find the vertex;*
b) *find the axis of symmetry;*
c) *determine whether there is a maximum or minimum value and find that value;*
d) *find the range; and*
e) *graph the function.*   [3.3]

43. $f(x) = -4x^2 + 3x - 1$

44. $f(x) = 5x^2 - 10x + 3$

In Exercises 45–48, match the equation with one of the figures (a)–(d), which follow. [3.3]

a)
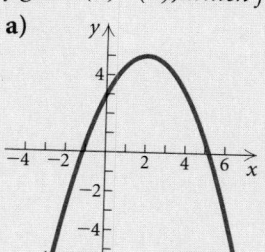

b)
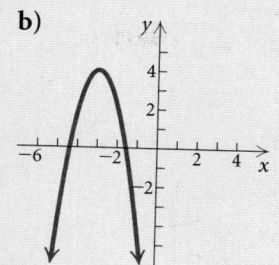

c)
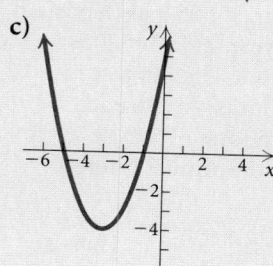

d)
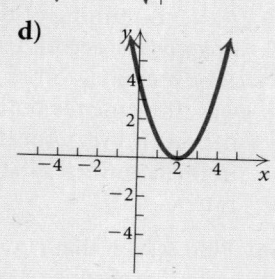

45. $y = (x - 2)^2$

46. $y = (x + 3)^2 - 4$

47. $y = -2(x + 3)^2 + 4$

48. $y = -\frac{1}{2}(x - 2)^2 + 5$

49. *Legs of a Right Triangle.* The hypotenuse of a right triangle is 50 ft. One leg is 10 ft longer than the other. What are the lengths of the legs? [3.2]

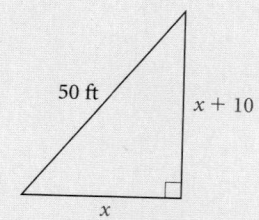

50. *Motion.* A Riverboat Cruise Line boat travels 8 mi upstream and 8 mi downstream. The total time for both parts of the trip is 3 hr. The speed of the stream is 2 mph. What is the speed of the boat in still water? [3.4]

51. *Motion.* Two freight trains leave the same city at right angles. The first train travels at a speed of 60 km/h. In 1 hr, the trains are 100 km apart. How fast is the second train traveling? [3.2]

52. *Sidewalk Width.* A 60-ft by 80-ft parking lot is torn up to install a sidewalk of uniform width around its perimeter. The new area of the parking lot is two-thirds of the old area. How wide is the sidewalk? [3.2]

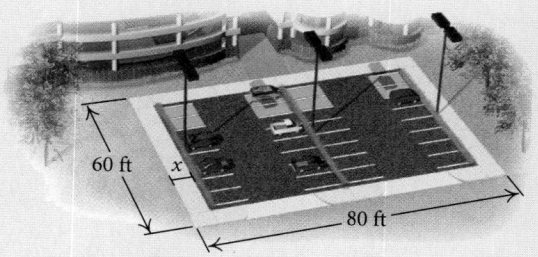

53. *Maximizing Volume.* The Berniers have 24 ft of flexible fencing with which to build a rectangular "toy corral." If the fencing is 2 ft high, what dimensions should the corral have in order to maximize its volume? [3.3]

54. *Dimensions of a Box.* An open box is made from a 10-cm by 20-cm piece of aluminum by cutting a square from each corner and folding up the edges. The area of the resulting base is 90 cm². What is the length of the sides of the squares? [3.2]

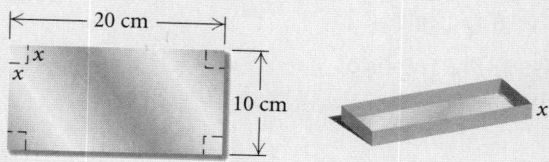

55. Find the zeros of $f(x) = 2x^2 - 5x + 1$. [3.2]

A. $\dfrac{5 \pm \sqrt{17}}{2}$ B. $\dfrac{5 \pm \sqrt{17}}{4}$

C. $\dfrac{5 \pm \sqrt{33}}{4}$ D. $\dfrac{-5 \pm \sqrt{17}}{4}$

56. Solve: $\sqrt{4x + 1} + \sqrt{2x} = 1$. [3.4]

A. There are two solutions.
B. There is only one solution. It is less than 1.
C. There is only one solution. It is greater than 1.
D. There is no solution.

57. The graph of $f(x) = (x - 2)^2 - 3$ is which of the following? [3.3]

A.

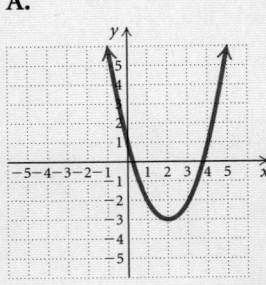

B.

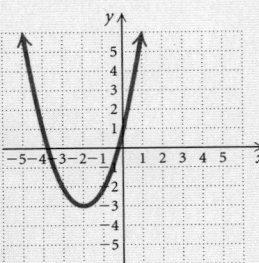

C.

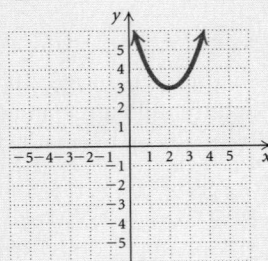

D.

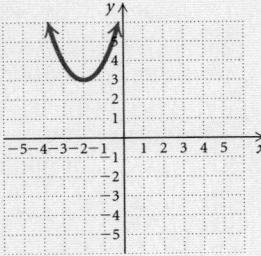

Collaborative Discussion and Writing

58. Explain how to write a quadratic equation that can be solved algebraically but not graphically. [3.2]

59. If the graphs of

$$f(x) = a_1(x - h_1)^2 + k_1$$

and

$$g(x) = a_2(x - h_2)^2 + k_2$$

have the same shape, what, if anything, can you conclude about the a's, the h's, and the k's? Explain your answer. [3.3]

Synthesis

Solve.

60. $\sqrt{\sqrt{\sqrt{\sqrt{x}}}} = 2$ [3.4]

61. $(t - 4)^{4/5} = 3$ [3.4]

62. $(x - 1)^{2/3} = 4$ [3.4]

63. $(2y - 2)^2 + y - 1 = 5$ [3.2]

64. $\sqrt{x + 2} + \sqrt[4]{x + 2} - 2 = 0$ [3.2]

65. At the beginning of the year, $3500 was deposited in a savings account. One year later, $4000 was deposited in another account. The interest rate was the same for both accounts. At the end of the second year, there was a total of $8518.35 in the accounts. What was the annual interest rate? [3.2]

66. Find b such that $f(x) = -3x^2 + bx - 1$ has a maximum value of 2. [3.3]

CHAPTER 3   Test

Solve. Find exact solutions.

1. $(2x - 1)(x + 5) = 0$

2. $6x^2 - 36 = 0$

3. $x^2 + 4 = 0$

4. $x^2 - 2x - 3 = 0$

5. $x^2 - 5x + 3 = 0$

6. $2t^2 - 3t + 4 = 0$

7. $x + 5\sqrt{x} - 36 = 0$

8. $\dfrac{3}{3x + 4} + \dfrac{2}{x - 1} = 2$

9. $\sqrt{x + 4} - 2 = 1$

10. $\sqrt{x + 4} - \sqrt{x - 4} = 2$

11. $|x + 4| = 7$

12. $|4y - 3| = 5$

Solve and write interval notation for the solution set. Then graph the solution set.

13. $|x + 3| \le 4$

14. $|2x - 1| < 5$

15. $|x + 5| > 2$

16. $|3x - 5| \ge 7$

17. Solve $\dfrac{1}{A} + \dfrac{1}{B} = \dfrac{1}{C}$ for B.

18. Solve $R = \sqrt{3np}$ for n.

19. Solve $x^2 + 4x = 1$ by completing the square. Find the exact solutions. Show your work.

20. *River Current.* Deke's boat travels 12 km/h in still water. Deke travels 45 km downstream and then returns 45 km upstream in a total time of 8 hr. Find the speed of the current.

Express in terms of i.

21. $\sqrt{-43}$

22. $-\sqrt{-25}$

Simplify.

23. $(5 - 2i) - (2 + 3i)$

24. $(3 + 4i)(2 - i)$

25. $\dfrac{1 - i}{6 + 2i}$

26. i^{33}

Find the zeros of each function.

27. $f(x) = 4x^2 - 11x - 3$

28. $f(x) = 2x^2 - x - 7$

29. For the graph of the function
$f(x) = -x^2 + 2x + 8$:

  a) Find the vertex.

  b) Find the axis of symmetry.

  c) State whether there is a maximum or minimum value and find that value.

  d) Find the range.

  e) Graph the function.

30. *Maximizing Area.* A homeowner wants to fence a rectangular play yard using 80 ft of fencing. The side of the house will be used as one side of the rectangle. Find the dimensions for which the area is a maximum.

31. The graph of $f(x) = x^2 - 2x - 1$ is which of the following?

A.

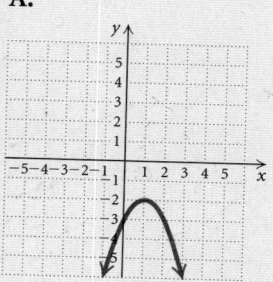

B.

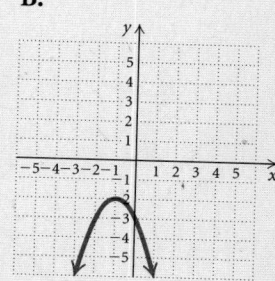

C.

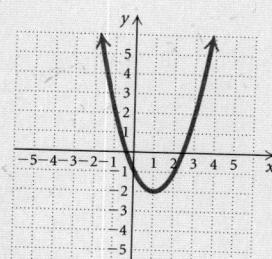

D.

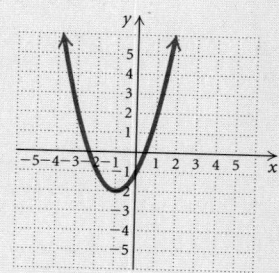

Synthesis

32. Find a such that $f(x) = ax^2 - 4x + 3$ has a maximum value of 12.

Polynomial and Rational Functions

APPLICATION The quartic function
$$f(x) = 0.2872960373x^4 - 5.04959855x^3$$
$$+ 31.68667444x^2 - 86.28334628x$$
$$+ 135.5819736,$$
where x is the number of years since 1998, can be used to estimate the number of tornado fatalities in the United States from 1998 to 2006. Using this function, estimate the number of deaths in 1998, in 2001, and in 2006.

This problem appears as Exercise 63 in Section 4.1.

4.1 Polynomial Functions and Modeling

4.2 Graphing Polynomial Functions

4.3 Polynomial Division; The Remainder and Factor Theorems

4.4 Theorems about Zeros of Polynomial Functions

4.5 Rational Functions

4.6 Polynomial and Rational Inequalities

4.1

Polynomial Functions and Modeling

❖ Determine the behavior of the graph of a polynomial function using the leading-term test.

❖ Factor polynomial functions and find their zeros and their multiplicities.

❖ Use a graphing calculator to graph a polynomial function and find its real-number zeros, its relative maximum and minimum values, and its domain and range.

❖ Solve applied problems using polynomial models; fit linear, quadratic, power, cubic, and quartic polynomial functions to data.

There are many different kinds of functions. The constant, linear, and quadratic functions that we studied in Chapters 1 and 3 are part of a larger group of functions called *polynomial functions.*

> ### Polynomial Function
> A **polynomial function** P is given by
> $$P(x) = a_n x^n + a_{n-1} x^{n-1} + a_{n-2} x^{n-2} + \cdots + a_1 x + a_0,$$
> where the coefficients $a_n, a_{n-1}, \ldots, a_1, a_0$ are real numbers and the exponents are whole numbers.

The first nonzero coefficient, a_n, is called the **leading coefficient**. The term $a_n x^n$ is called the **leading term**. The **degree** of the polynomial function is n. Some examples of polynomial functions follow.

POLYNOMIAL FUNCTION	EXAMPLE	DEGREE	LEADING TERM	LEADING COEFFICIENT
Constant	$f(x) = 3$ $(f(x) = 3 = 3x^0)$	0	3	3
Linear	$f(x) = \frac{2}{3}x + 5$ $\left(f(x) = \frac{2}{3}x + 5 = \frac{2}{3}x^1 + 5\right)$	1	$\frac{2}{3}x$	$\frac{2}{3}$
Quadratic	$f(x) = 4x^2 - x + 3$	2	$4x^2$	4
Cubic	$f(x) = x^3 + 2x^2 + x - 5$	3	x^3	1
Quartic	$f(x) = -x^4 - 1.1x^3 + 0.3x^2 - 2.8x - 1.7$	4	$-x^4$	-1

The function $f(x) = 0$ can be described in many ways:

$$f(x) = 0 = 0x^2 = 0x^{15} = 0x^{48},$$

and so on. For this reason, we say that the constant function $f(x) = 0$ has no degree.

Functions such as

$$f(x) = \frac{2}{x} + 5, \text{ or } 2x^{-1} + 5, \quad \text{and} \quad g(x) = \sqrt{x} - 6, \text{ or } x^{1/2} - 6,$$

are *not* polynomial functions because the exponents -1 and $\frac{1}{2}$ are *not* whole numbers.

From our study of functions in Chapters 1–3, we know how to find or at least estimate many characteristics of a polynomial function. Let's consider two examples for review.

Quadratic Function

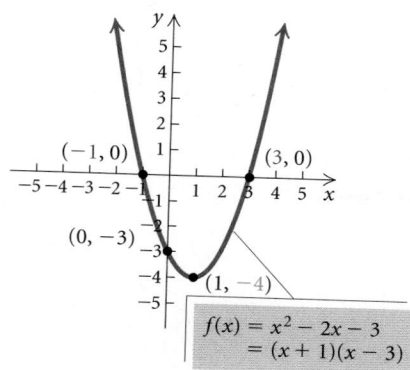

$f(x) = x^2 - 2x - 3$
$= (x + 1)(x - 3)$

Function: $f(x) = x^2 - 2x - 3$
$= (x + 1)(x - 3)$

Zeros: $-1, 3$

x-intercepts: $(-1, 0), (3, 0)$

y-intercept: $(0, -3)$

Minimum: -4 at $x = 1$

Maximum: None

Domain: All real numbers, $(-\infty, \infty)$

Range: $[-4, \infty)$

Cubic Function

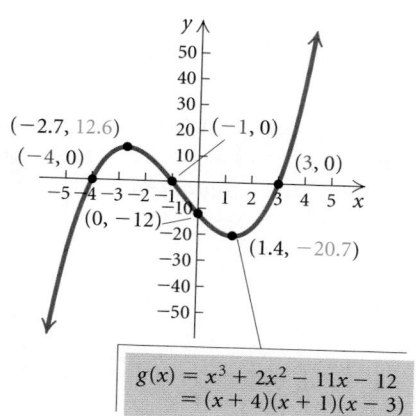

$g(x) = x^3 + 2x^2 - 11x - 12$
$= (x + 4)(x + 1)(x - 3)$

Function: $g(x) = x^3 + 2x^2 - 11x - 12$
$= (x + 4)(x + 1)(x - 3)$

Zeros: $-4, -1, 3$

x-intercepts: $(-4, 0), (-1, 0), (3, 0)$

y-intercept: $(0, -12)$

Relative minimum: -20.7 at $x = 1.4$

Relative maximum: 12.6 at $x = -2.7$

Domain: All real numbers, $(-\infty, \infty)$

Range: All real numbers, $(-\infty, \infty)$

All graphs of polynomial functions have some characteristics in common. Compare the following graphs. How do the graphs of polynomial functions differ from the graphs of nonpolynomial functions? Describe some characteristics of the graphs of polynomial functions that you observe.

Polynomial Functions

$f(x) = x^2 + 3x + 1$

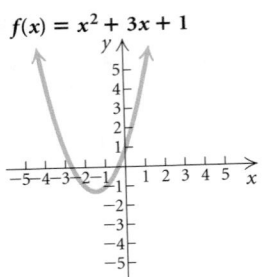

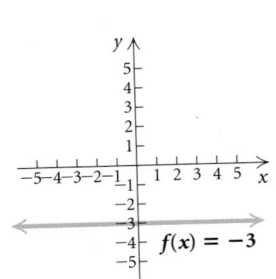

$f(x) = -3$

$f(x) = 2x^3 + x^2 + x - 1$

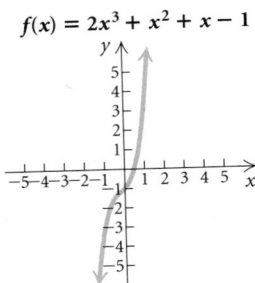

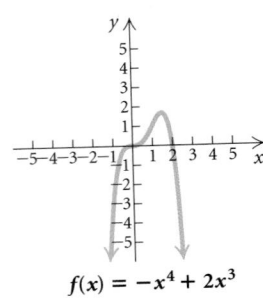

$f(x) = -x^4 + 2x^3$

Nonpolynomial Functions

$h(x) = |x + 2|$

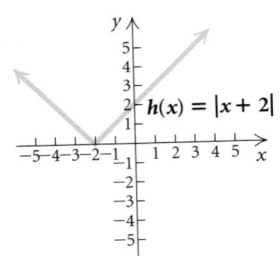

$h(x) = \sqrt{x + 1} - 1$

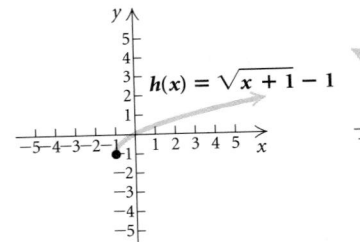

$h(x) = x^{4/5}$

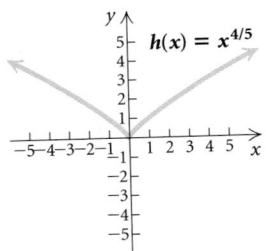

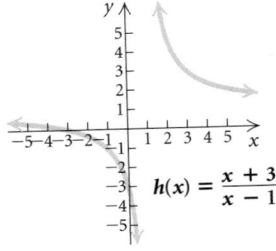

$h(x) = \dfrac{x + 3}{x - 1}$

You probably noted that the graph of a polynomial function is *continuous*; that is, it has no holes or breaks. It is also smooth; there are no sharp corners. Furthermore, the *domain* of a polynomial function is the set of all real numbers, $(-\infty, \infty)$.

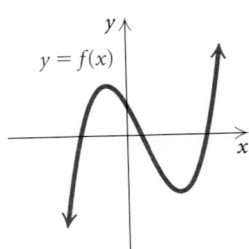

A continuous function

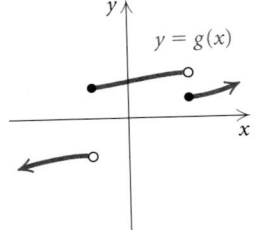

A discontinuous function

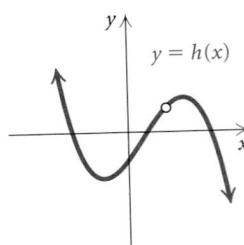

A discontinuous function

❋ The Leading-Term Test

The behavior of the graph of a polynomial function as x becomes very large ($x \to \infty$) or very small ($x \to -\infty$) is referred to as the end behavior of the graph. The leading term of a polynomial function determines its end behavior.

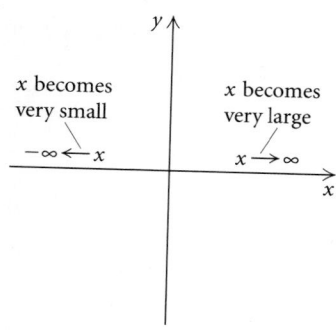

Using the graphs shown below, let's see if we can discover some general patterns by comparing the end behavior of even- and odd-degree functions. We also observe the effect of positive and negative leading coefficients.

Even Degree

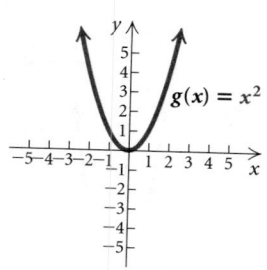

$g(x) = x^2$

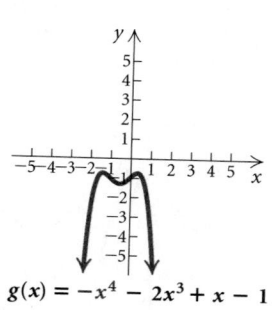

$g(x) = -x^4 - 2x^3 + x - 1$

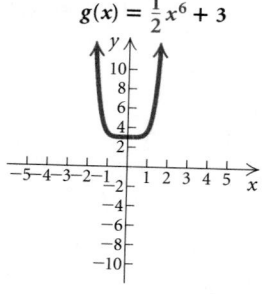

$g(x) = \frac{1}{2}x^6 + 3$

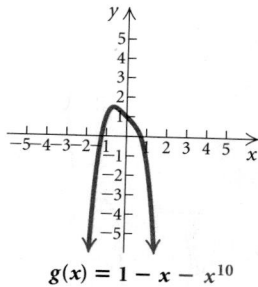

$g(x) = 1 - x - x^{10}$

Odd Degree

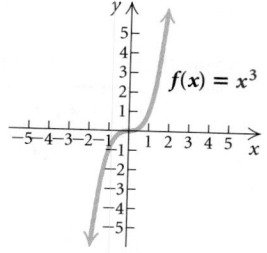

$f(x) = x^3$

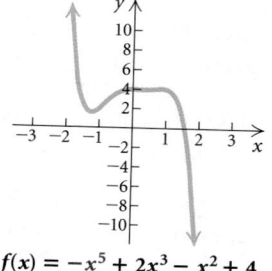

$f(x) = -x^5 + 2x^3 - x^2 + 4$

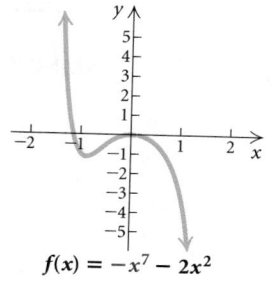

$f(x) = -x^7 - 2x^2$

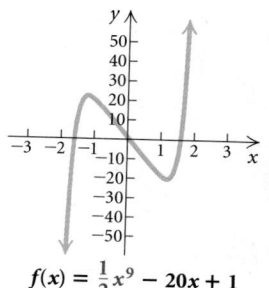

$f(x) = \frac{1}{2}x^9 - 20x + 1$

We can summarize our observations as follows.

The Leading-Term Test

If $a_n x^n$ is the leading term of a polynomial function, then the behavior of the graph as $x \to \infty$ or as $x \to -\infty$ can be described in one of the four following ways.

If n is even, and $a_n > 0$:   If n is even, and $a_n < 0$:

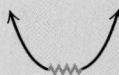   

If n is odd, and $a_n > 0$:   If n is odd, and $a_n < 0$:

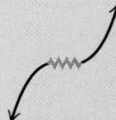   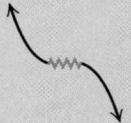

The ⌇⌇⌇ portion of the graph is not determined by this test.

EXAMPLE 1 Using the leading-term test, match each of the following functions with one of the graphs A–D, which follow.

a) $f(x) = 3x^4 - 2x^3 + 3$ **b)** $f(x) = -5x^3 - x^2 + 4x + 2$
c) $f(x) = x^5 + \frac{1}{4}x + 1$ **d)** $f(x) = -x^6 + x^5 - 4x^3$

A.
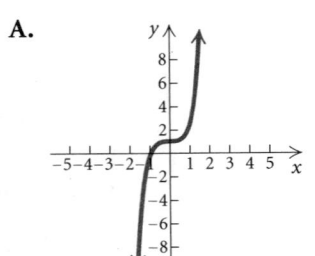

B.
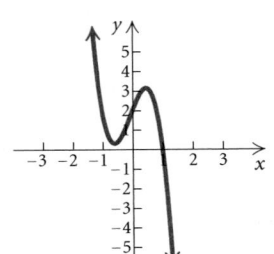

C.
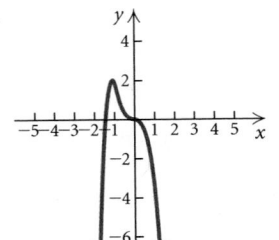

D.
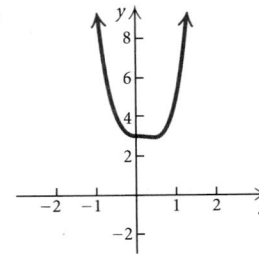

Solution

	LEADING TERM	DEGREE OF LEADING TERM	SIGN OF LEADING COEFFICIENT	GRAPH
a)	$3x^4$	4, even	Positive	D
b)	$-5x^3$	3, odd	Negative	B
c)	x^5	5, odd	Positive	A
d)	$-x^6$	6, even	Negative	C

Now Try Exercise 19. ◼

❇ Finding Zeros of Factored Polynomial Functions

Let's review the meaning of the real zeros of a function and their connection to the x-intercepts of the function's graph.

CONNECTING *the* CONCEPTS

Zeros, Solutions, and Intercepts

FUNCTION	ZEROS OF THE FUNCTION; SOLUTIONS OF THE EQUATION	ZEROS OF THE FUNCTION; X-INTERCEPTS OF THE GRAPH

Quadratic Polynomial
$g(x) = x^2 - 2x - 8$
$\quad = (x + 2)(x - 4),$
or
$\quad y = (x + 2)(x - 4)$

To find the **zeros** of $g(x)$, we solve $g(x) = 0$:
$$x^2 - 2x - 8 = 0$$
$$(x + 2)(x - 4) = 0$$
$$x + 2 = 0 \quad or \quad x - 4 = 0$$
$$x = -2 \quad or \quad x = 4.$$

The **solutions** of $x^2 - 2x - 8 = 0$ are -2 and 4. They are the zeros of the function $g(x)$. That is,
$$g(-2) = 0 \quad and \quad g(4) = 0.$$

The real-number zeros of $g(x)$ are the x-coordinates of the **x-intercepts** of the graph of $y = g(x)$.

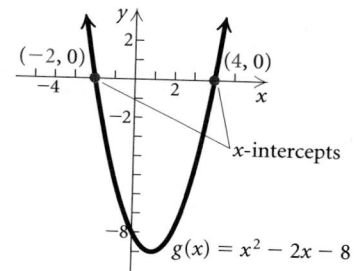

Cubic Polynomial
$h(x)$
$\quad = x^3 + 2x^2 - 5x - 6$
$\quad = (x + 3)(x + 1)(x - 2),$
or
$\quad y = (x + 3)(x + 1)(x - 2)$

To find the **zeros** of $h(x)$, we solve $h(x) = 0$:
$$x^3 + 2x^2 - 5x - 6 = 0$$
$$(x + 3)(x + 1)(x - 2) = 0$$
$$x + 3 = 0 \quad or \quad x + 1 = 0 \quad or \quad x - 2 = 0$$
$$x = -3 \quad or \quad x = -1 \quad or \quad x = 2.$$

The **solutions** of $x^3 + 2x^2 - 5x - 6 = 0$ are $-3, -1$, and 2. They are the zeros of the function $h(x)$. That is,
$$h(-3) = 0,$$
$$h(-1) = 0, \quad and$$
$$h(2) = 0.$$

The real-number zeros of $h(x)$ are the x-coordinates of the **x-intercepts** of the graph of $y = h(x)$.

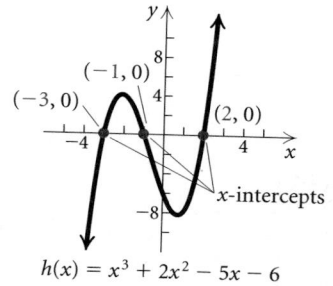

The connection between the real-number zeros of a function and the x-intercepts of the graph of the function is easily seen in the preceding examples. If c is a real zero of a function (that is, $f(c) = 0$), then $(c, 0)$ is an x-intercept of the graph of the function.

$y = x^3 + x^2 - 17x + 15$

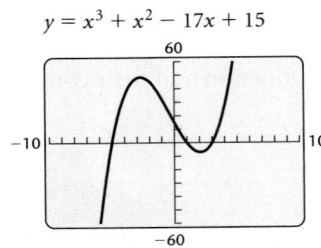

EXAMPLE 2   Consider $P(x) = x^3 + x^2 - 17x + 15$. Determine whether each of the numbers 2 and -5 is a zero of $P(x)$.

Solution   We first evaluate $P(2)$:

$$P(2) = (2)^3 + (2)^2 - 17(2) + 15 = -7.$$ Substituting 2 into the polynomial

Since $P(2) \neq 0$, we know that 2 is *not* a zero of the polynomial function. We then evaluate $P(-5)$:

$$P(-5) = (-5)^3 + (-5)^2 - 17(-5) + 15 = 0.$$ Substituting -5 into the polynomial

Since $P(-5) = 0$, we know that -5 is a zero of $P(x)$.

Now Try Exercise 23. ◼

Let's take a closer look at the polynomial function

$$h(x) = x^3 + 2x^2 - 5x - 6$$

(see Connecting the Concepts on p. 301). The factors of $h(x)$ are

$$x + 3, \qquad x + 1, \quad \text{and} \quad x - 2,$$

and the zeros are

$$-3, \qquad -1, \quad \text{and} \quad 2.$$

> **PRINCIPLE OF ZERO PRODUCTS**
>
> REVIEW SECTION **3.2.**

We note that when the polynomial is expressed as a product of linear factors, each factor determines a zero of the function. Thus if we know the linear factors of a polynomial function $f(x)$, we can easily find the zeros of $f(x)$ by solving the equation $f(x) = 0$ using the principle of zero products.

$y = 5(x - 2)^3 (x + 1)$

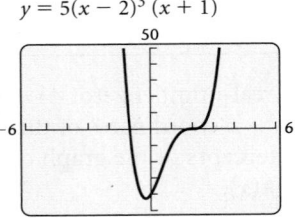

FIGURE 1

EXAMPLE 3   Find the zeros of

$$f(x) = 5(x - 2)(x - 2)(x - 2)(x + 1)$$
$$= 5(x - 2)^3(x + 1).$$

Solution   To solve the equation $f(x) = 0$, we use the principle of zero products, solving $x - 2 = 0$ and $x + 1 = 0$. The zeros of $f(x)$ are 2 and -1. (See Fig. 1.)   ◼

$y = -(x - 1)^2 (x + 2)^2$

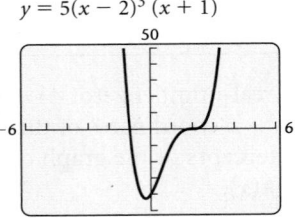

FIGURE 2

EXAMPLE 4   Find the zeros of

$$g(x) = -(x - 1)(x - 1)(x + 2)(x + 2)$$
$$= -(x - 1)^2(x + 2)^2.$$

Solution   To solve the equation $g(x) = 0$, we use the principle of zero products, solving $x - 1 = 0$ and $x + 2 = 0$. The zeros of $g(x)$ are 1 and -2. (See Fig. 2.)   ◼

Let's consider the occurrences of the zeros in the functions in Examples 3 and 4 and their relationship to the graphs of those functions. In Example 3, the factor $x - 2$ occurs three times. In a case like this, we say that the zero we obtain from this factor, 2, has a **multiplicity** of 3. The factor $x + 1$ occurs one time. The zero we obtain from this factor, -1, has a *multiplicity* of 1.

In Example 4, the factors $x - 1$ and $x + 2$ each occur two times. Thus both zeros, 1 and -2, have a *multiplicity* of 2.

Note, in Example 3, that the zeros have odd multiplicities and the graph crosses the x-axis at both -1 and 2. But in Example 4, the zeros have even multiplicities and the graph is tangent to (touches but does not cross) the x-axis at -2 and 1. This leads us to the following generalization.

> ### Even and Odd Multiplicity
> If $(x - c)^k, k \geq 1$, is a factor of a polynomial function $P(x)$ and $(x - c)^{k+1}$ is not a factor and:
>
> - k is odd, then the graph crosses the x-axis at $(c, 0)$;
> - k is even, then the graph is tangent to the x-axis at $(c, 0)$.

Some polynomials can be factored by grouping. Then we use the principle of zero products to find their zeros.

EXAMPLE 5 Find the zeros of
$$f(x) = x^3 - 2x^2 - 9x + 18.$$

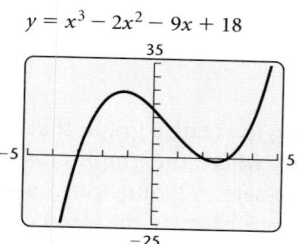

$y = x^3 - 2x^2 - 9x + 18$

Solution We factor by grouping, as follows:

$$
\begin{aligned}
f(x) &= x^3 - 2x^2 - 9x + 18 \\
&= x^2(x - 2) - 9(x - 2) && \text{Grouping } x^3 \text{ with } -2x^2 \text{ and } -9x \\
&&& \text{with 18 and factoring each group} \\
&= (x - 2)(x^2 - 9) && \text{Factoring out } x - 2 \\
&= (x - 2)(x + 3)(x - 3). && \text{Factoring } x^2 - 9
\end{aligned}
$$

Then, by the principle of zero products, the solutions of the equation $f(x) = 0$ are 2, -3, and 3. These are the zeros of $f(x)$.

Now Try Exercise 39. ◼

Other factoring techniques can also be used.

EXAMPLE 6 Find the zeros of
$$f(x) = x^4 + 4x^2 - 45.$$

Solution We factor as follows:

$$f(x) = x^4 + 4x^2 - 45 = (x^2 - 5)(x^2 + 9).$$

We now solve the equation $f(x) = 0$ to determine the zeros. We use the principle of zero products:

$$
\begin{aligned}
(x^2 - 5)(x^2 + 9) &= 0 \\
x^2 - 5 = 0 \quad &or \quad x^2 + 9 = 0 \\
x^2 = 5 \quad &or \quad x^2 = -9 \\
x = \pm\sqrt{5} \quad &or \quad x = \pm\sqrt{-9} = \pm 3i.
\end{aligned}
$$

The solutions are $\pm\sqrt{5}$ and $\pm 3i$. These are the zeros of $f(x)$.

Now Try Exercise 37. ◼

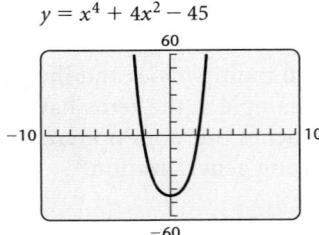

$y = x^4 + 4x^2 - 45$

Only the real-number zeros of a function correspond to the *x*-intercepts of its graph. For instance, the real-number zeros of the function in Example 6, $-\sqrt{5}$ and $\sqrt{5}$, can be seen on the graph of the function at left, but the nonreal zeros, $-3i$ and $3i$, cannot.

> Every polynomial function of degree *n*, with $n \geq 1$, has at least one zero and at most *n* zeros.

This is often stated as follows: "Every polynomial function of degree *n*, with $n \geq 1$, has *exactly n* zeros." This statement is compatible with the preceding statement, if one takes multiplicities into account.

❖ Finding Real Zeros on a Calculator

Finding exact values of the real zeros of a function can be difficult. We can find approximations using a graphing calculator.

EXAMPLE 7 Find the real zeros of the function *f* given by

$$f(x) = 0.1x^3 - 0.6x^2 - 0.1x + 2.$$

Approximate the zeros to three decimal places.

Solution We use a graphing calculator, trying to create a graph that clearly shows the curvature. Then we look for points where the graph crosses the *x*-axis. It appears that there are three zeros, one near -2, one near 2, and one near 6. We know that there are no more than 3 because the degree of the polynomial is 3. We use the ZERO feature to find them.

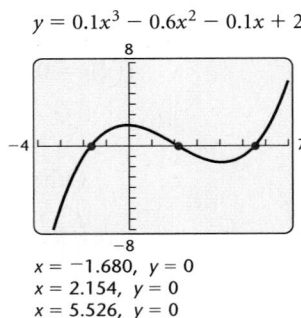

$y = 0.1x^3 - 0.6x^2 - 0.1x + 2$

$x = -1.680, \ y = 0$
$x = 2.154, \ y = 0$
$x = 5.526, \ y = 0$

The zeros are approximately -1.680, 2.154, and 5.526.

Now Try Exercise 43. ■

❖ Polynomial Models

Polynomial functions have many uses as models in science, engineering, and business. The simplest use of polynomial functions in applied problems occurs when we merely evaluate a polynomial function. In such cases, a model has already been developed.

EXAMPLE 8   *Ibuprofen in the Bloodstream.*   The polynomial function

$$M(t) = 0.5t^4 + 3.45t^3 - 96.65t^2 + 347.7t$$

can be used to estimate the number of milligrams of the pain relief medication ibuprofen in the bloodstream t hours after 400 mg of the medication has been taken.

a) Find the number of milligrams in the bloodstream at $t = 0, 0.5, 1, 1.5$, and so on, up to 6 hr. Round the function values to the nearest tenth.

b) Find the domain, the relative maximum and where it occurs, and the range.

Solution

a) We can evaluate the function with the TABLE feature of a graphing calculator set in AUTO mode. We start at 0 and use a step-value of 0.5.

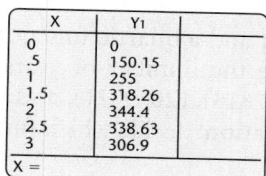

$M(0) = 0,$		$M(3.5) = 255.9,$
$M(0.5) = 150.2,$		$M(4) = 193.2,$
$M(1) = 255,$		$M(4.5) = 126.9,$
$M(1.5) = 318.3,$		$M(5) = 66,$
$M(2) = 344.4,$		$M(5.5) = 20.2,$
$M(2.5) = 338.6,$		$M(6) = 0.$
$M(3) = 306.9,$		

b) Recall that the domain of a polynomial function, unless restricted by a statement of the function, is $(-\infty, \infty)$. The implications of this application restrict the domain of the function. If we assume that a patient had not taken any of the medication before, it seems reasonable that $M(0) = 0$; that is, at time 0, there is 0 mg of the medication in the bloodstream. After the medication has been taken, $M(t)$ will be positive for a period of time and eventually decrease back to 0 when $t = 6$ and not increase again (unless another dose is taken). Thus the restricted domain is $[0, 6]$.

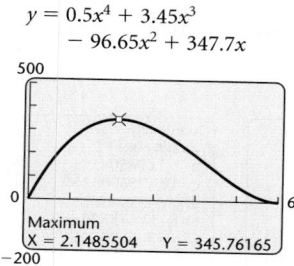

$y = 0.5x^4 + 3.45x^3 - 96.65x^2 + 347.7x$

To determine the range, we find the relative maximum value of the function using the MAXIMUM feature. The maximum is about 345.8 mg. It occurs approximately 2.15 hr, or 2 hr 9 min, after the initial dose has been taken. The range is about $[0, 345.8]$.   **Now Try Exercise 61.** ∎

In Chapter 1, we used regression to model data with linear functions. We now expand that procedure to include quadratic, cubic, and quartic models.

GCM   **EXAMPLE 9**   *Declining Number of Railroad Miles in the United States.* The greatest combined length of U.S.-owned operating railroad track existed in 1916, when industrial activity increased during World War I. The total length has decreased ever since.

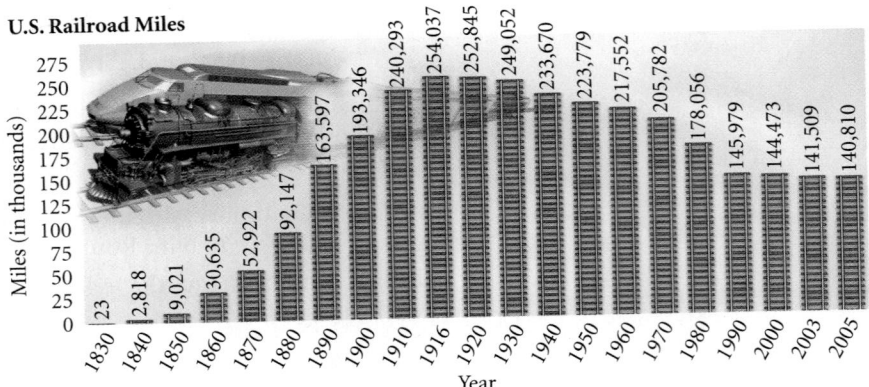

U.S. Railroad Miles

Note: The lengths exclude yard tracks, sidings, and parallel tracks.

Source: Association of American Railroads

Looking at the graph above, we note that the data, in miles, could be modeled with a cubic function or a quartic function.

a) Model the data with both a cubic function and a quartic function. Let the first coordinate of each data point be the number of years after 1830; that is, enter the data as (0, 23), (10, 2818), (20, 9021), and so on. Then using R^2, the **coefficient of determination**, decide which function is the better fit.

b) Graph the function with the scatterplot of the data.

c) Use the answer to part (a) to estimate the number of miles of railroad track in 1905, in 1965, in 2010, and in 2015.

Solution

a) Using the REGRESSION feature with DIAGNOSTIC turned on, we get the following.

```
CubicReg
y=ax³+bx²+cx+d
a=-.1085564619
b=5.473204363
c=3230.70885
d=-36548.65554
R²=.906657478
```

```
QuarticReg
y=ax⁴+bx³+...+e
a=.0044895151
b=-1.698508535
c=183.7560082
d=-3425.885898
↓e=8437.878526
```

```
QuarticReg
y=ax⁴+bx³+...+e
↑b=-1.698508535
c=183.7560082
d=-3425.885898
e=8437.878526
R²=.9845028016
```

Since the R^2-value for the quartic function, 0.9845028016, is closer to 1 than that for the cubic function, 0.906657478, the quartic function is the better fit.

$$f(x) = 0.0044895151x^4 - 1.698508535x^3$$
$$+ 183.7560082x^2 - 3425.885898x$$
$$+ 8437.878526$$

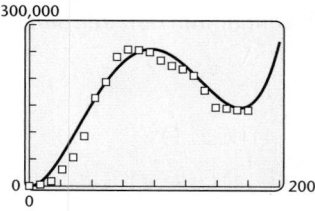

300,000

0
200

b) The scatterplot and the graph are shown at left.

c) We evaluate the function found in part (a).

X	Y₁
75	210617
135	207124
180	152685
185	168176

X =

   With this function, we can estimate that there were 210,617 mi of railroad track in 1905 and 207,124 mi in 1965. Looking at the bar graph shown on the preceding page, we see that these estimates appear to be fairly accurate.
   If we use the function to estimate the number of miles of track in 2010 and in 2015, we get about 152,685 mi and 168,176 mi, respectively. These estimates are probably not realistic since it is not reasonable to expect the number of miles of track to increase in the future. The quartic model has a higher value for R^2 than the cubic function over the domain of the data, but this number does not reflect the degree of accuracy for extended values. It is always important when using regression to evaluate predictions with common sense and knowledge of current trends.

Now Try Exercise 77. ■

4.1   Exercise Set

Determine the leading term, the leading coefficient, and the degree of the polynomial. Then classify the polynomial function as constant, linear, quadratic, cubic, or quartic.

1. $g(x) = \frac{1}{2}x^3 - 10x + 8$

2. $f(x) = 15x^2 - 10 + 0.11x^4 - 7x^3$

3. $h(x) = 0.9x - 0.13$

4. $f(x) = -6$

5. $g(x) = 305x^4 + 4021$

6. $h(x) = 2.4x^3 + 5x^2 - x + \frac{7}{8}$

7. $h(x) = -5x^2 + 7x^3 + x^4$

8. $f(x) = 2 - x^2$

9. $g(x) = 4x^3 - \frac{1}{2}x^2 + 8$

10. $f(x) = 12 + x$

In Exercises 11–18, select one of the following four sketches to describe the end behavior of the graph of the function.

a)

b)

c)

d)

11. $f(x) = -3x^3 - x + 4$

12. $f(x) = \frac{1}{4}x^4 + \frac{1}{2}x^3 - 6x^2 + x - 5$

13. $f(x) = -x^6 + \frac{3}{4}x^4$

14. $f(x) = \frac{2}{5}x^5 - 2x^4 + x^3 - \frac{1}{2}x + 3$

15. $f(x) = -3.5x^4 + x^6 + 0.1x^7$

16. $f(x) = -x^3 + x^5 - 0.5x^6$

17. $f(x) = 10 + \frac{1}{10}x^4 - \frac{2}{5}x^3$

18. $f(x) = 2x + x^3 - x^5$

In Exercises 19–22, use the leading-term test to match the function with one of the graphs (a)–(d), which follow.

a)

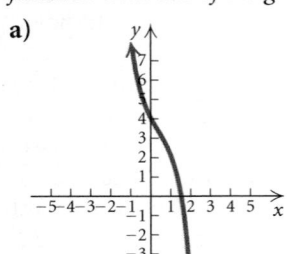

b)

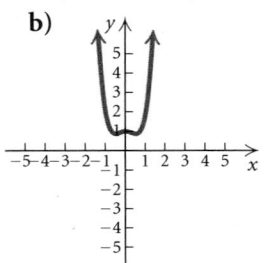

c)

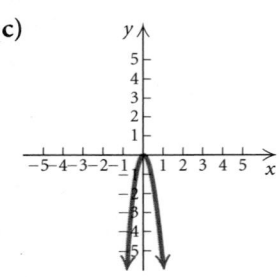

d)

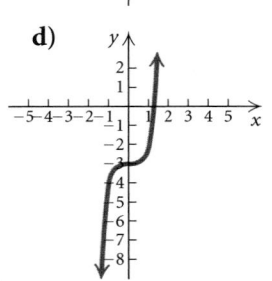

19. $f(x) = -x^6 + 2x^5 - 7x^2$

20. $f(x) = 2x^4 - x^2 + 1$

21. $f(x) = x^5 + \frac{1}{10}x - 3$

22. $f(x) = -x^3 + x^2 - 2x + 4$

23. Use substitution to determine whether 4, 5, and -2 are zeros of
$$f(x) = x^3 - 9x^2 + 14x + 24.$$

24. Use substitution to determine whether 2, 3, and -1 are zeros of
$$f(x) = 2x^3 - 3x^2 + x + 6.$$

25. Use substitution to determine whether 2, 3, and -1 are zeros of
$$g(x) = x^4 - 6x^3 + 8x^2 + 6x - 9.$$

26. Use substitution to determine whether 1, -2, and 3 are zeros of
$$g(x) = x^4 - x^3 - 3x^2 + 5x - 2.$$

Find the zeros of the polynomial function and state the multiplicity of each.

27. $f(x) = (x + 3)^2(x - 1)$

28. $f(x) = (x + 5)^3(x - 4)(x + 1)^2$

29. $f(x) = -2(x - 4)(x - 4)(x - 4)(x + 6)$

30. $f(x) = \left(x + \frac{1}{2}\right)(x + 7)(x + 7)(x + 5)$

31. $f(x) = (x^2 - 9)^3$

32. $f(x) = (x^2 - 4)^2$

33. $f(x) = x^3(x - 1)^2(x + 4)$

34. $f(x) = x^2(x + 3)^2(x - 4)(x + 1)^4$

35. $f(x) = -8(x - 3)^2(x + 4)^3x^4$

36. $f(x) = (x^2 - 5x + 6)^2$

37. $f(x) = x^4 - 4x^2 + 3$

38. $f(x) = x^4 - 10x^2 + 9$

39. $f(x) = x^3 + 3x^2 - x - 3$

40. $f(x) = x^3 - x^2 - 2x + 2$

41. $f(x) = 2x^3 - x^2 - 8x + 4$

42. $f(x) = 3x^3 + x^2 - 48x - 16$

Using a graphing calculator, find the real zeros of the function.

43. $f(x) = x^3 - 3x - 1$

44. $f(x) = x^3 + 3x^2 - 9x - 13$

45. $f(x) = x^4 - 2x^2$

46. $f(x) = x^4 - 2x^3 - 5.6$

47. $f(x) = x^3 - x$

48. $f(x) = 2x^3 - x^2 - 14x - 10$

49. $f(x) = x^8 + 8x^7 - 28x^6 - 56x^5 + 70x^4 + 56x^3 - 28x^2 - 8x + 1$

50. $f(x) = x^6 - 10x^5 + 13x^3 - 4x^2 - 5$

Using a graphing calculator, estimate the real zeros, the relative maxima and minima, and the range of the polynomial function.

51. $g(x) = x^3 - 1.2x + 1$

52. $h(x) = -\frac{1}{2}x^4 + 3x^3 - 5x^2 + 3x + 6$

53. $f(x) = x^6 - 3.8$

54. $h(x) = 2x^3 - x^4 + 20$

55. $f(x) = x^2 + 10x - x^5$

56. $f(x) = 2x^4 - 5.6x^2 + 10$

Determine whether the statement is true or false.

57. If $P(x) = (x - 3)^4(x + 1)^3$, then the graph of the polynomial function $y = P(x)$ crosses the x-axis at $(3, 0)$.

58. If $P(x) = (x + 2)^2(x - \frac{1}{4})^5$, then the graph of the polynomial function $y = P(x)$ crosses the x-axis at $(\frac{1}{4}, 0)$.

59. If $P(x) = (x - 2)^3(x + 5)^6$, then the graph of $y = P(x)$ is tangent to the x-axis at $(-5, 0)$.

60. If $P(x) = (x + 4)^2(x - 1)^2$, then the graph of $y = P(x)$ is tangent to the x-axis at $(4, 0)$.

61. *Milk Cows.* The number of milk cows on farms in the United States peaked at 24,940,000 in 1940 and decreased to 9,058,000 in 2006 (*Sources:* National Agricultural Statistics Service; U.S. Department of Agriculture).

The quartic function
$$f(x) = 0.001258507x^4 - 0.1862275538x^3$$
$$+ 1.865895038x^2 + 339.7440683x$$
$$+ 16{,}138.43225,$$
where x is the number of years since 1900, can be used to estimate the number of milk cows, in thousands, from 1900 to 2006. Estimate the number of milk cows in 1960 and in 1980.

62. *Projectile Motion.* A stone thrown downward with an initial velocity of 34.3 m/sec will travel a distance of s meters, where
$$s(t) = 4.9t^2 + 34.3t$$
and t is in seconds. If a stone is thrown downward at 34.3 m/sec from a height of 294 m, how long will it take the stone to hit the ground?

63. *Tornado Fatalities.* May and June are the worst months for tornado fatalities. The number of U.S. tornado deaths in 2006 was the highest since 1999 (*Source:* National Weather Service).

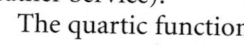

The quartic function
$$f(x) = 0.2872960373x^4$$
$$- 5.04959855x^3 + 31.68667444x^2$$
$$- 86.28334628x + 135.5819736,$$
where x is the number of years since 1998, can be used to estimate the number of tornado fatalities from 1998 to 2006. Using this function, estimate the number of deaths in 1998, in 2001, and in 2006.

64. *International Travel.* In 2003, the number of international visitors to the United States was 41.2 million, the lowest number since 1990 (*Sources:* Office of Travel and Tourism Industries, Department of Commerce; World Tourism Organization). The quartic function
$$f(x) = -0.1810606061x^4 + 2.433838384x^3$$
$$- 9.637121212x^2 + 9.457287157x$$
$$+ 48.5521645,$$
where x is the number of years since 1999, can be used to estimate the number of international visitors, in millions, from 1999 to 2005. Estimate the number of visitors in 2001 and in 2005.

65. *Games in a Sports League.*   If there are x teams in a sports league and all the teams play each other twice, a total of $N(x)$ games are played, where

$$N(x) = x^2 - x.$$

A softball league has 9 teams, each of which plays the others twice. If the league pays $110 per game for the field and umpires, how much will it cost to play the entire schedule?

66. *Windmill Power.*   Under certain conditions, the power P, in watts per hour, generated by a windmill with winds blowing v miles per hour is given by

$$P(v) = 0.015v^3.$$

a) Find the power generated by 15-mph winds.
b) How fast must the wind blow in order to generate 120 watts of power in 1 hr?

67. *Auto Insurance.*   The average cost of auto insurance rose only 1.5% from 2004 to 2005 after increasing almost 6% per year in the three years prior to 2004. The average cost from 1995 to 2005 can be modeled by the quartic function

$$f(x) = -0.2607808858x^4 + 5.281857032x^3$$
$$- 30.77185315x^2 + 60.89063714x$$
$$+ 663.8111888,$$

where x is the number of years since 1995 (*Source*: Insurance Information Institute). Estimate the average cost of auto insurance in 1996, in 1999, and in 2004.

68. *Debt Held by Foreigners.*   The percentage of the United States' public debt that is held by foreigners has greatly increased in recent years (*Source*: White House Office of Management and Budget). The data can be modeled by the quadratic function

$$f(x) = 0.0742857143x^2 - 0.76x + 17.14285714,$$

where x is the number of years since 1980. Find the percentage of the public debt held by foreigners in 1987, in 1996, and in 2006.

69. *Interest Compounded Annually.*   When P dollars is invested at interest rate i, compounded annually, for t years, the investment grows to A dollars, where

$$A = P(1 + i)^t.$$

a) Find the interest rate i if $4000 grows to $4368.10 in 2 yr.

b) Find the interest rate i if $10,000 grows to $13,310 in 3 yr.

70. *Threshold Weight.*   In a study performed by Alvin Shemesh, it was found that the **threshold weight** W, defined as the weight above which the risk of death rises dramatically, is given by

$$W(h) = \left(\frac{h}{12.3}\right)^3,$$

where W is in pounds and h is a person's height, in inches. Find the threshold weight of a person who is 5 ft 7 in. tall.

For the scatterplots and graphs in Exercises 71–76, determine which, if any, of the following functions might be used as a model for the data.

a) *Linear,* $f(x) = mx + b$
b) *Quadratic,* $f(x) = ax^2 + bx + c, a > 0$
c) *Quadratic,* $f(x) = ax^2 + bx + c, a < 0$
d) *Polynomial, not quadratic or linear*

71.

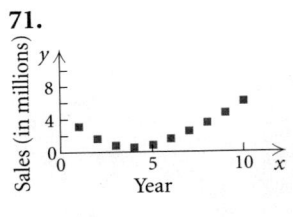

72.

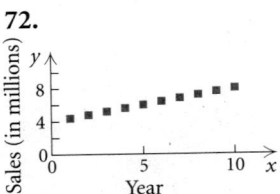

73.

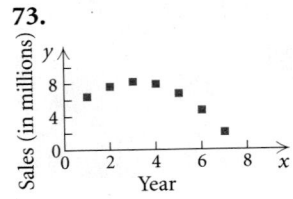

74.

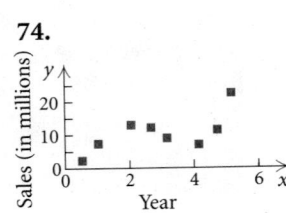

75.

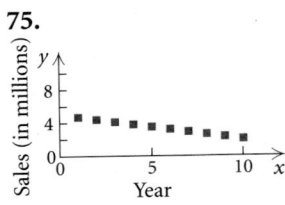

76.

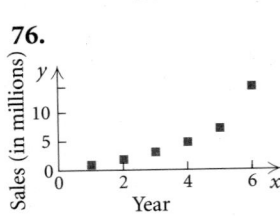

77. *Foreign-Born Population.*   The percentage of the U.S. population that is foreign-born has increased in recent years, as shown in the table on the following page.

a) Use a graphing calculator to fit cubic and quartic functions to the data. Let x represent the number of years since 1900.

Year	Percentage That Is Foreign-Born
1900	13.6%
1910	14.7
1920	13.2
1930	11.6
1940	8.8
1950	6.9
1960	5.4
1970	4.7
1980	6.2
1990	8.0
2000	10.4
2005	11.7

Sources: U.S. Census Bureau;
U.S. Department of Commerce

b) Use the functions found in part (a) to estimate what percentage of the U.S. population will be foreign-born in 2010. Compare the estimates and determine which model gives the more realistic estimate.

78. *U.S. Farm Acreage.*   As the number of farms has decreased in the United States, the average size of the remaining farms has grown larger, as shown in the following table.

Year	Average Acreage per Farm
1900	147
1910	139
1920	149
1930	157
1940	175
1950	213
1960	297
1970	374
1980	426
1990	460
1995	438
2000	436
2003	441
2005	444

Sources: Statistical History of the United States (1970); National Agricultural Statistics Service

a) Use a graphing calculator to fit quadratic, cubic, and quartic functions to the data. Let x represent the number of years since 1900.

b) With each function found in part (a), estimate the average acreage in 2010 and in 2015. Compare the estimates and determine which function gives the most realistic estimates.

79. *Unemployed.*   The table below shows the number of unemployed in the United States from 1996 through 2006.

Year	Number of Unemployed (in thousands)
1996	397
1997	350
1998	340
1999	275
2000	269
2001	310
2002	360
2003	457
2004	460
2005	435
2006	411

Source: U.S. Department of Labor

a) Use a graphing calculator to fit cubic and quartic functions to the data. Let x represent the number of years since 1996.

b) Use the functions found in part (a) to estimate the number of unemployed in 2008. Compare the estimates and determine which model gives the more realistic estimate.

80. *Dog Years.*   A dog's life span is typically much shorter than that of a human. Age equivalents for dogs and humans are listed in the table below.

Age of Dog, x (in years)	Human Age, $h(x)$ (in years)
0.25	5
0.5	10
1	15
2	24
4	32
6	40
8	48
10	56
14	72
18	91
21	106

Source: *Country*, December 1992, p. 60

a) Use a graphing calculator to fit linear and cubic functions to the data. Which function has the better fit?
b) Use the function from part (a) with the better fit to estimate the equivalent human age for dogs that are 5, 10, and 15 years old.

Collaborative Discussion and Writing

81. How is the range of a polynomial function related to the degree of the polynomial?

82. Polynomial functions are continuous. Discuss what "continuous" means in terms of the domains of the functions and the characteristics of their graphs.

Skill Maintenance

Find the distance between the pair of points.

83. $(3, -5)$ and $(0, -1)$

84. $(4, 2)$ and $(-2, -4)$

85. Find the center and the radius of the circle
$$(x - 3)^2 + (y + 5)^2 = 49.$$

86. The diameter of a circle connects the points $(-6, 5)$ and $(-2, 1)$ on the circle. Find the coordinates of the center of the circle and the length of the radius.

Solve.

87. $2y - 3 \geq 1 - y + 5$

88. $(x - 2)(x + 5) > x(x - 3)$

89. $|x + 6| \geq 7$

90. $\left|x + \frac{1}{4}\right| \leq \frac{2}{3}$

Synthesis

Determine the degree and the leading term of the polynomial function.

91. $f(x) = (x^5 - 1)^2(x^2 + 2)^3$

92. $f(x) = (10 - 3x^5)^2(5 - x^4)^3(x + 4)$

4.2 Graphing Polynomial Functions

❖ Graph polynomial functions.

❖ Use the intermediate value theorem to determine whether a function has a real zero between two given real numbers.

❖ Graphing Polynomial Functions

In addition to using the leading-term test and finding the zeros of the function, it is helpful to consider the following facts when graphing a polynomial function.

> If $P(x)$ is a polynomial function of degree n, the graph of the function has:
>
> - at most n real zeros, and thus at most n x-intercepts;
> - at most $n - 1$ turning points.
>
> (Turning points on a graph, also called relative maxima and minima, occur when the function changes from decreasing to increasing or from increasing to decreasing.)

EXAMPLE 1   Graph the polynomial function $h(x) = -2x^4 + 3x^3$.

Solution

1. First, we use the leading-term test to determine the end behavior of the graph. The leading term is $-2x^4$. The degree, 4, is even, and the coefficient, -2, is negative. Thus the end behavior of the graph as $x \to \infty$ and as $x \to -\infty$ can be sketched as follows.

2. The zeros of the function are the first coordinates of the x-intercepts of the graph. To find the zeros, we solve $h(x) = 0$ by factoring and using the principle of zero products.

$$-2x^4 + 3x^3 = 0$$
$$-x^3(2x - 3) = 0 \qquad \text{Factoring}$$
$$-x^3 = 0 \quad or \quad 2x - 3 = 0 \qquad \text{Using the principle of zero products}$$
$$x = 0 \quad or \qquad x = \frac{3}{2}.$$

The zeros of the function are 0 and $\frac{3}{2}$. Note that the multiplicity of 0 is 3 and the multiplicity of $\frac{3}{2}$ is 1. The x-intercepts are $(0,0)$ and $\left(\frac{3}{2},0\right)$.

3. The zeros divide the x-axis into three intervals:

$$(-\infty, 0), \qquad \left(0, \frac{3}{2}\right), \quad \text{and} \quad \left(\frac{3}{2}, \infty\right).$$

The sign of $h(x)$ is the same for all values of x in each of the three intervals. That is, $h(x)$ is positive for all x-values in an interval or $h(x)$ is negative for all x-values in an interval. To determine which, we choose a test value for x from each interval and find $h(x)$.

$y_1 = -2x^4 + 3x^3$

X	Y1
-1	-5
1	1
2	-8

X =

Interval	$(-\infty, 0)$	$\left(0, \frac{3}{2}\right)$	$\left(\frac{3}{2}, \infty\right)$
Test Value, x	-1	1	2
Function Value, $h(x)$	-5	1	-8
Sign of $h(x)$	$-$	$+$	$-$
Location of Points on Graph	Below x-axis	Above x-axis	Below x-axis

This test-point procedure also gives us three points to plot. In this case, we have $(-1, -5)$, $(1, 1)$, and $(2, -8)$.

4. To determine the y-intercept, we find $h(0)$:

$$h(x) = -2x^4 + 3x^3$$
$$h(0) = -2 \cdot 0^4 + 3 \cdot 0^3 = 0.$$

The y-intercept is $(0, 0)$.

5. A few additional points are helpful when completing the graph.

x	$h(x)$
-1.5	-20.25
-0.5	-0.5
0.5	0.25
2.5	-31.25

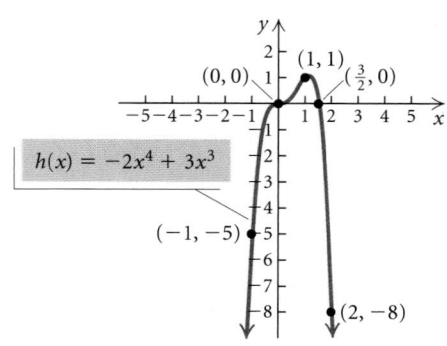

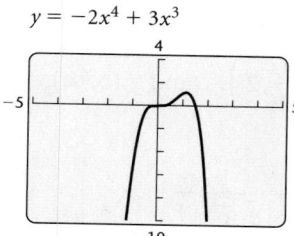

$y = -2x^4 + 3x^3$

6. The degree of h is 4. The graph of h can have at most 4 x-intercepts and at most 3 turning points. In fact, it has 2 x-intercepts and 1 turning point. The zeros, 0 and $\frac{3}{2}$, each have odd multiplicities: 3 for 0 and 1 for $\frac{3}{2}$. Since the multiplicities are odd, the graph crosses the x-axis at 0 and $\frac{3}{2}$. The end behavior of the graph is what we described in step (1). As $x \to \infty$ and also as $x \to -\infty$, $h(x) \to -\infty$. The graph appears to be correct.

Now Try Exercise 19. ▨

The following is a procedure for graphing polynomial functions.

To graph a polynomial function:

1. Use the leading-term test to determine the end behavior.
2. Find the zeros of the function by solving $f(x) = 0$. Any real zeros are the first coordinates of the x-intercepts.
3. Use the x-intercepts (zeros) to divide the x-axis into intervals and choose a test point in each interval to determine the sign of all function values in that interval.
4. Find $f(0)$. This gives the y-intercept of the function.
5. If necessary, find additional function values to determine the general shape of the graph and then draw the graph.
6. As a partial check, use the facts that the graph has at most n x-intercepts and at most $n - 1$ turning points. Multiplicity of zeros can also be considered in order to check where the graph crosses or is tangent to the x-axis. We can also check the graph with a graphing calculator.

EXAMPLE 2 Graph the polynomial function

$$f(x) = 2x^3 + x^2 - 8x - 4.$$

Solution

1. The leading term is $2x^3$. The degree, 3, is odd, and the coefficient, 2, is positive. Thus the end behavior of the graph will appear as follows.

2. To find the zeros, we solve $f(x) = 0$. Here we can use factoring by grouping.

$$2x^3 + x^2 - 8x - 4 = 0$$
$$x^2(2x + 1) - 4(2x + 1) = 0 \qquad \text{Factoring by grouping}$$
$$(2x + 1)(x^2 - 4) = 0$$
$$(2x + 1)(x + 2)(x - 2) = 0 \qquad \text{Factoring a difference of squares}$$

The zeros are $-\frac{1}{2}$, -2, and 2. Each is of multiplicity 1. The x-intercepts are $(-2, 0)$, $\left(-\frac{1}{2}, 0\right)$, and $(2, 0)$.

3. The zeros divide the x-axis into four intervals:

$$(-\infty, -2), \qquad \left(-2, -\frac{1}{2}\right), \qquad \left(-\frac{1}{2}, 2\right), \quad \text{and} \quad (2, \infty).$$

We choose a test value for x from each interval and find $f(x)$.

Interval	$(-\infty, -2)$	$\left(-2, -\frac{1}{2}\right)$	$\left(-\frac{1}{2}, 2\right)$	$(2, \infty)$
Test Value, x	-3	-1	1	3
Function Value, $f(x)$	-25	3	-9	35
Sign of $f(x)$	$-$	$+$	$-$	$+$
Location of Points on Graph	Below x-axis	Above x-axis	Below x-axis	Above x-axis

$y_1 = 2x^3 + x^2 - 8x - 4$

X	Y1
-3	-25
-1	3
1	-9
3	35

X =

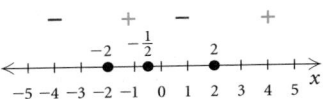

The test values and corresponding function values also give us four points on the graph: $(-3, -25), (-1, 3), (1, -9),$ and $(3, 35).$

4. To determine the y-intercept, we find $f(0)$:

$$f(x) = 2x^3 + x^2 - 8x - 4$$
$$f(0) = 2 \cdot 0^3 + 0^2 - 8 \cdot 0 - 4 = -4.$$

The y-intercept is $(0, -4).$

5. We find a few additional points and complete the graph.

x	f(x)
-2.5	-9
-1.5	3.5
0.5	-7.5
1.5	-7

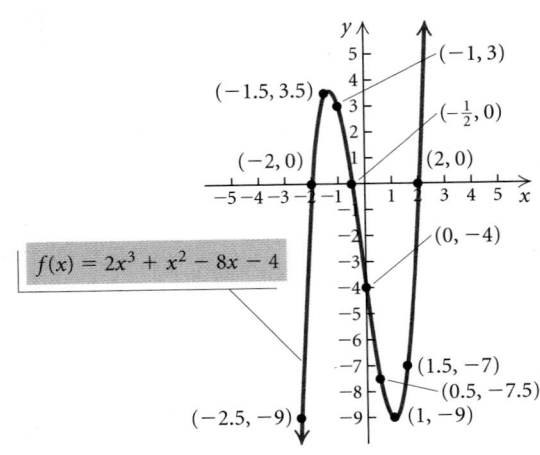

$f(x) = 2x^3 + x^2 - 8x - 4$

$y = 2x^3 + x^2 - 8x - 4$

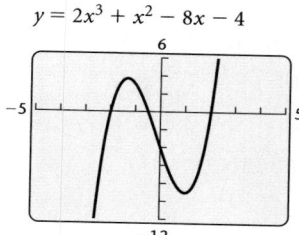

6. The degree of f is 3. The graph of f can have at most 3 x-intercepts and at most 2 turning points. It has 3 x-intercepts and 2 turning points. Each zero has a multiplicity of 1; thus the graph crosses the x-axis at -2, $-\frac{1}{2}$, and 2. The graph has the end behavior described in step (1). As $x \to -\infty$, $h(x) \to -\infty$, and as $x \to \infty$, $h(x) \to \infty$. The graph appears to be correct.

Now Try Exercise 29. ■

Some polynomials are difficult to factor. In the next example, the polynomial is given in factored form. In Sections 4.3 and 4.4, we will learn methods that facilitate determining factors of such polynomials.

EXAMPLE 3   Graph the polynomial function

$$g(x) = x^4 - 7x^3 + 12x^2 + 4x - 16$$
$$= (x + 1)(x - 2)^2(x - 4).$$

Solution

1. The leading term is x^4. The degree, 4, is even, and the coefficient, 1, is positive. The sketch below shows the end behavior.

2. To find the zeros, we solve $g(x) = 0$:

$$(x + 1)(x - 2)^2(x - 4) = 0.$$

The zeros are $-1, 2$, and 4; 2 is of multiplicity 2; the others are of multiplicity 1. The x-intercepts are $(-1, 0)$, $(2, 0)$, and $(4, 0)$.

$y_1 = x^4 - 7x^3 + 12x^2 + 4x - 16$

X	Y₁
−1.25	13.863
1	−6
3	−4
4.25	6.6445

X =

3. The zeros divide the x-axis into four intervals:

$$(-\infty, -1), \quad (-1, 2), \quad (2, 4), \quad \text{and} \quad (4, \infty).$$

We choose a test value for x from each interval and find $g(x)$.

Interval	$(-\infty, -1)$	$(-1, 2)$	$(2, 4)$	$(4, \infty)$
Test Value, x	-1.25	1	3	4.25
Function Value, $g(x)$	≈ 13.9	-6	-4	≈ 6.6
Sign of $g(x)$	$+$	$-$	$-$	$+$
Location of Points on Graph	Above x-axis	Below x-axis	Below x-axis	Above x-axis

The test values and corresponding function values also give us four points on the graph: $(-1.25, 13.9)$, $(1, -6)$, $(3, -4)$, and $(4.25, 6.6)$.

4. To determine the y-intercept, we find $g(0)$:

$$g(x) = x^4 - 7x^3 + 12x^2 + 4x - 16$$
$$g(0) = 0^4 - 7 \cdot 0^3 + 12 \cdot 0^2 + 4 \cdot 0 - 16 = -16.$$

The y-intercept is $(0, -16)$.

5. We find a few additional points and draw the graph.

$y = x^4 - 7x^3 + 12x^2 + 4x - 16$

x	$g(x)$
-0.5	-14.1
0.5	-11.8
1.5	-1.6
2.5	-1.3
3.5	-5.1

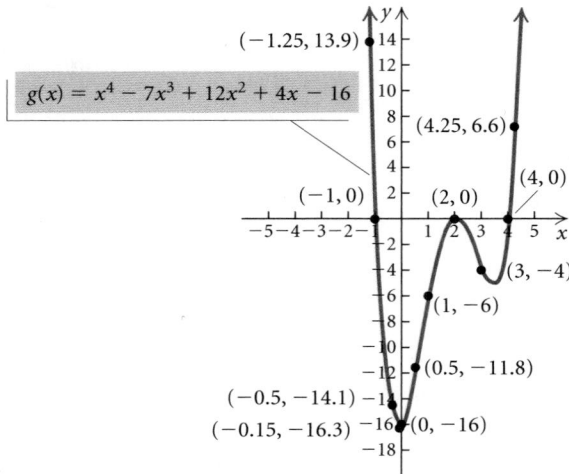

6. The degree of g is 4. The graph of g can have at most 4 x-intercepts and at most 3 turning points. It has 3 x-intercepts and 3 turning points. One of the zeros, 2, has a multiplicity of 2, so the graph is tangent to the x-axis at 2. The other zeros, -1 and 4, each have a multiplicity of 1 so the graph crosses the x-axis at -1 and 4. The graph has the end behavior described in step (1). As $x \to \infty$ and as $x \to -\infty$, $g(x) \to \infty$. The graph appears to be correct. **Now Try Exercise 17.** ■

❋ The Intermediate Value Theorem

Polynomial functions are continuous, hence their graphs are unbroken. The domain of a polynomial function, unless restricted by the statement of the function, is $(-\infty, \infty)$. Suppose two polynomial function values $P(a)$ and $P(b)$ have opposite signs. Since P is continuous, its graph must be a curve from $(a, P(a))$ to $(b, P(b))$ without a break. Then it follows that the curve must cross the x-axis at some point c between a and b; that is, the function has a zero at c between a and b.

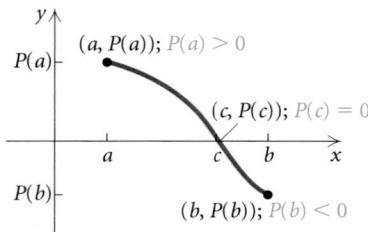

> ### The Intermediate Value Theorem
>
> For any polynomial function $P(x)$ with real coefficients, suppose that for $a \neq b$, $P(a)$ and $P(b)$ are of opposite signs. Then the function has a real zero between a and b.

EXAMPLE 4   Using the intermediate value theorem, determine, if possible, whether the function has a real zero between a and b.

a) $f(x) = x^3 + x^2 - 6x$; $a = -4, b = -2$

b) $f(x) = x^3 + x^2 - 6x$; $a = -1, b = 3$

c) $g(x) = \frac{1}{3}x^4 - x^3$; $a = -\frac{1}{2}, b = \frac{1}{2}$

d) $g(x) = \frac{1}{3}x^4 - x^3$; $a = 1, b = 2$

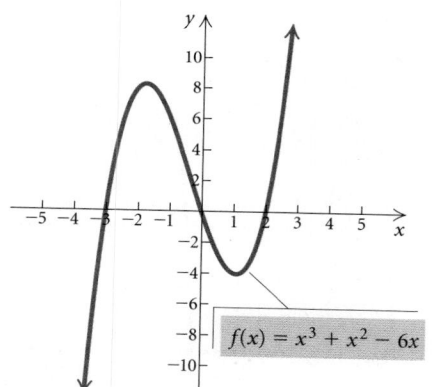

$f(x) = x^3 + x^2 - 6x$

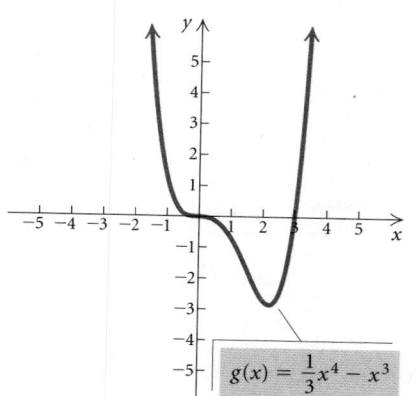

$g(x) = \frac{1}{3}x^4 - x^3$

Solution   We find $f(a)$ and $f(b)$ or $g(a)$ and $g(b)$ and determine whether they differ in sign. The graphs of $f(x)$ and $g(x)$ at left provide a visual check of the conclusions.

a) $f(-4) = (-4)^3 + (-4)^2 - 6(-4) = -24$,

$f(-2) = (-2)^3 + (-2)^2 - 6(-2) = 8$

Note that $f(-4)$ is negative and $f(-2)$ is positive. By the intermediate value theorem, since $f(-4)$ and $f(-2)$ have opposite signs, then $f(x)$ has a zero between -4 and -2. The graph confirms this.

b) $f(-1) = (-1)^3 + (-1)^2 - 6(-1) = 6$,

$f(3) = 3^3 + 3^2 - 6(3) = 18$

Both $f(-1)$ and $f(3)$ are positive. Thus the intermediate value theorem does not allow us to determine whether there is a real zero between -1 and 3. Note that the graph of $f(x)$ shows that there are two zeros between -1 and 3.

c) $g\left(-\frac{1}{2}\right) = \frac{1}{3}\left(-\frac{1}{2}\right)^4 - \left(-\frac{1}{2}\right)^3 = \frac{7}{48}$,

$g\left(\frac{1}{2}\right) = \frac{1}{3}\left(\frac{1}{2}\right)^4 - \left(\frac{1}{2}\right)^3 = -\frac{5}{48}$

Since $g\left(-\frac{1}{2}\right)$ and $g\left(\frac{1}{2}\right)$ have opposite signs, $g(x)$ has a zero between $-\frac{1}{2}$ and $\frac{1}{2}$. The graph confirms this.

d) $g(1) = \frac{1}{3}(1)^4 - 1^3 = -\frac{2}{3}$,

$g(2) = \frac{1}{3}(2)^4 - 2^3 = -\frac{8}{3}$

Both $g(1)$ and $g(2)$ are negative. Thus the intermediate value theorem does not allow us to determine whether there is a real zero between 1 and 2. Note that the graph of $g(x)$ shows that there are no zeros between 1 and 2.

Now Try Exercises 33 and 37. ■

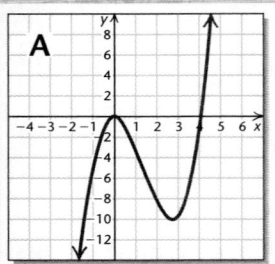

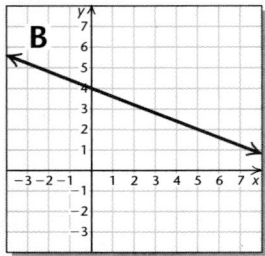

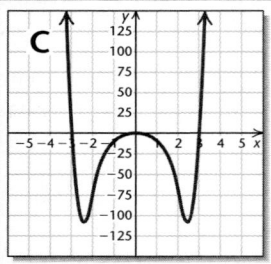

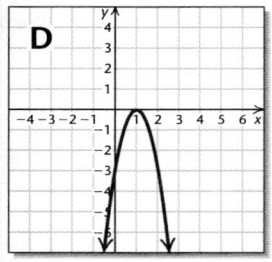

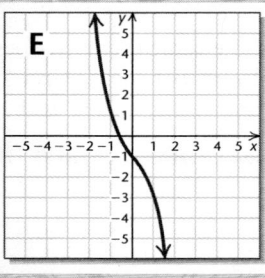

Visualizing the Graph

Match the function with its graph.

1. $f(x) = -x^4 - x + 5$

2. $f(x) = -3x^2 + 6x - 3$

3. $f(x) = x^4 - 4x^3 + 3x^2 + 4x - 4$

4. $f(x) = -\dfrac{2}{5}x + 4$

5. $f(x) = x^3 - 4x^2$

6. $f(x) = x^6 - 9x^4$

7. $f(x) = x^5 - 3x^3 + 2$

8. $f(x) = -x^3 - x - 1$

9. $f(x) = x^2 + 7x + 6$

10. $f(x) = \dfrac{7}{2}$

Answers on page A-23

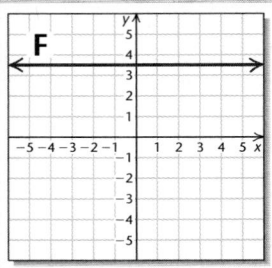

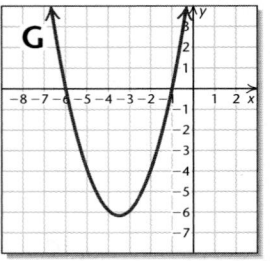

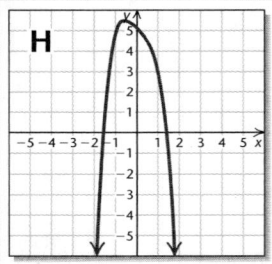

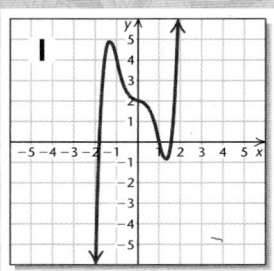

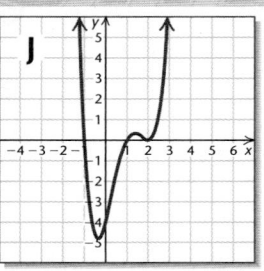

4.2 Exercise Set

For each function in Exercises 1–6, state:

a) the maximum number of real zeros that the function can have;

b) the maximum number of x-intercepts that the graph of the function can have; and

c) the maximum number of turning points that the graph of the function can have.

1. $f(x) = x^5 - x^2 + 6$

2. $f(x) = -x^2 + x^4 - x^6 + 3$

3. $f(x) = x^{10} - 2x^5 + 4x - 2$

4. $f(x) = \frac{1}{4}x^3 + 2x^2$

5. $f(x) = -x - x^3$

6. $f(x) = -3x^4 + 2x^3 - x - 4$

In Exercises 7–12, use the leading-term test and your knowledge of y-intercepts to match the function with one of the graphs (a)–(f), which follow.

a)

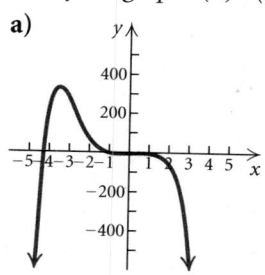

b)

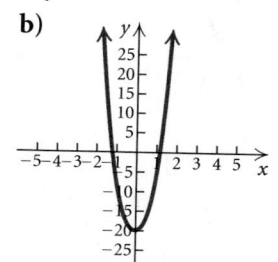

c)

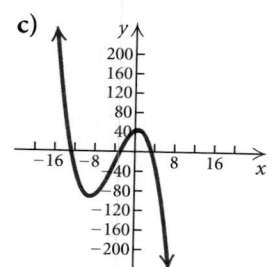

d)

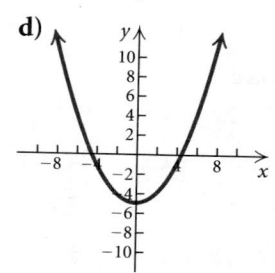

e)

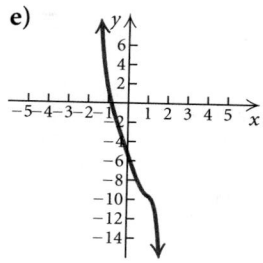

f)

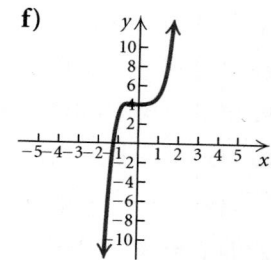

7. $f(x) = \frac{1}{4}x^2 - 5$

8. $f(x) = -0.5x^6 - x^5 + 4x^4 - 5x^3 - 7x^2 + x - 3$

9. $f(x) = x^5 - x^4 + x^2 + 4$

10. $f(x) = -\frac{1}{3}x^3 - 4x^2 + 6x + 42$

11. $f(x) = x^4 - 2x^3 + 12x^2 + x - 20$

12. $f(x) = -0.3x^7 + 0.11x^6 - 0.25x^5 + x^4 + x^3 - 6x - 5$

Graph the polynomial function. Follow the steps outlined in the procedure on p. 315.

13. $f(x) = -x^3 - 2x^2$

14. $g(x) = x^4 - 4x^3 + 3x^2$

15. $h(x) = x^5 - 4x^3$

16. $f(x) = x^3 - x$

17. $h(x) = x(x - 4)(x + 1)(x - 2)$

18. $f(x) = \frac{1}{2}x^3 + \frac{5}{2}x^2$

19. $g(x) = -x^4 - 2x^3$

20. $h(x) = x^3 - 3x^2$

21. $f(x) = -\frac{1}{2}(x - 2)(x + 1)^2(x - 1)$

22. $g(x) = (x - 2)^3(x + 3)$

23. $g(x) = -x(x - 1)^2(x + 4)^2$

24. $h(x) = -x(x - 3)(x - 3)(x + 2)$

25. $f(x) = (x - 2)^2(x + 1)^4$

26. $g(x) = x^4 - 9x^2$

27. $g(x) = -(x - 1)^4$

28. $h(x) = (x + 2)^3$

29. $h(x) = x^3 + 3x^2 - x - 3$

30. $g(x) = -x^3 + 2x^2 + 4x - 8$

31. $f(x) = 6x^3 - 8x^2 - 54x + 72$

32. $h(x) = x^5 - 5x^3 + 4x$

Using the intermediate value theorem, determine, if possible, whether the function f has a real zero between a and b.

33. $f(x) = x^3 + 3x^2 - 9x - 13;\ a = -5, b = -4$

34. $f(x) = x^3 + 3x^2 - 9x - 13;\ a = 1, b = 2$

35. $f(x) = 3x^2 - 2x - 11;\ a = -3, b = -2$

36. $f(x) = 3x^2 - 2x - 11;\ a = 2, b = 3$

37. $f(x) = x^4 - 2x^2 - 6;\ a = 2, b = 3$

38. $f(x) = 2x^5 - 7x + 1;\ a = 1, b = 2$

39. $f(x) = x^3 - 5x^2 + 4;\ a = 4, b = 5$

40. $f(x) = x^4 - 3x^2 + x - 1;\ a = -3, b = -2$

Collaborative Discussion and Writing

41. Explain why values of a function must be all positive or all negative between zeros.

42. Is it possible for the graph of a polynomial function to have no y-intercepts? no x-intercepts? Explain your answer.

Skill Maintenance

Match the equation with one of the graphs (a)–(f), which follow.

a)
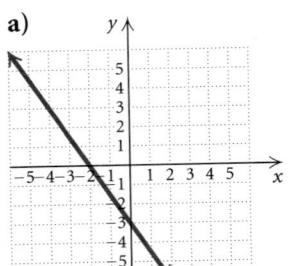

b)
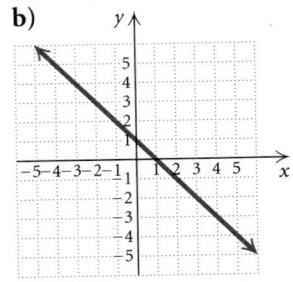

c)
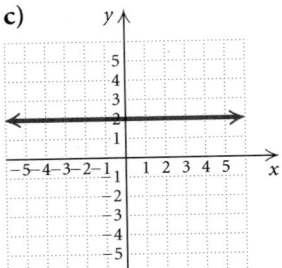

d)
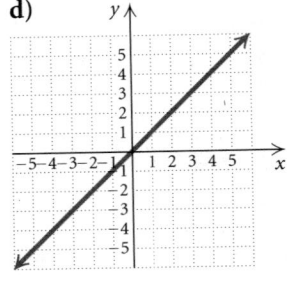

e)
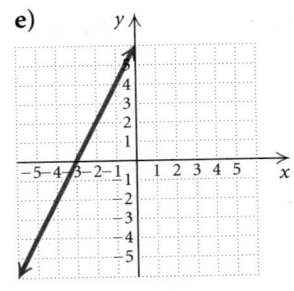

f)
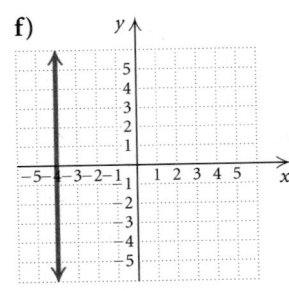
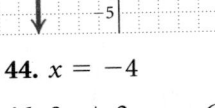

43. $y = x$

44. $x = -4$

45. $y - 2x = 6$

46. $3x + 2y = -6$

47. $y = 1 - x$

48. $y = 2$

Solve.

49. $2x - \frac{1}{2} = 4 - 3x$

50. $x^3 - x^2 - 12x = 0$

51. $6x^2 - 23x - 55 = 0$

52. $\frac{3}{4}x + 10 = \frac{1}{5} + 2x$

4.3

Polynomial Division; The Remainder and Factor Theorems

❖ Perform long division with polynomials and determine whether one polynomial is a factor of another.

❖ Use synthetic division to divide a polynomial by $x - c$.

❖ Use the remainder theorem to find a function value $f(c)$.

❖ Use the factor theorem to determine whether $x - c$ is a factor of $f(x)$.

In general, finding exact zeros of many polynomial functions is neither easy nor straightforward. In this section and the one that follows, we develop concepts that help us find exact zeros of certain polynomial functions with degree 3 or greater.

Consider the polynomial

$$h(x) = x^3 + 2x^2 - 5x - 6 = (x + 3)(x + 1)(x - 2).$$

The factors are

$$x + 3, \quad x + 1, \quad \text{and} \quad x - 2,$$

and the zeros are

$$-3, \quad -1, \quad \text{and} \quad 2.$$

When a polynomial is expressed in factored form, each factor determines a zero of the function. Thus if we know the factors of a polynomial, we can easily find the zeros. This idea can be "reversed" so that if we know the zeros of a polynomial function, we can find the factors of the polynomial.

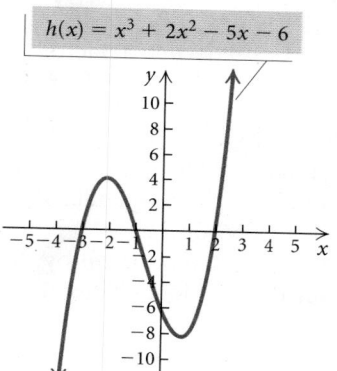

$h(x) = x^3 + 2x^2 - 5x - 6$

❖ Division and Factors

When we divide one polynomial by another, we obtain a quotient and a remainder. If the remainder is 0, then the divisor is a factor of the dividend.

EXAMPLE 1 Divide to determine whether $x + 1$ and $x - 3$ are factors of

$$x^3 + 2x^2 - 5x - 6.$$

Solution We divide $x^3 + 2x^2 - 5x - 6$ by $x + 1$.

$$
\begin{array}{r}
\overbrace{x^2 + x - 6}^{\text{Quotient}} \\
x + 1 \overline{)\, x^3 + 2x^2 - 5x - 6} \leftarrow \text{Dividend} \\
\underline{x^3 + x^2} \\
x^2 - 5x \\
\underline{x^2 + x} \\
-6x - 6 \\
\underline{-6x - 6} \\
0 \leftarrow \text{Remainder}
\end{array}
$$

Divisor

Since the remainder is 0, we know that $x + 1$ is a factor of $x^3 + 2x^2 - 5x - 6$. In fact, we know that

$$x^3 + 2x^2 - 5x - 6 = (x + 1)(x^2 + x - 6).$$

We divide $x^3 + 2x^2 - 5x - 6$ by $x - 3$.

$$
\begin{array}{r}
x^2 + 5x + 10 \\
x - 3{\overline{\smash{\big)}\,x^3 + 2x^2 - 5x - 6}} \\
\underline{x^3 - 3x^2} \\
5x^2 - 5x \\
\underline{5x^2 - 15x} \\
10x - 6 \\
\underline{10x - 30} \\
24 \leftarrow \text{Remainder}
\end{array}
$$

Since the remainder is not 0, we know that $x - 3$ is *not* a factor of $x^3 + 2x^2 - 5x - 6$.

Now Try Exercise 3. ■

When we divide a polynomial $P(x)$ by a divisor $d(x)$, a polynomial $Q(x)$ is the quotient and a polynomial $R(x)$ is the remainder. The quotient $Q(x)$ must have degree less than that of the dividend $P(x)$. The remainder $R(x)$ must either be 0 or have degree less than that of the divisor $d(x)$.

As in arithmetic, to check a division, we multiply the quotient by the divisor and add the remainder, to see if we get the dividend. Thus these polynomials are related as follows:

$$P(x) = d(x) \cdot Q(x) + R(x)$$

Dividend   Divisor   Quotient   Remainder

For example, if $P(x) = x^3 + 2x^2 - 5x - 6$ and $d(x) = x - 3$, as in Example 1, then $Q(x) = x^2 + 5x + 10$ and $R(x) = 24$, and

$$
\begin{aligned}
P(x) \quad &= \quad d(x) \; \cdot \quad Q(x) \quad + R(x) \\
x^3 + 2x^2 - 5x - 6 &= (x - 3) \cdot (x^2 + 5x + 10) + 24 \\
&= x^3 + 5x^2 + 10x - 3x^2 - 15x - 30 + 24 \\
&= x^3 + 2x^2 - 5x - 6.
\end{aligned}
$$

❊ The Remainder Theorem and Synthetic Division

Consider the function

$$h(x) = x^3 + 2x^2 - 5x - 6.$$

When we divided $h(x)$ by $x + 1$ and $x - 3$ in Example 1, the remainders were 0 and 24, respectively. Let's now find the function values $h(-1)$ and $h(3)$:

$$
\begin{aligned}
h(-1) &= (-1)^3 + 2(-1)^2 - 5(-1) - 6 = 0; \\
h(3) &= (3)^3 + 2(3)^2 - 5(3) - 6 = 24.
\end{aligned}
$$

Note that the function values are the same as the remainders. This suggests the following theorem.

> **The Remainder Theorem**
>
> If a number c is substituted for x in the polynomial $f(x)$, then the result $f(c)$ is the remainder that would be obtained by dividing $f(x)$ by $x - c$. That is, if $f(x) = (x - c) \cdot Q(x) + R$, then $f(c) = R$.

Proof (Optional). The equation $f(x) = d(x) \cdot Q(x) + R(x)$, where $d(x) = x - c$, is the basis of this proof. If we divide $f(x)$ by $x - c$, we obtain a quotient $Q(x)$ and a remainder $R(x)$ related as follows:

$$f(x) = (x - c) \cdot Q(x) + R(x).$$

The remainder $R(x)$ must either be 0 or have degree less than $x - c$. Thus, $R(x)$ must be a constant. Let's call this constant R. The equation above is true for any replacement of x, so we replace x with c. We get

$$
\begin{aligned}
f(c) &= (c - c) \cdot Q(c) + R \\
&= 0 \cdot Q(c) + R \\
&= R.
\end{aligned}
$$

Thus the function value $f(c)$ is the remainder obtained when we divide $f(x)$ by $x - c$. ■

The remainder theorem motivates us to find a rapid way of dividing by $x - c$ in order to find function values. To streamline division, we can arrange the work so that duplicate and unnecessary writing is avoided. Consider the following:

$$(4x^3 - 3x^2 + x + 7) \div (x - 2).$$

A.
$$
\begin{array}{r}
4x^2 + 5x + 11 \\
x - 2 \overline{)\,4x^3 - 3x^2 + x + 7} \\
\underline{4x^3 - 8x^2} \\
5x^2 + x \\
\underline{5x^2 - 10x} \\
11x + 7 \\
\underline{11x - 22} \\
29
\end{array}
$$

B.
$$
\begin{array}{r}
4 \quad 5 \quad 11 \\
1 - 2 \overline{)\,4 - 3 + 1 + 7} \\
\underline{4 - 8} \\
5 + 1 \\
\underline{5 - 10} \\
11 + 7 \\
\underline{11 - 22} \\
29
\end{array}
$$

The division in (B) is the same as that in (A), but we wrote only the coefficients. The red numerals are duplicated, so we look for an arrangement in which they are not duplicated. In place of the divisor in the form $x - c$, we can simply use c and then add rather than subtract. When the procedure is "collapsed," we have the algorithm known as **synthetic division**.

C. *Synthetic Division*

$$\begin{array}{r|rrrr} 2 & 4 & -3 & 1 & 7 \\ & & 8 & 10 & 22 \\ \hline & 4 & 5 & 11 & 29 \end{array}$$

The divisor is $x - 2$; thus we use 2 in synthetic division.

We "bring down" the 4. Then we multiply it by the 2 to get 8 and add to get 5. We then multiply 5 by 2 to get 10, add, and so on. The last number, 29, is the remainder. The others, 4, 5, and 11, are the coefficients of the quotient, $4x^2 + 5x + 11$. (Note that the degree of the quotient is 1 less than the degree of the dividend when the degree of the divisor is 1.)

When using synthetic division, we write a 0 for a missing term in the dividend.

EXAMPLE 2 Use synthetic division to find the quotient and the remainder:

$$(2x^3 + 7x^2 - 5) \div (x + 3).$$

Solution First, we note that $x + 3 = x - (-3)$.

$$\begin{array}{r|rrrr} -3 & 2 & 7 & 0 & -5 \\ & & -6 & -3 & 9 \\ \hline & 2 & 1 & -3 & 4 \end{array}$$

Note: We must write a 0 for the missing x-term.

The quotient is $2x^2 + x - 3$. The remainder is 4.　　**Now Try Exercise 11.** ◼

We can now use synthetic division to find polynomial function values.

EXAMPLE 3 Given that $f(x) = 2x^5 - 3x^4 + x^3 - 2x^2 + x - 8$, find $f(10)$.

Solution By the remainder theorem, $f(10)$ is the remainder when $f(x)$ is divided by $x - 10$. We use synthetic division to find that remainder.

$$\begin{array}{r|rrrrrr} 10 & 2 & -3 & 1 & -2 & 1 & -8 \\ & & 20 & 170 & 1710 & 17{,}080 & 170{,}810 \\ \hline & 2 & 17 & 171 & 1708 & 17{,}081 & 170{,}802 \end{array}$$

Thus, $f(10) = 170{,}802$.　　**Now Try Exercise 25.** ◼

Compare the computations in Example 3 with those in a direct substitution:

$$\begin{aligned} f(10) &= 2(10)^5 - 3(10)^4 + (10)^3 - 2(10)^2 + 10 - 8 \\ &= 2 \cdot 100{,}000 - 3 \cdot 10{,}000 + 1000 - 2 \cdot 100 + 10 - 8 \\ &= 200{,}000 - 30{,}000 + 1000 - 200 + 10 - 8 \\ &= 170{,}802. \end{aligned}$$

The computations in synthetic division are less complicated than those involved in substituting. The easiest way to find $f(10)$ is to use one of the methods for evaluating a function on a graphing calculator. In the figure at left, we show the result when we enter $y_1 = 2x^5 - 3x^4 + x^3 - 2x^2 + x - 8$ and then use function notation on the home screen.

Y1(10)

170802

EXAMPLE 4 Determine whether 5 is a zero of $g(x)$, where
$$g(x) = x^4 - 26x^2 + 25.$$

Solution We use synthetic division and the remainder theorem to find $g(5)$.

$$
\begin{array}{r|rrrrr}
5 & 1 & 0 & -26 & 0 & 25 \\
  &   & 5 & 25 & -5 & -25 \\
\hline
  & 1 & 5 & -1 & -5 & 0
\end{array}
$$
Writing 0's for missing terms:
$x^4 + 0x^3 - 26x^2 + 0x + 25$

Since $g(5) = 0$, the number 5 is a zero of $g(x)$.   **Now Try Exercise 31.** ■

EXAMPLE 5 Determine whether i is a zero of $f(x)$, where
$$f(x) = x^3 - 3x^2 + x - 3.$$

Solution We use synthetic division and the remainder theorem to find $f(i)$.

$$
\begin{array}{r|rrrr}
i & 1 & -3 & 1 & -3 \\
  &   & i & -3i-1 & -3i \\
\hline
  & 1 & -3+i & -3i & 0
\end{array}
$$

$i(-3+i) = -3i + i^2 = -3i - 1$
$i(-3i) = -3i^2 = 3$

Since $f(i) = 0$, the number i is a zero of $f(x)$.   **Now Try Exercise 35.** ■

❊ Finding Factors of Polynomials

We now consider a useful result that follows from the remainder theorem.

> **The Factor Theorem**
> For a polynomial $f(x)$, if $f(c) = 0$, then $x - c$ is a factor of $f(x)$.

Proof (Optional). If we divide $f(x)$ by $x - c$, we obtain a quotient and a remainder, related as follows:
$$f(x) = (x - c) \cdot Q(x) + f(c).$$
Then if $f(c) = 0$, we have
$$f(x) = (x - c) \cdot Q(x),$$
so $x - c$ is a factor of $f(x)$.   ■

The factor theorem is very useful in factoring polynomials and hence in solving polynomial equations and finding zeros of polynomial functions. If we know a zero of a polynomial function, we know a factor.

EXAMPLE 6 Let $f(x) = x^3 - 3x^2 - 6x + 8$. Factor $f(x)$ and solve the equation $f(x) = 0$.

Solution We look for linear factors of the form $x - c$. Let's try $x + 1$, or $x - (-1)$. (In the next section, we will learn a method for choosing the numbers to try for c.) We use synthetic division to determine whether $f(-1) = 0$.

$$
\begin{array}{r|rrrr}
-1 & 1 & -3 & -6 & 8 \\
   &   & -1 & 4 & 2 \\
\hline
   & 1 & -4 & -2 & 10
\end{array}
$$

Since $f(-1) \neq 0$, we know that $x + 1$ *is not a factor* of $f(x)$. We now try $x - 1$.

$$
\begin{array}{r|rrrr}
1 & 1 & -3 & -6 & 8 \\
  &   & 1 & -2 & -8 \\
\hline
  & 1 & -2 & -8 & 0
\end{array}
$$

Since $f(1) = 0$, we know that $x - 1$ *is one factor* of $f(x)$ and the quotient, $x^2 - 2x - 8$, is another. Thus,

$$f(x) = (x - 1)(x^2 - 2x - 8).$$

The trinomial $x^2 - 2x - 8$ is easily factored in this case, so we have

$$f(x) = (x - 1)(x - 4)(x + 2).$$

Our goal is to solve the equation $f(x) = 0$. To do so, we use the principle of zero products:

$$(x - 1)(x - 4)(x + 2) = 0$$

$$x - 1 = 0 \quad or \quad x - 4 = 0 \quad or \quad x + 2 = 0$$

$$x = 1 \quad or \quad x = 4 \quad or \quad x = -2.$$

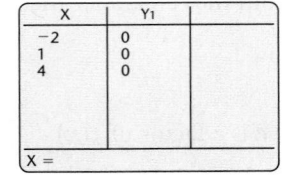

The solutions of the equation $x^3 - 3x^2 - 6x + 8 = 0$ are -2, 1, and 4. They are also the zeros of the function $f(x) = x^3 - 3x^2 - 6x + 8$. We can use a table set in ASK mode to check the solutions. (See the table at left.)

Now Try Exercise 41. ▨

CONNECTING
the CONCEPTS

Consider the function

$$f(x) = (x - 2)(x + 3)(x + 1), \quad or \quad f(x) = x^3 + 2x^2 - 5x - 6,$$

and its graph.

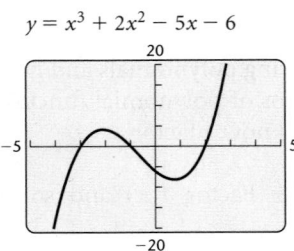

We can make the following statements:

- -3 is a zero of f.
- $f(-3) = 0$.
- -3 is a solution of $f(x) = 0$.
- $(-3, 0)$ is an x-intercept of the graph of f.
- 0 is the remainder when $f(x)$ is divided by $x - (-3)$.
- $x - (-3)$ is a factor of f.

Similar statements are also true for -1 and 2.

4.3 Exercise Set

1. For the function
$$f(x) = x^4 - 6x^3 + x^2 + 24x - 20,$$
use long division to determine whether each of the following is a factor of $f(x)$.
   a) $x + 1$ b) $x - 2$ c) $x + 5$

2. For the function
$$h(x) = x^3 - x^2 - 17x - 15,$$
use long division to determine whether each of the following is a factor of $h(x)$.
   a) $x + 5$ b) $x + 1$ c) $x + 3$

3. For the function
$$g(x) = x^3 - 2x^2 - 11x + 12,$$
use long division to determine whether each of the following is a factor of $g(x)$.
   a) $x - 4$ b) $x - 3$ c) $x - 1$

4. For the function
$$f(x) = x^4 + 8x^3 + 5x^2 - 38x + 24,$$
use long division to determine whether each of the following is a factor of $f(x)$.
   a) $x + 6$ b) $x + 1$ c) $x - 4$

In each of the following, a polynomial $P(x)$ and a divisor $d(x)$ are given. Use long division to find the quotient $Q(x)$ and the remainder $R(x)$ when $P(x)$ is divided by $d(x)$. Express $P(x)$ in the form $d(x) \cdot Q(x) + R(x)$.

5. $P(x) = x^3 - 8,$
   $d(x) = x + 2$

6. $P(x) = 2x^3 - 3x^2 + x - 1,$
   $d(x) = x - 3$

7. $P(x) = x^3 + 6x^2 - 25x + 18,$
   $d(x) = x + 9$

8. $P(x) = x^3 - 9x^2 + 15x + 25,$
   $d(x) = x - 5$

9. $P(x) = x^4 - 2x^2 + 3,$
   $d(x) = x + 2$

10. $P(x) = x^4 + 6x^3,$
    $d(x) = x - 1$

Use synthetic division to find the quotient and the remainder.

11. $(2x^4 + 7x^3 + x - 12) \div (x + 3)$

12. $(x^3 - 7x^2 + 13x + 3) \div (x - 2)$

13. $(x^3 - 2x^2 - 8) \div (x + 2)$

14. $(x^3 - 3x + 10) \div (x - 2)$

15. $(3x^3 - x^2 + 4x - 10) \div (x + 1)$

16. $(4x^4 - 2x + 5) \div (x + 3)$

17. $(x^5 + x^3 - x) \div (x - 3)$

18. $(x^7 - x^6 + x^5 - x^4 + 2) \div (x + 1)$

19. $(x^4 - 1) \div (x - 1)$

20. $(x^5 + 32) \div (x + 2)$

21. $(2x^4 + 3x^2 - 1) \div \left(x - \frac{1}{2}\right)$

22. $(3x^4 - 2x^2 + 2) \div \left(x - \frac{1}{4}\right)$

Use synthetic division to find the function values. Then check your work using a graphing calculator.

23. $f(x) = x^3 - 6x^2 + 11x - 6;$ find $f(1), f(-2),$ and $f(3)$.

24. $f(x) = x^3 + 7x^2 - 12x - 3;$ find $f(-3), f(-2),$ and $f(1)$.

25. $f(x) = x^4 - 3x^3 + 2x + 8;$ find $f(-1), f(4),$ and $f(-5)$.

26. $f(x) = 2x^4 + x^2 - 10x + 1;$ find $f(-10), f(2),$ and $f(3)$.

27. $f(x) = 2x^5 - 3x^4 + 2x^3 - x + 8;$ find $f(20)$ and $f(-3)$.

28. $f(x) = x^5 - 10x^4 + 20x^3 - 5x - 100;$ find $f(-10)$ and $f(5)$.

29. $f(x) = x^4 - 16;$ find $f(2), f(-2), f(3),$ and $f(1 - \sqrt{2})$.

30. $f(x) = x^5 + 32;$ find $f(2), f(-2), f(3),$ and $f(2 + 3i)$.

Using synthetic division, determine whether the numbers are zeros of the polynomial function.

31. $-3, 2; \ f(x) = 3x^3 + 5x^2 - 6x + 18$

32. $-4, 2; \ f(x) = 3x^3 + 11x^2 - 2x + 8$

33. $-3, 1; \ h(x) = x^4 + 4x^3 + 2x^2 - 4x - 3$

34. $2, -1; \ g(x) = x^4 - 6x^3 + x^2 + 24x - 20$

35. $i, -2i; \ g(x) = x^3 - 4x^2 + 4x - 16$

36. $\frac{1}{3}, 2; \ h(x) = x^3 - x^2 - \frac{1}{9}x + \frac{1}{9}$

37. $-3, \frac{1}{2}; \ f(x) = x^3 - \frac{7}{2}x^2 + x - \frac{3}{2}$

38. $i, -i, -2; \ f(x) = x^3 + 2x^2 + x + 2$

Factor the polynomial function $f(x)$. Then solve the equation $f(x) = 0$.

39. $f(x) = x^3 + 4x^2 + x - 6$

40. $f(x) = x^3 + 5x^2 - 2x - 24$

41. $f(x) = x^3 - 6x^2 + 3x + 10$

42. $f(x) = x^3 + 2x^2 - 13x + 10$

43. $f(x) = x^3 - x^2 - 14x + 24$

44. $f(x) = x^3 - 3x^2 - 10x + 24$

45. $f(x) = x^4 - 7x^3 + 9x^2 + 27x - 54$

46. $f(x) = x^4 - 4x^3 - 7x^2 + 34x - 24$

47. $f(x) = x^4 - x^3 - 19x^2 + 49x - 30$

48. $f(x) = x^4 + 11x^3 + 41x^2 + 61x + 30$

Sketch the graph of the polynomial function. Follow the procedure outlined on p. 315. Use synthetic division and the remainder theorem to find the zeros.

49. $f(x) = x^4 - x^3 - 7x^2 + x + 6$

50. $f(x) = x^4 + x^3 - 3x^2 - 5x - 2$

51. $f(x) = x^3 - 7x + 6$

52. $f(x) = x^3 - 12x + 16$

53. $f(x) = -x^3 + 3x^2 + 6x - 8$

54. $f(x) = -x^4 + 2x^3 + 3x^2 - 4x - 4$

Collaborative Discussion and Writing

55. Suppose a polynomial function $P(x)$ has a factor $p(x)$. If $p(2) = 0$, does it follow that $P(2) = 0$? Why or why not? If $P(2) = 0$, does it follow that $p(2) = 0$? Why or why not?

56. In synthetic division, why is the degree of the quotient 1 less than that of the dividend?

Skill Maintenance

Solve. Find exact solutions.

57. $2x^2 + 12 = 5x$

58. $7x^2 + 4x = 3$

Consider the function

$$g(x) = x^2 + 5x - 14$$

in Exercises 59–61.

59. What are the inputs if the output is -14?

60. What is the output if the input is 3?

61. Given an output of -20, find the corresponding inputs.

62. *Hours Online.* The average time spent online has grown linearly in recent years. The average number of hours spent online per week has risen from 9.4 hr in 2000 to 14.0 hr in 2006 (*Source:* USC Annenburg School for Communication, Center for the Digital Future). Using these two data points, find a linear function, $f(x) = mx + b$, that models the data. Let x represent the number of years since 2000. Then use this function to estimate the average number of hours spent online in 2005 and in 2010.

63. The sum of the base and the height of a triangle is 30 in. Find the dimensions for which the area is a maximum.

Synthesis

In Exercises 64 and 65, a graph of a polynomial function is given. On the basis of the graph:

a) *Find as many factors of the polynomial as you can.*
b) *Construct a polynomial function with the zeros shown in the graph.*
c) *Can you find any other polynomial functions with the given zeros?*
d) *Can you find any other polynomial functions with the given zeros and the same graph?*

64.

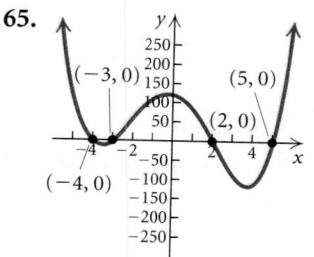

65.

66. For what values of k will the remainder be the same when $x^2 + kx + 4$ is divided by $x - 1$ and by $x + 1$?

67. Find k such that $x + 2$ is a factor of
$x^3 - kx^2 + 3x + 7k$.

68. *Beam Deflection.*   A beam rests at two points A and B and has a concentrated load applied to its center. Let $y =$ the deflection, in feet, of the beam at a distance of x feet from A. Under certain conditions, this deflection is given by

$$y = \frac{1}{13}x^3 - \frac{1}{14}x.$$

Find the zeros of the polynomial in the interval $[0, 2]$.

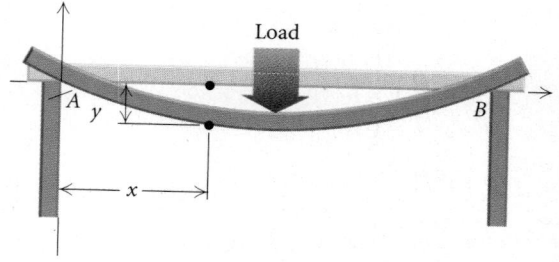

Solve.

69. $\dfrac{2x^2}{x^2 - 1} + \dfrac{4}{x + 3} = \dfrac{12x - 4}{x^3 + 3x^2 - x - 3}$

70. $\dfrac{6x^2}{x^2 + 11} + \dfrac{60}{x^3 - 7x^2 + 11x - 77} = \dfrac{1}{x - 7}$

71. Find a 15th-degree polynomial for which $x - 1$ is a factor. Answers may vary.

Use synthetic division to divide.

72. $(x^4 - y^4) \div (x - y)$

73. $(x^3 + 3ix^2 - 4ix - 2) \div (x + i)$

74. $(x^2 - 4x - 2) \div [x - (3 + 2i)]$

75. $(x^2 - 3x + 7) \div (x - i)$

4.4 Theorems about Zeros of Polynomial Functions

✹ Find a polynomial with specified zeros.

✹ For a polynomial function with integer coefficients, find the rational zeros and the other zeros, if possible.

✹ Use Descartes' rule of signs to find information about the number of real zeros of a polynomial function with real coefficients.

We will now allow the coefficients of a polynomial to be complex numbers. In certain cases, we will restrict the coefficients to be real numbers, rational numbers, or integers, as shown in the following examples.

Polynomial	Type of Coefficient
$5x^3 - 3x^2 + (2 + 4i)x + i$	Complex
$5x^3 - 3x^2 + \sqrt{2}x - \pi$	Real
$5x^3 - 3x^2 + \frac{2}{3}x - \frac{7}{4}$	Rational
$5x^3 - 3x^2 + 8x - 11$	Integer

✹ The Fundamental Theorem of Algebra

A linear, or first-degree, polynomial function $f(x) = mx + b$ (where $m \neq 0$) has just one zero, $-b/m$. It can be shown that any quadratic polynomial function $f(x) = ax^2 + bx + c$ with complex numbers for coefficients has at least one, and at most two, complex zeros. The following theorem is a generalization. No proof is given in this text.

> **The Fundamental Theorem of Algebra**
>
> Every polynomial function of degree n, with $n \geq 1$, has at least one zero in the set of complex numbers.

Note that although the fundamental theorem of algebra guarantees that a zero exists, it does not tell how to find it. Recall that the zeros of a polynomial function $f(x)$ are the solutions of the polynomial equation $f(x) = 0$. We now develop some concepts that can help in finding zeros. First, we consider one of the results of the fundamental theorem of algebra.

> Every polynomial function f of degree n, with $n \geq 1$, can be factored into n linear factors (not necessarily unique); that is,
>
> $$f(x) = a_n(x - c_1)(x - c_2)\cdots(x - c_n).$$

❉ Finding Polynomials with Given Zeros

Given several numbers, we can find a polynomial function with those numbers as its zeros.

EXAMPLE 1   Find a polynomial function of degree 3, having the zeros 1, $3i$, and $-3i$.

Solution   Such a function has factors $x - 1$, $x - 3i$, and $x + 3i$, so we have

$$f(x) = a_n(x - 1)(x - 3i)(x + 3i).$$

The number a_n can be any nonzero number. The simplest polynomial function will be obtained if we let it be 1. If we then multiply the factors, we obtain

$$f(x) = (x - 1)(x^2 + 9) \qquad \text{Multiplying } (x - 3i)(x + 3i)$$
$$= x^3 - x^2 + 9x - 9. \qquad \text{\textbf{Now Try Exercise 3.}} \blacksquare$$

$y = x^3 - x^2 + 9x - 9$

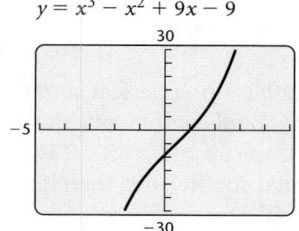

EXAMPLE 2   Find a polynomial function of degree 5 with -1 as a zero of multiplicity 3, 4 as a zero of multiplicity 1, and 0 as a zero of multiplicity 1.

Solution   Proceeding as in Example 1, letting $a_n = 1$, we obtain

$$f(x) = (x + 1)^3(x - 4)(x - 0)$$
$$= x^5 - x^4 - 9x^3 - 11x^2 - 4x. \qquad \text{\textbf{Now Try Exercise 13.}} \blacksquare$$

$y = x^5 - x^4 - 9x^3 - 11x^2 - 4x$

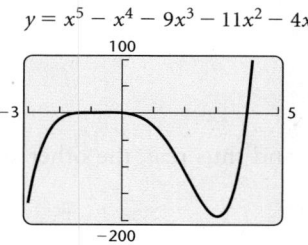

❉ Zeros of Polynomial Functions with Real Coefficients

Consider the quadratic equation $x^2 - 2x + 2 = 0$, with real coefficients. Its solutions are $1 + i$ and $1 - i$. Note that they are complex conjugates. This generalizes to any polynomial equation with real coefficients.

Nonreal Zeros: $a + bi$ and $a - bi$, $b \neq 0$

If a complex number $a + bi$, $b \neq 0$, is a zero of a polynomial function $f(x)$ with *real* coefficients, then its conjugate, $a - bi$, is also a zero. For example, if $2 + 7i$ is a zero of a polynomial function $f(x)$, with real coefficients, then its conjugate, $2 - 7i$, is also a zero. (Nonreal zeros occur in conjugate pairs.)

In order for the preceding to be true, it is essential that the coefficients be *real* numbers.

❋ Rational Coefficients

When a polynomial function has rational numbers for coefficients, certain irrational zeros also occur in pairs, as described in the following theorem.

Irrational Zeros: $a + c\sqrt{b}$ and $a - c\sqrt{b}$, b is not a perfect square

If $a + c\sqrt{b}$, where a, b, and c are rational and b is not a perfect square, is a zero of a polynomial function $f(x)$ with *rational* coefficients, then its conjugate, $a - c\sqrt{b}$, is also a zero. For example, if $-3 + 5\sqrt{2}$ is a zero of a polynomial function $f(x)$ with rational coefficients, then its conjugate, $-3 - 5\sqrt{2}$, is also a zero. (Irrational zeros occur in conjugate pairs.)

EXAMPLE 3   Suppose that a polynomial function of degree 6 with rational coefficients has

$$-2 + 5i, \qquad -2i, \quad \text{and} \quad 1 - \sqrt{3}$$

as three of its zeros. Find the other zeros.

Solution   Since the coefficients are rational, and thus real, the other zeros are the conjugates of the given zeros,

$$-2 - 5i, \qquad 2i, \quad \text{and} \quad 1 + \sqrt{3}.$$

There are no other zeros because a polynomial function of degree 6 can have at most 6 zeros.   **Now Try Exercise 17.** ■

EXAMPLE 4   Find a polynomial function of lowest degree with rational coefficients that has $1 - \sqrt{2}$ and $1 + 2i$ as two of its zeros.

Solution   The function must also have the zeros $1 + \sqrt{2}$ and $1 - 2i$. Because we want to find the polynomial function of lowest degree with the given zeros, we will not include additional zeros; that is, we will write a polynomial function of degree 4. Thus if we let $a_n = 1$, the polynomial function is

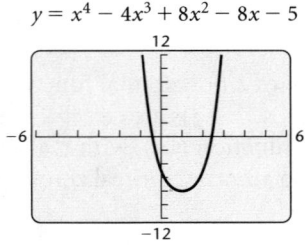

$$y = x^4 - 4x^3 + 8x^2 - 8x - 5$$

$$
\begin{aligned}
f(x) &= \left[x - \left(1 - \sqrt{2}\right)\right]\left[x - \left(1 + \sqrt{2}\right)\right]\left[x - (1 + 2i)\right]\left[x - (1 - 2i)\right] \\
&= (x^2 - 2x - 1)(x^2 - 2x + 5) \\
&= x^4 - 4x^3 + 8x^2 - 8x - 5. \qquad \text{\textbf{Now Try Exercise 39.}} ■
\end{aligned}
$$

❋ Integer Coefficients and the Rational Zeros Theorem

It is not always easy to find the zeros of a polynomial function. However, if a polynomial function has integer coefficients, there is a procedure that will yield all the rational zeros.

> ### The Rational Zeros Theorem
>
> Let
>
> $$P(x) = a_n x^n + a_{n-1} x^{n-1} + \cdots + a_1 x + a_0,$$
>
> where all the coefficients are integers. Consider a rational number denoted by p/q, where p and q are relatively prime (having no common factor besides -1 and 1). If p/q is a zero of $P(x)$, then p is a factor of a_0 and q is a factor of a_n.

EXAMPLE 5 Given $f(x) = 3x^4 - 11x^3 + 10x - 4$:

a) Find the rational zeros and then the other zeros; that is, solve $f(x) = 0$.
b) Factor $f(x)$ into linear factors.

Solution

a) Because the degree of $f(x)$ is 4, there are at most 4 distinct zeros. The rational zeros theorem says that if a rational number p/q is a zero of $f(x)$, then p must be a factor of -4 and q must be a factor of 3. Thus the possibilities for p/q are

$$\frac{\textit{Possibilities for } p}{\textit{Possibilities for } q}: \quad \frac{\pm 1, \pm 2, \pm 4}{\pm 1, \pm 3};$$

$$\textit{Possibilities for } p/q: \quad 1, -1, 2, -2, 4, -4, \tfrac{1}{3}, -\tfrac{1}{3}, \tfrac{2}{3}, -\tfrac{2}{3}, \tfrac{4}{3}, -\tfrac{4}{3}.$$

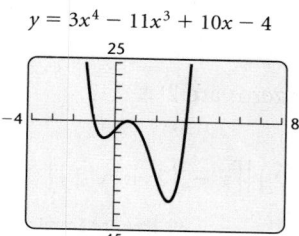

$y = 3x^4 - 11x^3 + 10x - 4$

We could use the TABLE feature or some other method to find function values. However, if we use synthetic division, the quotient polynomial becomes a beneficial by-product if a zero is found. Rather than use synthetic division to check *each* of these possibilities, we graph the function and inspect the graph for zeros that appear to be near any of the possible rational zeros. (See the graph at left.)

From the graph, we see that of the possibilities in the list, only the numbers $-1, \tfrac{1}{3}$, and $\tfrac{2}{3}$ might be rational zeros.

We try -1.

$$\begin{array}{r|rrrrr} -1 & 3 & -11 & 0 & 10 & -4 \\ & & -3 & 14 & -14 & 4 \\ \hline & 3 & -14 & 14 & -4 & \big|\ 0 \end{array}$$

We have $f(-1) = 0$, so -1 is a zero. Thus, $x + 1$ is a factor of $f(x)$. Using the results of the synthetic division, we can express $f(x)$ as

$$f(x) = (x + 1)(3x^3 - 14x^2 + 14x - 4).$$

We now use $3x^3 - 14x^2 + 14x - 4$ and check the other possible zeros. We try $\tfrac{1}{3}$.

$$\begin{array}{r|rrrr} 1/3 & 3 & -14 & 14 & -4 \\ & & 1 & -\tfrac{13}{3} & \tfrac{29}{9} \\ \hline & 3 & -13 & \tfrac{29}{3} & \big|\ -\tfrac{7}{9} \end{array}$$

Since $f\!\left(\tfrac{1}{3}\right) \neq 0$, we know that $\tfrac{1}{3}$ is not a zero.

Let's now try $\frac{2}{3}$.

$$\begin{array}{r|rrrr}
2/3 & 3 & -14 & 14 & -4 \\
 & & 2 & -8 & 4 \\
\hline
 & 3 & -12 & 6 & 0
\end{array}$$

Since the remainder is 0, we know that $x - \frac{2}{3}$ is a factor of $3x^3 - 14x^2 + 14x - 4$ and is also a factor of $f(x)$. Thus, $\frac{2}{3}$ is a zero of $f(x)$.

We can check the zeros with the TABLE feature. (See the window at left.) Note that the graphing calculator converts $\frac{1}{3}$ and $\frac{2}{3}$ to decimal notation. Since $f(-1) = 0$ and $f(\frac{2}{3}) = 0$, -1 and $\frac{2}{3}$ are zeros. Since $f(\frac{1}{3}) \neq 0$, $\frac{1}{3}$ is not a zero.

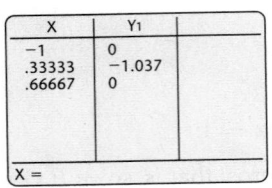

Using the results of the synthetic division, we can factor further:

$$f(x) = (x + 1)\left(x - \tfrac{2}{3}\right)(3x^2 - 12x + 6) \qquad \begin{array}{l}\text{Using the results of the}\\ \text{last synthetic division}\end{array}$$

$$= (x + 1)\left(x - \tfrac{2}{3}\right) \cdot 3 \cdot (x^2 - 4x + 2). \qquad \text{Removing a factor of 3}$$

The quadratic formula can be used to find the values of x for which $x^2 - 4x + 2 = 0$. Those values are also zeros of $f(x)$:

$$x = \frac{-b \pm \sqrt{b^2 - 4ac}}{2a}$$

$$= \frac{-(-4) \pm \sqrt{(-4)^2 - 4 \cdot 1 \cdot 2}}{2 \cdot 1} \qquad a = 1, b = -4, \text{ and } c = 2$$

$$= \frac{4 \pm \sqrt{8}}{2} = \frac{4 \pm 2\sqrt{2}}{2} = \frac{2(2 \pm \sqrt{2})}{2}$$

$$= 2 \pm \sqrt{2}.$$

The rational zeros are -1 and $\frac{2}{3}$. The other zeros are $2 \pm \sqrt{2}$.

b) The complete factorization of $f(x)$ is

$$f(x) = 3(x + 1)\left(x - \tfrac{2}{3}\right)\left[x - \left(2 - \sqrt{2}\right)\right]\left[x - \left(2 + \sqrt{2}\right)\right].$$

Now Try Exercise 55. ■

EXAMPLE 6 Given $f(x) = 2x^5 - x^4 - 4x^3 + 2x^2 - 30x + 15$:

a) Find the rational zeros and then the other zeros; that is, solve $f(x) = 0$.

b) Factor $f(x)$ into linear factors.

Solution

a) Because the degree of $f(x)$ is 5, there are at most 5 distinct zeros. According to the rational zeros theorem, any rational zero of f must be of the form p/q, where p is a factor of 15 and q is a factor of 2. The possibilities are

$$\frac{\textit{Possibilities for } p}{\textit{Possibilities for } q} : \quad \frac{\pm 1, \pm 3, \pm 5, \pm 15}{\pm 1, \pm 2};$$

$$\textit{Possibilities for } p/q: \quad 1, -1, 3, -3, 5, -5, 15, -15, \tfrac{1}{2}, -\tfrac{1}{2}, \tfrac{3}{2}, -\tfrac{3}{2},$$

$$\tfrac{5}{2}, -\tfrac{5}{2}, \tfrac{15}{2}, -\tfrac{15}{2}.$$

$y = 2x^5 - x^4 - 4x^3 + 2x^2$
$- 30x + 15$

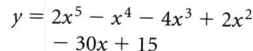

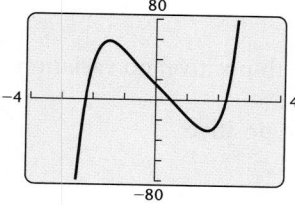

X	Y1	
-2.5	-69.38	
.5	0	
2.5	46.25	

X =

Rather than use synthetic division to check each of these possibilities, we graph $y = 2x^5 - x^4 - 4x^3 + 2x^2 - 30x + 15$. (See the graph at left.) We can then inspect the graph for zeros that appear to be near any of the possible rational zeros.

From the graph, we see that of the possibilities in the list, only the numbers $-\frac{5}{2}, \frac{1}{2}$, and $\frac{5}{2}$ might be rational zeros. By synthetic division or using the TABLE feature of a graphing calculator (see the table at left), we see that only $\frac{1}{2}$ is actually a rational zero.

$$
\begin{array}{r|rrrrrr}
1/2 & 2 & -1 & -4 & 2 & -30 & 15 \\
 & & 1 & 0 & -2 & 0 & -15 \\
\hline
 & 2 & 0 & -4 & 0 & -30 & 0
\end{array}
$$

This means that $x - \frac{1}{2}$ is a factor of $f(x)$. We write the factorization and try to factor further:

$$
\begin{aligned}
f(x) &= \left(x - \tfrac{1}{2}\right)(2x^4 - 4x^2 - 30) \\
&= \left(x - \tfrac{1}{2}\right) \cdot 2 \cdot (x^4 - 2x^2 - 15) \qquad \text{\textbf{Factoring out the 2}} \\
&= \left(x - \tfrac{1}{2}\right) \cdot 2 \cdot (x^2 - 5)(x^2 + 3). \qquad \text{\textbf{Factoring the trinomial}}
\end{aligned}
$$

We now solve the equation $f(x) = 0$ to determine the zeros. We use the principle of zero products:

$$
\left(x - \tfrac{1}{2}\right) \cdot 2 \cdot (x^2 - 5)(x^2 + 3) = 0
$$

$$
\begin{aligned}
x - \tfrac{1}{2} = 0 \quad &or \quad x^2 - 5 = 0 \quad &or \quad x^2 + 3 = 0 \\
x = \tfrac{1}{2} \quad &or \quad x^2 = 5 \quad &or \quad x^2 = -3 \\
x = \tfrac{1}{2} \quad &or \quad x = \pm\sqrt{5} \quad &or \quad x = \pm\sqrt{3}i.
\end{aligned}
$$

There is only one rational zero, $\frac{1}{2}$. The other zeros are $\pm\sqrt{5}$ and $\pm\sqrt{3}i$.

b) The factorization into linear factors is

$$
f(x) = 2\left(x - \tfrac{1}{2}\right)\left(x + \sqrt{5}\right)\left(x - \sqrt{5}\right)\left(x + \sqrt{3}i\right)\left(x - \sqrt{3}i\right).
$$

Now Try Exercise 61. ■

❉ Descartes' Rule of Signs

The development of a rule that helps determine the number of positive real zeros and the number of negative real zeros of a polynomial function is credited to the French mathematician René Descartes. To use the rule, we must have the polynomial arranged in descending or ascending order, with no zero terms written in and the constant term not 0. Then we determine the number of *variations of sign*, that is, the number of times, in reading through the polynomial, that successive coefficients are of different signs.

EXAMPLE 7 Determine the number of variations of sign in the polynomial function $P(x) = 2x^5 - 3x^2 + x + 4$.

Solution We have

$$P(x) = 2x^5 - 3x^2 + x + 4$$

From positive to negative; a variation

Both positive; no variation

From negative to positive; a variation

The number of variations of sign is 2.

Note the following:

$$P(-x) = 2(-x)^5 - 3(-x)^2 + (-x) + 4$$
$$= -2x^5 - 3x^2 - x + 4.$$

We see that the number of variations of sign in $P(-x)$ is 1. It occurs as we go from $-x$ to 4.

We now state Descartes' rule, without proof.

Descartes' Rule of Signs

Let $P(x)$ written in descending or ascending order be a polynomial function with real coefficients and a nonzero constant term. The number of positive real zeros of $P(x)$ is either:

1. The same as the number of variations of sign in $P(x)$, or
2. Less than the number of variations of sign in $P(x)$ by a positive even integer.

The number of negative real zeros of $P(x)$ is either:

3. The same as the number of variations of sign in $P(-x)$, or
4. Less than the number of variations of sign in $P(-x)$ by a positive even integer.

A zero of multiplicity m must be counted m times.

In each of Examples 8–10, what does Descartes' rule of signs tell you about the number of positive real zeros and the number of negative real zeros?

EXAMPLE 8 $P(x) = 2x^5 - 5x^2 - 3x + 6$

Solution The number of variations of sign in $P(x)$ is 2. Therefore, the number of positive real zeros is either 2 or less than 2 by 2, 4, 6, and so on. Thus the number of positive real zeros is either 2 or 0, since a negative number of zeros has no meaning.

$$P(-x) = -2x^5 - 5x^2 + 3x + 6$$

The number of variations of sign in $P(-x)$ is 1. Thus there is exactly 1 negative real zero. Since nonreal, complex conjugates occur in pairs, we also know the possible ways in which nonreal zeros might occur. The table shown at left summarizes all the possibilities for real zeros and nonreal zeros of $P(x)$.

Now Try Exercise 91. ■

Total Number of Zeros	5	
Positive Real	2	0
Negative Real	1	1
Nonreal	2	4

STUDY TIP

It is never too soon to begin reviewing for the final examination. The Skill Maintenance exercises found in each exercise set review and reinforce skills taught in earlier sections. Be sure to do these exercises as you do the homework assignment in each section. Answers to all of the skill maintenance exercises along with section references appear at the back of the book.

Total Number of Zeros	6	
0 as a Zero	1	1
Positive Real	1	1
Negative Real	2	0
Nonreal	2	4

EXAMPLE 9 $P(x) = 5x^4 - 3x^3 + 7x^2 - 12x + 4$

Solution There are 4 variations of sign. Thus the number of positive real zeros is either

$$4 \quad \text{or} \quad 4 - 2 \quad \text{or} \quad 4 - 4.$$

That is, the number of positive real zeros is 4, 2, or 0.

$$P(-x) = 5x^4 + 3x^3 + 7x^2 + 12x + 4$$

There are 0 changes in sign, so there are no negative real zeros.

Now Try Exercise 79. ■

EXAMPLE 10 $P(x) = 6x^6 - 2x^2 - 5x$

Solution As written, the polynomial does not satisfy the conditions of Descartes' rule of signs because the constant term is 0. But because x is a factor of every term, we know that the polynomial has 0 as a zero. We can then factor as follows:

$$P(x) = x(6x^5 - 2x - 5).$$

Now we analyze $Q(x) = 6x^5 - 2x - 5$ and $Q(-x) = -6x^5 + 2x - 5$. The number of variations of sign in $Q(x)$ is 1. Therefore, there is exactly 1 positive real zero. The number of variations of sign in $Q(-x)$ is 2. Thus the number of negative real zeros is 2 or 0. The same results apply to $P(x)$. Since nonreal, complex conjugates occur in pairs, we know the possible ways in which nonreal zeros might occur. The table at left summarizes all the possibilities for real zeros and nonreal zeros of $P(x)$. **Now Try Exercise 93.** ■

(4.4) Exercise Set

Find a polynomial function of degree 3 with the given numbers as zeros.

1. $-2, 3, 5$

2. $-1, 0, 4$

3. $-3, 2i, -2i$

4. $2, i, -i$

5. $\sqrt{2}, -\sqrt{2}, 3$

6. $-5, \sqrt{3}, -\sqrt{3}$

7. $1 - \sqrt{3}, 1 + \sqrt{3}, -2$

8. $-4, 1 - \sqrt{5}, 1 + \sqrt{5}$

9. $1 + 6i, 1 - 6i, -4$

10. $1 + 4i, 1 - 4i, -1$

11. $-\frac{1}{3}, 0, 2$

12. $-3, 0, \frac{1}{2}$

13. Find a polynomial function of degree 5 with -1 as a zero of multiplicity 3, 0 as a zero of multiplicity 1, and 1 as a zero of multiplicity 1.

14. Find a polynomial function of degree 4 with -2 as a zero of multiplicity 1, 3 as a zero of multiplicity 2, and -1 as a zero of multiplicity 1.

rational zeros
→ theorem

15. Find a polynomial function of degree 4 with -1 as a zero of multiplicity 3 and 0 as a zero of multiplicity 1.

16. Find a polynomial function of degree 5 with $-\frac{1}{2}$ as a zero of multiplicity 2, 0 as a zero of multiplicity 1, and 1 as a zero of multiplicity 2.

Suppose that a polynomial function of degree 4 with rational coefficients has the given numbers as zeros. Find the other zero(s).

17. $-1, \sqrt{3}, \frac{11}{3}$

18. $-\sqrt{2}, -1, \frac{4}{5}$

19. $-i, 2 - \sqrt{5}$

20. $i, -3 + \sqrt{3}$

21. $3i, 0, -5$

22. $3, 0, -2i$

23. $-4 - 3i, 2 - \sqrt{3}$

24. $6 - 5i, -1 + \sqrt{7}$

Suppose that a polynomial function of degree 5 with rational coefficients has the given numbers as zeros. Find the other zero(s).

25. $-\frac{1}{2}, \sqrt{5}, -4i$

26. $\frac{3}{4}, -\sqrt{3}, 2i$

27. $-5, 0, 2 - i, 4$

28. $-2, 3, 4, 1 - i$

29. $6, -3 + 4i, 4 - \sqrt{5}$

30. $-3 - 3i, 2 + \sqrt{13}, 6$

31. $-\frac{3}{4}, \frac{3}{4}, 0, 4 - i$

32. $-0.6, 0, 0.6, -3 + \sqrt{2}$

Find a polynomial function of lowest degree with rational coefficients that has the given numbers as some of its zeros.

33. $1 + i, 2$

34. $2 - i, -1$

35. $4i$

36. $-5i$

37. $-4i, 5$

38. $3, -i$

39. $1 - i, -\sqrt{5}$

40. $2 - \sqrt{3}, 1 + i$

41. $\sqrt{5}, -3i$

42. $-\sqrt{2}, 4i$

Given that the polynomial function has the given zero, find the other zeros.

43. $f(x) = x^3 + 5x^2 - 2x - 10; \ -5$

44. $f(x) = x^3 - x^2 + x - 1; \ 1$

45. $f(x) = x^4 - 5x^3 + 7x^2 - 5x + 6; \ -i$

46. $f(x) = x^4 - 16; \ 2i$

47. $f(x) = x^3 - 6x^2 + 13x - 20; \ 4$

48. $f(x) = x^3 - 8; \ 2$

List all possible rational zeros of the function.

49. $f(x) = x^5 - 3x^2 + 1$

50. $f(x) = x^7 + 37x^5 - 6x^2 + 12$

51. $f(x) = 2x^4 - 3x^3 - x + 8$

52. $f(x) = 3x^3 - x^2 + 6x - 9$

53. $f(x) = 15x^6 + 47x^2 + 2$

54. $f(x) = 10x^{25} + 3x^{17} - 35x + 6$

★ *For each polynomial function:*

a) *Find the rational zeros and then the other zeros; that is, solve $f(x) = 0$.*

b) *Factor $f(x)$ into linear factors.*

55. $f(x) = x^3 + 3x^2 - 2x - 6$

56. $f(x) = x^3 - x^2 - 3x + 3$

57. $f(x) = x^3 - 3x + 2$

58. $f(x) = x^3 - 2x + 4$

59. $f(x) = x^3 - 5x^2 + 11x + 17$

60. $f(x) = 2x^3 + 7x^2 + 2x - 8$

61. $f(x) = 5x^4 - 4x^3 + 19x^2 - 16x - 4$

62. $f(x) = 3x^4 - 4x^3 + x^2 + 6x - 2$

63. $f(x) = x^4 - 3x^3 - 20x^2 - 24x - 8$

64. $f(x) = x^4 + 5x^3 - 27x^2 + 31x - 10$

65. $f(x) = x^3 - 4x^2 + 2x + 4$

66. $f(x) = x^3 - 8x^2 + 17x - 4$

67. $f(x) = x^3 + 8$

68. $f(x) = x^3 - 8$

69. $f(x) = \frac{1}{3}x^3 - \frac{1}{2}x^2 - \frac{1}{6}x + \frac{1}{6}$

70. $f(x) = \frac{2}{3}x^3 - \frac{1}{2}x^2 + \frac{2}{3}x - \frac{1}{2}$

Find only the rational zeros of the function.

71. $f(x) = x^4 + 2x^3 - 5x^2 - 4x + 6$

72. $f(x) = x^4 - 3x^3 - 9x^2 - 3x - 10$

73. $f(x) = x^3 - x^2 - 4x + 3$

74. $f(x) = 2x^3 + 3x^2 + 2x + 3$

75. $f(x) = x^4 + 2x^3 + 2x^2 - 4x - 8$

76. $f(x) = x^4 + 6x^3 + 17x^2 + 36x + 66$

77. $f(x) = x^5 - 5x^4 + 5x^3 + 15x^2 - 36x + 20$

78. $f(x) = x^5 - 3x^4 - 3x^3 + 9x^2 - 4x + 12$

What does Descartes' rule of signs tell you about the number of positive real zeros and the number of negative real zeros of the function?

79. $f(x) = 3x^5 - 2x^2 + x - 1$

80. $g(x) = 5x^6 - 3x^3 + x^2 - x$

81. $h(x) = 6x^7 + 2x^2 + 5x + 4$

82. $P(x) = -3x^5 - 7x^3 - 4x - 5$

83. $F(p) = 3p^{18} + 2p^4 - 5p^2 + p + 3$

84. $H(t) = 5t^{12} - 7t^4 + 3t^2 + t + 1$

85. $C(x) = 7x^6 + 3x^4 - x - 10$

86. $g(z) = -z^{10} + 8z^7 + z^3 + 6z - 1$

87. $h(t) = -4t^5 - t^3 + 2t^2 + 1$

88. $P(x) = x^6 + 2x^4 - 9x^3 - 4$

89. $f(y) = y^4 + 13y^3 - y + 5$

90. $Q(x) = x^4 - 2x^2 + 12x - 8$

91. $r(x) = x^4 - 6x^2 + 20x - 24$

92. $f(x) = x^5 - 2x^3 - 8x$

93. $R(x) = 3x^5 - 5x^3 - 4x$

94. $f(x) = x^4 - 9x^2 - 6x + 4$

Sketch the graph of the polynomial function. Follow the procedure outlined on p. 315. Use the rational zeros theorem when finding the zeros.

95. $f(x) = 4x^3 + x^2 - 8x - 2$

96. $f(x) = 3x^3 - 4x^2 - 5x + 2$

97. $f(x) = 2x^4 - 3x^3 - 2x^2 + 3x$

98. $f(x) = 4x^4 - 37x^2 + 9$

Collaborative Discussion and Writing

99. Is it possible for a third-degree polynomial with rational coefficients to have no real zeros? Why or why not?

100. If $P(x)$ is an even function, and by Descartes' rule of signs, $P(x)$ has one positive real zero, how many negative real zeros does $P(x)$ have? Explain.

Skill Maintenance

For Exercises 101 and 102, complete the square to:
a) *find the vertex;*
b) *find the axis of symmetry; and*
c) *determine whether there is a maximum or minimum function value and find that value.*

101. $f(x) = x^2 - 8x + 10$

102. $f(x) = 3x^2 - 6x - 1$

Find the zeros of the function.

103. $f(x) = -\frac{4}{5}x + 8$

104. $g(x) = x^2 - 8x - 33$

Determine the leading term, the leading coefficient, and the degree of the polynomial. Then describe the end behavior of the function's graph and classify the polynomial function as constant, linear, quadratic, cubic, or quartic.

105. $g(x) = -x^3 - 2x^2$

106. $f(x) = -x^2 - 3x + 6$

107. $f(x) = -\frac{4}{9}$

108. $h(x) = x - 2$

109. $g(x) = x^4 - 2x^3 + x^2 - x + 2$

110. $h(x) = x^3 + \frac{1}{2}x^2 - 4x - 3$

Synthesis

111. Consider $f(x) = 2x^3 - 5x^2 - 4x + 3$. Find the solutions of each equation.
a) $f(x) = 0$ **b)** $f(x - 1) = 0$
c) $f(x + 2) = 0$ **d)** $f(2x) = 0$

112. Use the rational zeros theorem and the equation $x^4 - 12 = 0$ to show that $\sqrt[4]{12}$ is irrational.

Find the rational zeros of the function.

113. $P(x) = 2x^5 - 33x^4 - 84x^3 + 2203x^2 - 3348x - 10{,}080$

114. $P(x) = x^6 - 6x^5 - 72x^4 - 81x^2 + 486x + 5832$

4.5 Rational Functions

❖ For a rational function, find the domain and graph the function, identifying all of the asymptotes.

❖ Solve applied problems involving rational functions.

Now we turn our attention to functions that represent the quotient of two polynomials. Whereas the sum, difference, or product of two polynomials is a polynomial, in general the quotient of two polynomials is *not* itself a polynomial.

A *rational number* can be expressed as the quotient of two integers, p/q, where $q \neq 0$. A *rational function* is formed by the quotient of two polynomials, $p(x)/q(x)$, where $q(x) \neq 0$. Here are some examples of rational functions and their graphs.

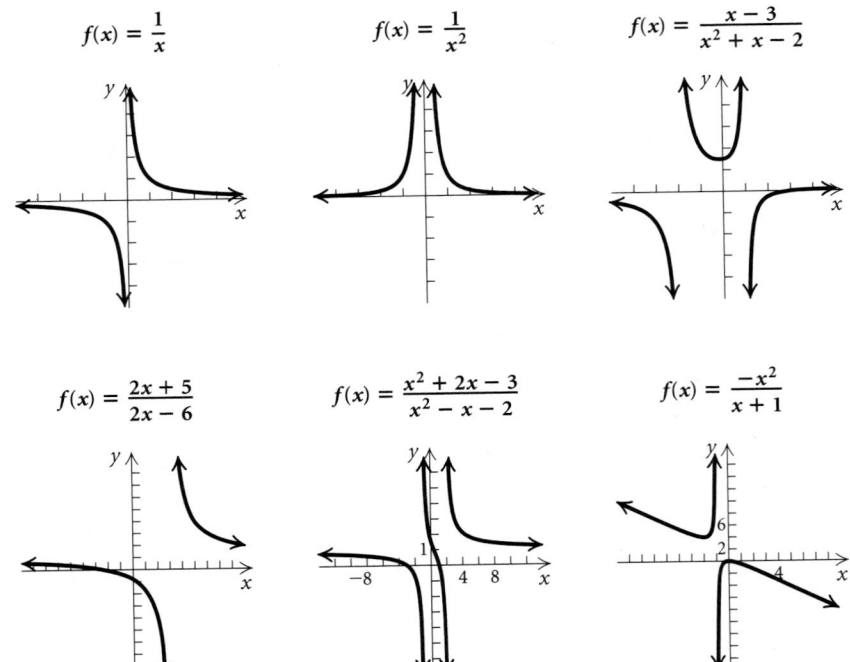

$$f(x) = \frac{1}{x} \qquad f(x) = \frac{1}{x^2} \qquad f(x) = \frac{x-3}{x^2+x-2}$$

$$f(x) = \frac{2x+5}{2x-6} \qquad f(x) = \frac{x^2+2x-3}{x^2-x-2} \qquad f(x) = \frac{-x^2}{x+1}$$

Rational Function

A **rational function** is a function f that is a quotient of two polynomials. That is,

$$f(x) = \frac{p(x)}{q(x)},$$

where $p(x)$ and $q(x)$ are polynomials and where $q(x)$ is not the zero polynomial. The domain of f consists of all inputs x for which $q(x) \neq 0$.

❋ The Domain of a Rational Function

GCM **EXAMPLE 1** Consider

$$f(x) = \frac{1}{x-3}.$$

Find the domain and graph f.

DOMAINS OF FUNCTIONS
REVIEW SECTION 1.2.

Solution When the denominator $x - 3$ is 0, we have $x = 3$, so the only input that results in a denominator of 0 is 3. Thus the domain is

$$\{x \mid x \neq 3\}, \text{ or } (-\infty, 3) \cup (3, \infty).$$

The graph of this function is the graph of $y = 1/x$ translated right 3 units. Two versions of the graph on a graphing calculator are shown below.

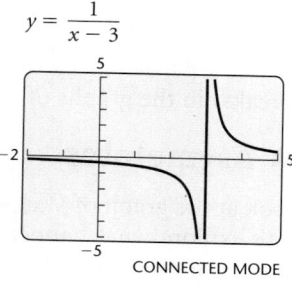

CONNECTED MODE

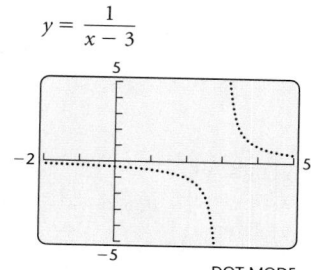

DOT MODE

Using CONNECTED mode can lead to an incorrect graph. In CONNECTED mode, a graphing calculator connects plotted points with line segments. In DOT mode, it simply plots unconnected points. In the first graph, the graphing calculator has connected the points plotted on either side of the x-value 3 with a line that appears to be the vertical line $x = 3$. (It is not actually vertical since it connects the last point to the left of $x = 3$ with the first point to the right of $x = 3$.) Since 3 is not in the domain of the function, the vertical line $x = 3$ cannot be part of the graph. We will see later in this section that vertical lines like $x = 3$, although not part of the graph, are important in the construction of graphs. If you have a choice when graphing rational functions, use DOT mode. ∎

EXAMPLE 2 Determine the domain of each of the functions illustrated at the beginning of this section.

Solution The domain of each rational function will be the set of all real numbers except those values that make the denominator 0. To determine those exceptions, we set the denominator equal to 0 and solve for x.

FUNCTION	DOMAIN
$f(x) = \dfrac{1}{x}$	$\{x \mid x \neq 0\}$, or $(-\infty, 0) \cup (0, \infty)$
$f(x) = \dfrac{1}{x^2}$	$\{x \mid x \neq 0\}$, or $(-\infty, 0) \cup (0, \infty)$
$f(x) = \dfrac{x-3}{x^2+x-2} = \dfrac{x-3}{(x+2)(x-1)}$	$\{x \mid x \neq -2 \text{ and } x \neq 1\}$, or $(-\infty, -2) \cup (-2, 1) \cup (1, \infty)$
$f(x) = \dfrac{2x+5}{2x-6} = \dfrac{2x+5}{2(x-3)}$	$\{x \mid x \neq 3\}$, or $(-\infty, 3) \cup (3, \infty)$
$f(x) = \dfrac{x^2+2x-3}{x^2-x-2} = \dfrac{x^2+2x-3}{(x+1)(x-2)}$	$\{x \mid x \neq -1 \text{ and } x \neq 2\}$, or $(-\infty, -1) \cup (-1, 2) \cup (2, \infty)$
$f(x) = \dfrac{-x^2}{x+1}$	$\{x \mid x \neq -1\}$, or $(-\infty, -1) \cup (-1, \infty)$

As a partial check of the domains, we can observe the discontinuities (breaks) in the graphs of these functions. (See p. 342.) ■

❈ Asymptotes

Look at the graph of $f(x) = 1/(x-3)$, shown at left. (Also see Example 1.) Let's explore what happens as x-values get closer and closer to 3 from the left. We then explore what happens as x-values get closer and closer to 3 from the right.

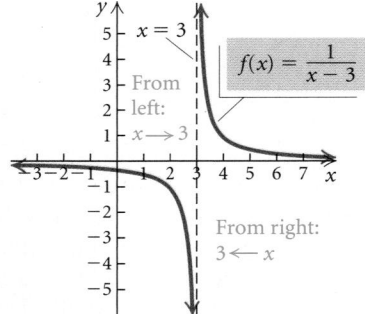

Vertical asymptote: $x = 3$

From the left:

x	2	$2\frac{1}{2}$	$2\frac{99}{100}$	$2\frac{9999}{10,000}$	$2\frac{999,999}{1,000,000}$	$\longrightarrow 3$
$f(x)$	-1	-2	-100	$-10,000$	$-1,000,000$	$\longrightarrow -\infty$

From the right:

x	4	$3\frac{1}{2}$	$3\frac{1}{100}$	$3\frac{1}{10,000}$	$3\frac{1}{1,000,000}$	$\longrightarrow 3$
$f(x)$	1	2	100	10,000	1,000,000	$\longrightarrow \infty$

We see that as x-values get closer and closer to 3 from the left, the function values (y-values) decrease without bound (that is, they approach negative infinity, $-\infty$). Similarly, as the x-values approach 3 from the right, the function values increase without bound (that is, they approach positive infinity, ∞). We write this as

$$f(x) \to -\infty \text{ as } x \to 3^- \quad \text{and} \quad f(x) \to \infty \text{ as } x \to 3^+.$$

We read "$f(x) \to -\infty$ as $x \to 3^-$" as "$f(x)$ decreases without bound as x approaches 3 from the left." We read "$f(x) \to \infty$ as $x \to 3^+$" as "$f(x)$ increases without bound as x approaches 3 from the right." The notation $x \to 3$ means that x gets as close to 3 as possible without being equal to 3. The vertical line $x = 3$ is said to be a *vertical asymptote* for this curve.

In general, the line $x = a$ is a **vertical asymptote** for the graph of f if any of the following is true:

$$f(x) \to \infty \text{ as } x \to a^-, \quad \text{or} \quad f(x) \to -\infty \text{ as } x \to a^-, \quad \text{or}$$
$$f(x) \to \infty \text{ as } x \to a^+, \quad \text{or} \quad f(x) \to -\infty \text{ as } x \to a^+.$$

The following figures show the four ways in which a vertical asymptote can occur.

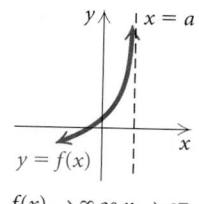
$f(x) \to \infty$ as $x \to a^-$

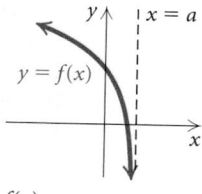
$f(x) \to -\infty$ as $x \to a^-$

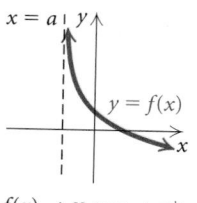
$f(x) \to \infty$ as $x \to a^+$

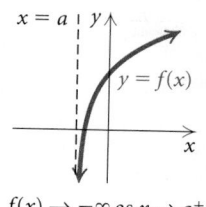
$f(x) \to -\infty$ as $x \to a^+$

The vertical asymptotes of a rational function $f(x) = p(x)/q(x)$ are found by determining the zeros of $q(x)$ that are not also zeros of $p(x)$. If $p(x)$ and $q(x)$ are polynomials with no common factors other than constants, we need determine only the zeros of the denominator $q(x)$.

> **Determining Vertical Asymptotes**
>
> For a rational function $f(x) = p(x)/q(x)$, where $p(x)$ and $q(x)$ are polynomials with no common factors other than constants, if a is a zero of the denominator, then the line $x = a$ is a vertical asymptote for the graph of the function.

EXAMPLE 3 Determine the vertical asymptotes for the graph of each of the following functions.

a) $f(x) = \dfrac{2x - 11}{x^2 + 2x - 8}$

b) $h(x) = \dfrac{x^2 - 4x}{x^3 - x}$

c) $g(x) = \dfrac{x - 2}{x^3 - 5x}$

Solution

a) First, we factor the denominator:

$$f(x) = \frac{2x - 11}{x^2 + 2x - 8} = \frac{2x - 11}{(x + 4)(x - 2)}.$$

The numerator and the denominator have no common factors. The zeros of the denominator are -4 and 2. Thus the vertical asymptotes for the graph of $f(x)$ are the lines $x = -4$ and $x = 2$. (See Fig. 1.)

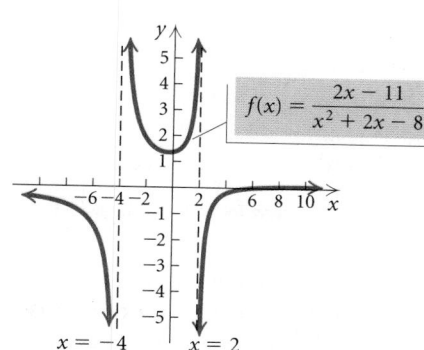
$f(x) = \dfrac{2x - 11}{x^2 + 2x - 8}$

$x = -4 \qquad x = 2$

FIGURE 1

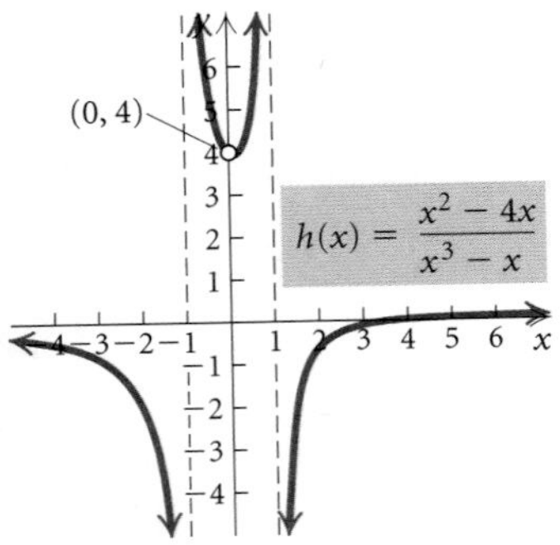

b) We factor the numerator and the denominator:

$$h(x) = \frac{x^2 - 4x}{x^3 - x} = \frac{x(x - 4)}{x(x^2 - 1)} = \frac{x(x - 4)}{x(x + 1)(x - 1)}.$$

The domain of the function is $\{x \mid x \neq -1 \text{ and } x \neq 0 \text{ and } x \neq 1\}$, or $(-\infty, -1) \cup (-1, 0) \cup (0, 1) \cup (1, \infty)$. Note that the numerator and the denominator share a common factor, x. The vertical asymptotes of $h(x)$ are found by determining the zeros of the denominator, $x(x + 1)(x - 1)$, that are *not* also zeros of the numerator, $x(x - 4)$. The zeros of $x(x + 1)(x - 1)$ are 0, −1, and 1. The zeros of $x(x - 4)$ are 0 and 4. Thus, although the denominator has three zeros, the graph of $h(x)$ has only two vertical asymptotes, $x = -1$ and $x = 1$. (See Fig. 2.)

The rational expression $[x(x - 4)]/[x(x + 1)(x - 1)]$ can be simplified. Thus,

$$h(x) = \frac{x(x - 4)}{x(x + 1)(x - 1)} = \frac{x - 4}{(x + 1)(x - 1)},$$

where $x \neq 0$, $x \neq -1$, and $x \neq 1$. The graph of $h(x)$ is the graph of

$$h(x) = \frac{x - 4}{(x + 1)(x - 1)}$$

with the point where $x = 0$ missing. To determine the y-coordinate of the "hole," we substitute 0 for x:

$$h(0) = \frac{0 - 4}{(0 + 1)(0 - 1)} = \frac{-4}{1 \cdot (-1)} = 4.$$

Thus the "hole" is located at $(0, 4)$.

c) We factor the denominator:

$$g(x) = \frac{x - 2}{x^3 - 5x} = \frac{x - 2}{x(x^2 - 5)}.$$

The numerator and the denominator have no common factors. We find the zeros of the denominator, $x(x^2 - 5)$. Solving $x(x^2 - 5) = 0$, we get

$$x = 0 \quad \text{or} \quad x^2 - 5 = 0$$
$$x = 0 \quad \text{or} \quad x^2 = 5$$
$$x = 0 \quad \text{or} \quad x = \pm\sqrt{5}.$$

The zeros of the denominator are 0, $\sqrt{5}$, and $-\sqrt{5}$. Thus the vertical asymptotes are the lines $x = 0$, $x = \sqrt{5}$, and $x = -\sqrt{5}$. (See Fig. 3.)

Now Try Exercises 15 and 19. ▦

Looking again at the graph of $f(x) = 1/(x - 3)$, shown on the following page (also see Example 1), let's explore what happens to $f(x) = 1/(x - 3)$ as x increases without bound (approaches positive infinity, ∞) and as x decreases without bound (approaches negative infinity, $-\infty$).

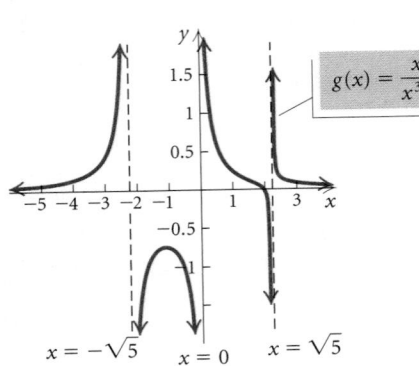

FIGURE 2

FIGURE 3

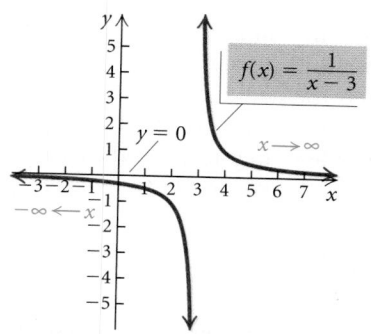

Horizontal asymptote: $y = 0$

x increases without bound:

x	100	5000	1,000,000	$\longrightarrow \infty$
$f(x)$	≈ 0.0103	≈ 0.0002	≈ 0.000001	$\longrightarrow 0$

x decreases without bound:

x	-300	-8000	$-1,000,000$	$\longrightarrow -\infty$
$f(x)$	≈ -0.0033	≈ -0.0001	≈ -0.000001	$\longrightarrow 0$

We see that

$$\frac{1}{x-3} \to 0 \text{ as } x \to \infty \quad \text{and} \quad \frac{1}{x-3} \to 0 \text{ as } x \to -\infty.$$

Since $y = 0$ is the equation of the *x*-axis, we say that the curve approaches the *x*-axis asymptotically and that the *x*-axis is a *horizontal asymptote* for the curve.

In general, the line $y = b$ is a **horizontal asymptote** for the graph of *f* if either or both of the following are true:

$$f(x) \to b \text{ as } x \to \infty \quad \text{or} \quad f(x) \to b \text{ as } x \to -\infty.$$

The following figures illustrate four ways in which horizontal asymptotes can occur. In each case, the curve gets close to the line $y = b$ either as $x \to \infty$ or as $x \to -\infty$. Keep in mind that the symbols ∞ and $-\infty$ convey the idea of increasing without bound and decreasing without bound, respectively.

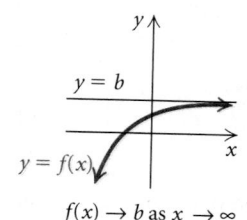
$f(x) \to b$ as $x \to \infty$

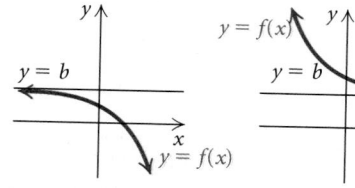
$f(x) \to b$ as $x \to -\infty$ ... $f(x) \to b$ as $x \to \infty$

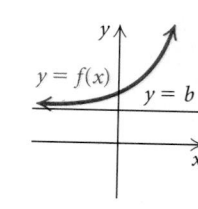
$f(x) \to b$ as $x \to -\infty$

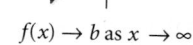

How can we determine a horizontal asymptote? As *x* gets very large or very small, the value of a polynomial function $p(x)$ is dominated by the function's leading term. Because of this, if $p(x)$ and $q(x)$ have the *same* degree, the value of $p(x)/q(x)$ as $x \to \infty$ or as $x \to -\infty$ is dominated by the ratio of the numerator's leading coefficient to the denominator's leading coefficient.

For $f(x) = (3x^2 + 2x - 4)/(2x^2 - x + 1)$, we see that the numerator, $3x^2 + 2x - 4$, is dominated by $3x^2$ and the denominator, $2x^2 - x + 1$, is dominated by $2x^2$, so $f(x)$ approaches $3x^2/2x^2$, or $3/2$ as *x* gets very large or very small:

$$\frac{3x^2 + 2x - 4}{2x^2 - x + 1} \to \frac{3}{2}, \text{ or } 1.5, \text{ as } x \to \infty, \quad \text{and}$$

$$\frac{3x^2 + 2x - 4}{2x^2 - x + 1} \to \frac{3}{2}, \text{ or } 1.5, \text{ as } x \to -\infty.$$

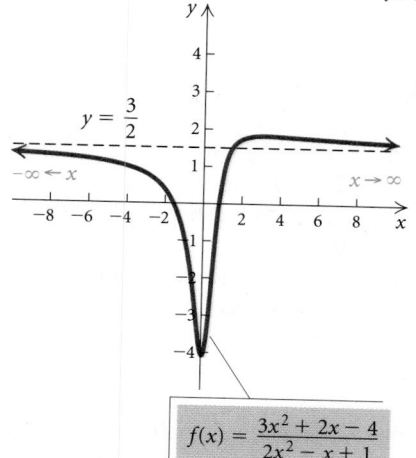

$$f(x) = \frac{3x^2 + 2x - 4}{2x^2 - x + 1}$$

We say that the curve approaches the horizontal line $y = \frac{3}{2}$ asymptotically and that $y = \frac{3}{2}$ is a *horizontal asymptote* for the curve.

It follows that when the numerator and the denominator of a rational function have the same degree, the line $y = a/b$ is the horizontal asymptote, where a and b are the leading coefficients of the numerator and the denominator, respectively.

EXAMPLE 4 Find the horizontal asymptote: $f(x) = \dfrac{-7x^4 - 10x^2 + 1}{11x^4 + x - 2}$.

Solution The numerator and the denominator have the same degree. The ratio of the leading coefficients is $-\frac{7}{11}$, so the line $y = -\frac{7}{11}$, or $-0.\overline{63}$, is the horizontal asymptote.

Now Try Exercise 21. ■

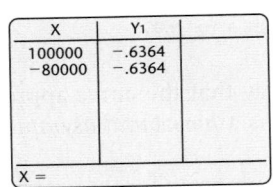

X	Y₁
100000	-.6364
-80000	-.6364

X =

As a partial check of Example 4, we could use a graphing calculator to evaluate the function for a very large and a very small value of x. (See the window at left.) It is useful in calculus to multiply by 1, using $(1/x^4)/(1/x^4)$:

$$f(x) = \frac{-7x^4 - 10x^2 + 1}{11x^4 + x - 2} \cdot \frac{\dfrac{1}{x^4}}{\dfrac{1}{x^4}} = \frac{\dfrac{-7x^4}{x^4} - \dfrac{10x^2}{x^4} + \dfrac{1}{x^4}}{\dfrac{11x^4}{x^4} + \dfrac{x}{x^4} - \dfrac{2}{x^4}}$$

$$= \frac{-7 - \dfrac{10}{x^2} + \dfrac{1}{x^4}}{11 + \dfrac{1}{x^3} - \dfrac{2}{x^4}}.$$

As $|x|$ becomes very large, each expression whose denominator is a power of x tends toward 0. Specifically, as $x \to \infty$ or as $x \to -\infty$, we have

$$f(x) \to \frac{-7 - 0 + 0}{11 + 0 - 0}, \quad \text{or} \quad f(x) \to -\frac{7}{11}.$$

The horizontal asymptote is $y = -\frac{7}{11}$, or $-0.\overline{63}$.

We now investigate the occurrence of a horizontal asymptote when the degree of the numerator is less than the degree of the denominator.

EXAMPLE 5 Find the horizontal asymptote: $f(x) = \dfrac{2x + 3}{x^3 - 2x^2 + 4}$.

Solution We let $p(x) = 2x + 3$, $q(x) = x^3 - 2x^2 + 4$, and $f(x) = p(x)/q(x)$. Note that as $x \to \infty$, the value of $q(x)$ grows much faster than the value of $p(x)$. Because of this, the ratio $p(x)/q(x)$ shrinks toward 0. As $x \to -\infty$, the ratio $p(x)/q(x)$ behaves in a similar manner. The horizontal asymptote is $y = 0$, the x-axis. This is the case for all rational functions for which the degree of the numerator is less than the degree of the denominator. Note in Example 1 that $y = 0$, the x-axis, is the horizontal asymptote of $f(x) = 1/(x - 3)$.

Now Try Exercise 23. ■

The following statements describe the two ways in which a horizontal asymptote occurs.

Determining a Horizontal Asymptote

- When the numerator and the denominator of a rational function have the same degree, the line $y = a/b$ is the horizontal asymptote, where a and b are the leading coefficients of the numerator and the denominator, respectively.
- When the degree of the numerator of a rational function is less than the degree of the denominator, the x-axis, or $y = 0$, is the horizontal asymptote.
- When the degree of the numerator of a rational function is greater than the degree of the denominator, there is no horizontal asymptote.

The following statements are also true.

Crossing an Asymptote

- The graph of a rational function *never crosses* a vertical asymptote.
- The graph of a rational function *might cross* a horizontal asymptote but does not necessarily do so.

EXAMPLE 6   Graph

$$g(x) = \frac{2x^2 + 1}{x^2}.$$

Include and label all asymptotes.

Solution   Since 0 is the zero of the denominator and not of the numerator, the y-axis, $x = 0$, is the vertical asymptote. Note also that the degree of the numerator is the same as the degree of the denominator. Thus, $y = 2/1$, or 2, is the horizontal asymptote.

To draw the graph, we first draw the asymptotes with dashed lines. Then we compute and plot some ordered pairs and draw the two branches of the curve. We can check the graph with a graphing calculator.

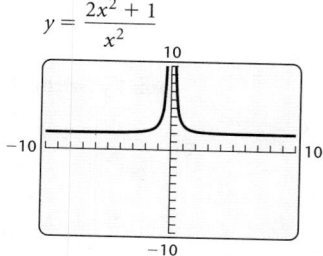

$y = \dfrac{2x^2 + 1}{x^2}$

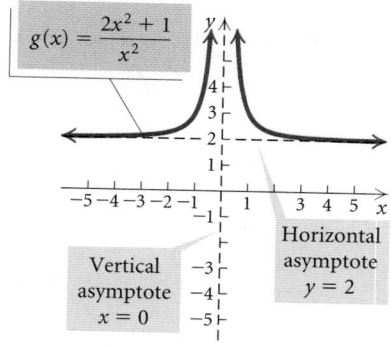

x	$g(x)$
-2	2.25
-1.5	$2.\overline{4}$
-1	3
-0.5	6
0.5	6
1	3
1.5	$2.\overline{4}$
2	2.25

Now Try Exercise 41. ■

Sometimes a line that is neither horizontal nor vertical is an asymptote. Such a line is called an **oblique asymptote**, or a **slant asymptote**.

EXAMPLE 7 Find all the asymptotes of

$$f(x) = \frac{2x^2 - 3x - 1}{x - 2}.$$

Solution The line $x = 2$ is the vertical asymptote because 2 is the zero of the denominator and is not a zero of the numerator. There is no horizontal asymptote because the degree of the numerator is greater than the degree of the denominator. When the degree of the numerator is 1 greater than the degree of the denominator, we divide to find an equivalent expression:

$$\frac{2x^2 - 3x - 1}{x - 2} = (2x + 1) + \frac{1}{x - 2}. \qquad \begin{array}{r} 2x + 1 \\ x - 2 \overline{)\, 2x^2 - 3x - 1} \\ \underline{2x^2 - 4x} \\ x - 1 \\ \underline{x - 2} \\ 1 \end{array}$$

Now we see that when $x \to \infty$ or $x \to -\infty$, $1/(x - 2) \to 0$ and the value of $f(x) \to 2x + 1$. This means that as $|x|$ becomes very large, the graph of $f(x)$ gets very close to the graph of $y = 2x + 1$. Thus the line $y = 2x + 1$ is the oblique asymptote.

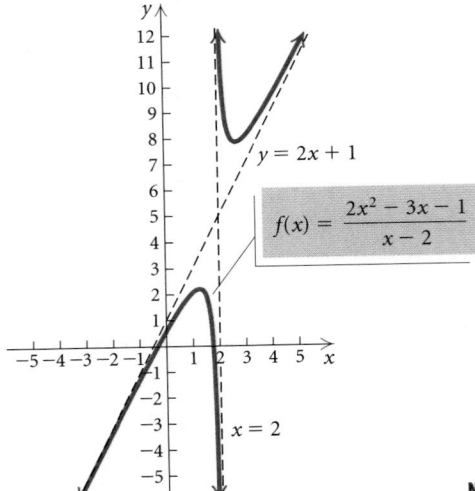

$y = 2x + 1$

$$f(x) = \frac{2x^2 - 3x - 1}{x - 2}$$

$x = 2$

Now Try Exercise 57. ■

Occurrence of Lines as Asymptotes of Rational Functions

For a rational function $f(x) = p(x)/q(x)$, where $p(x)$ and $q(x)$ have no common factors other than constants:

> **Vertical asymptotes** occur at any x-values that make the denominator 0.
>
> **The x-axis is the horizontal asymptote** when the degree of the numerator is less than the degree of the denominator.
>
> **A horizontal asymptote other than the x-axis** occurs when the numerator and the denominator have the same degree.
>
> **An oblique asymptote** occurs when the degree of the numerator is 1 greater than the degree of the denominator.

There can be only one horizontal asymptote or one oblique asymptote and never both.

An asymptote is *not* part of the graph of the function.

The following is an outline of a procedure that we can follow to create accurate graphs of rational functions.

To graph a rational function $f(x) = p(x)/q(x)$, where $p(x)$ and $q(x)$ have no common factor other than constants:

1. Find any real zeros of the denominator. Determine the domain of the function and sketch any vertical asymptotes.
2. Find the horizontal asymptote or the oblique asymptote, if there is one, and sketch it.
3. Find any zeros of the function. The zeros are found by determining the zeros of the numerator. These are the first coordinates of the x-intercepts of the graph.
4. Find $f(0)$. This gives the y-intercept $(0, f(0))$, of the function.
5. Find other function values to determine the general shape. Then draw the graph.

EXAMPLE 8   Graph: $f(x) = \dfrac{2x + 3}{3x^2 + 7x - 6}$.

Solution

1. We find the zeros of the denominator by solving $3x^2 + 7x - 6 = 0$. Since

$$3x^2 + 7x - 6 = (3x - 2)(x + 3),$$

the zeros are $\frac{2}{3}$ and -3. Thus the domain excludes $\frac{2}{3}$ and -3 and is

$$(-\infty, -3) \cup \left(-3, \tfrac{2}{3}\right) \cup \left(\tfrac{2}{3}, \infty\right).$$

Since neither zero of the denominator is a zero of the numerator, the graph has vertical asymptotes $x = -3$ and $x = \frac{2}{3}$. We sketch these as dashed lines.

2. Because the degree of the numerator is less than the degree of the denominator, the x-axis, $y = 0$, is the horizontal asymptote.

3. To find the zeros of the numerator, we solve $2x + 3 = 0$ and get $x = -\frac{3}{2}$. Thus, $-\frac{3}{2}$ is the zero of the function, and the pair $\left(-\frac{3}{2}, 0\right)$ is the x-intercept.

4. We find $f(0)$:

$$f(0) = \frac{2 \cdot 0 + 3}{3 \cdot 0^2 + 7 \cdot 0 - 6}$$

$$= \frac{3}{-6} = -\frac{1}{2}.$$

The point $\left(0, -\frac{1}{2}\right)$ is the y-intercept.

5. We find other function values to determine the general shape and then draw the graph. Note that the graph of this function crosses its horizontal asymptote at $x = -\frac{3}{2}$.

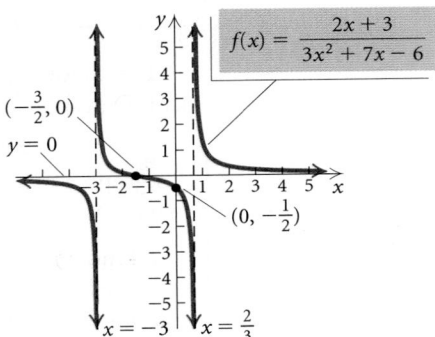

Now Try Exercise 63. ◼

EXAMPLE 9   Graph: $g(x) = \dfrac{x^2 - 1}{x^2 + x - 6}$.

Solution

1. We find the zeros of the denominator by solving $x^2 + x - 6 = 0$. Since

$$x^2 + x - 6 = (x + 3)(x - 2),$$

the zeros are -3 and 2. Thus the domain excludes the x-values -3 and 2 and is

$$(-\infty, -3) \cup (-3, 2) \cup (2, \infty).$$

Since neither zero of the denominator is a zero of the numerator, the graph has vertical asymptotes $x = -3$ and $x = 2$. We sketch these as dashed lines.

2. The numerator and the denominator have the same degree, so the horizontal asymptote is determined by the ratio of the leading coefficients: $1/1$, or 1. Thus, $y = 1$ is the horizontal asymptote. We sketch it with a dashed line.

3. To find the zeros of the numerator, we solve $x^2 - 1 = 0$. The solutions are -1 and 1. Thus, -1 and 1 are the zeros of the function and the pairs $(-1, 0)$ and $(1, 0)$ are the x-intercepts.

4. We find $g(0)$:

$$g(0) = \frac{0^2 - 1}{0^2 + 0 - 6} = \frac{-1}{-6} = \frac{1}{6}.$$

Thus, $\left(0, \frac{1}{6}\right)$ is the y-intercept.

5. We find other function values to determine the general shape and then draw the graph.

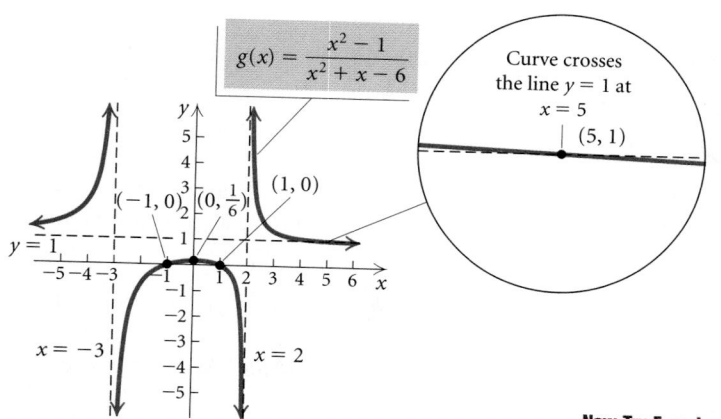

Now Try Exercise 73.

The magnified portion of the graph in Example 9 above shows another situation in which a graph can cross its horizontal asymptote. The point where $g(x)$ crosses $y = 1$ can be found by setting $g(x) = 1$ and solving for x:

$$\frac{x^2 - 1}{x^2 + x - 6} = 1$$

$$x^2 - 1 = x^2 + x - 6$$

$$-1 = x - 6 \qquad \text{Subtracting } x^2$$

$$5 = x. \qquad \text{Adding 6}$$

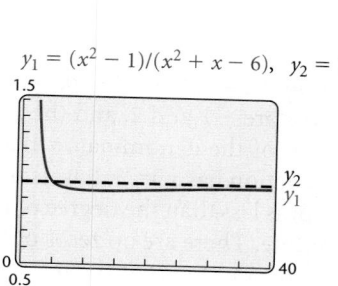

$y_1 = (x^2 - 1)/(x^2 + x - 6), \quad y_2 = 1$

The point of intersection is $(5, 1)$. Let's observe the behavior of the curve after it crosses the horizontal asymptote at $x = 5$. (See the graph at left.) It continues to decrease for a short interval and then begins to increase, getting closer and closer to $y = 1$ as $x \to \infty$.

Graphs of rational functions can also cross an oblique asymptote. The graph of

$$f(x) = \frac{2x^3}{x^2 + 1}$$

shown below crosses its oblique asymptote $y = 2x$. **Remember, graphs can cross horizontal asymptotes or oblique asymptotes, but they cannot cross vertical asymptotes.**

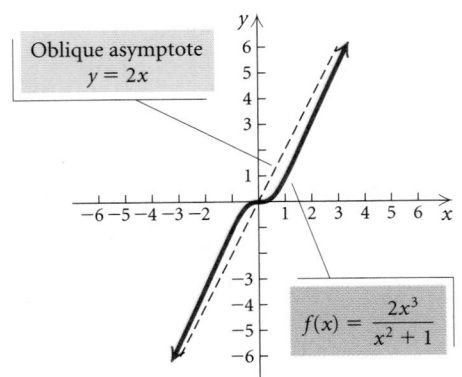

Oblique asymptote $y = 2x$

$f(x) = \dfrac{2x^3}{x^2 + 1}$

Let's now graph a rational function $f(x) = p(x)/q(x)$, where $p(x)$ and $q(x)$ have a common factor.

GCM **EXAMPLE 10**   Graph: $g(x) = \dfrac{x - 2}{x^2 - x - 2}$.

Solution   We first express the denominator in factored form:

$$g(x) = \frac{x - 2}{x^2 - x - 2} = \frac{x - 2}{(x + 1)(x - 2)}.$$

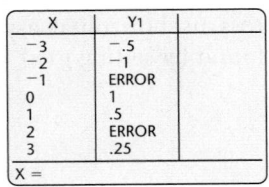

X	Y1
-3	-.5
-2	-1
-1	ERROR
0	1
1	.5
2	ERROR
3	.25
X =	

The domain of the function is $\{x \mid x \neq -1 \ and \ x \neq 2\}$, or $(-\infty, -1) \cup (-1, 2) \cup (2, \infty)$. The zeros of the denominator are -1 and 2, and the zero of the numerator is 2. Since -1 is the only zero of the denominator that is *not* a zero of the numerator, the graph of the function has $x = -1$ as its only vertical asymptote. The degree of the numerator is less than the degree of the denominator, so $y = 0$ is the horizontal asymptote. There are no zeros of the function and thus no x-intercepts, because 2 is the only zero of the numerator and 2 is not in the domain of the function. Since $g(0) = 1$, $(0, 1)$ is the y-intercept. We draw the graph indicating the "hole" when $x = 2$ with an open circle.

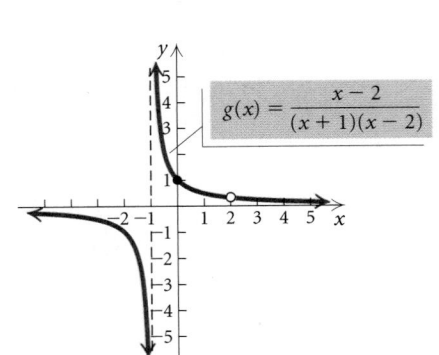

$g(x) = \dfrac{x - 2}{(x + 1)(x - 2)}$

The rational expression $(x - 2)/[(x + 1)(x - 2)]$ can be simplified. Thus,

$$g(x) = \frac{x - 2}{(x + 1)(x - 2)} = \frac{1}{x + 1}, \quad \text{where } x \neq -1 \text{ and } x \neq 2.$$

The graph of $g(x)$ is the graph of $y = 1/(x + 1)$ with the point where $x = 2$ missing. To determine the coordinates of the "hole," we substitute 2 for x in $g(x) = 1/(x + 1)$:

$$g(2) = \frac{1}{2 + 1} = \frac{1}{3}.$$

Thus the "hole" is located at $(2, \frac{1}{3})$. With certain window dimensions, the "hole" is visible on a graphing calculator. **Now Try Exercise 47.** ■

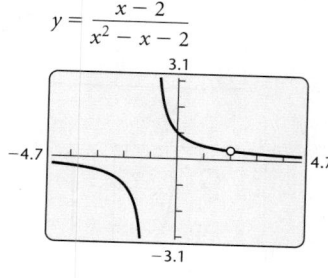

$y = \dfrac{x - 2}{x^2 - x - 2}$

❋ Applications

EXAMPLE 11 *Temperature During an Illness.* A person's temperature T, in degrees Fahrenheit, during an illness is given by the function

$$T(t) = \frac{4t}{t^2 + 1} + 98.6,$$

where time t is given in hours since the onset of the illness.

a) Graph the function on the interval $[0, 48]$.
b) Find the temperature at $t = 0, 1, 2, 5, 12,$ and 24.
c) Find the horizontal asymptote of the graph of $T(t)$. Complete:

$$T(t) \to \boxed{} \text{ as } t \to \infty.$$

d) Give the meaning of the answer to part (c) in terms of the application.
e) Find the maximum temperature during the illness.

Solution

a) The graph is shown at left.
b) We have

$$T(0) = 98.6, \quad T(1) = 100.6, \quad T(2) = 100.2,$$
$$T(5) \approx 99.369, \quad T(12) \approx 98.931, \quad \text{and} \quad T(24) \approx 98.766.$$

c) Since

$$T(t) = \frac{4t}{t^2 + 1} + 98.6$$
$$= \frac{98.6t^2 + 4t + 98.6}{t^2 + 1},$$

the horizontal asymptote is $y = 98.6/1$, or 98.6. Then it follows that $T(t) \to 98.6$ as $t \to \infty$.

d) As time goes on, the temperature returns to "normal," which is 98.6°F.
e) Using the MAXIMUM feature on a graphing calculator, we find the maximum temperature to be 100.6°F at $t = 1$ hr. **Now Try Exercise 79.** ■

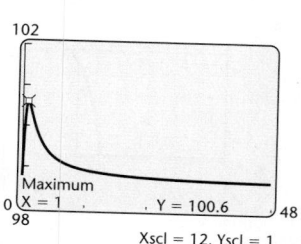

$T(x) = \dfrac{4x}{x^2 + 1} + 98.6$

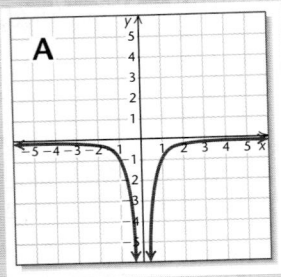

A

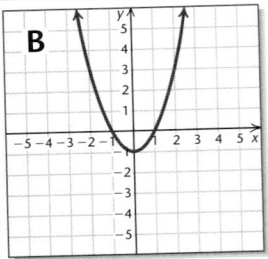

B

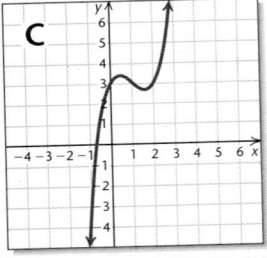

C

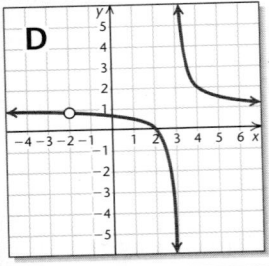

D

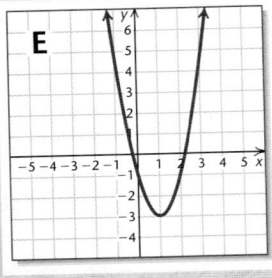

E

Visualizing the Graph

Match the function with its graph.

1. $f(x) = -\dfrac{1}{x^2}$

2. $f(x) = x^3 - 3x^2 + 2x + 3$

3. $f(x) = \dfrac{x^2 - 4}{x^2 - x - 6}$

4. $f(x) = -x^2 + 4x - 1$

5. $f(x) = \dfrac{x - 3}{x^2 + x - 6}$

6. $f(x) = \dfrac{3}{4}x + 2$

7. $f(x) = x^2 - 1$

8. $f(x) = x^4 - 2x^2 - 5$

9. $f(x) = \dfrac{8x - 4}{3x + 6}$

10. $f(x) = 2x^2 - 4x - 1$

Answers on page A-25

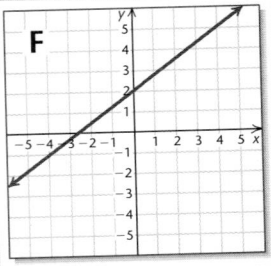

F

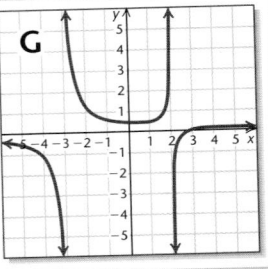

G

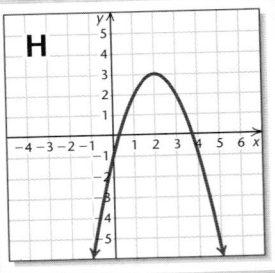

H

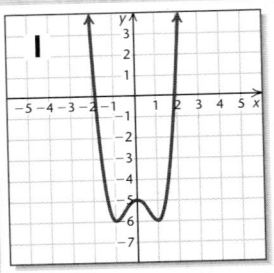

I

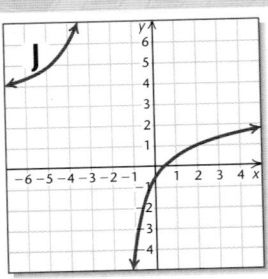

J

4.5 Exercise Set

Determine the domain of the function.

1. $f(x) = \dfrac{x^2}{2 - x}$

2. $f(x) = \dfrac{1}{x^3}$

3. $f(x) = \dfrac{x + 1}{x^2 - 6x + 5}$

4. $f(x) = \dfrac{(x + 4)^2}{4x - 3}$

5. $f(x) = \dfrac{3x - 4}{3x + 15}$

6. $f(x) = \dfrac{x^2 + 3x - 10}{x^2 + 2x}$

In Exercises 7–12, use your knowledge of asymptotes and intercepts to match the equation with one of the graphs (a)–(f), which follow. List all asymptotes. Check your work using a graphing calculator.

a)

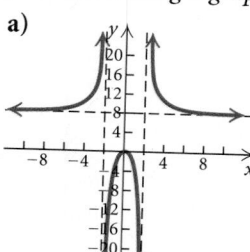

b)

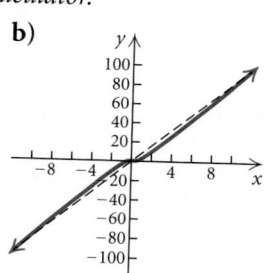

c)

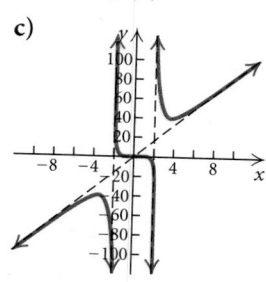

d)

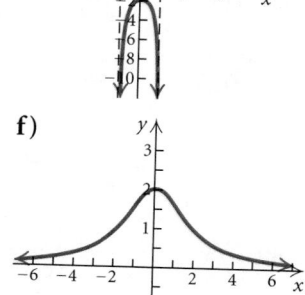

e)

f)

7. $f(x) = \dfrac{8}{x^2 - 4}$

8. $f(x) = \dfrac{8}{x^2 + 4}$

9. $f(x) = \dfrac{8x}{x^2 - 4}$

10. $f(x) = \dfrac{8x^2}{x^2 - 4}$

11. $f(x) = \dfrac{8x^3}{x^2 - 4}$

12. $f(x) = \dfrac{8x^3}{x^2 + 4}$

Determine the vertical asymptotes of the graph of the function.

13. $g(x) = \dfrac{1}{x^2}$

14. $f(x) = \dfrac{4x}{x^2 + 10x}$

15. $h(x) = \dfrac{x + 7}{2 - x}$

16. $g(x) = \dfrac{x^4 + 2}{x}$

17. $f(x) = \dfrac{3 - x}{(x - 4)(x + 6)}$

18. $h(x) = \dfrac{x^2 - 4}{x(x + 5)(x - 2)}$

19. $g(x) = \dfrac{x^3}{2x^3 - x^2 - 3x}$

20. $f(x) = \dfrac{x + 5}{x^2 + 4x - 32}$

Determine the horizontal asymptote of the graph of the function.

21. $f(x) = \dfrac{3x^2 + 5}{4x^2 - 3}$

22. $g(x) = \dfrac{x + 6}{x^3 + 2x^2}$

23. $h(x) = \dfrac{x^2 - 4}{2x^4 + 3}$

24. $f(x) = \dfrac{x^5}{x^5 + x}$

25. $g(x) = \dfrac{x^3 - 2x^2 + x - 1}{x^2 - 16}$

26. $h(x) = \dfrac{8x^4 + x - 2}{2x^4 - 10}$

Determine the oblique asymptote of the graph of the function.

27. $g(x) = \dfrac{x^2 + 4x - 1}{x + 3}$

28. $f(x) = \dfrac{x^2 - 6x}{x - 5}$

29. $h(x) = \dfrac{x^4 - 2}{x^3 + 1}$

30. $g(x) = \dfrac{12x^3 - x}{6x^2 + 4}$

31. $f(x) = \dfrac{x^3 - x^2 + x - 4}{x^2 + 2x - 1}$

32. $h(x) = \dfrac{5x^3 - x^2 + x - 1}{x^2 - x + 2}$

Make a hand-drawn graph for each of Exercises 38–74. Be sure to label all the asymptotes. List the domain and the x- and y-intercepts. Check your work using a graphing calculator.

33. $f(x) = \dfrac{1}{x}$

34. $g(x) = \dfrac{1}{x^2}$

35. $h(x) = -\dfrac{4}{x^2}$

36. $f(x) = -\dfrac{6}{x}$

37. $g(x) = \dfrac{x^2 - 4x + 3}{x + 1}$

38. $h(x) = \dfrac{2x^2 - x - 3}{x - 1}$

39. $f(x) = \dfrac{-2}{x - 5}$

40. $f(x) = \dfrac{1}{x - 5}$

41. $f(x) = \dfrac{2x + 1}{x}$

42. $f(x) = \dfrac{3x - 1}{x}$

43. $f(x) = \dfrac{x}{x^2 + 3x}$

44. $f(x) = \dfrac{3x}{3x - x^2}$

45. $f(x) = \dfrac{1}{(x - 2)^2}$

46. $f(x) = \dfrac{-2}{(x - 3)^2}$

47. $f(x) = \dfrac{x^2 + 2x - 3}{x^2 + 4x + 3}$

48. $f(x) = \dfrac{x^2 - x - 2}{x^2 - 5x - 6}$

49. $f(x) = \dfrac{1}{x^2 + 3}$

50. $f(x) = \dfrac{-1}{x^2 + 2}$

51. $f(x) = \dfrac{x^2 - 4}{x - 2}$

52. $f(x) = \dfrac{x^2 - 9}{x + 3}$

53. $f(x) = \dfrac{x - 1}{x + 2}$

54. $f(x) = \dfrac{x - 2}{x + 1}$

55. $f(x) = \dfrac{x^2 + 3x}{2x^3 - 5x^2 - 3x}$

56. $f(x) = \dfrac{3x}{x^2 + 5x + 4}$

57. $f(x) = \dfrac{x^2 - 9}{x + 1}$

58. $f(x) = \dfrac{x^3 - 4x}{x^2 - x}$

59. $f(x) = \dfrac{x^2 + x - 2}{2x^2 + 1}$

60. $f(x) = \dfrac{x^2 - 2x - 3}{3x^2 + 2}$

61. $g(x) = \dfrac{3x^2 - x - 2}{x - 1}$

62. $f(x) = \dfrac{2x + 1}{2x^2 - 5x - 3}$

63. $f(x) = \dfrac{x - 1}{x^2 - 2x - 3}$

64. $f(x) = \dfrac{x + 2}{x^2 + 2x - 15}$

65. $f(x) = \dfrac{x - 3}{(x + 1)^3}$

66. $f(x) = \dfrac{x + 2}{(x - 1)^3}$

67. $f(x) = \dfrac{x^3 + 1}{x}$

68. $f(x) = \dfrac{x^3 - 1}{x}$

69. $f(x) = \dfrac{x^3 + 2x^2 - 15x}{x^2 - 5x - 14}$

70. $f(x) = \dfrac{x^3 + 2x^2 - 3x}{x^2 - 25}$

71. $f(x) = \dfrac{5x^4}{x^4 + 1}$

72. $f(x) = \dfrac{x + 1}{x^2 + x - 6}$

73. $f(x) = \dfrac{x^2}{x^2 - x - 2}$

74. $f(x) = \dfrac{x^2 - x - 2}{x + 2}$

Find a rational function that satisfies the given conditions. Answers may vary, but try to give the simplest answer possible.

75. Vertical asymptotes $x = -4, x = 5$

76. Vertical asymptotes $x = -4, x = 5$; x-intercept $(-2, 0)$

77. Vertical asymptotes $x = -4, x = 5$; horizontal asymptote $y = \frac{3}{2}$; x-intercept $(-2, 0)$

78. Oblique asymptote $y = x - 1$

79. *Medical Dosage.* The function
$$N(t) = \dfrac{0.8t + 1000}{5t + 4}, \quad t \ge 15$$
gives the body concentration $N(t)$, in parts per million, of a certain dosage of medication after time t, in hours.

a) Graph the function on the interval $[15, \infty)$ and complete the following:
$$N(t) \rightarrow \boxed{} \text{ as } t \rightarrow \infty.$$

b) Explain the meaning of the answer to part (a) in terms of the application.

80. *Average Cost.* The average cost per DVD, in dollars, for a company to produce x fitness workout DVDs is given by the function
$$A(x) = \dfrac{2x + 100}{x}, \quad x > 0.$$

a) Graph the function on the interval $(0, \infty)$ and complete the following:
$$A(x) \rightarrow \boxed{} \text{ as } x \rightarrow \infty.$$

b) Explain the meaning of the answer to part (a) in terms of the application.

81. *Population Growth.* The population P, in thousands, of Lordsburg is given by
$$P(t) = \dfrac{500t}{2t^2 + 9},$$
where t is the time, in months.

a) Graph the function on the interval $[0, \infty)$.

b) Find the population at $t = 0, 1, 3$, and 8 months.

c) Find the horizontal asymptote of the graph and complete the following:
$$P(t) \rightarrow \boxed{} \text{ as } t \rightarrow \infty.$$

d) Explain the meaning of the answer to part (c) in terms of the application.

e) Find the maximum population and the value of t that will yield it.

82. *Minimizing Surface Area.* The Hold-It Container Co. is designing an open-top rectangular box, with a square base, that will hold 108 cubic centimeters.

a) Express the surface area S as a function of the length x of a side of the base.

b) Use a graphing calculator to graph the function on the interval $(0, \infty)$.

c) Estimate the minimum surface area and the value of x that will yield it.

83. Graph
$$y_1 = \dfrac{x^3 + 4}{x} \quad \text{and} \quad y_2 = x^2$$
using the same viewing window. Explain how the parabola $y_2 = x^2$ can be thought of as a nonlinear asymptote for y_1.

Collaborative Discussion and Writing

84. Explain why the graph of a rational function cannot have both a horizontal asymptote and an oblique asymptote.

85. Under what circumstances will a rational function have a domain consisting of all real numbers?

Skill Maintenance

In each of Exercises 80–88, fill in the blank with the correct term. Some of the given choices will not be used. Others will be used more than once.

x-intercept
y-intercept
odd function
even function
domain
range
slope
distance formula
midpoint formula
horizontal lines

vertical lines
point–slope equation
slope–intercept equation
difference quotient
$f(x) = f(-x)$
$f(-x) = -f(x)$

86. A function is a correspondence between a first set, called the _____, and a second set, called the _____, such that each member of the _____ corresponds to exactly one member of the _____.

87. The _____ of a line containing (x_1, y_1) and (x_2, y_2) is given by $(y_2 - y_1)/(x_2 - x_1)$.

88. The _____ of the line with slope m and y-intercept $(0, b)$ is $y = mx + b$.

89. The _____ of the line with slope m passing through (x_1, y_1) is $y - y_1 = m(x - x_1)$.

90. A(n) _____ is a point $(a, 0)$.

91. For each x in the domain of an odd function f, _____.

92. _____ are given by equations of the type $x = a$.

93. The _____ is $\left(\dfrac{x_1 + x_2}{2}, \dfrac{y_1 + y_2}{2} \right)$.

94. A(n) _____ is a point $(0, b)$.

Synthesis

Find the nonlinear asymptote of the function.

95. $f(x) = \dfrac{x^5 + 2x^3 + 4x^2}{x^2 + 2}$

96. $f(x) = \dfrac{x^4 + 3x^2}{x^2 + 1}$

Graph the function.

97. $f(x) = \dfrac{2x^3 + x^2 - 8x - 4}{x^3 + x^2 - 9x - 9}$

98. $f(x) = \dfrac{x^3 + 4x^2 + x - 6}{x^2 - x - 2}$

4.6

Polynomial and Rational Inequalities

❖ Solve polynomial inequalities and rational inequalities.

We will use a combination of algebraic and graphical methods to solve polynomial inequalities and rational inequalities.

❖ Polynomial Inequalities

Just as a quadratic equation can be written in the form $ax^2 + bx + c = 0$, a **quadratic inequality** can be written in the form $ax^2 + bx + c \ \blacksquare\ 0$, where $\blacksquare$ is $<$, $>$, $\leq$, or $\geq$. Here are some examples of quadratic inequalities:

$$x^2 - 4x - 5 < 0 \quad \text{and} \quad -\tfrac{1}{2}x^2 + 4x - 7 \geq 0.$$

When the inequality symbol in a polynomial inequality is replaced with an equals sign, a **related equation** is formed. Polynomial inequalities can be solved once the related equation has been solved.

EXAMPLE 1 Solve: $x^2 - 4x - 5 > 0$.

Solution We are asked to find all x-values for which $x^2 - 4x - 5 > 0$. To locate these values, we graph $f(x) = x^2 - 4x - 5$. Then we note that whenever the function changes sign, its graph passes through an x-intercept. Thus to solve $x^2 - 4x - 5 > 0$, we first solve the *related equation* $x^2 - 4x - 5 = 0$ to find all zeros of the function:

$$x^2 - 4x - 5 = 0$$
$$(x + 1)(x - 5) = 0.$$

The zeros are -1 and 5. Thus the x-intercepts of the graph are $(-1, 0)$ and $(5, 0)$, as shown below.

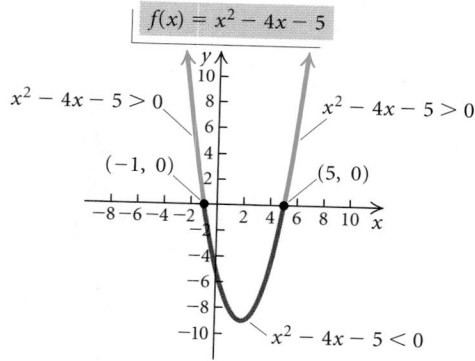

The zeros divide the x-axis into three intervals:

$$(-\infty, -1), \qquad (-1, 5), \quad \text{and} \quad (5, \infty).$$

The sign of $x^2 - 4x - 5$ is the same for all values of x in a given interval. Thus we choose a test value for x from each interval and find $f(x)$. We can also determine the sign of $f(x)$ in each interval by simply looking at the graph of the function.

Interval	$(-\infty, -1)$	$(-1, 5)$	$(5, \infty)$
Test Value	$f(-2) = 7$	$f(0) = -5$	$f(7) = 16$
Sign of $f(x)$	Positive	Negative	Positive

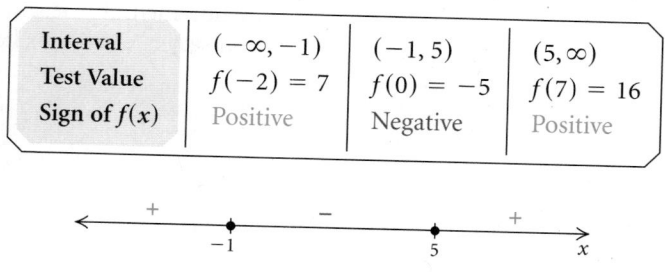

Since we are solving $x^2 - 4x - 5 > 0$, the solution set consists of only two of the three intervals, those in which the sign of $f(x)$ is positive. Since the inequality sign is $>$, we do not include the endpoints of the intervals in the solution set. The solution set is

$$(-\infty, -1) \cup (5, \infty), \quad \text{or} \quad \{x \,|\, x < -1 \text{ or } x > 5\}.$$

Now Try Exercise 27. ■

EXAMPLE 2   Solve: $x^2 + 3x - 5 < x + 3$.

Solution   By subtracting $x + 3$, we form an equivalent inequality:

$$x^2 + 3x - 5 - x - 3 < 0$$
$$x^2 + 2x - 8 < 0.$$

We need to find all x-values for which $x^2 + 2x - 8 < 0$. To visualize these values, we first graph $f(x) = x^2 + 2x - 8$ and then determine the zeros of the function.

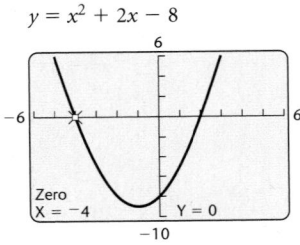
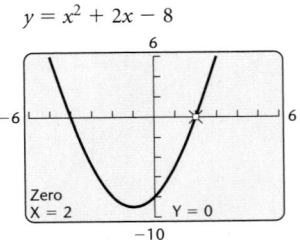

Using the ZERO feature, we see that the zeros are -4 and 2.

The intervals to be considered are $(-\infty, -4)$, $(-4, 2)$, and $(2, \infty)$. Using test values for $f(x)$, we determine the sign of $f(x)$ in each interval.

X	Y1	
-5	7	
0	-8	
4	16	
X =		

Function values are negative in the interval $(-4, 2)$. We can also note on the graph where the function values are negative. Since the inequality sign is $<$, we do not include the endpoints of the interval in the solution set. The solution set is $(-4, 2)$, or $\{x \,|\, -4 < x < 2\}$.

An alternative approach to solving the inequality

$$x^2 + 3x - 5 < x + 3$$

is to graph both sides: $y_1 = x^2 + 3x - 5$ and $y_2 = x + 3$ and determine where the graph of y_1 is below the graph of y_2. Using the INTERSECT feature, we determine the points of intersection, $(-4, -1)$ and $(2, 5)$.

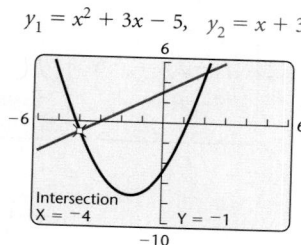

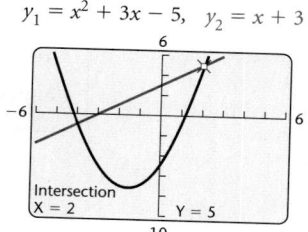

The graph of y_1 is below the graph of y_2 over the interval $(-4, 2)$. Thus the solution set of $x^2 + 3x - 5 < x + 3$ is $(-4, 2)$.   **Now Try Exercise 29.** ■

Quadratic inequalities are one type of **polynomial inequality**. Other examples of polynomial inequalities are

$$-2x^4 + x^2 - 3 < 7, \qquad \tfrac{2}{3}x + 4 \geq 0, \quad \text{and} \quad 4x^3 - 2x^2 > 5x + 7.$$

EXAMPLE 3   Solve: $x^3 - x > 0$.

Solution   We are asked to find all x-values for which $x^3 - x > 0$. To locate these values, we graph $f(x) = x^3 - x$. Then we note that whenever the function changes sign, its graph passes through an x-intercept. Thus to solve $x^3 - x > 0$, we first solve the related equation $x^3 - x = 0$ to find all zeros of the function:

$$x^3 - x = 0$$
$$x(x^2 - 1) = 0$$
$$x(x + 1)(x - 1) = 0.$$

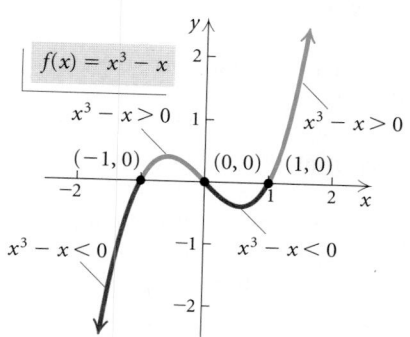

The zeros are -1, 0, and 1. Thus the x-intercepts of the graph are $(-1, 0)$, $(0, 0)$, and $(1, 0)$, as shown in the figure at left. The zeros divide the x-axis into four intervals:

$$(-\infty, -1), \qquad (-1, 0), \qquad (0, 1), \quad \text{and} \quad (1, \infty).$$

X	Y1	
-2	-6	
-.5	.375	
.5	-.375	
2	6	

X =

The sign of $x^3 - x$ is the same for all values of x in a given interval. Thus we choose a test value for x from each interval and find $f(x)$. We can use the TABLE feature set in ASK mode to determine the sign of $f(x)$ in each interval. (See the table at left.) We can also determine the sign of $f(x)$ in each interval by simply looking at the graph of the function.

Interval	$(-\infty, -1)$	$(-1, 0)$	$(0, 1)$	$(1, \infty)$
Test Value	$f(-2) = -6$	$f(-0.5) = 0.375$	$f(0.5) = -0.375$	$f(2) = 6$
Sign of $f(x)$	Negative	Positive	Negative	Positive

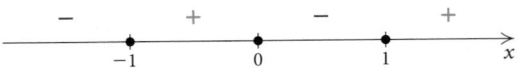

Since we are solving $x^3 - x > 0$, the solution set consists of only two of the four intervals, those in which the sign of $f(x)$ is *positive*. We see that the solution set is $(-1, 0) \cup (1, \infty)$, or $\{x \mid -1 < x < 0 \ or \ x > 1\}$.

Now Try Exercise 39. ■

To solve a polynomial inequality:

1. Find an equivalent inequality with $P(x)$ on one side and 0 on the other.
2. Change the inequality symbol to an equals sign and solve the related equation; that is, solve $P(x) = 0$.
3. Use the solutions to divide the x-axis into intervals. Then select a test value from each interval and determine the polynomial's sign on the interval.
4. Determine the intervals for which the inequality is satisfied and write interval notation or set-builder notation for the solution set. Include the endpoints of the intervals in the solution set if the inequality symbol is $\leq$ or $\geq$.

EXAMPLE 4   Solve: $3x^4 + 10x \leq 11x^3 + 4$.

Solution   By subtracting $11x^3 + 4$, we form the equivalent inequality

$$3x^4 - 11x^3 + 10x - 4 \leq 0.$$

To solve the related equation

$$3x^4 - 11x^3 + 10x - 4 = 0,$$

we need to use the theorems of Section 4.4. We solved this equation in Example 5 in Section 4.4. The solutions are

$$-1, \quad 2 - \sqrt{2}, \quad \tfrac{2}{3}, \quad \text{and} \quad 2 + \sqrt{2},$$

or approximately

$$-1, \quad 0.586, \quad 0.667, \quad \text{and} \quad 3.414.$$

These numbers divide the x-axis into five intervals: $(-\infty, -1), (-1, 2 - \sqrt{2}), (2 - \sqrt{2}, \tfrac{2}{3}),$ $(\tfrac{2}{3}, 2 + \sqrt{2})$, and $(2 + \sqrt{2}, \infty)$.

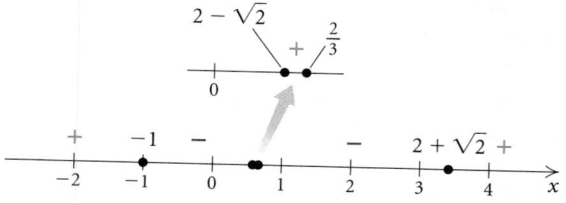

We then let $f(x) = 3x^4 - 11x^3 + 10x - 4$ and, using test values for x, determine the sign of $f(x)$ in each interval.

X	Y1
-2	112
0	-4
.6	.0128
1	-2
4	100

X =

Function values are negative in the intervals $(-1, 2 - \sqrt{2})$ and $(\tfrac{2}{3}, 2 + \sqrt{2})$. Since the inequality sign is $\le$, we include the endpoints of the intervals in the solution set. The solution set is

$$\left[-1, 2 - \sqrt{2}\right] \cup \left[\tfrac{2}{3}, 2 + \sqrt{2}\right], \quad \text{or}$$

$$\left\{x \mid -1 \le x \le 2 - \sqrt{2} \text{ or } \tfrac{2}{3} \le x \le 2 + \sqrt{2}\right\}.$$

We graph $y = 3x^4 - 11x^3 + 10x - 4$ using a viewing window that reveals the curvature of the graph.

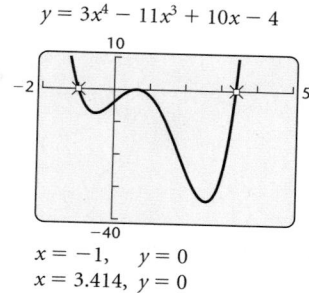

$$y = 3x^4 - 11x^3 + 10x - 4$$

$x = -1, \quad y = 0$
$x = 3.414, \quad y = 0$

Using the ZERO feature, we see that two of the zeros are -1 and approximately 3.414 $(2 + \sqrt{2} \approx 3.414)$. However, this window leaves us uncertain about the number of zeros of the function in the interval $[0, 1]$. The following window shows another view of the zeros in the interval $[0, 1]$. Those zeros are about 0.586 and 0.667 $(2 - \sqrt{2} \approx 0.586; \tfrac{2}{3} \approx 0.667)$.

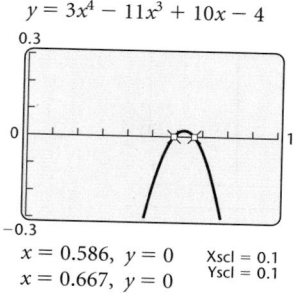

$$y = 3x^4 - 11x^3 + 10x - 4$$

$x = 0.586, \quad y = 0$ Xscl = 0.1
$x = 0.667, \quad y = 0$ Yscl = 0.1

The intervals to be considered are $(-\infty, -1)$, $(-1, 0.586), (0.586, 0.667), (0.667, 3.414)$, and $(3.414, \infty)$. We note on the graph where the function is negative. Then including appropriate endpoints, we find that the solution set is approximately

$$[-1, 0.586] \cup [0.667, 3.414], \quad \text{or}$$

$$\{x \mid -1 \le x \le 0.586 \text{ or } 0.667 \le x \le 3.414\}.$$

Now Try Exercise 45. ■

✿ Rational Inequalities

Some inequalities involve rational expressions and functions. These are called **rational inequalities**. To solve rational inequalities, we need to make some adjustments to the preceding method.

EXAMPLE 5   Solve: $\dfrac{3x}{x + 6} < 0$.

Solution   We look for all values of x for which the related function

$$f(x) = \frac{3x}{x + 6}$$

is not defined or is 0. These are called **critical values**.

The denominator tells us that $f(x)$ is not defined when $x = -6$. Next, we solve $f(x) = 0$:

$$\frac{3x}{x + 6} = 0$$

$$(x + 6) \cdot \frac{3x}{x + 6} = (x + 6) \cdot 0$$

$$3x = 0$$

$$x = 0.$$

The critical values are -6 and 0. These values divide the x-axis into three intervals:

$$(-\infty, -6), \qquad (-6, 0), \quad \text{and} \quad (0, \infty).$$

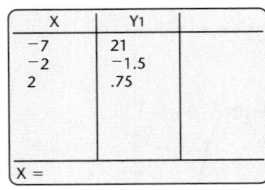

X	Y1
−7	21
−2	−1.5
2	.75

X =

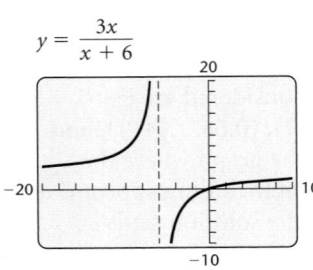

$y = \dfrac{3x}{x + 6}$

We then use a test value to determine the sign of $f(x)$ in each interval.

Function values are negative in only the interval $(-6, 0)$. Since $f(0) = 0$ and the inequality symbol is $<$, we know that 0 is not included in the solution set. Note that since -6 is not in the domain of f, -6 cannot be part of the solution set. The solution set is

$$(-6, 0), \quad \text{or} \quad \{x \mid -6 < x < 0\}.$$

The graph of $f(x)$ shows where $f(x)$ is positive and where it is negative.

Now Try Exercise 57. ■

EXAMPLE 6   Solve: $\dfrac{x - 3}{x + 4} \geq \dfrac{x + 2}{x - 5}$.

Solution   We first subtract $(x + 2)/(x - 5)$ on both sides in order to find an equivalent inequality with 0 on one side:

$$\frac{x - 3}{x + 4} - \frac{x + 2}{x - 5} \geq 0.$$

ALGEBRAIC SOLUTION

We look for all values of x for which the related function

$$f(x) = \frac{x-3}{x+4} - \frac{x+2}{x-5}$$

is not defined or is 0.

A look at the denominators shows that $f(x)$ is not defined for $x = -4$ and $x = 5$. Next, we solve $f(x) = 0$:

$$\frac{x-3}{x+4} - \frac{x+2}{x-5} = 0$$

$$(x+4)(x-5)\left(\frac{x-3}{x+4} - \frac{x+2}{x-5}\right) = (x+4)(x-5)\cdot 0$$

$$(x-5)(x-3) - (x+4)(x+2) = 0$$

$$x^2 - 8x + 15 - (x^2 + 6x + 8) = 0$$

$$-14x + 7 = 0$$

$$x = \tfrac{1}{2}.$$

The critical values are -4, $\frac{1}{2}$, and 5. These values divide the x-axis into four intervals:

$$(-\infty, -4), \quad \left(-4, \tfrac{1}{2}\right), \quad \left(\tfrac{1}{2}, 5\right), \quad \text{and} \quad (5, \infty).$$

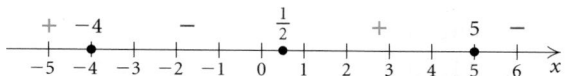

We then use a test value to determine the sign of $f(x)$ in each interval.

X	Y₁
−5	7.7
−2	−2.5
3	2.5
6	−7.7
X =	

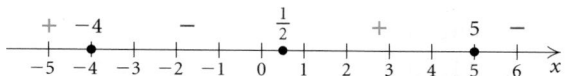

Function values are positive in the intervals $(-\infty, -4)$ and $\left(\tfrac{1}{2}, 5\right)$. Since $f\left(\tfrac{1}{2}\right) = 0$ and the inequality symbol is $\geq$, we know that $\tfrac{1}{2}$ must be in the solution set. Note that since neither -4 nor 5 is in the domain of f, they cannot be part of the solution set.

The solution set is $(-\infty, -4) \cup \left[\tfrac{1}{2}, 5\right)$.

GRAPHICAL SOLUTION

We graph

$$y = \frac{x-3}{x+4} - \frac{x+2}{x-5}$$

in the standard window, which shows the curvature of the function.

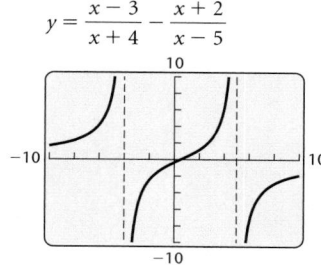

Using the ZERO feature, we find that 0.5 is a zero.

We then look for values where the function is not defined. By examining the denominators $x + 4$ and $x - 5$, we see that $f(x)$ is not defined for $x = -4$ and $x = 5$.

The critical values are -4, 0.5, and 5.

The graph shows where y is positive and where it is negative. Note that -4 and 5 cannot be in the solution set since y is not defined for these values. We do include 0.5, however, since the inequality symbol is $\geq$ and $f(0.5) = 0$. This solution set is

$$(-\infty, -4) \cup [0.5, 5).$$

The following is a method for solving rational inequalities.

> **To solve a rational inequality:**
> 1. Find an equivalent inequality with 0 on one side.
> 2. Change the inequality symbol to an equals sign and solve the related equation.
> 3. Find values of the variable for which the related rational function is not defined.
> 4. The numbers found in steps (2) and (3) are called critical values. Use the critical values to divide the x-axis into intervals. Then determine the function's sign in each interval using an x-value from the interval or the graph of the equation.
> 5. Select the intervals for which the inequality is satisfied and write interval notation or set-builder notation for the solution set. If the inequality symbol is $\leq$ or $\geq$, then the solutions to step (2) should be included in the solution set. The x-values found in step (3) are never included in the solution set.

It works well to use a combination of algebraic and graphical methods to solve polynomial inequalities and rational inequalities. The algebraic methods give exact numbers for the critical values, and the graphical methods usually allow us to see easily what intervals satisfy the inequality.

(4.6) Exercise Set

For the function $f(x) = x^2 + 2x - 15$, solve each of the following.

1. $f(x) = 0$ **2.** $f(x) < 0$

3. $f(x) \leq 0$ **4.** $f(x) > 0$

5. $f(x) \geq 0$

For the function $g(x) = \dfrac{x - 2}{x + 4}$, solve each of the following.

6. $g(x) = 0$ **7.** $g(x) > 0$

8. $g(x) \leq 0$ **9.** $g(x) \geq 0$

10. $g(x) < 0$

For the function

$$h(x) = \frac{7x}{(x - 1)(x + 5)},$$

solve each of the following.

11. $h(x) = 0$ **12.** $h(x) \leq 0$

13. $h(x) \geq 0$ **14.** $h(x) > 0$

15. $h(x) < 0$

For the function $g(x) = x^5 - 9x^3$, solve each of the following.

16. $g(x) = 0$ **17.** $g(x) < 0$

18. $g(x) \leq 0$

19. $g(x) > 0$

20. $g(x) \geq 0$

In Exercises 21–24, a related function is graphed. Solve the given inequality.

21. $x^3 + 6x^2 < x + 30$

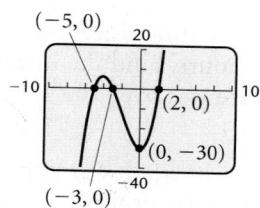

22. $x^4 - 27x^2 - 14x + 120 \geq 0$

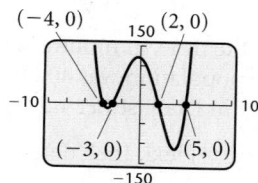

23. $\dfrac{8x}{x^2 - 4} \geq 0$

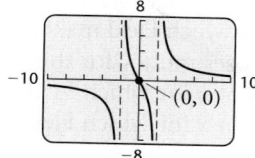

24. $\dfrac{8}{x^2 - 4} < 0$

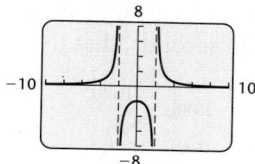

Solve.

25. $(x - 1)(x + 4) < 0$

26. $(x + 3)(x - 5) < 0$

27. $x^2 + x - 2 > 0$

28. $x^2 - x - 6 > 0$

29. $x^2 - x - 5 \geq x - 2$

30. $x^2 + 4x + 7 \geq 5x + 9$

31. $x^2 > 25$

32. $x^2 \leq 1$

33. $4 - x^2 \leq 0$

34. $11 - x^2 \geq 0$

35. $6x - 9 - x^2 < 0$

36. $x^2 + 2x + 1 \leq 0$

37. $x^2 + 12 < 4x$

38. $x^2 - 8 > 6x$

39. $4x^3 - 7x^2 \leq 15x$

40. $2x^3 - x^2 < 5x$

41. $x^3 + 3x^2 - x - 3 \geq 0$

42. $x^3 + x^2 - 4x - 4 \geq 0$

43. $x^3 - 2x^2 < 5x - 6$

44. $x^3 + x \leq 6 - 4x^2$

45. $x^5 + x^2 \geq 2x^3 + 2$

46. $x^5 + 24 > 3x^3 + 8x^2$

47. $2x^3 + 6 \leq 5x^2 + x$

48. $2x^3 + x^2 < 10 + 11x$

49. $x^3 + 5x^2 - 25x \leq 125$

50. $x^3 - 9x + 27 \geq 3x^2$

51. $0.1x^3 - 0.6x^2 - 0.1x + 2 < 0$

52. $19.2x^3 + 12.8x^2 + 144 \geq 172.8x + 3.2x^4$

53. $\dfrac{1}{x + 4} > 0$

54. $\dfrac{1}{x - 3} \leq 0$

55. $\dfrac{-4}{2x + 5} < 0$

56. $\dfrac{-2}{5 - x} \geq 0$

57. $\dfrac{2x}{x - 4} \geq 0$

58. $\dfrac{5x}{x + 1} < 0$

59. $\dfrac{x - 4}{x + 3} - \dfrac{x + 2}{x - 1} \leq 0$

60. $\dfrac{x + 1}{x - 2} + \dfrac{x - 3}{x - 1} < 0$

61. $\dfrac{2x - 1}{x + 3} \geq \dfrac{x + 1}{3x + 1}$

62. $\dfrac{x + 5}{x - 4} > \dfrac{3x + 2}{2x + 1}$

63. $\dfrac{x + 1}{x - 2} \geq 3$

64. $\dfrac{x}{x - 5} < 2$

65. $x - 2 > \dfrac{1}{x}$

66. $4 \geq \dfrac{4}{x} + x$

67. $\dfrac{2}{x^2 - 4x + 3} \leq \dfrac{5}{x^2 - 9}$

68. $\dfrac{3}{x^2 - 4} \leq \dfrac{5}{x^2 + 7x + 10}$

69. $\dfrac{3}{x^2 + 1} \geq \dfrac{6}{5x^2 + 2}$

70. $\dfrac{4}{x^2 - 9} < \dfrac{3}{x^2 - 25}$

71. $\dfrac{5}{x^2 + 3x} < \dfrac{3}{2x + 1}$

72. $\dfrac{2}{x^2 + 3} > \dfrac{3}{5 + 4x^2}$

73. $\dfrac{5x}{7x - 2} > \dfrac{x}{x + 1}$

74. $\dfrac{x^2 - x - 2}{x^2 + 5x + 6} < 0$

75. $\dfrac{x}{x^2 + 4x - 5} + \dfrac{3}{x^2 - 25} \leq \dfrac{2x}{x^2 - 6x + 5}$

76. $\dfrac{2x}{x^2 - 9} + \dfrac{x}{x^2 + x - 12} \geq \dfrac{3x}{x^2 + 7x + 12}$

77. *Temperature During an Illness.* A person's temperature T, in degrees Fahrenheit, during an illness is given by the function

$$T(t) = \dfrac{4t}{t^2 + 1} + 98.6,$$

where t is the time since the onset of the illness, in hours. Find the interval on which the temperature was over 100°F. (See Example 11 in Section 4.5.)

78. *Population Growth.* The population P, in thousands, of Lordsburg is given by

$$P(t) = \dfrac{500t}{2t^2 + 9},$$

where t is the time, in months. Find the interval on which the population was 40,000 or greater. (See Exercise 81 in Exercise Set 4.5.)

79. *Total Profit.* Flexl, Inc., determines that its total profit is given by the function

$$P(x) = -3x^2 + 630x - 6000.$$

a) Flexl makes a profit for those nonnegative values of x for which $P(x) > 0$. Find the values of x for which Flexl makes a profit.

b) Flexl loses money for those nonnegative values of x for which $P(x) < 0$. Find the values of x for which Flexl loses money.

80. *Height of a Thrown Object.* The function

$$S(t) = -16t^2 + 32t + 1920$$

gives the height S, in feet, of an object thrown upward from a cliff that is 1920 ft high. Here t is the time, in seconds, that the object is in the air.

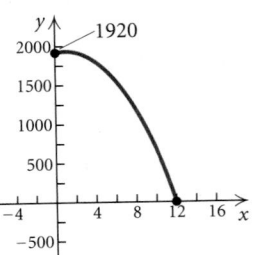

a) For what times is the height greater than 1920 ft?

b) For what times is the height less than 640 ft?

81. *Number of Diagonals.* A polygon with n sides has D diagonals, where D is given by the function

$$D(n) = \frac{n(n-3)}{2}.$$

Find the number of sides n if

$$27 \le D \le 230.$$

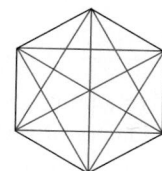

82. *Number of Handshakes.* If there are n people in a room, the number N of possible handshakes by all the people in the room is given by the function

$$N(n) = \frac{n(n-1)}{2}.$$

For what number n of people is

$$66 \le N \le 300?$$

Collaborative Discussion and Writing

83. Under what circumstances would a quadratic inequality have a solution set that is a closed interval? Under what circumstances would a quadratic inequality have an empty solution set?

84. Why, when solving rational inequalities, do we need to find values for which the function is undefined as well as zeros of the function?

Skill Maintenance

Find an equation for a circle satisfying the given conditions.

85. Center: $(-2, 4)$; radius of length 3

86. Center: $(0, -3)$; diameter of length $\frac{7}{2}$

In Exercises 87 and 88:
a) *Find the vertex.*
b) *Determine whether there is a maximum or minimum value and find that value.*
c) *Find the range.*

87. $h(x) = -2x^2 + 3x - 8$

88. $g(x) = x^2 - 10x + 2$

Synthesis

Solve.

89. $x^4 + 3x^2 > 4x - 15$

90. $x^4 - 6x^2 + 5 > 0$

91. $|x^2 - 5| = 5 - x^2$

92. $|x^2 - 6| = 5x$

93. $2|x|^2 - |x| + 2 \le 5$

94. $\left| \dfrac{x+3}{x-4} \right| < 2$

95. $\left| 1 + \dfrac{1}{x} \right| < 3$

96. $(7 - x)^{-2} < 0$

97. $|x^2 + 3x - 1| < 3$

98. $\left| 2 - \dfrac{1}{x} \right| \le 2 + \left| \dfrac{1}{x} \right|$

99. Write a quadratic inequality for which the solution set is $(-4, 3)$.

100. Write a polynomial inequality for which the solution set is $[-4, 3] \cup [7, \infty)$.

Find the domain of the function.

101. $f(x) = \sqrt{\dfrac{72}{x^2 - 4x - 21}}$

102. $f(x) = \sqrt{x^2 - 4x - 21}$

CHAPTER 4   Summary and Review

Important Properties and Formulas

Polynomial Function:

$$P(x) = a_n x^n + a_{n-1} x^{n-1} + a_{n-2} x^{n-2} + \cdots$$
$$+ a_1 x + a_0,$$

where the coefficients $a_n, a_{n-1}, \ldots, a_1, a_0$ are real numbers and the exponents are whole numbers.

The Leading-Term Test: If $a_n x^n$ is the leading term of a polynomial function, then the behavior of the graph as $x \to \infty$ and as $x \to -\infty$ can be described in one of the four following ways.

If n is even,
and $a_n > 0$:

If n is even,
and $a_n < 0$:

If n is odd,
and $a_n > 0$:

If n is odd,
and $a_n < 0$:

The Intermediate Value Theorem: For any polynomial function $P(x)$ with real coefficients, suppose that for $a \neq b$, $P(a)$ and $P(b)$ are of opposite signs. Then the function has a real zero between a and b.

Polynomial Division:

$$P(x) = d(x) \cdot Q(x) + R(x)$$

Dividend   Divisor   Quotient   Remainder

The Remainder Theorem: If a number c is substituted for x in the polynomial $f(x)$, then the result $f(c)$ is the remainder that would be obtained by dividing $f(x)$ by $x - c$. That is, if $f(x) = (x - c) \cdot Q(x) + R$, then $f(c) = R$.

The Factor Theorem: For a polynomial $f(x)$, if $f(c) = 0$, then $x - c$ is a factor of $f(x)$.

The Fundamental Theorem of Algebra: Every polynomial function of degree n, $n \geq 1$, with complex coefficients has at least one zero in the system of complex numbers.

The Rational Zeros Theorem: Consider the polynomial function

$$P(x) = a_n x^n + a_{n-1} x^{n-1} + a_{n-2} x^{n-2}$$
$$+ \cdots + a_1 x + a_0,$$

where all the coefficients are integers and $n \geq 1$. Also, consider a rational number p/q, where p and q have no common factor other than -1 and 1. If p/q is a zero of $P(x)$, then p is a factor of a_0 and q is a factor of a_n.

Descartes' Rule of Signs

Let $P(x)$, written in descending or ascending order, be a polynomial function with real coefficients and a nonzero constant term. The number of positive real zeros of $P(x)$ is either:

1. The same as the number of variations of sign in $P(x)$, or
2. Less than the number of variations of sign in $P(x)$ by a positive even integer.

The number of negative real zeros of $P(x)$ is either:

3. The same as the number of variations of sign in $P(-x)$, or
4. Less than the number of variations of sign in $P(-x)$ by a positive even integer.

A zero of multiplicity m must be counted m times.

Rational Function:

$$f(x) = \frac{p(x)}{q(x)},$$

where $p(x)$ and $q(x)$ are polynomials and where $q(x)$ is not the zero polynomial. The domain of $f(x)$ consists of all x for which $q(x) \neq 0$.

Occurrence of Lines as Asymptotes

For a rational function $f(x) = p(x)/q(x)$, where $p(x)$ and $q(x)$ have no common factors other than constants:

Vertical asymptotes occur at any x-values that make the denominator 0.

The x-axis is the horizontal asymptote when the degree of the numerator is less than the degree of the denominator.

A horizontal asymptote other than the x-axis occurs when the numerator and the denominator have the same degree.

An oblique asymptote occurs when the degree of the numerator is 1 greater than the degree of the denominator.

Review Exercises

Determine whether the statement is true or false.

1. If $f(x) = (x + a)(x + b)(x - c)$, then $f(-b) = 0$.   [4.3]

2. The graph of a rational function never crosses a vertical asymptote.   [4.5]

3. For the function $g(x) = x^4 - 8x^2 - 9$, the only possible rational zeros are 1, −1, 3, and −3.   [4.4]

4. The graph of $P(x) = x^6 - x^8$ has at most 6 x-intercepts.   [4.2]

5. The domain of the function

$$f(x) = \frac{x - 4}{(x + 2)(x - 3)}$$

is $(-\infty, -2) \cup (3, \infty)$.   [4.5]

*Use a graphing calculator to graph the polynomial function. Then estimate the function's **(a)** zeros, **(b)** relative maxima, **(c)** relative minima, and **(d)** domain and range.*   [4.1]

6. $f(x) = -2x^2 - 3x + 6$

7. $f(x) = x^3 + 3x^2 - 2x - 6$

8. $f(x) = x^4 - 3x^3 + 2x^2$

Determine the leading term, the leading coefficient, and the degree of the polynomial. Then classify the polynomial function as constant, linear, quadratic, cubic, or quartic.   [4.1]

9. $f(x) = 7x^2 - 5 + 0.45x^4 - 3x^3$

10. $h(x) = -25$

11. $g(x) = 6 - 0.5x$

12. $f(x) = \frac{1}{3}x^3 - 2x + 3$

Use the leading-term test to describe the end behavior of the graph of the function.   [4.1]

13. $f(x) = -\frac{1}{2}x^4 + 3x^2 + x - 6$

14. $f(x) = x^5 + 2x^3 - x^2 + 5x + 4$

Find the zeros of the polynomial function and state the multiplicity of each.   [4.1]

15. $g(x) = \left(x - \frac{2}{3}\right)(x + 2)^3(x - 5)^2$

16. $f(x) = x^4 - 26x^2 + 25$

17. $h(x) = x^3 + 4x^2 - 9x - 36$

18. *Interest Compounded Annually.* When P dollars is invested at interest rate i, compounded annually, for t years, the investment grows to A dollars, where

$$A = P(1 + i)^t.$$

a) Find the interest rate i if $6250 grows to $6760 in 2 yr.   [4.1]

b) Find the interest rate i if $1,000,000 grows to $1,215,506.25 in 4 yr.   [4.1]

19. *Cholesterol Level and the Risk of Heart Attack.* The table below lists data concerning the relationship of cholesterol level in men to the risk of a heart attack.

Cholesterol Level	Number of Men per 100,000 Who Suffer a Heart Attack
100	30
200	65
250	100
275	130

Source: Nutrition Action Healthletter

a) Use regression on a graphing calculator to fit linear, quadratic, and cubic functions to the data.   [4.1]

b) It is also known that 180 of 100,000 men with a cholesterol level of 300 have a heart attack. Which function in part (a) would best make this prediction?   [4.1]

c) Use the answer to part (b) to predict the heart attack rate for men with cholesterol levels of 350 and of 400.   [4.1]

Sketch the graph of the polynomial function.

20. $f(x) = -x^4 + 2x^3$ [4.2]

21. $g(x) = (x - 1)^3(x + 2)^2$ [4.2]

22. $h(x) = x^3 + 3x^2 - x - 3$ [4.2]

23. $f(x) = x^4 - 5x^3 + 6x^2 + 4x - 8$
[4.2], [4.3], [4.4]

24. $g(x) = 2x^3 + 7x^2 - 14x + 5$ [4.2], [4.4]

Using the intermediate value theorem, determine, if possible, whether the function f has a zero between a and b.   [4.2]

25. $f(x) = 4x^2 - 5x - 3$; $a = 1, b = 2$

26. $f(x) = x^3 - 4x^2 + \frac{1}{2}x + 2$; $a = -1, b = 1$

In each of the following, a polynomial $P(x)$ and a divisor $d(x)$ are given. Use long division to find the quotient $Q(x)$ and the remainder $R(x)$ when $P(x)$ is divided by $d(x)$. Express $P(x)$ in the form $d(x) \cdot Q(x) + R(x)$.   [4.3]

27. $P(x) = 6x^3 - 2x^2 + 4x - 1$,
   $d(x) = x - 3$

28. $P(x) = x^4 - 2x^3 + x + 5$,
   $d(x) = x + 1$

Use synthetic division to find the quotient and the remainder.   [4.3]

29. $(x^3 + 2x^2 - 13x + 10) \div (x - 5)$

30. $(x^4 + 3x^3 + 3x^2 + 3x + 2) \div (x + 2)$

31. $(x^5 - 2x) \div (x + 1)$

Use synthetic division to find the indicated function value.   [4.3]

32. $f(x) = x^3 + 2x^2 - 13x + 10$; $f(-2)$

33. $f(x) = x^4 - 16$; $f(-2)$

34. $f(x) = x^5 - 4x^4 + x^3 - x^2 + 2x - 100$;
   $f(-10)$

Using synthetic division, determine whether the given numbers are zeros of the polynomial function.   [4.3]

35. $-i, -5$; $f(x) = x^3 - 5x^2 + x - 5$

36. $-1, -2$; $f(x) = x^4 - 4x^3 - 3x^2 + 14x - 8$

37. $\frac{1}{3}, 1$; $f(x) = x^3 - \frac{4}{3}x^2 - \frac{5}{3}x + \frac{2}{3}$

38. $2, -\sqrt{3}$; $f(x) = x^4 - 5x^2 + 6$

Factor the polynomial $f(x)$. Then solve the equation $f(x) = 0$.   [4.3], [4.4]

39. $f(x) = x^3 + 2x^2 - 7x + 4$

40. $f(x) = x^3 + 4x^2 - 3x - 18$

41. $f(x) = x^4 - 4x^3 - 21x^2 + 100x - 100$

42. $f(x) = x^4 - 3x^2 + 2$

Find a polynomial function of degree 3 with the given numbers as zeros. [4.4]

43. $-4, -1, 2$

44. $-3, 1 - i, 1 + i$

45. $\frac{1}{2}, 1 - \sqrt{2}, 1 + \sqrt{2}$

46. Find a polynomial function of degree 4 with -5 as a zero of multiplicity 3 and $\frac{1}{2}$ as a zero of multiplicity 1. [4.4]

47. Find a polynomial function of degree 5 with -3 as a zero of multiplicity 2, 2 as a zero of multiplicity 1, and 0 as a zero of multiplicity 2. [4.4]

Suppose that a polynomial function of degree 5 with rational coefficients has the given zeros. Find the other zero(s). [4.4]

48. $-\frac{2}{3}, \sqrt{5}, i$

49. $0, 1 + \sqrt{3}, -\sqrt{3}$

50. $-\sqrt{2}, \frac{1}{2}, 1, 2$

Find a polynomial function of lowest degree with rational coefficients and the following as some of its zeros. [4.4]

51. $\sqrt{11}$

52. $-i, 6$

53. $-1, 4, 1 + i$

54. $\sqrt{5}, -2i$

55. $\frac{1}{3}, 0, -3$

List all possible rational zeros. [4.4]

56. $h(x) = 4x^5 - 2x^3 + 6x - 12$

57. $g(x) = 3x^4 - x^3 + 5x^2 - x + 1$

58. $f(x) = x^3 - 2x^2 + x - 24$

For each polynomial function:

a) *Find the rational zeros and then the other zeros; that is, solve $f(x) = 0$.* [4.4]

b) *Factor $f(x)$ into linear factors.* [4.4]

59. $f(x) = 3x^5 + 2x^4 - 25x^3 - 28x^2 + 12x$

60. $f(x) = x^3 - 2x^2 - 3x + 6$

61. $f(x) = x^4 - 6x^3 + 9x^2 + 6x - 10$

62. $f(x) = x^3 + 3x^2 - 11x - 5$

63. $f(x) = 3x^3 - 8x^2 + 7x - 2$

64. $f(x) = x^5 - 8x^4 + 20x^3 - 8x^2 - 32x + 32$

65. $f(x) = x^6 + x^5 - 28x^4 - 16x^3 + 192x^2$

66. $f(x) = 2x^5 - 13x^4 + 32x^3 - 38x^2 + 22x - 5$

What does Descartes' rule of signs tell you about the number of positive real zeros and the number of negative real zeros of each of the following polynomial functions? [4.4]

67. $f(x) = 2x^6 - 7x^3 + x^2 - x$

68. $h(x) = -x^8 + 6x^5 - x^3 + 2x - 2$

69. $g(x) = 5x^5 - 4x^2 + x - 1$

Graph the function. Be sure to label all the asymptotes. List the domain and the x- and y-intercepts. [4.5]

70. $f(x) = \dfrac{x^2 - 5}{x + 2}$

71. $f(x) = \dfrac{5}{(x - 2)^2}$

72. $f(x) = \dfrac{x^2 + x - 6}{x^2 - x - 20}$

73. $f(x) = \dfrac{x - 2}{x^2 - 2x - 15}$

In Exercises 74 and 75, find a rational function that satisfies the given conditions. Answers may vary, but try to give the simplest answer possible. [4.5]

74. Vertical asymptotes $x = -2, x = 3$

75. Vertical asymptotes $x = -2, x = 3$; horizontal asymptote $y = 4$; x-intercept $(-3, 0)$

76. *Medical Dosage.* The function

$$N(t) = \frac{0.7t + 2000}{8t + 9}, \quad t \geq 5$$

gives the body concentration $N(t)$, in parts per million, of a certain dosage of medication after time t, in hours.

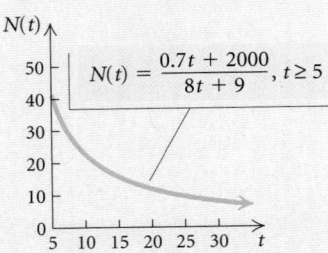

a) Find the horizontal asymptote of the graph and complete the following:

$$N(t) \rightarrow \boxed{} \text{ as } t \rightarrow \infty. \quad [4.5]$$

b) Explain the meaning of the answer to part (a) in terms of the application.   [4.5]

Solve.   [4.6]

77. $x^2 - 9 < 0$

78. $2x^2 > 3x + 2$

79. $(1 - x)(x + 4)(x - 2) \leq 0$

80. $\dfrac{x - 2}{x + 3} < 4$

81. *Height of a Rocket.* The function

$$S(t) = -16t^2 + 80t + 224$$

gives the height S, in feet, of a model rocket launched with a velocity of 80 ft/sec from a hill that is 224 ft high, where t is the time, in seconds.

a) Determine when the rocket reaches the ground.   [4.1]

b) On what interval is the height greater than 320 ft?   [4.1], [4.6]

82. *Population Growth.* The population P, in thousands, of Novi is given by

$$P(t) = \frac{8000t}{4t^2 + 10},$$

where t is the time, in months. Find the interval on which the population was 400,000 or greater.   [4.6]

83. Determine the domain of the function

$$g(x) = \frac{x^2 + 2x - 3}{x^2 - 5x + 6}. \quad [4.5]$$

A. $(-\infty, 2) \cup (2, 3) \cup (3, \infty)$
B. $(-\infty, -3) \cup (-3, 1) \cup (1, \infty)$
C. $(-\infty, 2) \cup (3, \infty)$
D. $(-\infty, -3) \cup (1, \infty)$

84. Determine the vertical asymptotes of the function

$$f(x) = \frac{x - 4}{(x + 1)(x - 2)(x + 4)}. \quad [4.5]$$

A. $x = 1, x = -2,$ and $x = 4$
B. $x = -1, x = 2, x = -4,$ and $x = 4$
C. $x = -1, x = 2,$ and $x = -4$
D. $x = 4$

85. The graph of $f(x) = -\dfrac{1}{2}x^4 + x^3 + 1$ is which of the following?   [4.2]

A.
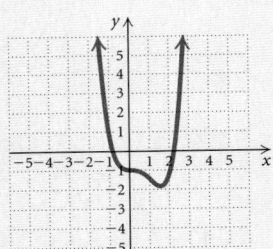

B.
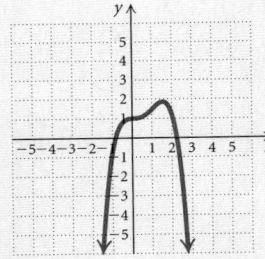

C.
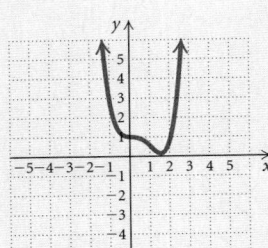

D.
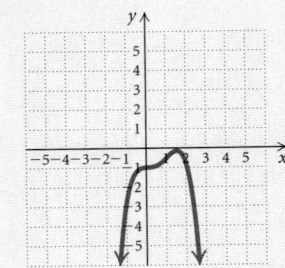

Collaborative Discussion and Writing

86. Explain the difference between a polynomial function and a rational function.   [4.1], [4.5]

87. Explain and contrast the three types of asymptotes considered for rational functions.   [4.5]

Synthesis

Solve.

88. $x^2 \geq 5 - 2x$ [4.6]

89. $\left|1 - \dfrac{1}{x^2}\right| < 3$ [4.6]

90. $x^4 - 2x^3 + 3x^2 - 2x + 2 = 0$ [4.4], [4.5]

91. $(x - 2)^{-3} < 0$ [4.6]

92. Express $x^3 - 1$ as a product of linear factors. [4.4]

93. Find k such that $x + 3$ is a factor of
$x^3 + kx^2 + kx - 15$. [4.3]

94. When $x^2 - 4x + 3k$ is divided by $x + 5$, the remainder is 33. Find the value of k. [4.3]

Find the domain of the function. [4.6]

95. $f(x) = \sqrt{x^2 + 3x - 10}$

96. $f(x) = \sqrt{x^2 - 3.1x + 2.2} + 1.75$

97. $f(x) = \dfrac{1}{\sqrt{5 - |7x + 2|}}$

CHAPTER 4   Test

Determine the leading term, the leading coefficient, and the degree of the polynomial. Then classify the polynomial as constant, linear, quadratic, cubic, or quartic.

1. $f(x) = 2x^3 + 6x^2 - x^4 + 11$

2. $h(x) = -4.7x + 29$

3. Find the zeros of the polynomial function and state the multiplicity of each:
$$f(x) = x(3x - 5)(x - 3)^2(x + 1)^3.$$

4. From 1980 to 2000, nearly 90,000 young people under age 20 were killed by gunfire in the United States. The quartic function
$$f(x) = 0.152x^4 - 9.786x^3 + 173.317x^2$$
$$-831.220x + 4053.134$$
can be used to estimate the number of people under age 20 who were killed x years after 1980. Using this function, estimate the number who were killed in 1986, in 1993, and in 1999.

Sketch the graph of the polynomial function.

5. $f(x) = x^3 - 5x^2 + 2x + 8$

6. $f(x) = -2x^4 + x^3 + 11x^2 - 4x - 12$

Using the intermediate value theorem, determine, if possible, whether the function has a zero between a and b.

7. $f(x) = -5x^2 + 3;\ a = 0, b = 2$

8. $g(x) = 2x^3 + 6x^2 - 3;\ a = -2, b = -1$

9. Use long division to find the quotient $Q(x)$ and the remainder $R(x)$ when $P(x)$ is divided by $d(x)$. Express $P(x)$ in the form $d(x) \cdot Q(x) + R(x)$. Show your work.
$$P(x) = x^4 + 3x^3 + 2x - 5,$$
$$d(x) = x - 1$$

10. Use synthetic division to find the quotient and the remainder. Show your work.
$$(3x^3 - 12x + 7) \div (x - 5)$$

11. Use synthetic division to find $P(-3)$ for $P(x) = 2x^3 - 6x^2 + x - 4$. Show your work.

12. Use synthetic division to determine whether -2 is a zero of $f(x) = x^3 + 4x^2 + x - 6$. Answer yes or no. Show your work.

13. Find a polynomial of degree 4 with -3 as a zero of multiplicity 2 and 0 and 6 as zeros of multiplicity 1.

14. Suppose that a polynomial function of degree 5 with rational coefficients has 1, $\sqrt{3}$, and $2 - i$ as zeros. Find the other zeros.

Find a polynomial function of lowest degree with rational coefficients and the following as some of its zeros.

15. $-10, 3i$

16. $0, -\sqrt{3}, 1 - i$

List all possible rational zeros.

17. $f(x) = 2x^3 + x^2 - 2x + 12$

18. $h(x) = 10x^4 - x^3 + 2x - 5$

For each polynomial function:

a) *Find the rational zeros and then the other zeros; that is, solve $f(x) = 0$.*

b) *Factor $f(x)$ into linear factors.*

19. $f(x) = x^3 + x^2 - 5x - 5$

20. $f(x) = 2x^4 - 11x^3 + 16x^2 - x - 6$

21. $f(x) = x^3 + 4x^2 + 4x + 16$

22. $f(x) = 3x^4 - 11x^3 + 15x^2 - 9x + 2$

23. What does Descartes' rule of signs tell you about the number of positive real zeros and the number of negative real zeros of the following function?
$$g(x) = -x^8 + 2x^6 - 4x^3 - 1$$

Graph the function. Be sure to label all the asymptotes. List the domain and the x- and y-intercepts.

24. $f(x) = \dfrac{2}{(x-3)^2}$

25. $f(x) = \dfrac{x+3}{x^2 - 3x - 4}$

26. Find a rational function that has vertical asymptotes $x = -1$ and $x = 2$ and x-intercept $(-4, 0)$.

Solve.

27. $2x^2 > 5x + 3$

28. $\dfrac{x+1}{x-4} \le 3$

29. The function $S(t) = -16t^2 + 64t + 192$ gives the height S, in feet, of a model rocket launched with a velocity of 64 ft/sec from a hill that is 192 ft high.

   a) Determine how long it will take the rocket to reach the ground.

   b) Find the interval on which the height of the rocket is greater than 240 ft.

30. The graph of $f(x) = x^3 - x^2 - 2$ is which of the following?

A.

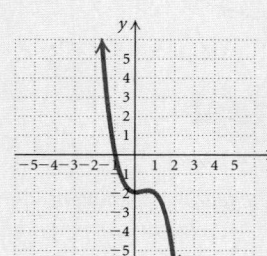

B.

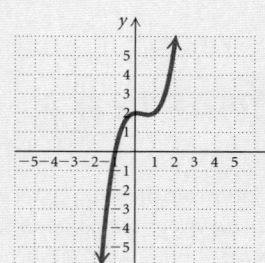

C.

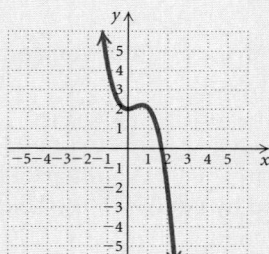

D.

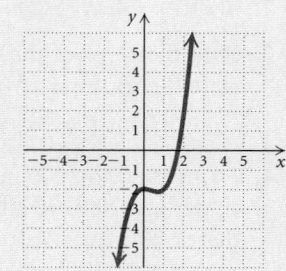

Synthesis

31. Find the domain of $f(x) = \sqrt{x^2 + x - 12}$.

Exponential and Logarithmic Functions

APPLICATION Total retail sales of coffee have increased exponentially from $8.3 billion in 2001 to $12.3 billion in 2006 (*Source*: Specialty Coffee Association of America). The following function models sales:

$$C(t) = 7.8867(1.0854)^t,$$

where t is the number of years since 2001 and $C(t)$ is in billions. Estimate the total retail sales of coffee in 2004 and in 2009.

This problem appears as Exercise 55 in Section 5.2.

5.1 Inverse Functions

5.2 Exponential Functions and Graphs

5.3 Logarithmic Functions and Graphs

5.4 Properties of Logarithmic Functions

5.5 Solving Exponential and Logarithmic Equations

5.6 Applications and Models: Growth and Decay; Compound Interest

5.1   Inverse Functions

❖ Determine whether a function is one-to-one, and if it is, find a formula for its inverse.

❖ Simplify expressions of the type $(f \circ f^{-1})(x)$ and $(f^{-1} \circ f)(x)$.

❖ Inverses

When we go from an output of a function back to its input or inputs, we get an inverse relation. When that relation is a function, we have an inverse function.

Consider the relation h given as follows:

$$h = \{(-8, 5), (4, -2), (-7, 1), (3.8, 6.2)\}.$$

Suppose we *interchange* the first and second coordinates. The relation we obtain is called the **inverse** of the relation h and is given as follows:

$$\text{Inverse of } h = \{(5, -8), (-2, 4), (1, -7), (6.2, 3.8)\}.$$

RELATIONS

REVIEW SECTION **1.2.**

> **Inverse Relation**
>
> Interchanging the first and second coordinates of each ordered pair in a relation produces the **inverse relation.**

EXAMPLE 1   Consider the relation g given by

$$g = \{(2, 4), (-1, 3), (-2, 0)\}.$$

Graph the relation in blue. Find the inverse and graph it in red.

Solution   The relation g is shown in blue in the figure at left. The inverse of the relation is

$$\{(4, 2), (3, -1), (0, -2)\}$$

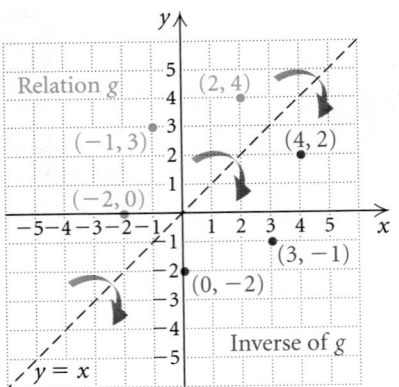

and is shown in red. The pairs in the inverse are reflections of the pairs in g across the line $y = x$.   **Now Try Exercise 1.** ■

> **Inverse Relation**
>
> If a relation is defined by an equation, interchanging the variables produces an equation of the **inverse relation.**

EXAMPLE 2 Find an equation for the inverse of the relation

$$y = x^2 - 5x.$$

Solution We interchange x and y and obtain an equation of the inverse:

$$x = y^2 - 5y.$$

Now Try Exercise 9. ■

If a relation is given by an equation, then the solutions of the inverse can be found from those of the original equation by interchanging the first and second coordinates of each ordered pair. Thus the graphs of a relation and its inverse are always reflections of each other across the line $y = x$. This is illustrated with the equations of Example 2 in the tables and graph below. We will explore inverses and their graphs later in this section.

$x = y^2 - 5y$	y
6	−1
0	0
−6	2
−4	4

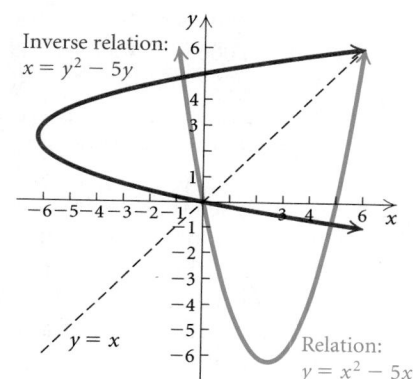

Inverse relation: $x = y^2 - 5y$

$y = x$

Relation: $y = x^2 - 5x$

x	$y = x^2 - 5x$
−1	6
0	0
2	−6
4	−4

❊ Inverses and One-to-One Functions

Let's consider the following two functions.

Year (domain)	First-Class Postage Cost, in cents (range)
1992	29
1994	29
1996	32
1998	32
2000	33
2002	37
2004	37
2006	39
2007	41

Number (domain)	Cube (range)
−3	−27
−2	−8
−1	−1
0	0
1	1
2	8
3	27

Source: U.S. Postal Service

Suppose we reverse the arrows. Are these inverse relations functions?

Year (range)	First-Class Postage Cost, in cents (domain)
1992 ←	
1994 ←	29
1996 ←	
1998 ←	32
2000 ←	33
2002 ←	
2004 ←	37
2006 ←	39
2007 ←	41

Number (range)	Cube (domain)
−3 ←	−27
−2 ←	−8
−1 ←	−1
0 ←	0
1 ←	1
2 ←	8
3 ←	27

Source: U.S. Postal Service

We see that the inverse of the postage function is not a function. Like all functions, each input in the postage function has exactly one output. However, the output for both 1996 and 1998 is 32. Thus in the inverse of the postage function, the input 32 has *two* outputs, 1996 and 1998. When two or more inputs of a function have the same output, the inverse relation cannot be a function. In the cubing function, each output corresponds to exactly one input, so its inverse is also a function. The cubing function is an example of a **one-to-one function**.

One-to-One Functions

A function f is **one-to-one** if different inputs have different outputs— that is,

$$\text{if } a \neq b, \quad \text{then} \quad f(a) \neq f(b).$$

Or a function f is **one-to-one** if when the outputs are the same, the inputs are the same—that is,

$$\text{if } f(a) = f(b), \quad \text{then} \quad a = b.$$

If the inverse of a function f is also a function, it is named f^{-1} (read "f-inverse").

The -1 in f^{-1} is *not* an exponent!

Do *not* misinterpret the -1 in f^{-1} as a negative exponent: f^{-1} does *not* mean the reciprocal of f and $f^{-1}(x)$ is *not* equal to $\dfrac{1}{f(x)}$.

One-to-One Functions and Inverses

- If a function f is one-to-one, then its inverse f^{-1} is a function.
- The domain of a one-to-one function f is the range of the inverse f^{-1}.
- The range of a one-to-one function f is the domain of the inverse f^{-1}.

$$D_f \quad D_{f^{-1}}$$
$$R_f \quad R_{f^{-1}}$$

- A function that is increasing over its domain or is decreasing over its domain is a one-to-one function.

EXAMPLE 3 Given the function f described by $f(x) = 2x - 3$, prove that f is one-to-one (that is, it has an inverse that is a function).

Solution To show that f is one-to-one, we show that if $f(a) = f(b)$, then $a = b$. Assume that $f(a) = f(b)$ for a and b in the domain of f. Since $f(a) = 2a - 3$ and $f(b) = 2b - 3$, we have

$$2a - 3 = 2b - 3$$
$$2a = 2b \qquad \text{Adding 3}$$
$$a = b. \qquad \text{Dividing by 2}$$

Thus, if $f(a) = f(b)$, then $a = b$. This shows that f is one-to-one.

Now Try Exercise 17. ■

EXAMPLE 4 Given the function g described by $g(x) = x^2$, prove that g is not one-to-one.

Solution We can prove that g is not one-to-one by finding two numbers a and b for which $a \neq b$ and $g(a) = g(b)$. Two such numbers are -3 and 3, because $-3 \neq 3$ and $g(-3) = g(3) = 9$. Thus g is not one-to-one.

Now Try Exercise 21. ■

The following graphs show a function, in blue, and its inverse, in red. To determine whether the inverse is a function, we can apply the vertical-line test to its graph. By reflecting each such vertical line back across the line $y = x$, we obtain an equivalent **horizontal-line test** for the original function.

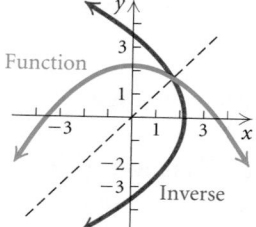

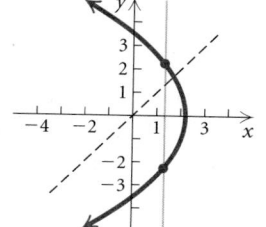

The vertical-line test shows that the inverse is not a function.

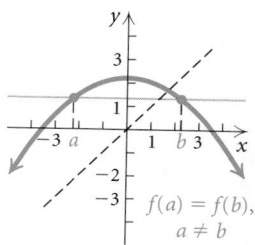

The horizontal-line test shows that the function is not one-to-one.

Horizontal-Line Test

If it is possible for a horizontal line to intersect the graph of a function more than once, then the function is *not* one-to-one and its inverse is *not* a function.

EXAMPLE 5 From the graph shown, determine whether each function is one-to-one and thus has an inverse that is a function.

a)
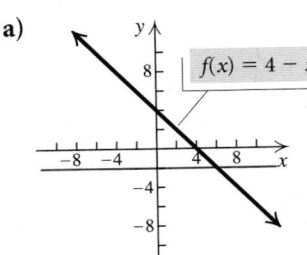

b)
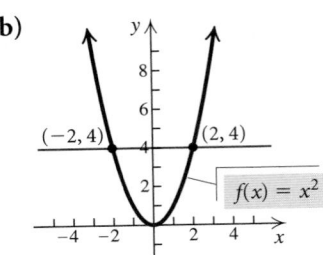

c)
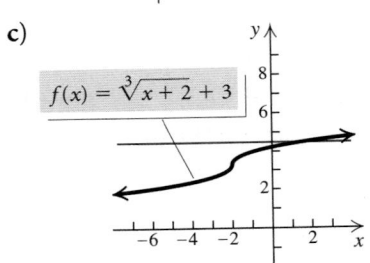

d)
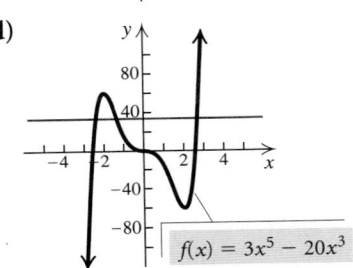

Solution For each function, we apply the horizontal-line test.

RESULT	REASON
a) One-to-one; inverse is a function	No horizontal line intersects the graph more than once.
b) Not one-to-one; inverse is not a function	There are many horizontal lines that intersect the graph more than once. Note that where the line $y = 4$ intersects the graph, the first coordinates are -2 and 2. Although these are different inputs, they have the same output, 4.
c) One-to-one; inverse is a function	No horizontal line intersects the graph more than once.
d) Not one-to-one; inverse is not a function	There are many horizontal lines that intersect the graph more than once.

Now Try Exercises 25 and 27. ■

❋ Finding Formulas for Inverses

Suppose that a function is described by a formula. If it has an inverse that is a function, we proceed as follows to find a formula for f^{-1}.

> ### Obtaining a Formula for an Inverse
> If a function f is one-to-one, a formula for its inverse can generally be found as follows:
>
> 1. Replace $f(x)$ with y.
> 2. Interchange x and y.
> 3. Solve for y.
> 4. Replace y with $f^{-1}(x)$.

EXAMPLE 6   Determine whether the function $f(x) = 2x - 3$ is one-to-one, and if it is, find a formula for $f^{-1}(x)$.

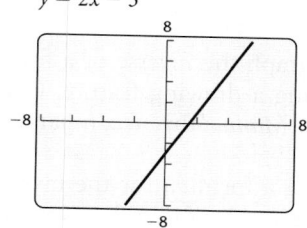

$y = 2x - 3$

Solution   The graph of f is shown at left. It passes the horizontal-line test. Thus it is one-to-one and its inverse is a function. We also proved that f is one-to-one in Example 3. We find a formula for $f^{-1}(x)$.

1. Replace $f(x)$ with y: $y = 2x - 3$
2. Interchange x and y: $x = 2y - 3$
3. Solve for y: $x + 3 = 2y$

$$\frac{x + 3}{2} = y$$

4. Replace y with $f^{-1}(x)$: $f^{-1}(x) = \dfrac{x + 3}{2}.$

Now Try Exercise 57. ■

Consider

$$f(x) = 2x - 3 \quad \text{and} \quad f^{-1}(x) = \frac{x + 3}{2}$$

from Example 6. For the input 5, we have

$$f(5) = 2 \cdot 5 - 3 = 10 - 3 = 7.$$

The output is 7. Now we use 7 for the input in the inverse:

$$f^{-1}(7) = \frac{7 + 3}{2} = \frac{10}{2} = 5.$$

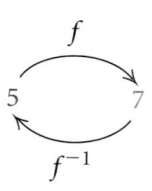

The function f takes the number 5 to 7. The inverse function f^{-1} takes the number 7 back to 5.

GCM | **EXAMPLE 7** Graph

$$f(x) = 2x - 3 \quad \text{and} \quad f^{-1}(x) = \frac{x + 3}{2}$$

using the same set of axes. Then compare the two graphs.

Solution The graphs of f and f^{-1} are shown at left. The solutions of the inverse function can be found from those of the original function by interchanging the first and second coordinates of each ordered pair.

$f^{-1}(x) = \dfrac{x + 3}{2}$

$f(x) = 2x - 3$

$y = x$

x	$f(x) = 2x - 3$
-1	-5
0	-3 ← y-intercept
2	1
3	3

x	$f^{-1}(x) = \dfrac{x + 3}{2}$
-5	-1
-3	0 ← x-intercept
1	2
3	3

On some graphing calculators, we can graph the inverse of a function after graphing the function itself by accessing a drawing feature. Consult your user's manual or the *Graphing Calculator Manual* that accompanies this text for the procedure.

When we interchange x and y in finding a formula for the inverse of $f(x) = 2x - 3$, we are in effect reflecting the graph of that function across the line $y = x$. For example, when the coordinates of the y-intercept, $(0, -3)$, of the graph of f are reversed, we get the x-intercept, $(-3, 0)$, of the graph of f^{-1}. If we were to graph $f(x) = 2x - 3$ in wet ink and fold along the line $y = x$, the graph of $f^{-1}(x) = (x + 3)/2$ would be formed by the ink transferred from f.

> The graph of f^{-1} is a reflection of the graph of f across the line $y = x$.

EXAMPLE 8 Consider $g(x) = x^3 + 2$.

a) Determine whether the function is one-to-one.

b) If it is one-to-one, find a formula for its inverse.

c) Graph the function and its inverse.

Solution

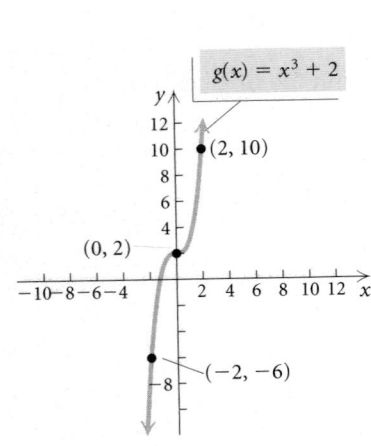

$g(x) = x^3 + 2$

$(2, 10)$

$(0, 2)$

$(-2, -6)$

a) The graph of $g(x) = x^3 + 2$ is shown at left. It passes the horizontal-line test and thus has an inverse that is a function. We also know that $g(x)$ is one-to-one because it is an increasing function over its entire domain.

b) We follow the procedure for finding an inverse.

1. Replace $g(x)$ with y: $\qquad y = x^3 + 2$
2. Interchange x and y: $\qquad x = y^3 + 2$
3. Solve for y: $\qquad x - 2 = y^3$
$$\sqrt[3]{x - 2} = y$$
4. Replace y with $g^{-1}(x)$: $\qquad g^{-1}(x) = \sqrt[3]{x - 2}.$

We can test a point as a partial check:

$$g(x) = x^3 + 2$$
$$g(3) = 3^3 + 2 = 27 + 2 = 29.$$

Will $g^{-1}(29) = 3$? We have

$$g^{-1}(x) = \sqrt[3]{x - 2}$$
$$g^{-1}(29) = \sqrt[3]{29 - 2} = \sqrt[3]{27} = 3.$$

Since $g(3) = 29$ and $g^{-1}(29) = 3$, we can be reasonably certain that the formula for $g^{-1}(x)$ is correct.

c) To find the graph, we reflect the graph of $g(x) = x^3 + 2$ across the line $y = x$. This can be done by plotting points or by using a graphing calculator.

x	$g(x)$
-2	-6
-1	1
0	2
1	3
2	10

x	$g^{-1}(x)$
-6	-2
1	-1
2	0
3	1
10	2

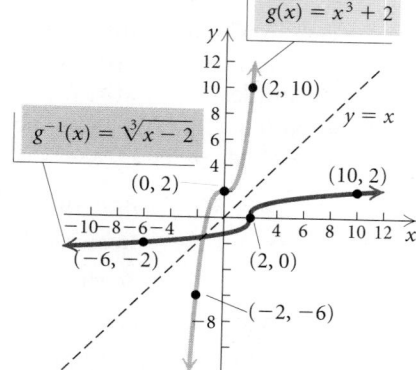

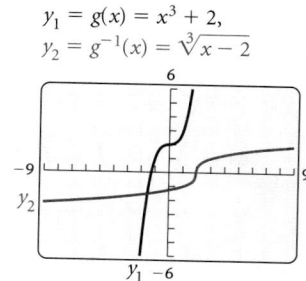

Now Try Exercise 79.

❋ Inverse Functions and Composition

Suppose that we were to use some input a for a one-to-one function f and find its output, $f(a)$. The function f^{-1} would then take that output back to a. Similarly, if we began with an input b for the function f^{-1} and found its output, $f^{-1}(b)$, the original function f would then take that output back to b. This is summarized as follows.

> If a function f is one-to-one, then f^{-1} is the unique function such that each of the following holds:
>
> $(f^{-1} \circ f)(x) = f^{-1}(f(x)) = x$,   for each x in the domain of f, and
>
> $(f \circ f^{-1})(x) = f(f^{-1}(x)) = x$,   for each x in the domain of f^{-1}.

EXAMPLE 9   Given that $f(x) = 5x + 8$, use composition of functions to show that

$$f^{-1}(x) = \frac{x - 8}{5}.$$

Solution   We find $(f^{-1} \circ f)(x)$ and $(f \circ f^{-1})(x)$ and check to see that each is x:

$$(f^{-1} \circ f)(x) = f^{-1}(f(x))$$

$$= f^{-1}(5x + 8) = \frac{(5x + 8) - 8}{5} = \frac{5x}{5} = x;$$

$$(f \circ f^{-1})(x) = f(f^{-1}(x))$$

$$= f\left(\frac{x - 8}{5}\right) = 5\left(\frac{x - 8}{5}\right) + 8 = x - 8 + 8 = x.$$

Now Try Exercise 87.

❋ Restricting a Domain

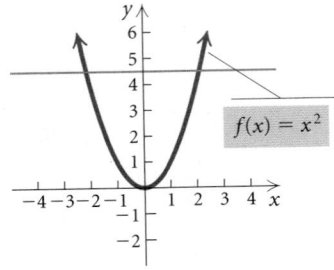

In the case in which the inverse of a function is not a function, the domain of the function can be restricted to allow the inverse to be a function. We saw in Examples 4 and 5(b) that $f(x) = x^2$ is not one-to-one. The graph is shown at left.

Suppose that we had tried to find a formula for the inverse as follows:

$$y = x^2 \qquad \text{Replacing } f(x) \text{ with } y$$
$$x = y^2 \qquad \text{Interchanging } x \text{ and } y$$
$$\pm\sqrt{x} = y. \qquad \text{Solving for } y$$

This is not the equation of a function. An input of, say, 4 would yield two outputs, -2 and 2. In such cases, it is convenient to consider "part" of the function by restricting the domain of $f(x)$. For example, if we restrict the domain of $f(x) = x^2$ to nonnegative numbers, then its inverse is a function, as shown with the graphs of $f(x) = x^2, x \geq 0$, and $f^{-1}(x) = \sqrt{x}$ below.

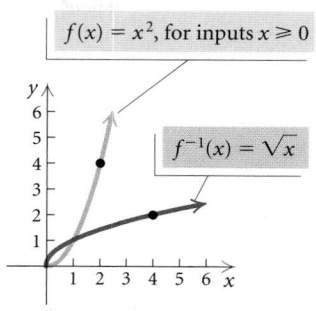

$f(x) = x^2$, for inputs $x \geq 0$

$f^{-1}(x) = \sqrt{x}$

5.1   Exercise Set

Find the inverse of the relation.

1. $\{(7,8),(-2,8),(3,-4),(8,-8)\}$

2. $\{(0,1),(5,6),(-2,-4)\}$

3. $\{(-1,-1),(-3,4)\}$

4. $\{(-1,3),(2,5),(-3,5),(2,0)\}$

Find an equation of the inverse relation.

5. $y = 4x - 5$

6. $2x^2 + 5y^2 = 4$

7. $x^3 y = -5$

8. $y = 3x^2 - 5x + 9$

9. $x = y^2 - 2y$

10. $x = \frac{1}{2}y + 4$

Graph the equation by substituting and plotting points. Then reflect the graph across the line $y = x$ to obtain the graph of its inverse.

11. $x = y^2 - 3$ **12.** $y = x^2 + 1$

13. $y = 3x - 2$ **14.** $x = -y + 4$

15. $y = |x|$ **16.** $x + 2 = |y|$

Given the function f, prove that f is one-to-one using the definition of a one-to-one function on p. 382.

17. $f(x) = \frac{1}{3}x - 6$ **18.** $f(x) = 4 - 2x$

19. $f(x) = x^3 + \frac{1}{2}$ **20.** $f(x) = \sqrt[3]{x}$

Given the function g, prove that g is not one-to-one using the definition of a one-to-one function on p. 382.

21. $g(x) = 1 - x^2$ **22.** $g(x) = 3x^2 + 1$

23. $g(x) = x^4 - x^2$ **24.** $g(x) = \dfrac{1}{x^6}$

Using the horizontal-line test, determine whether the function is one-to-one.

25. $f(x) = 2.7^x$ **26.** $f(x) = 2^{-x}$

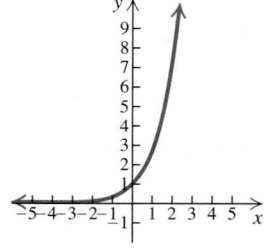

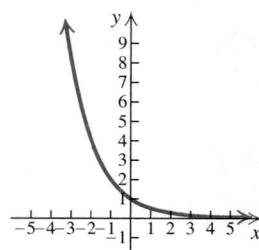

27. $f(x) = 4 - x^2$

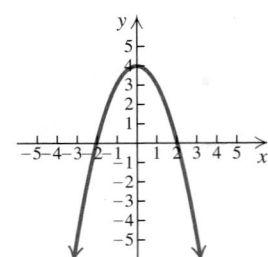

28. $f(x) = x^3 - 3x + 1$

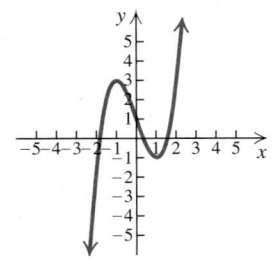

Graph the function and its inverse using a graphing calculator. Use an inverse drawing feature, if available. Find the domain and the range of f and of f^{-1}.

45. $f(x) = 0.8x + 1.7$

46. $f(x) = 2.7 - 1.08x$

47. $f(x) = \frac{1}{2}x - 4$

48. $f(x) = x^3 - 1$

49. $f(x) = \sqrt{x - 3}$

50. $f(x) = -\dfrac{2}{x}$

51. $f(x) = x^2 - 4, \; x \geq 0$

52. $f(x) = 3 - x^2, \; x \geq 0$

53. $f(x) = (3x - 9)^3$

54. $f(x) = \sqrt[3]{\dfrac{x - 3.2}{1.4}}$

29. $f(x) = \dfrac{8}{x^2 - 4}$

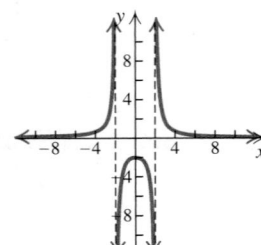

30. $f(x) = \sqrt{\dfrac{10}{4 + x}}$

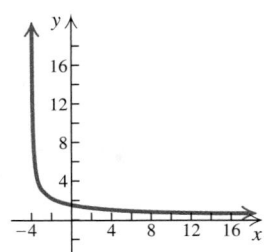

31. $f(x) = \sqrt[3]{x + 2} - 2$

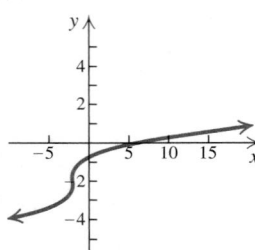

32. $f(x) = \dfrac{8}{x}$

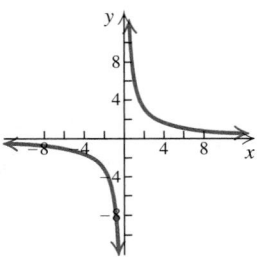

Graph the function and determine whether the function is one-to-one using the horizontal-line test.

33. $f(x) = 5x - 8$

34. $f(x) = 3 + 4x$

35. $f(x) = 1 - x^2$

36. $f(x) = |x| - 2$

37. $f(x) = |x + 2|$

38. $f(x) = -0.8$

39. $f(x) = -\dfrac{4}{x}$

40. $f(x) = \dfrac{2}{x + 3}$

41. $f(x) = \frac{2}{3}$

42. $f(x) = \frac{1}{2}x^2 + 3$

43. $f(x) = \sqrt{25 - x^2}$

44. $f(x) = -x^3 + 2$

In Exercises 55–70, for each function:
a) Determine whether it is one-to-one.
b) If the function is one-to-one, find a formula for the inverse.

55. $f(x) = x + 4$

56. $f(x) = 7 - x$

57. $f(x) = 2x - 1$

58. $f(x) = 5x + 8$

59. $f(x) = \dfrac{4}{x + 7}$

60. $f(x) = -\dfrac{3}{x}$

61. $f(x) = \dfrac{x + 4}{x - 3}$

62. $f(x) = \dfrac{5x - 3}{2x + 1}$

63. $f(x) = x^3 - 1$

64. $f(x) = (x + 5)^3$

65. $f(x) = x\sqrt{4 - x^2}$

66. $f(x) = 2x^2 - x - 1$

67. $f(x) = 5x^2 - 2, \; x \geq 0$

68. $f(x) = 4x^2 + 3, \; x \geq 0$

69. $f(x) = \sqrt{x + 1}$

70. $f(x) = \sqrt[3]{x - 8}$

Find the inverse by thinking about the operations of the function and then reversing, or undoing, them. Check your work algebraically.

Function	Inverse
71. $f(x) = 3x$	$f^{-1}(x) = $
72. $f(x) = \frac{1}{4}x + 7$	$f^{-1}(x) = $
73. $f(x) = -x$	$f^{-1}(x) = $
74. $f(x) = \sqrt[3]{x} - 5$	$f^{-1}(x) = $
75. $f(x) = \sqrt[3]{x - 5}$	$f^{-1}(x) = $
76. $f(x) = x^{-1}$	$f^{-1}(x) = $

Each graph in Exercises 77–82 is the graph of a one-to-one function f. Sketch the graph of the inverse function f^{-1}.

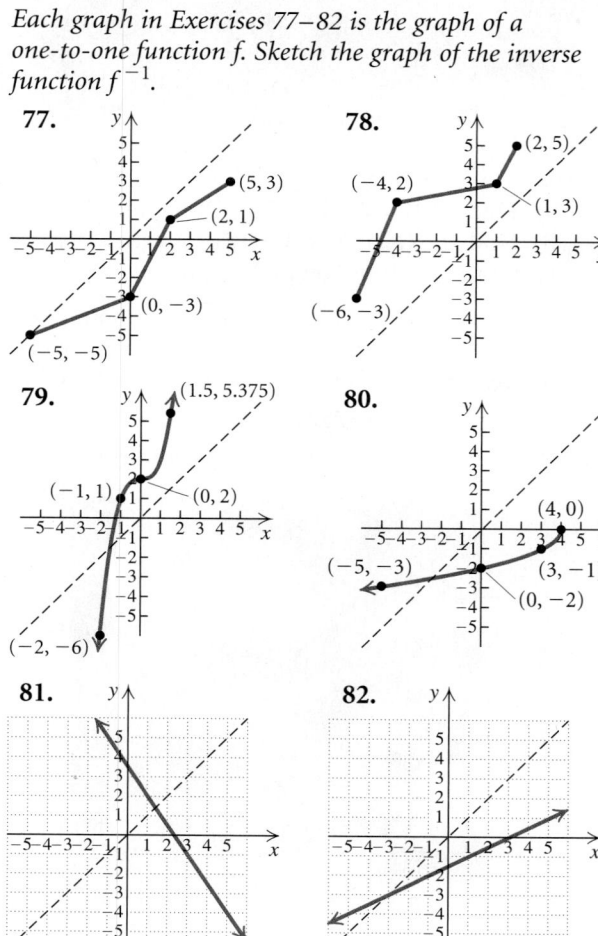

77. points $(5,3)$, $(2,1)$, $(0,-3)$, $(-5,-5)$

78. points $(2,5)$, $(-4,2)$, $(1,3)$, $(-6,-3)$

79. points $(1.5, 5.375)$, $(-1,1)$, $(0,2)$, $(-2,-6)$

80. points $(4,0)$, $(-5,-3)$, $(3,-1)$, $(0,-2)$

81.

82.

For the function f, use composition of functions to show that f^{-1} is as given.

83. $f(x) = \frac{7}{8}x$, $f^{-1}(x) = \frac{8}{7}x$

84. $f(x) = \frac{x+5}{4}$, $f^{-1}(x) = 4x - 5$

85. $f(x) = \frac{1-x}{x}$, $f^{-1}(x) = \frac{1}{x+1}$

86. $f(x) = \sqrt[3]{x+4}$, $f^{-1}(x) = x^3 - 4$

87. $f(x) = \frac{2}{5}x + 1$, $f^{-1}(x) = \frac{5x-5}{2}$

88. $f(x) = \frac{x+6}{3x-4}$, $f^{-1}(x) = \frac{4x+6}{3x-1}$

Find the inverse of the given one-to-one function f. Give the domain and the range of f and of f^{-1}, and then graph both f and f^{-1} on the same set of axes.

89. $f(x) = 5x - 3$

90. $f(x) = 2 - x$

91. $f(x) = \frac{2}{x}$

92. $f(x) = -\frac{3}{x+1}$

93. $f(x) = \frac{1}{3}x^3 - 2$

94. $f(x) = \sqrt[3]{x} - 1$

95. $f(x) = \frac{x+1}{x-3}$

96. $f(x) = \frac{x-1}{x+2}$

97. Find $f(f^{-1}(5))$ and $f^{-1}(f(a))$:
$$f(x) = x^3 - 4.$$

98. Find $f^{-1}(f(p))$ and $f(f^{-1}(1253))$:
$$f(x) = \sqrt[5]{\frac{2x-7}{3x+4}}.$$

99. *Women's Shoe Sizes.* A function that will convert women's shoe sizes in the United States to those in Australia is

$$s(x) = \frac{2x - 3}{2}$$

(*Source*: OnlineConversion.com).

a) Determine the women's shoe sizes in Australia that correspond to sizes 5, $7\frac{1}{2}$, and 8 in the United States.
b) Find a formula for the inverse of the function.
c) Use the inverse function to determine the women's shoe sizes in the United States that correspond to sizes 3, $5\frac{1}{2}$, and 7 in Australia.

100. *Bus Chartering.* An organization determines that the cost per person of chartering a bus is given by the formula

$$C(x) = \frac{100 + 5x}{x},$$

where x is the number of people in the group and $C(x)$ is in dollars. Determine $C^{-1}(x)$ and explain what it represents.

101. *Reaction Distance.* Suppose you are driving a car when a deer suddenly darts across the road in front of you. During the time it takes you to step on the brake, the car travels a distance D, in feet, where D is a function of the speed r, in miles per hour, that the car is traveling when you see the deer. That reaction distance D is a linear function given by

$$D(r) = \frac{11r + 5}{10}.$$

a) Find $D(0)$, $D(10)$, $D(20)$, $D(50)$, and $D(65)$.
b) Find $D^{-1}(r)$ and explain what it represents.
c) Graph the function and its inverse.

102. *Cheese Consumption.* The number of pounds of cheese consumed in the United States per person per year x years after 1990 is given by the function

$$N(x) = 0.4737x + 24.7702$$

(*Source*: U.S. Department of Agriculture).

a) Determine the cheese consumption per person in 2007 and in 2010.
b) Graph the function and its inverse.
c) Explain what the inverse represents.

Collaborative Discussion and Writing

103. Explain why an even function f does not have an inverse f^{-1}.

104. The following formulas for the conversion between Fahrenheit temperature and Celsius temperature have been considered several times in this text:

$$C = \tfrac{5}{9}(F - 32)$$

and

$$F = \tfrac{9}{5}C + 32.$$

Discuss these formulas from the standpoint of inverses.

Skill Maintenance

Consider the following quadratic functions. Without graphing them, answer the questions below.

a) $f(x) = 2x^2$
b) $f(x) = -x^2$
c) $f(x) = \tfrac{1}{4}x^2$
d) $f(x) = -5x^2 + 3$
e) $f(x) = \tfrac{2}{3}(x - 1)^2 - 3$
f) $f(x) = -2(x + 3)^2 + 1$
g) $f(x) = (x - 3)^2 + 1$
h) $f(x) = -4(x + 1)^2 - 3$

105. Which functions have a maximum value?

106. Which graphs open up?

107. Consider (a) and (c). Which graph is narrower?

108. Consider (d) and (e). Which graph is narrower?

109. Which graph has vertex $(-3, 1)$?

110. For which is the line of symmetry $x = 0$?

Synthesis

Using only a graphing calculator, determine whether the functions are inverses of each other.

111. $f(x) = \sqrt[3]{\dfrac{x - 3.2}{1.4}}$, $g(x) = 1.4x^3 + 3.2$

112. $f(x) = \dfrac{2x - 5}{4x + 7}$, $g(x) = \dfrac{7x - 4}{5x + 2}$

113. $f(x) = \dfrac{2}{3}$, $g(x) = \dfrac{3}{2}$

114. $f(x) = x^4, x \geq 0$; $g(x) = \sqrt[4]{x}$

115. The function $f(x) = x^2 - 3$ is not one-to-one. Restrict the domain of f so that its inverse is a function. Find the inverse and state the restriction on the domain of the inverse.

116. Consider the function f given by

$$f(x) = \begin{cases} x^3 + 2, & x \leq -1, \\ x^2, & -1 < x < 1, \\ x + 1, & x \geq 1. \end{cases}$$

Does f have an inverse that is a function? Why or why not?

117. Find three examples of functions that are their own inverses; that is, $f = f^{-1}$.

118. Given the function $f(x) = ax + b, a \neq 0$, find the values of a and b for which $f^{-1}(x) = f(x)$.

5.2

Exponential Functions and Graphs

❋ Graph exponential equations and exponential functions.

❋ Solve applied problems involving exponential functions and their graphs.

We now turn our attention to the study of a set of functions that are very rich in application. Consider the following graphs. Each one illustrates an *exponential function*. In this section, we consider such functions and some important applications.

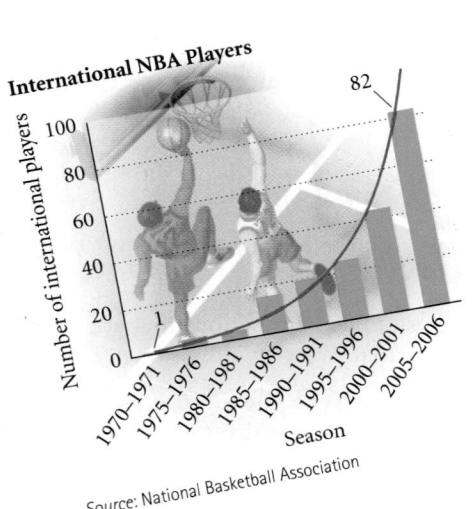

Source: National Basketball Association

iPod Sales

Source: SEC filings via 10(k) Wizard

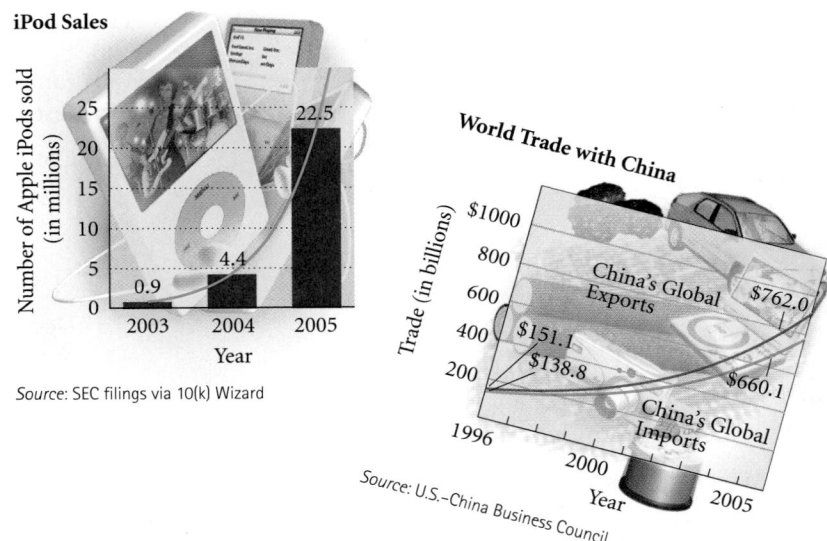

Source: U.S.–China Business Council

EXPONENTS

REVIEW SECTIONS **R.2** AND **R.7.**

❋ Graphing Exponential Functions

We now define exponential functions. We assume that a^x has meaning for any real number x and any positive real number a and that the laws of exponents still hold, though we will not prove them here.

> **Exponential Function**
>
> The function $f(x) = a^x$, where x is a real number, $a > 0$ and $a \neq 1$, is called the **exponential function**, base a.

We require the **base** to be positive in order to avoid the imaginary numbers that would occur by taking even roots of negative numbers—an example is $(-1)^{1/2}$, the square root of -1, which is not a real number. The restriction $a \neq 1$ is made to exclude the constant function $f(x) = 1^x = 1$, which does not have an inverse because it is not one-to-one.

The following are examples of exponential functions:

$$f(x) = 2^x, \qquad f(x) = \left(\frac{1}{2}\right)^x, \qquad f(x) = (3.57)^x.$$

Note that, in contrast to functions like $f(x) = x^5$ and $f(x) = x^{1/2}$ in which the variable is the base of an exponential expression, the variable in an exponential function is *in the exponent*.

Let's now consider graphs of exponential functions.

EXAMPLE 1 Graph the exponential function

$$y = f(x) = 2^x.$$

Solution We compute some function values and list the results in a table.

$$f(0) = 2^0 = 1; \qquad f(-1) = 2^{-1} = \frac{1}{2^1} = \frac{1}{2};$$
$$f(1) = 2^1 = 2;$$
$$f(2) = 2^2 = 4; \qquad f(-2) = 2^{-2} = \frac{1}{2^2} = \frac{1}{4};$$
$$f(3) = 2^3 = 8;$$
$$f(-3) = 2^{-3} = \frac{1}{2^3} = \frac{1}{8}.$$

x	$y = f(x) = 2^x$	(x, y)
0	1	$(0, 1)$
1	2	$(1, 2)$
2	4	$(2, 4)$
3	8	$(3, 8)$
-1	$\frac{1}{2}$	$\left(-1, \frac{1}{2}\right)$
-2	$\frac{1}{4}$	$\left(-2, \frac{1}{4}\right)$
-3	$\frac{1}{8}$	$\left(-3, \frac{1}{8}\right)$

Next, we plot these points and connect them with a smooth curve. Be sure to plot enough points to determine how steeply the curve rises.

The curve comes very close to the x-axis, but does not touch or cross it.

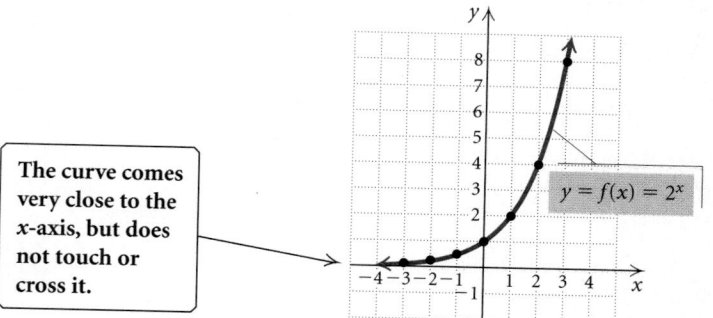

$y = f(x) = 2^x$

HORIZONTAL ASYMPTOTES

REVIEW SECTION **4.5.**

Note that as x increases, the function values increase without bound. As x decreases, the function values decrease, getting close to 0. That is, as $x \to -\infty$, $y \to 0$. Thus the x-axis, or the line $y = 0$, is a horizontal asymptote. As the x-inputs decrease, the curve gets closer and closer to this line, but does not cross it.

Now Try Exercise 11. ■

Exploring with Technology

Check the graph of $y = f(x) = 2^x$ using a graphing calculator. Then graph $y = f(x) = 5^x$. Use the TRACE and TABLE features to confirm that the graph never crosses the x-axis.

EXAMPLE 2   Graph the exponential function $y = f(x) = \left(\dfrac{1}{2}\right)^x$.

Solution   Before we plot points and draw the curve, note that

$$y = f(x) = \left(\frac{1}{2}\right)^x = (2^{-1})^x = 2^{-x}.$$

This tells us that this graph is a reflection of the graph of $y = 2^x$ across the y-axis. For example, if $(3, 8)$ is a point of the graph of $f(x) = 2^x$, then $(-3, 8)$ is a point of the graph of $f(x) = 2^{-x}$. Selected points are listed in the table at left.

Next, we plot points and connect them with a smooth curve.

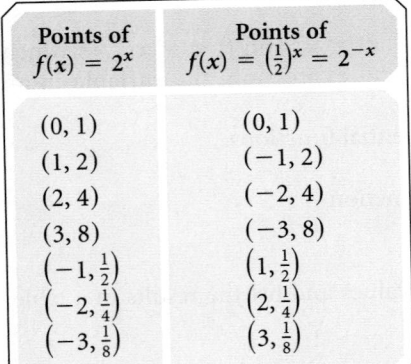

Points of $f(x) = 2^x$	Points of $f(x) = \left(\frac{1}{2}\right)^x = 2^{-x}$
$(0, 1)$	$(0, 1)$
$(1, 2)$	$(-1, 2)$
$(2, 4)$	$(-2, 4)$
$(3, 8)$	$(-3, 8)$
$\left(-1, \frac{1}{2}\right)$	$\left(1, \frac{1}{2}\right)$
$\left(-2, \frac{1}{4}\right)$	$\left(2, \frac{1}{4}\right)$
$\left(-3, \frac{1}{8}\right)$	$\left(3, \frac{1}{8}\right)$

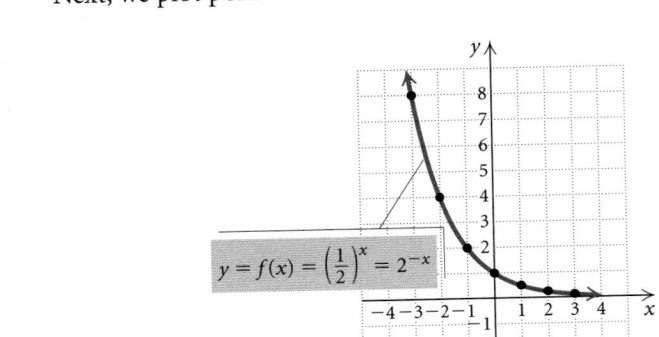

Note that as x increases, the function values decrease, getting close to 0. The x-axis, $y = 0$, is the horizontal asymptote. As x decreases, the function values increase without bound.

Observe the following graphs of exponential functions and look for patterns in them.

Now Try Exercise 15. ■

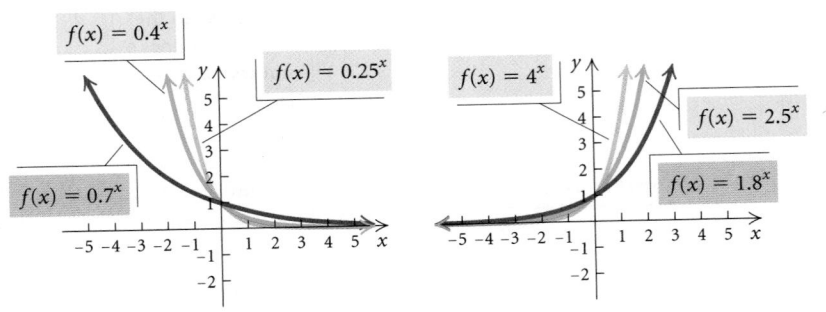

What relationship do you see between the base a and the shape of the resulting graph of $f(x) = a^x$? What do all the graphs have in common? How do they differ?

CONNECTING *the* CONCEPTS

Properties of Exponential Functions

Let's list and compare some characteristics of exponential functions, keeping in mind that the definition of an exponential function, $f(x) = a^x$, requires that a be positive and different from 1.

$f(x) = a^x, \ a > 0, a \neq 1$

Continuous

One-to-one

Domain: $(-\infty, \infty)$

Range: $(0, \infty)$

Increasing if $a > 1$

Decreasing if $0 < a < 1$

Horizontal asymptote is x-axis

y-intercept: $(0, 1)$

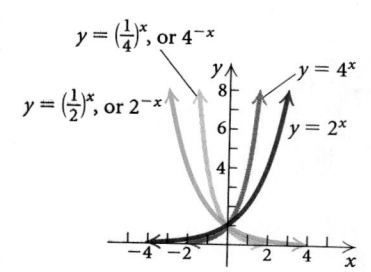

TRANSFORMATIONS OF FUNCTIONS

REVIEW SECTION **2.4.**

To graph other types of exponential functions, keep in mind the ideas of translation, stretching, and reflection. All these concepts allow us to visualize the graph before drawing it.

EXAMPLE 3 Graph each of the following by hand. Before doing so, describe how each graph can be obtained from the graph of $f(x) = 2^x$. Then check your graph with a graphing calculator.

a) $f(x) = 2^{x-2}$ **b)** $f(x) = 2^x - 4$ **c)** $f(x) = 5 - 0.5^x$

Solution

a) The graph of $f(x) = 2^{x-2}$ is the graph of $y = 2^x$ shifted *right* 2 units.

x	$f(x)$
-1	$\frac{1}{8}$
0	$\frac{1}{4}$
1	$\frac{1}{2}$
2	1
3	2
4	4
5	8

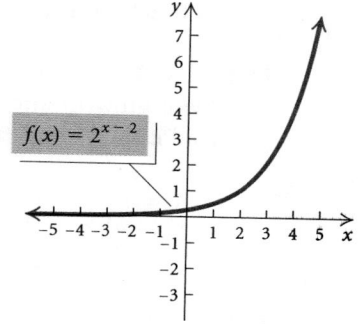

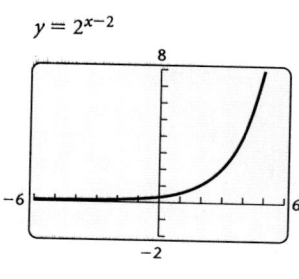

b) The graph of $f(x) = 2^x - 4$ is the graph of $y = 2^x$ shifted *down* 4 units.

x	$f(x)$
-2	$-3\frac{3}{4}$
-1	$-3\frac{1}{2}$
0	-3
1	-2
2	0
3	4

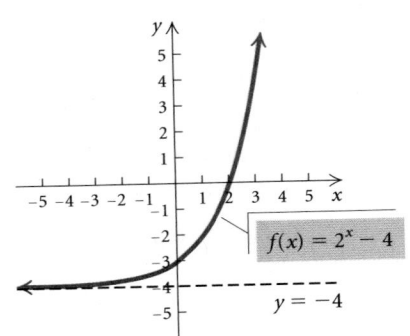

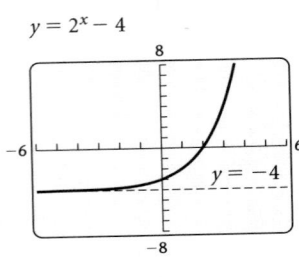

c) The graph of $f(x) = 5 - 0.5^x = 5 - \left(\frac{1}{2}\right)^x = 5 - 2^{-x}$ is a reflection of the graph of $y = 2^x$ across the y-axis, followed by a reflection across the x-axis and then a shift *up* 5 units.

x	$f(x)$
-3	-3
-2	1
-1	3
0	4
1	$4\frac{1}{2}$
2	$4\frac{3}{4}$

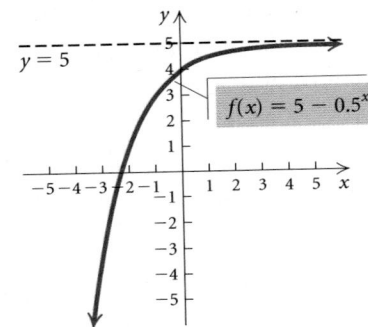

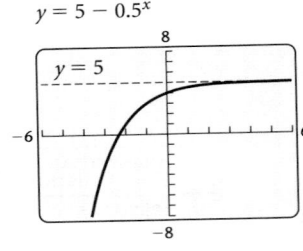

Now Try Exercises 27 and 31. ■

❋ Applications

Graphing calculators are especially helpful when we are working with exponential functions. They not only facilitate computations but they also allow us to visualize the functions. One of the most frequent applications of exponential functions occurs with compound interest.

EXAMPLE 4   *Compound Interest.*   The amount of money A that a principal P will grow to after t years at interest rate r (in decimal form), compounded n times per year, is given by the formula

$$A = P\left(1 + \frac{r}{n}\right)^{nt}.$$

Suppose that $100,000 is invested at 6.5% interest, compounded semiannually.

a) Find a function for the amount to which the investment grows after t years.

b) Graph the function.

c) Find the amount of money in the account at $t = 0, 4, 8,$ and 10 yr.

d) When will the amount of money in the account reach $400,000?

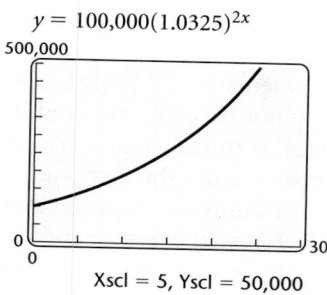

$$y = 100,000(1.0325)^{2x}$$

Y1(0)	
	100000
Y1(4)	
	129157.7535
Y1(8)	
	166817.253

Solution

a) Since $P = \$100,000$, $r = 6.5\% = 0.065$, and $n = 2$, we can substitute these values and write the following function:

$$A(t) = 100,000 \left(1 + \frac{0.065}{2}\right)^{2 \cdot t} = \$100,000(1.0325)^{2t}.$$

b) For the graph shown at left, we use the viewing window $[0, 30, 0, 500,000]$ because of the large numbers and the fact that negative time values have no meaning in this application.

c) We can compute function values using function notation on the home screen of a graphing calculator. (See the window at left.) We can also calculate the values directly on a graphing calculator by substituting in the expression for $A(t)$:

$$A(0) = 100,000(1.0325)^{2 \cdot 0} = \$100,000;$$
$$A(4) = 100,000(1.0325)^{2 \cdot 4} \approx \$129,157.75;$$
$$A(8) = 100,000(1.0325)^{2 \cdot 8} \approx \$166,817.25;$$
$$A(10) = 100,000(1.0325)^{2 \cdot 10} \approx \$189,583.79.$$

d) To find the amount of time it takes for the account to grow to $400,000, we set

$$100,000(1.0325)^{2t} = 400,000$$

and solve for t. One way we can do this is by graphing the equations

$$y_1 = 100,000(1.0325)^{2x} \quad \text{and} \quad y_2 = 400,000.$$

Then we can use the Intersect method to estimate the first coordinate of the point of intersection. (See Fig. 1 below.)

We can also use the Zero method to estimate the zero of the function $y = 100,000(1.0325)^{2x} - 400,000$. (See Fig. 2 below.)

Regardless of the method we use, we see that the account grows to $400,000 after about 21.67 yr, or about 21 yr, 8 mo, and 2 days.

INTERSECT METHOD
REVIEW SECTION **1.5.**

ZERO METHOD
REVIEW SECTION **1.5.**

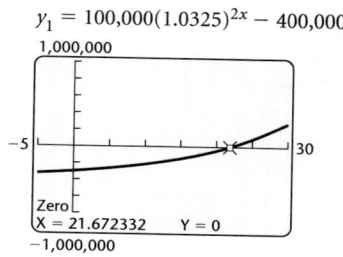

FIGURE 1

$$y_1 = 100,000(1.0325)^{2x} - 400,000$$

FIGURE 2

Now Try Exercise 43. ■

❊ The Number *e*

We now consider a very special number in mathematics. In 1741, Leonhard Euler named this number *e*. Though you may not have encountered it before, you will see here and in future mathematics courses that it has many important applications. To explain this number, we use the compound interest formula $A = P(1 + r/n)^{nt}$ discussed in Example 4. Suppose that \$1 is invested at 100% interest for 1 yr. Since $P = 1$, $r = 100\% = 1$, and $t = 1$, the formula above becomes a function A defined in terms of the number of compounding periods *n*:

$$A = P\left(1 + \frac{r}{n}\right)^{nt} = 1\left(1 + \frac{1}{n}\right)^{n \cdot 1} = \left(1 + \frac{1}{n}\right)^{n}.$$

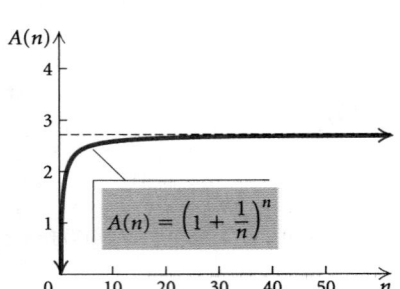

Let's visualize this function with its graph shown at left and explore the values of $A(n)$ as $n \to \infty$. Consider the graph for larger and larger values of *n*. Does this function have a horizontal asymptote?

Let's find some function values using a calculator.

n, Number of Compounding Periods	$A(n) = \left(1 + \dfrac{1}{n}\right)^{n}$
1 (compounded annually)	\$2.00
2 (compounded semiannually)	2.25
3	2.3704
4 (compounded quarterly)	2.4414
5	2.4883
100	2.7048
365 (compounded daily)	2.7146
8760 (compounded hourly)	2.7181

It appears from these values that the graph does have a horizontal asymptote, $y \approx 2.7$. As the values of *n* get larger and larger, the function values get closer and closer to the number Euler named *e*. Its decimal representation does not terminate or repeat; it is irrational.

$$e = 2.7182818284\ldots$$

GCM   **EXAMPLE 5**   Find each value of e^{x}, to four decimal places, using the $\boxed{e^x}$ key on a calculator.

a) e^{3} **b)** $e^{-0.23}$ **c)** e^{0} **d)** e^{1}

Solution

FUNCTION VALUE	READOUT	ROUNDED
a) e^3	e^(3) 20.08553692	20.0855
b) $e^{-0.23}$	e^(−.23) .7945336025	0.7945
c) e^0	e^(0) 1	1
d) e^1	e^(1) 2.718281828	2.7183

Now Try Exercises 1 and 3. ■

❈ Graphs of Exponential Functions, Base *e*

We demonstrate ways in which to graph exponential functions.

EXAMPLE 6 Graph $f(x) = e^x$ and $g(x) = e^{-x}$.

Solution We can compute points for each equation using the $\boxed{e^x}$ key on a calculator. Then we plot these points and draw the graphs of the functions.

$$y_1 = e^x, \quad y_2 = e^{-x}$$

X	Y1	Y2
−3	.04979	20.086
−2	.13534	7.3891
−1	.36788	2.7183
0	1	1
1	2.7183	.36788
2	7.3891	.13534
3	20.086	.04979

X =

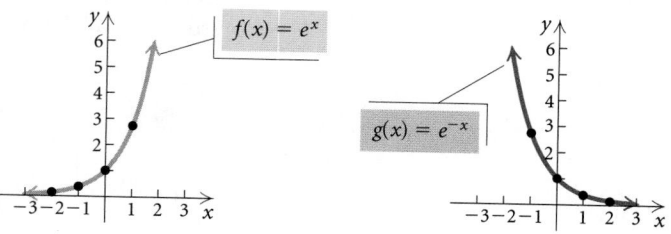

Note that the graph of *g* is a reflection of the graph of *f* across the *y*-axis.

Now Try Exercise 23. ■

EXAMPLE 7   Graph each of the following by hand. Before doing so, describe how each graph can be obtained from the graph of $y = e^x$.

a) $f(x) = e^{x+3}$ b) $f(x) = e^{-0.5x}$ c) $f(x) = 1 - e^{-2x}$

Solution

a) The graph of $f(x) = e^{x+3}$ is a translation of the graph of $y = e^x$ left 3 units.

x	$f(x)$
-7	0.018
-5	0.135
-3	1
-1	7.389
0	20.086

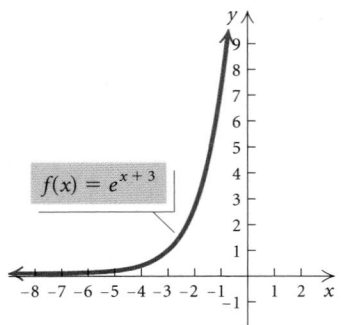

b) We note that the graph of $f(x) = e^{-0.5x}$ is a horizontal stretching of the graph of $y = e^x$ followed by a reflection across the y-axis.

x	$f(x)$
-2	2.718
-1	1.649
0	1
1	0.607
2	0.368

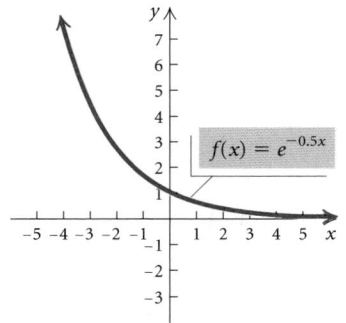

c) The graph of $f(x) = 1 - e^{-2x}$ is a horizontal shrinking of the graph of $y = e^x$, followed by a reflection across the y-axis, then across the x-axis, followed by a translation up 1 unit.

x	$f(x)$
-1	-6.389
0	0
1	0.865
2	0.982
3	0.998

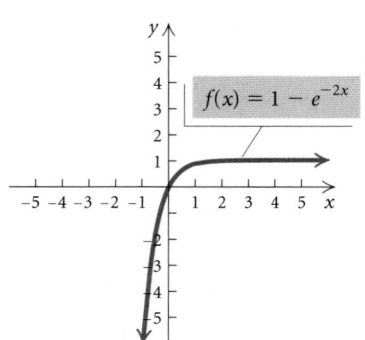

Now Try Exercises 37 and 41. ■

5.2 Exercise Set

Find each of the following, to four decimal places, using a calculator.

1. e^4

2. e^{10}

3. $e^{-2.458}$

4. $\left(\dfrac{1}{e^3}\right)^2$

In Exercises 5–10, match the function with one of the graphs (a)–(f), which follow.

5. $f(x) = -2^x - 1$

6. $f(x) = -\left(\dfrac{1}{2}\right)^x$

7. $f(x) = e^x + 3$

8. $f(x) = e^{x+1}$

9. $f(x) = 3^{-x} - 2$

10. $f(x) = 1 - e^x$

a)
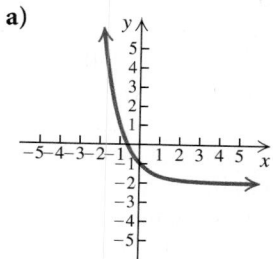

b)
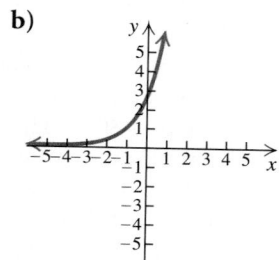

c)
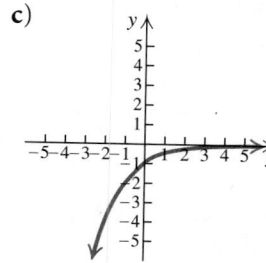

d)
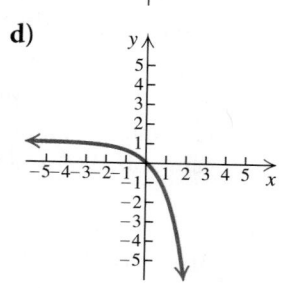

e)
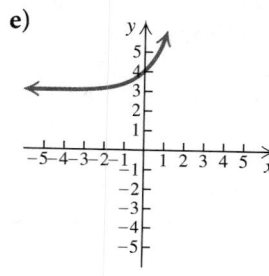

f)
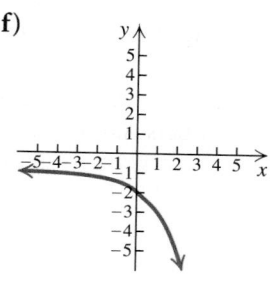

Graph the function by substituting and plotting points. Then check your work using a graphing calculator.

11. $f(x) = 3^x$

12. $f(x) = 5^x$

13. $f(x) = 6^x$

14. $f(x) = 3^{-x}$

15. $f(x) = \left(\dfrac{1}{4}\right)^x$

16. $f(x) = \left(\dfrac{2}{3}\right)^x$

17. $y = -2^x$

18. $y = 3 - 3^x$

19. $f(x) = -0.25^x + 4$

20. $f(x) = 0.6^x - 3$

21. $f(x) = 1 + e^{-x}$

22. $f(x) = 2 - e^{-x}$

23. $y = \dfrac{1}{4}e^x$

24. $y = 2e^{-x}$

25. $f(x) = 1 - e^{-x}$

26. $f(x) = e^x - 2$

Sketch the graph of the function and check the graph with a graphing calculator. Describe how each graph can be obtained from the graph of a basic exponential function.

27. $f(x) = 2^{x+1}$

28. $f(x) = 2^{x-1}$

29. $f(x) = 2^x - 3$

30. $f(x) = 2^x + 1$

31. $f(x) = 4 - 3^{-x}$

32. $f(x) = 2^{x-1} - 3$

33. $f(x) = \left(\dfrac{3}{2}\right)^{x-1}$

34. $f(x) = 3^{4-x}$

35. $f(x) = 2^{x+3} - 5$

36. $f(x) = -3^{x-2}$

37. $f(x) = e^{2x}$

38. $f(x) = e^{-0.2x}$

39. $y = e^{-x+1}$

40. $y = e^{2x} + 1$

41. $f(x) = 2(1 - e^{-x})$

42. $f(x) = 1 - e^{-0.01x}$

43. *Compound Interest.* Suppose that $82{,}000$ is invested at $4\frac{1}{2}\%$ interest, compounded quarterly.

a) Find the function for the amount to which the investment grows after t years.

b) Graph the function.

c) Find the amount of money in the account at $t = 0, 2, 5$, and 10 yr.

d) When will the amount of money in the account reach $100{,}000$?

44. *Compound Interest.* Suppose that $750 is invested at 7% interest, compounded semiannually.

a) Find the function for the amount to which the investment grows after t years.

b) Graph the function.

c) Find the amount of money in the account at $t = 1, 6, 10, 15$, and 25 yr.

d) When will the amount of money in the account reach 3000?

45. *Interest on a CD.* On Jacob's sixth birthday, his grandparents present him with a $3000 certificate of deposit (CD) that earns 5% interest, compounded quarterly. If the CD matures on his sixteenth birthday, what amount will be available then?

46. *Interest in a College Trust Fund.* Following the birth of his child, Juan deposits $10,000 in a college trust fund where interest is 6.4%, compounded semiannually.

    a) Find a function for the amount in the account after t years.

    b) Find the amount of money in the account at $t = 0, 4, 8, 10,$ and 18 yr.

In Exercises 47–54, use the compound-interest formula to find the account balance A with the given conditions:

P = principal,
r = interest rate,
n = number of compounding periods per year,
t = time, in years,
A = account balance.

	P	r	Compounded	n	t	A
47.	$3,000	4%	Semiannually		2	
48.	$12,500	3%	Quarterly		3	
49.	$120,000	2.5%	Annually		10	
50.	$120,000	2.5%	Quarterly		10	
51.	$53,500	$5\frac{1}{2}$%	Quarterly		$6\frac{1}{2}$	
52.	$6,250	$6\frac{3}{4}$%	Semiannually		$4\frac{1}{2}$	
53.	$17,400	8.1%	Daily		5	
54.	$900	7.3%	Daily		$7\frac{1}{4}$	

55. *Coffee Sales.* Retail sales of coffee and the number of coffeehouses and kiosks have steadily increased in recent years. Total retail sales of coffee have increased exponentially from $8.3 billion in 2001 to $12.3 billion in 2006 (*Source*: Specialty Coffee Association of America).

    The following function models sales:

$$C(t) = 7.8867(1.0854)^t,$$

where t is the number of years since 2001 and $C(t)$ is in billions. Estimate the total retail sales of coffee in 2004 and in 2009.

56. *Garlic Supply* The amount of garlic in the food supply has significantly increased in the past 80 yr. The average U.S. per capita consumption of garlic has increased from about 0.15 lb in the 1920s to 2.5 lb in the mid-2000s (*Source*: U.S. Department of Agriculture).

    The following exponential function models this growth in the consumption of garlic:

$$G(x) = 0.0882(1.0414)^x,$$

where x is the number of years since 1925. Estimate the amount of garlic consumed per capita in 1960 and in 1990.

57. *Corn-Based Ethanol.* The United States produced 3.9 billion gal of fuel ethanol in 2005. The production of corn-based ethanol has increased exponentially from only 1.1 billion gal in 1996. (*Sources*: Energy Information Administration; Renewable Fuels Association)

The following function can be used to model this growth:

$$C(t) = 1.0283(1.1483)^t,$$

where t is the number of years since 1996 and $C(t)$ is in billions.

a) Estimate the number of gallons that will be produced in 2008 and in 2010.
b) Graph the function.
c) When will the production of ethanol reach 5.0 billion gal?

58. *Growth of Bacteria* Escherichia coli. The bacteria *Escherichia coli* are commonly found in the human intestines. Suppose that 3000 of the bacteria are present at time $t = 0$. Then under certain conditions, t minutes later, the number of bacteria present is

$$N(t) = 3000(2)^{t/20}.$$

a) How many bacteria will be present after 10 min? 20 min? 30 min? 40 min? 60 min?
b) Graph the function.
c) These bacteria can cause intestinal infections in humans when the number of bacteria reaches 100,000,000. Find the length of time it takes for an intestinal infection to be possible.

59. *Storage of Data.* The amount of data storage in gigabytes (GB) needed by a typical family in the United States has increased dramatically since 2004 (*Source*: Coughlin Associates). The function

$$G(x) = 433.6(1.5)^x,$$

where x is the number of years since 2004, models the amount of storage for recent years. Estimate the number of gigabytes needed by a typical family in 2009 and in 2014.

60. *Bachelor's Degrees Earned by Women.* The function

$$D(t) = 73,630.7487(1.0415)^t$$

gives the number of bachelor's degrees conferred on women in the United States t years after 1940 (*Sources*: National Center for Educational Statistics; U.S. Department of Education). Find the number of bachelor's degrees earned by women in 1960, in 1985, and in 2000. Then estimate the number of bachelor's degrees earned by women in 2010.

61. *World Trade with China.* China has recently taken some major steps in expanding its trade with the world. Both exports and imports have increased exponentially since 1996 (*Source*: U.S.–China Business Council).

The following functions model these trends:

Exports from China: $E(x) = 139.76(1.194)^x,$
Imports to China: $I(x) = 130.67(1.187)^x,$

where t is the number of years since 1996 and $E(x)$ and $I(x)$ are in billions of dollars. Find the exports and the imports, in billions of dollars, for 2007, 2012, and 2020.

62. *Children as a Percentage of the Population.*
Children under 18 accounted for 26% of the U.S.
population in 1999. This percentage was down
from 36% in 1964 and is expected to continue to
decrease. (*Source*: www.childstats.gov) The
following function can be used to project the
percentage of children in the United States:

$$K(t) = 35.37(0.99)^t,$$

where t is the number of years since 1964.
Project the percentage of the U.S. population
under age 18 in 2010 and in 2015.

63. *Salvage Value.* A top-quality phone–fax
copying machine is purchased for $595. Its value
each year is about 80% of the value of the
preceding year. After t years, its value, in dollars,
is given by the exponential function

$$V(t) = 595(0.8)^t.$$

a) Graph the function.
b) Find the value of the machine after 0 yr, 1 yr,
2 yr, 5 yr, and 10 yr.
c) The company decides to replace the
machine when its value has declined to
$200. After how long will the machine be
replaced?

64. *Fiancé(e) Visas.* The number of foreign nationals
who came to the United States to marry an
American using a "fiancé(e) visa" has grown
exponentially in recent years. The total number
of fiancé(e) visas issued is given by the function

$$f(x) = 5728.98(1.1214)^x,$$

where x is the number of years since 1990 (*Source*:
U.S. Department of Homeland Security). Find
the total number of fiancé(e) visas issued in
2006 and in 2009.

65. *Online Sales.* As consumers become more
comfortable shopping using the Internet,
online retail sales will continue to increase.
Online sales increased 20% from 2005 to 2006
(*Source*: National Retail Federation). Online
retail sales, in billions of dollars, are given by
the function

$$S(t) = 29.0626(1.3438)^t,$$

where t is the number of years since 1999.

a) Graph the function.
b) Find the amount of online sales in 2000 and
in 2005. Then use the function to estimate
the total online sales in 2009.
c) After how many years will the amount of
online sales be $350.2 billion?

66. *Typing Speed.* Sarah is taking keyboarding at
a community college. After she practices for
t hours, her speed, in words per minute, is given
by the function

$$S(t) = 200[1 - (0.86)^t].$$

a) Graph the function.
b) What is Sarah's speed after practicing for
10 hr? 20 hr? 40 hr? 100 hr?
c) How much time passes before Sarah's speed
is 100 words per minute?
d) Does the graph have an asymptote? If so,
what is it, and what is its significance to
Sarah's learning?

67. *Advertising.* A company begins a radio
advertising campaign in New York City to
market a new video game. The percentage of
the target market that buys a game is generally
a function of the length of the advertising
campaign. The estimated percentage is given by

$$f(t) = 100(1 - e^{-0.04t}),$$

where t is the number of days of the campaign.

a) Graph the function.
b) Find $f(25)$, the percentage of the target
market that has bought the product after a
25-day advertising campaign.
c) After how long will 90% of the target market
have bought the product?

68. *Growth of a Stock.* The value of a stock is given by the function

$$V(t) = 58(1 - e^{-1.1t}) + 20,$$

where V is the value of the stock after time t, in months.

a) Graph the function.
b) Find $V(1)$, $V(2)$, $V(4)$, $V(6)$, and $V(12)$.
c) After how long will the value of the stock be $75?

In Exercises 69–82, use a graphing calculator to match the equation with one of the figures (a)–(n), which follow.

a)
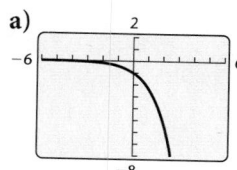

b)
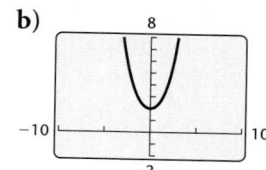

c)
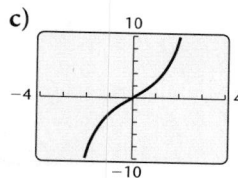

d)
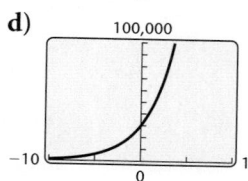

e)
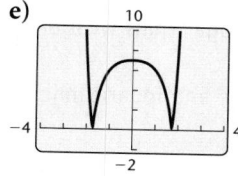

f)
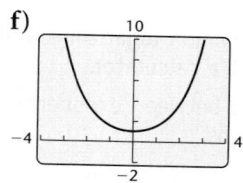

g)
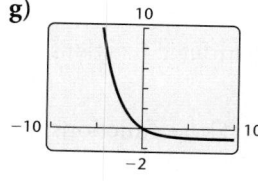

h)
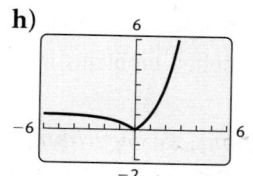

i)
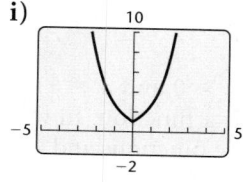

j)
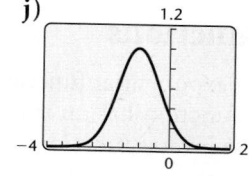

k)
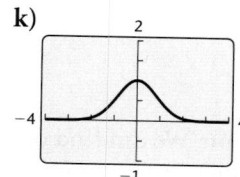

l)
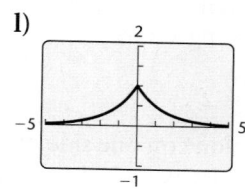

m)
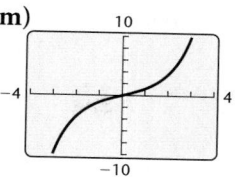

n)
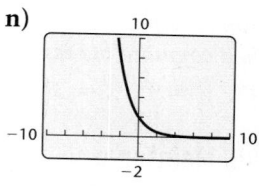

69. $y = 3^x - 3^{-x}$

70. $y = 3^{-(x+1)^2}$

71. $f(x) = -2.3^x$

72. $f(x) = 30,000(1.4)^x$

73. $y = 2^{-|x|}$

74. $y = 2^{-(x-1)}$

75. $f(x) = (0.58)^x - 1$

76. $y = 2^x + 2^{-x}$

77. $g(x) = e^{|x|}$

78. $f(x) = |2^x - 1|$

79. $y = 2^{-x^2}$

80. $y = |2^{x^2} - 8|$

81. $g(x) = \dfrac{e^x - e^{-x}}{2}$

82. $f(x) = \dfrac{e^x + e^{-x}}{2}$

Use a graphing calculator to find the point(s) of intersection of the graphs of each of the following pairs of equations.

83. $y = |1 - 3^x|$,
$y = 4 + 3^{-x^2}$

84. $y = 4^x + 4^{-x}$,
$y = 8 - 2x - x^2$

85. $y = 2e^x - 3$, $y = \dfrac{e^x}{x}$

86. $y = \dfrac{1}{e^x + 1}$, $y = 0.3x + \dfrac{7}{9}$

Solve graphically.

87. $5.3^x - 4.2^x = 1073$

88. $e^x = x^3$

89. $2^x > 1$

90. $3^x \leq 1$

91. $2^x + 3^x = x^2 + x^3$

92. $31,245e^{-3x} = 523,467$

Collaborative Discussion and Writing

93. Describe the differences between the graphs of $f(x) = x^3$ and $g(x) = 3^x$.

94. Suppose that $10,000 is invested for 8 yr at 6.4% interest, compounded annually. In what year will the most interest be earned? Why?

Graph the pair of equations using the same set of axes. Then compare the results.

95. $y = 3^x$, $x = 3^y$ **96.** $y = 1^x$, $x = 1^y$

Skill Maintenance

Simplify.

97. $(1 - 4i)(7 + 6i)$ **98.** $\dfrac{2 - i}{3 + i}$

Find the x-intercepts and the zeros of the function.

99. $f(x) = 2x^2 - 13x - 7$

100. $h(x) = x^3 - 3x^2 + 3x - 1$

101. $h(x) = x^4 - x^2$

102. $g(x) = x^3 + x^2 - 12x$

Solve.

103. $x^3 + 6x^2 - 16x = 0$ **104.** $3x^2 - 6 = 5x$

Synthesis

105. Which is larger, 7^{π} or π^7? 70^{80} or 80^{70}?

In Exercises 106 and 107:

a) *Graph using a graphing calculator.*
b) *Approximate the zeros.*
c) *Approximate the relative maximum and minimum values. If your graphing calculator has a* MAX–MIN *feature, use it.*

106. $f(x) = x^2 e^{-x}$ **107.** $f(x) = e^{-x^2}$

108. Graph $f(x) = x^{1/(x-1)}$ for $x > 0$. Use a graphing calculator and the TABLE feature to identify the horizontal asymptote.

Logarithmic Functions and Graphs

❖ Find common logarithms and natural logarithms with and without a calculator.

❖ Convert between exponential equations and logarithmic equations.

❖ Change logarithm bases.

❖ Graph logarithmic functions.

❖ Solve applied problems involving logarithmic functions.

We now consider *logarithmic*, or *logarithm*, *functions*. These functions are inverses of exponential functions and have many applications.

❖ Logarithmic Functions

We have noted that every exponential function (with $a > 0$ and $a \neq 1$) is one-to-one. Thus such a function has an inverse that is a function. In this section, we will name these inverse functions logarithmic functions and use them in applications. We can draw the graph of the inverse of an exponential function by interchanging x and y.

EXAMPLE 1 Graph: $x = 2^y$.

Solution Note that x is alone on one side of the equation. We can find ordered pairs that are solutions by choosing values for y and then computing the corresponding x-values.

For $y = 0, x = 2^0 = 1$.

For $y = 1, x = 2^1 = 2$.

For $y = 2, x = 2^2 = 4$.

For $y = 3, x = 2^3 = 8$.

For $y = -1, x = 2^{-1} = \dfrac{1}{2^1} = \dfrac{1}{2}$.

For $y = -2, x = 2^{-2} = \dfrac{1}{2^2} = \dfrac{1}{4}$.

For $y = -3, x = 2^{-3} = \dfrac{1}{2^3} = \dfrac{1}{8}$.

x		
$x = 2^y$	y	(x, y)
1	0	$(1, 0)$
2	1	$(2, 1)$
4	2	$(4, 2)$
8	3	$(8, 3)$
$\dfrac{1}{2}$	-1	$\left(\dfrac{1}{2}, -1\right)$
$\dfrac{1}{4}$	-2	$\left(\dfrac{1}{4}, -2\right)$
$\dfrac{1}{8}$	-3	$\left(\dfrac{1}{8}, -3\right)$

(1) Choose values for y.
(2) Compute values for x.

We plot the points and connect them with a smooth curve. Note that the curve does not touch or cross the y-axis. The y-axis is a vertical asymptote.

Note too that this curve looks just like the graph of $y = 2^x$, except that it is reflected across the line $y = x$, as we would expect for an inverse. The inverse of $y = 2^x$ is $x = 2^y$.

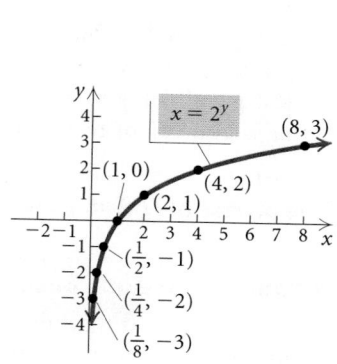

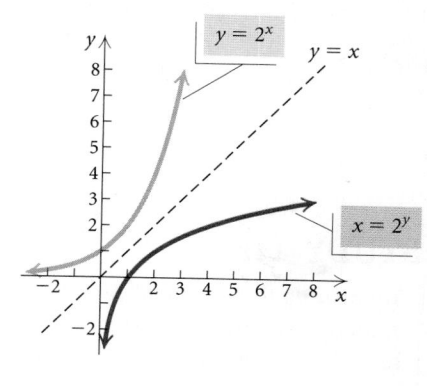

Now Try Exercise 1. ■

To find a formula for f^{-1} when $f(x) = 2^x$, we try to use the method of Section 5.1:

1. Replace $f(x)$ with y: $y = 2^x$
2. Interchange x and y: $x = 2^y$
3. Solve for y: $y =$ the power to which we raise 2 to get x
4. Replace y with $f^{-1}(x)$: $f^{-1}(x) =$ the power to which we raise 2 to get x.

Mathematicians have defined a new symbol to replace the words "the power to which we raise 2 to get x." That symbol is "$\log_2 x$," read "the logarithm, base 2, of x."

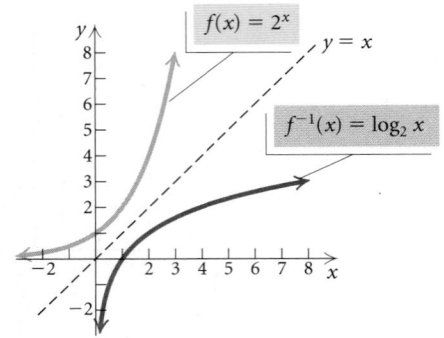

$f(x) = 2^x$

$y = x$

$f^{-1}(x) = \log_2 x$

Logarithmic Function, Base 2

"$\log_2 x$," read "the logarithm, base 2, of x," means "the power to which we raise 2 to get x."

Thus if $f(x) = 2^x$, then $f^{-1}(x) = \log_2 x$. For example,

$$f^{-1}(8) = \log_2 8 = 3,$$

because

3 is the power to which we raise 2 to get 8.

Similarly, $\log_2 13$ is the power to which we raise 2 to get 13. As yet, we have no simpler way to say this other than

"$\log_2 13$ is the power to which we raise 2 to get 13."

Later, however, we will learn how to approximate this expression using a calculator.

For any exponential function $f(x) = a^x$, its inverse is called a **logarithmic function, base a**. The graph of the inverse can be obtained by reflecting the graph of $y = a^x$ across the line $y = x$, to obtain $x = a^y$. Then $x = a^y$ is equivalent to $y = \log_a x$. We read $\log_a x$ as "the logarithm, base a, of x."

The inverse of $f(x) = a^x$ is given by $f^{-1}(x) = \log_a x$.

Logarithmic Function, Base a

We define $y = \log_a x$ as that number y such that $x = a^y$, where $x > 0$ and a is a positive constant other than 1.

STUDY TIP

Try being a tutor for a fellow student. Understanding and retention of concepts can be increased when you explain the material to someone else.

Let's look at the graphs of $f(x) = a^x$ and $f^{-1}(x) = \log_a x$ for $a > 1$ and $0 < a < 1$.

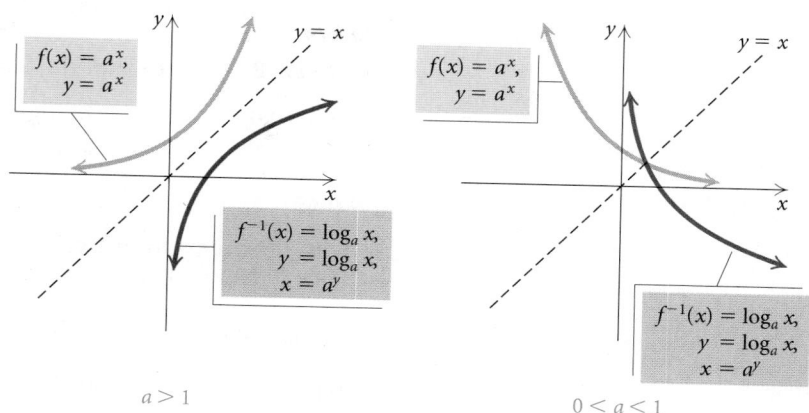

Note that the graphs of $f(x)$ and $f^{-1}(x)$ are reflections of each other across the line $y = x$.

CONNECTING the CONCEPTS

Comparing Exponential Functions and Logarithmic Functions

In the following table, we compare exponential functions and logarithmic functions with bases a greater than 1. Similar statements could be made for a, where $0 < a < 1$. It is helpful to visualize the differences by carefully observing the graphs.

EXPONENTIAL FUNCTION

$y = a^x$
$f(x) = a^x$
$a > 1$
Continuous
One-to-one
Domain: All real numbers, $(-\infty, \infty)$
Range: All positive real numbers, $(0, \infty)$
Increasing
Horizontal asymptote is x-axis: ($a^x \to 0$ as $x \to -\infty$)
y-intercept: $(0, 1)$
There is no x-intercept.

LOGARITHMIC FUNCTION

$x = a^y$
$f^{-1}(x) = \log_a x$
$a > 1$
Continuous
One-to-one
Domain: All positive real numbers, $(0, \infty)$
Range: All real numbers, $(-\infty, \infty)$
Increasing
Vertical asymptote is y-axis: ($\log_a x \to -\infty$ as $x \to 0^+$)
x-intercept: $(1, 0)$
There is no y-intercept.

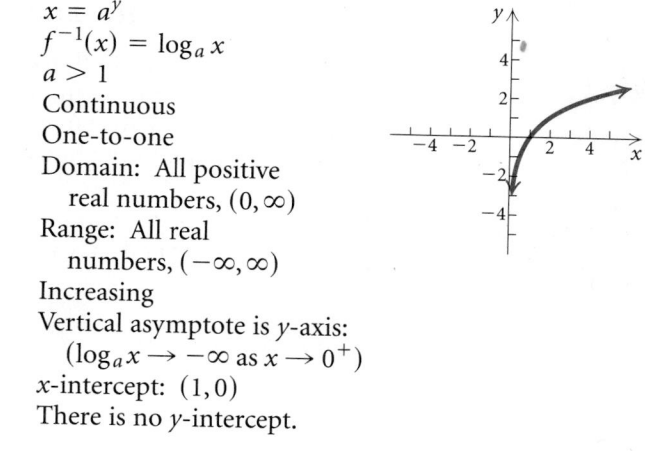

❊ Finding Certain Logarithms

Let's use the definition of logarithms to find some logarithmic values.

EXAMPLE 2 Find each of the following logarithms.

a) $\log_{10} 10{,}000$ **b)** $\log_{10} 0.01$ **c)** $\log_2 8$

d) $\log_9 3$ **e)** $\log_6 1$ **f)** $\log_8 8$

Solution

a) The exponent to which we raise 10 to obtain 10,000 is 4; thus $\log_{10} 10{,}000 = 4$.

b) We have $0.01 = \dfrac{1}{100} = \dfrac{1}{10^2} = 10^{-2}$. The exponent to which we raise 10 to get 0.01 is -2, so $\log_{10} 0.01 = -2$.

c) $8 = 2^3$. The exponent to which we raise 2 to get 8 is 3, so $\log_2 8 = 3$.

d) $3 = \sqrt{9} = 9^{1/2}$. The exponent to which we raise 9 to get 3 is $\frac{1}{2}$, so $\log_9 3 = \frac{1}{2}$.

e) $1 = 6^0$. The exponent to which we raise 6 to get 1 is 0, so $\log_6 1 = 0$.

f) $8 = 8^1$. The exponent to which we raise 8 to get 8 is 1, so $\log_8 8 = 1$.

Now Try Exercise 9. ▪

Examples 2(e) and 2(f) illustrate two important properties of logarithms. The property $\log_a 1 = 0$ follows from the fact that $a^0 = 1$. Thus, $\log_5 1 = 0, \log_{10} 1 = 0$, and so on. The property $\log_a a = 1$ follows from the fact that $a^1 = a$. Thus, $\log_5 5 = 1, \log_{10} 10 = 1$, and so on.

> $\log_a 1 = 0$ and $\log_a a = 1$,   for any logarithmic base a.

❊ Converting Between Exponential Equations and Logarithmic Equations

It is helpful in dealing with logarithmic functions to remember that a logarithm of a number is an *exponent*. It is the exponent y in $x = a^y$. You might think to yourself, "the logarithm, base a, of a number x is the power to which a must be raised to get x."

We are led to the following. (The symbol $\longleftrightarrow$ means that the two statements are equivalent; that is, when one is true, the other is true. The words "if and only if" can be used in place of $\longleftrightarrow$.)

> $\log_a x = y \longleftrightarrow x = a^y$ A logarithm is an exponent!

EXAMPLE 3   Convert each of the following to a logarithmic equation.

a) $16 = 2^x$ **b)** $10^{-3} = 0.001$ **c)** $e^t = 70$

Solution

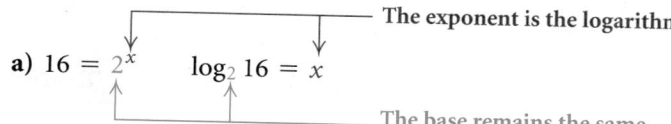

The exponent is the logarithm.

a) $16 = 2^x$ $\log_2 16 = x$

The base remains the same.

b) $10^{-3} = 0.001 \rightarrow \log_{10} 0.001 = -3$
c) $e^t = 70 \rightarrow \log_e 70 = t$

Now Try Exercise 37. ■

EXAMPLE 4   Convert each of the following to an exponential equation.

a) $\log_2 32 = 5$ **b)** $\log_a Q = 8$ **c)** $x = \log_t M$

Solution

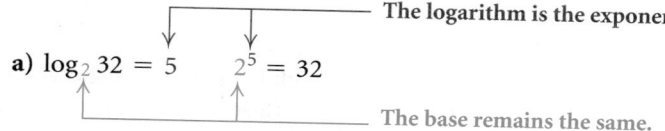

The logarithm is the exponent.

a) $\log_2 32 = 5$ $2^5 = 32$

The base remains the same.

b) $\log_a Q = 8 \rightarrow a^8 = Q$
c) $x = \log_t M \rightarrow t^x = M$

Now Try Exercise 45. ■

❋ Finding Logarithms on a Calculator

Before calculators became so widely available, base-10 logarithms, or **common logarithms**, were used extensively to simplify complicated calculations. In fact, that is why logarithms were invented. The abbreviation **log**, with no base written, is used to represent common logarithms, or base-10 logarithms. Thus,

$\log x$ means $\log_{10} x$.

For example, $\log 29$ means $\log_{10} 29$. Let's compare $\log 29$ with $\log 10$ and $\log 100$:

$$\left.\begin{array}{l} \log 10 = \log_{10} 10 = 1 \\ \log 29 = ? \\ \log 100 = \log_{10} 100 = 2 \end{array}\right\}$$

Since 29 is between 10 and 100, it seems reasonable that $\log 29$ is between 1 and 2.

On a calculator, the key for common logarithms is generally marked **LOG**. Using that key, we find that

$$\log 29 \approx 1.462397998 \approx 1.4624$$

rounded to four decimal places. Since $1 < 1.4624 < 2$, our answer seems reasonable. This also tells us that $10^{1.4624} \approx 29$.

EXAMPLE 5 Find each of the following common logarithms on a calculator. If you are using a graphing calculator, set the calculator in REAL mode. Round to four decimal places.

a) log 645,778   b) log 0.0000239   c) log (−3)

Solution

FUNCTION VALUE	READOUT	ROUNDED
a) log 645,778	log(645778) 5.810083246	5.8101
b) log 0.0000239	log(0.0000239) −4.621602099	−4.6216
c) log (−3)	ERR:NONREAL ANS *	Does not exist

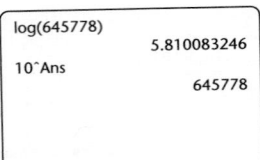

A check for part (a) is shown at left. Since 5.810083246 is the power to which we raise 10 to get 645,778, we can check part (a) by finding $10^{5.810083246}$. We can check part (b) in a similar manner. In part (c), log (−3) does not exist as a real number because there is no real-number power to which we can raise 10 to get −3. The number 10 raised to any real-number power is positive. The common logarithm of a negative number does not exist as a real number. Recall that logarithmic functions are inverses of exponential functions, and since the range of an exponential function is $(0, \infty)$, the domain of $f(x) = \log_a x$ is $(0, \infty)$.   **Now Try Exercises 57 and 61.** ■

❖ Natural Logarithms

Logarithms, base e, are called **natural logarithms**. The abbreviation "ln" is generally used for natural logarithms. Thus,

   ln x means $\log_e x$.

For example, ln 53 means $\log_e 53$. On a calculator, the key for natural logarithms is generally marked **LN**. Using that key, we find that

   ln 53 ≈ 3.970291914

   ≈ 3.9703

rounded to four decimal places. This also tells us that $e^{3.9703} \approx 53$.

GCM **EXAMPLE 6** Find each of the following natural logarithms on a calculator. If you are using a graphing calculator, set the calculator in REAL mode. Round to four decimal places.

a) ln 645,778   b) ln 0.0000239   c) ln (−5)

d) ln e e) ln 1

*If the graphing calculator is set in $a + bi$ mode, the readout is .4771212547 + 1.364376354i.

Solution

FUNCTION VALUE	READOUT	ROUNDED
a) ln 645,778	ln(645778) 13.37821107	13.3782
b) ln 0.0000239	ln(0.0000239) −10.6416321	−10.6416
c) ln (−5)	ERR:NONREAL ANS ✱	Does not exist
d) ln *e*	ln(e) 1	1
e) ln 1	ln(1) 0	0

ln(0.0000239)
            −10.6416321
e^Ans
            2.39E−5

Since 13.37821107 is the power to which we raise *e* to get 645,778, we can check part (a) by finding $e^{13.37821107}$. We can check parts (b), (d), and (e) in a similar manner. A check for part (b) is shown at left. In parts (d) and (e), note that ln *e* = $\log_e e$ = 1 and ln 1 = $\log_e 1$ = 0.

Now Try Exercises 65 and 67. ∎

> ln 1 = 0 and ln *e* = 1, for the logarithmic base *e*.

❈ Changing Logarithmic Bases

Most calculators give the values of both common logarithms and natural logarithms. To find a logarithm with a base other than 10 or *e*, we can use the following conversion formula.

The Change-of-Base Formula

For any logarithmic bases *a* and *b*, and any positive number *M*,

$$\log_b M = \frac{\log_a M}{\log_a b}.$$

We will prove this result in the next section.

---

*If the graphing calculator is set in *a* + *bi* mode, the readout is 1.609437912 + 3.141592654*i*.

GCM **EXAMPLE 7** Find $\log_5 8$ using common logarithms.

Solution First, we let $a = 10$, $b = 5$, and $M = 8$. Then we substitute into the change-of-base formula:

$$\log_5 8 = \frac{\log_{10} 8}{\log_{10} 5} \qquad \text{Substituting}$$

$$\approx 1.2920. \qquad \text{Using a calculator}$$

Since $\log_5 8$ is the power to which we raise 5 to get 8, we would expect this power to be greater than 1 $(5^1 = 5)$ and less than 2 $(5^2 = 25)$, so the result is reasonable. The check is shown in the window below.

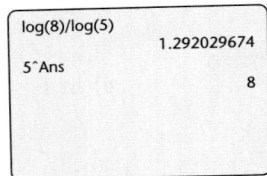

log(8)/log(5)
                                    1.292029674
5^Ans
                                              8

Now Try Exercise 69. ◼

We can also use base e for a conversion.

EXAMPLE 8 Find $\log_5 8$ using natural logarithms.

Solution Substituting e for a, 5 for b, and 8 for M, we have

$$\log_5 8 = \frac{\log_e 8}{\log_e 5}$$

$$= \frac{\ln 8}{\ln 5} \approx 1.2920.$$

The check is shown below. Note that we get the same value using base e for the conversion that we did using base 10 in Example 7.

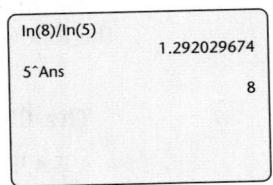

ln(8)/ln(5)
                                    1.292029674
5^Ans
                                              8

Now Try Exercise 75. ◼

❄ Graphs of Logarithmic Functions

We demonstrate several ways to graph logarithmic functions.

GCM **EXAMPLE 9** Graph: $y = f(x) = \log_5 x$.

Solution

Method 1. The equation $y = \log_5 x$ is equivalent to $x = 5^y$. We can find ordered pairs that are solutions by choosing values for y and computing the corresponding x-values. We then plot points, remembering that x is still the first coordinate.

For $y = 0, x = 5^0 = 1$.

For $y = 1, x = 5^1 = 5$.

For $y = 2, x = 5^2 = 25$.

For $y = 3, x = 5^3 = 125$.

For $y = -1, x = 5^{-1} = \dfrac{1}{5}$.

For $y = -2, x = 5^{-2} = \dfrac{1}{25}$.

x, or 5^y	y
1	0
5	1
25	2
125	3
$\dfrac{1}{5}$	-1
$\dfrac{1}{25}$	-2

(1) Select y.
(2) Compute x.

Method 2. To use a graphing calculator, we must first change the base. Here we change from base 5 to base e:

$$y = \log_5 x = \frac{\ln x}{\ln 5}. \qquad \text{Using } \log_b M = \frac{\log_a M}{\log_a b}$$

The graph is as shown in the window on the left below.

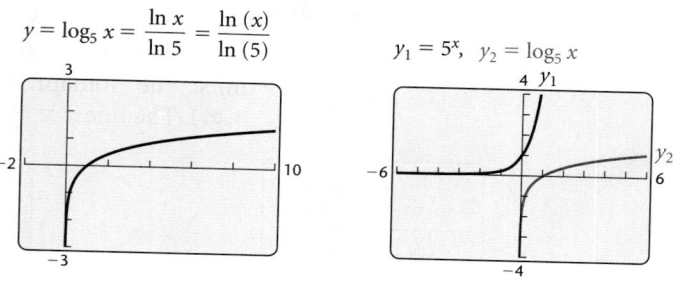

$$y = \log_5 x = \frac{\ln x}{\ln 5} = \frac{\ln (x)}{\ln (5)}$$

$$y_1 = 5^x, \quad y_2 = \log_5 x$$

Method 3. Some graphing calculators have a feature that graphs inverses automatically. If we begin with $y_1 = 5^x$, the graphs of both y_1 and its inverse $y_2 = \log_5 x$ will be drawn. (See the window on the right above.)

Now Try Exercises 5 and 79. ■

EXAMPLE 10   Graph: $g(x) = \ln x$.

Solution   To graph $y = g(x) = \ln x$, we select values for x and use the **LN** key on a calculator to find the corresponding values of $\ln x$. We then plot points and draw the curve. We can graph the function $y = \ln x$ directly using a graphing calculator.

x	$g(x)$ $g(x) = \ln x$
0.5	-0.7
1	0
2	0.7
3	1.1
4	1.4
5	1.6

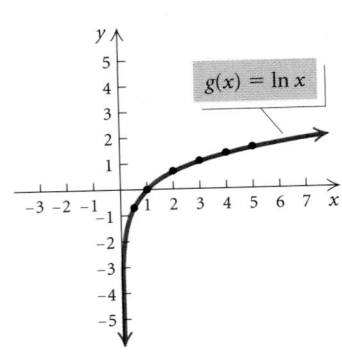

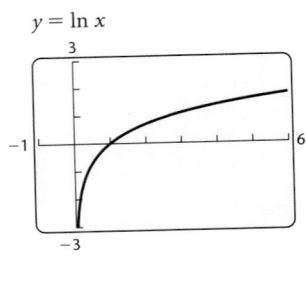

We could also write $g(x) = \ln x$, or $y = \ln x$, as $x = e^y$, select values for y, and use a calculator to find the corresponding values of x.

Now Try Exercise 7. ▪

Recall that the graph of $f(x) = \log_a x$, for any base a, has the x-intercept $(1, 0)$. The domain is the set of positive real numbers, and the range is the set of all real numbers. The y-axis is the vertical asymptote.

EXAMPLE 11   Graph each of the following by hand. Before doing so, describe how each graph can be obtained from the graph of $y = \ln x$. Give the domain and the vertical asymptote of each function. Then check your graph with a graphing calculator.

a) $f(x) = \ln (x + 3)$
b) $f(x) = 3 - \frac{1}{2} \ln x$
c) $f(x) = |\ln (x - 1)|$

Solution

a) The graph of $f(x) = \ln (x + 3)$ is a shift of the graph of $y = \ln x$ left 3 units. The domain is the set of all real numbers greater than -3, $(-3, \infty)$. The line $x = -3$ is the vertical asymptote.

x	$f(x)$
-2.9	-2.303
-2	0
0	1.099
2	1.609
4	1.946

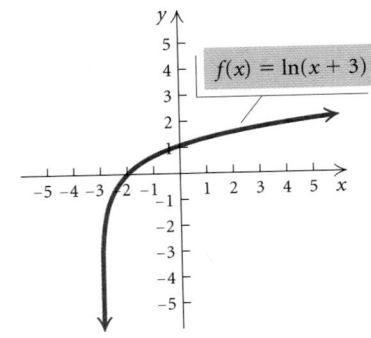

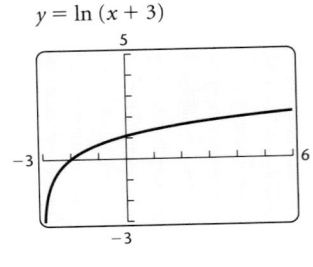

b) The graph of $f(x) = 3 - \frac{1}{2} \ln x$ is a vertical shrinking of the graph of $y = \ln x$, followed by a reflection across the x-axis, and then a translation up 3 units. The domain is the set of all positive real numbers, $(0, \infty)$. The y-axis is the vertical asymptote.

x	$f(x)$
0.1	4.151
1	3
3	2.451
6	2.104
9	1.901

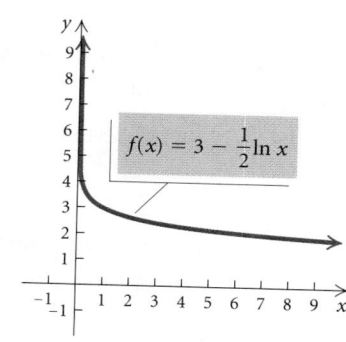

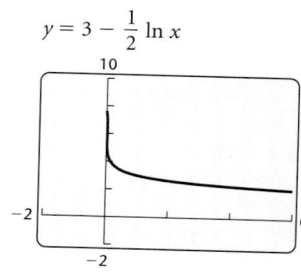

c) The graph of $f(x) = |\ln(x - 1)|$ is a translation of the graph of $y = \ln x$, right 1 unit. Then the absolute value has the effect of reflecting negative outputs across the x-axis. The domain is the set of all real numbers greater than 1, $(1, \infty)$. The line $x = 1$ is the vertical asymptote.

x	$f(x)$
1.1	2.303
2	0
4	1.099
6	1.609
8	1.946

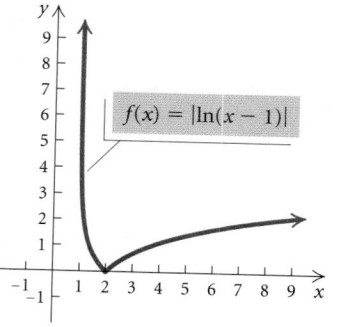

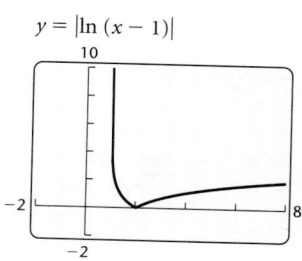

Now Try Exercise 89. ■

❋ Applications

EXAMPLE 12 *Walking Speed.* In a study by psychologists Bornstein and Bornstein, it was found that the average walking speed w, in feet per second, of a person living in a city of population P, in thousands, is given by the function

$$w(P) = 0.37 \ln P + 0.05$$

(*Source*: *International Journal of Psychology*).

a) The population of Hartford, Connecticut, is 124,848. Find the average walking speed of people living in Hartford.

b) The population of San Antonio, Texas, is 1,236,249. Find the average walking speed of people living in San Antonio.

c) Graph the function.

d) A sociologist computes the average walking speed in a city to be approximately 2.0 ft/sec. Use this information to estimate the population of the city.

Solution

a) Since P is in thousands and $124{,}848 = 124.848$ thousand, we substitute 124.848 for P:

$$w(124.848) = 0.37 \ln 124.848 + 0.05 \qquad \text{Substituting}$$
$$\approx 1.8. \qquad \text{Finding the natural logarithm and simplifying}$$

The average walking speed of people living in Hartford is about 1.8 ft/sec.

b) We substitute 1236.249 for P:

$$w(1236.249) = 0.37 \ln 1236.249 + 0.05 \qquad \text{Substituting}$$
$$\approx 2.7.$$

The average walking speed of people living in San Antonio is about 2.7 ft/sec.

c) We graph with a viewing window of $[0, 600, 0, 4]$ because inputs are very large and outputs are very small by comparison.

$$y = 0.37 \ln (x) + 0.05$$

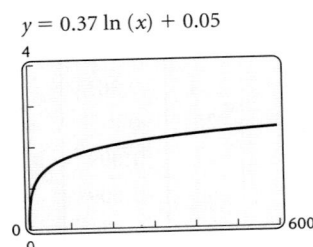

d) To find the population for which the average walking speed is 2.0 ft/sec, we substitute 2.0 for $w(P)$,

$$2.0 = 0.37 \ln P + 0.05,$$

and solve for P.

$y_1 = 0.37 \ln (x) + 0.05,\ y_2 = 2$

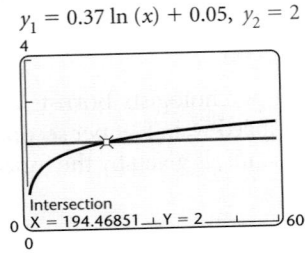

We will use the Intersect method. We graph the equations $y_1 = 0.37 \ln x + 0.05$ and $y_2 = 2$ and use the INTERSECT feature to approximate the point of intersection. (See the window at left.) We see that in a city in which the average walking speed is 2.0 ft/sec, the population is about 194.5 thousand, or 194,500. **Now Try Exercise 91(d).** ■

EXAMPLE 13 *Earthquake Magnitude.* The magnitude R, measured on the Richter scale, of an earthquake of intensity I is defined as

$$R = \log \frac{I}{I_0},$$

where I_0 is a minimum intensity used for comparison. We can think of I_0 as a threshold intensity that is the weakest earthquake that can be recorded on a seismograph. If one earthquake is 10 times as intense as another, its magnitude on the Richter scale is 1 greater than that of the other. If one earthquake is 100 times as intense as another, its magnitude on the Richter scale is 2 higher, and so on. Thus an earthquake whose magnitude is 7 on the Richter scale is 10 times as intense as an earthquake whose magnitude is 6. Earthquake intensities can be interpreted as multiples of the minimum intensity I_0.

The undersea earthquake off the west coast of northern Sumatra on December 26, 2004, had an intensity of $10^{9.3} \cdot I_0$ (*Sources:* U.S. Geological Survey; National Earthquake Information Center). It caused devastating tsunamis that hit twelve Indian Ocean countries. What was its magnitude on the Richter scale?

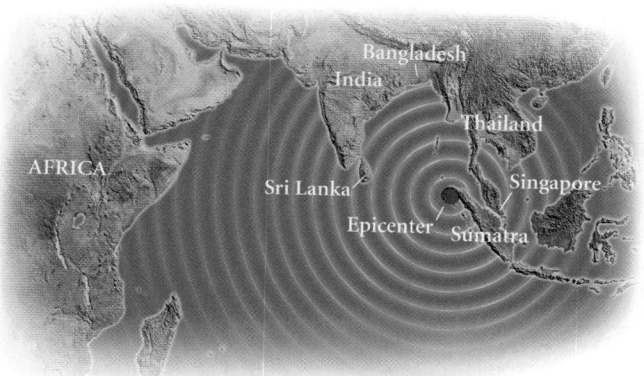

Solution We substitute into the formula:

$$R = \log \frac{I}{I_0} = \log \frac{10^{9.3} \cdot I_0}{I_0} = \log 10^{9.3} = 9.3.$$

The magnitude of the earthquake was 9.3 on the Richter scale.

Now Try Exercise 93(a). ■

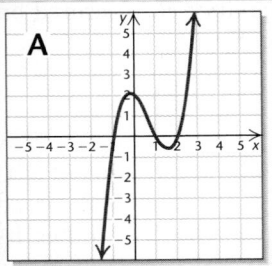

A

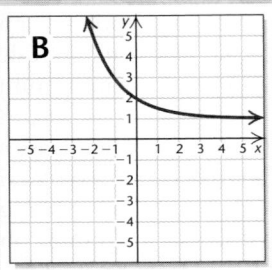

B

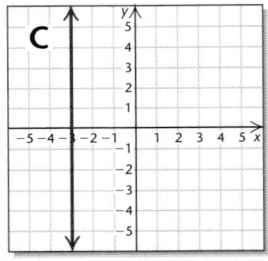

C

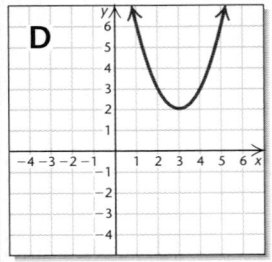

D

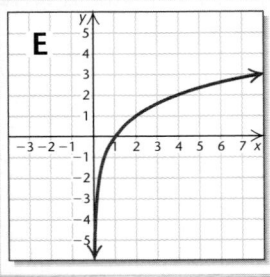

E

Visualizing the Graph

Match the equation or function with its graph.

1. $f(x) = 4^x$

2. $f(x) = \ln x - 3$

3. $(x + 3)^2 + y^2 = 9$

4. $f(x) = 2^{-x} + 1$

5. $f(x) = \log_2 x$

6. $f(x) = x^3 - 2x^2 - x + 2$

7. $x = -3$

8. $f(x) = e^x - 4$

9. $f(x) = (x - 3)^2 + 2$

10. $3x = 6 + y$

Answers on page A-35

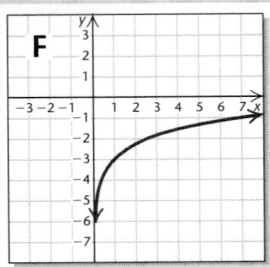

F

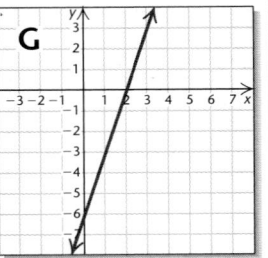

G

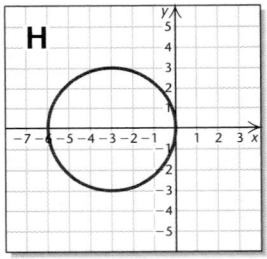

H

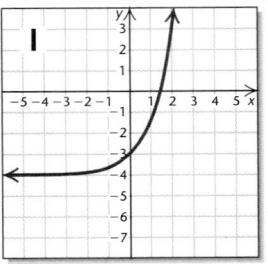

I

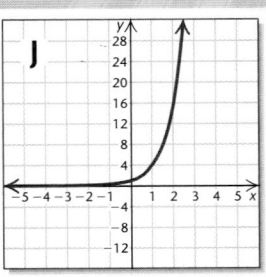

J

5.3 Exercise Set

Make a hand-drawn graph of each of the following. Then check your work using a graphing calculator.

1. $x = 3^y$

2. $x = 4^y$

3. $x = \left(\frac{1}{2}\right)^y$

4. $x = \left(\frac{4}{3}\right)^y$

5. $y = \log_3 x$

6. $y = \log_4 x$

7. $f(x) = \log x$

8. $f(x) = \ln x$

Find each of the following. Do not use a calculator.

9. $\log_2 16$

10. $\log_3 9$

11. $\log_5 125$

12. $\log_2 64$

13. $\log 0.001$

14. $\log 100$

15. $\log_2 \frac{1}{4}$

16. $\log_8 2$

17. $\ln 1$

18. $\ln e$

19. $\log 10$

20. $\log 1$

21. $\log_5 5^4$

22. $\log \sqrt{10}$

23. $\log_3 \sqrt[4]{3}$

24. $\log 10^{8/5}$

25. $\log 10^{-7}$

26. $\log_5 1$

27. $\log_{49} 7$

28. $\log_3 3^{-2}$

29. $\ln e^{3/4}$

30. $\log_2 \sqrt{2}$

31. $\log_4 1$

32. $\ln e^{-5}$

33. $\ln \sqrt{e}$

34. $\log_{64} 4$

Convert to a logarithmic equation.

35. $10^3 = 1000$

36. $5^{-3} = \frac{1}{125}$

37. $8^{1/3} = 2$

38. $10^{0.3010} = 2$

39. $e^3 = t$

40. $Q^t = x$

41. $e^2 = 7.3891$

42. $e^{-1} = 0.3679$

43. $p^k = 3$

44. $e^{-t} = 4000$

Convert to an exponential equation.

45. $\log_5 5 = 1$

46. $t = \log_4 7$

47. $\log 0.01 = -2$

48. $\log 7 = 0.845$

49. $\ln 30 = 3.4012$

50. $\ln 0.38 = -0.9676$

51. $\log_a M = -x$

52. $\log_t Q = k$

53. $\log_a T^3 = x$

54. $\ln W^5 = t$

Find each of the following using a calculator. Round to four decimal places.

55. $\log 3$

56. $\log 8$

57. $\log 532$

58. $\log 93{,}100$

59. $\log 0.57$

60. $\log 0.082$

61. $\log(-2)$

62. $\ln 50$

63. $\ln 2$

64. $\ln(-4)$

65. $\ln 809.3$

66. $\ln 0.00037$

67. $\ln(-1.32)$

68. $\ln 0$

Find the logarithm using common logarithms and the change-of-base formula.

69. $\log_4 100$

70. $\log_3 20$

71. $\log_{100} 0.3$

72. $\log_\pi 100$

73. $\log_{200} 50$

74. $\log_{5.3} 1700$

Find the logarithm using natural logarithms and the change-of-base formula.

75. $\log_3 12$

76. $\log_4 25$

77. $\log_{100} 15$

78. $\log_9 100$

Graph the function and its inverse using the same set of axes. Use any method.

79. $f(x) = 3^x,\ f^{-1}(x) = \log_3 x$

80. $f(x) = \log_4 x,\ f^{-1}(x) = 4^x$

81. $f(x) = \log x,\ f^{-1}(x) = 10^x$

82. $f(x) = e^x,\ f^{-1}(x) = \ln x$

For each of the following functions, briefly describe how the graph can be obtained from the graph of a basic logarithmic function. Then graph the function using a graphing calculator. Give the domain and the vertical asymptote of each function.

83. $f(x) = \log_2(x + 3)$ **84.** $f(x) = \log_3(x - 2)$

85. $y = \log_3 x - 1$ **86.** $y = 3 + \log_2 x$

87. $f(x) = 4 \ln x$ **88.** $f(x) = \frac{1}{2} \ln x$

89. $y = 2 - \ln x$ **90.** $y = \ln(x + 1)$

91. *Walking Speed.* Refer to Example 12. Various cities and their populations are given below. Find the average walking speed in each city.

    a) Indianapolis, Indiana: 784,242
    b) Albuquerque, New Mexico: 484,246
    c) St. Paul, Minnesota: 276,963
    d) Chicago, Illinois: 2,862,244
    e) Portland, Oregon: 533,492
    f) Toledo, Ohio: 304,973
    g) New York, New York: 8,104,079
    h) Cedar Rapids, Iowa: 122,206

92. *Forgetting.* Students in an accounting class took a final exam and then took equivalent forms of the exam at monthly intervals thereafter. The average score $S(t)$, as a percent, after t months was found to be given by the function

$$S(t) = 78 - 15 \log(t + 1), \quad t \geq 0.$$

    a) What was the average score when the students initially took the test, $t = 0$?
    b) What was the average score after 4 months? after 24 months?
    c) Graph the function.
    d) After what time t was the average score 50%?

93. *Earthquake Magnitude.* Refer to Example 13. Various locations of earthquakes and their intensities are given below. What was the magnitude on the Richter scale?

    a) Mexico City, 1978: $10^{7.85} \cdot I_0$
    b) San Francisco, 1906: $10^{8.25} \cdot I_0$
    c) Chile, 1960: $10^{9.6} \cdot I_0$
    d) Italy, 1980: $10^{7.85} \cdot I_0$
    e) San Francisco, 1989: $10^{6.9} \cdot I_0$

94. *pH of Substances in Chemistry.* In chemistry, the pH of a substance is defined as

$$pH = -\log[H^+],$$

where H^+ is the hydrogen ion concentration, in moles per liter. Find the pH of each substance.

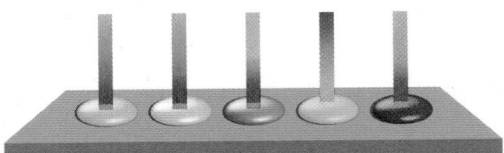

Litmus paper is used to test pH.

Substance	Hydrogen Ion Concentration
a) Pineapple juice	1.6×10^{-4}
b) Hair conditioner	0.0013
c) Mouthwash	6.3×10^{-7}
d) Eggs	1.6×10^{-8}
e) Tomatoes	6.3×10^{-5}

95. Find the hydrogen ion concentration of each substance, given the pH. (See Exercise 94.) Express the answer in scientific notation.

Substance	pH
a) Tap water	7
b) Rainwater	5.4
c) Orange juice	3.2
d) Wine	4.8

96. *Advertising.* A model for advertising response is given by the function

$$N(a) = 1000 + 200 \ln a, \quad a \geq 1,$$

where $N(a)$ is the number of units sold when a is the amount spent on advertising, in thousands of dollars.

a) How many units were sold after spending $1000 ($a = 1$) on advertising?

b) How many units were sold after spending $5000?

c) Graph the function.

d) How much would have to be spent in order to sell 2000 units?

97. *Loudness of Sound.* The **loudness L**, in bels (after Alexander Graham Bell), of a sound of intensity I is defined to be

$$L = \log \frac{I}{I_0},$$

where I_0 is the minimum intensity detectable by the human ear (such as the tick of a watch at 20 ft under quiet conditions). If a sound is 10 times as intense as another, its loudness is 1 bel greater than that of the other. If a sound is 100 times as intense as another, its loudness is 2 bels greater, and so on. The bel is a large unit, so a subunit, the **decibel**, is generally used. For L, in decibels, the formula is

$$L = 10 \log \frac{I}{I_0}.$$

Find the loudness, in decibels, of each sound with the given intensity.

Sound	Intensity
a) Library	$2510 \cdot I_0$
b) Dishwasher	$2{,}500{,}000 \cdot I_0$
c) Conversational speech	$10^6 \cdot I_0$
d) Heavy truck	$10^9 \cdot I_0$

Collaborative Discussion and Writing

98. Explain how the graph of $f(x) = \ln x$ can be used to obtain the graph of $g(x) = e^{x-2}$.

99. If $\log b < 0$, what can you say about b?

Skill Maintenance

Find the slope and the y-intercept of the line.

100. $3x - 10y = 14$

101. $y = 6$

102. $x = -4$

Use synthetic division to find the function values.

103. $g(x) = x^3 - 6x^2 + 3x + 10$; find $g(-5)$

104. $f(x) = x^4 - 2x^3 + x - 6$; find $f(-1)$

Find a polynomial function of degree 3 with the given numbers as zeros. Answers may vary.

105. $\sqrt{7}, -\sqrt{7}, 0$

106. $4i, -4i, 1$

Synthesis

Simplify.

107. $\dfrac{\log_5 8}{\log_5 2}$

108. $\dfrac{\log_3 64}{\log_3 16}$

Find the domain of the function.

109. $f(x) = \log_5 x^3$

110. $f(x) = \log_4 x^2$

111. $f(x) = \ln |x|$

112. $f(x) = \log (3x - 4)$

Solve.

113. $\log_2 (2x + 5) < 0$

114. $\log_2 (x - 3) \geq 4$

In Exercises 115–118, match the equation with one of the figures (a)–(d), which follow.

115. $f(x) = \ln|x|$

116. $f(x) = |\ln x|$

117. $f(x) = \ln x^2$

118. $g(x) = |\ln(x - 1)|$

a)
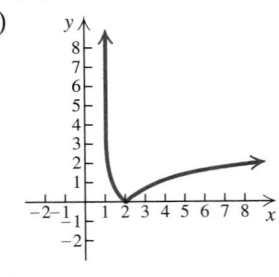

b)
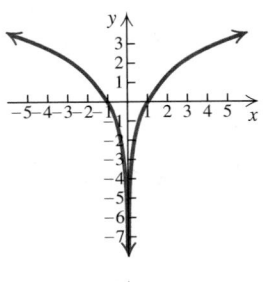

c)
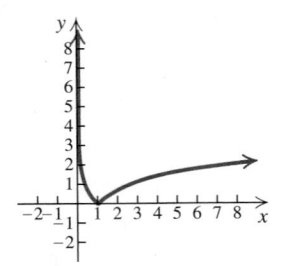

d)
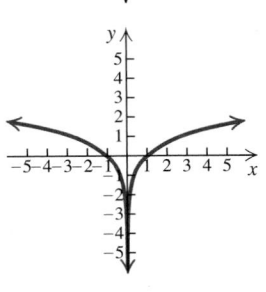

For Exercises 119–122:

a) *Graph the function.*

b) *Estimate the zeros.*

c) *Estimate the relative maximum values and the relative minimum values.*

119. $f(x) = x \ln x$

120. $f(x) = x^2 \ln x$

121. $f(x) = \dfrac{\ln x}{x^2}$

122. $f(x) = e^{-x} \ln x$

123. Using a graphing calculator, find the point(s) of intersection of the graphs of the following.

$$y = 4 \ln x, \qquad y = \frac{4}{e^x + 1}$$

5.4

Properties of Logarithmic Functions

❖ Convert from logarithms of products, powers, and quotients to expressions in terms of individual logarithms, and conversely.

❖ Simplify expressions of the type $\log_a a^x$ and $a^{\log_a x}$.

We now establish some properties of logarithmic functions. These properties are based on the corresponding rules for exponents.

❖ Logarithms of Products

The first property of logarithms corresponds to the product rule for exponents: $a^m \cdot a^n = a^{m+n}$.

> **The Product Rule**
>
> For any positive numbers M and N and any logarithmic base a,
>
> $$\log_a MN = \log_a M + \log_a N.$$
>
> (The logarithm of a product is the sum of the logarithms of the factors.)

EXAMPLE 1 Express as a sum of logarithms: $\log_3 (9 \cdot 27)$.

Solution We have

$$\log_3 (9 \cdot 27) = \log_3 9 + \log_3 27. \qquad \text{Using the product rule}$$

As a check, note that

$$\log_3 (9 \cdot 27) = \log_3 243 = 5 \qquad 3^5 = 243$$
and $\quad \log_3 9 + \log_3 27 = 2 + 3 = 5. \qquad 3^2 = 9; 3^3 = 27$

Now Try Exercise 1. ■

EXAMPLE 2 Express as a single logarithm: $\log_2 p^3 + \log_2 q$.

Solution We have

$$\log_2 p^3 + \log_2 q = \log_2 (p^3 q). \qquad \textit{Now Try Exercise 35.} ■$$

A Proof of the Product Rule: Let $\log_a M = x$ and $\log_a N = y$. Converting to exponential equations, we have $a^x = M$ and $a^y = N$. Then

$$MN = a^x \cdot a^y = a^{x+y}.$$

Converting back to a logarithmic equation, we get

$$\log_a MN = x + y.$$

Remembering what x and y represent, we know it follows that

$$\log_a MN = \log_a M + \log_a N.$$ ■

❈ Logarithms of Powers

The second property of logarithms corresponds to the power rule for exponents: $(a^m)^n = a^{mn}$.

> **The Power Rule**
>
> For any positive number M, any logarithmic base a, and any real number p,
>
> $$\log_a M^p = p \log_a M.$$
>
> (The logarithm of a power of M is the exponent times the logarithm of M.)

EXAMPLE 3 Express each of the following as a product.

a) $\log_a 11^{-3}$ b) $\log_a \sqrt[4]{7}$ c) $\ln x^6$

Solution

RATIONAL EXPONENTS
REVIEW SECTION **R.7.**

a) $\log_a 11^{-3} = -3 \log_a 11 \qquad$ Using the power rule

b) $\log_a \sqrt[4]{7} = \log_a 7^{1/4} \qquad$ Writing exponential notation

$\qquad = \frac{1}{4} \log_a 7 \qquad$ Using the power rule

c) $\ln x^6 = 6 \ln x \qquad$ Using the power rule

Now Try Exercises 13 and 15. ■

A Proof of the Power Rule: Let $x = \log_a M$. The equivalent exponential equation is $a^x = M$. Raising both sides to the power p, we obtain

$$(a^x)^p = M^p, \quad \text{or} \quad a^{xp} = M^p.$$

Converting back to a logarithmic equation, we get

$$\log_a M^p = xp.$$

But $x = \log_a M$, so substituting gives us

$$\log_a M^p = (\log_a M)p = p \log_a M. \qquad ∎$$

❁ Logarithms of Quotients

The third property of logarithms corresponds to the quotient rule for exponents: $a^m/a^n = a^{m-n}$.

> ### The Quotient Rule
>
> For any positive numbers M and N, and any logarithmic base a,
>
> $$\log_a \frac{M}{N} = \log_a M - \log_a N.$$
>
> (The logarithm of a quotient is the logarithm of the numerator minus the logarithm of the denominator.)

EXAMPLE 4 Express as a difference of logarithms: $\log_t \dfrac{8}{w}$.

Solution We have

$$\log_t \frac{8}{w} = \log_t 8 - \log_t w. \qquad \text{Using the quotient rule}$$

Now Try Exercise 17. ■

EXAMPLE 5 Express as a single logarithm: $\log_b 64 - \log_b 16$.

Solution We have

$$\log_b 64 - \log_b 16 = \log_b \frac{64}{16} = \log_b 4. \qquad \text{Now Try Exercise 37.} ■$$

A Proof of the Quotient Rule: The proof follows from both the product rule and the power rule:

$$\log_a \frac{M}{N} = \log_a MN^{-1}$$

$$= \log_a M + \log_a N^{-1} \qquad \text{Using the product rule}$$

$$= \log_a M + (-1)\log_a N \qquad \text{Using the power rule}$$

$$= \log_a M - \log_a N. \qquad\qquad ∎$$

Common Errors

$\log_a MN \neq (\log_a M)(\log_a N)$ The logarithm of a product is *not* the product of the logarithms.

$\log_a (M + N) \neq \log_a M + \log_a N$ The logarithm of a sum is *not* the sum of the logarithms.

$\log_a \dfrac{M}{N} \neq \dfrac{\log_a M}{\log_a N}$ The logarithm of a quotient is *not* the quotient of the logarithms.

$(\log_a M)^p \neq p \log_a M$ The power of a logarithm is *not* the exponent times the logarithm.

❋ Applying the Properties

EXAMPLE 6   Express each of the following in terms of sums and differences of logarithms.

a) $\log_a \dfrac{x^2 y^5}{z^4}$ b) $\log_a \sqrt[3]{\dfrac{a^2 b}{c^5}}$ c) $\log_b \dfrac{a y^5}{m^3 n^4}$

Solution

a) $\log_a \dfrac{x^2 y^5}{z^4} = \log_a (x^2 y^5) - \log_a z^4$ Using the quotient rule

$\qquad\qquad = \log_a x^2 + \log_a y^5 - \log_a z^4$ Using the product rule

$\qquad\qquad = 2 \log_a x + 5 \log_a y - 4 \log_a z$ Using the power rule

b) $\log_a \sqrt[3]{\dfrac{a^2 b}{c^5}} = \log_a \left(\dfrac{a^2 b}{c^5} \right)^{1/3}$ Writing exponential notation

$\qquad\qquad = \dfrac{1}{3} \log_a \dfrac{a^2 b}{c^5}$ Using the power rule

$\qquad\qquad = \dfrac{1}{3} (\log_a a^2 b - \log_a c^5)$ Using the quotient rule. The parentheses are necessary.

$\qquad\qquad = \dfrac{1}{3} (2 \log_a a + \log_a b - 5 \log_a c)$ Using the product and power rules

$\qquad\qquad = \dfrac{1}{3} (2 + \log_a b - 5 \log_a c)$ $\log_a a = 1$

$\qquad\qquad = \dfrac{2}{3} + \dfrac{1}{3} \log_a b - \dfrac{5}{3} \log_a c$ Multiplying to remove parentheses

c) $\log_b \dfrac{a y^5}{m^3 n^4} = \log_b a y^5 - \log_b m^3 n^4$ Using the quotient rule

$\qquad\qquad = (\log_b a + \log_b y^5) - (\log_b m^3 + \log_b n^4)$ Using the product rule

$\qquad\qquad = \log_b a + \log_b y^5 - \log_b m^3 - \log_b n^4$ Removing parentheses

$\qquad\qquad = \log_b a + 5 \log_b y - 3 \log_b m - 4 \log_b n$ Using the power rule

Now Try Exercises 25 and 31. ◼

EXAMPLE 7 Express as a single logarithm:

$$5 \log_b x - \log_b y + \frac{1}{4} \log_b z.$$

Solution We have

$$5 \log_b x - \log_b y + \frac{1}{4} \log_b z = \log_b x^5 - \log_b y + \log_b z^{1/4}$$

              Using the power rule

$$= \log_b \frac{x^5}{y} + \log_b z^{1/4} \qquad \text{Using the quotient rule}$$

$$= \log_b \frac{x^5 z^{1/4}}{y}, \text{ or } \log_b \frac{x^5 \sqrt[4]{z}}{y}.$$

              Using the product rule

              Now Try Exercise 41. ■

EXAMPLE 8 Express as a single logarithm:

$$\ln (3x + 1) - \ln (3x^2 - 5x - 2).$$

Solution We have

$$\ln (3x + 1) - \ln (3x^2 - 5x - 2)$$

$$= \ln \frac{3x + 1}{3x^2 - 5x - 2} \qquad \text{Using the quotient rule}$$

$$= \ln \frac{3x + 1}{(3x + 1)(x - 2)} \qquad \text{Factoring}$$

$$= \ln \frac{1}{x - 2}. \qquad \text{Simplifying} \qquad \text{Now Try Exercise 45.} ■$$

EXAMPLE 9 Given that $\log_a 2 \approx 0.301$ and $\log_a 3 \approx 0.477$, find each of the following, if possible.

a) $\log_a 6$ **b)** $\log_a \dfrac{2}{3}$ **c)** $\log_a 81$

d) $\log_a \dfrac{1}{4}$ **e)** $\log_a 5$ **f)** $\dfrac{\log_a 3}{\log_a 2}$

Solution

a) $\log_a 6 = \log_a (2 \cdot 3) = \log_a 2 + \log_a 3$ *Using the product rule*

$$\approx 0.301 + 0.477$$

$$\approx 0.778$$

b) $\log_a \frac{2}{3} = \log_a 2 - \log_a 3$ *Using the quotient rule*

$$\approx 0.301 - 0.477 \approx -0.176$$

c) $\log_a 81 = \log_a 3^4 = 4 \log_a 3$ *Using the power rule*

$$\approx 4(0.477) \approx 1.908$$

d) $\log_a \frac{1}{4} = \log_a 1 - \log_a 4$ Using the quotient rule

$\qquad\quad = 0 - \log_a 2^2$ $\log_a 1 = 0$

$\qquad\quad = -2 \log_a 2$ Using the power rule

$\qquad\quad \approx -2(0.301) \approx -0.602$

e) $\log_a 5$ *cannot* be found using these properties and the given information.

$$\log_a 5 \neq \log_a 2 + \log_a 3 \qquad \log_a 2 + \log_a 3 = \log_a (2 \cdot 3) = \log_a 6$$

f) $\dfrac{\log_a 3}{\log_a 2} \approx \dfrac{0.477}{0.301} \approx 1.585$ We simply divide, not using any of the properties.

Now Try Exercises 53 and 55. ■

❈ Simplifying Expressions of the Type $\log_a a^x$ and $a^{\log_a x}$

We have two final properties of logarithms to consider. The first follows from the product rule: Since $\log_a a^x = x \log_a a = x \cdot 1 = x$, we have $\log_a a^x = x$. This property also follows from the definition of a logarithm: x is the power to which we raise a in order to get a^x.

> **The Logarithm of a Base to a Power**
> For any base a and any real number x,
> $$\log_a a^x = x.$$
> (The logarithm, base a, of a to a power is the power.)

EXAMPLE 10 Simplify each of the following.

a) $\log_a a^8$ **b)** $\ln e^{-t}$ **c)** $\log 10^{3k}$

Solution

a) $\log_a a^8 = 8$ 8 is the power to which we raise a in order to get a^8.

b) $\ln e^{-t} = \log_e e^{-t} = -t$ $\ln e^x = x$

c) $\log 10^{3k} = \log_{10} 10^{3k} = 3k$ **Now Try Exercises 65 and 73.** ■

Let $M = \log_a x$. Then $a^M = x$. Substituting $\log_a x$ for M, we obtain $a^{\log_a x} = x$. This also follows from the definition of a logarithm: $\log_a x$ is the power to which a is raised in order to get x.

> **A Base to a Logarithmic Power**
> For any base a and any positive real number x,
> $$a^{\log_a x} = x.$$
> (The number a raised to the power $\log_a x$ is x.)

STUDY TIP

Immediately after each quiz or chapter test, write out a step-by-step solution to the questions you missed. Visit your instructor during office hours for help with problems that are still giving you trouble. When the week of the final examination arrives, you will be glad to have the excellent study guide these corrected tests provide.

EXAMPLE 11   Simplify each of the following.

a) $4^{\log_4 k}$ **b)** $e^{\ln 5}$ **c)** $10^{\log 7t}$

Solution

a) $4^{\log_4 k} = k$
b) $e^{\ln 5} = e^{\log_e 5} = 5$
c) $10^{\log 7t} = 10^{\log_{10} 7t} = 7t$

Now Try Exercises 69 and 71. ▨

A Proof of the Change-of-Base Formula: We close this section by proving the change-of-base formula and summarizing the properties of logarithms considered thus far in this chapter. In Section 5.3, we used the change-of-base formula,

$$\log_b M = \frac{\log_a M}{\log_a b},$$

to make base conversions in order to find logarithmic values using a calculator. Let $x = \log_b M$. Then

$b^x = M$	Definition of logarithm
$\log_a b^x = \log_a M$	Taking the logarithm on both sides
$x \log_a b = \log_a M$	Using the power rule
$x = \dfrac{\log_a M}{\log_a b},$	Dividing by $\log_a b$

so

$$x = \log_b M = \frac{\log_a M}{\log_a b}.$$

▪

CHANGE-OF-BASE FORMULA

REVIEW SECTION **5.3.**

Following is a summary of the properties of logarithms.

Summary of the Properties of Logarithms

The Product Rule:	$\log_a MN = \log_a M + \log_a N$
The Power Rule:	$\log_a M^p = p \log_a M$
The Quotient Rule:	$\log_a \dfrac{M}{N} = \log_a M - \log_a N$
The Change-of-Base Formula:	$\log_b M = \dfrac{\log_a M}{\log_a b}$
Other Properties:	$\log_a a = 1, \qquad \log_a 1 = 0,$
	$\log_a a^x = x, \qquad a^{\log_a x} = x$

5.4 Exercise Set

Express as a sum of logarithms.

1. $\log_3 (81 \cdot 27)$

2. $\log_2 (8 \cdot 64)$

3. $\log_5 (5 \cdot 125)$

4. $\log_4 (64 \cdot 4)$

5. $\log_t 8Y$

6. $\log 0.2x$

7. $\ln xy$

8. $\ln ab$

Express as a product.

9. $\log_b t^3$

10. $\log_a x^4$

11. $\log y^8$

12. $\ln y^5$

13. $\log_c K^{-6}$

14. $\log_b Q^{-8}$

15. $\ln \sqrt[3]{4}$

16. $\ln \sqrt{a}$

Express as a difference of logarithms.

17. $\log_t \dfrac{M}{8}$

18. $\log_a \dfrac{76}{13}$

19. $\log \dfrac{x}{y}$

20. $\ln \dfrac{a}{b}$

21. $\ln \dfrac{r}{s}$

22. $\log_b \dfrac{3}{w}$

Express in terms of sums and differences of logarithms.

23. $\log_a 6xy^5 z^4$

24. $\log_a x^3 y^2 z$

25. $\log_b \dfrac{p^2 q^5}{m^4 b^9}$

26. $\log_b \dfrac{x^2 y}{b^3}$

27. $\ln \dfrac{2}{3x^3 y}$

28. $\log \dfrac{5a}{4b^2}$

29. $\log \sqrt{r^3 t}$

30. $\ln \sqrt[3]{5x^5}$

31. $\log_a \sqrt{\dfrac{x^6}{p^5 q^8}}$

32. $\log_c \sqrt[3]{\dfrac{y^3 z^2}{x^4}}$

33. $\log_a \sqrt[4]{\dfrac{m^8 n^{12}}{a^3 b^5}}$

34. $\log_a \sqrt{\dfrac{a^6 b^8}{a^2 b^5}}$

Express as a single logarithm and, if possible, simplify.

35. $\log_a 75 + \log_a 2$

36. $\log 0.01 + \log 1000$

37. $\log 10{,}000 - \log 100$

38. $\ln 54 - \ln 6$

39. $\frac{1}{2} \log n + 3 \log m$

40. $\frac{1}{2} \log a - \log 2$

41. $\frac{1}{2} \log_a x + 4 \log_a y - 3 \log_a x$

42. $\frac{2}{5} \log_a x - \frac{1}{3} \log_a y$

43. $\ln x^2 - 2 \ln \sqrt{x}$

44. $\ln 2x + 3(\ln x - \ln y)$

45. $\ln (x^2 - 4) - \ln (x + 2)$

46. $\log (x^3 - 8) - \log (x - 2)$

47. $\log (x^2 - 5x - 14) - \log (x^2 - 4)$

48. $\log_a \dfrac{a}{\sqrt{x}} - \log_a \sqrt{ax}$

49. $\ln x - 3[\ln (x - 5) + \ln (x + 5)]$

50. $\frac{2}{3} [\ln (x^2 - 9) - \ln (x + 3)] + \ln (x + y)$

51. $\frac{3}{2} \ln 4x^6 - \frac{4}{5} \ln 2y^{10}$

52. $120 \left(\ln \sqrt[5]{x^3} + \ln \sqrt[3]{y^2} - \ln \sqrt[4]{16z^5} \right)$

Given that $\log_a 2 \approx 0.301$, $\log_a 7 \approx 0.845$, *and* $\log_a 11 \approx 1.041$, *find each of the following, if possible. Round the answer to the nearest thousandth.*

53. $\log_a \frac{2}{11}$

54. $\log_a 14$

55. $\log_a 98$

56. $\log_a \frac{1}{7}$

57. $\dfrac{\log_a 2}{\log_a 7}$

58. $\log_a 9$

Given that $\log_b 2 \approx 0.693$, $\log_b 3 \approx 1.099$, *and* $\log_b 5 \approx 1.609$, *find each of the following, if possible. Round the answer to the nearest thousandth.*

59. $\log_b 125$

60. $\log_b \frac{5}{3}$

61. $\log_b \frac{1}{6}$

62. $\log_b 30$

63. $\log_b \dfrac{3}{b}$

64. $\log_b 15b$

Simplify.

65. $\log_p p^3$

66. $\log_t t^{2713}$

67. $\log_e e^{|x-4|}$

68. $\log_q q^{\sqrt{3}}$

69. $3^{\log_3 4x}$

70. $5^{\log_5 (4x-3)}$

71. $10^{\log w}$

72. $e^{\ln x^3}$

73. $\ln e^{8t}$

74. $\log 10^{-k}$

75. $\log_b \sqrt{b}$

76. $\log_b \sqrt{b^3}$

Collaborative Discussion and Writing

77. Given that $f(x) = a^x$ and $g(x) = \log_a x$, find $(f \circ g)(x)$ and $(g \circ f)(x)$. These results are alternative proofs of which properties of logarithms already proven in this section? Explain.

78. Explain the errors, if any, in the following:
$$\log_a ab^3 = (\log_a a)(\log_a b^3) = 3 \log_a b.$$

Skill Maintenance

In each of Exercises 79–88, classify the function as linear, quadratic, cubic, quartic, rational, exponential, or logarithmic.

79. $f(x) = 5 - x^2 + x^4$ **80.** $f(x) = 2^x$

81. $f(x) = -\frac{3}{4}$ **82.** $f(x) = 4^x - 8$

83. $f(x) = -\dfrac{3}{x}$ **84.** $f(x) = \log x + 6$

85. $f(x) = -\frac{1}{3}x^3 - 4x^2 + 6x + 42$

86. $f(x) = \dfrac{x^2 - 1}{x^2 + x - 6}$

87. $f(x) = \frac{1}{2}x + 3$

88. $f(x) = 2x^2 - 6x + 3$

Synthesis

Solve for x.

89. $5^{\log_5 8} = 2x$ **90.** $\ln e^{3x-5} = -8$

Express as a single logarithm and, if possible, simplify.

91. $\log_a (x^2 + xy + y^2) + \log_a (x - y)$

92. $\log_a (a^{10} - b^{10}) - \log_a (a + b)$

Express as a sum or a difference of logarithms.

93. $\log_a \dfrac{x - y}{\sqrt{x^2 - y^2}}$ **94.** $\log_a \sqrt{9 - x^2}$

95. Given that $\log_a x = 2$, $\log_a y = 3$, and $\log_a z = 4$, find
$$\log_a \frac{\sqrt[4]{y^2 z^5}}{\sqrt[4]{x^3 z^{-2}}}.$$

Determine whether each of the following is true. Assume that a, x, M, and N are positive.

96. $\log_a M + \log_a N = \log_a (M + N)$

97. $\log_a M - \log_a N = \log_a \dfrac{M}{N}$

98. $\dfrac{\log_a M}{\log_a N} = \log_a M - \log_a N$

99. $\dfrac{\log_a M}{x} = \log_a M^{1/x}$

100. $\log_a x^3 = 3 \log_a x$

101. $\log_a 8x = \log_a x + \log_a 8$

102. $\log_N (MN)^x = x \log_N M + x$

Suppose that $\log_a x = 2$. Find each of the following.

103. $\log_a \left(\dfrac{1}{x} \right)$ **104.** $\log_{1/a} x$

105. Simplify:
$$\log_{10} 11 \cdot \log_{11} 12 \cdot \log_{12} 13 \cdots \log_{998} 999 \cdot \log_{999} 1000.$$

Write each of the following without using logarithms.

106. $\log_a x + \log_a y - mz = 0$

107. $\ln a - \ln b + xy = 0$

Prove each of the following for any base a and any positive number x.

108. $\log_a \left(\dfrac{1}{x} \right) = -\log_a x = \log_{1/a} x$

109. $\log_a \left(\dfrac{x + \sqrt{x^2 - 5}}{5} \right) = -\log_a \left(x - \sqrt{x^2 - 5} \right)$

5.5

Solving Exponential and Logarithmic Equations

❖ Solve exponential equations.
❖ Solve logarithmic equations.

❖ Solving Exponential Equations

Equations with variables in the exponents, such as

$$3^x = 20 \quad \text{and} \quad 2^{5x} = 64,$$

are called **exponential equations**.

Sometimes, as is the case with the equation $2^{5x} = 64$, we can write each side as a power of the same number:

$$2^{5x} = 2^6.$$

We can then set the exponents equal and solve:

$$5x = 6$$
$$x = \tfrac{6}{5}, \text{ or } 1.2.$$

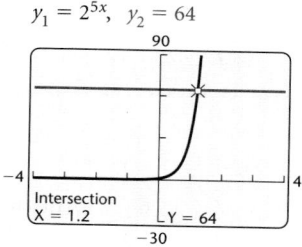

$y_1 = 2^{5x}, \quad y_2 = 64$

We use the following property to solve exponential equations.

Base–Exponent Property

For any $a > 0$, $a \neq 1$,

$$a^x = a^y \longleftrightarrow x = y.$$

ONE-TO-ONE FUNCTIONS

REVIEW SECTION **5.1.**

This property follows from the fact that for any $a > 0$, $a \neq 1$, $f(x) = a^x$ is a one-to-one function. If $a^x = a^y$, then $f(x) = f(y)$. Then since f is one-to-one, it follows that $x = y$. Conversely, if $x = y$, it follows that $a^x = a^y$, since we are raising a to the same power in each case.

EXAMPLE 1 Solve: $2^{3x-7} = 32$.

ALGEBRAIC SOLUTION

Note that $32 = 2^5$. Thus we can write each side as a power of the same number:

$$2^{3x-7} = 2^5.$$

Since the bases are the same number, 2, we can use the base–exponent property and set the exponents equal:

$$3x - 7 = 5$$
$$3x = 12$$
$$x = 4.$$

Check: $2^{3x-7} = 32$

$$\begin{array}{c|c} 2^{3(4)-7} \ ? \ 32 & \\ 2^{12-7} & \\ 2^5 & \\ 32 & 32 \quad \text{TRUE} \end{array}$$

The solution is 4.

GRAPHICAL SOLUTION

We will use the Intersect method. We graph

$$y_1 = 2^{3x-7} \quad \text{and} \quad y_2 = 32$$

to find the coordinates of the point of intersection. The first coordinate of this point is the solution of the equation $y_1 = y_2$, or $2^{3x-7} = 32$.

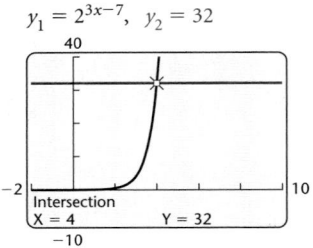

$y_1 = 2^{3x-7}, \ y_2 = 32$

The solution is 4.

We could also write the equation in the form $2^{3x-7} - 32 = 0$ and use the Zero method.

Now Try Exercise 7. ▪

Another property that is used when solving some exponential and logarithmic equations is as follows.

> **Property of Logarithmic Equality**
>
> For any $M > 0$, $N > 0$, $a > 0$, and $a \neq 1$,
>
> $$\log_a M = \log_a N \longleftrightarrow M = N.$$

This property follows from the fact that for any $a > 0$, $a \neq 1$, $f(x) = \log_a x$ is a one-to-one function. If $\log_a x = \log_a y$, then $f(x) = f(y)$. Then since f is one-to-one, it follows that $x = y$. Conversely, if $x = y$, it follows that $\log_a x = \log_a y$, since we are taking the logarithm of the same number in each case.

When it does not seem possible to write each side as a power of the same base, we can use the property of logarithmic equality and take the logarithm with any base on each side and then use the power rule for logarithms.

EXAMPLE 2 Solve: $3^x = 20$.

ALGEBRAIC SOLUTION

We have

$$3^x = 20$$

$$\log 3^x = \log 20 \quad \text{Taking the common logarithm on both sides}$$

$$x \log 3 = \log 20 \quad \text{Using the power rule}$$

$$x = \frac{\log 20}{\log 3}. \quad \text{Dividing by log 3}$$

This is an exact answer. We cannot simplify further, but we can approximate using a calculator:

$$x = \frac{\log 20}{\log 3} \approx 2.7268.$$

We can check this by finding $3^{2.7268}$:

$$3^{2.7268} \approx 20.$$

The solution is about 2.7268.

GRAPHICAL SOLUTION

We will use the Intersect method. We graph

$$y_1 = 3^x \quad \text{and} \quad y_2 = 20$$

to find the x-coordinate of the point of intersection. That x-coordinate is the value of x for which $3^x = 20$ and is thus the solution of the equation.

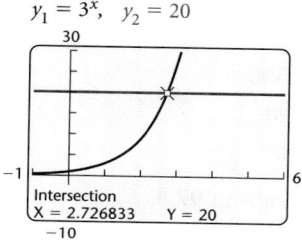

The solution is approximately 2.7268.

We could also write the equation in the form $3^x - 20 = 0$ and use the Zero method.

Now Try Exercise 11. ■

In Example 2, we took the common logarithm on both sides of the equation. Any base will give the same result. Let's try base 3. We have

$$3^x = 20$$

$$\log_3 3^x = \log_3 20$$

$$x = \log_3 20 \quad \log_a a^x = x$$

$$x = \frac{\log 20}{\log 3} \quad \text{Using the change-of-base formula}$$

$$x \approx 2.7268.$$

Note that we have to change the base to do the final calculation.

EXAMPLE 3 Solve: $e^{0.08t} = 2500$.

ALGEBRAIC SOLUTION	GRAPHICAL SOLUTION

ALGEBRAIC SOLUTION

It will make our work easier if we take the natural logarithm when working with equations that have e as a base.

We have

$$e^{0.08t} = 2500$$

$\ln e^{0.08t} = \ln 2500$ — Taking the natural logarithm on both sides

$0.08t = \ln 2500$ — Finding the logarithm of a base to a power: $\log_a a^x = x$

$t = \dfrac{\ln 2500}{0.08}$ — Dividing by 0.08

$\approx 97.8.$

The solution is about 97.8.

GRAPHICAL SOLUTION

Using the Intersect method, we graph the equations

$$y_1 = e^{0.08x} \quad \text{and} \quad y_2 = 2500$$

and determine the point of intersection. The first coordinate of the point of intersection is the solution of the equation $e^{0.08x} = 2500$.

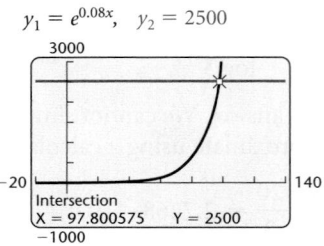

$y_1 = e^{0.08x}, \quad y_2 = 2500$

Intersection
X = 97.800575   Y = 2500

The solution is approximately 97.8.

Now Try Exercise 19.

EXAMPLE 4 Solve: $4^{x+3} = 3^{-x}$.

ALGEBRAIC SOLUTION	GRAPHICAL SOLUTION

ALGEBRAIC SOLUTION

We have

$$4^{x+3} = 3^{-x}$$

$\log 4^{x+3} = \log 3^{-x}$ — Taking the common logarithm on both sides

$(x + 3)\log 4 = -x \log 3$ — Using the power rule

$x \log 4 + 3 \log 4 = -x \log 3$ — Removing parentheses

$x \log 4 + x \log 3 = -3 \log 4$ — Adding $x \log 3$ and subtracting $3 \log 4$

$x(\log 4 + \log 3) = -3 \log 4$ — Factoring on the left

$x = \dfrac{-3 \log 4}{\log 4 + \log 3}$ — Dividing by $\log 4 + \log 3$

$x \approx -1.6737.$

GRAPHICAL SOLUTION

We will use the Intersect method. We graph

$$y_1 = 4^{x+3} \quad \text{and} \quad y_2 = 3^{-x}$$

to find the x-coordinate of the point of intersection. That x-coordinate is the value of x for which $4^{x+3} = 3^{-x}$ and is the solution of the equation.

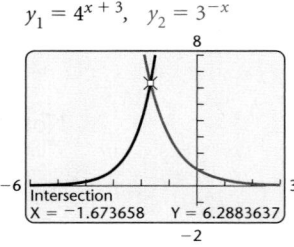

$y_1 = 4^{x+3}, \quad y_2 = 3^{-x}$

Intersection
X = -1.673658   Y = 6.2883637

The solution is approximately -1.6737.

We could also write the equation in the form $4^{x+3} - 3^{-x} = 0$ and use the Zero method.

Now Try Exercise 21.

EQUATIONS REDUCIBLE TO
QUADRATIC

REVIEW SECTION **3.2.**

EXAMPLE 5 Solve: $e^x + e^{-x} - 6 = 0$.

ALGEBRAIC SOLUTION

In this case, we have more than one term with x in the exponent:

$$e^x + e^{-x} - 6 = 0$$

$$e^x + \frac{1}{e^x} - 6 = 0 \qquad \text{Rewriting } e^{-x} \text{ with}$$
$$\text{a positive exponent}$$

$$e^{2x} + 1 - 6e^x = 0. \qquad \text{Multiplying by } e^x \text{ on both sides}$$

This equation is reducible to quadratic with $u = e^x$:

$$u^2 - 6u + 1 = 0.$$

We use the quadratic formula with $a = 1$, $b = -6$, and $c = 1$:

$$u = \frac{-b \pm \sqrt{b^2 - 4ac}}{2a}$$

$$u = \frac{-(-6) \pm \sqrt{(-6)^2 - 4 \cdot 1 \cdot 1}}{2 \cdot 1}$$

$$u = \frac{6 \pm \sqrt{32}}{2} = \frac{6 \pm 4\sqrt{2}}{2}$$

$$u = 3 \pm 2\sqrt{2}$$

$$e^x = 3 \pm 2\sqrt{2}. \qquad \text{Replacing } u \text{ with } e^x$$

We now take the natural logarithm on both sides:

$$\ln e^x = \ln\left(3 \pm 2\sqrt{2}\right)$$

$$x = \ln\left(3 \pm 2\sqrt{2}\right). \qquad \text{Using } \ln e^x = x$$

Approximating each of the solutions, we obtain 1.76 and -1.76.

GRAPHICAL SOLUTION

Using the Zero method, we begin by graphing the function

$$y = e^x + e^{-x} - 6.$$

Then we find the zeros of the function.

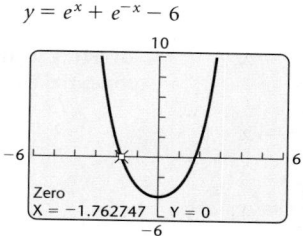

$y = e^x + e^{-x} - 6$

10

−6   6

Zero
X = −1.762747   Y = 0

−6

The leftmost zero is about -1.76. Using the ZERO feature one more time, we find that the other zero is about 1.76.

The solutions are about -1.76 and 1.76.

Now Try Exercise 25. ■

It is possible that when encountering an equation like the one in Example 5, you might not recognize that it could be solved in the algebraic manner shown. This points out the value of the graphical solution.

❄ Solving Logarithmic Equations

Equations containing variables in logarithmic expressions, such as $\log_2 x = 4$ and $\log x + \log(x + 3) = 1$, are called **logarithmic equations**. To solve logarithmic equations algebraically, we first try to obtain a single logarithmic expression on one side and then write an equivalent exponential equation.

EXAMPLE 6 Solve: $\log_3 x = -2$.

ALGEBRAIC SOLUTION

We have

$$\log_3 x = -2$$

$$3^{-2} = x \qquad \text{Converting to an exponential equation}$$

$$\frac{1}{3^2} = x$$

$$\frac{1}{9} = x.$$

Check:
$$\begin{array}{c|c} \log_3 x = -2 \\ \hline \log_3 \dfrac{1}{9} \ ? \ -2 \\ \log_3 3^{-2} \\ -2 \ \Big| \ -2 \quad \text{TRUE} \end{array}$$

The solution is $\frac{1}{9}$.

GRAPHICAL SOLUTION

We use the change-of-base formula and graph the equations

$$y_1 = \log_3 x = \frac{\ln x}{\ln 3} \quad \text{and} \quad y_2 = -2.$$

Then we use the Intersect method.

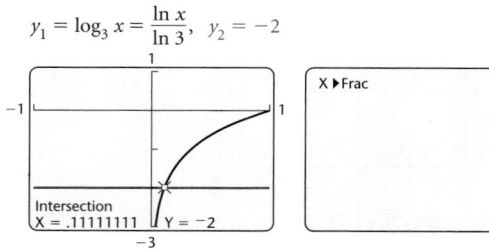

If the solution is a rational number, we can usually find fraction notation for the exact solution by using the FRAC feature from the MATH submenu of the MATH menu.

The solution is $\frac{1}{9}$.

Now Try Exercise 31. ■

EXAMPLE 7 Solve: $\log x + \log(x + 3) = 1$.

ALGEBRAIC SOLUTION

In this case, we have common logarithms. Writing the base of 10 will help us understand the problem:

$$\log_{10} x + \log_{10}(x + 3) = 1$$
$$\log_{10}[x(x + 3)] = 1 \quad \text{Using the product rule to obtain a single logarithm}$$
$$x(x + 3) = 10^1 \quad \text{Writing an equivalent exponential equation}$$
$$x^2 + 3x = 10$$
$$x^2 + 3x - 10 = 0$$
$$(x - 2)(x + 5) = 0 \quad \text{Factoring}$$
$$x - 2 = 0 \quad or \quad x + 5 = 0$$
$$x = 2 \quad or \quad x = -5.$$

Check: For 2:

$$\log x + \log(x + 3) = 1$$
$$\overline{\log 2 + \log(2 + 3)} \ ? \ 1$$
$$\log 2 + \log 5$$
$$\log(2 \cdot 5)$$
$$\log 10$$
$$1 \ | \ 1 \ \text{TRUE}$$

For -5:

$$\log x + \log(x + 3) = 1$$
$$\overline{\log(-5) + \log(-5 + 3)} \ ? \ 1 \ \text{FALSE}$$

The number -5 is not a solution because negative numbers do not have real-number logarithms. The solution is 2.

GRAPHICAL SOLUTION

We can graph the equations

$$y_1 = \log x + \log(x + 3)$$

and

$$y_2 = 1$$

and use the Intersect method. The first coordinate of the point of intersection is the solution of the equation.

We could also graph the function

$$y = \log x + \log(x + 3) - 1$$

and use the Zero method. The zero of the function is the solution of the equation.

The solution of the equation is 2. From the graph, we can easily see that there is only one solution.

Now Try Exercise 37.

EXAMPLE 8 Solve: $\log_3 (2x - 1) - \log_3 (x - 4) = 2$.

. We have

$$\log_3 (2x - 1) - \log_3 (x - 4) = 2$$

$$\log_3 \frac{2x - 1}{x - 4} = 2 \qquad \text{Using the quotient rule}$$

$$\frac{2x - 1}{x - 4} = 3^2 \qquad \text{Writing an equivalent exponential equation}$$

$$\frac{2x - 1}{x - 4} = 9$$

$$(x - 4) \cdot \frac{2x - 1}{x - 4} = 9(x - 4) \qquad \begin{array}{l}\text{Multiplying} \\ \text{by the LCD,} \\ x - 4\end{array}$$

$$2x - 1 = 9x - 36$$

$$35 = 7x$$

$$5 = x.$$

Check:

$$\begin{array}{c|c} \log_3 (2x - 1) - \log_3 (x - 4) = 2 \\ \hline \log_3 (2 \cdot 5 - 1) - \log_3 (5 - 4) \; ? \; 2 \\ \log_3 9 - \log_3 1 \\ 2 - 0 \\ 2 & 2 \quad \text{TRUE} \end{array}$$

The solution is 5.

Here we use the Intersect method to find the solutions of the equation. We use the change-of-base formula and graph the equations

$$y_1 = \frac{\ln (2x - 1)}{\ln 3} - \frac{\ln (x - 4)}{\ln 3}$$

and

$$y_2 = 2.$$

$$y_1 = \frac{\ln (2x - 1)}{\ln 3} - \frac{\ln (x - 4)}{\ln 3}, \; y_2 = 2$$

The solution is 5.

Now Try Exercise 41. ■

EXAMPLE 9 Solve: $\ln(4x + 6) - \ln(x + 5) = \ln x$.

ALGEBRAIC SOLUTION

We have

$$\ln(4x + 6) - \ln(x + 5) = \ln x$$

$$\ln \frac{4x + 6}{x + 5} = \ln x \qquad \text{Using the quotient rule}$$

$$\frac{4x + 6}{x + 5} = x \qquad \begin{array}{l}\text{Using the property}\\ \text{of logarithmic}\\ \text{equality}\end{array}$$

$$(x + 5) \cdot \frac{4x + 6}{x + 5} = x(x + 5) \qquad \text{Multiplying by } x + 5$$

$$4x + 6 = x^2 + 5x$$

$$0 = x^2 + x - 6$$

$$0 = (x + 3)(x - 2) \qquad \text{Factoring}$$

$$x + 3 = 0 \quad or \quad x - 2 = 0$$

$$x = -3 \quad or \quad x = 2.$$

The number -3 is not a solution because $4(-3) + 6 = -6$ and $\ln(-6)$ is not a real number. The value 2 checks and is the solution.

GRAPHICAL SOLUTION

The solution of the equation

$$\ln(4x + 6) - \ln(x + 5) = \ln x$$

is the zero of the function

$$f(x) = \ln(4x + 6) - \ln(x + 5) - \ln x.$$

The solution is also the first coordinate of the x-intercept of the graph of the function. Here we use the Zero method.

$y_1 = \ln(4x + 6) - \ln(x + 5) - \ln x$

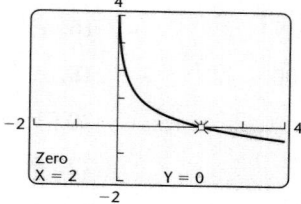

The solution of the equation is 2. From the graph, we can easily see that there is only one solution.

Now Try Exercise 39. ■

Sometimes we encounter equations for which an algebraic solution seems difficult or impossible.

EXAMPLE 10 Solve: $e^{0.5x} - 7.3 = 2.08x + 6.2$.

Graphical Solution We graph the equations

$$y_1 = e^{0.5x} - 7.3 \quad \text{and} \quad y_2 = 2.08x + 6.2$$

on a graphing calculator and use the Intersect method. (See the window at left.)

We can also consider the equation

$$y = e^{0.5x} - 7.3 - 2.08x - 6.2, \quad \text{or} \quad y = e^{0.5x} - 2.08x - 13.5,$$

and use the Zero method. The approximate solutions are -6.471 and 6.610.

$y_1 = e^{0.5x} - 7.3, \quad y_2 = 2.08x + 6.2$

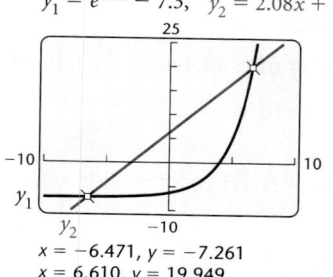

$x = -6.471, y = -7.261$
$x = 6.610, y = 19.949$

Now Try Exercise 53. ■

5.5 Exercise Set

Solve the exponential equation algebraically. Then check using a graphing calculator.

1. $3^x = 81$

2. $2^x = 32$

3. $2^{2x} = 8$

4. $3^{7x} = 27$

5. $2^x = 33$

6. $2^x = 40$

7. $5^{4x-7} = 125$

8. $4^{3x-5} = 16$

9. $27 = 3^{5x} \cdot 9^{x^2}$

10. $3^{x^2+4x} = \frac{1}{27}$

11. $84^x = 70$

12. $28^x = 10^{-3x}$

13. $10^{-x} = 5^{2x}$

14. $15^x = 30$

15. $e^{-c} = 5^{2c}$

16. $e^{4t} = 200$

17. $e^t = 1000$

18. $e^{-t} = 0.04$

19. $e^{-0.03t} = 0.08$

20. $1000e^{0.09t} = 5000$

21. $3^x = 2^{x-1}$

22. $5^{x+2} = 4^{1-x}$

23. $(3.9)^x = 48$

24. $250 - (1.87)^x = 0$

25. $e^x + e^{-x} = 5$

26. $e^x - 6e^{-x} = 1$

27. $3^{2x-1} = 5^x$

28. $2^{x+1} = 5^{2x}$

Solve the logarithmic equation algebraically. Then check using a graphing calculator.

29. $\log_5 x = 4$

30. $\log_2 x = -3$

31. $\log x = -4$

32. $\log x = 1$

33. $\ln x = 1$

34. $\ln x = -2$

35. $\log_2 (10 + 3x) = 5$

36. $\log_5 (8 - 7x) = 3$

37. $\log x + \log (x - 9) = 1$

38. $\log_2 (x + 1) + \log_2 (x - 1) = 3$

39. $\log_2 (x + 20) - \log_2 (x + 2) = \log_2 x$

40. $\log (x + 5) - \log (x - 3) = \log 2$

41. $\log_8 (x + 1) - \log_8 x = 2$

42. $\log x - \log (x + 3) = -1$

43. $\log x + \log (x + 4) = \log 12$

44. $\ln x - \ln (x - 4) = \ln 3$

45. $\log_4 (x + 3) + \log_4 (x - 3) = 2$

46. $\ln (x + 1) - \ln x = \ln 4$

47. $\log (2x + 1) - \log (x - 2) = 1$

48. $\log_5 (x + 4) + \log_5 (x - 4) = 2$

49. $2e^x = 5 - e^{-x}$

50. $e^x + e^{-x} = 4$

51. $\ln (x + 8) + \ln (x - 1) = 2 \ln x$

52. $\log_3 x + \log_3 (x + 1) = \log_3 2 + \log_3 (x + 3)$

Use a graphing calculator to find the approximate solutions of the equation.

53. $e^{7.2x} = 14.009$

54. $0.082e^{0.05x} = 0.034$

55. $2^x - 5 = 3x + 1$

56. $4x - 3^x = -6$

57. $xe^{3x} - 1 = 3$

58. $5e^{5x} + 10 = 3x + 40$

59. $4 \ln (x + 3.4) = 2.5$

60. $\ln x^2 = -x^2$

61. $\log_8 x + \log_8 (x + 2) = 2$

62. $\log_3 x + 7 = 4 - \log_5 x$

63. $\log_5 (x + 7) - \log_5 (2x - 3) = 1$

Approximate the point(s) of intersection of the pair of equations.

64. $y = \ln 3x, \ y = 3x - 8$

65. $2.3x + 3.8y = 12.4, \ y = 1.1 \ln (x - 2.05)$

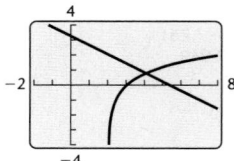

66. $y = 2.3 \ln(x + 10.7)$, $y = 10e^{-0.07x^2}$

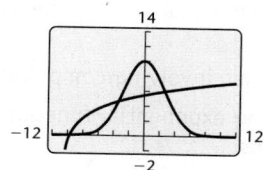

67. $y = 2.3 \ln(x + 10.7)$, $y = 10e^{-0.007x^2}$

Collaborative Discussion and Writing

68. In Example 3, we took the natural logarithm on both sides of the equation. What would have happened had we used the common logarithm? Explain which approach seems better to you and why.

69. Explain how Exercises 33 and 34 could be solved using the graph of $f(x) = \ln x$.

Skill Maintenance

In Exercises 70–73:

a) *Find the vertex.*
b) *Find the axis of symmetry.*
c) *Determine whether there is a maximum or minimum value and find that value.*

70. $f(x) = -x^2 + 6x - 8$

71. $g(x) = x^2 - 6$

72. $H(x) = 3x^2 - 12x + 16$

73. $G(x) = -2x^2 - 4x - 7$

Synthesis

Solve using any method.

74. $\dfrac{5^x - 5^{-x}}{5^x + 5^{-x}} = 8$

75. $\dfrac{e^x + e^{-x}}{e^x - e^{-x}} = 3$

76. $\ln(\ln x) = 2$

77. $\ln(\log x) = 0$

78. $\ln \sqrt[4]{x} = \sqrt{\ln x}$

79. $\sqrt{\ln x} = \ln \sqrt{x}$

80. $\log_3(\log_4 x) = 0$

81. $(\log_3 x)^2 - \log_3 x^2 = 3$

82. $(\log x)^2 - \log x^2 = 3$

83. $\ln x^2 = (\ln x)^2$

84. $e^{2x} - 9 \cdot e^x + 14 = 0$

85. $5^{2x} - 3 \cdot 5^x + 2 = 0$

86. $x\left(\ln \tfrac{1}{6}\right) = \ln 6$

87. $\log_3 |x| = 2$

88. $x^{\log x} = \dfrac{x^3}{100}$

89. $\ln x^{\ln x} = 4$

90. $\dfrac{(e^{3x+1})^2}{e^4} = e^{10x}$

91. $\dfrac{\sqrt{(e^{2x} \cdot e^{-5x})^{-4}}}{e^x \div e^{-x}} = e^7$

92. $e^x < \dfrac{4}{5}$

93. $|\log_5 x| + 3\log_5 |x| = 4$

94. $|2^{x^2} - 8| = 3$

95. Given that $a = \log_8 225$ and $b = \log_2 15$, express a as a function of b.

96. Given that $a = (\log_{125} 5)^{\log_5 125}$, find the value of $\log_3 a$.

97. Given that
$$\log_2[\log_3(\log_4 x)] = \log_3[\log_2(\log_4 y)]$$
$$= \log_4[\log_3(\log_2 z)]$$
$$= 0,$$
find $x + y + z$.

98. Given that $f(x) = e^x - e^{-x}$, find $f^{-1}(x)$ if it exists.

5.6

Applications and Models: Growth and Decay; Compound Interest

❖ Solve applied problems involving exponential growth and decay.

❖ Solve applied problems involving compound interest.

❖ Find models involving exponential functions and logarithmic functions.

Exponential functions and logarithmic functions with base e are rich in applications to many fields such as business, science, psychology, and sociology.

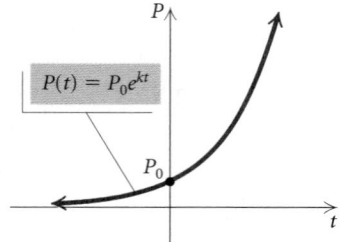

❖ Population Growth

The function

$$P(t) = P_0e^{kt}, \quad k > 0$$

is a model of many kinds of population growth, whether it be a population of people, bacteria, cellular phones, or money. In this function, P_0 is the population at time 0, P is the population after time t, and k is called the **exponential growth rate**. The graph of such an equation is shown at left.

EXAMPLE 1 *Population Growth of China.* In 2006, the population of China was about 1.314 billion, and the exponential growth rate was 0.6% per year (*Source*: *Time Almanac* 2007).

a) Find the exponential growth function.

b) Graph the exponential growth function.

c) Estimate the population in 2010.

d) After how long will the population be double what it was in 2006?

Solution

a) At $t = 0$ (2006), the population was 1.314 billion and the exponential growth rate was 0.6% per year. We substitute 1.314 for P_0 and 0.6%, or 0.006, for k to obtain the exponential growth function

$$P(t) = 1.314e^{0.006t},$$

where t is the number of years after 2006 and $P(t)$ is in billions.

b) Using a graphing calculator, we obtain the graph of the exponential growth function, shown at left.

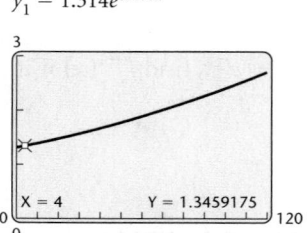

$y_1 = 1.314e^{0.006x}$

c) In 2010, $t = 4$; that is, 4 yr have passed since 2006. To find the population in 2010, we substitute 4 for t:

$$P(4) = 1.314e^{0.006(4)} = 1.314e^{0.024} \approx 1.346.$$

We can also use the VALUE feature from the CALC menu on a graphing calculator to find $P(4)$. (See the window at left.) The population will be about 1.346 billion, or 1,346,000,000, in 2010.

d) We are looking for the time T for which $P(T) = 2 \cdot 1.314$, or 2.628. The number T is called the **doubling time**. To find T, we solve the equation

$$2.628 = 1.314e^{0.006T}$$

using both an algebraic method and a graphical method.

ALGEBRAIC SOLUTION	GRAPHICAL SOLUTION

ALGEBRAIC SOLUTION

We have

$$2.628 = 1.314e^{0.006T}$$ Substituting 2.628 for $P(T)$

$$2 = e^{0.006T}$$ Dividing by 1.314

$$\ln 2 = \ln e^{0.006T}$$ Taking the natural logarithm on both sides

$$\ln 2 = 0.006T$$ $\ln e^x = x$

$$\frac{\ln 2}{0.006} = T$$ Dividing by 0.006

$$115.5 \approx T.$$

The population of China will be double what it was in 2006 about 115.5 yr after 2006.

GRAPHICAL SOLUTION

Using the Intersect method, we graph the equations

$$y_1 = 1.314e^{0.006x} \quad \text{and} \quad y_2 = 2.628$$

and find the first coordinate of their point of intersection.

$y_1 = 1.314e^{0.006x}, \quad y_2 = 2.628$

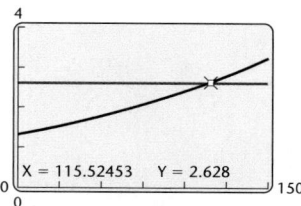

X = 115.52453   Y = 2.628

The solution is about 115.5, so the population of China will be double that of 2006 about 115.5 yr after 2006.

Now Try Exercise 1. ■

❊ Interest Compounded Continuously

When interest is paid on interest, we call it **compound interest**. Suppose that an amount P_0 is invested in a savings account at interest rate k **compounded continuously**. The amount $P(t)$ in the account after t years is given by the exponential function

$$P(t) = P_0 e^{kt}.$$

EXAMPLE 2 *Interest Compounded Continuously.* Suppose that $2000 is invested at interest rate k, compounded continuously, and grows to $2504.65 in 5 yr.

a) What is the interest rate?

b) Find the exponential growth function.

c) What will the balance be after 10 yr?

d) After how long will the $2000 have doubled?

STUDY TIP

Newspapers and magazines are full of mathematical applications, many of which follow the exponential model. Find an application and share it with your class. As you obtain higher math skills, you will more readily observe the world from a mathematical perspective. Math courses become more interesting when we connect the concepts to the real world.

Solution

a) At $t = 0, P(0) = P_0 = \$2000$. Thus the exponential growth function is of the form

$$P(t) = 2000e^{kt}.$$

We know that $P(5) = \$2504.65$. We substitute and solve for k:

$$2504.65 = 2000e^{k(5)} \qquad \text{Substituting 2504.65 for } P(t) \text{ and 5 for } t$$

$$2504.65 = 2000e^{5k}$$

$$\frac{2504.65}{2000} = e^{5k} \qquad \text{Dividing by 2000}$$

$$\ln \frac{2504.65}{2000} = \ln e^{5k} \qquad \text{Taking the natural logarithm}$$

$$\ln \frac{2504.65}{2000} = 5k \qquad \text{Using } \ln e^x = x$$

$$\frac{\ln \frac{2504.65}{2000}}{5} = k \qquad \text{Dividing by 5}$$

$$0.045 \approx k.$$

The interest rate is about 0.045, or 4.5%.

We can also find k by graphing the equations

$$y_1 = 2000e^{5x} \quad \text{and} \quad y_2 = 2504.65$$

and use the Intersect feature to approximate the first coordinate of the point of intersection.

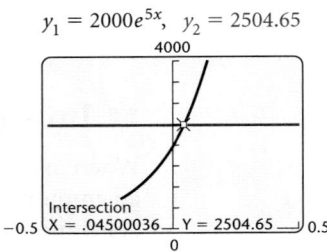

The interest rate is about 0.045, or 4.5%.

b) Substituting 0.045 for k in the function $P(t) = 2000e^{kt}$, we see that the exponential growth function is

$$P(t) = 2000e^{0.045t}.$$

c) The balance after 10 yr is

$$P(10) = 2000e^{0.045(10)} = 2000e^{0.45} \approx \$3136.62.$$

d) To find the doubling time T, we set $P(T) = 2 \cdot P_0 = 2 \cdot \$2000 = \$4000$ and solve for T. We solve

$$4000 = 2000e^{0.045T}$$

using both an algebraic method and a graphical method.

ALGEBRAIC SOLUTION

We have

$$4000 = 2000e^{0.045T}$$

$$2 = e^{0.045T} \qquad \text{Dividing by 2000}$$

$$\ln 2 = \ln e^{0.045T} \qquad \text{Taking the natural logarithm}$$

$$\ln 2 = 0.045T \qquad \ln e^x = x$$

$$\frac{\ln 2}{0.045} = T \qquad \text{Dividing by 0.045}$$

$$15.4 \approx T.$$

Thus the original investment of $2000 will double in about 15.4 yr.

GRAPHICAL SOLUTION

We use the Zero method. We graph the equation

$$y = 2000e^{0.045x} - 4000$$

and find the zero of the function. The zero of the function is the solution of the equation.

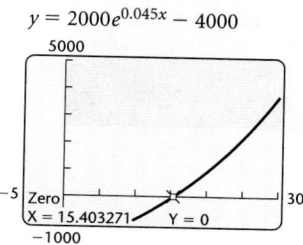

$$y = 2000e^{0.045x} - 4000$$

The solution is about 15.4, so the original investment of $2000 will double in about 15.4 yr.

Now Try Exercise 7. ▪

We can find a general expression relating the growth rate k and the doubling time T by solving the following equation:

$$2P_0 = P_0 e^{kT} \qquad \text{Substituting } 2P_0 \text{ for } P \text{ and } T \text{ for } t$$

$$2 = e^{kT} \qquad \text{Dividing by } P_0$$

$$\ln 2 = \ln e^{kT} \qquad \text{Taking the natural logarithm}$$

$$\ln 2 = kT \qquad \text{Using } \ln e^x = x$$

$$\frac{\ln 2}{k} = T.$$

Growth Rate and Doubling Time

The **growth rate k** and the **doubling time T** are related by

$$kT = \ln 2, \quad \text{or} \quad k = \frac{\ln 2}{T}, \quad \text{or} \quad T = \frac{\ln 2}{k}.$$

Note that the relationship between k and T does not depend on P_0.

EXAMPLE 3 *World Population Growth.* The population of the world is now doubling every 60.8 yr (*Source: Time Almanac* 2007). What is the exponential growth rate?

Solution We have

$$k = \frac{\ln 2}{T} = \frac{\ln 2}{60.8} \approx 0.0114 \approx 1.14\%.$$

The growth rate of the world population is about 1.14% per year.

Now Try Exercise 3(e). ■

❖ Models of Limited Growth

The model $P(t) = P_0 e^{kt}$, $k > 0$, has many applications involving unlimited population growth. However, in some populations, there can be factors that prevent a population from exceeding some limiting value—perhaps a limitation on food, living space, or other natural resources. One model of such growth is

$$P(t) = \frac{a}{1 + be^{-kt}}.$$

This is called a **logistic function**. This function increases toward a *limiting value a* as $t \leq \infty$. Thus, $y = a$ is the horizontal asymptote of the graph of $P(t)$.

EXAMPLE 4 *Limited Population Growth.* A ship carrying 1000 passengers has the misfortune to be shipwrecked on a small island from which the passengers are never rescued. The natural resources of the island limit the population to 5780. The population gets closer and closer to this limiting value, but never reaches it. The population of the island after time t, in years, is given by the logistic function

$$P(t) = \frac{5780}{1 + 4.78e^{-0.4t}}.$$

a) Graph the function.

b) Find the population after 0, 1, 2, 5, 10, and 20 yr.

Solution

a) We use a graphing calculator to graph the function. The graph is the S-shaped curve shown at left. Note that this function increases toward a limiting value of 5780. The graph has $y = 5780$ as a horizontal asymptote.

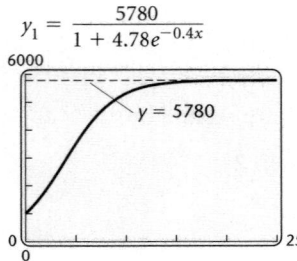

$$y_1 = \frac{5780}{1 + 4.78e^{-0.4x}}$$

b) We can use the TABLE feature on a graphing calculator set in ASK mode to find the function values. (See the window on the left below.) The VALUE feature can also be used. (See the window on the right below.)

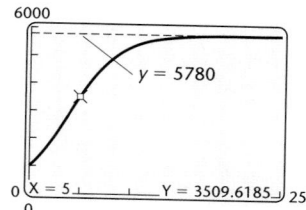

X	Y1
0	1000
1	1374.8
2	1836.2
5	3509.6
10	5314.7
20	5770.7

X =

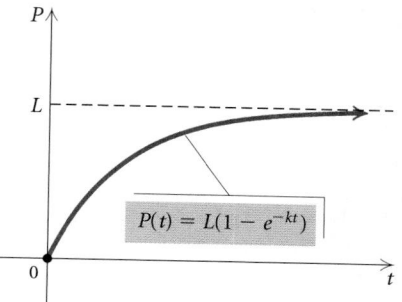

Thus the population will be about 1000 after 0 yr, 1375 after 1 yr, 1836 after 2 yr, 3510 after 5 yr, 5315 after 10 yr, and 5771 after 20 yr.

Now Try Exercise 17.

Another model of limited growth is provided by the function

$$P(t) = L(1 - e^{-kt}), \quad k > 0,$$

which is shown graphed at right. This function also increases toward a limiting value L, as $t \to \infty$, so $y = L$ is the horizontal asymptote of the graph of $P(t)$.

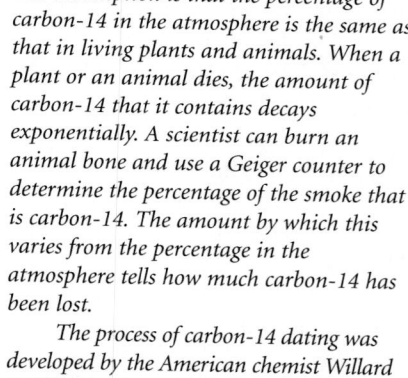

How can scientists determine that an animal bone has lost 30% of its carbon-14? The assumption is that the percentage of carbon-14 in the atmosphere is the same as that in living plants and animals. When a plant or an animal dies, the amount of carbon-14 that it contains decays exponentially. A scientist can burn an animal bone and use a Geiger counter to determine the percentage of the smoke that is carbon-14. The amount by which this varies from the percentage in the atmosphere tells how much carbon-14 has been lost.

The process of carbon-14 dating was developed by the American chemist Willard E. Libby in 1952. It is known that the radioactivity in a living plant is 16 disintegrations per gram per minute. Since the half-life of carbon-14 is 5750 years, an object with an activity of 8 disintegrations per gram per minute is 5750 years old, one with an activity of 4 disintegrations per gram per minute is 11,500 years old, and so on. Carbon-14 dating can be used to measure the age of objects up to 40,000 years old. Beyond such an age, it is too difficult to measure the radioactivity and some other method would have to be used.

Carbon-14 dating was used to find the age of the Dead Sea Scrolls. It was also used to refute the authenticity of the Shroud of Turin, presumed to have covered the body of Christ.

❄ Exponential Decay

The function

$$P(t) = P_0 e^{-kt}, \quad k > 0$$

is an effective model of the decline, or decay, of a population. An example is the decay of a radioactive substance. In this case, P_0 is the amount of the substance at time $t = 0$, and $P(t)$ is the amount of the substance left after time t, where k is a positive constant that depends on the situation. The constant k is called the **decay rate**.

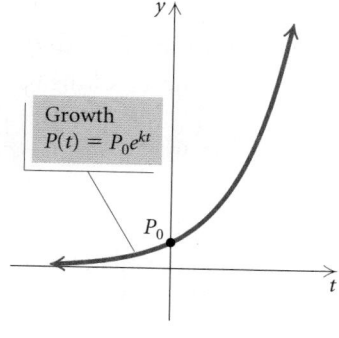

Growth
$P(t) = P_0 e^{kt}$

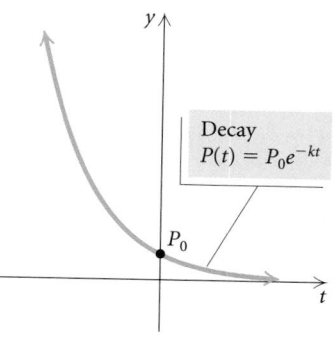

Decay
$P(t) = P_0 e^{-kt}$

The **half-life** of bismuth is 5 days. This means that half of an amount of bismuth will cease to be radioactive in 5 days. The effect of half-life T for nonnegative inputs is shown in the graph below. The exponential function gets close to 0, but never reaches 0, as t gets very large. Thus, according to an exponential decay model, a radioactive substance never completely decays.

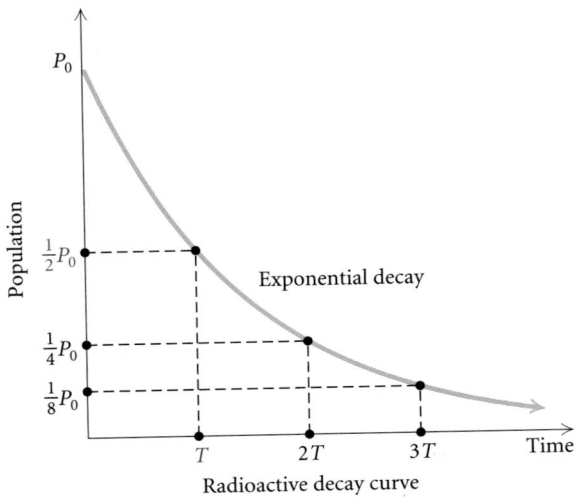

Radioactive decay curve

We can find a general expression relating the decay rate k and the half-life time T by solving the following equation:

$$\frac{1}{2}P_0 = P_0 e^{-kT} \qquad \text{Substituting } \frac{1}{2}P_0 \text{ for } P \text{ and } T \text{ for } t$$

$$\frac{1}{2} = e^{-kT} \qquad \text{Dividing by } P_0$$

$$\ln\frac{1}{2} = \ln e^{-kT} \qquad \text{Taking the natural logarithm}$$

$$\ln 2^{-1} = -kT \qquad \frac{1}{2} = 2^{-1}; \ln e^x = x$$

$$-\ln 2 = -kT \qquad \text{Using the power rule}$$

$$\frac{\ln 2}{k} = T. \qquad \text{Dividing by } -k$$

Decay Rate and Half-Life

The decay rate k and the half-life T are related by

$$kT = \ln 2, \quad \text{or} \quad k = \frac{\ln 2}{T}, \quad \text{or} \quad T = \frac{\ln 2}{k}.$$

Note that the relationship between decay rate and half-life is the same as that between growth rate and doubling time.

In 1947, a Bedouin youth looking for a stray goat climbed into a cave at Kirbet Qumran on the shores of the Dead Sea near Jericho and came upon earthenware jars containing an incalculable treasure of ancient manuscripts. Shown here are fragments of those Dead Sea Scrolls, a portion of some 600 or so texts found so far and which concern the Jewish books of the Bible. Officials date them before 70 A.D., making them the oldest Biblical manuscripts by 1000 years.

EXAMPLE 5 *Carbon Dating.* The radioactive element carbon-14 has a half-life of 5750 yr. The percentage of carbon-14 present in the remains of organic matter can be used to determine the age of that organic matter. Archaeologists discovered that the linen wrapping from one of the Dead Sea Scrolls had lost 22.3% of its carbon-14 at the time it was found. How old was the linen wrapping?

Solution   We first find k when the half-life T is 5750 yr:

$$k = \frac{\ln 2}{T}$$

$$k = \frac{\ln 2}{5750} \qquad \text{Substituting 5750 for } T$$

$$k = 0.00012.$$

Now we have the function

$$P(t) = P_0 e^{-0.00012t}.$$

(This function can be used for any subsequent carbon-dating problem.) If the linen wrapping has lost 22.3% of its carbon-14 from an initial amount P_0, then $77.7\% P_0$ is the amount present. To find the age t of the wrapping, we solve the following equation for t:

$$77.7\% P_0 = P_0 e^{-0.00012t} \qquad \text{Substituting } 77.7\% P_0 \text{ for } P$$

$$0.777 = e^{-0.00012t} \qquad \text{Dividing by } P_0 \text{ and writing 77.7\% as 0.777}$$

$$\ln 0.777 = \ln e^{-0.00012t} \qquad \text{Taking the natural logarithm on both sides}$$

$$\ln 0.777 = -0.00012t \qquad \ln e^x = x$$

$$\frac{\ln 0.777}{-0.00012} = t \qquad \text{Dividing by } -0.00012$$

$$2103 \approx t.$$

Thus the linen wrapping on the Dead Sea Scrolls was about 2103 yr old when it was found.

Now Try Exercise 9. ■

❋ Exponential and Logarithmic Curve Fitting

We have added several new functions that can be considered when we fit curves to data. Let's review some of them.

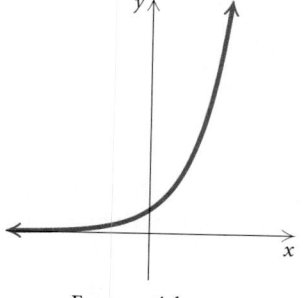

Exponential:
$f(x) = ab^x$, or ae^{kx}
$a > 0, b > 1, k > 0$

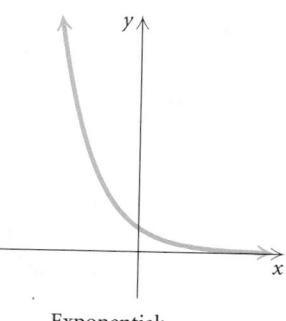

Exponential:
$f(x) = ab^{-x}$, or ae^{-kx}
$a > 0, b > 1, k > 0$

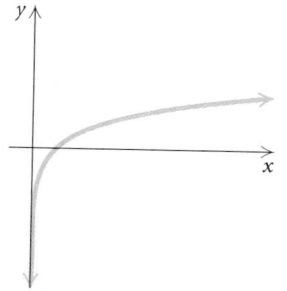

Logarithmic:
$f(x) = a + b \ln x$
$b > 0$

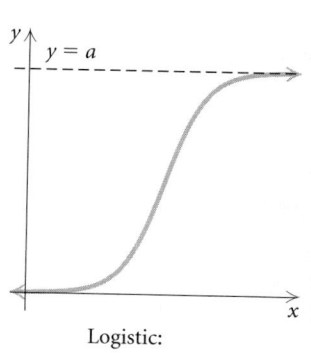

Logistic:
$f(x) = \dfrac{a}{1 + be^{-kx}}$
$a, b, k > 0$

Now, when we analyze a set of data for curve fitting, these models can be considered as well as polynomial functions (such as linear, quadratic, cubic, and quartic functions) and rational functions.

GCM | **EXAMPLE 6** *Surveillance Cameras.* The number of U.S. communities using surveillance cameras at intersections has greatly increased in recent years, as shown in the following table.

Year, *x*	Number of U.S. Communities Using Surveillance Cameras at Intersections
1999, 0	19
2001, 2	35
2003, 4	75
2005, 6	130
2007, 8	243

Source: Insurance Institute for Highway Safety

a) Use a graphing calculator to fit an exponential function to the data.

b) Graph the function with the scatterplot of the data.

c) Estimate the number of U.S. communities using surveillance cameras at intersections in 2010.

Solution

a) We will fit an equation of the type $y = a \cdot b^x$ to the data, where x is the number of years since 1999. Entering the data into the calculator and carrying out the regression procedure, we find that the equation is

$$y = 19.17654555(1.377777324)^x.$$

(See Fig. 1.) The correlation coefficient is very close to 1. This gives us an indication that the exponential function fits the data well.

b) The graph is shown in Fig. 2 on the left below.

```
ExpReg
y=a*b^x
a=19.17654555
b=1.377777324
r²=.9978104488
r=.9989046245
```

FIGURE 1

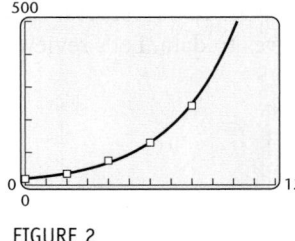

FIGURE 2

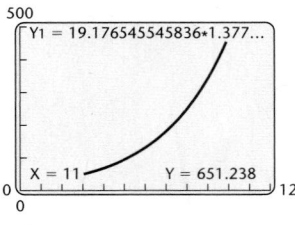

FIGURE 3

c) Using the VALUE feature in the CALC menu (see Fig. 3 on the right above), we evaluate the function found in part (a) for $x = 11$ ($2010 - 1999 = 11$), and estimate the number of communities using surveillance cameras at intersections in 2010 to be about 651. ·

Now Try Exercise 29. ▪

On some graphing calculators, there may be a REGRESSION feature that yields an exponential function, base *e*. If not, and you wish to find such a function, a conversion can be done using the following.

Converting from Base *b* to Base *e*

$$b^x = e^{x(\ln b)}$$

Then, for the equation in Example 6, we have

$$y = 19.17654555(1.377777324)^x$$
$$= 19.17654555e^{x(\ln 1.377777324)}$$
$$= 19.17654555e^{0.3204715659x}. \qquad \textbf{ln 1.377777324} \approx \textbf{0.3204715659}$$

We can prove this conversion formula using properties of logarithms, as follows:

$$e^{x(\ln b)} = e^{\ln b^x} = b^x.$$

5.6 Exercise Set

1. *World Population Growth.* In 2006, the world population was 6.5 billion. The exponential growth rate was 1.14% per year.

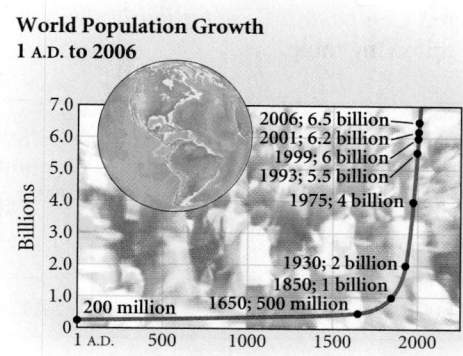

World Population Growth
1 A.D. to 2006

2006; 6.5 billion
2001; 6.2 billion
1999; 6 billion
1993; 5.5 billion
1975; 4 billion
1930; 2 billion
1850; 1 billion
1650; 500 million
200 million

Sources: U.S. Census Bureau; International Data Base; World Population Profile

a) Find the exponential growth function.
b) Estimate the population of the world in 2009 and in 2015.

c) When will the world population be 8 billion?
d) Find the doubling time.

2. *Population Growth of Rabbits.* Under ideal conditions, a population of rabbits has an exponential growth rate of 11.7% per day. Consider an initial population of 100 rabbits.

a) Find the exponential growth function.
b) Graph the function.
c) What will the population be after 7 days? after 2 weeks?
d) Find the doubling time.

3. *Population Growth.* Complete the following table.

	Population	Growth Rate, *k*	Doubling Time, *T*
a) Rwanda		2.4% per year	
b) Brazil			63 yr

(continued)

Population	Growth Rate, k	Doubling Time, T
c) India	1.4% per year	
d) Finland	0.2% per year	
e) Egypt		38.5 yr
f) Philippines		36.5 yr
g) United States	0.9% per year	
h) Japan		69.3 yr
i) Ireland	1.2% per year	
j) Kenya	2.6% per year	

4. *DVD Videos.* The total number of DVD videos produced and shipped in 1998 was 0.5 million. In 2004, the total number of units reached 29.01 million.

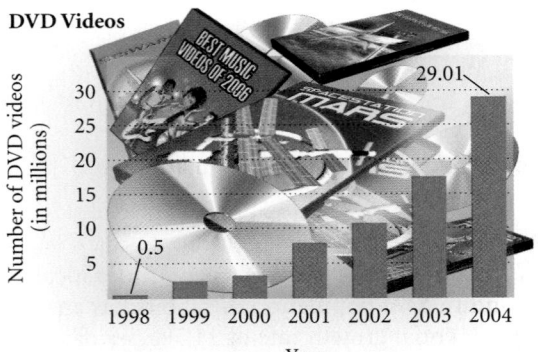

DVD Videos

Source: Recording Industry Association of America

Assuming the exponential growth model applies:
a) Find the value of k and write the function.
b) Estimate the number of DVD videos produced and shipped in 2005, in 2008, and in 2011.

5. *Population Growth of Israel.* The population of Israel has a growth rate of 1.3% per year. In 2005, the population was 6,276,883. The land area of Israel is 24,839,654,400 square yards. (*Source: Statistical Abstract of the United States*) Assuming this growth rate continues and is exponential,

after how long will there be one person for every square yard of land?

6. *Value of Manhattan Island.* In 1626, Peter Minuit of the Dutch West India Company purchased Manhattan Island from Native Americans for $24. Assuming an exponential rate of inflation of 6% per year, find the value of Manhattan Island in 2010.

7. *Interest Compounded Continuously.* Suppose that $10,000 is invested at an interest rate of 5.4% per year, compounded continuously.
a) Find the exponential function that describes the amount in the account after time t, in years.
b) What is the balance after 1 yr? 2 yr? 5 yr? 10 yr?
c) What is the doubling time?

8. *Interest Compounded Continuously.* Complete the following table.

Initial Investment at $t = 0$, P_0	Interest Rate, k	Doubling Time, T	Amount After 5 yr
a) $35,000	6.2%		
b) $5000			$7,130.90
c)	8.4%		$11,414.71
d)		11 yr	$17,539.32

9. *Carbon Dating.* A mummy discovered in the pyramid Khufu in Egypt has lost 46% of its carbon-14. Determine its age.

10. *Tomb in the Valley of the Kings.* In February 2006, in the Valley of the Kings in Egypt, a team of archaeologists uncovered the first tomb since King Tut's tomb was found in 1922. The tomb contained five wooden sarcophagi that contained mummies. The archaeologists believe that the mummies are from the 18th Dynasty, about 3300 to 3500 yr ago. Determine the amount of carbon-14 that the mummies have lost.

11. *Radioactive Decay.* Complete the following table.

Radioactive Substance	Decay Rate, k	Half-life, T
a) Polonium		3 min
b) Lead		22 yr
c) Iodine-131	9.6% per day	
d) Krypton-85	6.3% per year	
e) Strontium-90		25 yr
f) Uranium-238		4560 yr
g) Plutonium		23,105 yr

12. *Number of Farms.* The number N of farms in the United States has declined continually since

1950. In 1950, there were 5,650,000 farms, and in 2005 that number had decreased to 2,100,990 (*Sources:* U.S. Department of Agriculture; National Agricultural Statistics Service).

Assuming the number of farms decreased according to the exponential decay model:

a) Find the value of k, and write an exponential function that describes the number of farms after time t, in years, where t is the number of years since 1950.

b) Estimate the number of farms in 2009 and in 2015.

c) At this decay rate, when will only 1,000,000 farms remain?

13. *Cases of Tuberculosis.* The number of cases of tuberculosis in the United States has decreased continually since 1956, as shown in the graph below. In 1956 ($t = 0$), there were 69,895 cases. By 2006 ($t = 50$), this number had decreased over 80% to 13,767 cases.

Cases of Tuberculosis in the United States

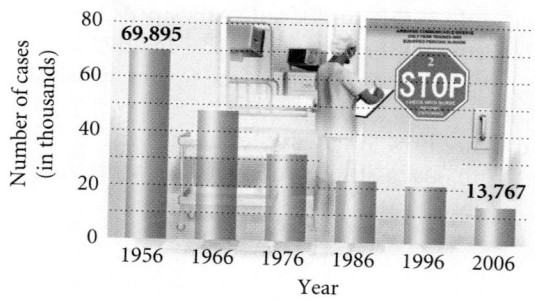

Source: U.S. Centers for Disease Control and Prevention

a) Find the value of k, and write an exponential function that describes the number of tuberculosis cases after time t, in years, where t is the number of years since 1956.

b) Estimate the number of cases in 2008 and in 2010.

c) At this decay rate, in what year will there be 5000 cases?

14. *1957 Studebaker Golden Hawk.*   The 1957 Studebaker Golden Hawk has become a car of interest to those investing in classic cars. In 1967, a 10-year-old Golden Hawk sold for only $800, and in 2006, this car, in top condition, was worth about $27,000. (*Source: 1957 Studebaker Golden Hawk,* George Mattar, *Hemmings Motor News,* June 2006, p. 36)

Assuming the value V_0 of the car has grown exponentially:

a) Find the value of k and determine the exponential growth function, assuming $V_0 = 800$ and t is the number of years since 1967.

b) Estimate the value of the car in 2008.

c) What is the doubling time for the value of the car?

d) After how long will the value of the car be $40,000, assuming there is no change in the growth rate?

15. *Norman Rockwell Painting.*   *Breaking Home Ties*, painted by Norman Rockwell, appeared on the cover of the *Saturday Evening Post* in 1954. In 1960, the original sold for only $900; in 2006, this painting was sold at Sotheby's auction house for $15.4 million (*Source: Associated Press, Indianapolis Star,* December 2, 2006, p. B5).

Assuming the value R_0 of the painting has grown exponentially:

a) Find the value of k and determine the exponential growth function, assuming $R_0 = 900$ and t is the number of years since 1960.

b) Estimate the value of the painting in 2010.

c) What is the doubling time for the value of the painting?

d) After how long will the value of the painting be $25 million, assuming there is no change in the growth rate?

16. *T206 Wagner Baseball Card.*   In 1909, the Pittsburgh Pirates shortstop Honus Wagner forced the American Tobacco Company to withdraw his baseball card that was packaged with cigarettes. Fewer than 60 of the Wagner cards still exist. In 1971, a Wagner card sold for $1000; and in September 2007, a card in near-mint condition was purchased for a record $2.8 million (*Source: USA Today,* 9/6/07; Kathy Willens/AP).

Assuming the value W_0 of the baseball card has grown exponentially:

a) Find the value of k and determine the exponential growth function, assuming $W_0 = 1000$ and t is the number of years since 1971.

b) Estimate the value of the Wagner card in 2011.

c) What is the doubling time for the value of the card?

d) After how long will the value of the Wagner card be \$3 million, assuming there is no change in the growth rate?

17. *Spread of an Epidemic.* In a town whose population is 3500, a disease creates an epidemic. The number of people N infected t days after the disease has begun is given by the function

$$N(t) = \frac{3500}{1 + 19.9e^{-0.6t}}.$$

a) Graph the function.

b) How many are initially infected with the disease $(t = 0)$?

c) Find the number infected after 2 days, 5 days, 8 days, 12 days, and 16 days.

d) Using this model, can you say whether all 3500 people will ever be infected? Explain.

18. *Limited Population Growth in a Lake.* A lake is stocked with 400 fish of a new variety. The size of the lake, the availability of food, and

the number of other fish restrict the growth of that type of fish in the lake to a limiting value of 2500. The population of fish in the lake after time t, in months, is given by the function

$$P(t) = \frac{2500}{1 + 5.25e^{-0.32t}}.$$

a) Graph the function.

b) Find the population after 0, 1, 5, 10, 15, and 20 months.

Newton's Law of Cooling. Suppose that a body with temperature T_1 is placed in surroundings with temperature T_0 different from that of T_1. The body will either cool or warm to temperature $T(t)$ after time t, in minutes, where

$$T(t) = T_0 + (T_1 - T_0)e^{-kt}.$$

Use this law in Exercises 19–22.

19. A cup of coffee with temperature 105°F is placed in a freezer with temperature 0°F. After 5 min, the temperature of the coffee is 70°F. What will its temperature be after 10 min?

20. A dish of lasagna baked at 375°F is taken out of the oven at 11:15 A.M. in a kitchen that is 72°F. After 3 min, the temperature of the lasagna is 365°F. What will the temperature of the lasagna be at 11:30 A.M.?

21. A chilled jello salad that has a temperature of 43°F is taken from the refrigerator and placed on the dining room table in a room that is 68°F. After 12 min, the temperature of the salad is 55°F. What will the temperature of the salad be after 20 min?

22. *When Was the Murder Committed?* The police discover the body of a murder victim. Critical to solving the crime is determining when the murder was committed. The coroner arrives at the murder scene at 12:00 P.M. She immediately takes the temperature of the body and finds it to be 94.6°F. She then takes the temperature 1 hr later and finds it to be 93.4°F. The temperature of the room is 70°F. When was the murder committed?

In Exercises 23–28, determine which, if any, of these functions might be used as a model for the data in the scatterplot.

a) *Quadratic,* $f(x) = ax^2 + bx + c$
b) *Polynomial, not quadratic*
c) *Exponential,* $f(x) = ab^x$, *or* $P_0 e^{kx}$, $k > 0$
d) *Exponential,* $f(x) = ab^{-x}$, *or* $P_0 e^{-kx}$, $k > 0$
e) *Logarithmic,* $f(x) = a + b \ln x$
f) *Logistic,* $f(x) = \dfrac{a}{1 + be^{-kx}}$

23.   **24.**

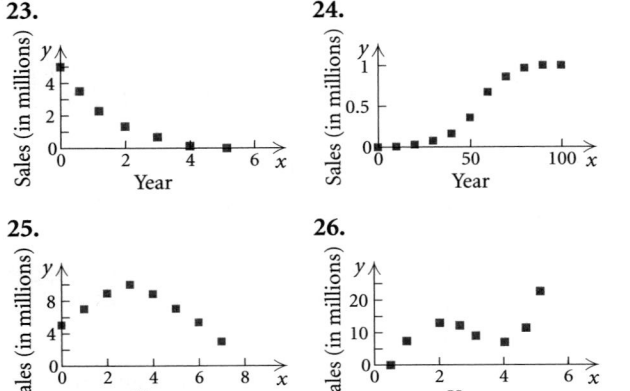

25.   **26.**

27.   **28.**

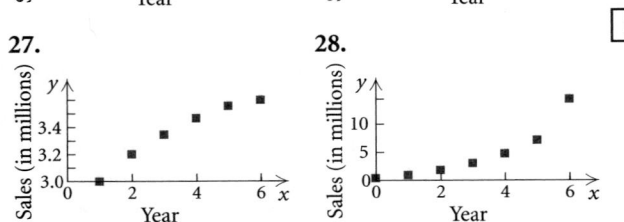

29. *Percent of Americans 85 and Older.* In 1900, 0.2% of the U.S. population, or 122,000 people, were 85 and older. The number of people 85 and older reached 5,096,000 in 2005. The table below lists data regarding the percentage of the U.S. population 85 and older in selected years from 1900 to 2005.

a) Use a graphing calculator to fit an exponential function to the data, where x is the number of years after 1900. Determine whether the function is a good fit.
b) Graph the function found in part (a) with a scatterplot of the data.
c) Estimate the percentage of the U.S. population 85 and older in 2007, in 2015, and in 2020.

Year, x	Percent of U.S. Population 85 and Older, y
1900, 0	0.2%
1910, 10	0.2
1920, 20	0.2
1930, 30	0.2
1940, 40	0.3
1950, 50	0.4
1960, 60	0.5
1970, 70	0.7
1980, 80	1.0
1990, 90	1.2
1995, 95	1.4
2000, 100	1.5
2001, 101	1.6
2002, 102	1.6
2003, 103	1.6
2004, 104	1.7
2005, 105	1.7

Sources: U.S. Census Bureau; U.S. Department of Commerce

GCM **30.** *Forgetting.* In an art class, students were given a final exam at the end of the course. Then they were retested with an equivalent test at subsequent time intervals. Their scores after time x, in months, are given in the following table.

Time, x (in months)	Score, y
1	84.9%
2	84.6
3	84.4
4	84.2
5	84.1
6	83.9

a) Use a graphing calculator to fit a logarithmic function $y = a + b \ln x$ to the data.
b) Use the function to predict test scores after 8, 10, 24, and 36 months.
c) After how long will the test scores fall below 82%?

31. *College Applications.*   Acceptance to the college of one's choice has become increasingly competitive. The table below lists the percent of college applicants who sent out 7 or more applications.

Year, x	Percent of College Applicants Who Sent Out 7 or More Applications, y
1967, 0	1.8%
1977, 10	4.0
1987, 20	7.9
1997, 30	10.8
2005, 38	17.4

Source: Higher Education Research Institute at UCLA

a) Create a scatterplot of the data. Let x = the number of years since 1967.
b) Use a graphing calculator to fit linear, quadratic, and exponential functions to the data, where x is the number of years after 1967. Determine which function has the best fit.
c) Graph all three functions found in part (b) with the scatterplot in part (a).
d) Use the functions found in part (b) to estimate the percent of college applicants who will send out 7 or more applications in 2008. Which function provides the most realistic prediction?

32. *Cost of Political Conventions.*   The total cost of the Democratic and the Republican national conventions has increased 638% over the 24-year period between 1980 and 2004. The table below lists the total cost, in millions of dollars, for selected years.

a) Use a graphing calculator to fit the data with an exponential function, where x is the number of years after 1980.
b) Use the function to estimate the cost in 2008 and in 2012.

c) In what year is total cost expected to exceed $546 million?

Year, x	Total Cost of Conventions, C(x) (in millions)
1980, 0	$ 23.1
1984, 4	31.8
1988, 8	44.4
1992, 12	58.8
1996, 16	90.6
2000, 20	160.8
2004, 24	170.5

Source: Campaign Finance Institute

33. *Obesity-Related Surgeries.*   The number of obesity-related surgeries in Indiana more than quadrupled from 1995 to 2002. The table below lists the number of surgeries for these years.

Year, x	Number of Obesity-Related Surgeries in Indiana, y
1995, 0	332
1996, 1	297
1997, 2	396
1998, 3	445
1999, 4	585
2000, 5	911
2001, 6	1146
2002, 7	1611

Source: Indiana State Department of Health

a) Use a graphing calculator to fit the data with an exponential function, where x is the number of years after 1995.
b) Use the function to estimate the number of obesity-related surgeries in Indiana in 2005 and in 2009.
c) In what year will the number of surgeries reach 12,000?

34. *Effect of Advertising.* A company introduced a new software product on a trial run in a city. They advertised the product on television and found the following data regarding the percent P of people who bought the product after x ads were run.

Number of Ads, x	Percentage Who Bought, P
0	0.2%
10	0.7
20	2.7
30	9.2
40	27.0
50	57.6
60	83.3
70	94.8
80	98.5
90	99.6

GCM

a) Use a graphing calculator to fit a logistic function

$$P(x) = \frac{a}{1 + be^{-kx}}$$

to the data.
b) What percent of people bought the product when 55 ads were run? 100 ads?
c) Find the horizontal asymptote for the graph. Interpret the asymptote in terms of the advertising situation.

Collaborative Discussion and Writing

35. Browse through some newspapers or magazines to find some data that appear to fit an exponential model. Make a case for why such a fit is appropriate. Then fit an exponential function to the data and make some predictions.

36. *Atmospheric Pressure.* Atmospheric pressure P at an altitude a is given by

$$P = P_0 e^{-0.00005a},$$

where P_0 is the pressure at sea level, approximately 14.7 lb/in^2 (pounds per square inch). Explain how a barometer, or some device for

measuring atmospheric pressure, can be used to find the height of a skyscraper.

Skill Maintenance

Fill in the blank with the correct name of the principle or rule from the given choices.

principle of zero products
multiplication principle for equations
product rule
addition principle for inequalities
power rule
multiplication principle for inequalities
principle of square roots
quotient rule

37. For any real numbers a, b, and c: If $a < b$ and $c > 0$ are true, then $ac < bc$ is true. If $a < b$ and $c < 0$ are true, then $ac > bc$ is true.

_____

38. For any positive numbers M and N and any logarithm base a, $\log_a MN = \log_a M + \log_a N$.

_____

39. If $ab = 0$ is true, then $a = 0$ or $b = 0$, and if $a = 0$ or $b = 0$, then $ab = 0$.

_____

40. If $x^2 = k$, then $x = \sqrt{k}$ or $x = -\sqrt{k}$.

_____

41. For any positive number M, any logarithm base a, and any real number p, $\log_a M^p = p \log_a M$.

_____

42. For any real numbers a, b, and c: If $a = b$ is true, then $ac = bc$ is true.

_____

Synthesis

43. *Supply and Demand.* The supply function and the demand function for the sale of a certain type of DVD player are given by

$$S(p) = 150e^{0.004p} \quad \text{and} \quad D(p) = 480e^{-0.003p},$$

where $S(p)$ is the number of DVD players that the company is willing to sell at price p and $D(p)$ is the quantity that the public is willing to buy at price p. Find p such that $D(p) = S(p)$. This is called the **equilibrium price**.

44. *Carbon Dating.* Recently, while digging in Chaco Canyon, New Mexico, archaeologists found corn pollen that was 4000 yr old (*Source: American Anthropologist*). This was evidence that Native Americans had been cultivating crops in the Southwest centuries earlier than scientists had thought. What percent of the carbon-14 had been lost from the pollen?

45. *Present Value.* Following the birth of a child, a parent wants to make an initial investment P_0 that will grow to $50,000 for the child's education at age 18. Interest is compounded continuously at 7%. What should the initial investment be? Such an amount is called the **present value** of $50,000 due 18 yr from now.

46. *Present Value.*
a) Solve $P = P_0 e^{kt}$ for P_0.
b) Referring to Exercise 45, find the present value of $50,000 due 18 yr from now at interest rate 6.4%, compounded continuously.

47. *Electricity.* The formula
$$i = \frac{V}{R}[1 - e^{-(R/L)t}]$$
occurs in the theory of electricity. Solve for t.

48. *The Beer–Lambert Law.* A beam of light enters a medium such as water or smog with initial intensity I_0. Its intensity decreases depending on the thickness (or concentration) of the medium. The intensity I at a depth (or concentration) of x units is given by
$$I = I_0 e^{-\mu x}.$$
The constant μ (the Greek letter "mu") is called the **coefficient of absorption**, and it varies with the medium. For sea water, $\mu = 1.4$.
a) What percentage of light intensity I_0 remains in sea water at a depth of 1 m? 3 m? 5 m? 50 m?
b) Plant life cannot exist below 10 m. What percentage of I_0 remains at 10 m?

49. Given that $y = ae^x$, take the natural logarithm on both sides. Let $Y = \ln y$. Consider Y as a function of x. What kind of function is Y?

50. Given that $y = ax^b$, take the natural logarithm on both sides. Let $Y = \ln y$ and $X = \ln x$. Consider Y as a function of X. What kind of function is Y?

CHAPTER 5 Summary and Review

Important Properties and Formulas

One-to-One Function: If $f(a) = f(b)$, then $a = b$.

Exponential Function: $f(x) = a^x$, for $a > 0$ and $a \neq 1$

The Number e: $e = 2.7182818284\ldots$

Logarithmic Function: $f(x) = \log_a x$, $a > 0$ and $a \neq 1$

A Logarithm Is an Exponent: $\log_a x = y \longleftrightarrow x = a^y$

The Change-of-Base Formula: $\log_b M = \dfrac{\log_a M}{\log_a b}$

The Product Rule: $\log_a MN = \log_a M + \log_a N$

The Power Rule: $\log_a M^p = p \log_a M$

The Quotient Rule: $\log_a \dfrac{M}{N} = \log_a M - \log_a N$

Other Properties: $\log_a a = 1, \quad \log_a 1 = 0,$

$\log_a a^x = x, \quad a^{\log_a x} = x$

Base–Exponent Property: $a^x = a^y \longleftrightarrow x = y$, for $a > 0$ and $a \neq 1$

Property of Logarithmic Equality: $\log_a M = \log_a N \longleftrightarrow M = N$, for $a > 0$ and $a \neq 1$

Exponential Growth Model: $P(t) = P_0 e^{kt}, \ k > 0$

Exponential Decay Model: $P(t) = P_0 e^{-kt}, \ k > 0$

Interest Compounded Continuously: $P(t) = P_0 e^{kt}, \ k > 0$

Doubling Time: $kT = \ln 2$, or $k = \dfrac{\ln 2}{T}$, or $T = \dfrac{\ln 2}{k}$

Half-Life: $kT = \ln 2$, or $k = \dfrac{\ln 2}{T}$, or $T = \dfrac{\ln 2}{k}$

Limited Growth Models: $P(t) = \dfrac{a}{1 + be^{-kt}}, \ k > 0$

$P(t) = L(1 - e^{-kt}), \ k > 0$

Review Exercises

Determine whether the statement is true or false.

1. The range of a one-to-one function f is the domain of the inverse f^{-1}. [5.1]

2. The y-intercept of $f(x) = e^{-x}$ is $(0, -1)$. [5.2]

3. The graph of f^{-1} is a reflection of the graph of f across $y = 0$. [5.1]

4. If it is not possible for a horizontal line to intersect the graph of a function more than once, then the function is one-to-one and its inverse is a function.   [5.1]

5. The domain of all logarithmic functions is $[1, \infty)$.   [5.3]

6. The horizontal asymptote of $y = 2^x$ is $y = 0$.   [5.2]

7. Find the inverse of the relation
$\{(1.3, -2.7), (8, -3), (-5, 3), (6, -3), (7, -5)\}$.
[5.1]

8. Find an equation of the inverse relation.   [5.1]
a) $y = -2x + 3$
b) $y = 3x^2 + 2x - 1$
c) $0.8x^3 - 5.4y^2 = 3x$

Graph the function and determine whether the function is one-to-one using the horizontal-line test.   [5.1]

9. $f(x) = -|x| + 3$

10. $f(x) = x^2 + 1$

11. $f(x) = 2x - \dfrac{3}{4}$

12. $f(x) = -\dfrac{6}{x + 1}$

In Exercises 13–18, given the function:
a) *Sketch the graph and determine whether the function is one-to-one.*   [5.1], [5.3]
b) *If it is one-to-one, find a formula for the inverse.*   [5.1], [5.3]

13. $f(x) = 2 - 3x$

14. $f(x) = \dfrac{x + 2}{x - 1}$

15. $f(x) = \sqrt{x - 6}$

16. $f(x) = x^3 - 8$

17. $f(x) = 3x^2 + 2x - 1$

18. $f(x) = e^x$

For the function f, use composition of functions to show that f^{-1} is as given.   [5.1]

19. $f(x) = 6x - 5$, $f^{-1}(x) = \dfrac{x + 5}{6}$

20. $f(x) = \dfrac{x + 1}{x}$, $f^{-1}(x) = \dfrac{1}{x - 1}$

Find the inverse of the given one-to-one function f. Give the domain and the range of f and of f^{-1} and then graph both f and f^{-1} on the same set of axes.   [5.1]

21. $f(x) = 2 - 5x$

22. $f(x) = \dfrac{x - 3}{x + 2}$

23. Find $f(f^{-1}(657))$: $f(x) = \dfrac{4x^5 - 16x^{37}}{119x}$, $x > 1$.   [5.1]

24. Find $f(f^{-1}(a))$: $f(x) = \sqrt[3]{3x - 4}$.   [5.1]

Graph the function.

25. $f(x) = \left(\frac{1}{3}\right)^x$
[5.2]

26. $f(x) = 1 + e^x$
[5.2]

27. $f(x) = -e^{-x}$
[5.2]

28. $f(x) = \log_2 x$
[5.3]

29. $f(x) = \frac{1}{2} \ln x$
[5.3]

30. $f(x) = \log x - 2$
[5.3]

In Exercises 31–36, match the equation with one of the figures (a)–(f), which follow

a)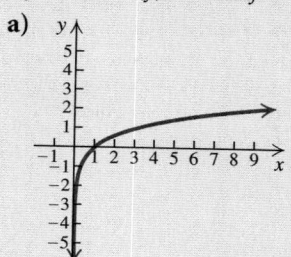

b)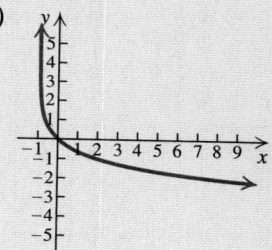

c)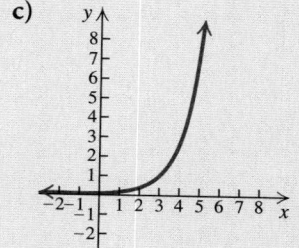

d)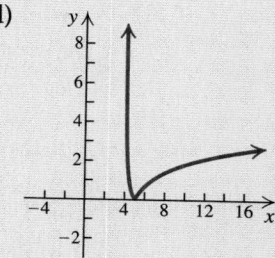

e)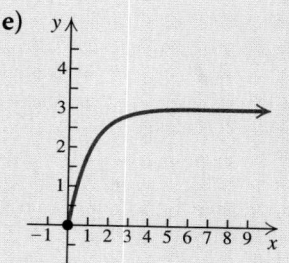

f)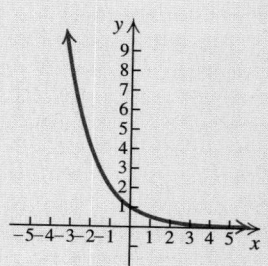

31. $f(x) = e^{x-3}$ [5.2]

32. $f(x) = \log_3 x$
[5.3]

33. $y = -\log_3 (x + 1)$
[5.3]

34. $y = \left(\frac{1}{2}\right)^x$
[5.2]

35. $f(x) = 3(1 - e^{-x})$, $x \geq 0$ [5.2]

36. $f(x) = |\ln(x - 4)|$ [5.3]

Find each of the following. Do not use a calculator. [5.3]

37. $\log_5 125$

38. $\log 100{,}000$

39. $\ln e$

40. $\ln 1$

41. $\log 10^{1/4}$

42. $\log_3 \sqrt{3}$

43. $\log 1$

44. $\log 10$

45. $\log_2 \sqrt[3]{2}$

46. $\log 0.01$

Convert to an exponential equation. [5.3]

47. $\log_4 x = 2$

48. $\log_a Q = k$

Convert to a logarithmic equation. [5.3]

49. $4^{-3} = \frac{1}{64}$

50. $e^x = 80$

Find each of the following using a calculator. Round to four decimal places. [5.3]

51. $\log 11$

52. $\log 0.234$

53. $\ln 3$

54. $\ln 0.027$

55. $\log(-3)$

56. $\ln 0$

Find the logarithm using the change-of-base formula. [5.3]

57. $\log_5 24$

58. $\log_8 3$

Express as a single logarithm and, if possible, simplify. [5.4]

59. $3 \log_b x - 4 \log_b y + \frac{1}{2} \log_b z$

60. $\ln(x^3 - 8) - \ln(x^2 + 2x + 4) + \ln(x + 2)$

Express in terms of sums and differences of logarithms. [5.4]

61. $\ln \sqrt[4]{wr^2}$

62. $\log \sqrt[3]{\dfrac{M^2}{N}}$

Given that $\log_a 2 = 0.301$, $\log_a 5 = 0.699$, *and* $\log_a 6 = 0.778$, *find each of the following.* [5.4]

63. $\log_a 3$

64. $\log_a 50$

65. $\log_a \frac{1}{5}$

66. $\log_a \sqrt[3]{5}$

Simplify. [5.4]

67. $\ln e^{-5k}$

68. $\log_5 5^{-6t}$

Solve. [5.5]

69. $\log_4 x = 2$

70. $3^{1-x} = 9^{2x}$

71. $e^x = 80$

72. $4^{2x-1} - 3 = 61$

73. $\log_{16} 4 = x$

74. $\log_x 125 = 3$

75. $\log_2 x + \log_2(x - 2) = 3$

76. $\log(x^2 - 1) - \log(x - 1) = 1$

77. $\log x^2 = \log x$

78. $e^{-x} = 0.02$

79. *Saving for College.* Following the birth of twins, the grandparents deposit $16,000 in a college trust fund that earns 4.2% interest, compounded quarterly.

a) Find a function for the amount in the account after t years. [5.2]

b) Find the amount in the account at $t = 0$, 6, 12, and 18 yr. [5.2]

80. *Whooping Cough.* In recent years, the number of new cases of whooping cough, a bacterial infection that induces a cough, has been increasing exponentially. The number of cases is estimated by the function

$$W(t) = 1665.945(1.087)^t,$$

where t is the number of years since 1980 (*Source*: U.S. Centers for Disease Control and Prevention). Find the number of new cases in 1990 and in 2000. Then use this function to estimate the number of cases in 2008. [5.2]

81. How long will it take an investment to double if it is invested at 8.6%, compounded continuously? [5.6]

82. The population of Murrayville doubled in 30 yr. What was the exponential growth rate? [5.6]

83. How old is a skeleton that has lost 27% of its carbon-14? [5.6]

84. The hydrogen ion concentration of milk is 2.3×10^{-6}. What is the pH? (See Exercise 94 in Exercise Set 5.3.) [5.3]

85. *Earthquake Magnitude.* The earthquake in Kashgar, China, on February 25, 2003, had an intensity of $10^{6.3} \cdot I_0$ (*Source*: U.S. Geological Survey). What is the magnitude on the Richter scale?   [5.3]

86. What is the loudness, in decibels, of a sound whose intensity is $1000I_0$? (See Exercise 97 in Exercise Set 5.3.)   [5.3]

87. *Walking Speed.*   The average walking speed w, in feet per second, of a person living in a city of population P, in thousands, is given by the function
$$w(P) = 0.37 \ln P + 0.05.$$

a) The population of Wichita, Kansas, is 353,823. Find the average walking speed.   [5.3]
b) A city's population has an average walking speed of 3.4 ft/sec. Find the population.   [5.6]

88. *Social Security Distributions.*   Cash Social Security distributions were $35 million, or $0.035 billion, in 1940. This amount has increased exponentially to $492 billion in 2004. (*Source*: Social Security Administration) Assuming the exponential growth model applies:

a) Find the exponential growth rate k.   [5.6]
b) Find the exponential growth function.   [5.6]
c) Estimate the total cash distributions in 1965, in 1995, and in 2015.   [5.6]
d) In what year will the cash benefits reach $1 trillion?   [5.6]

89. *The Population of Panama.*   The population of Panama was 3.039 million in 2005, and the exponential growth rate was 1.3% per year

(*Source*: U.S. Census Bureau, World Population Profile).

a) Find the exponential growth function.   [5.6]
b) What will the population be in 2009? in 2015?   [5.6]
c) When will the population be 10 million?   [5.6]
d) What is the doubling time?   [5.6]

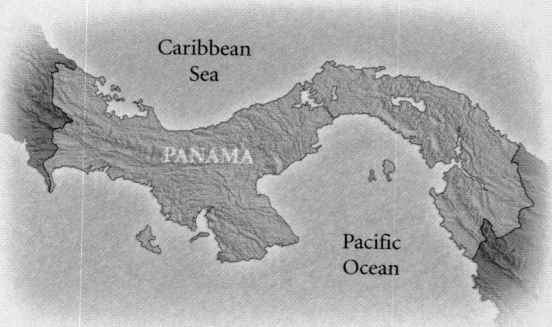

90. *Arson Damage to Churches.*   According to the federal government, 34.5% of church fires from 2001 through 2005 were caused by arson. The table below lists the dollar value (in millions) in arson damage.

Year, x	Arson Damage (in millions)
2001, 0	$2.8
2002, 1	3.3
2003, 2	3.6
2004, 3	5.0
2005, 4	8.5

Sources: U.S. Fire Administration; Bureau of Alcohol, Tobacco, Firearms, and Explosives

a) Use a graphing calculator to fit an exponential function to the data, where x is the number of years after 2001.   [5.6]
b) Graph the function with a scatterplot of the data.   [5.6]
c) Estimate the church arson damage (in millions of dollars) in 2010.   [5.6]

91. Using only a graphing calculator, determine whether the following functions are inverses of each other:

$$f(x) = \frac{4 + 3x}{x - 2}, \qquad g(x) = \frac{x + 4}{x - 3}. \quad [5.1]$$

92. a) Use a graphing calculator to graph $f(x) = 5e^{-x} \ln x$ in the viewing window $[-1, 10, -5, 5]$.   [5.2], [5.3]
  b) Estimate the relative maximum and minimum values of the function.   [5.2], [5.3]

93. Which of the following is the horizontal asymptote of the graph of $f(x) = e^{x-3} + 2$?   [5.2]

  A. $y = -2$ **B.** $y = -3$
  C. $y = 3$ **D.** $y = 2$

94. Which of the following is the domain of the logarithmic function $f(x) = \log(2x - 3)$?   [5.3]

  A. $\left(\frac{3}{2}, \infty\right)$ **B.** $\left(-\infty, \frac{3}{2}\right)$
  C. $(3, \infty)$ **D.** $(-\infty, \infty)$

95. The graph of $f(x) = 2^{x-2}$ is which of the following?   [5.2]

  A.
  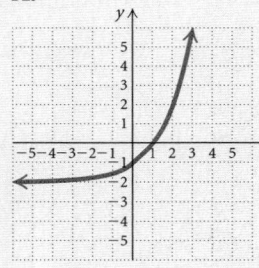

  B.
  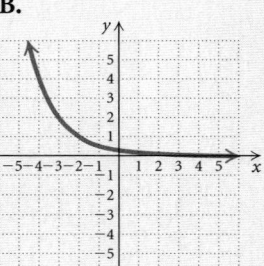

  C.
  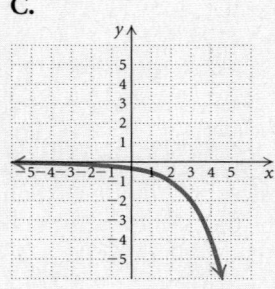

  D.
  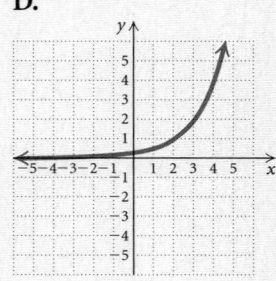

96. The graph of $f(x) = \log_2 x$ is which of the following?   [5.3]

  A.
  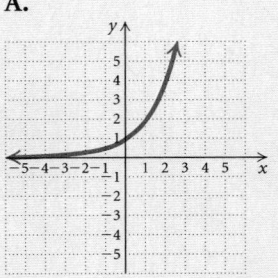

  B.
  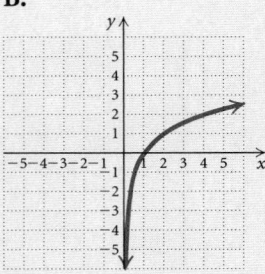

  C.
  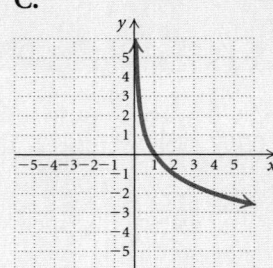

  D.
  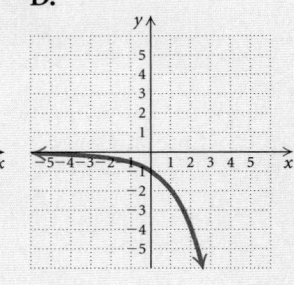

Collaborative Discussion and Writing

97. Suppose that you are trying to convince a fellow student that

$$\log_2(x + 5) \neq \log_2 x + \log_2 5.$$

Give as many explanations as you can.   [5.4]

98. Describe the difference between $f^{-1}(x)$ and $[f(x)]^{-1}$.   [5.1]

Synthesis

Solve.   [5.5]

99. $|\log_4 x| = 3$

100. $\log x = \ln x$

101. $5^{\sqrt{x}} = 625$

102. Find the domain: $f(x) = \log_3(\ln x)$.   [5.3]

CHAPTER 5   Test

1. Find the inverse of the relation
$$\{(-2,5),(4,3),(0,-1),(-6,-3)\}.$$

Determine whether the function is one-to-one. Answer yes or no.

2.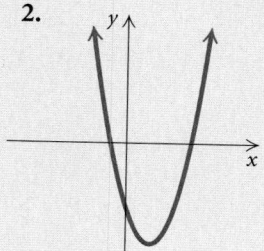

3.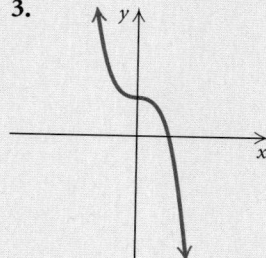

In Exercises 4–7, given the function:

a) *Sketch the graph and determine whether the function is one-to-one.*

b) *If it is one-to-one, find a formula for the inverse.*

4. $f(x) = x^3 + 1$

5. $f(x) = 1 - x$

6. $f(x) = \dfrac{x}{2-x}$

7. $f(x) = x^2 + x - 3$

8. Use composition of functions to show that f^{-1} is as given:
$$f(x) = -4x + 3, \qquad f^{-1}(x) = \frac{3-x}{4}.$$

9. Find the inverse of the one-to-one function
$$f(x) = \frac{1}{x-4}.$$
Give the domain and the range of f and of f^{-1} and then graph both f and f^{-1} on the same set of axes.

Graph the function.

10. $f(x) = 4^{-x}$

11. $f(x) = \log x$

12. $f(x) = e^x - 3$

13. $f(x) = \ln(x+2)$

Find each of the following. Do not use a calculator.

14. $\log 0.00001$

15. $\ln e$

16. $\ln 1$

17. $\log_4 \sqrt[5]{4}$

18. Convert to an exponential equation: $\ln x = 4$.

19. Convert to a logarithmic equation: $3^x = 5.4$.

Find each of the following using a calculator. Round to four decimal places.

20. $\ln 16$

21. $\log 0.293$

22. Find $\log_6 10$ using the change-of-base formula.

23. Express as a single logarithm:
$$2\log_a x - \log_a y + \tfrac{1}{2}\log_a z.$$

24. Express $\ln \sqrt[5]{x^2 y}$ in terms of sums and differences of logarithms.

25. Given that $\log_a 2 = 0.328$ and $\log_a 8 = 0.984$, find $\log_a 4$.

26. Simplify: $\ln e^{-4t}$.

Solve.

27. $\log_{25} 5 = x$

28. $\log_3 x + \log_3(x+8) = 2$

29. $3^{4-x} = 27^x$

30. $e^x = 65$

31. *Earthquake Magnitude.* The earthquake in Bam, in southeast Iran, on December 26, 2003, had an intensity of $10^{6.6} \cdot I_0$ (*Source:* U.S. Geological Survey). What was its magnitude on the Richter scale?

32. *Growth Rate.* A country's population doubled in 45 yr. What was the exponential growth rate?

33. *Compound Interest.* Suppose $1000 is invested at interest rate k, compounded continuously, and grows to $1144.54 in 3 yr.

a) Find the interest rate.
b) Find the exponential growth function.
c) Find the balance after 8 yr.
d) Find the doubling time.

34. The graph of $f(x) = 2^{x-1} + 1$ is which of the following?

A.

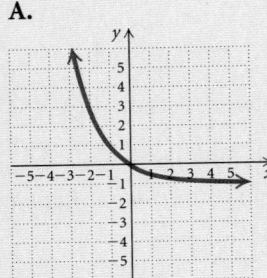

B.

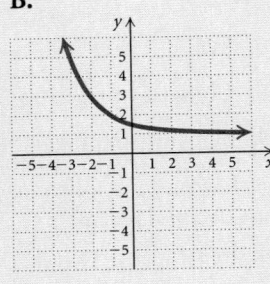

C.

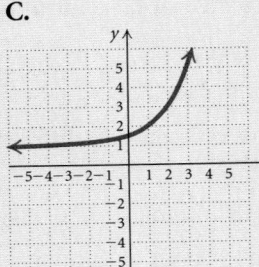

D.

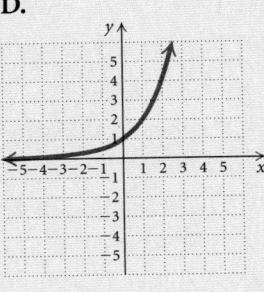

Synthesis

35. Solve: $4^{\sqrt[3]{x}} = 8$.

The Trigonometric Functions

APPLICATION In Telluride, Colorado, there is a free gondola ride that provides a spectacular view of the town and the surrounding mountains. The gondolas that begin in the town at an elevation of 8725 ft travel 5750 ft to Station St. Sophia, whose altitude is 10,550 ft. They then continue 3913 ft to Mountain Village, whose elevation is 9500 ft. What is the angle of elevation from the town to Station St. Sophia?

This problem appears as Example 5 in Section 6.2.

6.1 Trigonometric Functions of Acute Angles

6.2 Applications of Right Triangles

6.3 Trigonometric Functions of Any Angle

6.4 Radians, Arc Length, and Angular Speed

6.5 Circular Functions: Graphs and Properties

6.6 Graphs of Transformed Sine and Cosine Functions

6.1 Trigonometric Functions of Acute Angles

❖ Determine the six trigonometric ratios for a given acute angle of a right triangle.

❖ Determine the trigonometric function values of 30°, 45°, and 60°.

❖ Using a calculator, find function values for any acute angle, and given a function value of an acute angle, find the angle.

❖ Given the function values of an acute angle, find the function values of its complement.

❖ The Trigonometric Ratios

We begin our study of trigonometry by considering right triangles and acute angles measured in degrees. An **acute angle** is an angle with measure greater than 0° and less than 90°. Greek letters such as α (alpha), β (beta), γ (gamma), θ (theta), and ϕ (phi) are often used to denote an angle. Consider a right triangle with one of its acute angles labeled θ. The side opposite the right angle is called the **hypotenuse**. The other sides of the triangle are referenced by their position relative to the acute angle θ. One side is opposite θ and one is adjacent to θ.

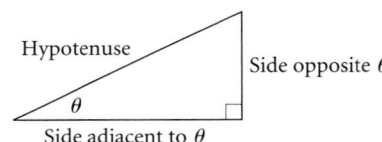

The *lengths* of the sides of the triangle are used to define the six trigonometric ratios:

sine (sin),   cosecant (csc),

cosine (cos),   secant (sec),

tangent (tan),   cotangent (cot).

The **sine of θ** is the *length* of the side opposite θ divided by the *length* of the hypotenuse (see Fig. 1):

$$\sin \theta = \frac{\text{length of side opposite } \theta}{\text{length of hypotenuse}}.$$

The ratio depends on the measure of angle θ and thus is a function of θ. The notation $\sin \theta$ actually means $\sin (\theta)$, where sin, or sine, is the name of the function.

The **cosine of θ** is the *length* of the side adjacent to θ divided by the *length* of the hypotenuse (see Fig. 2):

$$\cos \theta = \frac{\text{length of side adjacent to } \theta}{\text{length of hypotenuse}}.$$

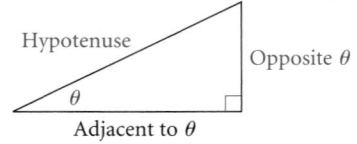

FIGURE 1

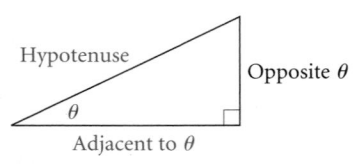

FIGURE 2

The six trigonometric ratios, or trigonometric functions, are defined as follows.

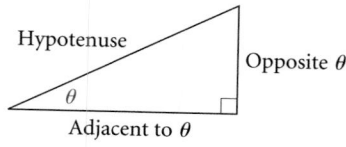

> ### Trigonometric Function Values of an Acute Angle θ
>
> Let θ be an acute angle of a right triangle. Then the six trigonometric functions of θ are as follows:
>
> $$\sin \theta = \frac{\text{side opposite } \theta}{\text{hypotenuse}}, \qquad \csc \theta = \frac{\text{hypotenuse}}{\text{side opposite } \theta},$$
>
> $$\cos \theta = \frac{\text{side adjacent to } \theta}{\text{hypotenuse}}, \qquad \sec \theta = \frac{\text{hypotenuse}}{\text{side adjacent to } \theta},$$
>
> $$\tan \theta = \frac{\text{side opposite } \theta}{\text{side adjacent to } \theta}, \qquad \cot \theta = \frac{\text{side adjacent to } \theta}{\text{side opposite } \theta}.$$

EXAMPLE 1 In the right triangle shown at left, find the six trigonometric function values of **(a)** θ and **(b)** α.

Solution We use the definitions.

a) $\sin \theta = \dfrac{\text{opp}}{\text{hyp}} = \dfrac{12}{13}, \qquad \csc \theta = \dfrac{\text{hyp}}{\text{opp}} = \dfrac{13}{12},$

$\quad\;\; \cos \theta = \dfrac{\text{adj}}{\text{hyp}} = \dfrac{5}{13}, \qquad \sec \theta = \dfrac{\text{hyp}}{\text{adj}} = \dfrac{13}{5},$

$\quad\;\; \tan \theta = \dfrac{\text{opp}}{\text{adj}} = \dfrac{12}{5}, \qquad \cot \theta = \dfrac{\text{adj}}{\text{opp}} = \dfrac{5}{12}$

The references to opposite, adjacent, and hypotenuse are relative to θ.

b) $\sin \alpha = \dfrac{\text{opp}}{\text{hyp}} = \dfrac{5}{13}, \qquad \csc \alpha = \dfrac{\text{hyp}}{\text{opp}} = \dfrac{13}{5},$

$\quad\;\; \cos \alpha = \dfrac{\text{adj}}{\text{hyp}} = \dfrac{12}{13}, \qquad \sec \alpha = \dfrac{\text{hyp}}{\text{adj}} = \dfrac{13}{12},$

$\quad\;\; \tan \alpha = \dfrac{\text{opp}}{\text{adj}} = \dfrac{5}{12}, \qquad \cot \alpha = \dfrac{\text{adj}}{\text{opp}} = \dfrac{12}{5}$

The references to opposite, adjacent, and hypotenuse are relative to α.

Now Try Exercise 1. ■

In Example 1(a), we note that the value of $\sin \theta$, $\frac{12}{13}$, is the reciprocal of $\frac{13}{12}$, the value of $\csc \theta$. Likewise, we see the same reciprocal relationship between the values of $\cos \theta$ and $\sec \theta$ and between the values of $\tan \theta$ and $\cot \theta$. For any angle, the cosecant, secant, and cotangent values are the reciprocals of the sine, cosine, and tangent function values, respectively.

> ### Reciprocal Functions
>
> $$\csc \theta = \frac{1}{\sin \theta}, \qquad \sec \theta = \frac{1}{\cos \theta}, \qquad \cot \theta = \frac{1}{\tan \theta}$$

If we know the values of the sine, cosine, and tangent functions of an angle, we can use these reciprocal relationships to find the values of the cosecant, secant, and cotangent functions of that angle.

EXAMPLE 2   Given that $\sin \phi = \frac{4}{5}$, $\cos \phi = \frac{3}{5}$, and $\tan \phi = \frac{4}{3}$, find $\csc \phi$, $\sec \phi$, and $\cot \phi$.

Solution   Using the reciprocal relationships, we have

$$\csc \phi = \frac{1}{\sin \phi} = \frac{1}{\frac{4}{5}} = \frac{5}{4}, \qquad \sec \phi = \frac{1}{\cos \phi} = \frac{1}{\frac{3}{5}} = \frac{5}{3},$$

and

$$\cot \phi = \frac{1}{\tan \phi} = \frac{1}{\frac{4}{3}} = \frac{3}{4}.$$

Now Try Exercise 7. ■

Triangles are said to be **similar** if their corresponding angles have the *same* measure. In similar triangles, the lengths of corresponding sides are in the same ratio. The right triangles shown below are similar. Note that the corresponding angles are equal and the length of each side of the second triangle is four times the length of the corresponding side of the first triangle.

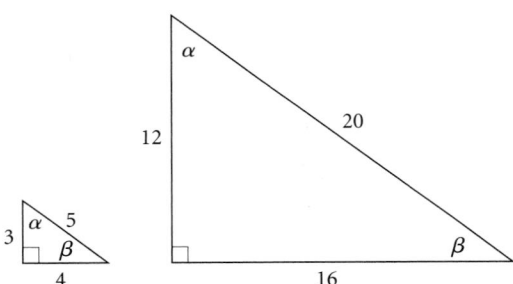

Let's observe the sine, cosine, and tangent values of β in each triangle. Can we expect corresponding function values to be the same?

FIRST TRIANGLE	SECOND TRIANGLE
$\sin \beta = \dfrac{3}{5}$	$\sin \beta = \dfrac{12}{20} = \dfrac{3}{5}$
$\cos \beta = \dfrac{4}{5}$	$\cos \beta = \dfrac{16}{20} = \dfrac{4}{5}$
$\tan \beta = \dfrac{3}{4}$	$\tan \beta = \dfrac{12}{16} = \dfrac{3}{4}$

For the two triangles, the corresponding values of $\sin \beta$, $\cos \beta$, and $\tan \beta$ are the same. The lengths of the sides are proportional—thus the *ratios* are the same. This must be the case because in order for the sine, cosine, and tangent to be functions, there must be only one output (the ratio) for each input (the angle β).

> The trigonometric function values of θ depend only on the measure of the angle, *not* on the size of the triangle.

❋ The Six Functions Related

We can find the other five trigonometric function values of an acute angle when one of the function-value ratios is known.

EXAMPLE 3 If $\sin \beta = \frac{6}{7}$ and β is an acute angle, find the other five trigonometric function values of β.

Solution We know from the definition of the sine function that the ratio

$$\frac{6}{7} \quad \text{is} \quad \frac{\text{opp}}{\text{hyp}}.$$

> **PYTHAGOREAN EQUATION**
> REVIEW SECTION **1.1.**

Using this information, let's consider a right triangle in which the hypotenuse has length 7 and the side opposite β has length 6. To find the length of the side adjacent to β, we recall the *Pythagorean equation*:

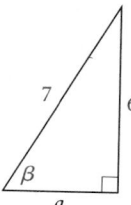

$$a^2 + b^2 = c^2$$
$$a^2 + 6^2 = 7^2$$
$$a^2 + 36 = 49$$
$$a^2 = 49 - 36 = 13$$
$$a = \sqrt{13}.$$

We now use the lengths of the three sides to find the other five ratios:

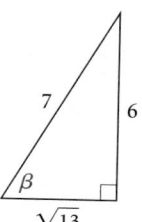

$$\sin \beta = \frac{6}{7}, \qquad\qquad\qquad \csc \beta = \frac{7}{6},$$

$$\cos \beta = \frac{\sqrt{13}}{7}, \qquad\qquad \sec \beta = \frac{7}{\sqrt{13}}, \quad \text{or} \quad \frac{7\sqrt{13}}{13},$$

$$\tan \beta = \frac{6}{\sqrt{13}}, \quad \text{or} \quad \frac{6\sqrt{13}}{13}, \qquad \cot \beta = \frac{\sqrt{13}}{6}.$$

Now Try Exercise 9. ■

❋ Function Values of 30°, 45°, and 60°

In Examples 1 and 3, we found the trigonometric function values of an acute angle of a right triangle when the lengths of the three sides were known. In most situations, we are asked to find the function values when the measure of the acute angle is given. For certain special angles such as

30°, 45°, and 60°, which are frequently seen in applications, we can use geometry to determine the function values.

A right triangle with a 45° angle actually has two 45° angles. Thus the triangle is *isosceles*, and the legs are the same length. Let's consider such a triangle whose legs have length 1. Then we can find the length of its hypotenuse, c, using the Pythagorean equation as follows:

$$1^2 + 1^2 = c^2, \quad \text{or} \quad c^2 = 2, \quad \text{or} \quad c = \sqrt{2}.$$

Such a triangle is shown below. From this diagram, we can easily determine the trigonometric function values of 45°.

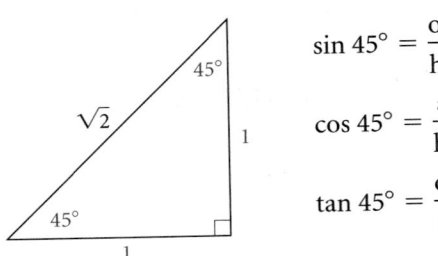

$$\sin 45° = \frac{\text{opp}}{\text{hyp}} = \frac{1}{\sqrt{2}} = \frac{\sqrt{2}}{2} \approx 0.7071,$$

$$\cos 45° = \frac{\text{adj}}{\text{hyp}} = \frac{1}{\sqrt{2}} = \frac{\sqrt{2}}{2} \approx 0.7071,$$

$$\tan 45° = \frac{\text{opp}}{\text{adj}} = \frac{1}{1} = 1$$

It is sufficient to find only the function values of the sine, cosine, and tangent, since the others are their reciprocals.

It is also possible to determine the function values of 30° and 60°. A right triangle with 30° and 60° acute angles is half of an equilateral triangle, as shown in the following figure. Thus if we choose an equilateral triangle whose sides have length 2 and take half of it, we obtain a right triangle that has a hypotenuse of length 2 and a leg of length 1. The other leg has length a, which can be found as follows:

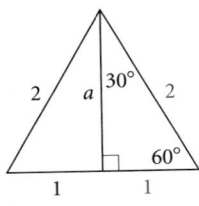

$$a^2 + 1^2 = 2^2$$
$$a^2 + 1 = 4$$
$$a^2 = 3$$
$$a = \sqrt{3}.$$

We can now determine the function values of 30° and 60°:

$$\sin 30° = \frac{1}{2} = 0.5, \qquad\qquad \sin 60° = \frac{\sqrt{3}}{2} \approx 0.8660,$$

$$\cos 30° = \frac{\sqrt{3}}{2} \approx 0.8660, \qquad \cos 60° = \frac{1}{2} = 0.5,$$

$$\tan 30° = \frac{1}{\sqrt{3}} = \frac{\sqrt{3}}{3} \approx 0.5774, \qquad \tan 60° = \frac{\sqrt{3}}{1} = \sqrt{3} \approx 1.7321.$$

Since we will often use the function values of 30°, 45°, and 60°, either the triangles that yield them or the values themselves should be memorized.

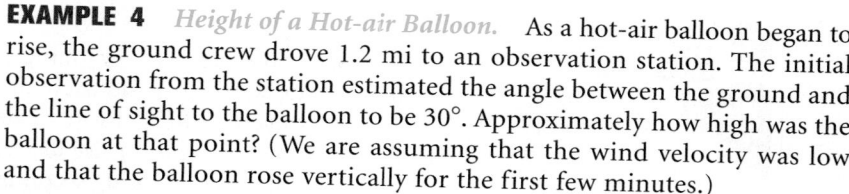

	30°	45°	60°
sin	$1/2$	$\sqrt{2}/2$	$\sqrt{3}/2$
cos	$\sqrt{3}/2$	$\sqrt{2}/2$	$1/2$
tan	$\sqrt{3}/3$	1	$\sqrt{3}$

Let's now use what we have learned about trigonometric functions of special angles to solve problems. We will consider such applications in greater detail in Section 6.2.

EXAMPLE 4   *Height of a Hot-air Balloon.*   As a hot-air balloon began to rise, the ground crew drove 1.2 mi to an observation station. The initial observation from the station estimated the angle between the ground and the line of sight to the balloon to be 30°. Approximately how high was the balloon at that point? (We are assuming that the wind velocity was low and that the balloon rose vertically for the first few minutes.)

Solution   We begin with a drawing of the situation. We know the measure of an acute angle and the length of its adjacent side.

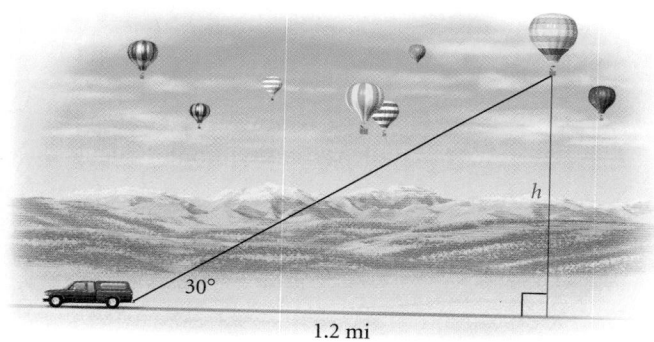

Since we want to determine the length of the opposite side, we can use the tangent ratio, or the cotangent ratio. Here we use the tangent ratio:

$$\tan 30° = \frac{\text{opp}}{\text{adj}} = \frac{h}{1.2}$$

$$1.2 \tan 30° = h$$

$$1.2\left(\frac{\sqrt{3}}{3}\right) = h \qquad \text{Substituting; } \tan 30° = \frac{\sqrt{3}}{3}$$

$$0.7 \approx h.$$

The balloon is approximately 0.7 mi, or 3696 ft, high.

Now Try Exercise 29. ■

❄ Function Values of Any Acute Angle

Historically, the measure of an angle has been expressed in degrees, minutes, and seconds. One minute, denoted $1'$, is such that $60' = 1°$, or $1' = \frac{1}{60} \cdot (1°)$. One second, denoted $1''$, is such that $60'' = 1'$, or $1'' = \frac{1}{60} \cdot (1')$. Then 61 degrees, 27 minutes, 4 seconds could be written as $61°27'4''$. This **D°M′S″ form** was common before the widespread use of scientific calculators. Now the preferred notation is to express fraction parts of degrees in **decimal degree form**. Although the D°M′S″ notation is still widely used in navigation, we will most often use the decimal form in this text.

Most calculators can convert D°M′S″ notation to decimal degree notation and vice versa. Procedures among calculators vary.

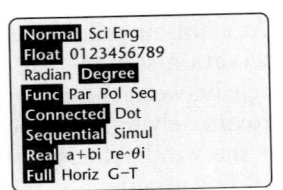

GCM **EXAMPLE 5** Convert $5°42'30''$ to decimal degree notation.

Solution We can use a graphing calculator set in DEGREE mode to convert between D°M′S″ form and decimal degree form. (See window at left.)

To convert D°M′S″ form to decimal degree form, we enter $5°42'30''$ using the ANGLE menu for the degree and minute symbols and **ALPHA** **+** for the symbol representing seconds. Pressing **ENTER** gives us

$$5°42'30'' \approx 5.71°,$$

rounded to the nearest hundredth of a degree.

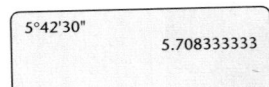

Without a calculator, we can convert as follows:

$$5°42'30'' = 5° + 42' + 30''$$

$$= 5° + 42' + \frac{30}{60}' \qquad 1'' = \frac{1}{60}'; 30'' = \frac{30}{60}'$$

$$= 5° + 42.5' \qquad \frac{30}{60}' = 0.5'$$

$$= 5° + \frac{42.5}{60}° \qquad 1' = \frac{1}{60}°; 42.5' = \frac{42.5}{60}°$$

$$\approx 5.71°. \qquad \frac{42.5}{60}° \approx 0.71°$$

Now Try Exercise 37. ■

GCM **EXAMPLE 6** Convert $72.18°$ to D°M′S″ notation.

Solution To convert decimal degree form to D°M′S″ form, we enter 72.18 and access the ▶DMS feature in the ANGLE menu. The result is

$$72.18° = 72°10'48''.$$

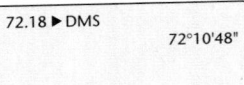

Without a calculator, we can convert as follows:

$$72.18° = 72° + 0.18 \times 1°$$
$$= 72° + 0.18 \times 60' \qquad 1° = 60'$$
$$= 72° + 10.8'$$
$$= 72° + 10' + 0.8 \times 1'$$
$$= 72° + 10' + 0.8 \times 60'' \qquad 1' = 60''$$
$$= 72° + 10' + 48''$$
$$= 72°10'48''.$$

Now Try Exercise 45. ■

So far we have measured angles using degrees. Another useful unit for angle measure is the radian, which we will study in Section 6.4. Calculators work with either degrees or radians. Be sure to use whichever mode is appropriate. In this section, we use the DEGREE mode.

Keep in mind the difference between an exact answer and an approximation. For example,

$$\sin 60° = \frac{\sqrt{3}}{2}. \qquad \textbf{This is exact!}$$

But using a calculator, you get an answer like

$$\sin 60° \approx 0.8660254038. \qquad \textbf{This is an approximation!}$$

Calculators generally provide values only of the sine, cosine, and tangent functions. You can find values of the cosecant, secant, and cotangent by taking reciprocals of the sine, cosine, and tangent functions, respectively.

GCM **EXAMPLE 7** Find the trigonometric function value, rounded to four decimal places, of each of the following.

a) tan 29.7°   b) sec 48°   c) sin 84°10'39"

Solution

a) We check to be sure that the calculator is in DEGREE mode. The function value is

$$\tan 29.7° \approx 0.5703899297$$
$$\approx 0.5704. \qquad \text{Rounded to four decimal places}$$

b) The secant function value can be found by taking the reciprocal of the cosine function value:

$$\sec 48° = \frac{1}{\cos 48°} \approx 1.49447655 \approx 1.4945.$$

c) We enter sin 84°10'39". The result is

$$\sin 84°10'39'' \approx 0.9948409474 \approx 0.9948.$$

Now Try Exercises 61 and 69. ■

We can use the TABLE feature on a graphing calculator to find an angle for which we know a trigonometric function value.

GCM **EXAMPLE 8**   Find the acute angle, to the nearest tenth of a degree, whose sine value is approximately 0.20113.

Solution   With a graphing calculator set in DEGREE mode, we first enter the equation $y = \sin x$. With a minimum value of 0 and a step-value of 0.1, we scroll through the table of values looking for the y-value closest to 0.20113.

X	Y₁
11.1	.19252
11.2	.19423
11.3	.19595
11.4	.19766
11.5	.19937
11.6	.20108 ◄
11.7	.20279
X = 11.6	

— sin 11.6° ≈ 0.20108

We find that 11.6° is the angle whose sine value is about 0.20113.

The quickest way to find the angle with a calculator is to use an inverse function key. (We first studied inverse functions in Section 5.1 and will consider inverse *trigonometric* functions in Section 7.4.) First check to be sure that your calculator is in DEGREE mode. Usually two keys must be pressed in sequence. For this example, if we press

2ND **SIN** .20113 **ENTER**,

we find that the acute angle whose sine is 0.20113 is approximately 11.60304613°, or 11.6°.   **Now Try Exercise 75.** ■

EXAMPLE 9   *Ladder Safety.*   A paint crew has purchased new 30-ft extension ladders. The manufacturer states that the safest placement on a wall is to extend the ladder to 25 ft and to position the base 6.5 ft from the wall (*Source*: R. D. Werner Co., Inc.). What angle does the ladder make with the ground in this position?

25 ft

θ

6.5 ft

Solution   We make a drawing and then use the most convenient trigonometric function. Because we know the length of the side adjacent to θ and the length of the hypotenuse, we choose the cosine function.

From the definition of the cosine function, we have

$$\cos \theta = \frac{\text{adj}}{\text{hyp}} = \frac{6.5 \text{ ft}}{25 \text{ ft}} = 0.26.$$

Using a calculator, we find the acute angle whose cosine is 0.26:

$\theta \approx 74.92993786°.$ Pressing **2ND** **COS** 0.26 **ENTER**

Thus when the ladder is in its safest position, it makes an angle of about 75° with the ground. ■

❖ Cofunctions and Complements

We recall that two angles are **complementary** whenever the sum of their measures is 90°. Each is the complement of the other. In a right triangle, the acute angles are complementary, since the sum of all three angle measures is 180° and the right angle accounts for 90° of this total. Thus if one acute angle of a right triangle is θ, the other is $90° - \theta$.

The six trigonometric function values of each of the acute angles in the triangle below are listed at the right. Note that 53° and 37° are complementary angles since 53° + 37° = 90°.

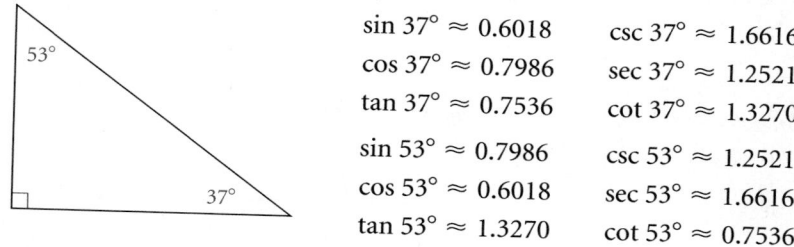

$$\sin 37° \approx 0.6018 \qquad \csc 37° \approx 1.6616$$
$$\cos 37° \approx 0.7986 \qquad \sec 37° \approx 1.2521$$
$$\tan 37° \approx 0.7536 \qquad \cot 37° \approx 1.3270$$

$$\sin 53° \approx 0.7986 \qquad \csc 53° \approx 1.2521$$
$$\cos 53° \approx 0.6018 \qquad \sec 53° \approx 1.6616$$
$$\tan 53° \approx 1.3270 \qquad \cot 53° \approx 0.7536$$

Try this with the acute, complementary angles 20.3° and 69.7° as well. What pattern do you observe? Look for this same pattern in Example 1 earlier in this section.

Note that the sine of an angle is also the cosine of the angle's complement. Similarly, the tangent of an angle is the cotangent of the angle's complement, and the secant of an angle is the cosecant of the angle's complement. These pairs of functions are called **cofunctions**. A list of cofunction identities follows.

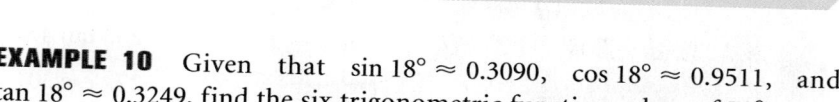

Cofunction Identities

$$\sin \theta = \cos(90° - \theta), \qquad \cos \theta = \sin(90° - \theta),$$
$$\tan \theta = \cot(90° - \theta), \qquad \cot \theta = \tan(90° - \theta),$$
$$\sec \theta = \csc(90° - \theta), \qquad \csc \theta = \sec(90° - \theta)$$

EXAMPLE 10 Given that $\sin 18° \approx 0.3090$, $\cos 18° \approx 0.9511$, and $\tan 18° \approx 0.3249$, find the six trigonometric function values of 72°.

Solution Using reciprocal relationships, we know that

$$\csc 18° = \frac{1}{\sin 18°} \approx 3.2361,$$

$$\sec 18° = \frac{1}{\cos 18°} \approx 1.0515,$$

and $\quad \cot 18° = \dfrac{1}{\tan 18°} \approx 3.0777.$

Since 72° and 18° are complementary, we have

$$\sin 72° = \cos 18° \approx 0.9511, \qquad \cos 72° = \sin 18° \approx 0.3090,$$
$$\tan 72° = \cot 18° \approx 3.0777, \qquad \cot 72° = \tan 18° \approx 0.3249,$$
$$\sec 72° = \csc 18° \approx 3.2361, \qquad \csc 72° = \sec 18° \approx 1.0515.$$

Now Try Exercise 97. ▧

6.1   Exercise Set

In Exercises 1–6, find the six trigonometric function values of the specified angle.

1.

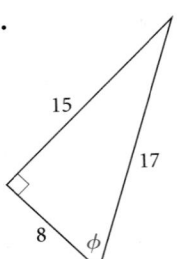

2.

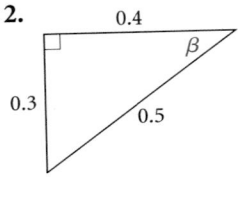

3.

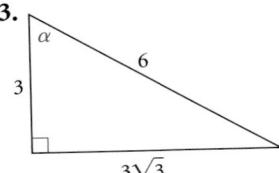

4.

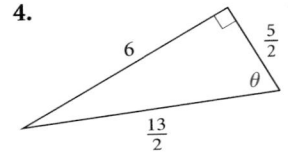

5.

6.

7. Given that $\sin \alpha = \dfrac{\sqrt{5}}{3}$, $\cos \alpha = \dfrac{2}{3}$, and $\tan \alpha = \dfrac{\sqrt{5}}{2}$, find $\csc \alpha$, $\sec \alpha$, and $\cot \alpha$.

8. Given that $\sin \beta = \dfrac{2\sqrt{2}}{3}$, $\cos \beta = \dfrac{1}{3}$, and $\tan \beta = 2\sqrt{2}$, find $\csc \beta$, $\sec \beta$, and $\cot \beta$.

Given a function value of an acute angle, find the other five trigonometric function values.

9. $\sin \theta = \dfrac{24}{25}$ **10.** $\cos \sigma = 0.7$

11. $\tan \phi = 2$ **12.** $\cot \theta = \dfrac{1}{3}$

13. $\csc \theta = 1.5$ **14.** $\sec \beta = \sqrt{17}$

15. $\cos \beta = \dfrac{\sqrt{5}}{5}$ **16.** $\sin \sigma = \dfrac{10}{11}$

Find the exact function value.

17. $\cos 45°$ **18.** $\tan 30°$

19. $\sec 60°$ **20.** $\sin 45°$

21. $\cot 60°$ **22.** $\csc 45°$

23. $\sin 30°$ **24.** $\cos 60°$

25. $\tan 45°$ **26.** $\sec 30°$

27. $\csc 30°$ **28.** $\tan 60°$

29. *Distance Across a River.* Find the distance a across the river.

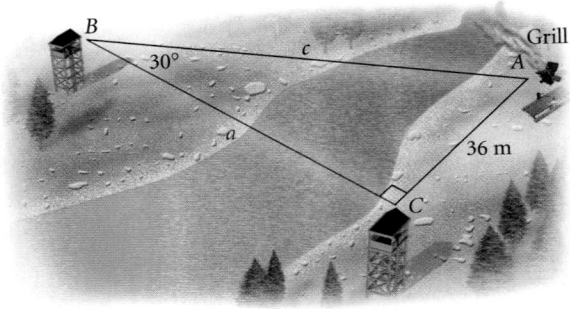

30. *Distance Between Bases.* A baseball diamond is actually a square 90 ft on a side. If a line is drawn from third base to first base, then a right triangle

QPH is formed, where ∠*QPH* is 45°. Using a trigonometric function, find the distance from third base to first base.

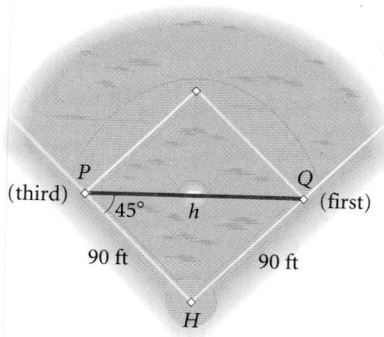

(third) *P* 45° *h* *Q* (first)

90 ft   90 ft

H

Convert to decimal degree notation. Round to two decimal places.

31. 9°43′   **32.** 52°15′

33. 35°50″   **34.** 64°53′

35. 3°2′   **36.** 19°47′23″

37. 49°38′46″   **38.** 76°11′34″

GCM **39.** 15′5″   **40.** 68°2″

GCM **41.** 5°53″   **42.** 44′10″

Convert to degrees, minutes, and seconds. Round to the nearest second.

43. 17.6°   **44.** 20.14°

45. 83.025°   **46.** 67.84°

47. 11.75°   **48.** 29.8°

49. 47.8268°   **50.** 0.253°

51. 0.9°   **52.** 30.2505°

53. 39.45°   **54.** 2.4°

Find the function value. Round to four decimal places.

55. cos 51°   **56.** cot 17°

57. tan 4°13′   **58.** sin 26.1°

59. sec 38.43°   **60.** cos 74°10′40″

61. cos 40.35°   **62.** csc 45.2°

63. sin 69°   **64.** tan 63°48′

65. tan 85.4°   **66.** cos 4°

67. csc 89.5°   **68.** sec 35.28°

69. cot 30°25′6″   **70.** sin 59.2°

Find the acute angle θ, to the nearest tenth of a degree, for the given function value.

71. sin θ = 0.5125   **72.** tan θ = 2.032

73. tan θ = 0.2226   **74.** cos θ = 0.3842

75. sin θ = 0.9022   **76.** tan θ = 3.056

77. cos θ = 0.6879   **78.** sin θ = 0.4005

GCM **79.** cot θ = 2.127   **80.** csc θ = 1.147

$$\left(\text{Hint: } \tan\theta = \frac{1}{\cot\theta}.\right)$$

81. sec θ = 1.279   **82.** cot θ = 1.351

Find the exact acute angle θ for the given function value.

83. $\sin\theta = \dfrac{\sqrt{2}}{2}$ **84.** $\cot\theta = \dfrac{\sqrt{3}}{3}$

85. $\cos\theta = \dfrac{1}{2}$ **86.** $\sin\theta = \dfrac{1}{2}$

87. tan θ = 1   **88.** $\cos\theta = \dfrac{\sqrt{3}}{2}$

89. $\csc\theta = \dfrac{2\sqrt{3}}{3}$ **90.** tan θ = $\sqrt{3}$

91. cot θ = $\sqrt{3}$ **92.** sec θ = $\sqrt{2}$

Use the cofunction and reciprocal identities to complete each of the following.

93. $\cos 20° = \underline{\hspace{1cm}} 70° = \dfrac{1}{\underline{\hspace{0.7cm}} 20°}$

94. $\sin 64° = \underline{\hspace{1cm}} 26° = \dfrac{1}{\underline{\hspace{0.7cm}} 64°}$

95. $\tan 52° = \cot\underline{\hspace{1cm}} = \dfrac{1}{\underline{\hspace{0.7cm}} 52°}$

96. $\sec 13° = \csc\underline{\hspace{1cm}} = \dfrac{1}{\underline{\hspace{0.7cm}} 13°}$

97. Given that

$$\sin 65° \approx 0.9063, \qquad \cos 65° \approx 0.4226,$$
$$\tan 65° \approx 2.1445, \qquad \cot 65° \approx 0.4663,$$
$$\sec 65° \approx 2.3662, \qquad \csc 65° \approx 1.1034,$$

find the six function values of 25°.

98. Given that

$$\sin 8° \approx 0.1392, \qquad \cos 8° \approx 0.9903,$$
$$\tan 8° \approx 0.1405, \qquad \cot 8° \approx 7.1154,$$
$$\sec 8° \approx 1.0098, \qquad \csc 8° \approx 7.1853,$$

find the six function values of 82°.

99. Given that $\sin 71°10'5'' \approx 0.9465,$
$\cos 71°10'5'' \approx 0.3228$, and
$\tan 71°10'5'' \approx 2.9321,$
find the six function values of $18°49'55''$.

100. Given that $\sin 38.7° \approx 0.6252,$
$\cos 38.7° \approx 0.7804$, and $\tan 38.7° \approx 0.8012,$
find the six function values of 51.3°.

101. Given that $\sin 82° = p$, $\cos 82° = q$, and
$\tan 82° = r$, find the six function values of $8°$ in
terms of p, q, and r.

Collaborative Discussion and Writing

102. Explain the difference between reciprocal
functions and cofunctions.

103. Explain why it is not necessary to memorize the
function values for both 30° and 60°.

Skill Maintenance

*Make a hand-drawn graph of the function. Then check
your work using a graphing calculator.*

104. $f(x) = 2^{-x}$ **105.** $f(x) = e^{x/2}$

106. $g(x) = \log_2 x$ **107.** $h(x) = \ln x$

Solve.

108. $e^t = 10{,}000$

109. $5^x = 625$

110. $\log (3x + 1) - \log (x - 1) = 2$

111. $\log_7 x = 3$

Synthesis

112. Given that $\cos \theta = 0.9651$, find $\csc (90° - \theta)$.

113. Given that $\sec \beta = 1.5304$, find $\sin (90° - \beta)$.

114. Find the six trigonometric function values of α.

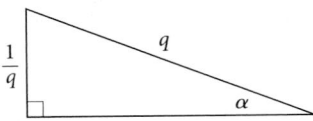

115. Show that the area of this right triangle is
$\frac{1}{2} bc \sin A$.

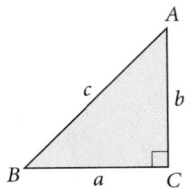

116. Show that the area of this triangle is
$\frac{1}{2} ab \sin \theta$.

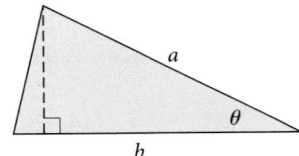

6.2 Applications of Right Triangles

❖ Solve right triangles.

❖ Solve applied problems involving right triangles and trigonometric functions.

❖ Solving Right Triangles

Now that we can find function values for any acute angle, it is possible to *solve* right triangles. To **solve** a triangle means to find the lengths of *all* sides and the measures of *all* angles.

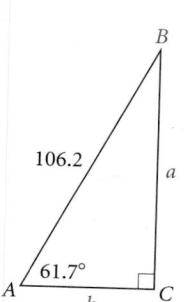

EXAMPLE 1 In $\triangle ABC$ (shown at left), find a, b, and B, where a and b represent lengths of sides and B represents the measure of $\angle B$. Here we use standard lettering for naming the sides and angles of a right triangle: Side a is opposite angle A, side b is opposite angle B, where a and b are the legs, and side c, the hypotenuse, is opposite angle C, the right angle.

Solution In $\triangle ABC$, we know three of the measures:

$$A = 61.7°, \qquad a = ?,$$
$$B = ?, \qquad b = ?,$$
$$C = 90°, \qquad c = 106.2.$$

Since the sum of the angle measures of any triangle is 180° and $C = 90°$, the sum of A and B is 90°. Thus,

$$B = 90° - A = 90° - 61.7° = 28.3°.$$

We are given an acute angle and the hypotenuse. This suggests that we can use the sine and cosine ratios to find a and b, respectively:

$$\sin 61.7° = \frac{\text{opp}}{\text{hyp}} = \frac{a}{106.2} \quad \text{and} \quad \cos 61.7° = \frac{\text{adj}}{\text{hyp}} = \frac{b}{106.2}.$$

Solving for a and b, we get

$$a = 106.2 \sin 61.7° \quad \text{and} \quad b = 106.2 \cos 61.7°$$
$$a \approx 93.5 \qquad\qquad\qquad b \approx 50.3.$$

Thus,

$$A = 61.7°, \qquad a \approx 93.5,$$
$$B = 28.3°, \qquad b \approx 50.3,$$
$$C = 90°, \qquad c = 106.2.$$

Now Try Exercise 1. ■

EXAMPLE 2 In $\triangle DEF$ (shown at left), find D and F. Then find d.

Solution In $\triangle DEF$, we know three of the measures:

$$D = ?, \qquad d = ?,$$
$$E = 90°, \qquad e = 23,$$
$$F = ?, \qquad f = 13.$$

We know the side adjacent to D and the hypotenuse. This suggests the use of the cosine ratio:

$$\cos D = \frac{\text{adj}}{\text{hyp}} = \frac{13}{23}.$$

We now find the angle whose cosine is $\frac{13}{23}$. To the nearest hundredth of a degree,

$$D \approx 55.58°. \qquad \text{Pressing } \boxed{\text{2ND}} \ \boxed{\text{COS}} \ (13/23) \ \boxed{\text{ENTER}}$$

Since the sum of D and F is $90°$, we can find F by subtracting:

$$F = 90° - D \approx 90° - 55.58° \approx 34.42°.$$

We could use the Pythagorean equation to find d, but we will use a trigonometric function here. We could use $\cos F$, $\sin D$, or the tangent or cotangent ratios for either D or F. Let's use $\tan D$:

$$\tan D = \frac{\text{opp}}{\text{adj}} = \frac{d}{13}, \quad \text{or} \quad \tan 55.58° \approx \frac{d}{13}.$$

Then

$$d \approx 13 \tan 55.58° \approx 19.$$

The six measures are

$$\begin{aligned} D &\approx 55.58°, & d &\approx 19, \\ E &= 90°, & e &= 23, \\ F &\approx 34.42°, & f &= 13. \end{aligned}$$

Now Try Exercise 9. ■

❁ Applications

Right triangles can be used to model and solve many applied problems in the real world.

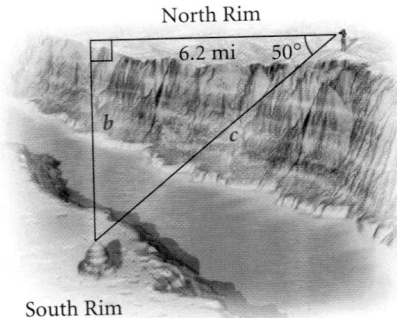

North Rim

6.2 mi   50°

b

c

South Rim

EXAMPLE 3 *Hiking at the Grand Canyon.* A backpacker hiking east along the North Rim of the Grand Canyon notices an unusual rock formation directly across the canyon. She decides to continue watching the landmark while hiking along the rim. In 2 hr, she has gone 6.2 mi due east and the landmark is still visible but at approximately a $50°$ angle to the North Rim. (See the figure at left.)

a) How many miles is she from the rock formation?

b) How far is it across the canyon from her starting point?

Solution

a) We know the side adjacent to the $50°$ angle and want to find the hypotenuse. We can use the cosine function:

$$\cos 50° = \frac{6.2 \text{ mi}}{c}$$

$$c = \frac{6.2 \text{ mi}}{\cos 50°} \approx 9.6 \text{ mi.}$$

After hiking 6.2 mi, she is approximately 9.6 mi from the rock formation.

b) We know the side adjacent to the 50° angle and want to find the opposite side. We can use the tangent function:

$$\tan 50° = \frac{b}{6.2 \text{ mi}}$$

$$b = 6.2 \text{ mi} \cdot \tan 50° \approx 7.4 \text{ mi}.$$

Thus it is approximately 7.4 mi across the canyon from her starting point.

Now Try Exercise 19. ■

EXAMPLE 4 *Rafters for a House.* House framers can use trigonometric functions to determine the lengths of rafters for a house. They first choose the pitch of the roof, or the ratio of the rise over the run. Then using a triangle with that ratio, they calculate the length of the rafter needed for the house. José is constructing rafters for a roof with a 10/12 pitch on a house that is 42 ft wide. Find the length x of the rafter of the house to the nearest tenth of a foot.

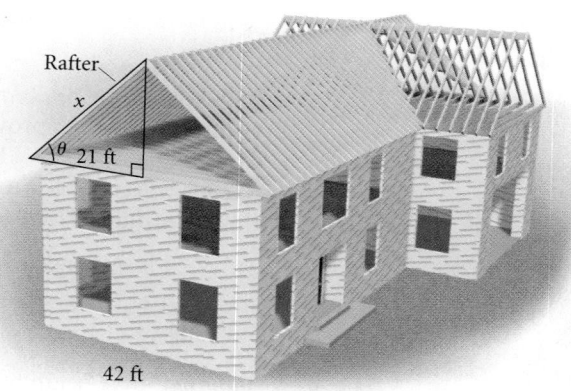
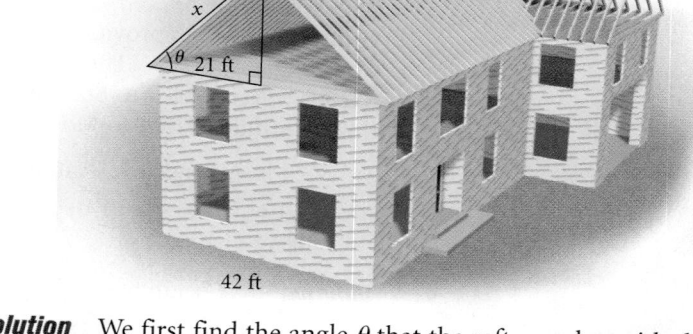

Solution We first find the angle θ that the rafter makes with the side wall. We know the rise, 10, and the run, 12, so we can use the tangent function to determine the angle that corresponds to the pitch of 10/12:

$$\tan \theta = \frac{10}{12} \approx 0.8333.$$

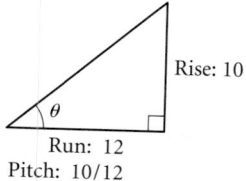

Rise: 10
Run: 12
Pitch: 10/12

Using a calculator, we find that $\theta \approx 39.8°$. Since trigonometric function values of θ depend only on the measure of the angle and not on the size of the triangle, the angle for the rafter is also 39.8°.

To determine the length x of the rafter, we can use the cosine function. (See the figure at left.) Note that the width of the house is 42 ft, and a leg of this triangle is half that length, 21 ft.

$$\cos 39.8° = \frac{21 \text{ ft}}{x}$$

$$x \cos 39.8° = 21 \text{ ft} \qquad \text{Multiplying by } x$$

$$x = \frac{21 \text{ ft}}{\cos 39.8°} \qquad \text{Dividing by } \cos 39.8°$$

$$x \approx 27.3 \text{ ft}$$

The length of the rafter for this house is approximately 27.3 ft.

Now Try Exercise 33. ■

Many applications with right triangles involve an *angle of elevation* or an *angle of depression*. The angle between the horizontal and a line of sight above the horizontal is called an **angle of elevation**. The angle between the horizontal and a line of sight below the horizontal is called an **angle of depression**. For example, suppose that you are looking straight ahead and then you move your eyes up to look at an approaching airplane. The angle that your eyes pass through is an angle of elevation. If the pilot of the plane is looking forward and then looks down, the pilot's eyes pass through an angle of depression.

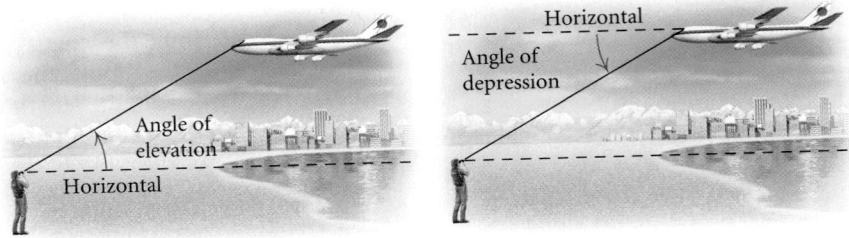

EXAMPLE 5   *Gondola Aerial Lift.*   In Telluride, Colorado, there is a free gondola ride that provides a spectacular view of the town and the surrounding mountains. The gondolas that begin in the town at an elevation of 8725 ft travel 5750 ft to Station St. Sophia, whose altitude is 10,550 ft. They then continue 3913 ft to Mountain Village, whose elevation is 9500 ft.

a) What is the angle of elevation from the town to Station St. Sophia?

b) What is the angle of depression from Station St. Sophia to Mountain Village?

Solution   We begin by labeling a drawing with the given information.

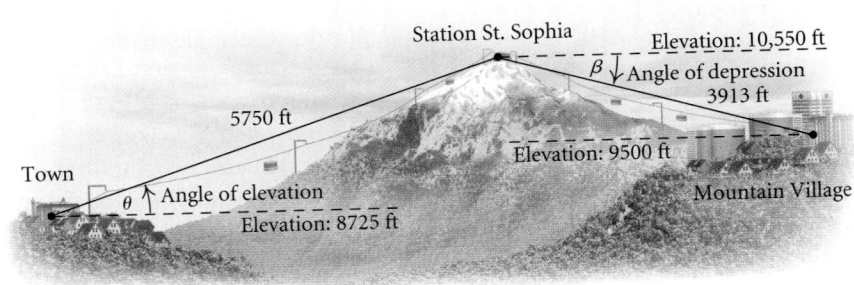

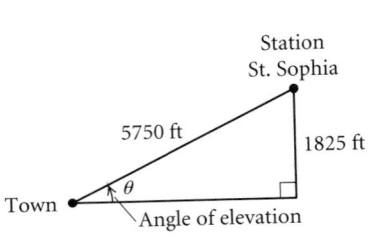

a) The difference in the elevation of Station St. Sophia and the elevation of the town is 10,550 ft − 8725 ft, or 1825 ft. This measure is the length of the side opposite the angle of elevation, θ, in the right triangle shown at left. Since we know the side opposite θ and the hypotenuse, we can find θ by using the sine function. We first find $\sin \theta$:

$$\sin \theta = \frac{1825 \text{ ft}}{5750 \text{ ft}} \approx 0.3174.$$

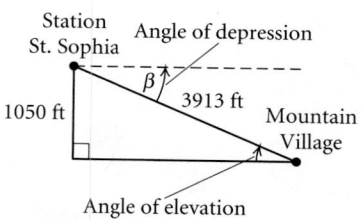

TRANSVERSALS

REVIEW THE
GEOMETRY APPENDIX.

Using a calculator, we find that

$\theta \approx 18.5°$. Pressing **2ND** **SIN** 0.3174 **ENTER**

Thus the angle of elevation from the town to Station St. Sophia is approximately 18.5°.

b) When parallel lines are cut by a transversal, alternate interior angles are equal. Thus the angle of depression, β, from Station St. Sophia to Mountain Village is equal to the angle of elevation from Mountain Village to Station St. Sophia, so we can use the right triangle shown at left.

The difference in the elevation of Station St. Sophia and the elevation of Mountain Village is 10,550 ft − 9500 ft, or 1050 ft. Since we know the side opposite the angle of elevation and the hypotenuse, we can again use the sine function:

$$\sin \beta = \frac{1050 \text{ ft}}{3913 \text{ ft}} \approx 0.2683.$$

Using a calculator, we find that

$\beta = 15.6°.$

The angle of depression from Station St. Sophia to Mountain Village is approximately 15.6°.

Now Try Exercise 17. ▨

EXAMPLE 6 *Cloud Height.* To measure cloud height at night, a vertical beam of light is directed on a spot on the cloud. From a point 135 ft away from the light source, the angle of elevation to the spot is found to be 67.35°. Find the height of the cloud.

Solution From the figure, we have

$$\tan 67.35° = \frac{h}{135 \text{ ft}}$$

$$h = 135 \text{ ft} \cdot \tan 67.35° \approx 324 \text{ ft.}$$

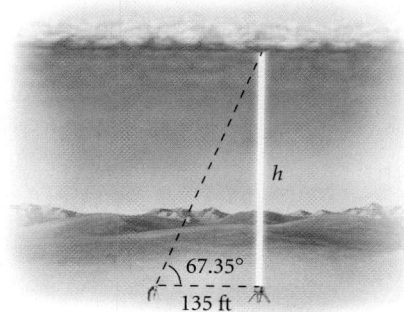

The height of the cloud is about 324 ft.

Now Try Exercise 23. ▨

Some applications of trigonometry involve the concept of direction, or bearing. In this text, we present two ways of giving direction, the first below and the second in Section 6.3.

Bearing: First-Type. One method of giving direction, or **bearing**, involves reference to a north–south line using an acute angle. For example, N55°W means 55° west of north and S67°E means 67° east of south.

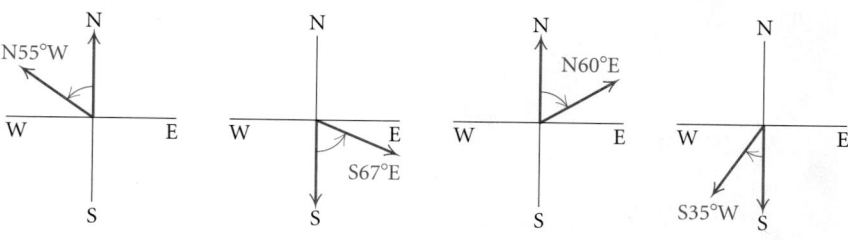

EXAMPLE 7 *Distance to a Forest Fire.* A forest ranger at point *A* sights a fire directly south. A second ranger at point *B*, 7.5 mi east, sights the same fire at a bearing of S27°23′W. How far from *A* is the fire?

Solution We first find the complement of 27°23′:

$$B = 90° - 27°23' \qquad \text{Angle } B \text{ is opposite side } d \text{ in the right triangle.}$$
$$= 62°37'$$
$$\approx 62.62°.$$

From the figure shown above, we see that the desired distance *d* is part of a right triangle. We have

$$\frac{d}{7.5 \text{ mi}} \approx \tan 62.62°$$
$$d \approx 7.5 \text{ mi} \cdot \tan 62.62° \approx 14.5 \text{ mi}.$$

The forest ranger at point *A* is about 14.5 mi from the fire.

Now Try Exercise 37. ■

EXAMPLE 8 *U.S. Cellular Field.* In U.S. Cellular Field, the home of the Chicago White Sox baseball team, the first row of seats in the upper deck is farther away from home plate than the last row of seats in the original Comiskey Park. Although there is no obstructed view in U.S. Cellular Field, some of the fans still complain about the present distance from home plate to the upper deck of seats. From a seat in the last row of the upper deck directly behind the batter, the angle of depression to home plate is 29.9°, and the angle of depression to the pitcher's mound is 24.2°. Find **(a)** the viewing distance to home plate and **(b)** the viewing distance to the pitcher's mound.

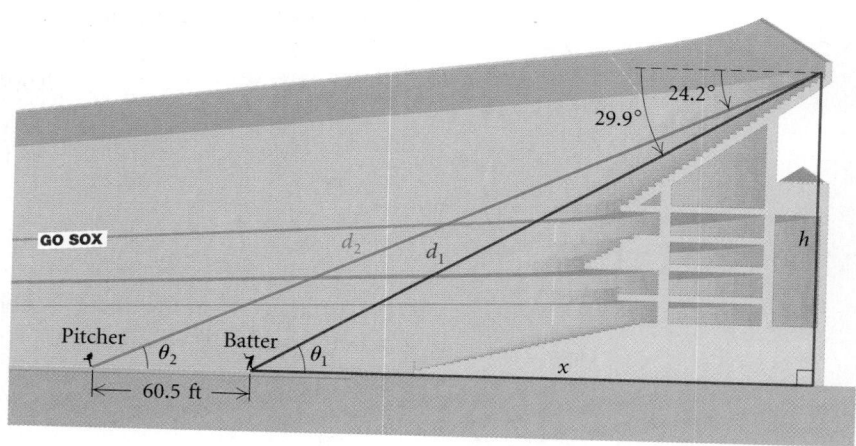

Solution From geometry we know that $\theta_1 = 29.9°$ and $\theta_2 = 24.2°$. The standard distance from home plate to the pitcher's mound is 60.5 ft. In the drawing, we let d_1 be the viewing distance to home plate, d_2 the viewing distance to the pitcher's mound, h the elevation of the last row, and x the horizontal distance from the batter to a point directly below the seat in the last row of the upper deck.

We begin by determining the distance x. We use the tangent function with $\theta_1 = 29.9°$ and $\theta_2 = 24.2°$:

$$\tan 29.9° = \frac{h}{x} \qquad \text{and} \qquad \tan 24.2° = \frac{h}{x + 60.5}$$

or

$$h = x \tan 29.9° \qquad \text{and} \qquad h = (x + 60.5) \tan 24.2°.$$

Then substituting $x \tan 29.9°$ for h in the second equation, we obtain

$$x \tan 29.9° = (x + 60.5) \tan 24.2°.$$

Solving for x, we get

$$x \tan 29.9° = x \tan 24.2° + 60.5 \tan 24.2°$$
$$x \tan 29.9° - x \tan 24.2° = x \tan 24.2° + 60.5 \tan 24.2° - x \tan 24.2°$$
$$x(\tan 29.9° - \tan 24.2°) = 60.5 \tan 24.2°$$

$$x = \frac{60.5 \tan 24.2°}{\tan 29.9° - \tan 24.2°}$$

$$x \approx 216.5.$$

We can then find d_1 and d_2 using the cosine function:

$$\cos 29.9° = \frac{216.5}{d_1} \quad \text{and} \quad \cos 24.2° = \frac{216.5 + 60.5}{d_2}$$

or

$$d_1 = \frac{216.5}{\cos 29.9°} \quad \text{and} \quad d_2 = \frac{277}{\cos 24.2°}$$

$$d_1 \approx 249.7 \qquad\qquad\qquad d_2 \approx 303.7.$$

The distance to home plate is about 250 ft,* and the distance to the pitcher's mound is about 304 ft. ■

$\left(\begin{array}{c} \textbf{6.2} \end{array}\right)$ # Exercise Set

In Exercises 1–6, solve the right triangle.

1.
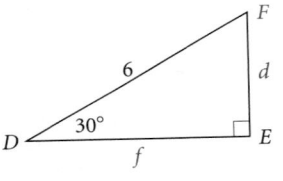

2.
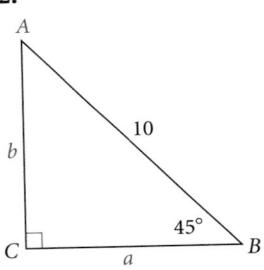

3.
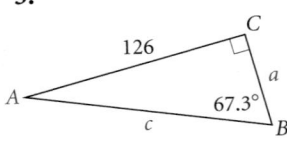

4.
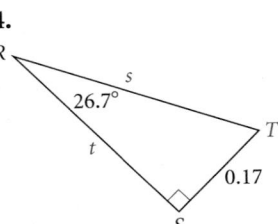

5.
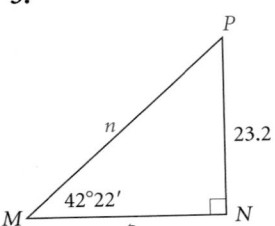

6.
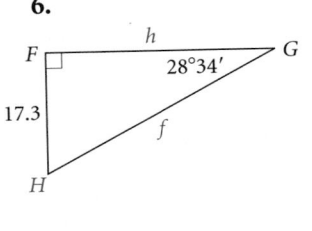

In Exercises 7–16, solve the right triangle. (Standard lettering has been used.)

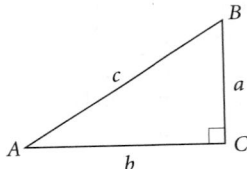

7. $A = 87°43'$, $a = 9.73$

8. $a = 12.5$, $b = 18.3$

9. $b = 100$, $c = 450$

10. $B = 56.5°$, $c = 0.0447$

11. $A = 47.58°$, $c = 48.3$

12. $B = 20.6°$, $a = 7.5$

13. $A = 35°$, $b = 40$

14. $B = 69.3°$, $b = 93.4$

15. $b = 1.86$, $c = 4.02$

16. $a = 10.2$, $c = 20.4$

---

*In the original Comiskey Park, the distance to home plate was only 150 ft.

17. *Aerial Photography.* An aerial photographer who photographs farm properties for a real estate company has determined from experience that the best photo is taken at a height of approximately 475 ft and a distance of 850 ft from the farmhouse. What is the angle of depression from the plane to the house?

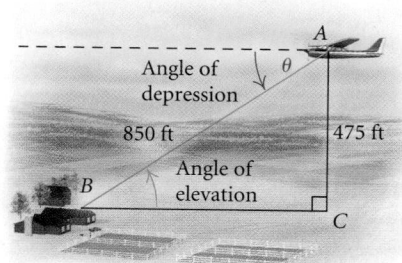

18. *Memorial Flag Case.* A tradition in the United States is to drape an American flag over the casket of a deceased U.S. Forces veteran. At the burial, the flag is removed, folded into a triangle, and presented to the family. The folded flag will fit in an isosceles right triangle case, as shown below. The inside dimension across the bottom is $21\frac{1}{2}$ in. (*Source*: Bruce Kieffer, *Woodworker's Journal*, August 2006). Using trigonometric functions, find the length x and round the answer to the nearest tenth of an inch.

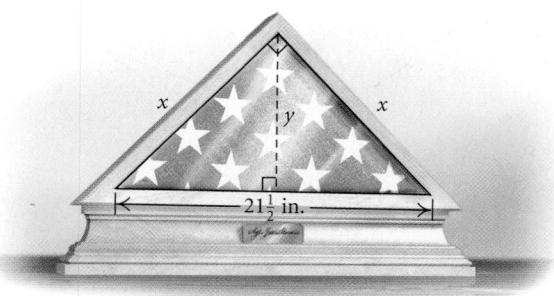

19. *Safety Line to Raft.* Each spring Bryan uses his vacation time to ready his lake property for the summer. He wants to run a new safety line from point B on the shore to the corner of the anchored diving raft. The current safety line,

which runs perpendicular to the shore line to point A, is 40 ft long. He estimates the angle from B to the corner of the raft to be 50°. Approximately how much rope does he need for the new safety line if he allows 5 ft of rope at each end to fasten the rope?

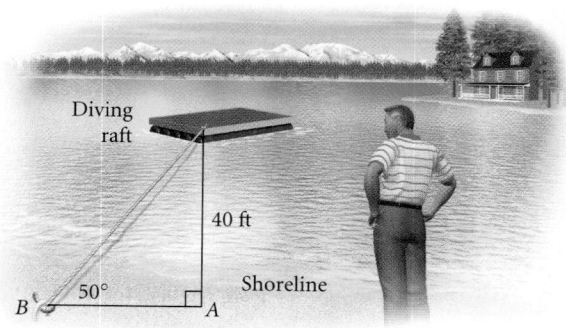

20. *Enclosing an Area.* Alicia is enclosing a triangular area in a corner of her fenced rectangular backyard for her Labrador retriever. In order for a certain tree to be included in this pen, one side needs to be 14.5 ft and make a 53° angle with the new side. How long is the new side?

21. *Setting a Fishing Reel Line Counter.* A fisherman who is fishing 50 ft directly out from a visible tree stump near the shore wants to position his line and bait approximately N35°W of the boat and west of the stump. Using the right triangle shown in the drawing, determine the reel's line counter setting, to the nearest foot, to position the line directly west of the stump.

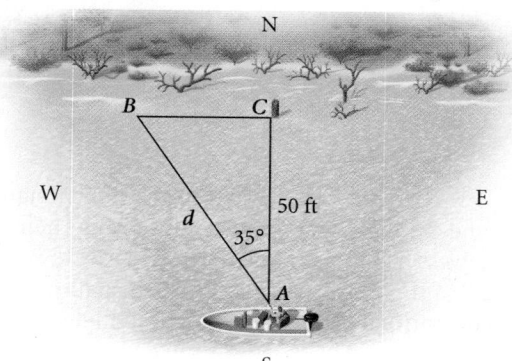

22. *Loading Ramp.* Kayden needs to purchase a custom ramp to use while loading and unloading a garden tractor. When down, the tailgate of his truck is 38 in. from the ground. If the recommended angle that the ramp makes with the ground is 28°, approximately how long does the ramp need to be?

23. *Height of a Tree.* A supervisor must train a new team of loggers to estimate the heights of trees. As an example, she walks off 40 ft from the base of a tree and estimates the angle of elevation to the tree's peak to be 70°. Approximately how tall is the tree?

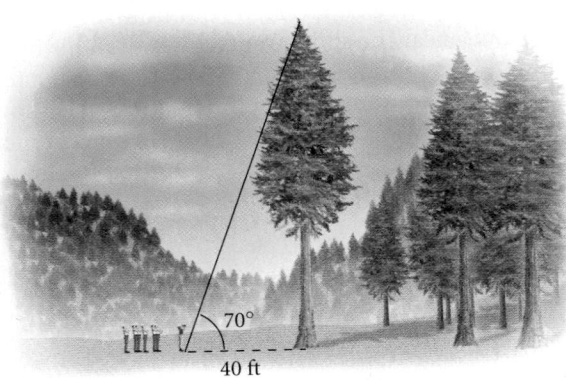

24. *Easel Display.* A marketing group is designing an easel to display posters advertising their newest products. They want the easel to be 6 ft tall and the back of it to fit flush against a wall. For optimal eye contact, the best angle between the front and back legs of the easel is 23°. How far from the wall should the front legs be placed in order to obtain this angle?

25. *Golden Gate Bridge.* The Golden Gate Bridge has two main towers of equal height that support the two main cables. A visitor on a tour boat passing through San Francisco Bay views the top of one of the towers and estimates the angle of elevation to be 30°. After sailing 670 ft closer, he estimates the angle of elevation to this same tower to be 50°. Approximate the height of the tower to the nearest foot.

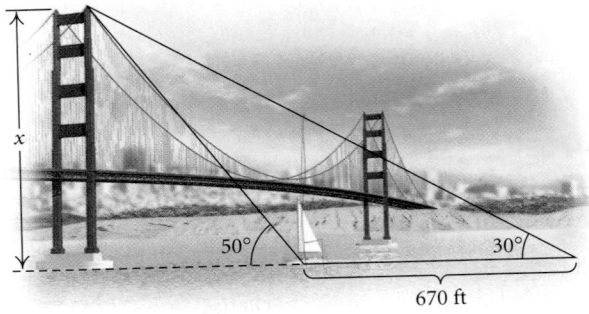

26. *Sand Dunes National Park.* While visiting the Sand Dunes National Park in Colorado, Cole approximated the angle of elevation to the top of a sand dune to be 20°. After walking 800 ft closer, he guessed that the angle of elevation had increased by 15°. Approximately how tall is the dune he was observing?

27. *Inscribed Pentagon.* A regular pentagon is inscribed in a circle of radius 15.8 cm. Find the perimeter of the pentagon.

28. *Height of a Weather Balloon.* A weather balloon is directly west of two observing stations that are 10 mi apart. The angles of elevation of the balloon from the two stations are 17.6° and 78.2°. How high is the balloon?

29. *Height of a Kite.* For a science fair project, a group of students tested different materials used to construct kites. Their instructor provided an instrument that accurately measures the angle of elevation. In one of the tests, the angle of elevation was 63.4° with 670 ft of string out. Assuming the string was taut, how high was the kite?

30. *Height of a Building.* A window washer on a ladder looks at a nearby building 100 ft away, noting that the angle of elevation to the top of the building is 18.7° and the angle of depression to the bottom of the building is 6.5°. How tall is the nearby building?

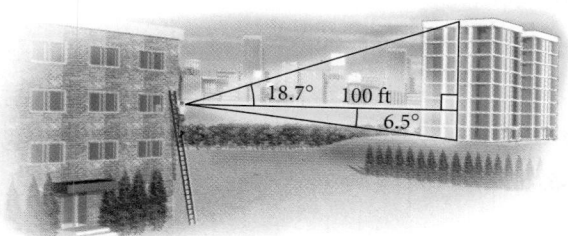

31. *Quilt Design.* Nancy is designing a quilt that she will enter in the quilt competition at the State Fair. The quilt consists of twelve identical squares with 4 rows of 3 squares each. Each square is to have a regular octagon inscribed in a circle, as shown in the figure. Each side of the octagon is to be 7 in. long. Find the radius of the circumscribed circle and the dimensions of the quilt. Round the answers to the nearest hundredth of an inch.

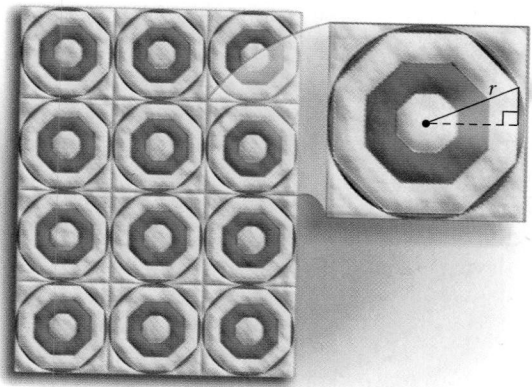

32. *Rafters for a House.* Blaise, an architect for luxury homes, is designing a house that is 46 ft wide with a roof whose pitch is 11/12. Determine the length of the rafters needed for this house. Round the answer to the nearest tenth of a foot.

33. *Rafters for a Medical Office.* The pitch of the roof for a medical office needs to be 5/12. If the building is 33 ft wide, how long must the rafters be?

34. *Angle of Elevation.* What is the angle of elevation of the sun when a 35-ft mast casts a 20-ft shadow?

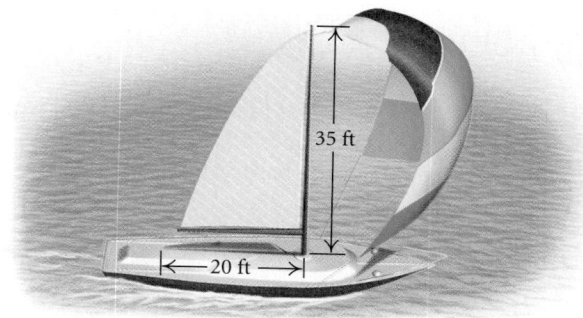

35. *Distance Between Towns.* From a hot-air balloon 2 km high, the angles of depression to two towns in line with the balloon are 81.2° and 13.5°. How far apart are the towns?

36. *Distance from a Lighthouse.* From the top of a lighthouse 55 ft above sea level, the angle of depression to a small boat is 11.3°. How far from the foot of the lighthouse is the boat?

37. *Lightning Detection.* In extremely large forests, it is not cost-effective to position forest rangers in towers or to use small aircraft to continually watch for fires. Since lightning is a frequent cause of fire, lightning detectors are now commonly used instead. These devices not only give a bearing on the location but also measure the intensity of the lightning. A detector at point Q is situated 15 mi west of a central fire station at point R. The bearing from Q to where lightning hits due south of R is S37.6°E. How far is the hit from point R?

38. *Length of an Antenna.*   A vertical antenna is mounted atop a 50-ft pole. From a point on level ground 75 ft from the base of the pole, the antenna subtends an angle of 10.5°. Find the length of the antenna.

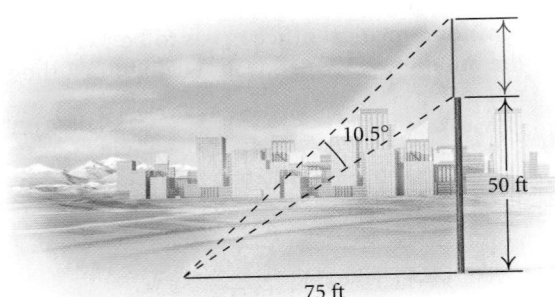

39. *Lobster Boat.*   A lobster boat is situated due west of a lighthouse. A barge is 12 km south of the lobster boat. From the barge, the bearing to the lighthouse is N63°20′E. How far is the lobster boat from the lighthouse?

Collaborative Discussion and Writing

40. Explain in your own words five ways in which length *c* can be determined in this triangle. Which way seems the most efficient?

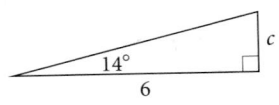

41. In this section, the trigonometric functions have been defined as functions of acute angles. Thus the set of angles whose measures are greater than 0° and less than 90° is the domain for each function. What appear to be the ranges for the sine, the cosine, and the tangent functions given this domain?

Skill Maintenance

Find the distance between the points.

42. $(-9, 3)$ and $(0, 0)$

43. $(8, -2)$ and $(-6, -4)$

44. Convert to a logarithmic equation: $e^4 = t$.

45. Convert to an exponential equation: $\log 0.001 = -3$.

Synthesis

46. Find *a*, to the nearest tenth.

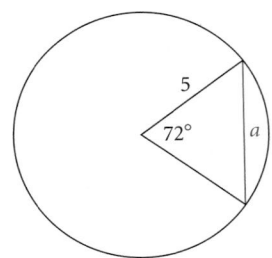

47. Find *h*, to the nearest tenth.

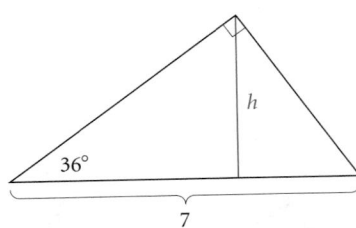

48. *Diameter of a Pipe.*   A V-gauge is used to find the diameter of a pipe. The advantage of such a device is that it is rugged, it is accurate, and it has no moving parts to break down. In the figure, the measure of angle *AVB* is 54°. A pipe is placed in the V-shaped slot and the distance *VP* is used to estimate the diameter. The line *VP* is calibrated by listing as its units the corresponding diameters. This, in effect, establishes a function between *VP* and *d*.

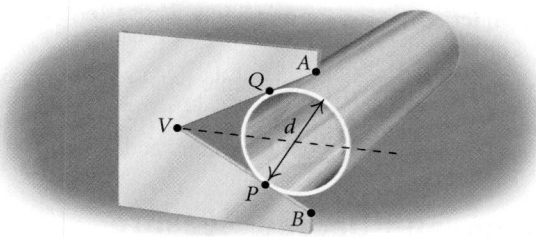

a) Suppose that the diameter of a pipe is 2 cm. What is the distance *VP*?

b) Suppose that the distance *VP* is 3.93 cm. What is the diameter of the pipe?

c) Find a formula for *d* in terms of *VP*.

d) Find a formula for *VP* in terms of *d*.

49. *Construction of Picnic Pavilions.* A construction company is mass-producing picnic pavilions for national parks, as shown in the figure. The rafter ends are to be sawed in such a way that they will be vertical when in place. The front is 8 ft high, the back is $6\frac{1}{2}$ ft high, and the distance between the front and back is 8 ft. At what angle should the rafters be cut?

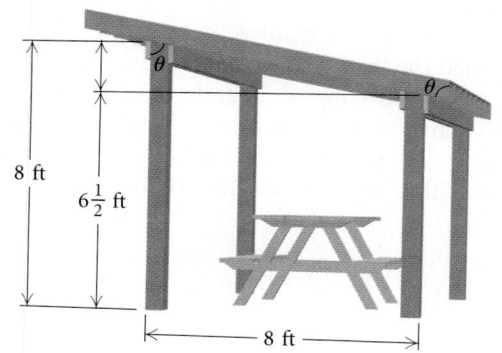

50. *Measuring the Radius of the Earth.* One way to measure the radius of the earth is to climb to the top of a mountain whose height above sea level is known and measure the angle between a vertical line to the center of the earth from the top of the mountain and a line drawn from the top of the mountain to the horizon, as shown in the figure. The height of Mt. Shasta in California is 14,162 ft. From the top of Mt. Shasta, one can see the horizon on the Pacific Ocean. The angle formed between a line to the horizon and the vertical is found to be 87°53'. Use this information to estimate the radius of the earth, in miles.

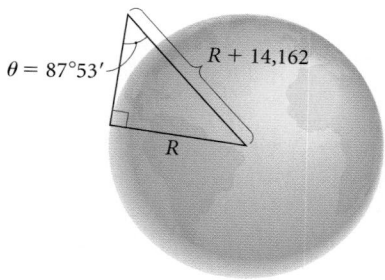

51. *Sound of an Airplane.* It is common experience to hear the sound of a low-flying airplane and look at the wrong place in the sky to see the plane. Suppose that a plane is traveling directly at you at a speed of 200 mph and an altitude of 3000 ft, and you hear the sound at what seems to be an angle of inclination of 20°. At what angle θ should you actually look in order to see the plane? Consider the speed of sound to be 1100 ft/sec.

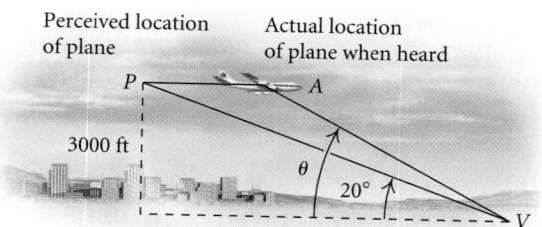

6.3 Trigonometric Functions of Any Angle

❀ Find angles that are coterminal with a given angle and find the complement and the supplement of a given angle.

❀ Determine the six trigonometric function values for any angle in standard position when the coordinates of a point on the terminal side are given.

❀ Find the function values for any angle whose terminal side lies on an axis.

❀ Find the function values for an angle whose terminal side makes an angle of 30°, 45°, or 60° with the *x*-axis.

❀ Use a calculator to find function values and angles.

❀ Angles, Rotations, and Degree Measure

An *angle* is a familiar figure in the world around us.

An **angle** is the union of two rays with a common endpoint called the **vertex**. In trigonometry, we often think of an angle as a **rotation**. To do so, think of locating a ray along the positive *x*-axis with its endpoint at the origin. This ray is called the **initial side** of the angle. Though we leave that ray fixed, think of making a copy of it and rotating it. A rotation *counterclockwise* is a **positive rotation**, and a rotation *clockwise* is a **negative rotation**. The ray at the end of the rotation is called the **terminal side** of the angle. The angle formed is said to be in **standard position**.

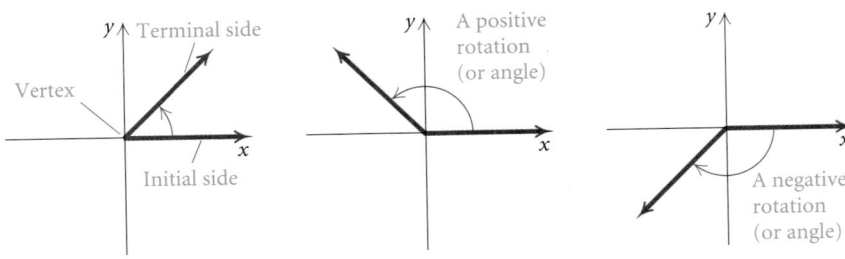

The measure of an angle or rotation may be given in degrees. The Babylonians developed the idea of dividing the circumference of a circle into 360 equal parts, or degrees. If we let the measure of one of these parts be 1°, then one complete positive revolution or rotation has a measure of 360°. One half of a revolution has a measure of 180°, one fourth of a revolution has a measure of 90°, and so on. We can also speak of an angle of measure 60°, 135°, 330°, or 420°. The terminal sides of these angles lie in quadrants I, II, IV, and I, respectively. The negative rotations −30°, −110°, and −225° represent angles with terminal sides in quadrants IV, III, and II, respectively.

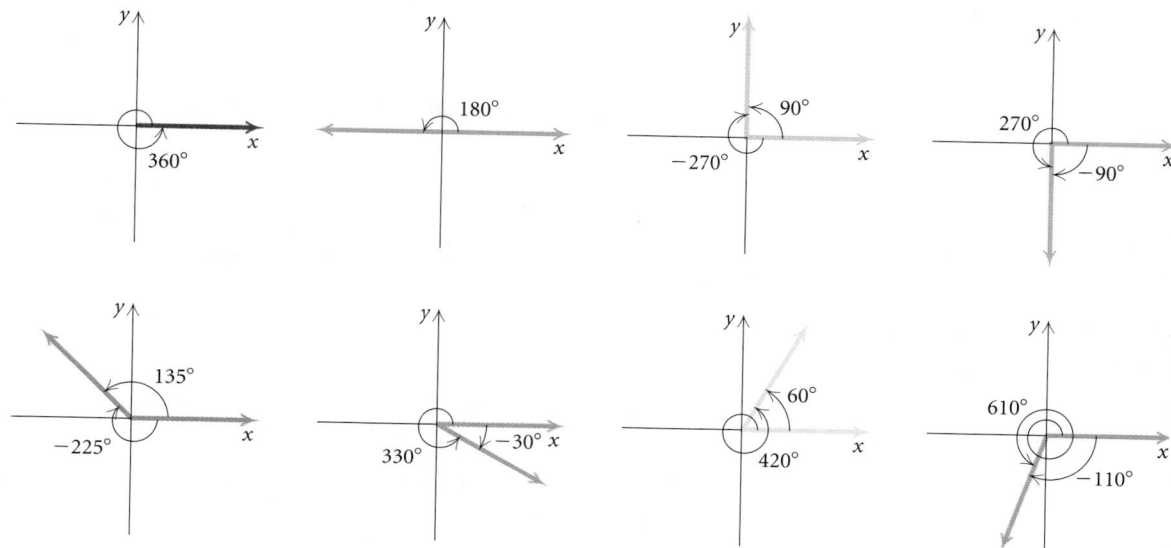

If two or more angles have the same terminal side, the angles are said to be **coterminal**. To find angles coterminal with a given angle, we add or subtract multiples of 360°. For example, 420°, shown above, has the same terminal side as 60°, since 420° = 360° + 60°. Thus we say that angles of measure 60° and 420° are coterminal. The negative rotation that measures −300° is also coterminal with 60° because 60° − 360° = −300°. The set of all angles coterminal with 60° can be expressed as 60° + n · 360°, where n is an integer. Other examples of coterminal angles shown above are 90° and −270°, −90° and 270°, 135° and −225°, −30° and 330°, and −110° and 610°.

EXAMPLE 1   Find two positive angles and two negative angles that are coterminal with (**a**) 51° and (**b**) −7°.

Solution

a) We add and subtract multiples of 360°. Many answers are possible.

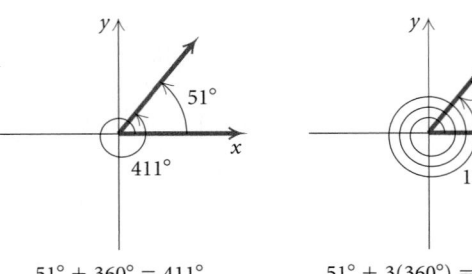

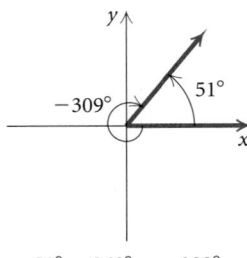

$$51° + 360° = 411°$$

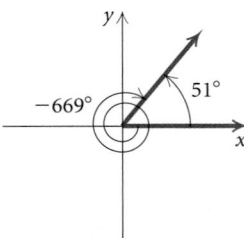

$$51° + 3(360°) = 1131°$$

$$51° − 360° = −309°$$

$$51° − 2(360°) = −669°$$

Thus angles of measure 411°, 1131°, −309°, and −669° are coterminal with 51°.

b) We have the following:

$$-7° + 360° = 353°, \qquad -7° + 2(360°) = 713°,$$
$$-7° - 360° = -367°, \qquad -7° - 10(360°) = -3607°.$$

Thus angles of measure 353°, 713°, −367°, and −3607° are coterminal with −7°.

Now Try Exercise 13. ■

Angles can be classified by their measures, as seen in the following figures.

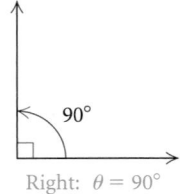

Right: $\theta = 90°$

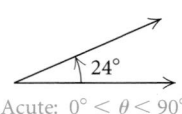

Acute: $0° < \theta < 90°$

Obtuse: $90° < \theta < 180°$

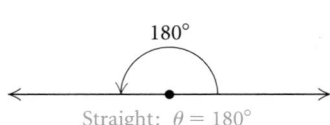

Straight: $\theta = 180°$

Recall that two acute angles are **complementary** if their sum is 90°. For example, angles that measure 10° and 80° are complementary because $10° + 80° = 90°$. Two positive angles are **supplementary** if their sum is 180°. For example, angles that measure 45° and 135° are supplementary because $45° + 135° = 180°$.

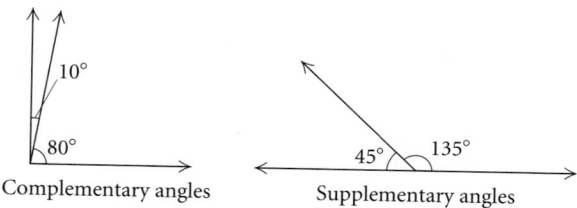

Complementary angles   Supplementary angles

EXAMPLE 2   Find the complement and the supplement of 71.46°.

Solution   We have

$$90° - 71.46° = 18.54°,$$
$$180° - 71.46° = 108.54°.$$

Thus the complement of 71.46° is 18.54° and the supplement is 108.54°.

Now Try Exercise 19. ■

❋ Trigonometric Functions of Angles or Rotations

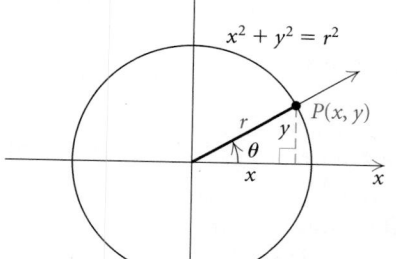

Many applied problems in trigonometry involve the use of angles that are not acute. Thus we need to extend the domains of the trigonometric functions defined in Section 6.1 to angles, or rotations, of *any* size. To do this, we first consider a right triangle with one vertex at the origin of a coordinate system and one vertex *on the positive x-axis*. (See the figure at left.) The other vertex is at P, a point on the circle whose center is at the origin and whose radius r is the length of the hypotenuse of the triangle. This triangle is a **reference triangle** for angle θ, which is in standard position. Note that y is the length of the side opposite θ and x is the length of the side adjacent to θ.

Recalling the definitions in Section 6.1, we note that three of the trigonometric functions of angle θ are defined as follows:

$$\sin\theta = \frac{\text{opp}}{\text{hyp}} = \frac{y}{r}, \qquad \cos\theta = \frac{\text{adj}}{\text{hyp}} = \frac{x}{r}, \qquad \tan\theta = \frac{\text{opp}}{\text{adj}} = \frac{y}{x}.$$

Since x and y are the coordinates of the point P and the length of the radius is the length of the hypotenuse, we can also define these functions as follows:

$$\sin\theta = \frac{y\text{-coordinate}}{\text{radius}},$$

$$\cos\theta = \frac{x\text{-coordinate}}{\text{radius}},$$

$$\tan\theta = \frac{y\text{-coordinate}}{x\text{-coordinate}}.$$

We will use these definitions for functions of angles of any measure. The following figures show angles whose terminal sides lie in quadrants II, III, and IV.

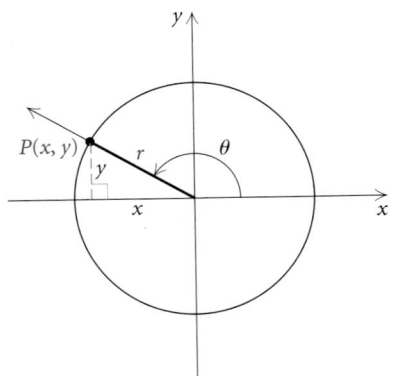

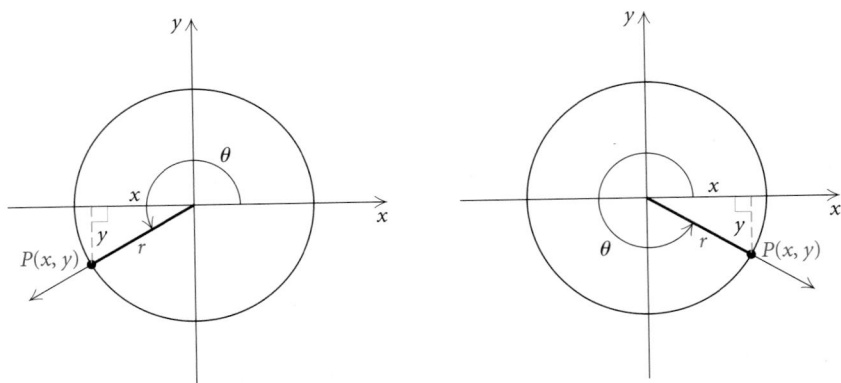

A reference triangle can be drawn for angles in any quadrant, as shown. Note that the angle is in standard position; that is, it is always measured from the positive half of the x-axis. The point $P(x, y)$ is a point, other than the vertex, on the terminal side of the angle. Each of its two coordinates may be positive, negative, or zero, depending on the location of the terminal side. *The length of the radius, which is also the length of the hypotenuse of the reference triangle, is always considered positive.* $\left(\text{Note that } x^2 + y^2 = r^2, \text{ or } r = \sqrt{x^2 + y^2}.\right)$ Regardless of the location of P, we have the following definitions.

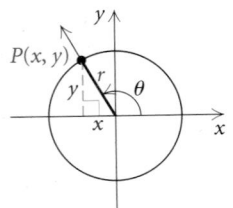

Trigonometric Functions of Any Angle θ

Suppose that $P(x, y)$ is any point other than the vertex on the terminal side of any angle θ in standard position, and r is the radius, or distance from the origin to $P(x, y)$. Then the trigonometric functions are defined as follows:

$$\sin \theta = \frac{y\text{-coordinate}}{\text{radius}} = \frac{y}{r}, \qquad \csc \theta = \frac{\text{radius}}{y\text{-coordinate}} = \frac{r}{y},$$

$$\cos \theta = \frac{x\text{-coordinate}}{\text{radius}} = \frac{x}{r}, \qquad \sec \theta = \frac{\text{radius}}{x\text{-coordinate}} = \frac{r}{x},$$

$$\tan \theta = \frac{y\text{-coordinate}}{x\text{-coordinate}} = \frac{y}{x}, \qquad \cot \theta = \frac{x\text{-coordinate}}{y\text{-coordinate}} = \frac{x}{y}.$$

Values of the trigonometric functions can be positive, negative, or zero, depending on where the terminal side of the angle lies. The length of the radius is always positive. Thus the signs of the function values depend only on the coordinates of the point P on the terminal side of the angle. In the first quadrant, all function values are positive because both coordinates are positive. In the second quadrant, first coordinates are negative and second coordinates are positive; thus only the sine and the cosecant values are positive. Similarly, we can determine the signs of the function values in the third and fourth quadrants. *Because of the reciprocal relationships, we need to learn only the signs for the sine, cosine, and tangent functions.*

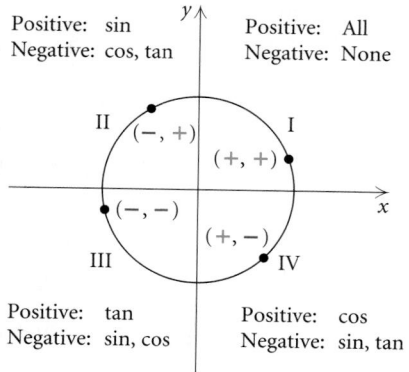

Positive: sin
Negative: cos, tan

Positive: All
Negative: None

II $(-, +)$

I $(+, +)$

$(-, -)$

$(+, -)$

III

IV

Positive: tan
Negative: sin, cos

Positive: cos
Negative: sin, tan

EXAMPLE 3 Find the six trigonometric function values for each angle shown.

a)

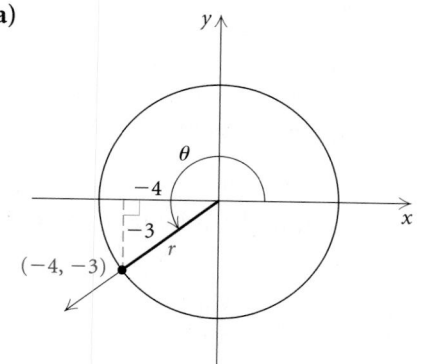

θ

-4

-3

r

$(-4, -3)$

b)

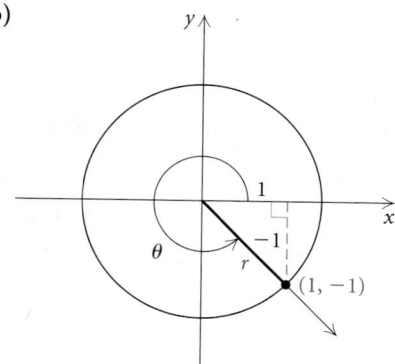

1

-1

r

θ

$(1, -1)$

c)

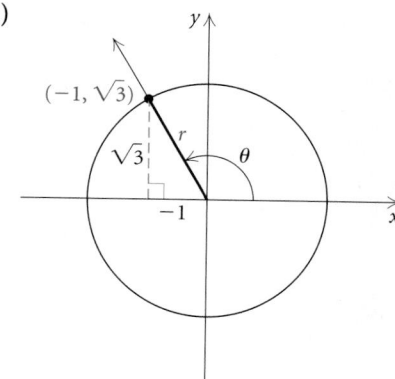

$(-1, \sqrt{3})$

r

$\sqrt{3}$

θ

-1

Solution

a) We first determine r, the distance from the origin $(0,0)$ to the point $(-4, -3)$. The distance between $(0,0)$ and any point (x, y) on the terminal side of the angle is

$$r = \sqrt{(x - 0)^2 + (y - 0)^2}$$
$$= \sqrt{x^2 + y^2}.$$

Substituting -4 for x and -3 for y, we find

$$r = \sqrt{(-4)^2 + (-3)^2}$$
$$= \sqrt{16 + 9} = \sqrt{25} = 5.$$

Using the definitions of the trigonometric functions, we can now find the function values of θ. We substitute -4 for x, -3 for y, and 5 for r:

$$\sin\theta = \frac{y}{r} = \frac{-3}{5} = -\frac{3}{5}, \qquad \csc\theta = \frac{r}{y} = \frac{5}{-3} = -\frac{5}{3},$$

$$\cos\theta = \frac{x}{r} = \frac{-4}{5} = -\frac{4}{5}, \qquad \sec\theta = \frac{r}{x} = \frac{5}{-4} = -\frac{5}{4},$$

$$\tan\theta = \frac{y}{x} = \frac{-3}{-4} = \frac{3}{4}, \qquad \cot\theta = \frac{x}{y} = \frac{-4}{-3} = \frac{4}{3}.$$

As expected, the tangent value and the cotangent value are positive and the other four are negative. This is true for all angles in quadrant III.

b) We first determine r, the distance from the origin to the point $(1, -1)$:

$$r = \sqrt{1^2 + (-1)^2} = \sqrt{1 + 1} = \sqrt{2}.$$

Substituting 1 for x, -1 for y, and $\sqrt{2}$ for r, we find

$$\sin\theta = \frac{y}{r} = \frac{-1}{\sqrt{2}} = -\frac{\sqrt{2}}{2}, \qquad \csc\theta = \frac{r}{y} = \frac{\sqrt{2}}{-1} = -\sqrt{2},$$

$$\cos\theta = \frac{x}{r} = \frac{1}{\sqrt{2}} = \frac{\sqrt{2}}{2}, \qquad \sec\theta = \frac{r}{x} = \frac{\sqrt{2}}{1} = \sqrt{2},$$

$$\tan\theta = \frac{y}{x} = \frac{-1}{1} = -1, \qquad \cot\theta = \frac{x}{y} = \frac{1}{-1} = -1.$$

c) We determine r, the distance from the origin to the point $\left(-1, \sqrt{3}\right)$:

$$r = \sqrt{(-1)^2 + \left(\sqrt{3}\right)^2} = \sqrt{1 + 3} = \sqrt{4} = 2.$$

Substituting -1 for x, $\sqrt{3}$ for y, and 2 for r, we find the trigonometric function values of θ are

$$\sin\theta = \frac{\sqrt{3}}{2}, \qquad \csc\theta = \frac{2}{\sqrt{3}} = \frac{2\sqrt{3}}{3},$$

$$\cos\theta = \frac{-1}{2} = -\frac{1}{2}, \qquad \sec\theta = \frac{2}{-1} = -2,$$

$$\tan\theta = \frac{\sqrt{3}}{-1} = -\sqrt{3}, \qquad \cot\theta = \frac{-1}{\sqrt{3}} = -\frac{\sqrt{3}}{3}.$$

Now Try Exercise 27. ■

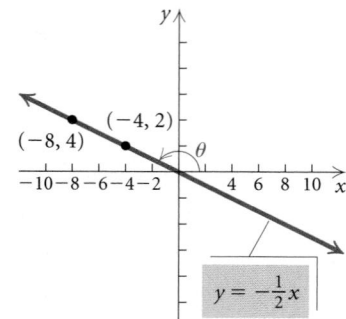

Any point other than the origin on the terminal side of an angle in standard position can be used to determine the trigonometric function values of that angle. The function values are the same regardless of which point is used. To illustrate this, let's consider an angle θ in standard position whose terminal side lies on the line $y = -\frac{1}{2}x$. We can determine two second-quadrant solutions of the equation, find the length r for each point, and then compare the sine, cosine, and tangent function values using each point.

If $x = -4$, then $y = -\frac{1}{2}(-4) = 2$.

If $x = -8$, then $y = -\frac{1}{2}(-8) = 4$.

For $(-4, 2)$, $r = \sqrt{(-4)^2 + 2^2} = \sqrt{20} = 2\sqrt{5}$.

For $(-8, 4)$, $r = \sqrt{(-8)^2 + 4^2} = \sqrt{80} = 4\sqrt{5}$.

Using $(-4, 2)$ and $r = 2\sqrt{5}$, we find that

$$\sin\theta = \frac{2}{2\sqrt{5}} = \frac{1}{\sqrt{5}} = \frac{\sqrt{5}}{5},$$

$$\cos\theta = \frac{-4}{2\sqrt{5}} = \frac{-2}{\sqrt{5}} = -\frac{2\sqrt{5}}{5},$$

and $\quad \tan\theta = \frac{2}{-4} = -\frac{1}{2}$.

Using $(-8, 4)$ and $r = 4\sqrt{5}$, we find that

$$\sin\theta = \frac{4}{4\sqrt{5}} = \frac{1}{\sqrt{5}} = \frac{\sqrt{5}}{5},$$

$$\cos\theta = \frac{-8}{4\sqrt{5}} = \frac{-2}{\sqrt{5}} = -\frac{2\sqrt{5}}{5},$$

and $\quad \tan\theta = \frac{4}{-8} = -\frac{1}{2}$.

We see that the function values are the same using either point. Any point other than the origin on the terminal side of an angle can be used to determine the trigonometric function values.

> The trigonometric function values of θ depend only on the angle, not on the choice of the point on the terminal side that is used to compute them.

❈ The Six Functions Related

When we know one of the function values of an angle, we can find the other five if we know the quadrant in which the terminal side lies. The procedure is to sketch a reference triangle in the appropriate quadrant, use the Pythagorean equation as needed to find the lengths of its sides, and then find the ratios of the sides.

EXAMPLE 4 Given that $\tan\theta = -\frac{2}{3}$ and θ is in the second quadrant, find the other function values.

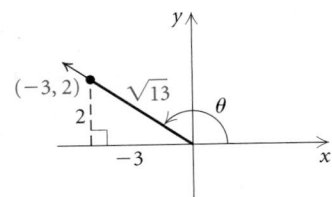

Solution We first sketch a second-quadrant angle. Since

$$\tan \theta = \frac{y}{x} = -\frac{2}{3} = \frac{2}{-3},$$

Expressing $-\frac{2}{3}$ as $\frac{2}{-3}$ since θ is in quadrant II

we make the legs lengths 2 and 3. The hypotenuse must then have length $\sqrt{2^2 + 3^2}$, or $\sqrt{13}$. Now we read off the appropriate ratios:

$$\sin \theta = \frac{2}{\sqrt{13}}, \quad \text{or} \quad \frac{2\sqrt{13}}{13}, \qquad \csc \theta = \frac{\sqrt{13}}{2},$$

$$\cos \theta = -\frac{3}{\sqrt{13}}, \quad \text{or} \quad -\frac{3\sqrt{13}}{13}, \qquad \sec \theta = -\frac{\sqrt{13}}{3},$$

$$\tan \theta = -\frac{2}{3}, \qquad \cot \theta = -\frac{3}{2}.$$

Now Try Exercise 33. ■

❖ Terminal Side on an Axis

An angle whose terminal side falls on one of the axes is a **quadrantal angle**. One of the coordinates of any point on that side is 0. The definitions of the trigonometric functions still apply, but in some cases, function values will not be defined because a denominator will be 0.

EXAMPLE 5 Find the sine, cosine, and tangent values for 90°, 180°, 270°, and 360°.

Solution We first make a drawing of each angle in standard position and label a point on the terminal side. Since the function values are the same for all points on the terminal side, we choose $(0, 1)$, $(-1, 0)$, $(0, -1)$, and $(1, 0)$ for convenience. Note that $r = 1$ for each choice.

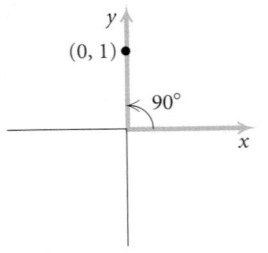 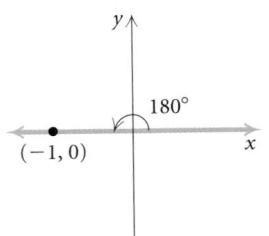 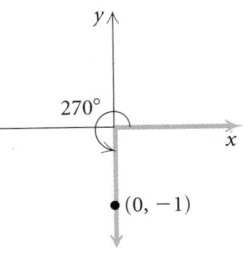 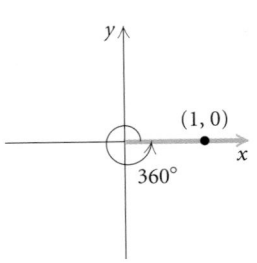

Then by the definitions we get

$$\sin 90° = \frac{1}{1} = 1, \qquad \sin 180° = \frac{0}{1} = 0, \qquad \sin 270° = \frac{-1}{1} = -1, \qquad \sin 360° = \frac{0}{1} = 0,$$

$$\cos 90° = \frac{0}{1} = 0, \qquad \cos 180° = \frac{-1}{1} = -1, \qquad \cos 270° = \frac{0}{1} = 0, \qquad \cos 360° = \frac{1}{1} = 1,$$

$$\tan 90° = \frac{1}{0}, \quad \text{Not defined} \qquad \tan 180° = \frac{0}{-1} = 0, \qquad \tan 270° = \frac{-1}{0}, \quad \text{Not defined} \qquad \tan 360° = \frac{0}{1} = 0.$$

■

In Example 5, all the values can be found using a calculator, but you will find that it is convenient to be able to compute them mentally. It is also helpful to note that coterminal angles have the same function values. For example, 0° and 360° are coterminal; thus, sin 0° = 0, cos 0° = 1, and tan 0° = 0.

EXAMPLE 6 Find each of the following.

a) sin (−90°)   **b)** csc 540°

Solution

a) We note that −90° is coterminal with 270°. Thus,

$$\sin(-90°) = \sin 270° = \frac{-1}{1} = -1.$$

b) Since 540° = 180° + 360°, 540° and 180° are coterminal. Thus,

$$\csc 540° = \csc 180° = \frac{1}{\sin 180°} = \frac{1}{0}, \quad \text{which is not defined.}$$

Now Try Exercises 43 and 53. ■

Trigonometric values can always be checked using a calculator. When the value is undefined, the calculator will display an ERROR message.

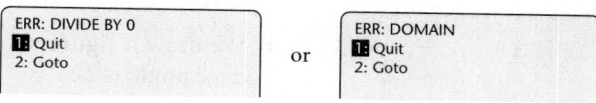

```
ERR: DIVIDE BY 0
1: Quit
2: Goto
```
or
```
ERR: DOMAIN
1: Quit
2: Goto
```

❋ Reference Angles: 30°, 45°, and 60°

We can also mentally determine trigonometric function values whenever the terminal side makes a 30°, 45°, or 60° angle with the x-axis. Consider, for example, an angle of 150°. The terminal side makes a 30° angle with the x-axis, since 180° − 150° = 30°.

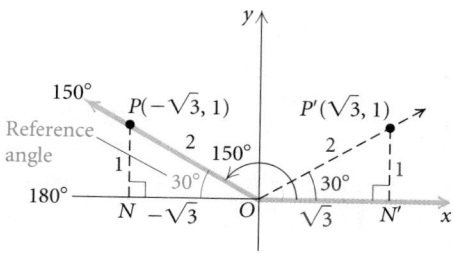

As the figure shows, △ONP is congruent to △ON′P′; therefore, the ratios of the sides of the two triangles are the same. Thus the trigonometric function values are the same except perhaps for the sign. We could determine the function values directly from △ONP, but this is not necessary. If we remember that in quadrant II, the sine is positive and the cosine and the tangent are negative, we can simply use the function values of 30° that we already know and prefix the appropriate sign. Thus,

$$\sin 150° = \sin 30° = \frac{1}{2},$$

$$\cos 150° = -\cos 30° = -\frac{\sqrt{3}}{2},$$

and $\quad \tan 150° = -\tan 30° = -\frac{1}{\sqrt{3}}, \quad \text{or} \quad -\frac{\sqrt{3}}{3}.$

Triangle ONP is the reference triangle and the acute angle $\angle NOP$ is called a *reference angle*.

Reference Angle
The **reference angle** for an angle is the acute angle formed by the terminal side of the angle and the x-axis.

EXAMPLE 7   Find the sine, cosine, and tangent function values for each of the following.

a) 225°   **b)** −780°

Solution

a) We draw a figure showing the terminal side of a 225° angle. The reference angle is 225° − 180°, or 45°.

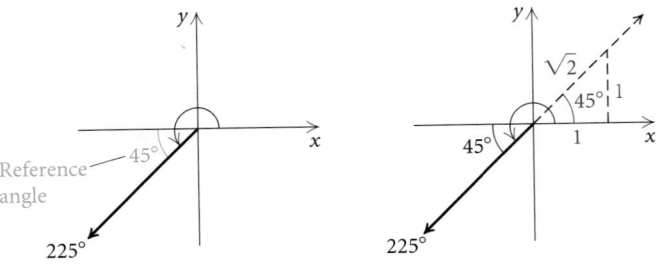

Recall from Section 6.1 that $\sin 45° = \sqrt{2}/2$, $\cos 45° = \sqrt{2}/2$, and $\tan 45° = 1$. Also note that in the third quadrant, the sine and the cosine are negative and the tangent is positive. Thus we have

$$\sin 225° = -\frac{\sqrt{2}}{2}, \quad \cos 225° = -\frac{\sqrt{2}}{2}, \quad \text{and} \quad \tan 225° = 1.$$

b) We draw a figure showing the terminal side of a $-780°$ angle. Since $-780° + 2(360°) = -60°$, we know that $-780°$ and $-60°$ are coterminal.

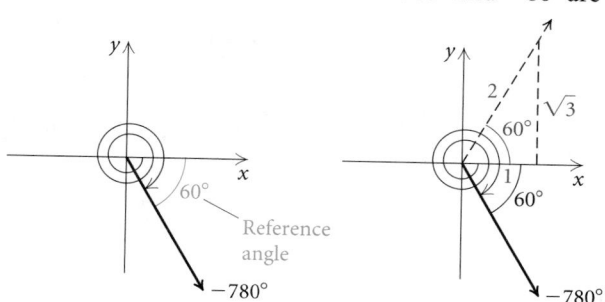

The reference angle for $-60°$ is the acute angle formed by the terminal side of the angle and the x-axis. Thus the reference angle for $-60°$ is $60°$. We know that since $-780°$ is a fourth-quadrant angle, the cosine is positive and the sine and the tangent are negative. Recalling that $\sin 60° = \sqrt{3}/2$, $\cos 60° = 1/2$, and $\tan 60° = \sqrt{3}$, we have

$$\sin(-780°) = -\frac{\sqrt{3}}{2}, \qquad \cos(-780°) = \frac{1}{2},$$

and $\qquad \tan(-780°) = -\sqrt{3}.$

Now Try Exercises 45 and 49. ■

❋ Function Values for Any Angle

When the terminal side of an angle falls on one of the axes or makes a $30°$, $45°$, or $60°$ angle with the x-axis, we can find exact function values without the use of a calculator. But this group is only a small subset of *all* angles. Using a calculator, we can approximate the trigonometric function values of *any* angle. In fact, we can approximate or find exact function values of all angles without using a reference angle.

EXAMPLE 8 Find each of the following function values using a calculator and round the answer to four decimal places, where appropriate.

a) $\cos 112°$ **b)** $\sec 500°$

c) $\tan(-83.4°)$ **d)** $\csc 351.75°$

e) $\cos 2400°$ **f)** $\sin 175°40'9''$

g) $\cot(-135°)$

Solution Using a calculator set in DEGREE mode, we find the values.

a) $\cos 112° \approx -0.3746$

b) $\sec 500° = \dfrac{1}{\cos 500°} \approx -1.3054$

c) $\tan(-83.4°) \approx -8.6427$

cos(112)
                −.3746065934
1/cos(500)
                −1.305407289
tan(−83.4)
                −8.642747461

d) $\csc 351.75° = \dfrac{1}{\sin 351.75°} \approx -6.9690$

e) $\cos 2400° = -0.5$

f) $\sin 175°40'9'' \approx 0.0755$

g) $\cot(-135°) = \dfrac{1}{\tan(-135°)} = 1$ **Now Try Exercises 87 and 93.** ◼

In many applications, we have a trigonometric function value and want to find the measure of a corresponding angle. When only acute angles are considered, there is only one angle for each trigonometric function value. This is not the case when we extend the domain of the trigonometric functions to the set of *all* angles. For a given function value, there is an infinite number of angles that have that function value. There can be two such angles for each value in the range from 0° to 360°. To determine a unique answer in the interval (0°, 360°), the quadrant in which the terminal side lies must be specified.

The calculator gives the reference angle as an output for each function value that is entered as an input. Knowing the reference angle and the quadrant in which the terminal side lies, we can find the specified angle.

EXAMPLE 9   Given the function value and the quadrant restriction, find θ.

a) $\sin \theta = 0.2812$, $90° < \theta < 180°$

b) $\cot \theta = -0.1611$, $270° < \theta < 360°$

Solution

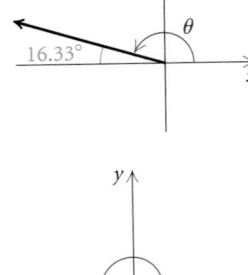

a) We first sketch the angle in the second quadrant. We use the calculator to find the acute angle (reference angle) whose sine is 0.2812. The reference angle is approximately 16.33°. We find the angle θ by subtracting 16.33° from 180°:

$$180° - 16.33° = 163.67°.$$

Thus, $\theta \approx 163.67°$.

b) We begin by sketching the angle in the fourth quadrant. Because the tangent and cotangent values are reciprocals, we know that

$$\tan \theta \approx \frac{1}{-0.1611} \approx -6.2073.$$

We use the calculator to find the acute angle (reference angle) whose tangent is 6.2073, ignoring the fact that $\tan \theta$ is negative. The reference angle is approximately 80.85°. We find angle θ by subtracting 80.85° from 360°:

$$360° - 80.85° = 279.15°.$$

Thus, $\theta \approx 279.15°$.   **Now Try Exercise 99.** ◼

Bearing: Second-Type. In aerial navigation, directions are given in degrees clockwise from north. Thus east is 90°, south is 180°, and west is 270°. Several aerial directions, or **bearings**, are given below.

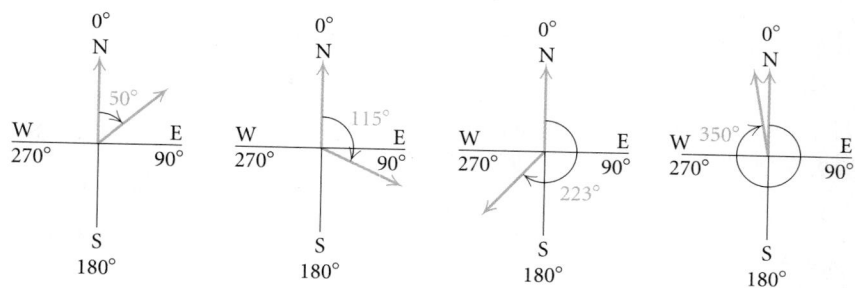

EXAMPLE 10 *Aerial Navigation.* An airplane flies 218 mi from an airport in a direction of 245°. How far south of the airport is the plane then? How far west?

Solution We first find the measure of $\angle ABC$:

$$B = 270° - 245° = 25°.$$ **Angle B is opposite side b in the right triangle.**

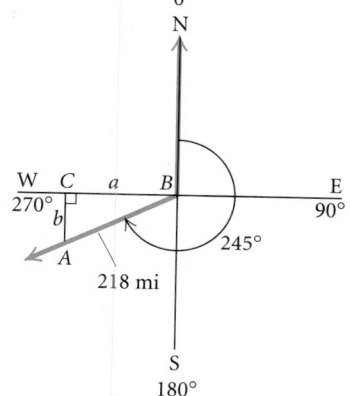

From the figure shown at left, we see that the distance south of the airport b and the distance west of the airport a are parts of a right triangle. We have

$$\frac{b}{218} = \sin 25°$$

$$b = 218 \sin 25° \approx 92 \text{ mi}$$

and

$$\frac{a}{218} = \cos 25°$$

$$a = 218 \cos 25° \approx 198 \text{ mi.}$$

The airplane is about 92 mi south and about 198 mi west of the airport.

Now Try Exercise 83. ▪

6.3 Exercise Set

For angles of the following measures, state in which quadrant the terminal side lies. It helps to sketch the angle in standard position.

1. 187°

2. −14.3°

3. 245°15′

4. −120°

5. 800°

6. 1075°

7. −460.5°

8. 315°

9. −912°

10. 13°15′58″

11. 537°

12. −345.14°

Find two positive angles and two negative angles that are coterminal with the given angle. Answers may vary.

13. 74° **14.** −81°

15. 115.3° **16.** 275°10′

17. −180° **18.** −310°

Find the complement and the supplement.

19. 17.11° **20.** 47°38′

21. 12°3′14″ **22.** 9.038°

23. 45.2° **24.** 67.31°

Find the six trigonometric function values for the angle shown.

25.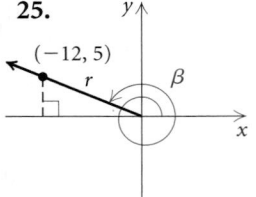

26.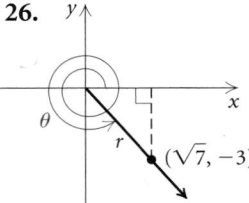

27.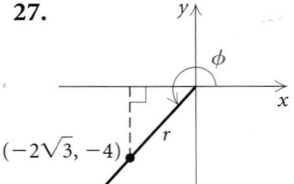

28.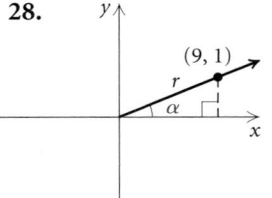

The terminal side of angle θ in standard position lies on the given line in the given quadrant. Find sin θ, cos θ, and tan θ.

29. $2x + 3y = 0$; quadrant IV

30. $4x + y = 0$; quadrant II

31. $5x - 4y = 0$; quadrant I

32. $y = 0.8x$; quadrant III

A function value and a quadrant are given. Find the other five function values. Give exact answers.

33. $\sin \theta = -\dfrac{1}{3}$; quadrant III

34. $\tan \beta = 5$; quadrant I

35. $\cot \theta = -2$; quadrant IV

36. $\cos \alpha = -\dfrac{4}{5}$; quadrant II

37. $\cos \phi = \dfrac{3}{5}$; quadrant IV

38. $\sin \theta = -\dfrac{5}{13}$; quadrant III

Find the reference angle and the exact function value if it exists.

39. cos 150° **40.** sec (−225°)

41. tan (−135°) **42.** sin (−45°)

43. sin 7560° **44.** tan 270°

45. cos 495° **46.** tan 675°

47. csc (−210°) **48.** sin 300°

49. cot 570° **50.** cos (−120°)

51. tan 330° **52.** cot 855°

53. sec (−90°) **54.** sin 90°

55. cos (−180°) **56.** csc 90°

57. tan 240° **58.** cot (−180°)

59. sin 495° **60.** sin 1050°

61. csc 225° **62.** sin (−450°)

63. cos 0° **64.** tan 480°

65. cot (−90°) **66.** sec 315°

67. cos 90° **68.** sin (−135°)

69. cos 270° **70.** tan 0°

Find the signs of the six trigonometric function values for the given angles.

71. 319° **72.** −57°

73. 194° **74.** −620°

75. −215° **76.** 290°

77. −272° **78.** 91°

Use a calculator in Exercises 79–82, but do not use the trigonometric function keys.

79. Given that
$$\sin 41° = 0.6561,$$
$$\cos 41° = 0.7547,$$
$$\tan 41° = 0.8693,$$
find the trigonometric function values for 319°.

80. Given that

$$\sin 27° = 0.4540,$$
$$\cos 27° = 0.8910,$$
$$\tan 27° = 0.5095,$$

find the trigonometric function values for 333°.

81. Given that

$$\sin 65° = 0.9063,$$
$$\cos 65° = 0.4226,$$
$$\tan 65° = 2.1445,$$

find the trigonometric function values for 115°.

82. Given that

$$\sin 35° = 0.5736,$$
$$\cos 35° = 0.8192,$$
$$\tan 35° = 0.7002,$$

find the trigonometric function values for 215°.

83. *Aerial Navigation.*   An airplane flies 150 km from an airport in a direction of 120°. How far east of the airport is the plane then? How far south?

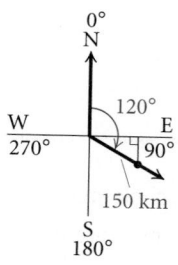

84. *Aerial Navigation.*   An airplane leaves an airport and travels for 100 mi in a direction of 300°. How far north of the airport is the plane then? How far west?

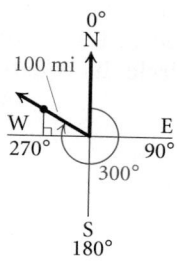

85. *Aerial Navigation.*   An airplane travels at 150 km/h for 2 hr in a direction of 138° from Omaha. At the end of this time, how far south of Omaha is the plane?

86. *Aerial Navigation.*   An airplane travels at 120 km/h for 2 hr in a direction of 319° from Chicago. At the end of this time, how far north of Chicago is the plane?

Find the function value. Round to four decimal places.

87. tan 310.8°   **88.** cos 205.5°

89. cot 146.15°   **90.** sin (−16.4°)

91. sin 118°42′   **92.** cos 273°45′

93. cos (−295.8°)   **94.** tan 1086.2°

95. cos 5417°   **96.** sec 240°55′

97. csc 520°   **98.** sin 3824°

Given the function value and the quadrant restriction, find θ.

Function Value	Interval	θ
99. sin θ = −0.9956	(270°, 360°)	
100. tan θ = 0.2460	(180°, 270°)	
101. cos θ = −0.9388	(180°, 270°)	
102. sec θ = −1.0485	(90°, 180°)	
103. tan θ = −3.0545	(270°, 360°)	
104. sin θ = −0.4313	(180°, 270°)	
105. csc θ = 1.0480	(0°, 90°)	
106. cos θ = −0.0990	(90°, 180°)	

Collaborative Discussion and Writing

107. Why do the function values of θ depend only on the angle and not on the choice of a point on the terminal side?

108. Why is the domain of the tangent function different from the domains of the sine function and the cosine function?

Skill Maintenance

Graph the function. Sketch and label any vertical asymptotes.

109. $f(x) = \dfrac{1}{x^2 - 25}$ **110.** $g(x) = x^3 - 2x + 1$

Determine the domain and the range of the function.

111. $f(x) = \dfrac{x - 4}{x + 2}$

112. $g(x) = \dfrac{x^2 - 9}{2x^2 - 7x - 15}$

Find the zeros of the function.

113. $f(x) = 12 - x$ **114.** $g(x) = x^2 - x - 6$

Find the x-intercepts of the graph of the function.

115. $f(x) = 12 - x$ **116.** $g(x) = x^2 - x - 6$

Synthesis

117. *Valve Cap on a Bicycle.*   The valve cap on a bicycle wheel is 12.5 in. from the center of the wheel. From the position shown, the wheel starts to roll. After the wheel has turned 390°, how far above the ground is the valve cap? Assume that the outer radius of the tire is 13.375 in.

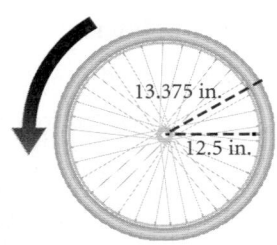

13.375 in.

12.5 in.

118. *Seats of a Ferris Wheel.*   The seats of a ferris wheel are 35 ft from the center of the wheel. When you board the wheel, you are 5 ft above the ground. After you have rotated through an angle of 765°, how far above the ground are you?

35 ft

5 ft

6.4

Radians, Arc Length, and Angular Speed

❖ Find points on the unit circle determined by real numbers.
❖ Convert between radian measure and degree measure; find coterminal, complementary, and supplementary angles.
❖ Find the length of an arc of a circle; find the measure of a central angle of a circle.
❖ Convert between linear speed and angular speed.

Another useful unit of angle measure is called a *radian*. To introduce radian measure, we use a circle centered at the origin with a radius of length 1. Such a circle is called a **unit circle**. Its equation is $x^2 + y^2 = 1$.

CIRCLES

REVIEW SECTION **1.1.**

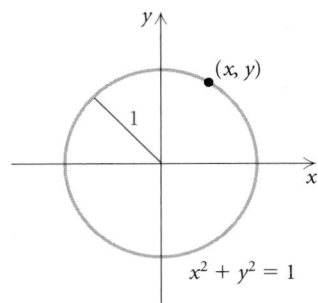

(x, y)

1

$x^2 + y^2 = 1$

❋ Distances on the Unit Circle

The circumference of a circle of radius r is $2\pi r$. Thus for the unit circle, where $r = 1$, the circumference is 2π. If a point starts at A and travels around the circle (Fig. 1), it will travel a distance of 2π. If it travels halfway around the circle (Fig. 2), it will travel a distance of $\frac{1}{2} \cdot 2\pi$, or π.

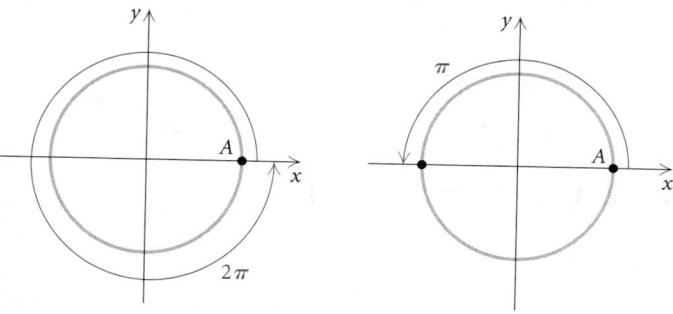

FIGURE 1   FIGURE 2

If a point C travels $\frac{1}{8}$ of the way around the circle (Fig. 3), it will travel a distance of $\frac{1}{8} \cdot 2\pi$, or $\pi/4$. Note that C is $\frac{1}{4}$ of the way from A to B. If a point D travels $\frac{1}{6}$ of the way around the circle (Fig. 4), it will travel a distance of $\frac{1}{6} \cdot 2\pi$, or $\pi/3$. Note that D is $\frac{1}{3}$ of the way from A to B.

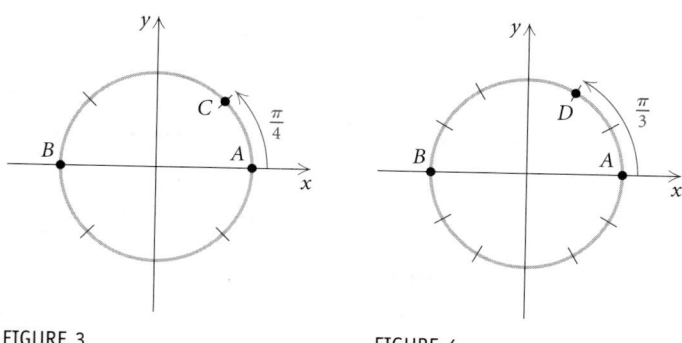

FIGURE 3   FIGURE 4

EXAMPLE 1 How far will a point travel if it goes (a) $\frac{1}{4}$, (b) $\frac{1}{12}$, (c) $\frac{3}{8}$, and (d) $\frac{5}{6}$ of the way around the unit circle?

Solution

a) $\frac{1}{4}$ of the total distance around the circle is $\frac{1}{4} \cdot 2\pi$, which is $\frac{1}{2} \cdot \pi$, or $\pi/2$.

b) The distance will be $\frac{1}{12} \cdot 2\pi$, which is $\frac{1}{6}\pi$, or $\pi/6$.

c) The distance will be $\frac{3}{8} \cdot 2\pi$, which is $\frac{3}{4}\pi$, or $3\pi/4$.

d) The distance will be $\frac{5}{6} \cdot 2\pi$, which is $\frac{5}{3}\pi$, or $5\pi/3$. Think of $5\pi/3$ as $\pi + \frac{2}{3}\pi$.

These distances are illustrated in the following figures.

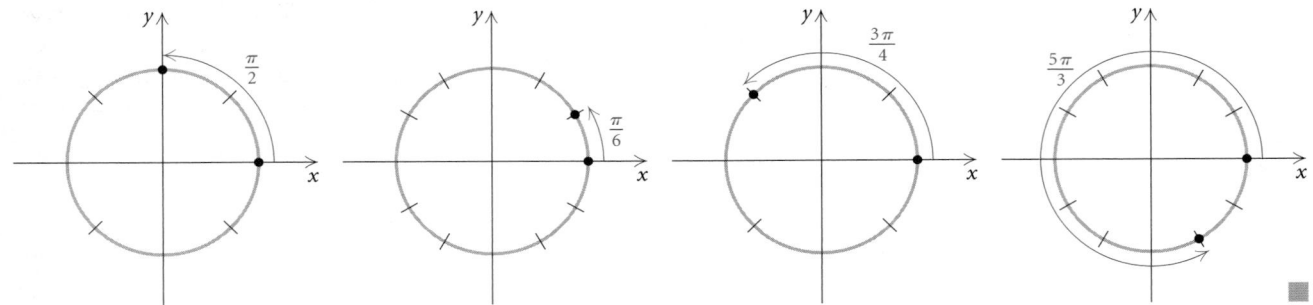

A point may travel completely around the circle and then continue. For example, if it goes around once and then continues $\frac{1}{4}$ of the way around, it will have traveled a distance of $2\pi + \frac{1}{4} \cdot 2\pi$, or $5\pi/2$ (Fig. 5). *Every* real number determines a point on the unit circle. For the positive number 10, for example, we start at A and travel counterclockwise a distance of 10. The point at which we stop is the point "determined" by the number 10. Note that $2\pi \approx 6.28$ and that $10 \approx 1.6(2\pi)$. Thus the point for 10 travels around the unit circle about $1\frac{3}{5}$ times (Fig. 6).

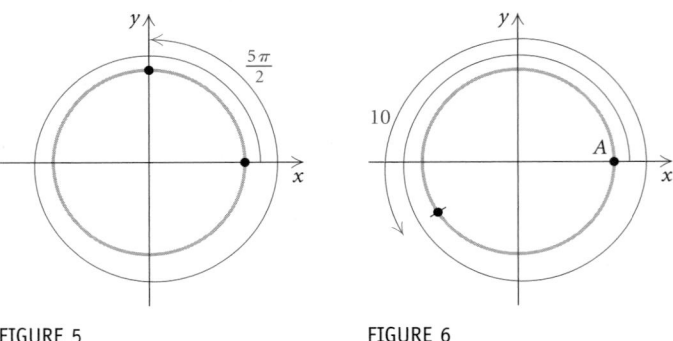

FIGURE 5   FIGURE 6

For a negative number, we move clockwise around the circle. Points for $-\pi/4$ and $-3\pi/2$ are shown in the figure below. The number 0 determines the point A.

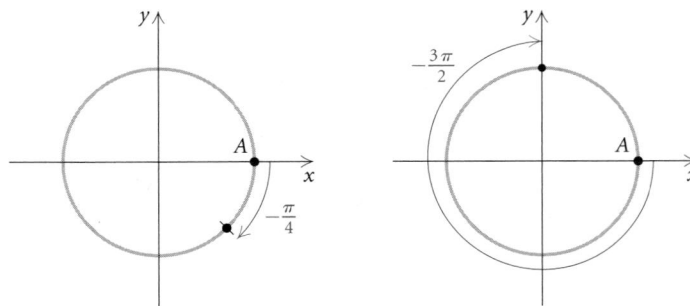

EXAMPLE 2 On the unit circle, mark the point determined by each of the following real numbers.

a) $\dfrac{9\pi}{4}$ b) $-\dfrac{7\pi}{6}$

Solution

a) Think of $9\pi/4$ as $2\pi + \frac{1}{4}\pi$. (See the figure below.) Since $9\pi/4 > 0$, the point moves counterclockwise. The point goes completely around once and then continues $\frac{1}{4}$ of the way from A to B.

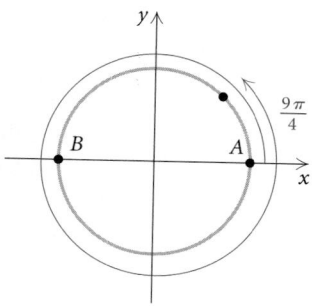

b) The number $-7\pi/6$ is negative, so the point moves clockwise. From A to B, the distance is π, or $\frac{6}{6}\pi$, so we need to go beyond B another distance of $\pi/6$, clockwise. (See the figure below.)

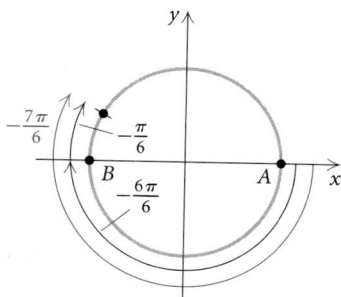

Now Try Exercise 1. ■

❖ Radian Measure

Degree measure is a common unit of angle measure in many everyday applications. But in many scientific fields and in mathematics (calculus, in particular), there is another commonly used unit of measure called the *radian*.

Consider the unit circle. Recall that this circle has radius 1. Suppose we measure, moving counterclockwise, an arc of length 1, and mark a point T on the circle.

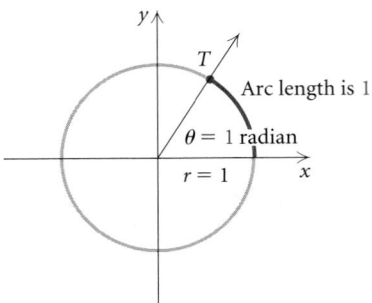

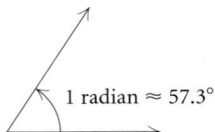

If we draw a ray from the origin through T, we have formed an angle. The measure of that angle is 1 **radian**. The word radian comes from the word *radius*. Thus measuring 1 "radius" along the circumference of the circle determines an angle whose measure is 1 *radian*. One radian is about 57.3°. Angles that measure 2 radians, 3 radians, and 6 radians are shown below.

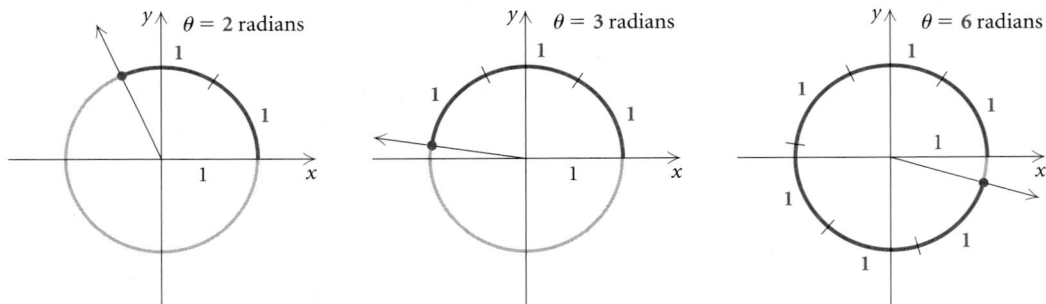

When we make a complete (counterclockwise) revolution, the terminal side coincides with the initial side on the positive x-axis. We then have an angle whose measure is 2π radians, or about 6.28 radians, which is the circumference of the circle:

$$2\pi r = 2\pi(1) = 2\pi.$$

Thus a rotation of 360° (1 revolution) has a measure of 2π radians. A half revolution is a rotation of 180°, or π radians. A quarter revolution is a rotation of 90°, or $\pi/2$ radians, and so on.

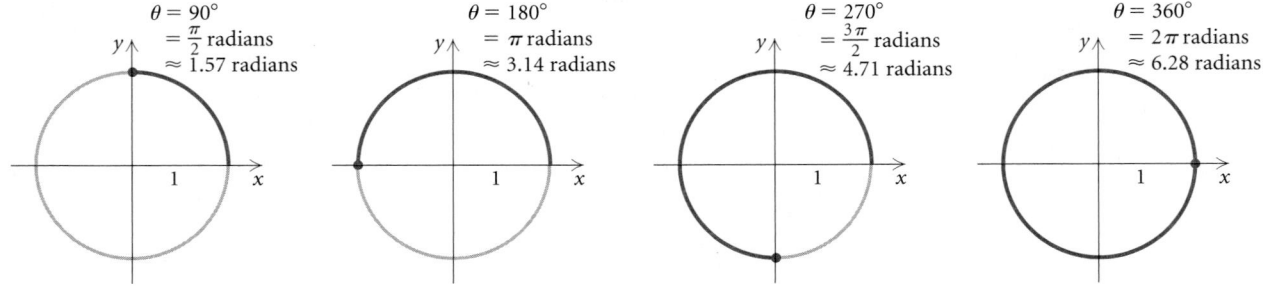

To convert between degrees and radians, we first note that

$$360° = 2\pi \text{ radians.}$$

It follows that

$$180° = \pi \text{ radians.}$$

To make conversions, we multiply by 1, noting that:

Converting Between Degree Measure and Radian Measure

$$\frac{\pi \text{ radians}}{180°} = \frac{180°}{\pi \text{ radians}} = 1.$$

To convert from degree to radian measure, multiply by $\dfrac{\pi \text{ radians}}{180°}$.

To convert from radian to degree measure, multiply by $\dfrac{180°}{\pi \text{ radians}}$.

GCM | **EXAMPLE 3**   Convert each of the following to radians.

a) 120°   **b)** −297.25°

Solution

a) $120° = 120° \cdot \dfrac{\pi \text{ radians}}{180°}$ **Multiplying by 1**

$= \dfrac{120°}{180°} \pi \text{ radians}$

$= \dfrac{2\pi}{3} \text{ radians, or about 2.09 radians}$

b) $-297.25° = -297.25° \cdot \dfrac{\pi \text{ radians}}{180°}$

$= -\dfrac{297.25°}{180°} \pi \text{ radians}$

$= -\dfrac{297.25\pi}{180} \text{ radians}$

$\approx -5.19 \text{ radians}$

```
120°
            2.094395102
-297.25°
           -5.187991202
```

We also can use a calculator set in RADIAN mode to convert the angle measures. We enter the angle measure followed by ° (degrees) from the ANGLE menu.

Now Try Exercises 23 and 35. ▪

GCM | **EXAMPLE 4**   Convert each of the following to degrees.

a) $\dfrac{3\pi}{4} \text{ radians}$ **b)** 8.5 radians

Solution

a) $\dfrac{3\pi}{4}$ radians $= \dfrac{3\pi}{4}$ radians $\cdot \dfrac{180°}{\pi \text{ radians}}$ Multiplying by 1

$$= \dfrac{3\pi}{4\pi} \cdot 180° = \dfrac{3}{4} \cdot 180° = 135°$$

b) 8.5 radians $= 8.5$ radians $\cdot \dfrac{180°}{\pi \text{ radians}}$

$$= \dfrac{8.5(180°)}{\pi} \approx 487.01°$$

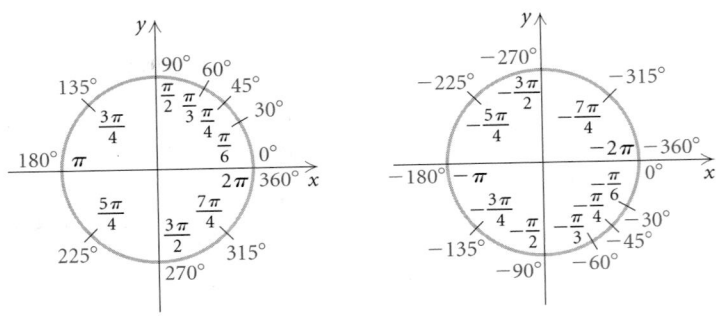

$(3\pi/4)^{\,r}$	
	135
$8.5^{\,r}$	
	487.0141259

With a calculator set in DEGREE mode, we can enter the angle measure followed by ^{r}(radians) from the ANGLE menu.   **Now Try Exercises 47 and 55.** ■

The radian–degree equivalents of the most commonly used angle measures are illustrated in the following figures.

When a rotation is given in radians, the word "radians" is optional and is most often omitted. **Thus if no unit is given for a rotation, the rotation is understood to be in radians.**

We can also find coterminal, complementary, and supplementary angles in radian measure just as we did for degree measure in Section 6.3.

EXAMPLE 5   Find a positive angle and a negative angle that are coterminal with $2\pi/3$. Many answers are possible.

Solution   To find angles coterminal with a given angle, we add or subtract multiples of 2π:

$$\dfrac{2\pi}{3} + 2\pi = \dfrac{2\pi}{3} + \dfrac{6\pi}{3} = \dfrac{8\pi}{3},$$

$$\dfrac{2\pi}{3} - 3(2\pi) = \dfrac{2\pi}{3} - \dfrac{18\pi}{3} = -\dfrac{16\pi}{3}.$$

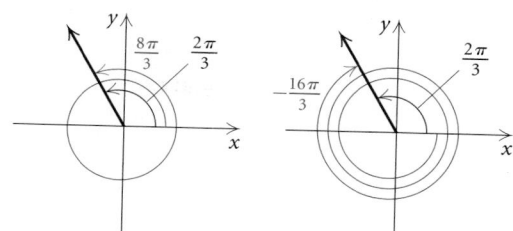

Thus, $8\pi/3$ and $-16\pi/3$ are two of the many angles coterminal with $2\pi/3$.

Now Try Exercise 11. ▩

EXAMPLE 6 Find the complement and the supplement of $\pi/6$.

Solution Since $90°$ equals $\pi/2$ radians, the complement of $\pi/6$ is

$$\frac{\pi}{2} - \frac{\pi}{6} = \frac{3\pi}{6} - \frac{\pi}{6} = \frac{2\pi}{6}, \quad \text{or} \quad \frac{\pi}{3}.$$

Since $180°$ equals π radians, the supplement of $\pi/6$ is

$$\pi - \frac{\pi}{6} = \frac{6\pi}{6} - \frac{\pi}{6} = \frac{5\pi}{6}.$$

Thus the complement of $\pi/6$ is $\pi/3$ and the supplement is $5\pi/6$.

Now Try Exercise 15. ▩

✿ Arc Length and Central Angles

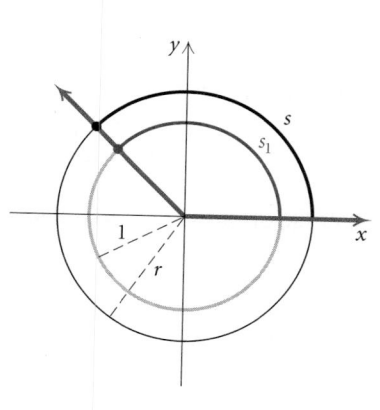

Radian measure can be determined using a circle other than a unit circle. In the figure at left, a unit circle (with radius 1) is shown along with another circle (with radius r, $r \neq 1$). The angle shown is a **central angle** of both circles.

From geometry, we know that the arcs that the angle subtends have their lengths in the same ratio as the radii of the circles. The radii of the circles are r and 1. The corresponding arc lengths are s and s_1. Thus we have the proportion

$$\frac{s}{s_1} = \frac{r}{1},$$

which also can be written as

$$\frac{s_1}{1} = \frac{s}{r}.$$

Now s_1 is the *radian measure* of the rotation in question. It is common to use a Greek letter, such as θ, for the measure of an angle or rotation and the letter s for arc length. Adopting this convention, we rewrite the proportion above as

$$\theta = \frac{s}{r}.$$

In any circle, the measure (in radians) of a central angle, the arc length the angle subtends, and the length of the radius are related in this fashion. Or, in general, the following is true.

Radian Measure

The **radian measure** θ of a rotation is the ratio of the distance s traveled by a point at a radius r from the center of rotation, to the length of the radius r:

$$\theta = \frac{s}{r}.$$

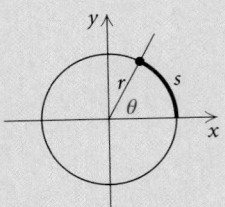

When we are using the formula $\theta = s/r$, θ must be in radians and s and r must be expressed in the same unit.

EXAMPLE 7   Find the measure of a rotation in radians when a point 2 m from the center of rotation travels 4 m.

Solution   We have

$$\theta = \frac{s}{r}$$

$$= \frac{4 \text{ m}}{2 \text{ m}} = 2. \qquad \text{The unit is understood to be radians.}$$

Now Try Exercise 65. ■

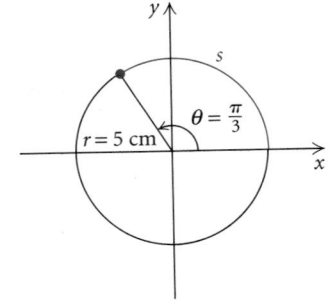

EXAMPLE 8   Find the length of an arc of a circle of radius 5 cm associated with an angle of $\pi/3$ radians.

Solution   We have

$$\theta = \frac{s}{r}, \quad \text{or} \quad s = r\theta.$$

Thus, $s = 5 \text{ cm} \cdot \pi/3$, or about 5.24 cm.   Now Try Exercise 63. ■

❖ Linear Speed and Angular Speed

Linear speed is defined to be distance traveled per unit of time. If we use v for linear speed, s for distance, and t for time, then

$$v = \frac{s}{t}.$$

Similarly, **angular speed** is defined to be amount of rotation per unit of time. For example, we might speak of the angular speed of a bicycle wheel as 150 revolutions per minute or the angular speed of the earth as 2π radians per day. The Greek letter ω (omega) is generally used for angular speed. Thus for a rotation θ and time t, angular speed is defined as

$$\omega = \frac{\theta}{t}.$$

As an example of how these definitions can be applied, let's consider the refurbished carousel at the Children's Museum in Indianapolis, Indiana. It consists of three circular rows of animals. All animals, regardless of the row, travel at the same angular speed. But the animals in the outer row travel at a greater linear speed than those in the inner rows. What is the relationship between the linear speed v and the angular speed ω?

To develop the relationship we seek, recall that, for rotations measured in radians, $\theta = s/r$. This is equivalent to

$$s = r\theta.$$

We divide by time, t, to obtain

$$\frac{s}{t} = \frac{r\theta}{t} \qquad \text{Dividing by } t$$

$$\frac{s}{t} = r \cdot \frac{\theta}{t}$$

$$\downarrow \qquad \downarrow$$
$$v \qquad \omega$$

Now s/t is linear speed v and θ/t is angular speed ω. Thus we have the relationship we seek,

$$v = r\omega.$$

Linear Speed in Terms of Angular Speed
The **linear speed** v of a point a distance r from the center of rotation is given by

$$v = r\omega,$$

where ω is the **angular speed** in radians per unit of time.

For the formula $v = r\omega$, the units of distance for v and r must be the same, ω must be in radians per unit of time, and the units of time for v and ω must be the same.

EXAMPLE 9 *Linear Speed of an Earth Satellite.* An earth satellite in circular orbit 1200 km high makes one complete revolution every 90 min. What is its linear speed? Use 6400 km for the length of a radius of the earth.

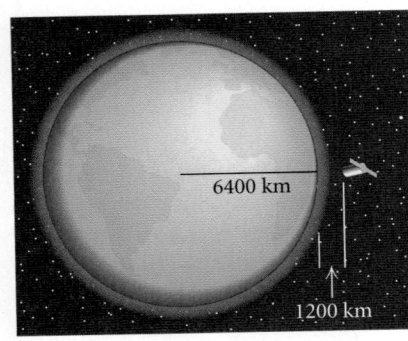

Solution To use the formula $v = r\omega$, we need to know r and ω:

$$r = 6400 \text{ km} + 1200 \text{ km}$$ Radius of earth plus height of satellite

$$= 7600 \text{ km};$$

$$\omega = \frac{\theta}{t} = \frac{2\pi}{90 \text{ min}} = \frac{\pi}{45 \text{ min}}.$$ We have, as usual, omitted the word radians.

Now, using $v = r\omega$, we have

$$v = 7600 \text{ km} \cdot \frac{\pi}{45 \text{ min}} = \frac{7600\pi}{45} \cdot \frac{\text{km}}{\text{min}} \approx 531 \frac{\text{km}}{\text{min}}.$$

Thus the linear speed of the satellite is approximately 531 km/min.

Now Try Exercise 71. ■

EXAMPLE 10 *Angular Speed of a Capstan.* An anchor is hoisted at a rate of 2 ft/sec as the chain is wound around a capstan with a 1.8-yd diameter. What is the angular speed of the capstan?

Solution We will use the formula $v = r\omega$ in the form $\omega = v/r$, taking care to use the proper units. Since v is given in feet per second, we need r in feet:

$$r = \frac{d}{2} = \frac{1.8}{2} \text{ yd} \cdot \frac{3 \text{ ft}}{1 \text{ yd}} = 2.7 \text{ ft}.$$

Then ω will be in radians per second:

$$\omega = \frac{v}{r} = \frac{2 \text{ ft/sec}}{2.7 \text{ ft}} = \frac{2 \text{ ft}}{\text{sec}} \cdot \frac{1}{2.7 \text{ ft}} \approx 0.741/\text{sec}.$$

Thus the angular speed is approximately 0.741 radian/sec.

Now Try Exercise 73. ■

The formulas $\theta = \omega t$ and $v = r\omega$ can be used in combination to find distances and angles in various situations involving rotational motion.

EXAMPLE 11   *Angle of Revolution.*   A 2006 Acura MDX is traveling at a speed of 70 mph. Its tires have an outside diameter of 28.56 in. Find the angle through which a tire turns in 10 sec.

28.56 in.

Solution   Recall that $\omega = \theta/t$, or $\theta = \omega t$. Thus we can find θ if we know ω and t. To find ω, we use the formula $v = r\omega$. The linear speed v of a point on the outside of the tire is the speed of the Acura, 70 mph. For convenience, we first convert 70 mph to feet per second:

$$v = 70 \ \frac{\text{mi}}{\text{hr}} \cdot \frac{1 \ \text{hr}}{60 \ \text{min}} \cdot \frac{1 \ \text{min}}{60 \ \text{sec}} \cdot \frac{5280 \ \text{ft}}{1 \ \text{mi}}$$

$$\approx 102.667 \ \frac{\text{ft}}{\text{sec}}.$$

The radius of the tire is half the diameter. Now $r = d/2 = 28.56/2 = 14.28$ in. We will convert to feet, since v is in feet per second:

$$r = 14.28 \ \text{in.} \cdot \frac{1 \ \text{ft}}{12 \ \text{in.}}$$

$$= \frac{14.28}{12} \ \text{ft.}$$

$$\approx 1.19 \ \text{ft.}$$

Using $v = r\omega$, we have

$$102.667 \ \frac{\text{ft}}{\text{sec}} = 1.19 \ \text{ft} \cdot \omega,$$

so

$$\omega = \frac{102.667 \ \text{ft/sec}}{1.19 \ \text{ft}} \approx \frac{86.27}{\text{sec}}.$$

Then in 10 sec,

$$\theta = \omega t = \frac{86.27}{\text{sec}} \cdot 10 \ \text{sec} \approx 863.$$

Thus the angle, in radians, through which a tire turns in 10 sec is 863.

Now Try Exercise 79. ◼

6.4 Exercise Set

For each of Exercises 1–4, sketch a unit circle and mark the points determined by the given real numbers.

1. a) $\dfrac{\pi}{4}$ **b)** $\dfrac{3\pi}{2}$ **c)** $\dfrac{3\pi}{4}$

   d) π **e)** $\dfrac{11\pi}{4}$ **f)** $\dfrac{17\pi}{4}$

2. a) $\dfrac{\pi}{2}$ **b)** $\dfrac{5\pi}{4}$ **c)** 2π

   d) $\dfrac{9\pi}{4}$ **e)** $\dfrac{13\pi}{4}$ **f)** $\dfrac{23\pi}{4}$

3. a) $\dfrac{\pi}{6}$ **b)** $\dfrac{2\pi}{3}$ **c)** $\dfrac{7\pi}{6}$

   d) $\dfrac{10\pi}{6}$ **e)** $\dfrac{14\pi}{6}$ **f)** $\dfrac{23\pi}{4}$

4. a) $-\dfrac{\pi}{2}$ **b)** $-\dfrac{3\pi}{4}$ **c)** $-\dfrac{5\pi}{6}$

   d) $-\dfrac{5\pi}{2}$ **e)** $-\dfrac{17\pi}{6}$ **f)** $-\dfrac{9\pi}{4}$

Find two real numbers between -2π and 2π that determine each of the points on the unit circle.

5.

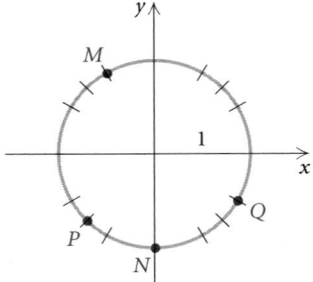

6.

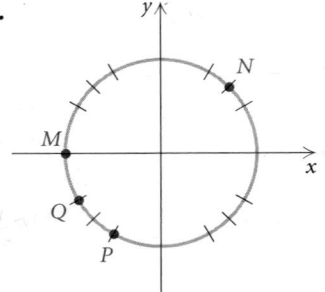

For Exercises 7 and 8, sketch a unit circle and mark the approximate location of the point determined by the given real number.

7. a) 2.4   **b)** 7.5
   c) 32   **d)** 320

8. a) 0.25   **b)** 1.8
   c) 47   **d)** 500

Find a positive angle and a negative angle that are coterminal with the given angle. Answers may vary.

9. $\dfrac{\pi}{4}$ **10.** $\dfrac{5\pi}{3}$

11. $\dfrac{7\pi}{6}$ **12.** π

13. $-\dfrac{2\pi}{3}$ **14.** $-\dfrac{3\pi}{4}$

Find the complement and the supplement.

15. $\dfrac{\pi}{3}$ **16.** $\dfrac{5\pi}{12}$ **17.** $\dfrac{3\pi}{8}$

18. $\dfrac{\pi}{4}$ **19.** $\dfrac{\pi}{12}$ **20.** $\dfrac{\pi}{6}$

Convert to radian measure. Leave the answer in terms of π.

21. $75°$ **22.** $30°$

23. $200°$ **24.** $-135°$

25. $-214.6°$ **26.** $37.71°$

27. $-180°$ **28.** $90°$

29. $12.5°$ **30.** $6.3°$

31. $-340°$ **32.** $-60°$

Convert to radian measure. Round the answer to two decimal places.

33. $240°$ **34.** $15°$

35. $-60°$ **36.** $145°$

37. $117.8°$ **38.** $-231.2°$

39. 1.354°

40. 584°

41. 345°

42. −75°

43. 95°

44. 24.8°

Convert to degree measure. Round the answer to two decimal places.

45. $-\dfrac{3\pi}{4}$

46. $\dfrac{7\pi}{6}$

47. 8π

48. $-\dfrac{\pi}{3}$

49. 1

50. −17.6

51. 2.347

52. 25

53. $\dfrac{5\pi}{4}$

54. -6π

55. −90

56. 37.12

57. $\dfrac{2\pi}{7}$

58. $\dfrac{\pi}{9}$

59. Certain positive angles are marked here in degrees. Find the corresponding radian measures.

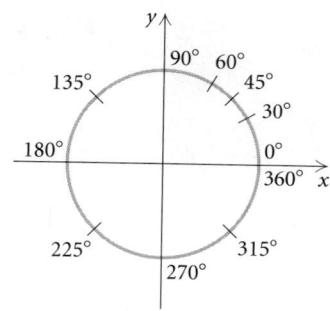

60. Certain negative angles are marked here in degrees. Find the corresponding radian measures.

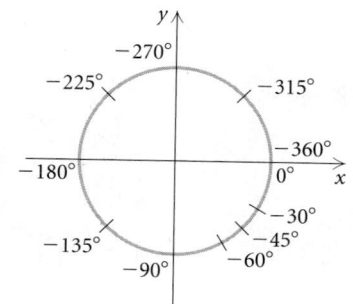

Arc Length and Central Angles. *Complete the table. Round the answers to two decimal places.*

Distance, *s* (arc length)	Radius, *r*	Angle, θ
61. 8 ft	$3\dfrac{1}{2}$ ft	_____
62. 200 cm	_____	45°
63. _____	4.2 in.	$\dfrac{5\pi}{12}$
64. 16 yd	_____	5

65. In a circle with a 120-cm radius, an arc 132 cm long subtends an angle of how many radians? how many degrees, to the nearest degree?

66. In a circle with a 10-ft diameter, an arc 20 ft long subtends an angle of how many radians? how many degrees, to the nearest degree?

67. In a circle with a 2-yd radius, how long is an arc associated with an angle of 1.6 radians?

68. In a circle with a 5-m radius, how long is an arc associated with an angle of 2.1 radians?

69. *Angle of Revolution.* Through how many radians does the minute hand of a clock rotate from 12:40 P.M. to 1:30 P.M.?

70. *Angle of Revolution.* A tire on a 2006 Dodge Ram truck has an outside diameter of 31.125 in. Through what angle (in radians) does the tire turn while traveling 1 mi?

31.125 in.

71. *Linear Speed.* A flywheel with a 15-cm diameter is rotating at a rate of 7 radians/sec. What is the linear speed of a point on its rim, in centimeters per minute?

72. *Linear Speed.* A wheel with a 30-cm radius is rotating at a rate of 3 radians/sec. What is the linear speed of a point on its rim, in meters per minute?

73. *Angular Speed of a Printing Press.* This text was printed on a four-color web heatset offset press. A cylinder on this press has a 21-in. diameter. The linear speed of a point on the cylinder's surface is 18.33 feet per second. What is the angular speed of the cylinder, in revolutions per hour? Printers often refer to the angular speed as impressions per hour (IPH). (*Source*: R. R. Donnelley, Willard, Ohio)

74. *Linear Speeds on a Carousel.* When Alicia and Zoe ride the carousel described earlier in this section, Alicia always selects a horse on the outside row, whereas Zoe prefers the row closest to the center. These rows are 19 ft 3 in. and 13 ft 11 in. from the center, respectively. The angular speed of the carousel is 2.4 revolutions per minute. What is the difference, in miles per hour, in the linear speeds of Alicia and Zoe? (*Source*: The Children's Museum, Indianapolis, IN)

75. *Linear Speed at the Equator.* The earth has a 4000-mi radius and rotates one revolution every 24 hr. What is the linear speed of a point on the equator, in miles per hour?

76. *Linear Speed of the Earth.* The earth is about 93,000,000 mi from the sun and traverses its orbit, which is nearly circular, every 365.25 days. What is the linear velocity of the earth in its orbit, in miles per hour?

77. *Tour of Flanders.* Tom Boonen of Belgium won the 2005 Tour of Flanders bicycle race. The wheel of his bicycle had a 67-cm diameter. His overall average linear speed during the race was 40.423 km/h. (*Source*: Toby Holsman, Bicycle Garage Indy, Indianapolis, Indiana; www.velonews.com) What was the angular speed of the wheel, in revolutions per hour?

78. *Determining the Speed of a River.* A water wheel has a 10-ft radius. To get a good approximation of the speed of the river, you count the revolutions of the wheel and find that it makes 14 revolutions per minute (rpm). What is the speed of the river, in miles per hour?

79. *John Deere Tractor.* A rear wheel on a John Deere 8300 farm tractor has a 23-in. radius. Find the angle (in radians) through which a wheel rotates in 12 sec if the tractor is traveling at a speed of 22 mph.

23 in.

Collaborative Discussion and Writing

80. Every real number determines a point on the unit circle. Does every point on the unit circle determine a real number? Explain.

81. In circular motion with a fixed angular speed, the length of the radius is directly proportional to the linear speed. Explain why with an example.

82. Two new cars are each driven at an average speed of 60 mph for an extended highway test drive of 2000 mi. The diameters of the wheels of the two cars are 15 in. and 16 in., respectively. If the cars use tires of equal durability and profile, differing only by the diameter, which car will probably need new tires first? Explain your answer.

Skill Maintenance

In each of Exercises 83–90, fill in the blanks with the correct terms. Some of the given choices will not be used.

inverse	relation
horizontal line	vertical asymptote
vertical line	horizontal asymptote
exponential function	even function
logarithmic function	odd function
natural	sine of θ
common	cosine of θ
logarithm	tangent of θ
one-to-one	

83. The domain of a(n) _____ function f is the range of the inverse f^{-1}.

84. The _____ is the length of the side adjacent to θ divided by the length of the hypotenuse.

85. The function $f(x) = a^x$, where x is a real number, $a > 0$ and $a \neq 1$, is called the _____, base a.

86. The graph of a rational function may or may not cross a(n) _____ .

87. If the graph of a function f is symmetric with respect to the origin, we say that it is a(n) _____ .

88. Logarithms, base e, are called _____ logarithms.

89. If it is possible for a(n) _____ to intersect the graph of a function more than once, then the function is not one-to-one and its _____ is not a function.

90. A(n) _____ is an exponent.

Synthesis

91. On the earth, one degree of latitude is how many kilometers? how many miles? (Assume that the radius of the earth is 6400 km, or 4000 mi, approximately.)

92. A point on the unit circle has y-coordinate $-\sqrt{21}/5$. What is its x-coordinate? Check using a calculator.

93. A **mil** is a unit of angle measure. A right angle has a measure of 1600 mils. Convert each of the following to degrees, minutes, and seconds.

   a) 100 mils   **b)** 350 mils

94. A **grad** is a unit of angle measure similar to a degree. A right angle has a measure of 100 grads. Convert each of the following to grads.

   a) 48°   **b)** $\dfrac{5\pi}{7}$

95. *Angular Speed of a Gear Wheel.* One gear wheel turns another, the teeth being on the rims. The wheels have 40-cm and 50-cm radii, and the smaller wheel rotates at 20 rpm. Find the angular speed of the larger wheel, in radians per second.

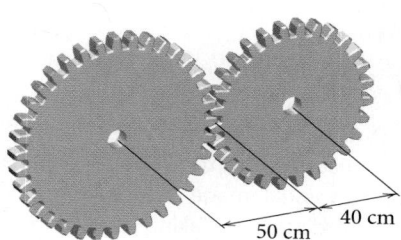

50 cm   40 cm

— larger wheel rotates slower

96. *Angular Speed of a Pulley.*   Two pulleys, 50 cm and 30 cm in diameter, respectively, are connected by a belt. The larger pulley makes 12 revolutions per minute. Find the angular speed of the smaller pulley, in radians per second.

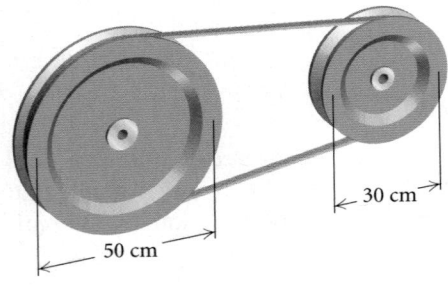

50 cm

30 cm

— same angular speed

97. *Distance Between Points on the Earth.*   To find the distance between two points on the earth when their latitude and longitude are known, we can use a right triangle for an excellent approximation if the points are not too far apart. Point *A* is at latitude 38°27′30″ N, longitude 82°57′15″ W; and point *B* is at latitude 38°28′45″ N, longitude 82°56′30″ W. Find the distance from *A* to *B* in nautical miles. (One minute of latitude is one nautical mile.)

98. *Hands of a Clock.*   At what time between noon and 1:00 P.M. are the hands of a clock perpendicular?

6.5

Circular Functions: Graphs and Properties

❖ Given the coordinates of a point on the unit circle, find its reflections across the *x*-axis, the *y*-axis, and the origin.

❖ Determine the six trigonometric function values for a real number when the coordinates of the point on the unit circle determined by that real number are given.

❖ Find function values for any real number using a calculator.

❖ Graph the six circular functions and state their properties.

The domains of the trigonometric functions, defined in Sections 6.1 and 6.3, have been sets of angles or rotations measured in a real number of degree units. We can also consider the domains to be sets of real numbers, or radians, introduced in Section 6.4. Many applications in calculus that use the trigonometric functions refer only to radians.

Let's again consider radian measure and the unit circle. We defined radian measure for θ as

$$\theta = \frac{s}{r}.$$

When $r = 1$,

$$\theta = \frac{s}{1}, \quad \text{or} \quad \theta = s.$$

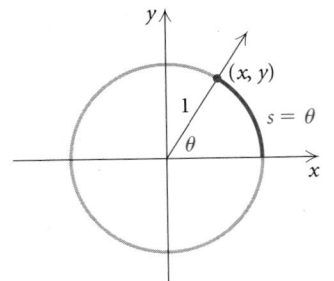

STUDY TIP

Take advantage of the numerous detailed art pieces in this text. They provide a visual image of the concept being discussed. Taking the time to study each figure is an efficient way to learn and retain the concepts.

The arc length *s* on the unit circle is the same as the radian measure of the angle *θ*.

In the figure on the preceding page, the point (x, y) is the point where the terminal side of the angle with radian measure *s* intersects the unit circle. We can now extend our definitions of the trigonometric functions using domains composed of real numbers, or radians.

In the definitions, *s can be considered the radian measure of an angle or the measure of an arc length on the unit circle. Either way, s is a real number.* To each real number *s*, there corresponds an arc length *s* on the unit circle. Trigonometric functions with domains composed of real numbers are called **circular functions**.

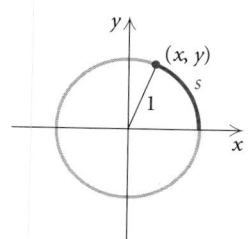

Basic Circular Functions

For a real number *s* that determines a point (x, y) on the unit circle:

$$\sin s = \text{second coordinate} = y,$$

$$\cos s = \text{first coordinate} = x,$$

$$\tan s = \frac{\text{second coordinate}}{\text{first coordinate}} = \frac{y}{x} \ (x \neq 0),$$

$$\csc s = \frac{1}{\text{second coordinate}} = \frac{1}{y} \ (y \neq 0),$$

$$\sec s = \frac{1}{\text{first coordinate}} = \frac{1}{x} \ (x \neq 0),$$

$$\cot s = \frac{\text{first coordinate}}{\text{second coordinate}} = \frac{x}{y} \ (y \neq 0).$$

We can consider the domains of trigonometric functions to be real numbers rather than angles. We can determine these values for a specific real number if we know the coordinates of the point on the unit circle determined by that number. As with degree measure, we can also find these function values directly using a calculator.

❋ Reflections on the Unit Circle

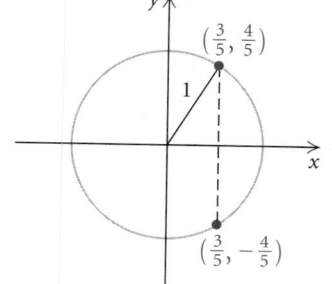

Let's consider the unit circle and a few of its points. For any point (x, y) on the unit circle, $x^2 + y^2 = 1$, we know that $-1 \leq x \leq 1$ and $-1 \leq y \leq 1$. If we know the *x*- or *y*-coordinate of a point on the unit circle, we can find the other coordinate. If $x = \frac{3}{5}$, then

$$\left(\tfrac{3}{5}\right)^2 + y^2 = 1$$

$$y^2 = 1 - \tfrac{9}{25} = \tfrac{16}{25}$$

$$y = \pm\tfrac{4}{5}.$$

Thus, $\left(\frac{3}{5}, \frac{4}{5}\right)$ and $\left(\frac{3}{5}, -\frac{4}{5}\right)$ are points on the unit circle. There are two points with an *x*-coordinate of $\frac{3}{5}$.

Now let's consider the radian measure $\pi/3$ and determine the coordinates of the point on the unit circle determined by $\pi/3$. We construct a right triangle by dropping a perpendicular segment from the point to the x-axis.

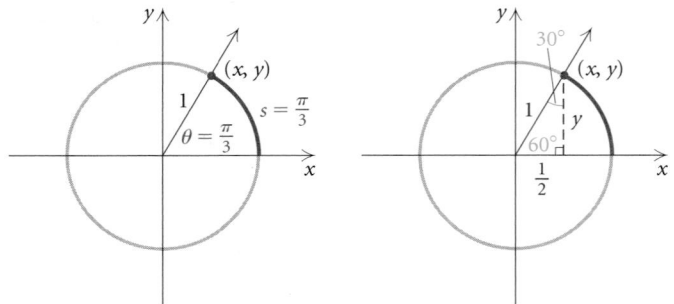

Since $\pi/3 = 60°$, we have a $30°-60°$ right triangle in which the side opposite the $30°$ angle is one half of the hypotenuse. The hypotenuse, or radius, is 1, so the side opposite the $30°$ angle is $\frac{1}{2} \cdot 1$, or $\frac{1}{2}$. Using the Pythagorean equation, we can find the other side:

$$\left(\frac{1}{2}\right)^2 + y^2 = 1$$

$$y^2 = 1 - \frac{1}{4} = \frac{3}{4}$$

$$y = \sqrt{\frac{3}{4}} = \frac{\sqrt{3}}{2}.$$

We know that y is positive since the point is in the first quadrant. Thus the coordinates of the point determined by $\pi/3$ are $x = 1/2$ and $y = \sqrt{3}/2$, or $\left(1/2, \sqrt{3}/2\right)$. We can always check to see if a point is on the unit circle by substituting into the equation $x^2 + y^2 = 1$:

$$\left(\frac{1}{2}\right)^2 + \left(\frac{\sqrt{3}}{2}\right)^2 = \frac{1}{4} + \frac{3}{4} = 1.$$

Because a unit circle is symmetric with respect to the x-axis, the y-axis, and the origin, we can use the coordinates of one point on the unit circle to find coordinates of its reflections.

EXAMPLE 1 Each of the following points lies on the unit circle. Find their reflections across the x-axis, the y-axis, and the origin.

a) $\left(\dfrac{3}{5}, \dfrac{4}{5}\right)$　　　　b) $\left(\dfrac{\sqrt{2}}{2}, \dfrac{\sqrt{2}}{2}\right)$　　　　c) $\left(\dfrac{1}{2}, \dfrac{\sqrt{3}}{2}\right)$

Solution

a)
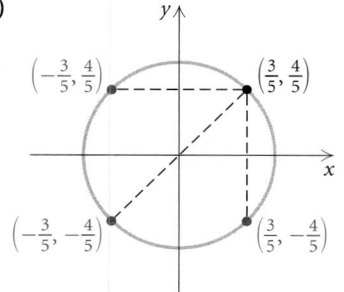

b)
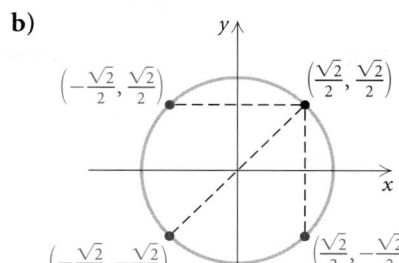

c)
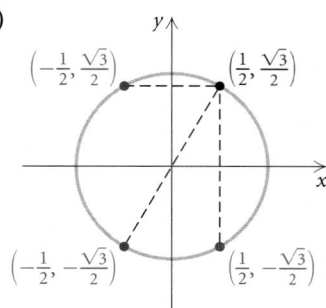

Now Try Exercise 1. ▨

❀ Finding Function Values

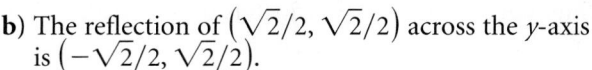

Knowing the coordinates of only a few points on the unit circle along with their reflections allows us to find trigonometric function values of the most frequently used real numbers, or radians.

EXAMPLE 2   Find each of the following function values.

a) $\tan \dfrac{\pi}{3}$

b) $\cos \dfrac{3\pi}{4}$

c) $\sin\left(-\dfrac{\pi}{6}\right)$

d) $\cos \dfrac{4\pi}{3}$

e) $\cot \pi$

f) $\csc\left(-\dfrac{7\pi}{2}\right)$

Solution   We locate the point on the unit circle determined by the rotation, and then find its coordinates using reflection if necessary.

a) The coordinates of the point determined by $\pi/3$ are $\left(1/2, \sqrt{3}/2\right)$.

b) The reflection of $\left(\sqrt{2}/2, \sqrt{2}/2\right)$ across the y-axis is $\left(-\sqrt{2}/2, \sqrt{2}/2\right)$.

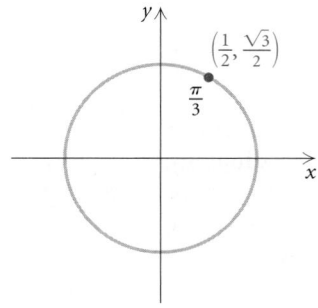

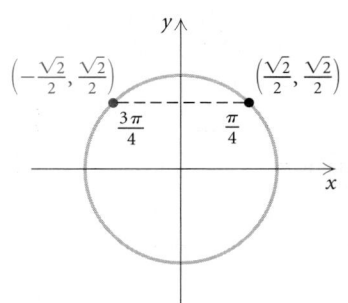

Thus, $\tan \dfrac{\pi}{3} = \dfrac{y}{x} = \dfrac{\sqrt{3}/2}{1/2} = \sqrt{3}.$

Thus, $\cos \dfrac{3\pi}{4} = x = -\dfrac{\sqrt{2}}{2}.$

c) The reflection of $\left(\sqrt{3}/2,\ 1/2\right)$ across the x-axis is $\left(\sqrt{3}/2,\ -1/2\right)$.

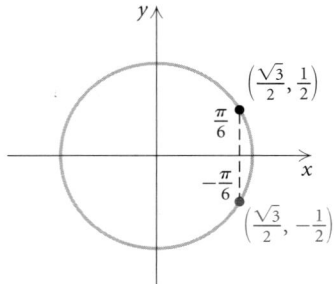

Thus, $\sin\left(-\dfrac{\pi}{6}\right) = y = -\dfrac{1}{2}$.

e) The coordinates of the point determined by π are $(-1, 0)$.

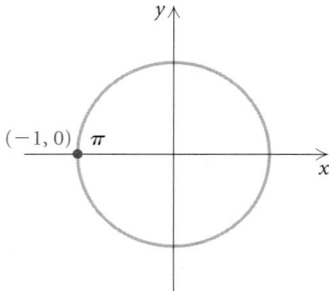

Thus, $\cot \pi = \dfrac{x}{y} = \dfrac{-1}{0}$, which is not defined.

We can also think of $\cot \pi$ as the reciprocal of $\tan \pi$. Since $\tan \pi = y/x = 0/-1 = 0$ and the reciprocal of 0 is not defined, we know that $\cot \pi$ is not defined.

d) The reflection of $\left(1/2,\ \sqrt{3}/2\right)$ across the origin is $\left(-1/2,\ -\sqrt{3}/2\right)$.

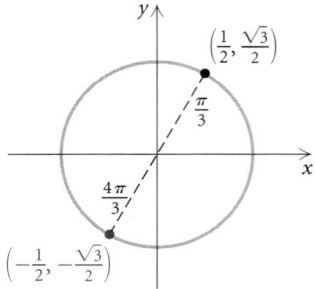

Thus, $\cos \dfrac{4\pi}{3} = x = -\dfrac{1}{2}$.

f) The coordinates of the point determined by $-7\pi/2$ are $(0, 1)$.

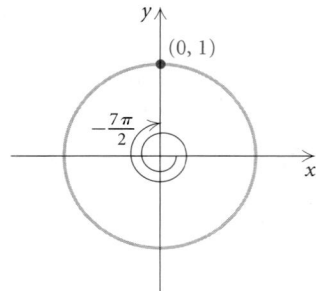

Thus, $\csc\left(-\dfrac{7\pi}{2}\right) = \dfrac{1}{y} = \dfrac{1}{1} = 1$.

Now Try Exercises 9 and 11. ▧

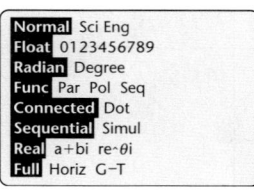

Using a calculator, we can find trigonometric function values of any real number without knowing the coordinates of the point that it determines on the unit circle. Most calculators have both degree and radian modes. When finding function values of radian measures, or real numbers, we *must* set the calculator in RADIAN mode. (See the window at left.)

EXAMPLE 3   Find each of the following function values of radian measures using a calculator. Round the answers to four decimal places.

a) $\cos \dfrac{2\pi}{5}$

b) $\tan(-3)$

c) $\sin 24.9$

d) $\sec \dfrac{\pi}{7}$

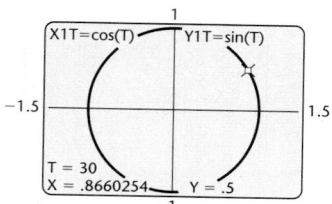

Solution Using a calculator set in RADIAN mode, we find the values.

a) $\cos \dfrac{2\pi}{5} \approx 0.3090$

b) $\tan(-3) \approx 0.1425$

c) $\sin 24.9 \approx -0.2306$

d) $\sec \dfrac{\pi}{7} = \dfrac{1}{\cos \dfrac{\pi}{7}} \approx 1.1099$

Note in part (d) that the secant function value can be found by taking the reciprocal of the cosine value. Thus we can enter cos $\pi/7$ and use the reciprocal key.   **Now Try Exercises 25 and 33.** ■

Exploring with Technology

We can graph the unit circle using a graphing calculator. We use PARAMETRIC mode with the following window and let $X_{1T} = \cos T$ and $Y_{1T} = \sin T$. Here we use DEGREE mode.

WINDOW

Tmin = 0
Tmax = 360
Tstep = 15
Xmin = −1.5
Xmax = 1.5
Xscl = 1
Ymin = −1
Ymax = 1
Yscl = 1

Using the trace key and an arrow key to move the cursor around the unit circle, we see the T, X, and Y values appear on the screen. What do they represent? Repeat this exercise in RADIAN mode. What do the T, X, and Y values represent? (For more on parametric equations, see Section 10.7.)

From the definitions on p. 531, we can relabel any point (x, y) on the unit circle as $(\cos s, \sin s)$, where s is any real number.

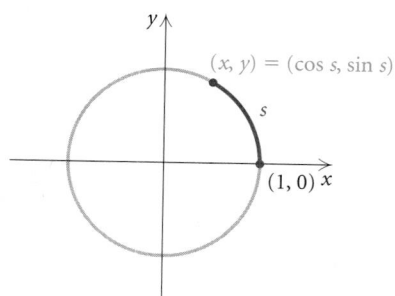

✾ Graphs of the Sine and Cosine Functions

Properties of functions can be observed from their graphs. We begin by graphing the sine and cosine functions. We make a table of values, plot the points, and then connect those points with a smooth curve. It is helpful to first draw a unit circle and label a few points with coordinates. We can either use the coordinates as the function values or find approximate sine and cosine values directly with a calculator.

s	sin s	cos s
0	0	1
$\pi/6$	0.5	0.8660
$\pi/4$	0.7071	0.7071
$\pi/3$	0.8660	0.5
$\pi/2$	1	0
$3\pi/4$	0.7071	−0.7071
π	0	−1
$5\pi/4$	−0.7071	−0.7071
$3\pi/2$	−1	0
$7\pi/4$	−0.7071	0.7071
2π	0	1

s	sin s	cos s
0	0	1
$-\pi/6$	−0.5	0.8660
$-\pi/4$	−0.7071	0.7071
$-\pi/3$	−0.8660	0.5
$-\pi/2$	−1	0
$-3\pi/4$	−0.7071	−0.7071
$-\pi$	0	−1
$-5\pi/4$	0.7071	−0.7071
$-3\pi/2$	1	0
$-7\pi/4$	0.7071	0.7071
-2π	0	1

The graphs are as follows.

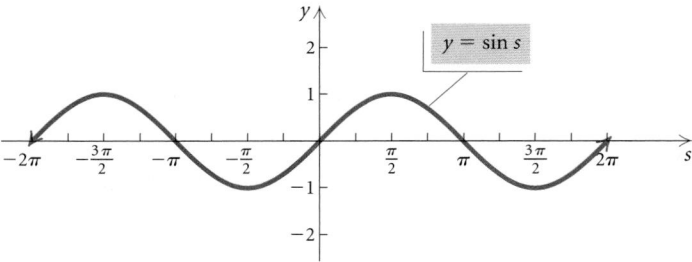

The sine function

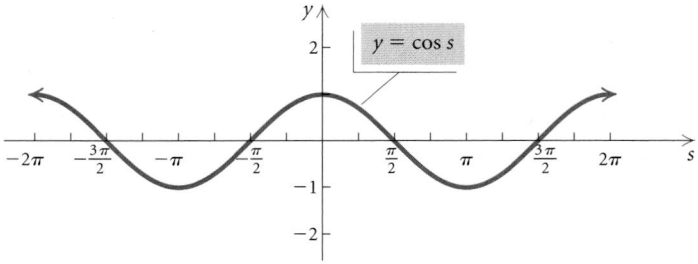

The cosine function

We can check these graphs using a graphing calculator.

$y = \sin x$

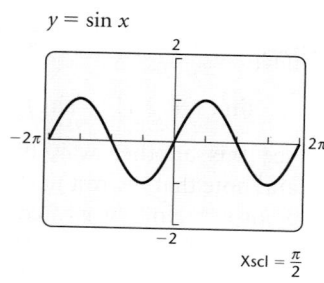

$\text{Xscl} = \frac{\pi}{2}$

$y = \cos x$

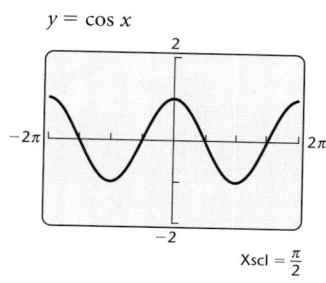

$\text{Xscl} = \frac{\pi}{2}$

The sine and cosine functions are continuous functions. Note in the graph of the sine function that function values increase from 0 at $s = 0$ to 1 at $s = \pi/2$, then decrease to 0 at $s = \pi$, decrease further to -1 at $s = 3\pi/2$, and increase to 0 at 2π. The reverse pattern follows when s decreases from 0 to -2π. Note in the graph of the cosine function that function values start at 1 when $s = 0$, and decrease to 0 at $s = \pi/2$. They decrease further to -1 at $s = \pi$, then increase to 0 at $s = 3\pi/2$, and increase further to 1 at $s = 2\pi$. An identical pattern follows when s decreases from 0 to -2π.

From the unit circle and the graphs of the functions, we know that the domain of both the sine and cosine functions is the entire set of real numbers, $(-\infty, \infty)$. The range of each function is the set of all real numbers from -1 to 1, $[-1, 1]$.

Domain and Range of Sine and Cosine Functions

The *domain* of the sine function and the cosine function is $(-\infty, \infty)$.

The *range* of the sine function and the cosine function is $[-1, 1]$.

Exploring with Technology

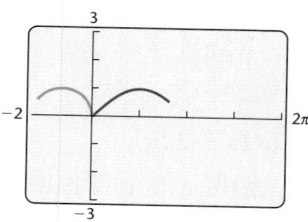

Another way to construct the sine and cosine graphs is by considering the unit circle and transferring vertical distances for the sine function and horizontal distances for the cosine function. Using a graphing calculator, we can visualize the transfer of these distances. We use the calculator set in PARAMETRIC and RADIAN modes and let $\text{X}_{1\text{T}} = \cos T - 1$ and $\text{Y}_{1\text{T}} = \sin T$ for the unit circle centered at $(-1, 0)$ and $\text{X}_{2\text{T}} = T$ and $\text{Y}_{2\text{T}} = \sin T$ for the sine curve. Use the following window settings.

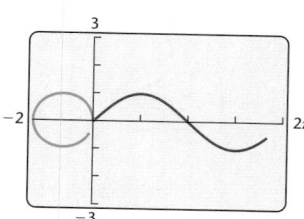

Tmin $= 0$	Xmin $= -2$	Ymin $= -3$
Tmax $= 2\pi$	Xmax $= 2\pi$	Ymax $= 3$
Tstep $= .1$	Xscl $= \pi/2$	Yscl $= 1$

With the calculator set in SIMULTANEOUS mode, we can actually watch the sine function (in red) "unwind" from the unit circle (in blue). In the two screens at left, we partially illustrate this animated procedure.

Consult your calculator's instruction manual for specific keystrokes and graph both the sine curve and the cosine curve in this manner. (For more on parametric equations, see Section 10.7.)

A function with a repeating pattern is called **periodic**. The sine function and the cosine function are examples of periodic functions. The values of these functions repeat themselves every 2π units. In other words, for any s, we have

$$\sin(s + 2\pi) = \sin s \quad \text{and} \quad \cos(s + 2\pi) = \cos s.$$

To see this another way, think of the part of the graph between 0 and 2π and note that the rest of the graph consists of copies of it. If we translate the graph of $y = \sin x$ or $y = \cos x$ to the left or right 2π units, we will obtain the original graph. We say that each of these functions has a period of 2π.

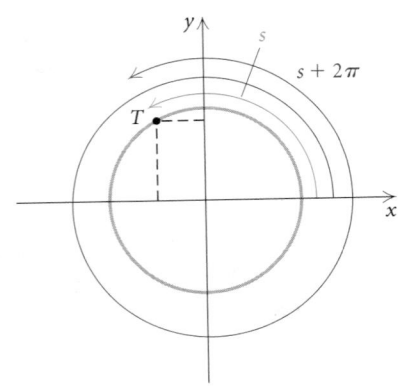

> **Periodic Function**
>
> A function f is said to be **periodic** if there exists a positive constant p such that
>
> $$f(s + p) = f(s)$$
>
> for all s in the domain of f. The smallest such positive number p is called the period of the function.

The period p can be thought of as the length of the shortest recurring interval.

We can also use the unit circle to verify that the period of the sine and cosine functions is 2π. Consider any real number s and the point T that it determines on a unit circle, as shown at left. If we increase s by 2π, the point determined by $s + 2\pi$ is again the point T. Hence for any real number s,

$$\sin(s + 2\pi) = \sin s \quad \text{and} \quad \cos(s + 2\pi) = \cos s.$$

It is also true that $\sin(s + 4\pi) = \sin s$, $\sin(s + 6\pi) = \sin s$, and so on. In fact, for *any* integer k, the following equations are identities:

$$\sin[s + k(2\pi)] = \sin s \quad \text{and} \quad \cos[s + k(2\pi)] = \cos s,$$

or

$$\sin s = \sin(s + 2k\pi) \quad \text{and} \quad \cos s = \cos(s + 2k\pi).$$

The **amplitude** of a periodic function is defined as one half of the distance between its maximum and minimum function values. It is always positive. Both the graphs and the unit circle verify that the maximum value of the sine and cosine functions is 1, whereas the minimum value of each is -1. Thus,

$$\text{the amplitude of the sine function} = \tfrac{1}{2}\left|1 - (-1)\right| = 1$$

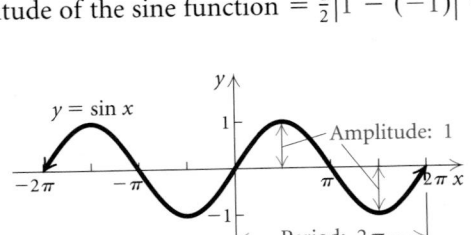

and

the amplitude of the cosine function is $\frac{1}{2}\left|1-(-1)\right|=1$.

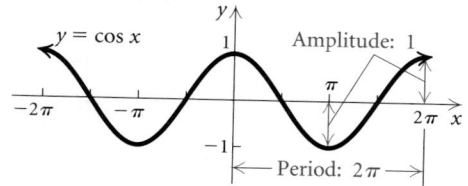

Exploring with Technology

Using the TABLE feature on a graphing calculator, compare the y-values for $y_1 = \sin x$ and $y_2 = \sin(-x)$ and for $y_3 = \cos x$ and $y_4 = \cos(-x)$. We set TblStart $= 0$ and $\triangle$Tbl $= \pi/12$.

X	Y1	Y2
0	0	0
.2618	.25882	−.2588
.5236	.5	−.5
.7854	.70711	−.7071
1.0472	.86603	−.866
1.309	.96593	−.9659
1.5708	1	−1

X = 0

X	Y3	Y4
0	1	1
.2618	.96593	.96593
.5236	.86603	.86603
.7854	.70711	.70711
1.0472	.5	.5
1.309	.25882	.25882
1.5708	0	0

X = 0

What appears to be the relationship between $\sin x$ and $\sin(-x)$ and between $\cos x$ and $\cos(-x)$?

Consider any real number s and its opposite, $-s$. These numbers determine points T and T_1 on a unit circle that are symmetric with respect to the x-axis.

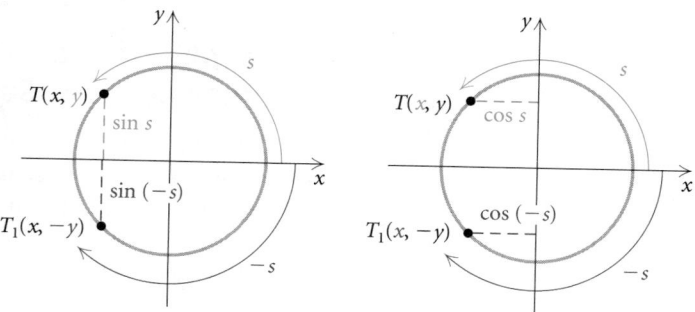

Because their second coordinates are opposites of each other, we know that for any number s,

$$\sin(-s) = -\sin s.$$

Because their first coordinates are the same, we know that for any number s,

$$\cos(-s) = \cos s.$$

Thus we have shown the following.

EVEN AND ODD FUNCTIONS

REVIEW SECTION **2.4.**

The sine function is *odd*.
The cosine function is *even*.

A summary of the properties of the sine and cosine functions follows.

CONNECTING *the* CONCEPTS

Comparing the Sine Function and the Cosine Function

SINE FUNCTION

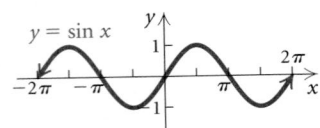

1. Continuous
2. Period: 2π
3. Domain: All real numbers
4. Range: $[-1, 1]$
5. Amplitude: 1
6. Odd: $\sin(-s) = -\sin s$

COSINE FUNCTION

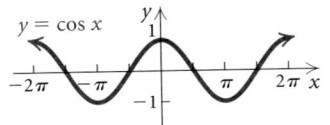

1. Continuous
2. Period: 2π
3. Domain: All real numbers
4. Range: $[-1, 1]$
5. Amplitude: 1
6. Even: $\cos(-s) = \cos s$

❋ Graphs of the Tangent, Cotangent, Cosecant, and Secant Functions

To graph the tangent function, we could make a table of values using a calculator, but in this case it is easier to begin with the definition of tangent and the coordinates of a few points on the unit circle. We recall that

$$\tan s = \frac{y}{x} = \frac{\sin s}{\cos s}.$$

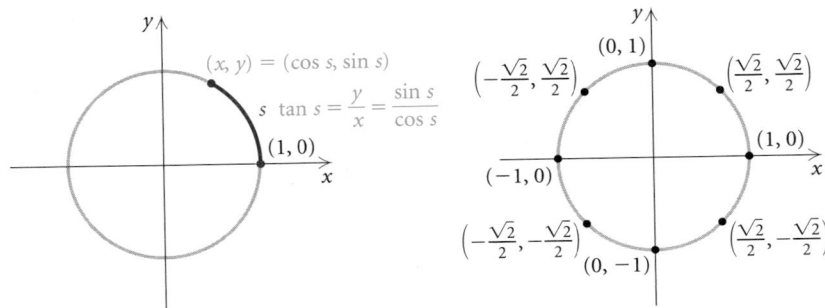

The tangent function is not defined when x, the first coordinate, is 0. That is, it is not defined for any number s whose cosine is 0:

$$s = \pm\frac{\pi}{2}, \pm\frac{3\pi}{2}, \pm\frac{5\pi}{2}, \ldots.$$

We draw vertical asymptotes at these locations (see Fig. 1 below).

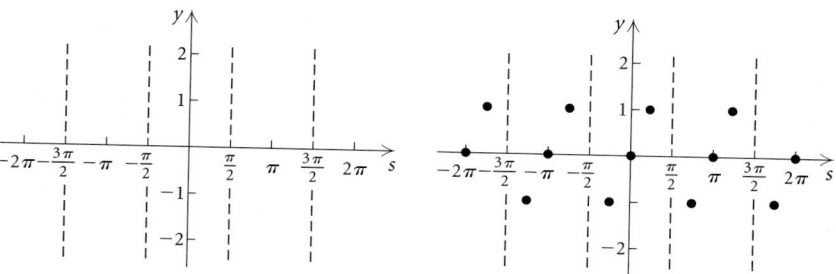

FIGURE 1   FIGURE 2

We also note that

$$\tan s = 0 \text{ at } s = 0, \pm\pi, \pm2\pi, \pm3\pi, \ldots,$$

$$\tan s = 1 \text{ at } s = \ldots -\frac{7\pi}{4}, -\frac{3\pi}{4}, \frac{\pi}{4}, \frac{5\pi}{4}, \frac{9\pi}{4}, \ldots,$$

$$\tan s = -1 \text{ at } s = \ldots -\frac{9\pi}{4}, -\frac{5\pi}{4}, -\frac{\pi}{4}, \frac{3\pi}{4}, \frac{7\pi}{4}, \ldots.$$

We can add these ordered pairs to the graph (see Fig. 2 above) and investigate the values in $(-\pi/2, \pi/2)$ using a calculator. Note that the function value is 0 when $s = 0$, and the values increase without bound as s increases toward $\pi/2$. The graph gets closer and closer to an asymptote as s gets closer to $\pi/2$, but it never touches the line. As s decreases from 0 to $-\pi/2$, the values decrease without bound. Again the graph gets closer and closer to an asymptote, but it never touches it. We now complete the graph.

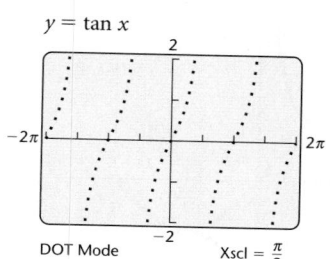

$y = \tan x$

DOT Mode $\text{Xscl} = \frac{\pi}{2}$

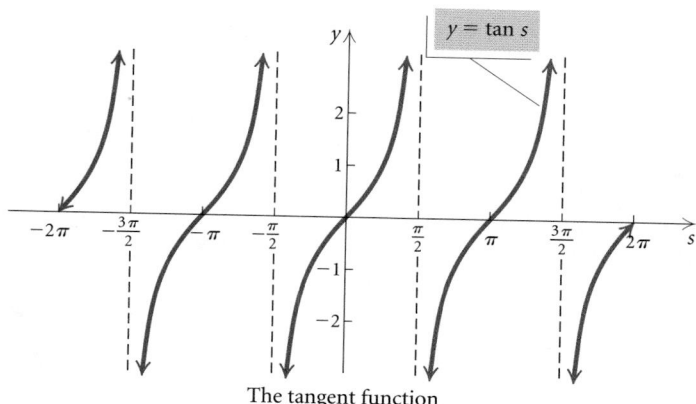

The tangent function

From the graph, we see that the tangent function is continuous except where it is not defined. The period of the tangent function is π. Note that although there is a period, there is no amplitude because there are no maximum and minimum values. When $\cos s = 0$, $\tan s$ is not defined ($\tan s = \sin s/\cos s$). Thus the domain of the tangent function is the set of all real numbers except $(\pi/2) + k\pi$, where k is an integer. The range of the function is the set of all real numbers.

The cotangent function ($\cot s = \cos s/\sin s$) is not defined when y, the second coordinate, is 0—that is, it is not defined for any number s whose sine is 0. Thus the cotangent is not defined for $s = 0, \pm\pi, \pm2\pi, \pm3\pi, \ldots$. The graph of the function is shown below.

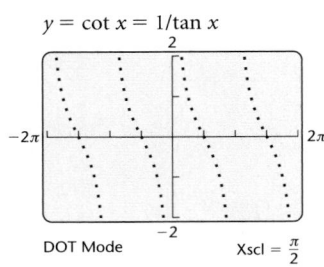

$y = \cot x = 1/\tan x$

DOT Mode $Xscl = \frac{\pi}{2}$

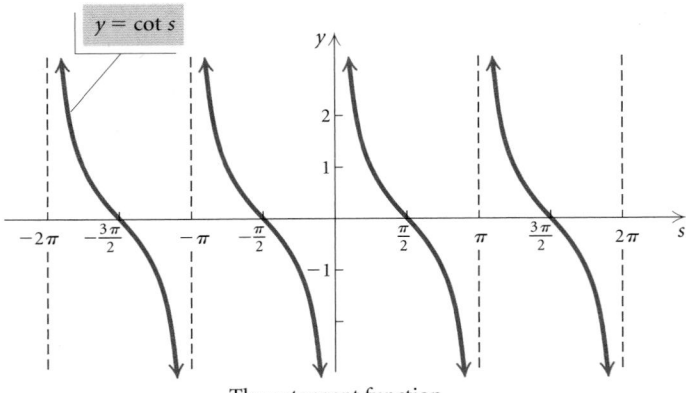

$y = \cot s$

The cotangent function

The cosecant and sine functions are reciprocal functions, as are the secant and cosine functions. The graphs of the cosecant and secant functions can be constructed by finding the reciprocals of the values of the sine and cosine functions, respectively. Thus the functions will be positive together and negative together. The cosecant function is not defined for those numbers s whose sine is 0. The secant function is not defined for those numbers s whose cosine is 0. In the graphs below, the sine and cosine functions are shown by the gray curves for reference.

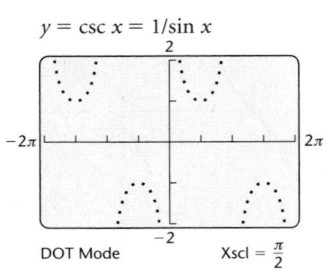

$y = \csc x = 1/\sin x$

DOT Mode $Xscl = \frac{\pi}{2}$

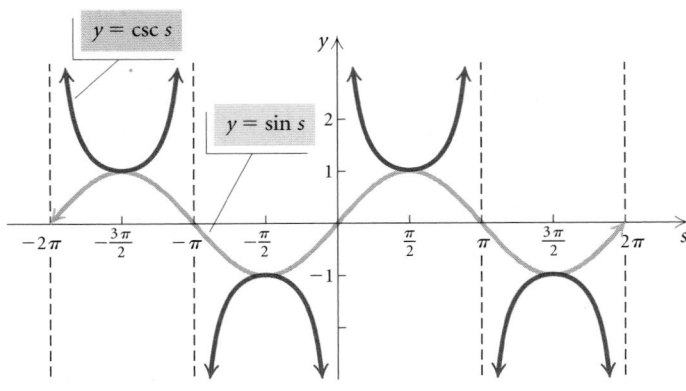

$y = \csc s$

$y = \sin s$

The cosecant function

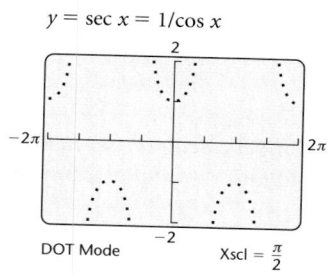

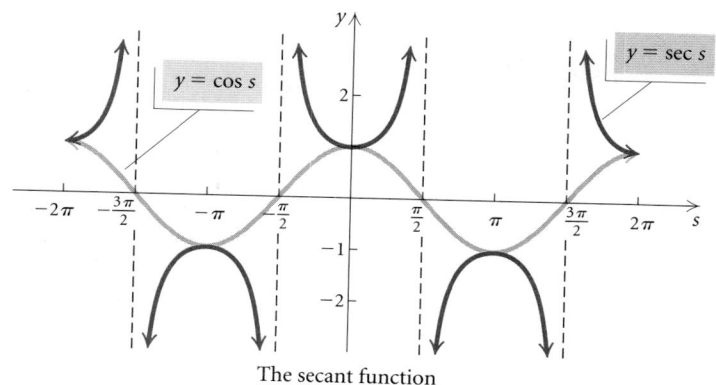

The secant function

The following is a summary of the basic properties of the tangent, cotangent, cosecant, and secant functions. These functions are continuous except where they are not defined.

CONNECTING the CONCEPTS

Comparing the Tangent, Cotangent, Cosecant, and Secant Functions

TANGENT FUNCTION

1. Period: π
2. Domain: All real numbers except $(\pi/2) + k\pi$, where k is an integer
3. Range: All real numbers

COTANGENT FUNCTION

1. Period: π
2. Domain: All real numbers except $k\pi$, where k is an integer
3. Range: All real numbers

COSECANT FUNCTION

1. Period: 2π
2. Domain: All real numbers except $k\pi$, where k is an integer
3. Range: $(-\infty, -1] \cup [1, \infty)$

SECANT FUNCTION

1. Period: 2π
2. Domain: All real numbers except $(\pi/2) + k\pi$, where k is an integer
3. Range: $(-\infty, -1] \cup [1, \infty)$

In this chapter, we have used the letter s for arc length and have avoided the letters x and y, which generally represent first and second coordinates. Nevertheless, we can represent the arc length on a unit circle by any variable, such as s, t, x, or θ. Each arc length determines a point that can be labeled with an ordered pair. The first coordinate of that ordered pair is the cosine of the arc length, and the second coordinate is the sine of the arc length. The identities we have developed hold no matter what symbols are used for variables—for example, $\cos(-s) = \cos s$, $\cos(-x) = \cos x$, $\cos(-\theta) = \cos \theta$, and $\cos(-t) = \cos t$.

6.5   Exercise Set

The following points are on the unit circle. Find the coordinates of their reflections across (a) the x-axis, (b) the y-axis, and (c) the origin.

1. $\left(-\dfrac{3}{4}, \dfrac{\sqrt{7}}{4}\right)$ **2.** $\left(\dfrac{2}{3}, \dfrac{\sqrt{5}}{3}\right)$

3. $\left(\dfrac{2}{5}, -\dfrac{\sqrt{21}}{5}\right)$ **4.** $\left(-\dfrac{\sqrt{3}}{2}, -\dfrac{1}{2}\right)$

5. The number $\pi/4$ determines a point on the unit circle with coordinates $\left(\sqrt{2}/2, \sqrt{2}/2\right)$. What are the coordinates of the point determined by $-\pi/4$?

6. A number β determines a point on the unit circle with coordinates $\left(-2/3, \sqrt{5}/3\right)$. What are the coordinates of the point determined by $-\beta$?

Find the function value using coordinates of points on the unit circle. Give exact answers.

7. $\sin \pi$ **8.** $\cos\left(-\dfrac{\pi}{3}\right)$

9. $\cot \dfrac{7\pi}{6}$ **10.** $\tan \dfrac{11\pi}{4}$

11. $\sin(-3\pi)$ **12.** $\csc \dfrac{3\pi}{4}$

13. $\cos \dfrac{5\pi}{6}$ **14.** $\tan\left(-\dfrac{\pi}{4}\right)$

15. $\sec \dfrac{\pi}{2}$ **16.** $\cos 10\pi$

17. $\cos \dfrac{\pi}{6}$ **18.** $\sin \dfrac{2\pi}{3}$

19. $\sin \dfrac{5\pi}{4}$ **20.** $\cos \dfrac{11\pi}{6}$

21. $\sin(-5\pi)$ **22.** $\tan \dfrac{3\pi}{2}$

23. $\cot \dfrac{5\pi}{2}$ **24.** $\tan \dfrac{5\pi}{3}$

Find the function value using a calculator set in RADIAN *mode. Round the answer to four decimal places, where appropriate.*

25. $\tan \dfrac{\pi}{7}$ **26.** $\cos\left(-\dfrac{2\pi}{5}\right)$

27. $\sec 37$ **28.** $\sin 11.7$

29. $\cot 342$ **30.** $\tan 1.3$

31. $\cos 6\pi$ **32.** $\sin \dfrac{\pi}{10}$

33. $\csc 4.16$ **34.** $\sec \dfrac{10\pi}{7}$

35. $\tan \dfrac{7\pi}{4}$ **36.** $\cos 2000$

37. $\sin\left(-\dfrac{\pi}{4}\right)$ **38.** $\cot 7\pi$

39. $\sin 0$ **40.** $\cos(-29)$

41. $\tan \dfrac{2\pi}{9}$ **42.** $\sin \dfrac{8\pi}{3}$

In Exercises 43–48, make hand-drawn graphs.

43. a) Sketch a graph of $y = \sin x$.
   b) By reflecting the graph in part (a), sketch a graph of $y = \sin(-x)$.
   c) By reflecting the graph in part (a), sketch a graph of $y = -\sin x$.
   d) How do the graphs in parts (b) and (c) compare?

44. a) Sketch a graph of $y = \cos x$.
   b) By reflecting the graph in part (a), sketch a graph of $y = \cos(-x)$.
   c) By reflecting the graph in part (a), sketch a graph of $y = -\cos x$.
   d) How do the graphs in parts (a) and (b) compare?

45. a) Sketch a graph of $y = \sin x$.
   b) By translating, sketch a graph of $y = \sin(x + \pi)$.

c) By reflecting the graph of part (a), sketch a graph of $y = -\sin x$.

d) How do the graphs of parts (b) and (c) compare?

46. a) Sketch a graph of $y = \sin x$.
   b) By translating, sketch a graph of $y = \sin(x - \pi)$.
   c) By reflecting the graph of part (a), sketch a graph of $y = -\sin x$.
   d) How do the graphs of parts (b) and (c) compare?

47. a) Sketch a graph of $y = \cos x$.
   b) By translating, sketch a graph of $y = \cos(x + \pi)$.
   c) By reflecting the graph of part (a), sketch a graph of $y = -\cos x$.
   d) How do the graphs of parts (b) and (c) compare?

48. a) Sketch a graph of $y = \cos x$.
   b) By translating, sketch a graph of $y = \cos(x - \pi)$.
   c) By reflecting the graph of part (a), sketch a graph of $y = -\cos x$.
   d) How do the graphs of parts (b) and (c) compare?

49. a) Sketch a graph of $y = \tan x$.
   b) By reflecting the graph of part (a), sketch a graph of $y = \tan(-x)$.
   c) By reflecting the graph of part (a), sketch a graph of $y = -\tan x$.
   d) How do the graphs of parts (b) and (c) compare?

50. a) Sketch a graph of $y = \sec x$.
   b) By reflecting the graph of part (a), sketch a graph of $y = \sec(-x)$.
   c) By reflecting the graph of part (a), sketch a graph of $y = -\sec x$.
   d) How do the graphs of parts (a) and (b) compare?

51. Of the six circular functions, which are even? Which are odd?

52. Of the six circular functions, which have period π? Which have period 2π?

Consider the coordinates on the unit circle for Exercises 53–56.

53. In which quadrants is the tangent function positive? negative?

54. In which quadrants is the sine function positive? negative?

55. In which quadrants is the cosine function positive? negative?

56. In which quadrants is the cosecant function positive? negative?

Collaborative Discussion and Writing

57. Describe how the graphs of the sine and cosine functions are related.

58. Explain why both the sine and cosine functions are continuous, but the tangent function, defined as sine/cosine, is not continuous.

Skill Maintenance

Graph both functions in the same viewing window and describe how g is a transformation of f.

59. $f(x) = x^2$, $g(x) = 2x^2 - 3$

60. $f(x) = x^2$, $g(x) = (x - 2)^2$

61. $f(x) = |x|$, $g(x) = \frac{1}{2}|x - 4| + 1$

62. $f(x) = x^3$, $g(x) = -x^3$

Write an equation for a function that has a graph with the given characteristics. Check using a graphing calculator.

63. The shape of $y = x^3$, but reflected across the x-axis, shifted right 2 units, and shifted down 1 unit

64. The shape of $y = 1/x$, but shrunk vertically by a factor of $\frac{1}{4}$ and shifted up 3 units

Synthesis

Complete. (For example, $\sin(x + 2\pi) = \sin x$.)

65. $\cos(-x) = $ _____

66. $\sin(-x) = $ _____

67. $\sin(x + 2k\pi)$, $k \in \mathbb{Z} = $ _____

68. $\cos(x + 2k\pi)$, $k \in \mathbb{Z} = $ _____

69. $\sin(\pi - x) = $ _____

70. $\cos(\pi - x) = $ _____

71. $\cos(x - \pi) = $ _____

72. $\cos(x + \pi) = $ _____

73. $\sin(x + \pi) = $ _____

74. $\sin(x - \pi) = $ _____

75. Find all numbers x that satisfy the following. Check using a graphing calculator.

   a) $\sin x = 1$
   b) $\cos x = -1$
   c) $\sin x = 0$

76. Find $f \circ g$ and $g \circ f$, where $f(x) = x^2 + 2x$ and $g(x) = \cos x$.

Use a graphing calculator to determine the domain, the range, the period, and the amplitude of the function.

77. $y = (\sin x)^2$ **78.** $y = |\cos x| + 1$

Determine the domain of the function.

79. $f(x) = \sqrt{\cos x}$ **80.** $g(x) = \dfrac{1}{\sin x}$

81. $f(x) = \dfrac{\sin x}{\cos x}$ **82.** $g(x) = \log(\sin x)$

Graph.

83. $y = 3\sin x$ **84.** $y = \sin|x|$

85. $y = \sin x + \cos x$ **86.** $y = |\cos x|$

87. One of the motivations for developing trigonometry with a unit circle is that you can actually "see" $\sin\theta$ and $\cos\theta$ on the circle. Note in the figure below that $AP = \sin\theta$ and $OA = \cos\theta$. It turns out that you can also "see" the other four trigonometric functions. Prove each of the following.

   a) $BD = \tan\theta$ **b)** $OD = \sec\theta$
   c) $OE = \csc\theta$ **d)** $CE = \cot\theta$

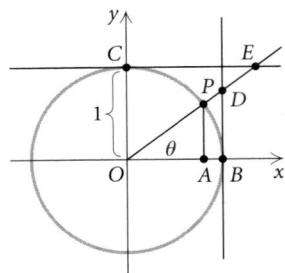

88. Using graphs, determine all numbers x that satisfy

$$\sin x < \cos x.$$

89. Using a calculator, consider $(\sin x)/x$, where x is between 0 and $\pi/2$. As x approaches 0, this function approaches a limiting value. What is it?

6.6

Graphs of Transformed Sine and Cosine Functions

❧ Graph transformations of $y = \sin x$ and $y = \cos x$ in the form

$$y = A \sin (Bx - C) + D$$

and

$$y = A \cos (Bx - C) + D$$

and determine the amplitude, the period, and the phase shift.

❧ Graph sums of functions.

❧ Graph functions (damped oscillations) found by multiplying trignometric functions by other functions.

❧ Variations of Basic Graphs

In Section 6.5, we graphed all six trigonometric functions. In this section, we will consider variations of the graphs of the sine and cosine functions. For example, we will graph equations like the following:

$$y = 5 \sin \tfrac{1}{2}x, \qquad y = \cos (2x - \pi), \quad \text{and} \quad y = \tfrac{1}{2} \sin x - 3.$$

In particular, we are interested in graphs of functions in the form

$$y = A \sin (Bx - C) + D$$

and

$$y = A \cos (Bx - C) + D,$$

where A, B, C, and D are constants. These constants have the effect of translating, reflecting, stretching, and shrinking the basic graphs. Let's first examine the effect of each constant individually. Then we will consider the combined effects of more than one constant.

> TRANSFORMATIONS
> OF FUNCTIONS
>
> REVIEW SECTION **2.4.**

The Constant D.   Let's observe the effect of the constant D in the graphs below.

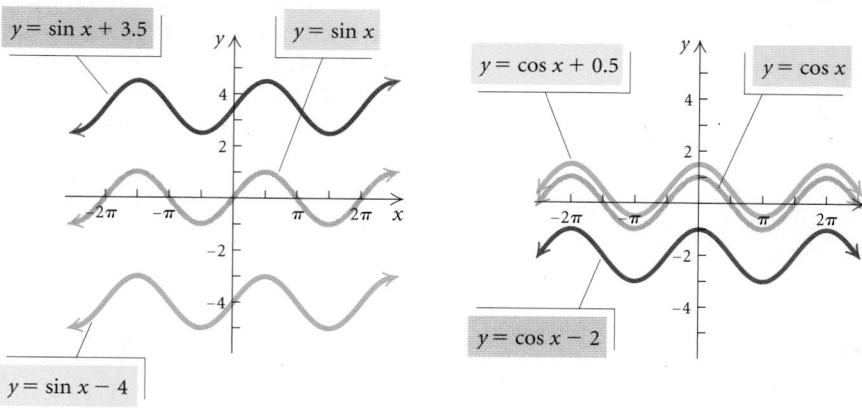

The constant D in

$$y = A \sin (Bx - C) + D \quad \text{and} \quad y = A \cos (Bx - C) + D$$

translates the graphs up D units if $D > 0$ or down $|D|$ units if $D < 0$.

EXAMPLE 1 Sketch a graph of $y = \sin x + 3$.

Solution The graph of $y = \sin x + 3$ is a *vertical* translation of the graph of $y = \sin x$ up 3 units. One way to sketch the graph is to first consider $y = \sin x$ on an interval of length 2π, say, $[0, 2\pi]$. The zeros of the function and the maximum and minimum values can be considered key points. These are

$$(0, 0), \quad \left(\frac{\pi}{2}, 1\right), \quad (\pi, 0), \quad \left(\frac{3\pi}{2}, -1\right), \quad (2\pi, 0).$$

These key points are transformed up 3 units to obtain the key points of the graph of $y = \sin x + 3$. These are

$$(0, 3), \quad \left(\frac{\pi}{2}, 4\right), \quad (\pi, 3), \quad \left(\frac{3\pi}{2}, 2\right), \quad (2\pi, 3).$$

The graph of $y = \sin x + 3$ can be sketched on the interval $[0, 2\pi]$ and extended to obtain the rest of the graph by repeating the graph on intervals of length 2π.

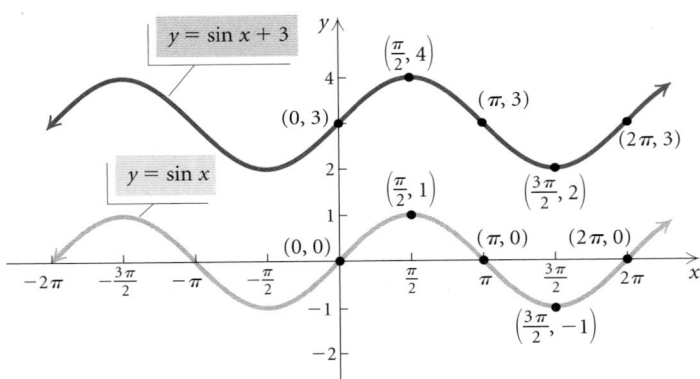

The Constant A. Next, we consider the effect of the constant A. What can we observe in the following graphs? What is the effect of the constant A on the graph of the basic function when **(a)** $0 < A < 1$? **(b)** $A > 1$? **(c)** $-1 < A < 0$? **(d)** $A < -1$?

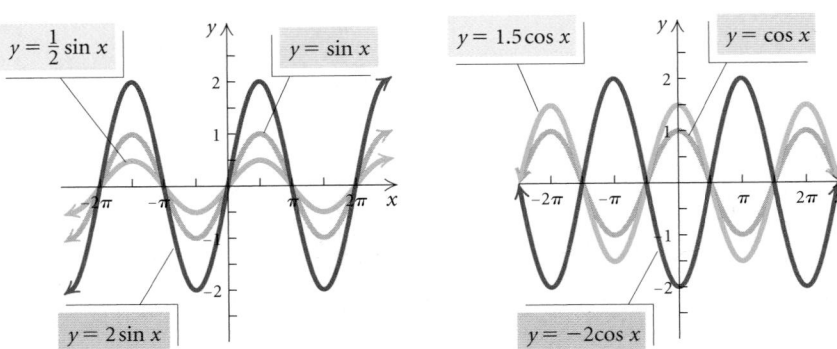

If $|A| > 1$, then there will be a vertical stretching. If $|A| < 1$, then there will be a vertical shrinking. If $A < 0$, the graph is also reflected across the x-axis.

> ### Amplitude
> The **amplitude** of the graphs of $y = A \sin(Bx - C) + D$ and $y = A \cos(Bx - C) + D$ is $|A|$.

EXAMPLE 2 Sketch a graph of $y = 2 \cos x$. What is the amplitude?

Solution The constant 2 in $y = 2 \cos x$ has the effect of stretching the graph of $y = \cos x$ vertically by a factor of 2 units. Since the function values of $y = \cos x$ are such that $-1 \le \cos x \le 1$, the function values of $y = 2 \cos x$ are such that $-2 \le 2 \cos x \le 2$. The maximum value of $y = 2 \cos x$ is 2, and the minimum value is -2. Thus the *amplitude*, A, is $\frac{1}{2}|2 - (-2)|$, or 2.

We draw the graph of $y = \cos x$ and consider its key points,

$$(0, 1), \quad \left(\frac{\pi}{2}, 0\right), \quad (\pi, -1), \quad \left(\frac{3\pi}{2}, 0\right), \quad (2\pi, 1),$$

on the interval $[0, 2\pi]$.

We then multiply the second coordinates by 2 to obtain the key points of $y = 2 \cos x$. These are

$$(0, 2), \quad \left(\frac{\pi}{2}, 0\right), \quad (\pi, -2), \quad \left(\frac{3\pi}{2}, 0\right), \quad (2\pi, 2).$$

We plot these points and sketch the graph on the interval $[0, 2\pi]$. Then we repeat this part of the graph on adjacent intervals of length 2π.

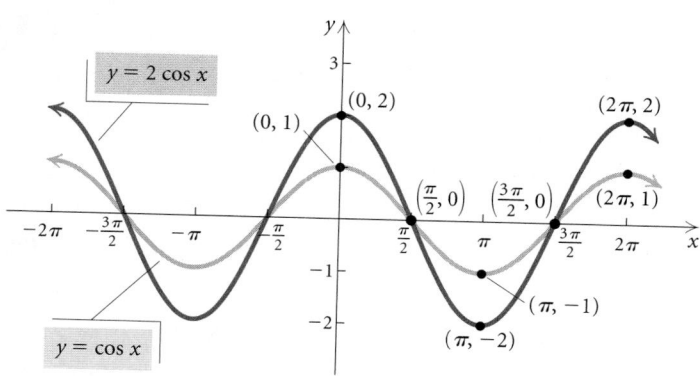

EXAMPLE 3 Sketch a graph of $y = -\frac{1}{2} \sin x$. What is the amplitude?

Solution The amplitude of the graph is $\left|-\frac{1}{2}\right|$, or $\frac{1}{2}$. The graph of $y = -\frac{1}{2} \sin x$ is a vertical shrinking and a reflection of the graph of $y = \sin x$ across the x-axis. In graphing, the key points of $y = \sin x$,

$$(0, 0), \quad \left(\frac{\pi}{2}, 1\right), \quad (\pi, 0), \quad \left(\frac{3\pi}{2}, -1\right), \quad (2\pi, 0),$$

are transformed to

$$(0, 0), \quad \left(\frac{\pi}{2}, -\frac{1}{2}\right), \quad (\pi, 0), \quad \left(\frac{3\pi}{2}, \frac{1}{2}\right), \quad (2\pi, 0).$$

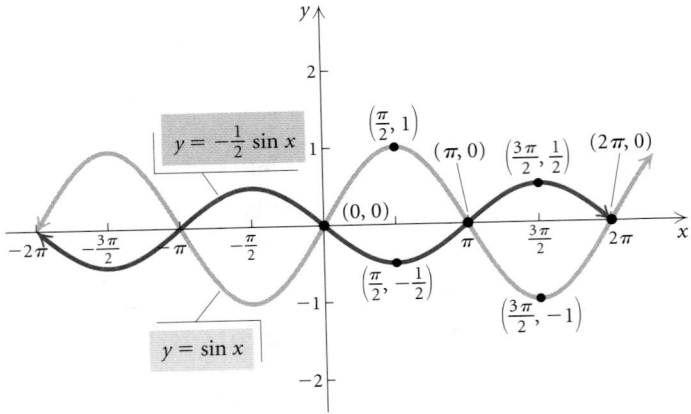

The Constant B.   Now, we consider the effect of the constant B. Changes in the constants A and D *do not* change the period. But what effect, if any, does a change in B have on the period of the function? Let's observe the period of each of the following graphs.

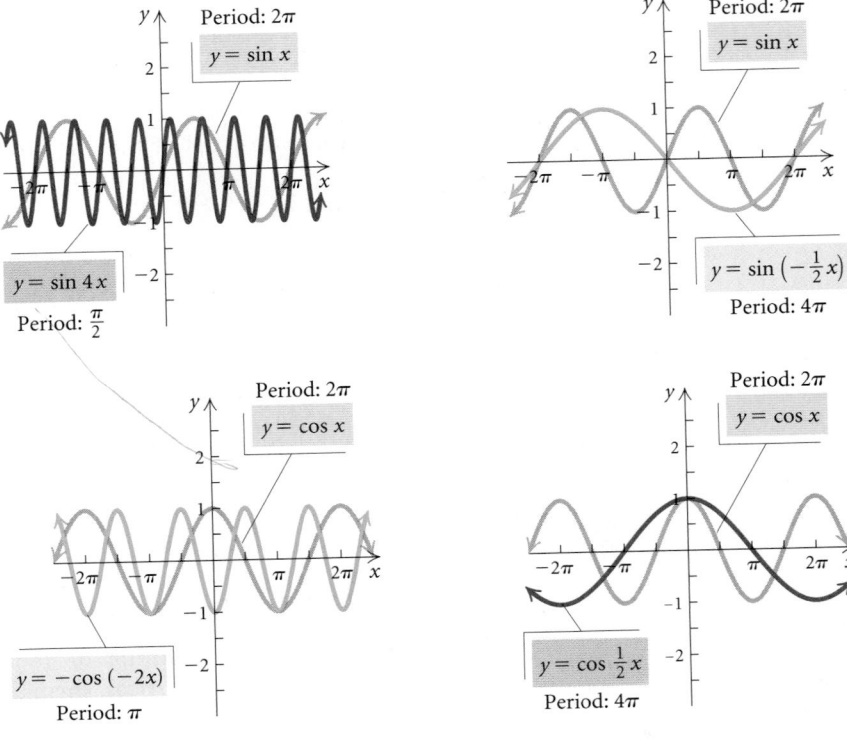

If $|B| < 1$, then there will be a horizontal stretching. If $|B| > 1$, then there will be a horizontal shrinking. If $B < 0$, the graph is also reflected across the y-axis.

Period

The **period** of the graphs of $y = A \sin(Bx - C) + D$ and $y = A \cos(Bx - C) + D$ is $\left|\dfrac{2\pi}{B}\right|$.*

EXAMPLE 4   Sketch a graph of $y = \sin 4x$. What is the period?

Solution   The constant B has the effect of changing the period. The graph of $y = f(4x)$ is obtained from the graph of $y = f(x)$ by shrinking the graph horizontally. The period of $y = \sin 4x$ is $|2\pi/4|$, or $\pi/2$. The new graph is obtained by dividing the first coordinate of each ordered-pair solution of $y = f(x)$ by 4. The key points of $y = \sin x$ are

$$(0, 0), \quad \left(\frac{\pi}{2}, 1\right), \quad (\pi, 0), \quad \left(\frac{3\pi}{2}, -1\right), \quad (2\pi, 0).$$

These are transformed to the key points of $y = \sin 4x$, which are

$$(0, 0), \quad \left(\frac{\pi}{8}, 1\right), \quad \left(\frac{\pi}{4}, 0\right), \quad \left(\frac{3\pi}{8}, -1\right), \quad \left(\frac{\pi}{2}, 0\right).$$

We plot these key points and sketch in the graph on the shortened interval $[0, \pi/2]$. Then we repeat the graph on other intervals of length $\pi/2$.

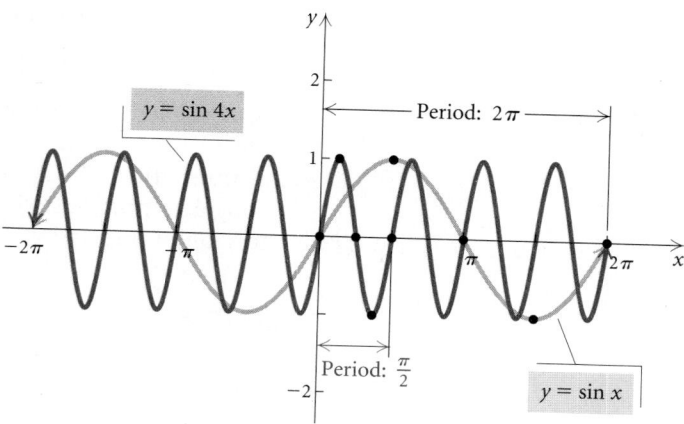

---

*The period of the graphs of $y = A \tan(Bx - C) + D$ and $y = A \cot(Bx - C) + D$ is $|\pi/B|$.

The period of the graphs of $y = A \sec(Bx - C) + D$ and $y = A \csc(Bx - C) + D$ is $|2\pi/B|$.

The Constant C.   Next, we examine the effect of the constant C. The curve in each of the following graphs has an amplitude of 1 and a period of 2π, but there are six distinct graphs. What is the effect of the constant C?

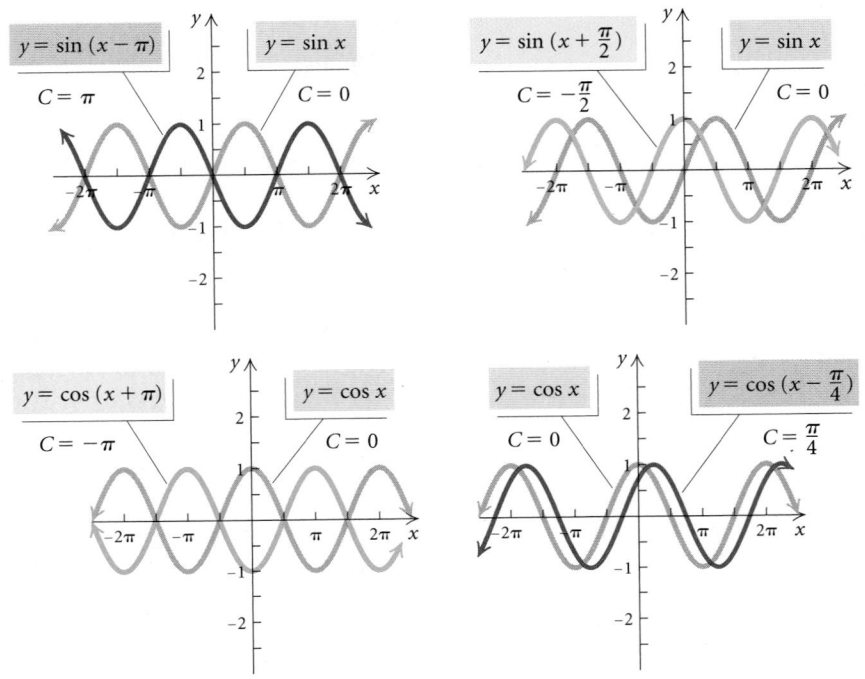

For each of the functions of the form

$$y = A \sin (Bx - C) + D \quad \text{and} \quad y = A \cos (Bx - C) + D$$

that are graphed above, the coefficient of x, which is B, is 1. In this case, the effect of the constant C on the graph of the basic function is a horizontal translation of $|C|$ units. In Example 5, which follows, $B = 1$. We will consider functions where $B \neq 1$ in Examples 6 and 7. When $B \neq 1$, the horizontal translation will be $|C/B|$.

EXAMPLE 5   Sketch a graph of $y = \sin \left(x - \dfrac{\pi}{2} \right)$.

Solution   The amplitude is 1, and the period is 2π. The graph of $y = f(x - c)$ is obtained from the graph of $y = f(x)$ by translating the graph horizontally—to the right c units if $c > 0$ and to the left $|c|$ units if $c < 0$. The graph of $y = \sin (x - \pi/2)$ is a translation of the graph of

$y = \sin x$ to the right $\pi/2$ units. The value $\pi/2$ is called the **phase shift**. The key points of $y = \sin x$,

$$(0, 0), \quad \left(\frac{\pi}{2}, 1\right), \quad (\pi, 0), \quad \left(\frac{3\pi}{2}, -1\right), \quad (2\pi, 0),$$

are transformed by adding $\pi/2$ to each of the first coordinates to obtain the following key points of $y = \sin(x - \pi/2)$:

$$\left(\frac{\pi}{2}, 0\right), \quad (\pi, 1), \quad \left(\frac{3\pi}{2}, 0\right), \quad (2\pi, -1), \quad \left(\frac{5\pi}{2}, 0\right).$$

We plot these key points and sketch the curve on the interval $[\pi/2, 5\pi/2]$. Then we repeat the graph on other intervals of length 2π.

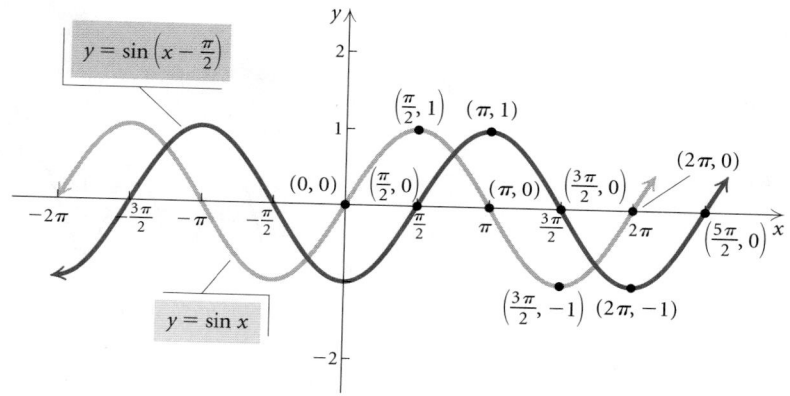

Combined Transformations.   Now we consider combined transformations of graphs. It is helpful to rewrite

$$y = A \sin(Bx - C) + D \qquad \text{and} \quad y = A \cos(Bx - C) + D$$

as

$$y = A \sin\left[B\left(x - \frac{C}{B}\right)\right] + D \quad \text{and} \quad y = A \cos\left[B\left(x - \frac{C}{B}\right)\right] + D.$$

EXAMPLE 6   Sketch a graph of $y = \cos(2x - \pi)$.

Solution   The graph of

$$y = \cos(2x - \pi)$$

is the same as the graph of

$$y = 1 \cdot \cos\left[2\left(x - \frac{\pi}{2}\right)\right] + 0.$$

The amplitude is 1. The factor 2 shrinks the period by half, making the period $|2\pi/2|$, or π. The phase shift $\pi/2$ translates the graph of $y = \cos 2x$

to the right $\pi/2$ units. Thus, to form the graph, we first graph $y = \cos x$, followed by $y = \cos 2x$ and then $y = \cos[2(x - \pi/2)]$.

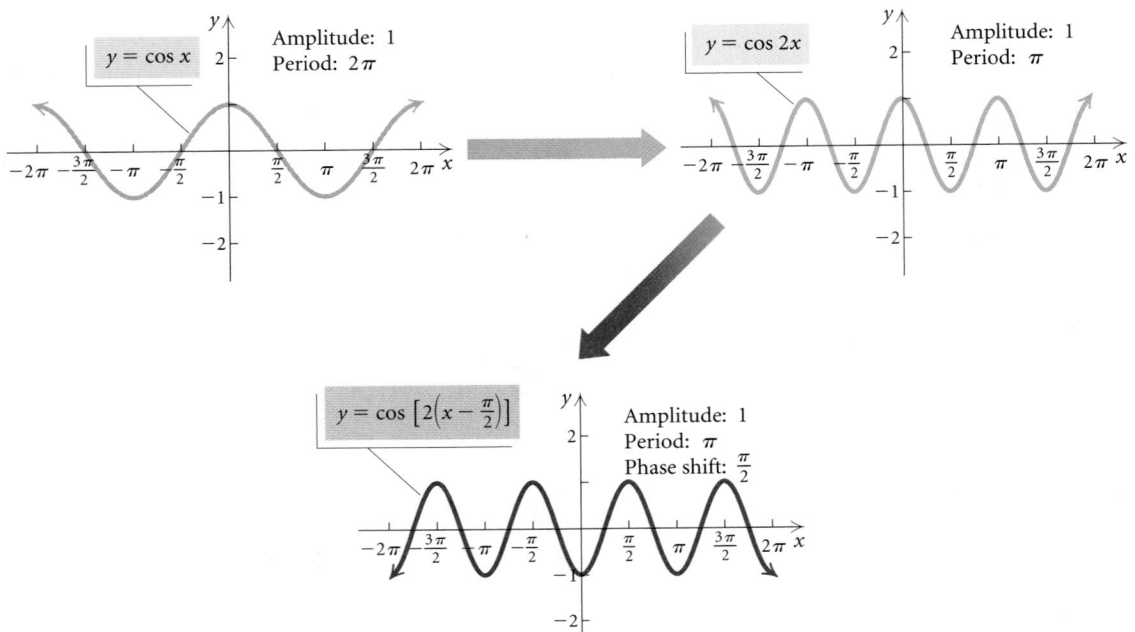

Phase Shift

The **phase shift** of the graphs

$$y = A \sin(Bx - C) + D = A \sin\left[B\left(x - \frac{C}{B}\right)\right] + D$$

and

$$y = A \cos(Bx - C) + D = A \cos\left[B\left(x - \frac{C}{B}\right)\right] + D$$

is the quantity $\dfrac{C}{B}$.

If $C/B > 0$, the graph is translated to the right C/B units. If $C/B < 0$, the graph is translated to the left $|C/B|$ units. Be sure that the horizontal stretching or shrinking based on the constant B is done before the translation based on the phase shift C/B.

Let's now summarize the effect of the constants. When graphing, we carry out the procedures in the order listed.

> **Transformations of Sine and Cosine Functions**
> To graph
>
> $$y = A \sin (Bx - C) + D = A \sin \left[B \left(x - \frac{C}{B} \right) \right] + D$$
>
> and
>
> $$y = A \cos (Bx - C) + D = A \cos \left[B \left(x - \frac{C}{B} \right) \right] + D,$$
>
> follow the steps listed below in the order in which they are listed.
>
> **1.** Stretch or shrink the graph horizontally according to B.
>
> $|B| < 1$ Stretch horizontally
>
> $|B| > 1$ Shrink horizontally
>
> $B < 0$ Reflect across the y-axis
>
> The *period* is $\left| \dfrac{2\pi}{B} \right|$.
>
> **2.** Stretch or shrink the graph vertically according to A.
>
> $|A| < 1$ Shrink vertically
>
> $|A| > 1$ Stretch vertically
>
> $A < 0$ Reflect across the x-axis
>
> The *amplitude* is $|A|$.
>
> **3.** Translate the graph horizontally according to C/B.
>
> $\dfrac{C}{B} < 0$ $\left| \dfrac{C}{B} \right|$ units to the left
>
> $\dfrac{C}{B} > 0$ $\dfrac{C}{B}$ units to the right
>
> The *phase shift* is $\dfrac{C}{B}$.
>
> **4.** Translate the graph vertically according to D.
>
> $D < 0$ $|D|$ units down
>
> $D > 0$ D units up

When graphing transformations of the tangent and cotangent functions, note that the period is $|\pi/B|$. When graphing transformations of the secant and cosecant functions, note that the period is $|2\pi/B|$.

EXAMPLE 7 Sketch a graph of $y = 3 \sin (2x + \pi/2) + 1$. Find the amplitude, the period, and the phase shift.

Solution We first note that

$$y = 3 \sin \left(2x + \frac{\pi}{2} \right) + 1 = 3 \sin \left[2 \left(x - \left(-\frac{\pi}{4} \right) \right) \right] + 1.$$

Then we have the following:

$$\text{Amplitude} = |A| = |3| = 3,$$

$$\text{Period} = \left|\frac{2\pi}{B}\right| = \left|\frac{2\pi}{2}\right| = \pi,$$

$$\text{Phase shift} = \frac{C}{B} = \frac{-\pi/2}{2} = -\frac{\pi}{4}.$$

To create the final graph, we begin with the basic sine curve, $y = \sin x$. Then we sketch graphs of each of the following equations in sequence.

1. $y = \sin 2x$

3. $y = 3 \sin\left[2\left(x - \left(-\frac{\pi}{4}\right)\right)\right]$

2. $y = 3 \sin 2x$

4. $y = 3 \sin\left[2\left(x - \left(-\frac{\pi}{4}\right)\right)\right] + 1$

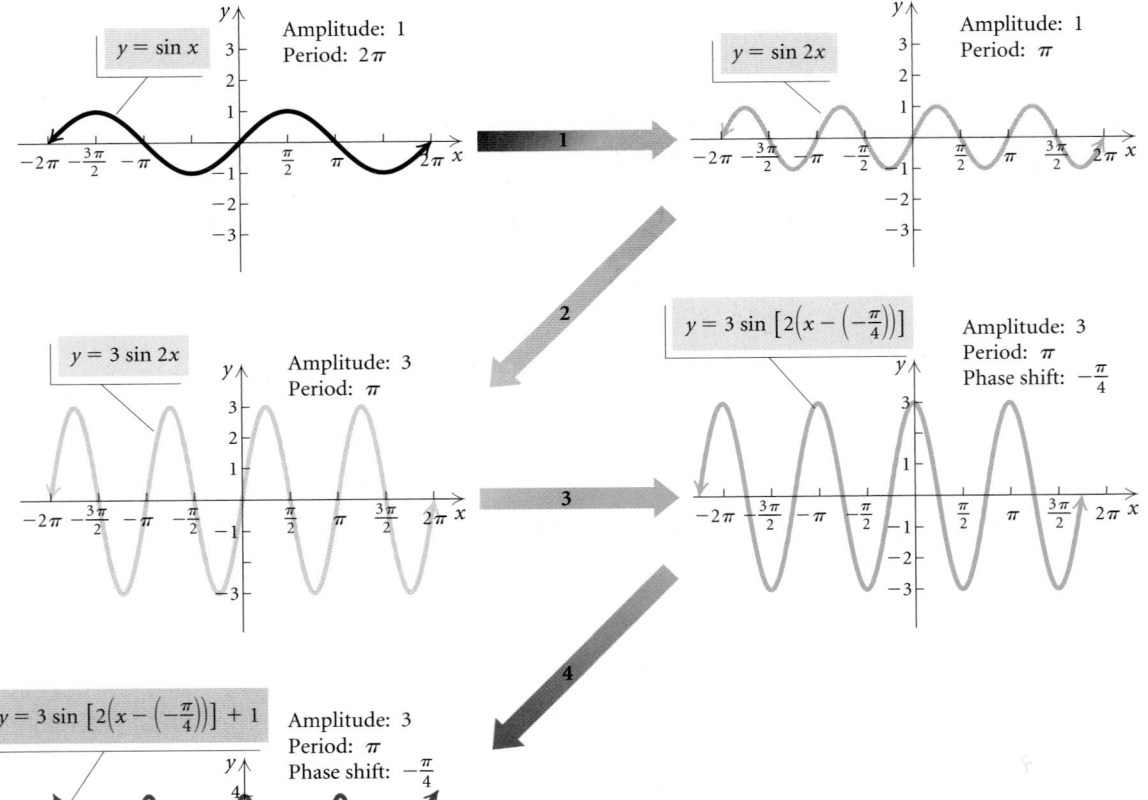

Now Try Exercise 27.

All the graphs in Examples 1–7 can be checked using a graphing calculator. Even though it is faster and more accurate to graph using a calculator, graphing by hand gives us a greater understanding of the effect of changing the constants A, B, C, and D.

Graphing calculators are especially convenient when a period or a phase shift is not a multiple of $\pi/4$.

EXAMPLE 8   Graph $y = 3 \cos 2\pi x - 1$. Find the amplitude, the period, and the phase shift.

Solution   First we note the following:

$$\text{Amplitude} = |A| = |3| = 3,$$

$$\text{Period} = \left|\frac{2\pi}{B}\right| = \left|\frac{2\pi}{2\pi}\right| = |1| = 1,$$

$$\text{Phase shift} = \frac{C}{B} = \frac{0}{2\pi} = 0.$$

There is no phase shift in this case because the constant $C = 0$. The graph has a vertical translation of the graph of the cosine function down 1 unit, an amplitude of 3, and a period of 1, so we can use $[-4, 4, -5, 5]$ as the viewing window.

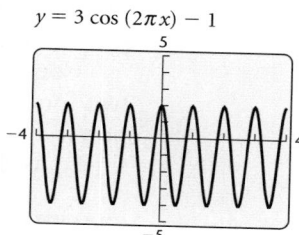

$y = 3 \cos (2\pi x) - 1$

Now Try Exercise 29. ■

The transformation techniques that we learned in this section for graphing the sine and cosine functions can also be applied in the same manner to the other trigonometric functions. Transformations of this type appear in the synthesis exercises in Exercise Set 6.6.

An **oscilloscope** is an electronic device that converts electrical signals into graphs like those in the preceding examples. These graphs are often called sine waves. By manipulating the controls, we can change the amplitude, the period, and the phase of sine waves. The oscilloscope has many applications, and the trigonometric functions play a major role in many of them.

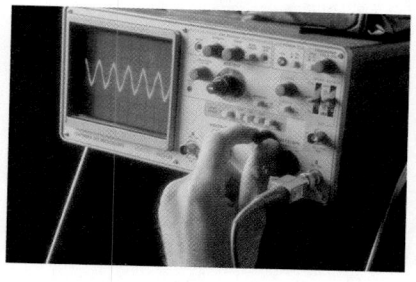

❖ Graphs of Sums: Addition of Ordinates

The output of an electronic synthesizer used in the recording and playing of music can be converted into sine waves by an oscilloscope. The following graphs illustrate simple tones of different frequencies. The frequency of a simple tone is the number of vibrations in the signal of the tone per second. The loudness or intensity of the tone is reflected in the height of the graph (its amplitude). The three tones in the diagrams below all have the same intensity but different frequencies.

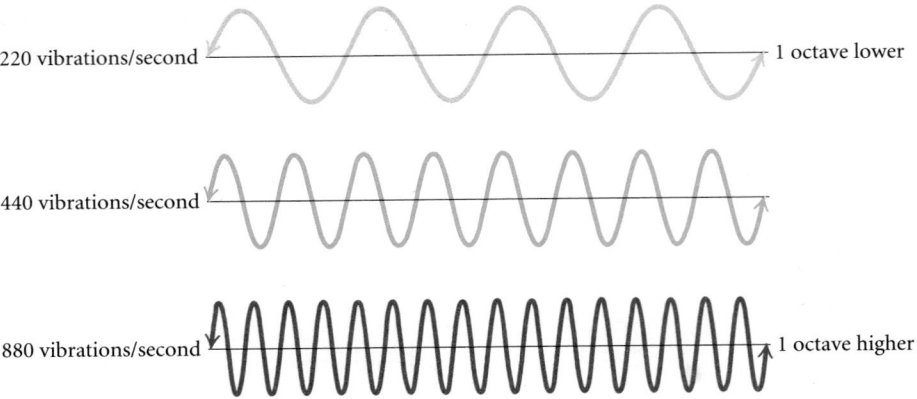

220 vibrations/second   1 octave lower

440 vibrations/second

880 vibrations/second   1 octave higher

Musical instruments can generate extremely complex sine waves. On a single instrument, overtones can become superimposed on a simple tone. When multiple notes are played simultaneously, graphs become very complicated. This can happen when multiple notes are played on a single instrument or a group of instruments, or even when the same simple note is played on different instruments.

Combinations of simple tones produce interesting curves. Consider two tones whose graphs are $y_1 = 2 \sin x$ and $y_2 = \sin 2x$. The combination of the two tones produces a new sound whose graph is $y = 2 \sin x + \sin 2x$, as shown in the following example.

EXAMPLE 9   Graph: $y = 2 \sin x + \sin 2x$.

Solution   We graph $y = 2 \sin x$ and $y = \sin 2x$ using the same set of axes.

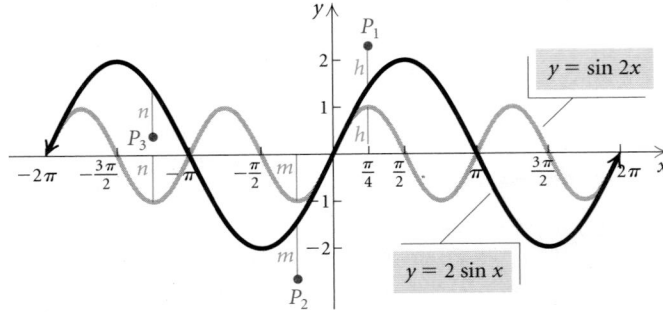

Now we graphically add some y-coordinates, or ordinates, to obtain points on the graph that we seek. At $x = \pi/4$, we transfer the distance h, which is the value of $\sin 2x$, up to add it to the value of $2 \sin x$. Point P_1 is on the graph that we seek. At $x = -\pi/4$, we use a similar procedure, but this time both ordinates are negative. Point P_2 is on the graph. At $x = -5\pi/4$, we add the negative ordinate of $\sin 2x$ to the positive ordinate of $2 \sin x$. Point P_3 is also on the graph. We continue to plot points in this fashion and then connect them to get the desired graph, shown below. This method is called **addition of ordinates**, because we add the y-values (ordinates) of $y = \sin 2x$ to the y-values (ordinates) of $y = 2 \sin x$. Note that the period of $2 \sin x$ is 2π and the period of $\sin 2x$ is π. The period of the sum $2 \sin x + \sin 2x$ is 2π, the least common multiple of 2π and π.

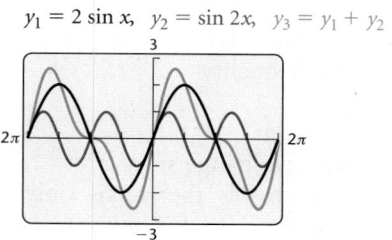

$y_1 = 2 \sin x, \quad y_2 = \sin 2x, \quad y_3 = y_1 + y_2$

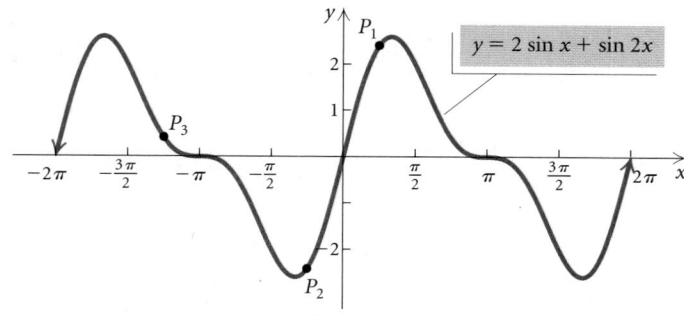

$y = 2 \sin x + \sin 2x$

Now Try Exercise 45. ■

Using a graphing calculator, we can quickly determine the period of a trigonometric function that is a combination of sine and cosine functions.

EXAMPLE 10   Graph $y = 2 \cos x - \sin 3x$ and determine its period.

Solution   We graph $y = 2 \cos x - \sin 3x$ with appropriate dimensions. The period appears to be 2π.

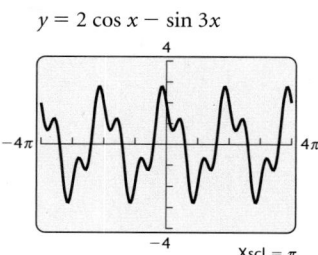

$y = 2 \cos x - \sin 3x$

Xscl = π

Now Try Exercise 53. ■

❖ Damped Oscillation: Multiplication of Ordinates

Suppose that a weight is attached to a spring and the spring is stretched and put into motion. The weight oscillates up and down. If we could assume falsely that the weight will bob up and down forever, then its height h after time t, in seconds, might be approximated by a function like

$$h(t) = 5 + 2 \sin (6\pi t).$$

Over a short time period, this might be a valid model, but experience tells us that eventually the spring will come to rest. A more appropriate model is provided by the following example, which illustrates **damped oscillation**.

EXAMPLE 11 Sketch a graph of $f(x) = e^{-x/2} \sin x$.

Solution The function f is the product of two functions g and h, where

$$g(x) = e^{-x/2} \quad \text{and} \quad h(x) = \sin x.$$

Thus, to find function values, we can **multiply ordinates.** Let's do more analysis before graphing. Note that for any real number x,

$$-1 \le \sin x \le 1.$$

Recall from Chapter 5 that all values of the exponential function are positive. Thus we can multiply by $e^{-x/2}$ and obtain the inequality

$$-e^{-x/2} \le e^{-x/2} \sin x \le e^{-x/2}.$$

The direction of the inequality symbols does not change since $e^{-x/2} > 0$. This also tells us that the original function crosses the x-axis only at values for which $\sin x = 0$. These are the numbers $k\pi$, for any integer k.

　　The inequality tells us that the function f is constrained between the graphs of $y = -e^{-x/2}$ and $y = e^{-x/2}$. We start by graphing these functions using dashed lines. Since we also know that $f(x) = 0$ when $x = k\pi$, k an integer, we mark these points on the graph. Then we use a calculator and compute other function values. The graph is as follows.

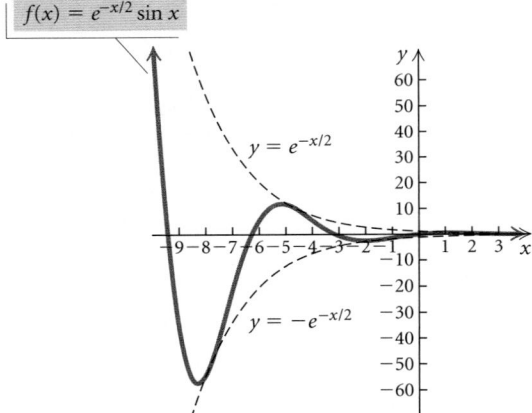

Now Try Exercise 61.

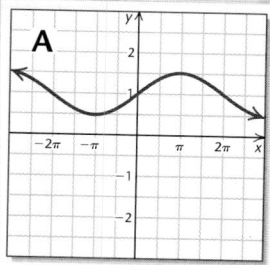

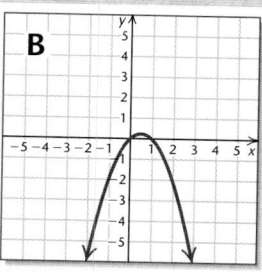

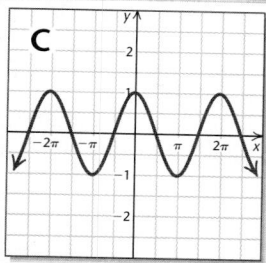

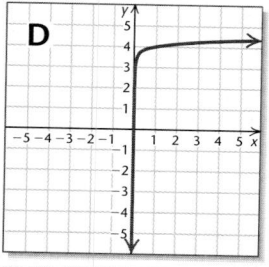

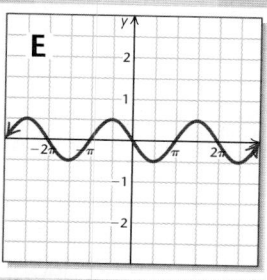

Visualizing the Graph

Match the function with its graph.

1. $f(x) = -\sin x$

2. $f(x) = 2x^3 - x + 1$

3. $y = \dfrac{1}{2}\cos\left(x + \dfrac{\pi}{2}\right)$

4. $f(x) = \cos\left(\dfrac{1}{2}x\right)$

5. $y = -x^2 + x$

6. $y = \dfrac{1}{2}\log x + 4$

7. $f(x) = 2^{x-1}$

8. $f(x) = \dfrac{1}{2}\sin\left(\dfrac{1}{2}x\right) + 1$

9. $f(x) = -\cos(x - \pi)$

10. $f(x) = -\dfrac{1}{2}x^4$

Answers on page A-43

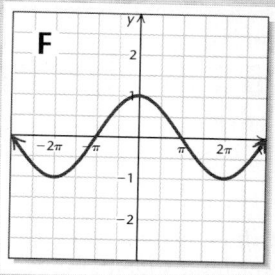

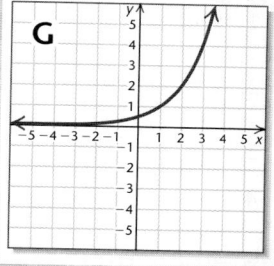

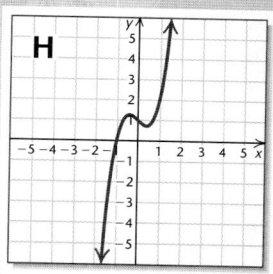

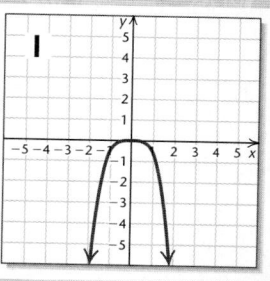

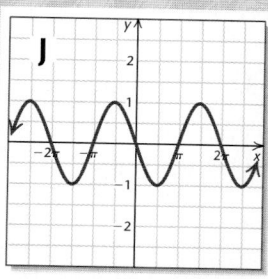

6.6 Exercise Set

Determine the amplitude, the period, and the phase shift of the function and, without a graphing calculator, sketch the graph of the function by hand. Then check the graph using a graphing calculator.

1. $y = \sin x + 1$

2. $y = \frac{1}{4} \cos x$

3. $y = -3 \cos x$

4. $y = \sin(-2x)$

5. $y = \frac{1}{2} \cos x$

6. $y = \sin\left(\frac{1}{2}x\right)$

7. $y = \sin(2x)$

8. $y = \cos x - 1$

9. $y = 2 \sin\left(\frac{1}{2}x\right)$

10. $y = \cos\left(x - \frac{\pi}{2}\right)$

11. $y = \frac{1}{2} \sin\left(x + \frac{\pi}{2}\right)$

12. $y = \cos x - \frac{1}{2}$

13. $y = 3 \cos(x - \pi)$

14. $y = -\sin\left(\frac{1}{4}x\right) + 1$

15. $y = \frac{1}{3} \sin x - 4$

16. $y = \cos\left(\frac{1}{2}x + \frac{\pi}{2}\right)$

17. $y = -\cos(-x) + 2$

18. $y = \frac{1}{2} \sin\left(2x - \frac{\pi}{4}\right)$

Determine the amplitude, the period, and the phase shift of the function. Then check by graphing the function using a graphing calculator. Try to visualize the graph before creating it.

19. $y = 2 \cos\left(\frac{1}{2}x - \frac{\pi}{2}\right)$

20. $y = 4 \sin\left(\frac{1}{4}x + \frac{\pi}{8}\right)$

21. $y = -\frac{1}{2} \sin\left(2x + \frac{\pi}{2}\right)$

22. $y = -3 \cos(4x - \pi) + 2$

23. $y = 2 + 3 \cos(\pi x - 3)$

24. $y = 5 - 2 \cos\left(\frac{\pi}{2}x + \frac{\pi}{2}\right)$

25. $y = -\frac{1}{2} \cos(2\pi x) + 2$

26. $y = -2 \sin(-2x + \pi) - 2$

27. $y = -\sin\left(\frac{1}{2}x - \frac{\pi}{2}\right) + \frac{1}{2}$

28. $y = \frac{1}{3} \cos(-3x) + 1$

29. $y = \cos(-2\pi x) + 2$

30. $y = \frac{1}{2} \sin(2\pi x + \pi)$

31. $y = -\frac{1}{4} \cos(\pi x - 4)$

32. $y = 2 \sin(2\pi x + 1)$

In Exercises 33–40, without a graphing calculator, match the function with one of the graphs (a)–(h), which follow. Then check your work using a graphing calculator.

a)

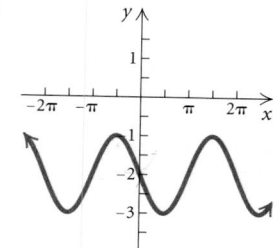

b)

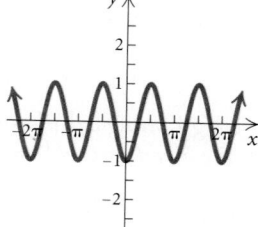

c)

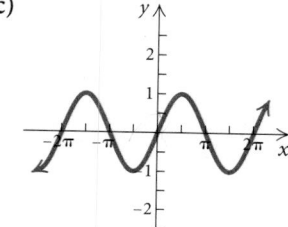

d)

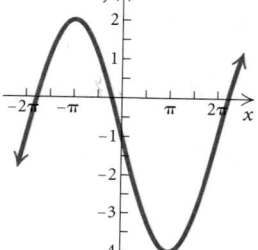

e)

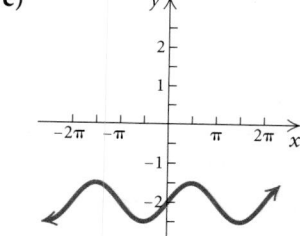

f)

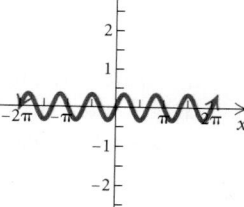

g)

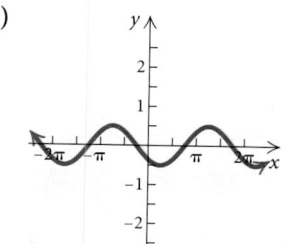

h)

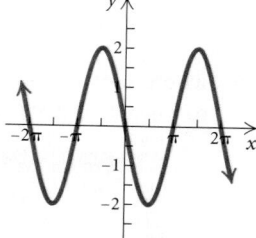

33. $y = -\cos 2x$

34. $y = \frac{1}{2} \sin x - 2$

35. $y = 2 \cos \left(x + \frac{\pi}{2} \right)$

36. $y = -3 \sin \frac{1}{2} x - 1$

37. $y = \sin (x - \pi) - 2$

38. $y = -\frac{1}{2} \cos \left(x - \frac{\pi}{4} \right)$

39. $y = \frac{1}{3} \sin 3x$

40. $y = \cos \left(x - \frac{\pi}{2} \right)$

In Exercises 41–44, determine the equation of the function that is graphed.

41.

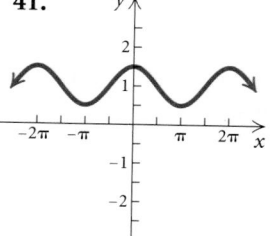

42.

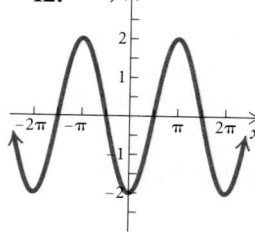

43.

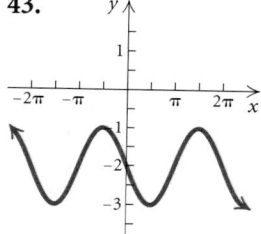

44.

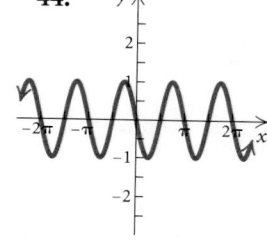

Graph using addition of ordinates. Then check your work using a graphing calculator.

45. $y = 2 \cos x + \cos 2x$

46. $y = 3 \cos x + \cos 3x$

47. $y = \sin x + \cos 2x$

48. $y = 2 \sin x + \cos 2x$

49. $y = \sin x - \cos x$

50. $y = 3 \cos x - \sin x$

51. $y = 3 \cos x + \sin 2x$

52. $y = 3 \sin x - \cos 2x$

Use a graphing calculator to graph the function.

53. $y = x + \sin x$

54. $y = -x - \sin x$

55. $y = \cos x - x$

56. $y = -(\cos x - x)$

57. $y = \cos 2x + 2x$

58. $y = \cos 3x + \sin 3x$

59. $y = 4 \cos 2x - 2 \sin x$

60. $y = 7.5 \cos x + \sin 2x$

Graph each of the following.

61. $f(x) = e^{-x/2} \cos x$

62. $f(x) = e^{-0.4x} \sin x$

63. $f(x) = 0.6x^2 \cos x$

64. $f(x) = e^{-x/4} \sin x$

65. $f(x) = x \sin x$

66. $f(x) = |x| \cos x$

67. $f(x) = 2^{-x} \sin x$

68. $f(x) = 2^{-x} \cos x$

Collaborative Discussion and Writing

69. In the equations $y = A \sin(Bx - C) + D$ and $y = A \cos(Bx - C) + D$, which constants translate the graphs and which constants stretch and shrink the graphs? Describe in your own words the effect of each constant.

70. In the transformation steps listed in this section, why must step (1) precede step (3)? Give an example that illustrates this.

Skill Maintenance

Classify the function as linear, quadratic, cubic, quartic, rational, exponential, logarithmic, or trigonometric.

71. $f(x) = \dfrac{x + 4}{x}$

72. $y = \dfrac{1}{2} \log x - 4$

73. $y = x^4 - x - 2$

74. $\dfrac{3}{4}x + \dfrac{1}{2}y = -5$

75. $f(x) = \sin x - 3$

76. $f(x) = 0.5e^{x-2}$

77. $y = \dfrac{2}{5}$

78. $y = \sin x + \cos x$

79. $y = x^2 - x^3$

80. $f(x) = \left(\dfrac{1}{2}\right)^x$

Synthesis

Find the maximum and minimum values of the function.

81. $y = 2 \cos \left[3\left(x - \dfrac{\pi}{2}\right)\right] + 6$

82. $y = \dfrac{1}{2} \sin(2x - 6\pi) - 4$

The transformation techniques that we learned in this section for graphing the sine and cosine functions can also be applied to the other trigonometric functions. Sketch a graph of each of the following. Then check your work using a graphing calculator.

83. $y = -\tan x$

84. $y = \tan(-x)$

85. $y = -2 + \cot x$

86. $y = -\dfrac{3}{2} \csc x$

87. $y = 2 \tan \dfrac{1}{2}x$

88. $y = \cot 2x$

89. $y = 2 \sec(x - \pi)$

90. $y = 4 \tan \left(\dfrac{1}{4}x + \dfrac{\pi}{8}\right)$

91. $y = 2 \csc \left(\dfrac{1}{2}x - \dfrac{3\pi}{4}\right)$

92. $y = 4 \sec(2x - \pi)$

Use a graphing calculator to graph each of the following on the given interval and approximate the zeros.

93. $f(x) = \dfrac{\sin x}{x}$; $[-12, 12]$

94. $f(x) = \dfrac{\cos x - 1}{x}$; $[-12, 12]$

95. $f(x) = x^3 \sin x$; $[-5, 5]$

96. $f(x) = \dfrac{(\sin x)^2}{x}$; $[-4, 4]$

97. *Temperature During an Illness.* The temperature T of a patient during a 12-day illness is given by

$$T(t) = 101.6° + 3° \sin \left(\dfrac{\pi}{8}t\right).$$

a) Graph the function on the interval $[0, 12]$.
b) What are the maximum and minimum temperatures during the illness?

98. *Periodic Sales.* A company in a northern climate has sales of skis as given by

$$S(t) = 10\left(1 - \cos \dfrac{\pi}{6}t\right),$$

where t is the time, in months ($t = 0$ corresponds to July 1), and $S(t)$ is in thousands of dollars.

a) Graph the function on a 12-month interval [0, 12].
b) What is the period of the function?
c) What is the minimum amount of sales and when does it occur?
d) What is the maximum amount of sales and when does it occur?

99. *Satellite Location.* A satellite circles the earth in such a way that it is y miles from the equator (north or south, height not considered) t minutes after its launch, where

$$y(t) = 3000\left[\cos\frac{\pi}{45}(t - 10)\right].$$

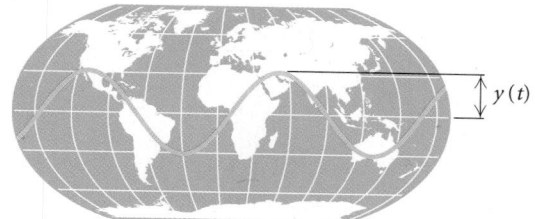

$$y = 3000\left[\cos\frac{\pi}{45}(x - 10)\right]$$

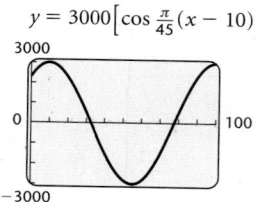

What are the amplitude, the period, and the phase shift?

100. *Water Wave.* The cross-section of a water wave is given by

$$y = 3\sin\left(\frac{\pi}{4}x + \frac{\pi}{4}\right),$$

where y is the vertical height of the water wave and x is the distance from the origin to the wave.

$$y = 3\sin\left(\frac{\pi}{4}x + \frac{\pi}{4}\right)$$

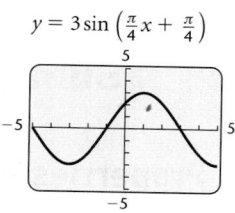

What are the amplitude, the period, and the phase shift?

101. *Damped Oscillations.* Suppose that the motion of a spring is given by

$$d(t) = 6e^{-0.8t}\cos(6\pi t) + 4,$$

where d is the distance, in inches, of a weight from the point at which the spring is attached to a ceiling, after t seconds. How far do you think the spring is from the ceiling when the spring stops bobbing?

102. *Rotating Beacon.* A police car is parked 10 ft from a wall. On top of the car is a beacon rotating in such a way that the light is at a distance $d(t)$ from point Q after t seconds, where

$$d(t) = 10\tan(2\pi t).$$

When d is positive, as shown in the figure, the light is pointing north of Q, and when d is negative, the light is pointing south of Q.

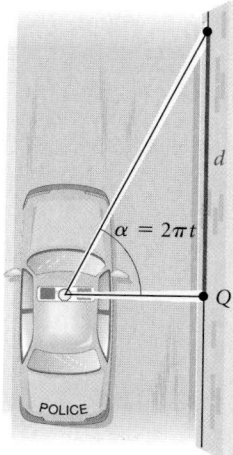

Explain the meaning of the values of t for which the function is not defined.

CHAPTER 6   Summary and Review

Important Properties and Formulas

Trigonometric Function Values of an Acute Angle θ

Let θ be an acute angle of a right triangle. The six trigonometric functions of θ are as follows:

$$\sin \theta = \frac{\text{opp}}{\text{hyp}}, \qquad \cos \theta = \frac{\text{adj}}{\text{hyp}}, \qquad \tan \theta = \frac{\text{opp}}{\text{adj}},$$

$$\csc \theta = \frac{\text{hyp}}{\text{opp}}, \qquad \sec \theta = \frac{\text{hyp}}{\text{adj}}, \qquad \cot \theta = \frac{\text{adj}}{\text{opp}}.$$

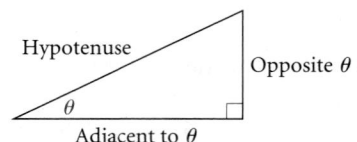

Reciprocal Functions

$$\csc \theta = \frac{1}{\sin \theta}, \qquad \sec \theta = \frac{1}{\cos \theta}, \qquad \cot \theta = \frac{1}{\tan \theta}$$

Function Values of Special Angles

	0°	30°	45°	60°	90°
sin	0	$1/2$	$\sqrt{2}/2$	$\sqrt{3}/2$	1
cos	1	$\sqrt{3}/2$	$\sqrt{2}/2$	$1/2$	0
tan	0	$\sqrt{3}/3$	1	$\sqrt{3}$	Not defined

Cofunction Identities

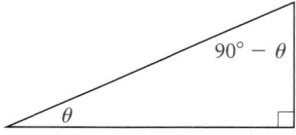

$$\sin \theta = \cos (90° - \theta), \qquad \cos \theta = \sin (90° - \theta),$$

$$\tan \theta = \cot (90° - \theta), \qquad \cot \theta = \tan (90° - \theta),$$

$$\sec \theta = \csc (90° - \theta), \qquad \csc \theta = \sec (90° - \theta)$$

Trigonometric Functions of Any Angle θ

If $P(x, y)$ is any point on the terminal side of any angle θ in standard position, and r is the distance from the origin to $P(x, y)$, where $r = \sqrt{x^2 + y^2}$, then

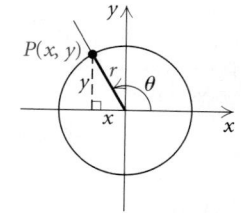

$$\sin \theta = \frac{y}{r}, \qquad \cos \theta = \frac{x}{r}, \qquad \tan \theta = \frac{y}{x},$$

$$\csc \theta = \frac{r}{y}, \qquad \sec \theta = \frac{r}{x}, \qquad \cot \theta = \frac{x}{y}.$$

Signs of Function Values

The signs of the function values depend only on the coordinates of the point P on the terminal side of an angle.

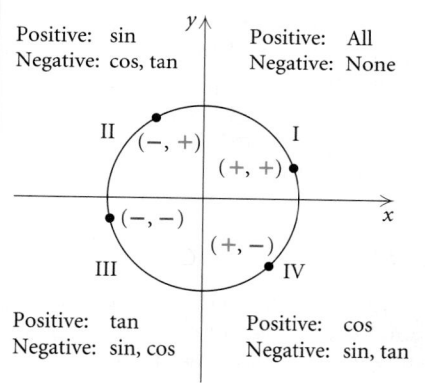

Positive: sin   Positive: All
Negative: cos, tan   Negative: None

Positive: tan   Positive: cos
Negative: sin, cos   Negative: sin, tan

Basic Circular Functions

For a real number s that determines a point (x, y) on the unit circle:

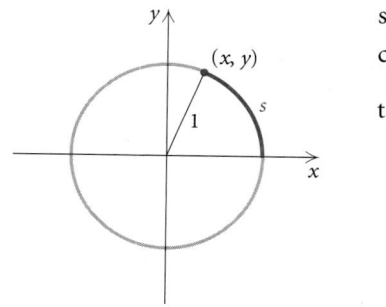

$$\sin s = y,$$
$$\cos s = x,$$
$$\tan s = \frac{y}{x}.$$

Sine is an odd function: $\sin(-s) = -\sin s$.

Cosine is an even function: $\cos(-s) = \cos s$.

Radian–Degree Equivalents

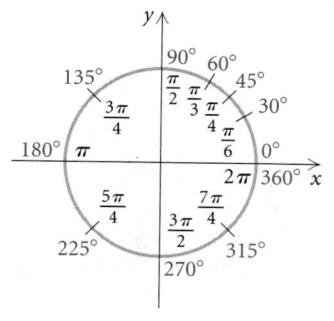

Transformations of Sine and Cosine Functions

To graph $y = A \sin(Bx - C) + D$ and $y = A \cos(Bx - C) + D$:

1. Stretch or shrink the graph horizontally according to B. $\left(\text{Period} = \left| \frac{2\pi}{B} \right| \right)$

2. Stretch or shrink the graph vertically according to A. (Amplitude $= |A|$)

3. Translate the graph horizontally according to C/B. $\left(\text{Phase shift} = \frac{C}{B} \right)$

4. Translate the graph vertically according to D.

Linear Speed in Terms of Angular Speed

$$v = r\omega$$

Review Exercises

Determine whether the statement is true or false.

1. Given that $(-a, b)$ is a point on the unit circle and θ is in the second quadrant, then $\cos \theta$ is a. [6.3]

2. The lengths of corresponding sides in similar triangles are in the same ratio. [6.1]

3. The measure $300°$ is greater than the measure 5 radians. [6.4]

4. If $\sec \theta > 0$ and $\cot \theta < 0$, then θ is in the fourth quadrant. [6.3]

5. The amplitude of $y = \frac{1}{2} \sin x$ is twice as large as the amplitude of $y = \sin \frac{1}{2} x$. [6.6]

6. The supplement of $\frac{9}{13} \pi$ is greater than the complement of $\frac{\pi}{6}$. [6.4]

7. Find the six trigonometric function values of θ. [6.1]

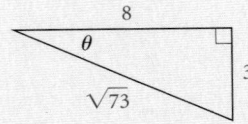

8. Given that β is acute and $\sin \beta = \frac{\sqrt{91}}{10}$, find the other five trigonometric function values. [6.1]

Find the exact function value, if it exists.

9. $\cos 45°$ [6.1]
10. $\cot 60°$ [6.1]
11. $\cos 495°$ [6.3]
12. $\sin 150°$ [6.3]
13. $\sec (-270°)$ [6.3]
14. $\tan (-600°)$ [6.3]
15. $\csc 60°$ [6.1]
16. $\cot (-45°)$ [6.3]

17. Convert $22.27°$ to degrees, minutes, and seconds. Round to the nearest second. [6.1]

18. Convert $47°33'27''$ to decimal degree notation. Round to two decimal places. [6.1]

Find the function value. Round to four decimal places. [6.3]

19. $\tan 2184°$
20. $\sec 27.9°$
21. $\cos 18°13'42''$
22. $\sin 245°24'$
23. $\cot (-33.2°)$
24. $\sin 556.13°$

Find θ in the interval indicated. Round the answer to the nearest tenth of a degree. [6.3]

25. $\cos \theta = -0.9041$, $(180°, 270°)$
26. $\tan \theta = 1.0799$, $(0°, 90°)$

Find the exact acute angle θ, in degrees, given the function value. [6.1]

27. $\sin \theta = \dfrac{\sqrt{3}}{2}$
28. $\tan \theta = \sqrt{3}$
29. $\cos \theta = \dfrac{\sqrt{2}}{2}$
30. $\sec \theta = \dfrac{2\sqrt{3}}{3}$

31. Given that $\sin 59.1° \approx 0.8581$, $\cos 59.1° \approx 0.5135$, and $\tan 59.1° \approx 1.6709$, find the six function values for $30.9°$. [6.1]

Solve each of the following right triangles. Standard lettering has been used. [6.2]

32. $a = 7.3$, $c = 8.6$
33. $a = 30.5$, $B = 51.17°$

34. One leg of a right triangle bears east. The hypotenuse is 734 m long and bears N57°23'E. Find the perimeter of the triangle.

35. An observer's eye is 6 ft above the floor. A mural is being viewed. The bottom of the mural is at floor level. The observer looks down 13° to see the bottom and up 17° to see the top. How tall is the mural?

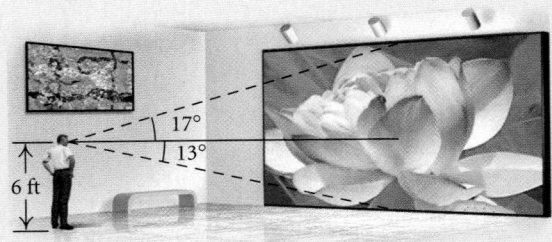

For angles of the following measures, state in which quadrant the terminal side lies. [6.3]

36. $142°11'5''$
37. $-635.2°$
38. $-392°$

Find a positive angle and a negative angle that are coterminal with the given angle. Answers may vary.

39. 65° [6.3]

40. $\dfrac{7\pi}{3}$ [6.4]

Find the complement and the supplement.

41. 13.4° [6.3]

42. $\dfrac{\pi}{6}$ [6.4]

43. Find the six trigonometric function values for the angle θ shown. [6.3]

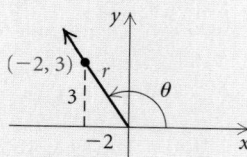

44. Given that $\tan \theta = 2/\sqrt{5}$ and that the terminal side is in quadrant III, find the other five function values. [6.3]

45. An airplane travels at 530 mph for $3\frac{1}{2}$ hr in a direction of 160° from Minneapolis, Minnesota. At the end of that time, how far south of Minneapolis is the airplane? [6.3]

46. On a unit circle, mark and label the points determined by $7\pi/6$, $-3\pi/4$, $-\pi/3$, and $9\pi/4$. [6.4]

For angles of the following measures, convert to radian measure in terms of π, and convert to radian measure not in terms of π. Round the answer to two decimal places. [6.4]

47. 145.2°

48. −30°

Convert to degree measure. Round the answer to two decimal places. [6.4]

49. $\dfrac{3\pi}{2}$

50. 3

51. −4.5

52. 11π

53. Find the length of an arc of a circle, given a central angle of $\pi/4$ and a radius of 7 cm. [6.4]

54. An arc 18 m long on a circle of radius 8 m subtends an angle of how many radians? how many degrees, to the nearest degree? [6.4]

55. At one time, inside La Madeleine French Bakery and Cafe in Houston, Texas, there was one of the few remaining working watermills in the world. The 300-year-old French-built waterwheel had a radius of 7 ft and made one complete revolution in 70 sec. (*Source*: La Madeleine French Bakery and Cafe, Houston, TX) What was the linear speed, in feet per minute, of a point on the rim? [6.4]

56. An automobile wheel has a diameter of 14 in. If the car travels at a speed of 55 mph, what is the angular velocity, in radians per hour, of a point on the edge of the wheel? [6.4]

57. The point $\left(\frac{3}{5}, -\frac{4}{5}\right)$ is on a unit circle. Find the coordinates of its reflections across the *x*-axis, the *y*-axis, and the origin. [6.5]

Find the exact function value, if it exists. [6.5]

58. $\cos \pi$

59. $\tan \dfrac{5\pi}{4}$

60. $\sin \dfrac{5\pi}{3}$

61. $\sin \left(-\dfrac{7\pi}{6}\right)$

62. $\tan \dfrac{\pi}{6}$

63. $\cos \left(-13\pi\right)$

Find the function value. Round to four decimal places. [6.5]

64. $\sin 24$

65. $\cos \left(-75\right)$

66. $\cot 16\pi$

67. $\tan \dfrac{3\pi}{7}$

68. $\sec 14.3$

69. $\cos \left(-\dfrac{\pi}{5}\right)$

70. Graph each of the six trigonometric functions from -2π to 2π. [6.5]

71. What is the period of each of the six trigonometric functions? [6.5]

72. Complete the following table. [6.5]

Function	Domain	Range
sine		
cosine		
tangent		

73. Complete the following table with the sign of the specified trigonometric function value in each of the four quadrants.   [6.3]

Function	I	II	III	IV
sine				
cosine				
tangent				

Determine the amplitude, the period, and the phase shift of the function, and sketch the graph of the function. Then check the graph using a graphing calculator.   [6.6]

74. $y = \sin\left(x + \dfrac{\pi}{2}\right)$

75. $y = 3 + \dfrac{1}{2}\cos\left(2x - \dfrac{\pi}{2}\right)$

In Exercises 76–79, without using a graphing calculator, match the function with one of the graphs (a)–(d), which follow. Then check your work using a graphing calculator.   [6.6]

a)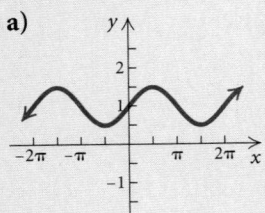

b)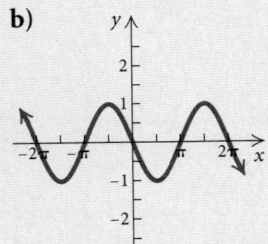

c)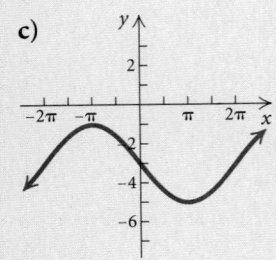

d)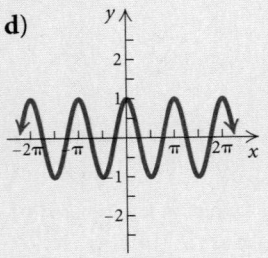

76. $y = \cos 2x$

77. $y = \dfrac{1}{2}\sin x + 1$

78. $y = -2\sin\dfrac{1}{2}x - 3$

79. $y = -\cos\left(x - \dfrac{\pi}{2}\right)$

80. Sketch a graph of $y = 3\cos x + \sin x$ for values of x between 0 and 2π.   [6.6]

81. Graph: $f(x) = e^{-0.7x}\cos x$.   [6.6]

82. Which of the following is the reflection of $\left(-\dfrac{1}{2}, \dfrac{\sqrt{3}}{2}\right)$ across the y-axis?   [6.5]

A. $\left(\dfrac{1}{2}, -\dfrac{\sqrt{3}}{2}\right)$ **B.** $\left(\dfrac{\sqrt{3}}{2}, \dfrac{1}{2}\right)$

C. $\left(\dfrac{1}{2}, \dfrac{\sqrt{3}}{2}\right)$ **D.** $\left(\dfrac{\sqrt{3}}{2}, -\dfrac{1}{2}\right)$

83. Which of the following is the domain of the cosine function?   [6.5]

A. $(-1, 1)$ **B.** $(-\infty, \infty)$
C. $[0, \infty)$ **D.** $[-1, 1]$

84. The graph of $f(x) = -\cos(-x)$ is which of the following?   [6.6]

A. 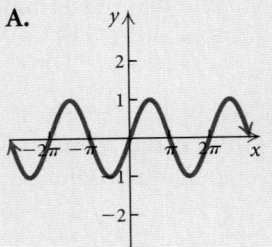   **B.**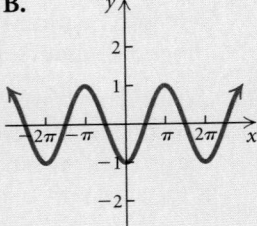

C. 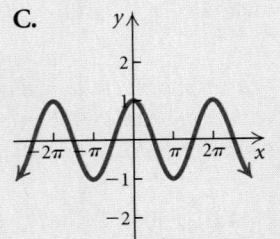   **D.**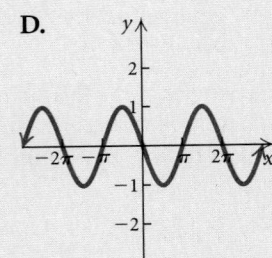

Collaborative Discussion and Writing

85. Describe the shape of the graph of the cosine function. How many maximum values are there of the cosine function? Where do they occur?   [6.5]

86. Compare the terms radian and degree.   [6.1], [6.4]

87. Does $5\sin x = 7$ have a solution for x? Why or why not?   [6.5]

88. Explain the disadvantage of a graphing calculator when graphing a function like

$$f(x) = \frac{\sin x}{x}. \quad [6.6]$$

Synthesis

89. Graph $y = 3 \sin (x/2)$, and determine the domain, the range, and the period.   [6.6]

90. In the graph below, $y_1 = \sin x$ is shown and y_2 is shown in red. Express y_2 as a transformation of the graph of y_1.   [6.6]

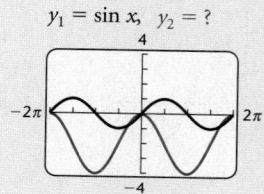

$y_1 = \sin x, \quad y_2 = ?$

91. Find the domain of $y = \log (\cos x)$.   [6.6]

92. Given that $\sin x = 0.6144$ and that the terminal side is in quadrant II, find the other basic circular function values.   [6.3]

CHAPTER 6   Test

1. Find the six trigonometric function values of θ.

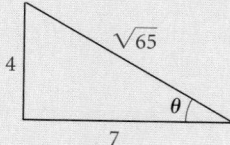

Find the exact function value, if it exists.

2. $\sin 120°$

3. $\tan (-45°)$

4. $\cos 3\pi$

5. $\sec \dfrac{5\pi}{4}$

6. Convert $38°27'56''$ to decimal degree notation. Round to two decimal places.

Find the function values. Round to four decimal places.

7. $\tan 526.4°$

8. $\sin (-12°)$

9. $\sec \dfrac{5\pi}{9}$

10. $\cos 76.07$

11. Find the exact acute angle θ, in degrees, for which $\sin \theta = \frac{1}{2}$.

12. Given that $\sin 28.4° \approx 0.4756$, $\cos 28.4° \approx 0.8796$, and $\tan 28.4° \approx 0.5407$, find the six trigonometric function values for $61.6°$.

13. Solve the right triangle with $b = 45.1$ and $A = 35.9°$. Standard lettering has been used.

14. Find a positive angle and a negative angle coterminal with a $112°$ angle.

15. Find the supplement of $\dfrac{5\pi}{6}$.

16. Given that $\sin \theta = -4/\sqrt{41}$ and that the terminal side is in quadrant IV, find the other five trigonometric function values.

17. Convert $210°$ to radian measure in terms of π.

18. Convert $\dfrac{3\pi}{4}$ to degree measure.

19. Find the length of an arc of a circle given a central angle of $\pi/3$ and a radius of 16 cm.

Consider the function $y = -\sin(x - \pi/2) + 1$ *for Exercises 20–23.*

20. Find the amplitude.

21. Find the period.

22. Find the phase shift.

23. Which is the graph of the function?

a)
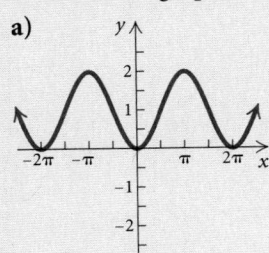

b)
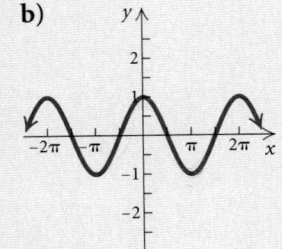

c)
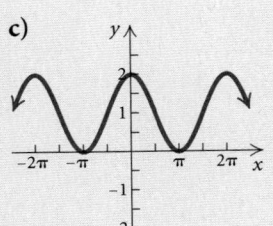

d)
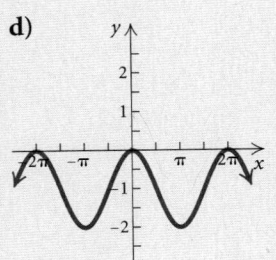

24. *Height of a Kite.*   The angle of elevation of a kite is 65° with 490 ft of string out. Assuming the string is taut, how high is the kite?

25. *Location.*   A pickup-truck camper travels at 50 mph for 6 hr in a direction of 115° from Buffalo, Wyoming. At the end of that time, how far east of Buffalo is the camper?

26. *Linear Speed.*   A ferris wheel has a radius of 6 m and revolves at 1.5 rpm. What is the linear speed, in meters per minute?

27. Graph: $f(x) = \frac{1}{2}x^2 \sin x$.

28. The graph of $f(x) = -\sin(-x)$ is which of the following?

A.
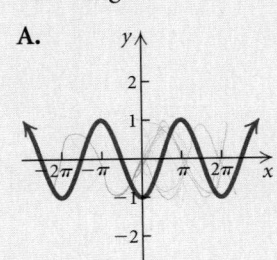

B.
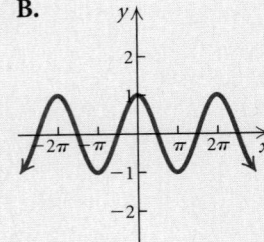

C.
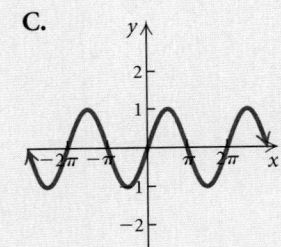

D.
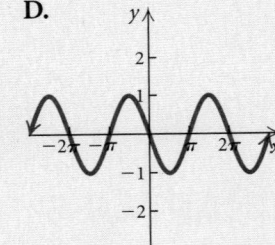

Synthesis

29. Determine the domain of $f(x) = \dfrac{-3}{\sqrt{\cos x}}$.

Trigonometric Identities, Inverse Functions, and Equations

APPLICATION For a rope course and climbing wall, a guy wire is attached 47 ft high on a vertical pole. Another guy wire is attached 40 ft above the ground on the same pole. (*Source*: Experiential Resources Inc., Todd Domeck, Owner) Find the angle between the wires if they are attached to the ground 50 ft from the pole.

This problem appears as Exercise 87 in Section 7.1.

7.1 Identities: Pythagorean and Sum and Difference

7.2 Identities: Cofunction, Double-Angle, and Half-Angle

7.3 Proving Trigonometric Identities

7.4 Inverses of the Trigonometric Functions

7.5 Solving Trigonometric Equations

7.1

Identities: Pythagorean and Sum and Difference

❖ State the Pythagorean identities.

❖ Simplify and manipulate expressions containing trigonometric expressions.

❖ Use the sum and difference identities to find function values.

An **identity** is an equation that is true for all *possible* replacements of the variables. The following is a list of the identities studied in Chapter 6.

Basic Identities

$$\sin x = \frac{1}{\csc x}, \qquad \csc x = \frac{1}{\sin x}, \qquad \sin(-x) = -\sin x,$$
$$\cos(-x) = \cos x,$$

$$\cos x = \frac{1}{\sec x}, \qquad \sec x = \frac{1}{\cos x}, \qquad \tan(-x) = -\tan x,$$

$$\tan x = \frac{1}{\cot x}, \qquad \cot x = \frac{1}{\tan x}, \qquad \tan x = \frac{\sin x}{\cos x},$$

$$\cot x = \frac{\cos x}{\sin x}$$

In this section, we will develop some other important identities.

❖ Pythagorean Identities

We now consider three other identities that are fundamental to a study of trigonometry. They are called the *Pythagorean identities*. Recall that the equation of a unit circle in the *xy*-plane is

$$x^2 + y^2 = 1.$$

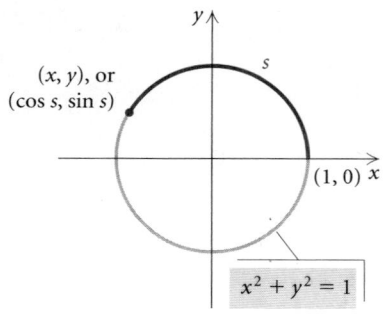

For any point on the unit circle, the coordinates *x* and *y* satisfy this equation. Suppose that a real number *s* determines a point on the unit circle with coordinates (x, y), or $(\cos s, \sin s)$. Then $x = \cos s$ and $y = \sin s$. Substituting $\cos s$ for *x* and $\sin s$ for *y* in the equation of the unit circle gives us the identity

$$(\cos s)^2 + (\sin s)^2 = 1, \qquad \begin{array}{l}\textbf{Substituting } \cos s \textbf{ for } x \\ \textbf{and } \sin s \textbf{ for } y\end{array}$$

which can be expressed as

$$\sin^2 s + \cos^2 s = 1.$$

It is conventional in trigonometry to use the notation $\sin^2 s$ rather than $(\sin s)^2$. Note that $\sin^2 s \neq \sin s^2$.

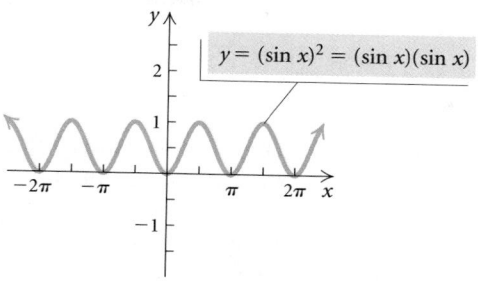

$y = (\sin x)^2 = (\sin x)(\sin x)$

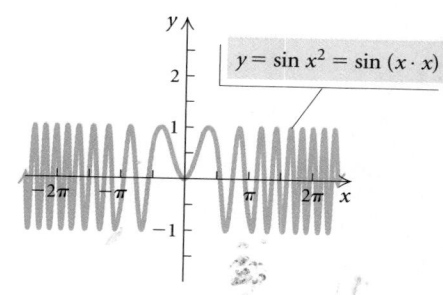

$y = \sin x^2 = \sin (x \cdot x)$

The identity $\sin^2 s + \cos^2 s = 1$ gives a relationship between the sine and the cosine of any real number s. It is an important **Pythagorean identity**.

Exploring with Technology

Addition of y-values provides a unique way of developing the identity $\sin^2 x + \cos^2 x = 1$. First, graph $y_1 = \sin^2 x$ and $y_2 = \cos^2 x$. By visually adding the y-values, sketch the graph of the sum, $y_3 = \sin^2 x + \cos^2 x$. Then graph y_3 using a graphing calculator and check your sketch. The resulting graph appears to be the line $y_4 = 1$, which is the graph of $\sin^2 x + \cos^2 x$. These graphs *do not* prove the identity, but they do provide a check in the interval shown.

$y_1 = \sin^2 x$

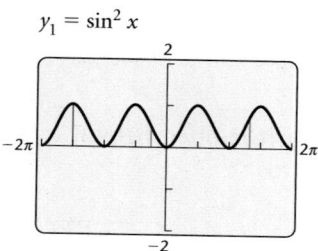

$y_2 = \cos^2 x$

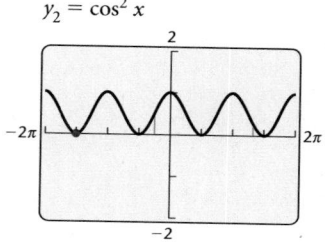

$y_3 = \sin^2 x + \cos^2 x, \quad y_4 = 1$

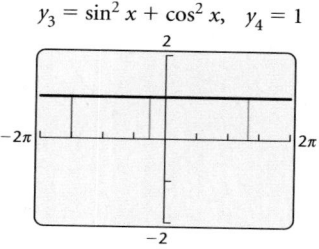

We can divide by $\sin^2 s$ on both sides of the preceding identity:

$$\frac{\sin^2 s}{\sin^2 s} + \frac{\cos^2 s}{\sin^2 s} = \frac{1}{\sin^2 s}. \qquad \text{Dividing by } \sin^2 s$$

Simplifying gives us a second Pythagorean identity:

$$1 + \cot^2 s = \csc^2 s.$$

This equation is true for any replacement of s with a real number for which $\sin^2 s \neq 0$, since we divided by $\sin^2 s$. But the numbers for which $\sin^2 s = 0$ (or $\sin s = 0$) are exactly the ones for which the cotangent function and the cosecant function are not defined. Hence our new equation holds for all real numbers s for which $\cot s$ and $\csc s$ are defined and is thus an identity.

The third Pythagorean identity can be obtained by dividing by $\cos^2 s$ on both sides of the first Pythagorean identity:

$$\frac{\sin^2 s}{\cos^2 s} + \frac{\cos^2 s}{\cos^2 s} = \frac{1}{\cos^2 s} \qquad \text{Dividing by } \cos^2 s$$

$$\tan^2 s + 1 = \sec^2 s. \qquad \text{Simplifying}$$

The identities we have developed hold no matter what symbols are used for the variables. For example, we could write $\sin^2 s + \cos^2 s = 1$, $\sin^2 \theta + \cos^2 \theta = 1$, or $\sin^2 x + \cos^2 x = 1$.

Pythagorean Identities

$\sin^2 x + \cos^2 x = 1$,

$1 + \cot^2 x = \csc^2 x$,

$1 + \tan^2 x = \sec^2 x$

It is often helpful to express the Pythagorean identities in equivalent forms.

Pythagorean Identities	Equivalent Forms
$\sin^2 x + \cos^2 x = 1$	$\sin^2 x = 1 - \cos^2 x$ $\cos^2 x = 1 - \sin^2 x$
$1 + \cot^2 x = \csc^2 x$	$1 = \csc^2 x - \cot^2 x$ $\cot^2 x = \csc^2 x - 1$
$1 + \tan^2 x = \sec^2 x$	$1 = \sec^2 x - \tan^2 x$ $\tan^2 x = \sec^2 x - 1$

❖ Simplifying Trigonometric Expressions

We can factor, simplify, and manipulate trigonometric expressions in the same way that we manipulate strictly algebraic expressions.

STUDY TIP

EXAMPLE 1 Multiply and simplify: $\cos x\,(\tan x - \sec x)$.

Solution

$\cos x\,(\tan x - \sec x)$

$= \cos x \tan x - \cos x \sec x$ **Multiplying**

$= \cos x\, \dfrac{\sin x}{\cos x} - \cos x\, \dfrac{1}{\cos x}$ **Recalling the identities** $\tan x = \dfrac{\sin x}{\cos x}$

                and $\sec x = \dfrac{1}{\cos x}$ **and substituting**

$= \sin x - 1$ **Simplifying**   **Now Try Exercise 3.** ▨

There is no general procedure for simplifying trigonometric expressions, but it is often helpful to write everything in terms of sines and cosines, as we did in Example 1. We also look for a Pythagorean identity within a trigonometric expression.

EXAMPLE 2 Factor and simplify: $\sin^2 x \cos^2 x + \cos^4 x$.

Solution

$\sin^2 x \cos^2 x + \cos^4 x$

$= \cos^2 x\,(\sin^2 x + \cos^2 x)$ **Removing a common factor**

$= \cos^2 x \cdot (1)$ **Using** $\sin^2 x + \cos^2 x = 1$

$= \cos^2 x$ **Now Try Exercise 5.** ▨

GCM   A graphing calculator can be used to perform a partial check of an identity. First, we graph the expression on the left side of the equals sign. Then we graph the expression on the right side using the same screen. If the two graphs are indistinguishable, then we have a partial verification that the equation is an identity. Of course, we can never see the entire graph, so there can always be some doubt. Also, the graphs may not overlap precisely, but you may not be able to tell because the difference between the graphs may be less than the width of a pixel. However, if the graphs are obviously different, we know that a mistake has been made.

Consider the identity in Example 1:

$$\cos x\,(\tan x - \sec x) = \sin x - 1.$$

Recalling that $\sec x = 1/\cos x$, we enter

$$y_1 = \cos x\,(\tan x - 1/\cos x) \quad \text{and} \quad y_2 = \sin x - 1.$$

To graph, we first select SEQUENTIAL mode. Then we select the "line"-graph style for y_1 and the "path"-graph style, denoted by $-\bigcirc$, for y_2. The calculator will graph y_1 first. Then it will graph y_2 as the circular cursor traces the leading edge of the graph, allowing us to determine whether the graphs coincide. As you can see in the graph screen at left, the graphs appear to be identical. Thus, $\cos x\,(\tan x - \sec x) = \sin x - 1$ is most likely an identity.

The TABLE feature can also be used to check identities. Note in the table at left that the function values are the same except for those values of x for which $\cos x = 0$. The domain of y_1 excludes these values. The domain of y_2

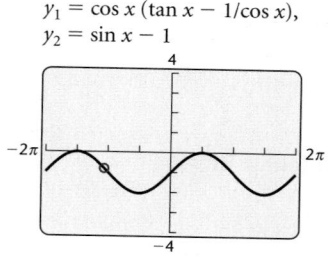

$y_1 = \cos x\,(\tan x - 1/\cos x),$
$y_2 = \sin x - 1$

X	Y1	Y2
−6.283	−1	−1
−5.498	−.2929	−.2929
−4.712	ERROR	0
−3.927	−.2929	−.2929
−3.142	−1	−1
−2.356	−1.707	−1.707
−1.571	ERROR	−2

X = −6.28318530718

TblStart = −2π
ΔTbl = π/4

is the set of all real numbers. Thus all real numbers except $\pm\pi/2$, $\pm3\pi/2$, $\pm5\pi/2,\ldots$ are possible replacements for x in the identity. Recall that an identity is an equation that is true for all *possible* replacements.

Suppose that we had simplified incorrectly in Example 1 and had gotten $\cos x - 1$ on the right instead of $\sin x - 1$. Then two different graphs would have appeared in the window. Thus we would have known that we did not have an identity and that $\cos x (\tan x - \sec x) \neq \cos x - 1$.

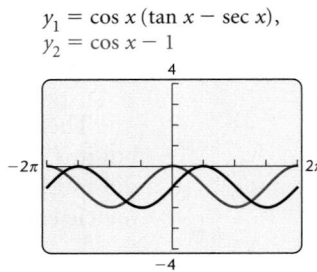

$$y_1 = \cos x (\tan x - \sec x),$$
$$y_2 = \cos x - 1$$

EXAMPLE 3 Simplify each of the following trigonometric expressions.

a) $\dfrac{\cot(-\theta)}{\csc(-\theta)}$

b) $\dfrac{2\sin^2 t + \sin t - 3}{1 - \cos^2 t - \sin t}$

Solution

a) $\dfrac{\cot(-\theta)}{\csc(-\theta)} = \dfrac{\dfrac{\cos(-\theta)}{\sin(-\theta)}}{\dfrac{1}{\sin(-\theta)}}$ Rewriting in terms of sines and cosines

$= \dfrac{\cos(-\theta)}{\sin(-\theta)} \cdot \sin(-\theta)$ Multiplying by the reciprocal, $\sin(-\theta)/1$

$= \cos(-\theta)$ Removing a factor of 1, $\sin(-\theta)/\sin(-\theta)$

$= \cos\theta$ The cosine function is even.

b) $\dfrac{2\sin^2 t + \sin t - 3}{1 - \cos^2 t - \sin t}$

$= \dfrac{2\sin^2 t + \sin t - 3}{\sin^2 t - \sin t}$ Substituting $\sin^2 t$ for $1 - \cos^2 t$

$= \dfrac{(2\sin t + 3)(\sin t - 1)}{\sin t(\sin t - 1)}$ Factoring in both the numerator and the denominator

$= \dfrac{2\sin t + 3}{\sin t}$ Simplifying

$= \dfrac{2\sin t}{\sin t} + \dfrac{3}{\sin t}$

$= 2 + \dfrac{3}{\sin t}$,   or $2 + 3\csc t$

> FACTORING
> REVIEW SECTION **R.4.**

Now Try Exercises 17 and 19. ■

We can add and subtract trigonometric rational expressions in the same way that we do algebraic expressions, writing expressions with a common denominator before adding and subtracting numerators.

EXAMPLE 4 Add and simplify: $\dfrac{\cos x}{1 + \sin x} + \tan x$.

Solution

$$\frac{\cos x}{1 + \sin x} + \tan x = \frac{\cos x}{1 + \sin x} + \frac{\sin x}{\cos x} \qquad \text{Using } \tan x = \frac{\sin x}{\cos x}$$

$$= \frac{\cos x}{1 + \sin x} \cdot \frac{\cos x}{\cos x} + \frac{\sin x}{\cos x} \cdot \frac{1 + \sin x}{1 + \sin x}$$

Multiplying by forms of 1

$$= \frac{\cos^2 x + \sin x + \sin^2 x}{\cos x \,(1 + \sin x)} \qquad \text{Adding}$$

$$= \frac{1 + \sin x}{\cos x \,(1 + \sin x)} \qquad \text{Using } \sin^2 x + \cos^2 x = 1$$

$$= \frac{1}{\cos x}, \quad \text{or} \quad \sec x \qquad \text{Simplifying}$$

Now Try Exercise 27. ■

When radicals occur, the use of absolute value is sometimes necessary, but it can be difficult to determine when to use it. In Examples 5 and 6, we will assume that all radicands are nonnegative. This means that the identities are meant to be confined to certain quadrants.

EXAMPLE 5 Multiply and simplify: $\sqrt{\sin^3 x \cos x} \cdot \sqrt{\cos x}$.

Solution

$$\sqrt{\sin^3 x \cos x} \cdot \sqrt{\cos x} = \sqrt{\sin^3 x \cos^2 x}$$

$$= \sqrt{\sin^2 x \cos^2 x \sin x}$$

$$= \sin x \cos x \sqrt{\sin x}$$

Now Try Exercise 31. ■

EXAMPLE 6 Rationalize the denominator: $\sqrt{\dfrac{2}{\tan x}}$.

Solution

$$\sqrt{\frac{2}{\tan x}} = \sqrt{\frac{2}{\tan x} \cdot \frac{\tan x}{\tan x}}$$

$$= \sqrt{\frac{2 \tan x}{\tan^2 x}}$$

$$= \frac{\sqrt{2 \tan x}}{\tan x}$$

Now Try Exercise 37. ■

RATIONAL EXPRESSIONS

REVIEW SECTION **R.6.**

Often in calculus, a substitution is a useful manipulation, as we show in the following example.

EXAMPLE 7   Express $\sqrt{9 + x^2}$ as a trigonometric function of θ without using radicals by letting $x = 3\tan\theta$. Assume that $0 < \theta < \pi/2$. Then find $\sin\theta$ and $\cos\theta$.

Solution   We have

$$
\begin{aligned}
\sqrt{9 + x^2} &= \sqrt{9 + (3\tan\theta)^2} && \text{Substituting } 3\tan\theta \text{ for } x \\
&= \sqrt{9 + 9\tan^2\theta} \\
&= \sqrt{9(1 + \tan^2\theta)} && \text{Factoring} \\
&= \sqrt{9\sec^2\theta} && \text{Using } 1 + \tan^2 x = \sec^2 x \\
&= 3|\sec\theta| = 3\sec\theta. && \text{For } 0 < \theta < \pi/2, \sec\theta > 0, \\
& && \text{so } |\sec\theta| = \sec\theta.
\end{aligned}
$$

We can express $\sqrt{9 + x^2} = 3\sec\theta$ as

$$\sec\theta = \frac{\sqrt{9 + x^2}}{3}.$$

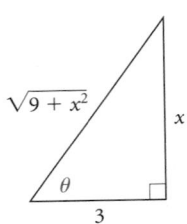

In a right triangle, we know that $\sec\theta$ is hypotenuse/adjacent, when θ is one of the acute angles. Using the Pythagorean theorem, we can determine that the side opposite θ is x. Then from the right triangle, we see that

$$\sin\theta = \frac{x}{\sqrt{9 + x^2}} \quad \text{and} \quad \cos\theta = \frac{3}{\sqrt{9 + x^2}}.$$

Now Try Exercise 45. ■

�֎ Sum and Difference Identities

We now develop some important identities involving sums or differences of two numbers (or angles), beginning with an identity for the cosine of the difference of two numbers. We use the letters u and v for these numbers.

Let's consider a real number u in the interval $[\pi/2, \pi]$ and a real number v in the interval $[0, \pi/2]$. These determine points A and B on the unit circle, as shown below. The arc length s is $u - v$, and we know that $0 \leq s \leq \pi$. Recall that the coordinates of A are $(\cos u, \sin u)$, and the coordinates of B are $(\cos v, \sin v)$.

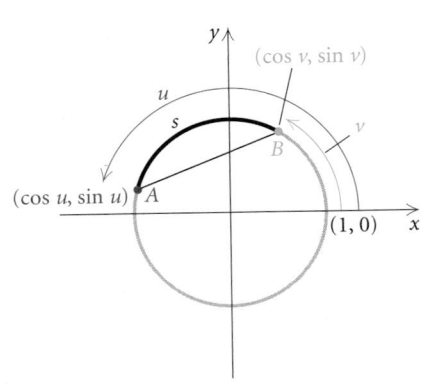

DISTANCE FORMULA

REVIEW SECTION **1.1.**

Using the distance formula, we can write an expression for the distance AB:

$$AB = \sqrt{(\cos u - \cos v)^2 + (\sin u - \sin v)^2}.$$

This can be simplified as follows:

$$
\begin{aligned}
AB &= \sqrt{\cos^2 u - 2\cos u \cos v + \cos^2 v + \sin^2 u - 2\sin u \sin v + \sin^2 v} \\
&= \sqrt{(\sin^2 u + \cos^2 u) + (\sin^2 v + \cos^2 v) - 2(\cos u \cos v + \sin u \sin v)} \\
&= \sqrt{2 - 2(\cos u \cos v + \sin u \sin v)}.
\end{aligned}
$$

Now let's imagine rotating the circle on p. 580 so that point B is at $(1, 0)$ as shown at left. Although the coordinates of point A are now $(\cos s, \sin s)$, the distance AB has not changed.

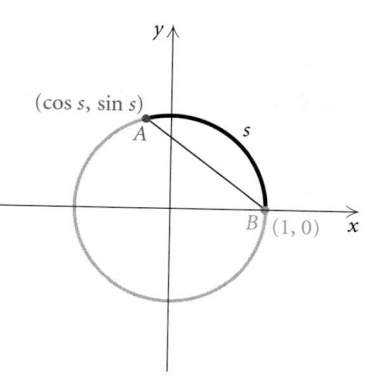

Again we use the distance formula to write an expression for the distance AB:

$$AB = \sqrt{(\cos s - 1)^2 + (\sin s - 0)^2}.$$

This can be simplified as follows:

$$
\begin{aligned}
AB &= \sqrt{\cos^2 s - 2\cos s + 1 + \sin^2 s} \\
&= \sqrt{(\sin^2 s + \cos^2 s) + 1 - 2\cos s} \\
&= \sqrt{2 - 2\cos s}.
\end{aligned}
$$

Equating our two expressions for AB, we obtain

$$\sqrt{2 - 2(\cos u \cos v + \sin u \sin v)} = \sqrt{2 - 2\cos s}.$$

Solving this equation for $\cos s$ gives

$$\cos s = \cos u \cos v + \sin u \sin v. \tag{1}$$

But $s = u - v$, so we have the equation

$$\cos(u - v) = \cos u \cos v + \sin u \sin v. \tag{2}$$

Formula (1) above holds when s is the length of the shortest arc from A to B. Given any real numbers u and v, the length of the shortest arc from A to B is not always $u - v$. In fact, it could be $v - u$. However, since $\cos(-x) = \cos x$, we know that $\cos(v - u) = \cos(u - v)$. Thus, $\cos s$ is always equal to $\cos(u - v)$. Formula (2) holds for all real numbers u and v. That formula is thus the identity we sought:

$$\cos(u - v) = \cos u \cos v + \sin u \sin v.$$

To illustrate this result using a graphing calculator, we replace u with x and v with 3 and graph

$$y_1 = \cos(x - 3)$$

and $y_2 = \cos x \cos 3 + \sin x \sin 3.$

$y_1 = \cos(x - 3),$
$y_2 = \cos x \cos 3 + \sin x \sin 3$

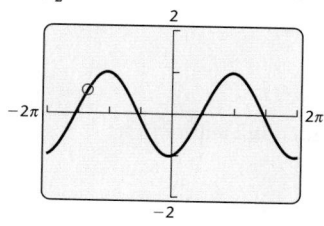

The cosine sum formula follows easily from the one we have just derived. Let's consider $\cos(u + v)$. This is equal to $\cos[u - (-v)]$, and by the identity above, we have

$$
\begin{aligned}
\cos(u + v) &= \cos[u - (-v)] \\
&= \cos u \cos(-v) + \sin u \sin(-v).
\end{aligned}
$$

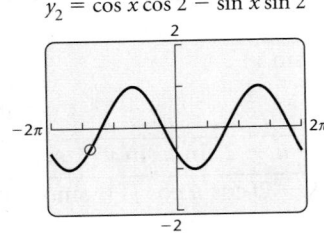

$y_1 = \cos(x + 2),$
$y_2 = \cos x \cos 2 - \sin x \sin 2$

But $\cos(-v) = \cos v$ and $\sin(-v) = -\sin v$, so the identity we seek is the following:

$$\cos(u + v) = \cos u \cos v - \sin u \sin v.$$

To illustrate this result using a graphing calculator, we replace u with x and v with 2 and graph

$$y_1 = \cos(x + 2)$$

and $$y_2 = \cos x \cos 2 - \sin x \sin 2.$$

EXAMPLE 8   Find $\cos(5\pi/12)$ exactly.

Solution   We can express $5\pi/12$ as a difference of two numbers whose exact sine and cosine values are known:

$$\frac{5\pi}{12} = \frac{9\pi}{12} - \frac{4\pi}{12}, \quad \text{or} \quad \frac{3\pi}{4} - \frac{\pi}{3}.$$

Then, using $\cos(u - v) = \cos u \cos v + \sin u \sin v$, we have

$$\cos\frac{5\pi}{12} = \cos\left(\frac{3\pi}{4} - \frac{\pi}{3}\right) = \cos\frac{3\pi}{4}\cos\frac{\pi}{3} + \sin\frac{3\pi}{4}\sin\frac{\pi}{3}$$

$$= -\frac{\sqrt{2}}{2} \cdot \frac{1}{2} + \frac{\sqrt{2}}{2} \cdot \frac{\sqrt{3}}{2}$$

$$= -\frac{\sqrt{2}}{4} + \frac{\sqrt{6}}{4}$$

$$= \frac{\sqrt{6} - \sqrt{2}}{4}.$$

cos(5π/12)
                    .2588190451
(√(6)−√(2))/4
                    .2588190451

We can check using a graphing calculator set in RADIAN mode.

Now Try Exercise 55. ■

Consider $\cos(\pi/2 - \theta)$. We can use the identity for the cosine of a difference to simplify as follows:

$$\cos\left(\frac{\pi}{2} - \theta\right) = \cos\frac{\pi}{2}\cos\theta + \sin\frac{\pi}{2}\sin\theta$$

$$= 0 \cdot \cos\theta + 1 \cdot \sin\theta$$

$$= \sin\theta.$$

Thus we have developed the identity

$$\sin\theta = \cos\left(\frac{\pi}{2} - \theta\right). \qquad \begin{array}{l}\text{This cofunction identity first}\\ \text{appeared in Section 6.1.}\end{array} \qquad (3)$$

This identity holds for any real number θ. From it we can obtain an identity for the cosine function. We first let α be any real number. Then we replace θ in $\sin\theta = \cos(\pi/2 - \theta)$ with $\pi/2 - \alpha$. This gives us

$$\sin\left(\frac{\pi}{2} - \alpha\right) = \cos\left[\frac{\pi}{2} - \left(\frac{\pi}{2} - \alpha\right)\right] = \cos\alpha,$$

which yields the identity

$$\cos \alpha = \sin \left(\frac{\pi}{2} - \alpha \right). \tag{4}$$

Using identities (3) and (4) and the identity for the cosine of a difference, we can obtain an identity for the sine of a sum. We start with identity (3) and substitute $u + v$ for θ:

$$\sin \theta = \cos \left(\frac{\pi}{2} - \theta \right) \qquad \text{Identity (3)}$$

$$\sin (u + v) = \cos \left[\frac{\pi}{2} - (u + v) \right] \qquad \text{Substituting } u + v \text{ for } \theta$$

$$= \cos \left[\left(\frac{\pi}{2} - u \right) - v \right]$$

$$= \cos \left(\frac{\pi}{2} - u \right) \cos v + \sin \left(\frac{\pi}{2} - u \right) \sin v$$

Using the identity for the cosine
of a difference

$$= \sin u \cos v + \cos u \sin v. \qquad \text{Using identities (3) and (4)}$$

Thus the identity we seek is

$$\sin (u + v) = \sin u \cos v + \cos u \sin v.$$

To find a formula for the sine of a difference, we can use the identity just derived, substituting $-v$ for v:

$$\sin (u + (-v)) = \sin u \cos (-v) + \cos u \sin (-v).$$

Simplifying gives us

$$\sin (u - v) = \sin u \cos v - \cos u \sin v.$$

EXAMPLE 9   Find sin 105° exactly.

Solution   We express 105° as the sum of two measures:

$$105° = 45° + 60°.$$

Then

$$\sin 105° = \sin (45° + 60°)$$

$$= \sin 45° \cos 60° + \cos 45° \sin 60°$$

Using $\sin (u + v) = \sin u \cos v + \cos u \sin v$

$$= \frac{\sqrt{2}}{2} \cdot \frac{1}{2} + \frac{\sqrt{2}}{2} \cdot \frac{\sqrt{3}}{2}$$

$$= \frac{\sqrt{2} + \sqrt{6}}{4}.$$

```
sin(105)
              .9659258263
(√(2)+√(6))/4
              .9659258263
```

We can check this result using a graphing calculator set in DEGREE mode.

Now Try Exercise 51. ■

Formulas for the tangent of a sum or a difference can be derived using identities already established. A summary of the sum and difference identities follows.

Sum and Difference Identities

$$\sin(u \pm v) = \sin u \cos v \pm \cos u \sin v,$$

$$\cos(u \pm v) = \cos u \cos v \mp \sin u \sin v,$$

$$\tan(u \pm v) = \frac{\tan u \pm \tan v}{1 \mp \tan u \tan v}$$

There are six identities here, half of them obtained by using the signs shown in color.

EXAMPLE 10   Find $\tan 15°$ exactly.

Solution   We rewrite $15°$ as $45° - 30°$ and use the identity for the tangent of a difference:

$$\tan 15° = \tan(45° - 30°) = \frac{\tan 45° - \tan 30°}{1 + \tan 45° \tan 30°}$$

$$= \frac{1 - \sqrt{3}/3}{1 + 1 \cdot \sqrt{3}/3} = \frac{3 - \sqrt{3}}{3 + \sqrt{3}}.$$

Now Try Exercise 53. ▉

EXAMPLE 11   Assume that $\sin \alpha = \frac{2}{3}$ and $\sin \beta = \frac{1}{3}$ and that α and β are between 0 and $\pi/2$. Then evaluate $\sin(\alpha + \beta)$.

Solution   Using the identity for the sine of a sum, we have

$$\sin(\alpha + \beta) = \sin \alpha \cos \beta + \cos \alpha \sin \beta$$

$$= \tfrac{2}{3} \cos \beta + \tfrac{1}{3} \cos \alpha.$$

To finish, we need to know the values of $\cos \beta$ and $\cos \alpha$. Using reference triangles and the Pythagorean theorem, we can determine these values from the diagrams:

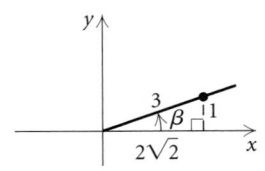

$$\cos \alpha = \frac{\sqrt{5}}{3} \quad \text{and} \quad \cos \beta = \frac{2\sqrt{2}}{3}.$$

Cosine values are positive in the first quadrant.

Substituting these values gives us

$$\sin(\alpha + \beta) = \frac{2}{3} \cdot \frac{2\sqrt{2}}{3} + \frac{1}{3} \cdot \frac{\sqrt{5}}{3}$$

$$= \frac{4}{9}\sqrt{2} + \frac{1}{9}\sqrt{5}, \quad \text{or} \quad \frac{4\sqrt{2} + \sqrt{5}}{9}.$$

Now Try Exercise 65. ▉

Exercise Set

Multiply and simplify. Check your result using a graphing calculator.

1. $(\sin x - \cos x)(\sin x + \cos x)$

2. $\tan x (\cos x - \csc x)$

3. $\cos y \sin y (\sec y + \csc y)$

4. $(\sin x + \cos x)(\sec x + \csc x)$

5. $(\sin \phi - \cos \phi)^2$

6. $(1 + \tan x)^2$

7. $(\sin x + \csc x)(\sin^2 x + \csc^2 x - 1)$

8. $(1 - \sin t)(1 + \sin t)$

Factor and simplify. Check your result using a graphing calculator.

9. $\sin x \cos x + \cos^2 x$

10. $\tan^2 \theta - \cot^2 \theta$

11. $\sin^4 x - \cos^4 x$

12. $4 \sin^2 y + 8 \sin y + 4$

13. $2 \cos^2 x + \cos x - 3$

14. $3 \cot^2 \beta + 6 \cot \beta + 3$

15. $\sin^3 x + 27$

16. $1 - 125 \tan^3 s$

Simplify and check using a graphing calculator.

17. $\dfrac{\sin^2 x \cos x}{\cos^2 x \sin x}$

18. $\dfrac{30 \sin^3 x \cos x}{6 \cos^2 x \sin x}$

19. $\dfrac{\sin^2 x + 2 \sin x + 1}{\sin x + 1}$

20. $\dfrac{\cos^2 \alpha - 1}{\cos \alpha + 1}$

21. $\dfrac{4 \tan t \sec t + 2 \sec t}{6 \tan t \sec t + 2 \sec t}$

22. $\dfrac{\csc (-x)}{\cot (-x)}$

23. $\dfrac{\sin^4 x - \cos^4 x}{\sin^2 x - \cos^2 x}$

24. $\dfrac{4 \cos^3 x}{\sin^2 x} \cdot \left(\dfrac{\sin x}{4 \cos x}\right)^2$

25. $\dfrac{5 \cos \phi}{\sin^2 \phi} \cdot \dfrac{\sin^2 \phi - \sin \phi \cos \phi}{\sin^2 \phi - \cos^2 \phi}$

26. $\dfrac{\tan^2 y}{\sec y} \div \dfrac{3 \tan^3 y}{\sec y}$

27. $\dfrac{1}{\sin^2 s - \cos^2 s} - \dfrac{2}{\cos s - \sin s}$

28. $\left(\dfrac{\sin x}{\cos x}\right)^2 - \dfrac{1}{\cos^2 x}$

29. $\dfrac{\sin^2 \theta - 9}{2 \cos \theta + 1} \cdot \dfrac{10 \cos \theta + 5}{3 \sin \theta + 9}$

30. $\dfrac{9 \cos^2 \alpha - 25}{2 \cos \alpha - 2} \cdot \dfrac{\cos^2 \alpha - 1}{6 \cos \alpha - 10}$

Simplify and check using a graphing calculator. Assume that all radicands are nonnegative.

31. $\sqrt{\sin^2 x \cos x} \cdot \sqrt{\cos x}$

32. $\sqrt{\cos^2 x \sin x} \cdot \sqrt{\sin x}$

33. $\sqrt{\cos \alpha \sin^2 \alpha} - \sqrt{\cos^3 \alpha}$

34. $\sqrt{\tan^2 x - 2 \tan x \sin x + \sin^2 x}$

35. $\left(1 - \sqrt{\sin y}\right)\left(\sqrt{\sin y} + 1\right)$

36. $\sqrt{\cos \theta}\left(\sqrt{2 \cos \theta} + \sqrt{\sin \theta \cos \theta}\right)$

Rationalize the denominator.

37. $\sqrt{\dfrac{\sin x}{\cos x}}$

38. $\sqrt{\dfrac{\cos x}{\tan x}}$

39. $\sqrt{\dfrac{\cos^2 y}{2 \sin^2 y}}$

40. $\sqrt{\dfrac{1 - \cos \beta}{1 + \cos \beta}}$

Rationalize the numerator.

41. $\sqrt{\dfrac{\cos x}{\sin x}}$

42. $\sqrt{\dfrac{\sin x}{\cot x}}$

43. $\sqrt{\dfrac{1 + \sin y}{1 - \sin y}}$

44. $\sqrt{\dfrac{\cos^2 x}{2 \sin^2 x}}$

Use the given substitution to express the given radical expression as a trigonometric function without radicals. Assume that $a > 0$ and $0 < \theta < \pi/2$. Then find expressions for the indicated trigonometric functions.

45. Let $x = a \sin \theta$ in $\sqrt{a^2 - x^2}$. Then find $\cos \theta$ and $\tan \theta$.

46. Let $x = 2 \tan \theta$ in $\sqrt{4 + x^2}$. Then find $\sin \theta$ and $\cos \theta$.

47. Let $x = 3 \sec \theta$ in $\sqrt{x^2 - 9}$. Then find $\sin \theta$ and $\cos \theta$.

48. Let $x = a \sec \theta$ in $\sqrt{x^2 - a^2}$. Then find $\sin \theta$ and $\cos \theta$.

Use the given substitution to express the given radical expression as a trigonometric function without radicals. Assume that $0 < \theta < \pi/2$.

49. Let $x = \sin \theta$ in $\dfrac{x^2}{\sqrt{1 - x^2}}$.

50. Let $x = 4 \sec \theta$ in $\dfrac{\sqrt{x^2 - 16}}{x^2}$.

Use the sum and difference identities to evaluate exactly. Then check using a graphing calculator.

51. $\sin \dfrac{\pi}{12}$

52. $\cos 75°$

53. $\tan 105°$

54. $\tan \dfrac{5\pi}{12}$

55. $\cos 15°$

56. $\sin \dfrac{7\pi}{12}$

First write each of the following as a trigonometric function of a single angle. Then evaluate.

57. $\sin 37° \cos 22° + \cos 37° \sin 22°$

58. $\cos 83° \cos 53° + \sin 83° \sin 53°$

59. $\cos 19° \cos 5° - \sin 19° \sin 5°$

60. $\sin 40° \cos 15° - \cos 40° \sin 15°$

61. $\dfrac{\tan 20° + \tan 32°}{1 - \tan 20° \tan 32°}$ DEGREE

62. $\dfrac{\tan 35° - \tan 12°}{1 + \tan 35° \tan 12°}$

63. Derive the formula for the tangent of a sum.

64. Derive the formula for the tangent of a difference.

Assuming that $\sin u = \frac{3}{5}$ and $\sin v = \frac{4}{5}$ and that u and v are between 0 and $\pi/2$, evaluate each of the following exactly.

65. $\cos (u + v)$

66. $\tan (u - v)$

67. $\sin (u - v)$

68. $\cos (u - v)$

Assuming that $\sin \theta = 0.6249$ and $\cos \phi = 0.1102$ and that both θ and ϕ are first-quadrant angles, evaluate each of the following.

69. $\tan (\theta + \phi)$

70. $\sin (\theta - \phi)$

71. $\cos (\theta - \phi)$

72. $\cos (\theta + \phi)$

Simplify.

73. $\sin (\alpha + \beta) + \sin (\alpha - \beta)$

74. $\cos (\alpha + \beta) - \cos (\alpha - \beta)$

75. $\cos (u + v) \cos v + \sin (u + v) \sin v$

76. $\sin (u - v) \cos v + \cos (u - v) \sin v$

Collaborative Discussion and Writing

77. What is the difference between a trigonometric equation that is an identity and a trigonometric equation that is not an identity? Give an example of each.

78. Explain why $\tan (x + 450°)$ cannot be simplified using the tangent sum formula, but can be simplified using the sine and cosine sum formulas.

Skill Maintenance

Solve.

79. $2x - 3 = 2\left(x - \frac{3}{2}\right)$

80. $x - 7 = x + 3.4$

Given that $\sin 31° = 0.5150$ and $\cos 31° = 0.8572$, find the specified function value.

81. $\sec 59°$

82. $\tan 59°$

Synthesis

Angles Between Lines. *One of the identities gives an easy way to find an angle formed by two lines. Consider two lines with equations $l_1: y = m_1x + b_1$ and $l_2: y = m_2x + b_2$.*

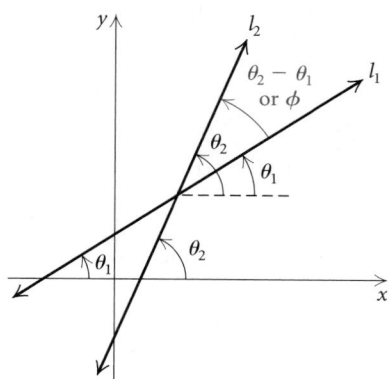

The slopes m_1 and m_2 are the tangents of the angles θ_1 and θ_2 that the lines form with the positive direction of the x-axis. Thus we have $m_1 = \tan \theta_1$ and $m_2 = \tan \theta_2$. To find the measure of $\theta_2 - \theta_1$, or ϕ, we proceed as follows:

$$\tan \phi = \tan(\theta_2 - \theta_1)$$
$$= \frac{\tan \theta_2 - \tan \theta_1}{1 + \tan \theta_2 \tan \theta_1}$$
$$= \frac{m_2 - m_1}{1 + m_2 m_1}.$$

This formula also holds when the lines are taken in the reverse order. When ϕ is acute, $\tan \phi$ will be positive. When ϕ is obtuse, $\tan \phi$ will be negative.

Find the measure of the angle from l_1 to l_2.

83. l_1: $2x = 3 - 2y$,
  l_2: $x + y = 5$

84. l_1: $3y = \sqrt{3}x + 3$,
  l_2: $y = \sqrt{3}x + 2$

85. l_1: $y = 3$,
  l_2: $x + y = 5$

86. l_1: $2x + y - 4 = 0$,
  l_2: $y - 2x + 5 = 0$

87. *Rope Course and Climbing Wall.* For a rope course and climbing wall, a guy wire R is attached 47 ft high on a vertical pole. Another guy wire S is attached 40 ft above the ground on the same pole. (*Source:* Experiential Resources, Inc., Todd Domeck, Owner) Find the angle α between the wires if they are attached to the ground 50 ft from the pole.

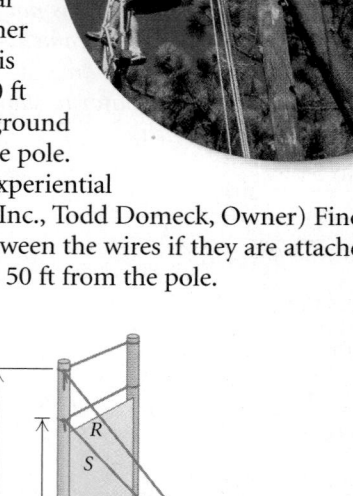

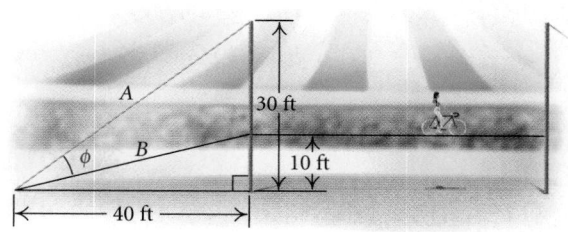

88. *Circus Guy Wire.* In a circus, a guy wire A is attached to the top of a 30-ft pole. Wire B is used for performers to walk up to the tight wire, 10 ft above the ground. Find the angle ϕ between the wires if they are attached to the ground 40 ft from the pole.

89. Given that $f(x) = \cos x$, show that

$$\frac{f(x + h) - f(x)}{h} = \cos x \left(\frac{\cos h - 1}{h} \right) - \sin x \left(\frac{\sin h}{h} \right).$$

90. Given that $f(x) = \sin x$, show that

$$\frac{f(x + h) - f(x)}{h} = \sin x \left(\frac{\cos h - 1}{h} \right) + \cos x \left(\frac{\sin h}{h} \right).$$

Show that each of the following is not an identity by finding a replacement or replacements for which the sides of the equation do not name the same number. Then use a graphing calculator to show that the equation is not an identity.

91. $\dfrac{\sin 5x}{x} = \sin 5$

92. $\sqrt{\sin^2 \theta} = \sin \theta$

93. $\cos (2\alpha) = 2 \cos \alpha$

94. $\sin (-x) = \sin x$

95. $\dfrac{\cos 6x}{\cos x} = 6$

96. $\tan^2 \theta + \cot^2 \theta = 1$

Find the slope of line l_1, where m_2 is the slope of line l_2 and ϕ is the smallest positive angle from l_1 to l_2.

97. $m_2 = \frac{2}{3}$, $\phi = 30°$

98. $m_2 = \frac{4}{3}$, $\phi = 45°$

99. Line l_1 contains the points $(-3, 7)$ and $(-3, -2)$. Line l_2 contains $(0, -4)$ and $(2, 6)$. Find the smallest positive angle from l_1 to l_2.

100. Line l_1 contains the points $(-2, 4)$ and $(5, -1)$. Find the slope of line l_2 such that the angle from l_1 to l_2 is $45°$.

101. Find an identity for $\cos 2\theta$. (*Hint:* $2\theta = \theta + \theta$.)

102. Find an identity for $\sin 2\theta$. (*Hint:* $2\theta = \theta + \theta$.)

Derive the identity. Check using a graphing calculator.

103. $\tan \left(x + \dfrac{\pi}{4} \right) = \dfrac{1 + \tan x}{1 - \tan x}$

104. $\sin \left(x - \dfrac{3\pi}{2} \right) = \cos x$

105. $\sin (\alpha + \beta) + \sin (\alpha - \beta) = 2 \sin \alpha \cos \beta$

106. $\dfrac{\sin (\alpha + \beta)}{\cos (\alpha - \beta)} = \dfrac{\tan \alpha + \tan \beta}{1 + \tan \alpha \tan \beta}$

7.2

Identities: Cofunction, Double-Angle, and Half-Angle

❖ Use cofunction identities to derive other identities.

❖ Use the double-angle identities to find function values of twice an angle when one function value is known for that angle.

❖ Use the half-angle identities to find function values of half an angle when one function value is known for that angle.

❖ Simplify trigonometric expressions using the double-angle identities and the half-angle identities.

❖ Cofunction Identities

Each of the identities listed on the following page yields a conversion to a *cofunction*. For this reason, we call them cofunction identities.

Cofunction Identities

$$\sin\left(\frac{\pi}{2} - x\right) = \cos x, \qquad \cos\left(\frac{\pi}{2} - x\right) = \sin x,$$

$$\tan\left(\frac{\pi}{2} - x\right) = \cot x, \qquad \cot\left(\frac{\pi}{2} - x\right) = \tan x,$$

$$\sec\left(\frac{\pi}{2} - x\right) = \csc x, \qquad \csc\left(\frac{\pi}{2} - x\right) = \sec x$$

We verified the first two of these identities in Section 7.1. The other four can be proved using the first two and the definitions of the trigonometric functions. These identities hold for all real numbers, and thus, for all angle measures, but if we restrict θ to values such that $0° < \theta < 90°$, or $0 < \theta < \pi/2$, then we have a special application to the acute angles of a right triangle.

Comparing graphs can lead to possible identities. On the left below, we see that the graph of $y = \sin(x + \pi/2)$ is a translation of the graph of $y = \sin x$ to the left $\pi/2$ units. On the right, we see the graph of $y = \cos x$.

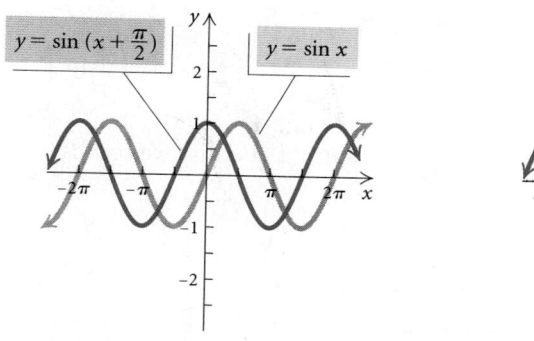
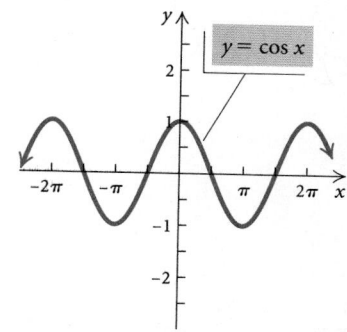

Comparing the graphs, we note a possible identity:

$$\sin\left(x + \frac{\pi}{2}\right) = \cos x.$$

The identity can be proved using the identity for the sine of a sum developed in Section 7.1.

EXAMPLE 1   Prove the identity $\sin(x + \pi/2) = \cos x$.

Solution

$$\sin\left(x + \frac{\pi}{2}\right) = \sin x \cos\frac{\pi}{2} + \cos x \sin\frac{\pi}{2} \qquad \text{Using } \sin(u + v) = \sin u \cos v + \cos u \sin v$$

$$= \sin x \cdot 0 + \cos x \cdot 1$$

$$= \cos x$$

We now state four more cofunction identities. These new identities that involve the sine and cosine functions can be verified using previously established identities as seen in Example 1.

> ### Cofunction Identities for the Sine and Cosine
>
> $$\sin\left(x \pm \frac{\pi}{2}\right) = \pm \cos x, \qquad \cos\left(x \pm \frac{\pi}{2}\right) = \mp \sin x$$

EXAMPLE 2   Find an identity for each of the following.

a) $\tan\left(x + \dfrac{\pi}{2}\right)$

b) $\sec\left(x - 90°\right)$

Solution

a) We have

$$\tan\left(x + \frac{\pi}{2}\right) = \frac{\sin\left(x + \dfrac{\pi}{2}\right)}{\cos\left(x + \dfrac{\pi}{2}\right)} \qquad \text{Using } \tan x = \frac{\sin x}{\cos x}$$

$$= \frac{\cos x}{-\sin x} \qquad\qquad \text{Using cofunction identities}$$

$$= -\cot x.$$

Thus the identity we seek is

$$\tan\left(x + \frac{\pi}{2}\right) = -\cot x.$$

b) We have

$$\sec\left(x - 90°\right) = \frac{1}{\cos\left(x - 90°\right)} = \frac{1}{\sin x} = \csc x.$$

Thus, $\sec\left(x - 90°\right) = \csc x$.   **Now Try Exercises 5 and 7.** ■

❖ Double-Angle Identities

If we double an angle of measure x, the new angle will have measure $2x$. **Double-angle identities** give trigonometric function values of $2x$ in terms of function values of x. To develop these identities, we will use the sum formulas from the preceding section. We first develop a formula for $\sin 2x$. Recall that

$$\sin\left(u + v\right) = \sin u \cos v + \cos u \sin v.$$

We will consider a number x and substitute it for both u and v in this identity. Doing so gives us

$$\sin(x + x) = \sin 2x$$
$$= \sin x \cos x + \cos x \sin x$$
$$= 2 \sin x \cos x.$$

Our first double-angle identity is thus

$$\sin 2x = 2 \sin x \cos x.$$

We can graph

$$y_1 = \sin 2x \quad \text{and} \quad y_2 = 2 \sin x \cos x$$

using the "line"-graph style for y_1 and the "path"-graph style for y_2 and see that they appear to have the same graph. We can also use the TABLE feature.

$$y_1 = \sin 2x, \quad y_2 = 2 \sin x \cos x$$

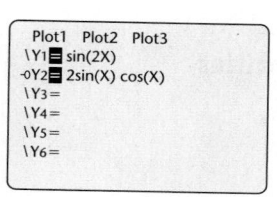

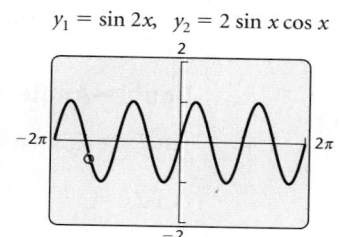

Double-angle identities for the cosine and tangent functions can be derived in much the same way as the identity above:

$$\cos 2x = \cos^2 x - \sin^2 x, \qquad \tan 2x = \frac{2 \tan x}{1 - \tan^2 x}.$$

EXAMPLE 3 Given that $\tan \theta = -\frac{3}{4}$ and θ is in quadrant II, find each of the following.

a) $\sin 2\theta$ **b)** $\cos 2\theta$

c) $\tan 2\theta$ **d)** The quadrant in which 2θ lies

Solution By drawing a reference triangle as shown, we find that

$$\sin \theta = \frac{3}{5}$$

and

$$\cos \theta = -\frac{4}{5}.$$

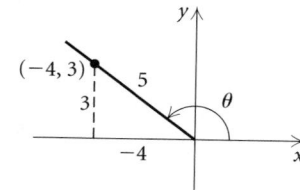

Thus we have the following.

a) $\sin 2\theta = 2 \sin \theta \cos \theta = 2 \cdot \dfrac{3}{5} \cdot \left(-\dfrac{4}{5}\right) = -\dfrac{24}{25}$

b) $\cos 2\theta = \cos^2 \theta - \sin^2 \theta = \left(-\dfrac{4}{5}\right)^2 - \left(\dfrac{3}{5}\right)^2 = \dfrac{16}{25} - \dfrac{9}{25} = \dfrac{7}{25}$

c) $\tan 2\theta = \dfrac{2\tan\theta}{1-\tan^2\theta} = \dfrac{2\cdot\left(-\frac{3}{4}\right)}{1-\left(-\frac{3}{4}\right)^2} = \dfrac{-\frac{3}{2}}{1-\frac{9}{16}} = -\dfrac{3}{2}\cdot\dfrac{16}{7} = -\dfrac{24}{7}$

Note that $\tan 2\theta$ could have been found more easily in this case by simply dividing:

$$\tan 2\theta = \frac{\sin 2\theta}{\cos 2\theta} = \frac{-\frac{24}{25}}{\frac{7}{25}} = -\frac{24}{7}.$$

d) Since $\sin 2\theta$ is negative and $\cos 2\theta$ is positive, we know that 2θ is in quadrant IV.   **Now Try Exercise 9.** ■

Two other useful identities for $\cos 2x$ can be derived easily, as follows.

$$\begin{aligned}\cos 2x &= \cos^2 x - \sin^2 x & \cos 2x &= \cos^2 x - \sin^2 x\\ &= (1-\sin^2 x) - \sin^2 x & &= \cos^2 x - (1-\cos^2 x)\\ &= 1 - 2\sin^2 x & &= 2\cos^2 x - 1\end{aligned}$$

Double-Angle Identities

$\sin 2x = 2\sin x \cos x,$

$\tan 2x = \dfrac{2\tan x}{1-\tan^2 x}$

$$\begin{aligned}\cos 2x &= \cos^2 x - \sin^2 x\\ &= 1 - 2\sin^2 x\\ &= 2\cos^2 x - 1\end{aligned}$$

Solving the last two cosine double-angle identities for $\sin^2 x$ and $\cos^2 x$, respectively, we obtain two more identities:

$$\sin^2 x = \frac{1-\cos 2x}{2} \quad\text{and}\quad \cos^2 x = \frac{1+\cos 2x}{2}.$$

Using division and these two identities gives us the following useful identity:

$$\tan^2 x = \frac{1-\cos 2x}{1+\cos 2x}.$$

EXAMPLE 4 Find an equivalent expression for each of the following.

a) $\sin 3\theta$ in terms of function values of θ

b) $\cos^3 x$ in terms of function values of x or $2x$, raised only to the first power

Solution

a) $\sin 3\theta = \sin(2\theta + \theta)$

$\qquad = \sin 2\theta \cos\theta + \cos 2\theta \sin\theta$

$\qquad = (2\sin\theta\cos\theta)\cos\theta + (2\cos^2\theta - 1)\sin\theta$

$\qquad\qquad$ Using $\sin 2\theta = 2\sin\theta\cos\theta$ and $\cos 2\theta = 2\cos^2\theta - 1$

$\qquad = 2\sin\theta\cos^2\theta + 2\sin\theta\cos^2\theta - \sin\theta$

$\qquad = 4\sin\theta\cos^2\theta - \sin\theta$

We could also substitute $\cos^2\theta - \sin^2\theta$ or $1 - 2\sin^2\theta$ for $\cos 2\theta$. Each substitution leads to a different result, but all results are equivalent.

b) $\cos^3 x = \cos^2 x \cos x$

$$= \frac{1 + \cos 2x}{2}\cos x$$

$$= \frac{\cos x + \cos x \cos 2x}{2}$$

Now Try Exercise 15. ■

❀ Half-Angle Identities

If we take half of an angle of measure x, the new angle will have measure $x/2$. **Half-angle identities** give trigonometric function values of $x/2$ in terms of function values of x. To develop these identities, we replace x with $x/2$ and take square roots. For example,

$$\sin^2 x = \frac{1 - \cos 2x}{2} \qquad \begin{array}{l}\text{Solving the identity}\\ \cos 2x = 1 - 2\sin^2 x \text{ for } \sin^2 x\end{array}$$

$$\sin^2 \frac{x}{2} = \frac{1 - \cos 2 \cdot \dfrac{x}{2}}{2} \qquad \text{Substituting } \frac{x}{2} \text{ for } x$$

$$\sin^2 \frac{x}{2} = \frac{1 - \cos x}{2}$$

$$\sin \frac{x}{2} = \pm\sqrt{\frac{1 - \cos x}{2}}. \qquad \text{Taking square roots}$$

The formula is called a *half-angle formula*. The use of $+$ and $-$ depends on the quadrant in which the angle $x/2$ lies. Half-angle identities for the cosine and tangent functions can be derived in a similar manner. Two additional formulas for the half-angle tangent identity are listed below.

Half-Angle Identities

$$\sin \frac{x}{2} = \pm\sqrt{\frac{1 - \cos x}{2}},$$

$$\cos \frac{x}{2} = \pm\sqrt{\frac{1 + \cos x}{2}},$$

$$\tan \frac{x}{2} = \pm\sqrt{\frac{1 - \cos x}{1 + \cos x}}$$

$$= \frac{\sin x}{1 + \cos x} = \frac{1 - \cos x}{\sin x}$$

EXAMPLE 5 Find $\tan(\pi/8)$ exactly. Then check the answer using a graphing calculator in RADIAN mode.

Solution We have

$$\tan \frac{\pi}{8} = \tan \frac{\frac{\pi}{4}}{2} = \frac{\sin \frac{\pi}{4}}{1 + \cos \frac{\pi}{4}} = \frac{\frac{\sqrt{2}}{2}}{1 + \frac{\sqrt{2}}{2}} = \frac{\frac{\sqrt{2}}{2}}{\frac{2 + \sqrt{2}}{2}}$$

$$= \frac{\sqrt{2}}{2 + \sqrt{2}} = \frac{\sqrt{2}}{2 + \sqrt{2}} \cdot \frac{2 - \sqrt{2}}{2 - \sqrt{2}}$$

$$= \sqrt{2} - 1.$$

Now Try Exercise 21. ▦

tan(π/8)	
	.4142135624
√(2)−1	
	.4142135624

The identities that we have developed are also useful for simplifying trigonometric expressions.

EXAMPLE 6 Simplify each of the following.

a) $\dfrac{\sin x \cos x}{\frac{1}{2} \cos 2x}$ **b)** $2 \sin^2 \dfrac{x}{2} + \cos x$

Solution

a) We can obtain $2 \sin x \cos x$ in the numerator by multiplying the expression by $\frac{2}{2}$:

$$\frac{\sin x \cos x}{\frac{1}{2} \cos 2x} = \frac{2}{2} \cdot \frac{\sin x \cos x}{\frac{1}{2} \cos 2x} = \frac{2 \sin x \cos x}{\cos 2x}$$

$$= \frac{\sin 2x}{\cos 2x} \qquad \text{Using } \sin 2x = 2 \sin x \cos x$$

$$= \tan 2x.$$

b) We have

$$2 \sin^2 \frac{x}{2} + \cos x = 2 \left(\frac{1 - \cos x}{2} \right) + \cos x$$

$$\text{Using } \sin \frac{x}{2} = \pm \sqrt{\frac{1 - \cos x}{2}}, \text{ or } \sin^2 \frac{x}{2} = \frac{1 - \cos x}{2}$$

$$= 1 - \cos x + \cos x$$

$$= 1.$$

We can check this result using a graph or a table.

$$y_1 = 2 \sin^2 \frac{x}{2} + \cos x, \quad y_2 = 1$$

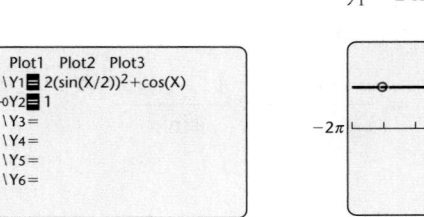

X	Y1	Y2
−6.283	1	1
−5.498	1	1
−4.712	1	1
−3.927	1	1
−3.142	1	1
−2.356	1	1
−1.571	1	1

X = −6.28318530718

ΔTbl = π/4

Plot1 Plot2 Plot3
\Y1◼ 2(sin(X/2))²+cos(X)
-◦Y2◼ 1
\Y3=
\Y4=
\Y5=
\Y6=

Now Try Exercise 33. ▦

7.2 Exercise Set

1. Given that $\sin (3\pi/10) \approx 0.8090$ and $\cos (3\pi/10) \approx 0.5878$, find each of the following.
   a) The other four function values for $3\pi/10$
   b) The six function values for $\pi/5$

2. Given that
   $$\sin \frac{\pi}{12} = \frac{\sqrt{2 - \sqrt{3}}}{2} \quad \text{and} \quad \cos \frac{\pi}{12} = \frac{\sqrt{2 + \sqrt{3}}}{2},$$
   find exact answers for each of the following.
   a) The other four function values for $\pi/12$
   b) The six function values for $5\pi/12$

3. Given that $\sin \theta = \frac{1}{3}$ and that the terminal side is in quadrant II, find exact answers for each of the following.
   a) The other function values for θ
   b) The six function values for $\pi/2 - \theta$
   c) The six function values for $\theta - \pi/2$

4. Given that $\cos \phi = \frac{4}{5}$ and that the terminal side is in quadrant IV, find exact answers for each of the following.
   a) The other function values for ϕ
   b) The six function values for $\pi/2 - \phi$
   c) The six function values for $\phi + \pi/2$

Find an equivalent expression for each of the following.

5. $\sec \left(x + \dfrac{\pi}{2} \right)$

6. $\cot \left(x - \dfrac{\pi}{2} \right)$

7. $\tan \left(x - \dfrac{\pi}{2} \right)$

8. $\csc \left(x + \dfrac{\pi}{2} \right)$

Find the exact value of $\sin 2\theta$, $\cos 2\theta$, $\tan 2\theta$, and the quadrant in which 2θ lies.

9. $\sin \theta = \dfrac{4}{5}$, θ in quadrant I

10. $\cos \theta = \dfrac{5}{13}$, θ in quadrant I

11. $\cos \theta = -\dfrac{3}{5}$, θ in quadrant III

12. $\tan \theta = -\dfrac{15}{8}$, θ in quadrant II

13. $\tan \theta = -\dfrac{5}{12}$, θ in quadrant II

14. $\sin \theta = -\dfrac{\sqrt{10}}{10}$, θ in quadrant IV

15. Find an equivalent expression for $\cos 4x$ in terms of function values of x.

16. Find an equivalent expression for $\sin^4 \theta$ in terms of function values of θ, 2θ, or 4θ, raised only to the first power.

Use the half-angle identities to evaluate exactly.

17. $\cos 15°$

18. $\tan 67.5°$

19. $\sin 112.5°$

20. $\cos \dfrac{\pi}{8}$

21. $\tan 75°$

22. $\sin \dfrac{5\pi}{12}$

Given that $\sin \theta = 0.3416$ and θ is in quadrant I, find each of the following using identities.

23. $\sin 2\theta$

24. $\cos \dfrac{\theta}{2}$

25. $\sin \dfrac{\theta}{2}$

26. $\sin 4\theta$

In Exercises 27–30, use a graphing calculator to determine which of the following expressions asserts an identity. Then derive the identity algebraically.

27. $\dfrac{\cos 2x}{\cos x - \sin x} = \cdots$

   a) $1 + \cos x$
   $y = 1 + \cos x$
   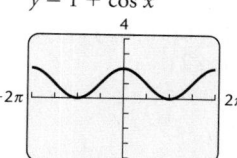

   b) $\cos x - \sin x$
   $y = \cos x - \sin x$
   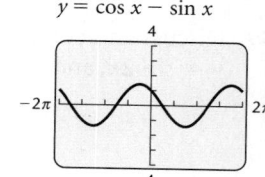

c) $-\cot x$

$y = -\cot x$

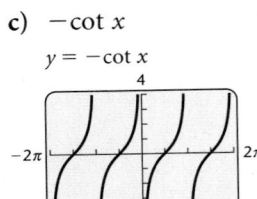

d) $\sin x (\cot x + 1)$

$y = \sin x (\cot x + 1)$

28. $2 \cos^2 \dfrac{x}{2} = \cdots$

a) $\sin x (\csc x + \tan x)$ b) $\sin x - 2 \cos x$
c) $2(\cos^2 x - \sin^2 x)$ d) $1 + \cos x$

29. $\dfrac{\sin 2x}{2 \cos x} = \cdots$

a) $\cos x$ b) $\tan x$
c) $\cos x + \sin x$ d) $\sin x$

30. $2 \sin \dfrac{\theta}{2} \cos \dfrac{\theta}{2} = \cdots$

a) $\cos^2 \theta$ b) $\sin \dfrac{\theta}{2}$

c) $\sin \theta$ d) $\sin \theta - \cos \theta$

Simplify. Check your results using a graphing calculator.

31. $2 \cos^2 \dfrac{x}{2} - 1$

32. $\cos^4 x - \sin^4 x$

33. $(\sin x - \cos x)^2 + \sin 2x$

34. $(\sin x + \cos x)^2$

35. $\dfrac{2 - \sec^2 x}{\sec^2 x}$

36. $\dfrac{1 + \sin 2x + \cos 2x}{1 + \sin 2x - \cos 2x}$

37. $(-4 \cos x \sin x + 2 \cos 2x)^2 +$
$(2 \cos 2x + 4 \sin x \cos x)^2$

38. $2 \sin x \cos^3 x - 2 \sin^3 x \cos x$

Collaborative Discussion and Writing

39. Discuss and compare the graphs of $y = \sin x$, $y = \sin 2x$, and $y = \sin (x/2)$.

40. Find all errors in the following:

$$2 \sin^2 2x + \cos 4x$$
$$= 2(2 \sin x \cos x)^2 + 2 \cos 2x$$
$$= 8 \sin^2 x \cos^2 x + 2(\cos^2 x + \sin^2 x)$$
$$= 8 \sin^2 x \cos^2 x + 2.$$

Skill Maintenance

Complete the identity.

41. $1 - \cos^2 x =$

42. $\sec^2 x - \tan^2 x =$

43. $\sin^2 x - 1 =$

44. $1 + \cot^2 x =$

45. $\csc^2 x - \cot^2 x$

46. $1 + \tan^2 x =$

47. $1 - \sin^2 x$

48. $\sec^2 x - 1$

Consider the following functions (a)–(f). Without graphing them, answer questions 49–52 below.

a) $f(x) = 2 \sin \left(\dfrac{1}{2} x - \dfrac{\pi}{2} \right)$

b) $f(x) = \dfrac{1}{2} \cos \left(2x - \dfrac{\pi}{4} \right) + 2$

c) $f(x) = -\sin \left[2 \left(x - \dfrac{\pi}{2} \right) \right] + 2$

d) $f(x) = \sin (x + \pi) - \dfrac{1}{2}$

e) $f(x) = -2 \cos (4x - \pi)$

f) $f(x) = -\cos \left[2 \left(x - \dfrac{\pi}{8} \right) \right]$

49. Which functions have a graph with an amplitude of 2?

50. Which functions have a graph with a period of π?

51. Which functions have a graph with a period of 2π?

52. Which functions have a graph with a phase shift of $\dfrac{\pi}{4}$?

Synthesis

53. Given that $\cos 51° \approx 0.6293$, find the six function values for $141°$.

Simplify. Check your results using a graphing calculator.

54. $\sin\left(\dfrac{\pi}{2} - x\right)[\sec x - \cos x]$

55. $\cos(\pi - x) + \cot x \sin\left(x - \dfrac{\pi}{2}\right)$

56. $\dfrac{\cos x - \sin\left(\dfrac{\pi}{2} - x\right)\sin x}{\cos x - \cos(\pi - x)\tan x}$

57. $\dfrac{\cos^2 y \sin\left(y + \dfrac{\pi}{2}\right)}{\sin^2 y \sin\left(\dfrac{\pi}{2} - y\right)}$

Find sin θ, cos θ, and tan θ under the given conditions.

58. $\cos 2\theta = \dfrac{7}{12}, \quad \dfrac{3\pi}{2} \le 2\theta \le 2\pi$

59. $\tan\dfrac{\theta}{2} = -\dfrac{5}{3}, \quad \pi < \theta \le \dfrac{3\pi}{2}$

60. *Nautical Mile.* Latitude is used to measure north–south location on the earth between the equator and the poles. For example, Chicago has latitude 42°N. (See the figure.) In Great Britain, the *nautical mile* is defined as the length of a minute of arc of the earth's radius. Since the earth is flattened slightly at the poles, a British nautical mile varies with latitude. In fact, it is given, in feet, by the function

$$N(\phi) = 6066 - 31 \cos 2\phi,$$

where ϕ is the latitude in degrees.

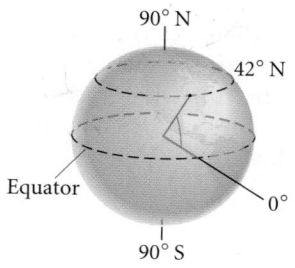

a) What is the length of a British nautical mile at Chicago?
b) What is the length of a British nautical mile at the North Pole?
c) Express $N(\phi)$ in terms of $\cos \phi$ only; that is, do not use the double angle.

61. *Acceleration Due to Gravity.* The acceleration due to gravity is often denoted by g in a formula such as $S = \frac{1}{2}gt^2$, where S is the distance that an object falls in time t. The number g relates to motion near the earth's surface and is usually considered constant. In fact, however, g is not constant, but varies slightly with latitude. If ϕ stands for latitude, in degrees, g is given with good approximation by the formula

$$g = 9.78049(1 + 0.005288 \sin^2 \phi - 0.000006 \sin^2 2\phi),$$

where g is measured in meters per second per second at sea level.

a) Chicago has latitude 42°N. Find g.
b) Philadelphia has latitude 40°N. Find g.
c) Express g in terms of $\sin \phi$ only. That is, eliminate the double angle.

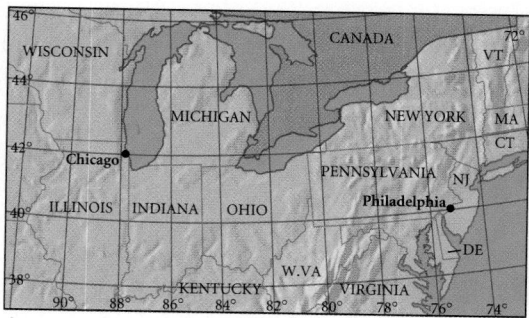

7.3 Proving Trigonometric Identities

❖ Prove identities using other identities.

❖ Use the product-to-sum identities and the sum-to-product identities to derive other identities.

❊ The Logic of Proving Identities

We outline two algebraic methods for proving identities.

Method 1. Start with either the left side or the right side of the equation and obtain the other side. For example, suppose you are trying to prove that the equation $P = Q$ is an identity. You might try to produce a string of statements $(R_1, R_2, \ldots$ or $T_1, T_2, \ldots)$ like the following, which start with P and end with Q or start with Q and end with P:

$$
\begin{array}{llll}
P &= R_1 & \text{or} \quad Q &= T_1 \\
&= R_2 & &= T_2 \\
&\ \ \vdots & &\ \ \vdots \\
&= Q & &= P.
\end{array}
$$

Method 2. Work with each side separately until you obtain the same expression. For example, suppose you are trying to prove that $P = Q$ is an identity. You might be able to produce two strings of statements like the following, each ending with the same statement S.

$$
\begin{array}{llll}
P &= R_1 & \quad Q &= T_1 \\
&= R_2 & &= T_2 \\
&\ \ \vdots & &\ \ \vdots \\
&= S & &= S.
\end{array}
$$

The number of steps in each string might be different, but in each case the result is S.

A first step in learning to prove identities is to have at hand a list of the identities that you have already learned. Such a list is on the inside back cover of this text. Ask your instructor which ones you are expected to memorize. The more identities you prove, the easier it will be to prove new ones. A list of helpful hints follows.

> **Hints for Proving Identities**
>
> 1. Use method 1 or 2 above.
> 2. Work with the more complex side first.
> 3. Carry out any algebraic manipulations, such as adding, subtracting, multiplying, or factoring.
> 4. Multiplying by 1 can be helpful when rational expressions are involved.
> 5. Converting all expressions to sines and cosines is often helpful.
> 6. Try something! Put your pencil to work and get involved. You will be amazed at how often this leads to success.

❖ Proving Identities

In what follows, method 1 is used in Examples 1, 3, and 4 and method 2 is used in Examples 2 and 5.

EXAMPLE 1 Prove the identity $1 + \sin 2\theta = (\sin \theta + \cos \theta)^2$.

Solution Let's use method 1. We begin with the right side and obtain the left side:

$$(\sin \theta + \cos \theta)^2 = \sin^2 \theta + 2 \sin \theta \cos \theta + \cos^2 \theta \qquad \text{Squaring}$$

$$= 1 + 2 \sin \theta \cos \theta \qquad \begin{array}{l}\text{Recalling the identity}\\ \sin^2 x + \cos^2 x = 1 \text{ and}\\ \text{substituting}\end{array}$$

$$= 1 + \sin 2\theta. \qquad \begin{array}{l}\text{Using } \sin 2x =\\ 2 \sin x \cos x\end{array}$$

We could also begin with the left side and obtain the right side:

$$1 + \sin 2\theta = 1 + 2 \sin \theta \cos \theta \qquad \text{Using } \sin 2x = 2 \sin x \cos x$$

$$= \sin^2 \theta + 2 \sin \theta \cos \theta + \cos^2 \theta \qquad \begin{array}{l}\text{Replacing 1 with}\\ \sin^2 \theta + \cos^2 \theta\end{array}$$

$$= (\sin \theta + \cos \theta)^2. \qquad \text{Factoring} \qquad \textbf{Now Try Exercise 19.} ■$$

EXAMPLE 2 Prove the identity

$$\sin^2 x \tan^2 x = \tan^2 x - \sin^2 x.$$

Solution For this proof, we are going to work with each side separately using method 2. We try to obtain the same expression on each side. In actual practice, you might work on one side for a while, then work on the other side, and then go back to the first side. In other words, you work back and forth until you arrive at the same expression. Let's start with the right side.

$$\tan^2 x - \sin^2 x = \frac{\sin^2 x}{\cos^2 x} - \sin^2 x$$

Recalling the identity
$\tan x = \dfrac{\sin x}{\cos x}$ and
substituting

$$= \frac{\sin^2 x}{\cos^2 x} - \sin^2 x \cdot \frac{\cos^2 x}{\cos^2 x}$$

Multiplying by 1 in
order to subtract

$$= \frac{\sin^2 x - \sin^2 x \cos^2 x}{\cos^2 x}$$

Carrying out
the subtraction

$$= \frac{\sin^2 x (1 - \cos^2 x)}{\cos^2 x}$$

Factoring

$$= \frac{\sin^2 x \sin^2 x}{\cos^2 x}$$

Recalling the identity
$1 - \cos^2 x = \sin^2 x$
and substituting

$$= \frac{\sin^4 x}{\cos^2 x}$$

At this point, we stop and work with the left side, $\sin^2 x \tan^2 x$, of the original identity and try to end with the same expression that we ended with on the right side:

$$\sin^2 x \tan^2 x = \sin^2 x \frac{\sin^2 x}{\cos^2 x}$$

Recalling the identity
$\tan x = \dfrac{\sin x}{\cos x}$ and substituting

$$= \frac{\sin^4 x}{\cos^2 x}.$$

We have obtained the same expression from each side, so the proof is complete.

Now Try Exercise 25. ■

EXAMPLE 3 Prove the identity

$$\frac{\sin 2x}{\sin x} - \frac{\cos 2x}{\cos x} = \sec x.$$

Solution

$$\frac{\sin 2x}{\sin x} - \frac{\cos 2x}{\cos x} = \frac{2 \sin x \cos x}{\sin x} - \frac{\cos^2 x - \sin^2 x}{\cos x}$$

Using double-angle identities

$$= 2 \cos x - \frac{\cos^2 x - \sin^2 x}{\cos x}$$

Simplifying

$$= \frac{2 \cos^2 x}{\cos x} - \frac{\cos^2 x - \sin^2 x}{\cos x}$$

Multiplying $2 \cos x$
by 1, or $\cos x / \cos x$

$$= \frac{2 \cos^2 x - \cos^2 x + \sin^2 x}{\cos x}$$

Subtracting

$$= \frac{\cos^2 x + \sin^2 x}{\cos x}$$

$$= \frac{1}{\cos x}$$

Using a Pythagorean identity

$$= \sec x$$

Recalling a basic identity

Now Try Exercise 15. ■

EXAMPLE 4 Prove the identity

$$\frac{\sec t - 1}{t \sec t} = \frac{1 - \cos t}{t}.$$

Solution We use method 1, starting with the left side. Note that the left side involves sec t, whereas the right side involves cos t, so it might be wise to make use of a basic identity that involves these two expressions: $\sec t = 1/\cos t$.

$$\frac{\sec t - 1}{t \sec t} = \frac{\dfrac{1}{\cos t} - 1}{t \dfrac{1}{\cos t}} \qquad \text{Substituting } 1/\cos t \text{ for sec } t$$

$$= \left(\frac{1}{\cos t} - 1\right) \cdot \frac{\cos t}{t}$$

$$= \frac{1}{t} - \frac{\cos t}{t} \qquad \text{Multiplying}$$

$$= \frac{1 - \cos t}{t}$$

We started with the left side and obtained the right side, so the proof is complete.

Now Try Exercise 5. ◼

EXAMPLE 5 Prove the identity

$$\cot \phi + \csc \phi = \frac{\sin \phi}{1 - \cos \phi}.$$

Solution We are again using method 2, beginning with the left side:

$$\cot \phi + \csc \phi = \frac{\cos \phi}{\sin \phi} + \frac{1}{\sin \phi} \qquad \text{Using basic identities}$$

$$= \frac{1 + \cos \phi}{\sin \phi}. \qquad \text{Adding}$$

At this point, we stop and work with the right side of the original identity:

$$\frac{\sin \phi}{1 - \cos \phi} = \frac{\sin \phi}{1 - \cos \phi} \cdot \frac{1 + \cos \phi}{1 + \cos \phi} \qquad \text{Multiplying by 1}$$

$$= \frac{\sin \phi (1 + \cos \phi)}{1 - \cos^2 \phi}$$

$$= \frac{\sin \phi (1 + \cos \phi)}{\sin^2 \phi} \qquad \text{Using } \sin^2 x = 1 - \cos^2 x$$

$$= \frac{1 + \cos \phi}{\sin \phi}. \qquad \text{Simplifying}$$

The proof is complete since we obtained the same expression from each side.

Now Try Exercise 29. ◼

❋ Product-to-Sum and Sum-to-Product Identities

On occasion, it is convenient to convert a product of trigonometric expressions to a sum, or the reverse. The following identities are useful in this connection.

Product-to-Sum Identities

$$\sin x \cdot \sin y = \frac{1}{2}[\cos(x - y) - \cos(x + y)] \tag{1}$$

$$\cos x \cdot \cos y = \frac{1}{2}[\cos(x - y) + \cos(x + y)] \tag{2}$$

$$\sin x \cdot \cos y = \frac{1}{2}[\sin(x + y) + \sin(x - y)] \tag{3}$$

$$\cos x \cdot \sin y = \frac{1}{2}[\sin(x + y) - \sin(x - y)] \tag{4}$$

We can derive product-to-sum identities (1) and (2) using the sum and difference identities for the cosine function:

$$\cos(x + y) = \cos x \cos y - \sin x \sin y, \qquad \text{Sum identity}$$
$$\cos(x - y) = \cos x \cos y + \sin x \sin y. \qquad \text{Difference identity}$$

Subtracting the sum identity from the difference identity, we have

$$\cos(x - y) - \cos(x + y) = 2 \sin x \sin y \qquad \text{Subtracting}$$

$$\frac{1}{2}[\cos(x - y) - \cos(x + y)] = \sin x \sin y. \qquad \text{Multiplying by } \tfrac{1}{2}$$

Thus, $\sin x \sin y = \frac{1}{2}[\cos(x - y) - \cos(x + y)]$.

Adding the cosine sum and difference identities, we have

$$\cos(x - y) + \cos(x + y) = 2 \cos x \cos y \qquad \text{Adding}$$

$$\frac{1}{2}[\cos(x - y) + \cos(x + y)] = \cos x \cos y. \qquad \text{Multiplying by } \tfrac{1}{2}$$

Thus, $\cos x \cos y = \frac{1}{2}[\cos(x - y) + \cos(x + y)]$.

Identities (3) and (4) can be derived in a similar manner using the sum and difference identities for the sine function.

EXAMPLE 6 Find an identity for $2 \sin 3\theta \cos 7\theta$.

Solution We will use the identity

$$\sin x \cdot \cos y = \frac{1}{2}[\sin(x+y) + \sin(x-y)].$$

Here $x = 3\theta$ and $y = 7\theta$. Thus,

$$2 \sin 3\theta \cos 7\theta = 2 \cdot \frac{1}{2}[\sin(3\theta + 7\theta) + \sin(3\theta - 7\theta)]$$

$$= \sin 10\theta + \sin(-4\theta)$$

$$= \sin 10\theta - \sin 4\theta. \qquad \text{Using } \sin(-\theta) = -\sin\theta$$

Now Try Exercise 37. ■

Sum-to-Product Identities

$$\sin x + \sin y = 2 \sin \frac{x+y}{2} \cos \frac{x-y}{2} \qquad\qquad (5)$$

$$\sin x - \sin y = 2 \cos \frac{x+y}{2} \sin \frac{x-y}{2} \qquad\qquad (6)$$

$$\cos y + \cos x = 2 \cos \frac{x+y}{2} \cos \frac{x-y}{2} \qquad\qquad (7)$$

$$\cos y - \cos x = 2 \sin \frac{x+y}{2} \sin \frac{x-y}{2} \qquad\qquad (8)$$

The sum-to-product identities (5)–(8) can be derived using the product-to-sum identities. Proofs are left to the exercises.

EXAMPLE 7 Find an identity for $\cos\theta + \cos 5\theta$.

Solution We will use the identity

$$\cos y + \cos x = 2 \cos \frac{x+y}{2} \cos \frac{x-y}{2}.$$

Here $x = 5\theta$ and $y = \theta$. Thus,

$$\cos\theta + \cos 5\theta = 2 \cos \frac{5\theta + \theta}{2} \cos \frac{5\theta - \theta}{2}$$

$$= 2 \cos 3\theta \cos 2\theta.$$

Now Try Exercise 35. ■

7.3 Exercise Set

Prove the identity.

1. $\sec x - \sin x \tan x = \cos x$

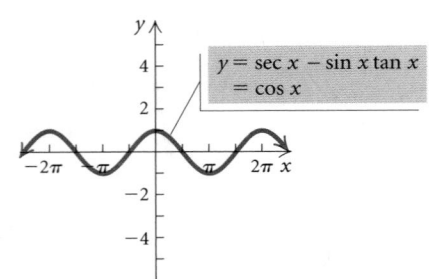

2. $\dfrac{1 + \cos \theta}{\sin \theta} + \dfrac{\sin \theta}{\cos \theta} = \dfrac{\cos \theta + 1}{\sin \theta \cos \theta}$

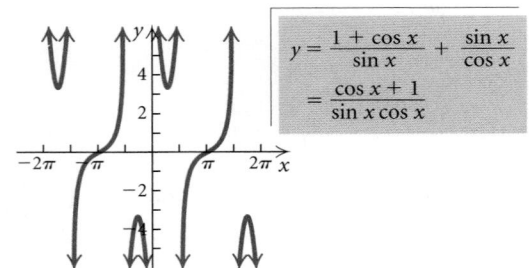

3. $\dfrac{1 - \cos x}{\sin x} = \dfrac{\sin x}{1 + \cos x}$

4. $\dfrac{1 + \tan y}{1 + \cot y} = \dfrac{\sec y}{\csc y}$

5. $\dfrac{1 + \tan \theta}{1 - \tan \theta} + \dfrac{1 + \cot \theta}{1 - \cot \theta} = 0$

6. $\dfrac{\sin x + \cos x}{\sec x + \csc x} = \dfrac{\sin x}{\sec x}$

7. $\dfrac{\cos^2 \alpha + \cot \alpha}{\cos^2 \alpha - \cot \alpha} = \dfrac{\cos^2 \alpha \tan \alpha + 1}{\cos^2 \alpha \tan \alpha - 1}$

8. $\sec 2\theta = \dfrac{\sec^2 \theta}{2 - \sec^2 \theta}$

9. $\dfrac{2 \tan \theta}{1 + \tan^2 \theta} = \sin 2\theta$

10. $\dfrac{\cos (u - v)}{\cos u \sin v} = \tan u + \cot v$

11. $1 - \cos 5\theta \cos 3\theta - \sin 5\theta \sin 3\theta = 2 \sin^2 \theta$

12. $\cos^4 x - \sin^4 x = \cos 2x$

13. $2 \sin \theta \cos^3 \theta + 2 \sin^3 \theta \cos \theta = \sin 2\theta$

14. $\dfrac{\tan 3t - \tan t}{1 + \tan 3t \tan t} = \dfrac{2 \tan t}{1 - \tan^2 t}$

15. $\dfrac{\tan x - \sin x}{2 \tan x} = \sin^2 \dfrac{x}{2}$

16. $\dfrac{\cos^3 \beta - \sin^3 \beta}{\cos \beta - \sin \beta} = \dfrac{2 + \sin 2\beta}{2}$

17. $\sin (\alpha + \beta) \sin (\alpha - \beta) = \sin^2 \alpha - \sin^2 \beta$

18. $\cos^2 x (1 - \sec^2 x) = -\sin^2 x$

19. $\tan \theta (\tan \theta + \cot \theta) = \sec^2 \theta$

20. $\dfrac{\cos \theta + \sin \theta}{\cos \theta} = 1 + \tan \theta$

21. $\dfrac{1 + \cos^2 x}{\sin^2 x} = 2 \csc^2 x - 1$

22. $\dfrac{\tan y + \cot y}{\csc y} = \sec y$

23. $\dfrac{1 + \sin x}{1 - \sin x} + \dfrac{\sin x - 1}{1 + \sin x} = 4 \sec x \tan x$

24. $\tan \theta - \cot \theta = (\sec \theta - \csc \theta)(\sin \theta + \cos \theta)$

25. $\cos^2 \alpha \cot^2 \alpha = \cot^2 \alpha - \cos^2 \alpha$

26. $\dfrac{\tan x + \cot x}{\sec x + \csc x} = \dfrac{1}{\cos x + \sin x}$

27. $2 \sin^2 \theta \cos^2 \theta + \cos^4 \theta = 1 - \sin^4 \theta$

28. $\dfrac{\cot \theta}{\csc \theta - 1} = \dfrac{\csc \theta + 1}{\cot \theta}$

29. $\dfrac{1 + \sin x}{1 - \sin x} = (\sec x + \tan x)^2$

30. $\sec^4 s - \tan^2 s = \tan^4 s + \sec^2 s$

31. Verify the product-to-sum identities (3) and (4) using the sine sum and difference identities.

32. Verify the sum-to-product identities (5)–(8) using the product-to-sum identities (1)–(4).

Use the product-to-sum identities and the sum-to-product identities to find identities for each of the following.

33. $\sin 3\theta - \sin 5\theta$

34. $\sin 7x - \sin 4x$

35. $\sin 8\theta + \sin 5\theta$

36. $\cos \theta - \cos 7\theta$

37. $\sin 7u \sin 5u$

38. $2 \sin 7\theta \cos 3\theta$

39. $7 \cos \theta \sin 7\theta$

40. $\cos 2t \sin t$

41. $\cos 55° \sin 25°$

42. $7 \cos 5\theta \cos 7\theta$

Use the product-to-sum identities and the sum-to-product identities to prove each of the following.

43. $\sin 4\theta + \sin 6\theta = \cot \theta (\cos 4\theta - \cos 6\theta)$

44. $\tan 2x (\cos x + \cos 3x) = \sin x + \sin 3x$

45. $\cot 4x (\sin x + \sin 4x + \sin 7x)$
$= \cos x + \cos 4x + \cos 7x$

46. $\tan \dfrac{x + y}{2} = \dfrac{\sin x + \sin y}{\cos x + \cos y}$

47. $\cot \dfrac{x + y}{2} = \dfrac{\sin y - \sin x}{\cos x - \cos y}$

48. $\tan \dfrac{\theta + \phi}{2} \tan \dfrac{\phi - \theta}{2} = \dfrac{\cos \theta - \cos \phi}{\cos \theta + \cos \phi}$

49. $\tan \dfrac{\theta + \phi}{2} (\sin \theta - \sin \phi)$
$= \tan \dfrac{\theta - \phi}{2} (\sin \theta + \sin \phi)$

50. $\sin 2\theta + \sin 4\theta + \sin 6\theta = 4 \cos \theta \cos 2\theta \sin 3\theta$

In Exercises 51–56, use a graphing calculator to determine which expression (A)–(F) on the right can be used to complete the identity. Then try to prove that identity algebraically.

51. $\dfrac{\cos x + \cot x}{1 + \csc x}$

52. $\cot x + \csc x$

53. $\sin x \cos x + 1$

54. $2 \cos^2 x - 1$

55. $\dfrac{1}{\cot x \sin^2 x}$

56. $(\cos x + \sin x)(1 - \sin x \cos x)$

A. $\dfrac{\sin^3 x - \cos^3 x}{\sin x - \cos x}$

B. $\cos x$

C. $\tan x + \cot x$

D. $\cos^3 x + \sin^3 x$

E. $\dfrac{\sin x}{1 - \cos x}$

F. $\cos^4 x - \sin^4 x$

Collaborative Discussion and Writing

What restrictions must be placed on the variable in each of the following identities? Why?

57. $\sin 2x = \dfrac{2 \tan x}{1 + \tan^2 x}$

58. $\dfrac{1 - \cos x}{\sin x} = \dfrac{\sin x}{1 + \cos x}$

Skill Maintenance

For each function:
a) *Graph the function.*
b) *Determine whether the function is one-to-one.*
c) *If the function is one-to-one, find an equation for its inverse.*
d) *Graph the inverse of the function.*

59. $f(x) = 3x - 2$

60. $f(x) = x^3 + 1$

61. $f(x) = x^2 - 4, \ x \geq 0$

62. $f(x) = \sqrt{x + 2}$

Solve.

63. $2x^2 = 5x$

64. $3x^2 + 5x - 10 = 18$

65. $x^4 + 5x^2 - 36 = 0$

66. $x^2 - 10x + 1 = 0$

67. $\sqrt{x - 2} = 5$

68. $x = \sqrt{x + 7} + 5$

Synthesis

Prove the identity.

69. $\ln |\tan x| = -\ln |\cot x|$

70. $\ln |\sec \theta + \tan \theta| = -\ln |\sec \theta - \tan \theta|$

71. $\log (\cos x - \sin x) + \log (\cos x + \sin x)$
$= \log \cos 2x$

72. *Mechanics.* The following equation occurs in the study of mechanics:

$$\sin \theta = \frac{I_1 \cos \phi}{\sqrt{(I_1 \cos \phi)^2 + (I_2 \sin \phi)^2}}.$$

It can happen that $I_1 = I_2$. Assuming that this happens, simplify the equation.

73. *Alternating Current.* In the theory of alternating current, the following equation occurs:

$$R = \frac{1}{\omega C(\tan \theta + \tan \phi)}.$$

Show that this equation is equivalent to

$$R = \frac{\cos \theta \cos \phi}{\omega C \sin (\theta + \phi)}.$$

74. *Electrical Theory.* In electrical theory, the following equations occur:

$$E_1 = \sqrt{2}E_t \cos \left(\theta + \frac{\pi}{P} \right)$$

and

$$E_2 = \sqrt{2}E_t \cos \left(\theta - \frac{\pi}{P} \right).$$

Assuming that these equations hold, show that

$$\frac{E_1 + E_2}{2} = \sqrt{2}E_t \cos \theta \cos \frac{\pi}{P}$$

and

$$\frac{E_1 - E_2}{2} = -\sqrt{2}E_t \sin \theta \sin \frac{\pi}{P}.$$

7.4

Inverses of the Trigonometric Functions

❖ Find values of the inverse trigonometric functions.

❖ Simplify expressions such as $\sin (\sin^{-1} x)$ and $\sin^{-1} (\sin x)$.

❖ Simplify expressions involving compositions such as $\sin \left(\cos^{-1} \frac{1}{2}\right)$ without using a calculator.

❖ Simplify expressions such as $\sin \arctan (a/b)$ by making a drawing and reading off appropriate ratios.

In this section, we develop inverse trigonometric functions. The graphs of the sine, cosine, and tangent functions follow. Do these functions have inverses that are functions? They do have inverses if they are one-to-one, which means that they pass the horizontal-line test.

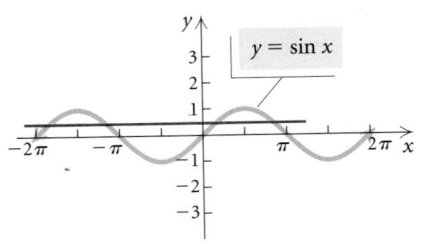

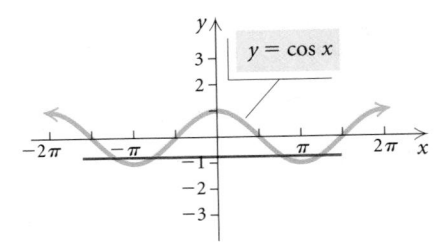

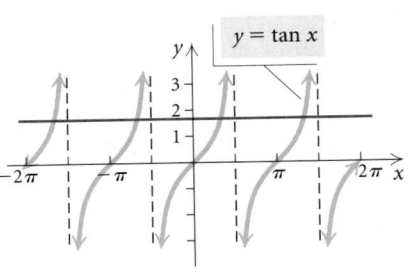

INVERSE FUNCTIONS

REVIEW SECTION **5.1.**

Note that for each function, a horizontal line (shown in red) crosses the graph more than once. Therefore, none of them has an inverse that is a function.

The graphs of an equation and its inverse are reflections of each other across the line $y = x$. Let's examine the graphs of the inverses of each of the three functions graphed above.

STUDY TIP

When you study a section of a mathematics text, read it slowly, observing all the details of the corresponding art pieces that are discussed in the paragraphs. Also note the precise color-coding in the art that enhances the learning of the concepts.

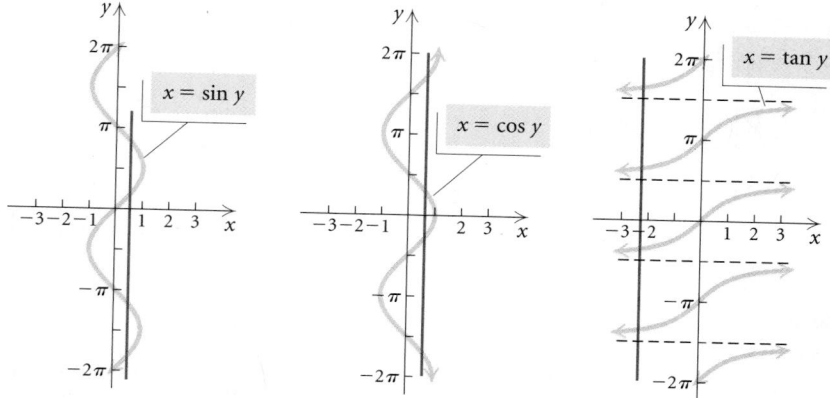

We can check again to see whether these are graphs of functions by using the vertical-line test. In each case, there is a vertical line (shown in red) that crosses the graph more than once, so each *fails* to be a function.

❋ Restricting Ranges to Define Inverse Functions

Recall that a function like $f(x) = x^2$ does not have an inverse that is a function, but by restricting the domain of f to nonnegative numbers, we have a new squaring function, $f(x) = x^2$, $x \geq 0$, that has an inverse, $f^{-1}(x) = \sqrt{x}$. This is equivalent to restricting the range of the inverse relation to exclude ordered pairs that contain negative numbers.

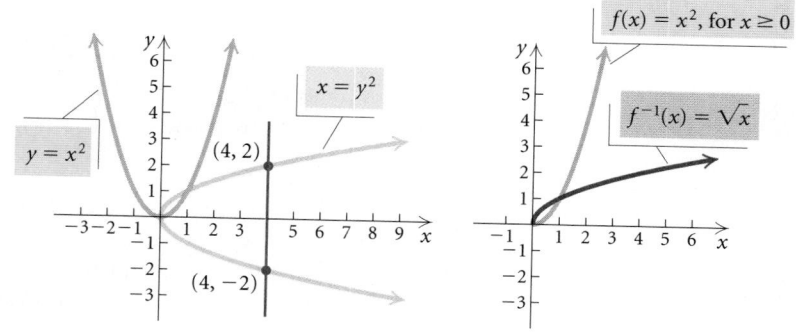

In a similar manner, we can define new trigonometric functions whose inverses are functions. We can do this by restricting either the domains of the basic trigonometric functions or the ranges of their inverse relations. This can be done in many ways, but the restrictions illustrated below with solid red curves are fairly standard in mathematics.

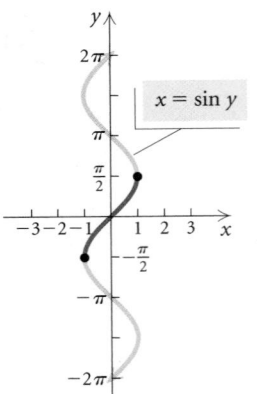

FIGURE 1

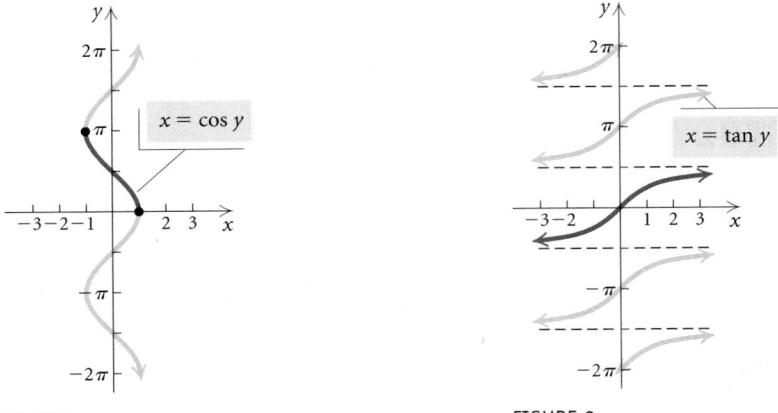

FIGURE 2

FIGURE 3

For the inverse sine function, we choose a range close to the origin that allows all inputs on the interval $[-1, 1]$ to have function values. Thus we choose the interval $[-\pi/2, \pi/2]$ for the range (Fig. 1). For the inverse cosine function, we choose a range close to the origin that allows all inputs on the interval $[-1, 1]$ to have function values. We choose the interval $[0, \pi]$ (Fig. 2). For the inverse tangent function, we choose a range close to the origin that allows all real numbers to have function values. The interval $(-\pi/2, \pi/2)$ satisfies this requirement (Fig. 3).

Inverse Trigonometric Functions

FUNCTION	DOMAIN	RANGE
$y = \sin^{-1} x$ $\quad = \arcsin x$, where $x = \sin y$	$[-1, 1]$	$[-\pi/2, \pi/2]$
$y = \cos^{-1} x$ $\quad = \arccos x$, where $x = \cos y$	$[-1, 1]$	$[0, \pi]$
$y = \tan^{-1} x$ $\quad = \arctan x$, where $x = \tan y$	$(-\infty, \infty)$	$(-\pi/2, \pi/2)$

The notation $\arcsin x$ arises because the function value, y, is the length of an arc on the unit circle for which the sine is x. Either of the two kinds of notation above can be read "the inverse sine of x" or "the arc sine of x" or "the number (or angle) whose sine is x."

CAUTION! The notation $\sin^{-1} x$ is *not* exponential notation.

It does *not* mean $\dfrac{1}{\sin x}$!

The graphs of the inverse trigonometric functions are as follows.

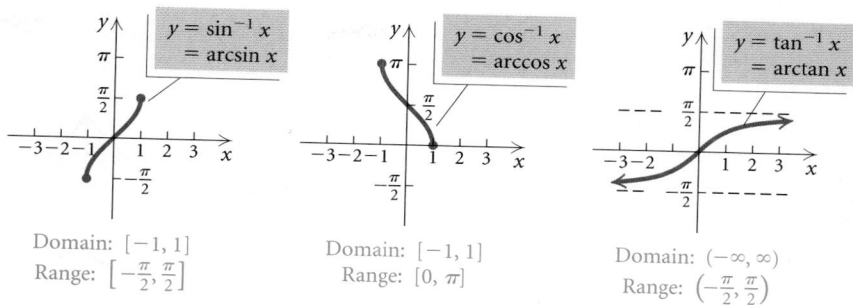

Domain: $[-1, 1]$
Range: $\left[-\frac{\pi}{2}, \frac{\pi}{2}\right]$

Domain: $[-1, 1]$
Range: $[0, \pi]$

Domain: $(-\infty, \infty)$
Range: $\left(-\frac{\pi}{2}, \frac{\pi}{2}\right)$

Exploring with Technology

Inverse trigonometric functions can be graphed using a graphing calculator. Graph $y = \sin^{-1} x$ using the viewing window $[-3, 3, -\pi, \pi]$, with Xscl $= 1$ and Yscl $= \pi/2$. Now try graphing $y = \cos^{-1} x$ and $y = \tan^{-1} x$. Then use the graphs to confirm the domain and the range of each inverse.

The following diagrams show the restricted ranges for the inverse trigonometric functions on a unit circle. Compare these graphs with the graphs above. The ranges of these functions should be memorized. The missing endpoints in the graph of the arctangent function indicate inputs that are not in the domain of the original function.

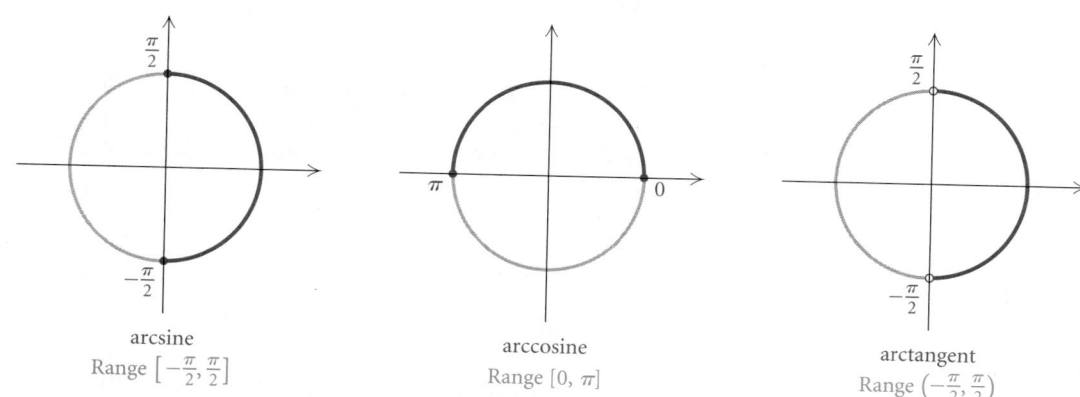

arcsine
Range $\left[-\frac{\pi}{2}, \frac{\pi}{2}\right]$

arccosine
Range $[0, \pi]$

arctangent
Range $\left(-\frac{\pi}{2}, \frac{\pi}{2}\right)$

EXAMPLE 1 Find each of the following function values.

a) $\sin^{-1} \dfrac{\sqrt{2}}{2}$ **b)** $\cos^{-1}\left(-\dfrac{1}{2}\right)$ **c)** $\tan^{-1}\left(-\dfrac{\sqrt{3}}{3}\right)$

Solution

a) Another way to state "find $\sin^{-1} \sqrt{2}/2$" is to say "find β such that $\sin \beta = \sqrt{2}/2$." In the restricted range $[-\pi/2, \pi/2]$, the only number with a sine of $\sqrt{2}/2$ is $\pi/4$. Thus, $\sin^{-1}\left(\sqrt{2}/2\right) = \pi/4$, or 45°. (See Fig. 4 below.)

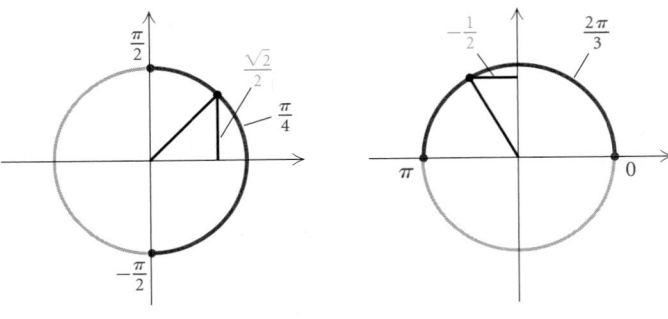

FIGURE 4   FIGURE 5

b) The only number with a cosine of $-\frac{1}{2}$ in the restricted range $[0, \pi]$ is $2\pi/3$. Thus, $\cos^{-1}\left(-\frac{1}{2}\right) = 2\pi/3$, or 120°. (See Fig. 5 above.)

c) The only number in the restricted range $(-\pi/2, \pi/2)$ with a tangent of $-\sqrt{3}/3$ is $-\pi/6$. Thus, $\tan^{-1}\left(-\sqrt{3}/3\right)$ is $-\pi/6$, or −30°. (See Fig. 6 at left.)

Now Try Exercises 1 and 5. ▪

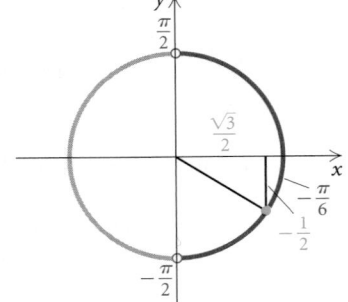

FIGURE 6

We can also use a calculator to find inverse trigonometric function values. On most graphing calculators, we can find inverse function values in either radians or degrees simply by selecting the appropriate mode. The keystrokes involved in finding inverse function values vary with the calculator. Be sure to read the instructions for the particular calculator that you are using.

GCM **EXAMPLE 2** Approximate each of the following function values in both radians and degrees. Round radian measure to four decimal places and degree measure to the nearest tenth of a degree.

a) $\cos^{-1}(-0.2689)$ b) $\tan^{-1}(-0.2623)$

c) $\sin^{-1} 0.20345$ d) $\cos^{-1} 1.318$

e) $\csc^{-1} 8.205$

Solution

FUNCTION VALUE	MODE	READOUT	ROUNDED
a) $\cos^{-1}(-0.2689)$	Radian	1.843047111	1.8430
	Degree	105.5988209	105.6°
b) $\tan^{-1}(-0.2623)$	Radian	−.2565212141	−0.2565
	Degree	−14.69758292	−14.7°
c) $\sin^{-1} 0.20345$	Radian	.2048803359	0.2049
	Degree	11.73877855	11.7°
d) $\cos^{-1} 1.318$	Radian	ERR:DOMAIN	
	Degree	ERR:DOMAIN	

The value 1.318 is not in $[-1, 1]$, the domain of the arccosine function.

e) The cosecant function is the reciprocal of the sine function:

$\csc^{-1} 8.205 =$
$\sin^{-1}(1/8.205)$

Radian	.1221806653	0.1222
Degree	7.000436462	7.0°

Now Try Exercises 21 and 25. ■

✱ memorize

CONNECTING
the CONCEPTS

Domains and Ranges

The following is a summary of the domains and the ranges of the trigonometric functions together with a summary of the domains and the ranges of the inverse trigonometric functions. For completeness, we have included the arccosecant, the arcsecant, and the arccotangent, though there is a lack of uniformity in their definitions in mathematical literature.

FUNCTION	DOMAIN	RANGE
sin	All reals, $(-\infty, \infty)$	$[-1, 1]$
cos	All reals, $(-\infty, \infty)$	$[-1, 1]$
tan	All reals except $k\pi/2$, k odd	All reals, $(-\infty, \infty)$
csc	All reals except $k\pi$	$(-\infty, -1] \cup [1, \infty)$
sec	All reals except $k\pi/2$, k odd	$(-\infty, -1] \cup [1, \infty)$
cot	All reals except $k\pi$	All reals, $(-\infty, \infty)$

INVERSE FUNCTION	DOMAIN	RANGE
$\sin^{-1}$	$[-1, 1]$	$\left[-\dfrac{\pi}{2}, \dfrac{\pi}{2}\right]$
$\cos^{-1}$	$[-1, 1]$	$[0, \pi]$
$\tan^{-1}$	All reals, or $(-\infty, \infty)$	$\left(-\dfrac{\pi}{2}, \dfrac{\pi}{2}\right)$
$\csc^{-1}$	$(-\infty, -1] \cup [1, \infty)$	$\left[-\dfrac{\pi}{2}, 0\right) \cup \left(0, \dfrac{\pi}{2}\right]$
$\sec^{-1}$	$(-\infty, -1] \cup [1, \infty)$	$\left[0, \dfrac{\pi}{2}\right) \cup \left(\dfrac{\pi}{2}, \pi\right]$
$\cot^{-1}$	All reals, or $(-\infty, \infty)$	$(0, \pi)$

❋ Composition of Trigonometric Functions and Their Inverses

Various compositions of trigonometric functions and their inverses often occur in practice. For example, we might want to try to simplify an expression such as

$$\sin(\sin^{-1} x) \quad \text{or} \quad \sin\left(\cot^{-1}\frac{x}{2}\right).$$

COMPOSITION OF FUNCTIONS

REVIEW SECTION **2.3.**

In the expression on the left, we are finding "the sine of a number whose sine is x." Recall from Section 5.1 that if a function f has an inverse that is also a function, then

$$f(f^{-1}(x)) = x, \quad \text{for all } x \text{ in the } \textit{domain} \text{ of } f^{-1},$$

and

$$f^{-1}(f(x)) = x, \quad \text{for all } x \text{ in the } \textit{domain} \text{ of } f.$$

Thus, if $f(x) = \sin x$ and $f^{-1}(x) = \sin^{-1} x$, then

$$\mathbf{\sin(\sin^{-1} x) = x, \quad \text{for all } x \text{ in the } \textit{domain} \text{ of } \sin^{-1},}$$

which is any number on the interval $[-1, 1]$. Similar results hold for the other trigonometric functions.

Composition of Trigonometric Functions

$\sin(\sin^{-1} x) = x, \quad$ for all x in the domain of $\sin^{-1}$.

$\cos(\cos^{-1} x) = x, \quad$ for all x in the domain of $\cos^{-1}$.

$\tan(\tan^{-1} x) = x, \quad$ for all x in the domain of $\tan^{-1}$.

EXAMPLE 3 Simplify each of the following.

a) $\cos\left(\cos^{-1}\dfrac{\sqrt{3}}{2}\right)$
b) $\sin(\sin^{-1} 1.8)$

Solution

a) Since $\sqrt{3}/2$ is in $[-1, 1]$, the domain of $\cos^{-1}$, it follows that

$$\cos\left(\cos^{-1}\frac{\sqrt{3}}{2}\right) = \frac{\sqrt{3}}{2}.$$

b) Since 1.8 is not in $[-1, 1]$, the domain of $\sin^{-1}$, we cannot evaluate this expression. We know that there is no number with a sine of 1.8. Since we cannot find $\sin^{-1} 1.8$, we state that $\sin(\sin^{-1} 1.8)$ does not exist. **Now Try Exercise 37.** ∎

Now let's consider an expression like $\sin^{-1}(\sin x)$. We might also suspect that this is equal to x for any x in the domain of $\sin x$, but this is not true unless x is in the range of the $\sin^{-1}$ function. Note that in order to define $\sin^{-1}$, we had to restrict the domain of the sine function. In doing so, we restricted the range of the inverse sine function. Thus,

$$\sin^{-1}(\sin x) = x, \quad \text{for all } x \text{ in the } range \text{ of } \sin^{-1}.$$

Similar results hold for the other trigonometric functions.

Special Cases

$\sin^{-1}(\sin x) = x, \quad$ for all x in the range of $\sin^{-1}$.
$\cos^{-1}(\cos x) = x, \quad$ for all x in the range of $\cos^{-1}$.
$\tan^{-1}(\tan x) = x, \quad$ for all x in the range of $\tan^{-1}$.

EXAMPLE 4 Simplify each of the following.

a) $\tan^{-1}\left(\tan \dfrac{\pi}{6}\right)$
b) $\sin^{-1}\left(\sin \dfrac{3\pi}{4}\right)$

Solution

a) Since $\pi/6$ is in $(-\pi/2, \pi/2)$, the range of the $\tan^{-1}$ function, we can use $\tan^{-1}(\tan x) = x$. Thus,

$$\tan^{-1}\left(\tan \frac{\pi}{6}\right) = \frac{\pi}{6}.$$

b) Note that $3\pi/4$ is not in $[-\pi/2, \pi/2]$, the range of the $\sin^{-1}$ function. Thus we *cannot* apply $\sin^{-1}(\sin x) = x$. Instead we first find $\sin(3\pi/4)$, which is $\sqrt{2}/2$, and substitute:

$$\sin^{-1}\left(\sin \frac{3\pi}{4}\right) = \sin^{-1}\left(\frac{\sqrt{2}}{2}\right) = \frac{\pi}{4}.$$

Now Try Exercise 43. ▨

Now we find some other function compositions.

EXAMPLE 5 Simplify each of the following.

a) $\sin\left[\tan^{-1}(-1)\right]$
b) $\cos^{-1}\left(\sin \dfrac{\pi}{2}\right)$

Solution

a) $\text{Tan}^{-1}(-1)$ is the number (or angle) θ in $(-\pi/2, \pi/2)$ whose tangent is -1. That is, $\tan \theta = -1$. Thus, $\theta = -\pi/4$ and

$$\sin\left[\tan^{-1}(-1)\right] = \sin\left[-\frac{\pi}{4}\right] = -\frac{\sqrt{2}}{2}.$$

b) $\cos^{-1}\left(\sin \dfrac{\pi}{2}\right) = \cos^{-1}(1) = 0 \qquad \sin \dfrac{\pi}{2} = 1$

Now Try Exercises 47 and 49. ▨

Next, let's consider

$$\cos\left(\sin^{-1}\frac{3}{5}\right).$$

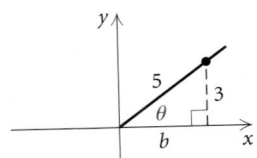

Without using a calculator, we cannot find $\sin^{-1}\frac{3}{5}$. However, we can still evaluate the entire expression by sketching a reference triangle. We are looking for angle θ such that $\sin^{-1}\frac{3}{5} = \theta$, or $\sin\theta = \frac{3}{5}$. Since $\sin^{-1}$ is defined in $[-\pi/2, \pi/2]$ and $\frac{3}{5} > 0$, we know that θ is in quadrant I. We sketch a reference right triangle, as shown at left. The angle θ in this triangle is an angle whose sine is $\frac{3}{5}$. We wish to find the cosine of this angle. Since the triangle is a right triangle, we can find the length of the base, b. It is 4. Thus we know that $\cos\theta = b/5$, or $\frac{4}{5}$. Therefore,

$$\cos\left(\sin^{-1}\frac{3}{5}\right) = \frac{4}{5}.$$

EXAMPLE 6 Find $\sin\left(\cot^{-1}\frac{x}{2}\right)$.

Solution Since $\cot^{-1}$ is defined in $(0, \pi)$, we consider quadrants I and II. We draw right triangles, as shown below, whose legs have lengths x and 2, so that $\cot\theta = x/2$.

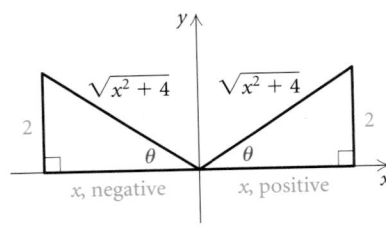

In each, we find the length of the hypotenuse and then read off the sine ratio. We get

$$\sin\left(\cot^{-1}\frac{x}{2}\right) = \frac{2}{\sqrt{x^2+4}}.$$

Now Try Exercise 55. ■

In the following example, we use a sum identity to evaluate an expression.

EXAMPLE 7 Evaluate:

$$\sin\left(\sin^{-1}\frac{1}{2} + \cos^{-1}\frac{5}{13}\right).$$

Solution Since $\sin^{-1}\frac{1}{2}$ and $\cos^{-1}\frac{5}{13}$ are both angles, the expression is the sine of a sum of two angles, so we use the identity

$$\sin(u + v) = \sin u \cos v + \cos u \sin v.$$

Thus,

$$\sin\left(\sin^{-1}\frac{1}{2} + \cos^{-1}\frac{5}{13}\right)$$

$$= \sin\left(\sin^{-1}\frac{1}{2}\right)\cdot\cos\left(\cos^{-1}\frac{5}{13}\right) + \cos\left(\sin^{-1}\frac{1}{2}\right)\cdot\sin\left(\cos^{-1}\frac{5}{13}\right)$$

$$= \frac{1}{2}\cdot\frac{5}{13} + \cos\left(\sin^{-1}\frac{1}{2}\right)\cdot\sin\left(\cos^{-1}\frac{5}{13}\right).$$ **Using composition identities**

Now since $\sin^{-1}\frac{1}{2} = \pi/6$, $\cos\left(\sin^{-1}\frac{1}{2}\right)$ simplifies to $\cos\pi/6$, or $\sqrt{3}/2$. We can illustrate this with a reference triangle in quadrant I.

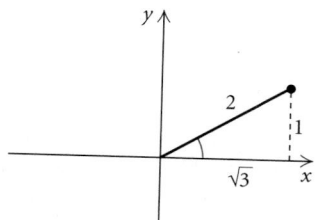

To find $\sin\left(\cos^{-1}\frac{5}{13}\right)$, we use a reference triangle in quadrant I and determine that the sine of the angle whose cosine is $\frac{5}{13}$ is $\frac{12}{13}$.

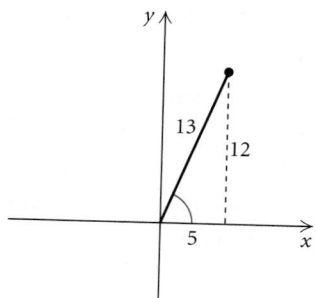

Our expression now simplifies to

$$\frac{1}{2}\cdot\frac{5}{13} + \frac{\sqrt{3}}{2}\cdot\frac{12}{13}, \quad\text{or}\quad \frac{5 + 12\sqrt{3}}{26}.$$

sin(sin−1(1/2)+cos−1(5/13))
.9917157573
(5+12√(3))/26
.9917157573

Thus,

$$\sin\left(\sin^{-1}\frac{1}{2} + \cos^{-1}\frac{5}{13}\right) = \frac{5 + 12\sqrt{3}}{26}.$$

Now Try Exercise 63. ■

7.4 Exercise Set

Find each of the following exactly in radians and degrees.

1. $\sin^{-1}\left(-\dfrac{\sqrt{3}}{2}\right)$

2. $\cos^{-1}\dfrac{1}{2}$

3. $\tan^{-1} 1$

4. $\sin^{-1} 0$

5. $\cos^{-1}\dfrac{\sqrt{2}}{2}$

6. $\sec^{-1}\sqrt{2}$

7. $\tan^{-1} 0$

8. $\tan^{-1}\dfrac{\sqrt{3}}{3}$

9. $\cos^{-1}\dfrac{\sqrt{3}}{2}$

10. $\cot^{-1}\left(-\dfrac{\sqrt{3}}{3}\right)$

11. $\csc^{-1} 2$

12. $\sin^{-1}\dfrac{1}{2}$

13. $\cot^{-1}\left(-\sqrt{3}\right)$

14. $\tan^{-1}(-1)$

15. $\sin^{-1}\left(-\dfrac{1}{2}\right)$

16. $\cos^{-1}\left(-\dfrac{\sqrt{2}}{2}\right)$

17. $\cos^{-1} 0$

18. $\sin^{-1}\dfrac{\sqrt{3}}{2}$

19. $\sec^{-1} 2$

20. $\csc^{-1}(-1)$

Use a calculator to find each of the following in radians, rounded to four decimal places, and in degrees, rounded to the nearest tenth of a degree.

21. $\tan^{-1} 0.3673$

22. $\cos^{-1}(-0.2935)$

23. $\sin^{-1} 0.9613$

24. $\sin^{-1}(-0.6199)$

25. $\cos^{-1}(-0.9810)$

26. $\tan^{-1} 158$

27. $\csc^{-1}(-6.2774)$

28. $\sec^{-1} 1.1677$

29. $\tan^{-1}(1.091)$

30. $\cot^{-1} 1.265$

31. $\sin^{-1}(-0.8192)$

32. $\cos^{-1}(-0.2716)$

33. State the domains of the inverse sine, inverse cosine, and inverse tangent functions.

34. State the ranges of the inverse sine, inverse cosine, and inverse tangent functions.

35. *Angle of Depression.*   An airplane is flying at an altitude of 2000 ft toward an island. The straight-line distance from the airplane to the island is *d* feet. Express θ, the angle of depression, as a function of *d*.

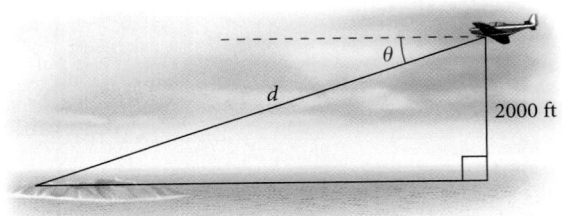

36. *Angle of Inclination.*   A guy wire is attached to the top of a 50-ft pole and stretched to a point that is *d* feet from the bottom of the pole. Express β, the angle of inclination, as a function of *d*.

Evaluate.

37. $\sin\left(\sin^{-1} 0.3\right)$

38. $\tan\left[\tan^{-1}(-4.2)\right]$

39. $\cos^{-1}\left[\cos\left(-\dfrac{\pi}{4}\right)\right]$

40. $\sin^{-1}\left(\sin\dfrac{2\pi}{3}\right)$

41. $\sin^{-1}\left(\sin\dfrac{\pi}{5}\right)$

42. $\cot^{-1}\left(\cot\dfrac{2\pi}{3}\right)$

43. $\tan^{-1}\left(\tan\dfrac{2\pi}{3}\right)$

44. $\cos^{-1}\left(\cos\dfrac{\pi}{7}\right)$

45. $\sin\left(\tan^{-1}\dfrac{\sqrt{3}}{3}\right)$

46. $\cos\left(\sin^{-1}\dfrac{\sqrt{3}}{2}\right)$

47. $\tan\left(\cos^{-1}\dfrac{\sqrt{2}}{2}\right)$

48. $\cos^{-1}(\sin\pi)$

49. $\sin^{-1}\left(\cos\dfrac{\pi}{6}\right)$

50. $\sin^{-1}\left[\tan\left(-\dfrac{\pi}{4}\right)\right]$

51. $\tan(\sin^{-1}0.1)$

52. $\cos\left(\tan^{-1}\dfrac{\sqrt{3}}{4}\right)$

53. $\sin^{-1}\left(\sin\dfrac{7\pi}{6}\right)$

54. $\tan^{-1}\left[\tan\left(-\dfrac{3\pi}{4}\right)\right]$

Find.

55. $\sin\left(\tan^{-1}\dfrac{a}{3}\right)$

56. $\tan\left(\cos^{-1}\dfrac{3}{x}\right)$

57. $\cot\left(\sin^{-1}\dfrac{p}{q}\right)$

58. $\sin(\cos^{-1}x)$

59. $\tan\left(\sin^{-1}\dfrac{p}{\sqrt{p^2+9}}\right)$

60. $\tan\left(\dfrac{1}{2}\sin^{-1}\dfrac{1}{2}\right)$

61. $\cos\left(\dfrac{1}{2}\sin^{-1}\dfrac{\sqrt{3}}{2}\right)$

62. $\sin\left(2\cos^{-1}\dfrac{3}{5}\right)$

Evaluate.

63. $\cos\left(\sin^{-1}\dfrac{\sqrt{2}}{2}+\cos^{-1}\dfrac{3}{5}\right)$

64. $\sin\left(\sin^{-1}\dfrac{1}{2}+\cos^{-1}\dfrac{3}{5}\right)$

65. $\sin(\sin^{-1}x+\cos^{-1}y)$

66. $\cos(\sin^{-1}x-\cos^{-1}y)$

67. $\sin(\sin^{-1}0.6032+\cos^{-1}0.4621)$

68. $\cos(\sin^{-1}0.7325-\cos^{-1}0.4838)$

Collaborative Discussion and Writing

69. Explain in your own words why the ranges of the inverse trigonometric functions are restricted.

70. How does the graph of $y=\sin^{-1}x$ differ from the graph of $y=\sin x$?

71. Why is it that
$$\sin\dfrac{5\pi}{6}=\dfrac{1}{2},\quad\text{but}\quad\sin^{-1}\left(\dfrac{1}{2}\right)\neq\dfrac{5\pi}{6}?$$

Skill Maintenance

In each of Exercises 72–80, fill in the blank with the correct term. Some of the given choices will not be used.

  linear speed
  angular speed
  angle of elevation
  angle of depression
  complementary
  supplementary
  similar
  congruent
  circular
  periodic
  period
  amplitude
  acute
  obtuse
  quadrantal
  radian measure

72. A function f is said to be _____ if there exists a positive constant p such that $f(s+p)=f(s)$ for all s in the domain of f.

73. The _____ of a rotation is the ratio of the distance s traveled by a point at a radius r from the center of rotation to the length of the radius r.

74. Triangles are _____ if their corresponding angles have the same measure.

75. The angle between the horizontal and a line of sight below the horizontal is called a(n) _____ .

76. _____ is the amount of rotation per unit of time.

77. Two positive angles are _____ if their sum is 180°.

78. The _____ of a periodic function is one half of the distance between its maximum and minimum function values.

79. A(n) _____ angle is an angle with measure greater than $0°$ and less than $90°$.

80. Trigonometric functions with domains composed of real numbers are called _____ functions.

Synthesis

Prove the identity.

81. $\sin^{-1} x + \cos^{-1} x = \dfrac{\pi}{2}$

82. $\tan^{-1} x + \cot^{-1} x = \dfrac{\pi}{2}$

83. $\sin^{-1} x = \tan^{-1} \dfrac{x}{\sqrt{1 - x^2}}$

84. $\tan^{-1} x = \sin^{-1} \dfrac{x}{\sqrt{x^2 + 1}}$

85. $\sin^{-1} x = \cos^{-1} \sqrt{1 - x^2}$, for $x \geq 0$

86. $\cos^{-1} x = \tan^{-1} \dfrac{\sqrt{1 - x^2}}{x}$, for $x > 0$

87. *Height of a Mural.* An art student's eye is at a point A, looking at a mural of height h, with the bottom of the mural y feet above the eye (see the figure). The eye is x feet from the wall. Write an expression for θ in terms of x, y, and h. Then evaluate the expression when $x = 20$ ft, $y = 7$ ft, and $h = 25$ ft.

88. Use a calculator to approximate the following expression:

$$16 \tan^{-1} \dfrac{1}{5} - 4 \tan^{-1} \dfrac{1}{239}.$$

What number does this expression seem to approximate?

7.5

Solving Trigonometric Equations

❖ Solve trigonometric equations.

When an equation contains a trigonometric expression with a variable, such as $\cos x$, it is called a trigonometric equation. Some trigonometric equations are identities, such as $\sin^2 x + \cos^2 x = 1$. Now we consider equations, such as $2 \cos x = -1$, that are usually not identities. As we have done for other types of equations, we will solve such equations by finding all values for x that make the equation true.

EXAMPLE 1 Solve: $2 \cos x = -1$.

Solution We first solve for $\cos x$:

$$2 \cos x = -1$$

$$\cos x = -\frac{1}{2}.$$

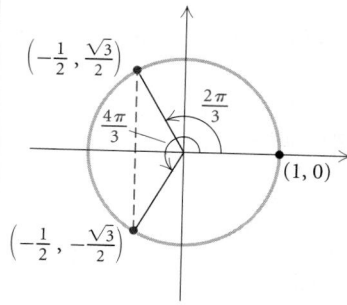

The solutions are numbers that have a cosine of $-\frac{1}{2}$. To find them, we use the unit circle (see Section 6.5).

There are just two points on the unit circle for which the cosine is $-\frac{1}{2}$, as shown in the figure at left. They are the points corresponding to $2\pi/3$ and $4\pi/3$. These numbers, plus any multiple of 2π, are the solutions:

$$\frac{2\pi}{3} + 2k\pi \quad \text{and} \quad \frac{4\pi}{3} + 2k\pi,$$

where k is any integer. In degrees, the solutions are

$$120° + k \cdot 360° \quad \text{and} \quad 240° + k \cdot 360°,$$

where k is any integer.

To check the solution to $2 \cos x = -1$, we can graph $y_1 = 2 \cos x$ and $y_2 = -1$ on the same set of axes and find the *first* coordinates of the points of intersection. Using $\pi/3$ as the Xscl facilitates our reading of the solutions. First, let's graph these equations on the interval from 0 to 2π, as shown in the figure on the left below. The only solutions in $[0, 2\pi)$ are $2\pi/3$ and $4\pi/3$.

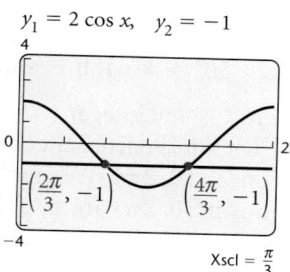

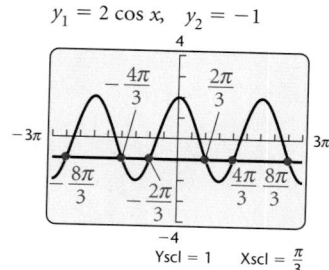

Next, let's change the viewing window to $[-3\pi, 3\pi, -4, 4]$ and graph again. Since the cosine function is periodic, there is an infinite number of solutions. A few of these appear in the graph on the right above. From the graph, we see that the solutions are $2\pi/3 + 2k\pi$ and $4\pi/3 + 2k\pi$, where k is any integer.

Now Try Exercise 1. ■

EXAMPLE 2   Solve: $4 \sin^2 x = 1$.

Solution   We begin by solving for $\sin x$:

$$4 \sin^2 x = 1$$

$$\sin^2 x = \frac{1}{4}$$

$$\sin x = \pm \frac{1}{2}.$$

Again, we use the unit circle to find those numbers having a sine of $\frac{1}{2}$ or $-\frac{1}{2}$. The solutions are

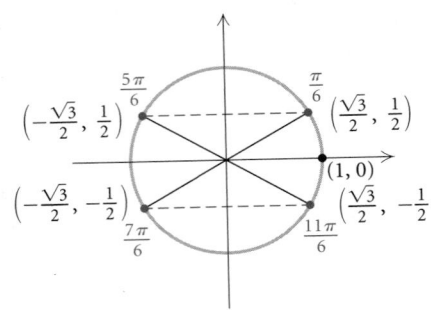

$$\frac{\pi}{6} + 2k\pi, \quad \frac{5\pi}{6} + 2k\pi, \quad \frac{7\pi}{6} + 2k\pi, \quad \text{and} \quad \frac{11\pi}{6} + 2k\pi,$$

where k is any integer. In degrees, the solutions are

$$30° + k \cdot 360°, \quad 150° + k \cdot 360°,$$
$$210° + k \cdot 360°, \quad \text{and} \quad 330° + k \cdot 360°,$$

where k is any integer.

The general solutions listed above could be condensed using odd as well as even multiples of π:

$$\frac{\pi}{6} + k\pi \quad \text{and} \quad \frac{5\pi}{6} + k\pi,$$

or, in degrees,

$$30° + k \cdot 180° \quad \text{and} \quad 150° + k \cdot 180°,$$

where k is any integer.

Let's do a partial check using a graphing calculator, checking only the solutions in $[0, 2\pi)$. We graph $y_1 = 4 \sin^2 x$ and $y_2 = 1$ and note that the solutions in $[0, 2\pi)$ are $\pi/6, 5\pi/6, 7\pi/6,$ and $11\pi/6$.

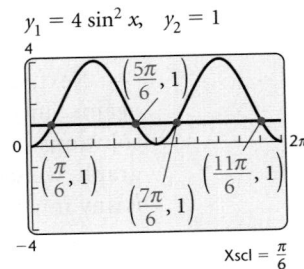

$$y_1 = 4 \sin^2 x, \quad y_2 = 1$$

Xscl $= \frac{\pi}{6}$ **Now Try Exercise 13.** ■

In most applications, it is sufficient to find just the solutions from 0 to 2π or from $0°$ to $360°$. We then remember that any multiple of 2π, or $360°$, can be added to obtain the rest of the solutions.

We must be careful to find all solutions in $[0, 2\pi)$ when solving trigonometric equations involving double angles.

EXAMPLE 3 Solve $3 \tan 2x = -3$ in the interval $[0, 2\pi)$.

Solution We first solve for $\tan 2x$:

$$3 \tan 2x = -3$$
$$\tan 2x = -1.$$

We are looking for solutions x to the equation for which

$$0 \leq x < 2\pi.$$

Multiplying by 2, we get

$$0 \leq 2x < 4\pi,$$

which is the interval we use when solving $\tan 2x = -1$.

Using the unit circle, we find points $2x$ in $[0, 4\pi)$ for which $\tan 2x = -1$. These values of $2x$ are as follows:

$$2x = \frac{3\pi}{4}, \quad \frac{7\pi}{4}, \quad \frac{11\pi}{4}, \quad \text{and} \quad \frac{15\pi}{4}.$$

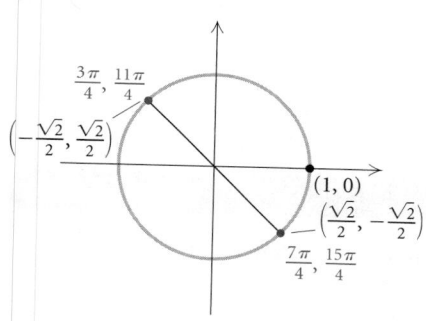

Thus the desired values of x in $[0, 2\pi)$ are each of these values divided by 2. Therefore,

$$x = \frac{3\pi}{8}, \quad \frac{7\pi}{8}, \quad \frac{11\pi}{8}, \quad \text{and} \quad \frac{15\pi}{8}.$$

Now Try Exercise 21. ■

Calculators are needed to solve some trigonometric equations. Answers can be found in radians or degrees, depending on the mode setting.

EXAMPLE 4 Solve $\frac{1}{2} \cos \phi + 1 = 1.2108$ in $[0, 360°)$.

Solution We have

$$\frac{1}{2} \cos \phi + 1 = 1.2108$$

$$\frac{1}{2} \cos \phi = 0.2108$$

$$\cos \phi = 0.4216.$$

Using a calculator set in DEGREE mode (see the window at left), we find that the reference angle, $\cos^{-1} 0.4216$, is

$$\phi \approx 65.06°.$$

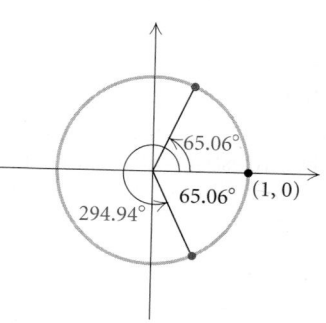

Since $\cos \phi$ is positive, the solutions are in quadrants I and IV. The solutions in $[0, 360°)$ are

$$65.06° \quad \text{and} \quad 360° - 65.06° = 294.94°.$$

Now Try Exercise 9. ■

EXAMPLE 5   Solve $2 \cos^2 u = 1 - \cos u$ in $[0°, 360°)$.

ALGEBRAIC SOLUTION

We use the principle of zero products:

$$2 \cos^2 u = 1 - \cos u$$
$$2 \cos^2 u + \cos u - 1 = 0$$
$$(2 \cos u - 1)(\cos u + 1) = 0$$
$$2 \cos u - 1 = 0 \quad or \quad \cos u + 1 = 0$$
$$2 \cos u = 1 \quad or \quad \cos u = -1$$
$$\cos u = \frac{1}{2} \quad or \quad \cos u = -1.$$

Thus,

$$u = 60°, 300° \quad or \quad u = 180°.$$

The solutions in $[0°, 360°)$ are $60°$, $180°$, and $300°$.

GRAPHICAL SOLUTION

We can use either the Intersect method or the Zero method to solve trigonometric equations. Here we illustrate by solving the equation using both methods. We set the calculator in DEGREE mode.

INTERSECT METHOD.   We graph the equations

$$y_1 = 2 \cos^2 x \quad \text{and} \quad y_2 = 1 - \cos x$$

and use the INTERSECT feature to find the first coordinates of the points of intersection.

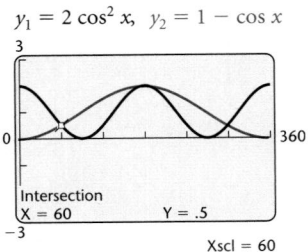

$y_1 = 2 \cos^2 x, \ y_2 = 1 - \cos x$

Intersection
X = 60   Y = .5

Xscl = 60

The leftmost solution is $60°$. Using the INTERSECT feature two more times, we find the other solutions, $180°$ and $300°$.

ZERO METHOD. We write the equation in the form

$$2 \cos^2 u + \cos u - 1 = 0.$$

Then we graph

$$y = 2 \cos^2 x + \cos x - 1$$

and use the ZERO feature to determine the zeros of the function.

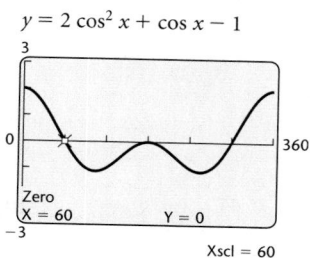

$$y = 2 \cos^2 x + \cos x - 1$$

The leftmost zero is 60°. Using the ZERO feature two more times, we find the other zeros, 180° and 300°. The solutions in [0°, 360°) are 60°, 180°, and 300°.

Now Try Exercise 15. ▧

EXAMPLE 6   Solve $\sin^2 \beta - \sin \beta = 0$ in $[0, 2\pi)$.

Solution   We factor and use the principle of zero products:

$$\sin^2 \beta - \sin \beta = 0$$
$$\sin \beta \,(\sin \beta - 1) = 0 \qquad \text{Factoring}$$
$$\sin \beta = 0 \quad \text{or} \quad \sin \beta - 1 = 0$$
$$\sin \beta = 0 \quad \text{or} \quad \sin \beta = 1$$
$$\beta = 0, \pi \quad \text{or} \qquad \beta = \frac{\pi}{2}.$$

The solutions in $[0, 2\pi)$ are 0, $\pi/2$, and π.

Now Try Exercise 17. ▧

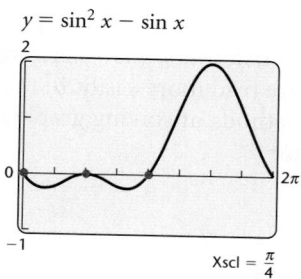

$$y = \sin^2 x - \sin x$$

If a trigonometric equation is quadratic but difficult or impossible to factor, we use the *quadratic formula*.

EXAMPLE 7   Solve $10 \sin^2 x - 12 \sin x - 7 = 0$ in $[0°, 360°)$.

Solution   This equation is quadratic in $\sin x$ with $a = 10$, $b = -12$, and $c = -7$. Substituting into the quadratic formula, we get

$$\sin x = \frac{-b \pm \sqrt{b^2 - 4ac}}{2a} \quad \begin{array}{l}\text{Using the quadratic} \\ \text{formula}\end{array}$$

$$= \frac{-(-12) \pm \sqrt{(-12)^2 - 4(10)(-7)}}{2 \cdot 10} \quad \text{Substituting}$$

$$= \frac{12 \pm \sqrt{144 + 280}}{20} = \frac{12 \pm \sqrt{424}}{20}$$

$$\approx \frac{12 \pm 20.5913}{20}$$

$$\sin x \approx 1.6296 \quad or \quad \sin x \approx -0.4296.$$

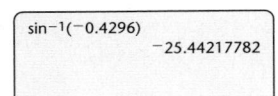

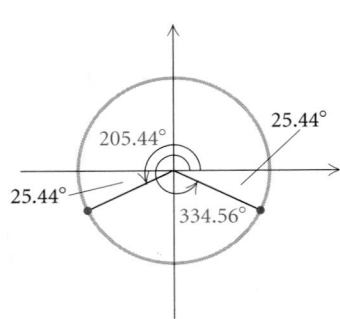

Since sine values are never greater than 1, the first of the equations has no solution. Using the other equation, we find the reference angle to be $25.44°$. Since $\sin x$ is negative, the solutions are in quadrants III and IV.

Thus the solutions in $[0°, 360°)$ are

$$180° + 25.44° = 205.44° \quad \text{and} \quad 360° - 25.44° = 334.56°.$$

Now Try Exercise 23. ■

Trigonometric equations can involve more than one function.

EXAMPLE 8   Solve $2 \cos^2 x \tan x = \tan x$ in $[0, 2\pi)$.

Solution   Using a graphing calculator, we can determine that there are six solutions. If we let Xscl $= \pi/4$, the solutions are read more easily. In the figures at left, we show the Intersect and Zero methods of solving graphically. Each illustrates that the solutions in $[0, 2\pi)$ are

$$0, \quad \frac{\pi}{4}, \quad \frac{3\pi}{4}, \quad \pi, \quad \frac{5\pi}{4}, \quad \text{and} \quad \frac{7\pi}{4}.$$

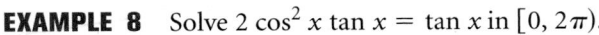

We can verify these solutions algebraically, as follows:

$$2 \cos^2 x \tan x = \tan x$$

$$2 \cos^2 x \tan x - \tan x = 0$$

$$\tan x (2 \cos^2 x - 1) = 0$$

$$\tan x = 0 \quad or \quad 2 \cos^2 x - 1 = 0$$

$$\cos^2 x = \frac{1}{2}$$

$$\cos x = \pm \frac{\sqrt{2}}{2}$$

$$x = 0, \pi \quad or \quad x = \frac{\pi}{4}, \frac{3\pi}{4}, \frac{5\pi}{4}, \frac{7\pi}{4}.$$

Thus, $x = 0, \pi/4, 3\pi/4, \pi, 5\pi/4$, and $7\pi/4$.

Now Try Exercise 29. ■

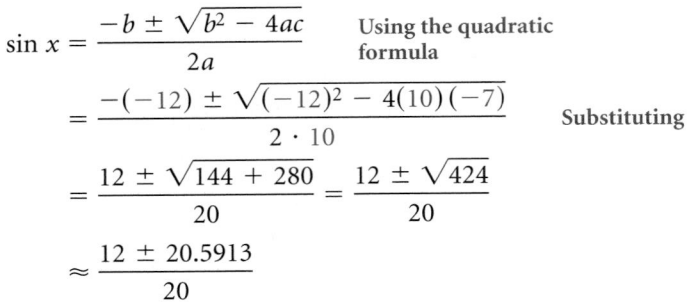

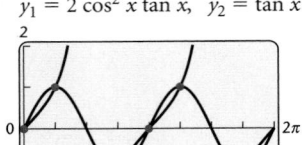

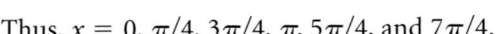

When a trigonometric equation involves more than one function, it is sometimes helpful to use identities to rewrite the equation in terms of a single function.

EXAMPLE 9 Solve $\sin x + \cos x = 1$ in $[0, 2\pi)$.

ALGEBRAIC SOLUTION

We have

$$\sin x + \cos x = 1$$
$$(\sin x + \cos x)^2 = 1^2 \qquad \text{Squaring both sides}$$
$$\sin^2 x + 2 \sin x \cos x + \cos^2 x = 1$$
$$2 \sin x \cos x + 1 = 1 \qquad \text{Using } \sin^2 x + \cos^2 x = 1$$
$$2 \sin x \cos x = 0$$
$$\sin 2x = 0. \qquad \text{Using } 2 \sin x \cos x = \sin 2x$$

We are looking for solutions x to the equation for which $0 \le x < 2\pi$. Multiplying by 2, we get $0 \le 2x < 4\pi$, which is the interval we consider to solve $\sin 2x = 0$. These values of $2x$ are 0, π, 2π, and 3π. Thus the desired values of x in $[0, 2\pi)$ satisfying this equation are 0, $\pi/2$, π, and $3\pi/2$. Now we check these in the original equation $\sin x + \cos x = 1$:

$$\sin 0 + \cos 0 = 0 + 1 = 1,$$
$$\sin \frac{\pi}{2} + \cos \frac{\pi}{2} = 1 + 0 = 1,$$
$$\sin \pi + \cos \pi = 0 + (-1) = -1,$$
$$\sin \frac{3\pi}{2} + \cos \frac{3\pi}{2} = (-1) + 0 = -1.$$

We find that π and $3\pi/2$ do not check, but the other values do. Thus the solutions in $[0, 2\pi)$ are

$$0 \quad \text{and} \quad \frac{\pi}{2}.$$

When the solution process involves squaring both sides, values are sometimes obtained that are not solutions of the original equation. As we saw in this example, it is important to check the possible solutions.

GRAPHICAL SOLUTION

We can graph the left side and then the right side of the equation as seen in the first window below. Then we look for points of intersection. We could also rewrite the equation as $\sin x + \cos x - 1 = 0$, graph the left side, and look for the zeros of the function, as illustrated in the second window below. In each window, we see the solutions in $[0, 2\pi)$ as 0 and $\pi/2$.

$y_1 = \sin x + \cos x, \quad y_2 = 1$

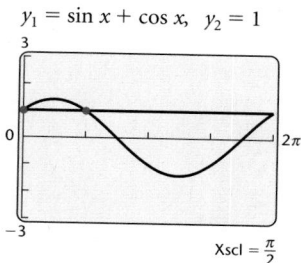

$Xscl = \frac{\pi}{2}$

$y = \sin x + \cos x - 1$

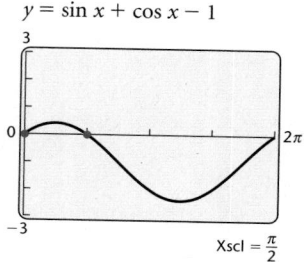

$Xscl = \frac{\pi}{2}$

This example illustrates a valuable advantage of the calculator—that is, with a graphing calculator, extraneous solutions do not appear.

Now Try Exercise 39. ■

EXAMPLE 10   Solve $\cos 2x + \sin x = 1$ in $[0, 2\pi)$.

ALGEBRAIC SOLUTION

We have

$$\cos 2x + \sin x = 1$$

$$1 - 2 \sin^2 x + \sin x = 1 \qquad \text{Using the identity}$$
$$\cos 2x = 1 - 2 \sin^2 x$$

$$-2 \sin^2 x + \sin x = 0$$

$$\sin x (-2 \sin x + 1) = 0 \qquad \text{Factoring}$$

$$\sin x = 0 \quad or \quad -2 \sin x + 1 = 0 \qquad \begin{array}{l}\text{Principle of}\\ \text{zero products}\end{array}$$

$$\sin x = 0 \quad or \qquad\qquad \sin x = \frac{1}{2}$$

$$x = 0, \pi \quad or \qquad\qquad x = \frac{\pi}{6}, \frac{5\pi}{6}.$$

All four values check. The solutions in $[0, 2\pi)$ are 0, $\pi/6$, $5\pi/6$, and π.

GRAPHICAL SOLUTION

We graph $y_1 = \cos 2x + \sin x - 1$ and look for the zeros of the function.

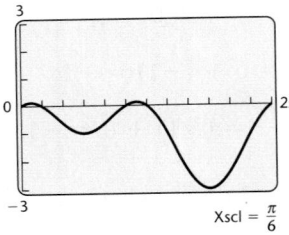

$\text{Xscl} = \frac{\pi}{6}$

The solutions in $[0, 2\pi)$ are 0, $\pi/6$, $5\pi/6$, and π.

EXAMPLE 11   Solve $\tan^2 x + \sec x - 1 = 0$ in $[0, 2\pi)$.

STUDY TIP

Check your solutions to the odd-numbered exercises in the exercise sets with the step-by-step annotated solutions in the *Student's Solutions Manual*. If you are still having difficulty with the concepts of this section, make time to view the content video that corresponds to the section.

ALGEBRAIC SOLUTION

We have

$$\tan^2 x + \sec x - 1 = 0$$
$$\sec^2 x - 1 + \sec x - 1 = 0 \qquad \text{Using the identity}$$
$$\qquad\qquad\qquad\qquad\qquad\qquad 1 + \tan^2 x = \sec^2 x, \text{ or}$$
$$\qquad\qquad\qquad\qquad\qquad\qquad \tan^2 x = \sec^2 x - 1$$
$$\sec^2 x + \sec x - 2 = 0$$
$$(\sec x + 2)(\sec x - 1) = 0 \qquad \text{Factoring}$$
$$\sec x = -2 \qquad or \quad \sec x = 1 \qquad \text{Principle of zero}$$
$$\qquad\qquad\qquad\qquad\qquad\qquad\qquad\qquad\qquad \text{products}$$
$$\cos x = -\frac{1}{2} \qquad or \quad \cos x = 1 \qquad \text{Using the identity}$$
$$\qquad\qquad\qquad\qquad\qquad\qquad\qquad\qquad\qquad \cos x = 1/\sec x$$
$$x = \frac{2\pi}{3}, \frac{4\pi}{3} \qquad or \qquad x = 0.$$

All these values check. The solutions in $[0, 2\pi)$ are 0, $2\pi/3$, and $4\pi/3$.

GRAPHICAL SOLUTION

We graph $y = \tan^2 x + \sec x - 1$, but we enter this equation in the form

$$y_1 = \tan^2 x + \frac{1}{\cos x} - 1.$$

We use the ZERO feature to find zeros of the function.

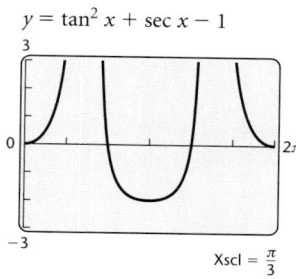

$y = \tan^2 x + \sec x - 1$

$\text{Xscl} = \frac{\pi}{3}$

The solutions in $[0, 2\pi)$ are 0, $2\pi/3$, and $4\pi/3$.

Now Try Exercise 27.

Sometimes we cannot find solutions algebraically, but we can approximate them with a graphing calculator.

EXAMPLE 12   Solve each of the following in $[0, 2\pi)$.

a) $x^2 - 1.5 = \cos x$

b) $\sin x - \cos x = \cot x$

Solution

a) In the screen on the left below, we graph $y_1 = x^2 - 1.5$ and $y_2 = \cos x$ and look for points of intersection. In the screen on the right, we graph $y_1 = x^2 - 1.5 - \cos x$ and look for the zeros of the function.

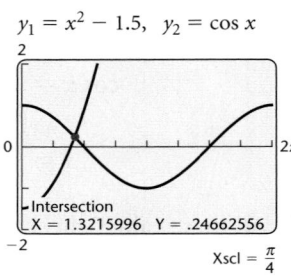

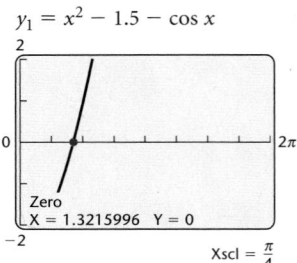

We determine the solution in $[0, 2\pi)$ to be approximately 1.32.

b) In the screen on the left below, we graph $y_1 = \sin x - \cos x$ and $y_2 = \cot x$ and determine the points of intersection. In the screen on the right, we graph the function $y_1 = \sin x - \cos x - \cot x$ and determine the zeros.

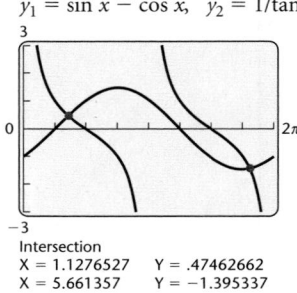

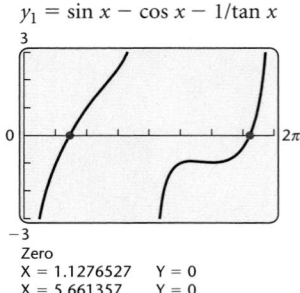

Each method leads to the approximate solutions 1.13 and 5.66 in $[0, 2\pi)$.

Now Try Exercise 47. ■

A

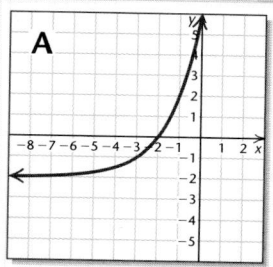

B

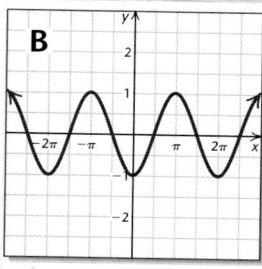

C

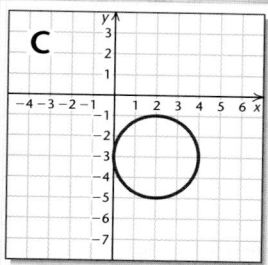

D

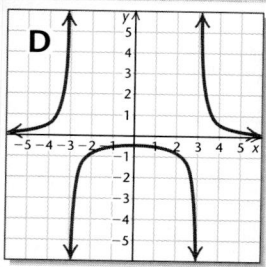

E

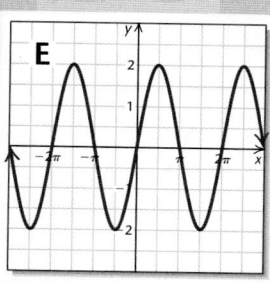

Visualizing the Graph

Match the equation with its graph.

1. $f(x) = \dfrac{4}{x^2 - 9}$

2. $f(x) = \dfrac{1}{2}\sin x - 1$

3. $(x - 2)^2 + (y + 3)^2 = 4$

4. $y = \sin^2 x + \cos^2 x$

5. $f(x) = 3 - \log x$

6. $f(x) = 2^{x+3} - 2$

7. $y = 2\cos\left(x - \dfrac{\pi}{2}\right)$

8. $y = -x^3 + 3x^2$

9. $f(x) = (x - 3)^2 + 2$

10. $f(x) = -\cos x$

Answers on page A-52

F

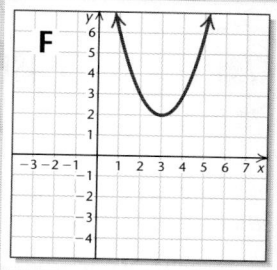

G

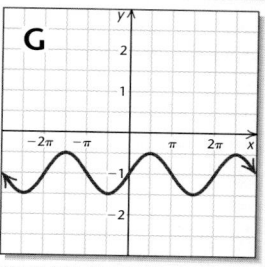

H

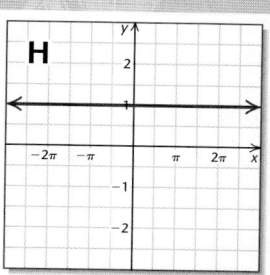

I

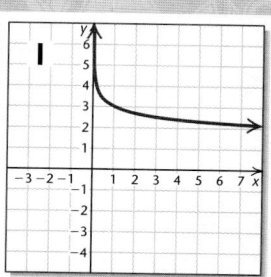

J

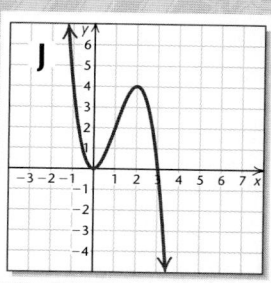

$$\boxed{7.5}\quad \textbf{Exercise Set}$$

Solve, finding all solutions. Express the solutions in both radians and degrees.

1. $\cos x = \dfrac{\sqrt{3}}{2}$

2. $\sin x = -\dfrac{\sqrt{2}}{2}$

3. $\tan x = -\sqrt{3}$

4. $\cos x = -\dfrac{1}{2}$

5. $\sin x = \dfrac{1}{2}$

6. $\tan x = -1$

7. $\cos x = -\dfrac{\sqrt{2}}{2}$

8. $\sin x = \dfrac{\sqrt{3}}{2}$

Solve, finding all solutions in $[0, 2\pi)$ or $[0°, 360°)$. Verify your answer using a graphing calculator.

9. $2 \cos x - 1 = -1.2814$

10. $\sin x + 3 = 2.0816$

11. $2 \sin x + \sqrt{3} = 0$

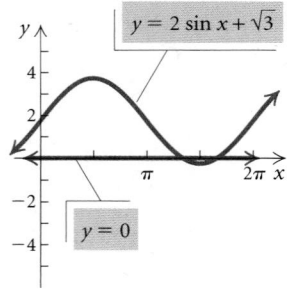

12. $2 \tan x - 4 = 1$

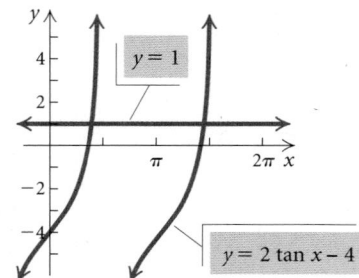

13. $2 \cos^2 x = 1$

14. $\csc^2 x - 4 = 0$

15. $2 \sin^2 x + \sin x = 1$

16. $\cos^2 x + 2 \cos x = 3$

17. $2 \cos^2 x - \sqrt{3} \cos x = 0$

18. $2 \sin^2 \theta + 7 \sin \theta = 4$

19. $6 \cos^2 \phi + 5 \cos \phi + 1 = 0$

20. $2 \sin t \cos t + 2 \sin t - \cos t - 1 = 0$

21. $\sin 2x \cos x - \sin x = 0$

22. $5 \sin^2 x - 8 \sin x = 3$

23. $\cos^2 x + 6 \cos x + 4 = 0$

24. $2 \tan^2 x = 3 \tan x + 7$

25. $7 = \cot^2 x + 4 \cot x$

26. $3 \sin^2 x = 3 \sin x + 2$

Solve, finding all solutions in $[0, 2\pi)$.

27. $\cos 2x - \sin x = 1$

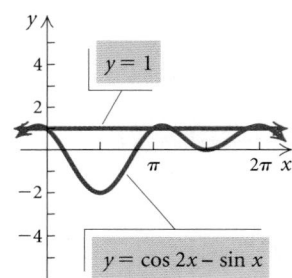

28. $2 \sin x \cos x + \sin x = 0$

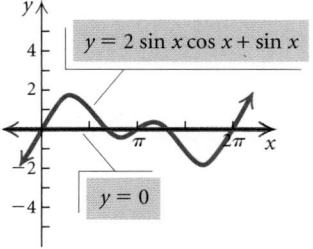

29. $\tan x \sin x - \tan x = 0$

30. $\sin 4x - 2 \sin 2x = 0$

31. $\sin 2x \cos x + \sin x = 0$

32. $\cos 2x \sin x + \sin x = 0$

33. $2 \sec x \tan x + 2 \sec x + \tan x + 1 = 0$

34. $\sin 2x \sin x - \cos 2x \cos x = -\cos x$

35. $\sin 2x + \sin x + 2 \cos x + 1 = 0$

36. $\tan^2 x + 4 = 2 \sec^2 x + \tan x$

37. $\sec^2 x - 2 \tan^2 x = 0$

38. $\cot x = \tan (2x - 3\pi)$

39. $2 \cos x + 2 \sin x = \sqrt{6}$

40. $\sqrt{3} \cos x - \sin x = 1$

41. $\sec^2 x + 2 \tan x = 6$

42. $5 \cos 2x + \sin x = 4$

43. $\cos (\pi - x) + \sin \left(x - \dfrac{\pi}{2} \right) = 1$

44. $\dfrac{\sin^2 x - 1}{\cos \left(\dfrac{\pi}{2} - x \right) + 1} = \dfrac{\sqrt{2}}{2} - 1$

Solve using a calculator, finding all solutions in $[0, 2\pi)$.

45. $x \sin x = 1$

46. $x^2 + 2 = \sin x$

47. $2 \cos^2 x = x + 1$

48. $x \cos x - 2 = 0$

49. $\cos x - 2 = x^2 - 3x$

50. $\sin x = \tan \dfrac{x}{2}$

GCM | *Some graphing calculators can use regression to fit a trigonometric function to a set of data.*

51. *Sales.* Sales of certain products fluctuate in cycles. The data in the following table show the total sales of skis per month for a business in a northern climate.

Month		Total Sales, y (in thousands)
August,	8	$ 0
November,	11	7
February,	2	14
May,	5	7
August,	8	0

a) Using the SINE REGRESSION feature on a graphing calculator, fit a sine function of the form $y = A \sin (Bx - C) + D$ to this set of data.

b) Approximate the total sales for December and for July.

52. *Daylight Hours.* The data in the following table give the number of daylight hours for certain days in Kajaani, Finland.

Day, x		Number of Daylight Hours, y
January 10,	10	5.0
February 19,	50	9.1
March 3,	62	10.4
April 28,	118	16.4
May 14,	134	18.2
June 11,	162	20.7
July 17,	198	19.5
August 22,	234	15.7
September 19,	262	12.7
October 1,	274	11.4
November 14,	318	6.7
December 28,	362	4.3

Source: The Astronomical Almanac, 1995, Washington: U.S. Government Printing Office

a) Using the SINE REGRESSION feature on a graphing calculator, model these data with an equation of the form $y = A \sin (Bx - C) + D$.

b) Approximate the number of daylight hours in Kajaani for April 22 ($x = 112$), July 4 ($x = 185$), and December 15 ($x = 349$).

c) Determine on which day of the year there will be about 12 hr of daylight.

Collaborative Discussion and Writing

53. Jan lists her answer to a problem as $\pi/6 + k\pi$, for any integer k, while Jacob lists his answer as $\pi/6 + 2k\pi$ and $7\pi/6 + 2\pi k$, for any integer k. Are their answers equivalent? Why or why not?

54. An identity is an equation that is true for all possible replacements of the variables. Explain the meaning of "possible" in this definition.

Skill Maintenance

Solve the right triangle.

55.

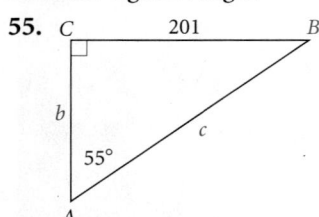

56.

T
3.8
14.2
S
t
R

Solve.

57. $\dfrac{x}{27} = \dfrac{4}{3}$

58. $\dfrac{0.01}{0.7} = \dfrac{0.2}{h}$

Synthesis

Solve in $[0, 2\pi)$.

59. $|\sin x| = \dfrac{\sqrt{3}}{2}$

60. $|\cos x| = \dfrac{1}{2}$

61. $\sqrt{\tan x} = \sqrt[4]{3}$

62. $12 \sin x - 7\sqrt{\sin x} + 1 = 0$

63. $\ln (\cos x) = 0$

64. $e^{\sin x} = 1$

65. $\sin (\ln x) = -1$

66. $e^{\ln (\sin x)} = 1$

67. *Temperature During an Illness.* The temperature T, in degrees Fahrenheit, of a patient t days into a 12-day illness is given by

$$T(t) = 101.6° + 3° \sin \left(\frac{\pi}{8} t \right).$$

Find the times t during the illness at which the patient's temperature was $103°$.

68. *Satellite Location.* A satellite circles the earth in such a manner that it is y miles from the equator (north or south, height from the surface not considered) t minutes after its launch, where

$$y = 5000 \left[\cos \frac{\pi}{45} (t - 10) \right].$$

At what times t in the interval $[0, 240]$, the first 4 hr, is the satellite 3000 mi north of the equator?

69. *Nautical Mile.* (See Exercise 60 in Exercise Set 7.2.) In Great Britain, the *nautical mile* is defined as the length of a minute of arc of the earth's radius. Since the earth is flattened at the poles, a British nautical mile varies with latitude. In fact, it is given, in feet, by the function

$$N(\phi) = 6066 - 31 \cos 2\phi,$$

where ϕ is the latitude in degrees. At what latitude north is the length of a British nautical mile found to be 6040 ft?

70. *Acceleration Due to Gravity.* (See Exercise 61 in Exercise Set 7.2.) The acceleration due to gravity is often denoted by g in a formula such as $S = \frac{1}{2}gt^2$, where S is the distance that an object falls in t seconds. The number g is generally considered constant, but in fact it varies slightly with latitude. If ϕ stands for latitude, in degrees, an excellent approximation of g is given by the formula

$$g = 9.78049(1 + 0.005288 \sin^2 \phi - 0.000006 \sin^2 2\phi),$$

where g is measured in meters per second per second at sea level. At what latitude north does $g = 9.8$?

Solve.

71. $\cos^{-1} x = \cos^{-1} \frac{3}{5} - \sin^{-1} \frac{4}{5}$

72. $\sin^{-1} x = \tan^{-1} \frac{1}{3} + \tan^{-1} \frac{1}{2}$

73. Suppose that $\sin x = 5 \cos x$. Find $\sin x \cos x$.

CHAPTER 7   Summary and Review

Important Properties and Formulas

Basic Identities

$$\sin x = \frac{1}{\csc x}, \qquad \tan x = \frac{\sin x}{\cos x},$$

$$\cos x = \frac{1}{\sec x}, \qquad \cot x = \frac{\cos x}{\sin x},$$

$$\tan x = \frac{1}{\cot x},$$

$$\sin(-x) = -\sin x,$$

$$\cos(-x) = \cos x,$$

$$\tan(-x) = -\tan x$$

Pythagorean Identities

$$\sin^2 x + \cos^2 x = 1,$$

$$1 + \cot^2 x = \csc^2 x,$$

$$1 + \tan^2 x = \sec^2 x$$

Sum and Difference Identities

$$\sin(u \pm v) = \sin u \cos v \pm \cos u \sin v,$$

$$\cos(u \pm v) = \cos u \cos v \mp \sin u \sin v,$$

$$\tan(u \pm v) = \frac{\tan u \pm \tan v}{1 \mp \tan u \tan v}$$

Cofunction Identities

$$\sin\left(\frac{\pi}{2} - x\right) = \cos x, \qquad \cos\left(\frac{\pi}{2} - x\right) = \sin x, \qquad \sin\left(x \pm \frac{\pi}{2}\right) = \pm\cos x,$$

$$\tan\left(\frac{\pi}{2} - x\right) = \cot x, \qquad \cot\left(\frac{\pi}{2} - x\right) = \tan x, \qquad \cos\left(x \pm \frac{\pi}{2}\right) = \mp\sin x$$

$$\sec\left(\frac{\pi}{2} - x\right) = \csc x, \qquad \csc\left(\frac{\pi}{2} - x\right) = \sec x,$$

Double-Angle Identities

$$\sin 2x = 2 \sin x \cos x,$$

$$\cos 2x = \cos^2 x - \sin^2 x$$

$$= 1 - 2\sin^2 x$$

$$= 2\cos^2 x - 1,$$

$$\tan 2x = \frac{2\tan x}{1 - \tan^2 x}$$

Half-Angle Identities

$$\sin\frac{x}{2} = \pm\sqrt{\frac{1 - \cos x}{2}},$$

$$\cos\frac{x}{2} = \pm\sqrt{\frac{1 + \cos x}{2}},$$

$$\tan\frac{x}{2} = \pm\sqrt{\frac{1 - \cos x}{1 + \cos x}}$$

$$= \frac{\sin x}{1 + \cos x}$$

$$= \frac{1 - \cos x}{\sin x}$$

(continued)

Product-to-Sum Identities

$$\sin x \cdot \sin y = \frac{1}{2}[\cos(x-y) - \cos(x+y)],$$

$$\cos x \cdot \cos y = \frac{1}{2}[\cos(x-y) + \cos(x+y)],$$

$$\sin x \cdot \cos y = \frac{1}{2}[\sin(x+y) + \sin(x-y)],$$

$$\cos x \cdot \sin y = \frac{1}{2}[\sin(x+y) - \sin(x-y)]$$

Sum-to-Product Identities

$$\sin x + \sin y = 2 \sin \frac{x+y}{2} \cos \frac{x-y}{2},$$

$$\sin x - \sin y = 2 \cos \frac{x+y}{2} \sin \frac{x-y}{2},$$

$$\cos y + \cos x = 2 \cos \frac{x+y}{2} \cos \frac{x-y}{2},$$

$$\cos y - \cos x = 2 \sin \frac{x+y}{2} \sin \frac{x-y}{2}$$

Inverse Trigonometric Functions

Function	Domain	Range
$y = \sin^{-1} x$	$[-1, 1]$	$\left[-\frac{\pi}{2}, \frac{\pi}{2}\right]$
$y = \cos^{-1} x$	$[-1, 1]$	$[0, \pi]$
$y = \tan^{-1} x$	$(-\infty, \infty)$	$\left(-\frac{\pi}{2}, \frac{\pi}{2}\right)$

Composition of Trigonometric Functions

The following are true for any x in the domain of the inverse function:

$$\sin(\sin^{-1} x) = x,$$
$$\cos(\cos^{-1} x) = x,$$
$$\tan(\tan^{-1} x) = x.$$

The following are true for any x in the range of the inverse function:

$$\sin^{-1}(\sin x) = x,$$
$$\cos^{-1}(\cos x) = x,$$
$$\tan^{-1}(\tan x) = x.$$

Review Exercises

Determine whether the statement is true or false.

1. $\sin^2 s \neq \sin s^2$. [7.1]

2. Given $0 < \alpha < \pi/2$ and $0 < \beta < \pi/2$ and that $\sin(\alpha + \beta) = 1$ and $\sin(\alpha - \beta) = 0$, then $\alpha = \pi/4$. [7.1]

3. If the terminal side of θ is in quadrant IV, then $\tan \theta < \cos \theta$. [7.1]

4. $\cos 5\pi/12 = \cos 7\pi/12$. [7.2]

5. Given that $\sin \theta = -\frac{2}{5}$, $\tan \theta < \cos \theta$. [7.1]

Complete the Pythagorean identity. [7.1]

6. $1 + \cot^2 x =$

7. $\sin^2 x + \cos^2 x =$

Multiply and simplify. Check using a graphing calculator. [7.1]

8. $(\tan y - \cot y)(\tan y + \cot y)$

9. $(\cos x + \sec x)^2$

Factor and simplify. Check using a graphing calculator. [7.1]

10. $\sec x \csc x - \csc^2 x$

11. $3 \sin^2 y - 7 \sin y - 20$

12. $1000 - \cos^3 u$

Simplify and check using a graphing calculator. [7.1]

13. $\dfrac{\sec^4 x - \tan^4 x}{\sec^2 x + \tan^2 x}$

14. $\dfrac{2 \sin^2 x}{\cos^3 x} \cdot \left(\dfrac{\cos x}{2 \sin x}\right)^2$

15. $\dfrac{3 \sin x}{\cos^2 x} \cdot \dfrac{\cos^2 x + \cos x \sin x}{\sin^2 x - \cos^2 x}$

16. $\dfrac{3}{\cos y - \sin y} - \dfrac{2}{\sin^2 y - \cos^2 y}$

17. $\left(\dfrac{\cot x}{\csc x}\right)^2 + \dfrac{1}{\csc^2 x}$

18. $\dfrac{4 \sin x \cos^2 x}{16 \sin^2 x \cos x}$

In Exercises 19–21, assume that all radicands are nonnegative.

19. Simplify:

$$\sqrt{\sin^2 x + 2 \cos x \sin x + \cos^2 x}. \quad [7.1]$$

20. Rationalize the denominator: $\sqrt{\dfrac{1 + \sin x}{1 - \sin x}}. \quad [7.1]$

21. Rationalize the numerator: $\sqrt{\dfrac{\cos x}{\tan x}}. \quad [7.1]$

22. Given that $x = 3 \tan \theta$, express $\sqrt{9 + x^2}$ as a trigonometric function without radicals. Assume that $0 < \theta < \pi/2$. [7.1]

Use the sum and difference formulas to write equivalent expressions. You need not simplify. [7.1]

23. $\cos\left(x + \dfrac{3\pi}{2}\right)$

24. $\tan(45° - 30°)$

25. Simplify: $\cos 27° \cos 16° + \sin 27° \sin 16°$. [7.1]

26. Find $\cos 165°$ exactly. [7.1]

27. Given that $\tan \alpha = \sqrt{3}$ and $\sin \beta = \sqrt{2}/2$ and that α and β are between 0 and $\pi/2$, evaluate $\tan(\alpha - \beta)$ exactly. [7.1]

28. Assume that $\sin \theta = 0.5812$ and $\cos \phi = 0.2341$ and that both θ and ϕ are first-quadrant angles. Evaluate $\cos(\theta + \phi)$. [7.1]

Complete the <u>cofunction</u> identity. [7.2]

29. $\cos\left(x + \dfrac{\pi}{2}\right) =$

30. $\cos\left(\dfrac{\pi}{2} - x\right) =$

31. $\sin\left(x - \dfrac{\pi}{2}\right) =$

32. Given that $\cos \alpha = -\frac{3}{5}$ and that the terminal side is in quadrant III:

  a) Find the other function values for α. [7.2]
  b) Find the six function values for $\pi/2 - \alpha$. [7.2]
  c) Find the six function values for $\alpha + \pi/2$. [7.2]

33. Find an equivalent expression for $\csc\left(x - \dfrac{\pi}{2}\right)$. [7.2]

34. Find $\tan 2\theta$, $\cos 2\theta$, and $\sin 2\theta$ and the quadrant in which 2θ lies, where $\cos \theta = -\frac{4}{5}$ and θ is in quadrant III. [7.2]

35. Find $\sin \dfrac{\pi}{8}$ exactly. [7.2]

36. Given that $\sin \beta = 0.2183$ and β is in quadrant I, find $\sin 2\beta$, $\cos \dfrac{\beta}{2}$, and $\cos 4\beta$. [7.2]

Simplify and check using a graphing calculator. [7.2]

37. $1 - 2 \sin^2 \dfrac{x}{2}$

38. $(\sin x + \cos x)^2 - \sin 2x$

39. $2 \sin x \cos^3 x + 2 \sin^3 x \cos x$

40. $\dfrac{2 \cot x}{\cot^2 x - 1}$

Prove the identity. [7.3]

41. $\dfrac{1 - \sin x}{\cos x} = \dfrac{\cos x}{1 + \sin x}$

42. $\dfrac{1 + \cos 2\theta}{\sin 2\theta} = \cot \theta$

43. $\dfrac{\tan y + \sin y}{2 \tan y} = \cos^2 \dfrac{y}{2}$

44. $\dfrac{\sin x - \cos x}{\cos^2 x} = \dfrac{\tan^2 x - 1}{\sin x + \cos x}$

Use the product-to-sum identities and the sum-to-product identities to find identities for each of the following. [7.3]

45. $3 \cos 2\theta \sin \theta$

46. $\sin \theta - \sin 4\theta$

In Exercises 47–50, use a graphing calculator to determine which expression (A)–(D) on the right can be used to complete the identity. Then prove the identity algebraically.   [7.3]

47. $\csc x - \cos x \cot x$

A. $\dfrac{\csc x}{\sec x}$

48. $\dfrac{1}{\sin x \cos x} - \dfrac{\cos x}{\sin x}$

B. $\sin x$

49. $\dfrac{\cot x - 1}{1 - \tan x}$

C. $\dfrac{2}{\sin x}$

50. $\dfrac{\cos x + 1}{\sin x} + \dfrac{\sin x}{\cos x + 1}$

D. $\dfrac{\sin x \cos x}{1 - \sin^2 x}$

Find each of the following exactly in both radians and degrees.   [7.4]

51. $\sin^{-1}\left(-\dfrac{1}{2}\right)$

52. $\cos^{-1}\dfrac{\sqrt{3}}{2}$

53. $\tan^{-1} 1$

54. $\sin^{-1} 0$

Use a calculator to find each of the following in radians, rounded to four decimal places, and in degrees, rounded to the nearest tenth of a degree.   [7.4]

55. $\cos^{-1}(-0.2194)$

56. $\cot^{-1} 2.381$

Evaluate.   [7.4]

57. $\cos\left(\cos^{-1}\dfrac{1}{2}\right)$

58. $\tan^{-1}\left(\tan \dfrac{\sqrt{3}}{3}\right)$

59. $\sin^{-1}\left(\sin \dfrac{\pi}{7}\right)$

60. $\cos\left(\sin^{-1}\dfrac{\sqrt{2}}{2}\right)$

Find.   [7.4]

61. $\cos\left(\tan^{-1}\dfrac{b}{3}\right)$

62. $\cos\left(2\sin^{-1}\dfrac{4}{5}\right)$

Solve, finding all solutions. Express the solutions in both radians and degrees.   [7.5]

63. $\cos x = -\dfrac{\sqrt{2}}{2}$

64. $\tan x = \sqrt{3}$

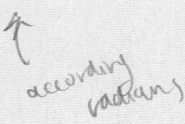
according radians

Solve, finding all solutions in $[0, 2\pi)$.   [7.5]

65. $4 \sin^2 x = 1$

all

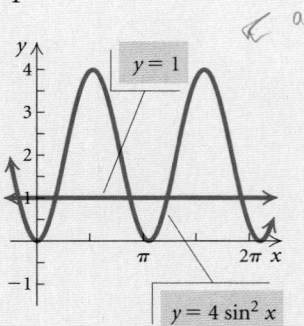

66. $\sin 2x \sin x - \cos x = 0$

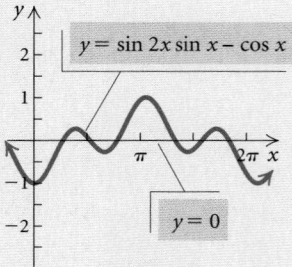

67. $2 \cos^2 x + 3 \cos x = -1$

68. $\sin^2 x - 7 \sin x = 0$

69. $\csc^2 x - 2 \cot^2 x = 0$

70. $\sin 4x + 2 \sin 2x = 0$

71. $2 \cos x + 2 \sin x = \sqrt{2}$

72. $6 \tan^2 x = 5 \tan x + \sec^2 x$

Solve using a graphing calculator, finding all solutions in $[0, 2\pi)$.   [7.5]

73. $x \cos x = 1$

74. $2 \sin^2 x = x + 1$

75. Determine the domain of the function $\cos^{-1} x$.   [7.4]

   A. $(0, \pi)$ **B.** $[-1, 1]$

   C. $[-\pi/2, \pi/2]$ **D.** $(-\infty, \infty)$

76. Simplify: $\sin^{-1}\left(\sin\dfrac{7\pi}{6}\right)$. [7.4]

    A. $-\pi/6$ **B.** $7\pi/6$

    C. $-1/2$ **D.** $11\pi/6$

77. The graph of $f(x) = \sin^{-1}x$ is which of the following? [7.4]

    A.
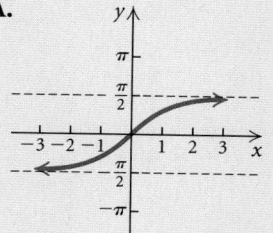
    B.
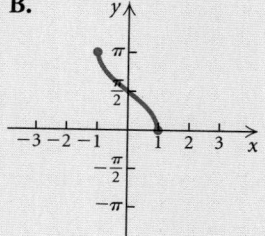

    C.
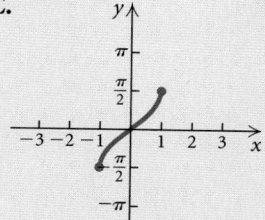
    D.
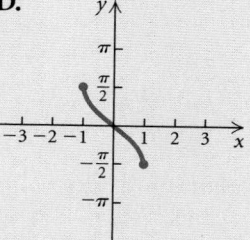

Collaborative Discussion and Writing

78. Prove the identity $2\cos^2 x - 1 = \cos^4 x - \sin^4 x$ in three ways:

    a) Start with the left side and deduce the right (method 1).

    b) Start with the right side and deduce the left (method 1).

    c) Work with each side separately until you deduce the same expression (method 2).

    Then determine the most efficient method and explain why you chose that method. [7.3]

79. Why are the ranges of the inverse trigonometric functions restricted? [7.4]

Synthesis

80. Find the measure of the angle from l_1 to l_2:

    l_1: $x + y = 3$ l_2: $2x - y = 5$. [7.1]

81. Find an identity for $\cos(u + v)$ involving only cosines. [7.1], [7.2]

82. Simplify: $\cos\left(\dfrac{\pi}{2} - x\right)[\csc x - \sin x]$. [7.2]

83. Find $\sin\theta$, $\cos\theta$, and $\tan\theta$ under the given conditions:

$$\sin 2\theta = \frac{1}{5}, \quad \frac{\pi}{2} \le 2\theta < \pi. \quad [7.2]$$

84. Prove the following equation to be an identity:

$$\ln e^{\sin t} = \sin t. \quad [7.3]$$

85. Graph: $y = \sec^{-1}x$. [7.4]

86. Show that

$$\tan^{-1}x = \frac{\sin^{-1}x}{\cos^{-1}x}$$

    is *not* an identity. [7.4]

87. Solve $e^{\cos x} = 1$ in $[0, 2\pi)$. [7.5]

CHAPTER 7   Test

Simplify.

1. $\dfrac{2\cos^2 x - \cos x - 1}{\cos x - 1}$

2. $\left(\dfrac{\sec x}{\tan x}\right)^2 - \dfrac{1}{\tan^2 x}$

3. Rationalize the denominator:
$$\sqrt{\dfrac{1 - \sin\theta}{1 + \sin\theta}}.$$
Assume the radicand is nonnegative.

4. Given that $x = 2\sin\theta$, express $\sqrt{4 - x^2}$ as a trigonometric function without radicals. Assume $0 < \theta < \pi/2$.

Use the sum or difference identities to evaluate exactly.

5. $\sin 75°$

6. $\tan\dfrac{\pi}{12}$

7. Assuming that $\cos u = \frac{5}{13}$ and $\cos v = \frac{12}{13}$ and that u and v are between 0 and $\pi/2$, evaluate $\cos(u - v)$ exactly.

8. Given that $\cos\theta = -\frac{2}{3}$ and that the terminal side is in quadrant II, find $\cos(\pi/2 - \theta)$.

9. Given that $\sin\theta = -\frac{4}{5}$ and θ is in quadrant III, find $\sin 2\theta$ and the quadrant in which 2θ lies.

10. Use a half-angle identity to evaluate $\cos\dfrac{\pi}{12}$ exactly.

11. Given that $\sin\theta = 0.6820$ and that θ is in quadrant I, find $\cos(\theta/2)$.

12. Simplify: $(\sin x + \cos x)^2 - 1 + 2\sin 2x$.

Prove each of the following identities.

13. $\csc x - \cos x \cot x = \sin x$

14. $(\sin x + \cos x)^2 = 1 + \sin 2x$

15. $(\csc\beta + \cot\beta)^2 = \dfrac{1 + \cos\beta}{1 - \cos\beta}$

16. $\dfrac{1 + \sin\alpha}{1 + \csc\alpha} = \dfrac{\tan\alpha}{\sec\alpha}$

Use the product-to-sum identities and the sum-to-product identities to find identities for each of the following.

17. $\cos 8\alpha - \cos\alpha$

18. $4\sin\beta\cos 3\beta$

19. Find $\sin^{-1}\left(-\dfrac{\sqrt{2}}{2}\right)$ exactly in degrees.

20. Find $\tan^{-1}\sqrt{3}$ exactly in radians.

21. Use a calculator to find $\cos^{-1}(-0.6716)$ in radians, rounded to four decimal places.

22. Evaluate $\cos\left(\sin^{-1}\dfrac{1}{2}\right)$.

23. Find $\tan\left(\sin^{-1}\dfrac{5}{x}\right)$.

24. Evaluate $\cos\left(\sin^{-1}\frac{1}{2} + \cos^{-1}\frac{1}{2}\right)$.

Solve, finding all solutions in $[0, 2\pi)$.

25. $4\cos^2 x = 3$

26. $2\sin^2 x = \sqrt{2}\sin x$

27. $\sqrt{3}\cos x + \sin x = 1$

28. The graph of $f(x) = \cos^{-1}x$ is which of the following?

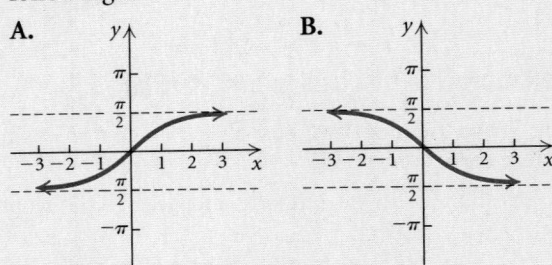

A.   B.

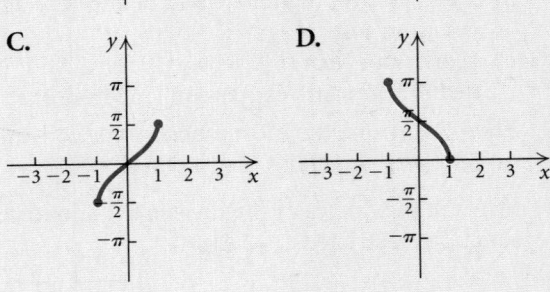

C.   D.

Synthesis

29. Find $\cos\theta$, given that $\cos 2\theta = \dfrac{5}{6}$,
$$\dfrac{3\pi}{2} < \theta < 2\pi.$$

Applications of Trigonometry

APPLICATION   A musician is constructing an octagonal recording studio in his home and needs to determine two distances for the electrician. The dimensions for the most acoustically perfect studio are shown in the figure on page 657 (*Source*: Tony Medeiros, Indianapolis, IN). Determine the distances from *D* to *F* and from *D* to *B* to the nearest tenth of an inch.

This problem appears as Example 2 in Section 8.2.

8.1   The Law of Sines

8.2   The Law of Cosines

8.3   Complex Numbers: Trigonometric Form

8.4   Polar Coordinates and Graphs

8.5   Vectors and Applications

8.6   Vector Operations

8.1   The Law of Sines

❀ Use the law of sines to solve triangles.

❀ Find the area of any triangle given the lengths of two sides and the measure of the included angle.

To **solve a triangle** means to find the lengths of all its sides and the measures of all its angles. We solved right triangles in Section 6.2. For review, let's solve the right triangle shown below. We begin by listing the known measures.

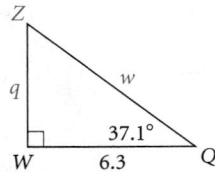

$$Q = 37.1° \qquad q = ?$$
$$W = 90° \qquad w = ?$$
$$Z = ? \qquad z = 6.3$$

Since the sum of the three angle measures of any triangle is 180°, we can immediately find the measure of the third angle:

$$Z = 180° - (90° + 37.1°)$$
$$= 52.9°.$$

Then using the tangent and cosine ratios, respectively, we can find q and w:

$$\tan 37.1° = \frac{q}{6.3}, \quad \text{or}$$
$$q = 6.3 \tan 37.1° \approx 4.8,$$

and $\cos 37.1° = \dfrac{6.3}{w}, \quad \text{or}$

$$w = \frac{6.3}{\cos 37.1°} \approx 7.9.$$

Now all six measures are known and we have solved triangle QWZ.

$$Q = 37.1° \qquad q \approx 4.8$$
$$W = 90° \qquad w \approx 7.9$$
$$Z = 52.9° \qquad z = 6.3$$

❀ Solving Oblique Triangles

The trigonometric functions can also be used to solve triangles that are not right triangles. Such triangles are called **oblique**. Any triangle, right or oblique, can be solved *if at least one side and any other two measures are known*. The five possible situations are illustrated on the next page.

1. AAS: Two angles of a triangle and a side opposite one of them are known.

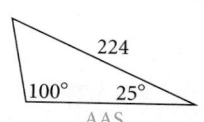

2. ASA: Two angles of a triangle and the included side are known.

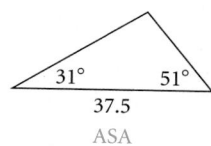

3. SSA: Two sides of a triangle and an angle opposite one of them are known. (In this case, there may be no solution, one solution, or two solutions. The latter is known as the ambiguous case.)

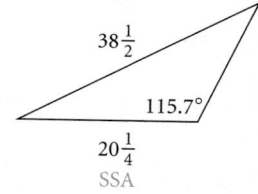

4. SAS: Two sides of a triangle and the included angle are known.

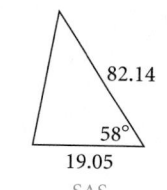

5. SSS: All three sides of the triangle are known.

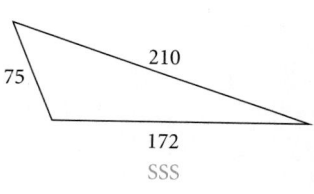

The list above does not include the situation in which only the three angle measures are given. The reason for this lies in the fact that the angle measures determine *only the shape* of the triangle and *not the size*, as shown with the following triangles. Thus we cannot solve a triangle when only the three angle measures are given.

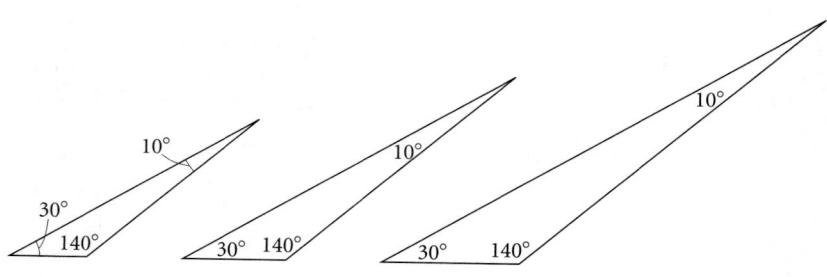

In order to solve oblique triangles, we need to derive the *law of sines* and the *law of cosines*. The law of sines applies to the first three situations listed above. The law of cosines, which we develop in Section 8.2, applies to the last two situations.

❋ The Law of Sines

We consider any oblique triangle. It may or may not have an obtuse angle. Although we look at only the acute-triangle case, the derivation of the obtuse-triangle case is essentially the same.

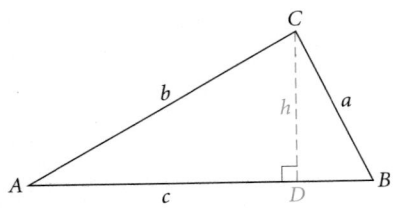

In acute △*ABC* at left, we have drawn an altitude from vertex *C*. It has length *h*. From △*ADC*, we have

$$\sin A = \frac{h}{b}, \quad \text{or} \quad h = b \sin A.$$

From △*BDC*, we have

$$\sin B = \frac{h}{a}, \quad \text{or} \quad h = a \sin B.$$

With $h = b \sin A$ and $h = a \sin B$, we now have

$$a \sin B = b \sin A$$

$$\frac{a \sin B}{\sin A \sin B} = \frac{b \sin A}{\sin A \sin B} \qquad \text{Dividing by } \sin A \sin B$$

$$\frac{a}{\sin A} = \frac{b}{\sin B}. \qquad \text{Simplifying}$$

There is no danger of dividing by 0 here because we are dealing with triangles whose angles are never 0° or 180°. Thus the sine value will never be 0.

If we were to consider altitudes from vertex *A* and vertex *B* in the triangle shown above, the same argument would give us

$$\frac{b}{\sin B} = \frac{c}{\sin C} \quad \text{and} \quad \frac{a}{\sin A} = \frac{c}{\sin C}.$$

We combine these results to obtain the law of sines.

The Law of Sines
In any triangle *ABC*,

$$\frac{a}{\sin A} = \frac{b}{\sin B} = \frac{c}{\sin C}.$$

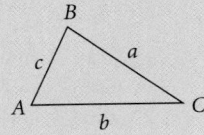

❋ Solving Triangles (AAS and ASA)

When two angles and a side of any triangle are known, the law of sines can be used to solve the triangle.

EXAMPLE 1   In $\triangle EFG$, $e = 4.56$, $E = 43°$, and $G = 57°$. Solve the triangle.

Solution   We first make a drawing. We know three of the six measures.

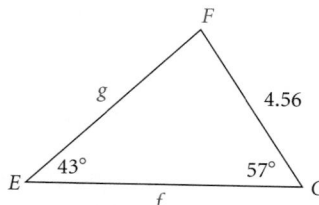

$$E = 43° \qquad e = 4.56$$
$$F = ? \qquad f = ?$$
$$G = 57° \qquad g = ?$$

From the figure, we see that we have the AAS situation. We begin by finding F:

$$F = 180° - (43° + 57°) = 80°.$$

We can now find the other two sides, using the law of sines:

$$\frac{f}{\sin F} = \frac{e}{\sin E}$$

$$\frac{f}{\sin 80°} = \frac{4.56}{\sin 43°} \qquad \text{Substituting}$$

$$f = \frac{4.56 \sin 80°}{\sin 43°} \qquad \text{Solving for } f$$

$$f \approx 6.58;$$

$$\frac{g}{\sin G} = \frac{e}{\sin E}$$

$$\frac{g}{\sin 57°} = \frac{4.56}{\sin 43°} \qquad \text{Substituting}$$

$$g = \frac{4.56 \sin 57°}{\sin 43°} \qquad \text{Solving for } g$$

$$g \approx 5.61.$$

Thus we have solved the triangle:

$$E = 43°, \qquad e = 4.56,$$
$$F = 80°, \qquad f \approx 6.58,$$
$$G = 57°, \qquad g \approx 5.61.$$

Now Try Exercise 1. ■

The law of sines is frequently used in determining distances.

EXAMPLE 2   *Rescue Mission.*   During a rescue mission, a Marine fighter pilot receives data on an unidentified aircraft from an AWACS plane and is instructed to intercept the aircraft. The diagram shown below appears on the screen, but before the distance to the point of interception appears on the screen, communications are jammed. Fortunately, the pilot remembers the law of sines. How far must the pilot fly?

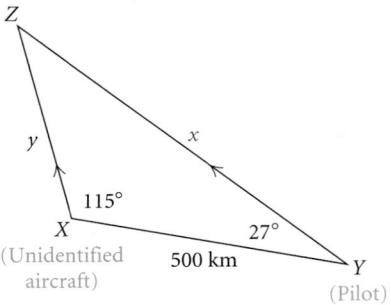

Solution   We let x represent the distance that the pilot must fly in order to intercept the aircraft and Z represent the point of interception. We first find angle Z:

$$Z = 180° - (115° + 27°)$$
$$= 38°.$$

Because this application involves the ASA situation, we use the law of sines to determine x:

$$\frac{x}{\sin X} = \frac{z}{\sin Z}$$

$$\frac{x}{\sin 115°} = \frac{500}{\sin 38°}   \qquad \textbf{Substituting}$$

$$x = \frac{500 \sin 115°}{\sin 38°}   \qquad \textbf{Solving for } x$$

$$x \approx 736.$$

Thus the pilot must fly approximately 736 km in order to intercept the unidentified aircraft.   **Now Try Exercise 23.** ■

❖ Solving Triangles (SSA)

When two sides of a triangle and an angle opposite one of them are known, the law of sines can be used to solve the triangle.

Suppose for $\triangle ABC$ that b, c, and B are given. The various possibilities are as shown in the eight cases on the following page: 5 cases when B is acute and 3 cases when B is obtuse. Note that $b < c$ in cases 1, 2, 3, and 6; $b = c$ in cases 4 and 7; and $b > c$ in cases 5 and 8.

Angle *B* Is Acute

Case 1: No solution
$b < c$; side b is too short to reach the base. No triangle is formed.

Case 2: One solution
$b < c$; side b just reaches the base and is perpendicular to it.

Case 3: Two solutions
$b < c$; an arc of radius b meets the base at two points. (This case is called the **ambiguous case**.)

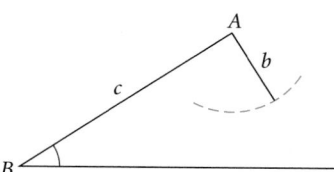

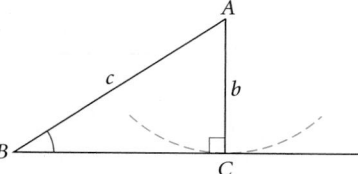

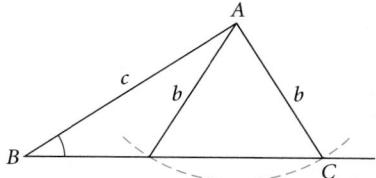

Case 4: One solution
$b = c$; an arc of radius b meets the base at just one point other than B.

Case 5: One solution
$b > c$; an arc of radius b meets the base at just one point.

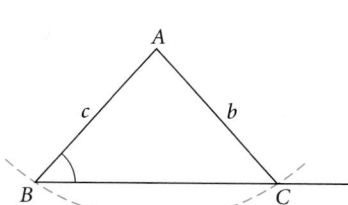

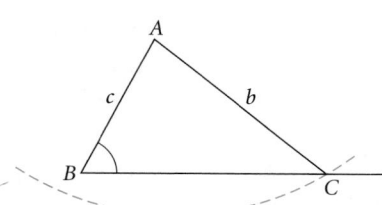

Angle *B* Is Obtuse

Case 6: No solution
$b < c$; side b is too short to reach the base. No triangle is formed.

Case 7: No solution
$b = c$; an arc of radius b meets the base only at point B. No triangle is formed.

Case 8: One solution
$b > c$; an arc of radius b meets the base at just one point.

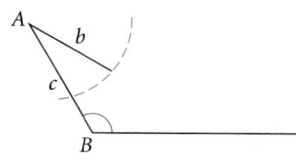

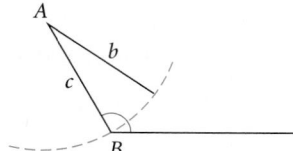

The eight cases above lead us to three possibilities in the SSA situation: *no* solution, *one* solution, or *two* solutions. Let's investigate these possibilities further, looking for ways to recognize the number of solutions.

EXAMPLE 3 *No solution.* In $\triangle QRS$, $q = 15$, $r = 28$, and $Q = 43.6°$. Solve the triangle.

Solution We make a drawing and list the known measures.

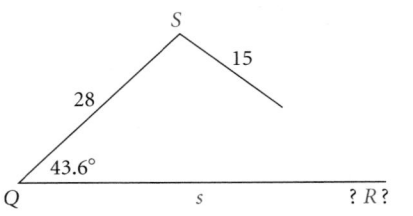

$$Q = 43.6° \qquad q = 15$$
$$R = ? \qquad r = 28$$
$$S = ? \qquad s = ?$$

We observe the SSA situation and use the law of sines to find R:

$$\frac{q}{\sin Q} = \frac{r}{\sin R}$$

$$\frac{15}{\sin 43.6°} = \frac{28}{\sin R} \qquad \text{Substituting}$$

$$\sin R = \frac{28 \sin 43.6°}{15} \qquad \text{Solving for } \sin R$$

$$\sin R \approx 1.2873.$$

Since there is no angle with a sine greater than 1, there is *no solution*.

Now Try Exercise 13. ▪

EXAMPLE 4 *One solution.* In $\triangle XYZ$, $x = 23.5$, $y = 9.8$, and $X = 39.7°$. Solve the triangle.

Solution We make a drawing and organize the given information.

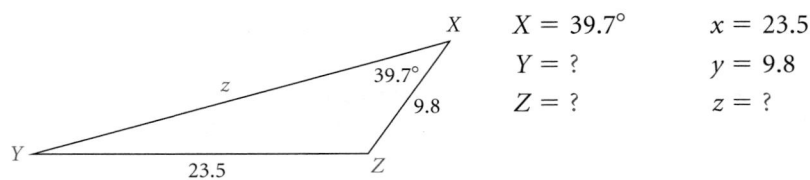

$$X = 39.7° \qquad x = 23.5$$
$$Y = ? \qquad y = 9.8$$
$$Z = ? \qquad z = ?$$

We see the SSA situation and begin by finding Y with the law of sines:

$$\frac{x}{\sin X} = \frac{y}{\sin Y}$$

$$\frac{23.5}{\sin 39.7°} = \frac{9.8}{\sin Y} \qquad \text{Substituting}$$

$$\sin Y = \frac{9.8 \sin 39.7°}{23.5} \qquad \text{Solving for } \sin Y$$

$$\sin Y \approx 0.2664.$$

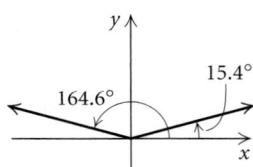

There are two angles less than 180° with a sine of 0.2664. They are 15.4° and 164.6°, to the nearest tenth of a degree. An angle of 164.6° cannot be an angle of this triangle because it already has an angle of 39.7° and these two angles would total more than 180°. Thus, 15.4° is the only possibility for Y. Therefore,

$$Z \approx 180° - (39.7° + 15.4°) \approx 124.9°.$$

We now find z:

$$\frac{z}{\sin Z} = \frac{x}{\sin X}$$

$$\frac{z}{\sin 124.9°} = \frac{23.5}{\sin 39.7°} \qquad \text{Substituting}$$

$$z = \frac{23.5 \sin 124.9°}{\sin 39.7°} \qquad \text{Solving for } z$$

$$z \approx 30.2.$$

We have now solved the triangle:

$X = 39.7°,$	$x = 23.5,$
$Y \approx 15.4°,$	$y = 9.8,$
$Z \approx 124.9°,$	$z \approx 30.2.$

Now Try Exercise 5. ▨

The next example illustrates the ambiguous case in which there are two possible solutions.

EXAMPLE 5 *Two solutions.* In $\triangle ABC$, $b = 15$, $c = 20$, and $B = 29°$. Solve the triangle.

Solution We make a drawing, list the known measures, and see that we again have the SSA situation.

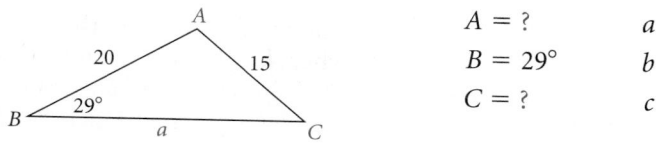

$A = ?$	$a = ?$
$B = 29°$	$b = 15$
$C = ?$	$c = 20$

We first find C:

$$\frac{b}{\sin B} = \frac{c}{\sin C}$$

$$\frac{15}{\sin 29°} = \frac{20}{\sin C} \qquad \text{Substituting}$$

$$\sin C = \frac{20 \sin 29°}{15} \approx 0.6464. \qquad \text{Solving for } \sin C$$

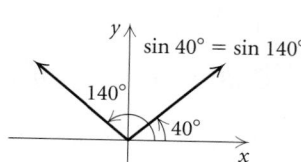

There are two angles less than 180° with a sine of 0.6464. They are 40° and 140°, to the nearest degree. This gives us two possible solutions.

Possible Solution I.

If $C = 40°$, then

$$A = 180° - (29° + 40°) = 111°.$$

Then we find a:

$$\frac{a}{\sin A} = \frac{b}{\sin B}$$

$$\frac{a}{\sin 111°} = \frac{15}{\sin 29°}$$

$$a = \frac{15 \sin 111°}{\sin 29°} \approx 29.$$

These measures make a triangle as shown below; thus we have a solution.

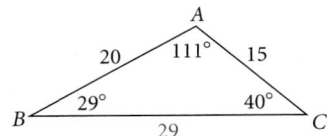

Possible Solution II.

If $C = 140°$, then

$$A = 180° - (29° + 140°) = 11°.$$

Then we find a:

$$\frac{a}{\sin A} = \frac{b}{\sin B}$$

$$\frac{a}{\sin 11°} = \frac{15}{\sin 29°}$$

$$a = \frac{15 \sin 11°}{\sin 29°} \approx 6.$$

These measures make a triangle as shown below; thus we have a second solution.

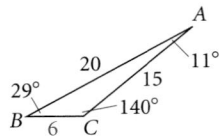

Now Try Exercise 3. ■

Examples 3–5 illustrate the SSA situation. Note that we need not memorize the eight cases or the procedures in finding no solution, one solution, or two solutions. When we are using the law of sines, the sine value leads us directly to the correct solution or solutions.

❖ The Area of a Triangle

The familiar formula for the area of a triangle, $A = \frac{1}{2}bh$, can be used only when h is known. However, we can use the method used to derive the law of sines to derive an area formula that does not involve the height.

Consider a general triangle $\triangle ABC$, with area K, as shown below.

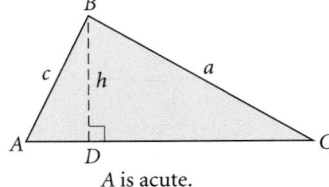

A is acute.

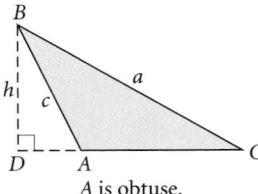

A is obtuse.

Note that in the triangle on the right, $\sin(\angle CAB) = \sin(\angle DAB)$, since $\sin A = \sin(180° - A)$. Then in each $\triangle ADB$,

$$\sin A = \frac{h}{c}, \quad \text{or} \quad h = c \sin A.$$

Substituting into the formula $K = \frac{1}{2}bh$, we get

$K = \frac{1}{2}bc \sin A.$

Any pair of sides and the included angle could have been used. Thus we also have

$K = \frac{1}{2}ab \sin C \quad \text{and} \quad K = \frac{1}{2}ac \sin B.$

The Area of a Triangle

The area K of any $\triangle ABC$ is one half the product of the lengths of two sides and the sine of the included angle:

$$K = \frac{1}{2}bc \sin A = \frac{1}{2}ab \sin C = \frac{1}{2}ac \sin B.$$

EXAMPLE 6 *Area of a Triangular Garden.* A university landscaping architecture department is designing a garden for a triangular area in a dormitory complex. Two sides of the garden, formed by the sidewalks in front of buildings A and B, measure 172 ft and 186 ft, respectively, and together form a 53° angle. The third side of the garden, formed by the sidewalk along Crossroads Avenue, measures 160 ft. What is the area of the garden to the nearest square foot?

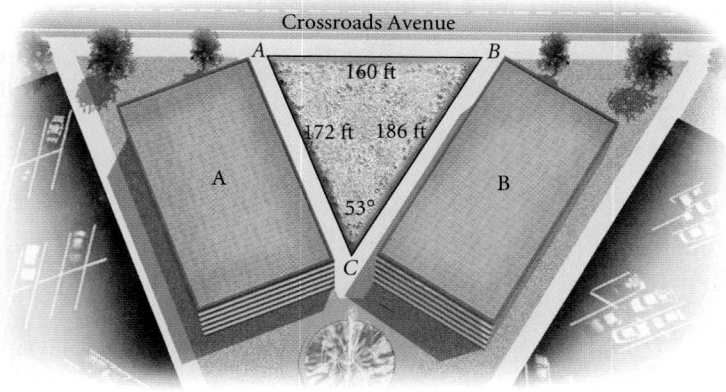

Solution Since we do not know a height of the triangle, we use the area formula:

$K = \frac{1}{2}ab \sin C$

$K = \frac{1}{2} \cdot 186 \text{ ft} \cdot 172 \text{ ft} \cdot \sin 53°$

$K \approx 12{,}775 \text{ ft}^2.$

The area of the garden is approximately 12,775 ft². **Now Try Exercise 25.** ■

8.1 Exercise Set

Solve the triangle, if possible.

1. $B = 38°, C = 21°, b = 24$

2. $A = 131°, C = 23°, b = 10$

3. $A = 36.5°, a = 24, b = 34$

4. $B = 118.3°, C = 45.6°, b = 42.1$

5. $C = 61°10', c = 30.3, b = 24.2$

6. $A = 126.5°, a = 17.2, c = 13.5$

7. $c = 3$ mi, $B = 37.48°, C = 32.16°$

8. $a = 2345$ mi, $b = 2345$ mi, $A = 124.67°$

9. $b = 56.78$ yd, $c = 56.78$ yd, $C = 83.78°$

10. $A = 129°32', C = 18°28', b = 1204$ in.

11. $a = 20.01$ cm, $b = 10.07$ cm, $A = 30.3°$

12. $b = 4.157$ km, $c = 3.446$ km, $C = 51°48'$

13. $A = 89°, a = 15.6$ in., $b = 18.4$ in.

14. $C = 46°32', a = 56.2$ m, $c = 22.1$ m

15. $a = 200$ m, $A = 32.76°, C = 21.97°$

16. $B = 115°, c = 45.6$ yd, $b = 23.8$ yd

Find the area of the triangle.

17. $B = 42°, a = 7.2$ ft, $c = 3.4$ ft

18. $A = 17°12', b = 10$ in., $c = 13$ in.

19. $C = 82°54', a = 4$ yd, $b = 6$ yd

20. $C = 75.16°, a = 1.5$ m, $b = 2.1$ m

21. $B = 135.2°, a = 46.12$ ft, $c = 36.74$ ft

22. $A = 113°, b = 18.2$ cm, $c = 23.7$ cm

Solve.

23. *Lunar Crater.* Points A and B are on opposite sides of a lunar crater. Point C is 50 m from A. The measure of $\angle BAC$ is determined to be 112° and the measure of $\angle ACB$ is determined to be 42°. What is the width of the crater?

24. *Rock Concert.* In preparation for an outdoor rock concert, a stage crew must determine how far apart to place the two large speaker columns on stage (see the figure below). What generally works best is to place them at 50° angles to the center of the front row. The distance from the center of the front row to each of the speakers is 10 ft. How far apart does the crew need to place the speakers on stage?

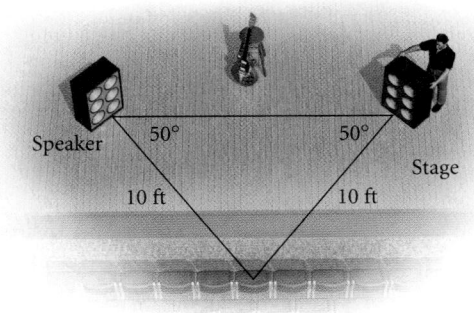

25. *Area of Back Yard.* A new homeowner has a triangular-shaped back yard. Two of the three

sides measure 53 ft and 42 ft and form an included angle of 135°. To determine the amount of fertilizer and grass seed to be purchased, the owner has to know, or at least approximate, the area of the yard. Find the area of the yard to the nearest square foot.

26. *Boarding Stable.* A rancher operates a boarding stable and temporarily needs to make an extra pen. He has a piece of rope 38 ft long and plans to tie the rope to one end of the barn (*S*) and run the rope around a tree (*T*) and back to the barn (*Q*). The tree is 21 ft from where the rope is first tied, and the rope from the barn to the tree makes an angle of 35° with the barn. Does the rancher have enough rope if he allows $4\frac{1}{2}$ ft at each end to fasten the rope?

27. *Length of Pole.* A pole leans away from the sun at an angle of 7° to the vertical. When the angle of elevation of the sun is 51°, the pole casts a shadow 47 ft long on level ground. How long is the pole?

In Exercises 28–31, keep in mind the two types of bearing considered in Sections 6.2 and 6.3.

28. *Reconnaissance Airplane.* A reconnaissance airplane leaves its airport on the east coast of the United States and flies in a direction of 85°. Because of bad weather, it returns to another airport 230 km due north of its home base. To get to the new airport, it flies in a direction of 283°. What is the total distance that the airplane flew?

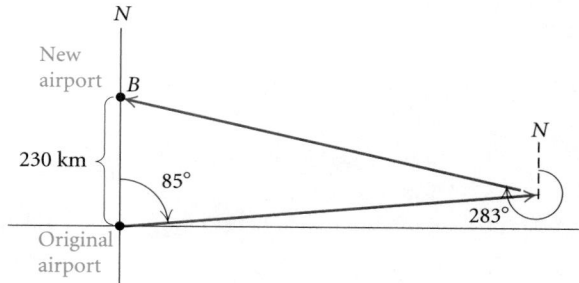

29. *Fire Tower.* A ranger in fire tower *A* spots a fire at a direction of 295°. A ranger in fire tower *B*, located 45 mi at a direction of 45° from tower *A*, spots the same fire at a direction of 255°. How far from tower *A* is the fire? from tower *B*?

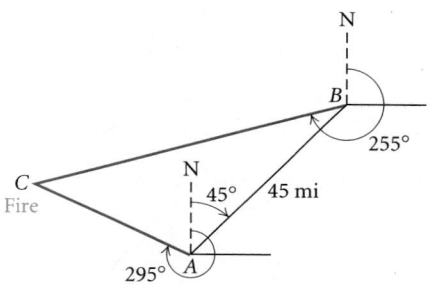

30. *Lighthouse.* A boat leaves lighthouse A and sails 5.1 km. At this time it is sighted from lighthouse B, 7.2 km west of A. The bearing of the boat from B is N65°10′E. How far is the boat from B?

31. *Mackinac Island.* Mackinac Island is located 18 mi N31°20′W of Cheboygan, Michigan, where the Coast Guard cutter Mackinaw is stationed. A freighter in distress radios the Coast Guard cutter for help. It radios its position as S78°40′E of Mackinac Island and N64°10′E of Cheboygan. How far is the freighter from Cheboygan?

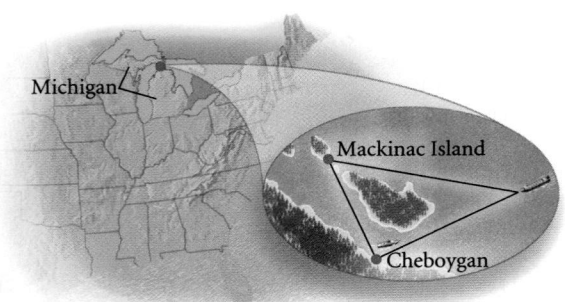

32. *Gears.* Three gears are arranged as shown in the figure below. Find the angle ϕ.

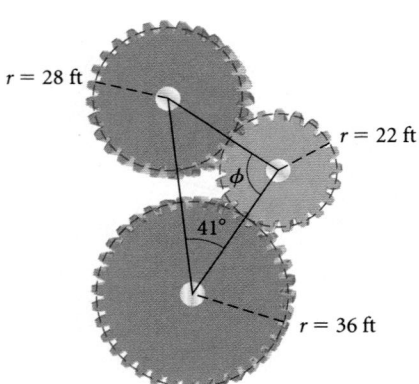

Collaborative Discussion and Writing

33. Explain why the law of sines cannot be used to find the first angle when solving a triangle given three sides.

34. We considered eight cases of solving triangles given two sides and an angle opposite one of them. Describe the relationship between side b and the height h in each.

Skill Maintenance

Find the acute angle A, in both radians and degrees, for the given function value.

35. $\cos A = 0.2213$

36. $\cos A = 1.5612$

Convert to decimal degree notation. Round to the nearest hundredth.

37. $18°14′20″$

38. $125°3′42″$

39. Find the absolute value: $|-5|$.

Find the values.

40. $\cos \dfrac{\pi}{6}$ **41.** $\sin 45°$

42. $\sin 300°$ **43.** $\cos\left(-\dfrac{2\pi}{3}\right)$

44. Multiply: $(1 - i)(1 + i)$.

Synthesis

45. Prove the following area formulas for a general triangle ABC with area represented by K.

$$K = \frac{a^2 \sin B \sin C}{2 \sin A}$$

$$K = \frac{c^2 \sin A \sin B}{2 \sin C}$$

$$K = \frac{b^2 \sin C \sin A}{2 \sin B}$$

46. *Area of a Parallelogram.* Prove that the area of a parallelogram is the product of two adjacent sides and the sine of the included angle.

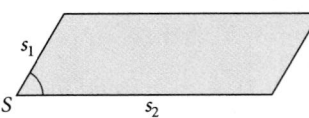

47. *Area of a Quadrilateral.* Prove that the area of a quadrilateral is one half the product of the lengths of its diagonals and the sine of the angle between the diagonals.

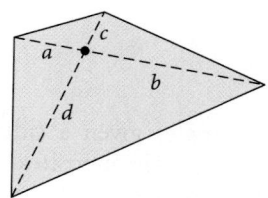

48. Find *d*.

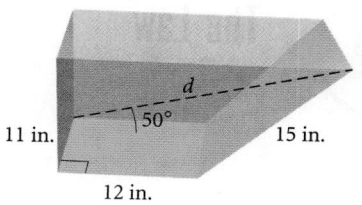

49. *Recording Studio.* A musician is constructing an octagonal recording studio in his home. The studio with dimensions shown at right is to be built within a rectangular 31′9″ by 29′9″ room. (*Source*: Tony Medeiros, Indianapolis, IN) Point *D* is 9″ from wall 2, and points *C* and *B* are each 9″ from wall 1. Using the law of sines and right triangles, determine to the nearest tenth of an inch how far point *A* is from wall 1 and from wall 4. (For more information on this studio, see Example 2 in Section 8.2.)

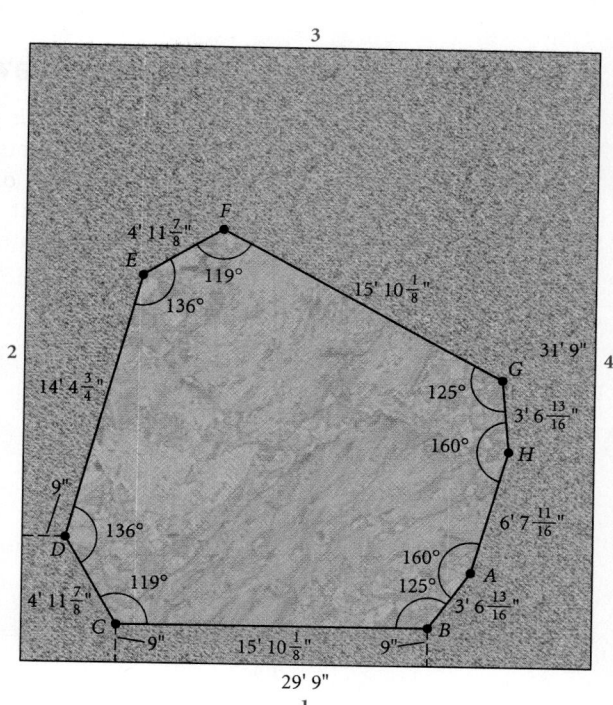

8.2 The Law of Cosines

❖ Use the law of cosines to solve triangles.

❖ Determine whether the law of sines or the law of cosines should be applied to solve a triangle.

The law of sines is used to solve triangles given a side and two angles (AAS and ASA) or given two sides and an angle opposite one of them (SSA). A second law, called the *law of cosines*, is needed to solve triangles given two sides and the included angle (SAS) or given three sides (SSS).

❖ The Law of Cosines

To derive this property, we consider any $\triangle ABC$ placed on a coordinate system. We position the origin at one of the vertices—say, C—and the positive half of the x-axis along one of the sides—say, CB. Let (x, y) be the coordinates of vertex A. Point B has coordinates $(a, 0)$ and point C has coordinates $(0, 0)$.

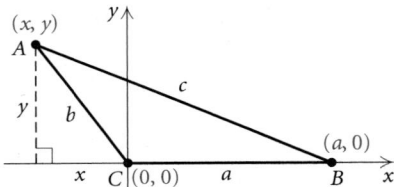

Then $\cos C = \dfrac{x}{b}$, so $x = b \cos C$

and $\sin C = \dfrac{y}{b}$, so $y = b \sin C$.

Thus point A has coordinates

$$(b \cos C, b \sin C).$$

Next, we use the distance formula to determine c^2:

$$c^2 = (x - a)^2 + (y - 0)^2,$$

or $$c^2 = (b \cos C - a)^2 + (b \sin C - 0)^2.$$

Now we multiply and simplify:

$$c^2 = b^2 \cos^2 C - 2ab \cos C + a^2 + b^2 \sin^2 C$$
$$= a^2 + b^2(\sin^2 C + \cos^2 C) - 2ab \cos C$$
$$= a^2 + b^2 - 2ab \cos C. \qquad \text{Using the identity } \sin^2 x + \cos^2 x = 1$$

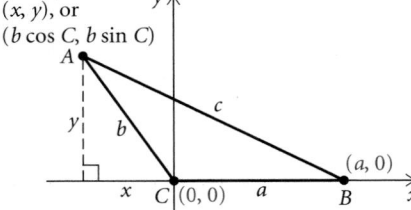

Had we placed the origin at one of the other vertices, we would have obtained

$$a^2 = b^2 + c^2 - 2bc \cos A$$
or $\quad b^2 = a^2 + c^2 - 2ac \cos B.$

The Law of Cosines

In any triangle ABC,

$$a^2 = b^2 + c^2 - 2bc \cos A,$$
$$b^2 = a^2 + c^2 - 2ac \cos B,$$
or $\quad c^2 = a^2 + b^2 - 2ab \cos C.$

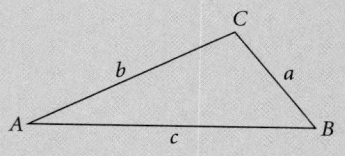

Thus, in any triangle, the square of a side is the sum of the squares of the other two sides, minus twice the product of those sides and the cosine of the included angle. When the included angle is 90°, the law of cosines reduces to the Pythagorean theorem.

❈ Solving Triangles (SAS)

When two sides of a triangle and the included angle are known, we can use the law of cosines to find the third side. The law of cosines or the law of sines can then be used to finish solving the triangle.

EXAMPLE 1 Solve $\triangle ABC$ if $a = 32$, $c = 48$, and $B = 125.2°$.

Solution We first label a triangle with the known and unknown measures.

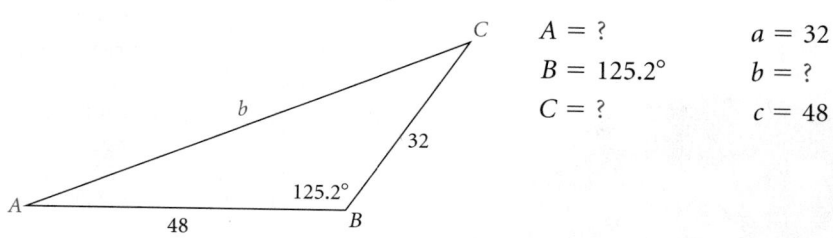

$$
\begin{array}{ll}
A = ? & a = 32 \\
B = 125.2° & b = ? \\
C = ? & c = 48
\end{array}
$$

We can find the third side using the law of cosines, as follows:

$$b^2 = a^2 + c^2 - 2ac \cos B$$
$$b^2 = 32^2 + 48^2 - 2 \cdot 32 \cdot 48 \cos 125.2° \qquad \textbf{Substituting}$$
$$b^2 \approx 5098.8$$
$$b \approx 71.$$

We now have $a = 32$, $b \approx 71$, and $c = 48$, and we need to find the other two angle measures. At this point, we can find them in two ways. One way uses the law of sines. The ambiguous case may arise, however, and we would have to be alert to this possibility. The advantage of using the law of cosines again is that if we solve for the cosine and find that its value is *negative*, then we know that the angle is obtuse. If the value of the cosine is *positive*, then the angle is acute. Thus we use the law of cosines to find a second angle.

Let's find angle A. We select the formula from the law of cosines that contains cos A and substitute:

$$a^2 = b^2 + c^2 - 2bc \cos A$$
$$32^2 = 71^2 + 48^2 - 2 \cdot 71 \cdot 48 \cos A \qquad \text{Substituting}$$
$$1024 = 5041 + 2304 - 6816 \cos A$$
$$-6321 = -6816 \cos A$$
$$\cos A \approx 0.9273768$$
$$A \approx 22.0°.$$

The third angle is now easy to find:

$$C \approx 180° - (125.2° + 22.0°)$$
$$\approx 32.8°.$$

Thus,

$$A \approx 22.0°, \qquad a = 32,$$
$$B = 125.2°, \qquad b \approx 71,$$
$$C \approx 32.8°, \qquad c = 48.$$

Now Try Exercise 1. ■

Due to errors created by rounding, answers may vary depending on the order in which they are found. Had we found the measure of angle C first in Example 1, the angle measures would have been $C \approx 34.1°$ and $A \approx 20.7°$. Variances in rounding also change the answers. Had we used 71.4 for b in Example 1, the angle measures would have been $A \approx 21.5°$ and $C \approx 33.3°$.

Suppose we used the law of sines at the outset in Example 1 to find b. We were given only three measures: $a = 32$, $c = 48$, and $B = 125.2°$. When substituting these measures into the proportions, we see that there is not enough information to use the law of sines:

$$\frac{a}{\sin A} = \frac{b}{\sin B} \rightarrow \frac{32}{\sin A} = \frac{b}{\sin 125.2°},$$
$$\frac{b}{\sin B} = \frac{c}{\sin C} \rightarrow \frac{b}{\sin 125.2°} = \frac{48}{\sin C},$$
$$\frac{a}{\sin A} = \frac{c}{\sin C} \rightarrow \frac{32}{\sin A} = \frac{48}{\sin C}.$$

In all three situations, the resulting equation, after the substitutions, still has two unknowns. Thus we cannot use the law of sines to find b.

EXAMPLE 2 *Recording Studio.* A musician is constructing an octagonal recording studio in his home and needs to determine two distances for the electrician. The dimensions for the most acoustically perfect studio are shown in the figure on the following page. (*Source:* Tony Medeiros, Indianapolis IN).

Determine the distances from D to F and from D to B to the nearest tenth of an inch.

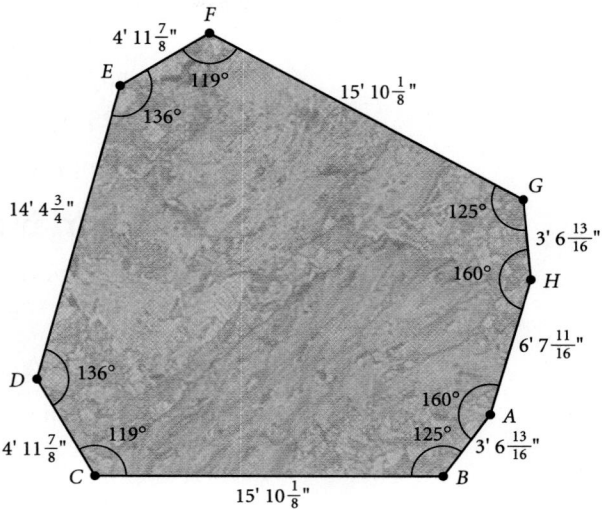

Solution   We begin by connecting points D and F and labeling the known measures of $\triangle DEF$. Converting the linear measures to decimal notation in inches, we have

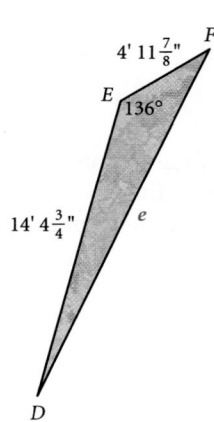

$$d = 4'11\frac{7}{8}'' = 59.875 \text{ in.,}$$

$$f = 14'4\frac{3}{4}'' = 172.75 \text{ in.,}$$

$$E = 136°.$$

We can find the measure of the third side, e, using the law of cosines:

$$e^2 = d^2 + f^2 - 2 \cdot d \cdot f \cdot \cos E \qquad \text{Using the law of cosines}$$
$$e^2 = (59.875 \text{ in.})^2 + (172.75 \text{ in.})^2$$
$$\qquad - 2(59.875 \text{ in.})(172.75 \text{ in.}) \cos 136° \qquad \text{Substituting}$$
$$e^2 \approx 48{,}308.4257 \text{ in}^2$$
$$e \approx 219.8 \text{ in.}$$

Thus it is approximately 219.8 in. from D to F.

We continue by connecting points D and B and labeling the known measures of $\triangle DCB$:

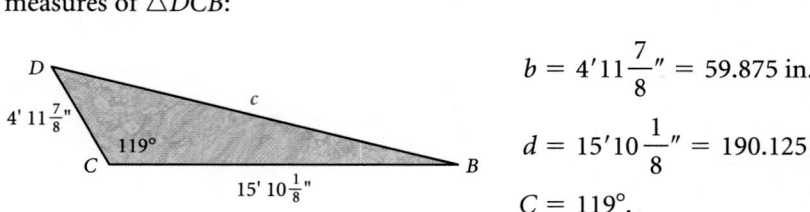

$$b = 4'11\frac{7}{8}'' = 59.875 \text{ in.,}$$

$$d = 15'10\frac{1}{8}'' = 190.125 \text{ in.,}$$

$$C = 119°.$$

Using the law of cosines, we can determine c, the length of the third side:

$$c^2 = b^2 + d^2 - 2 \cdot b \cdot d \cdot \cos C \qquad \text{Using the law of cosines}$$

$$c^2 = (59.875 \text{ in.})^2 + (190.125 \text{ in.})^2 - 2(59.875 \text{ in.})(190.125 \text{ in.}) \cos 119° \qquad \text{Substituting}$$

$$c^2 \approx 50{,}770.4191 \text{ in}^2$$

$$c \approx 225.3 \text{ in.}$$

The distance from D to B is approximately 225.3 in. **Now Try Exercise 25.**

Solving Triangles (SSS)

When all three sides of a triangle are known, the law of cosines can be used to solve the triangle.

EXAMPLE 3 Solve $\triangle RST$ if $r = 3.5$, $s = 4.7$, and $t = 2.8$.

Solution We sketch a triangle and label it with the given measures.

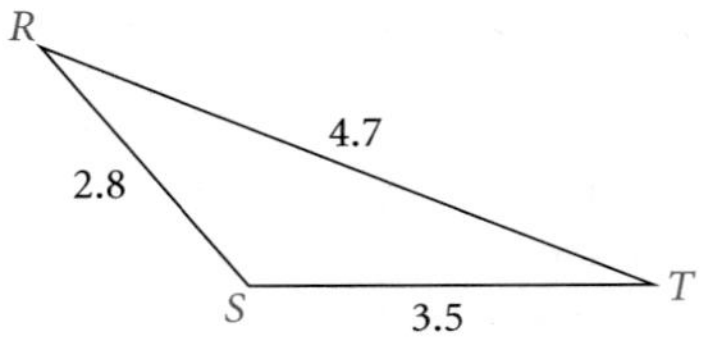

$R = ?$ $\quad r = 3.5$

$S = ?$ $\quad s = 4.7$

$T = ?$ $\quad t = 2.8$

Since we do not know any of the angle measures, we cannot use the law of sines. We begin instead by finding an angle with the law of cosines. We choose to find S first and select the formula that contains $\cos S$:

$$s^2 = r^2 + t^2 - 2rt \cos S$$

$$(4.7)^2 = (3.5)^2 + (2.8)^2 - 2(3.5)(2.8) \cos S \qquad \text{Substituting}$$

$$\cos S = \frac{(3.5)^2 + (2.8)^2 - (4.7)^2}{2(3.5)(2.8)}$$

$$\cos S \approx -0.1020408$$

$$S \approx 95.86°.$$

Similarly, we find angle R:

$$r^2 = s^2 + t^2 - 2st \cos R$$

$$(3.5)^2 = (4.7)^2 + (2.8)^2 - 2(4.7)(2.8) \cos R$$

$$\cos R = \frac{(4.7)^2 + (2.8)^2 - (3.5)^2}{2(4.7)(2.8)}$$

$$\cos R \approx 0.6717325$$

$$R \approx 47.80°.$$

Then

$$T \approx 180° - (95.86° + 47.80°) \approx 36.34°.$$

Thus,

$$R \approx 47.80°, \qquad r = 3.5,$$
$$S \approx 95.86°, \qquad s = 4.7,$$
$$T \approx 36.34°, \qquad t = 2.8.$$

Now Try Exercise 3. ■

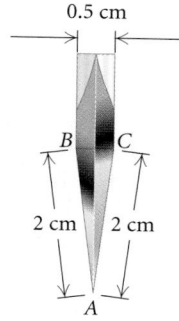

0.5 cm

B C

2 cm 2 cm

A

EXAMPLE 4 *Knife Bevel.* Knifemakers know that the *bevel* of the blade (the angle formed at the cutting edge of the blade) determines the cutting characteristics of the knife. A small bevel like that of a straight razor makes for a keen edge, but is impractical for heavy-duty cutting because the edge dulls quickly and is prone to chipping. A large bevel is suitable for heavy-duty work like chopping wood. Survival knives, being universal in application, are a compromise between small and large bevels. The diagram at left illustrates the blade of a hand-made Randall Model 18 survival knife. What is its bevel? (*Source*: Randall Made Knives, P.O. Box 1988, Orlando, FL 32802)

Solution   We know three sides of a triangle. We can use the law of cosines to find the bevel, angle A.

$$a^2 = b^2 + c^2 - 2bc \cos A$$
$$(0.5)^2 = 2^2 + 2^2 - 2 \cdot 2 \cdot 2 \cdot \cos A$$
$$0.25 = 4 + 4 - 8 \cos A$$
$$\cos A = \frac{4 + 4 - 0.25}{8}$$
$$\cos A = 0.96875$$
$$A \approx 14.36°.$$

Thus the bevel is approximately 14.36°.

Now Try Exercise 29. ■

CONNECTING
the CONCEPTS

Choosing the Appropriate Law

The following summarizes the situations in which to use the law of sines and the law of cosines.

To solve an oblique triangle:

Use the *law of sines* for: Use the *law of cosines* for:
    AAS   SAS
    ASA   SSS
    SSA

The law of cosines can also be used for the SSA situation, but since the process involves solving a quadratic equation, we do not include that option in the list above.

EXAMPLE 5 In $\triangle ABC$, three measures are given. Determine which law to use when solving the triangle. You need not solve the triangle.

a) $a = 14, b = 23, c = 10$
b) $a = 207, B = 43.8°, C = 57.6°$
c) $A = 112°, C = 37°, a = 84.7$
d) $B = 101°, a = 960, c = 1042$
e) $b = 17.26, a = 27.29, \ A = 39°$
f) $A = 61°, B = 39°, C = 80°$

Solution It is helpful to make a drawing of a triangle with the given information. The triangle need not be drawn to scale. The given parts are shown in color.

FIGURE		SITUATION	LAW TO USE
a)		SSS	Law of Cosines
b)		ASA	Law of Sines
c)		AAS	Law of Sines
d)		SAS	Law of Cosines
e)		SSA	Law of Sines
f)		AAA	Cannot be solved

Now Try Exercises 17 and 19. ■

STUDY TIP

The InterAct Math Tutorial software that accompanies this text provides practice exercises that correlate at the objective level to the odd-numbered exercises in the text. Each practice exercise is accompanied by an example and guided solution designed to involve students in the solution process. This software is available in your campus lab or on CD-ROM.

8.2 Exercise Set

Solve the triangle, if possible.

1. $A = 30°$, $b = 12$, $c = 24$

2. $B = 133°$, $a = 12$, $c = 15$

3. $a = 12$, $b = 14$, $c = 20$

4. $a = 22.3$, $b = 22.3$, $c = 36.1$

5. $B = 72°40'$, $c = 16$ m, $a = 78$ m

6. $C = 22.28°$, $a = 25.4$ cm, $b = 73.8$ cm

7. $a = 16$ m, $b = 20$ m, $c = 32$ m

8. $B = 72.66°$, $a = 23.78$ km, $c = 25.74$ km

9. $a = 2$ ft, $b = 3$ ft, $c = 8$ ft

10. $A = 96°13'$, $b = 15.8$ yd, $c = 18.4$ yd

11. $a = 26.12$ km, $b = 21.34$ km, $c = 19.25$ km

12. $C = 28°43'$, $a = 6$ mm, $b = 9$ mm

13. $a = 60.12$ mi, $b = 40.23$ mi, $C = 48.7°$

14. $a = 11.2$ cm, $b = 5.4$ cm, $c = 7$ cm

15. $b = 10.2$ in., $c = 17.3$ in., $A = 53.456°$

16. $a = 17$ yd, $b = 15.4$ yd, $c = 1.5$ yd

Determine which law applies. Then solve the triangle.

17. $A = 70°$, $B = 12°$, $b = 21.4$

18. $a = 15$, $c = 7$, $B = 62°$

19. $a = 3.3$, $b = 2.7$, $c = 2.8$

20. $a = 1.5$, $b = 2.5$, $A = 58°$

21. $A = 40.2°$, $B = 39.8°$, $C = 100°$

22. $a = 60$, $b = 40$, $C = 47°$

23. $a = 3.6$, $b = 6.2$, $c = 4.1$

24. $B = 110°30'$, $C = 8°10'$, $c = 0.912$

Solve.

25. *Poachers.* A park ranger establishes an observation post from which to watch for poachers. Despite losing her map, the ranger does have a compass and a rangefinder. She observes some poachers, and the rangefinder indicates that they are 500 ft from her position. They are headed toward big game that she knows to be 375 ft from her position. Using her compass, she finds that the poachers' azimuth (the direction measured as an angle from north) is 355° and that of the big game is 42°. What is the distance between the poachers and the game?

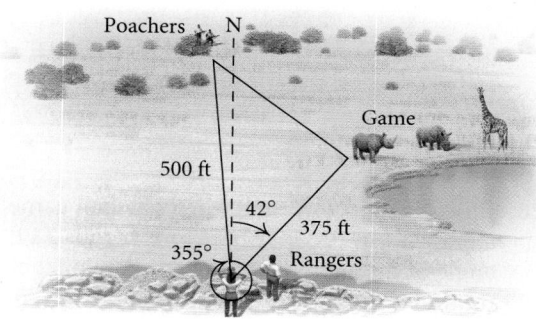

26. *Circus Highwire Act.* A circus highwire act walks up an approach wire to reach a highwire. The approach wire is 122 ft long and is currently anchored so that it forms the maximum allowable angle of 35° with the ground. A greater approach angle causes the aerialists to slip. However, the aerialists find that there is enough room to anchor the approach wire 30 ft back in order to make the approach angle less severe. When this is done, how much farther will they have to walk up the approach wire, and what will the new approach angle be?

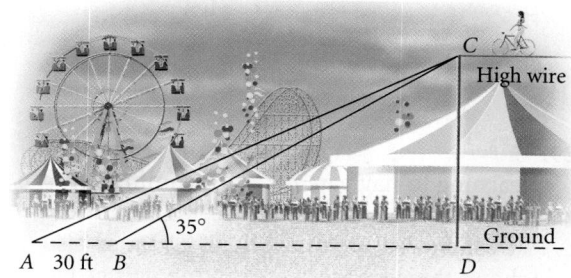

27. *In-line Skater.* An in-line skater skates on a fitness trail along the Pacific Ocean from point *A* to point *B*. As shown below, two streets inter-secting at point *C* also intersect the trail at *A* and *B*. In her car, the skater found the lengths of *AC* and *BC* to be approximately 0.5 mi and 1.3 mi, respectively. From a map, she estimates the included angle at *C* to be 110°. How far did she skate from *A* to *B*?

28. *Baseball Bunt.* A batter in a baseball game drops a bunt down the first-base line. It rolls 34 ft at an angle of 25° with the base path. The pitcher's mound is 60.5 ft from home plate. How far must the pitcher travel to pick up the ball? (*Hint:* A baseball diamond is a square.)

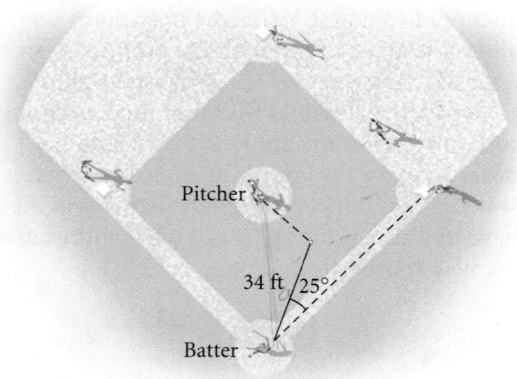

29. *Survival Trip.* A group of college students is learning to navigate for an upcoming survival trip. On a map, they have been given three points at which they are to check in. The map also shows the distances between the points. However,

to navigate they need to know the angle measurements. Calculate the angles for them.

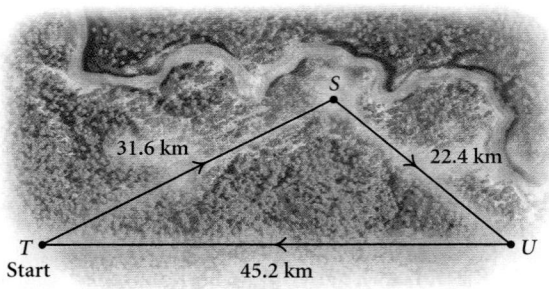

30. *Ships.* Two ships leave harbor at the same time. The first sails N15°W at 25 knots (a knot is one nautical mile per hour). The second sails N32°E at 20 knots. After 2 hr, how far apart are the ships?

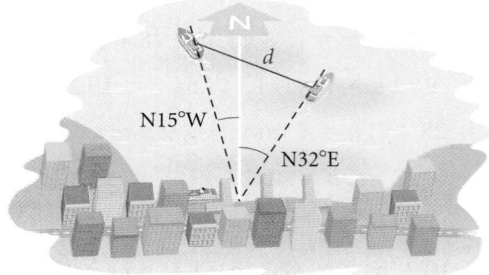

31. *Airplanes.* Two airplanes leave an airport at the same time. The first flies 150 km/h in a direction of 320°. The second flies 200 km/h in a direction of 200°. After 3 hr, how far apart are the planes?

32. *Slow-Pitch Softball.* A slow-pitch softball diamond is a square 65 ft on a side. The pitcher's mound is 46 ft from home plate. How far is it from the pitcher's mound to first base?

33. *Isosceles Trapezoid.* The longer base of an isosceles trapezoid measures 14 ft. The nonparallel sides measure 10 ft, and the base angles measure 80°.

a) Find the length of a diagonal.
b) Find the area.

34. *Dimensions of Sail.* A sail that is in the shape of an isosceles triangle has a vertex angle of 38°. The angle is included by two sides, each measuring 20 ft. Find the length of the other side of the sail.

35. Three circles are arranged as shown in the figure below. Find the length PQ.

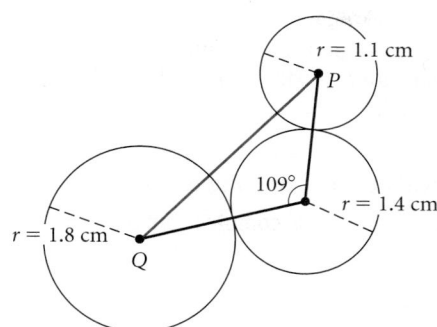

36. *Swimming Pool.* A triangular swimming pool measures 44 ft on one side and 32.8 ft on another side. These sides form an angle that measures 40.8°. How long is the other side?

Collaborative Discussion and Writing

37. Try to solve this triangle using the law of cosines. Then explain why it is easier to solve it using the law of sines.

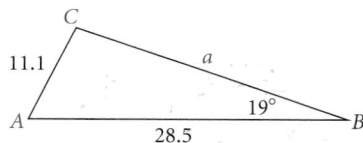

38. Explain why we cannot solve a triangle given SAS with the law of sines.

Skill Maintenance

Classify the function as linear, quadratic, cubic, quartic, rational, exponential, logarithmic, or trigonometric.

39. $f(x) = -\frac{3}{4}x^4$

40. $y - 3 = 17x$

41. $y = \sin^2 x - 3\sin x$

42. $f(x) = 2^{x-1/2}$

43. $f(x) = \dfrac{x^2 - 2x + 3}{x - 1}$

44. $f(x) = 27 - x^3$

45. $y = e^x + e^{-x} - 4$

46. $y = \log_2(x - 2) - \log_2(x + 3)$

47. $f(x) = -\cos(\pi x - 3)$

48. $y = \frac{1}{2}x^2 - 2x + 2$

Synthesis

49. *Canyon Depth.* A bridge is being built across a canyon. The length of the bridge is 5045 ft. From the deepest point in the canyon, the angles of elevation of the ends of the bridge are 78° and 72°. How deep is the canyon?

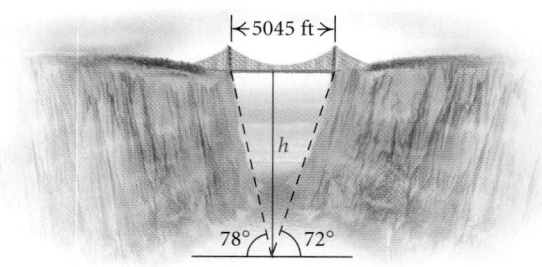

50. *Heron's Formula.* If a, b, and c are the lengths of the sides of a triangle, then the area K of the triangle is given by
$$K = \sqrt{s(s - a)(s - b)(s - c)},$$
where $s = \frac{1}{2}(a + b + c)$. The number s is called the *semiperimeter.* Prove Heron's formula. (*Hint:* Use the area formula $K = \frac{1}{2}bc\sin A$ developed in Section 8.1.) Then use Heron's formula to find the area of the triangular swimming pool described in Exercise 36.

51. *Area of Isosceles Triangle.* Find a formula for the area of an isosceles triangle in terms of the congruent sides and their included angle. Under what conditions will the area of a triangle with fixed congruent sides be maximum?

52. *Reconnaissance Plane.* A reconnaissance plane patrolling at 5000 ft sights a submarine at bearing 35° and at an angle of depression of 25°. A carrier is at bearing 105° and at an angle of depression of 60°. How far is the submarine from the carrier?

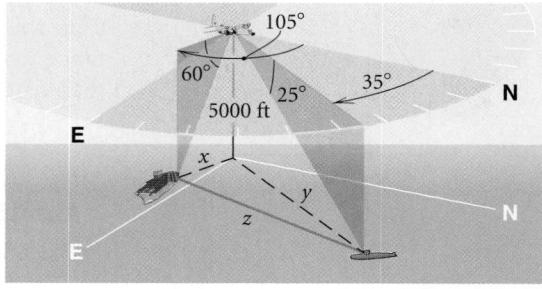

8.3

Complex Numbers: Trigonometric Form

❖ Graph complex numbers.

❖ Given a complex number in standard form, find trigonometric, or polar, notation; and given a complex number in trigonometric form, find standard notation.

❖ Use trigonometric notation to multiply and divide complex numbers.

❖ Use DeMoivre's theorem to raise complex numbers to powers.

❖ Find the *n*th roots of a complex number.

❖ Graphical Representation

Just as real numbers can be graphed on a line, complex numbers can be graphed on a plane. We graph a complex number $a + bi$ in the same way that we graph an ordered pair of real numbers (a, b). However, in place of an *x*-axis, we have a real axis, and in place of a *y*-axis, we have an imaginary axis. Horizontal distances correspond to the real part of a number. Vertical distances correspond to the imaginary part.

COMPLEX NUMBERS

REVIEW SECTION **3.1.**

EXAMPLE 1 Graph each of the following complex numbers.

a) $3 + 2i$ **b)** $-4 - 5i$ **c)** $-3i$

d) $-1 + 3i$ **e)** 2

Solution

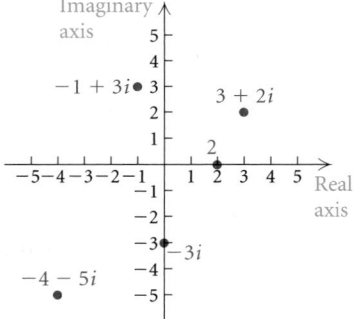

We recall that the absolute value of a real number is its distance from 0 on the number line. The absolute value of a complex number is its distance from the origin in the complex plane. For example, if $z = a + bi$, then using the distance formula, we have

$$|z| = |a + bi| = \sqrt{(a - 0)^2 + (b - 0)^2} = \sqrt{a^2 + b^2}.$$

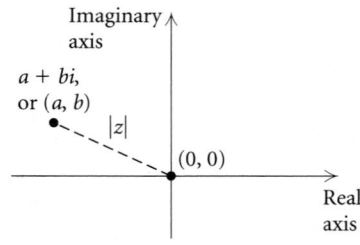

Absolute Value of a Complex Number

The **absolute value of a complex number** $a + bi$ is

$$|a + bi| = \sqrt{a^2 + b^2}.$$

EXAMPLE 2 Find the absolute value of each of the following.

a) $3 + 4i$ b) $-2 - i$ c) $\dfrac{4}{5}i$

Solution

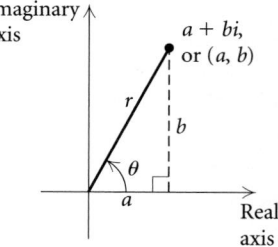

a) $|3 + 4i| = \sqrt{3^2 + 4^2} = \sqrt{9 + 16} = \sqrt{25} = 5$

b) $|-2 - i| = \sqrt{(-2)^2 + (-1)^2} = \sqrt{5}$

c) $\left|\dfrac{4}{5}i\right| = \left|0 + \dfrac{4}{5}i\right| = \sqrt{0^2 + \left(\dfrac{4}{5}\right)^2} = \dfrac{4}{5}$

We can check these results using a graphing calculator as shown at left. Note that $\sqrt{5} \approx 2.236067977$ and $4/5 = 0.8$.   **Now Try Exercises 3 and 5.** ◼

❋ Trigonometric Notation for Complex Numbers

Now let's consider a nonzero complex number $a + bi$. Suppose that its absolute value is r. If we let θ be an angle in standard position whose terminal side passes through the point (a, b), as shown in the figure, then

$$\cos \theta = \frac{a}{r}, \quad \text{or} \quad a = r \cos \theta$$

and

$$\sin \theta = \frac{b}{r}, \quad \text{or} \quad b = r \sin \theta.$$

Substituting these values for a and b into the $(a + bi)$ notation, we get

$$a + bi = r \cos \theta + (r \sin \theta)i$$
$$= r(\cos \theta + i \sin \theta).$$

This is **trigonometric notation** for a complex number $a + bi$. The number r is called the **absolute value** of $a + bi$, and θ is called the **argument** of $a + bi$. Trigonometric notation for a complex number is also called **polar notation**.

> **Trigonometric Notation for Complex Numbers**
> $$a + bi = r(\cos \theta + i \sin \theta)$$

To find trigonometric notation for a complex number given in **standard notation**, $a + bi$, we must find r and determine the angle θ for which $\sin \theta = b/r$ and $\cos \theta = a/r$.

GCM **EXAMPLE 3** Find trigonometric notation for each of the following complex numbers.

a) $1 + i$ **b)** $\sqrt{3} - i$

Solution

a) We note that $a = 1$ and $b = 1$. Then

$$r = \sqrt{a^2 + b^2} = \sqrt{1^2 + 1^2} = \sqrt{2},$$

$$\sin \theta = \frac{b}{r} = \frac{1}{\sqrt{2}}, \quad \text{or} \quad \frac{\sqrt{2}}{2},$$

and

$$\cos \theta = \frac{a}{r} = \frac{1}{\sqrt{2}}, \quad \text{or} \quad \frac{\sqrt{2}}{2}.$$

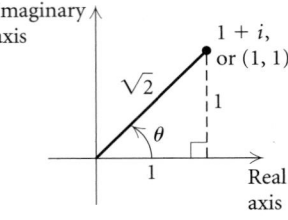

Since θ is in quadrant I, $\theta = \pi/4$, or $45°$, and we have

$$1 + i = \sqrt{2}\left(\cos \frac{\pi}{4} + i \sin \frac{\pi}{4}\right),$$

or

$$1 + i = \sqrt{2}(\cos 45° + i \sin 45°).$$

b) We see that $a = \sqrt{3}$ and $b = -1$. Then

$$r = \sqrt{(\sqrt{3})^2 + (-1)^2} = 2,$$

$$\sin \theta = \frac{-1}{2} = -\frac{1}{2},$$

and

$$\cos \theta = \frac{\sqrt{3}}{2}.$$

Since θ is in quadrant IV, $\theta = 11\pi/6$, or $330°$, and we have

$$\sqrt{3} - i = 2\left(\cos \frac{11\pi}{6} + i \sin \frac{11\pi}{6}\right),$$

or

$$\sqrt{3} - i = 2(\cos 330° + i \sin 330°).$$

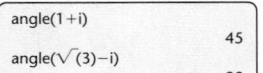

As shown at left, a graphing calculator (in DEGREE mode) can be used to determine angle values in degrees. **Now Try Exercise 13.** ■

In changing to trigonometric notation, note that there are many angles satisfying the given conditions. We ordinarily choose the *smallest positive* angle.

To change from trigonometric notation to standard notation, $a + bi$, we recall that $a = r \cos \theta$ and $b = r \sin \theta$.

EXAMPLE 4 Find standard notation, $a + bi$, for each of the following complex numbers.

a) $2(\cos 120° + i \sin 120°)$ **b)** $\sqrt{8}\left(\cos \dfrac{7\pi}{4} + i \sin \dfrac{7\pi}{4}\right)$

Solution

a) Rewriting, we have

$$2(\cos 120° + i \sin 120°) = 2 \cos 120° + (2 \sin 120°)i.$$

Thus,

$$a = 2 \cos 120° = 2 \cdot \left(-\frac{1}{2}\right) = -1$$

and

$$b = 2 \sin 120° = 2 \cdot \frac{\sqrt{3}}{2} = \sqrt{3},$$

so

$$2(\cos 120° + i \sin 120°) = -1 + \sqrt{3}i.$$

Degree Mode

```
2(cos(120)+isin(120))
              -1+1.732050808i
```

b) Rewriting, we have

$$\sqrt{8}\left(\cos \frac{7\pi}{4} + i \sin \frac{7\pi}{4}\right) = \sqrt{8} \cos \frac{7\pi}{4} + \left(\sqrt{8} \sin \frac{7\pi}{4}\right)i.$$

Thus,

$$a = \sqrt{8} \cos \frac{7\pi}{4} = \sqrt{8} \cdot \frac{\sqrt{2}}{2} = 2$$

and

$$b = \sqrt{8} \sin \frac{7\pi}{4} = \sqrt{8} \cdot \left(-\frac{\sqrt{2}}{2}\right) = -2,$$

Radian Mode

```
√(8)(cos(7π/4)+isin(7π/4))
                    2-2i
```

so

$$\sqrt{8}\left(\cos \frac{7\pi}{4} + i \sin \frac{7\pi}{4}\right) = 2 - 2i.$$

Now Try Exercises 23 and 27. ■

❋ Multiplication and Division with Trigonometric Notation

Multiplication of complex numbers is easier to manage with trigonometric notation than with standard notation. We simply multiply the absolute values and add the arguments. Let's state this in a more formal manner.

Complex Numbers: Multiplication
For any complex numbers $r_1(\cos \theta_1 + i \sin \theta_1)$ and $r_2(\cos \theta_2 + i \sin \theta_2)$,

$$r_1(\cos \theta_1 + i \sin \theta_1) \cdot r_2(\cos \theta_2 + i \sin \theta_2)$$
$$= r_1 r_2 [\cos (\theta_1 + \theta_2) + i \sin (\theta_1 + \theta_2)].$$

Proof

$$r_1(\cos \theta_1 + i \sin \theta_1) \cdot r_2(\cos \theta_2 + i \sin \theta_2) =$$
$$r_1 r_2(\cos \theta_1 \cos \theta_2 - \sin \theta_1 \sin \theta_2) + r_1 r_2(\sin \theta_1 \cos \theta_2 + \cos \theta_1 \sin \theta_2)i$$

Now, using identities for sums of angles, we simplify, obtaining

$$r_1 r_2 \cos (\theta_1 + \theta_2) + r_1 r_2 \sin (\theta_1 + \theta_2)i,$$

or

$$r_1 r_2 [\cos (\theta_1 + \theta_2) + i \sin (\theta_1 + \theta_2)],$$

which was to be shown. ∎

EXAMPLE 5 Multiply and express the answer to each of the following in standard notation.

a) $3(\cos 40° + i \sin 40°)$ and $4(\cos 20° + i \sin 20°)$

b) $2(\cos \pi + i \sin \pi)$ and $3\left[\cos \left(-\dfrac{\pi}{2} \right) + i \sin \left(-\dfrac{\pi}{2} \right) \right]$

Solution

a) $3(\cos 40° + i \sin 40°) \cdot 4(\cos 20° + i \sin 20°)$
$$= 3 \cdot 4 \cdot [\cos (40° + 20°) + i \sin (40° + 20°)]$$
$$= 12(\cos 60° + i \sin 60°)$$
$$= 12\left(\frac{1}{2} + \frac{\sqrt{3}}{2} i \right)$$
$$= 6 + 6\sqrt{3}i$$

Degree Mode

```
3(cos(40)+isin(40))*4(cos(20)+
isin(20))
                    6+10.39230485i
```

b) $2(\cos \pi + i \sin \pi) \cdot 3\left[\cos \left(-\dfrac{\pi}{2} \right) + i \sin \left(-\dfrac{\pi}{2} \right) \right]$

$$= 2 \cdot 3 \cdot \left[\cos \left(\pi + \left(-\frac{\pi}{2} \right) \right) + i \sin \left(\pi + \left(-\frac{\pi}{2} \right) \right) \right]$$
$$= 6\left(\cos \frac{\pi}{2} + i \sin \frac{\pi}{2} \right)$$
$$= 6(0 + i \cdot 1)$$
$$= 6i$$

Radian Mode

```
2(cos(π)+isin(π))*3(cos(-π/2)+
isin(-π/2))
                               6i
```

EXAMPLE 6 Convert to trigonometric notation and multiply:

$$(1 + i)(\sqrt{3} - i).$$

Solution We first find trigonometric notation:

$$1 + i = \sqrt{2}(\cos 45° + i \sin 45°), \qquad \text{See Example 3(a).}$$
$$\sqrt{3} - i = 2(\cos 330° + i \sin 330°). \qquad \text{See Example 3(b).}$$

Then we multiply:

$$\sqrt{2}(\cos 45° + i \sin 45°) \cdot 2(\cos 330° + i \sin 330°)$$
$$= 2\sqrt{2}[\cos (45° + 330°) + i \sin (45° + 330°)]$$
$$= 2\sqrt{2}(\cos 375° + i \sin 375°)$$
$$= 2\sqrt{2}(\cos 15° + i \sin 15°). \qquad \begin{array}{l}\text{375° has the same}\\ \text{terminal side as 15°.}\end{array}$$

Now Try Exercise 35. ■

To divide complex numbers, we divide the absolute values and subtract the arguments. We state this fact below, but omit the proof.

> **Complex Numbers: Division**
>
> For any complex numbers $r_1(\cos \theta_1 + i \sin \theta_1)$ and $r_2(\cos \theta_2 + i \sin \theta_2)$, $r_2 \neq 0$,
>
> $$\frac{r_1(\cos \theta_1 + i \sin \theta_1)}{r_2(\cos \theta_2 + i \sin \theta_2)} = \frac{r_1}{r_2}[\cos (\theta_1 - \theta_2) + i \sin (\theta_1 - \theta_2)].$$

EXAMPLE 7 Divide

$$2\left(\cos \frac{3\pi}{2} + i \sin \frac{3\pi}{2}\right) \quad \text{by} \quad 4\left(\cos \frac{\pi}{2} + i \sin \frac{\pi}{2}\right)$$

and express the solution in standard notation.

Solution We have

$$\frac{2\left(\cos \dfrac{3\pi}{2} + i \sin \dfrac{3\pi}{2}\right)}{4\left(\cos \dfrac{\pi}{2} + i \sin \dfrac{\pi}{2}\right)} = \frac{2}{4}\left[\cos \left(\frac{3\pi}{2} - \frac{\pi}{2}\right) + i \sin \left(\frac{3\pi}{2} - \frac{\pi}{2}\right)\right]$$

$$= \frac{1}{2}(\cos \pi + i \sin \pi)$$

$$= \frac{1}{2}(-1 + i \cdot 0)$$

$$= -\frac{1}{2}.$$

■

EXAMPLE 8 Convert to trigonometric notation and divide:

$$\frac{1 + i}{1 - i}.$$

Solution We first convert to trigonometric notation:

$$1 + i = \sqrt{2}(\cos 45° + i \sin 45°), \qquad \text{See Example 3(a).}$$
$$1 - i = \sqrt{2}(\cos 315° + i \sin 315°).$$

We now divide:

$$\frac{\sqrt{2}(\cos 45° + i \sin 45°)}{\sqrt{2}(\cos 315° + i \sin 315°)}$$

$$= 1[\cos (45° - 315°) + i \sin (45° - 315°)]$$
$$= \cos (-270°) + i \sin (-270°)$$
$$= 0 + i \cdot 1$$
$$= i. \qquad \text{\textbf{Now Try Exercise 39.}} ◼$$

(1 + i)/(1 − i)

i

❋ Powers of Complex Numbers

An important theorem about powers and roots of complex numbers is named for the French mathematician Abraham DeMoivre (1667–1754). Let's consider the square of a complex number $r(\cos \theta + i \sin \theta)$:

$$[r(\cos \theta + i \sin \theta)]^2 = [r(\cos \theta + i \sin \theta)] \cdot [r(\cos \theta + i \sin \theta)]$$
$$= r \cdot r \cdot [\cos (\theta + \theta) + i \sin (\theta + \theta)]$$
$$= r^2(\cos 2\theta + i \sin 2\theta).$$

Similarly, we see that

$$[r(\cos \theta + i \sin \theta)]^3$$
$$= r \cdot r \cdot r \cdot [\cos (\theta + \theta + \theta) + i \sin (\theta + \theta + \theta)]$$
$$= r^3(\cos 3\theta + i \sin 3\theta).$$

DeMoivre's theorem is the generalization of these results.

> **DeMoivre's Theorem**
>
> For any complex number $r(\cos \theta + i \sin \theta)$ and any natural number n,
>
> $$[r(\cos \theta + i \sin \theta)]^n = r^n(\cos n\theta + i \sin n\theta).$$

EXAMPLE 9 Find each of the following.

a) $(1 + i)^9$ **b)** $(\sqrt{3} - i)^{10}$

Solution

a) We first find trigonometric notation:

$$1 + i = \sqrt{2}(\cos 45° + i \sin 45°). \qquad \text{See Example 3(a).}$$

Then

$$(1 + i)^9 = \left[\sqrt{2}(\cos 45° + i \sin 45°)\right]^9$$

$$= \left(\sqrt{2}\right)^9\left[\cos(9 \cdot 45°) + i \sin(9 \cdot 45°)\right] \qquad \text{DeMoivre's theorem}$$

$$= 2^{9/2}(\cos 405° + i \sin 405°)$$

$$= 16\sqrt{2}(\cos 45° + i \sin 45°) \qquad \text{405° has the same terminal side as 45°.}$$

$$= 16\sqrt{2}\left(\frac{\sqrt{2}}{2} + i\frac{\sqrt{2}}{2}\right)$$

$$= 16 + 16i.$$

```
(1+i)^9
                16+16i
(√(3)−i)^10
        512+886.8100135i
```

b) We first convert to trigonometric notation:

$$\sqrt{3} - i = 2(\cos 330° + i \sin 330°). \qquad \text{See Example 3(b).}$$

Then

$$\left(\sqrt{3} - i\right)^{10} = [2(\cos 330° + i \sin 330°)]^{10}$$

$$= 2^{10}(\cos 3300° + i \sin 3300°)$$

$$= 1024(\cos 60° + i \sin 60°) \qquad \text{3300° has the same terminal side as 60°.}$$

$$= 1024\left(\frac{1}{2} + i\frac{\sqrt{3}}{2}\right)$$

$$= 512 + 512\sqrt{3}i. \qquad \text{Now Try Exercise 47.} \blacksquare$$

✢ Roots of Complex Numbers

As we will see, every nonzero complex number has two square roots. A nonzero complex number has three cube roots, four fourth roots, and so on. In general, a nonzero complex number has n different nth roots. They can be found using the formula that we now state but do not prove.

Roots of Complex Numbers

The nth roots of a complex number $r(\cos \theta + i \sin \theta)$, $r \neq 0$, are given by

$$r^{1/n}\left[\cos\left(\frac{\theta}{n} + k \cdot \frac{360°}{n}\right) + i \sin\left(\frac{\theta}{n} + k \cdot \frac{360°}{n}\right)\right],$$

where $k = 0, 1, 2, \ldots, n - 1$.

EXAMPLE 10 Find the square roots of $2 + 2\sqrt{3}i$.

Solution We first find trigonometric notation:

$$2 + 2\sqrt{3}i = 4(\cos 60° + i \sin 60°).$$

Then $n = 2, 1/n = 1/2$, and $k = 0, 1$; and

$$[4(\cos 60° + i \sin 60°)]^{1/2}$$

$$= 4^{1/2}\left[\cos\left(\frac{60°}{2} + k \cdot \frac{360°}{2}\right) + i \sin\left(\frac{60°}{2} + k \cdot \frac{360°}{2}\right)\right], \quad k = 0, 1$$

$$= 2[\cos(30° + k \cdot 180°) + i \sin(30° + k \cdot 180°), \quad k = 0, 1.$$

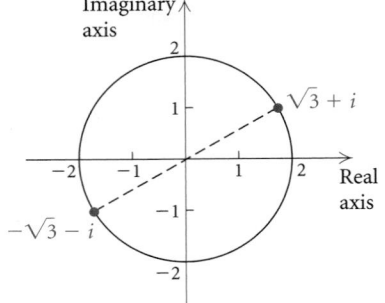

Thus the roots are

$$2(\cos 30° + i \sin 30°) \text{ for } k = 0$$

and $\quad 2(\cos 210° + i \sin 210°) \text{ for } k = 1,$

or $\quad \sqrt{3} + i \quad$ and $\quad -\sqrt{3} - i.$ **Now Try Exercise 57.** ■

In Example 10, we see that the two square roots of the number are opposites of each other. We can illustrate this graphically. We also note that the roots are equally spaced about a circle of radius r—in this case, $r = 2$. The roots are $360°/2$, or $180°$ apart.

EXAMPLE 11 Find the cube roots of 1. Then locate them on a graph.

Solution We begin by finding trigonometric notation:

$$1 = 1(\cos 0° + i \sin 0°).$$

Then $n = 3, 1/n = 1/3$, and $k = 0, 1, 2$; and

$$[1(\cos 0° + i \sin 0°)]^{1/3}$$

$$= 1^{1/3}\left[\cos\left(\frac{0°}{3} + k \cdot \frac{360°}{3}\right) + i \sin\left(\frac{0°}{3} + k \cdot \frac{360°}{3}\right)\right], \quad k = 0, 1, 2.$$

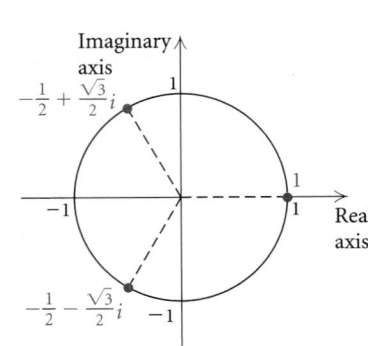

The roots are

$$1(\cos 0° + i \sin 0°), \quad 1(\cos 120° + i \sin 120°),$$

and $\quad 1(\cos 240° + i \sin 240°),$

or $\quad 1, \quad -\frac{1}{2} + \frac{\sqrt{3}}{2}i, \quad$ and $\quad -\frac{1}{2} - \frac{\sqrt{3}}{2}i.$

The graphs of the cube roots lie equally spaced about a circle of radius 1. The roots are $360°/3$, or $120°$ apart.   **Now Try Exercise 59.** ■

The nth roots of 1 are often referred to as the **nth roots of unity**. In Example 11, we found the cube roots of unity.

Exploring with Technology

Using a graphing calculator set in PARAMETRIC mode, we can approximate the nth roots of a number p. We use the following window and let

$$X_{1T} = (p^\wedge(1/n)) \cos T \quad \text{and} \quad Y_{1T} = (p^\wedge(1/n)) \sin T.$$

WINDOW

Tmin $= 0$

Tmax $= 360$, if in degree mode, or
    $= 2\pi$, if in radian mode

Tstep $= 360/n$, or $2\pi/n$

Xmin $= -3$, Xmax $= 3$, Xscl $= 1$

Ymin $= -2$, Ymax $= 2$, Yscl $= 1$

To find the fifth roots of 8, enter $X_{1T} = (8^\wedge(1/5)) \cos T$ and $Y_{1T} = (8^\wedge(1/5)) \sin T$. In this case, use DEGREE mode. After the graph has been generated, use the TRACE feature to locate the fifth roots. The T, X, and Y values appear on the screen. What do they represent?

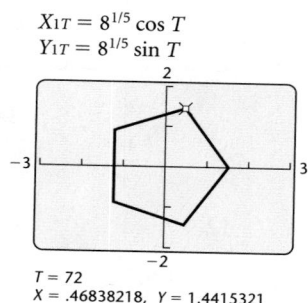

$X_{1T} = 8^{1/5} \cos T$
$Y_{1T} = 8^{1/5} \sin T$

T = 72
X = .46838218, Y = 1.4415321

Three of the fifth roots of 8 are approximately

$$1.5157, \qquad 0.46838 + 1.44153i, \quad \text{and} \quad -1.22624 + 0.89092i.$$

Find the other two. Then use a calculator to approximate the cube roots of unity that were found in Example 11. Also approximate the fourth roots of 5 and the tenth roots of unity.

(8.3) Exercise Set

Graph the complex number and find its absolute value.

1. $4 + 3i$

2. $-2 - 3i$

3. i

4. $-5 - 2i$

5. $4 - i$

6. $6 + 3i$

7. 3

8. $-2i$

Express the indicated number in both standard notation and trigonometric notation.

9.

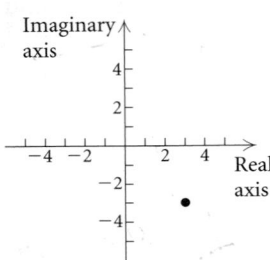

10.

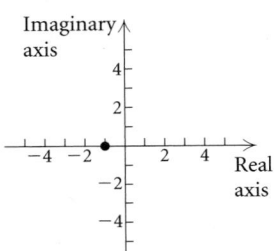

11.

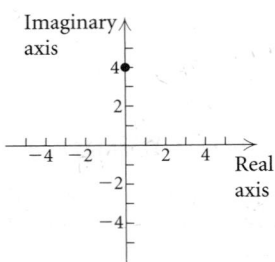

12.

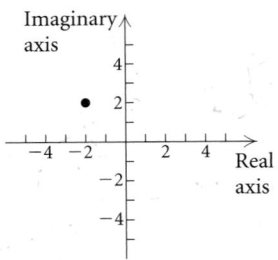

Find trigonometric notation.

13. $1 - i$

14. $-10\sqrt{3} + 10i$

15. $-3i$

16. $-5 + 5i$

17. $\sqrt{3} + i$

18. 4

19. $\dfrac{2}{5}$

20. $7.5i$

21. $-3\sqrt{2} - 3\sqrt{2}i$

22. $-\dfrac{9}{2} - \dfrac{9\sqrt{3}}{2}i$

Find standard notation, a + bi.

23. $3(\cos 30° + i \sin 30°)$

24. $6(\cos 120° + i \sin 120°)$

25. $10(\cos 270° + i \sin 270°)$

26. $3(\cos 0° + i \sin 0°)$

27. $\sqrt{8}\left(\cos \dfrac{\pi}{4} + i \sin \dfrac{\pi}{4} \right)$

28. $5\left(\cos \dfrac{\pi}{3} + i \sin \dfrac{\pi}{3} \right)$

29. $2\left(\cos \dfrac{\pi}{2} + i \sin \dfrac{\pi}{2} \right)$

30. $3\left[\cos\left(-\dfrac{3\pi}{4} \right) + i \sin\left(-\dfrac{3\pi}{4} \right) \right]$

31. $\sqrt{2}[\cos (-60°) + i \sin (-60°)]$

32. $4(\cos 135° + i \sin 135°)$

Multiply or divide and leave the answer in trigonometric notation.

33. $\dfrac{12(\cos 48° + i \sin 48°)}{3(\cos 6° + i \sin 6°)}$

34. $5\left(\cos \dfrac{\pi}{3} + i \sin \dfrac{\pi}{3} \right) \cdot 2\left(\cos \dfrac{\pi}{4} + i \sin \dfrac{\pi}{4} \right)$

35. $2.5(\cos 35° + i \sin 35°) \cdot 4.5(\cos 21° + i \sin 21°)$

36. $\dfrac{\dfrac{1}{2}\left(\cos \dfrac{2\pi}{3} + i \sin \dfrac{2\pi}{3} \right)}{\dfrac{3}{8}\left(\cos \dfrac{\pi}{6} + i \sin \dfrac{\pi}{6} \right)}$

Convert to trigonometric notation and then multiply or divide.

37. $(1 - i)(2 + 2i)$

38. $\left(1 + i\sqrt{3}\right)(1 + i)$

39. $\dfrac{1 - i}{1 + i}$

40. $\dfrac{1 - i}{\sqrt{3} - i}$

41. $\left(3\sqrt{3} - 3i\right)(2i)$ **42.** $\left(2\sqrt{3} + 2i\right)(2i)$

43. $\dfrac{2\sqrt{3} - 2i}{1 + \sqrt{3}i}$ **44.** $\dfrac{3 - 3\sqrt{3}i}{\sqrt{3} - i}$

Raise the number to the given power and write trigonometric notation for the answer.

45. $\left[2\left(\cos\dfrac{\pi}{3} + i\sin\dfrac{\pi}{3}\right)\right]^3$

46. $[2(\cos 120° + i\sin 120°)]^4$

47. $(1 + i)^6$

48. $\left(-\sqrt{3} + i\right)^5$

Raise the number to the given power and write standard notation for the answer.

49. $[3(\cos 20° + i\sin 20°)]^3$

50. $[2(\cos 10° + i\sin 10°)]^9$

51. $(1 - i)^5$ **52.** $(2 + 2i)^4$

53. $\left(\dfrac{1}{\sqrt{2}} - \dfrac{1}{\sqrt{2}}i\right)^{12}$ **54.** $\left(\dfrac{\sqrt{3}}{2} + \dfrac{1}{2}i\right)^{10}$

Find the square roots of the number.

55. $-i$ **56.** $1 + i$

57. $2\sqrt{2} - 2\sqrt{2}i$ **58.** $-\sqrt{3} - i$

Find the cube roots of the number.

59. i **60.** $-64i$

61. $2\sqrt{3} - 2i$ **62.** $1 - \sqrt{3}i$

63. Find and graph the fourth roots of 16.

64. Find and graph the fourth roots of i.

65. Find and graph the fifth roots of -1.

66. Find and graph the sixth roots of 1.

67. Find the tenth roots of 8.

68. Find the ninth roots of -4.

69. Find the sixth roots of -1.

70. Find the fourth roots of 12.

Find all the complex solutions of the equation.

71. $x^3 = 1$ **72.** $x^5 - 1 = 0$

73. $x^4 + i = 0$ **74.** $x^4 + 81 = 0$

75. $x^6 + 64 = 0$ **76.** $x^5 + \sqrt{3} + i = 0$

Collaborative Discussion and Writing

77. Explain why $x^6 - 2x^3 + 1 = 0$ has 3 distinct solutions, $x^6 - 2x^3 = 0$ has 4 distinct solutions, and $x^6 - 2x = 0$ has 6 distinct solutions.

78. Explain why trigonometric notation for a complex number is not unique, but rectangular, or standard, notation is unique.

Skill Maintenance

Convert to degree measure.

79. $\dfrac{\pi}{12}$ **80.** 3π

Convert to radian measure.

81. $330°$ **82.** $-225°$

83. Find r.

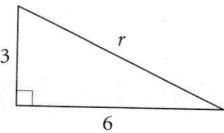

84. Graph these points in the rectangular coordinate system: $(2, -1)$, $(0, 3)$, and $\left(-\frac{1}{2}, -4\right)$.

Find the function value using coordinates of points on the unit circle.

85. $\sin\dfrac{2\pi}{3}$ **86.** $\cos\dfrac{\pi}{6}$

87. $\cos\dfrac{\pi}{4}$ **88.** $\sin\dfrac{5\pi}{6}$

Synthesis

Solve.

89. $x^2 + (1 - i)x + i = 0$

90. $3x^2 + (1 + 2i)x + 1 - i = 0$

91. Find polar notation for $(\cos\theta + i\sin\theta)^{-1}$.

92. Show that for any complex number z,
$$|z| = |-z|.$$

93. Show that for any complex number z and its conjugate $\bar{z}$,
$$|z| = |\bar{z}|.$$
(*Hint:* Let $z = a + bi$ and $\bar{z} = a - bi$.)

94. Show that for any complex number z and its conjugate $\bar{z}$,
$$|z\bar{z}| = |z^2|.$$
(*Hint*: Let $z = a + bi$ and $\bar{z} = a - bi$.)

95. Show that for any complex number z,
$$|z^2| = |z|^2.$$

96. Show that for any complex numbers z and w,
$$|z \cdot w| = |z| \cdot |w|.$$
(*Hint*: Let $z = r_1(\cos \theta_1 + i \sin \theta_1)$ and $w = r_2(\cos \theta_2 + i \sin \theta_2)$.)

97. Show that for any complex number z and any nonzero, complex number w,
$$\left| \frac{z}{w} \right| = \frac{|z|}{|w|}.$$
(Use the hint for Exercise 96.)

98. On a complex plane, graph $|z| = 1$.

99. On a complex plane, graph $z + \bar{z} = 3$.

100. Solve: $x^6 - 1 = 0$.

8.4 Polar Coordinates and Graphs

❖ Graph points given their polar coordinates.
❖ Convert from rectangular coordinates to polar coordinates and from polar coordinates to rectangular coordinates.
❖ Convert from rectangular equations to polar equations and from polar equations to rectangular equations.
❖ Graph polar equations.

❖ Polar Coordinates

All graphing throughout this text has been done with rectangular coordinates, (x, y), in the Cartesian coordinate system. We now introduce the polar coordinate system. As shown in the diagram at left, any point P has rectangular coordinates (x, y) and polar coordinates (r, θ). On a polar graph, the origin is called the **pole** and the positive half of the x-axis is called the **polar axis**. The point P can be plotted given the directed angle θ from the polar axis to the ray OP and the directed distance r from the pole to the point. The angle θ can be expressed in degrees or radians.

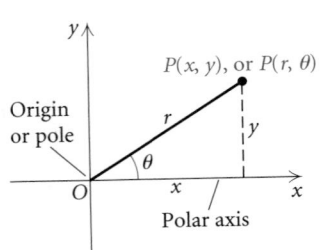

To plot points on a polar graph:

1. Locate the directed angle θ.
2. Move a directed distance r from the pole. If $r > 0$, move along ray OP. If $r < 0$, move in the opposite direction of ray OP.

Polar graph paper, shown below, facilitates plotting. Points B and G illustrate that θ may be in radians. Points E and F illustrate that the polar coordinates of a point are not unique.

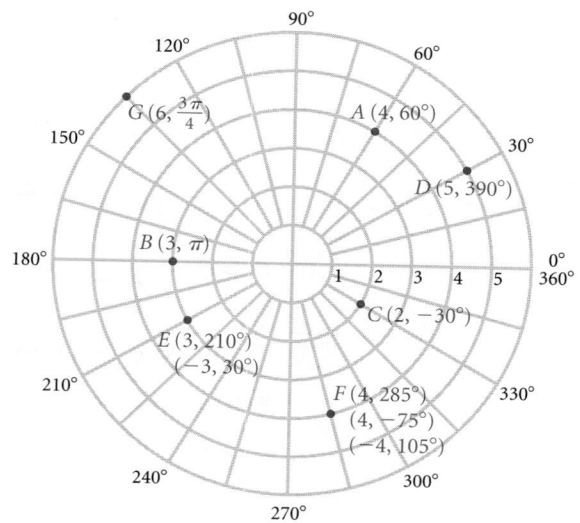

EXAMPLE 1 Graph each of the following points.

a) $A(3, 60°)$

b) $B(0, 10°)$

c) $C(-5, 120°)$

d) $D(1, -60°)$

e) $E\left(2, \dfrac{3\pi}{2}\right)$

f) $F\left(-4, \dfrac{\pi}{3}\right)$

Solution

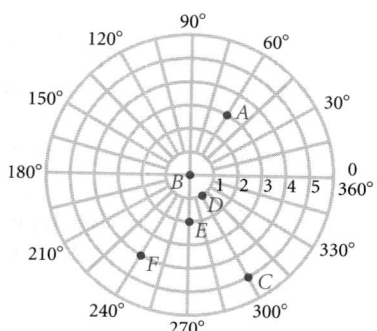

Now Try Exercises 3 and 7. ■

To convert from rectangular to polar coordinates and from polar to rectangular coordinates, we need to recall the following relationships.

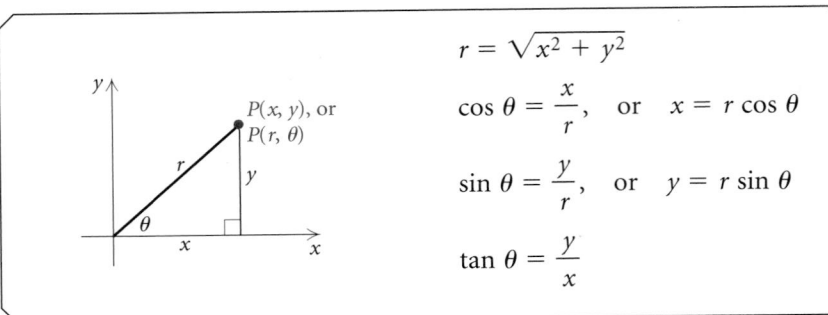

$$r = \sqrt{x^2 + y^2}$$

$$\cos \theta = \frac{x}{r}, \quad \text{or} \quad x = r \cos \theta$$

$$\sin \theta = \frac{y}{r}, \quad \text{or} \quad y = r \sin \theta$$

$$\tan \theta = \frac{y}{x}$$

GCM **EXAMPLE 2**   Convert each of the following to polar coordinates.

a) $(3, 3)$ **b)** $\left(2\sqrt{3}, -2\right)$

Solution

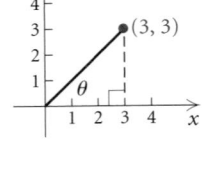

a) We first find r:

$$r = \sqrt{3^2 + 3^2} = \sqrt{18} = 3\sqrt{2}.$$

Then we determine θ:

$$\tan \theta = \frac{3}{3} = 1; \quad \text{therefore,} \quad \theta = 45°, \text{ or } \frac{\pi}{4}.$$

We know that for $r = 3\sqrt{2}$, $\theta = \pi/4$ and not $5\pi/4$ since $(3, 3)$ is in quadrant I. Thus, $(r, \theta) = \left(3\sqrt{2}, 45°\right)$, or $\left(3\sqrt{2}, \pi/4\right)$. Other possibilities for polar coordinates include $\left(3\sqrt{2}, -315°\right)$ and $\left(-3\sqrt{2}, 5\pi/4\right)$.

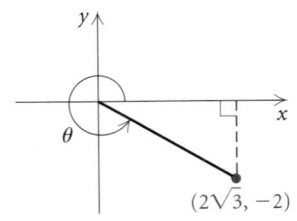

b) We first find r:

$$r = \sqrt{\left(2\sqrt{3}\right)^2 + (-2)^2} = \sqrt{12 + 4} = \sqrt{16} = 4.$$

Then we determine θ:

$$\tan \theta = \frac{-2}{2\sqrt{3}} = -\frac{1}{\sqrt{3}}; \quad \text{therefore,} \quad \theta = 330°, \text{ or } \frac{11\pi}{6}.$$

Thus, $(r, \theta) = (4, 330°)$, or $(4, 11\pi/6)$. Other possibilities for polar coordinates for this point include $(4, -\pi/6)$ and $(-4, 150°)$.

Now Try Exercise 19. ■

It is easier to convert from polar to rectangular coordinates than from rectangular to polar coordinates.

GCM **EXAMPLE 3** Convert each of the following to rectangular coordinates.

a) $\left(10, \dfrac{\pi}{3}\right)$

b) $(-5, 135°)$

Solution

a) The ordered pair $(10, \pi/3)$ gives us $r = 10$ and $\theta = \pi/3$. We now find x and y:

$$x = r \cos\theta = 10 \cos\frac{\pi}{3} = 10 \cdot \frac{1}{2} = 5$$

and

$$y = r \sin\theta = 10 \sin\frac{\pi}{3} = 10 \cdot \frac{\sqrt{3}}{2} = 5\sqrt{3}.$$

Thus, $(x, y) = \left(5, 5\sqrt{3}\right)$.

b) From the ordered pair $(-5, 135°)$, we know that $r = -5$ and $\theta = 135°$. We now find x and y:

$$x = -5 \cos 135° = -5 \cdot \left(-\frac{\sqrt{2}}{2}\right) = \frac{5\sqrt{2}}{2}$$

and

$$y = -5 \sin 135° = -5 \cdot \left(\frac{\sqrt{2}}{2}\right) = -\frac{5\sqrt{2}}{2}.$$

Thus, $(x, y) = \left(\dfrac{5\sqrt{2}}{2}, -\dfrac{5\sqrt{2}}{2}\right)$.

Now Try Exercises 31 and 37. ■

The conversions above can be easily made with some graphing calculators.

❖ Polar and Rectangular Equations

Some curves have simpler equations in polar coordinates than in rectangular coordinates. For others, the reverse is true.

EXAMPLE 4 Convert each of the following to a polar equation.

a) $x^2 + y^2 = 25$

b) $2x - y = 5$

Solution

a) We have

$$x^2 + y^2 = 25$$
$$(r \cos\theta)^2 + (r \sin\theta)^2 = 25 \qquad \text{Substituting for } x \text{ and } y$$
$$r^2 \cos^2\theta + r^2 \sin^2\theta = 25$$
$$r^2(\cos^2\theta + \sin^2\theta) = 25$$
$$r^2 = 25 \qquad \cos^2\theta + \sin^2\theta = 1$$
$$r = 5.$$

This example illustrates that the polar equation of a circle centered at the origin is much simpler than the rectangular equation.

b) We have

$$2x - y = 5$$
$$2(r \cos \theta) - (r \sin \theta) = 5$$
$$r(2 \cos \theta - \sin \theta) = 5.$$

In this example, we see that the rectangular equation is simpler than the polar equation. **Now Try Exercises 47 and 51.** ■

EXAMPLE 5 Convert each of the following to a rectangular equation.

a) $r = 4$

b) $r \cos \theta = 6$

c) $r = 2 \cos \theta + 3 \sin \theta$

Solution

a) We have

$$r = 4$$
$$\sqrt{x^2 + y^2} = 4 \qquad \text{Substituting for } r$$
$$x^2 + y^2 = 16. \qquad \text{Squaring}$$

In squaring, we must be careful not to introduce solutions of the equation that are not already present. In this case, we did not, because the graph of either equation is a circle of radius 4 centered at the origin.

b) We have

$$r \cos \theta = 6$$
$$x = 6. \qquad x = r \cos \theta$$

The graph of $r \cos \theta = 6$, or $x = 6$, is a vertical line.

c) We have

$$r = 2 \cos \theta + 3 \sin \theta$$
$$r^2 = 2r \cos \theta + 3r \sin \theta \qquad \text{Multiplying by } r \text{ on both sides}$$
$$x^2 + y^2 = 2x + 3y. \qquad \begin{array}{l}\text{Substituting } x^2 + y^2 \text{ for } r^2, \\ x \text{ for } r \cos \theta, \text{ and } y \text{ for } r \sin \theta\end{array}$$

Now Try Exercises 59 and 63. ■

❖ Graphing Polar Equations

To graph a polar equation, we can make a table of values, choosing values of θ and calculating corresponding values of r. We plot the points and complete the graph, as we do when graphing a rectangular equation. A difference occurs in the case of a polar equation however, because as θ increases sufficiently, points may begin to repeat and the curve will be traced again and again. When this happens, the curve is complete.

GCM **EXAMPLE 6** Graph: $r = 1 - \sin \theta$.

Solution We first make a table of values. The TABLE feature on a graphing calculator is the most efficient way to create this list. Note that the points begin to repeat at $\theta = 360°$. We plot these points and draw the curve, as shown below.

θ	r
0°	1
15°	0.7412
30°	0.5
45°	0.2929
60°	0.1340
75°	0.0341
90°	0
105°	0.0341
120°	0.1340
135°	0.2929
150°	0.5
165°	0.7412
180°	1

θ	r
195°	1.2588
210°	1.5
225°	1.7071
240°	1.8660
255°	1.9659
270°	2
285°	1.9659
300°	1.8660
315°	1.7071
330°	1.5
345°	1.2588
360°	1
375°	0.7412
390°	0.5

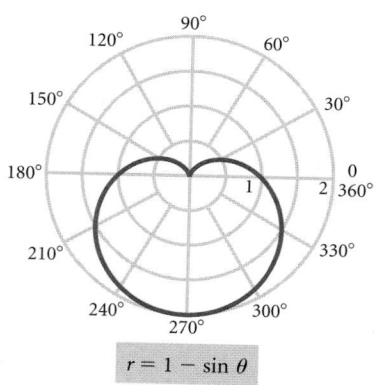

$r = 1 - \sin\theta$

Because of its heart shape, this curve is called a *cardioid*.

Now Try Exercise 77. ■

We plotted points in Example 6 because we feel that it is important to understand how these curves are developed. We also can graph polar equations using a graphing calculator. The equation usually must be written first in the form $r = f(\theta)$. It is necessary to decide on not only the best window dimensions but also the range of values for θ. Typically, we begin with a range of 0 to 2π for θ in radians and 0° to 360° for θ in degrees. Because most polar graphs are curved, it is important to square the window to minimize distortion.

Exploring with Technology

Graph $r = 4 \sin 3\theta$. Begin by setting the calculator in POLAR mode, and use either of the following windows:

WINDOW
(Radians)
   θmin = 0
   θmax = 2π
   θstep = $\pi/24$
   Xmin = −9
   Xmax = 9
   Xscl = 1
   Ymin = −6
   Ymax = 6
   Yscl = 1

WINDOW
(Degrees)
   θmin = 0
   θmax = 360
   θstep = 1
   Xmin = −9
   Xmax = 9
   Xscl = 1
   Ymin = −6
   Ymax = 6
   Yscl = 1

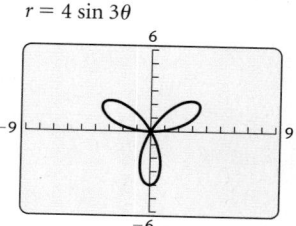

$r = 4 \sin 3\theta$

We observe the same graph in both windows. The calculator allows us to view the curve as it is formed.

Now graph each of the following equations and observe the effect of changing the coefficient of $\sin 3\theta$ and the coefficient of θ:

$$r = 2 \sin 3\theta, \qquad r = 6 \sin 3\theta, \qquad r = 4 \sin \theta,$$
$$r = 4 \sin 5\theta, \qquad r = 4 \sin 2\theta, \qquad r = 4 \sin 4\theta.$$

Polar equations of the form $r = a \cos n\theta$ and $r = a \sin n\theta$ have rose-shaped curves. The number a determines the length of the petals, and the number n determines the number of petals. If n is odd, there are n petals. If n is even, there are $2n$ petals.

EXAMPLE 7   Graph each of the following polar equations. Try to visualize the shape of the curve before graphing it.

a) $r = 3$
b) $r = 5 \sin \theta$
c) $r = 2 \csc \theta$

Solution   For each graph, we can begin with a table of values. Then we plot points and complete the graph.

a) $r = 3$

For all values of θ, r is 3. Thus the graph of $r = 3$ is a circle of radius 3 centered at the origin.

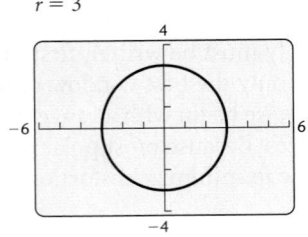

θ	r
0°	3
60°	3
135°	3
210°	3
300°	3
360°	3

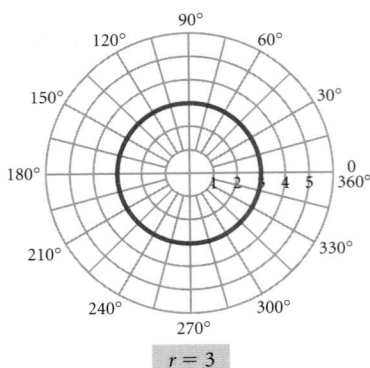

$r = 3$

We can verify our graph by converting to the equivalent rectangular equation. For $r = 3$, we substitute $\sqrt{x^2 + y^2}$ for r and square. The resulting equation,

$$x^2 + y^2 = 3^2,$$

is the equation of a circle with radius 3 centered at the origin.

b) $r = 5 \sin \theta$

$r = 5 \sin \theta$

θ	r
0°	0
15°	1.2941
30°	2.5
45°	3.5355
60°	4.3301
75°	4.8296
90°	5
105°	4.8296
120°	4.3301
135°	3.5355
150°	2.5
165°	1.2941
180°	0

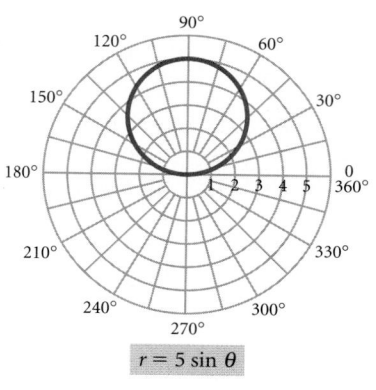

$r = 5 \sin \theta$

c) $r = 2 \csc \theta$

We can rewrite $r = 2 \csc \theta$ as $r = 2/\sin \theta$.

$r = 2 \csc \theta = 2/\sin \theta$

θ	r
0°	Not defined
15°	7.7274
30°	4
45°	2.8284
60°	2.3094
75°	2.0706
90°	2
105°	2.0706
120°	2.3094
135°	2.8284
150°	4
165°	7.7274
180°	Not defined

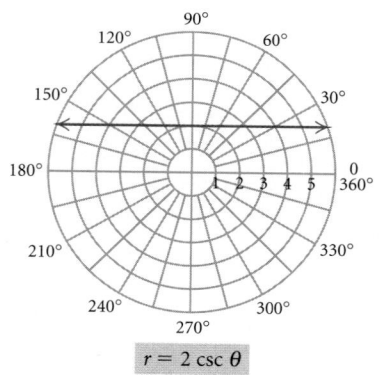

$r = 2 \csc \theta$

Now Try Exercise 71. ▪

We can check our graph in Example 7(c) by converting the polar equation to the equivalent rectangular equation:

$$r = 2 \csc \theta$$

$$r = \frac{2}{\sin \theta}$$

$$r \sin \theta = 2$$

$$y = 2. \qquad \text{Substituting } y \text{ for } r \sin \theta$$

The graph of $y = 2$ is a horizontal line passing through $(0, 2)$ on a rectangular grid.

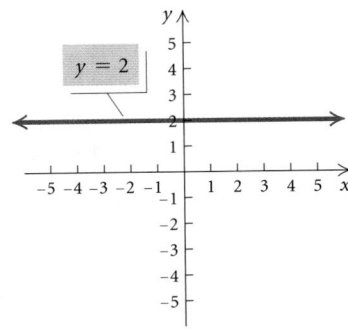

EXAMPLE 8 Graph the equation $r + 1 = 2 \cos 2\theta$ with a graphing calculator.

Solution We first solve for r:

$$r = 2 \cos 2\theta - 1.$$

We then obtain the following graph.

$$r = 2 \cos 2\theta - 1$$

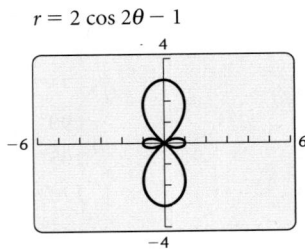

Now Try Exercise 91. ■

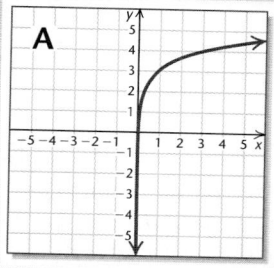

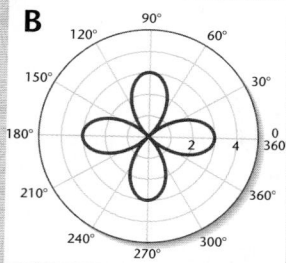

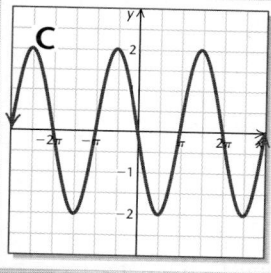

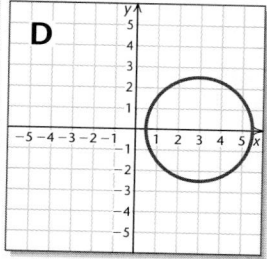

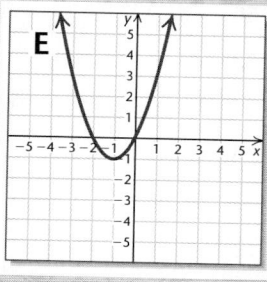

Visualizing the Graph

Match the equation with its graph.

1. $f(x) = 2^{(1/2)x}$

2. $y = -2 \sin x$

3. $y = (x + 1)^2 - 1$

4. $f(x) = \dfrac{x - 3}{x^2 + x - 6}$

5. $r = 1 + \sin \theta$

6. $f(x) = 2 \log x + 3$

7. $(x - 3)^2 + y^2 = \dfrac{25}{4}$

8. $y = -\cos\left(x - \dfrac{\pi}{2}\right)$

9. $r = 3 \cos 2\theta$

10. $f(x) = x^4 - x^3 + x^2 - x$

Answers on page A-57

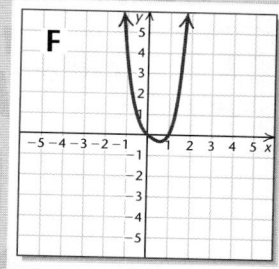

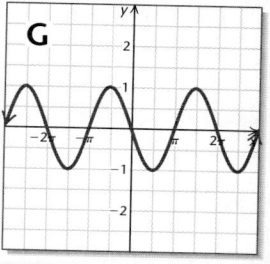

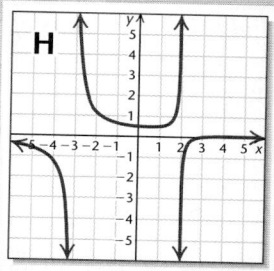

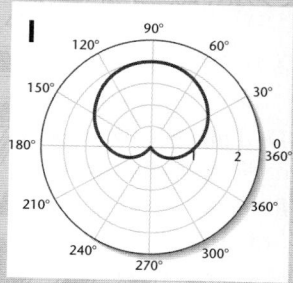

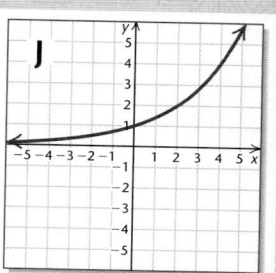

8.4 Exercise Set

Graph the point on a polar grid.

1. $(2, 45°)$

2. $(4, \pi)$

3. $(3.5, 210°)$

4. $(-3, 135°)$

5. $\left(1, \dfrac{\pi}{6}\right)$

6. $(2.75, 150°)$

7. $\left(-5, \dfrac{\pi}{2}\right)$

8. $(0, 15°)$

9. $(3, -315°)$

10. $\left(1.2, -\dfrac{2\pi}{3}\right)$

11. $(4.3, -60°)$

12. $(3, 405°)$

Find polar coordinates of points A, B, C, and D. Give three answers for each point.

13.

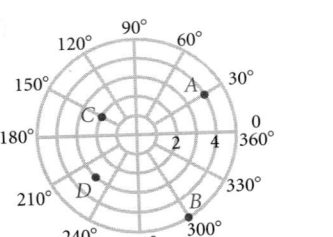

14.

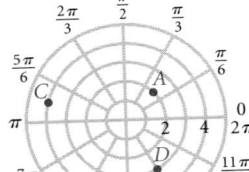

Find the polar coordinates of the point. Express the angle in degrees and then in radians, using the smallest possible positive angle.

15. $(0, -3)$

16. $(-4, 4)$

17. $\left(3, -3\sqrt{3}\right)$

18. $\left(-\sqrt{3}, 1\right)$

19. $\left(4\sqrt{3}, -4\right)$

20. $\left(2\sqrt{3}, 2\right)$

21. $\left(-\sqrt{2}, -\sqrt{2}\right)$

22. $\left(-3, 3\sqrt{3}\right)$

23. $\left(1, \sqrt{3}\right)$

24. $(0, -1)$

25. $\left(\dfrac{5\sqrt{2}}{2}, -\dfrac{5\sqrt{2}}{2}\right)$

26. $\left(-\dfrac{3}{2}, -\dfrac{3\sqrt{3}}{2}\right)$

Use a graphing calculator to convert from rectangular coordinates to polar coordinates. Express the answer in both degrees and radians, using the smallest possible positive angle.

27. $(3, 7)$

28. $\left(-2, -\sqrt{5}\right)$

29. $\left(-\sqrt{10}, 3.4\right)$

30. $(0.9, -6)$

Find the rectangular coordinates of the point.

31. $(5, 60°)$

32. $(0, -23°)$

33. $(-3, 45°)$

34. $(6, 30°)$

35. $(3, -120°)$

36. $\left(7, \dfrac{\pi}{6}\right)$

37. $\left(-2, \dfrac{5\pi}{3}\right)$

38. $(1.4, 225°)$

39. $(2, 210°)$

40. $\left(1, \dfrac{7\pi}{4}\right)$

41. $\left(-6, \dfrac{5\pi}{6}\right)$

42. $(4, 180°)$

Use a graphing calculator to convert from polar coordinates to rectangular coordinates. Round the coordinates to the nearest hundredth.

43. $(3, -43°)$

44. $\left(-5, \dfrac{\pi}{7}\right)$

45. $\left(-4.2, \dfrac{3\pi}{5}\right)$

46. $(2.8, 166°)$

Convert to a polar equation.

47. $3x + 4y = 5$

48. $5x + 3y = 4$

49. $x = 5$

50. $y = 4$

51. $x^2 + y^2 = 36$

52. $x^2 - 4y^2 = 4$

53. $x^2 = 25y$

54. $2x - 9y + 3 = 0$

55. $y^2 - 5x - 25 = 0$

56. $x^2 + y^2 = 8y$

57. $x^2 - 2x + y^2 = 0$

58. $3x^2y = 81$

Convert to a rectangular equation.

59. $r = 5$

60. $\theta = \dfrac{3\pi}{4}$

61. $r \sin \theta = 2$

62. $r = -3 \sin \theta$

63. $r + r \cos \theta = 3$

64. $r = \dfrac{2}{1 - \sin \theta}$

65. $r - 9 \cos \theta = 7 \sin \theta$

66. $r + 5 \sin \theta = 7 \cos \theta$

67. $r = 5 \sec \theta$

68. $r = 3 \cos \theta$

69. $\theta = \dfrac{5\pi}{3}$

70. $r = \cos \theta - \sin \theta$

Graph the equation by plotting points. Then check your work using a graphing calculator.

71. $r = \sin \theta$

72. $r = 1 - \cos \theta$

73. $r = 4 \cos 2\theta$

74. $r = 1 - 2 \sin \theta$

75. $r = \cos \theta$

76. $r = 2 \sec \theta$

77. $r = 2 - \cos 3\theta$

78. $r = \dfrac{1}{1 + \cos \theta}$

In Exercises 79–90, use a graphing calculator to match the equation with one of figures (a)–(l), which follow. Try matching the graphs mentally before using a calculator.

a)
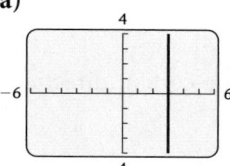

b)
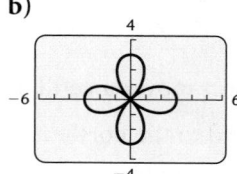

c)
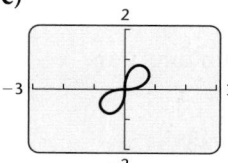

d)
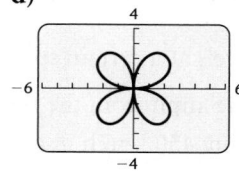

e)
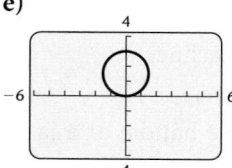

f)
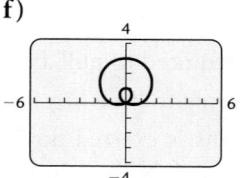

g)
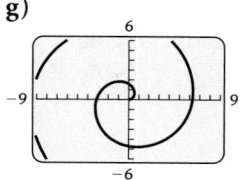

h)
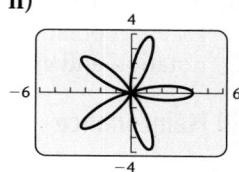

i)
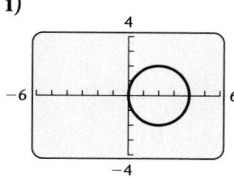

j)
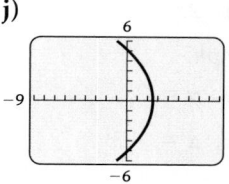

k)
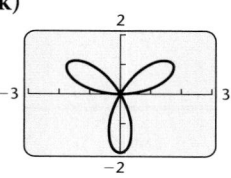

l)
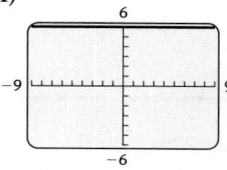

79. $r = 3 \sin 2\theta$

80. $r = 4 \cos \theta$

81. $r = \theta$

82. $r^2 = \sin 2\theta$

83. $r = \dfrac{5}{1 + \cos \theta}$

84. $r = 1 + 2 \sin \theta$

85. $r = 3 \cos 2\theta$

86. $r = 3 \sec \theta$

87. $r = 3 \sin \theta$

88. $r = 4 \cos 5\theta$

89. $r = 2 \sin 3\theta$

90. $r \sin \theta = 6$

Graph the equation using a graphing calculator.

91. $r = \sin \theta \tan \theta$ (Cissoid)

92. $r = 3\theta$ (Spiral of Archimedes)

93. $r = e^{\theta/10}$ (Logarithmic spiral)

94. $r = 10^{2\theta}$ (Logarithmic spiral)

95. $r = \cos 2\theta \sec \theta$ (Strophoid)

96. $r = \cos 2\theta - 2$ (Peanut)

97. $r = \frac{1}{4} \tan^2 \theta \sec \theta$ (Semicubical parabola)

98. $r = \sin 2\theta + \cos \theta$ (Twisted sister)

Collaborative Discussion and Writing

99. Explain why the rectangular coordinates of a point are unique and the polar coordinates of a point are not unique.

100. Give an example of an equation that is easier to graph in polar notation than in rectangular notation and explain why.

Skill Maintenance

Solve.

101. $2x - 4 = x + 8$ **102.** $4 - 5y = 3$

Graph.

103. $y = 2x - 5$ **104.** $4x - y = 6$

105. $x = -3$ **106.** $y = 0$

Synthesis

107. Convert to a rectangular equation:

$$r = \sec^2 \frac{\theta}{2}.$$

108. The center of a regular hexagon is at the origin, and one vertex is the point $(4, 0°)$. Find the coordinates of the other vertices.

Vectors and Applications

8.5

❖ Determine whether two vectors are equivalent.
❖ Find the sum, or resultant, of two vectors.
❖ Resolve a vector into its horizontal and vertical components.
❖ Solve applied problems involving vectors.

We measure some quantities using only their magnitudes. For example, we describe time, length, and mass using units like seconds, feet, and kilograms, respectively. However, to measure quantities like **displacement**, **velocity**, or **force**, we need to describe a *magnitude* and a *direction*. Together magnitude and direction describe a **vector**. The following are some examples.

Displacement.   An object moves a certain distance in a certain direction.

   A surveyor steps 20 yd to the northeast.

   A hiker follows a trail 5 mi to the west.

   A batter hits a ball 100 m along the left-field line.

Velocity.   An object travels at a certain speed in a certain direction.

   A breeze is blowing 15 mph from the northwest.

   An airplane is traveling 450 km/h in a direction of 243°.

Force.   A push or pull is exerted on an object in a certain direction.

   A force of 200 lb is required to pull a cart up a 30° incline.

   A 25-lb force is required to lift a box upward.

   A force of 15 newtons is exerted downward on the handle of a jack. (A newton, abbreviated N, is a unit of force used in physics, and 1 N ≈ 0.22 lb.)

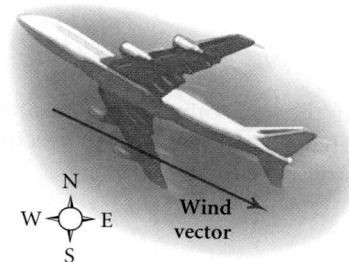

❋ Vectors

Vectors can be graphically represented by directed line segments. The length is chosen, according to some scale, to represent the **magnitude of the vector**, and the direction of the directed line segment represents the **direction of the vector**. For example, if we let 1 cm represent 5 km/h, then a 15-km/h wind from the northwest would be represented by a directed line segment 3 cm long, as shown in the figure at left.

> ### Vector
>
> A **vector** in the plane is a directed line segment. Two vectors are **equivalent** if they have the same *magnitude* and the same *direction*.

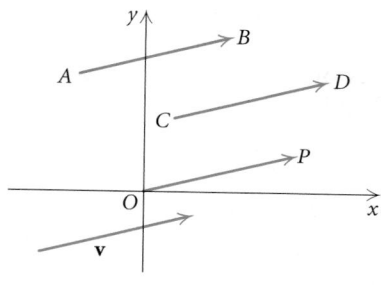

Consider a vector drawn from point A to point B. Point A is called the **initial point** of the vector, and point B is called the **terminal point**. Symbolic notation for this vector is $\overrightarrow{AB}$ (read "vector AB"). Vectors are also denoted by boldface letters such as **u**, **v**, and **w**. The four vectors in the figure at left have the *same* length and the *same* direction. Thus they represent **equivalent** vectors; that is,

$$\overrightarrow{AB} = \overrightarrow{CD} = \overrightarrow{OP} = \mathbf{v}.$$

In the context of vectors, we use = to mean equivalent.

The length, or **magnitude**, of $\overrightarrow{AB}$ is expressed as $|\overrightarrow{AB}|$. In order to determine whether vectors are equivalent, we find their magnitudes and directions.

EXAMPLE 1   The vectors **u**, $\overrightarrow{OR}$, and **w** are shown in the figure below. Show that $\mathbf{u} = \overrightarrow{OR} = \mathbf{w}$.

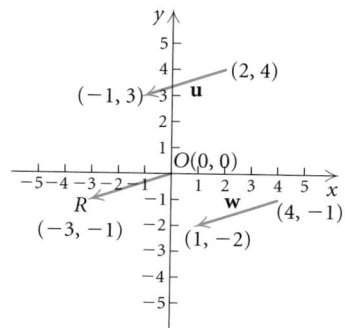

Solution   We first find the length of each vector using the distance formula:

$$|\mathbf{u}| = \sqrt{[2 - (-1)]^2 + (4 - 3)^2} = \sqrt{9 + 1} = \sqrt{10},$$
$$|\overrightarrow{OR}| = \sqrt{[0 - (-3)]^2 + [0 - (-1)]^2} = \sqrt{9 + 1} = \sqrt{10},$$
$$|\mathbf{w}| = \sqrt{(4 - 1)^2 + [-1 - (-2)]^2} = \sqrt{9 + 1} = \sqrt{10}.$$

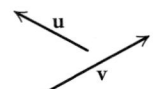

u ≠ **v** (not equivalent)
Different magnitudes;
different directions

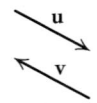

u ≠ **v**
Same magnitude;
different directions

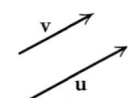

u ≠ **v**
Different magnitudes;
same direction

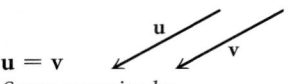

u = **v**
Same magnitude;
same direction

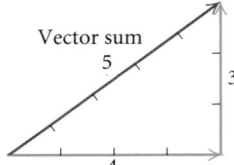

Thus,

$$|\mathbf{u}| = |\overrightarrow{OR}| = |\mathbf{w}|.$$

The vectors **u**, $\overrightarrow{OR}$, and **w** appear to go in the same direction so we check their slopes. If the lines that they are on all have the same slope, the vectors have the same direction. We calculate the slopes:

$$\text{Slope} = \overset{\mathbf{u}}{\frac{4-3}{2-(-1)}} = \overset{\overrightarrow{OR}}{\frac{0-(-1)}{0-(-3)}} = \overset{\mathbf{w}}{\frac{-1-(-2)}{4-1}} = \frac{1}{3}.$$

Since **u**, $\overrightarrow{OR}$, and **w** have the *same* magnitude and the *same* direction,

$$\mathbf{u} = \overrightarrow{OR} = \mathbf{w}.$$

Now Try Exercise 1. ■

Keep in mind that the equivalence of vectors requires only the same magnitude and the same direction—not the same location. In the illustrations at left, each of the first three pairs of vectors are not equivalent. The fourth set of vectors is an example of equivalence.

❖ Vector Addition

Suppose a person takes 4 steps east and then 3 steps north. He or she will then be 5 steps from the starting point in the direction shown at left. A vector 4 units long and pointing to the right represents 4 steps east and a vector 3 units long and pointing up represents 3 steps north. The **sum** of the two vectors is the vector 5 steps in magnitude and in the direction shown. The sum is also called the **resultant** of the two vectors.

In general, two nonzero vectors **u** and **v** can be added geometrically by placing the initial point of **v** at the terminal point of **u** and then finding the vector that has the same initial point as **u** and the same terminal point as **v**, as shown in the following figure.

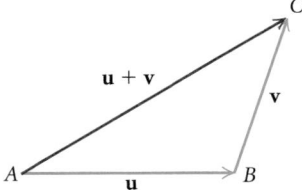

The sum **u** + **v** is the vector represented by the directed line segment from the initial point *A* of **u** to the terminal point *C* of **v**. That is, if

$$\mathbf{u} = \overrightarrow{AB} \quad \text{and} \quad \mathbf{v} = \overrightarrow{BC},$$

then

$$\mathbf{u} + \mathbf{v} = \overrightarrow{AB} + \overrightarrow{BC} = \overrightarrow{AC}.$$

We can also describe vector addition by placing the initial points of the vectors together, completing a parallelogram, and finding the diagonal of the parallelogram. (See the figure on the left below.) This description of addition is sometimes called the **parallelogram law** of vector addition. Vector addition is **commutative**. As shown in the figure on the right below, both **u** + **v** and **v** + **u** are represented by the same directed line segment.

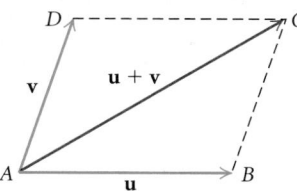
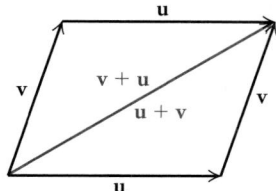

❋ Applications

If two forces F_1 and F_2 act on an object, the *combined* effect is the sum, or resultant, $F_1 + F_2$ of the separate forces.

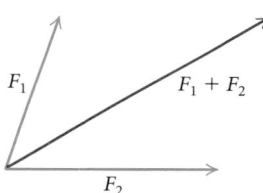

EXAMPLE 2 Forces of 15 newtons and 25 newtons act on an object at right angles to each other. Find their sum, or resultant, giving the magnitude of the resultant and the angle that it makes with the larger force.

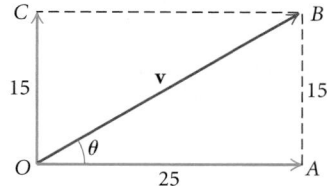

Solution We make a drawing—this time, a rectangle—using **v** or $\overrightarrow{OB}$ to represent the resultant. To find the magnitude, we use the Pythagorean equation:

$$|\mathbf{v}|^2 = 15^2 + 25^2 \quad \text{Here } |\mathbf{v}| \text{ denotes the length, or magnitude, of v.}$$
$$|\mathbf{v}| = \sqrt{15^2 + 25^2}$$
$$|\mathbf{v}| \approx 29.2.$$

To find the direction, we note that since OAB is a right triangle,

$$\tan \theta = \tfrac{15}{25} = 0.6.$$

Using a calculator, we find θ, the angle that the resultant makes with the larger force:

$$\theta = \tan^{-1}(0.6) \approx 31°.$$

The resultant $\overrightarrow{OB}$ has a magnitude of 29.2 and makes an angle of 31° with the larger force.   **Now Try Exercise 13.** ■

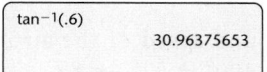
AERIAL BEARINGS

REVIEW SECTION **6.3.**

Pilots must adjust the direction of their flight when there is a crosswind. Both the wind and the aircraft velocities can be described by vectors.

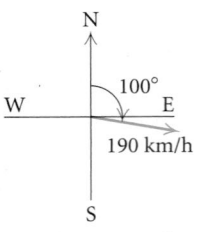

Airplane airspeed

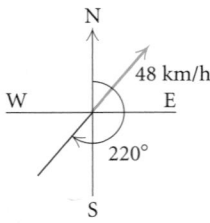

Windspeed

EXAMPLE 3 *Airplane Speed and Direction.* An airplane travels on a bearing of 100° at an airspeed of 190 km/h while a wind is blowing 48 km/h from 220°. Find the ground speed of the airplane and the direction of its track, or course, over the ground.

Solution We first make a drawing. The wind is represented by $\overrightarrow{OC}$ and the velocity vector of the airplane by $\overrightarrow{OA}$. The resultant velocity vector is **v**, the sum of the two vectors. The angle θ between **v** and $\overrightarrow{OA}$ is called a **drift angle**.

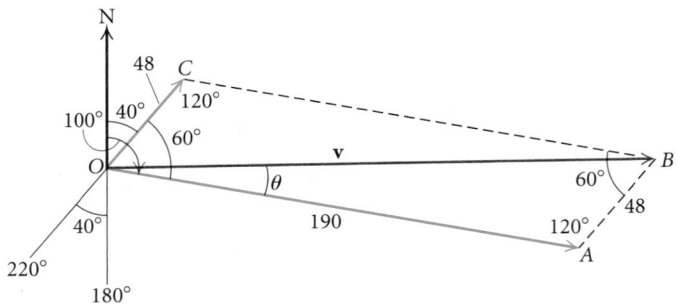

Note that the measure of $\angle COA = 100° - 40° = 60°$. Thus the measure of $\angle CBA$ is also 60° (opposite angles of a parallelogram are equal). Since the sum of all the angles of the parallelogram is 360° and $\angle OCB$ and $\angle OAB$ have the same measure, each must be 120°. By the *law of cosines* in $\triangle OAB$, we have

$$|\mathbf{v}|^2 = 48^2 + 190^2 - 2 \cdot 48 \cdot 190 \cos 120°$$
$$|\mathbf{v}|^2 = 47{,}524$$
$$|\mathbf{v}| = 218.$$

Thus, $|\mathbf{v}|$ is 218 km/h. By the *law of sines* in the same triangle,

$$\frac{48}{\sin \theta} = \frac{218}{\sin 120°},$$

or

$$\sin \theta = \frac{48 \sin 120°}{218} \approx 0.1907$$
$$\theta \approx 11°.$$

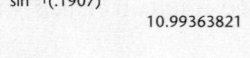

Thus, $\theta = 11°$, to the nearest degree. The ground speed of the airplane is 218 km/h, and its track is in the direction of $100° - 11°$, or 89°.

Now Try Exercise 27. ■

✻ Components

Given a vector **w**, we may want to find two other vectors **u** and **v** whose sum is **w**. The vectors **u** and **v** are called **components** of **w** and the process of finding them is called **resolving**, or **representing**, a vector into its vector components.

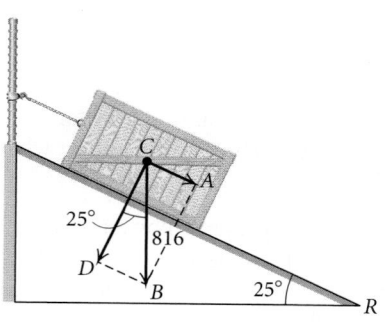

When we resolve a vector, we generally look for perpendicular components. Most often, one component will be parallel to the x-axis and the other will be parallel to the y-axis. For this reason, they are often called the **horizontal** and **vertical** components of a vector. In the figure at left, the vector $\mathbf{w} = \overrightarrow{AC}$ is resolved as the sum of $\mathbf{u} = \overrightarrow{AB}$ and $\mathbf{v} = \overrightarrow{BC}$. The horizontal component of $\mathbf{w}$ is $\mathbf{u}$ and the vertical component is $\mathbf{v}$.

EXAMPLE 4 A vector $\mathbf{w}$ has a magnitude of 130 and is inclined 40° with the horizontal. Resolve the vector into horizontal and vertical components.

Solution We first make a drawing showing horizontal and vertical vectors $\mathbf{u}$ and $\mathbf{v}$ whose sum is $\mathbf{w}$.

From $\triangle ABC$, we find $|\mathbf{u}|$ and $|\mathbf{v}|$ using the definitions of the cosine and sine functions:

$$\cos 40° = \frac{|\mathbf{u}|}{130}, \quad \text{or} \quad |\mathbf{u}| = 130 \cos 40° \approx 100,$$

$$\sin 40° = \frac{|\mathbf{v}|}{130}, \quad \text{or} \quad |\mathbf{v}| = 130 \sin 40° \approx 84.$$

Thus the horizontal component of $\mathbf{w}$ is 100 right, and the vertical component of $\mathbf{w}$ is 84 up. **Now Try Exercise 31.** ■

EXAMPLE 5 *Shipping Crate.* A wooden shipping crate that weighs 816 lb is placed on a loading ramp that makes an angle of 25° with the horizontal. To keep the crate from sliding, a chain is hooked to the crate and to a pole at the top of the ramp. Find the magnitude of the components of the crate's weight (disregarding friction) perpendicular and parallel to the incline.

Solution We first make a drawing illustrating the forces with a rectangle. We let

$|\overrightarrow{CB}|$ = the weight of the crate = 816 lb (force of gravity),
$|\overrightarrow{CD}|$ = the magnitude of the component of the crate's weight perpendicular to the incline (force against the ramp), and
$|\overrightarrow{CA}|$ = the magnitude of the component of the crate's weight parallel to the incline (force that pulls the crate down the ramp).

The angle at R is given to be 25° and $\angle BCD = \angle R = 25°$ because the sides of these angles are, respectively, perpendicular. Using the cosine and sine functions, we find that

$$\cos 25° = \frac{|\overrightarrow{CD}|}{816}, \quad \text{or} \quad |\overrightarrow{CD}| = 816 \cos 25° \approx 740 \text{ lb}, \quad \text{and}$$

$$\sin 25° = \frac{\overrightarrow{DB}}{816} = \frac{|\overrightarrow{CA}|}{816}, \quad \text{or} \quad |\overrightarrow{CA}| = 816 \sin 25° \approx 345 \text{ lb}.$$

Now Try Exercise 35. ■

8.5 Exercise Set

Sketch the pair of vectors and determine whether they are equivalent. Use the following ordered pairs for the initial and terminal points.

$A(-2, 2)$	$E(-4, 1)$	$I(-6, -3)$
$B(3, 4)$	$F(2, 1)$	$J(3, 1)$
$C(-2, 5)$	$G(-4, 4)$	$K(-3, -3)$
$D(-1, -1)$	$H(1, 2)$	$O(0, 0)$

1. $\overrightarrow{GE}, \overrightarrow{BJ}$

2. $\overrightarrow{DJ}, \overrightarrow{OF}$

3. $\overrightarrow{DJ}, \overrightarrow{AB}$

4. $\overrightarrow{CG}, \overrightarrow{FO}$

5. $\overrightarrow{DK}, \overrightarrow{BH}$

6. $\overrightarrow{BA}, \overrightarrow{DI}$

7. $\overrightarrow{EG}, \overrightarrow{BJ}$

8. $\overrightarrow{GC}, \overrightarrow{FO}$

9. $\overrightarrow{GA}, \overrightarrow{BH}$

10. $\overrightarrow{JD}, \overrightarrow{CG}$

11. $\overrightarrow{AB}, \overrightarrow{ID}$

12. $\overrightarrow{OF}, \overrightarrow{HB}$

13. Two forces of 32 N (newtons) and 45 N act on an object at right angles. Find the magnitude of the resultant and the angle that it makes with the smaller force.

14. Two forces of 50 N and 60 N act on an object at right angles. Find the magnitude of the resultant and the angle that it makes with the larger force.

15. Two forces of 410 N and 600 N act on an object. The angle between the forces is 47°. Find the magnitude of the resultant and the angle that it makes with the larger force.

16. Two forces of 255 N and 325 N act on an object. The angle between the forces is 64°. Find the magnitude of the resultant and the angle that it makes with the smaller force.

In Exercises 17–24, magnitudes of vectors $\mathbf{u}$ and $\mathbf{v}$ and the angle θ between the vectors are given. Find the sum of $\mathbf{u} + \mathbf{v}$. Give the magnitude to the nearest tenth and give the direction by specifying to the nearest degree the angle that the resultant makes with $\mathbf{u}$.

17. $|\mathbf{u}| = 45, \ |\mathbf{v}| = 35, \ \theta = 90°$

18. $|\mathbf{u}| = 54, \ |\mathbf{v}| = 43, \ \theta = 150°$

19. $|\mathbf{u}| = 10, \ |\mathbf{v}| = 12, \ \theta = 67°$

20. $|\mathbf{u}| = 25, \ |\mathbf{v}| = 30, \ \theta = 75°$

21. $|\mathbf{u}| = 20, \ |\mathbf{v}| = 20, \ \theta = 117°$

22. $|\mathbf{u}| = 30, \ |\mathbf{v}| = 30, \ \theta = 123°$

23. $|\mathbf{u}| = 23, \ |\mathbf{v}| = 47, \ \theta = 27°$

24. $|\mathbf{u}| = 32, \ |\mathbf{v}| = 74, \ \theta = 72°$

25. *Hot-Air Balloon.* A hot-air balloon is rising vertically 10 ft/sec while the wind is blowing horizontally 5 ft/sec. Find the speed $\mathbf{v}$ of the balloon and the angle θ that it makes with the horizontal.

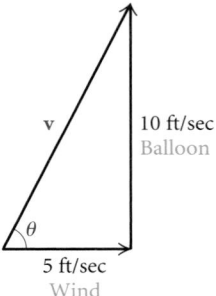

26. *Ship.* A ship sails first N80°E for 120 nautical mi, and then S20°W for 200 nautical mi. How far is the ship, then, from the starting point, and in what direction?

27. *Boat.* A boat heads 35°, propelled by a force of 750 lb. A wind from 320° exerts a force of 150 lb on the boat. How large is the resultant force $\mathbf{F}$, and in what direction is the boat moving?

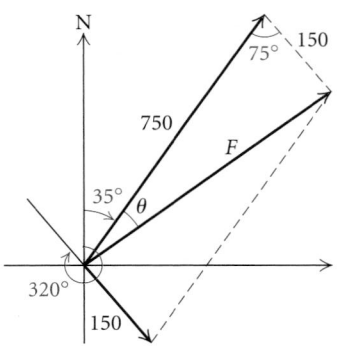

28. *Airplane.* An airplane flies 32° for 210 km, and then 280° for 170 km. How far is the airplane, then, from the starting point, and in what direction?

29. *Airplane.* An airplane has an airspeed of 150 km/h. It is to make a flight in a direction of 70° while there is a 25-km/h wind from 340°. What will the airplane's actual heading be?

30. *Wind.* A wind has an easterly component (*from* the east) of 10 km/h and a southerly component (*from* the south) of 16 km/h. Find the magnitude and the direction of the wind.

31. A vector **w** has magnitude 100 and points southeast. Resolve the vector into easterly and southerly components.

32. A vector **u** with a magnitude of 150 lb is inclined to the right and upward 52° from the horizontal. Resolve the vector into components.

33. *Airplane.* An airplane takes off at a speed **S** of 225 mph at an angle of 17° with the horizontal. Resolve the vector **S** into components.

34. *Wheelbarrow.* A wheelbarrow is pushed by applying a 97-lb force **F** that makes a 38° angle with the horizontal. Resolve **F** into its horizontal and vertical components. (The horizontal component is the effective force in the direction of motion and the vertical component adds weight to the wheelbarrow.)

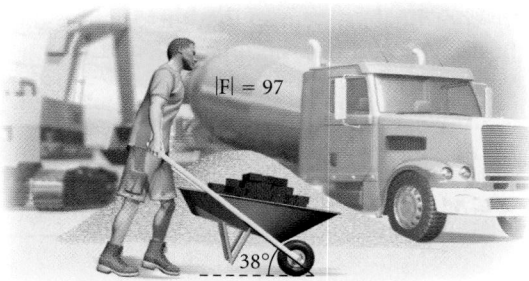

35. *Luggage Wagon.* A luggage wagon is being pulled with vector force **V**, which has a

magnitude of 780 lb at an angle of elevation of 60°. Resolve the vector **V** into components.

36. *Hot-air Balloon.* A hot-air balloon exerts a 1200-lb pull on a tether line at a 45° angle with the horizontal. Resolve the vector **B** into components.

37. *Airplane.* An airplane is flying at 200 km/h in a direction of 305°. Find the westerly and northerly components of its velocity.

38. *Baseball.* A baseball player throws a baseball with a speed **S** of 72 mph at an angle of 45° with the horizontal. Resolve the vector **S** into components.

39. A block weighing 100 lb rests on a 25° incline. Find the magnitude of the components of the block's weight perpendicular and parallel to the incline.

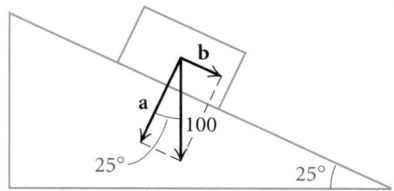

40. A shipping crate that weighs 450 kg is placed on a loading ramp that makes an angle of 30° with the horizontal. Find the magnitude of the components of the crate's weight perpendicular and parallel to the incline.

41. An 80-lb block of ice rests on a 37° incline. What force parallel to the incline is necessary in order to keep the ice from sliding down?

42. What force is necessary to pull a 3500-lb truck up a 9° incline?

Collaborative Discussion and Writing

43. Describe the concept of a vector as though you were explaining it to a classmate. Use the concept of an arrow shot from a bow in the explanation.

44. Explain why vectors $\overrightarrow{QR}$ and $\overrightarrow{RQ}$ are not equivalent.

Skill Maintenance

In each of Exercises 45–54, fill in the blank with the correct term. Some of the given choices will not be used.

angular speed
linear speed
acute
obtuse
secant of θ
cotangent of θ
identity
inverse
absolute value
sines
cosine
common
natural
horizontal line
vertical line
double-angle
half-angle
coterminal
reference angle

45. Logarithms, base e, are called _____ logarithms.

46. _____ identities give trigonometric function values of $x/2$ in terms of function values of x.

47. _____ is distance traveled per unit of time.

48. The sine of an angle is also the _____ of the angle's complement.

49. A(n) _____ is an equation that is true for all possible replacements of the variables.

50. The _____ is the length of the side adjacent to θ divided by the length of the side opposite θ.

51. If two or more angles have the same terminal side, the angles are said to be _____ .

52. In any triangle, the sides are proportional to the _____ of the opposite angles.

53. If it is possible for a(n) _____ to intersect the graph of a function more than once, then the function is not one-to-one and its _____ is not a function.

54. The _____ for an angle is the _____ angle formed by the terminal side of the angle and the x-axis.

Synthesis

55. *Eagle's Flight.* An eagle flies from its nest 7 mi in the direction northeast, where it stops to rest on a cliff. It then flies 8 mi in the direction S30°W to land on top of a tree. Place an xy-coordinate system so that the origin is the bird's nest, the x-axis points east, and the y-axis points north.

a) At what point is the cliff located?
b) At what point is the tree located?

8.6 Vector Operations

❉ Perform calculations with vectors in component form.

❉ Express a vector as a linear combination of unit vectors.

❉ Express a vector in terms of its magnitude and its direction.

❉ Find the angle between two vectors using the dot product.

❉ Solve applied problems involving forces in equilibrium.

❉ Position Vectors

Let's consider a vector **v** whose initial point is the *origin* in an *xy*-coordinate system and whose terminal point is (a, b). We say that the vector is in **standard position** and refer to it as a position vector. Note that the ordered pair (a, b) defines the vector uniquely. Thus we can use (a, b) to denote the vector. To emphasize that we are thinking of a vector and to avoid the confusion of notation with ordered-pair and interval notation, we generally write

$$\mathbf{v} = \langle a, b \rangle.$$

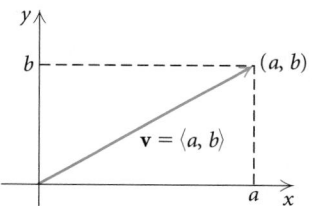

The coordinate a is the *scalar* **horizontal component** of the vector, and the coordinate b is the *scalar* **vertical component** of the vector. By **scalar**, we mean a *numerical* quantity rather than a *vector* quantity. Thus, $\langle a, b \rangle$ is considered to be the *component form* of **v**. Note that a and b are *not* vectors and should not be confused with the vector component definition given in Section 8.5.

Now consider $\overrightarrow{AC}$ with $A = (x_1, y_1)$ and $C = (x_2, y_2)$. Let's see how to find the position vector equivalent to $\overrightarrow{AC}$. As you can see in the figure below, the initial point A is relocated to the origin $(0, 0)$. The coordinates of P are found by subtracting the coordinates of A from the coordinates of C. Thus, $P = (x_2 - x_1, y_2 - y_1)$ and the position vector is $\overrightarrow{OP}$.

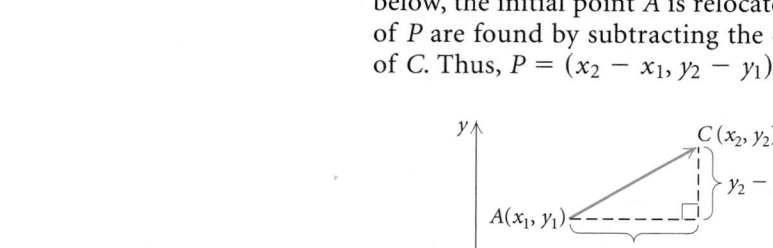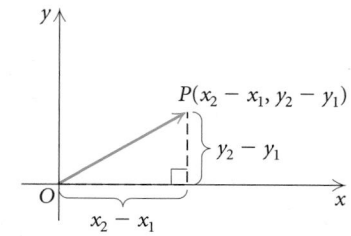

It can be shown that $\overrightarrow{OP}$ and $\overrightarrow{AC}$ have the same magnitude and direction and are therefore equivalent. Thus, $\overrightarrow{AC} = \overrightarrow{OP} = \langle x_2 - x_1, y_2 - y_1 \rangle$.

> **Component Form of a Vector**
>
> The **component form** of $\overrightarrow{AC}$ with $A = (x_1, y_1)$ and $C = (x_2, y_2)$ is
>
> $$\overrightarrow{AC} = \langle x_2 - x_1, y_2 - y_1 \rangle.$$

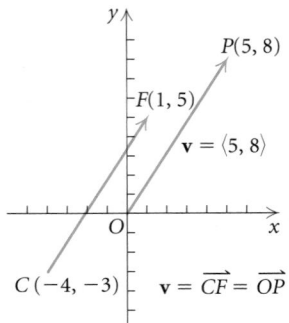

EXAMPLE 1 Find the component form of $\overrightarrow{CF}$ if $C = (-4, -3)$ and $F = (1, 5)$.

Solution We have

$$\overrightarrow{CF} = \langle 1 - (-4), 5 - (-3)\rangle = \langle 5, 8\rangle.$$

Note that vector $\overrightarrow{CF}$ is equivalent to *position vector* $\overrightarrow{OP}$ with $P = (5, 8)$ as shown in the figure at left. ∎

Now that we know how to write vectors in component form, let's restate some definitions that we first considered in Section 8.5.

The length of a vector **v** is easy to determine when the components of the vector are known. For $\mathbf{v} = \langle v_1, v_2\rangle$, we have

$$|\mathbf{v}|^2 = v_1^2 + v_2^2 \qquad \text{Using the Pythagorean equation}$$
$$|\mathbf{v}| = \sqrt{v_1^2 + v_2^2}.$$

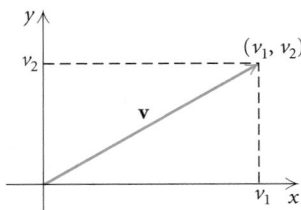

> **Length of a Vector**
> The **length**, or **magnitude**, of a vector $\mathbf{v} = \langle v_1, v_2\rangle$ is given by
> $$|\mathbf{v}| = \sqrt{v_1^2 + v_2^2}.$$

EXAMPLE 2 Find the length, or magnitude, of vector $\mathbf{v} = \langle 5, 8\rangle$, illustrated in Example 1.

Solution

$$|\mathbf{v}| = \sqrt{v_1^2 + v_2^2} \qquad \text{Length of vector } \mathbf{v} = \langle v_1, v_2\rangle$$
$$= \sqrt{5^2 + 8^2} \qquad \text{Substituting 5 for } v_1 \text{ and 8 for } v_2$$
$$= \sqrt{25 + 64}$$
$$= \sqrt{89} \qquad\qquad \textbf{Now Try Exercises 1 and 7.} \ ∎$$

Two vectors are **equivalent** if they have the *same* magnitude and the *same* direction.

> **Equivalent Vectors**
> Let $\mathbf{u} = \langle u_1, u_2\rangle$ and $\mathbf{v} = \langle v_1, v_2\rangle$. Then
> $$\langle u_1, u_2\rangle = \langle v_1, v_2\rangle \quad \text{if and only if} \quad u_1 = v_1 \quad \text{and} \quad u_2 = v_2.$$

❋ Operations on Vectors

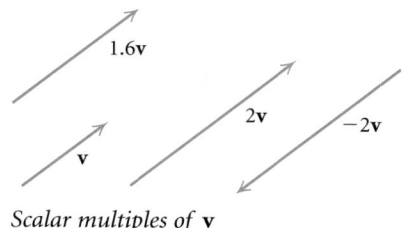

Scalar multiples of **v**

To multiply a vector **v** by a positive real number, we multiply its length by the number. Its direction stays the same. When a vector **v** is multiplied by 2, for instance, its length is doubled and its direction is not changed. When a vector is multiplied by 1.6, its length is increased by 60% and its direction stays the same. To multiply a vector **v** by a negative real number, we multiply its length by the number and reverse its direction. When a vector is multiplied by -2, its length is doubled and its direction is reversed. Since real numbers work like scaling factors in vector multiplication, we call them **scalars** and the products $k\mathbf{v}$ are called **scalar multiples** of **v**.

> ## Scalar Multiplication
>
> For a real number k and a vector $\mathbf{v} = \langle v_1, v_2 \rangle$, the **scalar product** of k and **v** is
>
> $$k\mathbf{v} = k\langle v_1, v_2 \rangle = \langle kv_1, kv_2 \rangle.$$
>
> The vector $k\mathbf{v}$ is a **scalar multiple** of the vector **v**.

EXAMPLE 3   Let $\mathbf{u} = \langle -5, 4 \rangle$ and $\mathbf{w} = \langle 1, -1 \rangle$. Find $-7\mathbf{w}$, $3\mathbf{u}$, and $-1\mathbf{w}$.

Solution

$$-7\mathbf{w} = -7\langle 1, -1 \rangle = \langle -7, 7 \rangle,$$
$$3\mathbf{u} = 3\langle -5, 4 \rangle = \langle -15, 12 \rangle,$$
$$-1\mathbf{w} = -1\langle 1, -1 \rangle = \langle -1, 1 \rangle$$ ■

In Section 8.5, we used the parallelogram law to add two vectors, but now we can add two vectors using components. To add two vectors given in component form, we add the corresponding components. Let $\mathbf{u} = \langle u_1, u_2 \rangle$ and $\mathbf{v} = \langle v_1, v_2 \rangle$. Then

$$\mathbf{u} + \mathbf{v} = \langle u_1 + v_1, u_2 + v_2 \rangle.$$

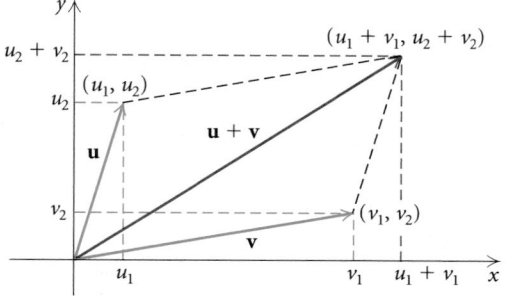

For example, if $\mathbf{v} = \langle -3, 2 \rangle$ and $\mathbf{w} = \langle 5, -9 \rangle$, then

$$\mathbf{v} + \mathbf{w} = \langle -3 + 5, 2 + (-9) \rangle = \langle 2, -7 \rangle.$$

> ### Vector Addition
> If $\mathbf{u} = \langle u_1, u_2 \rangle$ and $\mathbf{v} = \langle v_1, v_2 \rangle$, then
> $$\mathbf{u} + \mathbf{v} = \langle u_1 + v_1, u_2 + v_2 \rangle.$$

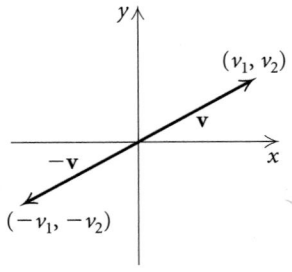

Before we define vector subtraction, we need to define $-\mathbf{v}$. The opposite of $\mathbf{v} = \langle v_1, v_2 \rangle$, shown at left, is

$$-\mathbf{v} = (-1)\mathbf{v} = (-1)\langle v_1, v_2 \rangle = \langle -v_1, -v_2 \rangle.$$

Vector subtraction such as $\mathbf{u} - \mathbf{v}$ involves subtracting corresponding components. We show this by rewriting $\mathbf{u} - \mathbf{v}$ as $\mathbf{u} + (-\mathbf{v})$. If $\mathbf{u} = \langle u_1, u_2 \rangle$ and $\mathbf{v} = \langle v_1, v_2 \rangle$, then

$$\begin{aligned}
\mathbf{u} - \mathbf{v} = \mathbf{u} + (-\mathbf{v}) &= \langle u_1, u_2 \rangle + \langle -v_1, -v_2 \rangle \\
&= \langle u_1 + (-v_1), u_2 + (-v_2) \rangle \\
&= \langle u_1 - v_1, u_2 - v_2 \rangle.
\end{aligned}$$

We can illustrate vector subtraction with parallelograms, just as we did vector addition.

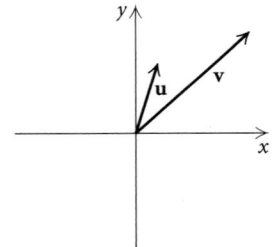

Sketch $\mathbf{u}$ and $\mathbf{v}$.

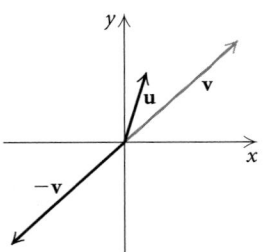

Sketch $-\mathbf{v}$.

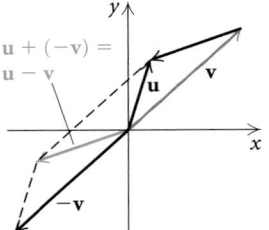

Sketch $\mathbf{u} + (-\mathbf{v})$, or $\mathbf{u} - \mathbf{v}$, using the parallelogram law.

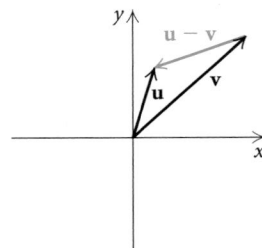

$\mathbf{u} - \mathbf{v}$ is the vector from the terminal point of $\mathbf{v}$ to the terminal point of $\mathbf{u}$.

> ### Vector Subtraction
> If $\mathbf{u} = \langle u_1, u_2 \rangle$ and $\mathbf{v} = \langle v_1, v_2 \rangle$, then
> $$\mathbf{u} - \mathbf{v} = \langle u_1 - v_1, u_2 - v_2 \rangle.$$

It is interesting to compare the sum of two vectors with the difference of the same two vectors in the same parallelogram. The vectors $\mathbf{u} + \mathbf{v}$ and $\mathbf{u} - \mathbf{v}$ are the diagonals of the parallelogram.

EXAMPLE 4 Do the following calculations, where $\mathbf{u} = \langle 7, 2 \rangle$ and $\mathbf{v} = \langle -3, 5 \rangle$.

a) $\mathbf{u} + \mathbf{v}$

b) $\mathbf{u} - 6\mathbf{v}$

c) $3\mathbf{u} + 4\mathbf{v}$

d) $|5\mathbf{v} - 2\mathbf{u}|$

Solution

a) $\mathbf{u} + \mathbf{v} = \langle 7, 2 \rangle + \langle -3, 5 \rangle = \langle 7 + (-3), 2 + 5 \rangle = \langle 4, 7 \rangle$

b) $\mathbf{u} - 6\mathbf{v} = \langle 7, 2 \rangle - 6\langle -3, 5 \rangle = \langle 7, 2 \rangle - \langle -18, 30 \rangle = \langle 25, -28 \rangle$

c) $3\mathbf{u} + 4\mathbf{v} = 3\langle 7, 2 \rangle + 4\langle -3, 5 \rangle = \langle 21, 6 \rangle + \langle -12, 20 \rangle = \langle 9, 26 \rangle$

d) $|5\mathbf{v} - 2\mathbf{u}| = |5\langle -3, 5 \rangle - 2\langle 7, 2 \rangle| = |\langle -15, 25 \rangle - \langle 14, 4 \rangle|$

$$= |\langle -29, 21 \rangle|$$
$$= \sqrt{(-29)^2 + 21^2}$$
$$= \sqrt{1282}$$
$$\approx 35.8$$

Now Try Exercises 9 and 11. ∎

Before we state the properties of vector addition and scalar multiplication, we need to define another special vector—the zero vector. The vector whose initial and terminal points are both $(0, 0)$ is the **zero vector**, denoted by $\mathbf{O}$, or $\langle 0, 0 \rangle$. Its magnitude is 0. In vector addition, the zero vector is the additive identity vector:

$$\mathbf{v} + \mathbf{O} = \mathbf{v}. \qquad \langle v_1, v_2 \rangle + \langle 0, 0 \rangle = \langle v_1, v_2 \rangle$$

Operations on vectors share many of the same properties as operations on real numbers.

Properties of Vector Addition and Scalar Multiplication

For all vectors $\mathbf{u}$, $\mathbf{v}$, and $\mathbf{w}$, and for all scalars b and c:

1. $\mathbf{u} + \mathbf{v} = \mathbf{v} + \mathbf{u}$.
2. $\mathbf{u} + (\mathbf{v} + \mathbf{w}) = (\mathbf{u} + \mathbf{v}) + \mathbf{w}$.
3. $\mathbf{v} + \mathbf{O} = \mathbf{v}$.
4. $1\mathbf{v} = \mathbf{v}; \quad 0\mathbf{v} = \mathbf{O}$.
5. $\mathbf{v} + (-\mathbf{v}) = \mathbf{O}$.
6. $b(c\mathbf{v}) = (bc)\mathbf{v}$.
7. $(b + c)\mathbf{v} = b\mathbf{v} + c\mathbf{v}$.
8. $b(\mathbf{u} + \mathbf{v}) = b\mathbf{u} + b\mathbf{v}$.

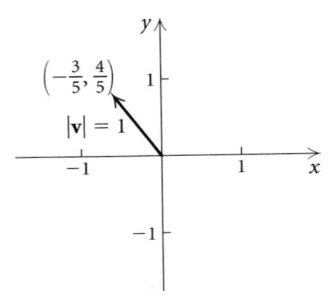

✳ Unit Vectors

A vector of magnitude, or length, 1 is called a **unit vector**. The vector $\mathbf{v} = \langle -\frac{3}{5}, \frac{4}{5} \rangle$ is a unit vector because

$$|\mathbf{v}| = \left|\left\langle -\tfrac{3}{5}, \tfrac{4}{5} \right\rangle\right| = \sqrt{\left(-\tfrac{3}{5}\right)^2 + \left(\tfrac{4}{5}\right)^2}$$
$$= \sqrt{\tfrac{9}{25} + \tfrac{16}{25}}$$
$$= \sqrt{\tfrac{25}{25}}$$
$$= \sqrt{1} = 1.$$

EXAMPLE 5   Find a unit vector that has the same direction as the vector $\mathbf{w} = \langle -3, 5 \rangle$.

Solution   We first find the length of $\mathbf{w}$:

$$|\mathbf{w}| = \sqrt{(-3)^2 + 5^2} = \sqrt{34}.$$

Thus we want a vector whose length is $1/\sqrt{34}$ of $\mathbf{w}$ and whose direction is the same as vector $\mathbf{w}$. That vector is

$$\mathbf{u} = \frac{1}{\sqrt{34}}\mathbf{w} = \frac{1}{\sqrt{34}}\langle -3, 5 \rangle = \left\langle \frac{-3}{\sqrt{34}}, \frac{5}{\sqrt{34}} \right\rangle.$$

The vector $\mathbf{u}$ is a *unit vector* because

$$|\mathbf{u}| = \left|\frac{1}{\sqrt{34}}\mathbf{w}\right| = \sqrt{\left(\frac{-3}{\sqrt{34}}\right)^2 + \left(\frac{5}{\sqrt{34}}\right)^2} = \sqrt{\frac{9}{34} + \frac{25}{34}}$$
$$= \sqrt{\frac{34}{34}} = \sqrt{1} = 1.$$

Now Try Exercise 33. ◼

Unit Vector

If $\mathbf{v}$ is a vector and $\mathbf{v} \neq \mathbf{O}$, then

$$\frac{1}{|\mathbf{v}|} \cdot \mathbf{v}, \quad \text{or} \quad \frac{\mathbf{v}}{|\mathbf{v}|},$$

is a **unit vector** in the direction of $\mathbf{v}$.

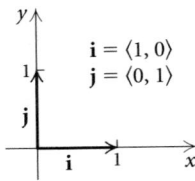

Although unit vectors can have any direction, the unit vectors parallel to the x- and y-axes are particularly useful. They are defined as

$$\mathbf{i} = \langle 1, 0 \rangle \quad \text{and} \quad \mathbf{j} = \langle 0, 1 \rangle.$$

Any vector can be expressed as a **linear combination** of unit vectors $\mathbf{i}$ and $\mathbf{j}$. For example, let $\mathbf{v} = \langle v_1, v_2 \rangle$. Then

$$\mathbf{v} = \langle v_1, v_2 \rangle = \langle v_1, 0 \rangle + \langle 0, v_2 \rangle$$
$$= v_1\langle 1, 0 \rangle + v_2\langle 0, 1 \rangle = v_1\mathbf{i} + v_2\mathbf{j}.$$

EXAMPLE 6   Express the vector $\mathbf{r} = \langle 2, -6 \rangle$ as a linear combination of $\mathbf{i}$ and $\mathbf{j}$.

Solution   We have

$$\mathbf{r} = \langle 2, -6 \rangle = 2\mathbf{i} + (-6)\mathbf{j} = 2\mathbf{i} - 6\mathbf{j}.$$ **Now Try Exercise 39.** ◾

EXAMPLE 7   Write the vector $\mathbf{q} = -\mathbf{i} + 7\mathbf{j}$ in component form.

Solution   We have

$$\mathbf{q} = -\mathbf{i} + 7\mathbf{j} = -1\mathbf{i} + 7\mathbf{j} = \langle -1, 7 \rangle.$$ ◾

Vector operations can also be performed when vectors are written as linear combinations of $\mathbf{i}$ and $\mathbf{j}$.

EXAMPLE 8   If $\mathbf{a} = 5\mathbf{i} - 2\mathbf{j}$ and $\mathbf{b} = -\mathbf{i} + 8\mathbf{j}$, find $3\mathbf{a} - \mathbf{b}$.

Solution   We have

$$
\begin{aligned}
3\mathbf{a} - \mathbf{b} &= 3(5\mathbf{i} - 2\mathbf{j}) - (-\mathbf{i} + 8\mathbf{j}) \\
&= 15\mathbf{i} - 6\mathbf{j} + \mathbf{i} - 8\mathbf{j} \\
&= 16\mathbf{i} - 14\mathbf{j}.
\end{aligned}
$$ **Now Try Exercise 45.** ◾

❈ Direction Angles

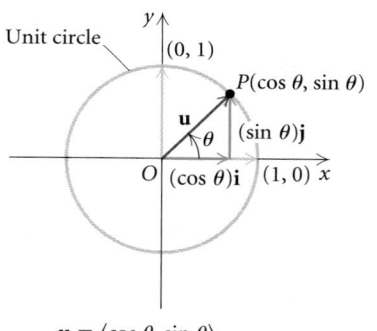

$$\mathbf{u} = \langle \cos\theta, \sin\theta \rangle$$
$$= (\cos\theta)\mathbf{i} + (\sin\theta)\mathbf{j}$$

The terminal point P of a unit vector in standard position is a point on the unit circle denoted by $(\cos\theta, \sin\theta)$. Thus the unit vector can be expressed in component form,

$$\mathbf{u} = \langle \cos\theta, \sin\theta \rangle,$$

or as a linear combination of the unit vectors $\mathbf{i}$ and $\mathbf{j}$,

$$\mathbf{u} = (\cos\theta)\mathbf{i} + (\sin\theta)\mathbf{j},$$

where the components of $\mathbf{u}$ are functions of the **direction angle** θ measured counterclockwise from the x-axis to the vector. As θ varies from 0 to 2π, the point P traces the circle $x^2 + y^2 = 1$. This takes in all possible directions for unit vectors so the equation $\mathbf{u} = (\cos\theta)\mathbf{i} + (\sin\theta)\mathbf{j}$ describes every possible unit vector in the plane.

> **UNIT CIRCLE**
>
> REVIEW SECTION **6.5.**

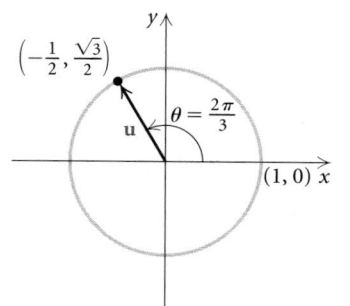

EXAMPLE 9   Calculate and sketch the unit vector $\mathbf{u} = (\cos\theta)\mathbf{i} + (\sin\theta)\mathbf{j}$ for $\theta = 2\pi/3$. Include the unit circle in your sketch.

Solution   We have

$$
\begin{aligned}
\mathbf{u} &= \left(\cos\frac{2\pi}{3}\right)\mathbf{i} + \left(\sin\frac{2\pi}{3}\right)\mathbf{j} \\
&= \left(-\frac{1}{2}\right)\mathbf{i} + \left(\frac{\sqrt{3}}{2}\right)\mathbf{j}.
\end{aligned}
$$ **Now Try Exercise 49.** ◾

Let $\mathbf{v} = \langle v_1, v_2 \rangle$ with direction angle θ. Using the definition of the tangent function, we can determine the direction angle from the components of $\mathbf{v}$:

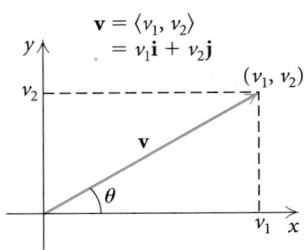

$$\tan \theta = \frac{v_2}{v_1}$$

$$\theta = \tan^{-1} \frac{v_2}{v_1}.$$

EXAMPLE 10   Determine the direction angle θ of the vector $\mathbf{w} = -4\mathbf{i} - 3\mathbf{j}$.

Solution   We know that

$$\mathbf{w} = -4\mathbf{i} - 3\mathbf{j} = \langle -4, -3 \rangle.$$

Thus we have

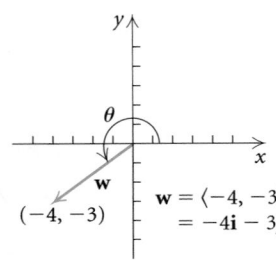

$$\tan \theta = \frac{-3}{-4} = \frac{3}{4} \quad \text{and} \quad \theta = \tan^{-1} \frac{3}{4}.$$

Since $\mathbf{w}$ is in the third quadrant, we know that θ is a third-quadrant angle. The reference angle is

$$\tan^{-1} \frac{3}{4} \approx 37°, \quad \text{and} \quad \theta \approx 180° + 37°, \text{ or } 217°.$$

Now Try Exercise 55. ■

It is convenient for work with applied problems and in subsequent courses, such as calculus, to have a way to express a vector so that both its magnitude and its direction can be determined, or read, easily. Let $\mathbf{v}$ be a vector. Then $\mathbf{v}/|\mathbf{v}|$ is a unit vector in the same direction as $\mathbf{v}$. Thus we have

$$\frac{\mathbf{v}}{|\mathbf{v}|} = (\cos \theta)\mathbf{i} + (\sin \theta)\mathbf{j}$$

$$\mathbf{v} = |\mathbf{v}|[(\cos \theta)\mathbf{i} + (\sin \theta)\mathbf{j}] \qquad \text{Multiplying by } |\mathbf{v}|$$

$$\mathbf{v} = |\mathbf{v}|(\cos \theta)\mathbf{i} + |\mathbf{v}|(\sin \theta)\mathbf{j}.$$

Let's revisit the applied problem in Example 3 of Section 8.5 and use this new notation.

EXAMPLE 11   *Airplane Speed and Direction.*   An airplane travels on a bearing of 100° at an airspeed of 190 km/h while a wind is blowing 48 km/h from 220°. Find the ground speed of the airplane and the direction of its track, or course, over the ground.

Solution We first make a drawing. The wind is represented by $\overrightarrow{OC}$ and the velocity vector of the airplane by $\overrightarrow{OA}$. The resultant velocity vector is **v**, the sum of the two vectors:

$$\mathbf{v} = \overrightarrow{OC} + \overrightarrow{OA}.$$

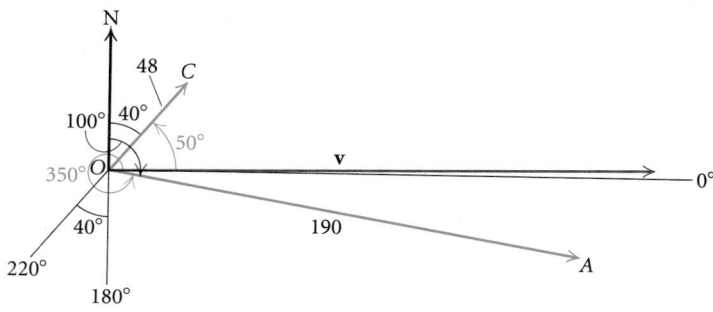

The bearing (measured from north) of the airspeed vector $\overrightarrow{OA}$ is 100°. Its *direction angle* (measured counterclockwise from the positive *x*-axis) is 350°. The bearing (measured from north) of the wind vector $\overrightarrow{OC}$ is 40°. Its direction angle (measured counterclockwise from the positive *x*-axis) is 50°. The magnitudes of $\overrightarrow{OA}$ and $\overrightarrow{OC}$ are 190 and 48, respectively. We have

$$\overrightarrow{OA} = 190(\cos 350°)\mathbf{i} + 190(\sin 350°)\mathbf{j}, \quad \text{and}$$
$$\overrightarrow{OC} = 48(\cos 50°)\mathbf{i} + 48(\sin 50°)\mathbf{j}.$$

Thus,

$$\mathbf{v} = \overrightarrow{OA} + \overrightarrow{OC}$$
$$= [190(\cos 350°)\mathbf{i} + 190(\sin 350°)\mathbf{j}] + [48(\cos 50°)\mathbf{i} + 48(\sin 50°)\mathbf{j}]$$
$$= [190(\cos 350°) + 48(\cos 50°)]\mathbf{i} + [190(\sin 350°) + 48(\sin 50°)]\mathbf{j}$$
$$\approx 217.97\mathbf{i} + 3.78\mathbf{j}.$$

From this form, we can determine the ground speed and the course:

$$\text{Ground speed} \approx \sqrt{(217.97)^2 + (3.78)^2}$$
$$\approx 218 \text{ km/h}.$$

We let α be the direction angle of **v**. Then

$$\tan \alpha = \frac{3.78}{217.97}$$

$$\alpha = \tan^{-1} \frac{3.78}{217.97} \approx 1°.$$

Thus the course of the airplane (the direction from north) is $90° - 1°$, or 89°.

❖ Angle Between Vectors

When a vector is multiplied by a scalar, the result is a vector. When two vectors are added, the result is also a vector. Thus we might expect the product of two vectors to be a vector as well, but it is not. The *dot product* of two vectors is a real number, or scalar. This product is useful in finding the angle between two vectors and in determining whether two vectors are perpendicular.

> ### Dot Product
>
> The **dot product** of two vectors $\mathbf{u} = \langle u_1, u_2 \rangle$ and $\mathbf{v} = \langle v_1, v_2 \rangle$ is
>
> $$\mathbf{u} \cdot \mathbf{v} = u_1 v_1 + u_2 v_2.$$
>
> (Note that $u_1 v_1 + u_2 v_2$ is a *scalar*, not a vector.)

EXAMPLE 12 Find the indicated dot product when

$$\mathbf{u} = \langle 2, -5 \rangle, \quad \mathbf{v} = \langle 0, 4 \rangle, \quad \text{and} \quad \mathbf{w} = \langle -3, 1 \rangle.$$

a) $\mathbf{u} \cdot \mathbf{w}$

b) $\mathbf{w} \cdot \mathbf{v}$

Solution

a) $\mathbf{u} \cdot \mathbf{w} = 2(-3) + (-5)1 = -6 - 5 = -11$

b) $\mathbf{w} \cdot \mathbf{v} = -3(0) + 1(4) = 0 + 4 = 4$ ∎

The dot product can be used to find the angle between two vectors. The angle *between* two vectors is the smallest positive angle formed by the two directed line segments. Thus the angle θ between $\mathbf{u}$ and $\mathbf{v}$ is the same angle as between $\mathbf{v}$ and $\mathbf{u}$, and $0 \le \theta \le \pi$.

> ### Angle Between Two Vectors
>
> If θ is the angle between two *nonzero* vectors $\mathbf{u}$ and $\mathbf{v}$, then
>
> $$\cos \theta = \frac{\mathbf{u} \cdot \mathbf{v}}{|\mathbf{u}||\mathbf{v}|}.$$

EXAMPLE 13 Find the angle between $\mathbf{u} = \langle 3, 7 \rangle$ and $\mathbf{v} = \langle -4, 2 \rangle$.

Solution We begin by finding $\mathbf{u} \cdot \mathbf{v}$, $|\mathbf{u}|$, and $|\mathbf{v}|$:

$$\mathbf{u} \cdot \mathbf{v} = 3(-4) + 7(2) = 2,$$
$$|\mathbf{u}| = \sqrt{3^2 + 7^2} = \sqrt{58}, \quad \text{and}$$
$$|\mathbf{v}| = \sqrt{(-4)^2 + 2^2} = \sqrt{20}.$$

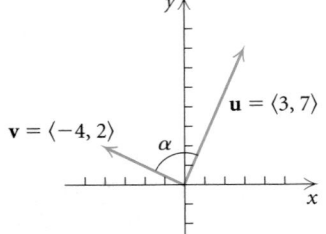

Then

$$\cos \alpha = \frac{\mathbf{u} \cdot \mathbf{v}}{|\mathbf{u}||\mathbf{v}|} = \frac{2}{\sqrt{58}\sqrt{20}}$$

$$\alpha = \cos^{-1} \frac{2}{\sqrt{58}\sqrt{20}}$$

$$\alpha \approx 86.6°.$$

Now Try Exercise 63. ∎

❈ Forces in Equilibrium

When several forces act through the same point on an object, their vector sum must be **O** in order for a balance to occur. When a balance occurs, then the object is either stationary or moving in a straight line without acceleration. The fact that the vector sum must be **O** for a balance, and vice versa, allows us to solve many applied problems involving forces.

EXAMPLE 14 *Suspended Block.* A 350-lb block is suspended by two cables, as shown at left. At point *A*, there are three forces acting: **W**, the block pulling down, and **R** and **S**, the two cables pulling upward and outward. Find the tension in each cable.

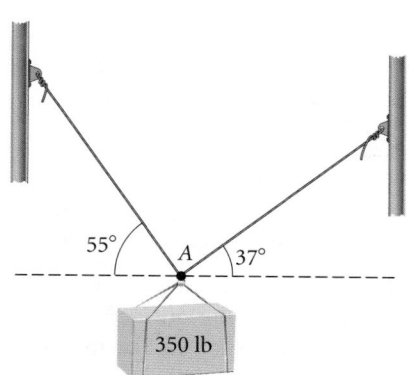

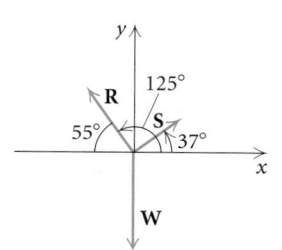

Solution We draw a force diagram with the initial points of each vector at the origin. For there to be a balance, the vector sum must be the vector **O**:

$$\mathbf{R} + \mathbf{S} + \mathbf{W} = \mathbf{O}.$$

We can express each vector in terms of its magnitude and its direction angle:

$$\mathbf{R} = |\mathbf{R}|[(\cos 125°)\mathbf{i} + (\sin 125°)\mathbf{j}],$$
$$\mathbf{S} = |\mathbf{S}|[(\cos 37°)\mathbf{i} + (\sin 37°)\mathbf{j}], \quad \text{and}$$
$$\mathbf{W} = |\mathbf{W}|[(\cos 270°)\mathbf{i} + (\sin 270°)\mathbf{j}]$$
$$= 350(\cos 270°)\mathbf{i} + 350(\sin 270°)\mathbf{j}$$
$$= -350\mathbf{j}. \qquad \cos 270° = 0; \sin 270° = -1$$

Substituting for **R**, **S**, and **W** in **R** + **S** + **W** = **O**, we have

$$[|\mathbf{R}|(\cos 125°) + |\mathbf{S}|(\cos 37°)]\mathbf{i} + [|\mathbf{R}|(\sin 125°) + |\mathbf{S}|(\sin 37°) - 350]\mathbf{j}$$
$$= 0\mathbf{i} + 0\mathbf{j}.$$

This gives us two equations:

$$|\mathbf{R}|(\cos 125°) + |\mathbf{S}|(\cos 37°) = 0 \quad \text{and} \qquad (1)$$
$$|\mathbf{R}|(\sin 125°) + |\mathbf{S}|(\sin 37°) - 350 = 0. \qquad (2)$$

Solving equation (1) for $|\mathbf{R}|$, we get

$$|\mathbf{R}| = -\frac{|\mathbf{S}|(\cos 37°)}{\cos 125°}. \qquad (3)$$

Substituting this expression for $|\mathbf{R}|$ in equation (2) gives us

$$-\frac{|\mathbf{S}|(\cos 37°)}{\cos 125°}(\sin 125°) + |\mathbf{S}|(\sin 37°) - 350 = 0.$$

Then solving this equation for $|\mathbf{S}|$, we get $|\mathbf{S}| \approx 201$, and substituting 201 for $|\mathbf{S}|$ in equation (3), we get $|\mathbf{R}| \approx 280$. The tensions in the cables are 280 lb and 201 lb.
Now Try Exercise 83. ■

8.6 Exercise Set

Find the component form of the vector given the initial and terminal points. Then find the length of the vector.

1. $\overrightarrow{MN}$; $M(6, -7)$, $N(-3, -2)$

2. $\overrightarrow{CD}$; $C(1, 5)$, $D(5, 7)$

3. $\overrightarrow{FE}$; $E(8, 4)$, $F(11, -2)$

4. $\overrightarrow{BA}$; $A(9, 0)$, $B(9, 7)$

5. $\overrightarrow{KL}$; $K(4, -3)$, $L(8, -3)$

6. $\overrightarrow{GH}$; $G(-6, 10)$, $H(-3, 2)$

7. Find the magnitude of vector **u** if $\mathbf{u} = \langle -1, 6 \rangle$.

8. Find the magnitude of vector $\overrightarrow{ST}$ if $\overrightarrow{ST} = \langle -12, 5 \rangle$.

Do the indicated calculations in Exercises 9–26 for the vectors

$\mathbf{u} = \langle 5, -2 \rangle$, $\quad \mathbf{v} = \langle -4, 7 \rangle$, $\quad$ and $\quad \mathbf{w} = \langle -1, -3 \rangle$.

9. $\mathbf{u} + \mathbf{w}$

10. $\mathbf{w} + \mathbf{u}$

11. $|3\mathbf{w} - \mathbf{v}|$

12. $6\mathbf{v} + 5\mathbf{u}$

13. $\mathbf{v} - \mathbf{u}$

14. $|2\mathbf{w}|$

15. $5\mathbf{u} - 4\mathbf{v}$

16. $-5\mathbf{v}$

17. $|3\mathbf{u}| - |\mathbf{v}|$

18. $|\mathbf{v}| + |\mathbf{u}|$

19. $\mathbf{v} + \mathbf{u} + 2\mathbf{w}$

20. $\mathbf{w} - (\mathbf{u} + 4\mathbf{v})$

21. $2\mathbf{v} + \mathbf{O}$

22. $10|7\mathbf{w} - 3\mathbf{u}|$

23. $\mathbf{u} \cdot \mathbf{w}$

24. $\mathbf{w} \cdot \mathbf{u}$

25. $\mathbf{u} \cdot \mathbf{v}$

26. $\mathbf{v} \cdot \mathbf{w}$

*The vectors **u**, **v**, and **w** are drawn below. Copy them on a sheet of paper. Then sketch each of the vectors in Exercises 27–30.*

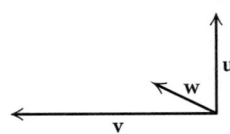

27. $\mathbf{u} + \mathbf{v}$

28. $\mathbf{u} - 2\mathbf{v}$

29. $\mathbf{u} + \mathbf{v} + \mathbf{w}$

30. $\frac{1}{2}\mathbf{u} - \mathbf{w}$

31. Vectors **u**, **v**, and **w** are determined by the sides of $\triangle ABC$ below.

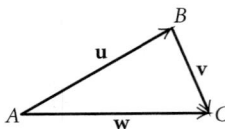

a) Find an expression for **w** in terms of **u** and **v**.
b) Find an expression for **v** in terms of **u** and **w**.

32. In $\triangle ABC$, vectors **u** and **w** are determined by the sides shown, where P is the midpoint of side BC. Find an expression for **v** in terms of **u** and **w**.

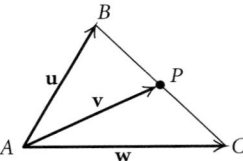

Find a unit vector that has the same direction as the given vector.

33. $\mathbf{v} = \langle -5, 12 \rangle$

34. $\mathbf{u} = \langle 3, 4 \rangle$

35. $\mathbf{w} = \langle 1, -10 \rangle$

36. $\mathbf{a} = \langle 6, -7 \rangle$

37. $\mathbf{r} = \langle -2, -8 \rangle$

38. $\mathbf{t} = \langle -3, -3 \rangle$

*Express the vector as a linear combination of the unit vectors **i** and **j**.*

39. $\mathbf{w} = \langle -4, 6 \rangle$

40. $\mathbf{r} = \langle -15, 9 \rangle$

41. $\mathbf{s} = \langle 2, 5 \rangle$

42. $\mathbf{u} = \langle 2, -1 \rangle$

*Express the vector as a linear combination of **i** and **j**.*

43.

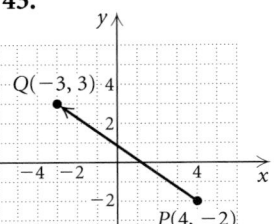

44.

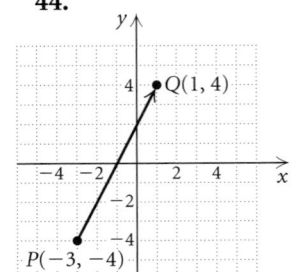

For Exercises 45–48, use the vectors

$$\mathbf{u} = 2\mathbf{i} + \mathbf{j}, \quad \mathbf{v} = -3\mathbf{i} - 10\mathbf{j}, \quad \text{and} \quad \mathbf{w} = \mathbf{i} - 5\mathbf{j}.$$

*Perform the indicated vector operations and state the answer in two forms: **(a)** as a linear combination of **i** and **j** and **(b)** in component form.*

45. $4\mathbf{u} - 5\mathbf{w}$ **46.** $\mathbf{v} + 3\mathbf{w}$

47. $\mathbf{u} - (\mathbf{v} + \mathbf{w})$ **48.** $(\mathbf{u} - \mathbf{v}) + \mathbf{w}$

Sketch (include the unit circle) and calculate the unit vector $\mathbf{u} = (\cos \theta)\mathbf{i} + (\sin \theta)\mathbf{j}$ for the given direction angle.

49. $\theta = \dfrac{\pi}{2}$ **50.** $\theta = \dfrac{\pi}{3}$

51. $\theta = \dfrac{4\pi}{3}$ **52.** $\theta = \dfrac{3\pi}{2}$

Determine the direction angle θ of the vector, to the nearest degree.

53. $\mathbf{u} = \langle -2, -5 \rangle$ **54.** $\mathbf{w} = \langle 4, -3 \rangle$

55. $\mathbf{q} = \mathbf{i} + 2\mathbf{j}$ **56.** $\mathbf{w} = 5\mathbf{i} - \mathbf{j}$

57. $\mathbf{t} = \langle 5, 6 \rangle$ **58.** $\mathbf{b} = \langle -8, -4 \rangle$

Find the magnitude and the direction angle θ of the vector.

59. $\mathbf{u} = 3[(\cos 45°)\mathbf{i} + (\sin 45°)\mathbf{j}]$

60. $\mathbf{w} = 6[(\cos 150°)\mathbf{i} + (\sin 150°)\mathbf{j}]$

61. $\mathbf{v} = \left\langle -\dfrac{1}{2}, \dfrac{\sqrt{3}}{2} \right\rangle$

62. $\mathbf{u} = -\mathbf{i} - \mathbf{j}$

Find the angle between the given vectors, to the nearest tenth of a degree.

63. $\mathbf{u} = \langle 2, -5 \rangle, \quad \mathbf{v} = \langle 1, 4 \rangle$

64. $\mathbf{a} = \langle -3, -3 \rangle, \quad \mathbf{b} = \langle -5, 2 \rangle$

65. $\mathbf{w} = \langle 3, 5 \rangle, \quad \mathbf{r} = \langle 5, 5 \rangle$

66. $\mathbf{v} = \langle -4, 2 \rangle, \quad \mathbf{t} = \langle 1, -4 \rangle$

67. $\mathbf{a} = \mathbf{i} + \mathbf{j}, \quad \mathbf{b} = 2\mathbf{i} - 3\mathbf{j}$

68. $\mathbf{u} = 3\mathbf{i} + 2\mathbf{j}, \quad \mathbf{v} = -\mathbf{i} + 4\mathbf{j}$

Express each vector in Exercises 69–72 in the form $a\mathbf{i} + b\mathbf{j}$ and sketch each in the coordinate plane.

69. The unit vectors $\mathbf{u} = (\cos \theta)\mathbf{i} + (\sin \theta)\mathbf{j}$ for $\theta = \pi/6$ and $\theta = 3\pi/4$. Include the unit circle $x^2 + y^2 = 1$ in your sketch.

70. The unit vectors $\mathbf{u} = (\cos \theta)\mathbf{i} + (\sin \theta)\mathbf{j}$ for $\theta = -\pi/4$ and $\theta = -3\pi/4$. Include the unit circle $x^2 + y^2 = 1$ in your sketch.

71. The unit vector obtained by rotating $\mathbf{j}$ counterclockwise $3\pi/4$ radians about the origin

72. The unit vector obtained by rotating $\mathbf{j}$ clockwise $2\pi/3$ radians about the origin

For the vectors in Exercises 73 and 74, find the unit vectors $\mathbf{u} = (\cos \theta)\mathbf{i} + (\sin \theta)\mathbf{j}$ in the same direction.

73. $-\mathbf{i} + 3\mathbf{j}$ **74.** $6\mathbf{i} - 8\mathbf{j}$

For the vectors in Exercises 75 and 76, express each vector in terms of its magnitude and its direction.

75. $2\mathbf{i} - 3\mathbf{j}$ **76.** $5\mathbf{i} + 12\mathbf{j}$

77. Use a sketch to show that

$$\mathbf{v} = 3\mathbf{i} - 6\mathbf{j} \quad \text{and} \quad \mathbf{u} = -\mathbf{i} + 2\mathbf{j}$$

have opposite directions.

78. Use a sketch to show that

$$\mathbf{v} = 3\mathbf{i} - 6\mathbf{j} \quad \text{and} \quad \mathbf{u} = \tfrac{1}{2}\mathbf{i} - \mathbf{j}$$

have the same direction.

Exercises 79–82 appeared first in Exercise Set 8.5, where we used the law of cosines and the law of sines to solve the applied problems. For this exercise set, solve the problem using the vector form

$$\mathbf{v} = |\mathbf{v}|[(\cos \theta)\mathbf{i} + (\sin \theta)\mathbf{j}].$$

79. *Ship.* A ship sails first N80°E for 120 nautical mi, and then S20°W for 200 nautical mi. How far is the ship, then, from the starting point, and in what direction is the ship moving?

80. *Boat.* A boat heads 35°, propelled by a force of 750 lb. A wind from 320° exerts a force of 150 lb on the boat. How large is the resultant force, and in what direction is the boat moving?

81. *Airplane.* An airplane has an airspeed of 150 km/h. It is to make a flight in a direction of 070° while there is a 25-km/h wind from 340°. What will the airplane's actual heading be?

82. *Airplane.* An airplane flies 032° for 210 mi, and then 280° for 170 mi. How far is the airplane, then, from the starting point, and in what direction is the plane moving?

83. Two cables support a 1000-lb weight, as shown. Find the tension in each cable.

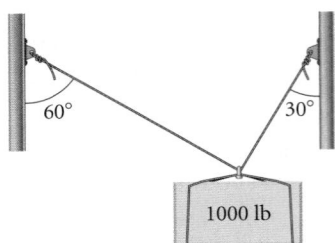

84. A 2500-kg block is suspended by two ropes, as shown. Find the tension in each rope.

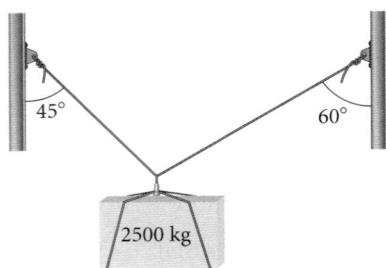

85. A 150-lb sign is hanging from the end of a hinged boom, supported by a cable inclined 42° with the horizontal. Find the tension in the cable and the compression in the boom.

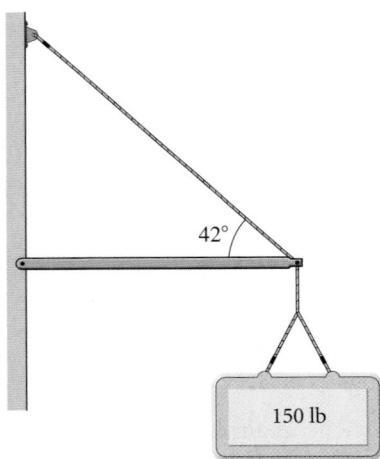

86. A weight of 200 lb is supported by a frame made of two rods and hinged at points A, B, and C. Find the forces exerted by the two rods.

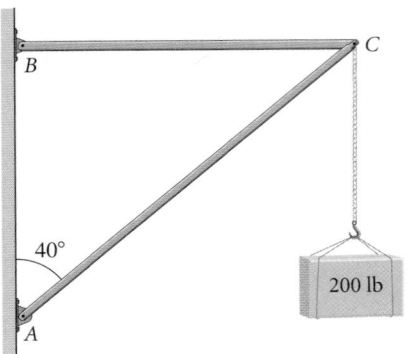

Let $\mathbf{u} = \langle u_1, u_2 \rangle$ *and* $\mathbf{v} = \langle v_1, v_2 \rangle$. *Prove each of the following properties.*

87. $\mathbf{u} + \mathbf{v} = \mathbf{v} + \mathbf{u}$

88. $\mathbf{u} \cdot \mathbf{v} = \mathbf{v} \cdot \mathbf{u}$

Collaborative Discussion and Writing

89. Explain how unit vectors are related to the unit circle.

90. Write a vector sum problem for a classmate for which the answer is $\mathbf{v} = 5\mathbf{i} - 8\mathbf{j}$.

Skill Maintenance

Find the slope and the y-intercept of the line with the given equation.

91. $-\frac{1}{5}x - y = 15$

92. $y = 7$

Find the zeros of the function.

93. $x^3 - 4x^2 = 0$

94. $6x^2 + 7x = 55$

Synthesis

95. If the dot product of two nonzero vectors **u** and **v** is 0, then the vectors are perpendicular (**orthogonal**). Let $\mathbf{u} = \langle u_1, u_2 \rangle$ and $\mathbf{v} = \langle v_1, v_2 \rangle$.

   a) Prove that if $\mathbf{u} \cdot \mathbf{v} = 0$, then **u** and **v** are perpendicular.

   b) Give an example of two perpendicular vectors and show that their dot product is 0.

96. If $\overrightarrow{PQ}$ is any vector, what is $\overrightarrow{PQ} + \overrightarrow{QP}$?

97. Find all the unit vectors that are parallel to the vector $\langle 3, -4 \rangle$.

98. Find a vector of length 2 whose direction is the opposite of the direction of the vector $\mathbf{v} = -\mathbf{i} + 2\mathbf{j}$. How many such vectors are there?

99. Given the vector $\overrightarrow{AB} = 3\mathbf{i} - \mathbf{j}$ and A is the point $(2, 9)$, find the point B.

100. Find vector **v** from point A to the origin, where $\overrightarrow{AB} = 4\mathbf{i} - 2\mathbf{j}$ and B is the point $(-2, 5)$.

CHAPTER 8   Summary and Review

Important Properties and Formulas

The Law of Sines

$$\frac{a}{\sin A} = \frac{b}{\sin B} = \frac{c}{\sin C}$$

The Area of a Triangle

$$K = \frac{1}{2} bc \sin A = \frac{1}{2} ab \sin C = \frac{1}{2} ac \sin B$$

The Law of Cosines

$$a^2 = b^2 + c^2 - 2bc \cos A,$$
$$b^2 = a^2 + c^2 - 2ac \cos B,$$
$$c^2 = a^2 + b^2 - 2ab \cos C$$

Complex Numbers

Absolute Value: $|a + bi| = \sqrt{a^2 + b^2}$

Trigonometric Notation: $a + bi = r(\cos \theta + i \sin \theta)$

Multiplication: $r_1(\cos \theta_1 + i \sin \theta_1) \cdot r_2(\cos \theta_2 + i \sin \theta_2)$
$$= r_1 r_2 [\cos (\theta_1 + \theta_2) + i \sin (\theta_1 + \theta_2)]$$

Division: $\dfrac{r_1(\cos \theta_1 + i \sin \theta_1)}{r_2(\cos \theta_2 + i \sin \theta_2)} = \dfrac{r_1}{r_2}[\cos (\theta_1 - \theta_2) + i \sin (\theta_1 - \theta_2)], \quad r_2 \neq 0$

DeMoivre's Theorem

$$[r(\cos \theta + i \sin \theta)]^n = r^n(\cos n\theta + i \sin n\theta)$$

Roots of Complex Numbers

The nth roots of $r(\cos \theta + i \sin \theta)$ are

$$r^{1/n}\left[\cos \left(\frac{\theta}{n} + k \cdot \frac{360°}{n}\right) + i \sin \left(\frac{\theta}{n} + k \cdot \frac{360°}{n}\right)\right], \quad r \neq 0, k = 0, 1, 2, \ldots, n - 1.$$

Vectors

If $\mathbf{u} = \langle u_1, u_2 \rangle$ and $\mathbf{v} = \langle v_1, v_2 \rangle$ and k is a scalar, then:

Length:	$\lvert \mathbf{v} \rvert = \sqrt{v_1^2 + v_2^2}$
Addition:	$\mathbf{u} + \mathbf{v} = \langle u_1 + v_1, u_2 + v_2 \rangle$
Subtraction:	$\mathbf{u} - \mathbf{v} = \langle u_1 - v_1, u_2 - v_2 \rangle$
Scalar Multiplication:	$k\mathbf{v} = \langle kv_1, kv_2 \rangle$
Dot Product:	$\mathbf{u} \cdot \mathbf{v} = u_1 v_1 + u_2 v_2$
Angle Between Two Vectors:	$\cos \theta = \dfrac{\mathbf{u} \cdot \mathbf{v}}{\lvert \mathbf{u} \rvert \lvert \mathbf{v} \rvert}$

Review Exercises

Determine whether the statement is true or false.

1. For any point (x, y) on the unit circle, $\langle x, y \rangle$ is a unit vector. [8.6]

2. The law of sines can be used to solve a triangle when all three sides are known. [8.1]

3. Two vectors are equivalent if they have the same magnitude and the lines that they are on have the same slope. [8.5]

4. Vectors $\langle 8, -2 \rangle$ and $\langle -8, 2 \rangle$ are equivalent. [8.6]

5. Any triangle, right or oblique, can be solved if at least one angle and any other two measures are known. [8.1]

6. When two angles and an included side of a triangle are known, the triangle cannot be solved using the law of cosines. [8.2]

Solve $\triangle ABC$, if possible. [8.1], [8.2]

7. $a = 23.4$ ft, $b = 15.7$ ft, $c = 8.3$ ft

8. $B = 27°$, $C = 35°$, $b = 19$ in.

9. $A = 133°28'$, $C = 31°42'$, $b = 890$ m

10. $B = 37°$, $b = 4$ yd, $c = 8$ yd

11. Find the area of $\triangle ABC$ if $b = 9.8$ m, $c = 7.3$ m, and $A = 67.3°$. [8.1]

12. A parallelogram has sides of lengths 3.21 ft and 7.85 ft. One of its angles measures 147°. Find the area of the parallelogram. [8.1]

13. *Sandbox.* A child-care center has a triangular-shaped sandbox. Two of the three sides measure 15 ft and 12.5 ft and form an included angle of 42°. To determine the amount of sand that is needed to fill the box, the director must determine the area of the floor of the box. Find the area of the floor of the box to the nearest square foot. [8.1]

14. *Flower Garden.* A triangular flower garden has sides of lengths 11 m, 9 m, and 6 m. Find the angles of the garden to the nearest degree. [8.2]

15. In an isosceles triangle, the base angles each measure 52.3° and the base is 513 ft long. Find the lengths of the other two sides to the nearest foot. [8.1]

16. *Airplanes.* Two airplanes leave an airport at the same time. The first flies 175 km/h in a direction of 305.6°. The second flies 220 km/h in a direction of 195.5°. After 2 hr, how far apart are the planes? [8.2]

Graph the complex number and find its absolute value. [8.3]

17. $2 - 5i$ **18.** 4

19. $2i$ **20.** $-3 + i$

Find trigonometric notation. [8.3]

21. $1 + i$ **22.** $-4i$

23. $-5\sqrt{3} + 5i$ **24.** $\dfrac{3}{4}$

Find standard notation, $a + bi$. [8.3]

25. $4(\cos 60° + i \sin 60°)$

26. $7(\cos 0° + i \sin 0°)$

27. $5\left(\cos \dfrac{2\pi}{3} + i \sin \dfrac{2\pi}{3}\right)$

28. $2\left[\cos\left(-\dfrac{\pi}{6}\right) + i \sin\left(-\dfrac{\pi}{6}\right)\right]$

Convert to trigonometric notation and then multiply or divide, expressing the answer in standard notation. [8.3]

29. $\left(1 + i\sqrt{3}\right)(1 - i)$ **30.** $\dfrac{2 - 2i}{2 + 2i}$

31. $\dfrac{2 + 2\sqrt{3}i}{\sqrt{3} - i}$ **32.** $i\left(3 - 3\sqrt{3}i\right)$

Raise the number to the given power and write trigonometric notation for the answer. [8.3]

33. $[2(\cos 60° + i \sin 60°)]^3$

34. $(1 - i)^4$

Raise the number to the given power and write standard notation for the answer. [8.3]

35. $(1 + i)^6$ **36.** $\left(\dfrac{1}{2} + \dfrac{\sqrt{3}}{2}i\right)^{10}$

37. Find the square roots of $-1 + i$. [8.3]

38. Find the cube roots of $3\sqrt{3} - 3i$. [8.3]

39. Find and graph the fourth roots of 81. [8.3]

40. Find and graph the fifth roots of 1. [8.3]

Find all the complex solutions of the equation. [8.3]

41. $x^4 - i = 0$ **42.** $x^3 + 1 = 0$

43. Find the polar coordinates of each of these points. Give three answers for each point. [8.4]

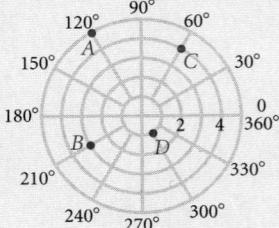

Find the polar coordinates of the point. Express the answer in degrees and then in radians. [8.4]

44. $\left(-4\sqrt{2}, 4\sqrt{2}\right)$ **45.** $(0, -5)$

Use a graphing calculator to convert from rectangular coordinates to polar coordinates. Express the answer in degrees and then in radians. [8.4]

46. $(-2, 5)$ **47.** $\left(-4.2, \sqrt{7}\right)$

Find the rectangular coordinates of the point. [8.4]

48. $\left(3, \dfrac{\pi}{4}\right)$ **49.** $(-6, -120°)$

Use a graphing calculator to convert from polar coordinates to rectangular coordinates. Round the coordinates to the nearest hundredth. [8.4]

50. $(2, -15°)$ **51.** $\left(-2.3, \dfrac{\pi}{5}\right)$

Convert to a polar equation. [8.4]

52. $5x - 2y = 6$ **53.** $y = 3$

54. $x^2 + y^2 = 9$ **55.** $y^2 - 4x - 16 = 0$

Convert to a rectangular equation. [8.4]

56. $r = 6$ **57.** $r + r \sin \theta = 1$

58. $r = \dfrac{3}{1 - \cos \theta}$ **59.** $r - 2 \cos \theta = 3 \sin \theta$

In Exercises 60–63, match the equation with one of figures (a)–(d), which follow. [8.4]

a)
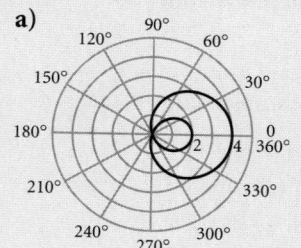

b)
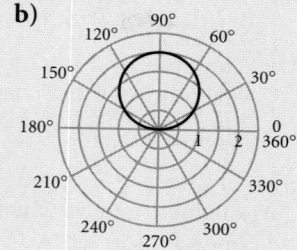

c)
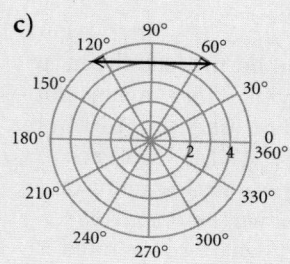

d)
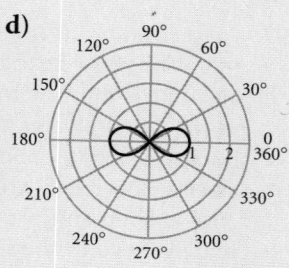

60. $r = 2 \sin \theta$

61. $r^2 = \cos 2\theta$

62. $r = 1 + 3 \cos \theta$

63. $r \sin \theta = 4$

Magnitudes of vectors **u** *and* **v** *and the angle* θ *between the vectors are given. Find the magnitude of the sum,* **u + v**, *to the nearest tenth and give the direction by specifying to the nearest degree the angle that it makes with the vector* **u**. [8.5]

64. $|\mathbf{u}| = 12$, $|\mathbf{v}| = 15$, $\theta = 120°$

65. $|\mathbf{u}| = 41$, $|\mathbf{v}| = 60$, $\theta = 25°$

The vectors **u**, **v**, *and* **w** *are drawn below. Copy them on a sheet of paper. Then sketch each of the vectors in Exercises 66 and 67.* [8.5]

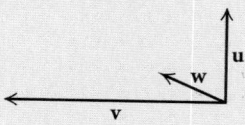

66. $\mathbf{u} - \mathbf{v}$

67. $\mathbf{u} + \frac{1}{2}\mathbf{w}$

68. Forces of 230 N and 500 N act on an object. The angle between the forces is 52°. Find the resultant, giving the angle that it makes with the smaller force. [8.5]

69. *Wind.* A wind has an easterly component of 15 km/h and a southerly component of 25 km/h. Find the magnitude and the direction of the wind. [8.5]

70. *Ship.* A ship sails N75°E for 90 nautical mi, and then S10°W for 100 nautical mi. How far is the ship, then, from the starting point, and in what direction? [8.5]

Find the component form of the vector given the initial and terminal points. [8.6]

71. $\overrightarrow{AB}$; $A(2, -8)$, $B(-2, -5)$

72. $\overrightarrow{TR}$; $R(0, 7)$, $T(-2, 13)$

73. Find the magnitude of vector **u** if $\mathbf{u} = \langle 5, -6 \rangle$. [8.6]

Do the calculations in Exercises 74–77 for the vectors $\mathbf{u} = \langle 3, -4 \rangle$, $\mathbf{v} = \langle -3, 9 \rangle$ *and* $\mathbf{w} = \langle -2, -5 \rangle$. [8.6]

74. $4\mathbf{u} + \mathbf{w}$

75. $2\mathbf{w} - 6\mathbf{v}$

76. $|\mathbf{u}| + |2\mathbf{w}|$

77. $\mathbf{u} \cdot \mathbf{w}$

78. Find a unit vector that has the same direction as $\mathbf{v} = \langle -6, -2 \rangle$. [8.6]

79. Express the vector $\mathbf{t} = \langle -9, 4 \rangle$ as a linear combination of the unit vectors **i** and **j**. [8.6]

80. Determine the direction angle θ of the vector $\mathbf{w} = \langle -4, -1 \rangle$ to the nearest degree. [8.6]

81. Find the magnitude and the direction angle θ of $\mathbf{u} = -5\mathbf{i} - 3\mathbf{j}$. [8.6]

82. Find the angle between $\mathbf{u} = \langle 3, -7 \rangle$ and $\mathbf{v} = \langle 2, 2 \rangle$ to the nearest tenth of a degree. [8.6]

83. *Airplane.* An airplane has an airspeed of 160 mph. It is to make a flight in a direction of 80° while there is a 20-mph wind from 310°. What will the airplane's actual heading be? [8.6]

Do the calculations in Exercises 84–87 for the vectors $\mathbf{u} = 2\mathbf{i} + 5\mathbf{j}$, $\mathbf{v} = -3\mathbf{i} + 10\mathbf{j}$, *and* $\mathbf{w} = 4\mathbf{i} + 7\mathbf{j}$. [8.6]

84. $5\mathbf{u} - 8\mathbf{v}$

85. $\mathbf{u} - (\mathbf{v} + \mathbf{w})$

86. $|\mathbf{u} - \mathbf{v}|$

87. $3|\mathbf{w}| + |\mathbf{v}|$

88. Express the vector $\overrightarrow{PQ}$ in the form $a\mathbf{i} + b\mathbf{j}$, if P is the point $(1, -3)$ and Q is the point $(-4, 2)$. [8.6]

Express each vector in Exercises 89 and 90 in the form $a\mathbf{i} + b\mathbf{j}$ *and sketch each in the coordinate plane.* [8.6]

89. The unit vectors $\mathbf{u} = (\cos \theta)\mathbf{i} + (\sin \theta)\mathbf{j}$ for $\theta = \pi/4$ and $\theta = 5\pi/4$. Include the unit circle $x^2 + y^2 = 1$ in your sketch.

90. The unit vector obtained by rotating **j** counter-clockwise $2\pi/3$ radians about the origin.

91. Express the vector $3\mathbf{i} - \mathbf{j}$ as a product of its magnitude and its direction.

92. Determine the trigonometric notation for $1 - i$.   [8.3]

   A. $\sqrt{2}\left(\cos\dfrac{5\pi}{4} + i\sin\dfrac{5\pi}{4}\right)$

   B. $\sqrt{2}\left(\cos\dfrac{7\pi}{4} - \sin\dfrac{7\pi}{4}\right)$

   C. $\cos\dfrac{7\pi}{4} + i\sin\dfrac{7\pi}{4}$

   D. $\sqrt{2}\left(\cos\dfrac{7\pi}{4} + i\sin\dfrac{7\pi}{4}\right)$

93. Convert the polar equation $r = 100$ to a rectangular equation.   [8.4]

   A. $x^2 + y^2 = 10{,}000$
   B. $x^2 + y^2 = 100$
   C. $\sqrt{x^2 + y^2} = 10$
   D. $\sqrt{x^2 + y^2} = 1000$

94. The graph of $r = 1 - 2\cos\theta$ is which of the following?   [8.4]

A.

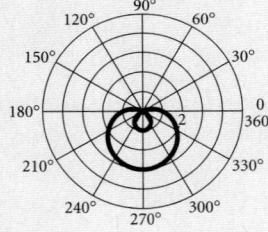

B.

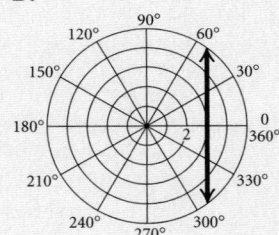

C.

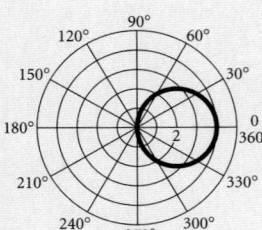

D.

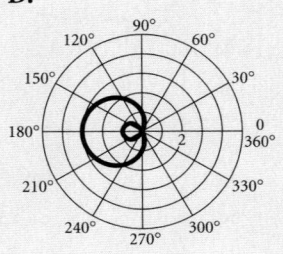

Collaborative Discussion and Writing

95. Explain why these statements are not contradictory:

   The number 1 has one real cube root.
   The number 1 has three complex cube roots.

   [8.3]

96. Summarize how you can tell algebraically when solving triangles whether there is no solution, one solution, or two solutions.   [8.1], [8.2]

97. *Golf: Distance versus Accuracy.*   It is often argued in golf that the farther you hit the ball, the more accurate it must be to stay safe. (Safe means not in the woods, water, or some other hazard.) In his book *Golf and the Spirit* (p. 54), M. Scott Peck asserts "Deviate 5° from your aiming point on a 150-yd shot, and your ball will land approximately 20 yd to the side of where you wanted it to be. Do the same on a 300-yd shot, and it will be 40 yd off target. Twenty yards may well be in the range of safety; 40 yards probably won't. This principle not infrequently allows a mediocre, short-hitting golfer like myself to score better than the long hitter." Check the accuracy of the mathematics in this statement, and comment on Peck's assertion.   [8.2]

Synthesis

98. Let $\mathbf{u} = 12\mathbf{i} + 5\mathbf{j}$. Find a vector that has the same direction as **u** but has length 3.   [8.6]

99. A parallelogram has sides of lengths 3.42 and 6.97. Its area is 18.4. Find the sizes of its angles.   [8.1]

CHAPTER 8   Test

Solve △ABC, if possible.

1. $a = 18$ ft, $B = 54°$, $C = 43°$

2. $b = 8$ m, $c = 5$ m, $C = 36°$

3. $a = 16.1$ in., $b = 9.8$ in., $c = 11.2$ in.

4. Find the area of △ABC if $C = 106.4°$, $a = 7$ cm, and $b = 13$ cm.

5. *Distance Across a Lake.* Points A and B are on opposite sides of a lake. Point C is 52 m from A. The measure of ∠BAC is determined to be 108°, and the measure of ∠ACB is determined to be 44°. What is the distance from A to B?

6. *Location of Airplanes.* Two airplanes leave an airport at the same time. The first flies 210 km/h in a direction of 290°. The second flies 180 km/h in a direction of 185°. After 3 hr, how far apart are the planes?

7. Graph: $-4 + i$.

8. Find the absolute value of $2 - 3i$.

9. Find trigonometric notation for $3 - 3i$.

10. Divide and express the result in standard notation $a + bi$:

$$\frac{2\left(\cos \frac{2\pi}{3} + i \sin \frac{2\pi}{3}\right)}{8\left(\cos \frac{\pi}{6} + i \sin \frac{\pi}{6}\right)}.$$

11. Find $(1 - i)^8$ and write standard notation for the answer.

12. Find the polar coordinates of $(-1, \sqrt{3})$. Express the angle in degrees using the smallest possible positive angle.

13. Convert $\left(-1, \frac{2\pi}{3}\right)$ to rectangular coordinates.

14. Convert to a polar equation: $x^2 + y^2 = 10$.

15. Graph: $r = 1 - \cos \theta$.

16. For vectors **u** and **v**, $|\mathbf{u}| = 8$, $|\mathbf{v}| = 5$, and the angle between the vectors is 63°. Find **u** + **v**. Give the magnitude to the nearest tenth, and give the direction by specifying the angle that the resultant makes with **u**, to the nearest degree.

17. For $\mathbf{u} = 2\mathbf{i} - 7\mathbf{j}$ and $\mathbf{v} = 5\mathbf{i} + \mathbf{j}$, find $2\mathbf{u} - 3\mathbf{v}$.

18. Find a unit vector in the same direction as $-4\mathbf{i} + 3\mathbf{j}$.

19. Which of the following is the graph of $r = 3 \cos \theta$?

A.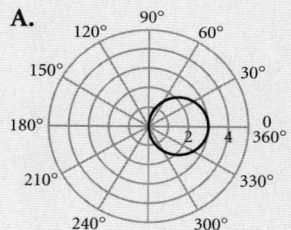

B.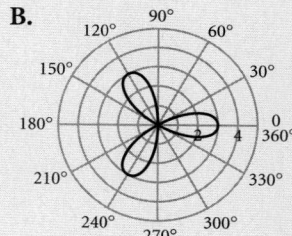

C.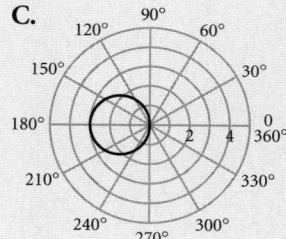

D.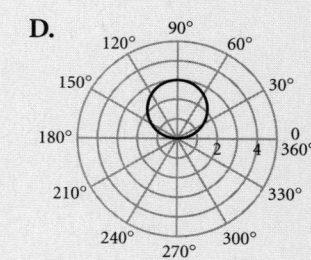

Synthesis

20. A parallelogram has sides of length 15.4 and 9.8. Its area is 72.9. Find the measures of the angles.

Systems of Equations and Matrices

APPLICATION Americans own a total of about 314 million fish, cats, and dogs as pets. The number of fish owned is 16 million less than the total number of cats and dogs owned. There are 17 million more cats than dogs. (*Source*: American Pet Products Manufacturers Association) How many of each type of pet do Americans own?

This problem appears as Exercise 17 in Section 9.2.

9.1 Systems of Equations in Two Variables

9.2 Systems of Equations in Three Variables

9.3 Matrices and Systems of Equations

9.4 Matrix Operations

9.5 Inverses of Matrices

9.6 Determinants and Cramer's Rule

9.7 Systems of Inequalities and Linear Programming

9.8 Partial Fractions

9.1

Systems of Equations in Two Variables

❄ Solve a system of two linear equations in two variables by graphing.

❄ Solve a system of two linear equations in two variables using the substitution method and the elimination method.

❄ Use systems of two linear equations to solve applied problems.

A **system of equations** is composed of two or more equations considered simultaneously. For example,

$$x - y = 5,$$
$$2x + y = 1$$

is a **system of two linear equations in two variables**. The solution set of this system consists of all ordered pairs that make *both* equations true. The ordered pair $(2, -3)$ is a solution of the system of equations above. We can verify this by substituting 2 for x and -3 for y in *each* equation.

$$\begin{array}{c|c}
x - y = 5 & 2x + y = 1 \\
\hline
2 - (-3)\ ?\ 5 & 2 \cdot 2 + (-3)\ ?\ 1 \\
2 + 3 & 4 - 3 \\
5 \;\big|\; 5 \quad \text{TRUE} & 1 \;\big|\; 1 \quad \text{TRUE}
\end{array}$$

❄ Solving Systems of Equations Graphically

Recall that the graph of a linear equation is a line that contains all the ordered pairs in the solution set of the equation. When we graph a system of linear equations, each point at which the graphs intersect is a solution of *both* equations and therefore a **solution of the system of equations**.

EXAMPLE 1 Solve the following system of equations graphically.

$$x - y = 5,$$
$$2x + y = 1$$

Solution We graph the equations on the same set of axes, as shown below.

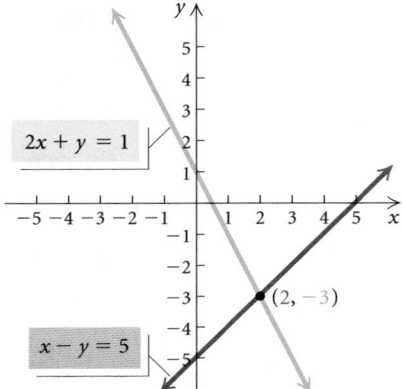

We see that the graphs intersect at a single point, $(2, -3)$, so $(2, -3)$ is the solution of the system of equations. To check this solution, we substitute 2 for x and -3 for y in both equations as we did above.

To use a graphing calculator to solve this system of equations, it might be necessary to write each equation in "$Y = \cdots$" form. If so, we would graph $y_1 = x - 5$ and $y_2 = -2x + 1$ and then use the INTERSECT feature. We see that the solution is $(2, -3)$.

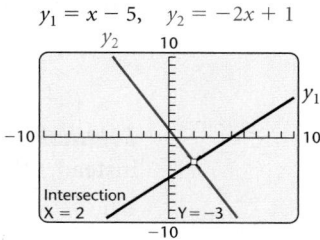

$$y_1 = x - 5, \quad y_2 = -2x + 1$$

Now Try Exercise 7. ▪

The graphs of most of the systems of equations that we use to model applications intersect at a single point, like the system above. However, it is possible that the graphs will have no points in common or infinitely many points in common. Each of these possibilities is illustrated below.

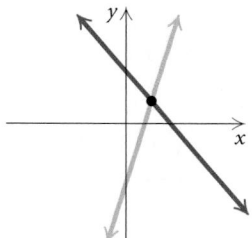
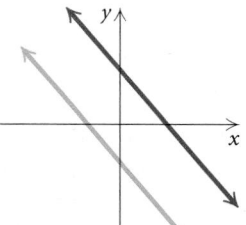
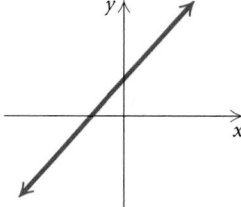

Exactly one common point	Parallel lines	Lines are identical.
One solution	No common points	Infinitely many common points
	No solution	Infinitely many solutions
Consistent	Inconsistent	Consistent
Independent	Independent	Dependent

If a system of equations has at least one solution, it is **consistent**. If the system has no solutions, it is **inconsistent**. In addition, for a system of two linear equations in two variables, if one equation can be obtained by multiplying both sides of the other equation by a constant, the equations are **dependent**. Otherwise, they are **independent**. A system of two dependent linear equations in two variables has an infinite number of solutions.

❋ The Substitution Method

Solving a system of equations graphically is not always accurate when the solutions are not integers. A solution like $\left(\frac{43}{27}, -\frac{19}{27}\right)$, for instance, will be difficult to determine from a hand-drawn graph.

Algebraic methods for solving systems of equations, when used correctly, always give accurate results. One such technique is the **substitution method**. It is used most often when a variable is alone on one side of an equation or when it is easy to solve for a variable. To apply the substitution method, we begin by using one of the equations to express one variable in terms of the other; then we substitute that expression in the other equation of the system.

EXAMPLE 2   Use the substitution method to solve the system

$$x - y = 5, \quad (1)$$
$$2x + y = 1. \quad (2)$$

Solution   First, we solve equation (1) for x. (We could have solved for y instead.) We have

$$x - y = 5, \qquad (1)$$
$$x = y + 5. \qquad \text{Solving for } x$$

Then we substitute $y + 5$ for x in equation (2). This gives an equation in one variable, which we know how to solve:

$$2x + y = 1 \qquad (2)$$
$$2(y + 5) + y = 1 \qquad \text{The parentheses are necessary.}$$
$$2y + 10 + y = 1 \qquad \text{Removing parentheses}$$
$$3y + 10 = 1 \qquad \text{Collecting like terms on the left}$$
$$3y = -9 \qquad \text{Subtracting 10 on both sides}$$
$$y = -3. \qquad \text{Dividing by 3 on both sides}$$

Now we substitute -3 for y in either of the original equations (this is called **back-substitution**) and solve for x. We choose equation (1):

$$x - y = 5, \qquad (1)$$
$$x - (-3) = 5 \qquad \text{Substituting } -3 \text{ for } y$$
$$x + 3 = 5$$
$$x = 2. \qquad \text{Subtracting 3 on both sides}$$

We have previously checked the pair $(2, -3)$ in both equations. The solution of the system of equations is $(2, -3)$.   **Now Try Exercise 17.** ■

❈ The Elimination Method

Another algebraic technique for solving systems of equations is the **elimination method**. With this method, we eliminate a variable by adding two equations. If the coefficients of a particular variable are opposites, we can eliminate that variable simply by adding the original equations. For example, if the x-coefficient is -3 in one equation and is 3 in the other equation, then the sum of the x-terms will be 0 and thus the variable x will be eliminated when we add the equations.

EXAMPLE 3   Use the elimination method to solve the system

$$2x + y = 2, \quad (1)$$
$$x - y = 7. \quad (2)$$

Solution   Since the y-coefficients, 1 and -1, are opposites, we can eliminate y by adding the equations:

$$
\begin{array}{ll}
2x + y = 2 & (1) \\
\underline{x - y = 7} & (2) \\
3x = 9 & \text{Adding} \\
x = 3.
\end{array}
$$

We then back-substitute 3 for x in either equation and solve for y. We choose equation (1):

$$
\begin{array}{ll}
2x + y = 2 & (1) \\
2 \cdot 3 + y = 2 & \text{Substituting 3 for } x \\
6 + y = 2 \\
y = -4.
\end{array}
$$

We can check the solution by substituting the pair $(3, -4)$ in both equations.

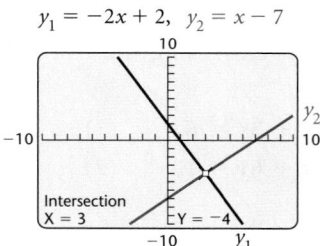

$y_1 = -2x + 2, \quad y_2 = x - 7$

$$
\begin{array}{c|c}
\underline{2x + y = 2} & \underline{x - y = 7} \\
2 \cdot 3 + (-4) \; ? \; 2 & 3 - (-4) \; ? \; 7 \\
6 - 4 \;\big|\; & 3 + 4 \;\big|\; \\
2 \;\big|\; 2 \quad \text{TRUE} & 7 \;\big|\; 7 \quad \text{TRUE}
\end{array}
$$

We can also check graphically using the INTERSECT feature, as shown at left. The solution is $(3, -4)$. Since there is exactly one solution, the system of equations is consistent and the equations are independent.

Now Try Exercise 33. ■

Before we add, it might be necessary to multiply one or both equations by suitable constants in order to find two equations in which coefficients are opposites.

EXAMPLE 4   Use the elimination method to solve the system

$$4x + 3y = 11, \quad (1)$$
$$-5x + 2y = 15. \quad (2)$$

Solution   We can obtain x-coefficients that are opposites by multiplying the first equation by 5 and the second equation by 4:

$$
\begin{array}{ll}
20x + 15y = 55 & \text{Multiplying equation (1) by 5} \\
\underline{-20x + 8y = 60} & \text{Multiplying equation (2) by 4} \\
23y = 115 & \text{Adding} \\
y = 5.
\end{array}
$$

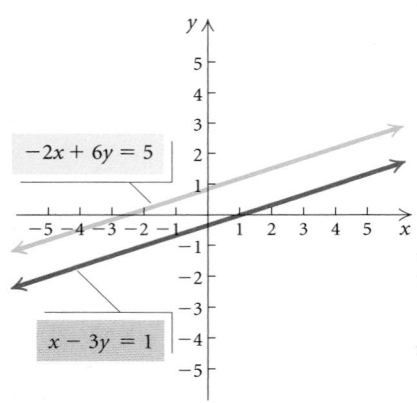

We then back-substitute 5 for y in either equation (1) or (2) and solve for x. We choose equation (1):

$$4x + 3y = 11 \quad (1)$$
$$4x + 3 \cdot 5 = 11 \quad \text{Substituting 5 for } y$$
$$4x + 15 = 11$$
$$4x = -4$$
$$x = -1.$$

We can check the pair $(-1, 5)$ by substituting in both equations or by graphing, as shown at left. The solution is $(-1, 5)$. The system of equations is consistent and the equations are independent. **Now Try Exercise 35.** ▪

In Example 4, the two systems

$$\begin{array}{cc} 4x + 3y = 11, & 20x + 15y = 55, \\ -5x + 2y = 15 & \text{and} \quad -20x + 8y = 60 \end{array}$$

are **equivalent** because they have exactly the same solutions. When we use the elimination method, we often multiply one or both equations by constants to find equivalent equations that allow us to eliminate a variable by adding.

EXAMPLE 5 Solve each of the following systems using the elimination method.

a) $x - 3y = 1,$ (1) $\qquad$ **b)** $2x + 3y = 6,$ (1)
$\quad\;\, -2x + 6y = 5$ (2) $\qquad\qquad\;\; 4x + 6y = 12$ (2)

Solution

a) We multiply equation (1) by 2 and add:

$$\begin{array}{ll} 2x - 6y = 2 & \text{Multiplying equation (1) by 2} \\ \underline{-2x + 6y = 5} & (2) \\ \quad\;\; 0 = 7. & \text{Adding} \end{array}$$

There are no values of x and y for which $0 = 7$ is true, so the system has *no solution*. The solution set is $\varnothing$. The system of equations is inconsistent and the equations are independent. The graphs of the equations are parallel lines, as shown in Fig. 1.

b) We multiply equation (1) by -2 and add:

$$\begin{array}{ll} -4x - 6y = -12 & \text{Multiplying equation (1) by } -2 \\ \underline{4x + 6y = 12} & (2) \\ \quad\;\; 0 = 0. & \text{Adding} \end{array}$$

We obtain the equation $0 = 0$, which is true for all values of x and y. This tells us that the equations are dependent, so there are *infinitely many solutions*. That is, any solution of one equation of the system is also a solution of the other. The system of equations is consistent. The graphs of the equations are identical, as shown in Fig. 2.

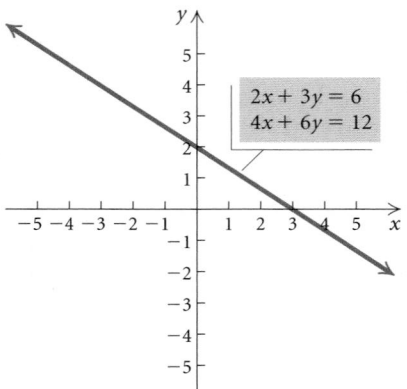

FIGURE 1

FIGURE 2

Solving either equation for y, we have $y = -\frac{2}{3}x + 2$, so we can write the solutions of the system as ordered pairs (x, y), where y is expressed as $-\frac{2}{3}x + 2$. Thus the solutions can be written in the form $\left(x, -\frac{2}{3}x + 2\right)$. Any real value that we choose for x then gives us a value for y and thus an ordered pair in the solution set. For example,

if $x = -3$, then $-\frac{2}{3}x + 2 = -\frac{2}{3}(-3) + 2 = 4$,

if $x = 0$, then $-\frac{2}{3}x + 2 = -\frac{2}{3} \cdot 0 + 2 = 2$, and

if $x = 6$, then $-\frac{2}{3}x + 2 = -\frac{2}{3} \cdot 6 + 2 = -2$.

Thus some of the solutions are $(-3, 4)$, $(0, 2)$, and $(6, -2)$.

Similarly, solving either equation for x, we have $x = -\frac{3}{2}y + 3$, so the solutions (x, y) can also be written, expressing x as $-\frac{3}{2}y + 3$, in the form $\left(-\frac{3}{2}y + 3, y\right)$.

Since the two forms of the solutions are equivalent, they yield the same solution set, as illustrated in the table at left. Note, for example, that when $y = 4$, we have the solution $(-3, 4)$; when $y = 2$, we have $(0, 2)$; and when $y = -2$, we have $(6, -2)$.

Now Try Exercise 37. ■

x	$-\frac{2}{3}x + 2$
$-\frac{3}{2}y + 3$	y
-3	4
0	2
6	-2

❖ Applications

Frequently the most challenging and time-consuming step in the problem-solving process is translating a situation to mathematical language. However, in many cases, this task is made easier if we translate to more than one equation in more than one variable.

EXAMPLE 6 *Snack Mixtures.* At Max's Munchies, caramel corn worth $2.50 per pound is mixed with honey roasted mixed nuts worth $7.50 per pound in order to get 20 lb of a mixture worth $4.50 per pound. How much of each snack is used?

Solution We use the five-step problem-solving process.

1. **Familiarize.** Let's begin by making a guess. Suppose 16 lb of caramel corn and 4 lb of nuts are used. Then the total weight of the mixture would be 16 lb + 4 lb, or 20 lb, the desired weight. The total values of these amounts of ingredients are found by multiplying the price per pound by the number of pounds used:

Caramel corn: $2.50(16) = $40

Nuts: $7.50(4) = \underline{$30}$

Total value: $70.

The desired value of the mixture is $4.50 per pound, so the value of 20 lb would be $4.50(20), or $90. Thus we see that our guess, which led to a total of $70, is incorrect. Nevertheless, these calculations will help us to translate.

2. **Translate.** We organize the information in a table. We let $x =$ the number of pounds of caramel corn in the mixture and $y =$ the number of pounds of nuts.

	Caramel Corn	Nuts	Mixture	
Price per Pound	$2.50	$7.50	$4.50	
Number of Pounds	x	y	20	$\longrightarrow x + y = 20$
Value of Mixture	$2.50x$	$7.50y$	4.50(20), or 90	$\longrightarrow 2.50x + 7.50y = 90$

From the second row of the table, we get one equation:

$$x + y = 20.$$

The last row of the table yields a second equation:

$$2.50x + 7.50y = 90, \quad \text{or} \quad 2.5x + 7.5y = 90.$$

We can multiply by 10 on both sides of the second equation to clear the decimals. This gives us the following system of equations:

$$x + y = 20, \qquad (1)$$
$$25x + 75y = 900. \qquad (2)$$

3. Carry out.   We carry out the solution as follows.

ALGEBRAIC SOLUTION

Using the elimination method, we multiply equation (1) by -25 and add it to equation (2):

$$\begin{array}{r} -25x - 25y = -500 \\ \underline{25x + 75y = 900} \\ 50y = 400 \\ y = 8. \end{array}$$

Then we back-substitute to find x:

$$x + y = 20 \qquad (1)$$
$$x + 8 = 20 \qquad \textbf{Substituting 8 for } y$$
$$x = 12.$$

GRAPHICAL SOLUTION

We solve each equation for y, getting

$$y = 20 - x,$$
$$y = \frac{900 - 25x}{75}.$$

Next, we graph these equations and find the point of intersection of the graphs.

$$y_1 = 20 - x, \quad y_2 = \frac{900 - 25x}{75}$$

4. **Check.** If 12 lb of caramel corn and 8 lb of nuts are used, the mixture weighs 12 + 8, or 20 lb. The value of the mixture is $2.50(12) + $7.50(8), or $30 + $60, or $90. Since the possible solution yields the desired weight and value of the mixture, our result checks.

5. **State.** The mixture should consist of 12 lb of caramel corn and 8 lb of honey roasted mixed nuts. **Now Try Exercise 71.** ■

EXAMPLE 7 *Boating.* Kerry's motorboat takes 3 hr to make a downstream trip with a 3-mph current. The return trip against the same current takes 5 hr. Find the speed of the boat in still water.

Solution

1. **Familiarize.** We first make a drawing, letting r = the speed of the boat in still water, in miles per hour, and d = the distance traveled, in miles. Note that the distance is the same for both parts of the trip. When the boat is traveling downstream, the current adds to its speed, so the downstream speed is $r + 3$. On the other hand, the current slows the boat down when it travels upstream, so the upstream speed is $r - 3$.

Downstream:
Speed: $r + 3$
Time: 3 hr
Distance: d

Upstream:
Speed: $r - 3$
Time: 5 hr
Distance: d

2. **Translate.** We organize the information in a table. Using the formula *Distance* = *Rate* (or *Speed*) · *Time*, we find that each row of the table yields an equation.

	Distance	Rate	Time	
Downstream	d	$r + 3$	3	⟶ $d = (r + 3)3$
Upstream	d	$r - 3$	5	⟶ $d = (r - 3)5$

We have a system of equations:

$$d = (r + 3)3, \quad (1)$$
$$d = (r - 3)5. \quad (2)$$

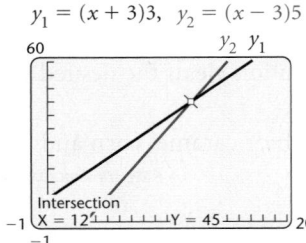

$y_1 = (x + 3)3,\ y_2 = (x - 3)5$

3. Carry out. We find r using substitution:

$(r + 3)3 = (r - 3)5$ Substituting $(r + 3)3$ for d in equation (2)

$3r + 9 = 5r - 15$ Multiplying

$9 = 2r - 15$ Subtracting $3r$ on both sides

$24 = 2r$ Adding 15 on both sides

$12 = r.$ Dividing by 2 on both sides

4. Check. If $r = 12$, then the boat's speed downstream is $r + 3 = 12 + 3$, or 15 mph, and the speed upstream is $r - 3 = 12 - 3$, or 9 mph. At 15 mph, in 3 hr the boat travels $15 \cdot 3 = 45$ mi; at 9 mph, in 5 hr the boat travels $9 \cdot 5 = 45$ mi. Since the distances are the same, the answer checks.

5. State. The speed of the boat in still water is 12 mph.

Now Try Exercise 67. ■

STUDY TIP

Make an effort to do your homework as soon as possible after each class. Make this part of your routine, choosing a time and a place where you can focus with a minimum of interruptions.

EXAMPLE 8 *Supply and Demand.* Suppose that the price and the supply of the Star Station satellite radio are related by the equation

$$y = 90 + 30x,$$

where y is the price, in dollars, at which the seller is willing to supply x thousand units. Also suppose that the price and the demand for the same model of satellite radio are related by the equation

$$y = 200 - 25x,$$

where y is the price, in dollars, at which the consumer is willing to buy x thousand units.

The **equilibrium point** for this radio is the pair (x, y) that is a solution of both equations. The **equilibrium price** is the price at which the amount of the product that the seller is willing to supply is the same as the amount demanded by the consumer. Find the equilibrium point for this radio.

Solution

1., 2. Familiarize and **Translate.** We are given a system of equations in the statement of the problem, so no further translation is necessary.

$y = 90 + 30x,$ (1)

$y = 200 - 25x.$ (2)

We substitute some values for x in each equation to get an idea of the corresponding prices. When $x = 1$,

$y = 90 + 30 \cdot 1 = 120,$ Substituting in equation (1)

$y = 200 - 25 \cdot 1 = 175.$ Substituting in equation (2)

This indicates that the price when 1 thousand units are supplied is lower than the price when 1 thousand units are demanded.

When $x = 4$,

$$y = 90 + 30 \cdot 4 = 210,$$ Substituting in equation (1)
$$y = 200 - 25 \cdot 4 = 100.$$ Substituting in equation (2)

In this case, the price related to supply is higher than the price related to demand. It would appear that the x-value we are looking for is between 1 and 4.

3. **Carry out.**   We use the substitution method:

$$y = 90 + 30x$$ Equation (1)
$$200 - 25x = 90 + 30x$$ Substituting $200 - 25x$ for y
$$110 = 55x$$ Adding $25x$ and subtracting 90 on both sides
$$2 = x.$$ Dividing by 55 on both sides

We now back-substitute 2 for x in either equation and find y:

$$y = 200 - 25x$$ (2)
$$y = 200 - 25 \cdot 2$$ Substituting 2 for x
$$y = 200 - 50$$
$$y = 150.$$

We can visualize the solution as the coordinates of the point of intersection of the graphs of the equations $y = 90 + 30x$ and $y = 200 - 25x$.

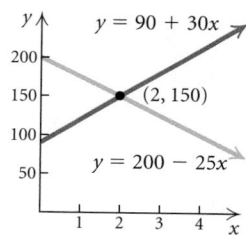

4. **Check.**   We can check by substituting 2 for x and 150 for y in both equations. Also note that 2 is between 1 and 4 as expected from the *Familiarize* and *Translate* steps.

5. **State.**   The equilibrium point is (2, $150). That is, the equilibrium quantity is 2 thousand units and the equilibrium price is $150.

Now Try Exercise 59. ▨

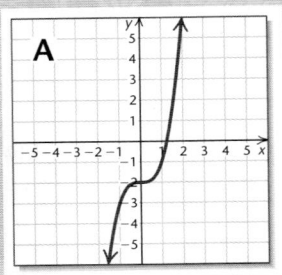

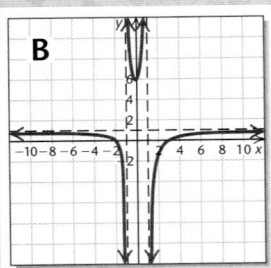

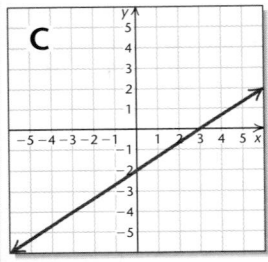

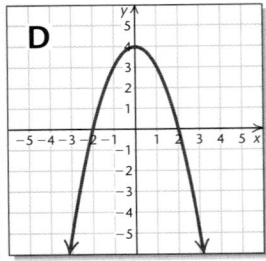

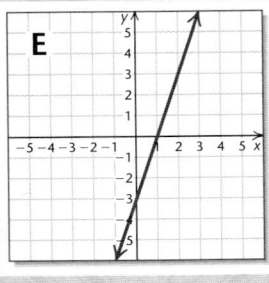

Visualizing the Graph

Match the equation or system of equations with its graph.

1. $2x - 3y = 6$

2. $f(x) = x^2 - 2x - 3$

3. $f(x) = -x^2 + 4$

4. $(x - 2)^2 + (y + 3)^2 = 9$

5. $f(x) = x^3 - 2$

6. $f(x) = -(x - 1)^2(x + 1)^2$

7. $f(x) = \dfrac{x - 1}{x^2 - 4}$

8. $f(x) = \dfrac{x^2 - x - 6}{x^2 - 1}$

9. $\begin{aligned} x - y &= -1, \\ 2x - y &= 2 \end{aligned}$

10. $\begin{aligned} 3x - y &= 3, \\ 2y &= 6x - 6 \end{aligned}$

Answers on page A-61

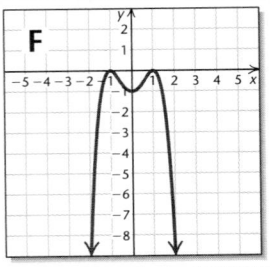

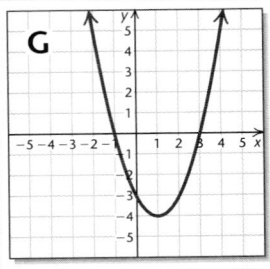

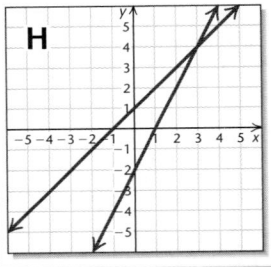

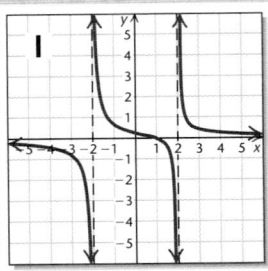

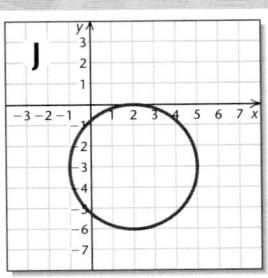

9.1 Exercise Set

In Exercises 1–6, match the system of equations with one of the graphs (a)–(f), which follow.

a)

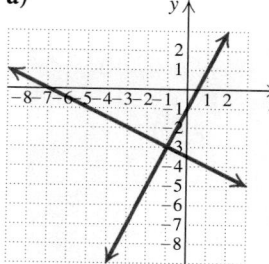

b)

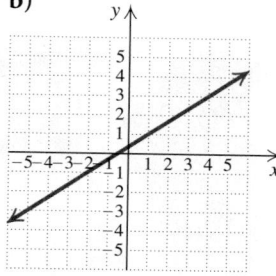

c)

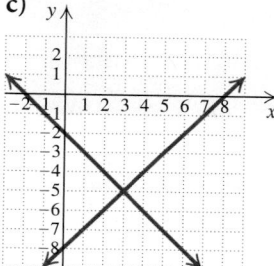

d)

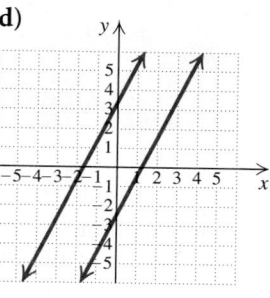

e)

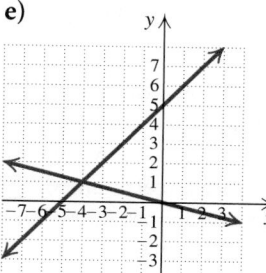

f)

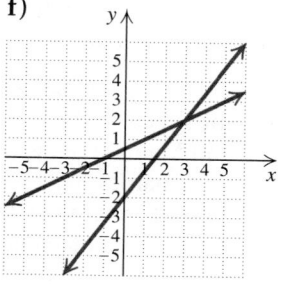

1. $x + y = -2,$
$y = x - 8$

2. $x - y = -5,$
$x = -4y$

3. $x - 2y = -1,$
$4x - 3y = 6$

4. $2x - y = 1,$
$x + 2y = -7$

5. $2x - 3y = -1,$
$-4x + 6y = 2$

6. $4x - 2y = 5,$
$6x - 3y = -10$

Solve graphically.

7. $x + y = 2,$
$3x + y = 0$

8. $x + y = 1,$
$3x + y = 7$

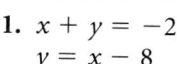

9. $x + 2y = 1,$
$x + 4y = 3$

10. $3x + 4y = 5,$
$x - 2y = 5$

11. $y + 1 = 2x,$
$y - 1 = 2x$

12. $2x - y = 1,$
$3y = 6x - 3$

13. $x - y = -6,$
$y = -2x$

14. $2x + y = 5,$
$x = -3y$

15. $2y = x - 1,$
$3x = 6y + 3$

16. $y = 3x + 2,$
$3x - y = -3$

Solve using the substitution method. Use a graphing calculator to check your answer.

17. $x + y = 9,$
$2x - 3y = -2$

18. $3x - y = 5,$
$x + y = \frac{1}{2}$

19. $x - 2y = 7,$
$x = y + 4$

20. $x + 4y = 6,$
$x = -3y + 3$

21. $y = 2x - 6,$
$5x - 3y = 16$

22. $3x + 5y = 2,$
$2x - y = -3$

23. $x + y = 3,$
$y = 4 - x$

24. $x - 2y = 3,$
$2x = 4y + 6$

25. $x - 5y = 4,$
$y = 7 - 2x$

26. $5x + 3y = -1,$
$x + y = 1$

27. $2x - 3y = 5,$
$5x + 4y = 1$

28. $3x + 4y = 6,$
$2x + 3y = 5$

29. $x + 2y = 2,$
$4x + 4y = 5$

30. $2x - y = 2,$
$4x + y = 3$

31. $3x - y = 5,$
$3y = 9x - 15$

32. $2x - y = 7,$
$y = 2x - 5$

Solve using the elimination method. Also determine whether each system is consistent or inconsistent and whether the equations are dependent or independent. Use a graphing calculator to check your answer.

33. $x + 2y = 7,$
$x - 2y = -5$

34. $3x + 4y = -2,$
$-3x - 5y = 1$

35. $x - 3y = 2,$
$6x + 5y = -34$

36. $x + 3y = 0,$
$20x - 15y = 75$

37. $3x - 12y = 6,$
$2x - 8y = 4$

38. $2x + 6y = 7,$
$3x + 9y = 10$

39. $2x = 5 - 3y,$
$4x = 11 - 7y$

40. $7(x - y) = 14,$
$2x = y + 5$

41. $0.3x - 0.2y = -0.9,$
$0.2x - 0.3y = -0.6$
(*Hint*: Since each coefficient has one decimal place, first multiply each equation by 10 to clear the decimals.)

42. $0.2x - 0.3y = 0.3,$
$0.4x + 0.6y = -0.2$
(*Hint*: Since each coefficient has one decimal place, first multiply each equation by 10 to clear the decimals.)

43. $\frac{1}{5}x + \frac{1}{2}y = 6,$
$\frac{3}{5}x - \frac{1}{2}y = 2$
(*Hint*: First multiply by the least common denominator to clear fractions.)

44. $\frac{2}{3}x + \frac{3}{5}y = -17,$
$\frac{1}{2}x - \frac{1}{3}y = -1$
(*Hint*: First multiply by the least common denominator to clear fractions.)

In Exercises 45–50, determine whether the statement is true or false.

45. If the graph of a system of equations is a pair of parallel lines, the system of equations is inconsistent.

46. If we obtain the equation $0 = 0$ when using the elimination method to solve a system of equations, the system has no solution.

47. If a system of two linear equations in two variables is consistent, it has exactly one solution.

48. If a system of two linear equations in two variables is dependent, it has infinitely many solutions.

49. It is possible for a system of two linear equations in two variables to be consistent and dependent.

50. It is possible for a system of two linear equations in two variables to be inconsistent and dependent.

51. *Winter Sports Injuries.* Skiers and snowboarders suffer about 288,400 injuries each winter, with skiing accounting for about 400 more injuries than snowboarding (*Source:* U.S. Consumer Product Safety Commission). How many injuries occur in each winter sport?

52. *Consumer Electronics.* A total of 1210 million mobile phones (cellphones and other hand-held devices) and PCs were shipped in 2006, with 752 million fewer PCs shipped than mobile phones (*Source*: IDC). How many mobile phones and how many PCs were shipped?

53. *Street Names.* The two most common names for streets in the United States are Second Street and Main Street, with 15,684 streets bearing one of these names. There are 260 more streets named Second Street than Main Street. (*Source*: 2006 Tele Atlas digital map data) How many streets bear each name?

54. *Museum Admission Prices.* Admission to the Indianapolis Children's Museum costs $5 more for an adult than for a child (*Source*: Indianapolis Children's Museum). Admission to the museum for two adults and two children costs $40. Find the cost of each adult's admission and each child's admission.

© 2006 Dale Chihuly; Photograph
© 2006 The Children's Museum of Indianapolis; photograph by Garry Chilluffo

55. *Visiting the Smithsonian.* The museums of the Smithsonian Institution and the Smithsonian website hosted a total of nearly 142 million visitors in 2005. The museums had about 94 million fewer visitors than the website. (*Source*: Smithsonian Institution) Find the number of visitors to the museums and to the website.

56. *Mail-Order Business.* A mail-order lacrosse equipment business shipped 120 packages one day. Customers are charged $3.50 for each standard-delivery package and $7.50 for each express-delivery package. Total shipping charges for the day were $596. How many of each kind of package were shipped?

57. *Sales Promotions.* During a one-month promotional campaign, Fran's Flix gave either a free DVD rental or a 12-serving box of microwave popcorn to new members. It cost the store $1 for each free rental and $2 for each box of popcorn. In all, 48 new members were signed up and the store's cost for the incentives was $86. How many of each incentive were given away?

58. *Concert Ticket Prices.* One evening 1500 concert tickets were sold for the Fairmont Summer Jazz

Festival. Tickets cost $25 for covered pavilion seats and $15 for lawn seats. Total receipts were $28,500. How many of each type of ticket were sold?

59. *Supply and Demand.* The supply and demand for a particular model of PDA (personal digital assistant) are related to price by the equations

$$y = 70 + 2x,$$
$$y = 175 - 5x,$$

respectively, where y is the price, in dollars, and x is the number of units, in thousands. Find the equilibrium point for this product.

60. *Supply and Demand.* The supply and demand for a particular model of treadmill are related to price by the equations

$$y = 240 + 40x,$$
$$y = 500 - 25x,$$

respectively, where y is the price, in dollars, and x is the number of units, in thousands. Find the equilibrium point for this product.

The point at which a company's costs equal its revenues is the **break-even point.** *In Exercises 61–64, C represents the production cost, in dollars, of x units of a product and R represents the revenue, in dollars, from the sale of x units. Find the number of units that must be produced and sold in order to break even. That is, find the value of x for which C = R.*

61. $C = 14x + 350,$
$R = 16.5x$

62. $C = 8.5x + 75,$
$R = 10x$

63. $C = 15x + 12,000,$
$R = 18x - 6000$

64. $C = 3x + 400,$
$R = 7x - 600$

65. *Nutrition.*   A one-cup serving of spaghetti with meatballs contains 260 Cal (calories) and 32 g of carbohydrates. A one-cup serving of chopped iceberg lettuce contains 5 Cal and 1 g of carbohydrates. (*Source: Home and Garden Bulletin No. 72*, U.S. Government Printing Office, Washington, D.C. 20402) How many servings of each would be required to obtain 400 Cal and 50 g of carbohydrates?

66. *Nutrition.*   One serving of tomato soup contains 100 Cal and 18 g of carbohydrates. One slice of whole wheat bread contains 70 Cal and 13 g of carbohydrates. (*Source: Home and Garden Bulletin No. 72*, U.S. Government Printing Office, Washington, D.C. 20402) How many servings of each would be required to obtain 230 Cal and 42 g of carbohydrates?

67. *Motion.*   A Leisure Time Cruises riverboat travels 46 km downstream in 2 hr. It travels 51 km upstream in 3 hr. Find the speed of the boat and the speed of the stream.

68. *Motion.*   A DC10 travels 3000 km with a tail wind in 3 hr. It travels 3000 km with a head wind in 4 hr. Find the speed of the plane and the speed of the wind.

69. *Investment.*   Bernadette inherited $15,000 and invested it in two municipal bonds, which pay 4% and 5% simple interest. The annual interest is $690. Find the amount invested at each rate.

70. *Tee Shirt Sales.*   Mack's Tee Shirt Shack sold 36 shirts one day. All short-sleeved tee shirts cost $12 each and all long-sleeved tee shirts cost $18 each. Total receipts for the day were $522. How many of each kind of shirt were sold?

71. *Coffee Mixtures.*   The owner of The Daily Grind coffee shop mixes French roast coffee worth $9.00 per pound with Kenyan coffee worth $7.50 per pound in order to get 10 lb of a mixture worth $8.40 per pound. How much of each type of coffee was used?

72. *Commissions.*   Jackson Manufacturing offers its sales representatives a choice between being paid a commission of 8% of sales or being paid a monthly salary of $1500 plus a commission of 1% of sales. For what monthly sales do the two plans pay the same amount?

73. *Motion.*   A Boeing 747 flies the 3000-mi distance from Los Angeles to New York, with a tail wind, in 5 hr. The return trip, against the wind, takes 6 hr. Find the speed of the plane and the speed of the wind.

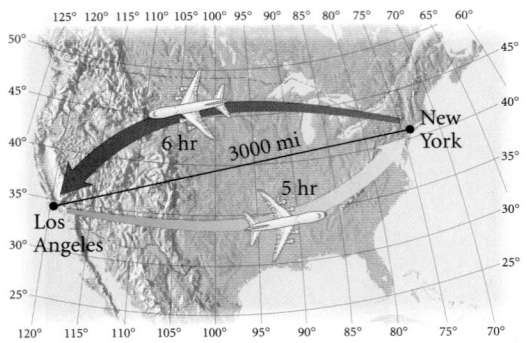

74. *Motion.*   Two private airplanes travel toward each other from cities that are 780 km apart at speeds of 190 km/h and 200 km/h. They left at the same time. In how many hours will they meet?

75. *Red Meat and Poultry Consumption.*   The amount of red meat consumed in the United States has remained fairly constant in recent years while the amount of poultry consumed has increased during those years, as shown by the data in the following table.

Year	Per Capita Red Meat Consumption (in pounds)	Per Capita Poultry Consumption (in pounds)
1995	113.6	62.1
2000	113.7	67.9
2003	111.9	71.2
2004	112.0	72.7

Source: U.S. Department of Agriculture, Economic Research Service, *Food Consumption, Prices, and Expenditures*

a) Find linear regression functions $r(x)$ and $p(x)$ that represent the per capita red meat consumption and the per capita poultry consumption, respectively, in number of pounds x years after 1995.

b) Use the functions found in part (a) to estimate when poultry consumption will equal red meat consumption.

76. *Prescription Drug Sales.* The table below lists the percentages of brand-name drugs and generic drugs sold in recent years.

Year	Percentage of Brand-Name Drugs Sold	Percentage of Generic Drugs Sold
1995	59.8	40.2
2000	57.6	42.4
2002	57.9	42.1
2004	51.9	48.1

Source: National Association of Chain Drug Stores

a) Find linear functions $b(x)$ and $g(x)$ that represent the percentages of brand-name drugs and generic drugs sold, respectively, x years after 1995.

b) Use the functions found in part (a) to estimate when the percentages of brand-name drugs and generic drugs sold will be equal.

Collaborative Discussion and Writing

77. Explain in your own words when the elimination method for solving a system of equations is preferable to the substitution method.

78. Cassidy solves the equation $2x + 5 = 3x - 7$ by finding the point of intersection of the graphs of $y_1 = 2x + 5$ and $y_2 = 3x - 7$. She finds the same point when she solves the system of equations

$$y = 2x + 5,$$
$$y = 3x - 7.$$

Explain the difference between the solution of the equation and the solution of the system of equations.

Skill Maintenance

79. *Book-Club Purchases.* Members of book clubs purchased a total of about 110 million books in 2005. Paperback sales exceeded sales of hardbacks by about 40 million books. (*Source:* Book Industry Study Group, Inc.) How many copies of each type of book were purchased?

80. *Soy-Food Sales.* Soy consumption has been growing in recent years. Soy-food sales in the United States totaled $4.1 billion in 2005. This was a 46% increase over soy-food sales in 2000. (*Source:* "Soyafoods: The U.S. Market 2006") Find the amount of soyfood sales in 2000.

Consider the function
$$f(x) = x^2 - 4x + 3$$
in Exercises 81–84.

81. What are the inputs if the output is 15?

82. Given an output of 8, find the corresponding inputs.

83. What is the output if the input is -2?

84. Find the zeros of the function.

Synthesis

85. *Motion.* Nancy jogs and walks to campus each day. She averages 4 km/h walking and 8 km/h jogging. The distance from home to the campus is 6 km and she makes the trip in 1 hr. How far does she jog on each trip?

86. *e-Commerce.* Shirts.com advertises a limited-time sale, offering 1 turtleneck for $15 and 2 turtlenecks for $25. A total of 1250 turtlenecks are sold and $16,750 is taken in. How many customers ordered 2 turtlenecks?

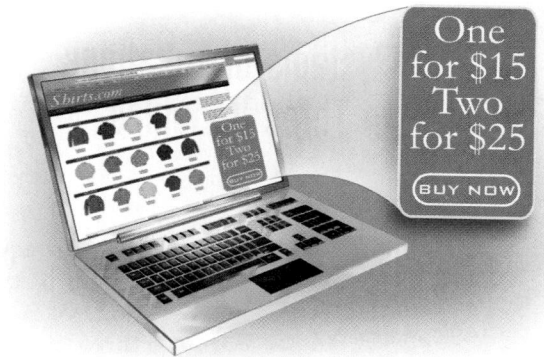

87. *Motion.* A train leaves Union Station for Central Station, 216 km away, at 9 A.M. One hour later, a train leaves Central Station for Union Station. They meet at noon. If the second train had started at 9 A.M. and the first train at 10:30 A.M., they would still have met at noon. Find the speed of each train.

88. *Antifreeze Mixtures.* An automobile radiator contains 16 L of antifreeze and water. This mixture is 30% antifreeze. How much of this mixture should be drained and replaced with pure antifreeze so that the final mixture will be 50% antifreeze?

89. Two solutions of the equation $Ax + By = 1$ are $(3, -1)$ and $(-4, -2)$. Find A and B.

90. *Ticket Line.* You are in line at a ticket window. There are 2 more people ahead of you in line than there are behind you. In the entire line, there are three times as many people as there are behind you. How many people are ahead of you?

91. *Gas Mileage.* The Honda Civic Hybrid vehicle, powered by gasoline–electric technology, gets 49 miles per gallon (mpg) in city driving and 51 mpg in highway driving (*Source*: Honda.com). The car is driven 447 mi on 9 gal of gasoline. How many miles were driven in the city and how many were driven on the highway?

92. *Motion.* Heather is standing on a railroad bridge, as shown in the figure below. A train is approaching from the direction shown by the yellow arrow. If Heather runs at a speed of 10 mph toward the train, she will reach point P on the bridge at the same moment that the train does. If she runs to point Q at the other end of the bridge at a speed of 10 mph, she will reach point Q also at the same moment that the train does. How fast, in miles per hour, is the train traveling?

<table>
<tr><td>**9.2**</td><td>**Systems of Equations in Three Variables**</td><td>❖ Solve systems of linear equations in three variables.
❖ Use systems of three equations to solve applied problems.
❖ Model a situation using a quadratic function.</td></tr>
</table>

A **linear equation in three variables** is an equation equivalent to one of the form $Ax + By + Cz = D$, where A, B, C, and D are real numbers and none of A, B, and C is 0. A **solution of a system of three equations in three variables** is an ordered triple that makes all three equations true. For example, the triple $(2, -1, 0)$ is a solution of the system of equations

$$4x + 2y + 5z = 6,$$
$$2x - y + z = 5,$$
$$3x + 2y - z = 4.$$

We can verify this by substituting 2 for x, -1 for y, and 0 for z in each equation.

❖ Solving Systems of Equations in Three Variables

We will solve systems of equations in three variables using an algebraic method called **Gaussian elimination,** named for the German mathematician Karl Friedrich Gauss (1777–1855). Our goal is to transform the original system to an equivalent system (one with the same solution set) of the form

$$Ax + By + Cz = D,$$
$$Ey + Fz = G,$$
$$Hz = K.$$

Then we solve the third equation for z and back-substitute to find y and then x.

Each of the following operations can be used to transform the original system to an equivalent system in the desired form.

1. Interchange any two equations.
2. Multiply both sides of one of the equations by a nonzero constant.
3. Add a nonzero multiple of one equation to another equation.

EXAMPLE 1 Solve the following system:

$$x - 2y + 3z = 11, \quad (1)$$
$$4x + 2y - 3z = 4, \quad (2)$$
$$3x + 3y - z = 4. \quad (3)$$

Solution First, we choose one of the variables to eliminate using two different pairs of equations. Let's eliminate x from equations (2) and (3). We multiply equation (1) by -4 and add it to equation (2). We also multiply equation (1) by -3 and add it to equation (3).

$$\begin{array}{rl} -4x + 8y - 12z = -44 & \text{Multiplying (1) by } -4 \\ \underline{4x + 2y - 3z = 4} & (2) \\ 10y - 15z = -40; & (4) \end{array}$$

$$\begin{array}{rl} -3x + 6y - 9z = -33 & \text{Multiplying (1) by } -3 \\ \underline{3x + 3y - z = 4} & (3) \\ 9y - 10z = -29. & (5) \end{array}$$

Now we have

$$x - 2y + 3z = 11, \quad (1)$$
$$10y - 15z = -40, \quad (4)$$
$$9y - 10z = -29. \quad (5)$$

Next, we multiply equation (5) by 10 to make the y-coefficient a multiple of the y-coefficient in the equation above it:

$$
\begin{aligned}
x - 2y + 3z &= 11, && (1)\\
10y - 15z &= -40, && (4)\\
90y - 100z &= -290. && (6)
\end{aligned}
$$

Next, we multiply equation (4) by -9 and add it to equation (6):

$$
\begin{array}{ll}
-90y + 135z = 360 & \textbf{Multiplying (4) by } -9\\
\underline{90y - 100z = -290} & (6)\\
35z = 70. & (7)
\end{array}
$$

We now have the system of equations

$$
\begin{aligned}
x - 2y + 3z &= 11, && (1)\\
10y - 15z &= -40, && (4)\\
35z &= 70. && (7)
\end{aligned}
$$

Now we solve equation (7) for z:

$$
\begin{aligned}
35z &= 70\\
z &= 2.
\end{aligned}
$$

Then we back-substitute 2 for z in equation (4) and solve for y:

$$
\begin{aligned}
10y - 15 \cdot 2 &= -40\\
10y - 30 &= -40\\
10y &= -10\\
y &= -1.
\end{aligned}
$$

Finally, we back-substitute -1 for y and 2 for z in equation (1) and solve for x:

$$
\begin{aligned}
x - 2(-1) + 3 \cdot 2 &= 11\\
x + 2 + 6 &= 11\\
x &= 3.
\end{aligned}
$$

We can check the triple $(3, -1, 2)$ in each of the three original equations. Since it makes all three equations true, the solution is $(3, -1, 2)$.

Now Try Exercise 1. ■

EXAMPLE 2   Solve the following system:

$$
\begin{aligned}
x + y + z &= 7, && (1)\\
3x - 2y + z &= 3, && (2)\\
x + 6y + 3z &= 25. && (3)
\end{aligned}
$$

Solution   We multiply equation (1) by -3 and add it to equation (2). We also multiply equation (1) by -1 and add it to equation (3).

$$
\begin{aligned}
x + y + z &= 7, && (1)\\
-5y - 2z &= -18, && (4)\\
5y + 2z &= 18 && (5)
\end{aligned}
$$

Next, we add equation (4) to equation (5):

$$x + y + z = 7, \qquad (1)$$
$$-5y - 2z = -18, \qquad (4)$$
$$0 = 0.$$

The equation $0 = 0$ tells us that equations (1), (2), and (3) are dependent. This means that the original system of three equations is equivalent to a system of two equations. One way to see this is to observe that four times equation (1) minus equation (2) is equation (3). Thus removing equation (3) from the system does not affect the solution of the system. We can say that the original system is equivalent to

$$x + y + z = 7, \qquad (1)$$
$$3x - 2y + z = 3. \qquad (2)$$

In this particular case, the original system has infinitely many solutions. (In some cases, a system containing dependent equations is inconsistent.) To find an expression for these solutions, we first solve equation (4) for either y or z. We choose to solve for y:

$$-5y - 2z = -18 \qquad (4)$$
$$-5y = 2z - 18$$
$$y = -\tfrac{2}{5}z + \tfrac{18}{5}.$$

Then we back-substitute in equation (1) to find an expression for x in terms of z:

$$x - \tfrac{2}{5}z + \tfrac{18}{5} + z = 7 \qquad \text{Substituting } -\tfrac{2}{5}z + \tfrac{18}{5} \text{ for } y$$
$$x + \tfrac{3}{5}z + \tfrac{18}{5} = 7$$
$$x + \tfrac{3}{5}z = \tfrac{17}{5}$$
$$x = -\tfrac{3}{5}z + \tfrac{17}{5}.$$

The solutions of the system of equations are ordered triples of the form $\left(-\tfrac{3}{5}z + \tfrac{17}{5}, -\tfrac{2}{5}z + \tfrac{18}{5}, z\right)$, where z can be any real number. Any real number that we use for z then gives us values for x and y and thus an ordered triple in the solution set. For example, if we choose $z = 0$, we have the solution $\left(\tfrac{17}{5}, \tfrac{18}{5}, 0\right)$. If we choose $z = -1$, we have $(4, 4, -1)$.   **Now Try Exercise 9.** ■

If we get a false equation, such as $0 = -5$, at some stage of the elimination process, we conclude that the original system is *inconsistent*; that is, it has no solutions.

Although systems of three linear equations in three variables do not lend themselves well to graphical solutions, it is of interest to picture some possible solutions. The graph of a linear equation in three variables is a plane. Thus the solution set of such a system is the intersection of three planes. Some possibilities are shown below.

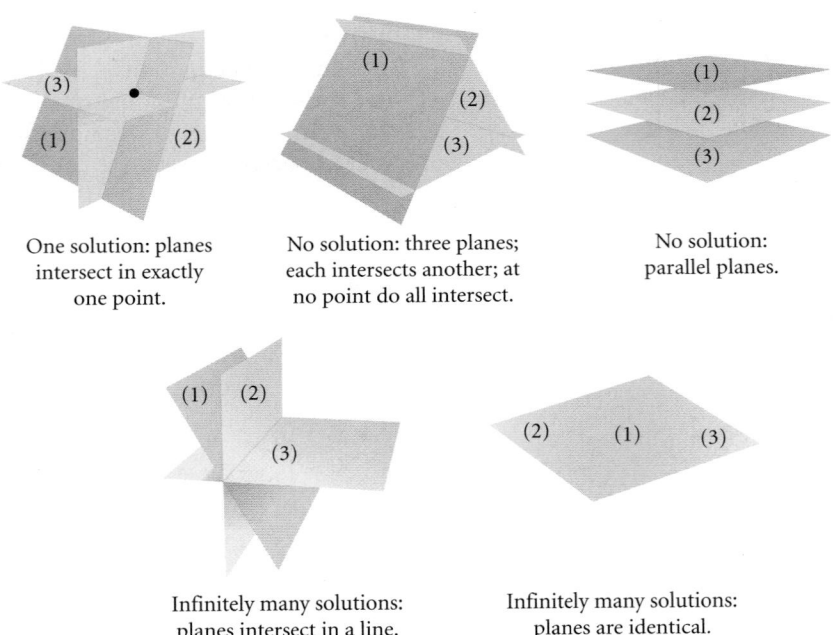

One solution: planes
intersect in exactly
one point.

No solution: three planes;
each intersects another; at
no point do all intersect.

No solution:
parallel planes.

Infinitely many solutions:
planes intersect in a line.

Infinitely many solutions:
planes are identical.

❖ Applications

Systems of equations in three or more variables allow us to solve many problems in fields such as business, the social and natural sciences, and engineering.

EXAMPLE 3   *Investment.*   Moira inherited $15,000 and invested part of it in a money market account, part in municipal bonds, and part in a mutual fund. After 1 yr, she received a total of $730 in simple interest from the three investments. The money market account paid 4% annually, the bonds paid 5% annually, and the mutual fund paid 6% annually. There was $2000 more invested in the mutual fund than in bonds. Find the amount that Moira invested in each category.

Solution

1. **Familiarize.**   We let x, y, and z represent the amounts invested in the money market account, the bonds, and the mutual fund, respectively. Then the amounts of income produced annually by each investment are given by 4%x, 5%y, and 6%z, or $0.04x$, $0.05y$, and $0.06z$.

2. **Translate.**   The fact that a total of $15,000 is invested gives us one equation:

$$x + y + z = 15,000.$$

Since the total interest is $730, we have a second equation:

$$0.04x + 0.05y + 0.06z = 730.$$

Another statement in the problem gives us a third equation.

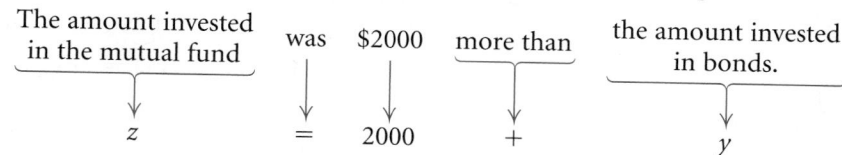

The amount invested in the mutual fund	was	$2000	more than	the amount invested in bonds.
z	$=$	2000	$+$	y

We now have a system of three equations:

$$x + y + z = 15,000,$$
$$0.04x + 0.05y + 0.06z = 730, \quad \text{or}$$
$$z = 2000 + y;$$

$$x + y + z = 15,000,$$
$$4x + 5y + 6z = 73,000,$$
$$-y + z = 2000.$$

3. Carry out. Solving the system of equations, we get

$$(7000, 3000, 5000).$$

4. Check. The sum of the numbers is 15,000. The income produced is

$$0.04(7000) + 0.05(3000) + 0.06(5000) = 280 + 150 + 300, \quad \text{or} \quad \$730.$$

Also the amount invested in the mutual fund, $5000, is $2000 more than the amount invested in bonds, $3000. Our solution checks in the original problem.

5. State. Moira invested $7000 in a money market account, $3000 in municipal bonds, and $5000 in a mutual fund.   **Now Try Exercise 27.** ■

❈ Mathematical Models and Applications

Recall that when we model a situation using a linear function $f(x) = mx + b$, we need to know two data points in order to determine m and b. For a quadratic model, $f(x) = ax^2 + bx + c$, we need three data points in order to determine a, b, and c.

EXAMPLE 4 *Country Music Stations.* The table below lists the number of commercial country music radio stations in the United States for three recent years. Use the data to find a quadratic function that gives the number of country music stations as a function of the number of years after 2004. Then use the function to estimate the number of country music stations in 2007.

YEAR, x	NUMBER OF COUNTRY MUSIC RADIO STATIONS
2004, 0	2047
2005, 1	2019
2006, 2	2097

Source: The M Street Radio Directory

Solution We let $x =$ the number of years after 2004 and $f(x) =$ the number of country music radio stations. Then $x = 0$ corresponds to 2004, $x = 1$ corresponds to 2005, and $x = 2$ corresponds to 2006. We use the three data points $(0, 2047)$, $(1, 2019)$, and $(2, 2097)$ to find a, b, and c in the function $f(x) = ax^2 + bx + c$. First, we substitute:

$$f(x) = ax^2 + bx + c$$

For $(0, 2047)$: $\quad 2047 = a \cdot 0^2 + b \cdot 0 + c;$
For $(1, 2019)$: $\quad 2019 = a \cdot 1^2 + b \cdot 1 + c;$
For $(2, 2097)$: $\quad 2097 = a \cdot 2^2 + b \cdot 2 + c.$

We now have a system of equations in the variables a, b, and c:

$$c = 2047,$$
$$a + b + c = 2019,$$
$$4a + 2b + c = 2097.$$

Solving this system, we get $(53, -81, 2047)$. Thus,

$$f(x) = 53x^2 - 81x + 2047.$$

To estimate the number of country music radio stations in 2007, we find $f(3)$, since 2007 is 3 yr after 2004:

$$f(3) = 53 \cdot 3^2 - 81 \cdot 3 + 2047$$
$$= 2281 \text{ country music stations.}$$

Now Try Exercise 33. ▪

```
QuadReg
y=ax²+bx+c
a=53
b=-81
c=2047
R²=1
```

```
Y₁(3)
              2281
```

The function in Example 4 can also be found using the QUADRATIC REGRESSION feature on a graphing calculator. Note that the method of Example 4 works when we have exactly three data points, whereas the QUADRATIC REGRESSION feature on a graphing calculator can be used for *three or more* points.

9.2 Exercise Set

Solve the system of equations.

1. $x + y + z = 2,$
$6x - 4y + 5z = 31,$
$5x + 2y + 2z = 13$

2. $x + 6y + 3z = 4,$
$2x + y + 2z = 3,$
$3x - 2y + z = 0$

3. $x - y + 2z = -3,$
$x + 2y + 3z = 4,$
$2x + y + z = -3$

4. $x + y + z = 6,$
$2x - y - z = -3,$
$x - 2y + 3z = 6$

5. $x + 2y - z = 5,$
$2x - 4y + z = 0,$
$3x + 2y + 2z = 3$

6. $2x + 3y - z = 1,$
$x + 2y + 5z = 4,$
$3x - y - 8z = -7$

7. $x + 2y - z = -8,$
$2x - y + z = 4,$
$8x + y + z = 2$

8. $x + 2y - z = 4,$
$4x - 3y + z = 8,$
$5x - y = 12$

9. $2x + y - 3z = 1,$
$x - 4y + z = 6,$
$4x - 7y - z = 13$

10. $x + 3y + 4z = 1,$
$3x + 4y + 5z = 3,$
$x + 8y + 11z = 2$

11. $4a + 9b = 8,$
$8a + 6c = -1,$
$6b + 6c = -1$

12. $3p + 2r = 11,$
$q - 7r = 4,$
$p - 6q = 1$

13. $2x \quad + \quad z = 1,$
$\qquad 3y - 2z = 6,$
$\quad x - 2y \qquad = -9$

14. $3x \qquad + 4z = -11,$
$\quad x - 2y \qquad = 5,$
$\qquad 4y - \quad z = -10$

15. $\quad w + \quad x + \quad y + \quad z = 2,$
$\quad w + 2x + 2y + 4z = 1,$
$-w + \quad x - \quad y - \quad z = -6,$
$-w + 3x + \quad y - \quad z = -2$

16. $\quad w + \quad x - \quad y + \quad z = 0,$
$-w + 2x + 2y + \quad z = 5,$
$-w + 3x + \quad y - \quad z = -4,$
$-2w + \quad x + \quad y - 3z = -7$

17. *Favorite Pets.* Americans own a total of about 314 million fish, cats, and dogs as pets. The number of fish owned is 16 million less than the total number of cats and dogs owned. There are 17 million more cats than dogs. (*Source*: American Pet Products Manufacturers Association) How many of each type of pet do Americans own?

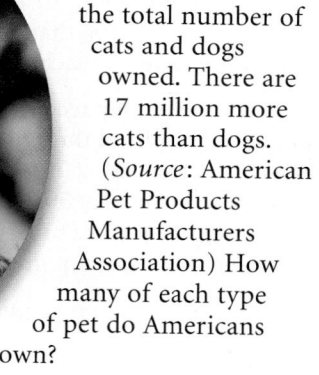

18. *Mother's Day Spending.* The top three Mother's Day gifts are flowers, jewelry, and gift certificates. The total of the average amounts spent on these gifts is $53.42. The average amount spent on jewelry is $4.40 more than the average amount spent on gift certificates. Together, the average amounts spent on flowers and gift certificates is $15.58 more than the average amount spent on jewelry. (*Source*: BIGresearch) What is the average amount spent on each type of gift?

19. *Jolts of Caffeine.* One 8-oz serving each of brewed coffee, Red Bull energy drink, and Mountain Dew soda contains 197 mg of caffeine. One serving of brewed coffee has 6 mg more caffeine than two servings of Mountain Dew. One serving of Red Bull contains 37 mg less caffeine than one serving each of brewed coffee and Mountain Dew. (*Source*: Australian Institute of Sport) Find the amount of caffeine in one serving of each beverage.

20. *Spring Cleaning.* In a group of 100 adults, 70 say they are most likely to do spring housecleaning in March, April, or May. Of these 70, the number who clean in April is 14 more than the total number who clean in March and May. The total number who clean in April and May is 2 more than three times the number who clean in March. (*Source*: Zoomerang online survey) Find the number who clean in each month.

21. *Low-Carb Gardening.* Gardeners on a low-carbohydrate diet are interested in knowing the carbohydrate content of the vegetables they plant. Together, 1 cup of raw lettuce, 6 raw asparagus spears, and 1 cup of raw tomatoes contain 12 grams of carbohydrates. One cup of raw lettuce and 6 raw asparagus spears have one-half the carbohydrates of 1 cup of raw tomatoes. One cup each of raw lettuce and raw tomatoes have three times the carbohydrate content of 6 raw asparagus spears. (*Source*: Burpee Seeds) Find the number of grams of carbohydrates in the given portion size of each vegetable.

22. *Toy Sales.* Three of the highest-grossing toy categories in 2005 were infant/preschool toys, dolls, and games/puzzles. Together they had gross sales of $8.2 billion. Sales of dolls and games/puzzles together were $2 billion more than sales of infant/preschool toys, and sales of games/puzzles were $0.3 billion less than doll sales. (*Source*: Toy Industry Association) Find the amount of sales of each type of toy.

23. *e-Commerce.* computerwarehouse.com charges $3 for shipping orders up to 10 lb, $5 for orders from 10 lb up to 15 lb, and $7.50 for orders of 15 lb or more. One day shipping charges for 150 orders totaled $680. The number of orders under 10 lb was three times the number of orders weighing 15 lb or more. Find the number of packages shipped at each rate.

24. *Mail-Order Business.* Natural Fibers Clothing charges $4 for shipping orders of $25 or less, $8 for orders from $25.01 to $75, and $10 for orders over $75. One week shipping charges for 600 orders totaled $4280. Eighty more orders for $25 or less were shipped than orders for more than $75. Find the number of orders shipped at each rate.

25. *Nutrition.* A hospital dietician must plan a lunch menu that provides 485 Cal, 41.5 g of carbohydrates, and 35 mg of calcium. A 3-oz serving of broiled ground beef contains 245 Cal, 0 g of carbohydrates, and 9 mg of calcium. One baked potato contains 145 Cal, 34 g of carbohydrates, and 8 mg of calcium. A one-cup serving of strawberries contains 45 Cal, 10 g of carbohydrates, and 21 mg of calcium. (*Source*: *Home and Garden Bulletin No. 72*, U.S. Government Printing Office, Washington, D.C. 20402) How many servings of each are required to provide the desired nutritional values?

26. *Nutrition.* A diabetic patient wishes to prepare a meal consisting of roasted chicken breast, mashed potatoes, and peas. A 3-oz serving of roasted skinless chicken breast contains 140 Cal, 27 g of protein, and 64 mg of sodium. A one-cup serving of mashed potatoes contains 160 Cal, 4 g of protein, and 636 mg of sodium, and a one-cup serving of peas contains 125 Cal, 8 g of protein, and 139 mg of sodium. (*Source*: *Home and Garden*

Bulletin No. 72, U.S. Government Printing Office, Washington, D.C. 20402) How many servings of each should be used if the meal is to contain 415 Cal, 50.5 g of protein, and 553 mg of sodium?

27. *Investment.* Jamal earns a year-end bonus of $5000 and puts it in 3 one-year investments that pay $243 in simple interest. Part is invested at 3%, part at 4%, and part at 6%. There is $1500 more invested at 6% than at 3%. Find the amount invested at each rate.

28. *Investment.* Casey receives $126 per year in simple interest from three investments. Part is invested at 2%, part at 3%, and part at 4%. There is $500 more invested at 3% than at 2%. The amount invested at 4% is three times the amount invested at 3%. Find the amount invested at each rate.

29. *Price Increases.* Orange juice, a raisin bagel, and a cup of coffee from Kelly's Koffee Kart cost a total of $5.35. Kelly posts a notice announcing that, effective the following week, the price of orange juice will increase 25% and the price of bagels will increase 20%. After the increase, the same purchase will cost a total of $6.20, and orange juice will cost 50¢ more than coffee. Find the price of each item before the increase.

30. *Cost of Snack Food.* Martin and Eva pool their loose change to buy snacks on their coffee break. One day, they spent $6.75 on 1 carton of milk, 2 donuts, and 1 cup of coffee. The next day, they spent $8.50 on 3 donuts and 2 cups of coffee. The third day, they bought 1 carton of milk, 1 donut, and 2 cups of coffee and spent $7.25. On the fourth day, they have a total of $6.45 left. Is this enough to buy 2 cartons of milk and 2 donuts?

31. *Golf.* On an 18-hole golf course, there are par-3 holes, par-4 holes, and par-5 holes. A golfer who shoots par on every hole has a score of 72. The sum of the number of par-3 holes and the number of par-5 holes is 8. How many of each type of hole are there on the golf course?

32. *Golf.* On an 18-hole golf course, there are par-3 holes, par-4 holes, and par-5 holes. A golfer who shoots par on every hole has a score of 70. There are twice as many par-4 holes as there are par-5 holes. How many of each type of hole are there on the golf course?

33. *Coffee Sales.* The table below lists retail sales of coffee, in billions of dollars, in the United States, represented in terms of the number of years since 2002.

Year, x	Retail Coffee Sales (in billions)
2002, 0	$ 8.4
2004, 2	9.6
2006, 4	12.3

Source: Specialty Coffee Association of America

a) Fit a quadratic function $f(x) = ax^2 + bx + c$ to the data.
b) Use the function to estimate the retail sales of coffee in 2008.

34. *Airport Security Violations.* The table below lists the number of fines levied on airline passengers for security violations by the Transportation Security Administration, represented as the number of years since 2002.

YEAR, x	NUMBER OF FINES
2002, 0	279
2004, 2	9741
2005, 3	4459

Source: Transportation Security Administration

a) Fit a quadratic function $f(x) = ax^2 + bx + c$ to the data.
b) Use the function to estimate the number of fines levied in 2003.

35. *Number of Marriages.* The table below lists the number of marriages, in thousands, in Texas, represented in terms of the number of years since 1990.

Year, x	Number of Marriages (in thousands)
1990, 0	179
2000, 10	196
2004, 14	179

Source: U.S. National Center for Health Statistics, *Vital Statistics of the United States*, annual, Monthly Vital Statistics Reports

a) Fit a quadratic function $f(x) = ax^2 + bx + c$ to the data.
b) Use the function to estimate the number of marriages in Texas in 2008.

36. *Female Smokers.* The table below lists the percentages of females age 18 to 24 who smoke, represented in terms of the number of years since 1995.

Year, x	Percent of Females Age 18 to 24 Who Smoke
1995, 0	22%
2000, 5	25
2004, 9	22

Source: U.S. National Center for Health Statistics

a) Fit a quadratic function $f(x) = ax^2 + bx + c$ to the data.
b) Use the function to estimate the percent of females age 18 to 24 who smoked in 2002.

37. *Morning Newspapers.*   The table below lists the number of morning newspapers published in the United States in various years.

GCM

Year	Number of Morning Newspapers
1920	437
1940	380
1960	312
1980	387
2000	766
2004	813

Source: Editor & Publisher

a) Use a graphing calculator to fit a quadratic function $f(x)$ to the data, where x is the number of years after 1920.
b) Use the function found in part (a) to estimate the number of morning newspapers published in 2008.

38. *Preprimary School Enrollment.*   The table below lists the number of children age 3 to 5 enrolled in nursery school and kindergarten, in millions, in various years.

Year	Number of Children Enrolled in Preprimary School (in millions)
1995	7.739
2000	7.592
2003	7.921
2004	7.968

Source: U.S. Census Bureau, Current Population Reports

a) Use a graphing calculator to fit a quadratic function $n(x)$ to the data, where x is the number of years after 1995.
b) Use the function found in part (a) to estimate the number of children enrolled in preprimary school in 2006.

Collaborative Discussion and Writing

39. Given two linear equations in three variables, $Ax + By + Cz = D$ and $Ex + Fy + Gz = H$, explain how you could find a third equation such that the system contains dependent equations.

40. Write a problem for a classmate to solve that can be translated to a system of three equations in three variables.

Skill Maintenance

In each of Exercises 41–48, fill in the blank with the correct term. Some of the given choices will not be used.

Descartes' rule of signs
the leading-term test
the intermediate value theorem
the fundamental theorem of algebra
a polynomial function
a rational function
a one-to-one function
a constant function
a horizontal asymptote
a vertical asymptote
an oblique asymptote
direct variation
inverse variation
a horizontal line
a vertical line
parallel
perpendicular

41. Two lines with slopes m_1 and m_2 are _____ if and only if the product of their slopes is -1.

42. We can use _____ to determine the behavior of the graph of a polynomial function as $x \to \infty$ or as $x \to -\infty$.

43. If it is possible for _____ to cross a graph more than once, then the graph is not the graph of a function.

44. A function is _____ if different inputs have different outputs.

45. _____ is a function that is a quotient of two polynomials.

46. If a situation gives rise to a function $f(x) = k/x$, or $y = k/x$, where k is a positive constant, we say that we have _____.

47. _____ of a rational function $p(x)/q(x)$, where $p(x)$ and $q(x)$ have no common factors other than constants, occurs at an x-value that makes the denominator 0.

48. When the numerator and the denominator of a rational function have the same degree, the graph of the function has _____.

Synthesis

In Exercises 49 and 50, let u represent $1/x$, v represent $1/y$, and w represent $1/z$. Solve first for u, v, and w. Then solve the system of equations.

49.
$$\frac{2}{x} - \frac{1}{y} - \frac{3}{z} = -1,$$
$$\frac{2}{x} - \frac{1}{y} + \frac{1}{z} = -9,$$
$$\frac{1}{x} + \frac{2}{y} - \frac{4}{z} = 17$$

50.
$$\frac{2}{x} + \frac{2}{y} - \frac{3}{z} = 3,$$
$$\frac{1}{x} - \frac{2}{y} - \frac{3}{z} = 9,$$
$$\frac{7}{x} - \frac{2}{y} + \frac{9}{z} = -39$$

51. Find the sum of the angle measures at the tips of the star.

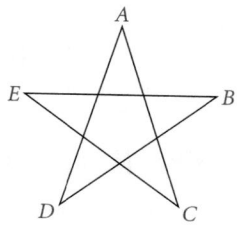

52. *Transcontinental Railroad.* Use the following facts to find the year in which the first U.S. transcontinental railroad was completed. The sum of the digits in the year is 24. The units digit is 1 more than the hundreds digit. Both the tens and the units digits are multiples of three.

In Exercises 53 and 54, three solutions of an equation are given. Use a system of three equations in three variables to find the constants and write the equation.

53. $Ax + By + Cz = 12$;
$\left(1, \frac{3}{4}, 3\right), \left(\frac{4}{3}, 1, 2\right)$, and $(2, 1, 1)$

54. $y = B - Mx - Nz$;
$(1, 1, 2), (3, 2, -6)$, and $\left(\frac{3}{2}, 1, 1\right)$

In Exercises 55 and 56, four solutions of the equation $y = ax^3 + bx^2 + cx + d$ are given. Use a system of four equations in four variables to find the constants a, b, c, and d and write the equation.

55. $(-2, 59), (-1, 13), (1, -1)$, and $(2, -17)$

56. $(-2, -39), (-1, -12), (1, -6)$, and $(3, 16)$

57. *Theater Attendance.* A performance at the Bingham Performing Arts Center was attended by 100 people. The audience consisted of adults, students, and children. The ticket prices were $10 for adults, $3 for students, and 50 cents for children. The total amount of money taken in was $100. How many adults, students, and children were in attendance? Does there seem to be some information missing? Do some careful reasoning.

9.3

Matrices and Systems of Equations

✥ Solve systems of equations using matrices.

✿ Matrices and Row-Equivalent Operations

In this section, we consider additional techniques for solving systems of equations. You have probably observed that when we solve a system of equations, we perform computations with the coefficients and the constants and continually rewrite the variables. We can streamline the solution process by omitting the variables until a solution is found. For example, the system

$$2x - 3y = 7,$$
$$x + 4y = -2$$

can be written more simply as

$$\begin{bmatrix} 2 & -3 & 7 \\ 1 & 4 & -2 \end{bmatrix}.$$

The vertical line replaces the equals signs.

A rectangular array of numbers like the one above is called a **matrix** (pl., **matrices**). The matrix above is called an **augmented matrix** for the given system of equations, because it contains not only the coefficients but also the constant terms. The matrix

$$\begin{bmatrix} 2 & -3 \\ 1 & 4 \end{bmatrix}$$

is called the **coefficient matrix** of the system.

The **rows** of a matrix are horizontal, and the **columns** are vertical. The augmented matrix above has 2 rows and 3 columns, and the coefficient matrix has 2 rows and 2 columns. A matrix with m rows and n columns is said to be of **order** $m \times n$. Thus the order of the augmented matrix above is 2×3, and the order of the coefficient matrix is 2×2. When $m = n$, a matrix is said to be **square**. The coefficient matrix above is a square matrix. The numbers 2 and 4 lie on the **main diagonal** of the coefficient matrix. The numbers in a matrix are called **entries** or **elements**.

✿ Gaussian Elimination with Matrices

In Section 9.2, we described a series of operations that can be used to transform a system of equations to an equivalent system. Each of these operations corresponds to one that can be used to produce *row-equivalent matrices*.

> **Row-Equivalent Operations**
>
> 1. Interchange any two rows.
> 2. Multiply each entry in a row by the same nonzero constant.
> 3. Add a nonzero multiple of one row to another row.

We can use these operations on the augmented matrix of a system of equations to solve the system.

GCM **EXAMPLE 1** Solve the following system:

$$
\begin{aligned}
2x - y + 4z &= -3, \\
x - 2y - 10z &= -6, \\
3x \phantom{{} - 2y} + 4z &= 7.
\end{aligned}
$$

Solution First, we write the augmented matrix, writing 0 for the missing y-term in the last equation:

$$
\left[\begin{array}{ccc|c}
2 & -1 & 4 & -3 \\
1 & -2 & -10 & -6 \\
3 & 0 & 4 & 7
\end{array}\right].
$$

Our goal is to find a row-equivalent matrix of the form

$$
\left[\begin{array}{ccc|c}
1 & a & b & c \\
0 & 1 & d & e \\
0 & 0 & 1 & f
\end{array}\right].
$$

The variables can then be reinserted to form equations from which we can complete the solution. This is done by working from the bottom equation to the top and using back-substitution.

The first step is to multiply and/or interchange rows so that each number in the first column below the first number is a multiple of that number. In this case, we interchange the first and second rows to obtain a 1 in the upper left-hand corner.

$$
\left[\begin{array}{ccc|c}
1 & -2 & -10 & -6 \\
2 & -1 & 4 & -3 \\
3 & 0 & 4 & 7
\end{array}\right]
\qquad
\begin{array}{l}
\text{New row 1 = row 2} \\
\text{New row 2 = row 1}
\end{array}
$$

Next, we multiply the first row by -2 and add it to the second row. We also multiply the first row by -3 and add it to the third row.

$$
\left[\begin{array}{ccc|c}
1 & -2 & -10 & -6 \\
0 & 3 & 24 & 9 \\
0 & 6 & 34 & 25
\end{array}\right]
\qquad
\begin{array}{l}
\text{Row 1 is unchanged.} \\
\text{New row 2} = -2(\text{row 1}) + \text{row 2} \\
\text{New row 3} = -3(\text{row 1}) + \text{row 3}
\end{array}
$$

Now we multiply the second row by $\frac{1}{3}$ to get a 1 in the second row, second column.

$$
\left[\begin{array}{ccc|c}
1 & -2 & -10 & -6 \\
0 & 1 & 8 & 3 \\
0 & 6 & 34 & 25
\end{array}\right]
\qquad
\text{New row 2} = \tfrac{1}{3}(\text{row 2})
$$

Then we multiply the second row by -6 and add it to the third row.

$$\left[\begin{array}{rrr|r} 1 & -2 & -10 & -6 \\ 0 & 1 & 8 & 3 \\ 0 & 0 & -14 & 7 \end{array}\right] \qquad \text{New row 3} = -6(\text{row 2}) + \text{row 3}$$

Finally, we multiply the third row by $-\frac{1}{14}$ to get a 1 in the third row, third column.

$$\left[\begin{array}{rrr|r} 1 & -2 & -10 & -6 \\ 0 & 1 & 8 & 3 \\ 0 & 0 & 1 & -\frac{1}{2} \end{array}\right] \qquad \text{New row 3} = -\frac{1}{14}(\text{row 3})$$

Now we can write the system of equations that corresponds to the last matrix above:

$$\begin{aligned} x - 2y - 10z &= -6, &&(1) \\ y + 8z &= 3, &&(2) \\ z &= -\tfrac{1}{2}. &&(3) \end{aligned}$$

We back-substitute $-\frac{1}{2}$ for z in equation (2) and solve for y:

$$\begin{aligned} y + 8\left(-\tfrac{1}{2}\right) &= 3 \\ y - 4 &= 3 \\ y &= 7. \end{aligned}$$

Next, we back-substitute 7 for y and $-\frac{1}{2}$ for z in equation (1) and solve for x:

$$\begin{aligned} x - 2 \cdot 7 - 10\left(-\tfrac{1}{2}\right) &= -6 \\ x - 14 + 5 &= -6 \\ x - 9 &= -6 \\ x &= 3. \end{aligned}$$

The triple $\left(3, 7, -\frac{1}{2}\right)$ checks in the original system of equations, so it is the solution.

Now Try Exercise 15. ■

Row-equivalent operations can be performed on a graphing calculator. For example, to interchange the first and second rows of the augmented matrix, as we did in the first step in Example 1, we enter the matrix as matrix **A** and select "rowSwap" from the MATRIX MATH menu. Some graphing calculators will not automatically store the matrix produced using a row-equivalent operation, so when several operations are to be performed in succession, it is helpful to store the result of each operation as it is produced. In the window at left, we see both the matrix produced by the rowSwap operation and the indication that this matrix is stored as matrix **B**.

The procedure followed in Example 1 is called **Gaussian elimination with matrices**. The last matrix in Example 1 is in **row-echelon form**. To be in this form, a matrix must have the following properties.

```
rowSwap([A],1,2)→[B]
[[1  -2  -10  -6]
 [2  -1   4   -3]
 [3   0   4    7 ]]
```

STUDY TIP

The *Graphing Calculator Manual* that accompanies this text contains the keystrokes for performing all the row-equivalent operations in Example 1.

> **Row-Echelon Form**
>
> 1. If a row does not consist entirely of 0's, then the first nonzero element in the row is a 1 (called a **leading 1**).
> 2. For any two successive nonzero rows, the leading 1 in the lower row is farther to the right than the leading 1 in the higher row.
> 3. All the rows consisting entirely of 0's are at the bottom of the matrix.
>
> If a fourth property is also satisfied, a matrix is said to be in **reduced row-echelon form**:
>
> 4. Each column that contains a leading 1 has 0's everywhere else.

EXAMPLE 2 Which of the following matrices are in row-echelon form? Which, if any, are in reduced row-echelon form?

a) $\begin{bmatrix} 1 & -3 & 5 & | & -2 \\ 0 & 1 & -4 & | & 3 \\ 0 & 0 & 1 & | & 10 \end{bmatrix}$ b) $\begin{bmatrix} 0 & -1 & | & 2 \\ 0 & 1 & | & 5 \end{bmatrix}$ c) $\begin{bmatrix} 1 & -2 & -6 & 4 & | & 7 \\ 0 & 3 & 5 & -8 & | & -1 \\ 0 & 0 & 1 & 9 & | & 2 \end{bmatrix}$

d) $\begin{bmatrix} 1 & 0 & 0 & | & -2.4 \\ 0 & 1 & 0 & | & 0.8 \\ 0 & 0 & 1 & | & 5.6 \end{bmatrix}$ e) $\begin{bmatrix} 1 & 0 & 0 & 0 & | & \frac{2}{3} \\ 0 & 1 & 0 & 0 & | & -\frac{1}{4} \\ 0 & 0 & 1 & 0 & | & \frac{6}{7} \\ 0 & 0 & 0 & 0 & | & 0 \end{bmatrix}$ f) $\begin{bmatrix} 1 & -4 & 2 & | & 5 \\ 0 & 0 & 0 & | & 0 \\ 0 & 1 & -3 & | & -8 \end{bmatrix}$

Solution The matrices in (a), (d), and (e) satisfy the row-echelon criteria and, thus, are in row-echelon form. In (b) and (c), the first nonzero elements of the first and second rows, respectively, are not 1. In (f), the row consisting entirely of 0's is not at the bottom of the matrix. Thus the matrices in (b), (c), and (f) are not in row-echelon form. In (d) and (e), not only are the row-echelon criteria met but each column that contains a leading 1 also has 0's elsewhere, so these matrices are in reduced row-echelon form. ∎

❖ Gauss–Jordan Elimination

We have seen that with Gaussian elimination we perform row-equivalent operations on a matrix to obtain a row-equivalent matrix in row-echelon form. When we continue to apply these operations until we have a matrix in *reduced* row-echelon form, we are using **Gauss–Jordan elimination**. This method is named for Karl Friedrich Gauss and Wilhelm Jordan (1842–1899).

GCM **EXAMPLE 3** Use Gauss–Jordan elimination to solve the system of equations in Example 1.

Solution Using Gaussian elimination in Example 1, we obtained the matrix

$$\begin{bmatrix} 1 & -2 & -10 & | & -6 \\ 0 & 1 & 8 & | & 3 \\ 0 & 0 & 1 & | & -\frac{1}{2} \end{bmatrix}.$$

We continue to perform row-equivalent operations until we have a matrix in reduced row-echelon form. We multiply the third row by 10 and add it to the first row. We also multiply the third row by -8 and add it to the second row.

$$\begin{bmatrix} 1 & -2 & 0 & | & -11 \\ 0 & 1 & 0 & | & 7 \\ 0 & 0 & 1 & | & -\frac{1}{2} \end{bmatrix} \qquad \begin{array}{l} \text{New row } 1 = 10(\text{row } 3) + \text{row } 1 \\ \text{New row } 2 = -8(\text{row } 3) + \text{row } 2 \end{array}$$

Next, we multiply the second row by 2 and add it to the first row.

$$\begin{bmatrix} 1 & 0 & 0 & | & 3 \\ 0 & 1 & 0 & | & 7 \\ 0 & 0 & 1 & | & -\frac{1}{2} \end{bmatrix} \qquad \text{New row } 1 = 2(\text{row } 2) + \text{row } 1$$

Writing the system of equations that corresponds to this matrix, we have

$$\begin{aligned} x \quad &= 3, \\ y \quad &= 7, \\ z &= -\tfrac{1}{2}. \end{aligned}$$

We can actually read the solution, $\left(3, 7, -\frac{1}{2}\right)$, directly from the last column of the reduced row-echelon matrix.

We can also use a graphing calculator to solve this system of equations. After the augmented matrix is entered, reduced row-echelon form can be found directly using the "rref" operation from the MATRIX MATH menu.

```
rref([A])▶Frac
   [[1 0 0 3    ]
    [0 1 0 7    ]
    [0 0 1 -1/2]]
```

Now Try Exercise 27. ▦

EXAMPLE 4 Solve the following system:

$$\begin{aligned} 3x - 4y - z &= 6, \\ 2x - y + z &= -1, \\ 4x - 7y - 3z &= 13. \end{aligned}$$

Solution We write the augmented matrix and use Gauss–Jordan elimination.

$$\begin{bmatrix} 3 & -4 & -1 & | & 6 \\ 2 & -1 & 1 & | & -1 \\ 4 & -7 & -3 & | & 13 \end{bmatrix}$$

We begin by multiplying the second and third rows by 3 so that each number in the first column below the first number, 3, is a multiple of that number.

$$\begin{bmatrix} 3 & -4 & -1 & | & 6 \\ 6 & -3 & 3 & | & -3 \\ 12 & -21 & -9 & | & 39 \end{bmatrix} \qquad \begin{array}{l} \text{New row } 2 = 3(\text{row } 2) \\ \text{New row } 3 = 3(\text{row } 3) \end{array}$$

Next, we multiply the first row by -2 and add it to the second row. We also multiply the first row by -4 and add it to the third row.

$$\begin{bmatrix} 3 & -4 & -1 & \bigm| & 6 \\ 0 & 5 & 5 & \bigm| & -15 \\ 0 & -5 & -5 & \bigm| & 15 \end{bmatrix}$$

New row 2 $= -2(\text{row } 1) + \text{row } 2$
New row 3 $= -4(\text{row } 1) + \text{row } 3$

Now we add the second row to the third row.

$$\begin{bmatrix} 3 & -4 & -1 & \bigm| & 6 \\ 0 & 5 & 5 & \bigm| & -15 \\ 0 & 0 & 0 & \bigm| & 0 \end{bmatrix}$$

New row 3 $= \text{row } 2 + \text{row } 3$

We can stop at this stage because we have a row consisting entirely of 0's. The last row of the matrix corresponds to the equation $0 = 0$, which is true for all values of x, y, and z. Consequently, the equations are dependent and the system is equivalent to

$$3x - 4y - z = 6,$$
$$5y + 5z = -15.$$

This particular system has infinitely many solutions. (A system containing dependent equations could be inconsistent.)

Solving the second equation for y gives us

$$y = -z - 3.$$

Substituting $-z - 3$ for y in the first equation and solving for x, we get

$$3x - 4(-z - 3) - z = 6$$
$$x = -z - 2.$$

Then the solutions of this system are of the form

$$(-z - 2, -z - 3, z),$$

where z can be any real number.   **Now Try Exercise 33.** ■

Similarly, if we obtain a row whose only nonzero entry occurs in the last column, we have an inconsistent system of equations. For example, in the matrix

$$\begin{bmatrix} 1 & 0 & 3 & \bigm| & -2 \\ 0 & 1 & 5 & \bigm| & 4 \\ 0 & 0 & 0 & \bigm| & 6 \end{bmatrix},$$

the last row corresponds to the false equation $0 = 6$, so we know the original system of equations has no solution.

9.3 Exercise Set

Determine the order of the matrix.

1. $\begin{bmatrix} 1 & -6 \\ -3 & 2 \\ 0 & 5 \end{bmatrix}$
2. $\begin{bmatrix} 7 \\ -5 \\ -1 \\ 3 \end{bmatrix}$

3. $[2 \quad -4 \quad 0 \quad 9]$
4. $[-8]$

5. $\begin{bmatrix} 1 & -5 & -8 \\ 6 & 4 & -2 \\ -3 & 0 & 7 \end{bmatrix}$
6. $\begin{bmatrix} 13 & 2 & -6 & 4 \\ -1 & 18 & 5 & -12 \end{bmatrix}$

Write the augmented matrix for the system of equations.

7. $2x - y = 7,$
   $x + 4y = -5$

8. $3x + 2y = 8,$
   $2x - 3y = 15$

9. $x - 2y + 3z = 12,$
   $2x \quad\quad - 4z = 8,$
   $\quad\quad 3y + z = 7$

10. $\quad x + y - z = 7,$
    $\quad\quad 3y + 2z = 1,$
    $-2x - 5y \quad\quad = 6$

Write the system of equations that corresponds to the augmented matrix.

11. $\left[\begin{array}{cc|c} 3 & -5 & 1 \\ 1 & 4 & -2 \end{array}\right]$

12. $\left[\begin{array}{cc|c} 1 & 2 & -6 \\ 4 & 1 & -3 \end{array}\right]$

13. $\left[\begin{array}{ccc|c} 2 & 1 & -4 & 12 \\ 3 & 0 & 5 & -1 \\ 1 & -1 & 1 & 2 \end{array}\right]$

14. $\left[\begin{array}{ccc|c} -1 & -2 & 3 & 6 \\ 0 & 4 & 1 & 2 \\ 2 & -1 & 0 & 9 \end{array}\right]$

Solve the system of equations using Gaussian elimination or Gauss–Jordan elimination. Use a graphing calculator to check your answer.

15. $4x + 2y = 11,$
    $3x - y = 2$

16. $2x + y = 1,$
    $3x + 2y = -2$

17. $5x - 2y = -3,$
    $2x + 5y = -24$

18. $2x + y = 1,$
    $3x - 6y = 4$

19. $3x + 4y = 7,$
    $-5x + 2y = 10$

20. $5x - 3y = -2,$
    $4x + 2y = 5$

21. $3x + 2y = 6,$
    $2x - 3y = -9$

22. $x - 4y = 9,$
    $2x + 5y = 5$

23. $x - 3y = 8,$
    $-2x + 6y = 3$

24. $4x - 8y = 12,$
    $-x + 2y = -3$

25. $-2x + 6y = 4,$
    $3x - 9y = -6$

26. $6x + 2y = -10,$
    $-3x - y = 6$

27. $x + 2y - 3z = 9,$
    $2x - y + 2z = -8,$
    $3x - y - 4z = 3$

28. $x - y + 2z = 0,$
    $x - 2y + 3z = -1,$
    $2x - 2y + z = -3$

29. $4x - y - 3z = 1,$
    $8x + y - z = 5,$
    $2x + y + 2z = 5$

30. $3x + 2y + 2z = 3,$
    $x + 2y - z = 5,$
    $2x - 4y + z = 0$

31. $x - 2y + 3z = -4,$
    $3x + y - z = 0,$
    $2x + 3y - 5z = 1$

32. $2x - 3y + 2z = 2,$
    $x + 4y - z = 9,$
    $-3x + y - 5z = 5$

33. $2x - 4y - 3z = 3,$
    $x + 3y + z = -1,$
    $5x + y - 2z = 2$

34. $x + y - 3z = 4,$
    $4x + 5y + z = 1,$
    $2x + 3y + 7z = -7$

35. $p + q + r = 1,$
    $p + 2q + 3r = 4,$
    $4p + 5q + 6r = 7$

36. $m + n + t = 9,$
    $m - n - t = -15,$
    $3m + n + t = 2$

37. $a + b - c = 7,$
    $a - b + c = 5,$
    $3a + b - c = -1$

38. $a - b + c = 3,$
    $2a + b - 3c = 5,$
    $4a + b - c = 11$

39. $-2w + 2x + 2y - 2z = -10,$
    $w + x + y + z = -5,$
    $3w + x - y + 4z = -2,$
    $w + 3x - 2y + 2z = -6$

40. $-w + 2x - 3y + z = -8,$
$\quad -w + x + y - z = -4,$
$\quad w + x + y + z = 22,$
$\quad -w + x - y - z = -14$

Use Gaussian elimination or Gauss–Jordan elimination in Exercises 41–44.

41. *Time of Return.* The Houlihans pay their babysitter $5 per hour before 11 P.M. and $7.50 per hour after 11 P.M. One evening they went out for 5 hr and paid the sitter $30. What time did they come home?

42. *Advertising Expense.* eAuction.com spent a total of $11 million on advertising in fiscal years 2006, 2007, and 2008. The amount spent in 2008 was three times the amount spent in 2006. The amount spent in 2007 was $3 million less than the amount spent in 2008. How much was spent on advertising each year?

43. *Borrowing.* Gonzalez Manufacturing borrowed $30,000 to buy a new piece of equipment. Part of the money was borrowed at 8%, part at 10%, and part at 12%. The annual interest was $3040, and the total amount borrowed at 8% and 10% was twice the amount borrowed at 12%. How much was borrowed at each rate?

44. *Stamp Purchase.* Ricardo spent $21 on 41¢ and 17¢ stamps. He bought a total of 60 stamps. How many of each type did he buy?

Collaborative Discussion and Writing

45. Solve the following system of equations using Gaussian elimination. Then solve it again using Gauss–Jordan elimination. Do you prefer one method over the other? Why or why not?

$$3x + 4y + 2z = 0,$$
$$x - y - z = 10,$$
$$2x + 3y + 3z = -10$$

46. Explain in your own words why the augmented matrix below represents a system of dependent equations.

$$\begin{bmatrix} 1 & -3 & 2 & | & -5 \\ 0 & 1 & -4 & | & 8 \\ 0 & 0 & 0 & | & 0 \end{bmatrix}$$

Skill Maintenance

Classify the function as linear, quadratic, cubic, quartic, rational, exponential, or logarithmic.

47. $f(x) = 3^{x-1}$

48. $f(x) = 3x - 1$

49. $f(x) = \dfrac{3x - 1}{x^2 + 4}$

50. $f(x) = -\frac{3}{4}x^4 + \frac{9}{2}x^3 + 2x^2 - 4$

51. $f(x) = \ln(3x - 1)$

52. $f(x) = \frac{3}{4}x^3 - x$

53. $f(x) = 3$

54. $f(x) = 2 - x - x^2$

Synthesis

In Exercises 55 and 56, three solutions of the equation $y = ax^2 + bx + c$ are given. Use a system of three equations in three variables and Gaussian elimination or Gauss–Jordan elimination to find the constants a, b, and c and write the equation.

55. $(-3, 12), (-1, -7),$ and $(1, -2)$

56. $(-1, 0), (1, -3),$ and $(3, -22)$

57. Find two different row-echelon forms of
$$\begin{bmatrix} 1 & 5 \\ 3 & 2 \end{bmatrix}.$$

58. Consider the system of equations
$$x - y + 3z = -8,$$
$$2x + 3y - z = 5,$$
$$3x + 2y + 2kz = -3k.$$

For what value(s) of k, if any, will the system have

a) no solution?
b) exactly one solution?
c) infinitely many solutions?

Solve using matrices.

59. $y = x + z,$
$3y + 5z = 4,$
$x + 4 = y + 3z$

60. $x + y = 2z,$
$2x - 5z = 4,$
$x - z = y + 8$

61. $x - 4y + 2z = 7,$
$3x + y + 3z = -5$

62. $x - y - 3z = 3,$
$-x + 3y + z = -7$

63. $4x + 5y = 3,$
$-2x + y = 9,$
$3x - 2y = -15$

64. $2x - 3y = -1,$
$-x + 2y = -2,$
$3x - 5y = 1$

9.4   **Matrix Operations**

❋ Add, subtract, and multiply matrices when possible.
❋ Write a matrix equation equivalent to a system of equations.

In Section 9.3, we used matrices to solve systems of equations. Matrices are useful in many other types of applications as well. In this section, we study matrices and some of their properties.

A capital letter is generally used to name a matrix, and lower-case letters with double subscripts generally denote its entries. For example, a_{47}, read "a sub four seven," indicates the entry in the fourth row and the seventh column. A general term is represented by a_{ij}. The notation a_{ij} indicates the entry in row i and column j. In general, we can write a matrix as

$$\mathbf{A} = [a_{ij}] = \begin{bmatrix} a_{11} & a_{12} & a_{13} & \cdots & a_{1n} \\ a_{21} & a_{22} & a_{23} & \cdots & a_{2n} \\ a_{31} & a_{32} & a_{33} & \cdots & a_{3n} \\ \vdots & \vdots & \vdots & & \vdots \\ a_{m1} & a_{m2} & a_{m3} & \cdots & a_{mn} \end{bmatrix}.$$

The matrix above has m rows and n columns; that is, its order is $m \times n$.

Two matrices are **equal** if they have the same order and corresponding entries are equal.

❋ Matrix Addition and Subtraction

To add or subtract matrices, we add or subtract their corresponding entries. The matrices must have the same order for this to be possible.

Addition and Subtraction of Matrices

Given two $m \times n$ matrices $\mathbf{A} = [a_{ij}]$ and $\mathbf{B} = [b_{ij}]$, their sum is

$$\mathbf{A} + \mathbf{B} = [a_{ij} + b_{ij}]$$

and their difference is

$$\mathbf{A} - \mathbf{B} = [a_{ij} - b_{ij}].$$

Addition of matrices is both commutative and associative.

GCM **EXAMPLE 1** Find $\mathbf{A} + \mathbf{B}$ for each of the following.

a) $\mathbf{A} = \begin{bmatrix} -5 & 0 \\ 4 & \frac{1}{2} \end{bmatrix}$, $\mathbf{B} = \begin{bmatrix} 6 & -3 \\ 2 & 3 \end{bmatrix}$

b) $\mathbf{A} = \begin{bmatrix} 1 & 3 \\ -1 & 5 \\ 6 & 0 \end{bmatrix}$, $\mathbf{B} = \begin{bmatrix} -1 & -2 \\ 1 & -2 \\ -3 & 1 \end{bmatrix}$

Solution We have a pair of 2×2 matrices in part (a) and a pair of 3×2 matrices in part (b). Since each pair of matrices has the same order, we can add the corresponding entries.

a) $\mathbf{A} + \mathbf{B} = \begin{bmatrix} -5 & 0 \\ 4 & \frac{1}{2} \end{bmatrix} + \begin{bmatrix} 6 & -3 \\ 2 & 3 \end{bmatrix}$

$= \begin{bmatrix} -5+6 & 0+(-3) \\ 4+2 & \frac{1}{2}+3 \end{bmatrix} = \begin{bmatrix} 1 & -3 \\ 6 & 3\frac{1}{2} \end{bmatrix}$

[A]+[B]
[[1 -3]
[6 3.5]]

We can also enter $\mathbf{A}$ and $\mathbf{B}$ in a graphing calculator and then find $\mathbf{A} + \mathbf{B}$.

b) $\mathbf{A} + \mathbf{B} = \begin{bmatrix} 1 & 3 \\ -1 & 5 \\ 6 & 0 \end{bmatrix} + \begin{bmatrix} -1 & -2 \\ 1 & -2 \\ -3 & 1 \end{bmatrix}$

$= \begin{bmatrix} 1+(-1) & 3+(-2) \\ -1+1 & 5+(-2) \\ 6+(-3) & 0+1 \end{bmatrix} = \begin{bmatrix} 0 & 1 \\ 0 & 3 \\ 3 & 1 \end{bmatrix}$

[A]+[B]
[[0 1]
[0 3]
[3 1]]

This sum can also be found on a graphing calculator after $\mathbf{A}$ and $\mathbf{B}$ have been entered. **Now Try Exercise 5.**

GCM **EXAMPLE 2** Find $\mathbf{C} - \mathbf{D}$ for each of the following.

a) $\mathbf{C} = \begin{bmatrix} 1 & 2 \\ -2 & 0 \\ -3 & -1 \end{bmatrix}$, $\mathbf{D} = \begin{bmatrix} 1 & -1 \\ 1 & 3 \\ 2 & 3 \end{bmatrix}$

b) $\mathbf{C} = \begin{bmatrix} 5 & -6 \\ -3 & 4 \end{bmatrix}$, $\mathbf{D} = \begin{bmatrix} -4 \\ 1 \end{bmatrix}$

Solution

a) Since the order of each matrix is 3×2, we can subtract corresponding entries:

$$\mathbf{C} - \mathbf{D} = \begin{bmatrix} 1 & 2 \\ -2 & 0 \\ -3 & -1 \end{bmatrix} - \begin{bmatrix} 1 & -1 \\ 1 & 3 \\ 2 & 3 \end{bmatrix}$$

$$= \begin{bmatrix} 1-1 & 2-(-1) \\ -2-1 & 0-3 \\ -3-2 & -1-3 \end{bmatrix} = \begin{bmatrix} 0 & 3 \\ -3 & -3 \\ -5 & -4 \end{bmatrix}.$$

[C]−[D]
```
[[0    3]
 [-3  -3]
 [-5  -4]]
```

This subtraction can also be done using a graphing calculator.

b) $\mathbf{C}$ is a 2×2 matrix and $\mathbf{D}$ is a 2×1 matrix. Since the matrices do not have the same order, we cannot subtract.   **Now Try Exercise 13.** ▣

The **opposite**, or **additive inverse**, of a matrix is obtained by replacing each entry with its opposite.

EXAMPLE 3   Find $-\mathbf{A}$ and $\mathbf{A} + (-\mathbf{A})$ for

$$\mathbf{A} = \begin{bmatrix} 1 & 0 & 2 \\ 3 & -1 & 5 \end{bmatrix}.$$

Solution   To find $-\mathbf{A}$, we replace each entry of $\mathbf{A}$ with its opposite.

$$-\mathbf{A} = \begin{bmatrix} -1 & 0 & -2 \\ -3 & 1 & -5 \end{bmatrix},$$

$$\mathbf{A} + (-\mathbf{A}) = \begin{bmatrix} 1 & 0 & 2 \\ 3 & -1 & 5 \end{bmatrix} + \begin{bmatrix} -1 & 0 & -2 \\ -3 & 1 & -5 \end{bmatrix}$$

$$= \begin{bmatrix} 0 & 0 & 0 \\ 0 & 0 & 0 \end{bmatrix}$$

[A]+(−[A])
```
[[0 0 0]
 [0 0 0]]
```

A matrix having 0's for all its entries is called a **zero matrix**. When a zero matrix is added to a second matrix of the same order, the second matrix is unchanged. Thus a zero matrix is an **additive identity**. For example,

$$\begin{bmatrix} 2 & 3 & -4 \\ 0 & 6 & 5 \end{bmatrix} + \begin{bmatrix} 0 & 0 & 0 \\ 0 & 0 & 0 \end{bmatrix} = \begin{bmatrix} 2 & 3 & -4 \\ 0 & 6 & 5 \end{bmatrix}.$$

The matrix

$$\begin{bmatrix} 0 & 0 & 0 \\ 0 & 0 & 0 \end{bmatrix}$$

is the additive identity for any 2×3 matrix.

❈ Scalar Multiplication

When we find the product of a number and a matrix, we obtain a **scalar product**.

> ### Scalar Product
>
> The **scalar product** of a number k and a matrix $\mathbf{A}$ is the matrix denoted $k\mathbf{A}$, obtained by multiplying each entry of $\mathbf{A}$ by the number k. The number k is called a **scalar**.

GCM **EXAMPLE 4** Find $3\mathbf{A}$ and $(-1)\mathbf{A}$ for

$$\mathbf{A} = \begin{bmatrix} -3 & 0 \\ 4 & 5 \end{bmatrix}.$$

Solution We have

$$3\mathbf{A} = 3\begin{bmatrix} -3 & 0 \\ 4 & 5 \end{bmatrix} = \begin{bmatrix} 3(-3) & 3 \cdot 0 \\ 3 \cdot 4 & 3 \cdot 5 \end{bmatrix} = \begin{bmatrix} -9 & 0 \\ 12 & 15 \end{bmatrix},$$

$$(-1)\mathbf{A} = -1\begin{bmatrix} -3 & 0 \\ 4 & 5 \end{bmatrix} = \begin{bmatrix} -1(-3) & -1 \cdot 0 \\ -1 \cdot 4 & -1 \cdot 5 \end{bmatrix} = \begin{bmatrix} 3 & 0 \\ -4 & -5 \end{bmatrix}.$$

These scalar products can also be found on a graphing calculator after $\mathbf{A}$ has been entered.

Now Try Exercise 9. ■

```
3[A]
          [[-9  0]
           [ 12 15]]
(−1)[A]
          [[3  0 ]
           [-4 -5]]
```

The properties of matrix addition and scalar multiplication are similar to the properties of addition and multiplication of real numbers.

> ### Properties of Matrix Addition and Scalar Multiplication
>
> For any $m \times n$ matrices $\mathbf{A}$, $\mathbf{B}$, and $\mathbf{C}$ and any scalars k and l:
>
> $\mathbf{A} + \mathbf{B} = \mathbf{B} + \mathbf{A}.$ Commutative property of addition
>
> $\mathbf{A} + (\mathbf{B} + \mathbf{C}) = (\mathbf{A} + \mathbf{B}) + \mathbf{C}.$ Associative property of addition
>
> $(kl)\mathbf{A} = k(l\mathbf{A}).$ Associative property of scalar multiplication
>
> $k(\mathbf{A} + \mathbf{B}) = k\mathbf{A} + k\mathbf{B}.$ Distributive property
> $(k + l)\mathbf{A} = k\mathbf{A} + l\mathbf{A}.$ Distributive property
>
> There exists a unique matrix $\mathbf{0}$ such that:
>
> $\mathbf{A} + \mathbf{0} = \mathbf{0} + \mathbf{A} = \mathbf{A}.$ Additive identity property
>
> There exists a unique matrix $-\mathbf{A}$ such that:
>
> $\mathbf{A} + (-\mathbf{A}) = -\mathbf{A} + \mathbf{A} = \mathbf{0}.$ Additive inverse property

EXAMPLE 5 *Production.* Mitchell Fabricators, Inc., manufactures three styles of bicycle frames in its two plants. The following table shows the number of each style produced at each plant in April.

	Mountain Bike	Racing Bike	Touring Bike
North Plant	150	120	100
South Plant	180	90	130

a) Write a 2 × 3 matrix **A** that represents the information in the table.

b) The manufacturer increased production by 20% in May. Find a matrix **M** that represents the increased production figures.

c) Find the matrix **A** + **M** and tell what it represents.

Solution

a) Write the entries in the table in a 2 × 3 matrix **A**.

$$\mathbf{A} = \begin{bmatrix} 150 & 120 & 100 \\ 180 & 90 & 130 \end{bmatrix}$$

b) The production in May will be represented by **A** + 20%**A**, or **A** + 0.2**A**, or 1.2**A**. Thus,

$$\mathbf{M} = (1.2)\begin{bmatrix} 150 & 120 & 100 \\ 180 & 90 & 130 \end{bmatrix} = \begin{bmatrix} 180 & 144 & 120 \\ 216 & 108 & 156 \end{bmatrix}.$$

c) $\mathbf{A} + \mathbf{M} = \begin{bmatrix} 150 & 120 & 100 \\ 180 & 90 & 130 \end{bmatrix} + \begin{bmatrix} 180 & 144 & 120 \\ 216 & 108 & 156 \end{bmatrix}$

$$= \begin{bmatrix} 330 & 264 & 220 \\ 396 & 198 & 286 \end{bmatrix}$$

The matrix **A** + **M** represents the total production of each of the three styles of frame at each plant in April and in May. **Now Try Exercise 29.** ■

❖ Products of Matrices

Matrix multiplication is defined in such a way that it can be used in solving systems of equations and in many applications.

Matrix Multiplication

For an $m \times n$ matrix $\mathbf{A} = [a_{ij}]$ and an $n \times p$ matrix $\mathbf{B} = [b_{ij}]$, the **product AB** $= [c_{ij}]$ is an $m \times p$ matrix, where

$$c_{ij} = a_{i1} \cdot b_{1j} + a_{i2} \cdot b_{2j} + a_{i3} \cdot b_{3j} + \cdots + a_{in} \cdot b_{nj}.$$

In other words, the entry c_{ij} in **AB** is obtained by multiplying the entries in row i of **A** by the corresponding entries in column j of **B** and adding the results.

> Note that we can multiply two matrices only when the number of columns in the first matrix is equal to the number of rows in the second matrix.

GCM **EXAMPLE 6** For

$$A = \begin{bmatrix} 3 & 1 & -1 \\ 2 & 0 & 3 \end{bmatrix}, \quad B = \begin{bmatrix} 1 & 6 \\ 3 & -5 \\ -2 & 4 \end{bmatrix}, \quad \text{and} \quad C = \begin{bmatrix} 4 & -6 \\ 1 & 2 \end{bmatrix},$$

find each of the following.

a) **AB**

b) **BA**

c) **BC**

d) **AC**

Solution

a) **A** is a 2 × 3 matrix and **B** is a 3 × 2 matrix, so **AB** will be a 2 × 2 matrix.

$$\mathbf{AB} = \begin{bmatrix} 3 & 1 & -1 \\ 2 & 0 & 3 \end{bmatrix} \begin{bmatrix} 1 & 6 \\ 3 & -5 \\ -2 & 4 \end{bmatrix}$$

$$= \begin{bmatrix} 3 \cdot 1 + 1 \cdot 3 + (-1)(-2) & 3 \cdot 6 + 1(-5) + (-1)(4) \\ 2 \cdot 1 + 0 \cdot 3 + 3(-2) & 2 \cdot 6 + 0(-5) + 3 \cdot 4 \end{bmatrix} = \begin{bmatrix} 8 & 9 \\ -4 & 24 \end{bmatrix}$$

b) **B** is a 3 × 2 matrix and **A** is a 2 × 3 matrix, so **BA** will be a 3 × 3 matrix.

$$\mathbf{BA} = \begin{bmatrix} 1 & 6 \\ 3 & -5 \\ -2 & 4 \end{bmatrix} \begin{bmatrix} 3 & 1 & -1 \\ 2 & 0 & 3 \end{bmatrix}$$

$$= \begin{bmatrix} 1 \cdot 3 + 6 \cdot 2 & 1 \cdot 1 + 6 \cdot 0 & 1(-1) + 6 \cdot 3 \\ 3 \cdot 3 + (-5)(2) & 3 \cdot 1 + (-5)(0) & 3(-1) + (-5)(3) \\ -2 \cdot 3 + 4 \cdot 2 & -2 \cdot 1 + 4 \cdot 0 & -2(-1) + 4 \cdot 3 \end{bmatrix} = \begin{bmatrix} 15 & 1 & 17 \\ -1 & 3 & -18 \\ 2 & -2 & 14 \end{bmatrix}$$

```
[A][B]
            [[8  9 ]
             [-4 24]]
[B][A]
          [[15  1  17 ]
           [-1  3 -18]
           [2  -2  14 ]]
```

> Note in parts (a) and (b) that **AB** ≠ **BA**. Multiplication of matrices is generally not commutative.

Matrix multiplication can be performed on a graphing calculator. The products in parts (a) and (b) are shown at left.

c) **B** is a 3×2 matrix and **C** is a 2×2 matrix, so **BC** will be a 3×2 matrix.

$$\mathbf{BC} = \begin{bmatrix} 1 & 6 \\ 3 & -5 \\ -2 & 4 \end{bmatrix} \begin{bmatrix} 4 & -6 \\ 1 & 2 \end{bmatrix}$$

$$= \begin{bmatrix} 1 \cdot 4 + 6 \cdot 1 & 1(-6) + 6 \cdot 2 \\ 3 \cdot 4 + (-5)(1) & 3(-6) + (-5)(2) \\ -2 \cdot 4 + 4 \cdot 1 & -2(-6) + 4 \cdot 2 \end{bmatrix}$$

$$= \begin{bmatrix} 10 & 6 \\ 7 & -28 \\ -4 & 20 \end{bmatrix}$$

d) The product **AC** is not defined because the number of columns of **A**, 3, is not equal to the number of rows of **C**, 2.

When the product **AC** is entered on a graphing calculator, an ERROR message is returned, indicating that the dimensions of the matrices are mismatched.

[A] [C]	ERR:DIM MISMATCH
	1: Quit
	2: Goto

Now Try Exercise 23. ▧

EXAMPLE 7   *Dairy Profit.*   Dalton's Dairy produces no-fat ice cream and frozen yogurt. The following table shows the number of gallons of each product that are sold at the dairy's three stores one week.

	Store 1	Store 2	Store 3
No-Fat Ice Cream (in gallons)	100	80	120
Frozen Yogurt (in gallons)	160	120	100

On each gallon of no-fat ice cream, the dairy's profit is $4, and on each gallon of frozen yogurt, it is $3. Use matrices to find the total profit on these items at each store for the given week.

Solution   We can write the table showing the distribution of the products as a 2 × 3 matrix:

$$\mathbf{D} = \begin{bmatrix} 100 & 80 & 120 \\ 160 & 120 & 100 \end{bmatrix}.$$

The profit per gallon for each product can also be written as a matrix:

$$\mathbf{P} = \begin{bmatrix} 4 & 3 \end{bmatrix}.$$

Then the total profit at each store is given by the matrix product **PD**:

$$\mathbf{PD} = \begin{bmatrix} 4 & 3 \end{bmatrix} \begin{bmatrix} 100 & 80 & 120 \\ 160 & 120 & 100 \end{bmatrix}$$
$$= \begin{bmatrix} 4 \cdot 100 + 3 \cdot 160 & 4 \cdot 80 + 3 \cdot 120 & 4 \cdot 120 + 3 \cdot 100 \end{bmatrix}$$
$$= \begin{bmatrix} 880 & 680 & 780 \end{bmatrix}.$$

The total profit on no-fat ice cream and frozen yogurt for the given week was $880 at store 1, $680 at store 2, and $780 at store 3.

Now Try Exercise 33. ■

A matrix that consists of a single row, like **P** in Example 7, is called a **row matrix**. Similarly, a matrix that consists of a single column, like

$$\begin{bmatrix} 8 \\ -3 \\ 5 \end{bmatrix},$$

is called a **column matrix**.

We have already seen that matrix multiplication is generally not commutative. Nevertheless, matrix multiplication does have some properties that are similar to those for multiplication of real numbers.

Properties of Matrix Multiplication

For matrices **A**, **B**, and **C**, assuming that the indicated operations are possible:

$\mathbf{A}(\mathbf{BC}) = (\mathbf{AB})\mathbf{C}.$	Associative property of multiplication
$\mathbf{A}(\mathbf{B} + \mathbf{C}) = \mathbf{AB} + \mathbf{AC}.$	Distributive property
$(\mathbf{B} + \mathbf{C})\mathbf{A} = \mathbf{BA} + \mathbf{CA}.$	Distributive property

---

❖ **Matrix Equations**

We can write a matrix equation equivalent to a system of equations.

STUDY TIP

The Addison-Wesley Math Tutor Center provides *free* tutoring to qualified students using this text. Assisted by mathematics instructors via telephone, fax, or e-mail, you can receive live tutoring on examples and exercises. This valuable resource provides you with immediate assistance when additional instruction is needed. Go to www.aw-bc.com/tutorcenter for more information.

EXAMPLE 8   Write a matrix equation equivalent to the following system of equations:

$$4x + 2y - z = 3,$$
$$9x \qquad + z = 5,$$
$$4x + 5y - 2z = 1.$$

Solution   We write the coefficients on the left in a matrix. We then write the product of that matrix and the column matrix containing the variables and set the result equal to the column matrix containing the constants on the right:

$$\begin{bmatrix} 4 & 2 & -1 \\ 9 & 0 & 1 \\ 4 & 5 & -2 \end{bmatrix} \begin{bmatrix} x \\ y \\ z \end{bmatrix} = \begin{bmatrix} 3 \\ 5 \\ 1 \end{bmatrix}.$$

If we let

$$A = \begin{bmatrix} 4 & 2 & -1 \\ 9 & 0 & 1 \\ 4 & 5 & -2 \end{bmatrix}, \quad X = \begin{bmatrix} x \\ y \\ z \end{bmatrix}, \quad \text{and} \quad B = \begin{bmatrix} 3 \\ 5 \\ 1 \end{bmatrix},$$

we can write this matrix equation as $AX = B$.   **Now Try Exercise 39.** ■

In the next section, we will solve systems of equations using a matrix equation like the one in Example 8.

9.4 Exercise Set

Find x and y.

1. $[5 \quad x] = [y \quad -3]$

2. $\begin{bmatrix} 6x \\ 25 \end{bmatrix} = \begin{bmatrix} -9 \\ 5y \end{bmatrix}$

3. $\begin{bmatrix} 3 & 2x \\ y & -8 \end{bmatrix} = \begin{bmatrix} 3 & -2 \\ 1 & -8 \end{bmatrix}$

4. $\begin{bmatrix} x - 1 & 4 \\ y + 3 & -7 \end{bmatrix} = \begin{bmatrix} 0 & 4 \\ -2 & -7 \end{bmatrix}$

For Exercises 5–20, let

$$A = \begin{bmatrix} 1 & 2 \\ 4 & 3 \end{bmatrix}, \quad B = \begin{bmatrix} -3 & 5 \\ 2 & -1 \end{bmatrix},$$

$$C = \begin{bmatrix} 1 & -1 \\ -1 & 1 \end{bmatrix}, \quad D = \begin{bmatrix} 1 & 1 \\ 1 & 1 \end{bmatrix},$$

$$E = \begin{bmatrix} 1 & 3 \\ 2 & 6 \end{bmatrix}, \quad F = \begin{bmatrix} 3 & 3 \\ -1 & -1 \end{bmatrix},$$

$$0 = \begin{bmatrix} 0 & 0 \\ 0 & 0 \end{bmatrix}, \quad I = \begin{bmatrix} 1 & 0 \\ 0 & 1 \end{bmatrix}.$$

Find each of the following.

5. $A + B$ **6.** $B + A$

7. $E + 0$ **8.** $2A$

9. $3F$ **10.** $(-1)D$

11. $3F + 2A$ **12.** $A - B$

13. $B - A$

14. AB

15. BA

16. $0F$

17. CD

18. EF

19. AI

20. IA

Find the product, if possible.

21. $\begin{bmatrix} -1 & 0 & 7 \\ 3 & -5 & 2 \end{bmatrix} \begin{bmatrix} 6 \\ -4 \\ 1 \end{bmatrix}$

22. $\begin{bmatrix} 6 & -1 & 2 \end{bmatrix} \begin{bmatrix} 1 & 4 \\ -2 & 0 \\ 5 & -3 \end{bmatrix}$

23. $\begin{bmatrix} -2 & 4 \\ 5 & 1 \\ -1 & -3 \end{bmatrix} \begin{bmatrix} 3 & -6 \\ -1 & 4 \end{bmatrix}$

24. $\begin{bmatrix} 2 & -1 & 0 \\ 0 & 5 & 4 \end{bmatrix} \begin{bmatrix} -3 & 1 & 0 \\ 0 & 2 & -1 \\ 5 & 0 & 4 \end{bmatrix}$

25. $\begin{bmatrix} 1 \\ -5 \\ 3 \end{bmatrix} \begin{bmatrix} -6 & 5 & 8 \\ 0 & 4 & -1 \end{bmatrix}$

26. $\begin{bmatrix} 2 & 0 & 0 \\ 0 & -1 & 0 \\ 0 & 0 & 3 \end{bmatrix} \begin{bmatrix} 0 & -4 & 3 \\ 2 & 1 & 0 \\ -1 & 0 & 6 \end{bmatrix}$

27. $\begin{bmatrix} 1 & -4 & 3 \\ 0 & 8 & 0 \\ -2 & -1 & 5 \end{bmatrix} \begin{bmatrix} 3 & 0 & 0 \\ 0 & -4 & 0 \\ 0 & 0 & 1 \end{bmatrix}$

28. $\begin{bmatrix} 4 \\ -5 \end{bmatrix} \begin{bmatrix} 2 & 0 \\ 6 & -7 \\ 0 & -3 \end{bmatrix}$

29. *Budget.* For the month of June, Nelia budgets $300 for food, $80 for clothes, and $40 for entertainment.

a) Write a 1×3 matrix **B** that represents the amounts budgeted for these items.

b) After receiving a raise, Nelia increases the amount budgeted for each item in July by 5%. Find a matrix **R** that represents the new amounts.

c) Find $B + R$ and tell what the entries represent.

30. *Produce.* The produce manager at Dugan's Market orders 40 lb of tomatoes, 20 lb of zucchini, and 30 lb of onions from a local farmer one week.

a) Write a 1×3 matrix **A** that represents the amount of each item ordered.

b) The following week the produce manager increases her order by 10%. Find a matrix **B** that represents this order.

c) Find $A + B$ and tell what the entries represent.

31. *Nutrition.* A 3-oz serving of roasted, skinless chicken breast contains 140 Cal, 27 g of protein, 3 g of fat, 13 mg of calcium, and 64 mg of sodium. One-half cup of potato salad contains 180 Cal, 4 g of protein, 11 g of fat, 24 mg of calcium, and 662 mg of sodium. One broccoli spear contains 50 Cal, 5 g of protein, 1 g of fat, 82 mg of calcium, and 20 mg of sodium. (*Source*: *Home and Garden Bulletin No. 72*, U.S. Government Printing Office, Washington, D.C. 20402)

a) Write 1×5 matrices **C**, **P**, and **B** that represent the nutritional values of each food.

b) Find $C + 2P + 3B$ and tell what the entries represent.

32. *Nutrition.* One slice of cheese pizza contains 290 Cal, 15 g of protein, 9 g of fat, and 39 g of carbohydrates. One-half cup of gelatin dessert contains 70 Cal, 2 g of protein, 0 g of fat, and 17 g of carbohydrates. One cup of whole milk contains 150 Cal, 8 g of protein, 8 g of fat, and 11 g of carbohydrates. (*Source*: *Home and Garden Bulletin No. 72*, U.S. Government Printing Office, Washington, D.C. 20402)

a) Write 1×4 matrices **P**, **G**, and **M** that represent the nutritional values of each food.

b) Find $3P + 2G + 2M$ and tell what the entries represent.

33. *Food Service Management.* The food service manager at a large hospital is concerned about maintaining reasonable food costs. The table below shows the cost per serving, in dollars, for items on four menus.

Menu	Meat	Potato	Vegetable	Salad	Dessert
1	1.03	0.15	0.26	0.23	0.27
2	1.10	0.14	0.24	0.21	0.25
3	1.06	0.22	0.31	0.28	0.34
4	1.21	0.20	0.29	0.33	0.31

On a particular day, a dietician orders 65 meals from menu 1, 48 from menu 2, 93 from menu 3, and 57 from menu 4.

a) Write the information in the table as a 4 × 5 matrix **M**.

b) Write a row matrix **N** that represents the number of each menu ordered.

c) Find the product **NM**.

d) State what the entries of **NM** represent.

34. *Food Service Management.* A college food service manager uses a table like the one below to show the number of units of ingredients, by weight, required for various menu items.

	White Cake	Bread	Coffee Cake	Sugar Cookies
Flour	1	2.5	0.75	0.5
Milk	0	0.5	0.25	0
Eggs	0.75	0.25	0.5	0.5
Butter	0.5	0	0.5	1

The cost per unit of each ingredient is 15 cents for flour, 28 cents for milk, 54 cents for eggs, and 83 cents for butter.

a) Write the information in the table as a 4 × 4 matrix **M**.

b) Write a row matrix **C** that represents the cost per unit of each ingredient.

c) Find the product **CM**.

d) State what the entries of **CM** represent.

35. *Production Cost.* Karin supplies two small campus coffee shops with homemade chocolate chip cookies, oatmeal cookies, and peanut butter cookies. The table below shows the number of each type of cookie, in dozens, that Karin sold in one week.

	Mugsy's Coffee Shop	The Coffee Club
Chocolate Chip	8	15
Oatmeal	6	10
Peanut Butter	4	3

Karin spends $3 for the ingredients for one dozen chocolate chip cookies, $1.50 for the ingredients for one dozen oatmeal cookies, and $2 for the ingredients for one dozen peanut butter cookies.

a) Write the information in the table as a 3 × 2 matrix **S**.

b) Write a row matrix **C** that represents the cost, per dozen, of the ingredients for each type of cookie.

c) Find the product **CS**.

d) State what the entries of **CS** represent.

36. *Profit.* A manufacturer produces exterior plywood, interior plywood, and fiberboard, which are shipped to two distributors. The table below shows the number of units of each type of product that are shipped to each warehouse.

	Distributor 1	Distributor 2
Exterior Plywood	900	500
Interior Plywood	450	1000
Fiberboard	600	700

The profits from each unit of exterior plywood, interior plywood, and fiberboard are $5, $8, and $4, respectively.

a) Write the information in the table as a 3 × 2 matrix **M**.
b) Write a row matrix **P** that represents the profit, per unit, of each type of product.
c) Find the product **PM**.
d) State what the entries of **PM** represent.

37. *Profit.*   In Exercise 35, suppose that Karin's profits on one dozen chocolate chip, oatmeal, and peanut butter cookies are $6, $4.50, and $5.20, respectively.

a) Write a row matrix **P** that represents this information.
b) Use the matrices **S** and **P** to find Karin's total profit from each coffee shop.

38. *Production Cost.*   In Exercise 36, suppose that the manufacturer's production costs for each unit of exterior plywood, interior plywood, and fiberboard are $20, $25, and $15, respectively.

a) Write a row matrix **C** that represents this information.
b) Use the matrices **M** and **C** to find the total production cost for the products shipped to each distributor.

Write a matrix equation equivalent to the system of equations.

39. $2x - 3y = 7,$
$x + 5y = -6$

40. $-x + y = 3,$
$5x - 4y = 16$

41. $x + y - 2z = 6,$
$3x - y + z = 7,$
$2x + 5y - 3z = 8$

42. $3x - y + z = 1,$
$x + 2y - z = 3,$
$4x + 3y - 2z = 11$

43. $3x - 2y + 4z = 17,$
$2x + y - 5z = 13$

44. $3x + 2y + 5z = 9,$
$4x - 3y + 2z = 10$

45. $-4w + x - y + 2z = 12,$
$w + 2x - y - z = 0,$
$-w + x + 4y - 3z = 1,$
$2w + 3x + 5y - 7z = 9$

46. $12w + 2x + 4y - 5z = 2,$
$-w + 4x - y + 12z = 5,$
$2w - x + 4y = 13,$
$ 2x + 10y + z = 5$

Collaborative Discussion and Writing

47. Is it true that if **AB** = **0**, for matrices **A** and **B**, then **A** = **0** or **B** = **0**? Why or why not?

48. Explain how Karin could use the matrix products found in Exercises 35 and 37 in making business decisions.

Skill Maintenance

In Exercises 49–52:
a) *Find the vertex.*
b) *Find the axis of symmetry.*
c) *Determine whether there is a maximum or minimum value and find that value.*
d) *Graph the function.*

49. $f(x) = x^2 - x - 6$

50. $f(x) = 2x^2 - 5x - 3$

51. $f(x) = -x^2 - 3x + 2$

52. $f(x) = -3x^2 + 4x + 4$

Synthesis

For Exercises 53–56, let
$$\mathbf{A} = \begin{bmatrix} -1 & 0 \\ 2 & 1 \end{bmatrix} \quad and \quad \mathbf{B} = \begin{bmatrix} 1 & -1 \\ 0 & 2 \end{bmatrix}.$$

53. Show that
$$(\mathbf{A} + \mathbf{B})(\mathbf{A} - \mathbf{B}) \neq \mathbf{A}^2 - \mathbf{B}^2,$$
where
$$\mathbf{A}^2 = \mathbf{A}\mathbf{A} \quad and \quad \mathbf{B}^2 = \mathbf{B}\mathbf{B}.$$

54. Show that
$$(\mathbf{A} + \mathbf{B})(\mathbf{A} + \mathbf{B}) \neq \mathbf{A}^2 + 2\mathbf{A}\mathbf{B} + \mathbf{B}^2.$$

55. Show that
$$(\mathbf{A} + \mathbf{B})(\mathbf{A} - \mathbf{B}) = \mathbf{A}^2 + \mathbf{B}\mathbf{A} - \mathbf{A}\mathbf{B} - \mathbf{B}^2.$$

56. Show that
$$(\mathbf{A} + \mathbf{B})(\mathbf{A} + \mathbf{B}) = \mathbf{A}^2 + \mathbf{B}\mathbf{A} + \mathbf{A}\mathbf{B} + \mathbf{B}^2.$$

In Exercises 57–61, let

$$A = \begin{bmatrix} a_{11} & a_{12} & a_{13} & \cdots & a_{1n} \\ a_{21} & a_{22} & a_{23} & \cdots & a_{2n} \\ a_{31} & a_{32} & a_{33} & \cdots & a_{3n} \\ \vdots & \vdots & \vdots & & \vdots \\ a_{m1} & a_{m2} & a_{m3} & \cdots & a_{mn} \end{bmatrix},$$

$$B = \begin{bmatrix} b_{11} & b_{12} & b_{13} & \cdots & b_{1n} \\ b_{21} & b_{22} & b_{23} & \cdots & b_{2n} \\ b_{31} & b_{32} & b_{33} & \cdots & b_{3n} \\ \vdots & \vdots & \vdots & & \vdots \\ b_{m1} & b_{m2} & b_{m3} & \cdots & b_{mn} \end{bmatrix},$$

$$and \quad C = \begin{bmatrix} c_{11} & c_{12} & c_{13} & \cdots & c_{1n} \\ c_{21} & c_{22} & c_{23} & \cdots & c_{2n} \\ c_{31} & c_{32} & c_{33} & \cdots & c_{3n} \\ \vdots & \vdots & \vdots & & \vdots \\ c_{m1} & c_{m2} & c_{m3} & \cdots & c_{mn} \end{bmatrix},$$

and let k and l be any scalars.

57. Prove that $A + B = B + A$.

58. Prove that $A + (B + C) = (A + B) + C$.

59. Prove that $(kl)A = k(lA)$.

60. Prove that $k(A + B) = kA + kB$.

61. Prove that $(k + l)A = kA + lA$.

9.5 **Inverses of Matrices**

❖ Find the inverse of a square matrix, if it exists.
❖ Use inverses of matrices to solve systems of equations.

In this section, we continue our study of matrix algebra, finding the **multiplicative inverse**, or simply **inverse**, of a square matrix, if it exists. Then we use such inverses to solve systems of equations.

❖ The Identity Matrix

Recall that, for real numbers, $a \cdot 1 = 1 \cdot a = a$; 1 is the multiplicative identity. A multiplicative identity matrix is very similar to the number 1.

> **Identity Matrix**
> For any positive integer n, the $n \times n$ **identity matrix** is an $n \times n$ matrix with 1's on the main diagonal and 0's elsewhere and is denoted by
>
> $$I = \begin{bmatrix} 1 & 0 & 0 & \cdots & 0 \\ 0 & 1 & 0 & \cdots & 0 \\ 0 & 0 & 1 & \cdots & 0 \\ \vdots & \vdots & \vdots & & \vdots \\ 0 & 0 & 0 & \cdots & 1 \end{bmatrix}.$$
>
> Then $AI = IA = A$, for any $n \times n$ matrix A.

EXAMPLE 1 For

$$\mathbf{A} = \begin{bmatrix} 4 & -7 \\ -3 & 2 \end{bmatrix} \quad \text{and} \quad \mathbf{I} = \begin{bmatrix} 1 & 0 \\ 0 & 1 \end{bmatrix},$$

find each of the following.

a) AI   **b) IA**

Solution

a) $\mathbf{AI} = \begin{bmatrix} 4 & -7 \\ -3 & 2 \end{bmatrix} \begin{bmatrix} 1 & 0 \\ 0 & 1 \end{bmatrix}$

$$= \begin{bmatrix} 4 \cdot 1 - 7 \cdot 0 & 4 \cdot 0 - 7 \cdot 1 \\ -3 \cdot 1 + 2 \cdot 0 & -3 \cdot 0 + 2 \cdot 1 \end{bmatrix} = \begin{bmatrix} 4 & -7 \\ -3 & 2 \end{bmatrix} = \mathbf{A}$$

b) $\mathbf{IA} = \begin{bmatrix} 1 & 0 \\ 0 & 1 \end{bmatrix} \begin{bmatrix} 4 & -7 \\ -3 & 2 \end{bmatrix}$

$$= \begin{bmatrix} 1 \cdot 4 + 0(-3) & 1(-7) + 0 \cdot 2 \\ 0 \cdot 4 + 1(-3) & 0(-7) + 1 \cdot 2 \end{bmatrix} = \begin{bmatrix} 4 & -7 \\ -3 & 2 \end{bmatrix} = \mathbf{A}$$

These products can also be found using a graphing calculator after **A** and **I** have been entered. ■

```
[A] [I]
              [[4  -7]
               [-3  2]]
[I] [A]
              [[4  -7]
               [-3  2]]
```

❋ The Inverse of a Matrix

Recall that for every nonzero real number a, there is a multiplicative inverse $1/a$, or a^{-1}, such that $a \cdot a^{-1} = a^{-1} \cdot a = 1$. The multiplicative inverse of a matrix behaves in a similar manner.

> **Inverse of a Matrix**
>
> For an $n \times n$ matrix **A**, if there is a matrix $\mathbf{A}^{-1}$ for which $\mathbf{A}^{-1} \cdot \mathbf{A} = \mathbf{I} = \mathbf{A} \cdot \mathbf{A}^{-1}$, then $\mathbf{A}^{-1}$ is the **inverse** of **A**.

We read $\mathbf{A}^{-1}$ as "**A** inverse." Note that not every matrix has an inverse.

EXAMPLE 2 Verify that

$$\mathbf{B} = \begin{bmatrix} 4 & -3 \\ 3 & -2 \end{bmatrix} \quad \text{is the inverse of} \quad \mathbf{A} = \begin{bmatrix} -2 & 3 \\ -3 & 4 \end{bmatrix}.$$

Solution   We show that $\mathbf{BA} = \mathbf{I} = \mathbf{AB}$.

$$\mathbf{BA} = \begin{bmatrix} 4 & -3 \\ 3 & -2 \end{bmatrix} \begin{bmatrix} -2 & 3 \\ -3 & 4 \end{bmatrix} = \begin{bmatrix} 1 & 0 \\ 0 & 1 \end{bmatrix}$$

$$\mathbf{AB} = \begin{bmatrix} -2 & 3 \\ -3 & 4 \end{bmatrix} \begin{bmatrix} 4 & -3 \\ 3 & -2 \end{bmatrix} = \begin{bmatrix} 1 & 0 \\ 0 & 1 \end{bmatrix}$$

Now Try Exercise 1. ■

We can find the inverse of a square matrix, if it exists, by using row-equivalent operations as in the Gauss–Jordan elimination method. For example, consider the matrix

$$\mathbf{A} = \begin{bmatrix} -2 & 3 \\ -3 & 4 \end{bmatrix}.$$

To find its inverse, we first form an **augmented matrix** consisting of **A** on the left side and the 2×2 identity matrix on the right side:

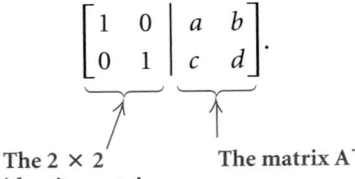

$$\begin{bmatrix} -2 & 3 & | & 1 & 0 \\ -3 & 4 & | & 0 & 1 \end{bmatrix}.$$

The 2×2 The 2×2
matrix A   identity matrix

Then we attempt to transform the augmented matrix to one of the form

$$\begin{bmatrix} 1 & 0 & | & a & b \\ 0 & 1 & | & c & d \end{bmatrix}.$$

The 2×2 The matrix $\mathbf{A}^{-1}$
identity matrix

If we can do this, the matrix on the right, $\begin{bmatrix} a & b \\ c & d \end{bmatrix}$, is $\mathbf{A}^{-1}$.

GCM **EXAMPLE 3** Find $\mathbf{A}^{-1}$, where

$$\mathbf{A} = \begin{bmatrix} -2 & 3 \\ -3 & 4 \end{bmatrix}.$$

Solution First, we write the augmented matrix. Then we transform it to the desired form.

$$\begin{bmatrix} -2 & 3 & | & 1 & 0 \\ -3 & 4 & | & 0 & 1 \end{bmatrix}$$

$$\begin{bmatrix} 1 & -\frac{3}{2} & | & -\frac{1}{2} & 0 \\ -3 & 4 & | & 0 & 1 \end{bmatrix} \qquad \text{New row } 1 = -\tfrac{1}{2}(\text{row } 1)$$

$$\begin{bmatrix} 1 & -\frac{3}{2} & | & -\frac{1}{2} & 0 \\ 0 & -\frac{1}{2} & | & -\frac{3}{2} & 1 \end{bmatrix} \qquad \text{New row } 2 = 3(\text{row } 1) + \text{row } 2$$

$$\begin{bmatrix} 1 & -\frac{3}{2} & | & -\frac{1}{2} & 0 \\ 0 & 1 & | & 3 & -2 \end{bmatrix} \qquad \text{New row } 2 = -2(\text{row } 2)$$

$$\begin{bmatrix} 1 & 0 & | & 4 & -3 \\ 0 & 1 & | & 3 & -2 \end{bmatrix} \qquad \text{New row } 1 = \tfrac{3}{2}(\text{row } 2) + \text{row } 1$$

Thus,

$$\mathbf{A}^{-1} = \begin{bmatrix} 4 & -3 \\ 3 & -2 \end{bmatrix},$$

which we verified in Example 2.

The $\boxed{x^{-1}}$ key on a graphing calculator can also be used to find the inverse of a matrix.

Now Try Exercise 5. ■

[A]⁻¹

[[4 -3]
[3 -2]]

EXAMPLE 4 Find $\mathbf{A}^{-1}$, where

$$\mathbf{A} = \begin{bmatrix} 1 & 2 & -1 \\ 3 & 5 & 3 \\ 2 & 4 & 3 \end{bmatrix}.$$

Solution First, we write the augmented matrix. Then we transform it to the desired form.

$$\begin{bmatrix} 1 & 2 & -1 & | & 1 & 0 & 0 \\ 3 & 5 & 3 & | & 0 & 1 & 0 \\ 2 & 4 & 3 & | & 0 & 0 & 1 \end{bmatrix}$$

$$\begin{bmatrix} 1 & 2 & -1 & | & 1 & 0 & 0 \\ 0 & -1 & 6 & | & -3 & 1 & 0 \\ 0 & 0 & 5 & | & -2 & 0 & 1 \end{bmatrix}$$
New row 2 = −3(row 1) + row 2
New row 3 = −2(row 1) + row 3

$$\begin{bmatrix} 1 & 2 & -1 & | & 1 & 0 & 0 \\ 0 & -1 & 6 & | & -3 & 1 & 0 \\ 0 & 0 & 1 & | & -\frac{2}{5} & 0 & \frac{1}{5} \end{bmatrix}$$
New row 3 = $\frac{1}{5}$(row 3)

$$\begin{bmatrix} 1 & 2 & 0 & | & \frac{3}{5} & 0 & \frac{1}{5} \\ 0 & -1 & 0 & | & -\frac{3}{5} & 1 & -\frac{6}{5} \\ 0 & 0 & 1 & | & -\frac{2}{5} & 0 & \frac{1}{5} \end{bmatrix}$$
New row 1 = row 3 + row 1
New row 2 = −6(row 3) + row 2

$$\begin{bmatrix} 1 & 0 & 0 & | & -\frac{3}{5} & 2 & -\frac{11}{5} \\ 0 & -1 & 0 & | & -\frac{3}{5} & 1 & -\frac{6}{5} \\ 0 & 0 & 1 & | & -\frac{2}{5} & 0 & \frac{1}{5} \end{bmatrix}$$
New row 1 = 2(row 2) + row 1

$$\begin{bmatrix} 1 & 0 & 0 & | & -\frac{3}{5} & 2 & -\frac{11}{5} \\ 0 & 1 & 0 & | & \frac{3}{5} & -1 & \frac{6}{5} \\ 0 & 0 & 1 & | & -\frac{2}{5} & 0 & \frac{1}{5} \end{bmatrix}$$
New row 2 = −1(row 2)

Thus,

$$\mathbf{A}^{-1} = \begin{bmatrix} -\frac{3}{5} & 2 & -\frac{11}{5} \\ \frac{3}{5} & -1 & \frac{6}{5} \\ -\frac{2}{5} & 0 & \frac{1}{5} \end{bmatrix}.$$

Now Try Exercise 9. ■

If a matrix has an inverse, we say that it is **invertible**, or **nonsingular**. When we cannot obtain the identity matrix on the left using the Gauss–Jordan method, then no inverse exists. This occurs when we obtain a row consisting entirely of 0's in either of the two matrices in the augmented matrix. In this case, we say that **A** is a **singular matrix**.

When we try to find the inverse of a noninvertible, or singular, matrix using a graphing calculator, the calculator returns an error message similar to ERR: SINGULAR MATRIX.

❈ Solving Systems of Equations

MATRIX EQUATIONS

REVIEW SECTION **9.4.**

We can write a system of n linear equations in n variables as a matrix equation $\mathbf{AX} = \mathbf{B}$. If **A** has an inverse, then the system of equations has a unique solution that can be found by solving for **X**, as follows:

$$\mathbf{AX} = \mathbf{B}$$
$$\mathbf{A}^{-1}(\mathbf{AX}) = \mathbf{A}^{-1}\mathbf{B} \qquad \text{Multiplying by } \mathbf{A}^{-1} \text{ on the left on both sides}$$
$$(\mathbf{A}^{-1}\mathbf{A})\mathbf{X} = \mathbf{A}^{-1}\mathbf{B} \qquad \text{Using the associative property of matrix multiplication}$$
$$\mathbf{IX} = \mathbf{A}^{-1}\mathbf{B} \qquad \mathbf{A}^{-1}\mathbf{A} = \mathbf{I}$$
$$\mathbf{X} = \mathbf{A}^{-1}\mathbf{B}. \qquad \mathbf{IX} = \mathbf{X}$$

> **Matrix Solutions of Systems of Equations**
>
> For a system of n linear equations in n variables, $\mathbf{AX} = \mathbf{B}$, if **A** is an invertible matrix, then the unique solution of the system is given by
>
> $$\mathbf{X} = \mathbf{A}^{-1}\mathbf{B}.$$

> Since matrix multiplication is not commutative in general, care must be taken to multiply *on the left* by $\mathbf{A}^{-1}$.

GCM | **EXAMPLE 5** Use an inverse matrix to solve the following system of equations:

$$-2x + 3y = 4,$$
$$-3x + 4y = 5.$$

Solution We write an equivalent matrix equation, $\mathbf{AX} = \mathbf{B}$:

$$\begin{bmatrix} -2 & 3 \\ -3 & 4 \end{bmatrix} \cdot \begin{bmatrix} x \\ y \end{bmatrix} = \begin{bmatrix} 4 \\ 5 \end{bmatrix}$$
$$\quad \mathbf{A} \qquad\quad \cdot \quad \mathbf{X} \;=\; \mathbf{B}$$

In Example 3, we found that

$$\mathbf{A}^{-1} = \begin{bmatrix} 4 & -3 \\ 3 & -2 \end{bmatrix}.$$

We also verified this in Example 2. Now we have

$$\mathbf{X} = \mathbf{A}^{-1}\mathbf{B}$$

$$\begin{bmatrix} x \\ y \end{bmatrix} = \begin{bmatrix} 4 & -3 \\ 3 & -2 \end{bmatrix} \begin{bmatrix} 4 \\ 5 \end{bmatrix} = \begin{bmatrix} 1 \\ 2 \end{bmatrix}.$$

The solution of the system of equations is $(1,2)$.

To use a graphing calculator to solve this system of equations, we enter **A** and **B** and then enter the notation $\mathbf{A}^{-1}\mathbf{B}$ on the home screen.

Now Try Exercise 25. ▪

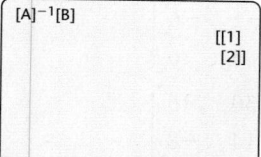

9.5 Exercise Set

Determine whether **B** *is the inverse of* **A**.

1. $\mathbf{A} = \begin{bmatrix} 1 & -3 \\ -2 & 7 \end{bmatrix}$, $\mathbf{B} = \begin{bmatrix} 7 & 3 \\ 2 & 1 \end{bmatrix}$

2. $\mathbf{A} = \begin{bmatrix} 3 & 2 \\ 4 & 3 \end{bmatrix}$, $\mathbf{B} = \begin{bmatrix} 3 & -2 \\ -4 & 3 \end{bmatrix}$

3. $\mathbf{A} = \begin{bmatrix} -1 & -1 & 6 \\ 1 & 0 & -2 \\ 1 & 0 & -3 \end{bmatrix}$, $\mathbf{B} = \begin{bmatrix} 2 & 3 & 2 \\ 3 & 3 & 4 \\ 1 & 1 & 1 \end{bmatrix}$

4. $\mathbf{A} = \begin{bmatrix} -2 & 0 & -3 \\ 5 & 1 & 7 \\ -3 & 0 & 4 \end{bmatrix}$, $\mathbf{B} = \begin{bmatrix} 4 & 0 & -3 \\ 1 & 1 & 1 \\ -3 & 0 & 2 \end{bmatrix}$

Use the Gauss–Jordan method to find $\mathbf{A}^{-1}$, *if it exists. Check your answers by using a graphing calculator to find* $\mathbf{A}^{-1}\mathbf{A}$ *and* $\mathbf{A}\mathbf{A}^{-1}$.

5. $\mathbf{A} = \begin{bmatrix} 3 & 2 \\ 5 & 3 \end{bmatrix}$

6. $\mathbf{A} = \begin{bmatrix} 3 & 5 \\ 1 & 2 \end{bmatrix}$

GCM 7. $\mathbf{A} = \begin{bmatrix} 6 & 9 \\ 4 & 6 \end{bmatrix}$

8. $\mathbf{A} = \begin{bmatrix} -4 & -6 \\ 2 & 3 \end{bmatrix}$

9. $A = \begin{bmatrix} 3 & 1 & 0 \\ 1 & 1 & 1 \\ 1 & -1 & 2 \end{bmatrix}$

10. $A = \begin{bmatrix} 1 & 0 & 1 \\ 2 & 1 & 0 \\ 1 & -1 & 1 \end{bmatrix}$

11. $A = \begin{bmatrix} 1 & -4 & 8 \\ 1 & -3 & 2 \\ 2 & -7 & 10 \end{bmatrix}$

12. $A = \begin{bmatrix} -2 & 5 & 3 \\ 4 & -1 & 3 \\ 7 & -2 & 5 \end{bmatrix}$

Use a graphing calculator to find A^{-1}, if it exists.

13. $A = \begin{bmatrix} 4 & -3 \\ 1 & -2 \end{bmatrix}$

14. $A = \begin{bmatrix} 0 & -1 \\ 1 & 0 \end{bmatrix}$

15. $A = \begin{bmatrix} 2 & 3 & 2 \\ 3 & 3 & 4 \\ -1 & -1 & -1 \end{bmatrix}$

16. $A = \begin{bmatrix} 1 & 2 & 3 \\ 2 & -1 & -2 \\ -1 & 3 & 3 \end{bmatrix}$

17. $A = \begin{bmatrix} 1 & 2 & -1 \\ -2 & 0 & 1 \\ 1 & -1 & 0 \end{bmatrix}$

18. $A = \begin{bmatrix} 7 & -1 & -9 \\ 2 & 0 & -4 \\ -4 & 0 & 6 \end{bmatrix}$

19. $A = \begin{bmatrix} 1 & 3 & -1 \\ 0 & 2 & -1 \\ 1 & 1 & 0 \end{bmatrix}$

20. $A = \begin{bmatrix} -1 & 0 & -1 \\ -1 & 1 & 0 \\ 0 & 1 & 1 \end{bmatrix}$

21. $A = \begin{bmatrix} 1 & 2 & 3 & 4 \\ 0 & 1 & 3 & -5 \\ 0 & 0 & 1 & -2 \\ 0 & 0 & 0 & -1 \end{bmatrix}$

22. $A = \begin{bmatrix} -2 & -3 & 4 & 1 \\ 0 & 1 & 1 & 0 \\ 0 & 4 & -6 & 1 \\ -2 & -2 & 5 & 1 \end{bmatrix}$

23. $A = \begin{bmatrix} 1 & -14 & 7 & 38 \\ -1 & 2 & 1 & -2 \\ 1 & 2 & -1 & -6 \\ 1 & -2 & 3 & 6 \end{bmatrix}$

24. $A = \begin{bmatrix} 10 & 20 & -30 & 15 \\ 3 & -7 & 14 & -8 \\ -7 & -2 & -1 & 2 \\ 4 & 4 & -3 & 1 \end{bmatrix}$

In Exercises 25–28, a system of equations is given, together with the inverse of the coefficient matrix. Use the inverse of the coefficient matrix to solve the system of equations.

25. $11x + 3y = -4,$
$7x + 2y = 5;$ $\quad A^{-1} = \begin{bmatrix} 2 & -3 \\ -7 & 11 \end{bmatrix}$

26. $8x + 5y = -6,$
$5x + 3y = 2;$ $\quad A^{-1} = \begin{bmatrix} -3 & 5 \\ 5 & -8 \end{bmatrix}$

27. $3x + y \quad\quad = 2,$
$2x - y + 2z = -5,$ $\quad A^{-1} = \frac{1}{9} \begin{bmatrix} 3 & 1 & -2 \\ 0 & -3 & 6 \\ -3 & 2 & 5 \end{bmatrix}$
$x + y + z = 5;$

28. $\quad\quad y - z = -4,$
$4x + y \quad\quad = -3,$ $\quad A^{-1} = \frac{1}{5} \begin{bmatrix} -3 & 2 & -1 \\ 12 & -3 & 4 \\ 7 & -3 & 4 \end{bmatrix}$
$3x - y + 3z = 1;$

Solve the system of equations using the inverse of the coefficient matrix of the equivalent matrix equation.

29. $4x + 3y = 2,$
$x - 2y = 6$

30. $2x - 3y = 7,$
$4x + y = -7$

31. $5x + y = 2,$
$3x - 2y = -4$

32. $x - 6y = 5,$
$-x + 4y = -5$

33. $x \quad\quad + z = 1,$
$2x + y \quad\quad = 3,$
$x - y + z = 4$

34. $\begin{aligned} x + 2y + 3z &= -1, \\ 2x - 3y + 4z &= 2, \\ -3x + 5y - 6z &= 4 \end{aligned}$

35. $\begin{aligned} 2x + 3y + 4z &= 2, \\ x - 4y + 3z &= 2, \\ 5x + y + z &= -4 \end{aligned}$

36. $\begin{aligned} x + y &= 2, \\ 3x + 2z &= 5, \\ 2x + 3y - 3z &= 9 \end{aligned}$

37. $\begin{aligned} 2w - 3x + 4y - 5z &= 0, \\ 3w - 2x + 7y - 3z &= 2, \\ w + x - y + z &= 1, \\ -w - 3x - 6y + 4z &= 6 \end{aligned}$

38. $\begin{aligned} 5w - 4x + 3y - 2z &= -6, \\ w + 4x - 2y + 3z &= -5, \\ 2w - 3x + 6y - 9z &= 14, \\ 3w - 5x + 2y - 4z &= -3 \end{aligned}$

39. *Sales.* Stefan sold a total of 145 Italian sausages and hot dogs from his curbside pushcart and collected $242.05. He sold 45 more hot dogs than sausages. How many of each did he sell?

40. *Price of School Supplies.* Miranda bought 4 lab record books and 3 highlighters for $17.83. Victor bought 3 lab record books and 2 highlighters for $13.05. Find the price of each item.

41. *Cost.* Evergreen Landscaping bought 4 tons of topsoil, 3 tons of mulch, and 6 tons of pea gravel for $2825. The next week the firm bought 5 tons of topsoil, 2 tons of mulch, and 5 tons of pea gravel for $2663. Pea gravel costs $17 less per ton than topsoil. Find the price per ton for each item.

42. *Investment.* Selena receives $230 per year in simple interest from three investments totaling $8500. Part is invested at 2.2%, part at 2.65%, and the rest at 3.05%. There is $1500 more invested at 3.05% than at 2.2%. Find the amount invested at each rate.

Collaborative Discussion and Writing

43. For square matrices **A** and **B**, is it true, in general, that $(\mathbf{A} + \mathbf{B})^{-1} = \mathbf{A}^{-1} + \mathbf{B}^{-1}$? Explain.

44. For square matrices **A** and **B**, is it true, in general, that $(\mathbf{AB})^{-1} = \mathbf{A}^{-1}\mathbf{B}^{-1}$? Explain.

Skill Maintenance

Use synthetic division to find the function values.

45. $f(x) = x^3 - 6x^2 + 4x - 8$; find $f(-2)$

46. $f(x) = 2x^4 - x^3 + 5x^2 + 6x - 4$; find $f(3)$

Solve.

47. $2x^2 + x = 7$

48. $\dfrac{1}{x + 1} - \dfrac{6}{x - 1} = 1$

49. $\sqrt{2x + 1} - 1 = \sqrt{2x - 4}$

50. $x - \sqrt{x - 6} = 0$

Factor the polynomial $f(x)$.

51. $f(x) = x^3 - 3x^2 - 6x + 8$

52. $f(x) = x^4 + 2x^3 - 16x^2 - 2x + 15$

Synthesis

State the conditions under which $\mathbf{A}^{-1}$ exists. Then find a formula for $\mathbf{A}^{-1}$.

53. $\mathbf{A} = [x]$

54. $\mathbf{A} = \begin{bmatrix} x & 0 \\ 0 & y \end{bmatrix}$

55. $\mathbf{A} = \begin{bmatrix} 0 & 0 & x \\ 0 & y & 0 \\ z & 0 & 0 \end{bmatrix}$

56. $\mathbf{A} = \begin{bmatrix} x & 1 & 1 & 1 \\ 0 & y & 0 & 0 \\ 0 & 0 & z & 0 \\ 0 & 0 & 0 & w \end{bmatrix}$

9.6 Determinants and Cramer's Rule

❖ Evaluate determinants of square matrices.

❖ Use Cramer's rule to solve systems of equations.

❖ Determinants of Square Matrices

With every square matrix, we associate a number called its *determinant*.

Determinant of a 2 × 2 Matrix

The **determinant** of the matrix $\begin{bmatrix} a & c \\ b & d \end{bmatrix}$ is denoted $\begin{vmatrix} a & c \\ b & d \end{vmatrix}$ and is defined as

$$\begin{vmatrix} a & c \\ b & d \end{vmatrix} = ad - bc.$$

EXAMPLE 1 Evaluate: $\begin{vmatrix} \sqrt{2} & -3 \\ -4 & -\sqrt{2} \end{vmatrix}$.

Solution

$$\begin{vmatrix} \sqrt{2} & -3 \\ -4 & -\sqrt{2} \end{vmatrix} \quad \text{The arrows indicate the products involved.}$$

$$= \sqrt{2}\left(-\sqrt{2}\right) - (-4)(-3)$$
$$= -2 - 12$$
$$= -14$$

Now Try Exercise 1. ■

We now consider a way to evaluate determinants of square matrices of order 3 × 3 or higher.

❖ Evaluating Determinants Using Cofactors

Often we first find minors and cofactors of matrices in order to evaluate determinants.

Minor

For a square matrix $\mathbf{A} = [a_{ij}]$, the **minor** M_{ij} of an entry a_{ij} is the determinant of the matrix formed by deleting the ith row and the jth column of $\mathbf{A}$.

EXAMPLE 2   For the matrix

$$\mathbf{A} = [a_{ij}] = \begin{bmatrix} -8 & 0 & 6 \\ 4 & -6 & 7 \\ -1 & -3 & 5 \end{bmatrix},$$

find each of the following.

a) M_{11} **b)** M_{23}

Solution

a) For M_{11}, we delete the first row and the first column and find the determinant of the 2×2 matrix formed by the remaining entries.

$$\begin{bmatrix} -8 & 0 & 6 \\ 4 & -6 & 7 \\ -1 & -3 & 5 \end{bmatrix}$$

$$\begin{aligned} M_{11} &= \begin{vmatrix} -6 & 7 \\ -3 & 5 \end{vmatrix} \\ &= (-6) \cdot 5 - (-3) \cdot 7 \\ &= -30 - (-21) \\ &= -30 + 21 \\ &= -9 \end{aligned}$$

b) For M_{23}, we delete the second row and the third column and find the determinant of the 2×2 matrix formed by the remaining entries.

$$\begin{bmatrix} -8 & 0 & 6 \\ 4 & -6 & 7 \\ -1 & -3 & 5 \end{bmatrix}$$

$$\begin{aligned} M_{23} &= \begin{vmatrix} -8 & 0 \\ -1 & -3 \end{vmatrix} \\ &= -8(-3) - (-1)0 \\ &= 24 \end{aligned}$$

Now Try Exercise 9. ◼

Cofactor

For a square matrix $\mathbf{A} = [a_{ij}]$, the **cofactor** A_{ij} of an entry a_{ij} is given by

$$A_{ij} = (-1)^{i+j}M_{ij},$$

where M_{ij} is the minor of a_{ij}.

EXAMPLE 3   For the matrix given in Example 2, find each of the following.

a) A_{11} **b)** A_{23}

Solution

a) In Example 2, we found that $M_{11} = -9$. Then

$$A_{11} = (-1)^{1+1}(-9) = (1)(-9) = -9.$$

b) In Example 2, we found that $M_{23} = 24$. Then

$$A_{23} = (-1)^{2+3}(24) = (-1)(24) = -24.$$

Now Try Exercise 11. ◼

Consider the matrix $\mathbf{A}$ given by

$$\mathbf{A} = \begin{bmatrix} a_{11} & a_{12} & a_{13} \\ a_{21} & a_{22} & a_{23} \\ a_{31} & a_{32} & a_{33} \end{bmatrix}.$$

The determinant of the matrix, denoted $|\mathbf{A}|$, can be found by multiplying each element of the first column by its cofactor and adding:

$$|\mathbf{A}| = a_{11}A_{11} + a_{21}A_{21} + a_{31}A_{31}.$$

Because

$$A_{11} = (-1)^{1+1}M_{11} = M_{11},$$
$$A_{21} = (-1)^{2+1}M_{21} = -M_{21},$$
and $$A_{31} = (-1)^{3+1}M_{31} = M_{31},$$

we can write

$$|\mathbf{A}| = a_{11} \cdot \begin{vmatrix} a_{22} & a_{23} \\ a_{32} & a_{33} \end{vmatrix} - a_{21} \cdot \begin{vmatrix} a_{12} & a_{13} \\ a_{32} & a_{33} \end{vmatrix} + a_{31} \cdot \begin{vmatrix} a_{12} & a_{13} \\ a_{22} & a_{23} \end{vmatrix}.$$

It can be shown that we can determine $|\mathbf{A}|$ by choosing *any* row or column, multiplying each element in that row or column by its cofactor, and adding. This is called *expanding* across a row or down a column. We just expanded down the first column. We now define the determinant of a square matrix of any order.

Determinant of Any Square Matrix

For any square matrix $\mathbf{A}$ of order $n \times n$ ($n > 1$), we define the **determinant** of $\mathbf{A}$, denoted $|\mathbf{A}|$, as follows. Choose any row or column. Multiply each element in that row or column by its cofactor and add the results. The determinant of a 1×1 matrix is simply the element of the matrix. The value of a determinant will be the same no matter which row or column is chosen.

EXAMPLE 4   Evaluate $|\mathbf{A}|$ by expanding across the third row.

$$\mathbf{A} = \begin{bmatrix} -8 & 0 & 6 \\ 4 & -6 & 7 \\ -1 & -3 & 5 \end{bmatrix}$$

Solution We have

$$|\mathbf{A}| = (-1)A_{31} + (-3)A_{32} + 5A_{33}$$

$$= (-1)(-1)^{3+1} \cdot \begin{vmatrix} 0 & 6 \\ -6 & 7 \end{vmatrix} + (-3)(-1)^{3+2} \cdot \begin{vmatrix} -8 & 6 \\ 4 & 7 \end{vmatrix}$$

$$+ 5(-1)^{3+3} \cdot \begin{vmatrix} -8 & 0 \\ 4 & -6 \end{vmatrix}$$

$$= (-1) \cdot 1 \cdot [0 \cdot 7 - (-6)6] + (-3)(-1)[-8 \cdot 7 - 4 \cdot 6]$$

$$+ 5 \cdot 1 \cdot [-8(-6) - 4 \cdot 0]$$

$$= -[36] + 3[-80] + 5[48]$$

$$= -36 - 240 + 240 = -36.$$

The value of this determinant is -36 no matter which row or column we expand on.

Now Try Exercise 13. ∎

Determinants can also be evaluated using a graphing calculator.

GCM **EXAMPLE 5** Use a graphing calculator to evaluate $|\mathbf{A}|$.

$$\mathbf{A} = \begin{bmatrix} 1 & 6 & -1 \\ -3 & -5 & 3 \\ 0 & 4 & 2 \end{bmatrix}$$

Solution First, we enter $\mathbf{A}$. Then we select the determinant operation, det, from the MATRIX MATH menu and enter the name of the matrix, $\mathbf{A}$. The calculator will return the value of the determinant of the matrix, 26.

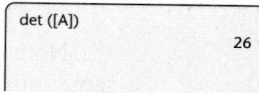

det ([A])
                  26

Now Try Exercise 17. ∎

❖ Cramer's Rule

Determinants can be used to solve systems of linear equations. Consider a system of two linear equations:

$$a_1x + b_1y = c_1,$$
$$a_2x + b_2y = c_2.$$

Solving this system using the elimination method, we obtain

$$x = \frac{c_1b_2 - c_2b_1}{a_1b_2 - a_2b_1} \quad \text{and} \quad y = \frac{a_1c_2 - a_2c_1}{a_1b_2 - a_2b_1}.$$

The numerators and the denominators of these expressions can be written as determinants:

$$x = \frac{\begin{vmatrix} c_1 & b_1 \\ c_2 & b_2 \end{vmatrix}}{\begin{vmatrix} a_1 & b_1 \\ a_2 & b_2 \end{vmatrix}} \quad \text{and} \quad y = \frac{\begin{vmatrix} a_1 & c_1 \\ a_2 & c_2 \end{vmatrix}}{\begin{vmatrix} a_1 & b_1 \\ a_2 & b_2 \end{vmatrix}}.$$

If we let

$$D = \begin{vmatrix} a_1 & b_1 \\ a_2 & b_2 \end{vmatrix}, \qquad D_x = \begin{vmatrix} c_1 & b_1 \\ c_2 & b_2 \end{vmatrix}, \quad \text{and} \quad D_y = \begin{vmatrix} a_1 & c_1 \\ a_2 & c_2 \end{vmatrix},$$

we have

$$x = \frac{D_x}{D} \quad \text{and} \quad y = \frac{D_y}{D}.$$

This procedure for solving systems of equations is known as *Cramer's rule*.

Cramer's Rule for 2 × 2 Systems

The solution of the system of equations

$$a_1x + b_1y = c_1,$$
$$a_2x + b_2y = c_2$$

is given by

$$x = \frac{D_x}{D}, \qquad y = \frac{D_y}{D},$$

where

$$D = \begin{vmatrix} a_1 & b_1 \\ a_2 & b_2 \end{vmatrix}, \qquad D_x = \begin{vmatrix} c_1 & b_1 \\ c_2 & b_2 \end{vmatrix},$$

$$D_y = \begin{vmatrix} a_1 & c_1 \\ a_2 & c_2 \end{vmatrix}, \quad \text{and} \quad D \neq 0.$$

Note that the denominator D contains the coefficients of x and y, in the same position as in the original equations. For x, the numerator is obtained by replacing the x-coefficients in D (the a's) with the c's. For y, the numerator is obtained by replacing the y-coefficients in D (the b's) with the c's.

GCM   **EXAMPLE 6**   Solve using Cramer's rule:

$$2x + 5y = 7,$$
$$5x - 2y = -3.$$

Solution   We have

$$x = \frac{\begin{vmatrix} 7 & 5 \\ -3 & -2 \end{vmatrix}}{\begin{vmatrix} 2 & 5 \\ 5 & -2 \end{vmatrix}} = \frac{7(-2) - (-3)5}{2(-2) - 5 \cdot 5} = \frac{1}{-29} = -\frac{1}{29},$$

$$y = \frac{\begin{vmatrix} 2 & 7 \\ 5 & -3 \end{vmatrix}}{\begin{vmatrix} 2 & 5 \\ 5 & -2 \end{vmatrix}} = \frac{2(-3) - 5 \cdot 7}{-29} = \frac{-41}{-29} = \frac{41}{29}.$$

The solution is $\left(-\frac{1}{29}, \frac{41}{29}\right)$.

To use Cramer's rule to solve this system of equations on a graphing calculator, we first enter the matrices corresponding to D, D_x, and D_y. We enter

$$\mathbf{A} = \begin{bmatrix} 2 & 5 \\ 5 & -2 \end{bmatrix}, \quad \mathbf{B} = \begin{bmatrix} 7 & 5 \\ -3 & -2 \end{bmatrix}, \quad \text{and} \quad \mathbf{C} = \begin{bmatrix} 2 & 7 \\ 5 & -3 \end{bmatrix}.$$

Then

$$x = \frac{\det(\mathbf{B})}{\det(\mathbf{A})} \quad \text{and} \quad y = \frac{\det(\mathbf{C})}{\det(\mathbf{A})}.$$

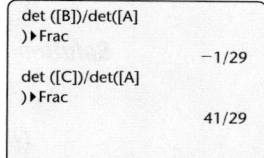

```
det ([B])/det([A]
)▶Frac
                      -1/29
det ([C])/det([A]
)▶Frac
                      41/29
```

Now Try Exercise 29. ▪

Cramer's rule works only when a system of equations has a unique solution. This occurs when $D \neq 0$. If $D = 0$ and D_x and D_y are also 0, then the equations are dependent. If $D = 0$ and D_x and/or D_y is not 0, then the system is inconsistent.

Cramer's rule can be extended to a system of n linear equations in n variables. We consider a 3×3 system.

Cramer's Rule for 3 × 3 Systems

The solution of the system of equations

$$\begin{aligned} a_1 x + b_1 y + c_1 z &= d_1, \\ a_2 x + b_2 y + c_2 z &= d_2, \\ a_3 x + b_3 y + c_3 z &= d_3 \end{aligned}$$

is given by

$$x = \frac{D_x}{D}, \quad y = \frac{D_y}{D}, \quad z = \frac{D_z}{D},$$

where

$$D = \begin{vmatrix} a_1 & b_1 & c_1 \\ a_2 & b_2 & c_2 \\ a_3 & b_3 & c_3 \end{vmatrix}, \quad D_x = \begin{vmatrix} d_1 & b_1 & c_1 \\ d_2 & b_2 & c_2 \\ d_3 & b_3 & c_3 \end{vmatrix},$$

$$D_y = \begin{vmatrix} a_1 & d_1 & c_1 \\ a_2 & d_2 & c_2 \\ a_3 & d_3 & c_3 \end{vmatrix}, \quad D_z = \begin{vmatrix} a_1 & b_1 & d_1 \\ a_2 & b_2 & d_2 \\ a_3 & b_3 & d_3 \end{vmatrix}, \quad \text{and} \quad D \neq 0.$$

Note that the determinant D_x is obtained from D by replacing the x-coefficients with d_1, d_2, and d_3. D_y and D_z are obtained in a similar manner. As with a system of two equations, Cramer's rule cannot be used if $D = 0$. If $D = 0$ and D_x, D_y, and D_z are 0, the equations are dependent. If $D = 0$ and one of D_x, D_y, or D_z is not 0, then the system is inconsistent.

EXAMPLE 7   Solve using Cramer's rule:

$$x - 3y + 7z = 13,$$
$$x + y + z = 1,$$
$$x - 2y + 3z = 4.$$

Solution   We have

$$D = \begin{vmatrix} 1 & -3 & 7 \\ 1 & 1 & 1 \\ 1 & -2 & 3 \end{vmatrix} = -10, \qquad D_x = \begin{vmatrix} 13 & -3 & 7 \\ 1 & 1 & 1 \\ 4 & -2 & 3 \end{vmatrix} = 20,$$

$$D_y = \begin{vmatrix} 1 & 13 & 7 \\ 1 & 1 & 1 \\ 1 & 4 & 3 \end{vmatrix} = -6, \qquad D_z = \begin{vmatrix} 1 & -3 & 13 \\ 1 & 1 & 1 \\ 1 & -2 & 4 \end{vmatrix} = -24.$$

Then

$$x = \frac{D_x}{D} = \frac{20}{-10} = -2, \qquad y = \frac{D_y}{D} = \frac{-6}{-10} = \frac{3}{5},$$

$$z = \frac{D_z}{D} = \frac{-24}{-10} = \frac{12}{5}.$$

The solution is $\left(-2, \frac{3}{5}, \frac{12}{5}\right)$.

In practice, it is not necessary to evaluate D_z. When we have found values for x and y, we can substitute them into one of the equations to find z.

Now Try Exercise 37. ▪

9.6   Exercise Set

Evaluate the determinant.

1. $\begin{vmatrix} 5 & 3 \\ -2 & -4 \end{vmatrix}$

2. $\begin{vmatrix} -8 & 6 \\ -1 & 2 \end{vmatrix}$

5. $\begin{vmatrix} -2 & -\sqrt{5} \\ -\sqrt{5} & 3 \end{vmatrix}$

6. $\begin{vmatrix} \sqrt{5} & -3 \\ 4 & 2 \end{vmatrix}$

3. $\begin{vmatrix} 4 & -7 \\ -2 & 3 \end{vmatrix}$

4. $\begin{vmatrix} -9 & -6 \\ 5 & 4 \end{vmatrix}$

7. $\begin{vmatrix} x & 4 \\ x & x^2 \end{vmatrix}$

8. $\begin{vmatrix} y^2 & -2 \\ y & 3 \end{vmatrix}$

Use the following matrix for Exercises 9–17:

$$A = \begin{bmatrix} 7 & -4 & -6 \\ 2 & 0 & -3 \\ 1 & 2 & -5 \end{bmatrix}.$$

9. Find M_{11}, M_{32}, and M_{22}.

10. Find M_{13}, M_{31}, and M_{23}.

11. Find A_{11}, A_{32}, and A_{22}.

12. Find A_{13}, A_{31}, and A_{23}.

13. Evaluate $|A|$ by expanding across the second row.

14. Evaluate $|A|$ by expanding down the second column.

15. Evaluate $|A|$ by expanding down the third column.

16. Evaluate $|A|$ by expanding across the first row.

17. Use a graphing calculator to evaluate $|A|$.

Use the following matrix for Exercises 18–24:

$$A = \begin{bmatrix} 1 & 0 & 0 & -2 \\ 4 & 1 & 0 & 0 \\ 5 & 6 & 7 & 8 \\ -2 & -3 & -1 & 0 \end{bmatrix}.$$

18. Find M_{12} and M_{44}.

19. Find M_{41} and M_{33}.

20. Find A_{22} and A_{34}.

21. Find A_{24} and A_{43}.

22. Evaluate $|A|$ by expanding down the third column.

23. Evaluate $|A|$ by expanding across the first row.

24. Use a graphing calculator to evaluate $|A|$.

Evaluate the determinant.

25. $\begin{vmatrix} 3 & 1 & 2 \\ -2 & 3 & 1 \\ 3 & 4 & -6 \end{vmatrix}$

26. $\begin{vmatrix} 3 & -2 & 1 \\ 2 & 4 & 3 \\ -1 & 5 & 1 \end{vmatrix}$

27. $\begin{vmatrix} x & 0 & -1 \\ 2 & x & x^2 \\ -3 & x & 1 \end{vmatrix}$

28. $\begin{vmatrix} x & 1 & -1 \\ x^2 & x & x \\ 0 & x & 1 \end{vmatrix}$

Solve using Cramer's rule.

29. $-2x + 4y = 3,$
$3x - 7y = 1$

30. $5x - 4y = -3,$
$7x + 2y = 6$

31. $2x - y = 5,$
$x - 2y = 1$

32. $3x + 4y = -2,$
$5x - 7y = 1$

33. $2x + 9y = -2,$
$4x - 3y = 3$

34. $2x + 3y = -1,$
$3x + 6y = -0.5$

35. $2x + 5y = 7,$
$3x - 2y = 1$

36. $3x + 2y = 7,$
$2x + 3y = -2$

37. $3x + 2y - z = 4,$
$3x - 2y + z = 5,$
$4x - 5y - z = -1$

38. $3x - y + 2z = 1,$
$x - y + 2z = 3,$
$-2x + 3y + z = 1$

39. $3x + 5y - z = -2,$
$x - 4y + 2z = 13,$
$2x + 4y + 3z = 1$

40. $3x + 2y + 2z = 1,$
$5x - y - 6z = 3,$
$2x + 3y + 3z = 4$

41. $x - 3y - 7z = 6,$
$2x + 3y + z = 9,$
$4x + y = 7$

42. $x - 2y - 3z = 4,$
$3x - 2z = 8,$
$2x + y + 4z = 13$

43. $6y + 6z = -1,$
$8x + 6z = -1,$
$4x + 9y = 8$

44. $3x + 5y = 2,$
$2x - 3z = 7,$
$4y + 2z = -1$

Collaborative Discussion and Writing

45. Given the system of equations

$$a_1 x + b_1 y = c_1,$$
$$a_2 x + b_2 y = c_2,$$

explain why the equations are dependent or the system is inconsistent when

$$\begin{vmatrix} a_1 & b_1 \\ a_2 & b_2 \end{vmatrix} = 0.$$

46. If the lines $a_1 x + b_1 y = c_1$ and $a_2 x + b_2 y = c_2$ are parallel, what can you say about the values of

$$\begin{vmatrix} a_1 & b_1 \\ a_2 & b_2 \end{vmatrix}, \quad \begin{vmatrix} c_1 & b_1 \\ c_2 & b_2 \end{vmatrix}, \quad \text{and} \quad \begin{vmatrix} a_1 & c_1 \\ a_2 & c_2 \end{vmatrix}?$$

Skill Maintenance

Determine whether the function is one-to-one, and if it is, find a formula for $f^{-1}(x)$.

47. $f(x) = 3x + 2$ **48.** $f(x) = x^2 - 4$

49. $f(x) = |x| + 3$ **50.** $f(x) = \sqrt[3]{x} + 1$

Simplify. Write answers in the form $a + bi$, where a and b are real numbers.

51. $(3 - 4i) - (-2 - i)$ **52.** $(5 + 2i) + (1 - 4i)$

53. $(1 - 2i)(6 + 2i)$ **54.** $\dfrac{3 + i}{4 - 3i}$

Synthesis

Solve.

55. $\begin{vmatrix} x & 5 \\ -4 & x \end{vmatrix} = 24$ **56.** $\begin{vmatrix} y & 2 \\ 3 & y \end{vmatrix} = y$

57. $\begin{vmatrix} x & -3 \\ -1 & x \end{vmatrix} \geq 0$ **58.** $\begin{vmatrix} y & -5 \\ -2 & y \end{vmatrix} < 0$

59. $\begin{vmatrix} x + 3 & 4 \\ x - 3 & 5 \end{vmatrix} = -7$

60. $\begin{vmatrix} m + 2 & -3 \\ m + 5 & -4 \end{vmatrix} = 3m - 5$

61. $\begin{vmatrix} 2 & x & 1 \\ 1 & 2 & -1 \\ 3 & 4 & -2 \end{vmatrix} = -6$

62. $\begin{vmatrix} x & 2 & x \\ 3 & -1 & 1 \\ 1 & -2 & 2 \end{vmatrix} = -10$

Rewrite the expression using a determinant. Answers may vary.

63. $2L + 2W$ **64.** $\pi r + \pi h$

65. $a^2 + b^2$ **66.** $\frac{1}{2}h(a + b)$

67. $2\pi r^2 + 2\pi rh$ **68.** $x^2y^2 - Q^2$

9.7 Systems of Inequalities and Linear Programming

❖ Graph linear inequalities.
❖ Graph systems of linear inequalities.
❖ Solve linear programming problems.

A graph of an inequality is a drawing that represents its solutions. We have already seen that an inequality in one variable can be graphed on a number line. An inequality in two variables can be graphed on a coordinate plane.

❖ Graphs of Linear Inequalities

A statement like $5x - 4y < 20$ is a linear inequality in two variables.

Linear Inequality in Two Variables

A **linear inequality in two variables** is an inequality that can be written in the form

$$Ax + By < C,$$

where A, B, and C are real numbers and A and B are not both zero. The symbol $<$ may be replaced with $\leq$, $>$, or $\geq$.

A solution of a linear inequality in two variables is an ordered pair (x, y) for which the inequality is true. For example, $(1, 3)$ is a solution of $5x - 4y < 20$ because $5 \cdot 1 - 4 \cdot 3 < 20$, or $-7 < 20$, is true. On the other hand, $(2, -6)$ is not a solution of $5x - 4y < 20$ because $5 \cdot 2 - 4 \cdot (-6) \not< 20$, or $34 \not< 20$.

The **solution set** of an inequality is the set of all ordered pairs that make it true. The **graph of an inequality** represents its solution set.

GCM **EXAMPLE 1**   Graph: $y < x + 3$.

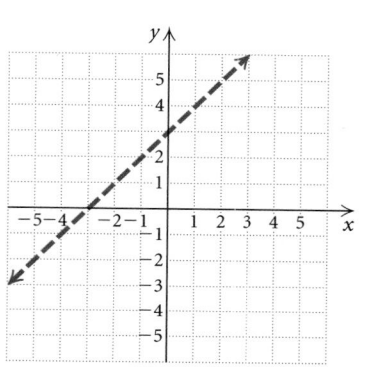

Solution   We begin by graphing the **related equation** $y = x + 3$. We use a dashed line because the inequality symbol is $<$. This indicates that the line itself is not in the solution set of the inequality.

Note that the line divides the coordinate plane into two regions called **half-planes**. One of these half-planes satisfies the inequality. Either *all* points in a half-plane are in the solution set of the inequality or *none* is.

To determine which half-plane satisfies the inequality, we try a test point in either region. The point $(0, 0)$ is usually a convenient choice so long as it does not lie on the line.

Since $(0, 0)$ satisfies the inequality, so do all points in the half-plane that contains $(0, 0)$. We shade this region to show the solution set of the inequality.

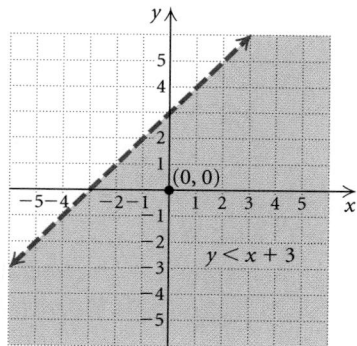

There are several ways to graph this inequality on a graphing calculator. One method is to first enter the related equation, $y = x + 3$. Then, after using a test point to determine which half-plane to shade as described above, we select the "shade below" graph style. Note that we must keep in mind that the line $y = x + 3$ is not included in the solution set. (Even if

DOT mode or the dot graph style is selected, the line appears to be solid rather than dashed.)

$$y = x + 3$$

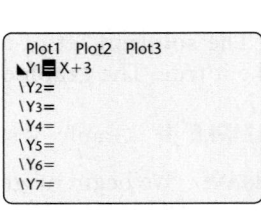
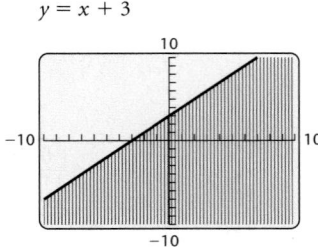

We could also use the Shade option from the DRAW menu to graph this inequality. This method is described in the *Graphing Calculator Manual* that accompanies the text.

Some calculators have a pre-loaded application that can be used to graph an inequality. This application, Inequalz, is found on the APPS menu. The keystrokes for using Inequalz are found in the *Graphing Calculator Manual* that accompanies the text. Note that when this application is used, the inequality $y < x + 3$ is entered directly and the graph of the related equation appears as a dashed line.

$$y < x + 3$$

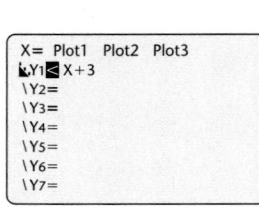
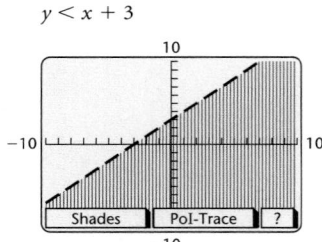

Now Try Exercise 13. ■

In general, we use the following procedure to graph linear inequalities in two variables by hand.

To graph a linear inequality in two variables:

1. Replace the inequality symbol with an equals sign and graph this related equation. If the inequality symbol is $<$ or $>$, draw the line dashed. If the inequality symbol is $\leq$ or $\geq$, draw the line solid.
2. The graph consists of a half-plane on one side of the line and, if the line is solid, the line as well. To determine which half-plane to shade, test a point not on the line in the original inequality. If that point is a solution, shade the half-plane containing that point. If not, shade the opposite half-plane.

EXAMPLE 2 Graph: $3x + 4y \geq 12$.

Solution

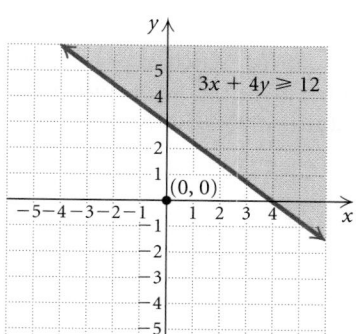

1. First, we graph the related equation $3x + 4y = 12$. We use a solid line because the inequality symbol is $\geq$. This indicates that the line is included in the solution set.

2. To determine which half-plane to shade, we test a point in either region. We choose $(0, 0)$.

$$
\begin{array}{c}
3x + 4y \geq 12 \\
\hline
3 \cdot 0 + 4 \cdot 0 \ ? \ 12 \\
0 \mid 12 \quad \text{FALSE} \qquad 0 \geq 12 \text{ is false.}
\end{array}
$$

Because $(0, 0)$ is *not* a solution, all the points in the half-plane that does *not* contain $(0, 0)$ are solutions. We shade that region, as shown in the figure at left.

To graph this inequality on a graphing calculator using either the "shade above" graph style or the Shade option from the DRAW menu, we must first solve the related equation for y and enter $y = \dfrac{-3x + 12}{4}$.

Similarly, to use the Inequalz application, we solve the inequality for y and enter $y \geq \dfrac{-3x + 12}{4}$.

Now Try Exercise 17. ▨

GCM **EXAMPLE 3** Graph $x > -3$ on a plane.

Solution

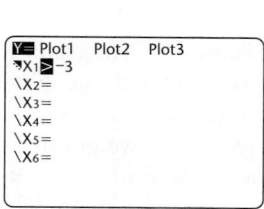

1. First, we graph the related equation $x = -3$. We use a dashed line because the inequality symbol is $>$. This indicates that the line is not included in the solution set.

2. The inequality tells us that all points (x, y) for which $x > -3$ are solutions. These are the points to the right of the line. We can also use a test point to determine the solutions. We choose $(5, 1)$.

$$
\begin{array}{c}
x > -3 \\
\hline
5 \ ? \ -3 \quad \text{TRUE} \qquad 5 > -3 \text{ is true.}
\end{array}
$$

Because $(5, 1)$ is a solution, we shade the region containing that point— that is, the region to the right of the dashed line.

We can also graph this inequality on a graphing calculator that has the Inequalz application on the APPS menu.

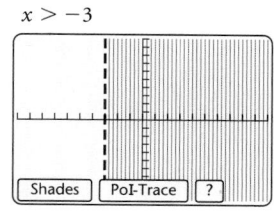

Now Try Exercise 23. ▨

EXAMPLE 4   Graph $y \leq 4$ on a plane.

Solution

1. First, we graph the related equation $y = 4$. We use a solid line because the inequality symbol is $\leq$.
2. The inequality tells us that all points (x, y) for which $y \leq 4$ are solutions of the inequality. These are the points on or below the line. We can also use a test point to determine the solutions. We choose $(-2, 5)$.

$$y \leq 4$$
$$\overline{}$$
$$5 \; ? \; 4 \quad \textsf{FALSE} \qquad 5 \leq 4 \text{ is false.}$$

Because $(-2, 5)$ is not a solution, we shade the half-plane that does not contain that point.

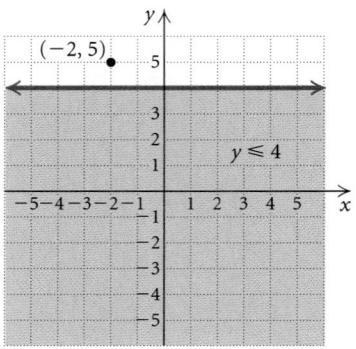

Now Try Exercise 25. ■

❋ Systems of Linear Inequalities

A system of inequalities consists of two or more inequalities considered simultaneously. For example,

$$x + y \leq 4,$$
$$x - y \geq 2$$

is a system of *two linear inequalities in two variables*.

A solution of a system of inequalities is an ordered pair that is a solution of each inequality in the system. To graph a system of linear inequalities, we graph each inequality and determine the region that is common to *all* the solution sets.

GCM   **EXAMPLE 5**   Graph the solution set of the system

$$x + y \leq 4,$$
$$x - y \geq 2.$$

Solution   We graph $x + y \leq 4$ by first graphing the equation $x + y = 4$ using a solid line. Next, we choose $(0, 0)$ as a test point and find that it is a solution of $x + y \leq 4$, so we shade the half-plane containing $(0, 0)$ using red. Next, we graph $x - y = 2$ using a solid line. We find that $(0, 0)$ is not a solution of $x - y \geq 2$, so we shade the half-plane that does not contain $(0, 0)$ using green. The arrows near the ends of each line help to indicate the half-plane that contains each solution set.

The solution set of the system of equations is the region shaded both red and green, or brown, including parts of the lines $x + y = 4$ and $x - y = 2$.

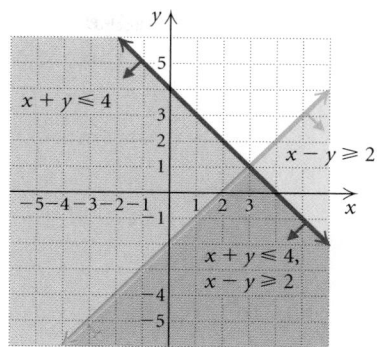

We can use different shading patterns on a graphing calculator to graph this system of inequalities. The solution set is the region shaded using both patterns.

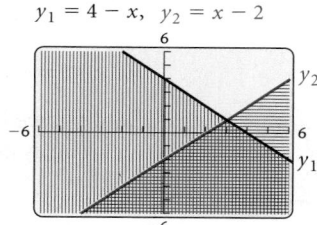

We can also use the Inequalz application to graph this system of inequalities. If we choose the Ineq Intersection option from the SHADES menu, only the solution set is shaded, as shown in the figure below.

$$y_1 \leq 4 - x, \quad y_2 \leq x - 2$$

A system of inequalities may have a graph that consists of a polygon and its interior. As we will see later in this section, in many applications we will need to know the vertices of such a polygon.

GCM **EXAMPLE 6** Graph the following system of inequalities and find the coordinates of any vertices formed:

$$3x - y \leq 6, \quad (1)$$
$$y - 3 \leq 0, \quad (2)$$
$$x + y \geq 0. \quad (3)$$

Solution   We graph the related equations $3x - y = 6$, $y - 3 = 0$, and $x + y = 0$ using solid lines. The half-plane containing the solution set for each inequality is indicated by the arrows near the ends of each line. We shade the region common to all three solution sets.

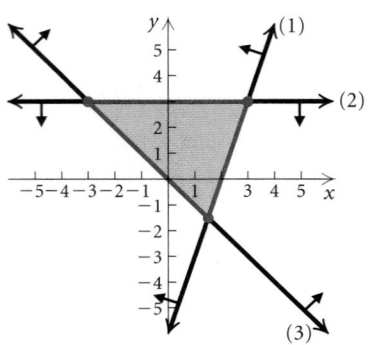

To find the vertices, we solve three systems of equations. The system of equations from inequalities (1) and (2) is

$$3x - y = 6,$$
$$y - 3 = 0.$$

Solving, we obtain the vertex $(3, 3)$.

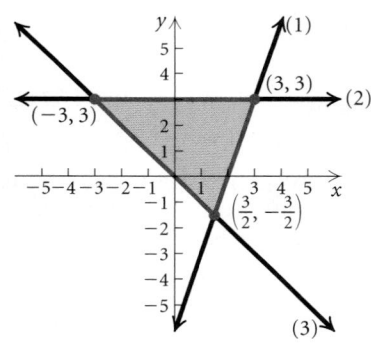

The system of equations from inequalities (1) and (3) is

$$3x - y = 6,$$
$$x + y = 0.$$

Solving, we obtain the vertex $\left(\frac{3}{2}, -\frac{3}{2}\right)$.
The system of equations from inequalities (2) and (3) is

$$y - 3 = 0,$$
$$x + y = 0.$$

Solving, we obtain the vertex $(-3, 3)$.

If a system of inequalities is graphed on a graphing calculator using a shading option, the coordinates of the vertices can be found using the INTERSECT feature. If the Inequalz application from the APPS menu is used to graph a system of inequalities, the PoI-Trace feature can be used to find the coordinates of the vertices. This method is covered in the *Graphing Calcula-tor Manual* that accompanies the text.   **Now Try Exercise 55.** ▪

❈ Applications: Linear Programming

In many applications, we want to find a maximum value or a minimum value. In business, for example, we might want to maximize profit and min-imize cost. **Linear programming** can tell us how to do this.

In our study of linear programming, we will consider linear functions of two variables that are to be maximized or minimized subject to several conditions, or **constraints**. These constraints are expressed as inequalities. The solution set of the system of inequalities made up of the constraints contains all the **feasible solutions** of a linear programming problem. The function that we want to maximize or minimize is called the **objective function**.

It can be shown that the maximum and minimum values of the objective function occur at a vertex of the region of feasible solutions. Thus we have the following procedure.

> **Linear Programming Procedure**
>
> To find the maximum or minimum value of a linear objective function subject to a set of constraints:
>
> **1.** Graph the region of feasible solutions.
> **2.** Determine the coordinates of the vertices of the region.
> **3.** Evaluate the objective function at each vertex. The largest and smallest of those values are the maximum and minimum values of the function, respectively.

EXAMPLE 7 *Maximizing Profit.* Dovetail Carpentry Shop makes bookcases and desks. Each bookcase requires 5 hr of woodworking and 4 hr of finishing. Each desk requires 10 hr of woodworking and 3 hr of finishing. Each month the shop has 600 hr of labor available for woodworking and 240 hr for finishing. The profit on each bookcase is $40 and on each desk is $75. How many of each product should be made each month in order to maximize profit?

Solution We let x = the number of bookcases to be produced and y = the number of desks. Then the profit P is given by the function

$$P = 40x + 75y.$$ To emphasize that P is a function of two variables, we sometimes write $P(x, y) = 40x + 75y$.

We know that x bookcases require $5x$ hr of woodworking and y desks require $10y$ hr of woodworking. Since there is no more than 600 hr of labor available for woodworking, we have one constraint:

$$5x + 10y \leq 600.$$

Similarly, the bookcases and desks require $4x$ hr and $3y$ hr of finishing, respectively. There is no more than 240 hr of labor available for finishing, so we have a second constraint:

$$4x + 3y \leq 240.$$

We also know that $x \geq 0$ and $y \geq 0$ because the carpentry shop cannot make a negative number of either product.

Thus we want to maximize the objective function

$$P = 40x + 75y$$

subject to the constraints

$$5x + 10y \leq 600,$$
$$4x + 3y \leq 240,$$
$$x \geq 0,$$
$$y \geq 0.$$

We graph the system of inequalities and determine the vertices.

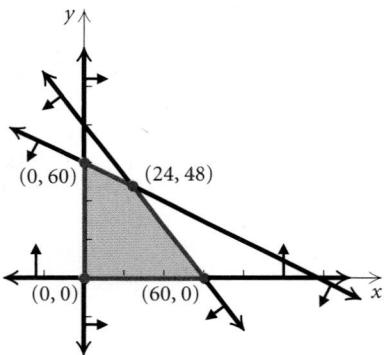

Next, we evaluate the objective function P at each vertex.

Vertices (x, y)	Profit $P = 40x + 75y$	
$(0, 0)$	$P = 40 \cdot 0 + 75 \cdot 0 = 0$	
$(60, 0)$	$P = 40 \cdot 60 + 75 \cdot 0 = 2400$	
$(24, 48)$	$P = 40 \cdot 24 + 75 \cdot 48 = 4560$	⟵ Maximum
$(0, 60)$	$P = 40 \cdot 0 + 75 \cdot 60 = 4500$	

The carpentry shop will make a maximum profit of $4560 when 24 bookcases and 48 desks are produced.   **Now Try Exercise 65.** ∎

(9.7) Exercise Set

In Exercises 1–8, match the inequality with one of the graphs (a)–(h), which follow.

a)

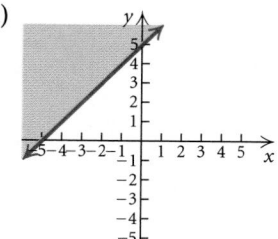

b)

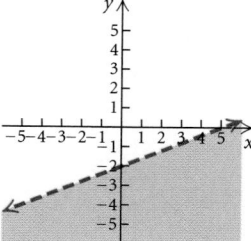

c)

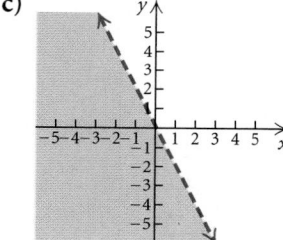

d)

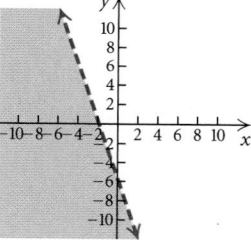

e)

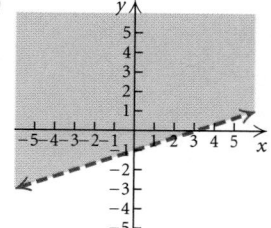

f)

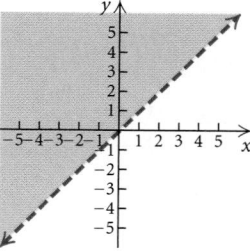

g)

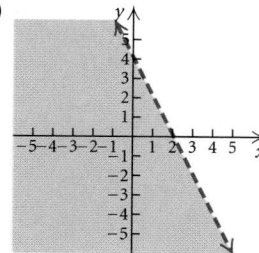

h)

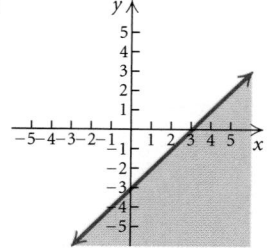

1. $y > x$

2. $y < -2x$

3. $y \leq x - 3$

4. $y \geq x + 5$

5. $2x + y < 4$

6. $3x + y < -6$

7. $2x - 5y > 10$

8. $3x - 9y < 9$

Graph.

9. $y > 2x$

10. $2y < x$

11. $y + x \geq 0$

12. $y - x < 0$

13. $y > x - 3$

14. $y \leq x + 4$

15. $x + y < 4$

16. $x - y \geq 5$

17. $3x - 2y \leq 6$

18. $2x - 5y < 10$

19. $3y + 2x \geq 6$

20. $2y + x \leq 4$

21. $3x - 2 \leq 5x + y$

22. $2x - 6y \geq 8 + 2y$

23. $x < -4$

24. $y > -3$

25. $y \geq 5$

26. $x \leq 5$

27. $-4 < y < -1$
   (*Hint*: Think of this as $-4 < y$ and $y < -1$.)

28. $-3 \leq x \leq 3$
   (*Hint*: Think of this as $-3 \leq x$ and $x \leq 3$.)

29. $y \geq |x|$

30. $y \leq |x + 2|$

In Exercises 31–36, match the system of inequalities with one of the graphs (a)–(f), which follow.

a)
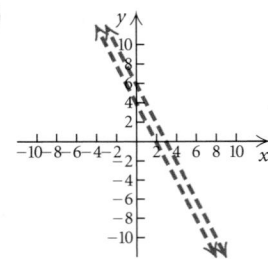

b)
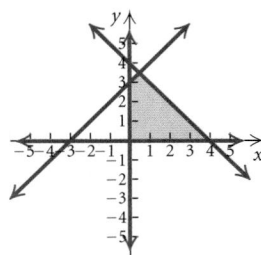

c)
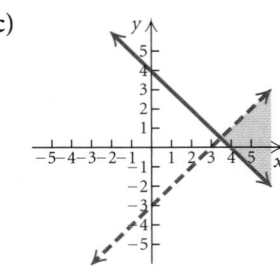

d)
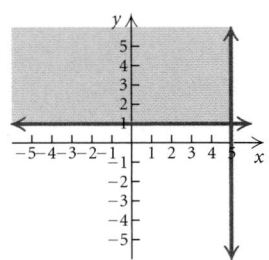

e)
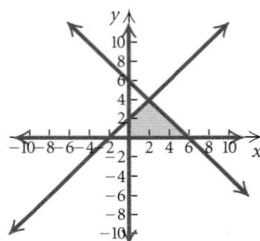

f)
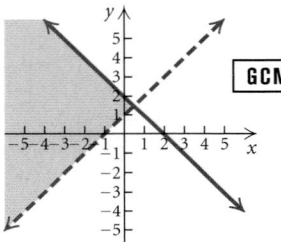

31. $y > x + 1,$
$y \leq 2 - x$

32. $y < x - 3,$
$y \geq 4 - x$

33. $2x + y < 4,$
$4x + 2y > 12$

34. $x \leq 5,$
$y \geq 1$

35. $x + y \leq 4,$
$x - y \geq -3,$
$x \geq 0,$
$y \geq 0$

36. $x - y \geq -2,$
$x + y \leq 6,$
$x \geq 0,$
$y \geq 0$

Find a system of inequalities with the given graph. Answers may vary.

37.
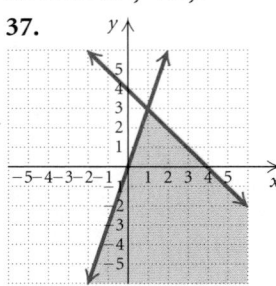

38.
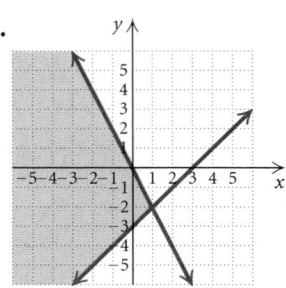

39.
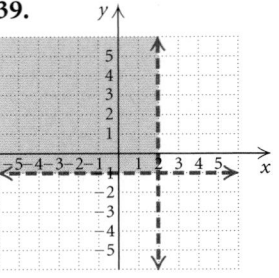

40.
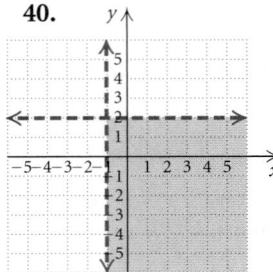

41.
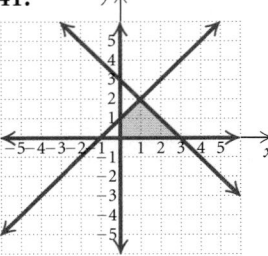

42.
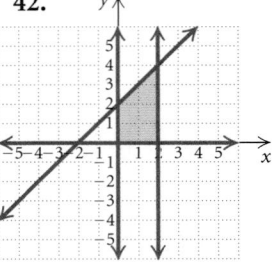

Graph the system of inequalities. Then find the coordinates of the vertices.

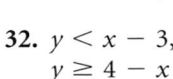

43. $y \leq x,$
$y \geq 3 - x$

44. $y \leq x,$
$y \geq 5 - x$

45. $y \geq x,$
$y \leq 4 - x$

46. $y \geq x,$
$y \leq 2 - x$

47. $y \geq -3,$
$x \geq 1$

48. $y \leq -2,$
$x \geq 2$

49. $x \leq 3,$
$y \geq 2 - 3x$

50. $x \geq -2,$
$y \leq 3 - 2x$

51. $x + y \leq 1,$
$x - y \leq 2$

52. $y + 3x \geq 0,$
$y + 3x \leq 2$

53. $2y - x \leq 2,$
$y + 3x \geq -1$

54. $y \leq 2x + 1,$
$y \geq -2x + 1,$
$x - 2 \leq 0$

55. $x - y \leq 2,$
$x + 2y \geq 8,$
$y - 4 \leq 0$

56. $x + 2y \leq 12,$
$2x + y \leq 12,$
$x \geq 0,$
$y \geq 0$

57. $4y - 3x \geq -12,$
$4y + 3x \geq -36,$
$y \leq 0,$
$x \leq 0$

58. $8x + 5y \leq 40,$
$x + 2y \leq 8,$
$x \geq 0,$
$y \geq 0$

59. $3x + 4y \geq 12,$
$5x + 6y \leq 30,$
$1 \leq x \leq 3$

60. $y - x \geq 1,$
$y - x \leq 3,$
$2 \leq x \leq 5$

Find the maximum value and the minimum value of the function and the values of x and y for which they occur.

61. $P = 17x - 3y + 60$, subject to

$$6x + 8y \le 48,$$
$$0 \le y \le 4,$$
$$0 \le x \le 7.$$

62. $Q = 28x - 4y + 72$, subject to

$$5x + 4y \ge 20,$$
$$0 \le y \le 4,$$
$$0 \le x \le 3.$$

63. $F = 5x + 36y$, subject to

$$5x + 3y \le 34,$$
$$3x + 5y \le 30,$$
$$x \ge 0,$$
$$y \ge 0.$$

64. $G = 16x + 14y$, subject to

$$3x + 2y \le 12,$$
$$7x + 5y \le 29,$$
$$x \ge 0,$$
$$y \ge 0.$$

65. *Maximizing Income.* Golden Harvest Foods makes jumbo biscuits and regular biscuits. The oven can cook at most 200 biscuits per day. Each jumbo biscuit requires 2 oz of flour, each regular biscuit requires 1 oz of flour, and there is 300 oz of flour available. The income from each jumbo biscuit is $0.10 and from each regular biscuit is $0.08. How many of each size biscuit should be made in order to maximize income? What is the maximum income?

66. *Maximizing Mileage.* Omar owns a car and a moped. He can afford 12 gal of gasoline to be split between the car and the moped. Omar's car gets 20 mpg and, with the fuel currently in the tank, can hold at most an additional 10 gal of gas. His moped gets 100 mpg and can hold at most 3 gal of gas. How many gallons of gasoline should each vehicle use if Omar wants to travel as far as

possible on the 12 gal of gas? What is the maximum number of miles that he can travel?

20 mpg

100 mpg

67. *Maximizing Profit.* Norris Mill can convert logs into lumber and plywood. In a given week, the mill can turn out 400 units of production, of which 100 units of lumber and 150 units of plywood are required by regular customers. The profit is $20 per unit of lumber and $30 per unit of plywood. How many units of each should the mill produce in order to maximize the profit?

68. *Maximizing Profit.* Sunnydale Farm includes 240 acres of cropland. The farm owner wishes to plant this acreage in corn and oats. The profit per acre in corn production is $40 and in oats is $30. A total of 320 hr of labor is available. Each acre of corn requires 2 hr of labor, whereas each acre of oats requires 1 hr of labor. How should the land be divided between corn and oats in order to yield the maximum profit? What is the maximum profit?

69. *Minimizing Cost.* An animal feed to be mixed from soybean meal and oats must contain at least 120 lb of protein, 24 lb of fat, and 10 lb of mineral ash. Each 100-lb sack of soybean meal costs $15 and contains 50 lb of protein, 8 lb of fat, and 5 lb of mineral ash. Each 100-lb sack of oats costs $5 and contains 15 lb of protein, 5 lb of fat, and 1 lb of mineral ash. How many sacks of each should be used to satisfy the minimum requirements at minimum cost?

70. *Minimizing Cost.* Suppose that in the preceding problem the oats were replaced by alfalfa, which costs $8 per 100 lb and contains 20 lb of protein, 6 lb of fat, and 8 lb of mineral ash. How much of each is now required in order to minimize the cost?

71. *Maximizing Income.*   Clayton is planning to invest up to $40,000 in corporate and municipal bonds. The least he is allowed to invest in corporate bonds is $6000, and he does not want to invest more than $22,000 in corporate bonds. He also does not want to invest more than $30,000 in municipal bonds. The interest is 8% on corporate bonds and $7\frac{1}{2}$% on municipal bonds. This is simple interest for one year. How much should he invest in each type of bond in order to maximize his income? What is the maximum income?

72. *Maximizing Income.*   Margaret is planning to invest up to $22,000 in certificates of deposit at City Bank and People's Bank. She wants to invest at least $2000 but no more than $14,000 at City Bank. People's Bank does not insure more than a $15,000 investment, so she will invest no more than that in People's Bank. The interest is 6% at City Bank and $6\frac{1}{2}$% at People's Bank. This is simple interest for one year. How much should she invest in each bank in order to maximize her income? What is the maximum income?

73. *Minimizing Transportation Cost.*   An airline with two types of airplanes, P_1 and P_2, has contracted with a tour group to provide transportation for a minimum of 2000 first-class, 1500 tourist-class, and 2400 economy-class passengers. For a certain trip, airplane P_1 costs $12 thousand to operate and can accommodate 40 first-class, 40 tourist-class, and 120 economy-class passengers, whereas airplane P_2 costs $10 thousand to operate and can accommodate 80 first-class, 30 tourist-class, and 40 economy-class passengers. How many of each type of airplane should be used in order to minimize the operating cost?

74. *Minimizing Transportation Cost.*   Suppose that in the preceding problem a new airplane P_3 becomes available, having an operating cost for the same trip of $15 thousand and accommodating 40 first-class, 40 tourist-class, and 80 economy-class passengers. If airplane P_1 were replaced by airplane P_3, how many of P_2 and P_3 should be used in order to minimize the operating cost?

75. *Maximizing Profit.*   It takes Just Sew 2 hr of cutting and 4 hr of sewing to make a knit suit. It takes 4 hr of cutting and 2 hr of sewing to make a worsted suit. At most 20 hr per day are available for cutting and at most 16 hr per day are available for sewing. The profit is $34 on a knit suit and $31 on a worsted suit. How many of each kind of suit should be made each day in order to maximize profit? What is the maximum profit?

76. *Maximizing Profit.*   Cambridge Metal Works manufactures two sizes of gears. The smaller gear requires 4 hr of machining and 1 hr of polishing and yields a profit of $25. The larger gear requires 1 hr of machining and 1 hr of polishing and yields a profit of $10. The firm has available at most 24 hr per day for machining and 9 hr per day for polishing. How many of each type of gear should be produced each day in order to maximize profit? What is the maximum profit?

77. *Minimizing Nutrition Cost.*   Suppose that it takes 12 units of carbohydrates and 6 units of protein to satisfy Jacob's minimum weekly requirements. A particular type of meat contains 2 units of carbohydrates and 2 units of protein per pound. A particular cheese contains 3 units of carbohydrates and 1 unit of protein per pound. The meat costs $3.50 per pound and the cheese costs $4.60 per pound. How many pounds of each are needed in order to minimize the cost and still meet the minimum requirements?

78. *Minimizing Salary Cost.* The Spring Hill school board is analyzing education costs for Hill Top School. It wants to hire teachers and teacher's aides to make up a faculty that satisfies its needs at minimum cost. The average annual salary for a teacher is $35,000 and for a teacher's aide is $18,000. The school building can accommodate a faculty of no more than 50 but needs at least 20 faculty members to function properly. The school must have at least 12 aides, but the number of teachers must be at least twice the number of aides in order to accommodate the expectations of the community. How many teachers and teacher's aides should be hired in order to minimize salary costs?

79. *Maximizing Animal Support in a Forest.* A certain area of forest is populated by two species of animal, which scientists refer to as A and B for simplicity. The forest supplies two kinds of food, referred to as F_1 and F_2. For one year, each member of species A requires 1 unit of F_1 and 0.5 unit of F_2. Each member of species B requires 0.2 unit of F_1 and 1 unit of F_2. The forest can normally supply at most 600 units of F_1 and 525 units of F_2 per year. What is the maximum total number of these animals that the forest can support?

80. *Maximizing Animal Support in a Forest.* Refer to Exercise 79. If there is a wet spring, then supplies of food increase to 1080 units of F_1 and 810 units of F_2. In this case, what is the maximum total number of these animals that the forest can support?

Collaborative Discussion and Writing

81. Write an applied linear programming problem for a classmate to solve. Devise your problem so that the answer is "The bakery will make a maximum profit when 5 dozen pies and 12 dozen cookies are baked."

82. Describe how the graph of a linear inequality differs from the graph of a linear equation.

Skill Maintenance

Solve.

83. $-5 \leq x + 2 < 4$

84. $|x - 3| \geq 2$

85. $x^2 - 2x \leq 3$

86. $\dfrac{x - 1}{x + 2} > 4$

Synthesis

Graph the system of inequalities.

87. $y \geq x^2 - 2,$
$\quad y \leq 2 - x^2$

88. $y < x + 1,$
$\quad y \geq x^2$

Graph the inequality.

89. $|x + y| \leq 1$

90. $|x| + |y| \leq 1$

91. $|x| > |y|$

92. $|x - y| > 0$

93. *Allocation of Resources.* Significant Sounds manufactures two types of speaker assemblies. The less expensive assembly, which sells for $350, consists of one midrange speaker and one tweeter. The more expensive speaker assembly, which sells for $600, consists of one woofer, one midrange speaker, and two tweeters. The manufacturer has in stock 44 woofers, 60 midrange speakers, and 90 tweeters. How many of each type of speaker assembly should be made in order to maximize income? What is the maximum income?

94. *Allocation of Resources.* Sitting Pretty Furniture produces chairs and sofas. Each chair requires 20 ft of wood, 1 lb of foam rubber, and 2 yd^2 of fabric. Each sofa requires 100 ft of wood, 50 lb of foam rubber, and 20 yd^2 of fabric. The manufacturer has in stock 1900 ft of wood, 500 lb of foam rubber, and 240 yd^2 of fabric. The chairs can be sold for $80 each and the sofas for $300 each. How many of each should be produced in order to maximize income? What is the maximum income?

9.8 ❖ Decompose rational expressions into partial fractions.

Partial Fractions

There are situations in calculus in which it is useful to write a rational expression as a sum of two or more simpler rational expressions. In the equation

$$\frac{4x - 13}{2x^2 + x - 6} = \frac{3}{x + 2} + \frac{-2}{2x - 3},$$

each fraction on the right side is called a **partial fraction**. The expression on the right side is the **partial fraction decomposition** of the rational expression on the left side. In this section, we learn how such decompositions are created.

❖ Partial Fraction Decompositions

The procedure for finding the partial fraction decomposition of a rational expression involves factoring its denominator into linear and quadratic factors.

Procedure for Decomposing a Rational Expression into Partial Fractions

Consider any rational expression $P(x)/Q(x)$ such that $P(x)$ and $Q(x)$ have no common factor other than 1 or -1.

1. If the degree of $P(x)$ is greater than or equal to the degree of $Q(x)$, divide to express $P(x)/Q(x)$ as a quotient $+$ remainder$/Q(x)$ and follow steps (2)–(5) to decompose the resulting rational expression.
2. If the degree of $P(x)$ is less than the degree of $Q(x)$, factor $Q(x)$ into linear factors of the form $(px + q)^n$ and/or quadratic factors of the form $(ax^2 + bx + c)^m$. Any quadratic factor $ax^2 + bx + c$ must be *irreducible*, meaning that it cannot be factored into linear factors with rational coefficients.
3. Assign to each linear factor $(px + q)^n$ the sum of n partial fractions:

$$\frac{A_1}{px + q} + \frac{A_2}{(px + q)^2} + \cdots + \frac{A_n}{(px + q)^n}.$$

4. Assign to each quadratic factor $(ax^2 + bx + c)^m$ the sum of m partial fractions:

$$\frac{B_1 x + C_1}{ax^2 + bx + c} + \frac{B_2 x + C_2}{(ax^2 + bx + c)^2} + \cdots + \frac{B_m x + C_m}{(ax^2 + bx + c)^m}.$$

5. Apply algebraic methods, as illustrated in the following examples, to find the constants in the numerators of the partial fractions.

EXAMPLE 1 Decompose into partial fractions:

$$\frac{4x - 13}{2x^2 + x - 6}.$$

Solution The degree of the numerator is less than the degree of the denominator. We begin by factoring the denominator: $(x + 2)(2x - 3)$. We find constants A and B such that

$$\frac{4x - 13}{(x + 2)(2x - 3)} = \frac{A}{x + 2} + \frac{B}{2x - 3}.$$

To determine A and B, we add the expressions on the right:

$$\frac{4x - 13}{(x + 2)(2x - 3)} = \frac{A(2x - 3) + B(x + 2)}{(x + 2)(2x - 3)}.$$

Next, we equate the numerators:

$$4x - 13 = A(2x - 3) + B(x + 2).$$

Since the last equation containing A and B is true for all x, we can substitute any value of x and still have a true equation. If we choose $x = \frac{3}{2}$, then $2x - 3 = 0$ and A will be eliminated when we make the substitution. This gives us

$$4\left(\tfrac{3}{2}\right) - 13 = A\left(2 \cdot \tfrac{3}{2} - 3\right) + B\left(\tfrac{3}{2} + 2\right)$$
$$-7 = 0 + \tfrac{7}{2}B.$$

Solving, we obtain $B = -2$.

If we choose $x = -2$, then $x + 2 = 0$ and B will be eliminated when we make the substitution. This gives us

$$4(-2) - 13 = A[2(-2) - 3] + B(-2 + 2)$$
$$-21 = -7A + 0.$$

Solving, we obtain $A = 3$.

The decomposition is as follows:

$$\frac{4x - 13}{2x^2 + x - 6} = \frac{3}{x + 2} + \frac{-2}{2x - 3}, \quad \text{or} \quad \frac{3}{x + 2} - \frac{2}{2x - 3}.$$

To check, we can add to see if we get the expression on the left. We can also use the TABLE feature on a graphing calculator, comparing values of

$$y_1 = \frac{4x - 13}{2x^2 + x - 6} \quad \text{and} \quad y_2 = \frac{3}{x + 2} - \frac{2}{2x - 3}$$

for the same values of x. Since $y_1 = y_2$ for the given values of x as we scroll through the table, the decomposition appears to be correct.

X	Y1	Y2
−1	3.4	3.4
0	2.1667	2.1667
1	3	3
2	−1.25	−1.25
3	−.0667	−.0667
4	.1	.1
5	.14286	.14286
X = −1		

Now Try Exercise 3. ∎

EXAMPLE 2   Decompose into partial fractions:

$$\frac{7x^2 - 29x + 24}{(2x - 1)(x - 2)^2}.$$

Solution   The degree of the numerator is 2 and the degree of the denominator is 3, so the degree of the numerator is less than the degree of the denominator. The denominator is given in factored form. The decomposition has the following form:

$$\frac{7x^2 - 29x + 24}{(2x - 1)(x - 2)^2} = \frac{A}{2x - 1} + \frac{B}{x - 2} + \frac{C}{(x - 2)^2}.$$

As in Example 1, we add the expressions on the right:

$$\frac{7x^2 - 29x + 24}{(2x - 1)(x - 2)^2} = \frac{A(x - 2)^2 + B(2x - 1)(x - 2) + C(2x - 1)}{(2x - 1)(x - 2)^2}.$$

Then we equate the numerators. This gives us

$$7x^2 - 29x + 24 = A(x - 2)^2 + B(2x - 1)(x - 2) + C(2x - 1).$$

Since the equation containing A, B, and C is true for all x, we can substitute any value of x and still have a true equation. In order to have $2x - 1 = 0$, we let $x = \frac{1}{2}$. This gives us

$$7\left(\tfrac{1}{2}\right)^2 - 29 \cdot \tfrac{1}{2} + 24 = A\left(\tfrac{1}{2} - 2\right)^2 + 0 + 0$$
$$\tfrac{45}{4} = \tfrac{9}{4}A.$$

Solving, we obtain $A = 5$.

In order to have $x - 2 = 0$, we let $x = 2$. Substituting gives us

$$7(2)^2 - 29(2) + 24 = 0 + 0 + C(2 \cdot 2 - 1)$$
$$-6 = 3C.$$

Solving, we obtain $C = -2$.

To find B, we choose any value for x except $\frac{1}{2}$ or 2 and replace A with 5 and C with -2. We let $x = 1$:

$$7 \cdot 1^2 - 29 \cdot 1 + 24 = 5(1 - 2)^2 + B(2 \cdot 1 - 1)(1 - 2)$$
$$+ (-2)(2 \cdot 1 - 1)$$
$$2 = 5 - B - 2$$
$$B = 1.$$

The decomposition is as follows:

$$\frac{7x^2 - 29x + 24}{(2x - 1)(x - 2)^2} = \frac{5}{2x - 1} + \frac{1}{x - 2} - \frac{2}{(x - 2)^2}.$$

We can check the result using a table of values. We let

$$y_1 = \frac{7x^2 - 29x + 24}{(2x - 1)(x - 2)^2} \quad \text{and} \quad y_2 = \frac{5}{2x - 1} + \frac{1}{x - 2} - \frac{2}{(x - 2)^2}.$$

Since $y_1 = y_2$ for given values of x as we scroll through the table, the decomposition appears to be correct.   **Now Try Exercise 7.** ■

X	Y1	Y2
−5	−.6382	−.6382
−4	−.7778	−.7778
−3	−.9943	−.9943
−2	−1.375	−1.375
−1	−2.222	−2.222
0	−6	−6
1	2	2

X = −5

EXAMPLE 3 Decompose into partial fractions:

$$\frac{6x^3 + 5x^2 - 7}{3x^2 - 2x - 1}.$$

Solution The degree of the numerator is greater than that of the denominator. Thus we divide and find an equivalent expression:

$$
\begin{array}{r}
2x + 3 \\
3x^2 - 2x - 1 \overline{)6x^3 + 5x^2 - 7} \\
\underline{6x^3 - 4x^2 - 2x} \\
9x^2 + 2x - 7 \\
\underline{9x^2 - 6x - 3} \\
8x - 4
\end{array}
$$

The original expression is thus equivalent to

$$2x + 3 + \frac{8x - 4}{3x^2 - 2x - 1}.$$

We decompose the fraction to get

$$\frac{8x - 4}{(3x + 1)(x - 1)} = \frac{5}{3x + 1} + \frac{1}{x - 1}.$$

The final result is

$$2x + 3 + \frac{5}{3x + 1} + \frac{1}{x - 1}.$$

Now Try Exercise 17. ■

Systems of equations can be used to decompose rational expressions. Let's reconsider Example 2.

EXAMPLE 4 Decompose into partial fractions:

$$\frac{7x^2 - 29x + 24}{(2x - 1)(x - 2)^2}.$$

Solution The decomposition has the following form:

$$\frac{A}{2x - 1} + \frac{B}{x - 2} + \frac{C}{(x - 2)^2}.$$

We first add as in Example 2:

$$\frac{7x^2 - 29x + 24}{(2x - 1)(x - 2)^2} = \frac{A}{2x - 1} + \frac{B}{x - 2} + \frac{C}{(x - 2)^2}$$

$$= \frac{A(x - 2)^2 + B(2x - 1)(x - 2) + C(2x - 1)}{(2x - 1)(x - 2)^2}.$$

Then we equate numerators:

$$7x^2 - 29x + 24$$
$$= A(x - 2)^2 + B(2x - 1)(x - 2) + C(2x - 1)$$
$$= A(x^2 - 4x + 4) + B(2x^2 - 5x + 2) + C(2x - 1)$$
$$= Ax^2 - 4Ax + 4A + 2Bx^2 - 5Bx + 2B + 2Cx - C,$$

or, combining like terms,

$$7x^2 - 29x + 24$$
$$= (A + 2B)x^2 + (-4A - 5B + 2C)x + (4A + 2B - C).$$

Next, we equate corresponding coefficients:

$7 = A + 2B,$	**The coefficients of the x^2-terms must be the same.**
$-29 = -4A - 5B + 2C,$	**The coefficients of the x-terms must be the same.**
$24 = 4A + 2B - C.$	**The constant terms must be the same.**

We now have a system of three equations. You should confirm that the solution of the system is

$$A = 5, \quad B = 1, \quad \text{and} \quad C = -2.$$

The decomposition is as follows:

$$\frac{7x^2 - 29x + 24}{(2x - 1)(x - 2)^2} = \frac{5}{2x - 1} + \frac{1}{x - 2} - \frac{2}{(x - 2)^2}.$$

Now Try Exercise 15. ■

SYSTEMS OF EQUATIONS IN THREE VARIABLES

REVIEW SECTION **9.2** OR **9.5.**

STUDY TIP

The review icons in the text margins provide references to earlier sections in which you can find content related to the concept at hand. Reviewing this earlier content will add to your understanding of the current concept.

EXAMPLE 5 Decompose into partial fractions:

$$\frac{11x^2 - 8x - 7}{(2x^2 - 1)(x - 3)}.$$

Solution The decomposition has the following form:

$$\frac{11x^2 - 8x - 7}{(2x^2 - 1)(x - 3)} = \frac{Ax + B}{2x^2 - 1} + \frac{C}{x - 3}.$$

Adding and equating numerators, we get

$$11x^2 - 8x - 7 = (Ax + B)(x - 3) + C(2x^2 - 1)$$
$$= Ax^2 - 3Ax + Bx - 3B + 2Cx^2 - C,$$

or $\quad 11x^2 - 8x - 7 = (A + 2C)x^2 + (-3A + B)x + (-3B - C).$

We then equate corresponding coefficients:

$11 = A + 2C,$	**The coefficients of the x^2-terms**
$-8 = -3A + B,$	**The coefficients of the x-terms**
$-7 = -3B - C.$	**The constant terms**

We solve this system of three equations and obtain

$$A = 3, \quad B = 1, \quad \text{and} \quad C = 4.$$

The decomposition is as follows:

$$\frac{11x^2 - 8x - 7}{(2x^2 - 1)(x - 3)} = \frac{3x + 1}{2x^2 - 1} + \frac{4}{x - 3}.$$

Now Try Exercise 13. ■

9.8 Exercise Set

Decompose into partial fractions. Check your answers using a graphing calculator.

1. $\dfrac{x + 7}{(x - 3)(x + 2)}$

2. $\dfrac{2x}{(x + 1)(x - 1)}$

3. $\dfrac{7x - 1}{6x^2 - 5x + 1}$

4. $\dfrac{13x + 46}{12x^2 - 11x - 15}$

5. $\dfrac{3x^2 - 11x - 26}{(x^2 - 4)(x + 1)}$

6. $\dfrac{5x^2 + 9x - 56}{(x - 4)(x - 2)(x + 1)}$

7. $\dfrac{9}{(x + 2)^2(x - 1)}$

8. $\dfrac{x^2 - x - 4}{(x - 2)^3}$

9. $\dfrac{2x^2 + 3x + 1}{(x^2 - 1)(2x - 1)}$

10. $\dfrac{x^2 - 10x + 13}{(x^2 - 5x + 6)(x - 1)}$

11. $\dfrac{x^4 - 3x^3 - 3x^2 + 10}{(x + 1)^2(x - 3)}$

12. $\dfrac{10x^3 - 15x^2 - 35x}{x^2 - x - 6}$

13. $\dfrac{-x^2 + 2x - 13}{(x^2 + 2)(x - 1)}$

14. $\dfrac{26x^2 + 208x}{(x^2 + 1)(x + 5)}$

15. $\dfrac{6 + 26x - x^2}{(2x - 1)(x + 2)^2}$

16. $\dfrac{5x^3 + 6x^2 + 5x}{(x^2 - 1)(x + 1)^3}$

17. $\dfrac{6x^3 + 5x^2 + 6x - 2}{2x^2 + x - 1}$

18. $\dfrac{2x^3 + 3x^2 - 11x - 10}{x^2 + 2x - 3}$

19. $\dfrac{2x^2 - 11x + 5}{(x - 3)(x^2 + 2x - 5)}$

20. $\dfrac{3x^2 - 3x - 8}{(x - 5)(x^2 + x - 4)}$

21. $\dfrac{-4x^2 - 2x + 10}{(3x + 5)(x + 1)^2}$

22. $\dfrac{26x^2 - 36x + 22}{(x - 4)(2x - 1)^2}$

23. $\dfrac{36x + 1}{12x^2 - 7x - 10}$

24. $\dfrac{-17x + 61}{6x^2 + 39x - 21}$

25. $\dfrac{-4x^2 - 9x + 8}{(3x^2 + 1)(x - 2)}$

26. $\dfrac{11x^2 - 39x + 16}{(x^2 + 4)(x - 8)}$

Collaborative Discussion and Writing

27. Describe the two methods for finding the constants in a partial fraction decomposition.

28. What would you say to a classmate who tells you that the partial fraction decomposition of

$$\frac{3x^2 - 8x + 9}{(x + 3)(x^2 - 5x + 6)}$$

is

$$\frac{2}{x + 3} + \frac{x - 1}{x^2 - 5x + 6}?$$

Explain.

29. Explain the error in the following process.

$$\frac{x^2 + 4}{(x + 2)(x + 1)} = \frac{A}{x + 2} + \frac{B}{x + 1}$$

$$= \frac{A(x + 1) + B(x + 2)}{(x + 2)(x + 1)}$$

Then

$$x^2 + 4 = A(x + 1) + B(x + 2).$$

When $x = -1$:

$$(-1)^2 + 4 = A(-1 + 1) + B(-1 + 2)$$
$$5 = B.$$

When $x = -2$:

$$(-2)^2 + 4 = A(-2 + 1) + B(-2 + 2)$$
$$8 = -A$$
$$-8 = A.$$

Thus,

$$\frac{x^2 + 4}{(x + 2)(x + 1)} = \frac{-8}{x + 2} + \frac{5}{x + 1}.$$

Skill Maintenance

Find the zeros of the polynomial function.

30. $f(x) = x^3 - 3x^2 + x - 3$

31. $f(x) = x^3 + x^2 - 3x - 2$

32. $f(x) = x^4 - x^3 - 5x^2 - x - 6$

33. $f(x) = x^3 + 5x^2 + 5x - 3$

Synthesis

Decompose into partial fractions.

34. $\dfrac{9x^3 - 24x^2 + 48x}{(x - 2)^4(x + 1)}$

[*Hint*: Let the expression equal

$$\frac{A}{x + 1} + \frac{P(x)}{(x - 2)^4}$$

and find $P(x)$].

35. $\dfrac{x}{x^4 - a^4}$

36. $\dfrac{1}{e^{-x} + 3 + 2e^x}$

37. $\dfrac{1 + \ln x^2}{(\ln x + 2)(\ln x - 3)^2}$

CHAPTER 9   Summary and Review

Important Properties and Formulas

Row-Equivalent Operations

1. Interchange any two rows.
2. Multiply each entry in a row by the same nonzero constant.
3. Add a nonzero multiple of one row to another row.

Row-Echelon Form

1. If a row does not consist entirely of 0's, then the first nonzero element in the row is a 1 (called a leading 1).
2. For any two successive nonzero rows, the leading 1 in the lower row is farther to the right than the leading 1 in the higher row.
3. All the rows consisting entirely of 0's are at the bottom of the matrix.

If a fourth property is also satisfied, a matrix is said to be in reduced row-echelon form:

4. Each column that contains a leading 1 has 0's everywhere else.

Properties of Matrix Addition and Scalar Multiplication

For matrices A, B, and C and any scalars k and l, assuming that the indicated operation is possible:

Commutative Property of Addition:

$$A + B = B + A.$$

Associative Property of Addition:

$$A + (B + C) = (A + B) + C.$$

Associative Property of Scalar Multiplication:

$$(kl)A = k(lA).$$

Additive Identity Property:

There exists a unique matrix 0 such that

$$A + 0 = 0 + A = A.$$

Additive Inverse Property:

There exists a unique matrix $-A$ such that

$$A + (-A) = -A + A = 0.$$

Distributive Properties:

$$k(A + B) = kA + kB,$$
$$(k + l)A = kA + lA.$$

Properties of Matrix Multiplication

For matrices A, B, and C, assuming that the indicated operation is possible:

Associative Property of Multiplication:

$$A(BC) = (AB)C.$$

Distributive Properties:

$$A(B + C) = AB + AC,$$
$$(B + C)A = BA + CA.$$

Matrix Solutions of Systems of Equations

For a system of n linear equations in n variables, $AX = B$, if A is an invertible matrix, then the unique solution of the system is given by $X = A^{-1}B$.

Determinant of a 2 × 2 Matrix

The **determinant** of the matrix $\begin{bmatrix} a & c \\ b & d \end{bmatrix}$ is

denoted $\begin{vmatrix} a & c \\ b & d \end{vmatrix}$ and is defined as

$$\begin{vmatrix} a & c \\ b & d \end{vmatrix} = ad - bc.$$

Determinant of Any Square Matrix

For any square matrix A of order $n \times n$ ($n > 1$), we define the **determinant** of A, denoted $|A|$, as follows. Choose any row or column. Multiply each element in that row or column by its cofactor and add the results. The determinant of a 1×1 matrix is simply the element of the matrix. The value of a determinant will be the same, no matter which row or column is chosen.

(continued)

806   **CHAPTER 9** ❖ Systems of Equations and Matrices

Cramer's Rule for 2 × 2 Systems

The solution of the system of equations

$$a_1x + b_1y = c_1,$$
$$a_2x + b_2y = c_2$$

is given by

$$x = \frac{D_x}{D}, \qquad y = \frac{D_y}{D},$$

where

$$D = \begin{vmatrix} a_1 & b_1 \\ a_2 & b_2 \end{vmatrix}, \qquad D_x = \begin{vmatrix} c_1 & b_1 \\ c_2 & b_2 \end{vmatrix},$$

$$D_y = \begin{vmatrix} a_1 & c_1 \\ a_2 & c_2 \end{vmatrix}, \quad \text{and} \quad D \neq 0.$$

Cramer's Rule for 3 × 3 Systems

The solution of the system of equations

$$a_1x + b_1y + c_1z = d_1,$$
$$a_2x + b_2y + c_2z = d_2,$$
$$a_3x + b_3y + c_3z = d_3$$

is given by

$$x = \frac{D_x}{D}, \qquad y = \frac{D_y}{D}, \qquad z = \frac{D_z}{D},$$

where

$$D = \begin{vmatrix} a_1 & b_1 & c_1 \\ a_2 & b_2 & c_2 \\ a_3 & b_3 & c_3 \end{vmatrix}, \qquad D_x = \begin{vmatrix} d_1 & b_1 & c_1 \\ d_2 & b_2 & c_2 \\ d_3 & b_3 & c_3 \end{vmatrix},$$

$$D_y = \begin{vmatrix} a_1 & d_1 & c_1 \\ a_2 & d_2 & c_2 \\ a_3 & d_3 & c_3 \end{vmatrix}, \qquad D_z = \begin{vmatrix} a_1 & b_1 & d_1 \\ a_2 & b_2 & d_2 \\ a_3 & b_3 & d_3 \end{vmatrix},$$

and $D \neq 0$.

To Graph a Linear Inequality in Two Variables:

1. Replace the inequality symbol with an equals sign and graph this related equation. If the inequality symbol is $<$ or $>$, draw the line dashed. If the inequality symbol is $\leq$ or $\geq$, draw the line solid.
2. The graph consists of a half-plane on one side of the line and, if the line is solid, the line as well. To determine which half-plane to shade, test a point not on the line in the original inequality. If that point is a solution, shade the half-plane containing that point. If not, shade the opposite half-plane.

Linear Programming Procedure

To find the maximum value or minimum value of a linear objective function subject to a set of constraints:

1. Graph the region of feasible solutions.
2. Determine the coordinates of the vertices of the region.
3. Evaluate the objective function at each vertex. The largest and smallest of those values are the maximum and minimum values of the function, respectively.

Procedure for Decomposing a Rational Expression into Partial Fractions

Consider any rational expression $P(x)/Q(x)$ such that $P(x)$ and $Q(x)$ have no common factor other than 1 or -1.

1. If the degree of $P(x)$ is greater than or equal to the degree of $Q(x)$, divide to express $P(x)/Q(x)$ as a quotient + remainder/$Q(x)$ and follow steps (2)–(5) to decompose the resulting rational expression.
2. If the degree of $P(x)$ is less than the degree of $Q(x)$, factor $Q(x)$ into linear factors of the form $(px + q)^n$ and/or quadratic factors of the form $(ax^2 + bx + c)^m$. Any quadratic factor $ax^2 + bx + c$ must be irreducible, meaning that it cannot be factored into linear factors with rational coefficients.
3. Assign to each linear factor $(px + q)^n$ the sum of n partial fractions:

$$\frac{A_1}{px + q} + \frac{A_2}{(px + q)^2} + \cdots + \frac{A_n}{(px + q)^n}.$$

4. Assign to each quadratic factor $(ax^2 + bx + c)^m$ the sum of m partial fractions:

$$\frac{B_1x + C_1}{ax^2 + bx + c} + \frac{B_2x + C_2}{(ax^2 + bx + c)^2} + \cdots +$$

$$\frac{B_mx + C_m}{(ax^2 + bx + c)^m}.$$

5. Apply algebraic methods to find the constants in the numerators of the partial fractions.

Review Exercises

Determine whether the statement is true or false.

1. A system of equations with exactly one solution is consistent and has independent equations.   [9.1]

2. A system of two linear equations in two variables can have exactly two solutions.   [9.1]

3. For any $m \times n$ matrices **A** and **B**, $\mathbf{A} + \mathbf{B} = \mathbf{B} + \mathbf{A}$.   [9.4]

4. In general, matrix multiplication is commutative. [9.4]

In Exercises 5–12, match the equations or inequalities with one of the graphs (a)–(h), which follow.

a)

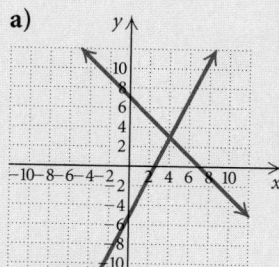

b)

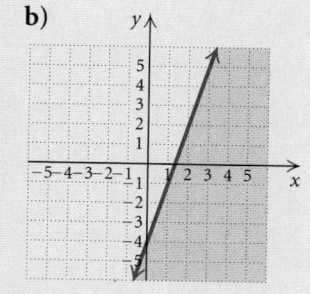

c)

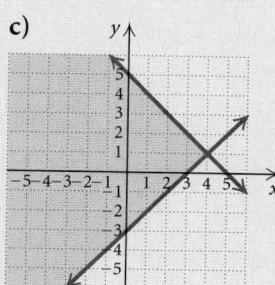

d)

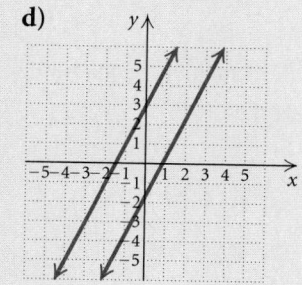

e)

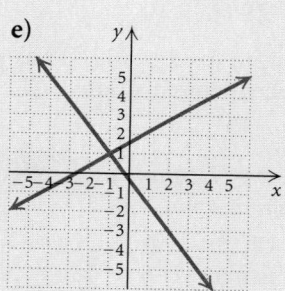

f)

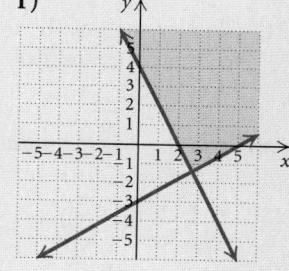

g)

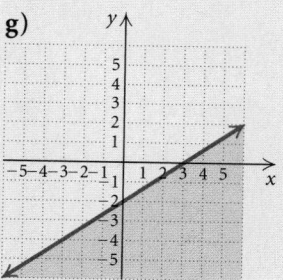

h)

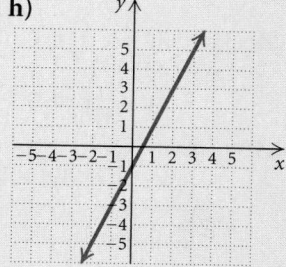

5. $x + y = 7$,
$2x - y = 5$ [9.1]

6. $3x - 5y = -8$,
$4x + 3y = -1$
[9.1]

7. $y = 2x - 1$,
$4x - 2y = 2$ [9.1]

8. $6x - 3y = 5$,
$y = 2x + 3$ [9.1]

9. $y \leq 3x - 4$ [9.7]

10. $2x - 3y \geq 6$ [9.7]

11. $x - y \leq 3$,
$x + y \leq 5$ [9.7]

12. $2x + y \geq 4$,
$3x - 5y \leq 15$
[9.7]

Solve.

13. $5x - 3y = -4$,
$3x - y = -4$ [9.1]

14. $2x + 3y = 2$,
$5x - y = -29$
[9.1]

15. $x + 5y = 12,$
$5x + 25y = 12$
[9.1]

16. $x + y = -2,$
$-3x - 3y = 6$
[9.1]

17. $x + 5y - 3z = 4,$
$3x - 2y + 4z = 3,$
$2x + 3y - z = 5$ [9.2]

18. $2x - 4y + 3z = -3,$
$-5x + 2y - z = 7,$
$3x + 2y - 2z = 4$ [9.2]

19. $x - y = 5,$
$y - z = 6,$
$z - w = 7,$
$x + w = 8$ [9.2]

20. Classify each of the systems in Exercises 13–19 as consistent or inconsistent.   [9.1], [9.2]

21. Classify each of the systems in Exercises 13–19 as having dependent or independent equations.   [9.1], [9.2]

Solve the system of equations using Gaussian elimination or Gauss–Jordan elimination.   [9.3]

22. $x + 2y = 5,$
$2x - 5y = -8$

23. $3x + 4y + 2z = 3,$
$5x - 2y - 13z = 3,$
$4x + 3y - 3z = 6$

24. $3x + 5y + z = 0,$
$2x - 4y - 3z = 0,$
$x + 3y + z = 0$

25. $w + x + y + z = -2,$
$-3w - 2x + 3y + 2z = 10,$
$2w + 3x + 2y - z = -12,$
$2w + 4x - y + z = 1$

26. *Coins.*   The value of 75 coins, consisting of nickels and dimes, is $5.95. How many of each kind are there?   [9.1]

27. *Investment.*   The Mendez family invested $5000, part at 3% and the remainder at 3.5%. The annual income from both investments is $167. What is the amount invested at each rate?   [9.1]

28. *Nutrition.*   A dietician must plan a breakfast menu that provides 460 Cal, 9 g of fat, and 55 mg of calcium. One plain bagel contains 200 Cal, 2 g of fat, and 29 mg of calcium. A one-tablespoon serving of cream cheese contains 100 Cal, 10 g of fat, and 24 mg of calcium. One banana contains 105 Cal, 1 g of fat, and 7 g of calcium. (*Source: Home and*

Garden Bulletin No. 72, U.S. Government Printing Office, Washington D.C. 20402) How many servings of each are required to provide the desired nutritional values?   [9.2]

29. *Test Scores.*   A student has a total of 226 on three tests. The sum of the scores on the first and second tests exceeds the score on the third test by 62. The first score exceeds the second by 6. Find the three scores.   [9.2]

30. *Trademarks.*   The table below lists the number of trademarks issued, in thousands, in the United States, represented in terms of the number of years since 2003.

Year, x	Number of Trademarks Issued (in thousands)
2003, 0	167
2004, 1	145
2005, 2	155

Source: U.S. Patent and Trademark Office

a) Use a system of equations to fit a quadratic function $f(x) = ax^2 + bx + c$ to the data. [9.2]

b) Use the function to estimate the number of trademarks issued in 2006.   [9.2]

For Exercises 31–38, let

$$A = \begin{bmatrix} 1 & -1 & 0 \\ 2 & 3 & -2 \\ -2 & 0 & 1 \end{bmatrix},$$

$$B = \begin{bmatrix} -1 & 0 & 6 \\ 1 & -2 & 0 \\ 0 & 1 & -3 \end{bmatrix},$$

and

$$C = \begin{bmatrix} -2 & 0 \\ 1 & 3 \end{bmatrix}.$$

Find each of the following, if possible.   [9.4]

31. $A + B$

32. $-3A$

33. $-A$

34. AB

35. $B + C$

36. $A - B$

37. BA

38. $A + 3B$

39. *Food Service Management.*   The table below lists the cost per serving, in dollars, for items on four menus that are served at an elder-care facility.

Menu	Meat	Potato	Vegetable	Salad	Dessert
1	0.98	0.23	0.30	0.28	0.45
2	1.03	0.19	0.27	0.34	0.41
3	1.01	0.21	0.35	0.31	0.39
4	0.99	0.25	0.29	0.33	0.42

On a particular day, a dietician orders 32 meals from menu 1, 19 from menu 2, 43 from menu 3, and 38 from menu 4.

a) Write the information in the table as a 4×5 matrix **M**.   [9.4]

b) Write a row matrix **N** that represents the number of each menu ordered.   [9.4]

c) Find the product **NM**.   [9.4]

d) State what the entries of **NM** represent.   [9.4]

Find $\mathbf{A}^{-1}$, *if it exists.*   [9.5]

40. $\mathbf{A} = \begin{bmatrix} -2 & 0 \\ 1 & 3 \end{bmatrix}$

41. $\mathbf{A} = \begin{bmatrix} 0 & 0 & 3 \\ 0 & -2 & 0 \\ 4 & 0 & 0 \end{bmatrix}$

42. $\mathbf{A} = \begin{bmatrix} 1 & 0 & 0 & 0 \\ 0 & 4 & -5 & 0 \\ 0 & 2 & 2 & 0 \\ 0 & 0 & 0 & 1 \end{bmatrix}$

43. Write a matrix equation equivalent to this system of equations:
$$3x - 2y + 4z = 13,$$
$$x + 5y - 3z = 7,$$
$$2x - 3y + 7z = -8. \quad [9.4]$$

Solve the system of equations using the inverse of the coefficient matrix of the equivalent matrix equation.   [9.5]

44. $2x + 3y = 5,$
$3x + 5y = 11$

45. $5x - y + 2z = 17,$
$3x + 2y - 3z = -16,$
$4x - 3y - z = 5$

46. $w - x - y + z = -1,$
$2w + 3x - 2y - z = 2,$
$-w + 5x + 4y - 2z = 3,$
$3w - 2x + 5y + 3z = 4$

Evaluate the determinant.   [9.6]

47. $\begin{vmatrix} 1 & -2 \\ 3 & 4 \end{vmatrix}$

48. $\begin{vmatrix} \sqrt{3} & -5 \\ -3 & -\sqrt{3} \end{vmatrix}$

49. $\begin{vmatrix} 2 & -1 & 1 \\ 1 & 2 & -1 \\ 3 & 4 & -3 \end{vmatrix}$

50. $\begin{vmatrix} 1 & -1 & 2 \\ -1 & 2 & 0 \\ -1 & 3 & 1 \end{vmatrix}$

Solve using Cramer's rule.   [9.6]

51. $5x - 2y = 19,$
$7x + 3y = 15$

52. $x + y = 4,$
$4x + 3y = 11$

53. $3x - 2y + z = 5,$
$4x - 5y - z = -1,$
$3x + 2y - z = 4$

54. $2x - y - z = 2,$
$3x + 2y + 2z = 10,$
$x - 5y - 3z = -2$

Graph.   [9.7]

55. $y \le 3x + 6$

56. $4x - 3y \ge 12$

57. Graph this system of inequalities and find the coordinates of any vertices formed.   [9.7]
$$2x + y \ge 9,$$
$$4x + 3y \ge 23,$$
$$x + 3y \ge 8,$$
$$x \ge 0,$$
$$y \ge 0$$

58. Find the maximum value and the minimum value of $T = 6x + 10y$ subject to
$$x + y \le 10,$$
$$5x + 10y \ge 50,$$
$$x \ge 2,$$
$$y \ge 0. \quad [9.7]$$

59. *Maximizing a Test Score.*   Marita is taking a test that contains questions in group A worth 7 points each and questions in group B worth 12 points each. The total number of questions answered must be at least 8. If Marita knows that group A questions take 8 min each and group B questions take 10 min each and the maximum time for the test is 80 min, how many questions from each group must she answer correctly in order to maximize her score? What is the maximum score?   [9.7]

Decompose into partial fractions.   [9.8]

60. $\dfrac{5}{(x + 2)^2(x + 1)}$

61. $\dfrac{-8x + 23}{2x^2 + 5x - 12}$

62. Solve: $2x + y = 7,$
$ x - 2y = 6.$ [9.1]

  A. x and y are both positive numbers.
  B. x and y are both negative numbers.
  C. x is positive and y is negative.
  D. x is negative and y is positive.

63. Which is *not* a row-equivalent operation on a matrix?   [9.3]

  A. Interchange any two columns.
  B. Interchange any two rows.
  C. Add two rows.
  D. Multiply each entry in a row by -3.

64. The graph of the given system of inequalities is which of the following?   [9.7]

$$x + y \leq 3,$$
$$x - y \leq 4$$

A.

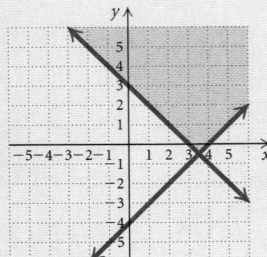

B.

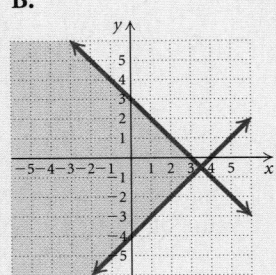

C.

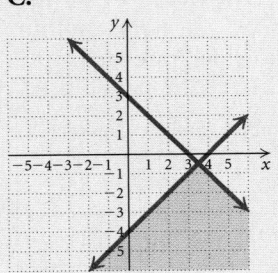

D.

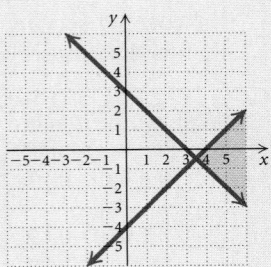

Collaborative Discussion and Writing

65. Write a problem for a classmate to solve that can be translated to a system of equations. Devise the problem so that the solution is "The caterer sold 20 cheese trays and 35 seafood trays."   [9.1]

66. For square matrices **A** and **B**, is it true, in general, that $(\mathbf{AB})^2 = \mathbf{A}^2\mathbf{B}^2$? Explain.   [9.4]

Synthesis

67. One year, Don invested a total of $40,000, part at 4%, part at 5%, and the rest at $5\frac{1}{2}$%. The total amount of interest received on the investments was $1990. The interest received on the $5\frac{1}{2}$% investment was $590 more than the interest received on the 4% investment. How much was invested at each rate?   [9.2]

Solve.

68. $\dfrac{2}{3x} + \dfrac{4}{5y} = 8,$
$ \dfrac{5}{4x} - \dfrac{3}{2y} = -6$
[9.1]

69. $\dfrac{3}{x} - \dfrac{4}{y} + \dfrac{1}{z} = -2,$
$ \dfrac{5}{x} + \dfrac{1}{y} - \dfrac{2}{z} = 1,$
$ \dfrac{7}{x} + \dfrac{3}{y} + \dfrac{2}{z} = 19$
[9.2]

Graph.   [9.7]

70. $|x| - |y| \leq 1$

71. $|xy| > 1$

CHAPTER 9   Test

Solve. Use any method. Also determine whether the system is consistent or inconsistent and whether the equations are dependent or independent.

1. $3x + 2y = 1,$
$2x - y = -11$

2. $2x - y = 3,$
$2y = 4x - 6$

3. $x - y = 4,$
$3y = 3x - 8$

4. $2x - 3y = 8,$
$5x - 2y = 9$

Solve.

5. $4x + 2y + z = 4,$
$3x - y + 5z = 4,$
$5x + 3y - 3z = -2$

6. *Ticket Sales.* One evening 750 tickets were sold for Shortridge Community College's spring musical. Tickets cost \$3 for students and \$5 for nonstudents. Total receipts were \$3066. How many of each type of ticket were sold?

7. Tricia, Maria, and Antonio can process 352 telephone orders per day. Tricia and Maria together can process 224 orders per day while Tricia and Antonio together can process 248 orders per day. How many orders can each of them process alone?

For Exercises 8–13, let

$$A = \begin{bmatrix} 1 & -1 & 3 \\ -2 & 5 & 2 \end{bmatrix}, \quad B = \begin{bmatrix} -5 & 1 \\ -2 & 4 \end{bmatrix},$$

and

$$C = \begin{bmatrix} 3 & -4 \\ -1 & 0 \end{bmatrix}.$$

Find each of the following, if possible.

8. $B + C$

9. $A - C$

10. CB

11. AB

12. $2A$

13. C^{-1}

14. *Food Service Management.* The table below lists the cost per serving, in dollars, for items on three lunch menus served at a senior citizens' center.

Menu	Main Dish	Side Dish	Dessert
1	0.95	0.40	0.39
2	1.10	0.35	0.41
3	1.05	0.39	0.36

On a particular day, 26 Menu 1 meals, 18 Menu 2 meals, and 23 Menu 3 meals are served.

a) Write the information in the table as a 3×3 matrix **M**.
b) Write a row matrix **N** that represents the number of each menu served.
c) Find the product **NM**.
d) State what the entries of **NM** represent.

15. Write a matrix equation equivalent to the system of equations

$$3x - 4y + 2z = -8,$$
$$2x + 3y + z = 7,$$
$$x - 5y - 3z = 3.$$

16. Solve the system of equations using the inverse of the coefficient matrix of the equivalent matrix equation.

$$3x + 2y + 6z = 2,$$
$$x + y + 2z = 1,$$
$$2x + 2y + 5z = 3$$

Evaluate the determinant.

17. $\begin{vmatrix} 3 & -5 \\ 8 & 7 \end{vmatrix}$

18. $\begin{vmatrix} 2 & -1 & 4 \\ -3 & 1 & -2 \\ 5 & 3 & -1 \end{vmatrix}$

19. Solve using Cramer's rule. Show your work.

$$5x + 2y = -1,$$
$$7x + 6y = 1$$

20. Graph: $3x + 4y \leq -12$.

21. Find the maximum value and the minimum value of $Q = 2x + 3y$ subject to

$$x + y \leq 6,$$
$$2x - 3y \geq -3,$$
$$x \geq 1,$$
$$y \geq 0.$$

22. *Maximizing Profit.*   Casey's Cakes prepares pound cakes and carrot cakes. In a given week, at most 100 cakes can be prepared, of which 25 pound cakes and 15 carrot cakes are required by regular customers. The profit from each pound cake is $3 and the profit from each carrot cake is $4. How many of each type of cake should be prepared in order to maximize the profit? What is the maximum profit?

23. Decompose into partial fractions:

$$\frac{3x - 11}{x^2 + 2x - 3}.$$

24. The graph of the given system of inequalities is which of the following?

$$x + 2y \geq 4,$$
$$x - y \leq 2$$

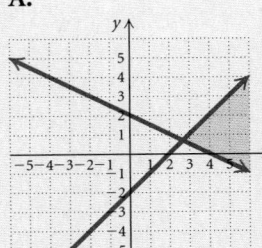

A.

B.

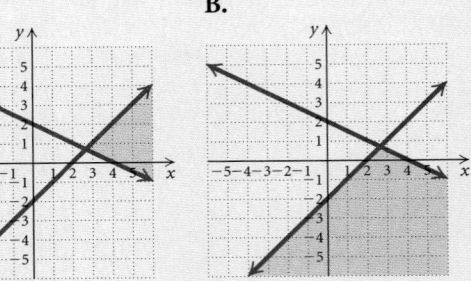

C.

D.

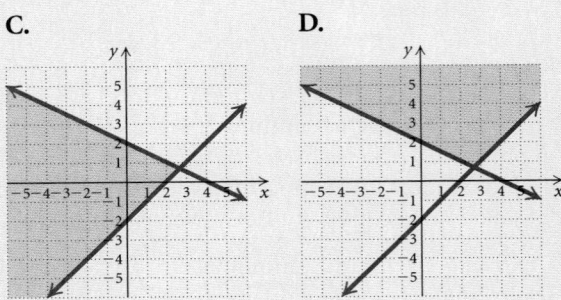

Synthesis

25. Three solutions of the equation $Ax - By = Cz - 8$ are $(2, -2, 2)$, $(-3, -1, 1)$, and $(4, 2, 9)$. Find A, B, and C.

Analytic Geometry Topics

APPLICATION The Burton Seed Company has two square test plots. The sum of their areas is 832 ft² and the difference of their areas is 320 ft². Find the length of a side of each plot.

This problem appears as Exercise 67 in Section 10.4.

10.1 The Parabola

10.2 The Circle and the Ellipse

10.3 The Hyperbola

10.4 Nonlinear Systems of Equations and Inequalities

10.5 Rotation of Axes

10.6 Polar Equations of Conics

10.7 Parametric Equations

10.1 The Parabola

❋ Given an equation of a parabola, complete the square, if necessary, and then find the vertex, the focus, and the directrix and graph the parabola.

A **conic section** is formed when a right circular cone with two parts, called *nappes*, is intersected by a plane. One of four types of curves can be formed: a parabola, a circle, an ellipse, or a hyperbola.

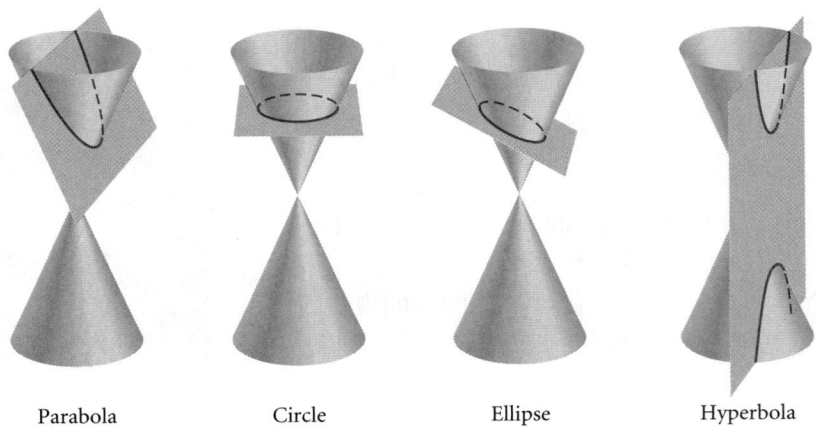

| Parabola | Circle | Ellipse | Hyperbola |

Conic Sections

Conic sections can be defined algebraically using second-degree equations of the form $Ax^2 + Bxy + Cy^2 + Dx + Ey + F = 0$. In addition, they can be defined geometrically as a set of points that satisfy certain conditions.

❋ Parabolas

In Section 3.3, we saw that the graph of the quadratic function $f(x) = ax^2 + bx + c$, $a \neq 0$, is a parabola. A parabola can be defined geometrically.

> **Parabola**
>
> A **parabola** is the set of all points in a plane equidistant from a fixed line (the **directrix**) and a fixed point not on the line (the **focus**).

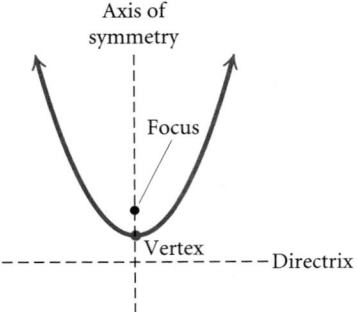

The line that is perpendicular to the directrix and contains the focus is the **axis of symmetry.** The **vertex** is the midpoint of the segment between the focus and the directrix. (See the figure at left.)

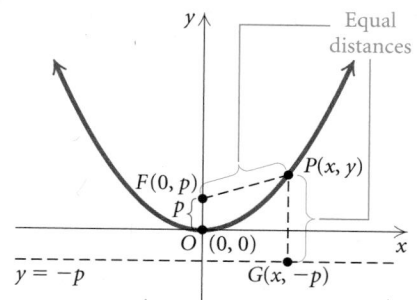

FIGURE 1

Let's derive the standard equation of a parabola with vertex $(0,0)$ and directrix $y = -p$, where $p > 0$. We place the coordinate axes as shown in Fig. 1. The y-axis is the axis of symmetry and contains the focus F. The distance from the focus to the vertex is the same as the distance from the vertex to the directrix. Thus the coordinates of F are $(0, p)$.

Let $P(x, y)$ be any point on the parabola and consider $\overline{PG}$ perpendicular to the line $y = -p$. The coordinates of G are $(x, -p)$. By the definition of a parabola,

$$PF = PG.$$ The distance from P to the focus is the same as the distance from P to the directrix.

Then using the distance formula, we have

$$\sqrt{(x - 0)^2 + (y - p)^2} = \sqrt{(x - x)^2 + [y - (-p)]^2}$$
$$x^2 + y^2 - 2py + p^2 = y^2 + 2py + p^2$$ Squaring both sides and squaring the binomials
$$x^2 = 4py.$$

We have shown that if $P(x, y)$ is on the parabola shown in Fig. 1, then its coordinates satisfy this equation. The converse is also true, but we will not prove it here.

Note that if $p > 0$, as above, the graph opens up. If $p < 0$, the graph opens down.

The equation of a parabola with vertex $(0,0)$ and directrix $x = -p$ is derived similarly. Such a parabola opens either to the right ($p > 0$), as shown in Fig. 2, or to the left ($p < 0$).

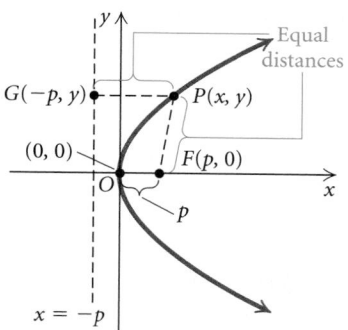

FIGURE 2

Standard Equation of a Parabola with Vertex at the Origin

The standard equation of a parabola with vertex $(0,0)$ and directrix $y = -p$ is

$$x^2 = 4py.$$

The focus is $(0, p)$ and the y-axis is the axis of symmetry.

The standard equation of a parabola with vertex $(0,0)$ and directrix $x = -p$ is

$$y^2 = 4px.$$

The focus is $(p, 0)$ and the x-axis is the axis of symmetry.

EXAMPLE 1   Find the focus and the directrix of the parabola $y = -\frac{1}{12}x^2$. Then graph the parabola.

Solution   We write $y = -\frac{1}{12}x^2$ in the form $x^2 = 4py$:

$$-\frac{1}{12}x^2 = y$$ Given equation
$$x^2 = -12y$$ Multiplying by -12 on both sides
$$x^2 = 4(-3)y.$$ Standard form

Thus, $p = -3$, so the focus is $(0, p)$, or $(0, -3)$. The directrix is $y = -p = -(-3) = 3$.

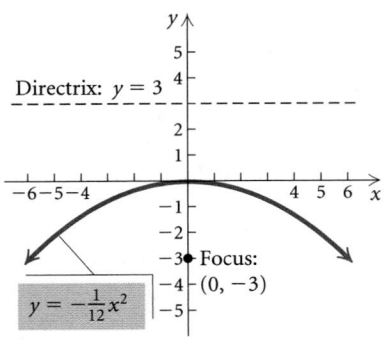

EXAMPLE 2   Find an equation of the parabola with vertex $(0, 0)$ and focus $(5, 0)$. Then graph the parabola.

Solution   The focus is on the x-axis so the line of symmetry is the x-axis. Thus the equation is of the type

$$y^2 = 4px.$$

Since the focus $(5, 0)$ is 5 units to the right of the vertex, $p = 5$ and the equation is

$$y^2 = 4(5)x, \quad \text{or} \quad y^2 = 20x.$$

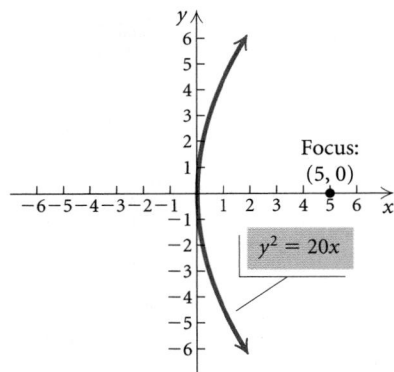

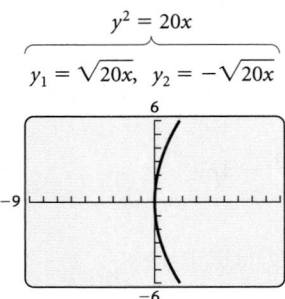

We can also use a graphing calculator to graph parabolas. It might be necessary to solve the equation for y before entering it in the calculator:

$$y^2 = 20x$$
$$y = \pm\sqrt{20x}.$$

We now graph $y_1 = \sqrt{20x}$ and $y_2 = -\sqrt{20x}$ or $y_1 = \sqrt{20x}$ and $y_2 = -y_1$ in a squared viewing window.

On some graphing calculators, the Conics application from the APPS menu can be used to graph parabolas. This method will be discussed in Example 4.

Now Try Exercise 15. ■

❋ Finding Standard Form by Completing the Square

If a parabola with vertex at the origin is translated horizontally $|h|$ units and vertically $|k|$ units, it has an equation as follows.

Standard Equation of a Parabola with Vertex (h, k) and Vertical Axis of Symmetry

The standard equation of a parabola with vertex (h, k) and vertical axis of symmetry is

$$(x - h)^2 = 4p(y - k),$$

where the vertex is (h, k), the focus is $(h, k + p)$, and the directrix is $y = k - p$.

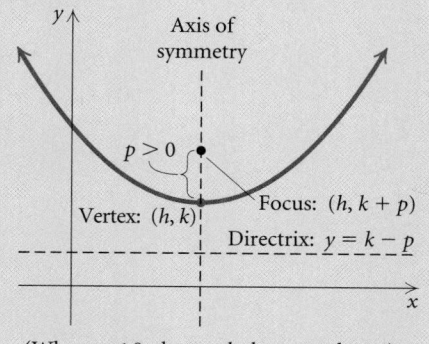

(When $p < 0$, the parabola opens down.)

Standard Equation of a Parabola with Vertex (h, k) and Horizontal Axis of Symmetry

The standard equation of a parabola with vertex (h, k) and horizontal axis of symmetry is

$$(y - k)^2 = 4p(x - h),$$

where the vertex is (h, k), the focus is $(h + p, k)$, and the directrix is $x = h - p$.

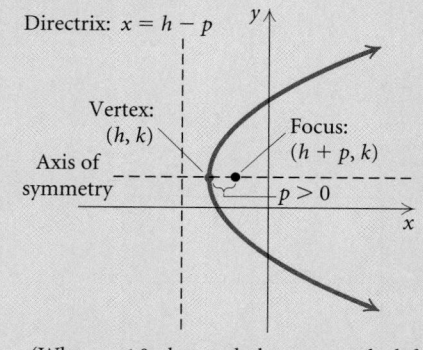

(When $p < 0$, the parabola opens to the left.)

STUDY TIP

Nearly every example in the textbook is worked out in a video presentation. Make time to visit your math lab or media center to view these presentations. Pause the video and take notes or work through the examples. You can proceed at your own pace, replaying all or part of a presentation as many times as you need to.

COMPLETING THE SQUARE

REVIEW SECTION **3.2.**

We can complete the square on equations of the form

$$y = ax^2 + bx + c \quad \text{or} \quad x = ay^2 + by + c$$

in order to write them in standard form.

EXAMPLE 3 For the parabola

$$x^2 + 6x + 4y + 5 = 0,$$

find the vertex, the focus, and the directrix. Then draw the graph.

Solution We first complete the square:

$$x^2 + 6x + 4y + 5 = 0$$

$$x^2 + 6x = -4y - 5 \qquad \text{Subtracting } 4y \text{ and } 5 \text{ on both sides}$$

$$x^2 + 6x + 9 = -4y - 5 + 9 \qquad \text{Adding } 9 \text{ on both sides to complete the square on the left side}$$

$$x^2 + 6x + 9 = -4y + 4$$

$$(x + 3)^2 = -4(y - 1) \qquad \text{Factoring}$$

$$[x - (-3)]^2 = 4(-1)(y - 1). \qquad \text{Writing standard form: } (x - h)^2 = 4p(y - k)$$

We see that $h = -3$, $k = 1$, and $p = -1$, so we have the following:

Vertex (h, k): $(-3, 1)$;

Focus $(h, k + p)$: $(-3, 1 + (-1))$, or $(-3, 0)$;

Directrix $y = k - p$: $y = 1 - (-1)$, or $y = 2$.

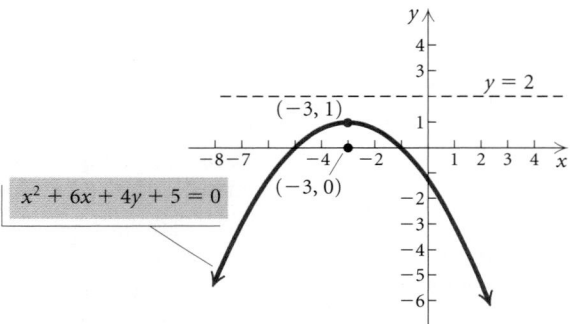

We can check the graph on a graphing calculator using a squared viewing window. It might be necessary to solve for y first:

$$x^2 + 6x + 4y + 5 = 0$$

$$4y = -x^2 - 6x - 5$$

$$y = \tfrac{1}{4}(-x^2 - 6x - 5).$$

The hand-drawn graph appears to be correct.   **Now Try Exercise 23.** ■

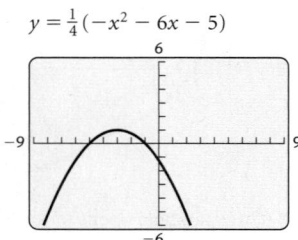

$y = \tfrac{1}{4}(-x^2 - 6x - 5)$

| GCM | **EXAMPLE 4** For the parabola

$$y^2 - 2y - 8x - 31 = 0,$$

find the vertex, the focus, and the directrix. Then draw the graph.

Solution We first complete the square:

$$y^2 - 2y - 8x - 31 = 0$$

$$y^2 - 2y = 8x + 31 \qquad \text{Adding } 8x \text{ and } 31 \text{ on both sides}$$

$$y^2 - 2y + 1 = 8x + 31 + 1 \qquad \text{Adding } 1 \text{ on both sides to complete the square on the left side}$$

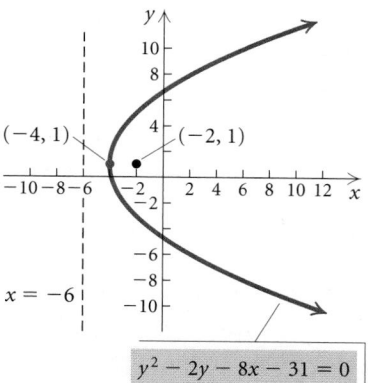

$$y^2 - 2y - 8x - 31 = 0$$

$$y^2 - 2y + 1 = 8x + 32$$
$$(y - 1)^2 = 8(x + 4) \qquad \text{Factoring}$$
$$(y - 1)^2 = 4(2)[x - (-4)]. \qquad \begin{array}{l}\text{Writing standard form:}\\(y - k)^2 = 4p(x - h)\end{array}$$

We see that $h = -4$, $k = 1$, and $p = 2$, so we have the following:

Vertex (h, k): $(-4, 1)$;

Focus $(h + p, k)$: $(-4 + 2, 1)$, or $(-2, 1)$;

Directrix $x = h - p$: $x = -4 - 2$, or $x = -6$.

When the equation of a parabola is written in standard form, we can use the Conics PARABOLA APP to graph it.

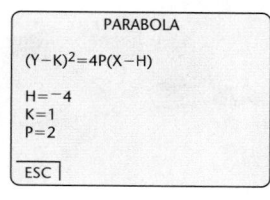
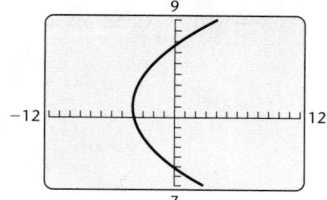

We can also draw the graph or check the hand-drawn graph on a graphing calculator by first solving the original equation for y using the quadratic formula:

$$y^2 - 2y - 8x - 31 = 0$$
$$y^2 - 2y + (-8x - 31) = 0$$
$$a = 1, \quad b = -2, \quad c = -8x - 31$$
$$y = \frac{-(-2) \pm \sqrt{(-2)^2 - 4 \cdot 1(-8x - 31)}}{2 \cdot 1}$$
$$y = \frac{2 \pm \sqrt{32x + 128}}{2}.$$

$$y^2 - 2y - 8x - 31 = 0$$

$$y_1 = \frac{2 + \sqrt{32x + 128}}{2},$$

$$y_2 = \frac{2 - \sqrt{32x + 128}}{2}$$

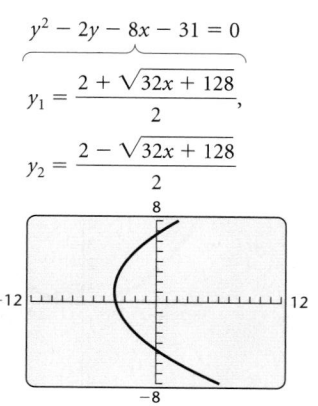

We now graph

$$y_1 = \frac{2 + \sqrt{32x + 128}}{2} \quad \text{and} \quad y_2 = \frac{2 - \sqrt{32x + 128}}{2}.$$

in a square viewing window.   **Now Try Exercise 29.** ■

❋ Applications

Parabolas have many applications. For example, cross sections of car headlights, flashlights, and searchlights are parabolas. The bulb is located at the focus and light from that point is reflected outward parallel to the axis of symmetry. Satellite dishes and field microphones used at sporting events often have parabolic cross sections. Incoming radio waves or sound waves parallel to the axis are reflected into the focus. Cables hung between structures in suspension bridges, such as the Golden Gate Bridge, form parabolas. When a cable supports only its own weight, however, it forms a curve called a *catenary* rather than a parabola.

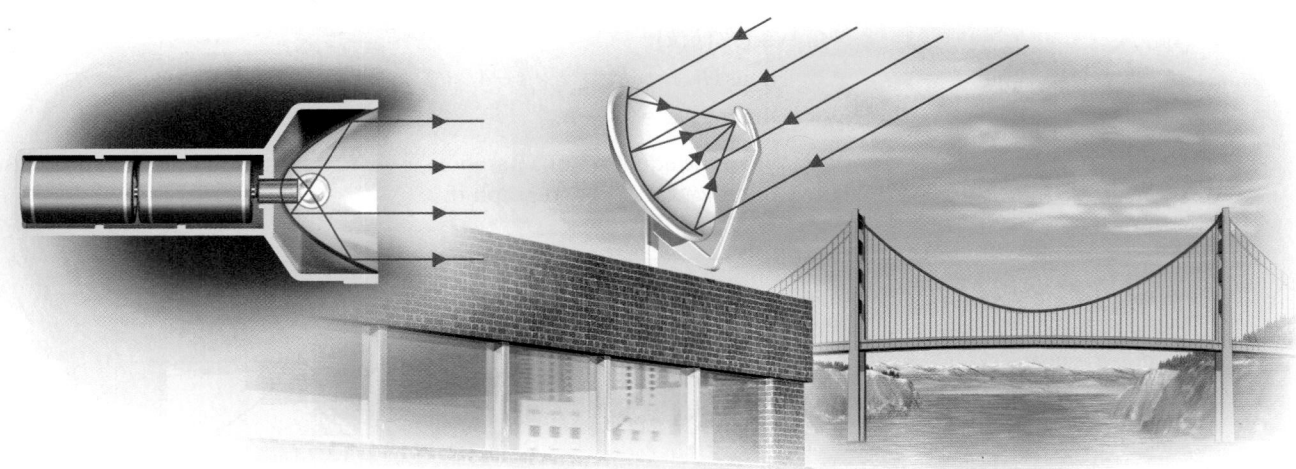

(10.1) Exercise Set

In Exercises 1–6, match the equation with one of the graphs (a)–(f), which follow.

a)

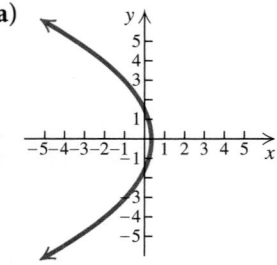

b)

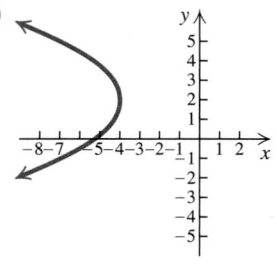

c)

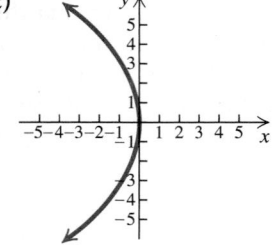

d)

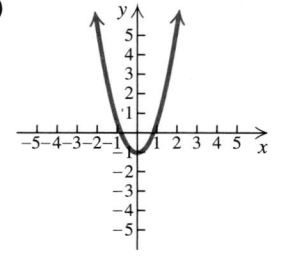

e)

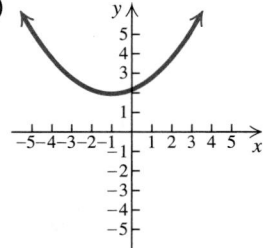

f)

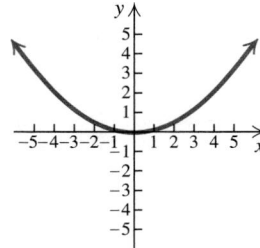

1. $x^2 = 8y$

2. $y^2 = -10x$

3. $(y - 2)^2 = -3(x + 4)$

4. $(x + 1)^2 = 5(y - 2)$

5. $13x^2 - 8y - 9 = 0$

6. $41x + 6y^2 = 12$

Find the vertex, the focus, and the directrix. Then draw the graph.

7. $x^2 = 20y$

8. $x^2 = 16y$

9. $y^2 = -6x$

10. $y^2 = -2x$

11. $x^2 - 4y = 0$

12. $y^2 + 4x = 0$

13. $x = 2y^2$

14. $y = \frac{1}{2}x^2$

Find an equation of a parabola satisfying the given conditions.

15. Focus $(4, 0)$, directrix $x = -4$

16. Focus $\left(0, \frac{1}{4}\right)$, directrix $y = -\frac{1}{4}$

17. Focus $(0, -\pi)$, directrix $y = \pi$

18. Focus $\left(-\sqrt{2}, 0\right)$, directrix $x = \sqrt{2}$

19. Focus $(3, 2)$, directrix $x = -4$

20. Focus $(-2, 3)$, directrix $y = -3$

Find the vertex, the focus, and the directrix. Then draw the graph.

21. $(x + 2)^2 = -6(y - 1)$

22. $(y - 3)^2 = -20(x + 2)$

23. $x^2 + 2x + 2y + 7 = 0$

24. $y^2 + 6y - x + 16 = 0$

25. $x^2 - y - 2 = 0$

26. $x^2 - 4x - 2y = 0$

27. $y = x^2 + 4x + 3$

28. $y = x^2 + 6x + 10$

29. $y^2 - y - x + 6 = 0$

30. $y^2 + y - x - 4 = 0$

31. *Satellite Dish.* An engineer designs a satellite dish with a parabolic cross section. The dish is 15 ft wide at the opening and the focus is placed 4 ft from the vertex.

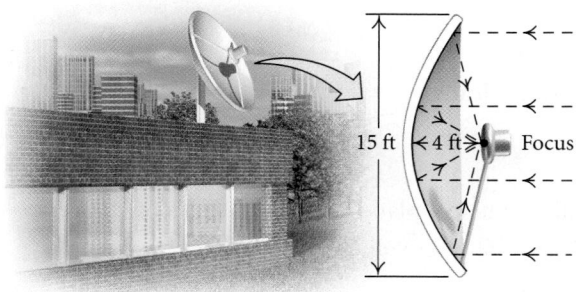

a) Position a coordinate system with the origin at the vertex and the x-axis on the parabola's axis of symmetry and find an equation of the parabola.

b) Find the depth of the satellite dish at the vertex.

32. *Headlight Mirror.* A car headlight mirror has a parabolic cross section with diameter 6 in. and depth 1 in.

a) Position a coordinate system with the origin at the vertex and the x-axis on the parabola's axis of symmetry and find an equation of the parabola.

b) How far from the vertex should the bulb be positioned if it is to be placed at the focus?

33. *Spotlight.* A spotlight has a parabolic cross section that is 4 ft wide at the opening and 1.5 ft deep at the vertex. How far from the vertex is the focus?

34. *Field Microphone.* A field microphone used at a football game has a parabolic cross section and is 18 in. deep. The focus is 4 in. from the vertex. Find the width of the microphone at the opening.

Collaborative Discussion and Writing

35. Is a parabola always the graph of a function? Why or why not?

36. Explain how the distance formula is used to find the standard equation of a parabola.

Skill Maintenance

Consider the following linear equations. Without graphing them, answer the questions below.

a) $y = 2x$

b) $y = \frac{1}{3}x + 5$

c) $y = -3x - 2$

d) $y = -0.9x + 7$

e) $y = -5x + 3$

f) $y = x + 4$

g) $8x - 4y = 7$

h) $3x + 6y = 2$

37. Which has/have x-intercept $\left(\frac{2}{3}, 0\right)$?

38. Which has/have y-intercept $(0, 7)$?

39. Which slant up from left to right?

40. Which has the least steep slant?

41. Which has/have slope $\frac{1}{3}$?

42. Which, if any, contain the point $(3,7)$?

43. Which, if any, are parallel?

44. Which, if any, are perpendicular?

Synthesis

45. Find an equation of the parabola with a vertical axis of symmetry and vertex $(-1,2)$ and containing the point $(-3,1)$.

46. Find an equation of a parabola with a horizontal axis of symmetry and vertex $(-2,1)$ and containing the point $(-3,5)$.

Use a graphing calculator to find the vertex, the focus, and the directrix of each of the following.

47. $4.5x^2 - 7.8x + 9.7y = 0$

48. $134.1y^2 + 43.4x - 316.6y - 122.4 = 0$

49. *Suspension Bridge.* The cables of a 200-ft portion of the roadbed of a suspension bridge are positioned as shown below. Vertical cables are to be spaced every 20 ft along this portion of the roadbed. Calculate the lengths of these vertical cables.

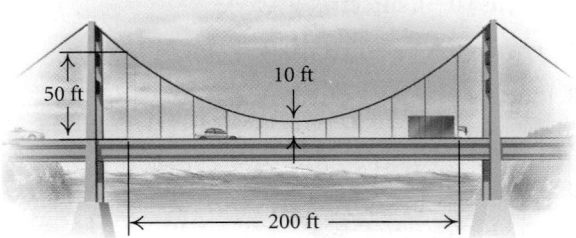

10.2

The Circle and the Ellipse

❖ Given an equation of a circle, complete the square, if necessary, and then find the center and the radius and graph the circle.

❖ Given an equation of an ellipse, complete the square, if necessary, and then find the center, the vertices, and the foci and graph the ellipse.

❖ Circles

We can define a circle geometrically.

CIRCLES
REVIEW SECTION 1.1.

Circle

A **circle** is the set of all points in a plane that are at a fixed distance from a fixed point (the **center**) in the plane.

Circles were introduced in Section 1.1. Recall the standard equation of a circle with center (h,k) and radius r.

Standard Equation of a Circle

The standard equation of a circle with center (h,k) and radius r is

$$(x-h)^2 + (y-k)^2 = r^2.$$

GCM **EXAMPLE 1** For the circle

$$x^2 + y^2 - 16x + 14y + 32 = 0,$$

find the center and the radius. Then graph the circle.

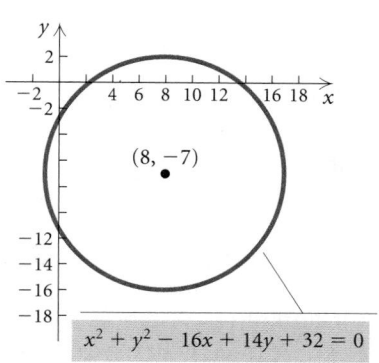

(8, −7)

$$x^2 + y^2 - 16x + 14y + 32 = 0$$

Solution First, we complete the square twice:

$$x^2 + y^2 - 16x + 14y + 32 = 0$$
$$x^2 - 16x \quad + y^2 + 14y \quad = -32$$
$$x^2 - 16x + 64 + y^2 + 14y + 49 = -32 + 64 + 49$$

$$\left[\tfrac{1}{2}(-16)\right]^2 = (-8)^2 = 64 \text{ and } \left(\tfrac{1}{2} \cdot 14\right)^2 = 7^2 = 49;$$
adding 64 and 49 on both sides to complete the square twice on the left side

$$(x - 8)^2 + (y + 7)^2 = 81$$
$$(x - 8)^2 + [y - (-7)]^2 = 9^2. \quad \textbf{Writing standard form}$$

The center is $(8, -7)$ and the radius is 9. We graph the circle as shown at left.

To use a graphing calculator to graph the circle, it might be necessary to solve for y first. The original equation can be solved using the quadratic formula, or the standard form of the equation can be solved using the principle of square roots. The second alternative is illustrated here:

$$(x - 8)^2 + (y + 7)^2 = 81$$
$$(y + 7)^2 = 81 - (x - 8)^2$$
$$y + 7 = \pm\sqrt{81 - (x - 8)^2} \quad \textbf{Using the principle of}$$
$$\textbf{square roots}$$
$$y = -7 \pm \sqrt{81 - (x - 8)^2}.$$

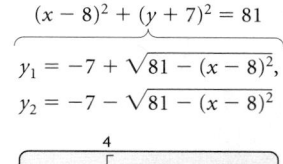

$$(x - 8)^2 + (y + 7)^2 = 81$$
$$y_1 = -7 + \sqrt{81 - (x - 8)^2},$$
$$y_2 = -7 - \sqrt{81 - (x - 8)^2}$$

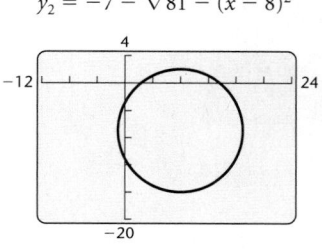

Then we graph

$$y_1 = -7 + \sqrt{81 - (x - 8)^2}$$

and

$$y_2 = -7 - \sqrt{81 - (x - 8)^2}$$

in a squared viewing window.

When we use the Conics CIRCLE APP to graph a circle, it is not necessary to write the equation in standard form or to solve it for y first. We enter the coefficients of x^2, y^2, x, and y and also the constant term when the equation is written in the form $ax^2 + ay^2 + bx + cy + d = 0$.

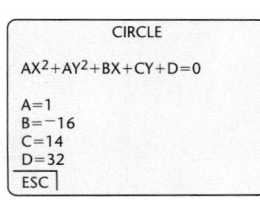

CIRCLE

$AX^2+AY^2+BX+CY+D=0$

A=1
B=−16
C=14
D=32
ESC

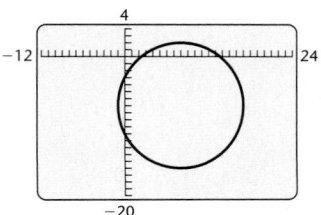

Some graphing calculators have a CIRCLE feature on a DRAW menu that provides a quick way to graph a circle when the center and the radius are known. This feature is described on p. 73. **Now Try Exercise 7.** ◼

❖ Ellipses

We have studied two conic sections, the parabola and the circle. Now we turn our attention to a third, the *ellipse*.

> ### Ellipse
>
> An **ellipse** is the set of all points in a plane, the sum of whose distances from two fixed points (the **foci**) is constant. The **center** of an ellipse is the midpoint of the segment between the foci.

We can draw an ellipse by first placing two thumbtacks in a piece of cardboard. These are the foci (singular, *focus*). We then attach a piece of string to the tacks. Its length is the constant sum of the distances $d_1 + d_2$ from the foci to any point on the ellipse. Next, we trace a curve with a pen held tight against the string. The figure traced is an ellipse.

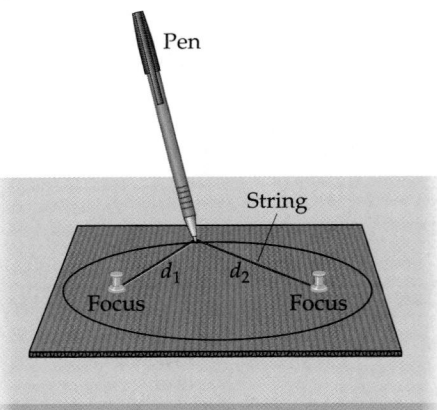

Let's first consider the ellipse shown below with center at the origin. The points F_1 and F_2 are the foci. The segment $\overline{A'A}$ is the **major axis**, and the points A' and A are the **vertices**. The segment $\overline{B'B}$ is the **minor axis**, and the points B' and B are the **y-intercepts**. Note that the major axis of an ellipse is longer than the minor axis.

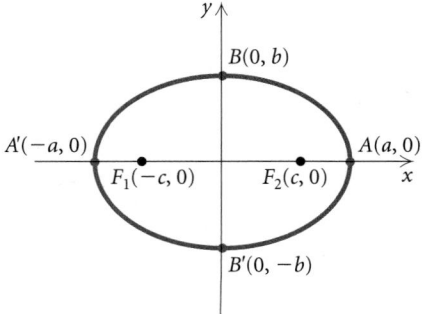

**Standard Equation of an Ellipse
with Center at the Origin**

Major Axis Horizontal

$$\frac{x^2}{a^2} + \frac{y^2}{b^2} = 1, \ a > b > 0$$

Vertices: $(-a, 0), (a, 0)$

y-intercepts: $(0, -b), (0, b)$

Foci: $(-c, 0), (c, 0)$, where $c^2 = a^2 - b^2$

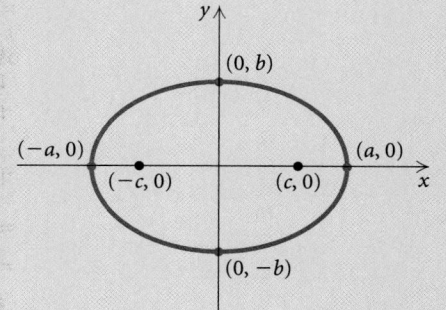

Major Axis Vertical

$$\frac{x^2}{b^2} + \frac{y^2}{a^2} = 1, \ a > b > 0$$

Vertices: $(0, -a), (0, a)$

x-intercepts: $(-b, 0), (b, 0)$

Foci: $(0, -c), (0, c)$, where $c^2 = a^2 - b^2$

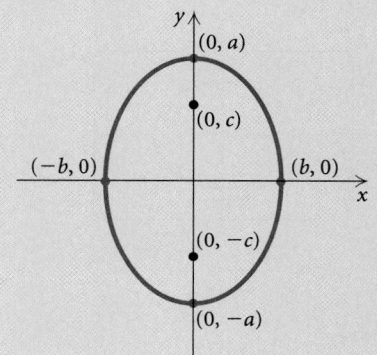

EXAMPLE 2 Find the standard equation of the ellipse with vertices $(-5, 0)$ and $(5, 0)$ and foci $(-3, 0)$ and $(3, 0)$. Then graph the ellipse.

Solution Since the foci are on the x-axis and the origin is the midpoint of the segment between them, the major axis is horizontal and $(0, 0)$ is the center of the ellipse. Thus the equation is of the form

$$\frac{x^2}{a^2} + \frac{y^2}{b^2} = 1.$$

Since the vertices are $(-5, 0)$ and $(5, 0)$ and the foci are $(-3, 0)$ and $(3, 0)$, we know that $a = 5$ and $c = 3$. These values can be used to find b^2:

$$c^2 = a^2 - b^2$$
$$3^2 = 5^2 - b^2$$
$$9 = 25 - b^2$$
$$b^2 = 16.$$

Thus the equation of the ellipse is

$$\frac{x^2}{5^2} + \frac{y^2}{4^2} = 1, \quad \text{or} \quad \frac{x^2}{25} + \frac{y^2}{16} = 1.$$

$$\frac{x^2}{25} + \frac{y^2}{16} = 1$$

$$y_1 = -\sqrt{\frac{400 - 16x^2}{25}},$$

$$y_2 = \sqrt{\frac{400 - 16x^2}{25}}$$

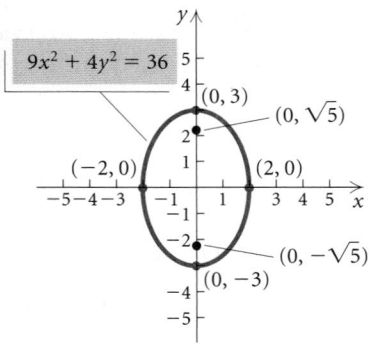

To graph the ellipse, we plot the vertices $(-5, 0)$ and $(5, 0)$. Since $b^2 = 16$, we know that $b = 4$ and the y-intercepts are $(0, -4)$ and $(0, 4)$. We plot these points as well and connect the four points we have plotted with a smooth curve.

To draw the graph using a graphing calculator, it might be necessary to solve for y first:

$$y = \pm\sqrt{\frac{400 - 16x^2}{25}}.$$

Then we graph

$$y_1 = -\sqrt{\frac{400 - 16x^2}{25}} \quad \text{and} \quad y_2 = \sqrt{\frac{400 - 16x^2}{25}}$$

or

$$y_1 = -\sqrt{\frac{400 - 16x^2}{25}} \quad \text{and} \quad y_2 = -y_1$$

in a squared viewing window.

On some graphing calculators, the Conics application from the APPS menu can be used to graph ellipses. This method will be discussed in Example 4.

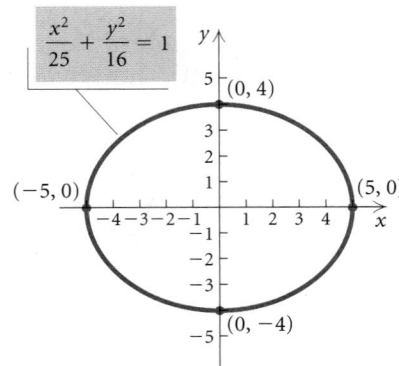

Now Try Exercise 31. ▪

EXAMPLE 3 For the ellipse

$$9x^2 + 4y^2 = 36,$$

find the vertices and the foci. Then draw the graph.

Solution We first find standard form:

$$9x^2 + 4y^2 = 36$$

$$\frac{9x^2}{36} + \frac{4y^2}{36} = \frac{36}{36} \qquad \text{Dividing by 36 on both sides to get 1 on the right side}$$

$$\frac{x^2}{4} + \frac{y^2}{9} = 1$$

$$\frac{x^2}{2^2} + \frac{y^2}{3^2} = 1. \qquad \text{Writing standard form}$$

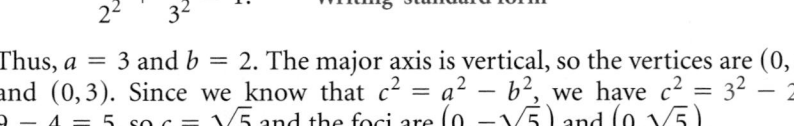

Thus, $a = 3$ and $b = 2$. The major axis is vertical, so the vertices are $(0, -3)$ and $(0, 3)$. Since we know that $c^2 = a^2 - b^2$, we have $c^2 = 3^2 - 2^2 = 9 - 4 = 5$, so $c = \sqrt{5}$ and the foci are $(0, -\sqrt{5})$ and $(0, \sqrt{5})$.

To graph the ellipse, we plot the vertices. Note also that since $b = 2$, the x-intercepts are $(-2, 0)$ and $(2, 0)$. We plot these points as well and connect the four points we have plotted with a smooth curve.

Now Try Exercise 25. ▪

If the center of an ellipse is not at the origin but at some point (h, k), then we can think of an ellipse with center at the origin being translated horizontally $|h|$ units and vertically $|k|$ units.

Standard Equation of an Ellipse with Center at (h, k)

Major Axis Horizontal

$$\frac{(x - h)^2}{a^2} + \frac{(y - k)^2}{b^2} = 1, \ a > b > 0$$

Vertices: $(h - a, k), (h + a, k)$

Length of minor axis: $2b$

Foci: $(h - c, k), (h + c, k)$, where $c^2 = a^2 - b^2$

Major Axis Vertical

$$\frac{(x - h)^2}{b^2} + \frac{(y - k)^2}{a^2} = 1, \ a > b > 0$$

Vertices: $(h, k - a), (h, k + a)$

Length of minor axis: $2b$

Foci: $(h, k - c), (h, k + c)$, where $c^2 = a^2 - b^2$

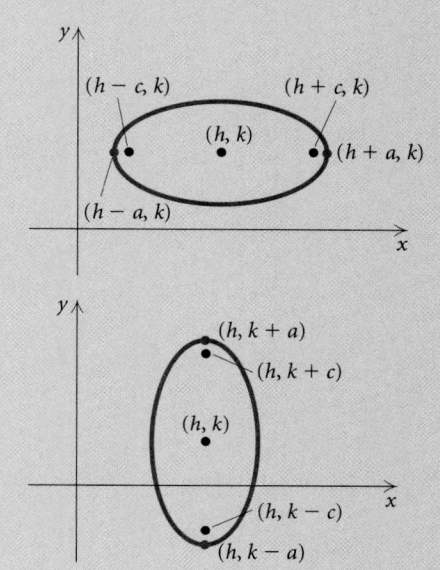

GCM **EXAMPLE 4** For the ellipse

$$4x^2 + y^2 + 24x - 2y + 21 = 0,$$

find the center, the vertices, and the foci. Then draw the graph.

Solution First, we complete the square twice to get standard form:

$$4x^2 + y^2 + 24x - 2y + 21 = 0$$

$$4(x^2 + 6x \quad) + (y^2 - 2y \quad) = -21$$

$$4(x^2 + 6x + 9) + (y^2 - 2y + 1) = -21 + 4 \cdot 9 + 1$$

Completing the square twice by adding $4 \cdot 9$ and 1 on both sides

$$4(x + 3)^2 + (y - 1)^2 = 16$$

$$\frac{1}{16}[4(x + 3)^2 + (y - 1)^2] = \frac{1}{16} \cdot 16$$

$$\frac{(x + 3)^2}{4} + \frac{(y - 1)^2}{16} = 1$$

$$\frac{[x - (-3)]^2}{2^2} + \frac{(y - 1)^2}{4^2} = 1.$$

Writing standard form:
$$\frac{(x - h)^2}{b^2} + \frac{(y - k)^2}{a^2} = 1$$

The center is $(-3, 1)$. Note that $a = 4$ and $b = 2$. The major axis is vertical, so the vertices are 4 units above and below the center:

$$(-3, 1 + 4) \text{ and } (-3, 1 - 4), \quad \text{or} \quad (-3, 5) \text{ and } (-3, -3).$$

We know that $c^2 = a^2 - b^2$, so $c^2 = 4^2 - 2^2 = 16 - 4 = 12$ and $c = \sqrt{12}$, or $2\sqrt{3}$. Then the foci are $2\sqrt{3}$ units above and below the center:

$$\left(-3, 1 + 2\sqrt{3}\right) \quad \text{and} \quad \left(-3, 1 - 2\sqrt{3}\right).$$

To graph the ellipse, we plot the vertices. Note also that since $b = 2$, two other points on the graph are the endpoints of the minor axis, 2 units right and left of the center:

$$(-3 + 2, 1) \quad \text{and} \quad (-3 - 2, 1),$$

or

$$(-1, 1) \quad \text{and} \quad (-5, 1).$$

We plot these points as well and connect the four points with a smooth curve.

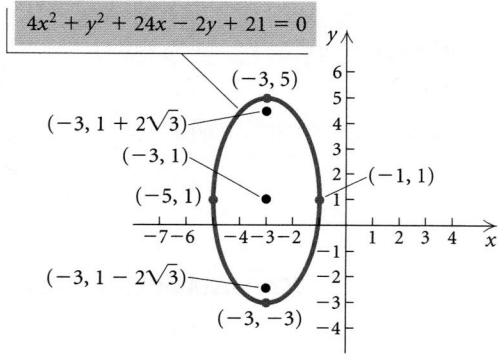

When the equation of an ellipse is written in standard form, we can use the Conics ELLIPSE APP to graph it.

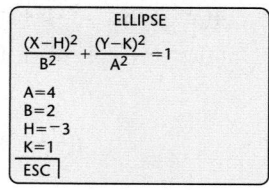
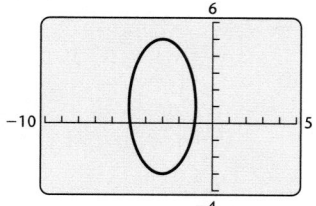

Now Try Exercise 43. ■

❊ Applications

An exciting medical application of an ellipse is a device called a *lithotripter*. One type of this device uses electromagnetic technology to generate a shock wave to pulverize kidney stones. The wave originates at one focus of an ellipse and is reflected to the kidney stone, which is positioned at the other focus. Recovery time following the use of this technique is much shorter than with conventional surgery and the mortality rate is far lower.

A room with an ellipsoidal ceiling is known as a *whispering gallery*. In such a room, a word whispered at one focus can be clearly heard at the other. Whispering galleries are found in the rotunda of the Capitol Building in Washington, D.C., and in the Mormon Tabernacle in Salt Lake City.

Ellipses have many other applications. Planets travel around the sun in elliptical orbits with the sun at one focus, for example, and satellites travel around the earth in elliptical orbits as well.

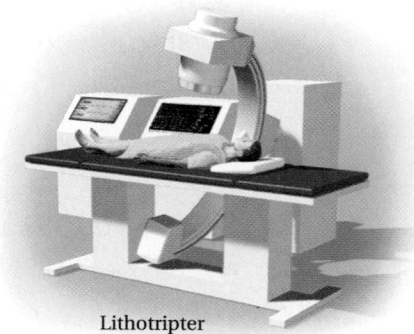

Lithotripter

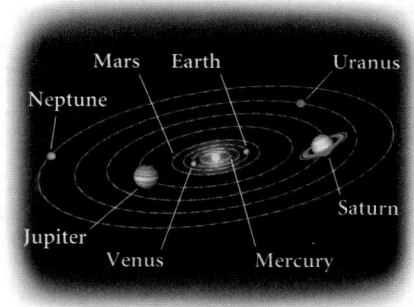

(10.2) Exercise Set

In Exercises 1–6, match the equation with one of the graphs (a)–(f), which follow.

a)
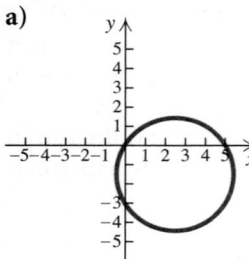

b)
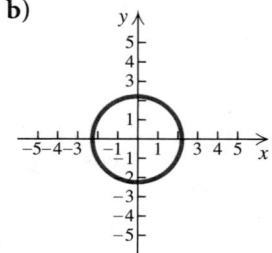

c)
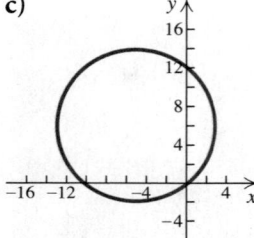

d)
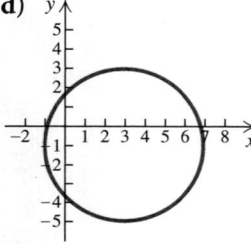

e)
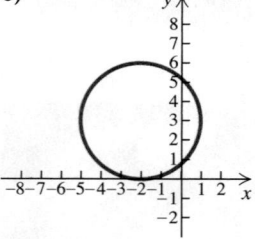

f)
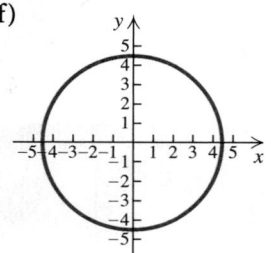

1. $x^2 + y^2 = 5$

2. $y^2 = 20 - x^2$

3. $x^2 + y^2 - 6x + 2y = 6$

4. $x^2 + y^2 + 10x - 12y = 3$

5. $x^2 + y^2 - 5x + 3y = 0$

6. $x^2 + 4x - 2 = 6y - y^2 - 6$

Find the center and the radius of the circle with the given equation. Then draw the graph.

7. $x^2 + y^2 - 14x + 4y = 11$

8. $x^2 + y^2 + 2x - 6y = -6$

9. $x^2 + y^2 + 6x - 2y = 6$

10. $x^2 + y^2 - 4x + 2y = 4$

11. $x^2 + y^2 + 4x - 6y - 12 = 0$

12. $x^2 + y^2 - 8x - 2y - 19 = 0$

13. $x^2 + y^2 - 6x - 8y + 16 = 0$

14. $x^2 + y^2 - 2x + 6y + 1 = 0$

15. $x^2 + y^2 + 6x - 10y = 0$

16. $x^2 + y^2 - 7x - 2y = 0$

17. $x^2 + y^2 - 9x = 7 - 4y$

18. $y^2 - 6y - 1 = 8x - x^2 + 3$

In Exercises 19–22, match the equation with one of the graphs (a)–(d), which follow.

a)
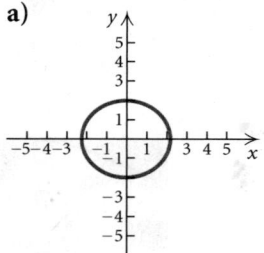

b)
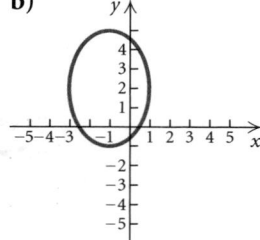

c)
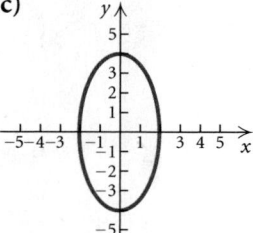

d)
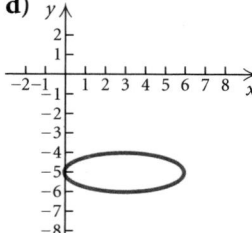

19. $16x^2 + 4y^2 = 64$

20. $4x^2 + 5y^2 = 20$

21. $x^2 + 9y^2 - 6x + 90y = -225$

22. $9x^2 + 4y^2 + 18x - 16y = 11$

Find the vertices and the foci of the ellipse with the given equation. Then draw the graph.

23. $\dfrac{x^2}{4} + \dfrac{y^2}{1} = 1$

24. $\dfrac{x^2}{25} + \dfrac{y^2}{36} = 1$

25. $16x^2 + 9y^2 = 144$

26. $9x^2 + 4y^2 = 36$

27. $2x^2 + 3y^2 = 6$

28. $5x^2 + 7y^2 = 35$

29. $4x^2 + 9y^2 = 1$

30. $25x^2 + 16y^2 = 1$

Find an equation of an ellipse satisfying the given conditions.

31. Vertices: $(-7,0)$ and $(7,0)$;
foci: $(-3,0)$ and $(3,0)$

32. Vertices: $(0,-6)$ and $(0,6)$;
foci: $(0,-4)$ and $(0,4)$

33. Vertices: $(0,-8)$ and $(0,8)$;
length of minor axis: 10

34. Vertices: $(-5,0)$ and $(5,0)$;
length of minor axis: 6

35. Foci: $(-2,0)$ and $(2,0)$;
length of major axis: 6

36. Foci: $(0,-3)$ and $(0,3)$;
length of major axis: 10

Find the center, the vertices, and the foci of the ellipse. Then draw the graph.

37. $\dfrac{(x-1)^2}{9} + \dfrac{(y-2)^2}{4} = 1$

38. $\dfrac{(x-1)^2}{1} + \dfrac{(y-2)^2}{4} = 1$

39. $\dfrac{(x+3)^2}{25} + \dfrac{(y-5)^2}{36} = 1$

40. $\dfrac{(x-2)^2}{16} + \dfrac{(y+3)^2}{25} = 1$

41. $3(x+2)^2 + 4(y-1)^2 = 192$

42. $4(x-5)^2 + 3(y-4)^2 = 48$

43. $4x^2 + 9y^2 - 16x + 18y - 11 = 0$

44. $x^2 + 2y^2 - 10x + 8y + 29 = 0$

45. $4x^2 + y^2 - 8x - 2y + 1 = 0$

46. $9x^2 + 4y^2 + 54x - 8y + 49 = 0$

*The **eccentricity** of an ellipse is defined as $e = c/a$. For an ellipse, $0 < c < a$, so $0 < e < 1$. When e is close to 0, an ellipse appears to be nearly circular. When e is close to 1, an ellipse is very flat.*

47. Observe the shapes of the ellipses in Examples 2 and 4. Which ellipse has the smaller eccentricity? Confirm your answer by computing the eccentricity of each ellipse.

48. Which ellipse has the smaller eccentricity? (Assume that the coordinate systems have the same scale.)

a)   b)

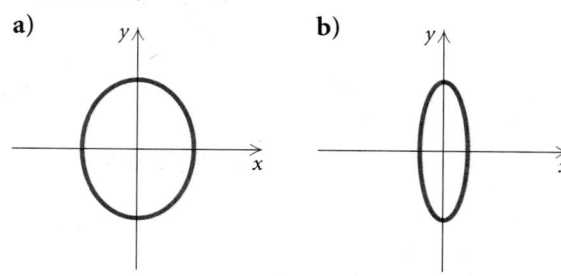

49. Find an equation of an ellipse with vertices $(0,-4)$ and $(0,4)$ and $e = \frac{1}{4}$.

50. Find an equation of an ellipse with vertices $(-3,0)$ and $(3,0)$ and $e = \frac{7}{10}$.

51. *Bridge Supports.* The bridge support shown in the figure below is the top half of an ellipse. Assuming that a coordinate system is superimposed on the drawing in such a way that point Q, the center of the ellipse, is at the origin, find an equation of the ellipse.

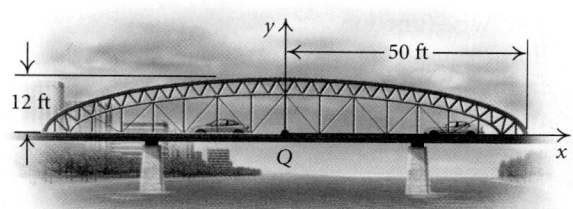

52. *The Ellipse.*   In Washington, D.C., there is a large grassy area south of the White House known as the Ellipse. It is actually an ellipse with major axis of length 1048 ft and minor axis of length 898 ft. Assuming that a coordinate system is super-imposed on the area in such a way that the center is at the origin and the major and minor axes are on the *x*- and *y*-axes of the coordinate system, respectively, find an equation of the ellipse.

53. *The Earth's Orbit.*   The maximum distance of the earth from the sun is 9.3×10^7 mi. The minimum distance is 9.1×10^7 mi. The sun is at one focus of the elliptical orbit. Find the distance from the sun to the other focus.

54. *Carpentry.*   A carpenter is cutting a 3-ft by 4-ft elliptical sign from a 3-ft by 4-ft piece of plywood. The ellipse will be drawn using a string attached to the board at the foci of the ellipse.

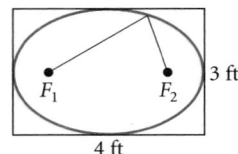

a) How far from the ends of the board should the string be attached?

b) How long should the string be?

Collaborative Discussion and Writing

55. Explain why function notation is not used in this section.

56. Is the center of an ellipse part of the graph of the ellipse? Why or why not?

Skill Maintenance

In each of Exercises 57–64, fill in the blank with the correct term. Some of the given choices will not be used.

piecewise function
linear equation
factor
remainder
solution
zero
x-intercept
y-intercept
parabola
circle
ellipse
midpoint
distance
one real-number solution
two different real-number solutions
two different imaginary-number solutions

57. The _____ between two points (x_1, y_1) and (x_2, y_2) is given by $\left(\dfrac{x_1 + x_2}{2}, \dfrac{y_1 + y_2}{2} \right)$.

58. An input *c* of a function *f* is a(n) _____ of the function if $f(c) = 0$.

59. A(n) _____ of the graph of an equation is a point $(0, b)$.

60. For a quadratic equation $ax^2 + bx + c = 0$, if $b^2 - 4ac > 0$, the equation has _____.

61. Given a polynomial $f(x)$, then $f(c)$ is the _____ that would be obtained by dividing $f(x)$ by $x - c$.

62. A(n) _____ is the set of all points in a plane the sum of whose distances from two fixed points is constant.

63. A(n) _____ is the set of all points in a plane equidistant from a fixed line and a fixed point not on the line.

64. A(n) _____ is the set of all points in a plane that are at a fixed distance from a fixed point in the plane.

Synthesis

Find an equation of an ellipse satisfying the given conditions.

65. Vertices: $(3, -4), (3, 6)$; endpoints of minor axis: $(1, 1), (5, 1)$

66. Vertices: $(-1, -1), (-1, 5)$; endpoints of minor axis: $(-3, 2), (1, 2)$

67. Vertices: $(-3, 0)$ and $(3, 0)$;
passing through $\left(2, \frac{22}{3}\right)$

68. Center: $(-2, 3)$; major axis vertical;
length of major axis: 4;
length of minor axis: 1

Use a graphing calculator to find the center and the
vertices of each of the following.

69. $4x^2 + 9y^2 - 16.025x + 18.0927y - 11.346 = 0$

70. $9x^2 + 4y^2 + 54.063x - 8.016y + 49.872 = 0$

71. *Bridge Arch.* A bridge with a semielliptical arch
spans a river as shown here. What is the clearance
6 ft from the riverbank?

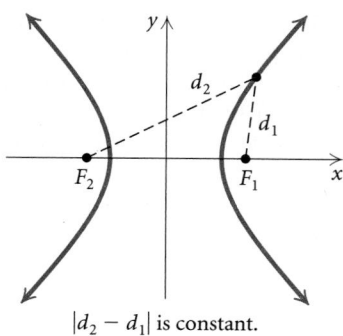

$|d_2 - d_1|$ is constant.

(**10.3**) **The Hyperbola**

❖ Given an equation of a hyperbola, complete the square, if
necessary, and then find the center, the vertices, and the
foci and graph the hyperbola.

The last type of conic section that we will study is the *hyperbola*.

> **Hyperbola**
>
> A **hyperbola** is the set of all points in a plane for which the absolute
> value of the difference of the distances from two fixed points (the **foci**)
> is constant. The midpoint of the segment between the foci is the
> **center** of the hyperbola.

❖ Standard Equations of Hyperbolas

We first consider the equation of a hyperbola with center at the origin. In the
figure below, F_1 and F_2 are the foci. The segment $\overline{V_2 V_1}$ is the **transverse axis**
and the points V_2 and V_1 are the **vertices**.

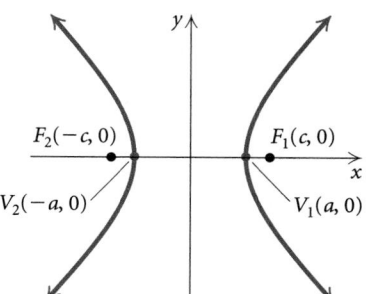

Standard Equation of a Hyperbola with Center at the Origin

Transverse Axis Horizontal

$$\frac{x^2}{a^2} - \frac{y^2}{b^2} = 1$$

Vertices: $(-a, 0), (a, 0)$

Foci: $(-c, 0), (c, 0)$,
where $c^2 = a^2 + b^2$

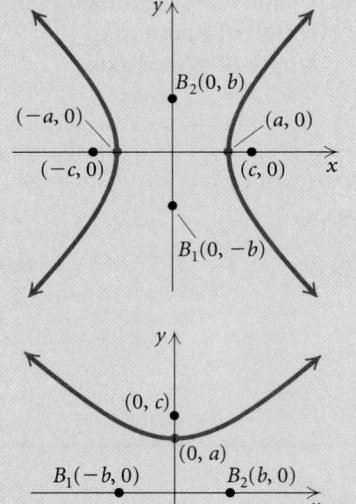

Transverse Axis Vertical

$$\frac{y^2}{a^2} - \frac{x^2}{b^2} = 1$$

Vertices: $(0, -a), (0, a)$

Foci: $(0, -c), (0, c)$,
where $c^2 = a^2 + b^2$

The segment $\overline{B_1 B_2}$ is the **conjugate axis** of the hyperbola.

To graph a hyperbola with a horizontal transverse axis, it is helpful to begin by graphing the lines $y = -(b/a)x$ and $y = (b/a)x$. These are the **asymptotes** of the hyperbola. For a hyperbola with a vertical transverse axis, the asymptotes are $y = -(a/b)x$ and $y = (a/b)x$. As $|x|$ gets larger and larger, the graph of the hyperbola gets closer and closer to the asymptotes.

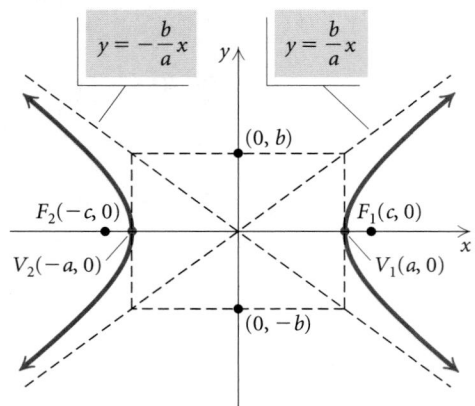

STUDY TIP

Take the time to include all the steps when working your homework problems. Doing so will help you organize your thinking and avoid computational errors. It will also give you complete, step-by-step solutions of the exercises that can be used to study for an exam.

EXAMPLE 1   Find an equation of the hyperbola with vertices $(0, -4)$ and $(0, 4)$ and foci $(0, -6)$ and $(0, 6)$.

Solution   We know that $a = 4$ and $c = 6$. We find b^2:

$$c^2 = a^2 + b^2$$
$$6^2 = 4^2 + b^2$$
$$36 = 16 + b^2$$
$$20 = b^2.$$

Since the vertices and the foci are on the y-axis, we know that the transverse axis is vertical. We can now write the equation of the hyperbola:

$$\frac{y^2}{a^2} - \frac{x^2}{b^2} = 1$$

$$\frac{y^2}{16} - \frac{x^2}{20} = 1.$$

Now Try Exercise 7. ■

GCM   **EXAMPLE 2**   For the hyperbola given by

$$9x^2 - 16y^2 = 144,$$

find the vertices, the foci, and the asymptotes. Then graph the hyperbola.

Solution   First, we find standard form:

$$9x^2 - 16y^2 = 144$$

$$\frac{1}{144}(9x^2 - 16y^2) = \frac{1}{144} \cdot 144 \qquad \text{Multiplying by } \tfrac{1}{144} \text{ to get 1 on the right side}$$

$$\frac{x^2}{16} - \frac{y^2}{9} = 1$$

$$\frac{x^2}{4^2} - \frac{y^2}{3^2} = 1. \qquad \text{Writing standard form}$$

The hyperbola has a horizontal transverse axis, so the vertices are $(-a, 0)$ and $(a, 0)$, or $(-4, 0)$ and $(4, 0)$. From the standard form of the equation, we know that $a^2 = 4^2$, or 16, and $b^2 = 3^2$, or 9. We find the foci:

$$c^2 = a^2 + b^2$$
$$c^2 = 16 + 9$$
$$c^2 = 25$$
$$c = 5.$$

Thus the foci are $(-5, 0)$ and $(5, 0)$.

Next, we find the asymptotes:

$$y = -\frac{b}{a}x = -\frac{3}{4}x \quad \text{and} \quad y = \frac{b}{a}x = \frac{3}{4}x.$$

To draw the graph, we sketch the asymptotes first. This is easily done by drawing the rectangle with horizontal sides passing through $(0, 3)$ and $(0, -3)$ and vertical sides through $(4, 0)$ and $(-4, 0)$. Then we draw and extend the diagonals of this rectangle. The two extended diagonals are the asymptotes of the hyperbola. Next, we plot the vertices and draw the branches of the hyperbola outward from the vertices toward the asymptotes.

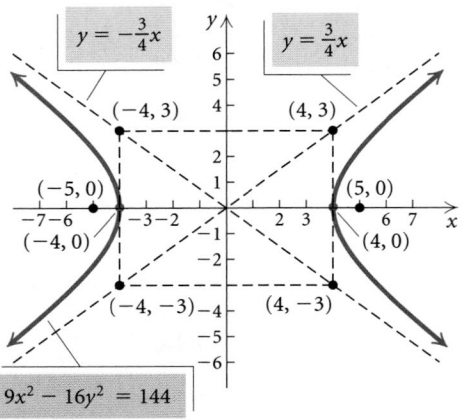

To graph this hyperbola on a graphing calculator, it might be necessary to solve for y first and then graph the top and bottom halves of the hyperbola in the same squared viewing window.

$$9x^2 - 16y^2 = 144$$

$$y_1 = \sqrt{\frac{9x^2 - 144}{16}}, \quad y_2 = -\sqrt{\frac{9x^2 - 144}{16}}$$

On some graphing calculators, the Conics HYPERBOLA APP can be used to graph hyperbolas. This method will be discussed in Example 3.

Now Try Exercise 17. ■

If a hyperbola with center at the origin is translated horizontally $|h|$ units and vertically $|k|$ units, the center is at the point (h, k).

Standard Equation of a Hyperbola with Center (h, k)

Transverse Axis Horizontal

$$\frac{(x - h)^2}{a^2} - \frac{(y - k)^2}{b^2} = 1$$

Vertices: $(h - a, k), (h + a, k)$

Asymptotes: $y - k = \dfrac{b}{a}(x - h), y - k = -\dfrac{b}{a}(x - h)$

Foci: $(h - c, k), (h + c, k)$, where $c^2 = a^2 + b^2$

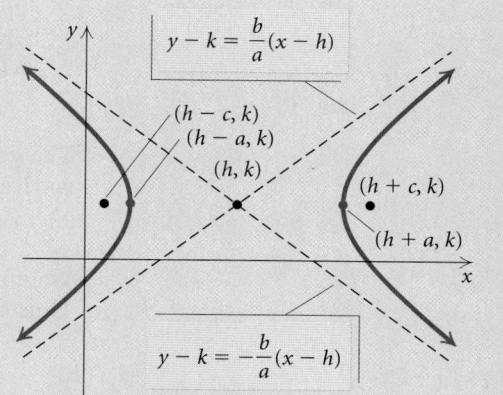

Transverse Axis Vertical

$$\frac{(y - k)^2}{a^2} - \frac{(x - h)^2}{b^2} = 1$$

Vertices: $(h, k - a), (h, k + a)$

Asymptotes: $y - k = \dfrac{a}{b}(x - h), y - k = -\dfrac{a}{b}(x - h)$

Foci: $(h, k - c), (h, k + c)$, where $c^2 = a^2 + b^2$

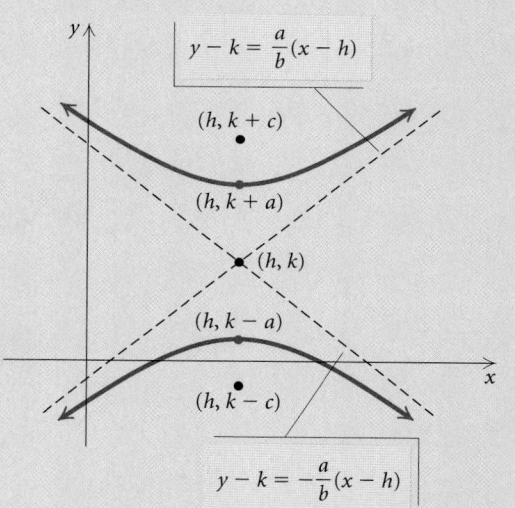

GCM | **EXAMPLE 3**   For the hyperbola given by

$$4y^2 - x^2 + 24y + 4x + 28 = 0,$$

find the center, the vertices, and the foci. Then draw the graph.

Solution   First, we complete the square to get standard form:

$$4y^2 - x^2 + 24y + 4x + 28 = 0$$
$$4(y^2 + 6y \qquad) - (x^2 - 4x \qquad) = -28$$
$$4(y^2 + 6y + 9 - 9) - (x^2 - 4x + 4 - 4) = -28$$
$$4(y^2 + 6y + 9) + 4(-9) - (x^2 - 4x + 4) - (-4) = -28$$
$$4(y^2 + 6y + 9) - 36 - (x^2 - 4x + 4) + 4 = -28$$
$$4(y^2 + 6y + 9) - (x^2 - 4x + 4) = -28 + 36 - 4$$
$$4(y + 3)^2 - (x - 2)^2 = 4$$

$$\frac{(y + 3)^2}{1} - \frac{(x - 2)^2}{4} = 1 \qquad \text{Dividing by 4}$$

$$\frac{[y - (-3)]^2}{1^2} - \frac{(x - 2)^2}{2^2} = 1. \qquad \text{Standard form}$$

The center is $(2, -3)$. Note that $a = 1$ and $b = 2$. The transverse axis is vertical, so the vertices are 1 unit below and above the center:

$$(2, -3 - 1) \text{ and } (2, -3 + 1), \quad \text{or} \quad (2, -4) \text{ and } (2, -2).$$

We know that $c^2 = a^2 + b^2$, so $c^2 = 1^2 + 2^2 = 1 + 4 = 5$ and $c = \sqrt{5}$. Thus the foci are $\sqrt{5}$ units below and above the center:

$$\left(2, -3 - \sqrt{5}\right) \quad \text{and} \quad \left(2, -3 + \sqrt{5}\right).$$

The asymptotes are

$$y - (-3) = \frac{1}{2}(x - 2) \quad \text{and} \quad y - (-3) = -\frac{1}{2}(x - 2),$$

or

$$y + 3 = \frac{1}{2}(x - 2) \quad \text{and} \quad y + 3 = -\frac{1}{2}(x - 2).$$

We sketch the asymptotes, plot the vertices, and draw the graph.

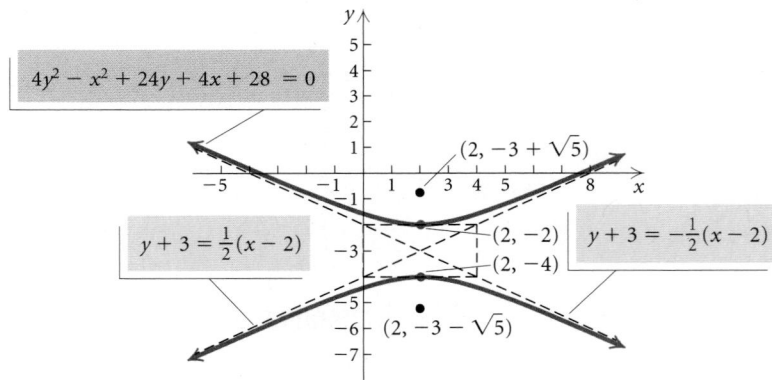

When the equation of a hyperbola is written in standard form, we can use the Conics HYPERBOLA APP to graph it.

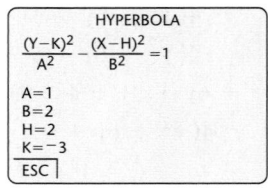

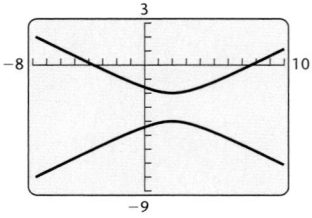

Now Try Exercise 29. ■

CONNECTING
the CONCEPTS

Classifying Equations of Conic Sections

EQUATION	TYPE OF CONIC SECTION	GRAPH

$x - 4 + 4y = y^2$

Only one variable is squared, so this cannot be a circle, an ellipse, or a hyperbola. Find an equivalent equation:

$$x = (y - 2)^2.$$

This is an equation of a parabola.

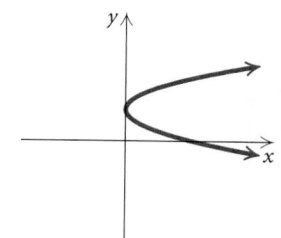

$3x^2 + 3y^2 = 75$

Both variables are squared, so this cannot be a parabola. The squared terms are added, so this cannot be a hyperbola. Divide by 3 on both sides to find an equivalent equation:

$$x^2 + y^2 = 25.$$

This is an equation of a circle.

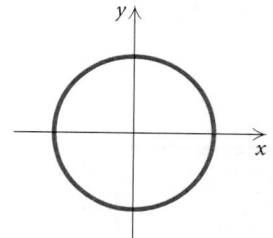

$y^2 = 16 - 4x^2$

Both variables are squared, so this cannot be a parabola. Add $4x^2$ on both sides to find an equivalent equation: $4x^2 + y^2 = 16$. The squared terms are added, so this cannot be a hyperbola. The coefficients of x^2 and y^2 are not the same, so this is not a circle. Divide by 16 on both sides to find an equivalent equation:

$$\frac{x^2}{4} + \frac{y^2}{16} = 1.$$

This is an equation of an ellipse.

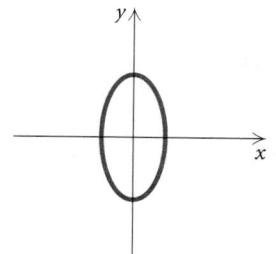

$x^2 = 4y^2 + 36$

Both variables are squared, so this cannot be a parabola. Subtract $4y^2$ on both sides to find an equivalent equation: $x^2 - 4y^2 = 36$. The squared terms are not added, so this cannot be a circle or an ellipse. Divide by 36 on both sides to find an equivalent equation:

$$\frac{x^2}{36} - \frac{y^2}{9} = 1.$$

This is an equation of a hyperbola.

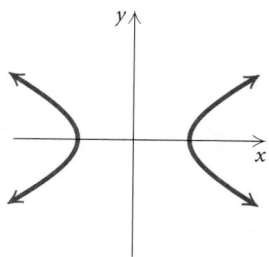

❋ Applications

Some comets travel in hyperbolic paths with the sun at one focus. Such comets pass by the sun only one time, unlike those with elliptical orbits, which reappear at intervals. A cross section of an amphitheater might be one branch of a hyperbola. A cross section of a nuclear cooling tower might also be a hyperbola.

Another application of hyperbolas is in the long-range navigation system LORAN. This system uses transmitting stations in three locations to send out simultaneous signals to a ship or aircraft. The difference in the arrival times of the signals from one pair of transmitters is recorded on the ship or aircraft. This difference is also recorded for signals from another pair of transmitters. For each pair, a computation is performed to determine the difference in the distances from each member of the pair to the ship or aircraft. If each pair of differences is kept constant, two hyperbolas can be drawn. Each has one of the pairs of transmitters as foci, and the ship or aircraft lies on the intersection of two of their branches.

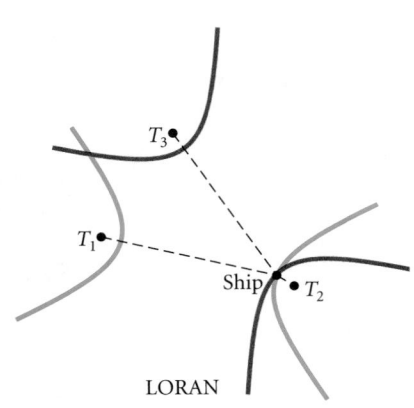

LORAN

(10.3)   ## Exercise Set

In Exercises 1–6, match the equation with one of the graphs (a)–(f), which follow.

a)
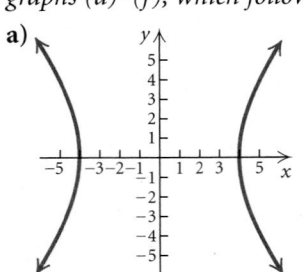

b)
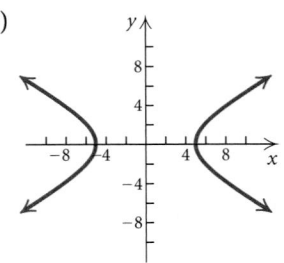

c)
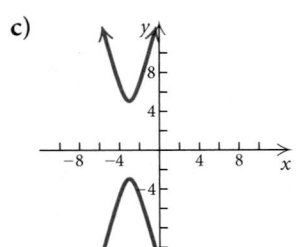

d)
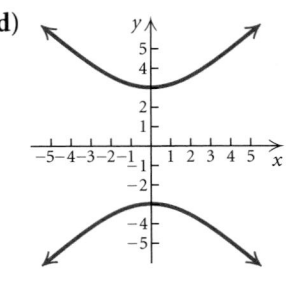

e)
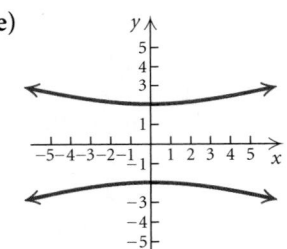

f)
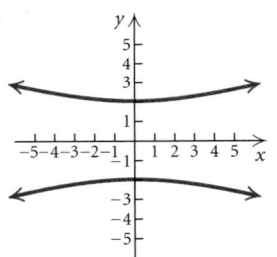

1. $\dfrac{x^2}{25} - \dfrac{y^2}{9} = 1$

2. $\dfrac{y^2}{4} - \dfrac{x^2}{36} = 1$

3. $\dfrac{(y-1)^2}{16} - \dfrac{(x+3)^2}{1} = 1$

4. $\dfrac{(x+4)^2}{100} - \dfrac{(y-2)^2}{81} = 1$

5. $25x^2 - 16y^2 = 400$

6. $y^2 - x^2 = 9$

Find an equation of a hyperbola satisfying the given conditions.

7. Vertices at $(0, 3)$ and $(0, -3)$;
foci at $(0, 5)$ and $(0, -5)$

8. Vertices at $(1, 0)$ and $(-1, 0)$;
foci at $(2, 0)$ and $(-2, 0)$

9. Asymptotes $y = \frac{3}{2}x$, $y = -\frac{3}{2}x$;
one vertex $(2, 0)$

10. Asymptotes $y = \frac{5}{4}x$, $y = -\frac{5}{4}x$;
one vertex $(0, 3)$

Find the center, the vertices, the foci, and the asymptotes. Then draw the graph.

11. $\dfrac{x^2}{4} - \dfrac{y^2}{4} = 1$

12. $\dfrac{x^2}{1} - \dfrac{y^2}{9} = 1$

13. $\dfrac{(x-2)^2}{9} - \dfrac{(y+5)^2}{1} = 1$

14. $\dfrac{(x-5)^2}{16} - \dfrac{(y+2)^2}{9} = 1$

15. $\dfrac{(y+3)^2}{4} - \dfrac{(x+1)^2}{16} = 1$

16. $\dfrac{(y+4)^2}{25} - \dfrac{(x+2)^2}{16} = 1$

17. $x^2 - 4y^2 = 4$

18. $4x^2 - y^2 = 16$

19. $9y^2 - x^2 = 81$

20. $y^2 - 4x^2 = 4$

21. $x^2 - y^2 = 2$

22. $x^2 - y^2 = 3$

23. $y^2 - x^2 = \frac{1}{4}$

24. $y^2 - x^2 = \frac{1}{9}$

Find the center, the vertices, the foci, and the asymptotes of the hyperbola. Then draw the graph.

25. $x^2 - y^2 - 2x - 4y - 4 = 0$

26. $4x^2 - y^2 + 8x - 4y - 4 = 0$

27. $36x^2 - y^2 - 24x + 6y - 41 = 0$

28. $9x^2 - 4y^2 + 54x + 8y + 41 = 0$

29. $9y^2 - 4x^2 - 18y + 24x - 63 = 0$

30. $x^2 - 25y^2 + 6x - 50y = 41$

31. $x^2 - y^2 - 2x - 4y = 4$

32. $9y^2 - 4x^2 - 54y - 8x + 41 = 0$

33. $y^2 - x^2 - 6x - 8y - 29 = 0$

34. $x^2 - y^2 = 8x - 2y - 13$

*The **eccentricity** of a hyperbola is defined as $e = c/a$. For a hyperbola, $c > a > 0$, so $e > 1$. When e is close to 1, a hyperbola appears to be very narrow. As the eccentricity increases, the hyperbola becomes "wider."*

35. Observe the shapes of the hyperbolas in Examples 2 and 3. Which hyperbola has the larger eccentricity? Confirm your answer by computing the eccentricity of each hyperbola.

36. Which hyperbola has the larger eccentricity? (Assume that the coordinate systems have the same scale.)

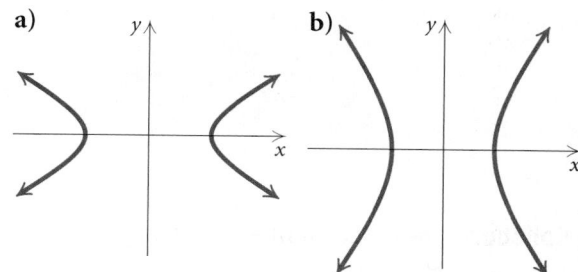

a)   **b)**

37. Find an equation of a hyperbola with vertices $(3, 7)$ and $(-3, 7)$ and $e = \frac{5}{3}$.

38. Find an equation of a hyperbola with vertices $(-1, 3)$ and $(-1, 7)$ and $e = 4$.

39. *Hyperbolic Mirror.* Certain telescopes contain both a parabolic mirror and a hyperbolic mirror. In the telescope shown in the figure, the parabola and the hyperbola share focus F_1, which is 14 m above the vertex of the parabola. The hyperbola's second focus F_2 is 2 m above the parabola's vertex. The vertex of the hyperbolic mirror is 1 m below F_1. Position a coordinate system with the origin at the center of the hyperbola and with the foci on the y-axis. Then find the equation of the hyperbola.

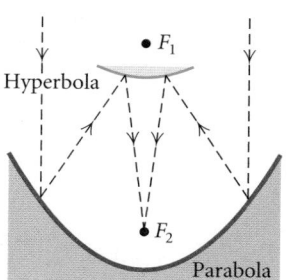

40. *Nuclear Cooling Tower.*   A cross section of a nuclear cooling tower is a hyperbola with equation

$$\frac{x^2}{90^2} - \frac{y^2}{130^2} = 1.$$

The tower is 450 ft tall and the distance from the top of the tower to the center of the hyperbola is half the distance from the base of the tower to the center of the hyperbola. Find the diameter of the top and the base of the tower.

Collaborative Discussion and Writing

41. How does the graph of a parabola differ from the graph of one branch of a hyperbola?

42. Are the asymptotes of a hyperbola part of the graph of the hyperbola? Why or why not?

Skill Maintenance

In Exercises 43–46, given the function:

a) *Determine whether it is one-to-one.*
b) *If it is one-to-one, find a formula for the inverse.*

43. $f(x) = 2x - 3$

44. $f(x) = x^3 + 2$

45. $f(x) = \dfrac{5}{x - 1}$

46. $f(x) = \sqrt{x + 4}$

Solve.

47. $x + y = 5,$
$\quad x - y = 7$

48. $3x - 2y = 5,$
$\quad 5x + 2y = 3$

49. $2x - 3y = 7,$
$\quad 3x + 5y = 1$

50. $3x + 2y = -1,$
$\quad 2x + 3y = 6$

Synthesis

Find an equation of a hyperbola satisfying the given conditions.

51. Vertices at $(3, -8)$ and $(3, -2)$; asymptotes $y = 3x - 14, y = -3x + 4$

52. Vertices at $(-9, 4)$ and $(-5, 4)$; asymptotes $y = 3x + 25, y = -3x - 17$

Use a graphing calculator to find the center, the vertices, and the asymptotes.

53. $5x^2 - 3.5y^2 + 14.6x - 6.7y + 3.4 = 0$

54. $x^2 - y^2 - 2.046x - 4.088y - 4.228 = 0$

55. *Navigation.*   Two radio transmitters positioned 300 mi apart along the shore send simultaneous signals to a ship that is 200 mi offshore, sailing parallel to the shoreline. The signal from transmitter S reaches the ship 200 microseconds later than the signal from transmitter T. The signals travel at a speed of 186,000 miles per second, or 0.186 mile per microsecond. Find the equation of the hyperbola with foci S and T on which the ship is located. (*Hint:* For any point on the hyperbola, the absolute value of the difference of its distances from the foci is $2a$.)

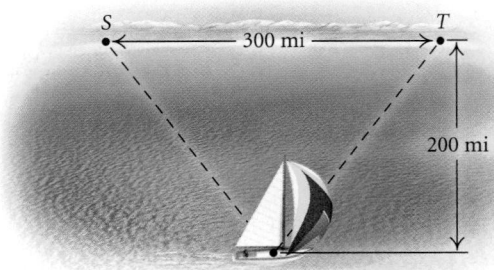

Nonlinear Systems of Equations and Inequalities

10.4

❋ Solve a nonlinear system of equations.
❋ Use nonlinear systems of equations to solve applied problems.
❋ Graph nonlinear systems of inequalities.

The systems of equations that we have studied so far have been composed of linear equations. Now we consider systems of two equations in two variables in which at least one equation is not linear.

❋ Nonlinear Systems of Equations

The graphs of the equations in a nonlinear system of equations can have no point of intersection or one or more points of intersection. The coordinates of each point of intersection represent a solution of the system of equations. When no point of intersection exists, the system of equations has no real-number solution.

Solutions of nonlinear systems of equations can be found using the substitution method or the elimination method. The substitution method is preferable for a system consisting of one linear and one nonlinear equation. The elimination method is preferable in most, but not all, cases when both equations are nonlinear.

EXAMPLE 1 Solve the following system of equations:

$$x^2 + y^2 = 25, \quad (1) \qquad \text{The graph is a circle.}$$
$$3x - 4y = 0. \quad (2) \qquad \text{The graph is a line.}$$

ALGEBRAIC SOLUTION

We use the substitution method. First, we solve equation (2) for x:

$$x = \tfrac{4}{3}y. \quad (3) \qquad \text{We could have solved for } y \text{ instead.}$$

Next, we substitute $\tfrac{4}{3}y$ for x in equation (1) and solve for y:

$$\left(\tfrac{4}{3}y\right)^2 + y^2 = 25$$
$$\tfrac{16}{9}y^2 + y^2 = 25$$
$$\tfrac{25}{9}y^2 = 25$$
$$y^2 = 9 \qquad \text{Multiplying by } \tfrac{9}{25}$$
$$y = \pm 3.$$

Now we substitute these numbers for y in equation (3) and solve for x:

$$x = \tfrac{4}{3}(3) = 4, \qquad (4, 3) \text{ appears to be a solution.}$$
$$x = \tfrac{4}{3}(-3) = -4. \qquad (-4, -3) \text{ appears to be a solution.}$$

(continued)

Check: For $(4, 3)$:

$$x^2 + y^2 = 25$$

$$4^2 + 3^2 \;?\; 25$$
$$16 + 9$$
$$\qquad 25 \;\bigg|\; 25 \quad \text{TRUE}$$

$$3x - 4y = 0$$

$$3(4) - 4(3) \;?\; 0$$
$$12 - 12$$
$$\qquad 0 \;\bigg|\; 0 \quad \text{TRUE}$$

For $(-4, -3)$:

$$x^2 + y^2 = 25$$

$$(-4)^2 + (-3)^2 \;?\; 25$$
$$16 + 9$$
$$\qquad 25 \;\bigg|\; 25 \quad \text{TRUE}$$

$$3x - 4y = 0$$

$$3(-4) - 4(-3) \;?\; 0$$
$$-12 + 12$$
$$\qquad 0 \;\bigg|\; 0 \quad \text{TRUE}$$

The pairs $(4, 3)$ and $(-4, -3)$ check, so they are the solutions.

GRAPHICAL SOLUTION

We graph both equations in the same viewing window. Note that there are two points of intersection. We can find their coordinates using the INTERSECT feature.

$$x^2 + y^2 = 25$$
$$y_1 = \sqrt{25 - x^2}, \quad y_2 = -\sqrt{25 - x^2}, \quad y_3 = \tfrac{3}{4}x$$

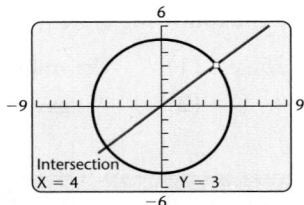

$$x^2 + y^2 = 25$$
$$y_1 = \sqrt{25 - x^2}, \quad y_2 = -\sqrt{25 - x^2}, \quad y_3 = \tfrac{3}{4}x$$

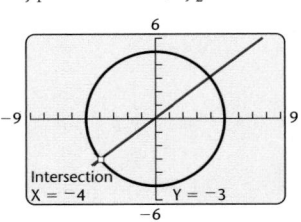

The solutions are $(4, 3)$ and $(-4, -3)$.

Now Try Exercise 7. ■

In the algebraic solution in Example 1, suppose that to find x we had substituted 3 and -3 in equation (1) rather than equation (3). If $y = 3$, $y^2 = 9$, and if $y = -3$, $y^2 = 9$, so both substitutions can be performed at the same time:

$$x^2 + y^2 = 25 \qquad (1)$$
$$x^2 + (\pm 3)^2 = 25$$
$$x^2 + 9 = 25$$
$$x^2 = 16$$
$$x = \pm 4.$$

Each y-value produces two values for x. Thus, if $y = 3$, $x = 4$ or $x = -4$, and if $y = -3$, $x = 4$ or $x = -4$. The possible solutions are $(4, 3)$, $(-4, 3)$, $(4, -3)$, and $(-4, -3)$. A check reveals that $(4, -3)$ and $(-4, 3)$ are not solutions of equation (2). Since a circle and a line can intersect in at most two points, it is clear that there can be at most two real-number solutions.

EXAMPLE 2 Solve the following system of equations:

$$x + y = 5, \quad (1) \qquad \text{The graph is a line.}$$
$$y = 3 - x^2. \quad (2) \qquad \text{The graph is a parabola.}$$

ALGEBRAIC SOLUTION

We use the substitution method, substituting $3 - x^2$ for y in equation (1):

$$x + 3 - x^2 = 5$$
$$-x^2 + x - 2 = 0 \qquad \text{Subtracting 5 and rearranging}$$
$$x^2 - x + 2 = 0. \qquad \text{Multiplying by } -1$$

Next, we use the quadratic formula:

$$x = \frac{-b \pm \sqrt{b^2 - 4ac}}{2a} = \frac{-(-1) \pm \sqrt{(-1)^2 - 4(1)(2)}}{2(1)}$$

$$= \frac{1 \pm \sqrt{1 - 8}}{2} = \frac{1 \pm \sqrt{-7}}{2} = \frac{1 \pm i\sqrt{7}}{2} = \frac{1}{2} \pm \frac{\sqrt{7}}{2}i.$$

Now, we substitute these values for x in equation (1) and solve for y:

$$\frac{1}{2} + \frac{\sqrt{7}}{2}i + y = 5$$

$$y = 5 - \frac{1}{2} - \frac{\sqrt{7}}{2}i = \frac{9}{2} - \frac{\sqrt{7}}{2}i$$

and $\quad \dfrac{1}{2} - \dfrac{\sqrt{7}}{2}i + y = 5$

$$y = 5 - \frac{1}{2} + \frac{\sqrt{7}}{2}i = \frac{9}{2} + \frac{\sqrt{7}}{2}i.$$

The solutions are

$$\left(\frac{1}{2} + \frac{\sqrt{7}}{2}i, \frac{9}{2} - \frac{\sqrt{7}}{2}i \right) \quad \text{and} \quad \left(\frac{1}{2} - \frac{\sqrt{7}}{2}i, \frac{9}{2} + \frac{\sqrt{7}}{2}i \right).$$

There are no real-number solutions.

GRAPHICAL SOLUTION

We graph both equations in the same viewing window.

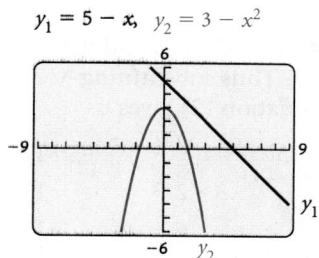

$y_1 = 5 - x, \quad y_2 = 3 - x^2$

Note that there are no points of intersection. This indicates that there are no real-number solutions. Algebra must be used, as at left, to find the imaginary-number solutions.

Now Try Exercise 17. ■

EXAMPLE 3 Solve the following system of equations:

$$2x^2 + 5y^2 = 39, \quad (1) \qquad \text{The graph is an ellipse.}$$
$$3x^2 - y^2 = -1. \quad (2) \qquad \text{The graph is a hyperbola.}$$

ALGEBRAIC SOLUTION

We use the elimination method. First, we multiply equation (2) by 5 and add to eliminate the y^2-term:

$$
\begin{array}{ll}
2x^2 + 5y^2 = 39 & (1) \\
\underline{15x^2 - 5y^2 = -5} & \text{Multiplying (2) by 5} \\
17x^2 \quad\quad = 34 & \text{Adding} \\
\quad\quad x^2 = 2 \\
\quad\quad x = \pm\sqrt{2}.
\end{array}
$$

If $x = \sqrt{2}$, $x^2 = 2$, and if $x = -\sqrt{2}$, $x^2 = 2$. Thus substituting $\sqrt{2}$ or $-\sqrt{2}$ for x in equation (2) gives us

$$
\begin{aligned}
3(\pm\sqrt{2})^2 - y^2 &= -1 \\
3 \cdot 2 - y^2 &= -1 \\
6 - y^2 &= -1 \\
-y^2 &= -7 \\
y^2 &= 7 \\
y &= \pm\sqrt{7}.
\end{aligned}
$$

Each x-value produces two values for y. Thus, for $x = \sqrt{2}$, we have $y = \sqrt{7}$ or $y = -\sqrt{7}$, and for $x = -\sqrt{2}$, we have $y = \sqrt{7}$ or $y = -\sqrt{7}$. The possible solutions are $(\sqrt{2}, \sqrt{7})$, $(\sqrt{2}, -\sqrt{7})$, $(-\sqrt{2}, \sqrt{7})$, and $(-\sqrt{2}, -\sqrt{7})$. All four pairs check, so they are the solutions.

GRAPHICAL SOLUTION

We graph both equations in the same viewing window. There are four points of intersection. We can use the INTERSECT feature to find their coordinates.

$$y_1 = \sqrt{(39 - 2x^2)/5}, \quad y_2 = -\sqrt{(39 - 2x^2)/5},$$
$$y_3 = \sqrt{3x^2 + 1}, \quad y_4 = -\sqrt{3x^2 + 1}$$

$(-1.414, 2.646)$ $(1.414, 2.646)$

$(-1.414, -2.646)$ $(1.414, -2.646)$

Note that the algebraic solution yields exact solutions, whereas the graphical solution yields decimal approximations of the solutions on most graphing calculators.

The solutions are approximately $(1.414, 2.646)$, $(1.414, -2.646)$, $(-1.414, 2.646)$, and $(-1.414, -2.646)$.

Now Try Exercise 27. ■

EXAMPLE 4 Solve the following system of equations:

$$x^2 - 3y^2 = 6, \quad (1)$$
$$xy = 3. \quad (2)$$

ALGEBRAIC SOLUTION

We use the substitution method. First, we solve equation (2) for y:

$$xy = 3 \quad (2)$$

$$y = \frac{3}{x}. \quad (3) \qquad \text{Dividing by } x$$

Next, we substitute $3/x$ for y in equation (1) and solve for x:

$$x^2 - 3\left(\frac{3}{x}\right)^2 = 6$$

$$x^2 - 3 \cdot \frac{9}{x^2} = 6$$

$$x^2 - \frac{27}{x^2} = 6$$

$$x^4 - 27 = 6x^2 \qquad \text{Multiplying by } x^2$$

$$x^4 - 6x^2 - 27 = 0$$

$$u^2 - 6u - 27 = 0 \qquad \text{Letting } u = x^2$$

$$(u - 9)(u + 3) = 0 \qquad \text{Factoring}$$

$$u = 9 \quad or \quad u = -3 \qquad \text{Principle of zero products}$$

$$x^2 = 9 \quad or \quad x^2 = -3 \qquad \text{Substituting } x^2 \text{ for } u$$

$$x = \pm 3 \quad or \quad x = \pm i\sqrt{3}.$$

Since $y = 3/x$,

when $x = 3$, $\qquad y = \dfrac{3}{3} = 1$;

when $x = -3$, $\qquad y = \dfrac{3}{-3} = -1$;

when $x = i\sqrt{3}$, $\qquad y = \dfrac{3}{i\sqrt{3}} = \dfrac{3}{i\sqrt{3}} \cdot \dfrac{-i\sqrt{3}}{-i\sqrt{3}} = -i\sqrt{3}$;

when $x = -i\sqrt{3}$, $\quad y = \dfrac{3}{-i\sqrt{3}} = \dfrac{3}{-i\sqrt{3}} \cdot \dfrac{i\sqrt{3}}{i\sqrt{3}} = i\sqrt{3}$.

The pairs $(3, 1)$, $(-3, -1)$, $\left(i\sqrt{3}, -i\sqrt{3}\right)$, and $\left(-i\sqrt{3}, i\sqrt{3}\right)$ check, so they are the solutions.

GRAPHICAL SOLUTION

We graph both equations in the same viewing window and find the coordinates of their points of intersection.

$$x^2 - 3y^2 = 6$$

$$y_1 = \sqrt{(x^2 - 6)/3}, \quad y_2 = -\sqrt{(x^2 - 6)/3}, \quad y_3 = 3/x$$

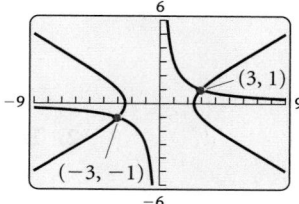

Again, note that the graphical method yields only the real-number solutions of the system of equations. The algebraic method must be used in order to find *all* the solutions.

Now Try Exercise 19. ▓

❋ Modeling and Problem Solving

EXAMPLE 5   *Dimensions of a Piece of Land.*   For a student recreation building at Southport Community College, an architect wants to lay out a rectangular piece of land that has a perimeter of 204 m and an area of 2565 m². Find the dimensions of the piece of land.

Solution

1. **Familiarize.**   We make a drawing and label it, letting l = the length of the piece of land, in meters, and w = the width, in meters.

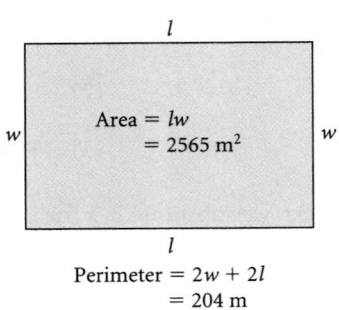

$$\text{Perimeter} = 2w + 2l$$
$$= 204 \text{ m}$$

2. **Translate.**   We now have the following:

   Perimeter: $2w + 2l = 204$,   (1)

   Area: $lw = 2565$.   (2)

3. **Carry out.**   We solve the system of equations both algebraically and graphically.

ALGEBRAIC SOLUTION

We solve the system of equations

$$2w + 2l = 204,$$
$$lw = 2565.$$

Solving the second equation for l gives us $l = 2565/w$. We then substitute $2565/w$ for l in equation (1) and solve for w:

$$2w + 2\left(\frac{2565}{w}\right) = 204$$

$$2w^2 + 2(2565) = 204w \qquad \text{Multiplying by } w$$

$$2w^2 - 204w + 2(2565) = 0$$

$$w^2 - 102w + 2565 = 0 \qquad \text{Multiplying by } \tfrac{1}{2}$$

$$(w - 57)(w - 45) = 0$$

$$w = 57 \quad or \quad w = 45. \qquad \begin{array}{l}\text{Principle of zero}\\ \text{products}\end{array}$$

If $w = 57$, then $l = 2565/w = 2565/57 = 45$. If $w = 45$, then $l = 2565/w = 2565/45 = 57$. Since length is generally considered to be longer than width, we have the solution $l = 57$ and $w = 45$, or $(57, 45)$.

GRAPHICAL SOLUTION

We replace l with x and w with y, graph $y_1 = (204 - 2x)/2$ and $y_2 = 2565/x$, and find the point(s) of intersection of the graphs.

$$y_1 = (204 - 2x)/2, \ y_2 = 2565/x$$

As in the algebraic solution, we have two possible solutions: $(45, 57)$ and $(57, 45)$. Since length, x, is generally considered to be longer than width, y, we have the solution $(57, 45)$.

4. **Check.**   If $l = 57$ and $w = 45$, the perimeter is $2 \cdot 45 + 2 \cdot 57$, or 204. The area is $57 \cdot 45$, or 2565. The numbers check.

5. **State.**   The length of the piece of land is 57 m and the width is 45 m.

Now Try Exercise 65. ■

❊ Nonlinear Systems of Inequalities

SYSTEMS OF INEQUALITIES

REVIEW SECTION **9.7.**

Recall that a solution of a system of inequalities is an ordered pair that is a solution of each inequality in the system. We graphed systems of linear inequalities in Section 9.7. Now we graph a nonlinear system of inequalities.

EXAMPLE 6   Graph the solution set of the system

$$x^2 + y^2 \le 25,$$
$$3x - 4y > 0.$$

Solution   We graph $x^2 + y^2 \le 25$ by first graphing the equation of the circle $x^2 + y^2 = 25$. We use a solid line since the inequality symbol is $\le$. Next, we choose $(0,0)$ as a test point and find that it is a solution of $x^2 + y^2 \le 25$, so we shade the region that contains $(0,0)$ using red. This is the region inside the circle. Now we graph the line $3x - 4y = 0$ using a dashed line since the inequality symbol is $>$. The point $(0,0)$ is on the line, so we choose another test point, say, $(0,2)$. We find that this point is not a solution of $3x - 4y > 0$, so we shade the half-plane that does not contain $(0,2)$ using green. The solution set of the system of inequalities is the region shaded both red and green, or brown, including part of the circle $x^2 + y^2 = 25$.

To find the points of intersection of the graphs, we solve the system of equations

$$x^2 + y^2 = 25,$$
$$3x - 4y = 0.$$

In Example 1 we found that these points are $(4,3)$ and $(-4,-3)$.

Now Try Exercise 75. ■

EXAMPLE 7   Use a graphing calculator to graph the system

$$y \le 4 - x^2,$$
$$x + y \ge 2.$$

Solution   We graph $y_1 = 4 - x^2$ and $y_2 = 2 - x$. Using the test point $(0,0)$ for each inequality, we find that we should shade below y_1 and above y_2. We can find the points of intersection of the graphs, $(-1,3)$ and $(2,0)$, using the INTERSECT feature.

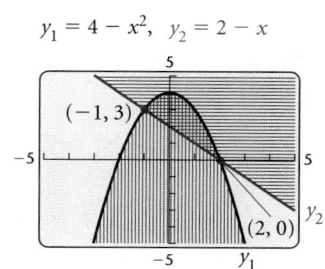

Now Try Exercise 81. ■

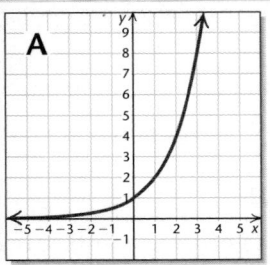

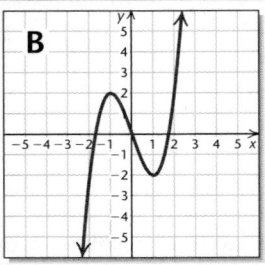

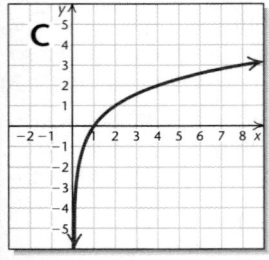

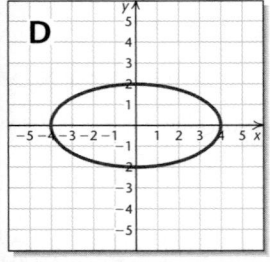

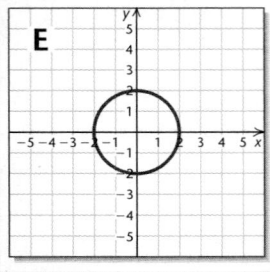

Visualizing the Graph

Match the equation or system of equations with its graph.

1. $y = x^3 - 3x$

2. $y = x^2 + 2x - 3$

3. $y = \dfrac{x - 1}{x^2 - x - 2}$

4. $y = -3x + 2$

5. $x + y = 3,$
   $2x + 5y = 3$

6. $9x^2 - 4y^2 = 36,$
   $x^2 + y^2 = 9$

7. $5x^2 + 5y^2 = 20$

8. $4x^2 + 16y^2 = 64$

9. $y = \log_2 x$

10. $y = 2^x$

Answers on page A-72

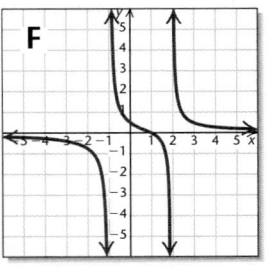

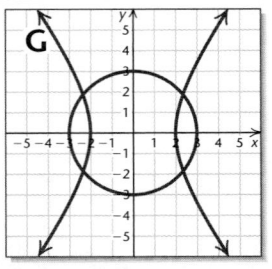

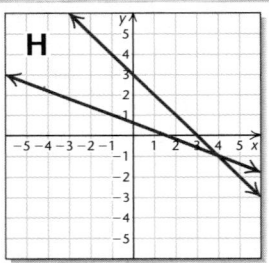

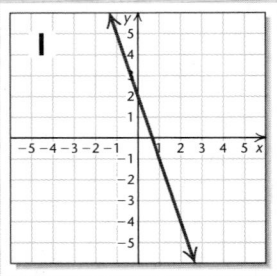

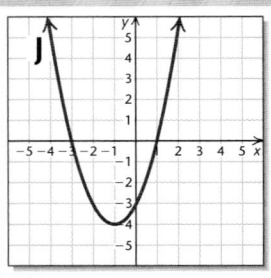

Exercise Set

In Exercises 1–6, match the system of equations with one of the graphs (a)–(f), which follow.

a)

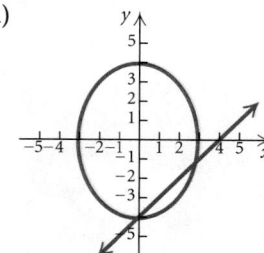

b)

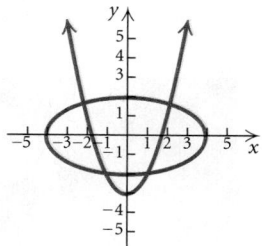

c)

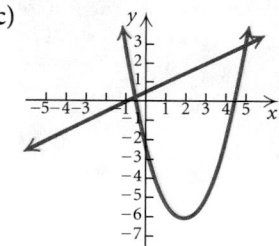

d)

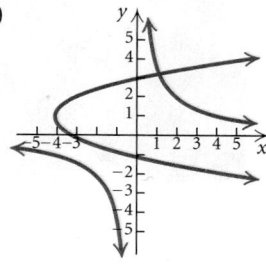

e)

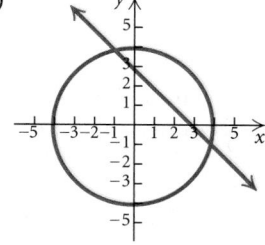

f)

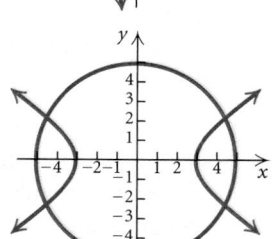

1. $x^2 + y^2 = 16,$
$x + y = 3$

2. $16x^2 + 9y^2 = 144,$
$x - y = 4$

3. $y = x^2 - 4x - 2,$
$2y - x = 1$

4. $4x^2 - 9y^2 = 36,$
$x^2 + y^2 = 25$

5. $y = x^2 - 3,$
$x^2 + 4y^2 = 16$

6. $y^2 - 2y = x + 3,$
$xy = 4$

Solve.

7. $x^2 + y^2 = 25,$
$y - x = 1$

8. $x^2 + y^2 = 100,$
$y - x = 2$

9. $4x^2 + 9y^2 = 36,$
$3y + 2x = 6$

10. $9x^2 + 4y^2 = 36,$
$3x + 2y = 6$

11. $x^2 + y^2 = 25,$
$y^2 = x + 5$

12. $y = x^2,$
$x = y^2$

13. $x^2 + y^2 = 9,$
$x^2 - y^2 = 9$

14. $y^2 - 4x^2 = 4,$
$4x^2 + y^2 = 4$

15. $y^2 - x^2 = 9,$
$2x - 3 = y$

16. $x + y = -6,$
$xy = -7$

17. $y^2 = x + 3,$
$2y = x + 4$

18. $y = x^2,$
$3x = y + 2$

19. $x^2 + y^2 = 25,$
$xy = 12$

20. $x^2 - y^2 = 16,$
$x + y^2 = 4$

21. $x^2 + y^2 = 4,$
$16x^2 + 9y^2 = 144$

22. $x^2 + y^2 = 25,$
$25x^2 + 16y^2 = 400$

23. $x^2 + 4y^2 = 25,$
$x + 2y = 7$

24. $y^2 - x^2 = 16,$
$2x - y = 1$

25. $x^2 - xy + 3y^2 = 27,$
$x - y = 2$

26. $2y^2 + xy + x^2 = 7,$
$x - 2y = 5$

27. $x^2 + y^2 = 16,$
$y^2 - 2x^2 = 10$

28. $x^2 + y^2 = 14,$
$x^2 - y^2 = 4$

29. $x^2 + y^2 = 5,$
$xy = 2$

30. $x^2 + y^2 = 20,$
$xy = 8$

31. $3x + y = 7,$
$4x^2 + 5y = 56$

32. $2y^2 + xy = 5,$
$4y + x = 7$

33. $a + b = 7,$
$ab = 4$

34. $p + q = -4,$
$pq = -5$

35. $x^2 + y^2 = 13,$
$xy = 6$

36. $x^2 + 4y^2 = 20,$
$xy = 4$

37. $x^2 + y^2 + 6y + 5 = 0,$
$x^2 + y^2 - 2x - 8 = 0$

38. $2xy + 3y^2 = 7,$
$3xy - 2y^2 = 4$

39. $2a + b = 1,$
$b = 4 - a^2$

40. $4x^2 + 9y^2 = 36,$
$x + 3y = 3$

41. $a^2 + b^2 = 89,$
$a - b = 3$

42. $xy = 4,$
$x + y = 5$

43. $xy - y^2 = 2,$
$2xy - 3y^2 = 0$

44. $4a^2 - 25b^2 = 0$,
$2a^2 - 10b^2 = 3b + 4$

45. $m^2 - 3mn + n^2 + 1 = 0$,
$3m^2 - mn + 3n^2 = 13$

46. $ab - b^2 = -4$,
$ab - 2b^2 = -6$

47. $x^2 + y^2 = 5$,
$x - y = 8$

48. $4x^2 + 9y^2 = 36$,
$y - x = 8$

49. $a^2 + b^2 = 14$,
$ab = 3\sqrt{5}$

50. $x^2 + xy = 5$,
$2x^2 + xy = 2$

51. $x^2 + y^2 = 25$,
$9x^2 + 4y^2 = 36$

52. $x^2 + y^2 = 1$,
$9x^2 - 16y^2 = 144$

53. $5y^2 - x^2 = 1$,
$xy = 2$

54. $x^2 - 7y^2 = 6$,
$xy = 1$

In Exercises 55–58, determine whether the statement is true or false.

55. A nonlinear system of equations can have both real-number solutions and imaginary-number solutions.

56. If the graph of a nonlinear system of equations consists of a line and a parabola, the system has two real-number solutions.

57. If the graph of a nonlinear system of equations consists of a line and a circle, the system has at most two real-number solutions.

58. If the graph of a nonlinear system of equations consists of a line and an ellipse, it is possible for the system to have exactly one real-number solution.

59. *Picture Frame Dimensions.* Frank's Frame Shop is building a frame for a rectangular oil painting with a perimeter of 28 cm and a diagonal of 10 cm. Find the dimensions of the painting.

60. *Sign Dimensions.* Peden's Advertising is building a rectangular sign with an area of 2 yd^2 and a perimeter of 6 yd. Find the dimensions of the sign.

61. *Banner Design.* A rectangular banner with an area of $\sqrt{3}$ m^2 is being designed to advertise an exhibit at the Davis Gallery. The length of a diagonal is 2 m. Find the dimensions of the banner.

62. *Landscaping.* Green Leaf Landscaping is planting a rectangular wildflower garden with a perimeter of 6 m and a diagonal of $\sqrt{5}$ m. Find the dimensions of the garden.

63. *Fencing.* It will take 210 yd of fencing to enclose a rectangular dog pen. The area of the pen is 2250 yd^2. What are the dimensions of the pen?

64. *Carpentry.* Ted Hansen of Hansen Woodworking Designs has been commissioned to make a rectangular tabletop with an area of $\sqrt{2}$ m^2 and a diagonal of $\sqrt{3}$ m for the Decorators' Show House. Find the dimensions of the tabletop.

65. *Graphic Design.* Marcia Graham, owner of Graham's Graphics, is designing an advertising brochure for the Art League's spring show. Each page of the brochure is rectangular with an area of 20 in^2 and a perimeter of 18 in. Find the dimensions of the brochure.

66. *Investment.*   Jenna made an investment for 1 yr that earned $7.50 simple interest. If the principal had been $25 more and the interest rate 1% less, the interest would have been the same. Find the principal and the interest rate.

67. *Seed Test Plots.*   The Burton Seed Company has two square test plots. The sum of their areas is 832 ft^2 and the difference of their areas is 320 ft^2. Find the length of a side of each plot.

68. *Office Dimensions.*   The diagonal of the floor of a rectangular office cubicle is 1 ft longer than the length of the cubicle and 3 ft longer than twice the width. Find the dimensions of the cubicle.

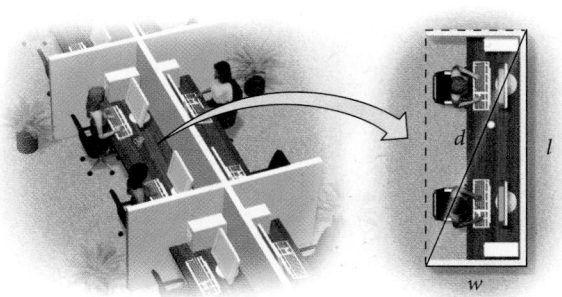

In Exercises 69–74, match the system of inequalities with one of the graphs (a)–(f), which follow.

a)
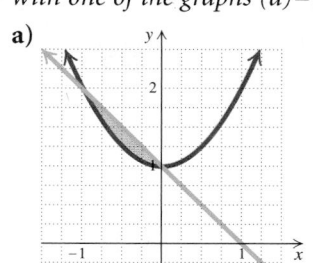

b)
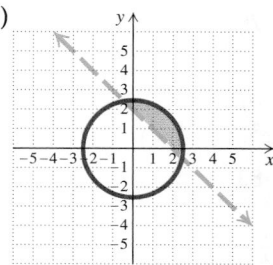

c)
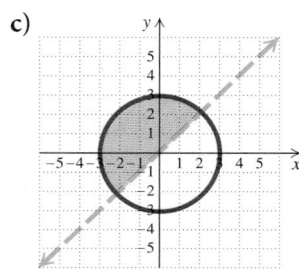

d)
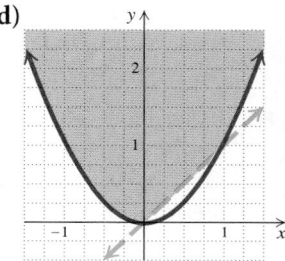

e)
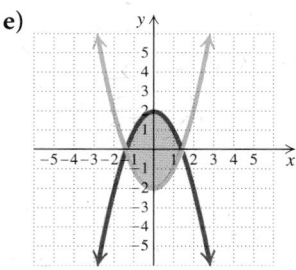

f)
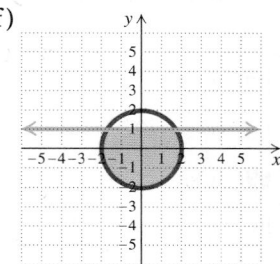

69. $x^2 + y^2 \leq 5,$
     $x + y > 2$

70. $y \leq 2 - x^2,$
     $y \geq x^2 - 2$

71. $y \geq x^2,$
     $y > x$

72. $x^2 + y^2 \leq 4,$
     $y \leq 1$

73. $y \geq x^2 + 1,$
     $x + y \leq 1$

74. $x^2 + y^2 \leq 9,$
     $y > x$

Graph the system of inequalities. Then find the coordinates of the points of intersection of the graphs.

75. $x^2 + y^2 \leq 16,$
     $y < x$

76. $x^2 + y^2 \leq 10,$
     $y > x$

77. $x^2 \leq y,$
$\quad x + y \geq 2$

78. $x \geq y^2,$
$\quad x - y \leq 2$

79. $x^2 + y^2 \leq 25,$
$\quad x - y > 5$

80. $x^2 + y^2 \geq 9,$
$\quad x - y > 3$

81. $y \geq x^2 - 3,$
$\quad y \leq 2x$

82. $y \leq 3 - x^2,$
$\quad y \geq x + 1$

83. $y \geq x^2,$
$\quad y < x + 2$

84. $y \leq 1 - x^2,$
$\quad y > x - 1$

Collaborative Discussion and Writing

85. What would you say to a classmate who tells you that any nonlinear system of equations can be solved graphically?

86. Write a problem that can be translated to a nonlinear system of equations, and ask a classmate to solve it. Devise the problem so that the solution is "The dimensions of the rectangle are 6 ft by 8 ft."

Skill Maintenance

Solve.

87. $2^{3x} = 64$

88. $5^x = 27$

89. $\log_3 x = 4$

90. $\log (x - 3) + \log x = 1$

Synthesis

91. Find an equation of the circle that passes through the points $(2, 4)$ and $(3, 3)$ and whose center is on the line $3x - y = 3$.

92. Find an equation of the circle that passes through the points $(-2, 3)$ and $(-4, 1)$ and whose center is on the line $5x + 8y = -2$.

93. Find an equation of an ellipse centered at the origin that passes through the points $\left(1, \sqrt{3}/2\right)$ and $\left(\sqrt{3}, 1/2\right)$.

94. Find an equation of a hyperbola of the type
$$\frac{x^2}{b^2} - \frac{y^2}{a^2} = 1$$
that passes through the points $\left(-3, -3\sqrt{5}/2\right)$ and $(-3/2, 0)$.

95. Find an equation of the circle that passes through the points $(4, 6), (6, 2),$ and $(1, -3)$.

96. Find an equation of the circle that passes through the points $(2, 3), (4, 5),$ and $(0, -3)$.

97. Show that a hyperbola does not intersect its asymptotes. That is, solve the system of equations
$$\frac{x^2}{a^2} - \frac{y^2}{b^2} = 1,$$
$$y = \frac{b}{a}x \left(\text{or } y = -\frac{b}{a}x\right).$$

98. *Numerical Relationship.* Find two numbers whose product is 2 and the sum of whose reciprocals is $\frac{33}{8}$.

99. *Numerical Relationship.* The square of a number exceeds twice the square of another number by $\frac{1}{8}$. The sum of their squares is $\frac{5}{16}$. Find the numbers.

100. *Box Dimensions.* Four squares with sides 5 in. long are cut from the corners of a rectangular metal sheet that has an area of 340 in². The edges are bent up to form an open box with a volume of 350 in³. Find the dimensions of the box.

101. *Numerical Relationship.* The sum of two numbers is 1, and their product is 1. Find the sum of their cubes. There is a method to solve this problem that is easier than solving a nonlinear system of equations. Can you discover it?

102. Solve for x and y:
$$x^2 - y^2 = a^2 - b^2,$$
$$x - y = a - b.$$

Solve.

103. $x^3 + y^3 = 72,$
$\quad x + y = 6$

104. $a + b = \dfrac{5}{6},$
$\quad \dfrac{a}{b} + \dfrac{b}{a} = \dfrac{13}{6}$

105. $p^2 + q^2 = 13,$
$\quad \dfrac{1}{pq} = -\dfrac{1}{6}$

106. $x^2 + y^2 = 4,$
$\quad (x - 1)^2 + y^2 = 4$

107. $5^{x+y} = 100$,
$3^{2x-y} = 1000$

108. $e^x - e^{x+y} = 0$,
$e^y - e^{x-y} = 0$

Solve using a graphing calculator.

109. $y - \ln x = 2$,
$y = x^2$

110. $y = \ln(x + 4)$,
$x^2 + y^2 = 6$

111. $e^x - y = 1$,
$3x + y = 4$

112. $y - e^{-x} = 1$,
$y = 2x + 5$

113. $y = e^x$,
$x - y = -2$

114. $y = e^{-x}$,
$x + y = 3$

115. $x^2 + y^2 = 19{,}380{,}510.36$,
$27{,}942.25x - 6.125y = 0$

116. $2x + 2y = 1660$,
$xy = 35{,}325$

117. $14.5x^2 - 13.5y^2 - 64.5 = 0$,
$5.5x - 6.3y - 12.3 = 0$

118. $13.5xy + 15.6 = 0$,
$5.6x - 6.7y - 42.3 = 0$

119. $0.319x^2 + 2688.7y^2 = 56{,}548$,
$0.306x^2 - 2688.7y^2 = 43{,}452$

120. $18.465x^2 + 788.723y^2 = 6408$,
$106.535x^2 - 788.723y^2 = 2692$

(10.5) Rotation of Axes

❖ Use rotation of axes to graph conic sections.

❖ Use the discriminant to determine the type of conic represented by a given equation.

CONIC SECTIONS

REVIEW SECTIONS **10.1–10.3.**

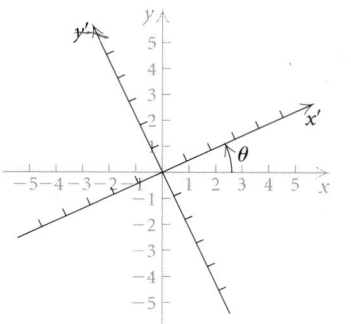

In Section 10.1, we saw that conic sections can be defined algebraically using a second-degree equation of the form $Ax^2 + Bxy + Cy^2 + Dx + Ey + F = 0$. Up to this point, we have considered only equations of this form for which $B = 0$. Now we turn our attention to equations of conics that contain an xy-term.

❖ Rotation of Axes

When B is nonzero, the graph of $Ax^2 + Bxy + Cy^2 + Dx + Ey + F = 0$ is a conic section with an axis that is parallel to neither the x-axis nor the y-axis. We use a technique called **rotation of axes** when we graph such an equation. The goal is to rotate the x- and y-axes through a positive angle θ to yield an $x'y'$-coordinate system, as shown at left. For the appropriate choice of θ, the graph of any conic section with an xy-term will have its axis parallel to the x'-axis or the y'-axis.

Algebraically we want to rewrite an equation

$$Ax^2 + Bxy + Cy^2 + Dx + Ey + F = 0$$

in the xy-coordinate system in the form

$$A'(x')^2 + C'(y')^2 + D'x' + E'y' + F' = 0$$

in the $x'y'$-coordinate system. Equations of this second type were graphed in Sections 10.1–10.3.

To achieve our goal, we find formulas relating the xy-coordinates of a point and the $x'y'$-coordinates of the same point. We begin by letting P be a point with coordinates (x, y) in the xy-coordinate system and (x', y') in the $x'y'$-coordinate system.

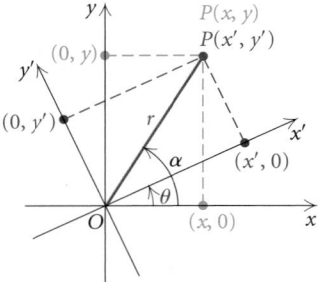

We let r represent the distance OP, and we let α represent the angle from the x-axis to OP. Then

$$\cos \alpha = \frac{x}{r} \quad \text{and} \quad \sin \alpha = \frac{y}{r},$$

so

$$x = r \cos \alpha \quad \text{and} \quad y = r \sin \alpha.$$

We also see from the figure above that

$$\cos (\alpha - \theta) = \frac{x'}{r} \quad \text{and} \quad \sin (\alpha - \theta) = \frac{y'}{r},$$

so

$$x' = r \cos (\alpha - \theta) \quad \text{and} \quad y' = r \sin (\alpha - \theta).$$

Then

$$x' = r \cos \alpha \cos \theta + r \sin \alpha \sin \theta$$

and

$$y' = r \sin \alpha \cos \theta - r \cos \alpha \sin \theta.$$

Substituting x for $r \cos \alpha$ and y for $r \sin \alpha$ gives us

$$x' = x \cos \theta + y \sin \theta \tag{1}$$

and

$$y' = y \cos \theta - x \sin \theta. \tag{2}$$

We can use these formulas to find the $x'y'$-coordinates of any point given that point's xy-coordinates and an angle of rotation θ. To express xy-coordinates in terms of $x'y'$-coordinates and an angle of rotation θ, we solve the system composed of equations (1) and (2) above for x and y. (See Exercise 45.) We get

$$x = x' \cos \theta - y' \sin \theta$$

and

$$y = x' \sin \theta + y' \cos \theta.$$

Rotation of Axes Formulas

If the x- and y-axes are rotated about the origin through a positive acute angle θ, then the coordinates (x, y) and (x', y') of a point P in the xy- and $x'y'$-coordinate systems are related by the following formulas:

$$x' = x \cos \theta + y \sin \theta, \qquad y' = -x \sin \theta + y \cos \theta;$$
$$x = x' \cos \theta - y' \sin \theta, \qquad y = x' \sin \theta + y' \cos \theta.$$

STUDY TIP

If you are finding it difficult to master a particular topic or concept, talk about it with a classmate. Verbalizing your questions about the material might help clarify it for you.

EXAMPLE 1   Suppose that the xy-axes are rotated through an angle of 45°. Write the equation $xy = 1$ in the $x'y'$-coordinate system.

Solution   We substitute 45° for θ in the rotation of axes formulas for x and y:

$$x = x' \cos 45° - y' \sin 45°,$$
$$y = x' \sin 45° + y' \cos 45°.$$

We know that

$$\sin 45° = \frac{\sqrt{2}}{2} \quad \text{and} \quad \cos 45° = \frac{\sqrt{2}}{2},$$

so we have

$$x = x' \left(\frac{\sqrt{2}}{2} \right) - y' \left(\frac{\sqrt{2}}{2} \right) = \frac{\sqrt{2}}{2}(x' - y')$$

and

$$y = x' \left(\frac{\sqrt{2}}{2} \right) + y' \left(\frac{\sqrt{2}}{2} \right) = \frac{\sqrt{2}}{2}(x' + y').$$

Next, we substitute these expressions for x and y in the equation $xy = 1$:

$$\frac{\sqrt{2}}{2}(x' - y') \cdot \frac{\sqrt{2}}{2}(x' + y') = 1$$

$$\frac{1}{2}[(x')^2 - (y')^2] = 1$$

$$\frac{(x')^2}{2} - \frac{(y')^2}{2} = 1, \quad \text{or} \quad \frac{(x')^2}{(\sqrt{2})^2} - \frac{(y')^2}{(\sqrt{2})^2} = 1.$$

We have the equation of a hyperbola in the $x'y'$-coordinate system with its transverse axis on the x'-axis and with vertices $\left(-\sqrt{2},0\right)$ and $\left(\sqrt{2},0\right)$. Its asymptotes are $y' = -x'$ and $y' = x'$. These correspond to the axes of the xy-coordinate system.

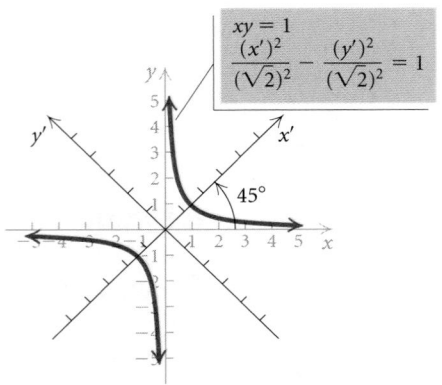

$$xy = 1$$
$$\frac{(x')^2}{(\sqrt{2})^2} - \frac{(y')^2}{(\sqrt{2})^2} = 1$$

Now let's substitute the rotation of axes formulas for x and y in the equation

$$Ax^2 + Bxy + Cy^2 + Dx + Ey + F = 0.$$

We have

$$A(x' \cos \theta - y' \sin \theta)^2 + B(x' \cos \theta - y' \sin \theta)(x' \sin \theta + y' \cos \theta)$$
$$+ C(x' \sin \theta + y' \cos \theta)^2 + D(x' \cos \theta - y' \sin \theta)$$
$$+ E(x' \sin \theta + y' \cos \theta) + F = 0.$$

Performing the operations indicated and collecting like terms yields the equation

$$A'(x')^2 + B'x'y' + C'(y')^2 + D'x' + E'y' + F' = 0, \qquad \textbf{(3)}$$

where

$$A' = A \cos^2 \theta + B \sin \theta \cos \theta + C \sin^2 \theta,$$
$$B' = 2(C - A) \sin \theta \cos \theta + B(\cos^2 \theta - \sin^2 \theta),$$
$$C' = A \sin^2 \theta - B \sin \theta \cos \theta + C \cos^2 \theta,$$
$$D' = D \cos \theta + E \sin \theta,$$
$$E' = -D \sin \theta + E \cos \theta, \quad \text{and}$$
$$F' = F.$$

Recall that our goal is to produce an equation without an $x'y'$-term, or with $B' = 0$. Then we must have

$$2(C - A) \sin \theta \cos \theta + B(\cos^2 \theta - \sin^2 \theta) = 0$$

$$(C - A) \sin 2\theta + B \cos 2\theta = 0 \qquad \text{Using double-angle formulas}$$

$$B \cos 2\theta = (A - C) \sin 2\theta$$

$$\frac{\cos 2\theta}{\sin 2\theta} = \frac{A - C}{B}$$

$$\cot 2\theta = \frac{A - C}{B}.$$

Thus, when θ is chosen so that

$$\cot 2\theta = \frac{A - C}{B},$$

equation (3) will have no $x'y'$-term. Although we will not do so here, it can be shown that we can always find θ such that $0° < 2\theta < 180°$, or $0° < \theta < 90°$.

> **Eliminating the xy-Term**
>
> To eliminate the xy-term from the equation
>
> $$Ax^2 + Bxy + Cy^2 + Dx + Ey + F = 0, \quad B \ne 0,$$
>
> select an angle θ such that
>
> $$\cot 2\theta = \frac{A - C}{B}, \quad 0° < 2\theta < 180°,$$
>
> and use the rotation of axes formulas.

EXAMPLE 2   Graph the equation

$$3x^2 - 2\sqrt{3}xy + y^2 + 2x + 2\sqrt{3}y = 0.$$

Solution   We have

$$A = 3, \quad B = -2\sqrt{3}, \quad C = 1, \quad D = 2, \quad E = 2\sqrt{3}, \quad \text{and} \quad F = 0.$$

To select the angle of rotation θ, we must have

$$\cot 2\theta = \frac{A - C}{B} = \frac{3 - 1}{-2\sqrt{3}} = \frac{2}{-2\sqrt{3}} = -\frac{1}{\sqrt{3}}.$$

Thus, $2\theta = 120°$, and $\theta = 60°$. We substitute this value for θ in the rotation of axes formulas for x and y:

$$x = x' \cos 60° - y' \sin 60°,$$
$$y = x' \sin 60° + y' \cos 60°.$$

This gives us

$$x = x' \cdot \frac{1}{2} - y' \cdot \frac{\sqrt{3}}{2} = \frac{x'}{2} - \frac{y'\sqrt{3}}{2}$$

and

$$y = x' \cdot \frac{\sqrt{3}}{2} + y' \cdot \frac{1}{2} = \frac{x'\sqrt{3}}{2} + \frac{y'}{2}.$$

Now we substitute these expressions for x and y in the given equation:

$$3\left(\frac{x'}{2} - \frac{y'\sqrt{3}}{2}\right)^2 - 2\sqrt{3}\left(\frac{x'}{2} - \frac{y'\sqrt{3}}{2}\right)\left(\frac{x'\sqrt{3}}{2} + \frac{y'}{2}\right) +$$

$$\left(\frac{x'\sqrt{3}}{2} + \frac{y'}{2}\right)^2 + 2\left(\frac{x'}{2} - \frac{y'\sqrt{3}}{2}\right) + 2\sqrt{3}\left(\frac{x'\sqrt{3}}{2} + \frac{y'}{2}\right) = 0.$$

After simplifying, we get

$$4(y')^2 + 4x' = 0, \quad \text{or}$$
$$(y')^2 = -x'.$$

This is the equation of a parabola with its vertex at $(0,0)$ of the $x'y'$-coordinate system and axis of symmetry $y' = 0$. We sketch the graph.

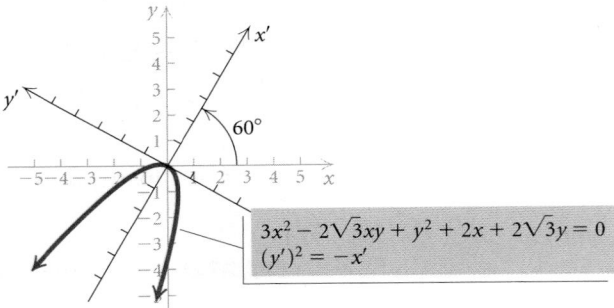

Now Try Exercise 23. ■

❖ The Discriminant

It is possible to determine the type of conic represented by the equation $Ax^2 + Bxy + Cy^2 + Dx + Ey + F = 0$ before rotating the axes. Using the expressions for A', B', and C' in terms of A, B, C, and θ developed earlier, it can be shown that

$$(B')^2 - 4A'C' = B^2 - 4AC.$$

Now when θ is chosen so that

$$\cot 2\theta = \frac{A - C}{B},$$

rotation of axes gives us an equation

$$A'(x')^2 + C'(y')^2 + D'x' + E'y' + F' = 0.$$

If A' and C' have the same sign, or $A'C' > 0$, then the graph of this equation is an ellipse or a circle. If A' and C' have different signs, or $A'C' < 0$, then the graph is a hyperbola. And, if either $A' = 0$ or $C' = 0$, or $A'C' = 0$, the graph is a parabola.

Since $B' = 0$ and $(B')^2 - 4A'C' = B^2 - 4AC$, it follows that $B^2 - 4AC = -4A'C'$. Then the graph is an ellipse or a circle if $B^2 - 4AC < 0$, a hyperbola if $B^2 - 4AC > 0$, or a parabola if $B^2 - 4AC = 0$. (There are certain special cases, called *degenerate conics*, where these statements do not hold, but we will not concern ourselves with these here.) The expression $B^2 - 4AC$ is the **discriminant** of the equation $Ax^2 + Bxy + Cy^2 + Dx + Ey + F = 0$.

The graph of the equation

$$Ax^2 + Bxy + Cy^2 + Dx + Ey + F = 0$$

is, except in degenerate cases,

1. an ellipse or a circle if $B^2 - 4AC < 0$,
2. a hyperbola if $B^2 - 4AC > 0$, and
3. a parabola if $B^2 - 4AC = 0$.

EXAMPLE 3 Graph the equation $3x^2 + 2xy + 3y^2 = 16$.

Solution We have

$$A = 3, \quad B = 2, \quad \text{and} \quad C = 3, \quad \text{so}$$
$$B^2 - 4AC = 2^2 - 4 \cdot 3 \cdot 3 = 4 - 36 = -32.$$

Since the discriminant is negative, the graph is an ellipse or a circle. Now, to rotate the axes, we begin by determining θ:

$$\cot 2\theta = \frac{A - C}{B} = \frac{3 - 3}{2} = \frac{0}{2} = 0.$$

Then $2\theta = 90°$ and $\theta = 45°$, so

$$\sin \theta = \frac{\sqrt{2}}{2} \quad \text{and} \quad \cos \theta = \frac{\sqrt{2}}{2}.$$

As we saw in Example 1, substituting these values for $\sin \theta$ and $\cos \theta$ in the rotation of axes formulas gives

$$x = \frac{\sqrt{2}}{2}(x' - y') \quad \text{and} \quad y = \frac{\sqrt{2}}{2}(x' + y').$$

Now we substitute for x and y in the given equation:

$$3\left[\frac{\sqrt{2}}{2}(x' - y')\right]^2 + 2\left[\frac{\sqrt{2}}{2}(x' - y')\right]\left[\frac{\sqrt{2}}{2}(x' + y')\right] +$$
$$3\left[\frac{\sqrt{2}}{2}(x' + y')\right]^2 = 16.$$

(calculator display:)
```
22-4*3*3
                    -32
```

After simplifying, we have

$$4(x')^2 + 2(y')^2 = 16, \quad \text{or}$$

$$\frac{(x')^2}{4} + \frac{(y')^2}{8} = 1.$$

This is the equation of an ellipse with vertices $\left(0, -\sqrt{8}\right)$ and $\left(0, \sqrt{8}\right)$, or $\left(0, -2\sqrt{2}\right)$ and $\left(0, 2\sqrt{2}\right)$, on the y'-axis. The x'-intercepts are $(-2, 0)$ and $(2, 0)$. We sketch the graph.

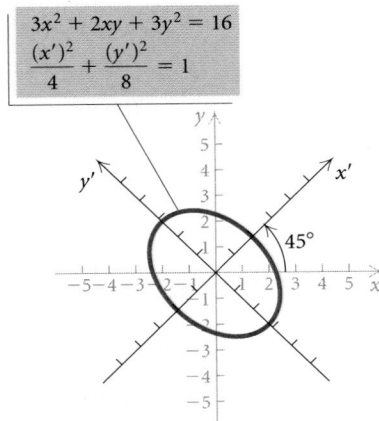

Now Try Exercise 21.

EXAMPLE 4   Graph the equation $4x^2 - 24xy - 3y^2 - 156 = 0$.

Solution   We have

$$A = 4, \qquad B = -24, \quad \text{and} \quad C = -3, \qquad \text{so}$$
$$B^2 - 4AC = (-24)^2 - 4 \cdot 4(-3) = 576 + 48 = 624.$$

Since the discriminant is positive, the graph is a hyperbola. To rotate the axes, we begin by determining θ:

$$\cot 2\theta = \frac{A - C}{B} = \frac{4 - (-3)}{-24} = -\frac{7}{24}.$$

Since $\cot 2\theta < 0$, we have $90° < 2\theta < 180°$. From the triangle at left, we see that $\cos 2\theta = -\frac{7}{25}$.

Using half-angle formulas, we have

$$\sin \theta = \sqrt{\frac{1 - \cos 2\theta}{2}} = \sqrt{\frac{1 - \left(-\frac{7}{25}\right)}{2}} = \frac{4}{5}$$

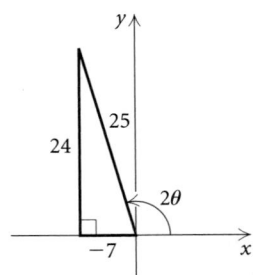

and

$$\cos \theta = \sqrt{\frac{1 + \cos 2\theta}{2}} = \sqrt{\frac{1 + \left(-\frac{7}{25}\right)}{2}} = \frac{3}{5}.$$

Substituting in the rotation of axes formulas gives us

$$x = x' \cos \theta - y' \sin \theta = \tfrac{3}{5}x' - \tfrac{4}{5}y'$$

and

$$y = x' \sin \theta + y' \cos \theta = \tfrac{4}{5}x' + \tfrac{3}{5}y'.$$

Now we substitute for x and y in the given equation:

$$4\left(\tfrac{3}{5}x' - \tfrac{4}{5}y'\right)^2 - 24\left(\tfrac{3}{5}x' - \tfrac{4}{5}y'\right)\left(\tfrac{4}{5}x' + \tfrac{3}{5}y'\right) - 3\left(\tfrac{4}{5}x' + \tfrac{3}{5}y'\right)^2 - 156 = 0.$$

After simplifying, we have

$$13(y')^2 - 12(x')^2 - 156 = 0$$
$$13(y')^2 - 12(x')^2 = 156$$
$$\frac{(y')^2}{12} - \frac{(x')^2}{13} = 1.$$

The graph of this equation is a hyperbola with vertices $\left(0, -\sqrt{12}\right)$ and $\left(0, \sqrt{12}\right)$, or $\left(0, -2\sqrt{3}\right)$ and $\left(0, 2\sqrt{3}\right)$, on the y'-axis. Since we know that $\sin \theta = \tfrac{4}{5}$ and $0° < \theta < 90°$, we can use a calculator to find that $\theta \approx 53.1°$. Thus the xy-axes are rotated through an angle of about $53.1°$ in order to obtain the $x'y'$-axes. We sketch the graph.

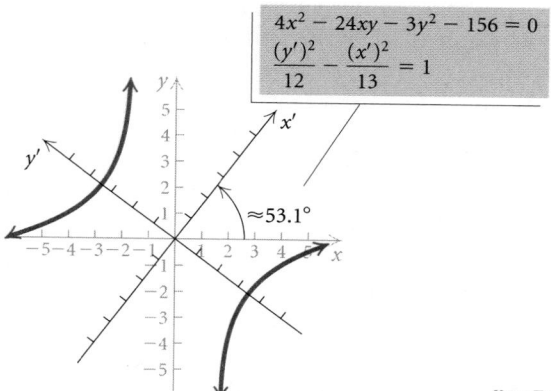

Now Try Exercise 35. ■

(10.5) Exercise Set

For the given angle of rotation and coordinates of a point in the xy-coordinate system, find the coordinates of the point in the x'y'-coordinate system.

1. $\theta = 45°, \left(\sqrt{2}, -\sqrt{2}\right)$ 2. $\theta = 45°, (-1, 3)$

3. $\theta = 30°, (0, 2)$ 4. $\theta = 60°, \left(0, \sqrt{3}\right)$

For the given angle of rotation and coordinates of a point in the x'y'-coordinate system, find the coordinates of the point in the xy-coordinate system.

5. $\theta = 45°, (1, -1)$

6. $\theta = 45°, \left(-3\sqrt{2}, \sqrt{2}\right)$

7. $\theta = 30°, (2, 0)$

8. $\theta = 60°, \left(-1, -\sqrt{3}\right)$

Use the discriminant to determine whether the graph of the equation is an ellipse (or a circle), a hyperbola, or a parabola.

9. $3x^2 - 5xy + 3y^2 - 2x + 7y = 0$

10. $5x^2 + 6xy - 4y^2 + x - 3y + 4 = 0$

11. $x^2 - 3xy - 2y^2 + 12 = 0$

12. $4x^2 + 7xy + 2y^2 - 3x + y = 0$

13. $4x^2 - 12xy + 9y^2 - 3x + y = 0$

14. $6x^2 + 5xy + 6y^2 + 15 = 0$

15. $2x^2 - 8xy + 7y^2 + x - 2y + 1 = 0$

16. $x^2 + 6xy + 9y^2 - 3x + 4y = 0$

17. $8x^2 - 7xy + 5y^2 - 17 = 0$

18. $x^2 + xy - y^2 - 4x + 3y - 2 = 0$

Graph the equation.

19. $4x^2 + 2xy + 4y^2 = 15$

20. $3x^2 + 10xy + 3y^2 + 8 = 0$

21. $x^2 - 10xy + y^2 + 36 = 0$

22. $x^2 + 2xy + y^2 + 4\sqrt{2}x - 4\sqrt{2}y = 0$

23. $x^2 - 2\sqrt{3}xy + 3y^2 - 12\sqrt{3}x - 12y = 0$

24. $13x^2 + 6\sqrt{3}xy + 7y^2 - 16 = 0$

25. $7x^2 + 6\sqrt{3}xy + 13y^2 - 32 = 0$

26. $x^2 + 4xy + y^2 - 9 = 0$

27. $11x^2 + 10\sqrt{3}xy + y^2 = 32$

28. $5x^2 - 8xy + 5y^2 = 81$

29. $\sqrt{2}x^2 + 2\sqrt{2}xy + \sqrt{2}y^2 - 8x + 8y = 0$

30. $x^2 + 2\sqrt{3}xy + 3y^2 - 8x + 8\sqrt{3}y = 0$

31. $x^2 + 6\sqrt{3}xy - 5y^2 + 8x - 8\sqrt{3}y - 48 = 0$

32. $3x^2 - 2xy + 3y^2 - 6\sqrt{2}x + 2\sqrt{2}y - 26 = 0$

33. $x^2 + xy + y^2 = 24$

34. $4x^2 + 3\sqrt{3}xy + y^2 = 55$

35. $4x^2 - 4xy + y^2 - 8\sqrt{5}x - 16\sqrt{5}y = 0$

36. $9x^2 - 24xy + 16y^2 - 400x - 300y = 0$

37. $11x^2 + 7xy - 13y^2 = 621$

38. $3x^2 + 4xy + 6y^2 = 28$

Collaborative Discussion and Writing

39. Explain how the procedure you would follow for graphing an equation of the form $Ax^2 + Bxy + Cy^2 + Dx + Ey + F = 0$ when $B \neq 0$ differs from the procedure you would follow when $B = 0$.

40. Discuss some circumstances under which you might use rotation of axes.

Skill Maintenance

Convert to radian measure.

41. $120°$ 42. $-315°$

Convert to degree measure.

43. $\dfrac{\pi}{3}$ 44. $\dfrac{3\pi}{4}$

Synthesis

45. Solve this system of equations for x and y:

$$x' = x \cos \theta + y \sin \theta,$$
$$y' = y \cos \theta - x \sin \theta.$$

Show your work.

46. Show that substituting $x' \cos \theta - y' \sin \theta$ for x and $x' \sin \theta + y' \cos \theta$ for y in the equation

$$Ax^2 + Bxy + Cy^2 + Dx + Ey + F = 0$$

yields the equation

$$A'(x')^2 + B'x'y' + C'(y')^2 + D'x' + E'y' + F' = 0,$$

where

$$A' = A \cos^2 \theta + B \sin \theta \cos \theta + C \sin^2 \theta,$$
$$B' = 2(C - A) \sin \theta \cos \theta + B(\cos^2 \theta - \sin^2 \theta),$$
$$C' = A \sin^2 \theta - B \sin \theta \cos \theta + C \cos^2 \theta,$$
$$D' = D \cos \theta + E \sin \theta,$$
$$E' = -D \sin \theta + E \cos \theta, \quad \text{and}$$
$$F' = F.$$

47. Show that $A + C = A' + C'$.

48. Show that for any angle θ, the equation $x^2 + y^2 = r^2$ becomes $(x')^2 + (y')^2 = r^2$ when the rotation of axes formulas are applied.

10.6 Polar Equations of Conics

❖ Graph polar equations of conics.
❖ Convert from polar equations of conics to rectangular equations of conics.
❖ Find polar equations of conics.

In Sections 10.1–10.3, we saw that the parabola, the ellipse, and the hyperbola have different definitions in rectangular coordinates. When polar coordinates are used, we can give a single definition that applies to all three conics.

CONIC SECTIONS

REVIEW SECTIONS 10.1–10.3.

An Alternative Definition of Conics

Let L be a fixed line (the **directrix**); let F be a fixed point (the **focus**) not on L; and let e be a positive constant (the **eccentricity**). A **conic** is the set of all points P in the plane such that

$$\frac{PF}{PL} = e,$$

where PF is the distance from P to F and PL is the distance from P to L. The conic is a parabola if $e = 1$, an ellipse if $e < 1$, and a hyperbola if $e > 1$.

Note that if $e = 1$, then $PF = PL$ and the alternative definition of a parabola is identical to the definition presented in Section 10.1.

❇ Polar Equations of Conics

To derive equations for the conics in polar coordinates, we position the focus F at the pole and position the directrix L either perpendicular to the polar axis or parallel to it. In the figure below, we place L perpendicular to the polar axis and p units to the right of the focus, or pole.

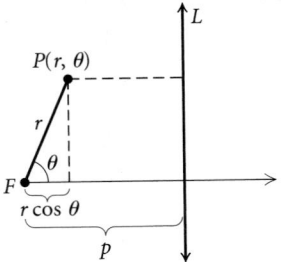

Note that $PL = p - r\cos\theta$. Then if P is any point on the conic, we have

$$\frac{PF}{PL} = e$$

$$\frac{r}{p - r\cos\theta} = e$$

$$r = ep - er\cos\theta$$

$$r + er\cos\theta = ep$$

$$r(1 + e\cos\theta) = ep$$

$$r = \frac{ep}{1 + e\cos\theta}.$$

Thus we see that the polar equation of a conic with focus at the pole and directrix perpendicular to the polar axis and p units to the right of the pole is

$$r = \frac{ep}{1 + e\cos\theta},$$

where e is the eccentricity of the conic.

For an ellipse and a hyperbola, we can make the following statement regarding eccentricity.

For an ellipse and a hyperbola, the **eccentricity e** is given by

$$e = \frac{c}{a},$$

where c is the distance from the center to a focus and a is the distance from the center to a vertex.

EXAMPLE 1   Describe and graph the conic $r = \dfrac{18}{6 + 3\cos\theta}$.

Solution   We begin by dividing the numerator and the denominator by 6 to obtain a constant term of 1 in the denominator:

$$r = \frac{3}{1 + 0.5\cos\theta}.$$

This equation is in the form

$$r = \frac{ep}{1 + e\cos\theta}$$

with $e = 0.5$. Since $e < 1$, the graph is an ellipse. Also, since $e = 0.5$ and $ep = 0.5p = 3$, we have $p = 6$. Thus the ellipse has a vertical directrix that lies 6 units to the right of the pole. We graph the equation in the square window $[-10, 5, -5, 5]$.

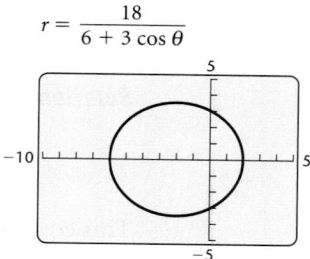

$$r = \frac{18}{6 + 3\cos\theta}$$

It follows that the major axis is horizontal and lies on the polar axis. The vertices are found by letting $\theta = 0$ and $\theta = \pi$. They are $(2, 0)$ and $(6, \pi)$. The center of the ellipse is at the midpoint of the segment connecting the vertices, or at $(2, \pi)$.

The length of the major axis is 8, so we have $2a = 8$, or $a = 4$. From the equation of the conic, we know that $e = 0.5$. Using the equation $e = c/a$, we can find that $c = 2$. Finally, using $a = 4$ and $c = 2$ in $b^2 = a^2 - c^2$ gives us

$$b^2 = 4^2 - 2^2 = 16 - 4 = 12$$
$$b = \sqrt{12}, \text{ or } 2\sqrt{3},$$

so the length of the minor axis is $2\sqrt{12}$, or $4\sqrt{3}$. This is useful to know when sketching a hand-drawn graph of the conic.   **Now Try Exercise 7.** ▪

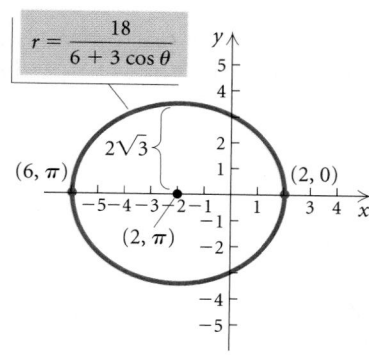

Other derivations similar to the one on p. 866 lead to the following result.

> **Polar Equations of Conics**
>
> A polar equation of any of the four forms
>
> $$r = \frac{ep}{1 \pm e\cos\theta}, \qquad r = \frac{ep}{1 \pm e\sin\theta}$$
>
> is a conic section. The conic is a parabola if $e = 1$, an ellipse if $0 < e < 1$, and a hyperbola if $e > 1$.

The table below describes the polar equations of conics with a focus at the pole and the directrix either perpendicular to or parallel to the polar axis.

Equation	Description
$r = \dfrac{ep}{1 + e \cos \theta}$	Vertical directrix p units to the right of the pole (or focus)
$r = \dfrac{ep}{1 - e \cos \theta}$	Vertical directrix p units to the left of the pole (or focus)
$r = \dfrac{ep}{1 + e \sin \theta}$	Horizontal directrix p units above the pole (or focus)
$r = \dfrac{ep}{1 - e \sin \theta}$	Horizontal directrix p units below the pole (or focus)

EXAMPLE 2 Describe and graph the conic $r = \dfrac{10}{5 - 5 \sin \theta}$.

Solution We first divide the numerator and the denominator by 5:

$$r = \frac{2}{1 - \sin \theta}.$$

This equation is in the form

$$r = \frac{ep}{1 - e \sin \theta}$$

with $e = 1$, so the graph is a parabola. Since $e = 1$ and $ep = 1 \cdot p = 2$, we have $p = 2$. Thus the parabola has a horizontal directrix 2 units below the pole. We graph the equation in the square window $[-9, 9, -4, 8]$.

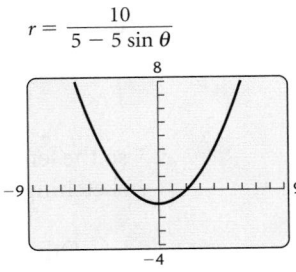

$$r = \frac{10}{5 - 5 \sin \theta}$$

It follows that the parabola has a vertical axis of symmetry. Since the directrix lies below the focus, or pole, the parabola opens up. The vertex is the midpoint of the segment of the axis of symmetry from the focus to the directrix. We find it by letting $\theta = 3\pi/2$. It is $(1, 3\pi/2)$.

Now Try Exercise 17. ◼

EXAMPLE 3 Describe and graph the conic $r = \dfrac{4}{2 + 6 \sin \theta}$.

Solution We first divide the numerator and the denominator by 2:

$$r = \frac{2}{1 + 3 \sin \theta}.$$

$r = \dfrac{4}{2 + 6 \sin \theta}$

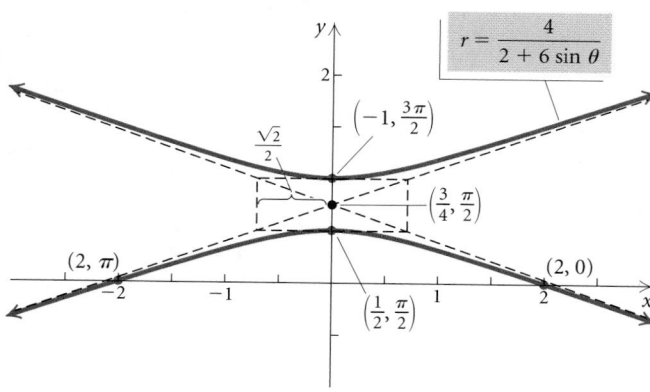

This equation is in the form

$$r = \frac{ep}{1 + e \sin \theta}$$

with $e = 3$. Since $e > 1$, the graph is a hyperbola. We have $e = 3$ and $ep = 3p = 2$, so $p = \frac{2}{3}$. Thus the hyperbola has a horizontal directrix that lies $\frac{2}{3}$ unit above the pole. We graph the equation in the square window $[-6, 6, -3, 5]$ using DOT mode.

It follows that the transverse axis is vertical. To find the vertices, we let $\theta = \pi/2$ and $\theta = 3\pi/2$. The vertices are $(1/2, \pi/2)$ and $(-1, 3\pi/2)$. The center of the hyperbola is the midpoint of the segment connecting the vertices, or $(3/4, \pi/2)$. Thus the distance c from the center to a focus is $3/4$. Using $c = 3/4$, $e = 3$, and $e = c/a$, we have $a = 1/4$. Then since $c^2 = a^2 + b^2$, we have

$$b^2 = \left(\frac{3}{4}\right)^2 - \left(\frac{1}{4}\right)^2 = \frac{9}{16} - \frac{1}{16} = \frac{1}{2}$$

$$b = \frac{1}{\sqrt{2}}, \text{ or } \frac{\sqrt{2}}{2}.$$

Knowing the values of a and b allows us to sketch the asymptotes if we are graphing the hyperbola by hand. We can also easily plot the points $(2, 0)$ and $(2, \pi)$ on the polar axis.

Now Try Exercise 13. ■

❋ Converting from Polar Equations to Rectangular Equations

We can use the relationships between polar coordinates and rectangular coordinates that were developed in Section 8.4 to convert polar equations of conics to rectangular equations.

EXAMPLE 4   Convert to a rectangular equation: $r = \dfrac{2}{1 - \sin \theta}$.

Solution   We have

$$r = \frac{2}{1 - \sin \theta}$$

$$r - r \sin \theta = 2 \qquad \text{Multiplying by } 1 - \sin \theta$$

$$r = r \sin \theta + 2$$

$$\sqrt{x^2 + y^2} = y + 2 \qquad \begin{array}{l}\text{Substituting } \sqrt{x^2 + y^2} \text{ for } r \text{ and} \\ y \text{ for } r \sin \theta\end{array}$$

$$x^2 + y^2 = y^2 + 4y + 4 \qquad \text{Squaring both sides}$$

$$x^2 = 4y + 4, \quad \text{or}$$

$$x^2 - 4y - 4 = 0.$$

This is the equation of a parabola, as we should have anticipated, since $e = 1$.   **Now Try Exercise 23.** ◼

❋ Finding Polar Equations of Conics

We can find the polar equation of a conic with a focus at the pole if we know its eccentricity and the equation of the directrix.

EXAMPLE 5   Find a polar equation of the conic with a focus at the pole, eccentricity $\frac{1}{3}$, and directrix $r = 2 \csc \theta$.

Solution   The equation of the directrix can be written

$$r = \frac{2}{\sin \theta}, \quad \text{or} \quad r \sin \theta = 2.$$

This corresponds to the equation $y = 2$ in rectangular coordinates, so the directrix is a horizontal line 2 units above the polar axis. Using the table on p. 868, we see that the equation is of the form

$$r = \frac{ep}{1 + e \sin \theta}.$$

Substituting $\frac{1}{3}$ for e and 2 for p gives us

$$r = \frac{\frac{1}{3} \cdot 2}{1 + \frac{1}{3} \sin \theta} = \frac{\frac{2}{3}}{1 + \frac{1}{3} \sin \theta} = \frac{2}{3 + \sin \theta}. \qquad \text{\textbf{Now Try Exercise 39.}} \ ◼$$

(10.6) Exercise Set

In Exercises 1–6, match the equation with one of the graphs (a)–(f), which follow.

a)

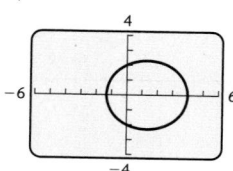

b)

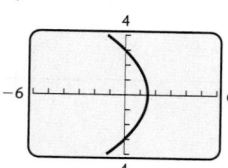

c)

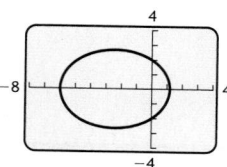

d)

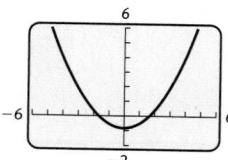

e)

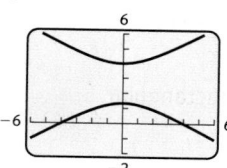

f)

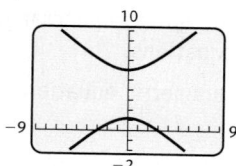

1. $r = \dfrac{3}{1 + \cos \theta}$

2. $r = \dfrac{4}{1 + 2 \sin \theta}$

3. $r = \dfrac{8}{4 - 2 \cos \theta}$

4. $r = \dfrac{12}{4 + 6 \sin \theta}$

5. $r = \dfrac{5}{3 - 3 \sin \theta}$

6. $r = \dfrac{6}{3 + 2 \cos \theta}$

For each equation:

a) *Tell whether the equation describes a parabola, an ellipse, or a hyperbola.*
b) *State whether the directrix is vertical or horizontal and give its location in relation to the pole.*
c) *Find the vertex or vertices.*
d) *Graph the equation.*

7. $r = \dfrac{1}{1 + \cos \theta}$

8. $r = \dfrac{4}{2 + \cos \theta}$

9. $r = \dfrac{15}{5 - 10 \sin \theta}$

10. $r = \dfrac{12}{4 + 8 \sin \theta}$

11. $r = \dfrac{8}{6 - 3 \cos \theta}$

12. $r = \dfrac{6}{2 + 2 \sin \theta}$

13. $r = \dfrac{20}{10 + 15 \sin \theta}$

14. $r = \dfrac{10}{8 - 2 \cos \theta}$

15. $r = \dfrac{9}{6 + 3 \cos \theta}$

16. $r = \dfrac{4}{3 - 9 \sin \theta}$

17. $r = \dfrac{3}{2 - 2 \sin \theta}$

18. $r = \dfrac{12}{3 + 9 \cos \theta}$

19. $r = \dfrac{4}{2 - \cos \theta}$

20. $r = \dfrac{5}{1 - \sin \theta}$

21. $r = \dfrac{7}{2 + 10 \sin \theta}$

22. $r = \dfrac{3}{8 - 4 \cos \theta}$

23.–38. Convert the equations in Exercises 7–22 to rectangular equations.

Find a polar equation of the conic with a focus at the pole and the given eccentricity and directrix.

39. $e = 2, r = 3 \csc \theta$

40. $e = \frac{2}{3}, r = -\sec \theta$

41. $e = 1, r = 4 \sec \theta$

42. $e = 3, r = 2 \csc \theta$

43. $e = \frac{1}{2}, r = -2 \sec \theta$

44. $e = 1, r = 4 \csc \theta$

45. $e = \frac{3}{4}, r = 5 \csc \theta$

46. $e = \frac{4}{5}, r = 2 \sec \theta$

47. $e = 4, r = -2 \csc \theta$

48. $e = 3, r = 3 \csc \theta$

Collaborative Discussion and Writing

49. Consider the graphs of

$$r = \frac{e}{1 - e \sin \theta}$$

for $e = 0.2, 0.4, 0.6,$ and 0.8. Explain the effect of the value of e on the graph.

50. When using a graphing calculator, would you prefer to graph a conic in rectangular form or in polar form? Why?

Skill Maintenance

For $f(x) = (x - 3)^2 + 4$, find each of the following.

51. $f(t)$

52. $f(2t)$

53. $f(t - 1)$

54. $f(t + 2)$

Synthesis

Parabolic Orbit.   Suppose that a comet travels in a parabolic orbit with the sun as its focus. Position a polar coordinate system with the pole at the sun and the axis of the orbit perpendicular to the polar axis. When the comet is the given distance from the sun, the segment from the comet to the sun makes the given angle with the polar axis. Find a polar equation of the orbit, assuming that the directrix lies above the pole.

55. 100 million miles, $\dfrac{\pi}{6}$

56. 120 million miles, $\dfrac{\pi}{4}$

Parametric Equations

❋ Graph parametric equations.

❋ Determine an equivalent rectangular equation for parametric equations.

❋ Determine parametric equations for a rectangular equation.

❋ Solve applied problems involving projectile motion.

❋ Graphing Parametric Equations

We have graphed *plane curves* that are composed of sets of ordered pairs (x, y) in the rectangular coordinate plane. Now we discuss a way to represent plane curves in which x and y are functions of a third variable, t.

EXAMPLE 1   Graph the curve represented by the equations

$$x = \tfrac{1}{2}t, \qquad y = t^2 - 3; \quad -3 \le t \le 3.$$

Solution   We can choose values for t between -3 and 3 and find the corresponding values of x and y. When $t = -3$, we have

$$x = \tfrac{1}{2}(-3) = -\tfrac{3}{2}, \qquad y = (-3)^2 - 3 = 6.$$

The table below lists other ordered pairs. We plot these points and then draw the curve.

t	x	y	(x, y)
-3	$-\frac{3}{2}$	6	$\left(-\frac{3}{2}, 6\right)$
-2	-1	1	$(-1, 1)$
-1	$-\frac{1}{2}$	-2	$\left(-\frac{1}{2}, -2\right)$
0	0	-3	$(0, -3)$
1	$\frac{1}{2}$	-2	$\left(\frac{1}{2}, -2\right)$
2	1	1	$(1, 1)$
3	$\frac{3}{2}$	6	$\left(\frac{3}{2}, 6\right)$

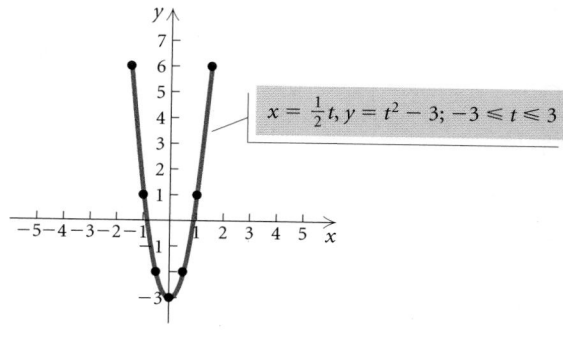

$x = \frac{1}{2}t, y = t^2 - 3; -3 \leq t \leq 3$

The curve above appears to be part of a parabola. Let's verify this by finding the equivalent rectangular equation. Solving $x = \frac{1}{2}t$ for t, we get $t = 2x$. Substituting $2x$ for t in $y = t^2 - 3$, we have

$$y = (2x)^2 - 3 = 4x^2 - 3.$$

This is a quadratic equation. Hence its graph is a parabola. The curve is part of the parabola $y = 4x^2 - 3$. Since $-3 \leq t \leq 3$ and $x = \frac{1}{2}t$, we must include the restriction $-\frac{3}{2} \leq x \leq \frac{3}{2}$ when we write the equivalent rectangular equation:

$$y = 4x^2 - 3, \quad -\frac{3}{2} \leq x \leq \frac{3}{2}.$$

The equations $x = \frac{1}{2}t$ and $y = t^2 - 3$ are **parametric equations** for the curve. The variable t is the **parameter**. When we write the corresponding rectangular equation, we say that we **eliminate the parameter**.

> **Parametric Equations**
>
> If f and g are continuous functions of t on an interval I, then the set of ordered pairs (x, y) such that $x = f(t)$ and $y = g(t)$ is a **plane curve**. The equations $x = f(t)$ and $y = g(t)$ are **parametric equations** for the curve. The variable t is the **parameter**.

❖ Determining a Rectangular Equation for Given Parametric Equations

GCM **EXAMPLE 2** Using a graphing calculator, graph each of the following plane curves given their respective parametric equations and the restriction on the parameter. Then find the equivalent rectangular equation.

a) $x = t^2,\ y = t - 1;\ -1 \leq t \leq 4$
b) $x = \sqrt{t},\ y = 2t + 3;\ 0 \leq t \leq 3$

Solution

a) To graph the curve, we set the graphing calculator in PARAMETRIC mode, enter the equations, and select minimum and maximum values for x, y, and t.

WINDOW

Tmin = −1

Tmax = 4

Tstep = .1

Xmin = −2

Xmax = 18

Xscl = 1

Ymin = −4

Ymax = 4

Yscl = 1

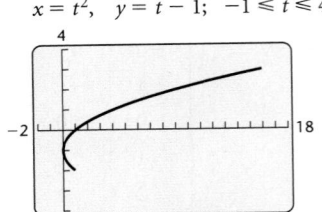

$x = t^2, \quad y = t - 1; \quad -1 \leq t \leq 4$

To find an equivalent rectangular equation, we first solve either equation for t. We choose the equation $y = t - 1$:

$$y = t - 1$$
$$y + 1 = t.$$

We then substitute $y + 1$ for t in $x = t^2$:

$$x = t^2$$
$$x = (y + 1)^2. \qquad \textbf{Substituting}$$

This is an equation of a parabola that opens to the right. Given that $-1 \leq t \leq 4$, we have the corresponding restrictions on x and y: $0 \leq x \leq 16$ and $-2 \leq y \leq 3$. Thus the equivalent rectangular equation is

$$x = (y + 1)^2, \quad -2 \leq y \leq 3.$$

b) To graph the curve, we use PARAMETRIC mode and enter the equations and window settings.

WINDOW

Tmin = 0

Tmax = 3

Tstep = .1

Xmin = −3

Xmax = 3

Xscl = 1

Ymin = −2

Ymax = 10

Yscl = 1

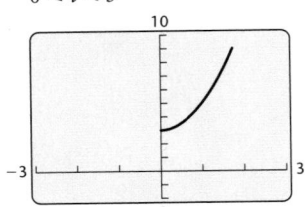

$x = \sqrt{t}, \quad y = 2t + 3;$
$0 \leq t \leq 3$

To find an equivalent rectangular equation, we first solve $x = \sqrt{t}$ for t:

$$x = \sqrt{t}$$
$$x^2 = t.$$

Then we substitute x^2 for t in $y = 2t + 3$:

$$y = 2t + 3$$
$$y = 2x^2 + 3. \qquad \text{Substituting}$$

When $0 \le t \le 3$, we have $0 \le x \le \sqrt{3}$. The equivalent rectangular equation is

$$y = 2x^2 + 3, \quad 0 \le x \le \sqrt{3}. \qquad \text{\textbf{Now Try Exercise 3.}} \ ▇$$

We first graphed in parametric mode in Section 6.5. There we used an angle measure as the parameter as we do in the next example.

EXAMPLE 3   Graph the plane curve represented by $x = \cos t$ and $y = \sin t$, with t in $[0, 2\pi]$. Then determine an equivalent rectangular equation.

Solution   Using a squared window and a Tstep of $\pi/48$, we obtain the graph at left. It appears to be the unit circle.

The equivalent rectangular equation can be obtained by squaring both sides of each parametric equation:

$$x^2 = \cos^2 t \quad \text{and} \quad y^2 = \sin^2 t.$$

This allows us to use the trigonometric identity $\sin^2 \theta + \cos^2 \theta = 1$. Substituting, we get

$$x^2 + y^2 = 1.$$

As expected, this is an equation of the unit circle.   **Now Try Exercise 13.** ▇

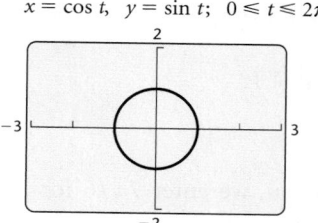

$x = \cos t, \ \ y = \sin t; \ \ 0 \le t \le 2\pi$

EXAMPLE 4   Graph the plane curve represented by

$$x = 5 \cos t \quad \text{and} \quad y = 3 \sin t; \quad 0 \le t \le 2\pi.$$

Then eliminate the parameter to find the rectangular equation.

Solution

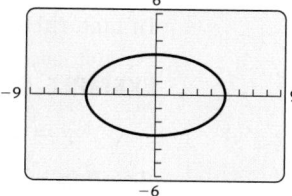

$x = 5 \cos t, \ \ y = 3 \sin t; \ \ 0 \le t \le 2\pi$

This appears to be the graph of an ellipse. To find the rectangular equation, we first solve for $\cos t$ and $\sin t$ in the parametric equations:

$$x = 5 \cos t \qquad y = 3 \sin t$$
$$\frac{x}{5} = \cos t, \qquad \frac{y}{3} = \sin t.$$

Using the identity $\sin^2 \theta + \cos^2 \theta = 1$, we can substitute to eliminate the parameter:

$$\sin^2 t + \cos^2 t = 1$$

$$\left(\frac{y}{3}\right)^2 + \left(\frac{x}{5}\right)^2 = 1 \qquad \text{Substituting}$$

$$\frac{x^2}{25} + \frac{y^2}{9} = 1. \qquad \text{Ellipse}$$

The rectangular form of the equation confirms that the graph is an ellipse centered at the origin with vertices at $(5, 0)$ and $(-5, 0)$.

Now Try Exercise 15. ■

One advantage of graphing the unit circle parametrically, as we did in Example 3, is that it provides a method of finding trigonometric function values.

EXAMPLE 5 Using the VALUE feature from the CALC menu and the parametric graph of the unit circle, find each of the following function values.

a) $\cos \dfrac{7\pi}{6}$
b) $\sin 4.13$

Solution

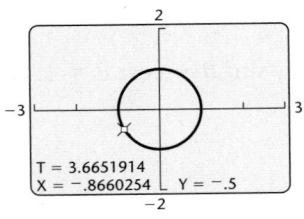

T = 3.6651914
X = −.8660254   Y = −.5

a) Using the VALUE feature from the CALC menu, we enter $7\pi/6$ for t. The value of x, which is $\cos t$, appears, as shown at left. Thus,

$$\cos \frac{7\pi}{6} \approx -0.8660.$$

b) We enter 4.13 for t. The value of y, which is $\sin t$, will appear on the screen. The calculator will show that $\sin 4.13 \approx -0.8352$.

Now Try Exercise 21. ■

❈ Determining Parametric Equations for a Given Rectangular Equation

Many sets of parametric equations can represent the same plane curve. In fact, there are infinitely many such equations.

EXAMPLE 6 Find three sets of parametric equations for the parabola

$$y = 4 - (x + 3)^2.$$

Solution

If $x = t$, then $y = 4 - (t + 3)^2$, or $-t^2 - 6t - 5$.

If $x = t - 3$, then $y = 4 - (t - 3 + 3)^2$, or $4 - t^2$.

If $x = \dfrac{t}{3}$, then $y = 4 - \left(\dfrac{t}{3} + 3\right)^2$, or $-\dfrac{t^2}{9} - 2t - 5$.

Now Try Exercise 29. ■

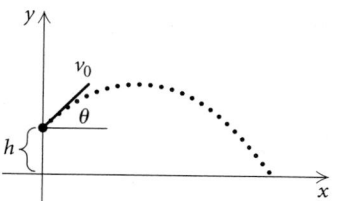

❊ Applications

The motion of an object that is propelled upward can be described with parametric equations. Such motion is called **projectile motion**. It can be shown using more advanced mathematics that, neglecting air resistance, the following equations describe the path of a projectile propelled upward at an angle θ with the horizontal from a height h, in feet, at an initial speed v_0, in feet per second:

$$x = (v_0 \cos \theta)t, \qquad y = h + (v_0 \sin \theta)t - 16t^2.$$

We can use these equations to determine the location of the object at time t, in seconds.

EXAMPLE 7   *Projectile Motion.*   A baseball is thrown from a height of 6 ft with an initial speed of 100 ft/sec at an angle of 45° with the horizontal.

a) Find parametric equations that give the position of the ball at time t, in seconds.

b) Graph the plane curve represented by the equations found in part (a).

c) Find the height of the ball after 1 sec, 2 sec, and 3 sec.

d) Determine how long the ball is in the air.

e) Determine the horizontal distance that the ball travels.

f) Find the maximum height of the ball.

Solution

a) We substitute 6 for h, 100 for v_0, and 45° for θ in the equations above:

$$
\begin{aligned}
x &= (v_0 \cos \theta)t \\
&= (100 \cos 45°)t \\
&= \left(100 \cdot \frac{\sqrt{2}}{2}\right)t = 50\sqrt{2}\,t; \\[4pt]
y &= h + (v_0 \sin \theta)t - 16t^2 \\
&= 6 + (100 \sin 45°)t - 16t^2 \\
&= 6 + \left(100 \cdot \frac{\sqrt{2}}{2}\right)t - 16t^2 \\
&= 6 + 50\sqrt{2}\,t - 16t^2.
\end{aligned}
$$

b) $x = 50\sqrt{2}\,t, \quad y = 6 + 50\sqrt{2}\,t - 16t^2$

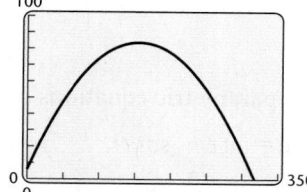

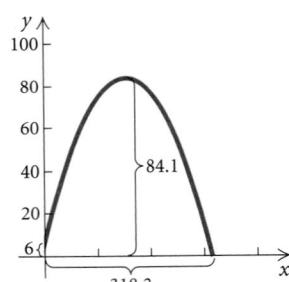

T	X₁ₜ	Y₁ₜ
1	70.711	60.711
2	141.42	83.421
3	212.13	74.132

T =

c) The height of the ball at time t is represented by y. We can use a table set in ASK mode to find the desired values of y as shown at left, or we can substitute in the equation for y as shown below.

When $t = 1$, $y = 6 + 50\sqrt{2}(1) - 16(1)^2 \approx 60.7$ ft.

When $t = 2$, $y = 6 + 50\sqrt{2}(2) - 16(2)^2 \approx 83.4$ ft.

When $t = 3$, $y = 6 + 50\sqrt{2}(3) - 16(3)^2 \approx 74.1$ ft.

d) The ball hits the ground when $y = 0$. Thus, in order to determine how long the ball is in the air, we solve the equation $y = 0$:

$$6 + 50\sqrt{2}t - 16t^2 = 0$$

$$-16t^2 + 50\sqrt{2}t + 6 = 0 \qquad \text{Standard form}$$

$$t = \frac{-50\sqrt{2} \pm \sqrt{(50\sqrt{2})^2 - 4(-16)(6)}}{2(-16)}$$

$$\text{Using the quadratic formula}$$

$$t \approx -0.1 \quad \text{or} \quad t \approx 4.5.$$

The negative value for t has no meaning in this application. Thus we determine that the ball is in the air for about 4.5 sec.

e) Since the ball is in the air for about 4.5 sec, the horizontal distance that it travels is given by

$$x = 50\sqrt{2}(4.5) \approx 318.2 \text{ ft.}$$

f) To find the maximum height of the ball, we find the maximum value of y. This occurs at the vertex of the quadratic function represented by y. At the vertex, we have

$$t = -\frac{b}{2a} = -\frac{50\sqrt{2}}{2(-16)} \approx 2.2.$$

When $t = 2.2$,

$$y = 6 + 50\sqrt{2}(2.2) - 16(2.2)^2 \approx 84.1 \text{ ft.} \qquad \text{\small\textbf{Now Try Exercise 33.}} \blacksquare$$

The path of a fixed point on the circumference of a circle as it rolls along a line is called a **cycloid**. For example, a point on the rim of a bicycle wheel traces a cycloid curve.

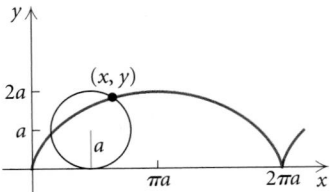

The parametric equations of a cycloid are

$$x = a(t - \sin t), \qquad y = a(1 - \cos t),$$

where a is the radius of the circle that traces the curve and t is in radian measure.

STUDY TIP

Prepare for each chapter test by rereading the text, reviewing your homework, reading the important properties and formulas in the Chapter Summary and Review, and doing the review exercises at the end of the chapter. Then take the chapter test at the end of the chapter.

EXAMPLE 8 Graph the cycloid described by the parametric equations

$$x = 3(t - \sin t), \quad y = 3(1 - \cos t); \quad 0 \le t \le 6\pi.$$

Solution

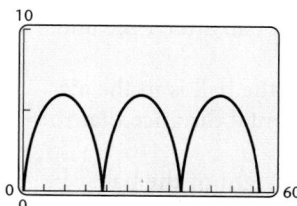

$x = 3(t - \sin t), \; y = 3(1 - \cos t);$
$0 \le t \le 6\pi$

Now Try Exercise 35. ■

(10.7) Exercise Set

Graph the plane curve given by the parametric equations. Then find an equivalent rectangular equation.

1. $x = \frac{1}{2}t, \; y = 6t - 7; \; -1 \le t \le 6$

2. $x = t, \; y = 5 - t; \; -2 \le t \le 3$

3. $x = 4t^2, \; y = 2t; \; -1 \le t \le 1$

4. $x = \sqrt{t}, \; y = 2t + 3; \; 0 \le t \le 8$

5. $x = t^2, \; y = \sqrt{t}; \; 0 \le t \le 4$

6. $x = t^3 + 1, \; y = t; \; -3 \le t \le 3$

7. $x = t + 3, \; y = \dfrac{1}{t + 3}; \; -2 \le t \le 2$

8. $x = 2t^3 + 1, \; y = 2t^3 - 1; \; -4 \le t \le 4$

9. $x = 2t - 1, \; y = t^2; \; -3 \le t \le 3$

10. $x = \frac{1}{3}t, \; y = t; \; -5 \le t \le 5$

11. $x = e^{-t}, \; y = e^t; \; -\infty < t < \infty$

12. $x = 2\ln t, \; y = t^2; \; 0 < t < \infty$

13. $x = 3\cos t, \; y = 3\sin t; \; 0 \le t \le 2\pi$

14. $x = 2\cos t, \; y = 4\sin t; \; 0 \le t \le 2\pi$

15. $x = \cos t, \; y = 2\sin t; \; 0 \le t \le 2\pi$

16. $x = 2\cos t, \; y = 2\sin t; \; 0 \le t \le 2\pi$

17. $x = \sec t, \; y = \cos t; \; -\dfrac{\pi}{2} < t < \dfrac{\pi}{2}$

18. $x = \sin t, \; y = \csc t; \; 0 < t < \pi$

19. $x = 1 + 2\cos t, \; y = 2 + 2\sin t; \; 0 \le t \le 2\pi$

20. $x = 2 + \sec t, \; y = 1 + 3\tan t; \; 0 < t < \dfrac{\pi}{2}$

Using a parametric graph of the unit circle, find the function value.

21. $\sin \dfrac{\pi}{4}$

22. $\cos \dfrac{2\pi}{3}$

23. $\cos \dfrac{17\pi}{12}$

24. $\sin \dfrac{4\pi}{5}$

25. $\tan \dfrac{\pi}{5}$

26. $\tan \dfrac{2\pi}{7}$

27. $\cos 5.29$

28. $\sin 1.83$

Find two sets of parametric equations for the rectangular equation.

29. $y = 4x - 3$

30. $y = x^2 - 1$

31. $y = (x - 2)^2 - 6x$

32. $y = x^3 + 3$

33. *Projectile Motion.*   A ball is thrown from a height of 7 ft with an initial speed of 80 ft/sec at an angle of 30° with the horizontal.

a) Find parametric equations that give the position of the ball at time t, in seconds.
b) Graph the plane curve represented by the equations found in part (a).
c) Find the height of the ball after 1 sec and after 2 sec.
d) Determine how long the ball is in the air.
e) Determine the horizontal distance that the ball travels.
f) Find the maximum height of the ball.

34. *Projectile Motion.*   A projectile is launched from the ground with an initial speed of 200 ft/sec at an angle of 60° with the horizontal.

a) Find parametric equations that give the position of the projectile at time t, in seconds.
b) Graph the plane curve represented by the equations found in part (a).
c) Find the height of the projectile after 4 sec and after 8 sec.
d) Determine how long the projectile is in the air.
e) Determine the horizontal distance that the projectile travels.
f) Find the maximum height of the projectile.

Graph the cycloid.

35. $x = 2(t - \sin t), \ y = 2(1 - \cos t); \ 0 \le t \le 4\pi$

36. $x = 4t - 4 \sin t, \ y = 4 - 4 \cos t; \ 0 \le t \le 6\pi$

37. $x = t - \sin t, \ y = 1 - \cos t; \ -2\pi \le t \le 2\pi$

38. $x = 5(t - \sin t), \ y = 5(1 - \cos t); \ -4\pi \le t \le 4\pi$

Collaborative Discussion and Writing

39. Show that $x = a \cos t + h$ and $y = b \sin t + k$, $a \ne b, 0 \le t \le 2\pi$, are parametric equations of an ellipse with center (h, k).

40. Consider the graph in Example 5. Explain how the values of T, X, and Y displayed relate to both the parametric and rectangular equations of the unit circle.

Skill Maintenance

Graph.

41. $y = x^3$

42. $x = y^3$

43. $f(x) = \sqrt{x - 2}$

44. $f(x) = \dfrac{3}{x^2 - 1}$

Synthesis

45. Graph the curve described by

$$x = 3 \cos t, \quad y = 3 \sin t; \quad 0 \le t \le 2\pi.$$

As t increases, the path of the curve is generated in the counterclockwise direction. How can this set of equations be changed so that the curve is generated in the clockwise direction?

46. Graph the plane curve described by

$$x = \cos^3 t, \quad y = \sin^3 t; \quad 0 \le t \le 2\pi.$$

Then find the equivalent rectangular equation.

CHAPTER 10   Summary and Review

Important Properties and Formulas

Standard Equation of a Parabola with Vertex at the Origin

The standard equation of a parabola with vertex $(0, 0)$ and directrix $y = -p$ is

$$x^2 = 4py.$$

The focus is $(0, p)$ and the y-axis is the axis of symmetry.

The standard equation of a parabola with vertex $(0, 0)$ and directrix $x = -p$ is

$$y^2 = 4px.$$

The focus is $(p, 0)$ and the x-axis is the axis of symmetry.

Standard Equation of a Parabola with Vertex (h, k) and Vertical Axis of Symmetry

The standard equation of a parabola with vertex (h, k) and vertical axis of symmetry is

$$(x - h)^2 = 4p(y - k),$$

where the vertex is (h, k), the focus is $(h, k + p)$, and the directrix is $y = k - p$.

Standard Equation of a Parabola with Vertex (h, k) and Horizontal Axis of Symmetry

The standard equation of a parabola with vertex (h, k) and horizontal axis of symmetry is

$$(y - k)^2 = 4p(x - h),$$

where the vertex is (h, k), the focus is $(h + p, k)$, and the directrix is $x = h - p$.

Standard Equation of a Circle

The standard equation of a circle with center (h, k) and radius r is

$$(x - h)^2 + (y - k)^2 = r^2.$$

Standard Equation of an Ellipse with Center at the Origin

Major axis horizontal

$$\frac{x^2}{a^2} + \frac{y^2}{b^2} = 1, \ a > b > 0$$

Vertices: $(-a, 0), (a, 0)$
y-intercepts: $(0, -b), (0, b)$
Foci: $(-c, 0), (c, 0)$, where $c^2 = a^2 - b^2$

Major axis vertical

$$\frac{x^2}{b^2} + \frac{y^2}{a^2} = 1, \ a > b > 0$$

Vertices: $(0, -a), (0, a)$
x-intercepts: $(-b, 0), (b, 0)$
Foci: $(0, -c), (0, c)$, where $c^2 = a^2 - b^2$

Standard Equation of an Ellipse with Center at (h, k)

Major axis horizontal

$$\frac{(x - h)^2}{a^2} + \frac{(y - k)^2}{b^2} = 1, \ a > b > 0$$

Vertices: $(h - a, k), (h + a, k)$
Length of minor axis: $2b$
Foci: $(h - c, k), (h + c, k)$, where $c^2 = a^2 - b^2$

(continued)

Major axis vertical

$$\frac{(x-h)^2}{b^2} + \frac{(y-k)^2}{a^2} = 1, \ a > b > 0$$

Vertices: $(h, k - a), (h, k + a)$

Length of minor axis: $2b$

Foci: $(h, k - c), (h, k + c)$, where $c^2 = a^2 - b^2$

Standard Equation of a Hyperbola with Center at the Origin

Transverse axis horizontal

$$\frac{x^2}{a^2} - \frac{y^2}{b^2} = 1$$

Vertices: $(-a, 0), (a, 0)$

Asymptotes: $y = -\frac{b}{a}x, y = \frac{b}{a}x$

Foci: $(-c, 0), (c, 0)$, where $c^2 = a^2 + b^2$

Transverse axis vertical

$$\frac{y^2}{a^2} - \frac{x^2}{b^2} = 1$$

Vertices: $(0, -a), (0, a)$

Asymptotes: $y = -\frac{a}{b}x, y = \frac{a}{b}x$

Foci: $(0, -c), (0, c)$, where $c^2 = a^2 + b^2$

Standard Equation of a Hyperbola with Center at (h, k)

Transverse axis horizontal

$$\frac{(x-h)^2}{a^2} - \frac{(y-k)^2}{b^2} = 1$$

Vertices: $(h - a, k), (h + a, k)$

Asymptotes: $y - k = \frac{b}{a}(x - h),$

$$y - k = -\frac{b}{a}(x - h)$$

Foci: $(h - c, k), (h + c, k)$, where $c^2 = a^2 + b^2$

Transverse axis vertical

$$\frac{(y-k)^2}{a^2} - \frac{(x-h)^2}{b^2} = 1$$

Vertices: $(h, k - a), (h, k + a)$

Asymptotes: $y - k = \frac{a}{b}(x - h),$

$$y - k = -\frac{a}{b}(x - h)$$

Foci: $(h, k - c), (h, k + c)$, where $c^2 = a^2 + b^2$

Rotation of Axes Formulas

$x' = x \cos\theta + y \sin\theta,$

$y' = -x \sin\theta + y \cos\theta;$

$x = x' \cos\theta - y' \sin\theta,$

$y = x' \sin\theta + y' \cos\theta$

Eliminating the xy-Term

To eliminate the xy-term from the equation

$$Ax^2 + Bxy + Cy^2 + Dx + Ey + F = 0, \quad B \neq 0,$$

select an angle θ such that

$$\cot 2\theta = \frac{A - C}{B}, \quad 0 < 2\theta < 180°,$$

and use the rotation of axes formulas.

The Discriminant

The graph of the equation $Ax^2 + Bxy + Cy^2 + Dx + Ey + F = 0$ is, except in degenerate cases,

1. an ellipse or a circle if $B^2 - 4AC < 0$,
2. a hyperbola if $B^2 - 4AC > 0$, and
3. a parabola if $B^2 - 4AC = 0$.

Polar Equations of Conics

A polar equation of any of the four forms

$$r = \frac{ep}{1 \pm e \cos\theta}, \qquad r = \frac{ep}{1 \pm e \sin\theta}$$

is a conic section. The conic is a parabola if $e = 1$, an ellipse if $0 < e < 1$, and a hyperbola if $e > 1$.

Review Exercises

Determine whether the statement is true or false.

1. The graph of $x + y^2 = 1$ is a parabola that opens to the left. [10.1]

2. The graph of $\dfrac{(x-2)^2}{4} + \dfrac{(y+3)^2}{9}$ is an ellipse with center $(-2, 3)$. [10.2]

3. The hyperbola $\dfrac{x^2}{5} - \dfrac{y^2}{10} = 1$ has a horizontal transverse axis. [10.3]

4. Every nonlinear system of equations has at least one real-number solution. [10.4]

5. The graph of $2x^2 + xy + 2y^2 = 10$ is a parabola. [10.5]

In Exercises 6–13, match the equation with one of the graphs (a)–(h), which follow.

a)
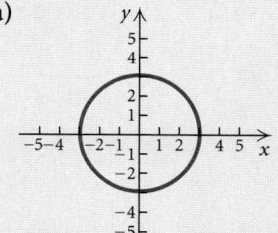

b)
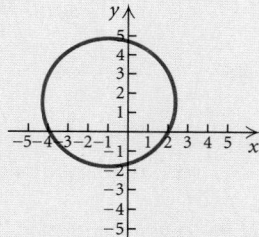

c)
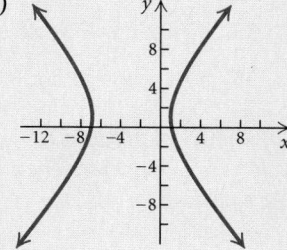

d)

e)
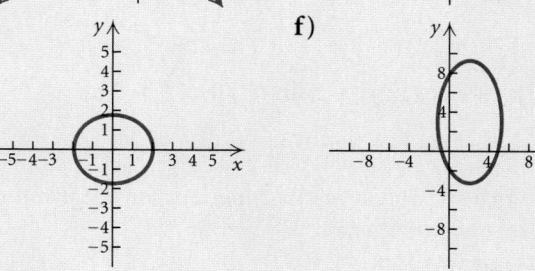

f)

g)
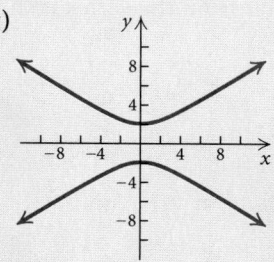

h)
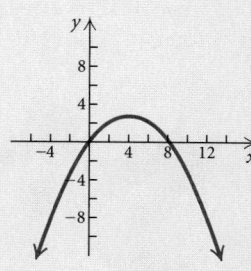

6. $y^2 = 5x$ [10.1]

7. $y^2 = 9 - x^2$ [10.2]

8. $3x^2 + 4y^2 = 12$ [10.2]

9. $9y^2 - 4x^2 = 36$ [10.3]

10. $x^2 + y^2 + 2x - 3y = 8$ [10.2]

11. $4x^2 + y^2 - 16x - 6y = 15$ [10.2]

12. $x^2 - 8x + 6y = 0$ [10.1]

13. $\dfrac{(x+3)^2}{16} - \dfrac{(y-1)^2}{25} = 1$ [10.3]

14. Find an equation of the parabola with directrix $y = \frac{3}{2}$ and focus $\left(0, -\frac{3}{2}\right)$. [10.1]

15. Find the focus, the vertex, and the directrix of the parabola given by
$$y^2 = -12x.$$ [10.1]

16. Find the vertex, the focus, and the directrix of the parabola given by
$$x^2 + 10x + 2y + 9 = 0.$$ [10.1]

17. Find the center, the vertices, and the foci of the ellipse given by
$$16x^2 + 25y^2 - 64x + 50y - 311 = 0.$$
Then draw the graph. [10.2]

18. Find an equation of the ellipse having vertices $(0, -4)$ and $(0, 4)$ with minor axis of length 6. [10.2]

19. Find the center, the vertices, the foci, and the asymptotes of the hyperbola given by
$$x^2 - 2y^2 + 4x + y - \tfrac{1}{8} = 0.$$ [10.3]

20. *Spotlight.*   A spotlight has a parabolic cross section that is 2 ft wide at the opening and 1.5 ft deep at the vertex. How far from the vertex is the focus?   [10.1]

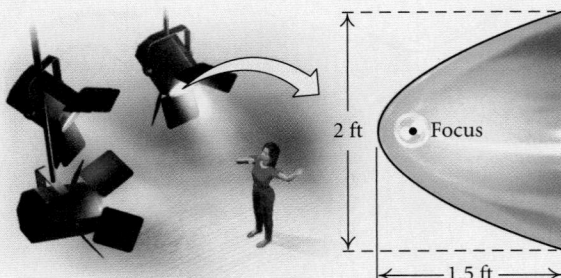

Solve.   [10.4]

21. $x^2 - 16y = 0,$
$x^2 - y^2 = 64$

22. $4x^2 + 4y^2 = 65,$
$6x^2 - 4y^2 = 25$

23. $x^2 - y^2 = 33,$
$x + y = 11$

24. $x^2 - 2x + 2y^2 = 8,$
$2x + y = 6$

25. $x^2 - y = 3,$
$2x - y = 3$

26. $x^2 + y^2 = 25,$
$x^2 - y^2 = 7$

27. $x^2 - y^2 = 3,$
$y = x^2 - 3$

28. $x^2 + y^2 = 18,$
$2x + y = 3$

29. $x^2 + y^2 = 100,$
$2x^2 - 3y^2 = -120$

30. $x^2 + 2y^2 = 12,$
$xy = 4$

31. *Numerical Relationship.*   The sum of two numbers is 11 and the sum of their squares is 65. Find the numbers.   [10.4]

32. *Dimensions of a Rectangle.*   A rectangle has a perimeter of 38 m and an area of 84 m². What are the dimensions of the rectangle?   [10.4]

33. *Numerical Relationship.*   Find two positive integers whose sum is 12 and the sum of whose reciprocals is $\frac{3}{8}$.   [10.4]

34. *Perimeter.*   The perimeter of a square is 12 cm more than the perimeter of another square. The area of the first square exceeds the area of the other by 39 cm². Find the perimeter of each square.   [10.4]

35. *Radius of a Circle.*   The sum of the areas of two circles is 130π ft². The difference of the areas is 112π ft². Find the radius of each circle.   [10.4]

Graph the system of inequalities. Then find the coordinates of the points of intersection of the graphs.   [10.4]

36. $y \le 4 - x^2,$
$x - y \le 2$

37. $x^2 + y^2 \le 16,$
$x + y < 4$

38. $y \ge x^2 - 1,$
$y < 1$

39. $x^2 + y^2 \le 9,$
$x \le -1$

Graph the equation.   [10.5]

40. $5x^2 - 2xy + 5y^2 - 24 = 0$

41. $x^2 - 10xy + y^2 + 12 = 0$

42. $5x^2 + 6\sqrt{3}xy - y^2 = 16$

43. $x^2 + 2xy + y^2 - \sqrt{2}x + \sqrt{2}y = 0$

Graph the equation. State whether the directrix is vertical or horizontal, describe its location in relation to the pole, and find the vertex or vertices.   [10.6]

44. $r = \dfrac{6}{3 - 3\sin\theta}$

45. $r = \dfrac{8}{2 + 4\cos\theta}$

46. $r = \dfrac{4}{2 - \cos\theta}$

47. $r = \dfrac{18}{9 + 6\sin\theta}$

48.–51. Convert the equations in Exercises 44–47 to rectangular equations.   [10.6]

Find a polar equation of the conic with a focus at the pole and the given eccentricity and directrix.   [10.6]

52. $e = \frac{1}{2}, r = 2\sec\theta$

53. $e = 3, r = -6\csc\theta$

54. $e = 1, r = -4\sec\theta$

55. $e = 2, r = 3\csc\theta$

Graph the plane curve given by the set of parametric equations and the restrictions for the parameter. Then find the equivalent rectangular equation.   [10.7]

56. $x = t, y = 2 + t;\quad -3 \le t \le 3$

57. $x = \sqrt{t}, y = t - 1;\quad 0 \le t \le 9$

58. $x = 2\cos t, y = 2\sin t;\quad 0 \le t \le 2\pi$

59. $x = 3\sin t, y = \cos t;\quad 0 \le t \le 2\pi$

Find two sets of parametric equations for the given rectangular equation.   [10.7]

60. $y = 2x - 3$

61. $y = x^2 + 4$

62. *Projectile Motion.* A projectile is launched from the ground with an initial speed of 150 ft/sec at an angle of 45° with the horizontal.   [10.7]

　a) Find parametric equations that give the position of the projectile at time *t*, in seconds.

　b) Find the height of the projectile after 3 sec and after 6 sec.

　c) Determine how long the projectile is in the air.

　d) Determine the horizontal distance that the projectile travels.

　e) Find the maximum height of the projectile.

63. The vertex of the parabola $y^2 - 4y - 12x - 8 = 0$ is which of the following?   [10.1]

　A. $(1, -2)$
　B. $(-1, 2)$
　C. $(2, -1)$
　D. $(-2, 1)$

64. Which of the following cannot be a number of solutions possible for a system of equations representing an ellipse and a straight line?   [10.4]

　A. 0
　B. 1
　C. 2
　D. 4

65. The graph of $x^2 + 4y^2 = 4$ is which of the following?   [10.2], [10.3]

A.
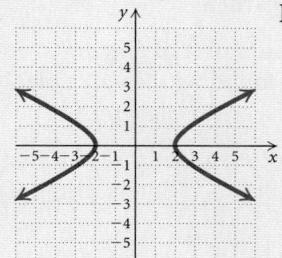

B.
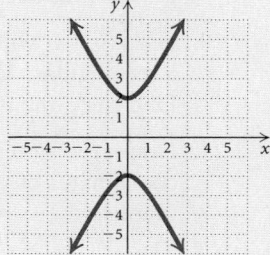

C.
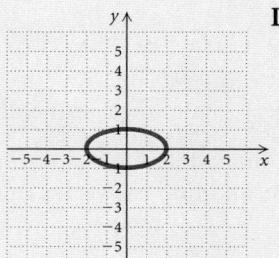

D.
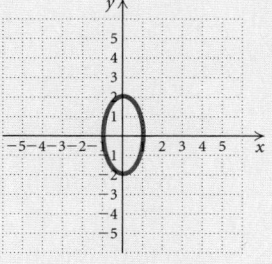

Collaborative Discussion and Writing

66. What would you say to a classmate who tells you that it is always possible to visualize all the solutions of a nonlinear system of equations? [10.4]

67. Is a circle a special type of ellipse? Why or why not?   [10.2]

Synthesis

68. Find two numbers whose product is 4 and the sum of whose reciprocals is $\frac{65}{56}$.   [10.4]

69. Find an equation of the circle that passes through the points $(10, 7), (-6, 7)$, and $(-8, 1)$. [10.2], [10.4]

70. Find an equation of the ellipse containing the point $\left(-1/2, 3\sqrt{3}/2\right)$ and with vertices $(0, -3)$ and $(0, 3)$.   [10.2]

71. *Navigation.* Two radio transmitters positioned 400 mi apart along the shore send simultaneous signals to a ship that is 250 mi offshore, sailing parallel to the shoreline. The signal from transmitter *A* reaches the ship 300 microseconds before the signal from transmitter *B*. The signals travel at a speed of 186,000 miles per second, or 0.186 mile per microsecond. Find the equation of the hyperbola with foci *A* and *B* on which the ship is located. (*Hint*: For any point on the hyperbola, the absolute value of the difference of its distances from the foci is 2*a*.)   [10.3]

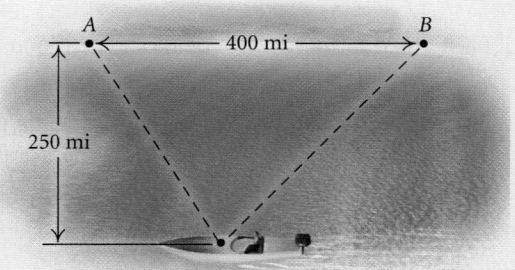

CHAPTER 10   Test

In Exercises 1–4, match the equation with one of the graphs (a)–(d), which follow.

a)

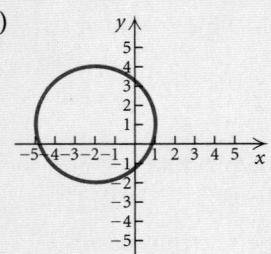

b)

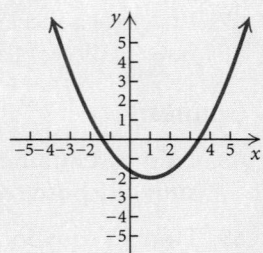

c)

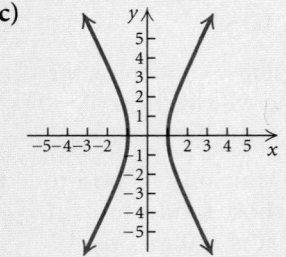

d)

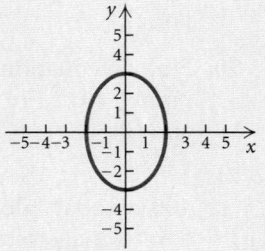

1. $4x^2 - y^2 = 4$

2. $x^2 - 2x - 3y = 5$

3. $x^2 + 4x + y^2 - 2y - 4 = 0$

4. $9x^2 + 4y^2 = 36$

Find the vertex, the focus, and the directrix of the parabola. Then draw the graph.

5. $x^2 = 12y$

6. $y^2 + 2y - 8x - 7 = 0$

7. Find an equation of the parabola with focus $(0, 2)$ and directrix $y = -2$.

8. Find the center and the radius of the circle given by $x^2 + y^2 + 2x - 6y - 15 = 0$. Then draw the graph.

Find the center, the vertices, and the foci of the ellipse. Then draw the graph.

9. $9x^2 + 16y^2 = 144$

10. $\dfrac{(x+1)^2}{4} + \dfrac{(y-2)^2}{9} = 1$

11. Find an equation of the ellipse having vertices $(0, -5)$ and $(0, 5)$ and with minor axis of length 4.

Find the center, the vertices, the foci, and the asymptotes of the hyperbola. Then draw the graph.

12. $4x^2 - y^2 = 4$

13. $\dfrac{(y-2)^2}{4} - \dfrac{(x+1)^2}{9} = 1$

14. Find the asymptotes of the hyperbola given by $2y^2 - x^2 = 18$.

15. *Satellite Dish.* A satellite dish has a parabolic cross section that is 18 in. wide at the opening and 6 in. deep at the vertex. How far from the vertex is the focus?

Solve.

16. $2x^2 - 3y^2 = -10,$
$x^2 + 2y^2 = 9$

17. $x^2 + y^2 = 13,$
$x + y = 1$

18. $x + y = 5,$
$xy = 6$

19. *Landscaping.* Leisurescape is planting a rectangular flower garden with a perimeter of 18 ft and a diagonal of $\sqrt{41}$ ft. Find the dimensions of the garden.

20. *Fencing.* It will take 210 ft of fencing to enclose a rectangular playground with an area of 2700 ft^2. Find the dimensions of the playground.

21. Graph the system of inequalities. Then find the coordinates of the points of intersection of the graphs.

$y \geq x^2 - 4,$
$y < 2x - 1$

22. Graph: $5x^2 - 8xy + 5y^2 = 9$.

23. Graph $r = \dfrac{2}{1 - \sin \theta}$. State whether the directrix is vertical or horizontal, describe its location in relation to the pole, and find the vertex or vertices.

24. Find a polar equation of the conic with a focus at the pole, eccentricity 2, and directrix $r = 3 \sec \theta$.

25. Graph the plane curve given by the parametric equations $x = \sqrt{t}, y = t + 2; 0 \le t \le 16$.

26. Find a rectangular equation equivalent to $x = 3 \cos \theta, y = 3 \sin \theta; 0 \le \theta \le 2\pi$.

27. Find two sets of parametric equations for the rectangular equation $y = x - 5$.

28. *Projectile Motion.*   A projectile is launched from a height of 10 ft with an initial speed of 250 ft/sec at an angle of 30° with the horizontal.

   a) Find parametric equations that give the position of the projectile at time t, in seconds.

   b) Find the height of the projectile after 1 sec and after 3 sec.

   c) Determine how long the projectile is in the air.

   d) Determine the horizontal distance that the projectile travels.

   e) Find the maximum height of the projectile.

29. The graph of $(y - 1)^2 = 4(x + 1)$ is which of the following?

A.

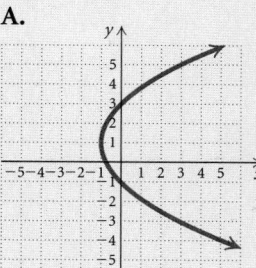

B.

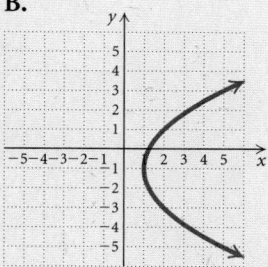

C.

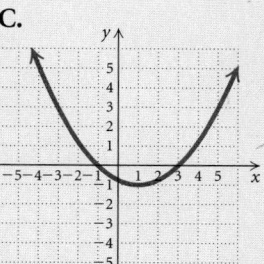

D.

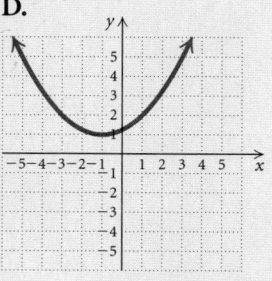

Synthesis

30. Find an equation of the circle for which the endpoints of a diameter are $(1, 1)$ and $(5, -3)$.

Sequences, Series, and Combinatorics

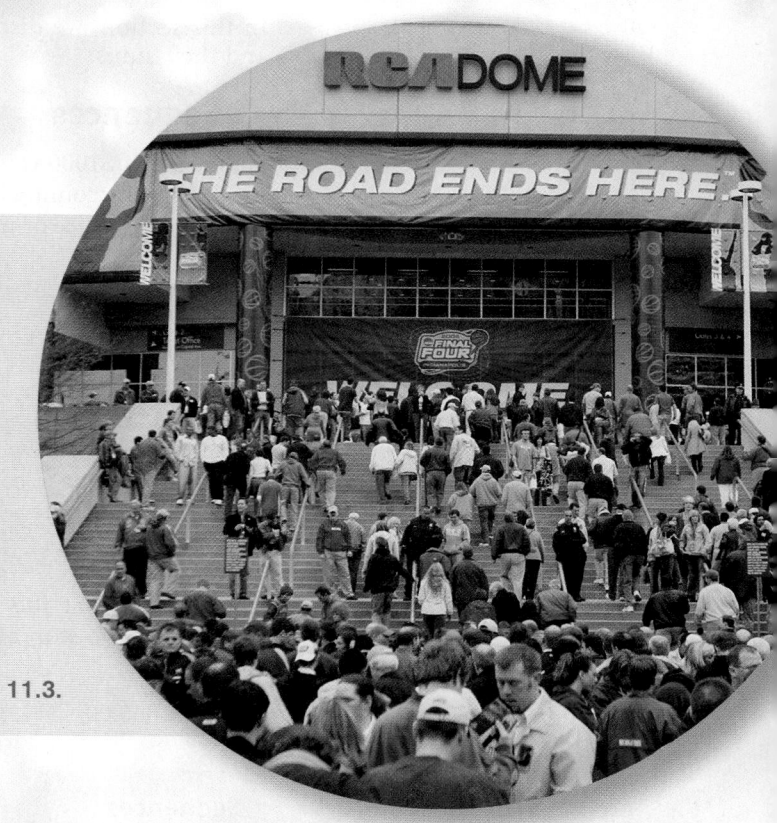

APPLICATION   Large sporting events have a significant impact on the economy of the host city. Those attending the 2006 NCAA men's Final Four basketball competition in Indianapolis poured $45 million into the local economy (*Source*: WISH-TV). Assume that 60% of that amount is spent again in the city, and then 60% of that amount is spent again, and so on. This is known as the *economic multiplier effect*. Find the total effect on the economy.

This problem appears as Example 10 in Section 11.3.

11.1   Sequences and Series

11.2   Arithmetic Sequences and Series

11.3   Geometric Sequences and Series

11.4   Mathematical Induction

11.5   Combinatorics: Permutations

11.6   Combinatorics: Combinations

11.7   The Binomial Theorem

11.8   Probability

11.1

Sequences and Series

❖ Find terms of sequences given the *n*th term.
❖ Look for a pattern in a sequence and try to determine a general term.
❖ Convert between sigma notation and other notation for a series.
❖ Construct the terms of a recursively defined sequence.

In this section, we discuss sets or lists of numbers, considered in order, and their sums.

❖ Sequences

Suppose that $1000 is invested at 6%, compounded annually. The amounts to which the account will grow after 1 yr, 2 yr, 3 yr, 4 yr, and so on, form the following sequence of numbers:

$$
\begin{array}{cccc}
(1) & (2) & (3) & (4) \\
\downarrow & \downarrow & \downarrow & \downarrow \\
\$1060.00, & \$1123.60, & \$1191.02, & \$1262.48, \ldots.
\end{array}
$$

We can think of this as a function that pairs 1 with $1060.00, 2 with $1123.60, 3 with $1191.02, and so on. A **sequence** is thus a *function*, where the domain is a set of consecutive positive integers beginning with 1.

If we continue to compute the amounts of money in the account forever, we obtain an **infinite sequence** with function values

$1060.00, $1123.60, $1191.02, $1262.48, $1338.23, $1418.52, ...

The dots "..." at the end indicate that the sequence goes on without stopping. If we stop after a certain number of years, we obtain a **finite sequence**:

$1060.00, $1123.60, $1191.02, $1262.48.

Sequences

An **infinite sequence** is a function having for its domain the set of positive integers, $\{1, 2, 3, 4, 5, \ldots\}$.

A **finite sequence** is a function having for its domain a set of positive integers, $\{1, 2, 3, 4, 5, \ldots, n\}$, for some positive integer n.

Consider the sequence given by the formula

$$a(n) = 2^n, \quad \text{or} \quad a_n = 2^n.$$

Some of the function values, also known as the **terms** of the sequence, follow:

$$a_1 = 2^1 = 2,$$
$$a_2 = 2^2 = 4,$$
$$a_3 = 2^3 = 8,$$
$$a_4 = 2^4 = 16,$$
$$a_5 = 2^5 = 32.$$

The first term of the sequence is denoted as a_1, the fifth term as a_5, and the nth term, or **general term**, as a_n. This sequence can also be denoted as

$$2, 4, 8, \ldots, \quad \text{or as} \quad 2, 4, 8, \ldots, 2^n, \ldots.$$

EXAMPLE 1 Find the first 4 terms and the 23rd term of the sequence whose general term is given by $a_n = (-1)^n n^2$.

Solution We have $a_n = (-1)^n n^2$, so

$$a_1 = (-1)^1 \cdot 1^2 = -1,$$
$$a_2 = (-1)^2 \cdot 2^2 = 4,$$
$$a_3 = (-1)^3 \cdot 3^2 = -9,$$
$$a_4 = (-1)^4 \cdot 4^2 = 16,$$
$$a_{23} = (-1)^{23} \cdot 23^2 = -529.$$

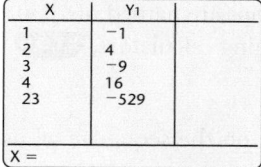

We can also use a graphing calculator to find the desired terms of this sequence. We enter $y_1 = (-1)^x x^2$. We then set up a table in ASK mode and enter 1, 2, 3, 4, and 23 as values for x.

Now Try Exercise 1. ■

Note in Example 1 that the power $(-1)^n$ causes the signs of the terms to alternate between positive and negative, depending on whether n is even or odd. This kind of sequence is called an **alternating sequence**.

[GCM] **EXAMPLE 2** Use a graphing calculator to find the first 5 terms of the sequence whose general term is given by $a_n = n/(n + 1)$.

STUDY TIP

Refer to the *Graphing Calculator Manual* that accompanies this text to find the keystrokes for finding the terms of a sequence.

Solution We can use a table or the SEQ feature, as shown here. We select SEQ from the LIST OPS menu and enter the general term, the variable, and the numbers of the first and last terms desired. The calculator will write the terms horizontally as a list. The list can also be written in fraction notation.

```
seq(X/(X+1),X,1,5)▶Frac
{1/2 2/3 3/4 4/...
```

```
seq(X/(X+1),X,1,5)▶Frac
.../3 3/4 4/5 5/6}
```

We use the ▶ key to view the two items that do not initially appear on the screen. The first 5 terms of the sequence are 1/2, 2/3, 3/4, 4/5, and 5/6. ■

We can graph a sequence just as we graph other functions. Consider the function given by $f(x) = x + 1$ and the sequence whose general term is given by $a_n = n + 1$. The graph of $f(x) = x + 1$ is shown on the left below. Since the domain of a sequence is a set of positive integers, the graph of a sequence is a set of points that are not connected. Thus if we use only positive integers for inputs of $f(x) = x + 1$, we have the graph of the sequence $a_n = n + 1$ as shown on the right below.

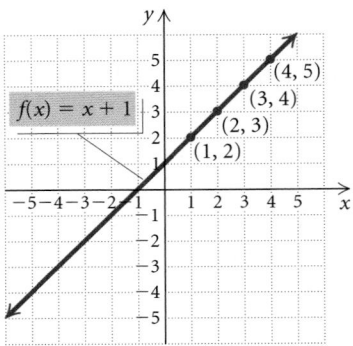
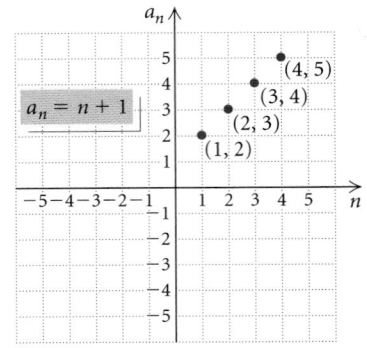

We can also use a graphing calculator to graph a sequence. Since we are graphing a set of unconnected points, we use DOT mode. We also select SEQUENCE mode. In this mode, the variable is n and functions are named $u(n)$, $v(n)$, and $w(n)$ rather than y_1, y_2, and y_3. On many graphing calculators, $\boxed{\text{X,T,}\Theta\text{,n}}$ is used to enter n in SEQUENCE mode.

EXAMPLE 3   Use a graphing calculator to graph the sequence whose general term is given by $a_n = n/(n + 1)$.

Solution   With the calculator set in DOT mode and SEQUENCE mode, we enter $u(n) = n/(n + 1)$. All the function values will be positive numbers that are less than 1, so we choose the window $[0, 10, 0, 1]$ and we also choose $n\text{Min} = 1$, $n\text{Max} = 10$, PlotStart $= 1$, and PlotStep $= 1$.

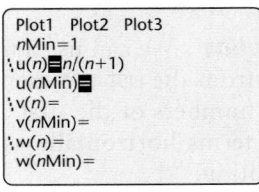
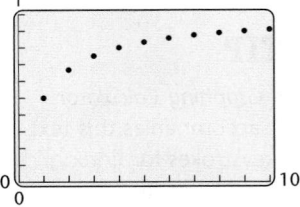

Now Try Exercise 19. ■

❋ Finding the General Term

When only the first few terms of a sequence are known, we do not know for sure what the general term is, but we might be able to make a prediction by looking for a pattern.

EXAMPLE 4 For each of the following sequences, predict the general term.

a) $1, \sqrt{2}, \sqrt{3}, 2, \ldots$ **b)** $-1, 3, -9, 27, -81, \ldots$

c) $2, 4, 8, \ldots$

Solution

a) These are square roots of consecutive integers, so the general term might be $\sqrt{n}$.

b) These are powers of 3 with alternating signs, so the general term might be $(-1)^n 3^{n-1}$.

c) If we see the pattern of powers of 2, we will see 16 as the next term and guess 2^n for the general term. Then the sequence could be written with more terms as

$$2, 4, 8, 16, 32, 64, 128, \ldots .$$

If we see that we can get the second term by adding 2, the third term by adding 4, and the next term by adding 6, and so on, we will see 14 as the next term. A general term for the sequence is $n^2 - n + 2$, and the sequence can be written with more terms as

$$2, 4, 8, 14, 22, 32, 44, 58, \ldots . \qquad \text{\small\textbf{Now Try Exercise 23.}} ∎$$

Example 4(c) illustrates that, in fact, you can never be certain about the general term when only a few terms are given. The fewer the given terms, the greater the uncertainty.

❋ Sums and Series

> **Series**
>
> Given the infinite sequence
>
> $$a_1, a_2, a_3, a_4, \ldots, a_n, \ldots,$$
>
> the sum of the terms
>
> $$a_1 + a_2 + a_3 + \cdots + a_n + \cdots$$
>
> is called an **infinite series**. A **partial sum** is the sum of the first n terms:
>
> $$a_1 + a_2 + a_3 + \cdots + a_n.$$
>
> A partial sum is also called a **finite series**, or **nth partial sum**, and is denoted S_n.

EXAMPLE 5 For the sequence $-2, 4, -6, 8, -10, 12, -14, \ldots$, find each of the following.

a) S_1 b) S_4 c) S_5

Solution

a) $S_1 = -2$
b) $S_4 = -2 + 4 + (-6) + 8 = 4$
c) $S_5 = -2 + 4 + (-6) + 8 + (-10) = -6$ **Now Try Exercise 33.** ■

We can also use a graphing calculator to find partial sums of a sequence when a formula for the general term is known.

GCM **EXAMPLE 6** Use a graphing calculator to find S_1, S_2, S_3, and S_4 for the sequence whose general term is given by $a_n = n^2 - 3$.

Solution We can use the CUMSUM feature from the LIST OPS menu. The calculator will write the partial sums as a list. (Note that the calculator can be set in either FUNCTION mode or SEQUENCE mode. Here we show SEQUENCE mode.)
     We have $S_1 = -2$, $S_2 = -1$, $S_3 = 5$, and $S_4 = 18$. ■

```
cumSum(seq(n²−3,n,1,4))
            {−2 −1 5 18}
```

❊ Sigma Notation

The Greek letter Σ (sigma) can be used to denote a sum when the general term of a sequence is a formula. For example, the sum of the first four terms of the sequence $3, 5, 7, 9, \ldots, 2k + 1, \ldots$ can be named as follows, using what is called **sigma notation**, or **summation notation**:

$$\sum_{k=1}^{4} (2k + 1).$$

This is read "the sum as k goes from 1 to 4 of $2k + 1$." The letter k is called the **index of summation**. The index of summation might start at a number other than 1, and letters other than k can be used.

GCM **EXAMPLE 7** Find and evaluate each of the following sums.

a) $\displaystyle\sum_{k=1}^{5} k^3$ b) $\displaystyle\sum_{k=0}^{4} (-1)^k 5^k$ c) $\displaystyle\sum_{i=8}^{11} \left(2 + \frac{1}{i} \right)$

Solution

a) We replace k with 1, 2, 3, 4, and 5. Then we add the results.

$$\sum_{k=1}^{5} k^3 = 1^3 + 2^3 + 3^3 + 4^3 + 5^3$$
$$= 1 + 8 + 27 + 64 + 125$$
$$= 225$$

We can also combine the SUM and SEQ features on a graphing calculator to add the terms of this sequence.

$$
\boxed{\begin{array}{l} \text{sum(seq}(n\text{^}3,n,1,5)) \\ \hspace{4cm} 225 \end{array}}
$$

b) $\displaystyle\sum_{k=0}^{4}(-1)^{k}5^{k} = (-1)^{0}5^{0} + (-1)^{1}5^{1} + (-1)^{2}5^{2} + (-1)^{3}5^{3} + (-1)^{4}5^{4}$

$$= 1 - 5 + 25 - 125 + 625 = 521$$

c) $\displaystyle\sum_{i=8}^{11}\left(2 + \frac{1}{i}\right) = \left(2 + \frac{1}{8}\right) + \left(2 + \frac{1}{9}\right) + \left(2 + \frac{1}{10}\right) + \left(2 + \frac{1}{11}\right)$

$$= 8\frac{1691}{3960}$$

Now Try Exercise 37. ■

EXAMPLE 8 Write sigma notation for each sum.

a) $1 + 2 + 4 + 8 + 16 + 32 + 64$

b) $-2 + 4 - 6 + 8 - 10$

c) $x + \dfrac{x^2}{2} + \dfrac{x^3}{3} + \dfrac{x^4}{4} + \cdots$

Solution

a) $1 + 2 + 4 + 8 + 16 + 32 + 64$

This is the sum of powers of 2, beginning with 2^0, or 1, and ending with 2^6, or 64. Sigma notation is $\Sigma_{k=0}^{6} 2^k$.

b) $-2 + 4 - 6 + 8 - 10$

Disregarding the alternating signs, we see that this is the sum of the first 5 even integers. Note that $2k$ is a formula for the kth positive even integer, and $(-1)^k = -1$ when k is odd and $(-1)^k = 1$ when k is even. Thus the general term is $(-1)^k(2k)$. The sum begins with $k = 1$ and ends with $k = 5$, so sigma notation is $\Sigma_{k=1}^{5}(-1)^k(2k)$.

c) $x + \dfrac{x^2}{2} + \dfrac{x^3}{3} + \dfrac{x^4}{4} + \cdots$

The general term is x^k/k, beginning with $k = 1$. This is also an infinite series. We use the symbol ∞ for infinity and write the series using sigma notation: $\Sigma_{k=1}^{\infty}(x^k/k)$.

Now Try Exercise 55. ■

❈ Recursive Definitions

A sequence may be defined **recursively** or by using a **recursion formula**. Such a definition lists the first term, or the first few terms, and then describes how to determine the remaining terms from the given terms.

GCM **EXAMPLE 9**   Find the first 5 terms of the sequence defined by

$$a_1 = 5, \qquad a_{n+1} = 2a_n - 3, \quad \text{for } n \geq 1.$$

Solution   We have

$$a_1 = 5,$$

$$a_2 = 2a_1 - 3 = 2 \cdot 5 - 3 = 7,$$

$$a_3 = 2a_2 - 3 = 2 \cdot 7 - 3 = 11,$$

$$a_4 = 2a_3 - 3 = 2 \cdot 11 - 3 = 19,$$

$$a_5 = 2a_4 - 3 = 2 \cdot 19 - 3 = 35.$$

Many graphing calculators have the capability to work with recursively defined sequences when they are set in SEQUENCE mode. For this sequence, for instance, the function could be entered as $u(n) = 2 * u(n - 1) - 3$ with $u(n\text{Min}) = 5$. We can read the terms of the sequence from a table.

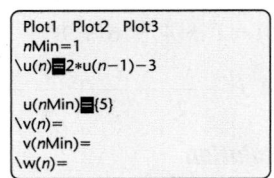

Now Try Exercise 65. ■

(11.1)   Exercise Set

In each of the following, the nth term of a sequence is given. Find the first 4 terms, a_{10}, and a_{15}.

1. $a_n = 4n - 1$

2. $a_n = (n - 1)(n - 2)(n - 3)$

3. $a_n = \dfrac{n}{n - 1},\ n \geq 2$

4. $a_n = n^2 - 1,\ n \geq 3$

5. $a_n = \dfrac{n^2 - 1}{n^2 + 1}$

6. $a_n = \left(-\dfrac{1}{2}\right)^{n-1}$

7. $a_n = (-1)^n n^2$

8. $a_n = (-1)^{n-1}(3n - 5)$

9. $a_n = 5 + \dfrac{(-2)^{n+1}}{2^n}$

10. $a_n = \dfrac{2n - 1}{n^2 + 2n}$

Find the indicated term of the given sequence.

11. $a_n = 5n - 6;\ a_8$

12. $a_n = (3n - 4)(2n + 5);\ a_7$

13. $a_n = (2n + 3)^2;\ a_6$

14. $a_n = (-1)^{n-1}(4.6n - 18.3);\ a_{12}$

15. $a_n = 5n^2(4n - 100);\ a_{11}$

16. $a_n = \left(1 + \dfrac{1}{n}\right)^2;\ a_{80}$

17. $a_n = \ln e^n;\ a_{67}$

18. $a_n = 2 - \dfrac{1000}{n};\ a_{100}$

Use a graphing calculator to construct a table of values and a graph for the first 10 terms of the sequence.

19. $a_n = \left(1 + \dfrac{1}{n}\right)^n$

20. $a_n = \sqrt{n + 1} - \sqrt{n}$

21. $a_1 = 2,\ a_{n+1} = \sqrt{1 + \sqrt{a_n}}$

22. $a_1 = 2,\ a_{n+1} = \dfrac{1}{2}\left(a_n + \dfrac{2}{a_n}\right)$

Predict the general term, or nth term, a_n, of the sequence. Answers may vary.

23. $2, 4, 6, 8, 10, \ldots$

24. $3, 9, 27, 81, 243, \ldots$

25. $-2, 6, -18, 54, \ldots$

26. $-2, 3, 8, 13, 18, \ldots$

27. $\frac{2}{3}, \frac{3}{4}, \frac{4}{5}, \frac{5}{6}, \frac{6}{7}, \ldots$

28. $\sqrt{2}, 2, \sqrt{6}, 2\sqrt{2}, \sqrt{10}, \ldots$

29. $1 \cdot 2, 2 \cdot 3, 3 \cdot 4, 4 \cdot 5, \ldots$

30. $-1, -4, -7, -10, -13, \ldots$

31. $0, \log 10, \log 100, \log 1000, \ldots$

32. $\ln e^2, \ln e^3, \ln e^4, \ln e^5, \ldots$

Find the indicated partial sums for the sequence.

33. $1, 2, 3, 4, 5, 6, 7, \ldots;\ S_3$ and S_7

34. $1, -3, 5, -7, 9, -11, \ldots;\ S_2$ and S_5

35. $2, 4, 6, 8, \ldots;\ S_4$ and S_5

36. $1, \frac{1}{4}, \frac{1}{9}, \frac{1}{16}, \frac{1}{25}, \ldots;\ S_1$ and S_5

Find and evaluate the sum.

37. $\displaystyle\sum_{k=1}^{5} \frac{1}{2k}$

38. $\displaystyle\sum_{i=1}^{6} \frac{1}{2i + 1}$

39. $\displaystyle\sum_{i=0}^{6} 2^i$

40. $\displaystyle\sum_{k=4}^{7} \sqrt{2k - 1}$

41. $\displaystyle\sum_{k=7}^{10} \ln k$

42. $\displaystyle\sum_{k=1}^{4} \pi k$

43. $\displaystyle\sum_{k=1}^{8} \frac{k}{k + 1}$

44. $\displaystyle\sum_{i=1}^{5} \frac{i - 1}{i + 3}$

45. $\displaystyle\sum_{i=1}^{5} (-1)^i$

46. $\displaystyle\sum_{k=0}^{5} (-1)^{k+1}$

47. $\displaystyle\sum_{k=1}^{8} (-1)^{k+1}3k$

48. $\displaystyle\sum_{k=0}^{7} (-1)^k 4^{k+1}$

49. $\displaystyle\sum_{k=0}^{6} \frac{2}{k^2 + 1}$

50. $\displaystyle\sum_{i=1}^{10} i(i + 1)$

51. $\displaystyle\sum_{k=0}^{5} (k^2 - 2k + 3)$

52. $\displaystyle\sum_{k=1}^{10} \frac{1}{k(k + 1)}$

53. $\displaystyle\sum_{i=0}^{10} \frac{2^i}{2^i + 1}$

54. $\displaystyle\sum_{k=0}^{3} (-2)^{2k}$

Write sigma notation. Answers may vary.

55. $5 + 10 + 15 + 20 + 25 + \cdots$

56. $7 + 14 + 21 + 28 + 35 + \cdots$

57. $2 - 4 + 8 - 16 + 32 - 64$

58. $3 + 6 + 9 + 12 + 15$

59. $-\dfrac{1}{2} + \dfrac{2}{3} - \dfrac{3}{4} + \dfrac{4}{5} - \dfrac{5}{6} + \dfrac{6}{7}$

60. $\dfrac{1}{1^2} + \dfrac{1}{2^2} + \dfrac{1}{3^2} + \dfrac{1}{4^2} + \dfrac{1}{5^2}$

61. $4 - 9 + 16 - 25 + \cdots + (-1)^n n^2$

62. $9 - 16 + 25 + \cdots + (-1)^{n+1} n^2$

63. $\dfrac{1}{1 \cdot 2} + \dfrac{1}{2 \cdot 3} + \dfrac{1}{3 \cdot 4} + \dfrac{1}{4 \cdot 5} + \cdots$

64. $\dfrac{1}{1 \cdot 2^2} + \dfrac{1}{2 \cdot 3^2} + \dfrac{1}{3 \cdot 4^2} + \dfrac{1}{4 \cdot 5^2} + \cdots$

Find the first 4 terms of the recursively defined sequence.

65. $a_1 = 4$, $a_{n+1} = 1 + \dfrac{1}{a_n}$

66. $a_1 = 256$, $a_{n+1} = \sqrt{a_n}$

67. $a_1 = 6561$, $a_{n+1} = (-1)^n \sqrt{a_n}$

68. $a_1 = e^Q$, $a_{n+1} = \ln a_n$

69. $a_1 = 2$, $a_2 = 3$, $a_{n+1} = a_n + a_{n-1}$

70. $a_1 = -10$, $a_2 = 8$, $a_{n+1} = a_n - a_{n-1}$

71. *Compound Interest.*   Suppose that $1000 is invested at 6.2%, compounded annually. The value of the investment after n years is given by the sequence model

$$a_n = \$1000(1.062)^n, \quad n = 1, 2, 3, \dots.$$

a) Find the first 10 terms of the sequence.
b) Find the value of the investment after 20 yr.

72. *Salvage Value.*   The value of an office machine is $5200. Its salvage value each year is 75% of its value the year before. Give a sequence that lists the salvage value of the machine for each year of a 10-yr period.

73. *Bacteria Growth.*   Suppose a single cell of bacteria divides into two every 15 min. Suppose that the same rate of division is maintained for 4 hr. Give a sequence that lists the number of cells after successive 15-min periods.

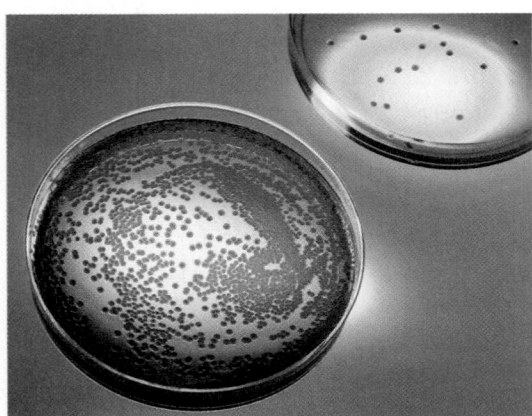

74. *Wage Sequence.*   Torrey is paid $8.30 per hour for working at Red Freight Limited. Each year he receives a $0.30 hourly raise. Give a sequence that lists Torrey's hourly wage over a 10-yr period.

75. *Fibonacci Sequence: Rabbit Population Growth.* One of the most famous recursively defined sequences is the **Fibonacci sequence**. In 1202, the Italian mathematician Leonardo da Pisa, also called Fibonacci, proposed the following model for rabbit population growth. Suppose that every month each mature pair of rabbits in the population produces a new pair that begins reproducing after two months, and also suppose that no rabbits die. Beginning with one pair of newborn rabbits, the population can be modeled by the following recursively defined sequence:

$$a_1 = 1, \quad a_2 = 1, \quad a_n = a_{n-1} + a_{n-2}, \text{ for } n \geq 3,$$

where a_n is the total number of pairs of rabbits in month n. Find the first 7 terms of the Fibonacci sequence.

76. *Prescription Drug Sales.*   The table below lists the retail sales of prescription drugs in the United States in recent years.

Year	Retail Sales of Prescription Drugs (in billions)
2000	$146
2001	164
2002	183
2003	203
2004	221
2005	230

Source: National Association of Chain Drug Stores

a) Use a graphing calculator to fit a linear sequence regression function

$$a_n = an + b$$

to the data, where n is the number of years after 2000.
b) Estimate the retail sales of prescription drugs in 2010, in 2012, and in 2015. Round to the nearest billion.

77. *Patents Issued.* The table below lists the number of patents issued in the United States in recent years.

Year	Patents Issued (in thousands)
2000	176.0
2001	184.0
2002	184.4
2003	187.0
2004	181.3
2005	157.7

Source: U.S. Patent and Trademark Office

a) Use a graphing calculator to fit a quadratic sequence regression function

$$a_n = an^2 + bn + c$$

to the data, where n is the number of years after 2000.

b) Estimate the number of patents issued in 2006, in 2007, and in 2008. Round to the nearest tenth of a thousand.

Collaborative Discussion and Writing

78. a) Find the first few terms of the sequence $a_n = n^2 - n + 41$ and describe the pattern you observe.

b) Does the pattern you found in part (a) hold for all choices of n? Why or why not?

79. The Fibonacci sequence has intrigued mathematicians for centuries. In fact, a journal called the *Fibonacci Quarterly* is devoted to publishing results pertaining to such sequences. Do some research on the connection of the Fibonacci sequence to the idea of the "Golden Section."

Skill Maintenance

Solve.

80. $3x - 2y = 3,$
$2x + 3y = -11$

81. *Dining Habits.* The average American ordered a total of 208 takeout meals and meals to be eaten in restaurants in 2006. The number of takeout meals ordered exceeded the number of meals eaten in restaurants by 46. (*Source:* NPD Group) Find the number of each type of meal ordered.

Find the center and the radius of the circle with the given equation.

82. $x^2 + y^2 - 6x + 4y = 3$

83. $x^2 + y^2 + 5x - 8y = 2$

Synthesis

Find the first 5 terms of the sequence, and then find S_5.

84. $a_n = \frac{1}{2^n} \log 1000^n$

85. $a_n = i^n, \ i = \sqrt{-1}$

86. $a_n = \ln(1 \cdot 2 \cdot 3 \cdots n)$

For each sequence, find a formula for S_n.

87. $a_n = \ln n$

88. $a_n = \frac{1}{n} - \frac{1}{n+1}$

Arithmetic Sequences and Series

❖ For any arithmetic sequence, find the nth term when n is given and n when the nth term is given, and given two terms, find the common difference and construct the sequence.

❖ Find the sum of the first n terms of an arithmetic sequence.

A sequence in which each term after the first is found by adding the same number to the preceding term is an **arithmetic sequence**.

❖ Arithmetic Sequences

The sequence 2, 5, 8, 11, 14, 17,... is arithmetic because adding 3 to any term produces the next term. In other words, the difference between any term and the preceding one is 3. Arithmetic sequences are also called *arithmetic progressions*.

> **Arithmetic Sequence**
>
> A sequence is **arithmetic** if there exists a number d, called the **common difference**, such that $a_{n+1} = a_n + d$ for any integer $n \geq 1$.

EXAMPLE 1   For each of the following arithmetic sequences, identify the first term, a_1, and the common difference, d.

a) 4, 9, 14, 19, 24,...

b) 34, 27, 20, 13, 6, -1, -8,...

c) 2, $2\frac{1}{2}$, 3, $3\frac{1}{2}$, 4, $4\frac{1}{2}$,...

Solution   The first term, a_1, is the first term listed. To find the common difference, d, we choose any term beyond the first and subtract the preceding term from it.

SEQUENCE	FIRST TERM, a_1	COMMON DIFFERENCE, d
a) 4, 9, 14, 19, 24,...	4	5 $(9 - 4 = 5)$
b) 34, 27, 20, 13, 6, -1, -8,...	34	-7 $(27 - 34 = -7)$
c) 2, $2\frac{1}{2}$, 3, $3\frac{1}{2}$, 4, $4\frac{1}{2}$,...	2	$\frac{1}{2}$ $\left(2\frac{1}{2} - 2 = \frac{1}{2}\right)$

We obtained the common difference by subtracting a_1 from a_2. Had we subtracted a_2 from a_3 or a_3 from a_4, we would have obtained the same values for d. Thus we can check by adding d to each term in a sequence to see if we progress correctly to the next term.

Check:

a) $4 + 5 = 9$, $9 + 5 = 14$, $14 + 5 = 19$, $19 + 5 = 24$

b) $34 + (-7) = 27$, $27 + (-7) = 20$, $20 + (-7) = 13$,
   $13 + (-7) = 6$, $6 + (-7) = -1$, $-1 + (-7) = -8$

STUDY TIP

The best way to prepare for a final exam is to do so over a period of at least two weeks. First review each chapter, studying the formulas, theorems, properties, and procedures in the sections and in the Chapter Summary and Review. Then take each of the Chapter Tests again. If you miss any questions, spend extra time reviewing the corresponding topics. Watch the videos that accompany the text or use the InterAct Math Tutorial Software. Also consider participating in a study group or attending a tutoring or review session.

c) $2 + \frac{1}{2} = 2\frac{1}{2}$, $2\frac{1}{2} + \frac{1}{2} = 3$, $3 + \frac{1}{2} = 3\frac{1}{2}$, $3\frac{1}{2} + \frac{1}{2} = 4$,
   $4 + \frac{1}{2} = 4\frac{1}{2}$

Now Try Exercise 1. ■

To find a formula for the general, or nth, term of any arithmetic sequence, we denote the common difference by d, write out the first few terms, and look for a pattern:

a_1,
$a_2 = a_1 + d$,
$a_3 = a_2 + d = (a_1 + d) + d = a_1 + 2d$,   Substituting for a_2
$a_4 = a_3 + d = (a_1 + 2d) + d = a_1 + 3d$.   Substituting for a_3

Note that the coefficient of d in each case is 1 less than the subscript.

Generalizing, we obtain the following formula.

> ### nth Term of an Arithmetic Sequence
> The **nth term** of an arithmetic sequence is given by
> $a_n = a_1 + (n - 1)d$, for any integer $n \geq 1$.

EXAMPLE 2   Find the 14th term of the arithmetic sequence $4, 7, 10, 13, \ldots$.

Solution   We first note that $a_1 = 4$, $d = 7 - 4$, or 3, and $n = 14$. Then using the formula for the nth term, we obtain

$$a_n = a_1 + (n - 1)d$$
$$a_{14} = 4 + (14 - 1) \cdot 3 \quad \text{Substituting}$$
$$= 4 + 13 \cdot 3 = 4 + 39$$
$$= 43.$$

The 14th term is 43.

Now Try Exercise 9. ■

EXAMPLE 3   In the sequence of Example 2, which term is 301? That is, find n if $a_n = 301$.

Solution   We substitute 301 for a_n, 4 for a_1, and 3 for d in the formula for the nth term and solve for n:

$$a_n = a_1 + (n - 1)d$$
$$301 = 4 + (n - 1) \cdot 3 \quad \text{Substituting}$$
$$301 = 4 + 3n - 3$$
$$301 = 3n + 1 \qquad \text{Solving for } n$$
$$300 = 3n$$
$$100 = n.$$

The term 301 is the 100th term of the sequence.

Now Try Exercise 15. ■

Given two terms and their places in an arithmetic sequence, we can construct the sequence.

EXAMPLE 4 The 3rd term of an arithmetic sequence is 8, and the 16th term is 47. Find a_1 and d and construct the sequence.

Solution We know that $a_3 = 8$ and $a_{16} = 47$. Thus we would have to add d 13 times to get from 8 to 47. That is,

$$8 + 13d = 47. \qquad a_3 \text{ and } a_{16} \text{ are } 16 - 3, \text{ or } 13, \text{ terms apart.}$$

Solving $8 + 13d = 47$, we obtain

$$13d = 39$$
$$d = 3.$$

Since $a_3 = 8$, we subtract d twice to get a_1. Thus,

$$a_1 = 8 - 2 \cdot 3 = 2. \qquad a_1 \text{ and } a_3 \text{ are } 3 - 1, \text{ or } 2, \text{ terms apart.}$$

The sequence is 2, 5, 8, 11, ... Note that we could also subtract d 15 times from a_{16} in order to find a_1. **Now Try Exercise 23.** ■

In general, d should be subtracted $n - 1$ times from a_n in order to find a_1.

Exploring with Technology

Graph the first 10 terms of each arithmetic sequence. What pattern do you observe?

$$a_n = 2 + (n - 1)(4),$$
$$a_n = -5 + (n - 1)(1.2),$$
$$a_n = -4 + (n - 1)(-3),$$
$$a_n = 3 + (n - 1)\left(-\tfrac{5}{2}\right)$$

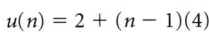
$u(n) = 2 + (n - 1)(4)$

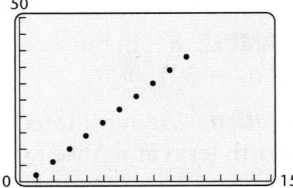

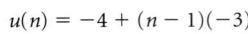
$u(n) = -4 + (n - 1)(-3)$

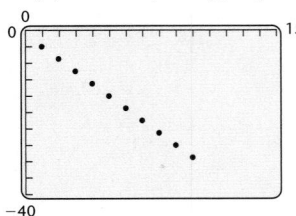

The pattern shown above holds in general. *The graph of an arithmetic sequence is a set of points that lie on the graph of a linear function.*

❈ Sum of the First *n* Terms of an Arithmetic Sequence

Consider the arithmetic sequence

$$3, 5, 7, 9, \ldots .$$

When we add the first 4 terms of the sequence, we get S_4, which is

$$3 + 5 + 7 + 9, \quad \text{or} \quad 24.$$

This sum is called an **arithmetic series**. To find a formula for the sum of the first n terms, S_n, of an arithmetic sequence, we first denote an arithmetic sequence, as follows:

> This term is two terms back from the last. If you add *d* to this term, the result is the next-to-last term, $a_n - d$.

$$a_1, \quad (a_1 + d), \quad (a_1 + 2d), \quad \ldots, \quad \overbrace{(a_n - 2d)}, \quad \underbrace{(a_n - d)}, \quad a_n.$$

> This is the next-to-last term. If you add *d* to this term, the result is a_n.

Then S_n is given by

$$S_n = a_1 + (a_1 + d) + (a_1 + 2d) + \cdots + (a_n - 2d)$$
$$+ (a_n - d) + a_n. \tag{1}$$

Reversing the order of the addition gives us

$$S_n = a_n + (a_n - d) + (a_n - 2d) + \cdots + (a_1 + 2d)$$
$$+ (a_1 + d) + a_1. \tag{2}$$

If we add corresponding terms of each side of equations (1) and (2), we get

$$2S_n = [a_1 + a_n] + [(a_1 + d) + (a_n - d)] + [(a_1 + 2d) + (a_n - 2d)]$$
$$+ \cdots + [(a_n - 2d) + (a_1 + 2d)]$$
$$+ [(a_n - d) + (a_1 + d)] + [a_n + a_1].$$

In the expression for $2S_n$, there are n expressions in square brackets. Each of these expressions is equivalent to $a_1 + a_n$. Thus the expression for $2S_n$ can be written in simplified form as

$$2S_n = [a_1 + a_n] + [a_1 + a_n] + [a_1 + a_n] + \cdots + [a_n + a_1]$$
$$+ [a_n + a_1] + [a_n + a_1].$$

Since $a_1 + a_n$ is being added n times, it follows that

$$2S_n = n(a_1 + a_n),$$

from which we get the following formula.

> **Sum of the First n Terms**
> The sum of the first n terms of an arithmetic sequence is given by
> $$S_n = \frac{n}{2}(a_1 + a_n).$$

EXAMPLE 5 Find the sum of the first 100 natural numbers.

Solution The sum is

$$1 + 2 + 3 + \cdots + 99 + 100.$$

This is the sum of the first 100 terms of the arithmetic sequence for which

$$a_1 = 1, \quad a_n = 100, \quad \text{and} \quad n = 100.$$

Thus substituting into the formula

$$S_n = \frac{n}{2}(a_1 + a_n),$$

we get

$$S_{100} = \frac{100}{2}(1 + 100) = 50(101) = 5050.$$

The sum of the first 100 natural numbers is 5050. **Now Try Exercise 27.** ▪

EXAMPLE 6 Find the sum of the first 15 terms of the arithmetic sequence 4, 7, 10, 13, ...

Solution Note that $a_1 = 4$, $d = 3$, and $n = 15$. Before using the formula

$$S_n = \frac{n}{2}(a_1 + a_n),$$

we find the last term, a_{15}:

$$a_{15} = 4 + (15 - 1)3 \qquad \text{Substituting into the formula } a_n = a_1 + (n-1)d$$
$$= 4 + 14 \cdot 3 = 46.$$

Thus,

$$S_{15} = \frac{15}{2}(4 + 46) = \frac{15}{2}(50) = 375.$$

The sum of the first 15 terms is 375. **Now Try Exercise 25.** ▪

EXAMPLE 7 Find the sum: $\displaystyle\sum_{k=1}^{130}(4k + 5)$.

Solution It is helpful to first write out a few terms:

$$9 + 13 + 17 + \cdots.$$

It appears that this is an arithmetic series coming from an arithmetic sequence with $a_1 = 9$, $d = 4$, and $n = 130$. Before using the formula

$$S_n = \frac{n}{2}(a_1 + a_n),$$

we find the last term, a_{130}:

$$a_{130} = 4 \cdot 130 + 5 \qquad \text{The } k\text{th term is } 4k + 5.$$
$$= 520 + 5$$
$$= 525.$$

Thus,

$$S_{130} = \frac{130}{2}(9 + 525) \qquad \text{Substituting into } S_n = \frac{n}{2}(a_1 + a_n)$$
$$= 34{,}710.$$

This sum can also be found on a graphing calculator. It is not necessary to have the calculator set in SEQUENCE mode to do this.

```
sum(seq(4X+5,X,1,130))
                    34710
```

Now Try Exercise 33. ■

❀ Applications

The translation of some applications and problem-solving situations may involve arithmetic sequences or series. We consider some examples.

EXAMPLE 8 *Hourly Wages.* Gloria accepts a job, starting with an hourly wage of \$14.25, and is promised a raise of 15¢ per hour every 2 months for 5 yr. At the end of 5 yr, what will Gloria's hourly wage be?

Solution It helps to first write down the hourly wage for several 2-month time periods:

Beginning: \$14.25,

After 2 months: \$14.40,

After 4 months: \$14.55,

and so on.

What appears is a sequence of numbers: 14.25, 14.40, 14.55,.... . This sequence is arithmetic, because adding 0.15 each time gives us the next term.

We want to find the last term of an arithmetic sequence, so we use the formula $a_n = a_1 + (n - 1)d$. We know that $a_1 = 14.25$ and $d = 0.15$, but what is n? That is, how many terms are in the sequence? Each year there are 12/2, or 6 raises, since Gloria gets a raise every 2 months. There are 5 yr, so the total number of raises will be $5 \cdot 6$, or 30. Thus there will be 31 terms: the original wage and 30 increased rates.

Substituting in the formula $a_n = a_1 + (n - 1)d$ gives us

$$a_{31} = 14.25 + (31 - 1) \cdot 0.15$$
$$= 18.75.$$

Thus, at the end of 5 yr, Gloria's hourly wage will be $18.75.

Now Try Exercise 43. ▪

The calculations in Example 8 could be done in a number of ways. There is often a variety of ways in which a problem can be solved. In this chapter, we concentrate on the use of sequences and series and their related formulas.

EXAMPLE 9 *Total in a Stack.* A stack of telephone poles has 30 poles in the bottom row. There are 29 poles in the second row, 28 in the next row, and so on. How many poles are in the stack if there are 5 poles in the top row?

Solution A picture will help in this case. The following figure shows the ends of the poles and the way in which they stack.

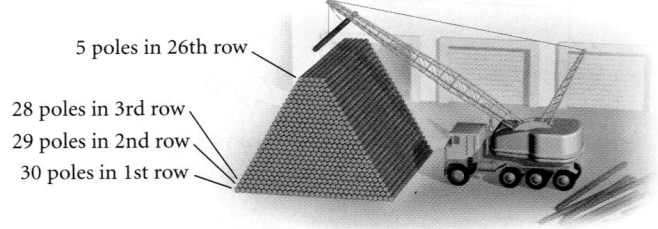

5 poles in 26th row
28 poles in 3rd row
29 poles in 2nd row
30 poles in 1st row

Since the number of poles goes from 30 in a row up to 5 in the top row, there must be 26 rows. We want the sum

$$30 + 29 + 28 + \cdots + 5.$$

Thus we have an arithmetic series. We use the formula

$$S_n = \frac{n}{2}(a_1 + a_n),$$

with $n = 26$, $a_1 = 30$, and $a_{26} = 5$.
    Substituting, we get

$$S_{26} = \frac{26}{2}(30 + 5) = 455.$$

There are 455 poles in the stack.

Now Try Exercise 39. ▪

(11.2) Exercise Set

Find the first term and the common difference.

1. $3, 8, 13, 18, \ldots$

2. $\$1.08, \$1.16, \$1.24, \$1.32, \ldots$

3. $9, 5, 1, -3, \ldots$

4. $-8, -5, -2, 1, 4, \ldots$

5. $\frac{3}{2}, \frac{9}{4}, 3, \frac{15}{4}, \ldots$

6. $\frac{3}{5}, \frac{1}{10}, -\frac{2}{5}, \ldots$

7. $\$316, \$313, \$310, \$307, \ldots$

8. Find the 11th term of the arithmetic sequence $0.07, 0.12, 0.17, \ldots$.

9. Find the 12th term of the arithmetic sequence $2, 6, 10, \ldots$.

10. Find the 17th term of the arithmetic sequence $7, 4, 1, \ldots$.

11. Find the 14th term of the arithmetic sequence $3, \frac{7}{3}, \frac{5}{3}, \ldots$.

12. Find the 13th term of the arithmetic sequence $\$1200, \$964.32, \$728.64, \ldots$.

13. Find the 10th term of the arithmetic sequence $\$2345.78, \$2967.54, \$3589.30, \ldots$.

14. In the sequence of Exercise 9, what term is the number 106?

15. In the sequence of Exercise 8, what term is the number 1.67?

16. In the sequence of Exercise 10, what term is -296?

17. In the sequence of Exercise 11, what term is -27?

18. Find a_{20} when $a_1 = 14$ and $d = -3$.

19. Find a_1 when $d = 4$ and $a_8 = 33$.

20. Find d when $a_1 = 8$ and $a_{11} = 26$.

21. Find n when $a_1 = 25$, $d = -14$, and $a_n = -507$.

22. In an arithmetic sequence, $a_{17} = -40$ and $a_{28} = -73$. Find a_1 and d. Write the first 5 terms of the sequence.

23. In an arithmetic sequence, $a_{17} = \frac{25}{3}$ and $a_{32} = \frac{95}{6}$. Find a_1 and d. Write the first 5 terms of the sequence.

24. Find the sum of the first 14 terms of the series $11 + 7 + 3 + \cdots$.

25. Find the sum of the first 20 terms of the series $5 + 8 + 11 + 14 + \cdots$.

26. Find the sum of the first 300 natural numbers.

27. Find the sum of the first 400 even natural numbers.

28. Find the sum of the odd numbers 1 to 199, inclusive.

29. Find the sum of the multiples of 7 from 7 to 98, inclusive.

30. Find the sum of all multiples of 4 that are between 14 and 523.

31. If an arithmetic series has $a_1 = 2$, $d = 5$, and $n = 20$, what is S_n?

32. If an arithmetic series has $a_1 = 7$, $d = -3$, and $n = 32$, what is S_n?

Find the sum.

33. $\displaystyle\sum_{k=1}^{40} (2k + 3)$

34. $\displaystyle\sum_{k=5}^{20} 8k$

35. $\displaystyle\sum_{k=0}^{19} \frac{k - 3}{4}$

36. $\displaystyle\sum_{k=2}^{50} (2000 - 3k)$

37. $\displaystyle\sum_{k=12}^{57} \frac{7 - 4k}{13}$

38. $\displaystyle\sum_{k=101}^{200} (1.14k - 2.8) - \sum_{k=1}^{5} \left(\frac{k + 4}{10}\right)$

39. *Stacking Poles.* How many poles will be in a stack of telephone poles if there are 50 in the first layer, 49 in the second, and so on, with 6 in the top layer?

40. *Investment Return.* Max, an investment counselor, sets up an investment situation for a client that will return $5000 the first year, $6125 the second year, $7250 the third year, and so on, for 25 yr. How much is received from the investment altogether?

41. *Total Savings.* If 10¢ is saved on October 1, 20¢ is saved on October 2, 30¢ on October 3, and so on, how much is saved during the 31 days of October?

42. *Theater Seating.* Theaters are often built with more seats per row as the rows move toward the back. Suppose that the first balcony of a theater has 28 seats in the first row, 32 in the second, 36 in the third, and so on, for 20 rows. How many seats are in the first balcony altogether?

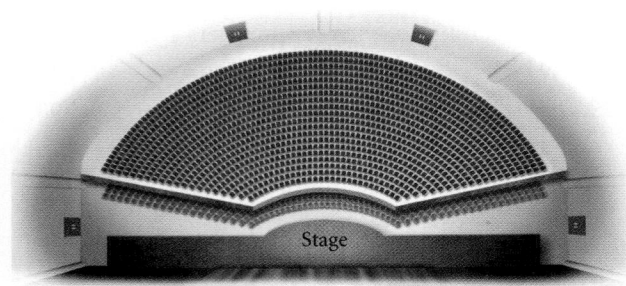

43. *Parachutist Free Fall.* When a parachutist jumps from an airplane, the distances, in feet, that the parachutist falls in each successive second before pulling the ripcord to release the parachute are as follows:

$$16, 48, 80, 112, 144, \ldots .$$

Is this sequence arithmetic? What is the common difference? What is the total distance fallen in 10 sec?

44. *Small Group Interaction.* In a social science study, Stephan found the following data regarding an interaction measurement r_n for groups of size n.

n	r_n
3	0.5908
4	0.6080
5	0.6252
6	0.6424
7	0.6596
8	0.6768
9	0.6940
10	0.7112

Source: *American Sociological Review*, 17 (1952)

Is this sequence arithmetic? What is the common difference?

45. *Garden Plantings.* A gardener is making a planting in the shape of a trapezoid. It will have 35 plants in the front row, 31 in the second row, 27 in the third row, and so on. If the pattern is consistent, how many plants will there be in the last row? How many plants are there altogether?

46. *Band Formation.* A formation of a marching band has 10 marchers in the front row, 12 in the second row, 14 in the third row, and so on, for 8 rows. How many marchers are in the last row? How many marchers are there altogether?

47. *Raw Material Production.* In a manufacturing process, it took 3 units of raw materials to produce 1 unit of a product. The raw material needs thus formed the sequence

$$3, 6, 9, \ldots, 3n, \ldots .$$

Is this sequence arithmetic? What is the common difference?

Collaborative Discussion and Writing

48. The sum of the first n terms of an arithmetic sequence can be given by

$$S_n = \frac{n}{2}[2a_1 + (n - 1)d].$$

Compare this formula to

$$S_n = \frac{n}{2}(a_1 + a_n).$$

Discuss the reasons for the use of one formula over the other.

49. It is said that as a young child, the mathematician Karl F. Gauss (1777–1855) was able to compute the sum $1 + 2 + 3 + \cdots + 100$ very quickly in his head to the amazement of a teacher. Explain how Gauss might have done this had he possessed some knowledge of arithmetic sequences and series. Then give a formula for the sum of the first n natural numbers.

Skill Maintenance

Solve.

50. $7x - 2y = 4,$
   $x + 3y = 17$

51. $2x + y + 3z = 12,$
   $x - 3y + 2z = 11,$
   $5x + 2y - 4z = -4$

52. Find the vertices and the foci of the ellipse with the equation $9x^2 + 16y^2 = 144.$

53. Find an equation of the ellipse with vertices $(0, -5)$ and $(0, 5)$ and minor axis of length 4.

Synthesis

54. Find three numbers in an arithmetic sequence such that the sum of the first and third is 10 and the product of the first and second is 15.

55. Find a formula for the sum of the first n odd natural numbers:

$$1 + 3 + 5 + \cdots + (2n - 1).$$

56. Find the first 10 terms of the arithmetic sequence for which

$$a_1 = \$8760 \quad \text{and} \quad d = -\$798.23.$$

Then find the sum of the first 10 terms.

57. Find the first term and the common difference for the arithmetic sequence for which

$$a_2 = 40 - 3q \quad \text{and} \quad a_4 = 10p + q.$$

58. The zeros of this polynomial function form an arithmetic sequence. Find them.

$$f(x) = x^4 + 4x^3 - 84x^2 - 176x + 640$$

*If p, m, and q form an arithmetic sequence, it can be shown that $m = (p + q)/2$. (See Exercise 65.) The number m is the **arithmetic mean**, or **average**, of p and q. Given two numbers p and q, if we find k other numbers $m_1, m_2, \ldots, m_k$ such that*

$$p, m_1, m_2, \ldots, m_k, q$$

forms an arithmetic sequence, we say that we have "inserted k arithmetic means between p and q."

59. Insert three arithmetic means between 4 and 12.

60. Insert three arithmetic means between -3 and 5.

61. Insert four arithmetic means between 4 and 13.

62. Insert ten arithmetic means between 27 and 300.

63. Insert enough arithmetic means between 1 and 50 so that the sum of the resulting series will be 459.

64. *Straight-Line Depreciation.* A company buys an office machine for $5200 on January 1 of a given year. The machine is expected to last for 8 yr, at the end of which time its **trade-in value**, or **salvage value**, will be $1100. If the company's accountant figures the decline in value to be the same each year, then its **book values**, or **salvage values**, after t years, $0 \le t \le 8$, form an arithmetic sequence given by

$$a_t = C - t\left(\frac{C - S}{N}\right),$$

where C is the original cost of the item ($5200), N is the number of years of expected life (8), and S is the salvage value ($1100).

a) Find the formula for a_t for the straight-line depreciation of the office machine.

b) Find the salvage value after 0 yr, 1 yr, 2 yr, 3 yr, 4 yr, 7 yr, and 8 yr.

65. Prove that if p, m, and q form an arithmetic sequence, then

$$m = \frac{p + q}{2}.$$

Geometric Sequences and Series

11.3

❋ Identify the common ratio of a geometric sequence, and find a given term and the sum of the first n terms.

❋ Find the sum of an infinite geometric series, if it exists.

A sequence in which each term after the first is found by multiplying the preceding term by the same number is a **geometric sequence**.

❋ Geometric Sequences

Consider the sequence:

$$2, \ 6, \ 18, \ 54, \ 162, \ldots .$$

Note that multiplying each term by 3 produces the next term. We call the number 3 the **common ratio** because it can be found by dividing any term by the preceding term. A geometric sequence is also called a *geometric progression*.

> ### Geometric Sequence
>
> A sequence is **geometric** if there is a number r, called the **common ratio**, such that
>
> $$\frac{a_{n+1}}{a_n} = r, \quad \text{or} \quad a_{n+1} = a_n r, \quad \text{for any integer } n \geq 1.$$

EXAMPLE 1 For each of the following geometric sequences, identify the common ratio.

a) 3, 6, 12, 24, 48,...

b) $1, \ -\dfrac{1}{2}, \ \dfrac{1}{4}, \ -\dfrac{1}{8}, \ldots$

c) \$5200, \$3900, \$2925, \$2193.75,...

d) \$1000, \$1060, \$1123.60,...

Solution

SEQUENCE	COMMON RATIO
a) 3, 6, 12, 24, 48,...	2 $\left(\frac{6}{3} = 2, \frac{12}{6} = 2, \text{and so on}\right)$
b) $1, \ -\dfrac{1}{2}, \ \dfrac{1}{4}, \ -\dfrac{1}{8}, \ldots$	$-\dfrac{1}{2}$ $\left(\dfrac{-\frac{1}{2}}{1} = -\dfrac{1}{2}, \dfrac{\frac{1}{4}}{-\frac{1}{2}} = -\dfrac{1}{2}, \text{and so on}\right)$
c) \$5200, \$3900, \$2925, \$2193.75,...	0.75 $\left(\dfrac{\$3900}{\$5200} = 0.75, \dfrac{\$2925}{\$3900} = 0.75, \text{and so on}\right)$
d) \$1000, \$1060, \$1123.60,...	1.06 $\left(\dfrac{\$1060}{\$1000} = 1.06, \dfrac{\$1123.60}{\$1060} = 1.06, \text{and so on}\right)$

Now Try Exercise 1. ■

We now find a formula for the general, or nth, term of a geometric sequence. Let a_1 be the first term and r the common ratio. The first few terms are as follows:

$a_1,$

$a_2 = a_1 r,$

$a_3 = a_2 r = (a_1 r)r = a_1 r^2,$ Substituting $a_1 r$ for a_2

$a_4 = a_3 r = (a_1 r^2)r = a_1 r^3.$ Substituting $a_1 r^2$ for a_3

Note that the exponent is 1 less than the subscript.

Generalizing, we obtain the following.

> **nth Term of a Geometric Sequence**
>
> The **nth term** of a geometric sequence is given by
>
> $$a_n = a_1 r^{n-1}, \quad \text{for any integer } n \geq 1.$$

EXAMPLE 2 Find the 7th term of the geometric sequence $4, \ 20, \ 100, \ldots$.

Solution We first note that

$$a_1 = 4 \quad \text{and} \quad n = 7.$$

To find the common ratio, we can divide any term (other than the first) by the preceding term. Since the second term is 20 and the first is 4, we get

$$r = \frac{20}{4}, \quad \text{or} \quad 5.$$

Then using the formula $a_n = a_1 r^{n-1}$, we have

$$a_7 = 4 \cdot 5^{7-1} = 4 \cdot 5^6 = 4 \cdot 15{,}625 = 62{,}500.$$

Thus the 7th term is 62,500.   **Now Try Exercise 11.** ■

EXAMPLE 3 Find the 10th term of the geometric sequence $64, \ -32, \ 16, \ -8, \ldots$.

Solution We first note that

$$a_1 = 64, \quad n = 10, \quad \text{and} \quad r = \frac{-32}{64}, \text{or} -\frac{1}{2}.$$

Then using the formula $a_n = a_1 r^{n-1}$, we have

$$a_{10} = 64 \cdot \left(-\frac{1}{2}\right)^{10-1} = 64 \cdot \left(-\frac{1}{2}\right)^9 = 2^6 \cdot \left(-\frac{1}{2^9}\right) = -\frac{1}{2^3} = -\frac{1}{8}.$$

Thus the 10th term is $-\frac{1}{8}$.   **Now Try Exercise 15.** ■

Exploring with Technology

Graph the first 7 terms of each geometric sequence. What pattern do you observe?

$$a_n = 5 \cdot 3^{n-1}, \qquad a_n = 0.2 \cdot (2.25)^{n-1},$$

$$a_n = \frac{1}{3} \cdot \left(\frac{3}{5}\right)^{n-1}, \qquad a_n = (0.95)^{n-1}$$

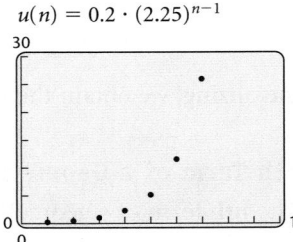

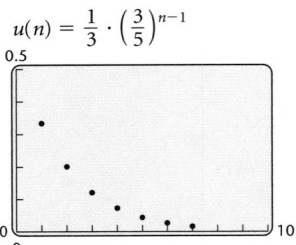

The pattern shown above holds in general. *The graph of a geometric sequence is a set of points that lie on the graph of an exponential function.*

❊ Sum of the First n Terms of a Geometric Sequence

Next, we develop a formula for the sum S_n of the first n terms of a geometric sequence:

$$a_1, \ a_1r, \ a_1r^2, \ a_1r^3, \ldots, \ a_1r^{n-1}, \ldots.$$

The associated **geometric series** is given by

$$S_n = a_1 + a_1r + a_1r^2 + a_1r^3 + \cdots + a_1r^{n-1}. \tag{1}$$

We want to find a formula for this sum. If we multiply on both sides of equation (1) by r, we have

$$rS_n = a_1r + a_1r^2 + a_1r^3 + a_1r^4 + \cdots + a_1r^n. \tag{2}$$

Subtracting equation (2) from equation (1), we see that the differences of the red terms are 0, leaving

$$S_n - rS_n = a_1 - a_1r^n,$$

or

$$S_n(1 - r) = a_1(1 - r^n). \qquad \text{Factoring}$$

Dividing on both sides by $1 - r$ gives us the following formula.

Sum of the First n Terms

The sum of the first n terms of a geometric sequence is given by

$$S_n = \frac{a_1(1 - r^n)}{1 - r}, \quad \text{for any } r \neq 1.$$

EXAMPLE 4 Find the sum of the first 7 terms of the geometric sequence $3, 15, 75, 375, \ldots$.

Solution We first note that

$$a_1 = 3, \quad n = 7, \quad \text{and} \quad r = \frac{15}{3}, \text{ or } 5.$$

Then using the formula

$$S_n = \frac{a_1(1 - r^n)}{1 - r},$$

we have

$$S_7 = \frac{3(1 - 5^7)}{1 - 5}$$

$$= \frac{3(1 - 78{,}125)}{-4}$$

$$= 58{,}593.$$

Thus the sum of the first 7 terms is 58,593.   Now Try Exercise 23. ■

EXAMPLE 5 Find the sum: $\displaystyle\sum_{k=1}^{11} (0.3)^k$.

Solution This is a geometric series with $a_1 = 0.3$, $r = 0.3$, and $n = 11$. Thus,

$$S_{11} = \frac{0.3(1 - 0.3^{11})}{1 - 0.3}$$

$$\approx 0.42857.$$

We can also find this sum using a graphing calculator set in either FUNCTION mode or SEQUENCE mode.

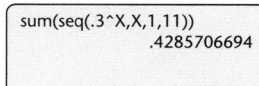

```
sum(seq(.3^X,X,1,11))
                .4285706694
```

Now Try Exercise 41. ■

❖ Infinite Geometric Series

The sum of the terms of an infinite geometric sequence is an **infinite geometric series**. For some geometric sequences, S_n gets close to a specific number as n gets large. For example, consider the infinite series

$$\frac{1}{2} + \frac{1}{4} + \frac{1}{8} + \frac{1}{16} + \cdots + \frac{1}{2^n} + \cdots.$$

We can visualize S_n by considering the area of a square. For S_1, we shade half the square. For S_2, we shade half the square plus half the remaining half, or $\frac{1}{4}$. For S_3, we shade the parts shaded in S_2 plus half the remaining part. We see that the values of S_n will continue to get close to 1 (shading the complete square).

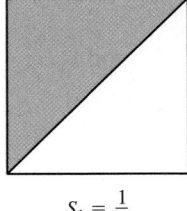 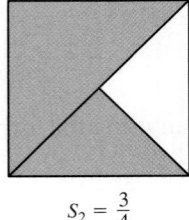 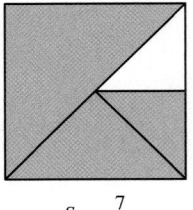 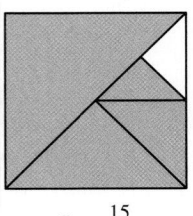

$S_1 = \dfrac{1}{2}$ $S_2 = \dfrac{3}{4}$ $S_3 = \dfrac{7}{8}$ $S_4 = \dfrac{15}{16}$

We examine some partial sums. Note that each of the partial sums is less than 1, but S_n gets very close to 1 as n gets large.

n	S_n
1	0.5
5	0.96875
10	0.9990234375
20	0.9999990463
30	0.9999999991

```
sum(seq(1/2^X,X,1,20))
                 .9999990463
sum(seq(1/2^X,X,1,30))
                 .9999999991
```

We say that 1 is the **limit** of S_n and also that 1 is the **sum of the infinite geometric sequence**. The sum of an infinite geometric sequence is denoted S_∞. In this case, $S_\infty = 1$.

Some infinite sequences do not have sums. Consider the infinite geometric series

$$2 + 4 + 8 + 16 + \cdots + 2^n + \cdots.$$

We again examine some partial sums. Note that as n gets large, S_n gets large without bound. This sequence does not have a sum.

n	S_n
1	2
5	62
10	2,046
20	2,097,150
30	2,147,483,646

```
sum(seq(2^X,X,1,20))
                 2097150
sum(seq(2^X,X,1,30))
              2147483646
```

It can be shown (but we will not do so here) that the sum of an infinite geometric series exists if and only if $|r| < 1$ (that is, the absolute value of the common ratio is less than 1).

To find a formula for the sum of an infinite geometric series, we first consider the sum of the first n terms:

$$S_n = \frac{a_1(1 - r^n)}{1 - r} = \frac{a_1 - a_1 r^n}{1 - r}. \qquad \text{Using the distributive law}$$

For $|r| < 1$, values of r^n get close to 0 as n gets large. As r^n gets close to 0, so does $a_1 r^n$. Thus, S_n gets close to $a_1/(1 - r)$.

Limit or Sum of an Infinite Geometric Series

When $|r| < 1$, the limit or sum of an infinite geometric series is given by

$$S_\infty = \frac{a_1}{1 - r}.$$

EXAMPLE 6 Determine whether each of the following infinite geometric series has a limit. If a limit exists, find it.

a) $1 + 3 + 9 + 27 + \cdots$

b) $-2 + 1 - \frac{1}{2} + \frac{1}{4} - \frac{1}{8} + \cdots$

Solution

a) Here $r = 3$, so $|r| = |3| = 3$. Since $|r| > 1$, the series *does not* have a limit.

b) Here $r = -\frac{1}{2}$, so $|r| = \left|-\frac{1}{2}\right| = \frac{1}{2}$. Since $|r| < 1$, the series *does* have a limit. We find the limit:

$$S_\infty = \frac{a_1}{1 - r} = \frac{-2}{1 - \left(-\frac{1}{2}\right)} = \frac{-2}{\frac{3}{2}} = -\frac{4}{3}.$$

Now Try Exercises 33 and 37. ■

EXAMPLE 7 Find fraction notation for $0.78787878\ldots$, or $0.\overline{78}$.

Solution We can express this as

$$0.78 + 0.0078 + 0.000078 + \cdots.$$

Then we see that this is an infinite geometric series, where $a_1 = 0.78$ and $r = 0.01$. Since $|r| < 1$, this series has a limit:

$$S_\infty = \frac{a_1}{1 - r} = \frac{0.78}{1 - 0.01} = \frac{0.78}{0.99} = \frac{78}{99}, \quad \text{or} \quad \frac{26}{33}.$$

Thus fraction notation for $0.78787878\ldots$ is $\frac{26}{33}$. You can check this on your calculator.

Now Try Exercise 51. ■

❈ Applications

The translation of some applications and problem-solving situations may involve geometric sequences or series. Examples 9 and 10, in particular, show applications in business and economics.

EXAMPLE 8   *A Daily Doubling Salary.*   Suppose someone offered you a job for the month of September (30 days) under the following conditions. You will be paid $0.01 for the first day, $0.02 for the second, $0.04 for the third, and so on, doubling your previous day's salary each day. How much would you earn? (Would you take the job? Make a conjecture before reading further.)

Solution   You earn $0.01 the first day, $0.01(2)$ the second day, $0.01(2)(2)$ the third day, and so on. The amount earned is the geometric series

$$\$0.01 + \$0.01(2) + \$0.01(2^2) + \$0.01(2^3) + \cdots + \$0.01(2^{29}),$$

where $a_1 = \$0.01$, $r = 2$, and $n = 30$. Using the formula

$$S_n = \frac{a_1(1 - r^n)}{1 - r},$$

we have

$$S_{30} = \frac{\$0.01(1 - 2^{30})}{1 - 2} = \$10{,}737{,}418.23.$$

The pay exceeds $10.7 million for the month.   **Now Try Exercise 57.** ▨

EXAMPLE 9   *The Amount of an Annuity.*   An **annuity** is a sequence of equal payments, made at equal time intervals, that earn interest. Fixed deposits in a savings account are an example of an annuity. Suppose that to save money to buy a car, Andrea deposits $1000 at the *end* of each of 5 yr in an account that pays 8% interest, compounded annually. The total amount in the account at the end of 5 yr is called the **amount of the annuity**. Find that amount.

Solution   The following time diagram can help visualize the problem. Note that no deposit is made until the end of the first year.

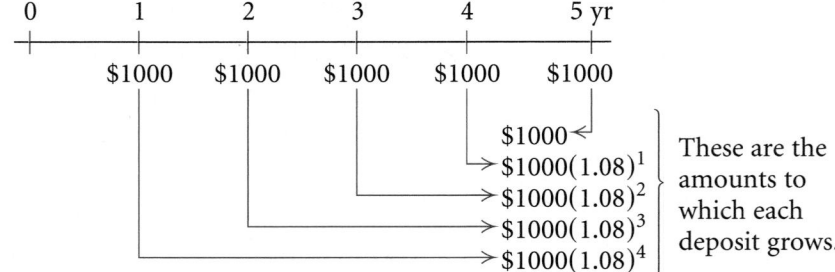

The amount of the annuity is the geometric series

$$\$1000 + \$1000(1.08)^1 + \$1000(1.08)^2 + \$1000(1.08)^3 + \$1000(1.08)^4,$$

where $a_1 = \$1000$, $n = 5$, and $r = 1.08$. Using the formula

$$S_n = \frac{a_1(1 - r^n)}{1 - r},$$

we have

$$S_5 = \frac{\$1000(1 - 1.08^5)}{1 - 1.08} \approx \$5866.60.$$

The amount of the annuity is \$5866.60.

Now Try Exercise 61.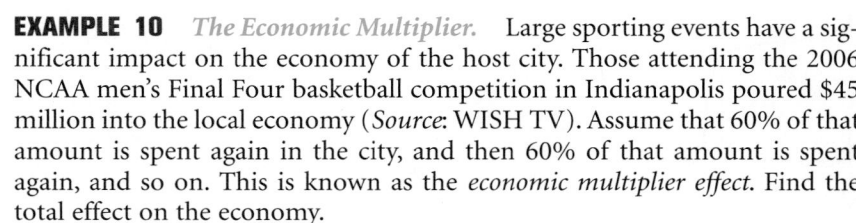

EXAMPLE 10 *The Economic Multiplier.* Large sporting events have a significant impact on the economy of the host city. Those attending the 2006 NCAA men's Final Four basketball competition in Indianapolis poured \$45 million into the local economy (*Source*: WISH TV). Assume that 60% of that amount is spent again in the city, and then 60% of that amount is spent again, and so on. This is known as the *economic multiplier effect*. Find the total effect on the economy.

Solution The total economic effect is given by the infinite series

$$\$45,000,000 + \$45,000,000(0.6) + \$45,000,000(0.6)^2 + \cdots.$$

Since $|r| = |0.6| = 0.6 < 1$, the series has a sum. Using the formula for the sum of an infinite geometric series, we have

$$S_\infty = \frac{a_1}{1 - r} = \frac{\$45,000,000}{1 - 0.6} = \$112,500,000.$$

The total effect of the spending on the economy is \$112,500,000.

Now Try Exercise 65.

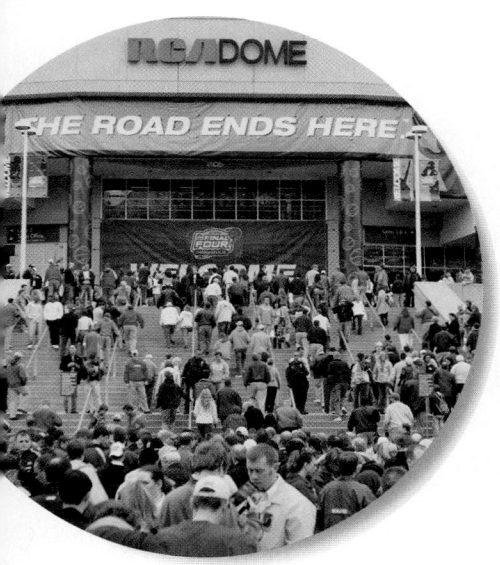

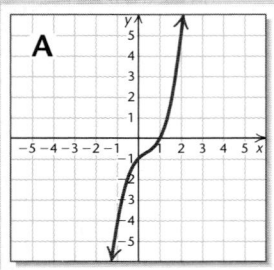

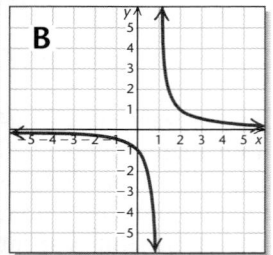

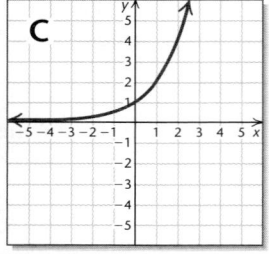

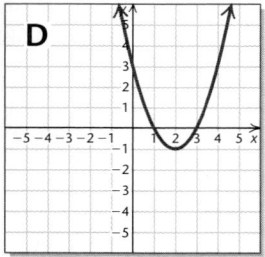

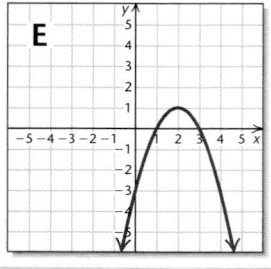

Visualizing the Graph

Match the equation with its graph.

1. $(x - 1)^2 + (y + 2)^2 = 9$

2. $y = x^3 - x^2 + x - 1$

3. $f(x) = 2^x$

4. $f(x) = x$

5. $a_n = n$

6. $y = \log(x + 3)$

7. $f(x) = -(x - 2)^2 + 1$

8. $f(x) = (x - 2)^2 - 1$

9. $y = \dfrac{1}{x - 1}$

10. $y = -3x + 4$

Answers on page A-81

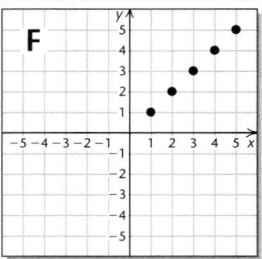

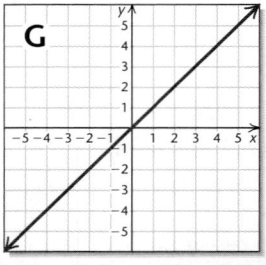

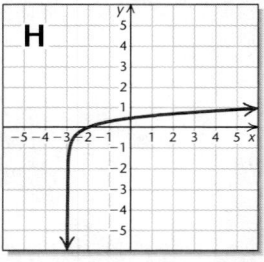

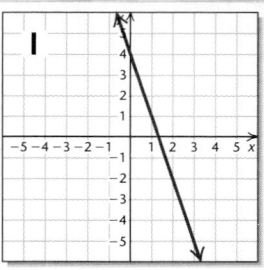

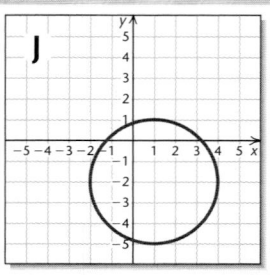

(11.3) Exercise Set

Find the common ratio.

1. $2, 4, 8, 16, \ldots$

2. $18, -6, 2, -\frac{2}{3}, \ldots$

3. $-1, 1, -1, 1, \ldots$

4. $-8, -0.8, -0.08, -0.008, \ldots$

5. $\frac{2}{3}, -\frac{4}{3}, \frac{8}{3}, -\frac{16}{3}, \ldots$

6. $75, 15, 3, \frac{3}{5}, \ldots$

7. $6.275, 0.6275, 0.06275, \ldots$

8. $\frac{1}{x}, \frac{1}{x^2}, \frac{1}{x^3}, \ldots$

9. $5, \frac{5a}{2}, \frac{5a^2}{4}, \frac{5a^3}{8}, \ldots$

10. $\$780, \$858, \$943.80, \$1038.18, \ldots$

Find the indicated term.

11. $2, 4, 8, 16, \ldots$; the 7th term

12. $2, -10, 50, -250, \ldots$; the 9th term

13. $2, 2\sqrt{3}, 6, \ldots$; the 9th term

14. $1, -1, 1, -1, \ldots$; the 57th term

15. $\frac{7}{625}, -\frac{7}{25}, \ldots$; the 23rd term

16. $\$1000, \$1060, \$1123.60, \ldots$; the 5th term

Find the nth, or general, term.

17. $1, 3, 9, \ldots$

18. $25, 5, 1, \ldots$

19. $1, -1, 1, -1, \ldots$

20. $-2, 4, -8, \ldots$

21. $\frac{1}{x}, \frac{1}{x^2}, \frac{1}{x^3}, \ldots$

22. $5, \frac{5a}{2}, \frac{5a^2}{4}, \frac{5a^3}{8}, \ldots$

23. Find the sum of the first 7 terms of the geometric series
$$6 + 12 + 24 + \cdots.$$

24. Find the sum of the first 10 terms of the geometric series
$$16 - 8 + 4 - \cdots.$$

25. Find the sum of the first 9 terms of the geometric series
$$\tfrac{1}{18} - \tfrac{1}{6} + \tfrac{1}{2} - \cdots.$$

26. Find the sum of the geometric series
$$-8 + 4 + (-2) + \cdots + \left(-\tfrac{1}{32}\right).$$

Determine whether the statement is true or false.

27. The sequence $2, -2\sqrt{2}, 4, -4\sqrt{2}, 8, \ldots$ is geometric.

28. The sequence with general term $3n$ is geometric.

29. The sequence with general term 2^n is geometric.

30. Multiplying a term of a geometric sequence by the common ratio produces the next term of the sequence.

31. An infinite geometric series with common ratio -0.75 has a sum.

32. Every infinite geometric series has a limit.

Find the sum, if it exists.

33. $4 + 2 + 1 + \cdots$

34. $7 + 3 + \frac{9}{7} + \cdots$

35. $25 + 20 + 16 + \cdots$

36. $100 - 10 + 1 - \frac{1}{10} + \cdots$

37. $8 + 40 + 200 + \cdots$

38. $-6 + 3 - \frac{3}{2} + \frac{3}{4} - \cdots$

39. $0.6 + 0.06 + 0.006 + \cdots$

40. $\displaystyle\sum_{k=0}^{10} 3^k$

41. $\displaystyle\sum_{k=1}^{11} 15\left(\frac{2}{3}\right)^k$

42. $\displaystyle\sum_{k=0}^{50} 200(1.08)^k$

43. $\displaystyle\sum_{k=1}^{\infty} \left(\frac{1}{2}\right)^{k-1}$

44. $\displaystyle\sum_{k=1}^{\infty} 2^k$

45. $\displaystyle\sum_{k=1}^{\infty} 12.5^k$

46. $\displaystyle\sum_{k=1}^{\infty} 400(1.0625)^k$

47. $\displaystyle\sum_{k=1}^{\infty} \$500(1.11)^{-k}$

48. $\displaystyle\sum_{k=1}^{\infty} \$1000(1.06)^{-k}$

49. $\displaystyle\sum_{k=1}^{\infty} 16(0.1)^{k-1}$

50. $\displaystyle\sum_{k=1}^{\infty} \frac{8}{3}\left(\frac{1}{2}\right)^{k-1}$

Find fraction notation.

51. $0.131313\ldots$, or $0.\overline{13}$

52. $0.2222\ldots$, or $0.\overline{2}$

53. $8.999\overline{9}$

54. $6.161\overline{616}$

55. $3.4125\overline{125}$

56. $12.7809\overline{809}$

57. *Daily Doubling Salary.*   Suppose someone offered you a job for the month of February (28 days) under the following conditions. You will be paid $0.01 the 1st day, $0.02 the 2nd, $0.04 the 3rd, and so on, doubling your previous day's salary each day. How much would you earn altogether?

58. *Bouncing Ping-Pong Ball.*   A ping-pong ball is dropped from a height of 16 ft and always rebounds $\frac{1}{4}$ of the distance fallen.

a) How high does it rebound the 6th time?
b) Find the total sum of the rebound heights of the ball.

59. *Bungee Jumping.*   A bungee jumper always rebounds 60% of the distance fallen. A bungee jump is made using a cord that stretches to 200 ft.

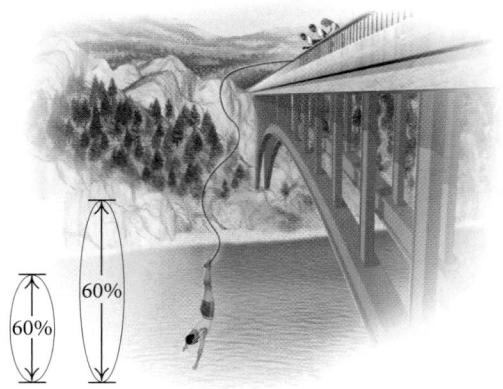

a) After jumping and then rebounding 9 times, how far has a bungee jumper traveled upward (the total rebound distance)?
b) About how far will a jumper have traveled upward (bounced) before coming to rest?

60. *Population Growth.*   Hadleytown has a present population of 100,000, and the population is increasing by 3% each year.

a) What will the population be in 15 yr?
b) How long will it take for the population to double?

61. *Amount of an Annuity.*   To create a college fund, a parent makes a sequence of 18 yearly deposits of $1000 each in a savings account on which interest is compounded annually at 3.2%. Find the amount of the annuity.

62. *Amount of an Annuity.*   A sequence of yearly payments of P dollars is invested at the end of each of N years at interest rate i, compounded annually. The total amount in the account, or the amount of the annuity, is V.

a) Show that
$$V = \frac{P[(1 + i)^N - 1]}{i}.$$

b) Suppose that interest is compounded n times per year and deposits are made every compounding period. Show that the formula for V is then given by
$$V = \frac{P\left[\left(1 + \dfrac{i}{n}\right)^{nN} - 1\right]}{i/n}.$$

63. *Loan Repayment.*   A family borrows $120,000. The loan is to be repaid in 13 yr at 12% interest, compounded annually. How much will have been repaid at the end of 13 yr?

64. *Doubling the Thickness of Paper.*   A piece of paper is 0.01 in. thick. It is cut and stacked repeatedly in such a way that its thickness is doubled each time for 20 times. How thick is the result?

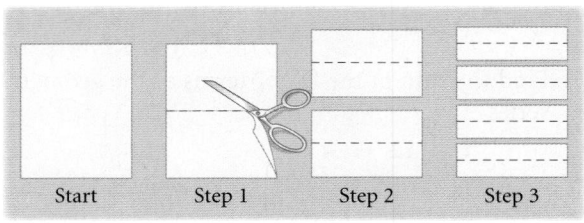

Start   Step 1   Step 2   Step 3

65. *The Economic Multiplier.* Suppose the government is making a $13,000,000,000 expenditure for educational improvement. If 85% of this is spent again, and so on, what is the total effect on the economy?

66. *Advertising Effect.* Great Grains Cereal Company is about to market a new low-carbohydrate cereal in a city of 5,000,000 people. They plan an advertising campaign that they think will induce 30% of the people to buy the product. They estimate that if those people like the product, they will induce $30\% \cdot 30\% \cdot 5{,}000{,}000$ more to buy the product, and those will induce $30\% \cdot 30\% \cdot 30\% \cdot 5{,}000{,}000$, and so on. In all, how many people will buy the product as a result of the advertising campaign? What percentage of the population is this?

Collaborative Discussion and Writing

67. Write a problem for a classmate to solve. Devise the problem so that a geometric series is involved and the solution is "The total amount in the bank is $900(1.08)^{40}$, or about $19,552."

68. The infinite series
$$S_\infty = 2 + \frac{1}{2} + \frac{1}{2 \cdot 3} + \frac{1}{2 \cdot 3 \cdot 4} + \frac{1}{2 \cdot 3 \cdot 4 \cdot 5} + \cdots$$

is not geometric, but it does have a sum. Consider S_1, S_2, S_3, S_4, S_5, and S_6. Construct a table and a graph of the sequence. Expand the sequence of sums, if needed. Make a conjecture about the value of S_∞ and explain your reasoning.

Skill Maintenance

For each pair of functions, find $(f \circ g)(x)$ and $(g \circ f)(x)$.

69. $f(x) = x^2$, $g(x) = 4x + 5$

70. $f(x) = x - 1$, $g(x) = x^2 + x + 3$

Solve.

71. $5^x = 35$ **72.** $\log_2 x = -4$

Synthesis

73. Prove that
$$\sqrt{3} - \sqrt{2}, \quad 4 - \sqrt{6}, \quad \text{and} \quad 6\sqrt{3} - 2\sqrt{2}$$
form a geometric sequence.

74. Consider the sequence
$$4, \ 20.4, \ 104.04, \ 531.6444, \ldots.$$
What is the error in using $a_{277} = 4(5.1)^{276}$ to find the 277th term?

75. Consider the sequence
$$x + 3, \ x + 7, \ 4x - 2, \ldots.$$
a) If the sequence is arithmetic, find x and then determine each of the 3 terms and the 4th term.
b) If the sequence is geometric, find x and then determine each of the 3 terms and the 4th term.

76. Find the sum of the first n terms of
$$1 + x + x^2 + \cdots.$$

77. Find the sum of the first n terms of
$$x^2 - x^3 + x^4 - x^5 + \cdots.$$

In Exercises 78 and 79, assume that $a_1, a_2, a_3, \ldots$ is a geometric sequence.

78. Prove that $a_1^2, a_2^2, a_3^2, \ldots$ is a geometric sequence.

79. Prove that $\ln a_1, \ln a_2, \ln a_3, \ldots$ is an arithmetic sequence.

80. Prove that $5^{a_1}, 5^{a_2}, 5^{a_3}, \ldots$ is a geometric sequence, if $a_1, a_2, a_3, \ldots$ is an arithmetic sequence.

81. The sides of a square are 16 cm long. A second square is inscribed by joining the midpoints of the sides, successively. In the second square, we repeat the process, inscribing a third square. If this process is continued indefinitely, what is the sum of all the areas of all the squares? (*Hint*: Use an infinite geometric series.)

11.4 Mathematical Induction

❖ List the statements of an infinite sequence that is defined by a formula.

❖ Do proofs by mathematical induction.

In this section, we learn to prove a sequence of mathematical statements using a procedure called *mathematical induction.*

❖ Sequences of Statements

Infinite sequences of statements occur often in mathematics. In an infinite sequence of statements, there is a statement for each natural number. For example, consider the sequence of statements represented by the following:

"For each x between 0 and 1, $0 < x^n < 1$."

Let's think of this as $S(n)$, or S_n. Substituting natural numbers for n gives a sequence of statements. We list a few of them.

Statement 1, S_1:   For each x between 0 and 1, $0 < x^1 < 1$.
Statement 2, S_2:   For each x between 0 and 1, $0 < x^2 < 1$.
Statement 3, S_3:   For each x between 0 and 1, $0 < x^3 < 1$.
Statement 4, S_4:   For each x between 0 and 1, $0 < x^4 < 1$.

In this context, the symbols S_1, S_2, S_3, and so on, do not represent sums.

EXAMPLE 1   List the first four statements in the sequence that can be obtained from each of the following.

a) $\log n < n$
b) $1 + 3 + 5 + \cdots + (2n - 1) = n^2$

Solution

a) This time, S_n is "$\log n < n$."

S_1: $\log 1 < 1$
S_2: $\log 2 < 2$
S_3: $\log 3 < 3$
S_4: $\log 4 < 4$

b) This time, S_n is "$1 + 3 + 5 + \cdots + (2n - 1) = n^2$."

S_1: $1 = 1^2$
S_2: $1 + 3 = 2^2$
S_3: $1 + 3 + 5 = 3^2$
S_4: $1 + 3 + 5 + 7 = 4^2$

Now Try Exercise 1. ■

❀ Proving Infinite Sequences of Statements

We now develop a method of proof, called **mathematical induction**, which we can use to try to prove that all statements in an infinite sequence of statements are true. The statements usually have the form:

"For all natural numbers n, S_n",

where S_n is some mathematical sentence such as those of the preceding examples. Of course, we cannot prove each statement of an infinite sequence individually. Instead, we try to show that "whenever S_k holds, then S_{k+1} must hold." We abbreviate this as $S_k \rightarrow S_{k+1}$. (This is also read "If S_k, then S_{k+1}," or "S_k implies S_{k+1}.") Suppose that we could somehow establish that this holds for all natural numbers k. Then we would have the following:

$S_1 \rightarrow S_2$ meaning "if S_1 is true, then S_2 is true";

$S_2 \rightarrow S_3$ meaning "if S_2 is true, then S_3 is true";

$S_3 \rightarrow S_4$ meaning "if S_3 is true, then S_4 is true";

and so on, indefinitely.

Even knowing that $S_k \rightarrow S_{k+1}$, we would still not be certain whether there is *any* k for which S_k is true. All we would know is that "if S_k is true, then S_{k+1} is true." Suppose now that S_k is true for some k, say, $k = 1$. We then must have the following.

S_1 is true.   **We have verified, or proved, this.**

$S_1 \rightarrow S_2$ **This means that whenever S_1 is true, S_2 is true.**

Therefore, S_2 is true.

$S_2 \rightarrow S_3$ **This means that whenever S_2 is true, S_3 is true.**

Therefore, S_3 is true.

and so on.

We conclude that S_n is true for all natural numbers n.

This leads us to the principle of mathematical induction, which we use to prove the types of statements considered here.

The Principle of Mathematical Induction

We can prove an infinite sequence of statements S_n by showing the following.

(1) *Basis step.* S_1 is true.

(2) *Induction step.* For all natural numbers k, $S_k \rightarrow S_{k+1}$.

Mathematical induction is analogous to lining up a sequence of dominoes. The induction step tells us that if any one domino is knocked over, then the one next to it will be hit and knocked over. The basis step tells us that the first domino can indeed be knocked over. Note that in order for all dominoes to fall, *both* conditions must be satisfied.

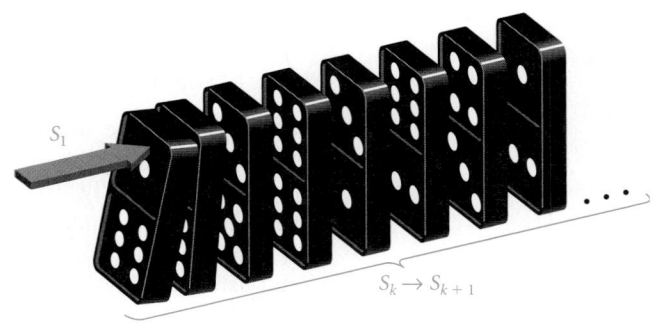

When you are learning to do proofs by mathematical induction, it is helpful to first write out S_n, S_1, S_k, and S_{k+1}. This helps to identify what is to be assumed and what is to be deduced.

EXAMPLE 2 Prove: For every natural number n,

$$1 + 3 + 5 + \cdots + (2n - 1) = n^2.$$

Proof. We first list S_n, S_1, S_k, and S_{k+1}.

S_n: $1 + 3 + 5 + \cdots + (2n - 1) = n^2$

S_1: $1 = 1^2$

S_k: $1 + 3 + 5 + \cdots + (2k - 1) = k^2$

S_{k+1}: $1 + 3 + 5 + \cdots + (2k - 1) + [2(k + 1) - 1] = (k + 1)^2$.

(1) *Basis step.* S_1, as listed, is true since $1 = 1^2$, or $1 = 1$.

(2) *Induction step.* We let k be any natural number. We assume S_k to be true and try to show that it implies that S_{k+1} is true. Now S_k is

$$1 + 3 + 5 + \cdots + (2k - 1) = k^2.$$

Starting with the left side of S_{k+1} and substituting k^2 for $1 + 3 + 5 + \cdots + (2k - 1)$, we have

$$\underbrace{1 + 3 + \cdots + (2k - 1)}_{} + [2(k + 1) - 1]$$

$$\begin{aligned} &= k^2 + [2(k + 1) - 1] \quad \text{We assume } S_k \text{ is true.} \\ &= k^2 + 2k + 2 - 1 \\ &= k^2 + 2k + 1 \\ &= (k + 1)^2. \end{aligned}$$

We have shown that for all natural numbers k, $S_k \to S_{k+1}$. This completes the induction step. It and the basis step tell us that the proof is complete.

Now Try Exercise 5. ■

EXAMPLE 3 Prove: For every natural number n,

$$\frac{1}{2} + \frac{1}{4} + \frac{1}{8} + \cdots + \frac{1}{2^n} = \frac{2^n - 1}{2^n}.$$

Proof. We first list S_n, S_1, S_k, and S_{k+1}.

S_n: $\dfrac{1}{2} + \dfrac{1}{4} + \dfrac{1}{8} + \cdots + \dfrac{1}{2^n} = \dfrac{2^n - 1}{2^n}$

S_1: $\dfrac{1}{2^1} = \dfrac{2^1 - 1}{2^1}$

S_k: $\dfrac{1}{2} + \dfrac{1}{4} + \dfrac{1}{8} + \cdots + \dfrac{1}{2^k} = \dfrac{2^k - 1}{2^k}$

S_{k+1}: $\dfrac{1}{2} + \dfrac{1}{4} + \dfrac{1}{8} + \cdots + \dfrac{1}{2^k} + \dfrac{1}{2^{k+1}} = \dfrac{2^{k+1} - 1}{2^{k+1}}$

(1) *Basis step.* We show S_1 to be true as follows:

$$\frac{2^1 - 1}{2^1} = \frac{2 - 1}{2} = \frac{1}{2}.$$

(2) *Induction step.* We let k be any natural number. We assume S_k to be true and try to show that it implies that S_{k+1} is true. Now S_k is

$$\frac{1}{2} + \frac{1}{4} + \frac{1}{8} + \cdots + \frac{1}{2^k} = \frac{2^k - 1}{2^k}.$$

We start with the left side of S_{k+1}. Since we assume S_k is true, we can substitute

$$\frac{2^k - 1}{2^k} \quad \text{for} \quad \frac{1}{2} + \frac{1}{4} + \cdots + \frac{1}{2^k}.$$

We have

$$\underbrace{\frac{1}{2} + \frac{1}{4} + \frac{1}{8} + \cdots + \frac{1}{2^k}} + \frac{1}{2^{k+1}}$$

$$= \frac{2^k - 1}{2^k} + \frac{1}{2^{k+1}} = \frac{2^k - 1}{2^k} \cdot \frac{2}{2} + \frac{1}{2^{k+1}}$$

$$= \frac{(2^k - 1) \cdot 2 + 1}{2^{k+1}}$$

$$= \frac{2^{k+1} - 2 + 1}{2^{k+1}}$$

$$= \frac{2^{k+1} - 1}{2^{k+1}}.$$

We have shown that for all natural numbers k, $S_k \rightarrow S_{k+1}$. This completes the induction step. It and the basis step tell us that the proof is complete.

Now Try Exercise 15. ■

EXAMPLE 4   Prove: For every natural number n, $n < 2^n$.

Proof.   We first list S_n, S_1, S_k, and S_{k+1}.

$$S_n: \quad n < 2^n$$
$$S_1: \quad 1 < 2^1$$
$$S_k: \quad k < 2^k$$
$$S_{k+1}: \quad k + 1 < 2^{k+1}$$

(1) *Basis step.* S_1, as listed, is true since $2^1 = 2$ and $1 < 2$.

(2) *Induction step.* We let k be any natural number. We assume S_k to be true and try to show that it implies that S_{k+1} is true. Now

$$k < 2^k \qquad \text{This is } S_k.$$
$$2k < 2 \cdot 2^k \qquad \text{Multiplying by 2 on both sides}$$
$$2k < 2^{k+1} \qquad \text{Adding exponents on the right}$$
$$k + k < 2^{k+1}. \qquad \text{Rewriting } 2k \text{ as } k + k$$

Since k is any natural number, we know that $1 \le k$. Thus,

$$k + 1 \le k + k. \qquad \text{Adding } k \text{ on both sides of } 1 \le k$$

Putting the results $k + 1 \le k + k$ and $k + k < 2^{k+1}$ together gives us

$$k + 1 < 2^{k+1}. \qquad \text{This is } S_{k+1}.$$

We have shown that for all natural numbers k, $S_k \rightarrow S_{k+1}$. This completes the induction step. It and the basis step tell us that the proof is complete.   **Now Try Exercise 11.** ▪

(11.4) Exercise Set

List the first five statements in the sequence that can be obtained from each of the following. Determine whether each of the statements is true or false.

1. $n^2 < n^3$

2. $n^2 - n + 41$ is prime. Find a value for n for which the statement is false.

3. A polygon of n sides has $[n(n - 3)]/2$ diagonals.

4. The sum of the angles of a polygon of n sides is $(n - 2) \cdot 180°$.

Use mathematical induction to prove each of the following.

5. $2 + 4 + 6 + \cdots + 2n = n(n + 1)$

6. $4 + 8 + 12 + \cdots + 4n = 2n(n + 1)$

7. $1 + 5 + 9 + \cdots + (4n - 3) = n(2n - 1)$

8. $3 + 6 + 9 + \cdots + 3n = \dfrac{3n(n + 1)}{2}$

9. $2 + 4 + 8 + \cdots + 2^n = 2(2^n - 1)$

10. $2 \le 2^n$

11. $n < n + 1$

12. $3^n < 3^{n+1}$

13. $2n \le 2^n$

14. $\dfrac{1}{1 \cdot 2} + \dfrac{1}{2 \cdot 3} + \cdots + \dfrac{1}{n(n + 1)} = \dfrac{n}{n + 1}$

15. $\dfrac{1}{1\cdot 2\cdot 3} + \dfrac{1}{2\cdot 3\cdot 4} + \dfrac{1}{3\cdot 4\cdot 5} + \cdots$

$\qquad + \dfrac{1}{n(n+1)(n+2)} = \dfrac{n(n+3)}{4(n+1)(n+2)}$

16. If x is any real number greater than 1, then for any natural number n, $x \le x^n$.

The following formulas can be used to find sums of powers of natural numbers. Use mathematical induction to prove each formula.

17. $1 + 2 + 3 + \cdots + n = \dfrac{n(n+1)}{2}$

18. $1^2 + 2^2 + 3^2 + \cdots + n^2 = \dfrac{n(n+1)(2n+1)}{6}$

19. $1^3 + 2^3 + 3^3 + \cdots + n^3 = \dfrac{n^2(n+1)^2}{4}$

20. $1^4 + 2^4 + 3^4 + \cdots + n^4$

$\qquad = \dfrac{n(n+1)(2n+1)(3n^2+3n-1)}{30}$

21. $1^5 + 2^5 + 3^5 + \cdots + n^5$

$\qquad = \dfrac{n^2(n+1)^2(2n^2+2n-1)}{12}$

Use mathematical induction to prove each of the following.

22. $\displaystyle\sum_{i=1}^{n}(3i-1) = \dfrac{n(3n+1)}{2}$

23. $\displaystyle\sum_{i=1}^{n}i(i+1) = \dfrac{n(n+1)(n+2)}{3}$

24. $\left(1+\dfrac{1}{1}\right)\left(1+\dfrac{1}{2}\right)\left(1+\dfrac{1}{3}\right)\cdots\left(1+\dfrac{1}{n}\right)$

$\qquad = n+1$

25. The sum of n terms of an arithmetic sequence:

$a_1 + (a_1+d) + (a_1+2d) + \cdots + [a_1+(n-1)d]$

$\qquad = \dfrac{n}{2}[2a_1+(n-1)d]$

Collaborative Discussion and Writing

26. Write an explanation of the idea behind mathematical induction for a fellow student.

27. Find two statements not considered in this section that are not true for all natural numbers. Then try to find where a proof by mathematical induction fails.

Skill Maintenance

Solve.

28. $2x - 3y = 1,$
$3x - 4y = 3$

29. $x + y + z = 3,$
$2x - 3y - 2z = 5,$
$3x + 2y + 2z = 8$

30. *e-Commerce.* ebooks.com ran a one-day promotion offering a hardback title for $24.95 and a paperback title for $9.95. A total of 80 books were sold and $1546 was taken in. How many of each type of book were sold?

31. *Investment.* Martin received $104 in simple interest one year from three investments. Part is invested at 1.5%, part at 2%, and part at 3%. The amount invested at 2% is twice the amount invested at 1.5%. There is $400 more invested at 3% than at 2%. Find the amount invested at each rate.

Synthesis

Use mathematical induction to prove each of the following.

32. The sum of n terms of a geometric sequence:

$a_1 + a_1 r + a_1 r^2 + \cdots + a_1 r^{n-1} = \dfrac{a_1 - a_1 r^n}{1-r}$

33. $x + y$ is a factor of $x^{2n} - y^{2n}$.

Prove each of the following using mathematical induction. Do the basis step for $n = 2$.

34. For every natural number $n \ge 2$,

$2n + 1 < 3^n.$

35. For every natural number $n \ge 2$,

$\log_a(b_1 b_2 \cdots b_n)$

$\qquad = \log_a b_1 + \log_a b_2 + \cdots + \log_a b_n.$

Prove each of the following for any complex numbers $z_1, z_2, \ldots, z_n$, where $i^2 = -1$ and $\bar{z}$ is the conjugate of z. (See Section 3.1.)

36. $\overline{z^n} = \bar{z}^n$

37. $\overline{z_1 + z_2 + \cdots + z_n} = \bar{z_1} + \bar{z_2} + \cdots + \bar{z_n}$

38. $\overline{z_1 z_2 \cdots z_n} = \bar{z_1} \cdot \bar{z_2} \cdots \bar{z_n}$

39. i^n is either $1, -1, i,$ or $-i$.

For any integers a and b, b is a factor of a if there exists an integer c such that a = bc. Prove each of the following for any natural number n.

40. 2 is a factor of $n^2 + n$.

41. 3 is a factor of $n^3 + 2n$.

42. *The Tower of Hanoi Problem.*   There are three pegs on a board. On one peg are n disks, each smaller than the one on which it rests. The problem is to move this pile of disks to another peg. The final order must be the same, but you can move only one disk at a time and can never place a larger disk on a smaller one.

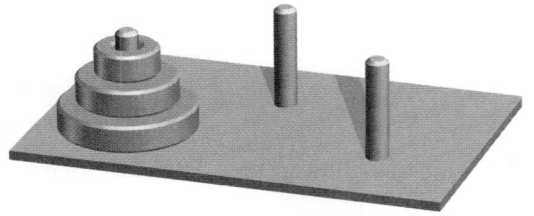

a) What is the *smallest* number of moves needed to move 3 disks? 4 disks? 2 disks? 1 disk?

b) Conjecture a formula for the *smallest* number of moves needed to move n disks. Prove it by mathematical induction.

11.5 Combinatorics: Permutations

❋ Evaluate factorial notation and permutation notation and solve related applied problems.

In order to study probability, it is first necessary to learn about **combinatorics**, the theory of counting.

❋ Permutations

In this section, we will consider the part of combinatorics called *permutations*.

> The study of permutations involves *order* and *arrangements*.

EXAMPLE 1   How many 3-letter code symbols can be formed with the letters A, B, C *without* repetition (that is, using each letter only once)?

Solution   Consider placing the letters in these boxes.

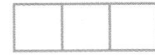

We can select any of the 3 letters for the first letter in the symbol. Once this letter has been selected, the second must be selected from the 2 remaining letters. After this, the third letter is already determined, since only 1 possibility is left. That is, we can place any of the 3 letters in the first box, either of the remaining 2 letters in the second box, and the only remaining letter in the third box. The possibilities can be determined using a **tree diagram**, as shown below.

TREE DIAGRAM			OUTCOMES	

We see that there are 6 possibilities. The set of all the possibilities is

$$\{ABC, ACB, BAC, BCA, CAB, CBA\}.$$

This is the set of all *permutations* of the letters A, B, C.

Suppose that we perform an experiment such as selecting letters (as in the preceding example), flipping a coin, or drawing a card. The results are called **outcomes**. An **event** is a set of outcomes. The following principle enables us to count actions that are combined to form an event.

The Fundamental Counting Principle

Given a combined action, or *event*, in which the first action can be performed in n_1 ways, the second action can be performed in n_2 ways, and so on, the total number of ways in which the combined action can be performed is the product

$$n_1 \cdot n_2 \cdot n_3 \cdot \cdots \cdot n_k.$$

Thus, in Example 1, there are 3 choices for the first letter, 2 for the second letter, and 1 for the third letter, making a total of $3 \cdot 2 \cdot 1$, or 6 possibilities.

EXAMPLE 2 How many 3-letter code symbols can be formed with the letters A, B, C, D, E *with* repetition (that is, allowing letters to be repeated)?

Solution Since repetition is allowed, there are 5 choices for the first letter, 5 choices for the second, and 5 for the third. Thus, by the fundamental counting principle, there are $5 \cdot 5 \cdot 5$, or 125 code symbols.

Permutation

A **permutation** of a set of n objects is an ordered arrangement of all n objects.

We can use the fundamental counting principle to count the number of permutations of the objects in a set. Consider, for example, a set of 4 objects

$$\{A, B, C, D\}.$$

To find the number of ordered arrangements of the set, we select a first letter: There are 4 choices. Then we select a second letter: There are 3 choices. Then we select a third letter: There are 2 choices. Finally, there is 1 choice for the last selection. Thus, by the fundamental counting principle, there are $4 \cdot 3 \cdot 2 \cdot 1$, or 24, permutations of a set of 4 objects.

We can find a formula for the total number of permutations of all objects in a set of n objects. We have n choices for the first selection, $n - 1$ choices for the second, $n - 2$ for the third, and so on. For the nth selection, there is only 1 choice.

The Total Number of Permutations of n Objects

The total number of permutations of n objects, denoted $_nP_n$, is given by

$$_nP_n = n(n - 1)(n - 2) \cdots 3 \cdot 2 \cdot 1.$$

| GCM | **EXAMPLE 3** Find each of the following.

a) $_4P_4$ **b)** $_7P_7$

Solution

Start with 4.

a) $_4P_4 = 4 \cdot 3 \cdot 2 \cdot 1 = 24$

4 factors

b) $_7P_7 = 7 \cdot 6 \cdot 5 \cdot 4 \cdot 3 \cdot 2 \cdot 1 = 5040$

We can also find the total number of permutations of n objects using the $_nP_r$ operation from the MATH PRB (probability) menu on a graphing calculator.

```
4 nPr 4
                24
7 nPr 7
              5040
```

Now Try Exercise 1. ■

EXAMPLE 4 In how many ways can 9 packages be placed in 9 mailboxes, one package in a box?

Solution We have

$$_9P_9 = 9 \cdot 8 \cdot 7 \cdot 6 \cdot 5 \cdot 4 \cdot 3 \cdot 2 \cdot 1 = 362,880.$$

Now Try Exercise 23. ■

❋ Factorial Notation

We will use products such as $7 \cdot 6 \cdot 5 \cdot 4 \cdot 3 \cdot 2 \cdot 1$ so often that it is convenient to adopt a notation for them. For the product

$$7 \cdot 6 \cdot 5 \cdot 4 \cdot 3 \cdot 2 \cdot 1,$$

we write 7!, read "7 factorial."

We now define factorial notation for natural numbers and for 0.

> **Factorial Notation**
>
> For any natural number n,
> $$n! = n(n-1)(n-2) \cdots 3 \cdot 2 \cdot 1.$$
> For the number 0,
> $$0! = 1.$$

We define 0! as 1 so that certain formulas can be stated concisely and with a consistent pattern.

Here are some examples of factorial notation.

GCM

$$
\begin{aligned}
7! &= 7 \cdot 6 \cdot 5 \cdot 4 \cdot 3 \cdot 2 \cdot 1 = 5040 \\
6! &= 6 \cdot 5 \cdot 4 \cdot 3 \cdot 2 \cdot 1 = 720 \\
5! &= 5 \cdot 4 \cdot 3 \cdot 2 \cdot 1 = 120 \\
4! &= 4 \cdot 3 \cdot 2 \cdot 1 = 24 \\
3! &= 3 \cdot 2 \cdot 1 = 6 \\
2! &= 2 \cdot 1 = 2 \\
1! &= 1 = 1 \\
0! &= 1 = 1
\end{aligned}
$$

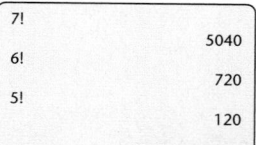

Factorial notation can also be evaluated using the ! operation from the MATH PRB (probability) menu, as shown at left.

We now see that the following statement is true.

> $$_nP_n = n!$$

We will often need to manipulate factorial notation. For example, note that

$$
\begin{aligned}
8! &= 8 \cdot 7 \cdot 6 \cdot 5 \cdot 4 \cdot 3 \cdot 2 \cdot 1 \\
&= 8 \cdot (7 \cdot 6 \cdot 5 \cdot 4 \cdot 3 \cdot 2 \cdot 1) = 8 \cdot 7!.
\end{aligned}
$$

Generalizing, we get the following.

> For any natural number n, $n! = n(n-1)!$.

By using this result repeatedly, we can further manipulate factorial notation.

EXAMPLE 5 Rewrite 7! with a factor of 5!.

Solution We have

$$7! = 7 \cdot 6! = 7 \cdot 6 \cdot 5!.$$

In general, we have the following.

For any natural numbers k and n, with $k < n$,

$$n! = \underbrace{n(n-1)(n-2) \cdots [n-(k-1)]}_{k \text{ factors}} \cdot \underbrace{(n-k)!}_{n-k \text{ factors}}$$

STUDY TIP

Take time to prepare for class. Review the material that was covered in the previous class and read the portion of the text that will be covered in the next class. When you are prepared, you will be able to follow the lecture more easily and derive the greatest benefit from the time you spend in class.

❖ Permutations of n Objects Taken k at a Time

Consider a set of 5 objects

$$\{A, B, C, D, E\}.$$

How many ordered arrangements can be formed using 3 objects from the set without repetition? Examples of such an arrangement are EBA, CAB, and BCD. There are 5 choices for the first object, 4 choices for the second, and 3 choices for the third. By the fundamental counting principle, there are

$$5 \cdot 4 \cdot 3,$$

or

60 *permutations* of a set of 5 objects taken 3 at a time.

Note that

$$5 \cdot 4 \cdot 3 = \frac{5 \cdot 4 \cdot 3 \cdot 2 \cdot 1}{2 \cdot 1}, \quad \text{or} \quad \frac{5!}{2!}.$$

Permutation of n Objects Taken k at a Time

A **permutation** of a set of n objects taken k at a time is an ordered arrangement of k objects taken from the set.

Consider a set of n objects and the selection of an ordered arrangement of k of them. There would be n choices for the first object. Then there would remain $n-1$ choices for the second, $n-2$ choices for the third, and so on. We make k choices in all, so there are k factors in the product. By the fundamental counting principle, the total number of permutations is

$$\underbrace{n(n-1)(n-2) \cdots [n-(k-1)]}_{k \text{ factors}}.$$

We can express this in another way by multiplying by 1, as follows:

$$n(n-1)(n-2)\cdots[n-(k-1)]\cdot\frac{(n-k)!}{(n-k)!}$$

$$=\frac{n(n-1)(n-2)\cdots[n-(k-1)](n-k)!}{(n-k)!}$$

$$=\frac{n!}{(n-k)!}.$$

This gives us the following.

The Number of Permutations of n Objects Taken k at a Time

The number of permutations of a set of n objects taken k at a time, denoted $_nP_k$, is given by

$$_nP_k = \underbrace{n(n-1)(n-2)\cdots[n-(k-1)]}_{k\text{ factors}} \tag{1}$$

$$= \frac{n!}{(n-k)!}. \tag{2}$$

GCM **EXAMPLE 6** Compute $_8P_4$ using both forms of the formula.

Solution Using form (1), we have

The 8 tells where to start.

$$_8P_4 = \underbrace{8 \cdot 7 \cdot 6 \cdot 5} = 1680.$$

The 4 tells how many factors.

Using form (2), we have

$$_8P_4 = \frac{8!}{(8-4)!}$$

$$= \frac{8!}{4!}$$

$$= \frac{8 \cdot 7 \cdot 6 \cdot 5 \cdot 4!}{4!} = \frac{8 \cdot 7 \cdot 6 \cdot 5 \cdot \cancel{4!}}{\cancel{4!}}$$

$$= 8 \cdot 7 \cdot 6 \cdot 5 = 1680.$$

We can also evaluate $_8P_4$ using the $_nP_r$ operation from the MATH PRB menu on a graphing calculator.

```
8 nPr 4
                    1680
```

Now Try Exercise 3. ■

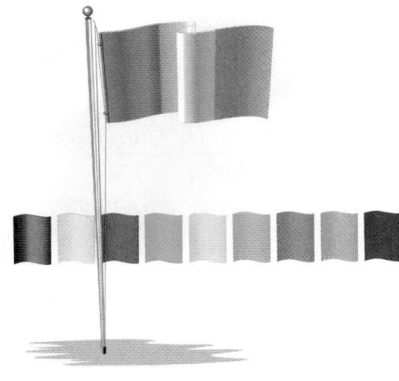

EXAMPLE 7 *Flags of Nations.* The flags of many nations consist of three vertical stripes. For example, the flag of Ireland, shown here, has its first stripe green, second white, and third orange.

Suppose that the following 9 colors are available:

{black, yellow, red, blue, white, gold, orange, pink, purple}.

How many different flags of 3 colors can be made without repetition of colors in a flag? This assumes that the order in which the stripes appear is considered.

Solution We are determining the number of permutations of 9 objects taken 3 at a time. There is no repetition of colors. Using form (1), we get

$$_9P_3 = 9 \cdot 8 \cdot 7 = 504.$$ **Now Try Exercise 37(a).** ■

EXAMPLE 8 *Batting Orders.* A baseball manager arranges the batting order as follows: The 4 infielders will bat first. Then the 3 outfielders, the catcher, and the pitcher will follow, not necessarily in that order. How many different batting orders are possible?

Solution The infielders can bat in $_4P_4$ different ways, the rest in $_5P_5$ different ways. Then by the fundamental counting principle, we have

$$_4P_4 \cdot _5P_5 = 4! \cdot 5!, \quad \text{or} \quad 2880 \text{ possible batting orders.}$$

Now Try Exercise 31. ■

If we allow repetition, a situation like the following can occur.

EXAMPLE 9 How many 5-letter code symbols can be formed with the letters A, B, C, and D if we allow a letter to occur more than once?

Solution We can select each of the 5 letters in 4 ways. That is, we can select the first letter in 4 ways, the second in 4 ways, and so on. Thus there are 4^5, or 1024 arrangements. **Now Try Exercise 37(b).** ■

> The number of distinct arrangements of n objects taken k at a time, allowing repetition, is n^k.

❖ Permutations of Sets with Nondistinguishable Objects

Consider a set of 7 marbles, 4 of which are blue and 3 of which are red. When they are lined up, one red marble will look just like any other red marble. In this sense, we say that the red marbles are nondistinguishable and, similarly, the blue marbles are nondistinguishable.

We know that there are 7! permutations of this set. Many of them will look alike, however. We develop a formula for finding the number of distinguishable permutations.

Consider a set of n objects in which n_1 are of one kind, n_2 are of a second kind, ..., and n_k are of a kth kind. The total number of permutations of the set is $n!$, but this includes many that are nondistinguishable. Let N be the total number of distinguishable permutations. For each of these N permutations, there are $n_1!$ actual, nondistinguishable permutations, obtained by permuting the objects of the first kind. For each of these $N \cdot n_1!$ permutations, there are $n_2!$ nondistinguishable permutations, obtained by permuting the objects of the second kind, and so on. By the fundamental counting principle, the total number of permutations, including those that are nondistinguishable, is

$$N \cdot n_1! \cdot n_2! \cdot \cdots \cdot n_k!.$$

Then we have $N \cdot n_1! \cdot n_2! \cdot \cdots \cdot n_k! = n!$. Solving for N, we obtain

$$N = \frac{n!}{n_1! \cdot n_2! \cdot \cdots \cdot n_k!}.$$

Now, to finish our problem with the marbles, we have

$$N = \frac{7!}{4!\,3!}$$

$$= \frac{7 \cdot 6 \cdot 5 \cdot 4!}{4! \cdot 3 \cdot 2 \cdot 1} = \frac{7 \cdot \cancel{3} \cdot \cancel{2} \cdot 5 \cdot \cancel{4!}}{\cancel{4!} \cdot \cancel{3} \cdot \cancel{2} \cdot 1}$$

$$= \frac{7 \cdot 5}{1}, \quad \text{or} \quad 35$$

distinguishable permutations of the marbles.

In general:

> For a set of n objects in which n_1 are of one kind, n_2 are of another kind, ..., and n_k are of a kth kind, the number of distinguishable permutations is
>
> $$\frac{n!}{n_1! \cdot n_2! \cdot \cdots \cdot n_k!}.$$

EXAMPLE 10 In how many distinguishable ways can the letters of the word CINCINNATI be arranged?

Solution There are 2 C's, 3 I's, 3 N's, 1 A, and 1 T for a total of 10 letters. Thus,

$$N = \frac{10!}{2! \cdot 3! \cdot 3! \cdot 1! \cdot 1!}, \quad \text{or} \quad 50{,}400.$$

The letters of the word CINCINNATI can be arranged in 50,400 distinguishable ways.

Now Try Exercise 35. ■

(11.5) Exercise Set

Evaluate.

1. $_6P_6$

2. $_4P_3$

3. $_{10}P_7$

4. $_{10}P_3$

5. $5!$ GCM **6.** $7!$

7. $0!$

8. $1!$

9. $\dfrac{9!}{5!}$ GCM

10. $\dfrac{9!}{4!}$

11. $(8-3)!$

12. $(8-5)!$

13. $\dfrac{10!}{7!\,3!}$

14. $\dfrac{7!}{(7-2)!}$

15. $_8P_0$

16. $_{13}P_1$

17. $_{52}P_4$

18. $_{52}P_5$

19. $_nP_3$

20. $_nP_2$

21. $_nP_1$

22. $_nP_0$

In each of Exercises 23–41, give your answer using permutation notation, factorial notation, or other operations. Then evaluate.

How many permutations are there of the letters in each of the following words, if all the letters are used without repetition?

23. MARVIN

24. JUDY

25. UNDERMOST

26. COMBINES

27. How many permutations are there of the letters of the word UNDERMOST if the letters are taken 4 at a time?

28. How many permutations are there of the letters of the word COMBINES if the letters are taken 5 at a time?

29. How many 5-digit numbers can be formed using the digits 2, 4, 6, 8, and 9 without repetition? with repetition?

30. In how many ways can 7 athletes be arranged in a straight line?

31. *Program Planning.* A program is planned to have 5 musical numbers and 4 speeches. In how many ways can this be done if a musical number and a speech are to alternate and a musical number is to come first?

32. A professor is going to grade her 24 students on a curve. She will give 3 A's, 5 B's, 9 C's, 4 D's, and 3 F's. In how many ways can she do this?

33. *Phone Numbers.* How many 7-digit phone numbers can be formed with the digits 0, 1, 2, 3, 4, 5, 6, 7, 8, and 9, assuming that the first number cannot be 0 or 1? Accordingly, how many telephone numbers can there be within a given area code, before the area needs to be split with a new area code?

34. How many distinguishable code symbols can be formed from the letters of the word BUSINESS? BIOLOGY? MATHEMATICS?

35. Suppose the expression $a^2b^3c^4$ is rewritten without exponents. In how many ways can this be done?

36. *Coin Arrangements.* A penny, a nickel, a dime, and a quarter are arranged in a straight line.

a) Considering just the coins, in how many ways can they be lined up?

b) Considering the coins and heads and tails, in how many ways can they be lined up?

37. How many code symbols can be formed using 5 out of 6 letters of A, B, C, D, E, F if the letters:

   a) are not repeated?
   b) can be repeated?
   c) are not repeated but must begin with D?
   d) are not repeated but must begin with DE?

38. *License Plates.* A state forms its license plates by first listing a number that corresponds to the county in which the owner of the car resides. (The names of the counties are alphabetized and the number is its location in that order.) Then the plate lists a letter of the alphabet, and this is followed by a number from 1 to 9999. How many such plates are possible if there are 80 counties?

39. *Zip Codes.* A U.S. postal zip code is a five-digit number.

   a) How many zip codes are possible if any of the digits 0 to 9 can be used?
   b) If each post office has its own zip code, how many possible post offices can there be?

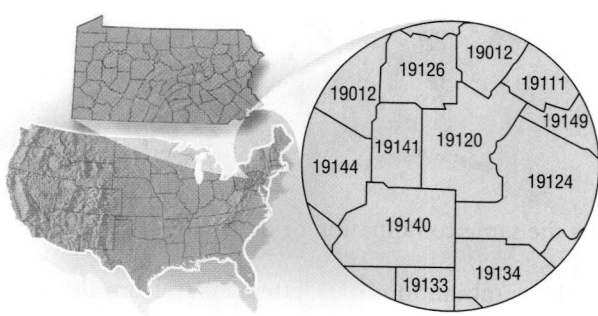

40. *Zip-Plus-4 Codes.* A zip-plus-4 postal code uses a 9-digit number like 75247-5456. How many 9-digit zip-plus-4 postal codes are possible?

41. *Social Security Numbers.* A social security number is a 9-digit number like 243-47-0825.

   a) How many different social security numbers can there be?
   b) There are about 303 million people in the United States. Can each person have a unique social security number?

Collaborative Discussion and Writing

42. How "long" is 15? Suppose you own 15 books and decide to make up all the possible arrangements of the books on a shelf. About how long, in years, would it take you if you make one arrangement per second? Write out the reasoning you used for this problem in the form of a paragraph.

43. *Circular Arrangements.* In how many ways can the numbers on a clock face be arranged? See if you can derive a formula for the number of distinct circular arrangements of n objects. Explain your reasoning.

Skill Maintenance

Find the zero(s) of the function.

44. $f(x) = 4x - 9$

45. $f(x) = x^2 + x - 6$

46. $f(x) = 2x^2 - 3x - 1$

47. $f(x) = x^3 - 4x^2 - 7x + 10$

Synthesis

Solve for n.

48. $_nP_5 = 7 \cdot {}_nP_4$ **49.** $_nP_4 = 8 \cdot {}_{n-1}P_3$

50. $_nP_5 = 9 \cdot {}_{n-1}P_4$ **51.** $_nP_4 = 8 \cdot {}_nP_3$

52. Show that $n! = n(n-1)(n-2)(n-3)!$.

53. *Single-Elimination Tournaments.* In a single-elimination sports tournament consisting of n teams, a team is eliminated when it loses one game. How many games are required to complete the tournament?

54. *Double-Elimination Tournaments.* In a double-elimination softball tournament consisting of n teams, a team is eliminated when it loses two games. At most, how many games are required to complete the tournament?

11.6 Combinatorics: Combinations

❖ Evaluate combination notation and solve related applied problems.

We now consider counting techniques in which order is not considered.

❖ Combinations

We sometimes make a selection from a set *without regard to order*. Such a selection is called a *combination*. If you play cards, for example, you know that in most situations the *order* in which you hold cards is not important. That is,

The hand

is "equivalent" to these hands.

Each hand contains the same combination of three cards.

EXAMPLE 1 Find all the combinations of 3 letters taken from the set of 5 letters {A, B, C, D, E}.

Solution The combinations are

{A, B, C},	{A, B, D},
{A, B, E},	{A, C, D},
{A, C, E},	{A, D, E},
{B, C, D},	{B, C, E},
{B, D, E},	{C, D, E}.

There are 10 combinations of the 5 letters taken 3 at a time.

When we find all the combinations from a set of 5 objects taken 3 at a time, we are finding all the 3-element subsets. When a set is named, the order of the elements is *not* considered. Thus,

{A, C, B}   names the same set as   {A, B, C}.

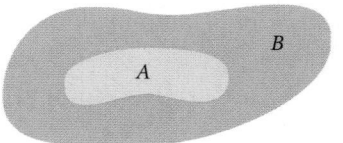

Subset

Set A is a subset of set B, denoted $A \subseteq B$, if every element of A is an element of B.

The elements of a set are not ordered. When thinking of *combinations*, do *not* think about order!

Combination

A **combination** containing k objects is a subset containing k objects.

We want to develop a formula for computing the number of combinations of n objects taken k at a time without actually listing the combinations or subsets.

Combination Notation

The number of combinations of n objects taken k at a time is denoted $_nC_k$.

We call $_nC_k$ **combination notation.** We want to derive a general formula for $_nC_k$ for any $k \leq n$. First, it is true that $_nC_n = 1$, because a set with n objects has only 1 subset with n objects, the set itself. Second, $_nC_1 = n$, because a set with n objects has n subsets with 1 element each. Finally, $_nC_0 = 1$, because a set with n objects has only one subset with 0 elements, namely, the empty set $\varnothing$. To consider other possibilities, let's return to Example 1 and compare the number of combinations with the number of permutations.

COMBINATIONS			PERMUTATIONS			
$\{A, B, C\} \longrightarrow$	ABC	BCA	CAB	CBA	BAC	ACB
$\{A, B, D\} \longrightarrow$	ABD	BDA	DAB	DBA	BAD	ADB
$\{A, B, E\} \longrightarrow$	ABE	BEA	EAB	EBA	BAE	AEB
$\{A, C, D\} \longrightarrow$	ACD	CDA	DAC	DCA	CAD	ADC
$\{A, C, E\} \longrightarrow$	ACE	CEA	EAC	ECA	CAE	AEC
$\{A, D, E\} \longrightarrow$	ADE	DEA	EAD	EDA	DAE	AED
$\{B, C, D\} \longrightarrow$	BCD	CDB	DBC	DCB	CBD	BDC
$\{B, C, E\} \longrightarrow$	BCE	CEB	EBC	ECB	CBE	BEC
$\{B, D, E\} \longrightarrow$	BDE	DEB	EBD	EDB	DBE	BED
$\{C, D, E\} \longrightarrow$	CDE	DEC	ECD	EDC	DCE	CED

$_5C_3$ of these (left bracket)

$3! \cdot {}_5C_3$ of these (right bracket)

Note that each combination of 3 objects yields 6, or 3!, permutations.

$$3! \cdot {}_5C_3 = 60 = {}_5P_3 = 5 \cdot 4 \cdot 3,$$

so

$${}_5C_3 = \frac{{}_5P_3}{3!} = \frac{5 \cdot 4 \cdot 3}{3 \cdot 2 \cdot 1} = 10.$$

In general, the number of combinations of n objects taken k at a time, ${}_nC_k$, times the number of permutations of these objects, $k!$, must equal the number of permutations of n objects taken k at a time:

$$k! \cdot {}_nC_k = {}_nP_k$$

$$\begin{aligned}{}_nC_k &= \frac{{}_nP_k}{k!}\\ &= \frac{1}{k!} \cdot {}_nP_k\\ &= \frac{1}{k!} \cdot \frac{n!}{(n-k)!} = \frac{n!}{k!\,(n-k)!}.\end{aligned}$$

> ## Combinations of n Objects Taken k at a Time
> The total number of combinations of n objects taken k at a time, denoted ${}_nC_k$, is given by
>
> $${}_nC_k = \frac{n!}{k!\,(n-k)!}, \tag{1}$$
>
> or
>
> $${}_nC_k = \frac{{}_nP_k}{k!} = \frac{n(n-1)(n-2)\cdots[n-(k-1)]}{k!}. \tag{2}$$

Another kind of notation for ${}_nC_k$ is **binomial coefficient notation**. The reason for such terminology will be seen later.

> ## Binomial Coefficient Notation
>
> $$\binom{n}{k} = {}_nC_k$$

You should be able to use either notation and either form of the formula.

GCM **EXAMPLE 2** Evaluate $\binom{7}{5}$, using forms (1) and (2).

Solution

a) By form (1),

$$\binom{7}{5} = \frac{7!}{5!\,(7-5)!} = \frac{7!}{5!\,2!}$$

$$= \frac{7 \cdot 6 \cdot 5!}{5! \cdot 2!} = \frac{7 \cdot 6 \cdot \cancel{5}!}{\cancel{5}! \cdot 2!} = \frac{7 \cdot 6}{2 \cdot 1} = 21.$$

b) By form (2),

The 7 tells where to start.

$$\binom{7}{5} = \frac{7 \cdot 6 \cdot 5 \cdot 4 \cdot 3}{5 \cdot 4 \cdot 3 \cdot 2 \cdot 1} = \frac{7 \cdot 6}{2 \cdot 1} = 21.$$

The 5 tells how many factors there are in both the numerator and the denominator and where to start the denominator.

```
7 nCr 5
                  21
```

We can also find combinations using the $_nC_r$ operation from the MATH PRB (probability) menu on a graphing calculator.   **Now Try Exercise 11.** ■

Be sure to keep in mind that $\binom{n}{k}$ does not mean $n \div k$, or n/k.

EXAMPLE 3 Evaluate $\binom{n}{0}$ and $\binom{n}{2}$.

Solution We use form (1) for the first expression and form (2) for the second. Then

$$\binom{n}{0} = \frac{n!}{0!\,(n-0)!} = \frac{n!}{1 \cdot n!} = 1,$$

using form (1), and

$$\binom{n}{2} = \frac{n(n-1)}{2!} = \frac{n(n-1)}{2}, \quad \text{or} \quad \frac{n^2 - n}{2},$$

using form (2).   **Now Try Exercise 19.** ■

Note that

$$\binom{7}{2} = \frac{7 \cdot 6}{2 \cdot 1} = 21.$$

Using the result of Example 2 gives us

$$\binom{7}{5} = \binom{7}{2}.$$

This says that the number of 5-element subsets of a set of 7 objects is the same as the number of 2-element subsets of a set of 7 objects. When 5 elements are chosen from a set, one also chooses *not* to include 2 elements. To see this, consider the set $\{A, B, C, D, E, F, G\}$:

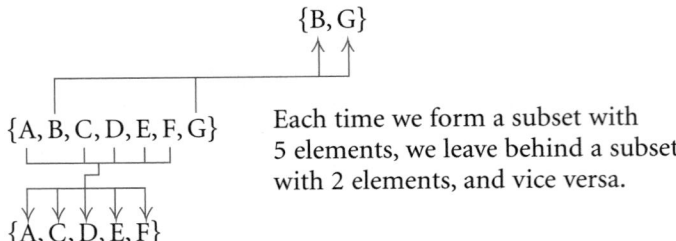

$\{B, G\}$

$\{A, B, C, D, E, F, G\}$ Each time we form a subset with 5 elements, we leave behind a subset with 2 elements, and vice versa.

$\{A, C, D, E, F\}$

In general, we have the following. This result provides an alternative way to compute combinations.

Subsets of Size *k* and of Size *n* − *k*

$$\binom{n}{k} = \binom{n}{n-k} \quad \text{and} \quad {}_nC_k = {}_nC_{n-k}$$

The number of subsets of size k of a set with n objects is the same as the number of subsets of size $n - k$. The number of combinations of n objects taken k at a time is the same as the number of combinations of n objects taken $n - k$ at a time.

We now solve problems involving combinations.

EXAMPLE 4 *Michigan Lottery.* Run by the state of Michigan, Classic Lotto 47 is a twice-weekly lottery game with jackpots starting at $1 million. For a wager of $1, a player can choose 6 numbers from 1 through 47. If the numbers match those drawn by the state, the player wins. (*Source:* www.michigan.gov/lottery)

a) How many 6-number combinations are there?

b) Suppose it takes you 10 min to pick your numbers and buy a game ticket. How many tickets can you buy in 4 days?

c) How many people would you have to hire for 4 days to buy tickets with all the possible combinations and ensure that you win?

Solution

a) No order is implied here. You pick any 6 different numbers from 1 through 47. Thus the number of combinations is

$$_{47}C_6 = \binom{47}{6} = \frac{47!}{6!\,(47-6)!} = \frac{47!}{6!\,41!}$$
$$= \frac{47 \cdot 46 \cdot 45 \cdot 44 \cdot 43 \cdot 42}{6 \cdot 5 \cdot 4 \cdot 3 \cdot 2 \cdot 1} = 10{,}737{,}573.$$

b) First we find the number of minutes in 4 days:

$$4 \text{ days} = 4 \text{ days} \cdot \frac{24 \text{ hr}}{1 \text{ day}} \cdot \frac{60 \text{ min}}{1 \text{ hr}} = 5760 \text{ min.}$$

Thus you could buy 5760/10, or 576 tickets in 4 days.

c) You would need to hire 10,737,573/576, or about 18,642 people, to buy tickets with all the possible combinations and ensure a win. (This presumes lottery tickets can be bought 24 hours a day.)

Now Try Exercise 23. ◼

EXAMPLE 5 How many committees can be formed from a group of 5 governors and 7 senators if each committee consists of 3 governors and 4 senators?

Solution The 3 governors can be selected in $_5C_3$ ways and the 4 senators can be selected in $_7C_4$ ways. If we use the fundamental counting principle, it follows that the number of possible committees is

$$_5C_3 \cdot {_7C_4} = \frac{5!}{3!\,2!} \cdot \frac{7!}{4!\,3!} = \frac{5 \cdot 4 \cdot 3!}{3! \cdot 2 \cdot 1} \cdot \frac{7 \cdot 6 \cdot 5 \cdot 4!}{4! \cdot 3 \cdot 2 \cdot 1}$$
$$= \frac{5 \cdot \cancel{2} \cdot 2 \cdot \cancel{3!}}{\cancel{3!} \cdot \cancel{2} \cdot 1} \cdot \frac{7 \cdot \cancel{3} \cdot \cancel{2} \cdot 5 \cdot \cancel{4!}}{\cancel{4!} \cdot \cancel{3} \cdot \cancel{2} \cdot 1}$$
$$= 10 \cdot 35$$
$$= 350.$$

Now Try Exercise 27. ◼

```
5 nCr 3*7 nCr 4
                    350
```

CONNECTING
the **CONCEPTS**

Permutations and Combinations

PERMUTATIONS

Permutations involve order and arrangements of objects.

Given 5 books, we can arrange 3 of them on a shelf in $_5P_3$, or 60 ways.

Placing the books in different orders produces different arrangements.

COMBINATIONS

Combinations do not involve order or arrangements of objects.

Given 5 books, we can select 3 of them in $_5C_3$, or 10 ways.

The order in which the books are chosen does not matter.

(11.6) **Exercise Set**

Evaluate.

1. $_{13}C_2$

2. $_9C_6$

3. $\binom{13}{11}$

4. $\binom{9}{3}$

5. $\binom{7}{1}$

6. $\binom{8}{8}$

7. $\dfrac{_5P_3}{3!}$

8. $\dfrac{_{10}P_5}{5!}$

9. $\binom{6}{0}$

10. $\binom{6}{1}$

11. $\binom{6}{2}$

12. $\binom{6}{3}$

13. $\binom{7}{0} + \binom{7}{1} + \binom{7}{2} + \binom{7}{3} + \binom{7}{4} + \binom{7}{5}$
$+ \binom{7}{6} + \binom{7}{7}$

14. $\binom{6}{0} + \binom{6}{1} + \binom{6}{2} + \binom{6}{3} + \binom{6}{4}$
$+ \binom{6}{5} + \binom{6}{6}$

15. $_{52}C_4$

16. $_{52}C_5$

17. $\binom{27}{11}$

18. $\binom{37}{8}$

19. $\binom{n}{1}$

20. $\binom{n}{3}$

21. $\binom{m}{m}$

22. $\binom{t}{4}$

In each of the following exercises, give an expression for the answer using permutation notation, combination notation, factorial notation, or other operations. Then evaluate.

23. *Fraternity Officers.* There are 23 students in a fraternity. How many sets of 4 officers can be selected?

24. *League Games.* How many games can be played in a 9-team sports league if each team plays all other teams once? twice?

25. *Test Options.* On a test, a student is to select 10 out of 13 questions. In how many ways can this be done?

26. *Senate Committees.* Suppose the Senate of the United States consists of 58 Republicans and 42 Democrats. How many committees can be formed consisting of 6 Republicans and 4 Democrats?

27. *Test Options.* Of the first 10 questions on a test, a student must answer 7. Of the second 5 questions, the student must answer 3. In how many ways can this be done?

28. *Lines and Triangles from Points.* How many lines are determined by 8 points, no 3 of which are collinear? How many triangles are determined by the same points?

29. *Poker Hands.* How many 5-card poker hands are possible with a 52-card deck?

30. *Bridge Hands.* How many 13-card bridge hands are possible with a 52-card deck?

31. *Baskin-Robbins Ice Cream.* Burt Baskin and Irv Robbins began making ice cream in 1945. Initially they developed 31 flavors—one for each day of the month. (*Source*: Baskin-Robbins)

a) How many 2-dip cones are possible using the 31 original flavors if order of flavors is to be considered and no flavor is repeated?

b) How many 2-dip cones are possible if order is to be considered and a flavor can be repeated?

c) How many 2-dip cones are possible if order is not considered and no flavor is repeated?

Collaborative Discussion and Writing

32. Give an explanation that you might use with a fellow student to explain that
$$\binom{n}{k} = \binom{n}{n-k}.$$

33. Explain why a "combination" lock should really be called a "permutation" lock.

Skill Maintenance

Solve.

34. $3x - 7 = 5x + 10$

35. $2x^2 - x = 3$

36. $x^2 + 5x + 1 = 0$

37. $x^3 + 3x^2 - 10x = 24$

Synthesis

38. *Full House.* A full house in poker consists of three of a kind and a pair (two of a kind). How many full houses are there that consist of 3 aces and 2 queens? (See Section 11.8 for a description of a 52-card deck.)

39. *Flush.* A flush in poker consists of a 5-card hand with all cards of the same suit. How many 5-card hands (flushes) are there that consist of all diamonds?

40. There are n points on a circle. How many quadrilaterals can be inscribed with these points as vertices?

41. *League Games.* How many games are played in a league with n teams if each team plays each other team once? twice?

Solve for n.

42. $\dbinom{n+1}{3} = 2 \cdot \dbinom{n}{2}$

43. $\dbinom{n}{n-2} = 6$

44. $\dbinom{n}{3} = 2 \cdot \dbinom{n-1}{2}$

45. $\dbinom{n+2}{4} = 6 \cdot \dbinom{n}{2}$

46. Prove that
$$\dbinom{n}{k-1} + \dbinom{n}{k} = \dbinom{n+1}{k}$$
for any natural numbers n and k, $k \le n$.

47. How many line segments are determined by the n vertices of an n-gon? Of these, how many are diagonals? Use mathematical induction to prove the result for the diagonals.

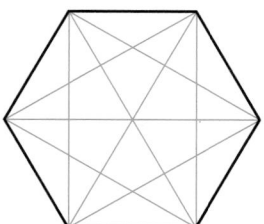

The Binomial Theorem

❖ Expand a power of a binomial using Pascal's triangle or factorial notation.

❖ Find a specific term of a binomial expansion.

❖ Find the total number of subsets of a set of n objects.

In this section, we consider ways of expanding a binomial $(a + b)^n$.

❖ Binomial Expansions Using Pascal's Triangle

Consider the following expanded powers of $(a + b)^n$, where $a + b$ is any binomial and n is a whole number. Look for patterns.

$$(a + b)^0 = 1$$
$$(a + b)^1 = a + b$$
$$(a + b)^2 = a^2 + 2ab + b^2$$
$$(a + b)^3 = a^3 + 3a^2b + 3ab^2 + b^3$$
$$(a + b)^4 = a^4 + 4a^3b + 6a^2b^2 + 4ab^3 + b^4$$
$$(a + b)^5 = a^5 + 5a^4b + 10a^3b^2 + 10a^2b^3 + 5ab^4 + b^5$$

Each expansion is a polynomial. There are some patterns to be noted.

1. There is one more term than the power of the exponent, n. That is, there are $n + 1$ terms in the expansion of $(a + b)^n$.

2. In each term, the sum of the exponents is n, the power to which the binomial is raised.

3. The exponents of a start with n, the power of the binomial, and decrease to 0. The last term has no factor of a. The first term has no factor of b, so powers of b start with 0 and increase to n.

4. The coefficients start at 1 and increase through certain values about "half"-way and then decrease through these same values back to 1.

Let's explore the coefficients further. Suppose that we want to find an expansion of $(a + b)^6$. The patterns we just noted indicate that there are 7 terms in the expansion:

$$a^6 + c_1a^5b + c_2a^4b^2 + c_3a^3b^3 + c_4a^2b^4 + c_5ab^5 + b^6.$$

How can we determine the value of each coefficient, c_i? We can do so in two ways. The first method involves writing the coefficients in a triangular array, as follows. This is known as **Pascal's triangle**:

$$(a + b)^0: \qquad\qquad\qquad 1$$
$$(a + b)^1: \qquad\qquad\quad 1 \quad 1$$
$$(a + b)^2: \qquad\qquad 1 \quad 2 \quad 1$$
$$(a + b)^3: \qquad\quad 1 \quad 3 \quad 3 \quad 1$$
$$(a + b)^4: \qquad 1 \quad 4 \quad 6 \quad 4 \quad 1$$
$$(a + b)^5: \quad 1 \quad 5 \quad 10 \quad 10 \quad 5 \quad 1$$

There are many patterns in the triangle. Find as many as you can.

Perhaps you discovered a way to write the next row of numbers, given the numbers in the row above it. There are always 1's on the outside. Each remaining number is the sum of the two numbers above it. Let's try to find an expansion for $(a + b)^6$ by adding another row using the patterns we have discovered:

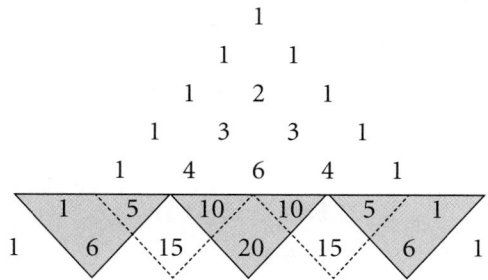

We see that in the last row

    the 1st and last numbers are **1**;

    the 2nd number is $1 + 5$, or **6**;

    the 3rd number is $5 + 10$, or **15**;

    the 4th number is $10 + 10$, or **20**;

    the 5th number is $10 + 5$, or **15**; and

    the 6th number is $5 + 1$, or **6**.

Thus the expansion for $(a + b)^6$ is

$$(a + b)^6 = 1a^6 + 6a^5b + 15a^4b^2 + 20a^3b^3 + 15a^2b^4 + 6ab^5 + 1b^6.$$

To find an expansion for $(a + b)^8$, we complete two more rows of Pascal's triangle:

```
                    1
                 1     1
              1     2     1
           1     3     3     1
        1     4     6     4     1
     1     5    10    10     5     1
  1     6    15    20    15     6     1
1     7    21    35    35    21     7     1
1  8    28    56    70    56    28     8     1
```

Thus the expansion of $(a + b)^8$ is

$$(a + b)^8 = a^8 + 8a^7b + 28a^6b^2 + 56a^5b^3 + 70a^4b^4 + 56a^3b^5$$
$$+ 28a^2b^6 + 8ab^7 + b^8.$$

We can generalize our results as follows.

The Binomial Theorem Using Pascal's Triangle

For any binomial $a + b$ and any natural number n,

$$(a + b)^n = c_0 a^n b^0 + c_1 a^{n-1} b^1 + c_2 a^{n-2} b^2 + \cdots$$
$$+ c_{n-1} a^1 b^{n-1} + c_n a^0 b^n,$$

where the numbers $c_0, c_1, c_2, \ldots, c_{n-1}, c_n$ are from the $(n + 1)$st row of Pascal's triangle.

EXAMPLE 1   Expand: $(u - v)^5$.

Solution   We have $(a + b)^n$, where $a = u$, $b = -v$, and $n = 5$. We use the 6th row of Pascal's triangle:

$$1 \quad 5 \quad 10 \quad 10 \quad 5 \quad 1$$

Then we have

$$
\begin{aligned}
(u - v)^5 &= [u + (-v)]^5 \\
&= 1(u)^5 + 5(u)^4(-v)^1 + 10(u)^3(-v)^2 + 10(u)^2(-v)^3. \\
&\quad + 5(u)(-v)^4 + 1(-v)^5 \\
&= u^5 - 5u^4 v + 10u^3 v^2 - 10u^2 v^3 + 5uv^4 - v^5.
\end{aligned}
$$

Note that the signs of the terms alternate between $+$ and $-$. When the power of $-v$ is odd, the sign is $-$.   **Now Try Exercise 5.** ◼

EXAMPLE 2   Expand: $\left(2t + \dfrac{3}{t} \right)^4$.

Solution   We have $(a + b)^n$, where $a = 2t$, $b = 3/t$, and $n = 4$. We use the 5th row of Pascal's triangle:

$$1 \quad 4 \quad 6 \quad 4 \quad 1$$

Then we have

$$
\begin{aligned}
\left(2t + \frac{3}{t} \right)^4 &= 1(2t)^4 + 4(2t)^3 \left(\frac{3}{t} \right)^1 + 6(2t)^2 \left(\frac{3}{t} \right)^2 + 4(2t)^1 \left(\frac{3}{t} \right)^3 + 1 \left(\frac{3}{t} \right)^4 \\
&= 1(16t^4) + 4(8t^3) \left(\frac{3}{t} \right) + 6(4t^2) \left(\frac{9}{t^2} \right) + 4(2t) \left(\frac{27}{t^3} \right) + 1 \left(\frac{81}{t^4} \right) \\
&= 16t^4 + 96t^2 + 216 + 216t^{-2} + 81t^{-4}.
\end{aligned}
$$

Now Try Exercise 9. ◼

❊ Binomial Expansion Using Factorial Notation

Suppose that we want to find the expansion of $(a + b)^{11}$. The disadvantage in using Pascal's triangle is that we must compute all the preceding rows of the triangle to obtain the row needed for the expansion. The following method avoids this. It also enables us to find a specific term— say, the 8th term—without computing all the other terms of the expansion. This method is useful in such courses as finite mathematics, calculus, and statistics, and it uses the *binomial coefficient notation* $\binom{n}{k}$ developed in Section 8.6.

We can restate the binomial theorem as follows.

The Binomial Theorem Using Factorial Notation

For any binomial $a + b$ and any natural number n,

$$(a + b)^n = \binom{n}{0}a^n b^0 + \binom{n}{1}a^{n-1}b^1 + \binom{n}{2}a^{n-2}b^2 + \cdots$$

$$+ \binom{n}{n-1}a^1 b^{n-1} + \binom{n}{n}a^0 b^n$$

$$= \sum_{k=0}^{n} \binom{n}{k}a^{n-k}b^k.$$

The binomial theorem can be proved by mathematical induction. (See Exercise 61.) This form shows why $\binom{n}{k}$ is called a *binomial coefficient*.

EXAMPLE 3 Expand: $(x^2 - 2y)^5$.

Solution We have $(a + b)^n$, where $a = x^2$, $b = -2y$, and $n = 5$. Then using the binomial theorem, we have

$$(x^2 - 2y)^5 = \binom{5}{0}(x^2)^5 + \binom{5}{1}(x^2)^4(-2y) + \binom{5}{2}(x^2)^3(-2y)^2$$

$$+ \binom{5}{3}(x^2)^2(-2y)^3 + \binom{5}{4}x^2(-2y)^4 + \binom{5}{5}(-2y)^5$$

$$= \frac{5!}{0!\,5!}x^{10} + \frac{5!}{1!\,4!}x^8(-2y) + \frac{5!}{2!\,3!}x^6(4y^2) + \frac{5!}{3!\,2!}x^4(-8y^3)$$

$$+ \frac{5!}{4!\,1!}x^2(16y^4) + \frac{5!}{5!\,0!}(-32y^5)$$

$$= 1 \cdot x^{10} + 5x^8(-2y) + 10x^6(4y^2) + 10x^4(-8y^3)$$

$$+ 5x^2(16y^4) + 1 \cdot (-32y^5)$$

$$= x^{10} - 10x^8 y + 40x^6 y^2 - 80x^4 y^3 + 80x^2 y^4 - 32y^5.$$

Now Try Exercise 11. ■

EXAMPLE 4 Expand: $\left(\dfrac{2}{x} + 3\sqrt{x}\right)^4$.

Solution We have $(a + b)^n$, where $a = 2/x$, $b = 3\sqrt{x}$, and $n = 4$. Then using the binomial theorem, we have

$$\left(\dfrac{2}{x} + 3\sqrt{x}\right)^4 = \binom{4}{0}\left(\dfrac{2}{x}\right)^4 + \binom{4}{1}\left(\dfrac{2}{x}\right)^3(3\sqrt{x}) + \binom{4}{2}\left(\dfrac{2}{x}\right)^2(3\sqrt{x})^2$$

$$+ \binom{4}{3}\left(\dfrac{2}{x}\right)(3\sqrt{x})^3 + \binom{4}{4}(3\sqrt{x})^4$$

$$= \dfrac{4!}{0!\,4!}\left(\dfrac{16}{x^4}\right) + \dfrac{4!}{1!\,3!}\left(\dfrac{8}{x^3}\right)(3x^{1/2})$$

$$+ \dfrac{4!}{2!\,2!}\left(\dfrac{4}{x^2}\right)(9x) + \dfrac{4!}{3!\,1!}\left(\dfrac{2}{x}\right)(27x^{3/2})$$

$$+ \dfrac{4!}{4!\,0!}(81x^2)$$

$$= \dfrac{16}{x^4} + \dfrac{96}{x^{5/2}} + \dfrac{216}{x} + 216x^{1/2} + 81x^2.$$

Now Try Exercise 13. ◾

❖ Finding a Specific Term

Suppose that we want to determine only a particular term of an expansion. The method we have developed will allow us to find such a term without computing all the rows of Pascal's triangle or all the preceding coefficients.

Note that in the binomial theorem, $\binom{n}{0}a^n b^0$ gives us the 1st term, $\binom{n}{1}a^{n-1}b^1$ gives us the 2nd term, $\binom{n}{2}a^{n-2}b^2$ gives us the 3rd term, and so on. This can be generalized as follows.

> **Finding the $(k + 1)$st Term**
>
> The $(k + 1)$st term of $(a + b)^n$ is $\binom{n}{k}a^{n-k}b^k$.

EXAMPLE 5 Find the 5th term in the expansion of $(2x - 5y)^6$.

Solution First, we note that $5 = 4 + 1$. Thus, $k = 4$, $a = 2x$, $b = -5y$, and $n = 6$. Then the 5th term of the expansion is

$$\binom{6}{4}(2x)^{6-4}(-5y)^4, \quad \text{or} \quad \dfrac{6!}{4!\,2!}(2x)^2(-5y)^4, \quad \text{or} \quad 37{,}500x^2y^4.$$

Now Try Exercise 21. ◾

EXAMPLE 6 Find the 8th term in the expansion of $(3x - 2)^{10}$.

Solution First, we note that $8 = 7 + 1$. Thus, $k = 7$, $a = 3x$, $b = -2$, and $n = 10$. Then the 8th term of the expansion is

$$\binom{10}{7}(3x)^{10-7}(-2)^7, \quad \text{or} \quad \frac{10!}{7!\,3!}(3x)^3(-2)^7, \quad \text{or} \quad -414,720x^3.$$

Now Try Exercise 27. ◼

❈ Total Number of Subsets

Suppose that a set has n objects. The number of subsets containing k elements is $\binom{n}{k}$ by a result of Section 11.6. The total number of subsets of a set is the number of subsets with 0 elements, plus the number of subsets with 1 element, plus the number of subsets with 2 elements, and so on. The total number of subsets of a set with n elements is

$$\binom{n}{0} + \binom{n}{1} + \binom{n}{2} + \cdots + \binom{n}{n}.$$

Now consider the expansion of $(1 + 1)^n$:

$$(1 + 1)^n = \binom{n}{0} \cdot 1^n + \binom{n}{1} \cdot 1^{n-1} \cdot 1^1 + \binom{n}{2} \cdot 1^{n-2} \cdot 1^2$$

$$+ \cdots + \binom{n}{n} \cdot 1^n$$

$$= \binom{n}{0} + \binom{n}{1} + \binom{n}{2} + \cdots + \binom{n}{n}.$$

Thus the total number of subsets is $(1 + 1)^n$, or 2^n. We have proved the following.

Total Number of Subsets

The total number of subsets of a set with n elements is 2^n.

EXAMPLE 7 The set $\{A, B, C, D, E\}$ has how many subsets?

Solution The set has 5 elements, so the number of subsets is 2^5, or 32.

Now Try Exercise 31. ◼

EXAMPLE 8 Wendy's, a national restaurant chain, offers the following toppings for its hamburgers:

{catsup, mustard, mayonnaise, tomato, lettuce, onions, pickle}.

How many different kinds of hamburgers can Wendy's serve, excluding size of hamburger or number of patties?

Solution The toppings on each hamburger are the elements of a subset of the set of all possible toppings, the empty set being a plain hamburger. The total number of possible hamburgers is

$$\binom{7}{0} + \binom{7}{1} + \binom{7}{2} + \cdots + \binom{7}{7} = 2^7 = 128.$$

Thus Wendy's serves hamburgers in 128 different ways.

Now Try Exercise 33.

(11.7) Exercise Set

Expand.

1. $(x + 5)^4$

2. $(x - 1)^4$

3. $(x - 3)^5$

4. $(x + 2)^9$

5. $(x - y)^5$

6. $(x + y)^8$

7. $(5x + 4y)^6$

8. $(2x - 3y)^5$

9. $\left(2t + \dfrac{1}{t}\right)^7$

10. $\left(3y - \dfrac{1}{y}\right)^4$

11. $(x^2 - 1)^5$

12. $(1 + 2q^3)^8$

13. $\left(\sqrt{5} + t\right)^6$

14. $\left(x - \sqrt{2}\right)^6$

15. $\left(a - \dfrac{2}{a}\right)^9$

16. $(1 + 3)^n$

17. $\left(\sqrt{2} + 1\right)^6 - \left(\sqrt{2} - 1\right)^6$

18. $\left(1 - \sqrt{2}\right)^4 + \left(1 + \sqrt{2}\right)^4$

19. $(x^{-2} + x^2)^4$

20. $\left(\dfrac{1}{\sqrt{x}} - \sqrt{x}\right)^6$

Find the indicated term of the binomial expansion.

21. 3rd; $(a + b)^7$

22. 6th; $(x + y)^8$

23. 6th; $(x - y)^{10}$

24. 5th; $(p - 2q)^9$

25. 12th; $(a - 2)^{14}$

26. 11th; $(x - 3)^{12}$

27. 5th; $\left(2x^3 - \sqrt{y}\right)^8$

28. 4th; $\left(\dfrac{1}{b^2} + \dfrac{b}{3}\right)^7$

29. Middle; $(2u - 3v^2)^{10}$

30. Middle two; $\left(\sqrt{x} + \sqrt{3}\right)^5$

Determine the number of subsets of each of the following.

31. A set of 7 elements

32. A set of 6 members

33. The set of letters of the Greek alphabet, which contains 24 letters

34. The set of letters of the English alphabet, which contains 26 letters

35. What is the degree of $(x^5 + 3)^4$?

36. What is the degree of $(2 - 5x^3)^7$?

Expand each of the following, where $i^2 = -1$.

37. $(3 + i)^5$

38. $(1 + i)^6$

39. $\left(\sqrt{2} - i\right)^4$

40. $\left(\dfrac{\sqrt{3}}{2} - \dfrac{1}{2}i\right)^{11}$

41. Find a formula for $(a - b)^n$. Use sigma notation.

42. Expand and simplify:
$$\frac{(x + h)^{13} - x^{13}}{h}.$$

43. Expand and simplify:
$$\frac{(x + h)^n - x^n}{h}.$$
Use sigma notation.

Collaborative Discussion and Writing

44. Discuss the advantages and disadvantages of each method of finding a binomial expansion. Give examples of when you might use one method rather than the other.

45. Blaise Pascal (1623–1662) was a French scientist and philosopher who founded the modern theory of probability. Do some research on Pascal and see if you can find out how he discovered his famous "triangle of numbers."

Skill Maintenance

Given that $f(x) = x^2 + 1$ and $g(x) = 2x - 3$, find each of the following.

46. $(f + g)(x)$

47. $(fg)(x)$

48. $(f \circ g)(x)$

49. $(g \circ f)(x)$

Synthesis

Solve for x.

50. $\displaystyle\sum_{k=0}^{8} \binom{8}{k} x^{8-k} 3^k = 0$

51. $\displaystyle\sum_{k=0}^{4} \binom{4}{k} (-1)^k x^{4-k} 6^k = 81$

52. Find the term of
$$\left(\frac{3x^2}{2} - \frac{1}{3x}\right)^{12}$$
that does not contain x.

53. Find the middle term of $(x^2 - 6y^{3/2})^6$.

54. Find the ratio of the 4th term of
$$\left(p^2 - \frac{1}{2} p \sqrt[3]{q}\right)^5$$
to the 3rd term.

55. Find the term of
$$\left(\sqrt[3]{x} - \frac{1}{\sqrt{x}}\right)^7$$
containing $1/x^{1/6}$.

56. *Money Combinations.* A money clip contains one each of the following bills: \$1, \$2, \$5, \$10, \$20, \$50, and \$100. How many different sums of money can be formed using the bills?

Find the sum.

57. $_{100}C_0 + {}_{100}C_1 + \cdots + {}_{100}C_{100}$

58. $_nC_0 + {}_nC_1 + \cdots + {}_nC_n$

Simplify.

59. $\displaystyle\sum_{k=0}^{23} \binom{23}{k} (\log_a x)^{23-k} (\log_a t)^k$

60. $\displaystyle\sum_{k=0}^{15} \binom{15}{k} i^{30-2k}$

61. Use mathematical induction and the property
$$\binom{n}{r-1} + \binom{n}{r} = \binom{n+1}{r}$$
to prove the binomial theorem.

✤   Compute the probability of a simple event.

11.8   Probability

When a coin is tossed, we can reason that the chance, or likelihood, that it will fall heads is 1 out of 2, or the **probability** that it will fall heads is $\frac{1}{2}$. Of course, this does not mean that if a coin is tossed 10 times it will necessarily fall heads 5 times. If the coin is a "fair coin" and it is tossed a great many times, however, it will fall heads very nearly half of the time. Here we give an introduction to two kinds of probability, **experimental** and **theoretical**.

✤ Experimental and Theoretical Probability

If we toss a coin a great number of times—say, 1000—and count the number of times it falls heads, we can determine the probability that it will fall heads. If it falls heads 503 times, we would calculate the probability of its falling heads to be

$$\frac{503}{1000}, \quad \text{or} \quad 0.503.$$

This is an **experimental** determination of probability. Such a determination of probability is discovered by the observation and study of data and is quite common and very useful. Here, for example, are some probabilities that have been determined *experimentally*:

1. The probability that a woman will get breast cancer in her lifetime is $\frac{1}{11}$.
2. If you kiss someone who has a cold, the probability of your catching a cold is 0.07.
3. A person who has just been released from prison has an 80% probability of returning.

If we consider a coin and reason that it is just as likely to fall heads as tails, we would calculate the probability that it will fall heads to be $\frac{1}{2}$. This is a **theoretical** determination of probability. Here are some other probabilities that have been determined *theoretically*, using mathematics:

1. If there are 30 people in a room, the probability that two of them have the same birthday (excluding year) is 0.706.
2. While on a trip, you meet someone and, after a period of conversation, discover that you have a common acquaintance. The typical reaction, "It's a small world!", is actually not appropriate, because the probability of such an occurrence is quite high—just over 22%.

In summary, experimental probabilities are determined by making observations and gathering data. Theoretical probabilities are determined by reasoning mathematically. Examples of experimental and theoretical probability like those above, especially those we do not expect, lead us to see the value of a study of probability. You might ask, "What is the *true* probability?" In fact, there is none. Experimentally, we can determine probabilities within certain limits. These may or may not agree with the probabilities that we obtain theoretically. There are situations in which it is much easier to determine one of these types of probabilities than the other. For example, it would be quite difficult to arrive at the probability of catching a cold using theoretical probability.

❋ Computing Experimental Probabilities

We first consider experimental determination of probability. The basic principle we use in computing such probabilities is as follows.

> ### Principle *P* (Experimental)
> Given an experiment in which *n* observations are made, if a situation, or event, *E* occurs *m* times out of *n* observations, then we say that the *experimental probability* of the event, $P(E)$, is given by
>
> $$P(E) = \frac{m}{n}.$$

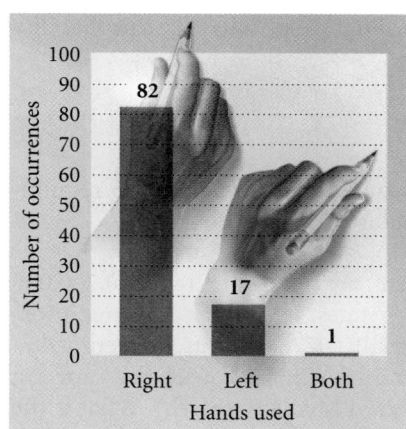

EXAMPLE 1 *Sociological Survey.* The authors of this text conducted an experimental survey to determine the number of people who are left-handed, right-handed, or both. The results are shown in the graph at left.

a) Determine the probability that a person is right-handed.

b) Determine the probability that a person is left-handed.

c) Determine the probability that a person is ambidextrous (uses both hands with equal ability).

d) There are 120 bowlers in most tournaments held by the Professional Bowlers Association. On the basis of the data in this experiment, how many of the bowlers would you expect to be left-handed?

Solution

a) The number of people who are right-handed is 82, the number who are left-handed is 17, and the number who are ambidextrous is 1. The total number of observations is 82 + 17 + 1, or 100. Thus the probability that a person is right-handed is *P*, where

$$P = \frac{82}{100}, \quad \text{or} \quad 0.82, \quad \text{or} \quad 82\%.$$

b) The probability that a person is left-handed is *P*, where

$$P = \frac{17}{100}, \quad \text{or} \quad 0.17, \quad \text{or} \quad 17\%.$$

c) The probability that a person is ambidextrous is *P*, where

$$P = \frac{1}{100}, \quad \text{or} \quad 0.01, \quad \text{or} \quad 1\%.$$

d) There are 120 bowlers, and from part (b) we can expect 17% to be left-handed. Since

$$17\% \text{ of } 120 = 0.17 \cdot 120 = 20.4,$$

we can expect that about 20 of the bowlers will be left-handed.

Now Try Exercise 1. ▢

EXAMPLE 2 *Quality Control.*   It is very important for a manufacturer to maintain the quality of its products. In fact, companies hire quality control inspectors to ensure this process. The goal is to produce as few defective products as possible. But since a company is producing thousands of products every day, it cannot afford to check every product to see if it is defective. To find out what percentage of its products are defective, the company checks a smaller sample.

The U.S. Department of Agriculture requires that 80% of the seeds that a company produces sprout. To determine the quality of the seeds it produces, a company takes 500 seeds from those it has produced and plants them. It finds that 417 of the seeds sprout.

a) What is the probability that a seed will sprout?

b) Did the seeds meet government standards?

Solution

a) We know that 500 seeds were planted and 417 sprouted. The probability of a seed sprouting is P, where

$$P = \frac{417}{500} = 0.834, \quad \text{or} \quad 83.4\%.$$

b) Since the percentage of seeds that sprouted exceeded the 80% requirement, the seeds meet government standards.   ▢

EXAMPLE 3 *Television Ratings.*   There are an estimated 110,200,000 households with televisions in the United States. Each week, viewing information is collected and reported. One week, 12,077,000 households tuned in to the news show "60 Minutes" on CBS and 10,672,000 households tuned in to the drama "Lost" on ABC (*Source*: Nielsen Media Research). What is the probability that a television household tuned in to "60 Minutes" during the given week? to "Lost"?

Solution   The probability that a television household was tuned in to "60 Minutes" is P, where

$$P = \frac{12,077,000}{110,200,000} \approx 0.110 \approx 11.0\%.$$

The probability that a television household was tuned in to "Lost" is P, where

$$P = \frac{10,672,000}{110,200,000} \approx 0.097 \approx 9.7\%.$$ ▢

❋ Theoretical Probability

Suppose that we perform an experiment such as flipping a coin, throwing a dart, drawing a card from a deck, or checking an item off an assembly line for quality. Each possible result of such an experiment is called an **outcome**. The set of all possible outcomes is called the **sample space**. An **event** is a set of outcomes, that is, a subset of the sample space.

EXAMPLE 4 *Dart Throwing.*   Consider this dartboard. Assume that the experiment is "throwing a dart" and that the dart hits the board. Find each of the following.

a) The outcomes

b) The sample space

Solution

a) The outcomes are *hitting black* (B), *hitting red* (R), and *hitting white* (W).

b) The sample space is {*hitting black*, *hitting red*, *hitting white*}, which can be simply stated as {B, R, W}.   ◼

EXAMPLE 5 *Die Rolling.*   A die (pl., dice) is a cube, with six faces, each containing a number of dots from 1 to 6 on each side.

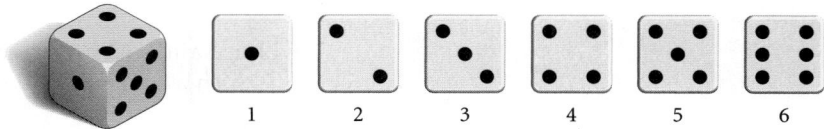

Suppose a die is rolled. Find each of the following.

a) The outcomes   **b)** The sample space

Solution

a) The outcomes are 1, 2, 3, 4, 5, 6.

b) The sample space is {1, 2, 3, 4, 5, 6}.   ◼

We denote the probability that an event E occurs as $P(E)$. For example, "a coin falling heads" may be denoted H. Then $P(H)$ represents the probability of the coin falling heads. When all the outcomes of an experiment have the same probability of occurring, we say that they are *equally likely*. To see the distinction between events that are equally likely and those that are not, consider the dartboards shown below.

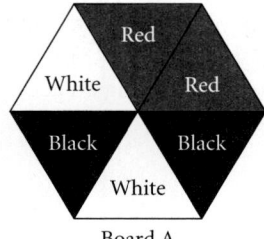

Board A

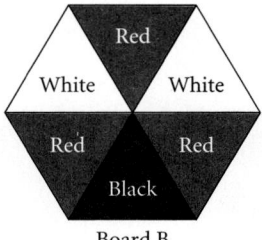

Board B

For board A, the events *hitting black*, *hitting red*, and *hitting white* are equally likely, because the black, red, and white areas are the same. However, for board B the areas are not the same so these events are not equally likely.

> ### Principle *P* (Theoretical)
> If an event *E* can occur *m* ways out of *n* possible equally likely outcomes of a sample space *S*, then the **theoretical probability** of the event, $P(E)$, is given by
>
> $$P(E) = \frac{m}{n}.$$

EXAMPLE 6   What is the probability of rolling a 3 on a die?

Solution   On a fair die, there are 6 equally likely outcomes and there is 1 way to roll a 3. By Principle *P*, $P(3) = \frac{1}{6}$.   **Now Try Exercise 7(a).** ■

EXAMPLE 7   What is the probability of rolling an even number on a die?

Solution   The event is rolling an *even* number. It can occur 3 ways (rolling 2, 4, or 6). The number of equally likely outcomes is 6. By Principle *P*, $P(\text{even}) = \frac{3}{6}$, or $\frac{1}{2}$.   **Now Try Exercise 7(d).** ■

We will use a number of examples related to a standard bridge deck of 52 cards. Such a deck is made up as shown in the following figure.

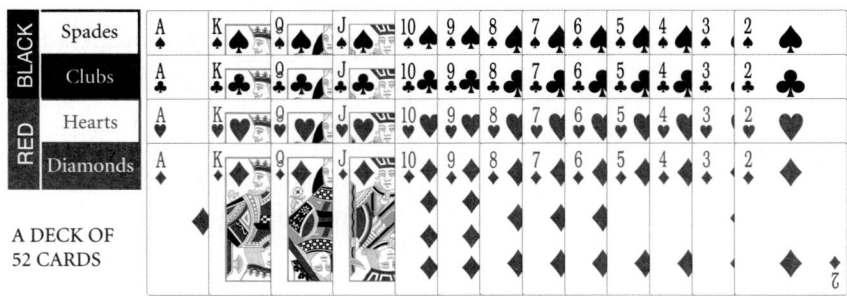

A DECK OF 52 CARDS

EXAMPLE 8   What is the probability of drawing an ace from a well-shuffled deck of cards?

Solution   There are 52 outcomes (the number of cards in the deck), they are equally likely (from a well-shuffled deck), and there are 4 ways to obtain an ace, so by Principle *P*, we have

$$P(\text{drawing an ace}) = \frac{4}{52}, \quad \text{or} \quad \frac{1}{13}.$$

■

EXAMPLE 9   Suppose that we select, without looking, one marble from a bag containing 3 red marbles and 4 green marbles. What is the probability of selecting a red marble?

Solution   There are 7 equally likely ways of selecting any marble, and since the number of ways of getting a red marble is 3, we have

$$P(\text{selecting a red marble}) = \frac{3}{7}.$$

Now Try Exercise 7(b). ▣

The following are some results that follow from Principle P.

> **Probability Properties**
>
> **a)** If an event E cannot occur, then $P(E) = 0$.
> **b)** If an event E is certain to occur, then $P(E) = 1$.
> **c)** The probability that an event E will occur is a number from 0 to 1: $0 \le P(E) \le 1$.

For example, in coin tossing, the event that a coin will land on its edge has probability 0. The event that a coin falls either heads or tails has probability 1.

In the following examples, we use the combinatorics that we studied in Sections 11.5 and 11.6 to calculate theoretical probabilities.

EXAMPLE 10   Suppose that 2 cards are drawn from a well-shuffled deck of 52 cards. What is the probability that both of them are spades?

Solution   The number of ways n of drawing 2 cards from a well-shuffled deck of 52 cards is $_{52}C_2$. Since 13 of the 52 cards are spades, the number of ways m of drawing 2 spades is $_{13}C_2$. Thus,

$$P(\text{drawing 2 spades}) = \frac{m}{n} = \frac{_{13}C_2}{_{52}C_2} = \frac{78}{1326} = \frac{1}{17}.$$

```
13 nCr 2/52 nCr 2►Frac
                  1/17
```

Now Try Exercise 13. ▣

EXAMPLE 11   Suppose that 3 people are selected at random from a group that consists of 6 men and 4 women. What is the probability that 1 man and 2 women are selected?

Solution   The number of ways of selecting 3 people from a group of 10 is $_{10}C_3$. One man can be selected in $_6C_1$ ways, and 2 women can be selected in $_4C_2$ ways. By the fundamental counting principle, the number of ways of selecting 1 man and 2 women is $_6C_1 \cdot {_4C_2}$. Thus the probability that 1 man and 2 women are selected is

$$P = \frac{_6C_1 \cdot {_4C_2}}{_{10}C_3} = \frac{3}{10}.$$

```
6 nCr 1*4 nCr 2/10 nCr
3►Frac
                  3/10
```

Now Try Exercise 9. ▣

EXAMPLE 12   *Rolling Two Dice.*   What is the probability of getting a total of 8 on a roll of a pair of dice?

Solution   On each die, there are 6 possible outcomes. The outcomes are paired so there are 6 · 6, or 36, possible ways in which the two can fall. (Assuming that the dice are different—say, one red and one blue—can help in visualizing this.)

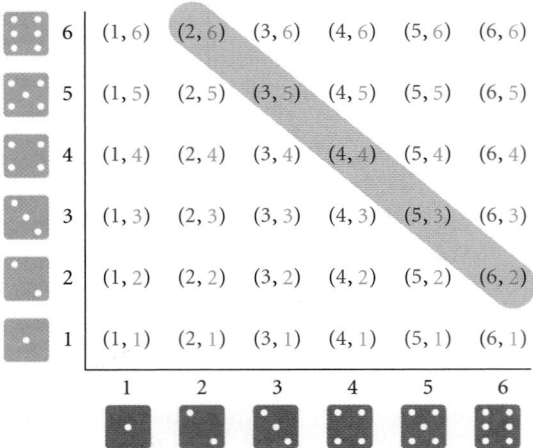

The pairs that total 8 are as shown in the figure above. There are 5 possible ways of getting a total of 8, so the probability is $\frac{5}{36}$.

Now Try Exercise 15. ▇

Exercise Set

1. *Select a Number.*   In a survey conducted by the authors, 100 people were polled and asked to select a number from 1 to 5. The results are shown in the following table.

Number Chosen	1	2	3	4	5
Number Who Chose That Number	18	24	23	23	12

a) What is the probability that the number chosen is 1? 2? 3? 4? 5?
b) What general conclusion might be made from the results of the experiment?

2. *Mason Dots®.*   Made by the Tootsie Industries of Chicago, Illinois, Mason Dots® is a gumdrop candy. A box was opened by the authors and was found to contain the following number of gumdrops:

Orange	9
Lemon	8
Strawberry	7
Grape	6
Lime	5
Cherry	4

If we take one gumdrop out of the box, what is the probability of getting lemon? lime? orange? grape? strawberry? licorice?

3. *Junk Mail.* In experimental studies, the U.S. Postal Service has found that the probability that a piece of advertising is opened and read is 78%. A business sends out 15,000 pieces of advertising. How many of these can the company expect to be opened and read?

4. *Linguistics.* An experiment was conducted by the authors to determine the relative occurrence of various letters of the English alphabet. The front page of a newspaper was considered. In all, there were 9136 letters. The number of occurrences of each letter of the alphabet is listed in the following table.

Letter	Number of Occurrences	Probability
A	853	$853/9136 \approx 9.3\%$
B	136	
C	273	
D	286	
E	1229	
F	173	
G	190	
H	399	
I	539	
J	21	
K	57	
L	417	
M	231	
N	597	
O	705	
P	238	
Q	4	
R	609	
S	745	
T	789	
U	240	
V	113	
W	127	
X	20	
Y	124	
Z	21	$21/9136 \approx 0.2\%$

a) Complete the table of probabilities with the percentage, to the nearest tenth of a percent, of the occurrence of each letter.

b) What is the probability of a vowel occurring?
c) What is the probability of a consonant occurring?

5. *"Wheel of Fortune®."* The results of the experiment in Exercise 4 can be quite useful to a person playing the popular television game show "Wheel of Fortune." Players guess letters to spell out a phrase, a person, or a thing.

a) What 5 consonants have the greatest probability of occurring?
b) What vowel has the greatest probability of occurring?
c) The daily winner of each show plays for a grand prize and at one time was allowed to guess 5 consonants and a vowel in order to discover the secret wording. The 5 consonants R, S, T, L, N, and the vowel E seemed to be chosen most often. Do the results in parts (a) and (b) support such a choice?

6. *Card Drawing.* Suppose we draw a card from a well-shuffled deck of 52 cards.

a) How many equally likely outcomes are there?
What is the probability of drawing each of the following?

b) A queen   **c)** A heart
d) A 7   **e)** A red card
f) A 9 or a king   **g)** A black ace

7. *Marbles.* Suppose we select, without looking, one marble from a bag containing 4 red marbles and 10 green marbles. What is the probability of selecting each of the following?

a) A red marble
b) A green marble
c) A purple marble
d) A red marble or a green marble

8. *Production Unit.* The sales force of a business consists of 10 men and 10 women. A production unit of 4 people is set up at random. What is the probability that 2 men and 2 women are chosen?

9. *Coin Drawing.* A sack contains 7 dimes, 5 nickels, and 10 quarters. Eight coins are drawn at random. What is the probability of getting 4 dimes, 3 nickels, and 1 quarter?

10. *Michigan Lottery.*   Run by the state of Michigan, Classic Lotto 47 is a twice-weekly lottery game with jackpots starting at $1 million. For a wager of $1, a player can choose 6 numbers from 1 through 47. If the numbers match those drawn by the state, the player wins. (*Source*: www.michigan.gov/lottery) Ava buys 1 game ticket. What is her probability of winning?

Five-Card Poker Hands.   *Suppose that 5 cards are drawn from a deck of 52 cards. What is the probability of drawing each of the following?*

11. 3 sevens and 2 kings

12. 5 aces

13. 5 spades

14. 4 aces and 1 five

15. *Tossing Three Coins.*   Three coins are flipped. An outcome might be HTH.

  a) Find the sample space.

  What is the probability of getting each of the following?

  b) Exactly one head
  c) At most two tails
  d) At least one head
  e) Exactly two tails

Roulette.   *An American roulette wheel contains 38 slots numbered 00, 0, 1, 2, 3, . . . , 35, 36. Eighteen of the slots numbered 1–36 are colored red and 18 are colored black. The 00 and 0 slots are considered to be uncolored. The wheel is spun, and a ball is rolled around the rim until it falls into a slot. What is the probability that the ball falls in each of the following?*

16. A red slot

17. A black slot

18. The 00 slot

19. The 0 slot

20. Either the 00 or the 0 slot

21. A red slot or a black slot

22. The number 24

23. An odd-numbered slot

24. *Dartboard.*   The figure below shows a dartboard. A dart is thrown and hits the board. Find the probabilities

$$P(\text{red}), \quad P(\text{green}), \quad P(\text{blue}), \quad P(\text{yellow}).$$

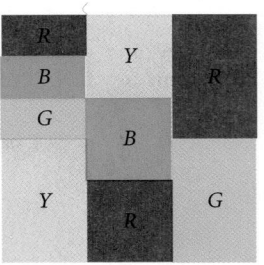

25. *Random-Number Generator.*   Many graphing calculators have a **random-number generator**. This feature produces a random number in the interval $[0, 1]$. (Consult your user's manual.) We can use such a feature to simulate coin flipping. A number r such that $0 \le r \le 0.5$ would indicate heads, H. A number r such that $0.5 < r \le 1.0$ would indicate tails, T. Use a random-number generator 100 times.

  a) What is the experimental probability of getting heads?
  b) What is the experimental probability of getting tails?

Collaborative Discussion and Writing

26. *Random Best-Selling Novels.*   Sir Arthur Stanley Eddington, an astronomer, once wrote in a satirical essay that if a monkey were left alone long enough with a typewriter and typed randomly, any great novel could be replicated. What is the probability that the following passage could have been written by a monkey? Ignore capital letters and punctuation and consider only letters and spaces.

  "*It was the best of times, it was the worst of times,* . . ." (Charles Dickens, 1859). Explain your answer.

27. Find at least one use of probability in today's newspaper. Make a report.

Skill Maintenance

In each of Exercises 28–35, fill in the blank with the correct term. Some of the given choices will be used more than once. Others will not be used.

    range
    domain
    function
    an inverse function
    a composite function
    direct variation
    inverse variation
    factor
    solution
    zero
    y-intercept
    one-to-one
    rational
    permutation
    combination
    arithmetic sequence
    geometric sequence

28. A(n) _____ of a function is an input for which the output is 0.

29. A function is _____ if different inputs have different outputs.

30. A(n) _____ is a correspondence between a first set, called the _____, and a second set, called the _____, such that each member of the _____ corresponds to exactly one member of the _____.

31. The first coordinate of an *x*-intercept of a function is a(n) _____ of the function.

32. A selection made from a set without regard to order is a(n) _____ .

33. If we have a function $f(x) = k/x$, where k is a positive constant, we have _____ .

34. For a polynomial function $f(x)$, if $f(c) = 0$, then $x - c$ is a(n) _____ of the polynomial.

35. We have $\dfrac{a_{n+1}}{a_n} = r$, for any integer $n \geq 1$, in a(n) _____ .

Synthesis

Five-Card Poker Hands. *Suppose that 5 cards are drawn from a deck of 52 cards. For the following exercises, give both a reasoned expression and an answer.*

36. *Royal Flush.* A *royal flush* consists of a 5-card hand with A-K-Q-J-10 of the same suit.

  a) How many royal flushes are there?
  b) What is the probability of getting a royal flush?

37. *Straight Flush.* A *straight flush* consists of 5 cards in sequence in the same suit, but excludes royal flushes. An ace can be used low, before a two, or high, following a king.

  a) How many straight flushes are there?
  b) What is the probability of getting a straight flush?

38. *Four of a Kind.* A *four-of-a-kind* is a 5-card hand in which 4 of the cards are of the same denomination, such as J-J-J-J-6, 7-7-7-7-A, or 2-2-2-2-5.

  a) How many four-of-a-kind hands are there?
  b) What is the probability of getting four of a kind?

39. *Full House.* A *full house* consists of 3 of a kind and a pair such as Q-Q-Q-4-4.

  a) How many full houses are there?
  b) What is the probability of getting a full house?

40. *Three of a Kind.* A *three-of-a-kind* is a 5-card hand in which exactly 3 of the cards are of the same denomination and the other 2 are not a pair, such as Q-Q-Q-10-7.

  a) How many three-of-a-kind hands are there?
  b) What is the probability of getting three of a kind?

41. *Flush.* An ordinary *flush* is a 5-card hand in which all the cards are of the same suit, but not all in sequence (not a straight or royal flush).

a) How many flushes are there?
b) What is the probability of getting a flush?

42. *Two Pairs.* A hand with *two pairs* is a hand like Q-Q-3-3-A.

a) How many are there?
b) What is the probability of getting two pairs?

43. *Straight.* An ordinary *straight* is any 5 cards in sequence, but not of the same suit—for example, 4 of spades, 5 of hearts, 6 of diamonds, 7 of hearts, and 8 of clubs.

a) How many straights are there?
b) What is the probability of getting a straight?

CHAPTER 11 Summary and Review

Important Properties and Formulas

Arithmetic Sequences and Series

General term: $a_{n+1} = a_n + d$

$a_n = a_1 + (n-1)d$

Common difference: d

Sum of the first n terms: $S_n = \dfrac{n}{2}(a_1 + a_n)$

Geometric Sequences and Series

General term: $a_{n+1} = a_n r$

$a_n = a_1 r^{n-1}$

Common ratio: r

Sum of the first n terms: $S_n = \dfrac{a_1(1 - r^n)}{1 - r}$

Sum of an infinite geometric series:

$S_\infty = \dfrac{a_1}{1 - r}, \quad |r| < 1$

The Principle of Mathematical Induction

(1) *Basis step*: Prove S_1 is true.
(2) *Induction step*: Prove for all numbers k,
$S_k \rightarrow S_{k+1}$.

The Fundamental Counting Principle

The total number of ways in which k actions can be performed together is $n_1 \cdot n_2 \cdot n_3 \cdots n_k$, where the first action can be performed in n_1 ways, the second in n_2 ways, and so on.

Factorial Notation

For any natural number n,

$n! = n(n-1)(n-2) \cdots 3 \cdot 2 \cdot 1$

and $0! = 1$.

Permutations of n Objects Taken n at a Time

$$_nP_n = n! = n(n-1)(n-2) \cdots 3 \cdot 2 \cdot 1$$

Permutations of n Objects Taken k at a Time

$$_nP_k = \underbrace{n(n-1)(n-2) \cdots [n-(k-1)]}_{k \text{ factors}}$$

$$= \dfrac{n!}{(n-k)!}$$

Permutations of Sets with Some Nondistinguishable Objects

$$\dfrac{n!}{n_1! \cdot n_2! \cdots \cdots n_k!},$$

where n_1 objects are of one kind, n_2 are of another kind, and so on.

Combinations of n Objects Taken k at a Time

$$_nC_k = \binom{n}{k} = \dfrac{_nP_k}{k!} = \dfrac{n!}{k!(n-k)!}$$

$$= \dfrac{n(n-1)(n-2) \cdots [n-(k-1)]}{k!}$$

The Binomial Theorem

$$(a+b)^n = \sum_{k=0}^{n} \binom{n}{k} a^{n-k} b^k$$

(continued)

The $(k + 1)$st Term of Binomial Expansion

The $(k + 1)$st term of $(a + b)^n$ is $\binom{n}{k}a^{n-k}b^k$.

Total Number of Subsets

The total number of subsets of a set with n elements is 2^n.

Probability Principle P (Experimental)

$P(E) = \dfrac{m}{n}$, where an event E occurs m times out of n observations.

Probability Principle P (Theoretical)

$P(E) = \dfrac{m}{n}$, where an event E can occur m ways out of n possible equally likely outcomes.

Review Exercises

Determine whether the statement is true or false.

1. A sequence is a function.   [11.1]

2. An infinite geometric series with $r = -1$ has a limit.   [11.3]

3. Permutations involve order and arrangements of objects.   [11.5]

4. The total number of subsets of a set with n elements is n^2.   [11.7]

5. Find the first 4 terms, a_{11}, and a_{23}:
$$a_n = (-1)^n \left(\frac{n^2}{n^4 + 1} \right).\quad [11.1]$$

6. Predict the general, or nth, term. Answers may vary.
$$2, -5, 10, -17, 26, \ldots \quad [11.1]$$

7. Find and evaluate:
$$\sum_{k=1}^{4} \frac{(-1)^{k+1}3^k}{3^k - 1}.\quad [11.1]$$

8. Use a graphing calculator to construct a table of values and a graph for the first 10 terms of this sequence.
$$a_1 = 0.3, \quad a_{k+1} = 5a_k + 1 \quad [11.1]$$

9. Write sigma notation. Answers may vary.
$$0 + 3 + 8 + 15 + 24 + 35 + 48 \quad [11.1]$$

10. Find the 10th term of the arithmetic sequence
$$\tfrac{3}{4}, \tfrac{13}{12}, \tfrac{17}{12}, \ldots . \quad [11.2]$$

11. Find the 6th term of the arithmetic sequence
$$a - b, a, a + b, \ldots . \quad [11.2]$$

12. Find the sum of the first 18 terms of the arithmetic sequence
$$4, 7, 10, \ldots . \quad [11.2]$$

13. Find the sum of the first 200 natural numbers. [11.2]

14. The 1st term in an arithmetic sequence is 5, and the 17th term is 53. Find the 3rd term.   [11.2]

15. The common difference in an arithmetic sequence is 3. The 10th term is 23. Find the first term.   [11.2]

16. For a geometric sequence, $a_1 = -2$, $r = 2$, and $a_n = -64$. Find n and S_n. [11.3]

17. For a geometric sequence, $r = \frac{1}{2}$ and $S_5 = \frac{31}{2}$. Find a_1 and a_5.   [11.3]

Find the sum of each infinite geometric series, if it exists. [11.3]

18. $25 + 27.5 + 30.25 + 33.275 + \cdots$

19. $0.27 + 0.0027 + 0.000027 + \cdots$

20. $\frac{1}{2} - \frac{1}{6} + \frac{1}{18} - \cdots$

21. Find fraction notation for $2.\overline{43}$.   [11.3]

22. Insert four arithmetic means between 5 and 9. [11.2]

23. *Bouncing Golfball.* A golfball is dropped from a height of 30 ft to the pavement. It always rebounds three fourths of the distance that it drops. How far (up and down) will the ball have traveled when it hits the pavement for the 6th time? [11.3]

24. *The Amount of an Annuity.* To create a college fund, a parent makes a sequence of 18 yearly deposits of $2000 each in a savings account on which interest is compounded annually at 2.8%. Find the amount of the annuity. [11.3]

25. *Total Gift.* Suppose you receive 10¢ on the first day of the year, 12¢ on the 2nd day, 14¢ on the 3rd day, and so on.

   a) How much will you receive on the 365th day? [11.2]

   b) What is the sum of these 365 gifts? [11.2]

26. *The Economic Multiplier.* Suppose the government is making a $24,000,000,000 expenditure for travel to Mars. If 73% of this amount is spent again, and so on, what is the total effect on the economy? [11.3]

Use mathematical induction to prove each of the following. [11.4]

27. For every natural number n,

$$1 + 4 + 7 + \cdots + (3n - 2) = \frac{n(3n - 1)}{2}.$$

28. For every natural number n,

$$1 + 3 + 3^2 + \cdots + 3^{n-1} = \frac{3^n - 1}{2}.$$

29. For every natural number $n \geq 2$,

$$\left(1 - \frac{1}{2}\right)\left(1 - \frac{1}{3}\right) \cdots \left(1 - \frac{1}{n}\right) = \frac{1}{n}.$$

30. *Book Arrangements.* In how many ways can 6 books be arranged on a shelf? [11.5]

31. *Flag Displays.* If 9 different signal flags are available, how many different displays are possible using 4 flags in a row? [11.5]

32. *Prize Choices.* The winner of a contest can choose any 8 of 15 prizes. How many different sets of prizes can be chosen? [11.6]

33. *Fraternity–Sorority Names.* The Greek alphabet contains 24 letters. How many fraternity or sorority names can be formed using 3 different letters? [11.5]

34. *Letter Arrangements.* In how many distinguishable ways can the letters of the word TENNESSEE be arranged? [11.5]

35. *Floor Plans.* A manufacturer of houses has 1 floor plan but achieves variety by having 3 different roofs, 4 different ways of attaching the garage, and 3 different types of entrances. Find the number of different houses that can be produced. [11.5]

36. *Code Symbols.* How many code symbols can be formed using 5 out of 6 of the letters of G, H, I, J, K, L if the letters:

   a) cannot be repeated? [11.5]

   b) can be repeated? [11.5]

   c) cannot be repeated but must begin with K? [11.5]

   d) cannot be repeated but must end with IGH? [11.5]

37. Determine the number of subsets of a set containing 8 members. [11.7]

Expand. [11.7]

38. $(m + n)^7$

39. $\left(x - \sqrt{2}\right)^5$

40. $(x^2 - 3y)^4$

41. $\left(a + \dfrac{1}{a}\right)^8$

42. $(1 + 5i)^6$, where $i^2 = -1$

43. Find the 4th term of $(a + x)^{12}$. [11.7]

44. Find the 12th term of $(2a - b)^{18}$. Do not multiply out the factorials. [11.7]

45. *Rolling Dice.* What is the probability of getting a 10 on a roll of a pair of dice? on a roll of 1 die? [11.8]

46. *Drawing a Card.* From a deck of 52 cards, 1 card is drawn at random. What is the probability that it is a club? [11.8]

47. *Drawing Three Cards.* From a deck of 52 cards, 3 are drawn at random without replacement.

What is the probability that 2 are aces and 1 is a king? [11.8]

48. *Election Poll.* Three people were running for mayor in an election campaign. A poll was conducted to see which candidate was favored. During the polling, 86 favored candidate A, 97 favored B, and 23 favored C. Assuming that the poll is a valid indicator of the election results, what is the probability that the election will be won by A? B? C? [11.8]

49. *Self-Employed Workers.* The table below lists the number of self-employed workers in the United States in various years.

Year	Self-Employed Workers (in millions)
2002	9.926
2003	10.295
2004	10.431
2005	10.464

Source: U.S. Bureau of Labor Statistics

a) Find a linear sequence function $a_n = an + b$ that models the data. Let n represent the number of years after 2002. [11.1]

b) Use the sequence found in part (a) to estimate the number of self-employed workers in 2010. [11.1]

50. Which of the following is the 25th term of the arithmetic sequence 12, 10, 8, 6, ... ? [11.2]

A. -38 B. -36

C. 32 D. 60

51. What is the probability of getting a total of 4 on a roll of a pair of dice? [11.8]

A. $\frac{1}{12}$ B. $\frac{1}{9}$

C. $\frac{1}{6}$ D. $\frac{5}{36}$

52. The graph of the sequence whose general term is $a_n = n - 1$ is which of the following? [11.1]

A.   B.

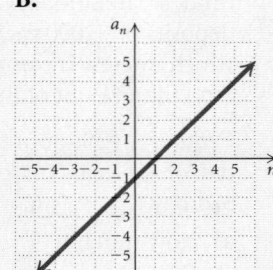

C.   D.

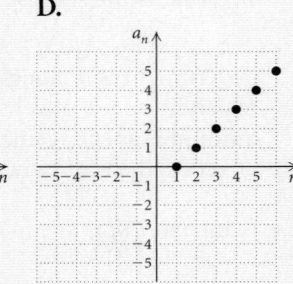

Collaborative Discussion and Writing

53. Write an exercise for a classmate to solve. Design it so that the solution is $_9C_4$. [11.6]

54. *Chain Business Deals.* Chain letters have been outlawed by the U.S. government. Nevertheless, "chain" business deals still exist and they can be fraudulent. Suppose that a salesperson is charged with the task of hiring 4 new salespersons. Each of them gives half of his or her profits to the person who hires them. Each of these people hires 4 new salespersons. Each of these gives half of his or her profits to the person who hired them. Half of these profits then go back to the original hiring person. Explain the lure of this business to someone who has managed several sequences of hirings. Explain the fallacy of such a business as well. Keep in mind that there are about 303 million people in the United States. [11.3]

Synthesis

55. Explain why the following cannot be proved by mathematical induction: For every natural number n,
a) $3 + 5 + \cdots + (2n + 1) = (n + 1)^2$. [11.4]
b) $1 + 3 + \cdots + (2n - 1) = n^2 + 3$. [11.4]

56. Suppose that $a_1, a_2, \ldots, a_n$ and $b_1, b_2, \ldots, b_n$ are geometric sequences. Prove that $c_1, c_2, \ldots, c_n$ is a geometric sequence, where $c_n = a_n b_n$. [11.3]

57. Suppose that $a_1, a_2, \ldots, a_n$ is an arithmetic sequence. Is $b_1, b_2, \ldots, b_n$ an arithmetic sequence if:
a) $b_n = |a_n|$? [11.2] **b)** $b_n = a_n + 8$? [11.2]
c) $b_n = 7a_n$? [11.2] **d)** $b_n = \dfrac{1}{a_n}$? [11.2]
e) $b_n = \log a_n$? [11.2] **f)** $b_n = a_n^3$? [11.2]

58. The zeros of this polynomial function form an arithmetic sequence. Find them. [11.2]
$$f(x) = x^4 - 4x^3 - 4x^2 + 16x$$

59. Write the first 3 terms of the infinite geometric series with $r = -\frac{1}{3}$ and $S_\infty = \frac{3}{8}$. [11.3]

60. Simplify:
$$\sum_{k=0}^{10} (-1)^k \binom{10}{k} (\log x)^{10-k} (\log y)^k. \quad [11.6]$$

Solve for n. [11.6]

61. $\binom{n}{6} = 3 \cdot \binom{n-1}{5}$ **62.** $\binom{n}{n-1} = 36$

63. Solve for a:
$$\sum_{k=0}^{5} \binom{5}{k} 9^{5-k} a^k = 0. \quad [11.7]$$

CHAPTER 11   Test

1. For the sequence whose nth term is $a_n = (-1)^n(2n + 1)$, find a_{21}.

2. Find the first 5 terms of the sequence with general term
$$a_n = \frac{n + 1}{n + 2}.$$

3. Find and evaluate:
$$\sum_{k=1}^{4} (k^2 + 1).$$

4. Use a graphing calculator to construct a table of values and a graph for the first 10 terms of the sequence with general term
$$a_n = \frac{n + 1}{n + 2}.$$

Write sigma notation. Answers may vary.

5. $4 + 8 + 12 + 16 + 20 + 24$

6. $2 + 4 + 8 + 16 + 32 + \cdots$

7. Find the first 4 terms of the recursively defined sequence
$$a_1 = 3, \quad a_{n+1} = 2 + \frac{1}{a_n}.$$

8. Find the 15th term of the arithmetic sequence $2, 5, 8, \ldots$.

9. The 1st term of an arithmetic sequence is 8 and the 21st term is 108. Find the 7th term.

10. Find the sum of the first 20 terms of the series $17 + 13 + 9 + \cdots$.

11. Find the sum: $\displaystyle\sum_{k=1}^{25}(2k+1)$.

12. Find the 11th term of the geometric sequence $10, -5, \frac{5}{2}, -\frac{5}{4}, \dots$.

13. For a geometric sequence, $r = 0.2$ and $S_4 = 1248$. Find a_1.

Find the sum, if it exists.

14. $\displaystyle\sum_{k=1}^{8} 2^k$

15. $18 + 6 + 2 + \cdots$

16. Find fraction notation for $0.\overline{56}$.

17. *Salvage Value.*   The value of an office machine is $10,000. Its salvage value each year is 80% of its value the year before. Give a sequence that lists the salvage value of the machine for each year of a 6-yr period.

18. *Hourly Wage.*   Tamika accepts a job, starting with an hourly wage of $8.50, and is promised a raise of 25¢ per hour every three months for 4 yr. What will Tamika's hourly wage be at the end of the 4-yr period?

19. *Amount of an Annuity.*   To create a college fund, a parent makes a sequence of 18 equal yearly deposits of $2500 in a savings account on which interest is compounded annually at 5.6%. Find the amount of the annuity.

20. Use mathematical induction to prove that, for every natural number n,

$$2 + 5 + 8 + \cdots + (3n - 1) = \frac{n(3n+1)}{2}.$$

Evaluate.

21. $_{15}P_6$ 22. $_{21}C_{10}$ 23. $\dbinom{n}{4}$

24. How many 4-digit numbers can be formed using the digits 1, 3, 5, 6, 7, and 9 without repetition?

25. How many code symbols can be formed using 4 of the 6 letters A, B, C, X, Y, Z if the letters:
   a) can be repeated?
   b) are not repeated and must begin with Z?

26. *Scuba Club Officers.*   The Bay Woods Scuba Club has 28 members. How many sets of 4 officers can be selected from this group?

27. *Test Options.*   On a test with 20 questions, a student must answer 8 of the first 12 questions and 4 of the last 8. In how many ways can this be done?

28. Expand: $(x + 1)^5$.

29. Find the 5th term of the binomial expansion $(x - y)^7$.

30. Determine the number of subsets of a set containing 9 members.

31. *Marbles.*   Suppose we select, without looking, one marble from a bag containing 6 red marbles and 8 blue marbles. What is the probability of selecting a blue marble?

32. *Drawing Coins.*   Ethan has 6 pennies, 5 dimes, and 4 quarters in his pocket. Six coins are drawn at random. What is the probability of getting 1 penny, 2 dimes, and 3 quarters?

33. The graph of the sequence whose general term is $a_n = 2n - 2$ is which of the following?

A.

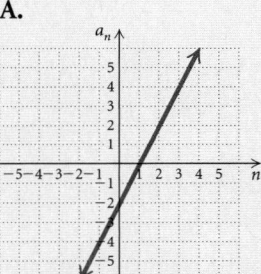

B.

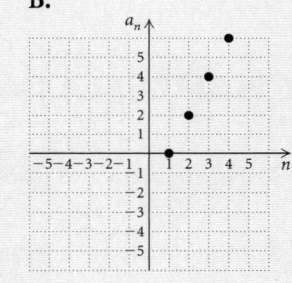

C.

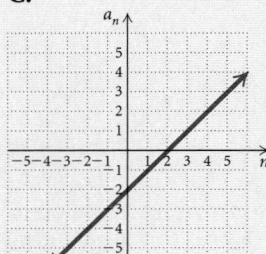

D.

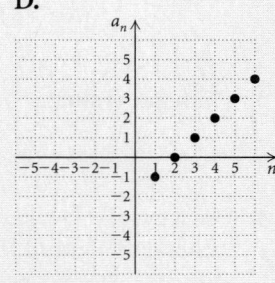

Synthesis

34. Solve for n: $_nP_7 = 9 \cdot {}_nP_6$.

Appendix

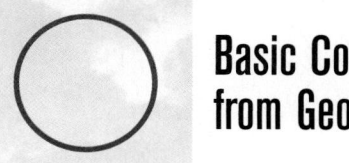

Basic Concepts from Geometry

❀ Classify an angle as right, straight, acute, or obtuse.

❀ Identify complementary angles and supplementary angles and find the measure of a complement or a supplement of a given angle.

❀ Determine whether angles are congruent.

❀ Use the vertical angle property to find measures of angles.

❀ Identify pairs of corresponding angles, interior angles, and alternate interior angles and apply properties of transversals and parallel lines to find measures of angles.

❀ Find the lengths of sides of similar triangles.

❀ Given the lengths of any two sides of a right triangle, find the length of the third side.

In this appendix, we present a review of some basic concepts from geometry. A summary of formulas from geometry is found near the back of the book.

❀ Classifying Angles

The following are ways in which we classify angles.

> **Types of Angles**
>
> ***Right angle:*** An angle whose measure is 90°.
>
> ***Straight angle:*** An angle whose measure is 180°.
>
> ***Acute angle:*** An angle whose measure is greater than 0° and less than 90°.
>
> ***Obtuse angle:*** An angle whose measure is greater than 90° and less than 180°.

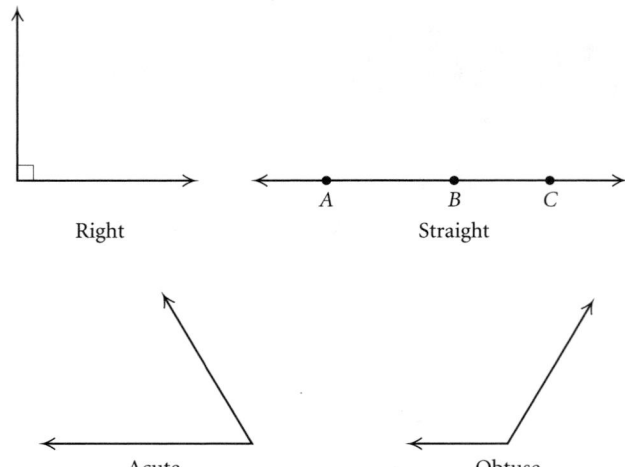

Right   Straight

Acute   Obtuse

EXAMPLE 1 Classify the angle as right, straight, acute, or obtuse.

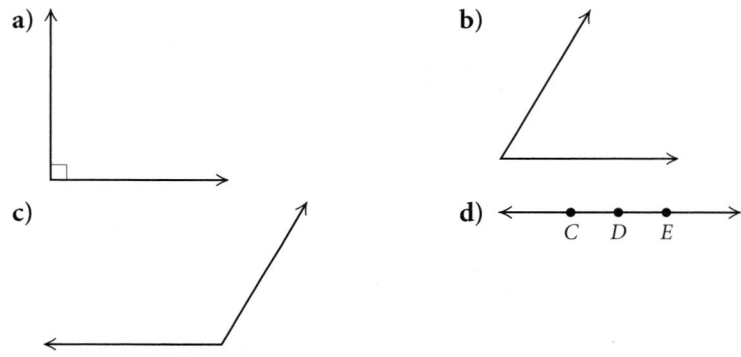

a)   b)

c)   d)

Solution

a) We observe that the measure of this angle is 90°, so it is a right angle. The symbol at the vertex is used to denote a right angle. (We could also use a protractor to measure the angle.)

b) We observe that the measure of this angle is greater than 0° and less than 90°. It is an acute angle.

c) We observe that the measure of this angle is greater than 90° and less than 180°. It is an obtuse angle.

d) We observe that the measure of this angle is 180°, so it is a straight angle.

Now Try Exercise 1. ◾

❊ Complementary and Supplementary Angles

We can describe the relationship between certain pairs of angles on the basis of the sum of their measures.

Two angles are **complementary** if the sum of their measures is 90°. Each angle is called a **complement** of the other.

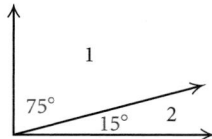

$\angle 1$ and $\angle 2$ above are **complementary** angles.

$$m\angle 1 + m\angle 2 = 90°$$
$$75° + 15° = 90°$$

If two angles are complementary, each is an acute angle. When complementary angles are adjacent to each other, they form a right angle.

EXAMPLE 2 Identify each pair of complementary angles.

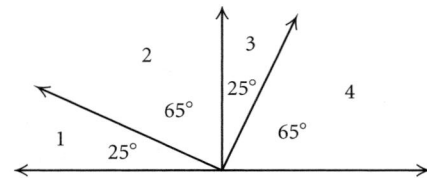

Solution We look for pairs of angles for which the sum of the measures is 90°. They are

$\angle 1$ and $\angle 2$,

$\angle 1$ and $\angle 4$,

$\angle 2$ and $\angle 3$,

$\angle 3$ and $\angle 4$.

EXAMPLE 3 Find the measure of a complement of an angle of 39°.

Solution

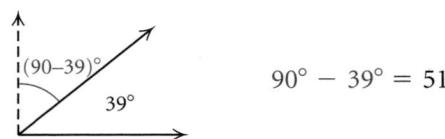

$$90° - 39° = 51°$$

The measure of a complement of an angle of 39° is 51°.

Now Try Exercise 7. ▩

Next, consider ∠1 and ∠2 as shown below. Because the sum of their measures is 180°, ∠1 and ∠2 are said to be **supplementary**. Note that when supplementary angles are adjacent, they form a straight angle.

$m\angle 1 + m\angle 2 = 180°;$
$30° + 150° = 180°$

> Two angles are **supplementary** if the sum of their measures is 180°. Each angle is called a **supplement** of the other.

EXAMPLE 4   Identify each pair of supplementary angles.

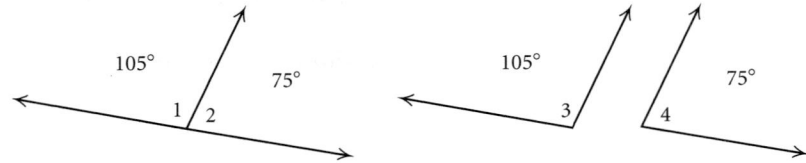

Solution   We look for pairs of angles for which the sum of the measures is 180°. They are

∠1 and ∠2,
∠1 and ∠4,
∠2 and ∠3,
∠3 and ∠4.

EXAMPLE 5   Find the measure of a supplement of an angle of 112°.

Solution

$180° - 112° = 68°$

The measure of a supplement of an angle of 112° is 68°.

Now Try Exercise 15.

❄ Congruent Angles

Congruent figures have the same shape and same size.

> Two angles are **congruent** if and only if they have the same measure.

EXAMPLE 6   Use a protractor to show that $\angle P$ and $\angle Q$ are congruent.

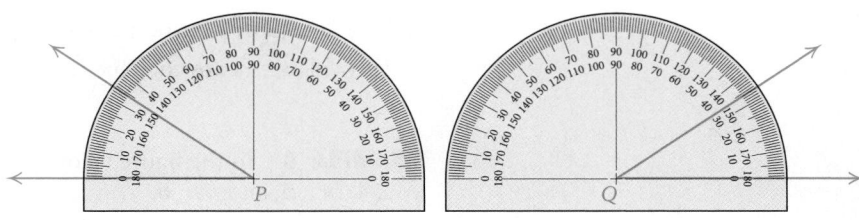

Solution   Since $m\angle P = m\angle Q = 34°$, $\angle P$ and $\angle Q$ are congruent. To say that $\angle P$ and $\angle Q$ are congruent, we write

$$\angle P \cong \angle Q.$$

EXAMPLE 7   Which pairs of angles are congruent? Use a protractor.

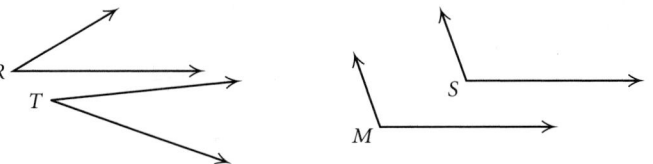

Solution $m\angle R = 31°$ and $m\angle T = 25°$. Since these angles do not have the same measure, they are not congruent.

$m\angle M = m\angle S = 108°$, so $\angle M \cong \angle S$.   **Now Try Exercise 23.** ■

If two angles are congruent, then their supplements are congruent and their complements are congruent.

❄ Vertical Angles

When $\overleftrightarrow{RT}$ intersects $\overleftrightarrow{SQ}$ at P, four angles are formed:

$\angle SPT$

$\angle RPQ$

$\angle SPR$

$\angle QPT$

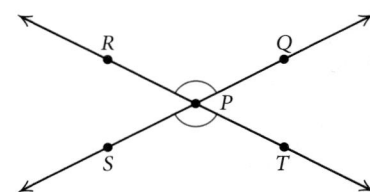

Pairs of angles such as $\angle RPQ$ and $\angle SPT$ are called **vertical angles**. $\angle RPS$ and $\angle QPT$ are also vertical angles.

> Two nonstraight angles are **vertical angles** if and only if their sides form two pairs of opposite rays.

Vertical angles are supplements of the same angle. Thus they are congruent.

> **The Vertical Angle Property**
> Vertical angles are congruent.

EXAMPLE 8 In the figure below, $m\angle 1 = 23°$ and $m\angle 3 = 34°$. Find $m\angle 2$, $m\angle 4$, $m\angle 5$, and $m\angle 6$.

Solution Since $\angle 1$ and $\angle 4$ are vertical angles, $m\angle 4 = 23°$. Likewise, $\angle 3$ and $\angle 6$ are vertical angles, so $m\angle 6 = 34°$.

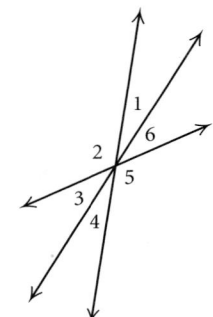

$$m\angle 1 + m\angle 2 + m\angle 3 = 180°$$
$$23° + m\angle 2 + 34° = 180° \qquad \text{Substituting}$$
$$m\angle 2 = 180° - 57°$$
$$m\angle 2 = 123°$$

Since $\angle 2$ and $\angle 5$ are vertical angles, $m\angle 5 = 123°$.

Now Try Exercise 25. ▧

❀ Transversals and Angles

> A **transversal** is a line that intersects two or more coplanar lines in different points.

In the figure below, the blue line is a transversal.

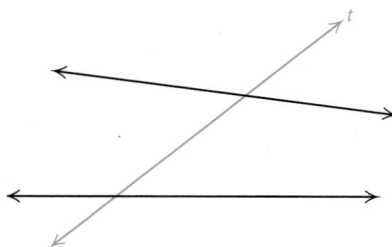

When a transversal intersects a pair of lines, eight angles are formed. Certain pairs of these angles have special names.

Corresponding Angles

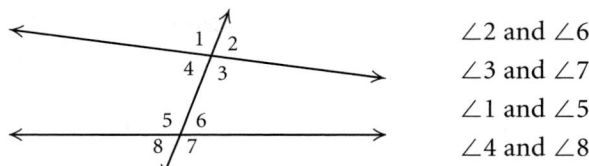

∠2 and ∠6
∠3 and ∠7
∠1 and ∠5
∠4 and ∠8

Interior Angles

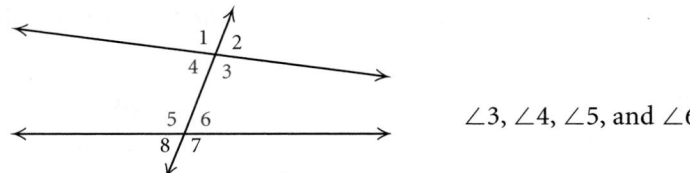

∠3, ∠4, ∠5, and ∠6

Alternate Interior Angles

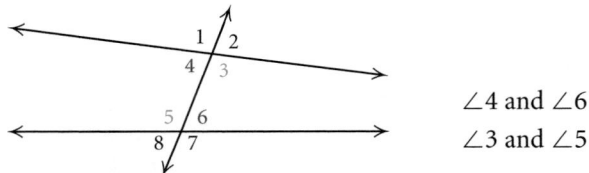

∠4 and ∠6
∠3 and ∠5

EXAMPLE 9 Identify all pairs of corresponding angles, all interior angles, and all pairs of alternate interior angles.

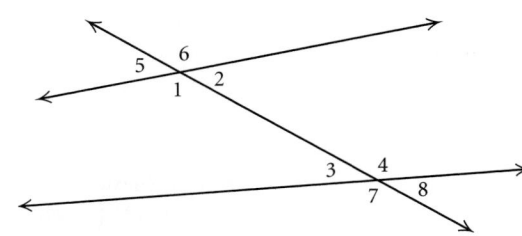

Solution

Corresponding angles: ∠6 and ∠4, ∠2 and ∠8, ∠5 and ∠3, ∠1 and ∠7

Interior angles: ∠1, ∠2, ∠3, ∠4

Alternate interior angles: ∠1 and ∠4, ∠2 and ∠3

Now Try Exercise 27. ▪

Given a line l and a point P not on l, there is exactly one line that contains P and is parallel to l.

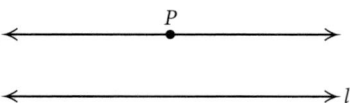

We use the symbol ‖ to indicate that two lines are parallel. For example, $l \parallel m$ indicates that lines l and m are parallel to each other.

If two lines are parallel, the following relations hold.

Properties of Parallel Lines

1. If a transversal intersects two parallel lines, then the corresponding angles are congruent.

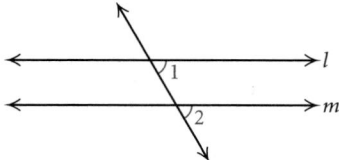

If $l \parallel m$, then $\angle 1 \cong \angle 2$.

2. If a transversal intersects two parallel lines, then the alternate interior angles are congruent.

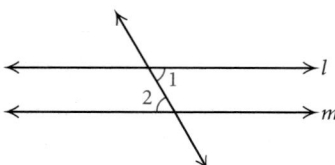

If $l \parallel m$, then $\angle 1 \cong \angle 2$.

3. In a plane, if two lines are parallel to a third line, then the two lines are parallel to each other.

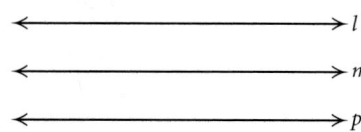

If $l \parallel p$ and $m \parallel p$, then $l \parallel m$.

APPENDIX ❖ Basic Concepts from Geometry   **979**

4. If a transversal intersects two parallel lines, then the interior angles on the same side of the transversal are supplementary.

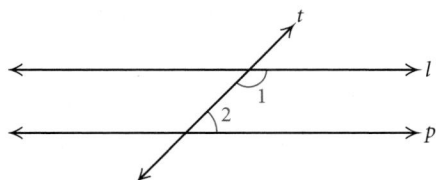

If $l \parallel p$, then $m \angle 1 + m \angle 2 = 180°$.

5. If a transversal is perpendicular to one of two parallel lines, then it is perpendicular to the other. (We use the symbol $\perp$ to indicate that two lines are perpendicular.)

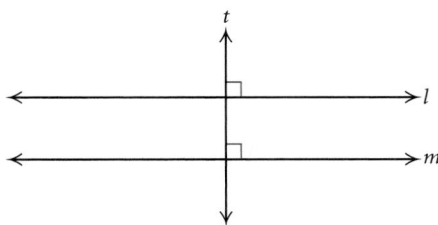

If $l \parallel m$ and $t \perp l$, then $t \perp m$.

EXAMPLE 10   If $l \parallel m$ and $m \angle 1 = 40°$, what are the measures of the other angles?

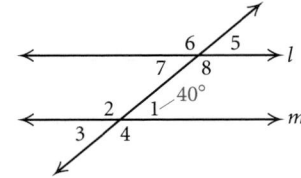

Solution

$$m \angle 7 = 40° \qquad \text{Using Property 2}$$
$$m \angle 5 = 40° \qquad \text{Using Property 1}$$
$$m \angle 8 = 140° \qquad \text{Using Property 4}$$
$$m \angle 3 = 40° \qquad \angle 1 \text{ and } \angle 3 \text{ are vertical angles}$$
$$m \angle 4 = 140° \qquad \text{Using Property 1 and } m \angle 8 = 140°$$
$$m \angle 2 = 140° \qquad \angle 2 \text{ and } \angle 4 \text{ are vertical angles and } m \angle 4 = 140°$$
$$m \angle 6 = 140° \qquad \angle 6 \text{ and } \angle 8 \text{ are vertical angles and } m \angle 8 = 140°$$

Now Try Exercise 29. ■

EXAMPLE 11 If $\overline{PT} \parallel \overline{SR}$, which pairs of angles are congruent?

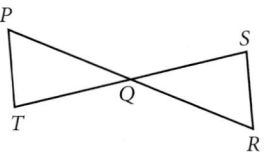

Solution

$$\angle TPQ \cong \angle SRQ \quad \text{and} \quad \angle PTQ \cong \angle RSQ \qquad \text{Using Property 2}$$
$$\angle PQT \cong \angle RQS \quad \text{and} \quad \angle PQS \cong \angle RQT \qquad \text{Vertical angles}$$

Now Try Exercise 31. ■

EXAMPLE 12 If $\overline{DE} \parallel \overline{BC}$, which pairs of angles are congruent?

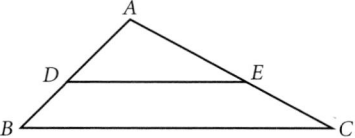

Solution

$$\angle ADE \cong \angle ABC \quad \text{and} \quad \angle AED \cong \angle ACB \qquad \text{Using Property 1}$$

■

✳ Similar Triangles

Similar figures have the same shape but are not necessarily the same size.

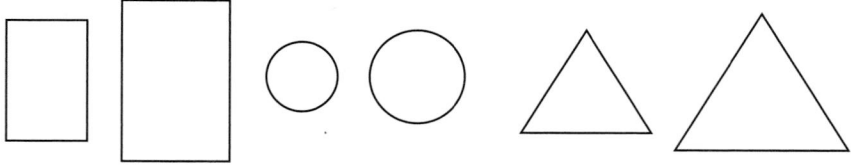

SIMILAR FIGURES

EXAMPLE 13 Which pairs of triangles appear to be similar?

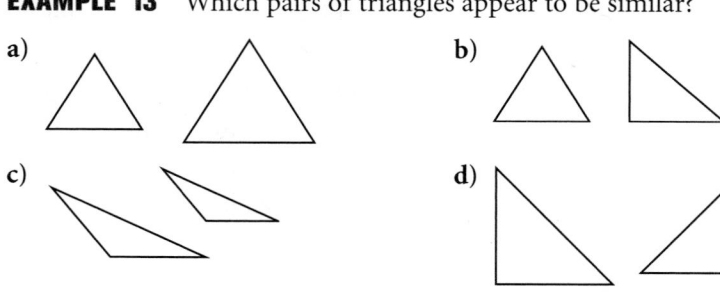

a)

b)

c)

d)

Solution   Pairs (a), (c), and (d) appear to be similar because they appear to have the same shape. ■

Similar triangles have corresponding sides and angles.

EXAMPLE 14 $\triangle ABC$ and $\triangle DEF$ are similar. Name their corresponding sides and angles.

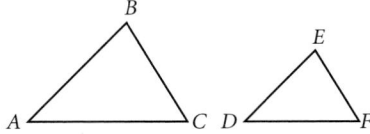

Solution

$\overline{AB} \leftrightarrow \overline{DE}$	$\angle A \leftrightarrow \angle D$	The symbol $\leftrightarrow$ means "corresponds to."
$\overline{AC} \leftrightarrow \overline{DF}$	$\angle B \leftrightarrow \angle E$	
$\overline{BC} \leftrightarrow \overline{EF}$	$\angle C \leftrightarrow \angle F$	**Now Try Exercise 35.** ■

> Two triangles are **similar** if and only if their vertices can be matched so that the corresponding angles have the same measure and the lengths of corresponding sides are proportional.

To say that $\triangle ABC$ and $\triangle DEF$ are similar, we write "$\triangle ABC \sim \triangle DEF$." We will agree that this symbol also tells us the way in which the vertices are matched.

$$\triangle ABC \sim \triangle DEF$$

Thus, $\triangle ABC \sim \triangle DEF$ means that

$$\begin{array}{l} \angle A \leftrightarrow \angle D \\ \angle B \leftrightarrow \angle E \quad \text{and} \quad \dfrac{AB}{DE} = \dfrac{AC}{DF} = \dfrac{BC}{EF}. \\ \angle C \leftrightarrow \angle F \end{array}$$

EXAMPLE 15   Suppose that $\triangle PQR \sim \triangle STV$. Name the corresponding angles. Which sides are proportional?

Solution

$$\begin{array}{l} \angle P \leftrightarrow \angle S \\ \angle Q \leftrightarrow \angle T \quad \text{and} \quad \dfrac{PQ}{ST} = \dfrac{PR}{SV} = \dfrac{QR}{TV}. \\ \angle R \leftrightarrow \angle V \end{array}$$

Now Try Exercise 39. ■

EXAMPLE 16   These triangles are similar. Which sides are proportional?

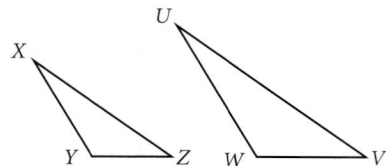

Solution   It appears that if we match X with U, Y with W, and Z with V, the corresponding angles have the same measure. Thus,

$$\frac{XY}{UW} = \frac{XZ}{UV} = \frac{YZ}{WV}.$$

Now Try Exercise 43. ▇

❀ Proportions and Similar Triangles

We can find lengths of sides in similar triangles.

EXAMPLE 17   If $\triangle RAE \sim \triangle GQL$, find QL and GL.

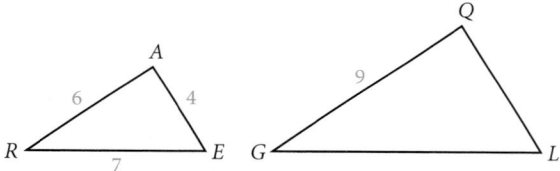

Solution   Since $\triangle RAE \sim \triangle GQL$, the corresponding sides are proportional. Thus,

$$\frac{6}{9} = \frac{4}{QL}$$

$6(QL) = 4 \cdot 9$ **Multiplying by $9(QL)$ on both sides**

$6(QL) = 36$

$QL = 6$ **Dividing by 6 on both sides**

and

$$\frac{6}{9} = \frac{7}{GL}$$

$6(GL) = 7 \cdot 9$

$6(GL) = 63$

$GL = 10\frac{1}{2}.$

Now Try Exercise 47. ▇

Similar triangles and proportions can often be used to find lengths that would ordinarily be difficult to measure. For example, we could find the height of a flagpole without climbing it or the distance across a river without crossing it.

EXAMPLE 18 *Height of a Flagpole.* How high is a flagpole that casts a 56-ft shadow at the same time that a 6-ft man casts a 5-ft shadow?

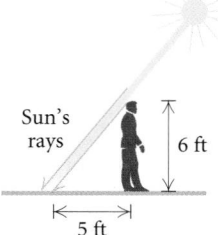
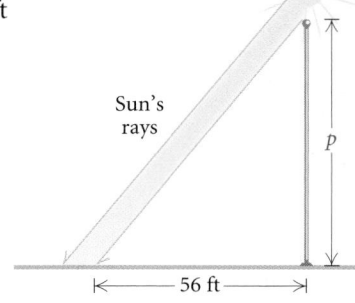

Solution If we use the sun's rays to represent the third side of the triangle in our drawing of the situation, we see that we have similar triangles. Let p = the height of the flagpole. Then the ratio of 6 to p is the same as the ratio of 5 to 56. Thus we have the proportion

$$\text{Height of man} \longrightarrow \frac{6}{p} = \frac{5}{56} \longleftarrow \text{Length of man's shadow} \\ \text{Height of pole} \longrightarrow \quad\;\; \longleftarrow \text{Length of pole's shadow}$$

$$6 \cdot 56 = 5 \cdot p \qquad \text{Multiplying by } 56p \text{ on both sides}$$

$$\frac{6 \cdot 56}{5} = p \qquad \text{Dividing by 5 on both sides}$$

$$67.2 = p \qquad \text{Simplifying}$$

The height of the flagpole is 67.2 ft. **Now Try Exercise 49.** ■

EXAMPLE 19 *F-106 Blueprint.* A blueprint for an F-106 Delta Dart fighter plane is a scale drawing. Each wing of the plane has a triangular shape. Find the length of side a of the wing.

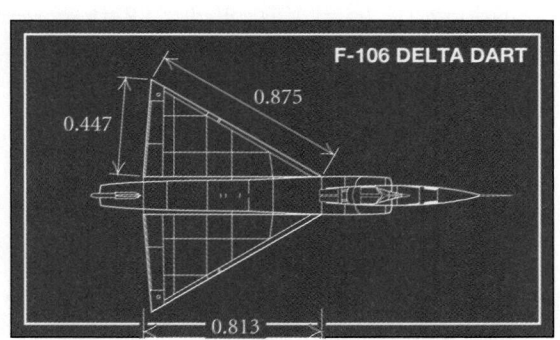
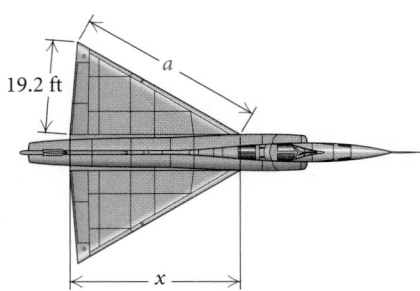

Solution   We let a = the length of the wing. Thus we have the proportion

Length on the blueprint → $\dfrac{0.447}{19.2}$ = $\dfrac{0.875}{a}$ ← Length on the blueprint
Length of the wing →   ← Length of the wing

$$0.447 \cdot a = 0.875 \cdot 19.2 \qquad \textbf{Multiplying by 19.2}a$$

$$a = \frac{0.875 \cdot 19.2}{0.447} \qquad \textbf{Dividing by 0.447}$$

$$a \approx 37.6 \text{ ft}$$

The length of side a of the wing is about 37.6 ft.   **Now Try Exercise 51.** ■

❀ Right Triangles

A **right triangle** is a triangle with a 90° angle, as shown in the figure below.

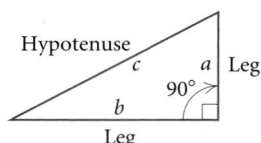

In a right triangle, the longest side is called the **hypotenuse**. It is the side opposite the right angle. The other two sides are called **legs**. We generally use the letters a and b for the lengths of the legs and c for the length of the hypotenuse. They are related as follows.

The Pythagorean Theorem

In any right triangle, if a and b are the lengths of the legs and c is the length of the hypotenuse, then

$$a^2 + b^2 = c^2.$$

The equation $a^2 + b^2 = c^2$ is called the **Pythagorean equation.**

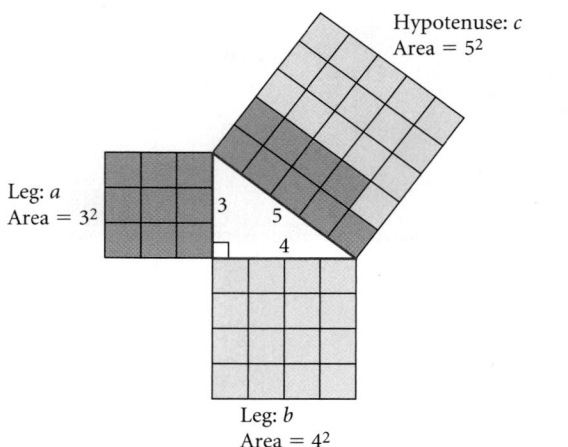

$$a^2 + b^2 = c^2$$
$$3^2 + 4^2 = 5^2$$
$$9 + 16 = 25$$

The Pythagorean theorem is named after the ancient Greek mathematician Pythagoras (569?–500? B.C.). It is uncertain who actually proved this result the first time. A proof can be found in most geometry books.

If we know the lengths of any two sides of a right triangle, we can find the length of the third side.

EXAMPLE 20 Find the length of the hypotenuse of this right triangle. Give an exact answer and an approximation to three decimal places.

Solution

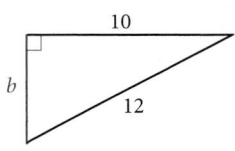

$$4^2 + 5^2 = c^2 \qquad \text{Substituting in the Pythagorean equation}$$
$$16 + 25 = c^2$$
$$41 = c^2$$
$$\sqrt{41} = c \qquad \text{Exact answer}$$
$$6.403 \approx c \qquad \text{Using a calculator to find an approximation}$$

Now Try Exercise 53. ■

EXAMPLE 21 Find the length b of the leg of this right triangle. Give an exact answer and an approximation to three decimal places.

Solution

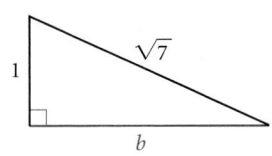

$$10^2 + b^2 = 12^2 \qquad \text{Substituting in the Pythagorean equation}$$
$$100 + b^2 = 144$$
$$b^2 = 144 - 100$$
$$b^2 = 44$$
$$b = \sqrt{44} \qquad \text{Exact answer}$$
$$b \approx 6.633 \qquad \text{Using a calculator} \qquad \text{Now Try Exercise 57.} ■$$

EXAMPLE 22 Find the length b of the leg of this right triangle. Give an exact answer and an approximation to three decimal places.

Solution

$$1^2 + b^2 = \left(\sqrt{7}\right)^2 \qquad \text{Substituting in the Pythagorean equation}$$
$$1 + b^2 = 7$$
$$b^2 = 7 - 1 = 6$$
$$b = \sqrt{6} \qquad \text{Exact answer}$$
$$b \approx 2.449 \qquad \text{Using a calculator} \qquad \text{Now Try Exercise 59.} ■$$

EXAMPLE 23   In a right triangle with $b = 10$ and $c = 15$, find the length of the third side. Give an exact answer and an approximation to three decimal places.

Solution   The third side is leg a. We make a drawing of the triangle, labeling the sides with the known and the unknown information.

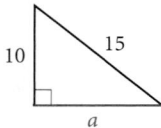

Now we substitute in the Pythagorean equation and solve for a:

$$a^2 + 10^2 = 15^2$$
$$a^2 + 100 = 225$$
$$a^2 = 225 - 100$$
$$a^2 = 125$$
$$a = \sqrt{125} \qquad \text{Exact answer}$$
$$a \approx 11.180. \qquad \text{Using a calculator} \qquad \text{Now Try Exercise 63.} \ \blacksquare$$

❋ An Application

EXAMPLE 24   *Dimensions of a Softball Diamond.*   A slow-pitch softball diamond is actually a square 65 ft on a side. How far is it from home plate to second base? Give an exact answer and an approximation to three decimal places. (This can be helpful information when lining up the bases.)

Solution

We first make a drawing. We note that the first and second base lines, together with a line from home plate to second base, form a right triangle. We label the unknown distance d.

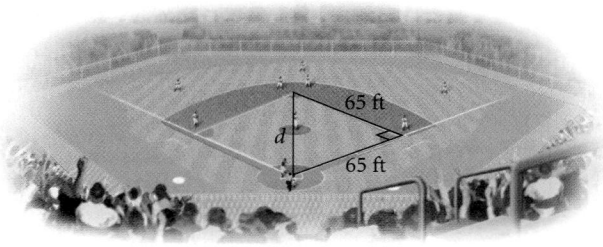

We know that $65^2 + 65^2 = d^2$. We solve this equation:

$$4225 + 4225 = d^2$$
$$8450 = d^2.$$

Exact answer: $\sqrt{8450}$ ft $= d$
Approximation: 91.924 ft $\approx d$ Now Try Exercise 73. ▨

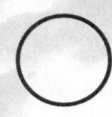

Exercise Set

Classify the angle as right, straight, acute, or obtuse.

1.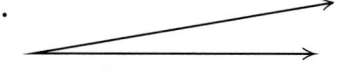

2.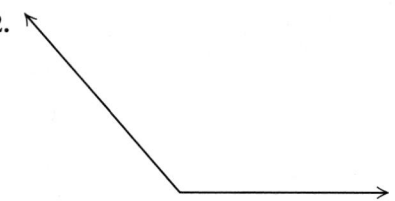

3. 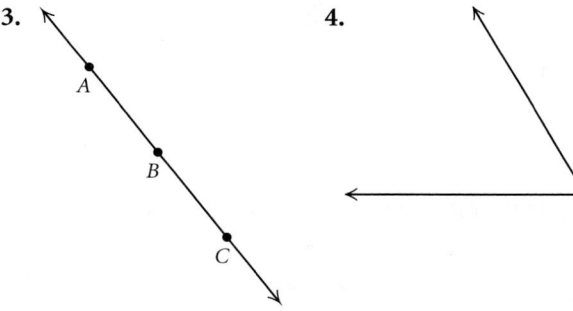 **4.**

5.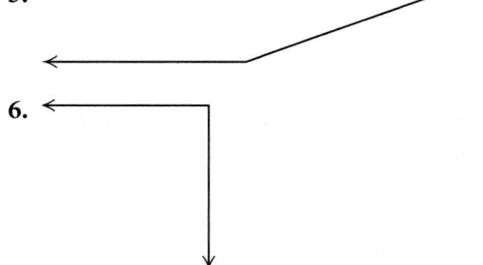

6.

Find the measure of a complement of an angle with the given measure.

7. 11°   **8.** 83°   **9.** 67°   **10.** 5°

11. 58°   **12.** 32°   **13.** 29°   **14.** 54°

Find the measure of a supplement of an angle with the given measure.

15. 3°   **16.** 54°   **17.** 139°   **18.** 13°

19. 85°   **20.** 129°   **21.** 102°   **22.** 45°

Determine whether the pair of angles is congruent. Use a protractor.

23.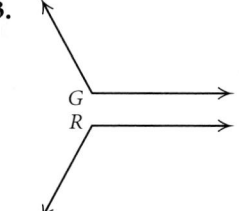

24.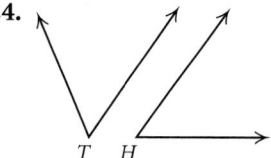

25. In the figure, $m \angle 1 = 80°$ and $m \angle 5 = 67°$. Find $m \angle 2$, $m \angle 3$, $m \angle 4$, and $m \angle 6$.

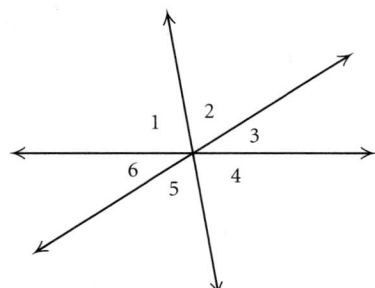

26. In the figure, $m \angle 2 = 42°$ and $m \angle 4 = 56°$. Find $m \angle 1$, $m \angle 3$, $m \angle 5$, and $m \angle 6$.

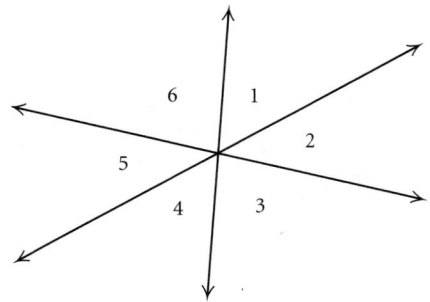

In Exercises 27 and 28, (a) identify all pairs of corresponding angles, (b) identify all interior angles, and (c) identify all pairs of alternate interior angles.

27.

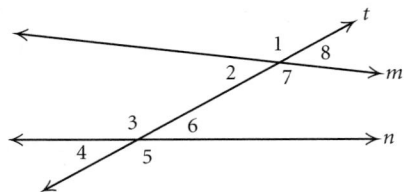

Lines *m* and *n*
Transversal *t*

28.

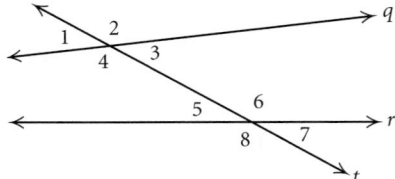

Lines *q* and *r*
Transversal *t*

29. If $m \parallel n$ and $m \angle 4 = 125°$, what are the measures of the other angles?

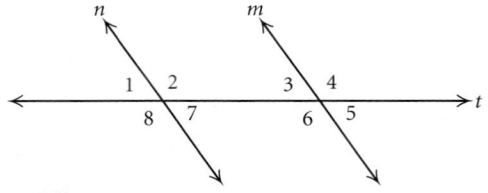

30. If $m \parallel n$ and $m \angle 8 = 34°$, what are the measures of the other angles?

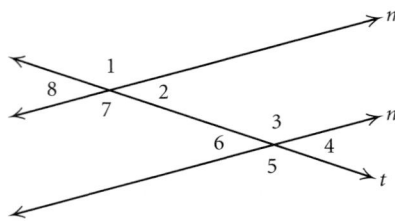

In each figure, $\overline{AB} \parallel \overline{CD}$. Identify pairs of congruent angles. Where possible, give the measures of the angles.

31.

32.

33.

34. If $\overline{PQ} \parallel \overline{RS}$, which pairs of angles are congruent?

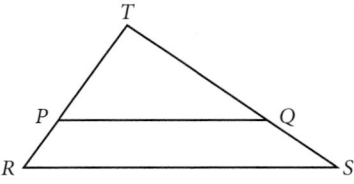

For each pair of similar triangles, name the corresponding angles and sides.

35.

36.

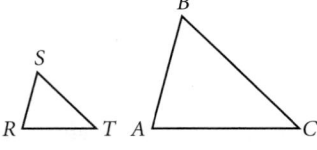

37.

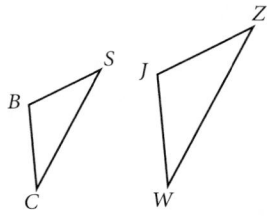

38.

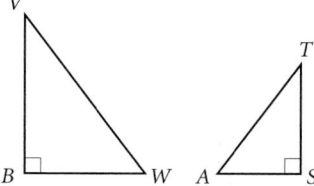

For each pair of similar triangles, name the angles with the same measure and name the proportional sides.

39. $\triangle ABC \sim \triangle RST$ **40.** $\triangle PQR \sim \triangle STV$

41. $\triangle MES \sim \triangle CLF$ **42.** $\triangle SMH \sim \triangle WLK$

Name the proportional sides in these similar triangles.

43.

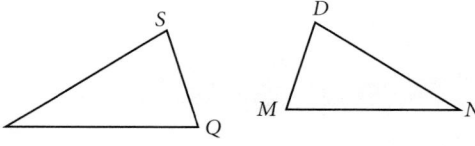

44.

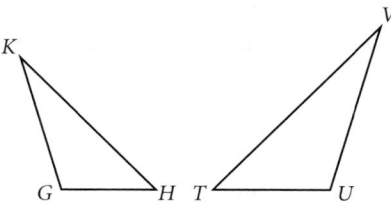

45.

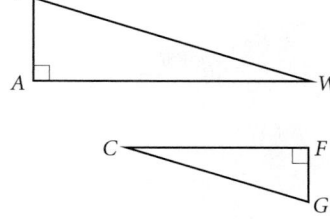

46.

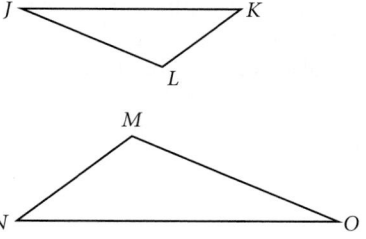

47. If $\triangle ABC \sim \triangle PQR$, find QR and PR.

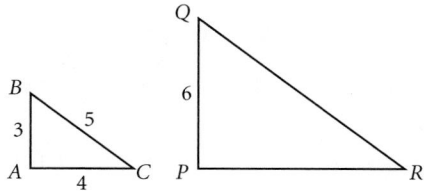

48. If $\triangle MAC \sim \triangle GET$, find AM and GT.

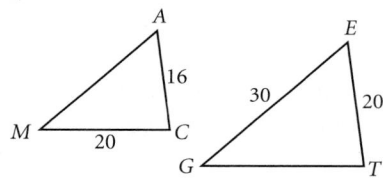

49. How high is a tree that casts a 27-ft shadow at the same time that a 4-ft fence post casts a 3-ft shadow?

50. How high is a flagpole that casts a 42-ft shadow at the same time that a $5\frac{1}{2}$-ft woman casts a 7-ft shadow?

51. Find the distance across the river. Assume that the ratio of d to 25 ft is the same as the ratio of 40 ft to 10 ft.

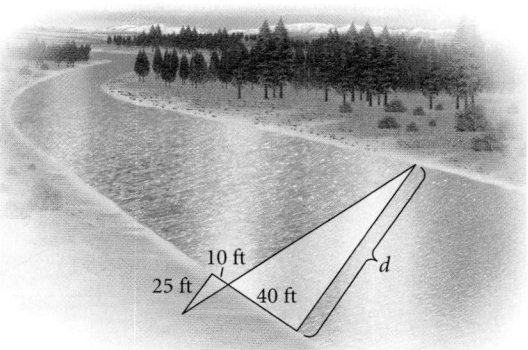

52. To measure the height of a cliff, a string is drawn tight from level ground to the top of the cliff. A 3-ft yardstick is placed under the string, touching it at point P, a distance of 5 ft from point G, where the string touches the ground. The string is then detached and found to be 120 ft long. How high is the cliff?

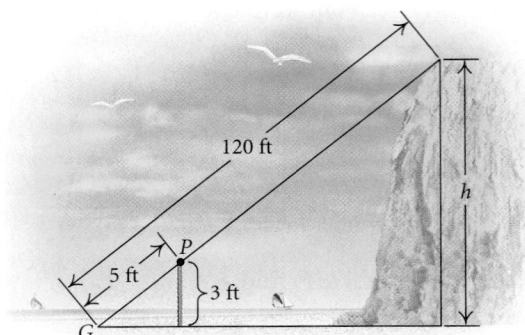

Find the length of the third side of the right triangle. Give an exact answer and an approximation to three decimal places.

53.

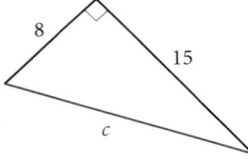

54.

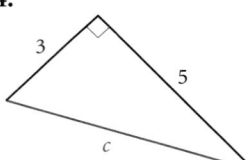

55.

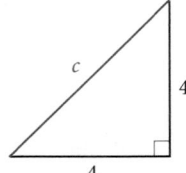

56.

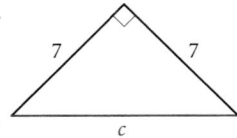

57.

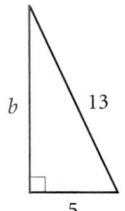

58.

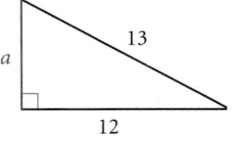

59.

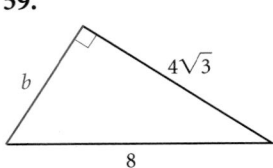

60.

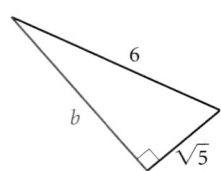

In a right triangle, find the length of the side not given. Give an exact answer and an approximation to three decimal places.

61. $a = 10$, $b = 24$ **62.** $a = 5$, $b = 12$

63. $a = 9$, $c = 15$ **64.** $a = 18$, $c = 30$

65. $b = 1$, $c = \sqrt{5}$ **66.** $b = 1$, $c = \sqrt{2}$

67. $a = 1$, $c = \sqrt{3}$ **68.** $a = \sqrt{3}$, $b = \sqrt{5}$

69. $c = 10$, $b = 5\sqrt{3}$ **70.** $a = 5$, $b = 5$

71. $a = \sqrt{2}$, $b = \sqrt{7}$ **72.** $c = \sqrt{7}$, $a = \sqrt{2}$

Solve. Give an exact answer and an approximation to three decimal places.

73. *Airport Distance.*   An airplane is flying at an altitude of 4100 ft. The slanted distance directly to the airport is 15,100 ft. How far is the airplane horizontally from the airport?

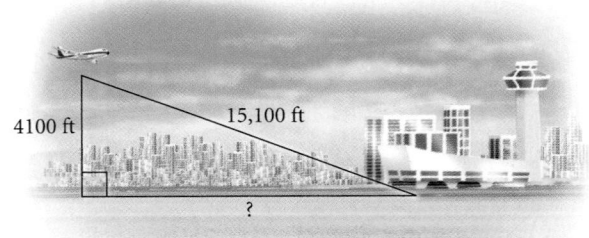

74. *Surveying Distance.*   A surveyor had poles located at points P, Q, and R. The distances that the surveyor was able to measure are marked on the drawing. What is the distance from P to R?

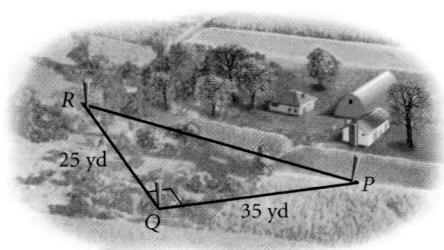

75. *Cordless Telephone.* Becky's cordless telephone has clear reception up to 300 ft from its base. Her base is located near a window in her apartment, 180 ft above ground level. How far into her backyard can Becky use her phone?

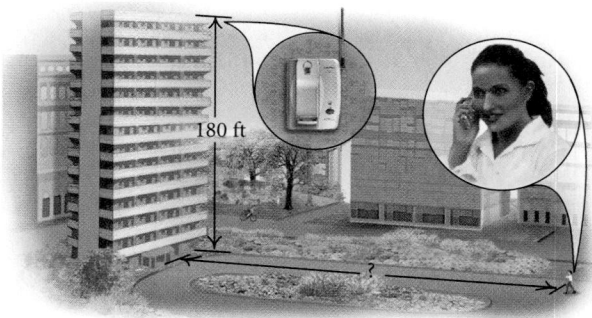

76. *Rope Course.* An outdoor rope course consists of a cable that slopes downward from a height of 37 ft to a resting place 30 ft above the ground. The trees that the cable connects are 24 ft apart. How long is the cable?

77. *Diagonal of a Square.* Find the length of a diagonal of a square whose sides are 3 cm long.

78. *Ladder Height.* A 10-m ladder is leaning against a building. The bottom of the ladder is 5 m from the building. How high is the top of the ladder?

79. *Guy Wire.* How long is a guy wire reaching from the top of a 12-ft pole to a point on the ground 8 ft from the base of the pole?

80. *Diagonal of a Soccer Field.* The largest regulation soccer field is 100 yd wide and 130 yd long. Find the length of a diagonal of such a field.

Photo Credits

Answers

CHAPTER R

Exercise Set R.1

1. $\frac{2}{3}, 6, -2.45, 18.\overline{4}, -11, \sqrt[3]{27}, 5\frac{1}{6}, -\frac{8}{7}, 0, \sqrt{16}$
3. $\sqrt{3}, \sqrt[6]{26}, 7.151551555\ldots, -\sqrt{35}, \sqrt[5]{3}$
5. $6, \sqrt[3]{27}, 0, \sqrt{16}$ **7.** $-11, 0$
9. $\frac{2}{3}, -2.45, 18.\overline{4}, 5\frac{1}{6}, -\frac{8}{7}$
11. $[-5, 5]$;
13. $(-3, -1]$;
15. $(-\infty, -2]$;
17. $(3.8, \infty)$;
19. $(7, \infty)$;   **21.** $(0, 5)$
23. $[-9, -4)$ **25.** $[x, x + h]$ **27.** (p, ∞) **29.** True
31. False   **33.** True   **35.** False   **37.** False
39. True   **41.** True   **43.** True   **45.** False
47. Commutative property of addition
49. Multiplicative identity property
51. Commutative property of multiplication
53. Commutative property of multiplication
55. Commutative property of addition
57. Multiplicative inverse property   **59.** 8.15
61. 295   **63.** $\sqrt{97}$ **65.** 0   **67.** $\frac{5}{4}$ **69.** 22
71. 6   **73.** 5.4   **75.** $\frac{21}{8}$ **77.** 7
79. Discussion and Writing
81. Answers may vary; $0.124124412444\ldots$
83. Answers may vary; -0.00999

85.

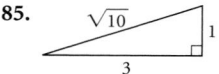

Exercise Set R.2

1. $\frac{1}{3^7}$ **3.** $\frac{y^4}{x^5}$ **5.** $\frac{t^6}{mn^{12}}$ **7.** 1   **9.** z^7
11. 5^2, or 25   **13.** 1   **15.** y^{-4}, or $\frac{1}{y^4}$
17. 3^6, or 729   **19.** $6x^5$ **21.** $-15a^{-12}$, or $-\frac{15}{a^{12}}$
23. $15a^{-1}b^5$, or $\frac{15b^5}{a}$ **25.** $-42x^{-1}y^{-4}$, or $-\frac{42}{xy^4}$
27. $432x^7$ **29.** $-200n^5$ **31.** y^4 **33.** b^{-19}, or $\frac{1}{b^{19}}$
35. x^3y^{-3}, or $\frac{x^3}{y^3}$ **37.** $8xy^{-5}$, or $\frac{8x}{y^5}$ **39.** $16x^8y^4$
41. $-32x^{15}$ **43.** $\frac{c^2d^4}{25}$ **45.** $432m^{-8}$, or $\frac{432}{m^8}$
47. $\frac{8x^{-9}y^{21}}{z^{-3}}$, or $\frac{8y^{21}z^3}{x^9}$ **49.** $2^{-5}a^{-20}b^{25}c^{-10}$, or $\frac{b^{25}}{32a^{20}c^{10}}$
51. 1.65×10^7 **53.** 4.37×10^{-7} **55.** 2.346×10^{11}
57. 1.04×10^{-3} **59.** 1.6×10^{-5} **61.** 760,000
63. 0.000000109   **65.** 34,960,000,000
67. 0.0000000541   **69.** 231,900,000
71. 1.344×10^6 **73.** 2.21×10^{-10}
75. 8×10^{-14} **77.** 2.5×10^5 **79.** 3.6×10^{-7} m
81. $\$1.19 \times 10^7$ **83.** 3.627×10^9 mi
85. 1.332×10^{14} disintegrations   **87.** 103
89. 2048   **91.** 5   **93.** $\$3944.71$ **95.** $\$5299.49$

97. Discussion and Writing **99.** $170,797.30
101. $309.79 **103.** x^{8t} **105.** t^{8x} **107.** $9x^{2a}y^{2b}$

Exercise Set R.3

1. $7x^3, -4x^2, 8x, 5; 3$ **3.** $3a^4b, -7a^3b^3, 5ab, -2; 6$
5. $2ab^2 - 9a^2b + 6ab + 10$ **7.** $3x + 2y - 2z - 3$
9. $-2x^2 + 6x - 2$ **11.** $x^4 - 3x^3 - 4x^2 + 9x - 3$
13. $2a^4 - 2a^3b - a^2b + 4ab^2 - 3b^3$
15. $y^2 + 2y - 15$ **17.** $x^2 + 9x + 18$
19. $2a^2 + 13a + 15$ **21.** $4x^2 + 8xy + 3y^2$
23. $x^2 + 6x + 9$ **25.** $y^2 - 10y + 25$
27. $25x^2 - 30x + 9$ **29.** $4x^2 + 12xy + 9y^2$
31. $4x^4 - 12x^2y + 9y^2$ **33.** $n^2 - 36$
35. $9y^2 - 16$ **37.** $9x^2 - 4y^2$
39. $4x^2 + 12xy + 9y^2 - 16$ **41.** $x^4 - 1$
43. Discussion and Writing **45.** $a^{2n} - b^{2n}$
47. $a^{2n} + 2a^nb^n + b^{2n}$ **49.** $x^6 - 1$ **51.** $x^{a^2-b^2}$
53. $a^2 + b^2 + c^2 + 2ab + 2ac + 2bc$

Exercise Set R.4

1. $3(x + 6)$ **3.** $2z^2(z - 4)$ **5.** $4(a^2 - 3a + 4)$
7. $(b - 2)(a + c)$ **9.** $(3x - 1)(x^2 + 6)$
11. $(y - 1)(y^2 + 2)$ **13.** $12(2x - 3)(x^2 + 3)$
15. $(x - 1)(x^2 - 5)$ **17.** $(a - 3)(a^2 - 2)$
19. $(w - 5)(w - 2)$ **21.** $(x + 1)(x + 5)$
23. $(t + 3)(t + 5)$ **25.** $(x + 3y)(x - 9y)$
27. $2(n - 12)(n + 2)$ **29.** $(y - 7)(y + 3)$
31. $y^2(y - 2)(y - 7)$ **33.** $2x(x + 3y)(x - 4y)$
35. $(2n - 7)(n + 8)$ **37.** $(3x + 2)(4x + 1)$
39. $(4x + 3)(x + 3)$ **41.** $(2y - 3)(y + 2)$
43. $(3a - 4b)(2a - 7b)$ **45.** $4(3a - 4)(a + 1)$
47. $(z + 9)(z - 9)$ **49.** $(4x + 3)(4x - 3)$
51. $6(x + y)(x - y)$ **53.** $4x(y^2 + z)(y^2 - z)$
55. $7p(q^2 + y^2)(q + y)(q - y)$ **57.** $(x + 6)^2$
59. $(3z - 2)^2$ **61.** $(1 - 4x)^2$ **63.** $a(a + 12)^2$
65. $4(p - q)^2$ **67.** $(x + 4)(x^2 - 4x + 16)$
69. $(m - 6)(m^2 + 6m + 36)$
71. $8(t + 1)(t^2 - t + 1)$
73. $3a^2(a - 2)(a^2 + 2a + 4)$
75. $(t^2 + 1)(t^4 - t^2 + 1)$ **77.** $3ab(6a - 5b)$
79. $(x - 4)(x^2 + 5)$ **81.** $8(x + 2)(x - 2)$
83. Prime **85.** $(m + 3n)(m - 3n)$
87. $(x + 4)(x + 5)$ **89.** $(y - 5)(y - 1)$
91. $(2a + 1)(a + 4)$ **93.** $(3x - 1)(2x + 3)$
95. $(y - 9)^2$ **97.** $(3z - 4)^2$ **99.** $(xy - 7)^2$
101. $4a(x + 7)(x - 2)$ **103.** $3(z - 2)(z^2 + 2z + 4)$
105. $2ab(2a^2 + 3b^2)(4a^4 - 6a^2b^2 + 9b^4)$
107. $(y - 3)(y + 2)(y - 2)$ **109.** $(x - 1)(x^2 + 1)$
111. $5(m^2 + 2)(m^2 - 2)$
113. $2(x + 3)(x + 2)(x - 2)$ **115.** $(2c - d)^2$
117. $(m^3 + 10)(m^3 - 2)$
119. $p(1 - 4p)(1 + 4p + 16p^2)$

121. Discussion and Writing **123.** $(y^2 + 12)(y^2 - 7)$
125. $\left(y + \frac{4}{7}\right)\left(y - \frac{2}{7}\right)$ **127.** $\left(x + \frac{3}{2}\right)^2$ **129.** $\left(x - \frac{1}{2}\right)^2$
131. $h(3x^2 + 3xh + h^2)$ **133.** $(y + 4)(y - 7)$
135. $(x^n + 8)(x^n - 3)$ **137.** $(x + a)(x + b)$
139. $(5y^m + x^n - 1)(5y^m - x^n + 1)$
141. $y(y - 1)^2(y - 2)$

Exercise Set R.5

1. 12 **3.** -4 **5.** 3 **7.** 10 **9.** 11 **11.** -1
13. -12 **15.** 2 **17.** -1 **19.** $\frac{18}{5}$ **21.** -3
23. 1 **25.** 0 **27.** $-\frac{1}{10}$ **29.** 5 **31.** $-\frac{3}{2}$
33. $\frac{20}{7}$ **35.** $-7, 4$ **37.** $-5, 0$ **39.** -3
41. 10 **43.** $-4, 8$ **45.** $-2, -\frac{2}{3}$ **47.** $-\frac{3}{4}, \frac{2}{3}$
49. $-\frac{4}{3}, \frac{7}{4}$ **51.** $-2, 7$ **53.** $-6, 6$ **55.** $-12, 12$
57. $-\sqrt{10}, \sqrt{10}$ **59.** $-\sqrt{3}, \sqrt{3}$
61. Discussion and Writing **63.** $\frac{23}{66}$ **65.** 8
67. $-\frac{6}{5}, -\frac{1}{4}, 0, \frac{2}{3}$ **69.** $-3, -2, 3$

Exercise Set R.6

1. $\{x \,|\, x \text{ is a real number}\}$
3. $\{x \,|\, x \text{ is a real number } and \, x \neq 0 \text{ and } x \neq 1\}$
5. $\{x \,|\, x \text{ is a real number } and \, x \neq -5 \text{ and } x \neq 1\}$
7. $\{x \,|\, x \text{ is a real number } and \, x \neq -2 \text{ and } x \neq 2 \text{ and } x \neq -5\}$
9. $\dfrac{x + 2}{x - 2}$ **11.** $\dfrac{x - 3}{x}$ **13.** $\dfrac{2(y + 4)}{y - 1}$
15. $-\dfrac{1}{x + 8}$ **17.** 1 **19.** $\dfrac{(x - 5)(3x + 2)}{7x}$
21. $\dfrac{a + 2}{a - 5}$ **23.** $m + n$ **25.** $\dfrac{3(x - 4)}{2(x + 4)}$
27. $\dfrac{1}{x + y}$ **29.** $\dfrac{x - y - z}{x + y + z}$ **31.** $\dfrac{2}{x}$ **33.** 1
35. $\dfrac{7}{8z}$ **37.** $\dfrac{3x - 4}{(x + 2)(x - 2)}$ **39.** $\dfrac{-y + 10}{(y + 4)(y - 5)}$
41. $\dfrac{4x - 8y}{(x + y)(x - y)}$ **43.** $\dfrac{y - 2}{y - 1}$ **45.** $\dfrac{x + y}{2x - 3y}$
47. $\dfrac{3x - 4}{(x - 2)(x - 1)}$ **49.** $\dfrac{5a^2 + 10ab - 4b^2}{(a + b)(a - b)}$
51. $\dfrac{11x^2 - 18x + 8}{(2 + x)(2 - x)^2}$, or $\dfrac{11x^2 - 18x + 8}{(x + 2)(x - 2)^2}$ **53.** 0
55. $\dfrac{a}{a + b}$ **57.** $x - y$ **59.** $\dfrac{c^2 - 2c + 4}{c}$
61. $\dfrac{xy}{x - y}$ **63.** $\dfrac{a^2 - 1}{a^2 + 1}$
65. $\dfrac{3(x - 1)^2(x + 2)}{(x - 3)(x + 3)(-x + 10)}$ **67.** $\dfrac{1 + a}{1 - a}$
69. $\dfrac{b + a}{b - a}$ **71.** Discussion and Writing

73. $2x + h$ **75.** $3x^2 + 3xh + h^2$ **77.** x^5

79. $\dfrac{(n + 1)(n + 2)(n + 3)}{2 \cdot 3}$

81. $\dfrac{x^3 + 2x^2 + 11x + 20}{2(x + 1)(2 + x)}$

Exercise Set R.7

1. 21   **3.** $3|y|$ **5.** $|a - 2|$ **7.** $-3x$ **9.** $3x^2$
11. 2   **13.** $6\sqrt{5}$ **15.** $6\sqrt{2}$ **17.** $3\sqrt[3]{2}$
19. $8\sqrt{2}|c|d^2$ **21.** $2|x||y|\sqrt[4]{3x^2}$ **23.** $|x - 2|$
25. $5\sqrt{21}$ **27.** $4\sqrt{5}$ **29.** $2x^2y\sqrt{6}$ **31.** $3x\sqrt[3]{4y}$
33. $2(x + 4)\sqrt[3]{(x + 4)^2}$ **35.** $\dfrac{m^2n^3}{2}$ **37.** $\sqrt{5y}$
39. $\dfrac{1}{2x}$ **41.** $\dfrac{4a\sqrt[3]{a}}{3b}$ **43.** $\dfrac{x\sqrt{7x}}{6y^3}$ **45.** $17\sqrt{2}$
47. $4\sqrt{5}$ **49.** $-2x\sqrt{2} - 12\sqrt{5x}$ **51.** -12
53. $-9 - 5\sqrt{15}$ **55.** $27 - 10\sqrt{2}$
57. $11 - 2\sqrt{30}$ **59.** About 13,709.5 ft
61. (a) $h = \dfrac{a}{2}\sqrt{3}$; (b) $A = \dfrac{a^2}{4}\sqrt{3}$ **63.** 8
65. $\dfrac{\sqrt{21}}{7}$ **67.** $\dfrac{\sqrt[3]{35}}{5}$ **69.** $\dfrac{2\sqrt[3]{6}}{3}$ **71.** $\sqrt{3} + 1$
73. $-\dfrac{\sqrt{6}}{6}$ **75.** $\dfrac{6\sqrt{m} + 6\sqrt{n}}{m - n}$ **77.** $\dfrac{10}{3\sqrt{2}}$
79. $\dfrac{2}{\sqrt[3]{20}}$ **81.** $\dfrac{11}{\sqrt{33}}$
83. $\dfrac{76}{27 + 3\sqrt{5} - 9\sqrt{3} - \sqrt{15}}$ **85.** $\dfrac{a - b}{3a\sqrt{a} - 3a\sqrt{b}}$
87. $\sqrt[6]{y^5}$ **89.** 8   **91.** $\dfrac{1}{5}$ **93.** $\dfrac{a\sqrt[4]{a}}{\sqrt[4]{b^3}}$, or $a\sqrt[4]{\dfrac{a}{b^3}}$
95. $mn^2\sqrt[3]{m^2n}$ **97.** $17^{3/5}$ **99.** $12^{4/5}$
101. $11^{1/6}$ **103.** $5^{5/6}$ **105.** 4   **107.** $8a^2$
109. $\dfrac{x^3}{3b^{-2}}$, or $\dfrac{x^3b^2}{3}$ **111.** $x\sqrt[3]{y}$ **113.** $n\sqrt[3]{mn^2}$
115. $a\sqrt[12]{a^5} + a^2\sqrt[12]{a}$ **117.** $\sqrt[6]{288}$ **119.** $\sqrt[12]{x^{11}y^7}$
121. $a\sqrt[6]{a^5}$ **123.** $(a + x)\sqrt[12]{(a + x)^{11}}$
125. Discussion and Writing   **127.** $\dfrac{(2 + x^2)\sqrt{1 + x^2}}{1 + x^2}$
129. $a^{a/2}$

Review Exercises: Chapter R

1. True   **2.** False   **3.** True   **4.** True
5. $-7, 43, -\dfrac{4}{9}, 0, \sqrt[3]{64}, 4\dfrac{3}{4}, \dfrac{12}{7}, 102$ **6.** $43, 0, \sqrt[3]{64}, 102$
7. $-7, 43, 0, \sqrt[3]{64}, 102$ **8.** All of them
9. $43, \sqrt[3]{64}, 102$
10. $\sqrt{17}, 2.191191119\ldots, -\sqrt{2}, \sqrt[5]{5}$
11. $(-4, 7]$ **12.** 24   **13.** $\dfrac{7}{8}$ **14.** 10   **15.** 2

16. -10 **17.** 0.000083   **18.** 20,700,000
19. 4.05×10^5 **20.** 3.9×10^{-7} **21.** 1.395×10^3
22. 7.8125×10^{-22} **23.** $-12x^2y^{-4}$, or $\dfrac{-12x^2}{y^4}$
24. $8a^{-6}b^3c$, or $\dfrac{8b^3c}{a^6}$ **25.** 3   **26.** -2 **27.** $\dfrac{b}{a}$
28. $\dfrac{x + y}{xy}$ **29.** -4 **30.** $27 - 10\sqrt{2}$
31. $13\sqrt{5}$ **32.** $x^3 + t^3$ **33.** $10a^2 - 7ab - 12b^2$
34. $10x^2y + xy^2 + 2xy - 11$ **35.** $8x(4x^3 - 5y^3)$
36. $(y + 3)(y^2 - 2)$ **37.** $(x + 12)^2$
38. $x(9x - 1)(x + 4)$ **39.** $(3x - 5)^2$
40. $(2x - 1)(4x^2 + 2x + 1)$ **41.** $3(6x^2 - x + 2)$
42. $(x - 1)(2x + 3)(2x - 3)$
43. $6(x + 2)(x^2 - 2x + 4)$ **44.** $(ab - 3)(ab + 2)$
45. $(2x - 1)(x + 3)$ **46.** 7   **47.** -1 **48.** 3
49. $-\dfrac{1}{13}$ **50.** -8 **51.** $-4, 5$ **52.** $-6, \dfrac{1}{2}$
53. $-1, 3$ **54.** $-4, 4$ **55.** $-\sqrt{7}, \sqrt{7}$ **56.** 3
57. $\dfrac{x - 5}{(x + 5)(x + 3)}$ **58.** $y^3\sqrt[6]{y}$ **59.** $\sqrt[3]{(a + b)^2}$
60. $b\sqrt[5]{b^2}$ **61.** $\dfrac{m^4n^2}{3}$ **62.** $\dfrac{23 - 9\sqrt{3}}{22}$
63. About 18.8 ft   **64.** B   **65.** C
66. Discussion and Writing: Anya is probably not following the rules for order of operations. She is subtracting 6 from 15 first, then dividing the difference by 3, and finally multiplying the quotient by 4. The correct answer is 7.
67. Discussion and Writing: When the number 4 is raised to a positive integer power, the last digit of the result is 4 or 6. Since the calculator returns $4.398046511 \times 10^{12}$, or 4,398,046,511,000, we can conclude that this result is an approximation.   **68.** $553.67   **69.** $606.92
70. $942.54   **71.** $857.57   **72.** $x^{2n} + 6x^n - 40$
73. $t^{2a} + 2 + t^{-2a}$ **74.** $y^{2b} - z^{2c}$
75. $a^{3n} - 3a^{2n}b^n + 3a^nb^{2n} - b^{3n}$
76. $(y^n + 8)^2$ **77.** $(x^t - 7)(x^t + 4)$
78. $m^{3n}(m^n - 1)(m^{2n} + m^n + 1)$

Test: Chapter R

1. [R.1] (a) $0, \sqrt[3]{8}, 29$; (b) $\sqrt{12}$; (c) $0, -5$; (d) $6\dfrac{6}{7}, -\dfrac{13}{4}, -1.2$
2. [R.1] 17.6   **3.** [R.1] $\dfrac{15}{11}$ **4.** [R.1] $5|x||y|$
5. [R.1] $(-3, 6]$;

6. [R.1] 15   **7.** [R.2] -5 **8.** [R.2] 4.509×10^6
9. [R.2] 0.000086   **10.** [R.2] 7.5×10^6
11. [R.2] x^{-3}, or $\dfrac{1}{x^3}$ **12.** [R.2] $72y^{14}$
13. [R.2] $-15a^4b^{-1}$, or $-\dfrac{15a^4}{b}$
14. [R.3] $8xy^4 - 9xy^2 + 4x^2 + 2y - 7$
15. [R.3] $3y^2 - 2y - 8$ **16.** [R.3] $16x^2 - 24x + 9$
17. [R.6] $\dfrac{x - y}{xy}$ **18.** [R.7] $3\sqrt{5}$ **19.** [R.7] $2\sqrt[3]{7}$

20. [R.7] $21\sqrt{3}$ **21.** [R.7] $6\sqrt{5}$
22. [R.7] $4 + \sqrt{3}$ **23.** [R.4] $2(2x + 3)(2x - 3)$
24. [R.4] $(y + 3)(y - 6)$ **25.** [R.4] $(2n - 3)(n + 4)$
26. [R.4] $x(x + 5)^2$
27. [R.4] $(m - 2)(m^2 + 2m + 4)$ **28.** [R.5] 4
29. [R.5] $\frac{15}{4}$ **30.** [R.5] $-\frac{3}{2}, -1$
31. [R.5] $-\sqrt{11}, \sqrt{11}$ **32.** [R.6] $\dfrac{x - 5}{x - 2}$
33. [R.6] $\dfrac{x + 3}{(x + 1)(x + 5)}$ **34.** [R.7] $\dfrac{35 + 5\sqrt{3}}{46}$
35. [R.7] $\sqrt[8]{m^3}$ **36.** [R.7] $3^{5/6}$ **37.** [R.7] 13 ft
38. [R.3] $x^2 - 2xy + y^2 - 2x + 2y + 1$

CHAPTER 1

Visualizing the Graph

1. H   **2.** B   **3.** D   **4.** A   **5.** G   **6.** I   **7.** C
8. J   **9.** F   **10.** E

Exercise Set 1.1

1. A: $(-5, 4)$; B: $(2, -2)$; C: $(0, -5)$;
D: $(3, 5)$; E: $(-5, -4)$; F: $(3, 0)$

3. 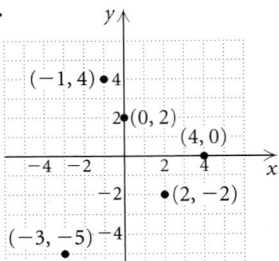   **5.**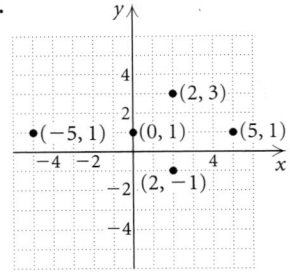

7. $(2002, 5.9), (2003, 5.9), (2004, 5.8), (2005, 5.6),$
$(2006, 5.2), (2007, 4.9)$ **9.** Yes; no
11. Yes; no   **13.** No; yes   **15.** No; yes
17. x-intercept: $(-3, 0)$;
y-intercept: $(0, 5)$;

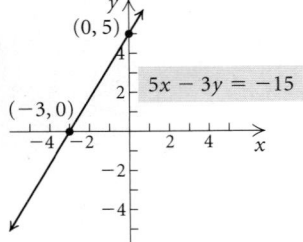

19. x-intercept: $(2, 0)$;
y-intercept: $(0, 4)$;

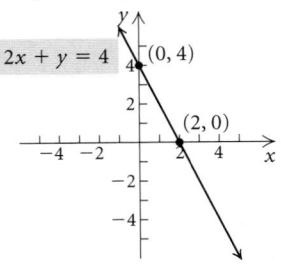

21. x-intercept: $(-4, 0)$;
y-intercept: $(0, 3)$;

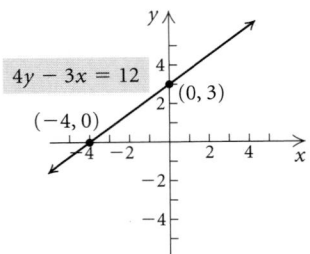

23. 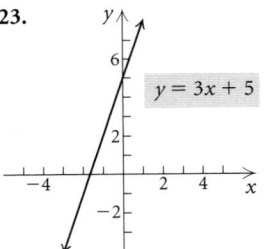   **25.**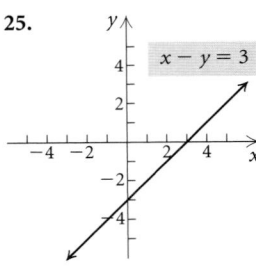

27. 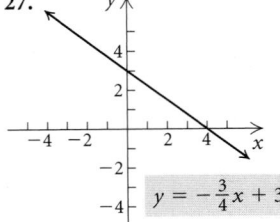   **29.**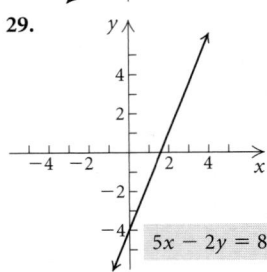

31. 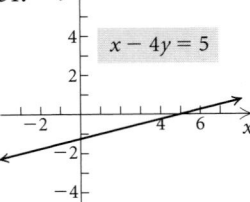   **33.**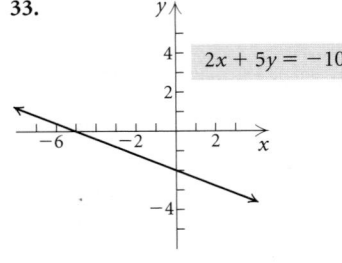

35.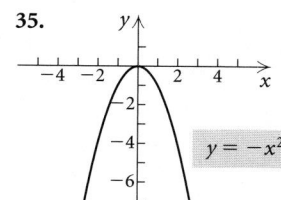
$y = -x^2$

37.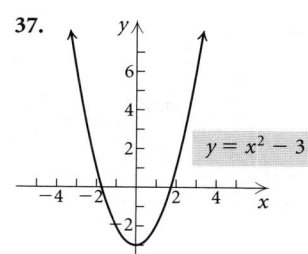
$y = x^2 - 3$

39.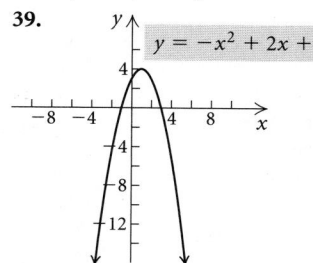
$y = -x^2 + 2x + 3$

41. (b)   **43.** (a)

45. $y = 2x + 1$ **47.** $4x + y = 7$

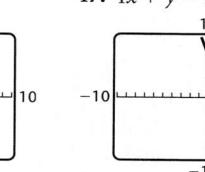

49. $y = \frac{1}{3}x + 2$ **51.** $2x + 3y = -5$

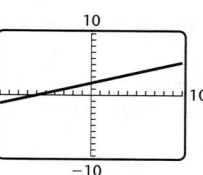   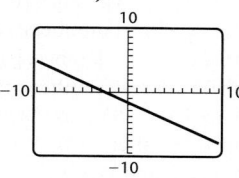

53. $y = x^2 + 6$ **55.** $y = 2 - x^2$

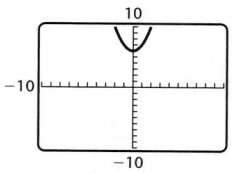   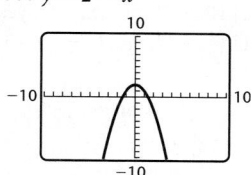

57. $y = x^2 + 4x - 2$ **59.** Standard window

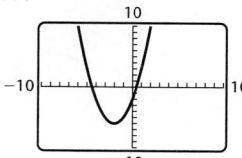

61. $[-1, 1, -0.3, 0.3]$ **63.** $\sqrt{10}, 3.162$ **65.** 13
67. $\sqrt{45}, 6.708$ **69.** 16   **71.** $\frac{14}{3}$ **73.** $\sqrt{128.05}, 11.316$

75. $\sqrt{a^2 + b^2}$ **77.** 6.5   **79.** Yes   **81.** No
83. $(-4, -6)$ **85.** $\left(-\frac{1}{5}, \frac{1}{4}\right)$ **87.** $(4.95, -4.95)$
89. $\left(-6, \frac{13}{2}\right)$ **91.** $\left(-\frac{5}{12}, \frac{13}{40}\right)$
93.

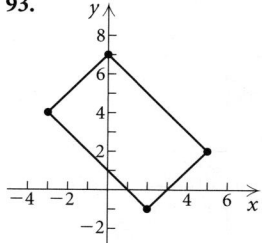

$\left(-\frac{1}{2}, \frac{3}{2}\right), \left(\frac{7}{2}, \frac{1}{2}\right), \left(\frac{5}{2}, \frac{9}{2}\right), \left(-\frac{3}{2}, \frac{11}{2}\right)$; no
95. $\left(\dfrac{\sqrt{7} + \sqrt{2}}{2}, -\dfrac{1}{2}\right)$ **97.** Square the window; for
example, use $[-12, 9, -4, 10]$.
99. $(x - 2)^2 + (y - 3)^2 = \dfrac{25}{9}$
101. $(x + 1)^2 + (y - 4)^2 = 25$
103. $(x - 2)^2 + (y - 1)^2 = 169$
105. $(x + 2)^2 + (y - 3)^2 = 4$

107. $(0, 0)$; 2;   **109.** $(0, 3)$; 4;

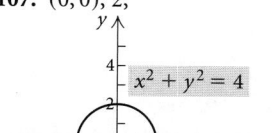   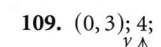 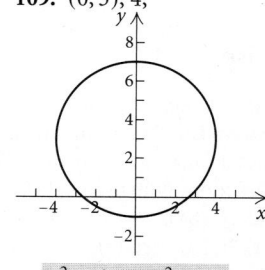

$x^2 + y^2 = 4$ $x^2 + (y - 3)^2 = 16$

111. $(1, 5)$; 6;   **113.** $(-4, -5)$; 3;

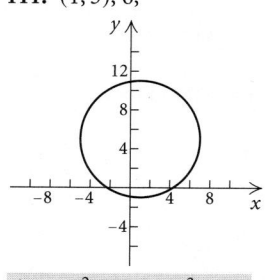   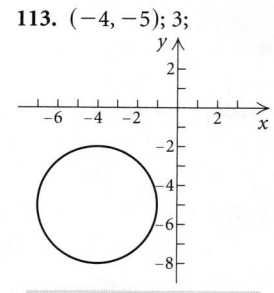

$(x - 1)^2 + (y - 5)^2 = 36$ $(x + 4)^2 + (y + 5)^2 = 9$
115. $(x + 2)^2 + (y - 1)^2 = 3^2$
117. $(x - 5)^2 + (y + 5)^2 = 15^2$
119. Discussion and Writing   **121.** Third
123. $\sqrt{h^2 + h + 2a - 2\sqrt{a^2 + ah}}$,
$\left(\dfrac{2a + h}{2}, \dfrac{\sqrt{a} + \sqrt{a + h}}{2}\right)$
125. $(x - 2)^2 + (y + 7)^2 = 36$ **127.** $(0, 4)$

129. $a_1 \approx 2.7$ ft, $a_2 \approx 37.3$ ft   **131.** Yes   **133.** Yes
135. Let $P_1 = (x_1, y_1)$, $P_2 = (x_2, y_2)$, and

$M = \left(\dfrac{x_1 + x_2}{2}, \dfrac{y_1 + y_2}{2} \right)$. Let $d(AB)$ denote the distance

from point A to point B.

(a) $d(P_1M) = \sqrt{\left(\dfrac{x_1 + x_2}{2} - x_1 \right)^2 + \left(\dfrac{y_1 + y_2}{2} - y_1 \right)^2}$

$= \dfrac{1}{2} \sqrt{(x_2 - x_1)^2 + (y_2 - y_1)^2}$;

$d(P_2M) = \sqrt{\left(\dfrac{x_1 + x_2}{2} - x_2 \right)^2 + \left(\dfrac{y_1 + y_2}{2} - y_2 \right)^2}$

$= \dfrac{1}{2} \sqrt{(x_1 - x_2)^2 + (y_1 - y_2)^2}$

$= \dfrac{1}{2} \sqrt{(x_2 - x_1)^2 + (y_2 - y_1)^2} = d(P_1M)$.

(b) $d(P_1M) + d(P_2M) = \dfrac{1}{2} \sqrt{(x_2 - x_1)^2 + (y_2 - y_1)^2}$

$+ \dfrac{1}{2} \sqrt{(x_2 - x_1)^2 + (y_2 - y_1)^2}$

$= \sqrt{(x_2 - x_1)^2 + (y_2 - y_1)^2}$

$= d(P_1P_2)$.

Exercise Set 1.2

1. Yes   **3.** Yes   **5.** No   **7.** Yes   **9.** Yes   **11.** Yes
13. No   **15.** Function; domain: $\{2, 3, 4\}$; range: $\{10, 15, 20\}$
17. Not a function; domain: $\{-7, -2, 0\}$; range: $\{3, 1, 4, 7\}$
19. Function; domain: $\{-2, 0, 2, 4, -3\}$; range: $\{1\}$
21. (a) 1; **(b)** 6; **(c)** 22; **(d)** $3x^2 + 2x + 1$; **(e)** $3t^2 - 4t + 2$
23. (a) 8; **(b)** -8; **(c)** $-x^3$; **(d)** $27y^3$;

(e) $8 + 12h + 6h^2 + h^3$ **25. (a)** $\dfrac{1}{8}$; **(b)** 0; **(c)** does not

exist; **(d)** $\dfrac{81}{53}$, or approximately 1.5283; **(e)** $\dfrac{x + h - 4}{x + h + 3}$

27. 0; does not exist; does not exist as a real number; $\dfrac{1}{\sqrt{3}}$,

or $\dfrac{\sqrt{3}}{3}$ **29.** $g(-2.1) \approx -21.8$; $g(5.08) \approx -130.4$;

$g(10.003) \approx -468.3$
31.

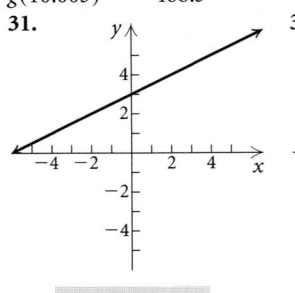

$f(x) = \frac{1}{2}x + 3$

33.

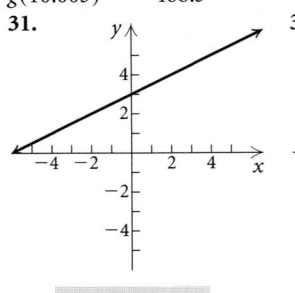

$f(x) = -x^2 + 4$

35.

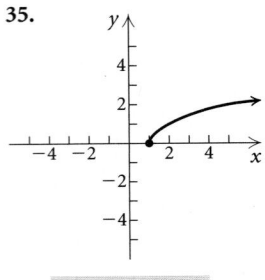

$f(x) = \sqrt{x - 1}$

37. $h(1) = -2$; $h(3) = 2$; $h(4) = 1$
39. $s(-4) = 3$; $s(-2) = 0$; $s(0) = -3$
41. $f(-1) = 2$; $f(0) = 0$; $f(1) = -2$
43. No   **45.** Yes   **47.** Yes   **49.** No
51. All real numbers, or $(-\infty, \infty)$
53. All real numbers, or $(-\infty, \infty)$
55. $\{x | x \neq 0\}$, or $(-\infty, 0) \cup (0, \infty)$
57. $\{x | x \neq 2\}$, or $(-\infty, 2) \cup (2, \infty)$
59. $\{x | x \neq -1 \text{ and } x \neq 5\}$, or $(-\infty, -1) \cup (-1, 5) \cup (5, \infty)$
61. $\{x | x \neq 0 \text{ and } x \neq 7\}$, or $(-\infty, 0) \cup (0, 7) \cup (7, \infty)$
63. All real numbers, or $(-\infty, \infty)$
65. Domain: $[0, 5]$; range: $[0, 3]$
67. Domain: $[-2\pi, 2\pi]$; range: $[-1, 1]$
69. Domain: $(-\infty, \infty)$; range: $\{-3\}$
71. Domain: $[-5, 3]$; range: $[-2, 2]$
73. Domain: all real numbers; range: $[0, \infty)$
75. Domain: $[-3, 3]$; range: $[0, 3]$
77. Domain: all real numbers; range: all real numbers
79. Domain: $(-\infty, 7]$; range: $[0, \infty)$
81. Domain: all real numbers; range: $(-\infty, 3]$
83. 645 m; 0 m   **85. (a)** 2008: \$20.68; 2015: \$23.57;
(b) about 41 yr after 1990, or in 2031
87. Discussion and Writing
88. [1.1] $(-3, -2)$, yes; $(2, -3)$, no
89. [1.1] $(0, -7)$, no; $(8, 11)$, yes
90. [1.1] $\left(\frac{4}{5}, -2\right)$, yes; $\left(\frac{11}{5}, \frac{1}{10}\right)$, yes
91. [1.1]   **92.** [1.1]

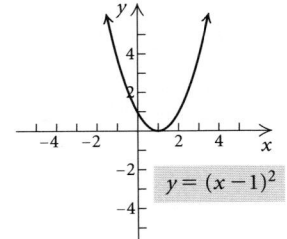

$y = (x - 1)^2$

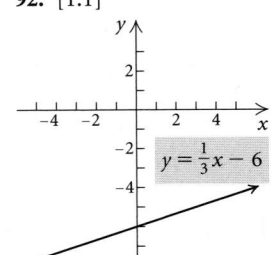

$y = \frac{1}{3}x - 6$

93. [1.1]

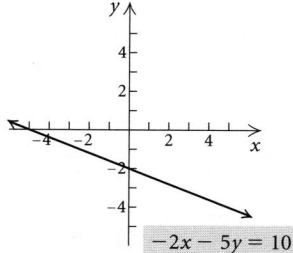

$-2x - 5y = 10$

94. [1.1]

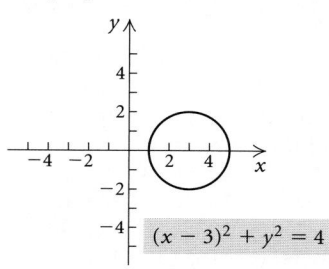

$(x - 3)^2 + y^2 = 4$

95. All real numbers, or $(-\infty, \infty)$
97. $\{x \mid x \le 8\}$, or $(-\infty, 8]$
99. $\{x \mid x \ge -6 \ and \ x \ne -2 \ and \ x \ne 3\}$, or $[-6, -2) \cup (-2, 3) \cup (3, \infty)$
101. $\{x \mid -5 \le x \le 3\}$, or $[-5, 3]$
103. $f(x) = x, g(x) = x + 1$
105.

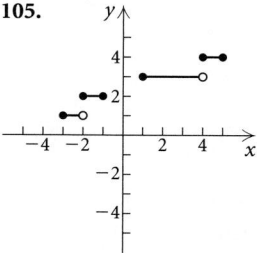

107. -7 **109.** (a) $g(x) = -2x + 1$; (b) $g(x) = 1$;
(c) $g(x) = 2x - 1$

Visualizing the Graph

1. E   **2.** D   **3.** A   **4.** J   **5.** C   **6.** F   **7.** H
8. G   **9.** B   **10.** I

Exercise Set 1.3

1. (a) Yes; (b) yes; (c) yes   **3.** (a) Yes; (b) no; (c) no
5. $\frac{6}{5}$ **7.** $-\frac{3}{5}$ **9.** 0   **11.** $\frac{1}{5}$ **13.** Not defined   **15.** 0.3
17. 0   **19.** $-\frac{6}{5}$ **21.** $-\frac{1}{3}$ **23.** Not defined   **25.** -2

27. 5   **29.** 1.3   **31.** Not defined   **33.** $-\frac{1}{2}$
35. -1 **37.** 0   **39.** The average rate of change in honey production from 1997 to 2006 was about -4.8 million lb per year.   **41.** The average rate of change in attendance was 1705 per race.   **43.** The average rate of change over the 5-yr period was -3.8 million per year.   **45.** $\frac{75}{457}$ mi per minute, or about 0.16 mi per minute   **47.** $\frac{3}{5}$; $(0, -7)$
49. Slope is not defined; there is no y-intercept.
51. $-\frac{1}{2}$; $(0, 5)$ **53.** $-\frac{3}{2}$; $(0, 5)$ **55.** 0; $(0, -6)$
57. $\frac{4}{5}$; $\left(0, \frac{8}{5}\right)$ **59.** $\frac{1}{4}$; $\left(0, -\frac{1}{2}\right)$
61.

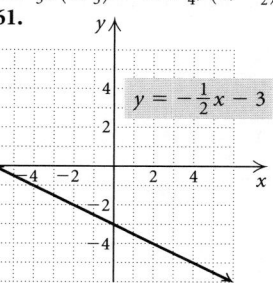

$y = -\frac{1}{2}x - 3$

63.

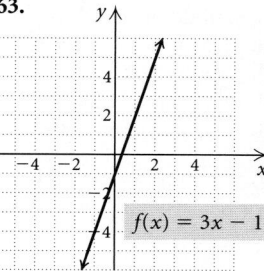

$f(x) = 3x - 1$

65.

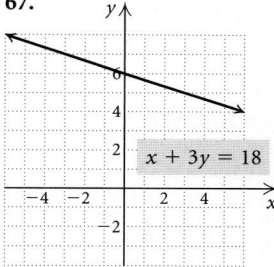

$3x - 4y = 20$

67.

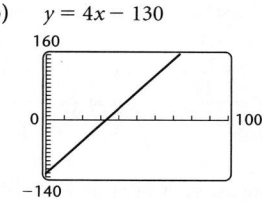

$x + 3y = 18$

69. (a) $W(h) = 4h - 130$;   (b) $y = 4x - 130$

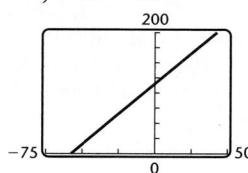

(c) 118 lb   **71.** (a) $y = 2x + 115$

(b) 115 ft, 75 ft, 135 ft, 179 ft;   (c) Below $-57.5°$, stopping distance is negative; above 32°, ice doesn't form.

73. (a) $\frac{11}{10}$. For each mile per hour faster that the car travels, it takes $\frac{11}{10}$ ft longer to stop; **(b)** $y = \frac{11}{10}x + \frac{1}{2}$

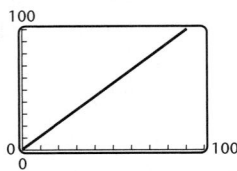

(c) 6 ft, 11.5 ft, 22.5 ft, 55.5 ft, 72 ft; **(d)** $\{r \mid r > 0\}$, or $(0, \infty)$. If r is allowed to be 0, the function says that a stopped car has a reaction distance of $\frac{1}{2}$ ft.
75. $C(t) = 60 + 29t$; $C(6) = \$234$
77. $C(x) = 800 + 3x$; $C(75) = \$1025$ **79.** Discussion and Writing **81.** [1.2] $-\frac{5}{4}$ **82.** [1.2] 10 **83.** [1.2] 40
84. [1.2] $a^2 + 3a$ **85.** [1.2] $a^2 + 2ah + h^2 - 3a - 3h$
87. $-\dfrac{d}{10c}$ **89.** 0 **91.** $2a + h$ **93.** False **95.** False
97. $f(x) = x + b$

Exercise Set 1.4

1. 4; $(0, -2)$; $y = 4x - 2$ **3.** -1, $(0, 0)$; $y = -x$
5. 0, $(0, -3)$; $y = -3$ **7.** $y = \frac{2}{9}x + 4$
9. $y = -4x - 7$ **11.** $y = -4.2x + \frac{3}{4}$
13. $y = \frac{2}{9}x + \frac{19}{3}$ **15.** $y = 8$ **17.** $y = -\frac{3}{5}x - \frac{17}{5}$
19. $y = -3x + 2$ **21.** $y = -\frac{1}{2}x + \frac{7}{2}$ **23.** $y = \frac{2}{3}x - 6$
25. $y = 7.3$ **27.** Horizontal: $y = -3$; vertical: $x = 0$
29. Horizontal: $y = -1$; vertical: $x = \frac{2}{11}$
31. $h(x) = -3x + 7$; 1 **33.** $f(x) = \frac{2}{5}x - 1$; -1
35. Perpendicular **37.** Neither parallel nor perpendicular
39. Parallel **41.** Perpendicular **43.** $y = \frac{2}{7}x + \frac{29}{7}$;
$y = -\frac{7}{2}x + \frac{31}{2}$ **45.** $y = -0.3x - 2.1$; $y = \frac{10}{3}x + \frac{70}{3}$
47. $y = -\frac{3}{4}x + \frac{1}{4}$; $y = \frac{4}{3}x - 6$ **49.** $x = 3$; $y = -3$
51. True **53.** True **55.** False **57.** No **59.** Yes
61. (a) Using $(5, 96.7)$ and $(13, 128.7)$, $y = 4x + 76.7$, where x is the number of years after 1990 and y is in thousands;
(b) 2009: about 152,700 twin births; 2012: about 164,700 twin births **63.** Using $(2, 300)$ and $(9, 328)$, $y = 4x + 292$, where x is the number of years after 1995 and y is in millions; 2006: 336 million; 2010: 352 million **65.** Using $(10, 321)$ and $(35, 955)$, $y = 25.36x + 67.4$, where x is the number of years after 1970; 2009: \$1056; 2012: \$1133; 2020: \$1335
67. (a) $M = 0.2H + 156$; **(b)** 164, 169, 171, 173; **(c)** $r = 1$; the regression line fits the data perfectly and should be a good predictor. **69. (a)** $y = 5.6524x + 8.731$, where x is the number of years since 1995 and y is in millions;
(b) 2010: 93,517,000 returns; 2014: 116,127,000 returns;
(c) $r \approx 0.9931$; the line fits the data well.
71. (a) $y = 3.176981132x + 88.15660377$, where x is the number of years after 1990 and y is in thousands; **(b)** about 158,050 twin births; this value is 6650 lower than the value found in Exercise 61; **(c)** $r \approx 0.9584$; the line fits the data fairly well.

73. Discussion and Writing **74.** [1.3] Not defined
75. [1.3] -1 **76.** [1.1] $x^2 + (y - 3)^2 = 6.25$
77. [1.1] $(x + 7)^2 + (y + 1)^2 = \frac{81}{25}$ **79.** -7.75

Exercise Set 1.5

1. 4 **3.** All real numbers, or $(-\infty, \infty)$ **5.** $-\frac{3}{4}$ **7.** -9
9. 6 **11.** No solution **13.** $\frac{11}{5}$ **15.** $\frac{35}{6}$ **17.** 8
19. -4 **21.** 6 **23.** -1 **25.** $\frac{4}{5}$ **27.** $-\frac{3}{2}$ **29.** $-\frac{2}{3}$
31. $\frac{1}{2}$ **33.** About \$6.9 billion **35.** \$2966 **37.** 406 GB
39. Great Smoky Mountains: 9.2 million visitors; Grand Canyon: 4.4 million visitors **41.** CBS: 11.4 million viewers; ABC: 9.7 million viewers; NBC: 8.0 million viewers
43. \$1300 **45.** \$9800 **47.** 12 mi **49.** 26°, 130°, 24°
51. Length: 93 m; width: 68 m **53.** Length: 100 yd; width: 65 yd **55.** 67.5 lb **57.** 3 hr **59.** 4.5 hr
61. 2.5 hr **63.** \$2400 at 3%; \$2600 at 4%
65. Yogurt: 452 mg; cheese: 224 mg **67.** 709 ft
69. 0.6336 in. **71.** -5 **73.** $\frac{11}{2}$ **75.** 16 **77.** -12
79. 6 **81.** 20 **83.** 25 **85.** 15 **87. (a)** $(4, 0)$; **(b)** 4
89. (a) $(-2, 0)$; **(b)** -2 **91. (a)** $(-4, 0)$; **(b)** -4
93. $b = \dfrac{2A}{h}$ **95.** $w = \dfrac{P - 2l}{2}$
97. $b_2 = \dfrac{2A}{h} - b_1$, or $\dfrac{2A - b_1 h}{h}$ **99.** $\pi = \dfrac{3V}{4r^3}$
101. $C = \dfrac{5}{9}(F - 32)$ **103.** $A = \dfrac{C - By}{x}$
105. $h = \dfrac{p - l - 2w}{2}$ **107.** $y = \dfrac{2x - 6}{3}$
109. $b = \dfrac{a}{1 + cd}$ **111.** $x = \dfrac{z}{y - y^2}$
113. Discussion and Writing **115.** [1.4] $y = -\frac{3}{4}x + \frac{13}{4}$
116. [1.4] $y = -\frac{3}{4}x + \frac{1}{4}$ **117.** [1.1] 13
118. [1.1] $\left(-1, \frac{1}{2}\right)$ **119.** [1.2] $f(-3) = \frac{1}{2}$; $f(0) = 0$; $f(3)$ does not exist. **120.** [1.3] $m = 7$; y-intercept: $\left(0, -\frac{1}{2}\right)$
121. Yes **123.** No **125.** $-\frac{2}{3}$ **127.** No; the 6-oz cup costs about 6.4% more per ounce. **129.** 11.25 mi

Exercise Set 1.6

1. $\{x \mid x > 5\}$, or $(5, \infty)$;
3. $\{x \mid x > 3\}$, or $(3, \infty)$;
5. $\{x \mid x \geq -3\}$, or $[-3, \infty)$;
7. $\left\{y \mid y \geq \frac{22}{13}\right\}$, or $\left[\frac{22}{13}, \infty\right)$;
9. $\{x \mid x > 6\}$, or $(6, \infty)$;

11. $\left\{x \,\middle|\, x \ge -\frac{5}{12}\right\}$, or $\left[-\frac{5}{12}, \infty\right)$;

13. $\left\{x \,\middle|\, x \le \frac{15}{34}\right\}$, or $\left(-\infty, \frac{15}{34}\right]$;

15. $\{x \mid x < 1\}$, or $(-\infty, 1)$;

17. $[-3, 3)$;

19. $[8, 10]$;

21. $[-7, -1]$;

23. $\left(-\frac{3}{2}, 2\right)$;

25. $(1, 5]$;

27. $\left(-\frac{11}{3}, \frac{13}{3}\right)$;

29. $(-\infty, -2] \cup (1, \infty)$;

31. $\left(-\infty, -\frac{7}{2}\right] \cup \left[\frac{1}{2}, \infty\right)$;

33. $(-\infty, 9.6) \cup (10.4, \infty)$;

35. $\left(-\infty, -\frac{57}{4}\right] \cup \left[-\frac{55}{4}, \infty\right)$;

37. More than 4 yr after 2002   **39.** Less than 4 hr
41. \$5000   **43.** More than 20 checks   **45.** Sales greater
than \$18,000   **47.** Discussion and Writing   **49.** [1.5] 3
50. [1.5] $\frac{5}{2}$, or 2.5   **51.** [1.5] $\frac{7}{5}$, or 1.4   **52.** [1.5] -5
53. [1.5] Mass transit: 14 million passengers; airlines: 1.8
million passengers   **54.** [1.5] \$14 million   **55.** $\left(-\frac{1}{4}, \frac{5}{9}\right]$
57. $\left(-\frac{1}{8}, \frac{1}{2}\right)$

Review Exercises: Chapter 1

1. False   **2.** True   **3.** True   **4.** False   **5.** True
6. False   **7.** Yes; no   **8.** Yes; no
9. x-intercept: $(3, 0)$;
y-intercept: $(0, -2)$;

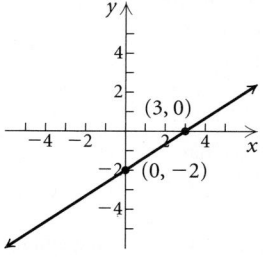

$2x - 3y = 6$

10. x-intercept: $(2, 0)$;
y-intercept: $(0, 5)$;

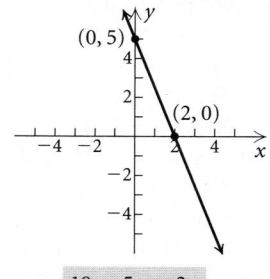

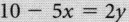

$10 - 5x = 2y$

11.

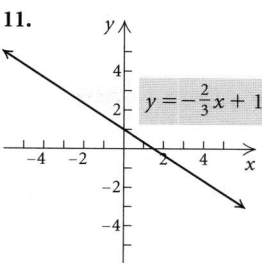

$y = -\frac{2}{3}x + 1$

12.

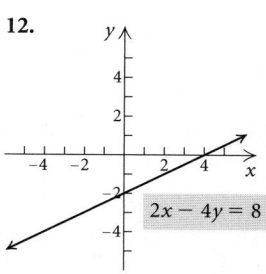

$2x - 4y = 8$

13.

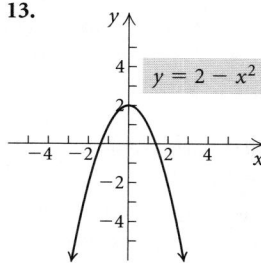

$y = 2 - x^2$

14. $\sqrt{34} \approx 5.831$ **15.** $\left(\frac{1}{2}, \frac{11}{2}\right)$ **16.** Center: $(-1, 3)$; radius: 3;

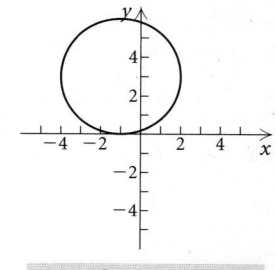

$(x + 1)^2 + (y - 3)^2 = 9$

17. $x^2 + (y + 4)^2 = \frac{9}{4}$ **18.** $(x + 2)^2 + (y - 6)^2 = 13$
19. $(x - 2)^2 + (y - 4)^2 = 26$ **20.** No   **21.** Yes
22. Not a function; domain: $\{3, 5, 7\}$; range: $\{1, 3, 5, 7\}$
23. Function; domain: $\{-2, 0, 1, 2, 7\}$; range: $\{-7, -4, -2, 2, 7\}$
24. (a) -3; (b) 9; (c) $a^2 - 3a - 1$; (d) $x^2 + x - 3$
25. (a) 0; (b) $\dfrac{x - 6}{x + 6}$; (c) does not exist; (d) $-\frac{5}{3}$
26. $f(2) = -1; f(-4) = -3; f(0) = -1$ **27.** No
28. Yes   **29.** No   **30.** Yes   **31.** All real numbers,
or $(-\infty, \infty)$ **32.** $\{x \mid x \ne 0\}$, or $(-\infty, 0) \cup (0, \infty)$
33. $\{x \mid x \ne 5 \text{ and } x \ne 1\}$, or $(-\infty, 1) \cup (1, 5) \cup (5, \infty)$
34. $\{x \mid x \ne -4 \text{ and } x \ne 4\}$, or $(-\infty, -4) \cup (-4, 4) \cup (4, \infty)$
35. Domain: $[-4, 4]$; range: $[0, 4]$ **36.** Domain: $(-\infty, \infty)$;
range: $[0, \infty)$ **37.** Domain: $(-\infty, \infty)$; range: $(-\infty, \infty)$
38. Domain: $(-\infty, \infty)$; range: $[0, \infty)$ **39.** (a) Yes; (b) no;
(c) no, strictly speaking, but data might be modeled by a
linear regression function.   **40.** (a) Yes; (b) yes; (c) yes
41. $\frac{5}{3}$ **42.** 0   **43.** Not defined   **44.** The average rate
of change over the 26-yr period was about \$0.08 per year.
45. $m = -\frac{7}{11}$; y-intercept: $(0, -6)$ **46.** $m = -2$;
y-intercept: $(0, -7)$

47.

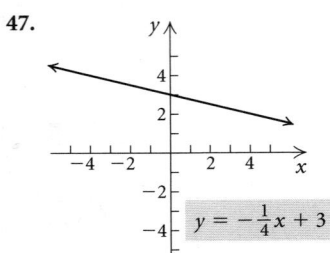

$y = -\frac{1}{4}x + 3$

48. $C(t) = 60 + 44t$; \$588   **49. (a)** 70°C, 220°C, 10,020°C;

(b) $[0, 5600]$ **50.** $y = -\frac{2}{3}x - 4$ **51.** $y = 3x + 5$
52. $y = \frac{1}{3}x - \frac{1}{3}$ **53.** Horizontal: $y = \frac{2}{5}$; vertical: $x = -4$
54. $h(x) = 2x - 5$; -5 **55.** Parallel   **56.** Neither
57. Perpendicular   **58.** $y = -\frac{2}{3}x - \frac{1}{3}$ **59.** $y = \frac{3}{2}x - \frac{5}{2}$
60. (a) Using $(3, 837)$ and $(15, 648)$: $y = -15.75x + 884.25$;
2009: 585 sites; **(b)** $y = -18.72380952x + 902.0952381$;
546 sites; $r \approx -0.9587$; the line fits the data well.   **61.** $\frac{3}{2}$

62. -6 **63.** -1 **64.** -21 **65.** $\frac{95}{24}$ **66.** No solution
67. All real numbers, or $(-\infty, \infty)$ **68.** China: 6493
adoptions; Russia: 3706 adoptions   **69.** \$2300   **70.** 3.4 hr

71. 3   **72.** 4   **73.** 0.2, or $\frac{1}{5}$ **74.** 4   **75.** $h = \dfrac{V}{lw}$

76. $s = \dfrac{M - n}{0.3}$ **77.** $t = \dfrac{A - P}{Pr}$

78. $(-\infty, 12)$;

79. $(-\infty, -4]$;

80. $\left[-\frac{4}{3}, \frac{4}{3}\right]$;

81. $\left(\frac{2}{5}, 2\right]$;

82. $\left(-\infty, -\frac{1}{2}\right) \cup (3, \infty)$;

83. $\left(-\infty, -\frac{5}{3}\right] \cup [1, \infty)$;

84. Years after 2004
85. Fahrenheit temperatures less than 113°   **86.** B   **87.** B
88. C   **89.** Discussion and Writing: If an equation contains
no fractions, using the addition principle before using the
multiplication principle eliminates the need to add or subtract
fractions.   **90.** Discussion and Writing: Think of the slopes
as $\dfrac{-3/5}{1}$ and $\dfrac{1/2}{1}$. The graph of $f(x)$ changes $\frac{3}{5}$ unit vertically
for each unit of horizontal change while the graph of $g(x)$
changes $\frac{1}{2}$ unit vertically for each unit of horizontal change.
Since $\frac{3}{5} > \frac{1}{2}$, the graph of $f(x) = -\frac{3}{5}x + 4$ is steeper than
the graph of $g(x) = \frac{1}{2}x - 6$.   **91.** $\left(\frac{5}{2}, 0\right)$

92. $\{x \mid x < 0\}$, or $(-\infty, 0)$
93. $\{x \mid x \neq -3 \text{ and } x \neq 0 \text{ and } x \neq 3\}$, or
$(-\infty, -3) \cup (-3, 0) \cup (0, 3) \cup (3, \infty)$

Test: Chapter 1

1. [1.1] Yes
2. [1.1] x-intercept: $(-2, 0)$; y-intercept: $(0, 5)$;

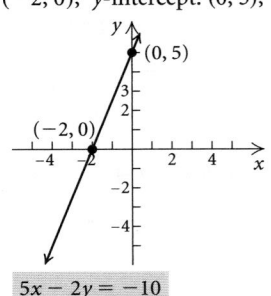

$5x - 2y = -10$

3. [1.1] $\sqrt{45} \approx 6.708$ **4.** [1.1] $\left(-3, \frac{9}{2}\right)$ **5.** [1.1] Center:
$(-4, 5)$; radius: 6   **6.** [1.1] $(x + 1)^2 + (y - 2)^2 = 5$
7. [1.2] **(a)** Yes; **(b)** $\{-4, 3, 1, 0\}$; **(c)** $\{7, 0, 5\}$ **8.** [1.2] **(a)** 8;
(b) $2a^2 + 7a + 11$ **9.** [1.2] **(a)** Does not exist; **(b)** 0
10. [1.2] 0   **11.** [1.2] **(a)** No; **(b)** yes
12. [1.2] $\{x \mid x \neq 4\}$, or $(-\infty, 4) \cup (4, \infty)$
13. [1.2] All real numbers, or $(-\infty, \infty)$
14. [1.2] $\{x \mid -5 \leq x \leq 5\}$, or $[-5, 5]$
15. [1.2] **(a)**

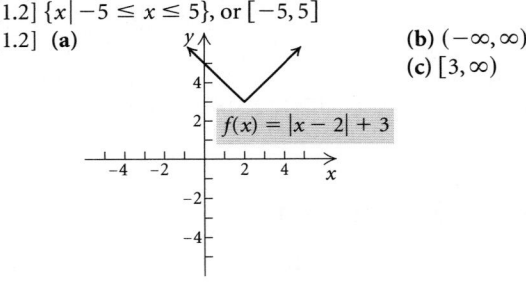

$f(x) = |x - 2| + 3$

(b) $(-\infty, \infty)$;
(c) $[3, \infty)$

16. [1.3] Not defined   **17.** [1.3] $-\frac{11}{6}$ **18.** [1.3] 0
19. [1.3] The average rate of change in weekend attendance
from 1995 to 2004 was about 0.167 million, or 167,000,
per year.   **20.** [1.3] Slope: $\frac{3}{2}$; y-intercept: $\left(0, \frac{5}{2}\right)$
21. [1.3] $C(t) = 80 + 39.95t$; \$1038.80
22. [1.4] $y = -\frac{5}{8}x - 5$ **23.** [1.4] $y - 4 = -\frac{3}{4}(x - (-5))$,
or $y - (-2) = -\frac{3}{4}(x - 3)$, or $y = -\frac{3}{4}x + \frac{1}{4}$
24. [1.4] $x = -\frac{3}{8}$ **25.** [1.4] Perpendicular
26. [1.4] $y - 3 = -\frac{1}{2}(x + 1)$, or $y = -\frac{1}{2}x + \frac{5}{2}$
27. [1.4] $y - 3 = 2(x + 1)$, or $y = 2x + 5$
28. [1.4] **(a)** Using $(0, 203)$ and $(3, 212)$, $y = 3x + 203$, where
x is the number of years after 2002 and y is in billions; 2010:
227 billion pieces of mail; **(b)** $y = 3x + 201.2$, where x is the
number of years after 2002 and y is in billions; 2010: 225.2
billion pieces of mail; $r \approx 0.9357$ **29.** [1.5] -1

30. [1.5] All real numbers, or $(-\infty, \infty)$ **31.** [1.5] -60
32. [1.5] $\frac{21}{11}$ **33.** [1.5] Length: 60 m; width: 45 m
34. [1.5] \$1.80   **35.** [1.5] -3 **36.** [1.5] $h = \frac{3V}{2\pi r^2}$
37. [1.5] $s = \dfrac{r}{1-t}$

38. [1.6] $(-\infty, -3]$;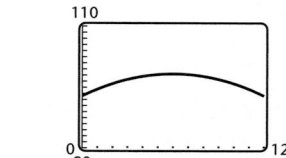

39. [1.6] $(-5, 3)$;

40. [1.6] $(-\infty, 2] \cup [4, \infty)$;

41. [1.6] More than 6 hr   **42.** [1.3] B   **43.** [1.2] -2

CHAPTER 2

Exercise Set 2.1

1. (a) $(-5, 1)$; (b) $(3, 5)$; (c) $(1, 3)$
3. (a) $(-3, -1), (3, 5)$; (b) $(1, 3)$; (c) $(-5, -3)$
5. (a) $(-\infty, -8), (-3, -2)$; (b) $(-8, -6)$;
(c) $(-6, -3), (-2, \infty)$ **7.** Domain: $[-5, 5]$; range: $[-3, 3]$
9. Domain: $[-5, -1] \cup [1, 5]$; range: $[-4, 6]$
11. Domain: $(-\infty, \infty)$; range: $(-\infty, 3]$
13. Relative maximum: 3.25 at $x = 2.5$; increasing: $(-\infty, 2.5)$;
decreasing: $(2.5, \infty)$
15. Relative maximum: 2.370 at $x = -0.667$; relative
minimum: 0 at $x = 2$; increasing: $(-\infty, -0.667), (2, \infty)$;
decreasing: $(-0.667, 2)$
17. Increasing: $(0, \infty)$; decreasing: $(-\infty, 0)$; relative minimum:
0 at $x = 0$
19. Increasing: $(-\infty, 0)$; decreasing: $(0, \infty)$; relative maximum:
5 at $x = 0$
21. Increasing: $(3, \infty)$; decreasing: $(-\infty, 3)$; relative minimum:
1 at $x = 3$
23. Increasing: $(1, 3)$; decreasing: $(-\infty, 1), (3, \infty)$; relative
maximum: -4 at $x = 3$; relative minimum: -8 at $x = 1$
25. Increasing: $(-1.552, 0), (1.552, \infty)$; decreasing:
$(-\infty, -1.552), (0, 1.552)$; relative maximum: 4.07 at $x = 0$;
relative minima: -2.314 at $x = -1.552$, -2.314 at $x = 1.552$
27. (a) $y = -0.1x^2 + 1.2x + 98.6$ (b) 6 days after the illness

began, the
temperature is 102.2°F.

29. Increasing: $(-1, 1)$; decreasing: $(-\infty, -1), (1, \infty)$
31. Increasing: $(-1.414, 1.414)$; decreasing: $(-2, -1.414)$,
$(1.414, 2)$ **33.** $A(x) = 30x - x^2$

35. $d(t) = \sqrt{(120t)^2 + 400^2}$ **37.** $A(w) = 10w - \dfrac{w^2}{2}$

39. $d(s) = \dfrac{14}{s}$ **41.** (a) $A(x) = x(30 - x)$, or $30x - x^2$;
(b) $\{x \,|\, 0 < x < 30\}$; (c) 15 ft by 15 ft
43. (a) $V(x) = x(12 - 2x)(12 - 2x)$, or $4x(6 - x)^2$;
(b) $\{x \,|\, 0 < x < 6\}$; (c) $y = 4x(6 - x)^2$

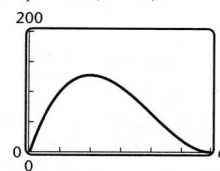

(d) 8 cm by 8 cm by 2 cm   **45.** (a) $A(x) = x\sqrt{256 - x^2}$;
(b) $\{x \,|\, 0 < x < 16\}$; (c) $y = x\sqrt{256 - x^2}$

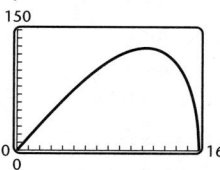

(d) 11.314 ft by 11.314 ft
47. $g(-4) = 0$; $g(0) = 4$; $g(1) = 5$; $g(3) = 5$
49. $h(-5) = 1$; $h(0) = 1$; $h(1) = 3$; $h(4) = 6$
51.

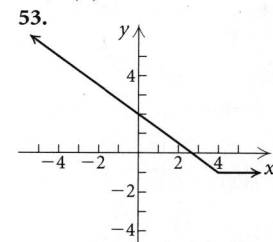

53.

55.

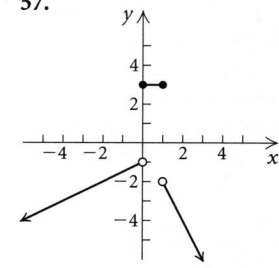

57.

59.

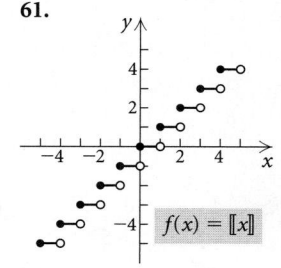

61.

$f(x) = [\![x]\!]$

63.

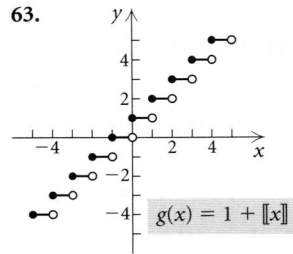

$g(x) = 1 + [\![x]\!]$

65. Domain: $(-\infty, \infty)$; range: $(-\infty, 0) \cup [3, \infty)$
67. Domain: $(-\infty, \infty)$; range: $[-1, \infty)$
69. Domain: $(-\infty, \infty)$; range: $\{y \mid y \le -2 \text{ or } y = -1 \text{ or } y \ge 2\}$
71. Domain: $(-\infty, \infty)$; range: $\{-5, -2, 4\}$;

$$f(x) = \begin{cases} -2, & \text{for } x < 2, \\ -5, & \text{for } x = 2, \\ 4, & \text{for } x > 2 \end{cases}$$

73. Domain: $(-\infty, \infty)$; range: $(-\infty, -1] \cup [2, \infty)$;

$$g(x) = \begin{cases} x, & \text{for } x \le -1, \\ 2, & \text{for } -1 < x < 2, \\ x, & \text{for } x \ge 2 \end{cases}$$

or

$$g(x) = \begin{cases} x, & \text{for } x \le -1, \\ 2, & \text{for } -1 < x \le 2, \\ x, & \text{for } x > 2 \end{cases}$$

75. Domain: $[-5, 3]$; range: $(-3, 5)$;

$$h(x) = \begin{cases} x + 8, & \text{for } -5 \le x < -3, \\ 3, & \text{for } -3 \le x \le 1, \\ 3x - 6, & \text{for } 1 < x \le 3 \end{cases}$$

77. Discussion and Writing   **79.** [1.2] Function; domain; range; domain; exactly one; range   **80.** [1.1] Midpoint formula   **81.** [1.1] x-intercept   **82.** [1.3] Constant; identity   **83.** Increasing: $(-5, -2)$, $(4, \infty)$; decreasing: $(-\infty, -5)$, $(-2, 4)$; relative maximum: 560 at $x = -2$; relative minima: 425 at $x = -5$, -304 at $x = 4$
85. (a) C **(b)** $C(t) = 2([\![t]\!] + 1)$, $t > 0$

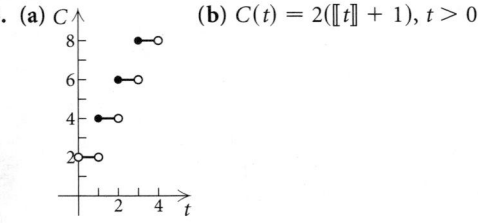

87. $\{x \mid -5 \le x < -4 \text{ or } 5 \le x < 6\}$

89. (a) $h(r) = \dfrac{30 - 5r}{3}$; **(b)** $V(r) = \pi r^2 \left(\dfrac{30 - 5r}{3} \right)$;

(c) $V(h) = \pi h \left(\dfrac{30 - 3h}{5} \right)^2$

Exercise Set 2.2

1. 33   **3.** -1 **5.** Does not exist   **7.** 0   **9.** 1
11. Does not exist   **13.** 0   **15.** 5
17. (a) Domain of f, g, $f + g$, $f - g$, fg, and ff: $(-\infty, \infty)$; domain of f/g: $\left(-\infty, \frac{3}{5}\right) \cup \left(\frac{3}{5}, \infty\right)$;
domain of g/f: $\left(-\infty, -\frac{3}{2}\right) \cup \left(-\frac{3}{2}, \infty\right)$;
(b) $(f + g)(x) = -3x + 6$; $(f - g)(x) = 7x$;
$(fg)(x) = -10x^2 - 9x + 9$; $(ff)(x) = 4x^2 + 12x + 9$;
$(f/g)(x) = \dfrac{2x + 3}{3 - 5x}$; $(g/f)(x) = \dfrac{3 - 5x}{2x + 3}$
19. (a) Domain of f: $(-\infty, \infty)$; domain of g: $[-4, \infty)$;
domain of $f + g$, $f - g$, and fg: $[-4, \infty)$;
domain of ff: $(-\infty, \infty)$; domain of f/g: $(-4, \infty)$;
domain of g/f: $[-4, 3) \cup (3, \infty)$;
(b) $(f + g)(x) = x - 3 + \sqrt{x + 4}$;
$(f - g)(x) = x - 3 - \sqrt{x + 4}$; $(fg)(x) = (x - 3)\sqrt{x + 4}$;
$(ff)(x) = x^2 - 6x + 9$; $(f/g)(x) = \dfrac{x - 3}{\sqrt{x + 4}}$;
$(g/f)(x) = \dfrac{\sqrt{x + 4}}{x - 3}$
21. (a) Domain of f, g, $f + g$, $f - g$, fg, and ff: $(-\infty, \infty)$;
domain of f/g: $(-\infty, 0) \cup (0, \infty)$;
domain of g/f: $\left(-\infty, \frac{1}{2}\right) \cup \left(\frac{1}{2}, \infty\right)$
(b) $(f + g)(x) = -2x^2 + 2x - 1$;
$(f - g)(x) = 2x^2 + 2x - 1$; $(fg)(x) = -4x^3 + 2x^2$;
$(ff)(x) = 4x^2 - 4x + 1$; $(f/g)(x) = \dfrac{2x - 1}{-2x^2}$;
$(g/f)(x) = \dfrac{-2x^2}{2x - 1}$
23. (a) Domain of f: $[3, \infty)$; domain of g: $[-3, \infty)$;
domain of $f + g$, $f - g$, fg, and ff: $[3, \infty)$;
domain of f/g: $[3, \infty)$; domain of g/f: $(3, \infty)$;
(b) $(f + g)(x) = \sqrt{x - 3} + \sqrt{x + 3}$;
$(f - g)(x) = \sqrt{x - 3} - \sqrt{x + 3}$; $(fg)(x) = \sqrt{x^2 - 9}$;
$(ff)(x) = |x - 3|$; $(f/g)(x) = \dfrac{\sqrt{x - 3}}{\sqrt{x + 3}}$; $(g/f)(x) = \dfrac{\sqrt{x + 3}}{\sqrt{x - 3}}$
25. (a) Domain of f, g, $f + g$, $f - g$, fg, and ff: $(-\infty, \infty)$;
domain of f/g: $(-\infty, 0) \cup (0, \infty)$;
domain of g/f: $(-\infty, -1) \cup (-1, \infty)$;
(b) $(f + g)(x) = x + 1 + |x|$; $(f - g)(x) = x + 1 - |x|$;
$(fg)(x) = (x + 1)|x|$; $(ff)(x) = x^2 + 2x + 1$;
$(f/g)(x) = \dfrac{x + 1}{|x|}$; $(g/f)(x) = \dfrac{|x|}{x + 1}$
27. (a) Domain of f, g, $f + g$, $f - g$, fg, and ff: $(-\infty, \infty)$;
domain of f/g: $(-\infty, -3) \cup \left(-3, \frac{1}{2}\right) \cup \left(\frac{1}{2}, \infty\right)$;
domain of g/f: $(-\infty, 0) \cup (0, \infty)$;
(b) $(f + g)(x) = x^3 + 2x^2 + 5x - 3$;
$(f - g)(x) = x^3 - 2x^2 - 5x + 3$;
$(fg)(x) = 2x^5 + 5x^4 - 3x^3$; $(ff)(x) = x^6$;
$(f/g)(x) = \dfrac{x^3}{2x^2 + 5x - 3}$; $(g/f)(x) = \dfrac{2x^2 + 5x - 3}{x^3}$

29. (a) Domain of f: $(-\infty, -1) \cup (-1, \infty)$;
domain of g: $(-\infty, 6) \cup (6, \infty)$; domain of $f + g, f - g$, and
fg: $(-\infty, -1) \cup (-1, 6) \cup (6, \infty)$;
domain of ff: $(-\infty, -1) \cup (-1, \infty)$;
domain of f/g and g/f: $(-\infty, -1) \cup (-1, 6) \cup (6, \infty)$;
(b) $(f + g)(x) = \dfrac{4}{x + 1} + \dfrac{1}{6 - x}$;

$(f - g)(x) = \dfrac{4}{x + 1} - \dfrac{1}{6 - x}$; $(fg)(x) = \dfrac{4}{(x + 1)(6 - x)}$;

$(ff)(x) = \dfrac{16}{(x + 1)^2}$; $(f/g)(x) = \dfrac{4(6 - x)}{x + 1}$;

$(g/f)(x) = \dfrac{x + 1}{4(6 - x)}$
31. (a) Domain of f: $(-\infty, 0) \cup (0, \infty)$;
domain of g: $(-\infty, \infty)$; domain of $f + g, f - g, fg$, and ff:
$(-\infty, 0) \cup (0, \infty)$; domain of f/g: $(-\infty, 0) \cup (0, 3) \cup (3, \infty)$;
domain of g/f: $(-\infty, 0) \cup (0, \infty)$;
(b) $(f + g)(x) = \dfrac{1}{x} + x - 3$; $(f - g)(x) = \dfrac{1}{x} - x + 3$;

$(fg)(x) = 1 - \dfrac{3}{x}$; $(ff)(x) = \dfrac{1}{x^2}$; $(f/g)(x) = \dfrac{1}{x(x - 3)}$;

$(g/f)(x) = x(x - 3)$
33. Domain of F: $[2, 11]$; domain of G: $[1, 9]$; domain of
$F + G$: $[2, 9]$ **35.** $[2, 3) \cup (3, 9]$
37.

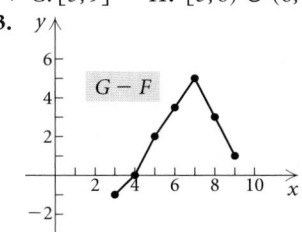

39. Domain of F: $[0, 9]$; domain of G: $[3, 10]$; domain of
$F + G$: $[3, 9]$ **41.** $[3, 6) \cup (6, 8) \cup (8, 9]$
43.

45. (a) $P(x) = -0.4x^2 + 57x - 13$; **(b)** $R(100) = 2000$;
$C(100) = 313$; $P(100) = 1687$; **(c)** Left to the student

47. 3   **49.** 6   **51.** $\frac{1}{3}$ **53.** $\dfrac{-1}{3x(x + h)}$, or $-\dfrac{1}{3x(x + h)}$

55. $\dfrac{1}{4x(x + h)}$ **57.** $2x + h$ **59.** $-2x - h$

61. $6x + 3h - 2$ **63.** $\dfrac{5|x + h| - 5|x|}{h}$

65. $3x^2 + 3xh + h^2$ **67.** $\dfrac{7}{(x + h + 3)(x + 3)}$
69. Discussion and Writing
71. [1.4] (c)   **72.** [1.4] None   **73.** [1.3] (b), (d), (f), and (h)
74. [1.3] (b)   **75.** [1.4] (a)   **76.** [1.4] (c) and (g)
77. [1.4] (c) and (g)   **78.** [1.4] (a) and (f)

79. $f(x) = \dfrac{1}{x + 7}, g(x) = \dfrac{1}{x - 3}$; answers may vary

81. $(-\infty, -1) \cup (-1, 1) \cup (1, \tfrac{7}{3}) \cup (\tfrac{7}{3}, 3) \cup (3, \infty)$

Exercise Set 2.3

1. -8 **3.** 64   **5.** 218   **7.** -80
9. $(f \circ g)(x) = (g \circ f)(x) = x$;
domain of $f \circ g$ and $g \circ f$: $(-\infty, \infty)$
11. $(f \circ g)(x) = 3x^2 - 2x$; $(g \circ f)(x) = 3x^2 + 4x$; domain of
$f \circ g$ and $g \circ f$: $(-\infty, \infty)$
13. $(f \circ g)(x) = 16x^2 - 24x + 6$; $(g \circ f)(x) = 4x^2 - 15$;
domain of $f \circ g$ and $g \circ f$: $(-\infty, \infty)$

15. $(f \circ g)(x) = \dfrac{4x}{x - 5}$; $(g \circ f)(x) = \dfrac{1 - 5x}{4}$;

domain of $f \circ g$: $(-\infty, 0) \cup (0, 5) \cup (5, \infty)$;
domain of $g \circ f$: $\left(-\infty, \tfrac{1}{5}\right) \cup \left(\tfrac{1}{5}, \infty\right)$
17. $(f \circ g)(x) = (g \circ f)(x) = x$;
domain of $f \circ g$ and $g \circ f$: $(-\infty, \infty)$
19. $(f \circ g)(x) = 2\sqrt{x} + 1$; $(g \circ f)(x) = \sqrt{2x + 1}$;
domain of $f \circ g$: $[0, \infty)$; domain of $g \circ f$: $\left[-\tfrac{1}{2}, \infty\right)$
21. $(f \circ g)(x) = 20$; $(g \circ f)(x) = 0.05$; domain of $f \circ g$ and
$g \circ f$: $(-\infty, \infty)$
23. $(f \circ g)(x) = |x|$; $(g \circ f)(x) = x$;
domain of $f \circ g$: $(-\infty, \infty)$; domain of $g \circ f$: $[-5, \infty)$
25. $(f \circ g)(x) = 5 - x$; $(g \circ f)(x) = \sqrt{1 - x^2}$;
domain of $f \circ g$: $(-\infty, 3]$; domain of $g \circ f$: $[-1, 1]$
27. $(f \circ g)(x) = (g \circ f)(x) = x$;
domain of $f \circ g$: $(-\infty, -1) \cup (-1, \infty)$;
domain of $g \circ f$: $(-\infty, 0) \cup (0, \infty)$
29. $(f \circ g)(x) = x^3 - 2x^2 - 4x + 6$;
$(g \circ f)(x) = x^3 - 5x^2 + 3x + 8$;
domain of $f \circ g$ and $g \circ f$: $(-\infty, \infty)$
31. $f(x) = x^5$; $g(x) = 4 + 3x$

33. $f(x) = \dfrac{1}{x}$; $g(x) = (x - 2)^4$

35. $f(x) = \dfrac{x - 1}{x + 1}$; $g(x) = x^3$

37. $f(x) = x^6$; $g(x) = \dfrac{2 + x^3}{2 - x^3}$

39. $f(x) = \sqrt{x}$; $g(x) = \dfrac{x - 5}{x + 2}$

41. $f(x) = x^3 - 5x^2 + 3x - 1$; $g(x) = x + 2$
43. $f(x) = x + 1$
45. Discussion and Writing

47. $[1.1], [1.3]$

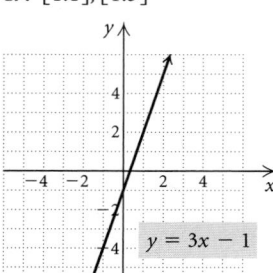

$y = 3x - 1$

48. $[1.1], [1.3]$

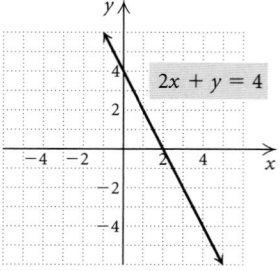

$2x + y = 4$

49. Start with the graph of $f(x) = x^2$. Shift it right 3 units.

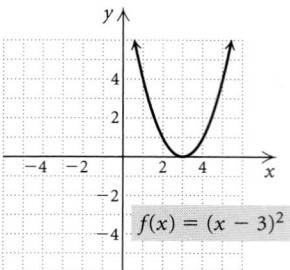

$f(x) = (x - 3)^2$

49. $[1.1], [1.3]$

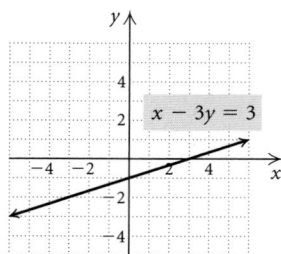

$x - 3y = 3$

50. $[1.1]$

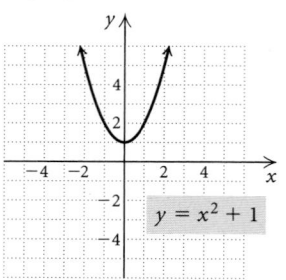

$y = x^2 + 1$

51. Only $(c \circ p)(a)$ makes sense. It represents the cost of the grass seed required to seed a lawn with area a.

Visualizing the Graph

1. C   **2.** B   **3.** A   **4.** E   **5.** G   **6.** D   **7.** H
8. I   **9.** F

Exercise Set 2.4

1. x-axis, no; y-axis, yes; origin, no
3. x-axis, yes; y-axis, no; origin, no
5. x-axis, no; y-axis, no; origin, yes
7. x-axis, no; y-axis, yes; origin, no
9. x-axis, no; y-axis, no; origin, no
11. x-axis, no; y-axis, yes; origin, no
13. x-axis, no; y-axis, no; origin, yes
15. x-axis, no; y-axis, no; origin, yes
17. x-axis, yes; y-axis, yes; origin, yes
19. x-axis, no; y-axis, yes; origin, no
21. x-axis, yes; y-axis, yes; origin, yes
23. x-axis, no; y-axis, no; origin, no
25. x-axis, no; y-axis, no; origin, yes
27. x-axis: $(-5, -6)$; y-axis: $(5, 6)$; origin: $(5, -6)$
29. x-axis: $(-10, 7)$; y-axis: $(10, -7)$; origin: $(10, 7)$
31. x-axis: $(0, 4)$; y-axis: $(0, -4)$; origin: $(0, 4)$
33. Even   **35.** Odd   **37.** Neither   **39.** Odd   **41.** Even
43. Odd   **45.** Neither   **47.** Even

51. Start with the graph of $g(x) = x$. Shift it down 3 units.

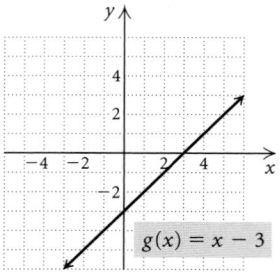

$g(x) = x - 3$

53. Start with the graph of $h(x) = \sqrt{x}$. Reflect it across the x-axis.

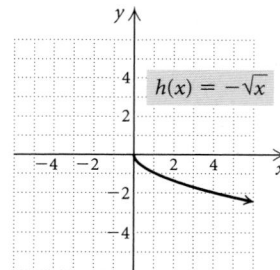

$h(x) = -\sqrt{x}$

55. Start with the graph of $h(x) = \dfrac{1}{x}$. Shift it up 4 units.

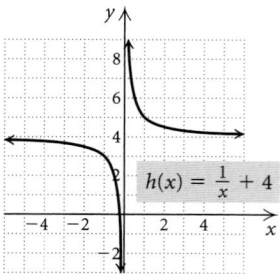

$h(x) = \frac{1}{x} + 4$

57. Start with the graph of $h(x) = x$. Stretch it vertically by multiplying each y-coordinate by 3. Then reflect it across the x-axis and shift it up 3 units.

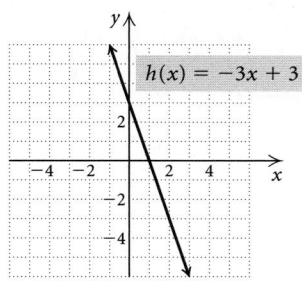

$h(x) = -3x + 3$

59. Start with the graph of $h(x) = |x|$. Shrink it vertically by multiplying each y-coordinate by $\frac{1}{2}$. Then shift it down 2 units.

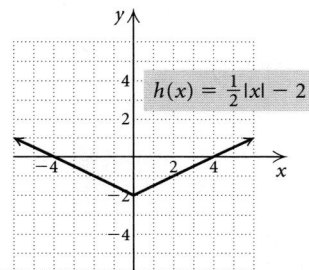

$h(x) = \frac{1}{2}|x| - 2$

61. Start with the graph of $g(x) = x^3$. Shift it right 2 units. Then reflect it across the x-axis.

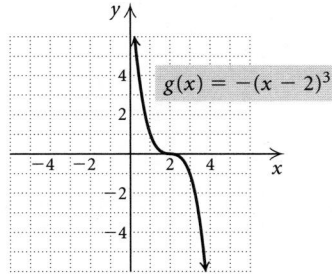

$g(x) = -(x - 2)^3$

63. Start with the graph of $g(x) = x^2$. Shift it left 1 unit. Then shift it down 1 unit.

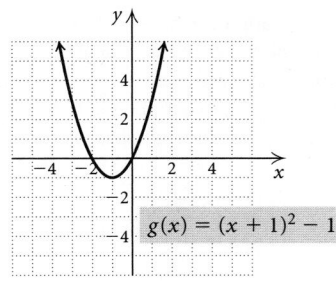

$g(x) = (x + 1)^2 - 1$

65. Start with the graph of $g(x) = x^3$. Shrink it vertically by multiplying each y-coordinate by $\frac{1}{3}$. Then shift it up 2 units.

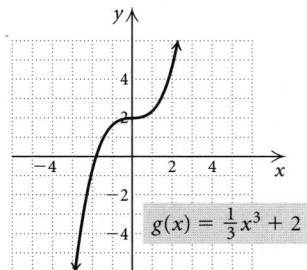

$g(x) = \frac{1}{3}x^3 + 2$

67. Start with the graph of $f(x) = \sqrt{x}$. Shift it left 2 units.

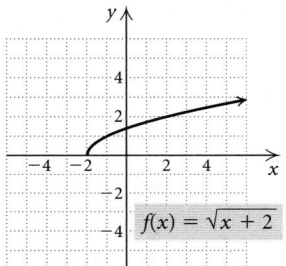

$f(x) = \sqrt{x + 2}$

69. Start with the graph of $f(x) = \sqrt[3]{x}$. Shift it down 2 units.

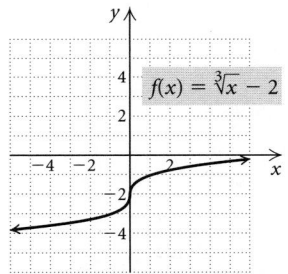

$f(x) = \sqrt[3]{x} - 2$

71. Start with the graph of $f(x) = |x|$. Shrink it horizontally by multiplying each x-coordinate by $\frac{1}{3}$ (or dividing each x-coordinate by 3).

73. Start with the graph of $h(x) = \dfrac{1}{x}$. Stretch it vertically by multiplying each y-coordinate by 2.

75. Start with the graph of $g(x) = \sqrt{x}$. Stretch it vertically by multiplying each y-coordinate by 3. Then shift it down 5 units.

77. Start with the graph of $f(x) = |x|$. Stretch it horizontally by multiplying each x-coordinate by 3. Then shift it down 4 units.

79. Start with the graph of $g(x) = x^2$. Shift it right 5 units, shrink it vertically by multiplying each y-coordinate by $\frac{1}{4}$, and then reflect it across the x-axis.

81. Start with the graph of $g(x) = \dfrac{1}{x}$. Shift it left 3 units, then up 2 units.

83. Start with the graph of $h(x) = x^2$. Shift it right 3 units. Then reflect it across the x-axis and shift it up 5 units.
85. $(-12, 2)$ **87.** $(12, 4)$ **89.** $(-12, 2)$ **91.** $(-12, 16)$
93. B   **95.** A   **97.** $f(x) = -(x - 8)^2$
99. $f(x) = |x + 7| + 2$ **101.** $f(x) = \dfrac{1}{2x} - 3$
103. $f(x) = -(x - 3)^2 + 4$ **105.** $f(x) = \sqrt{-(x + 2)} - 1$

107.

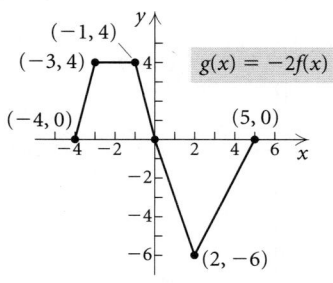

109.

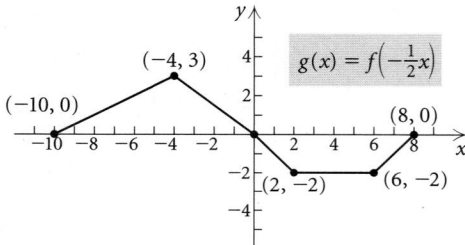

111.

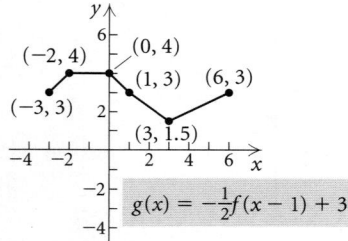

113.

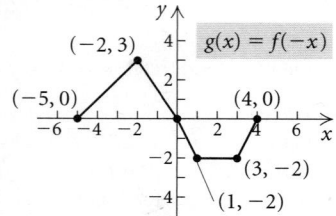

115.

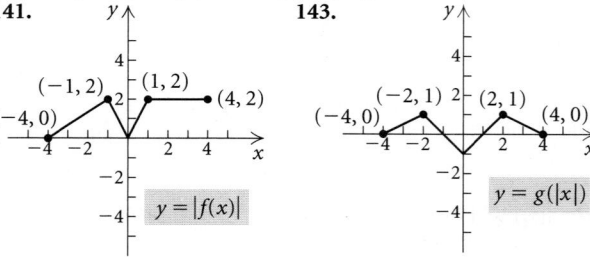

117.

119. (f)   **121.** (f)   **123.** (d)   **125.** (c)
127. $f(-x) = 2(-x)^4 - 35(-x)^3 + 3(-x) - 5 = 2x^4 + 35x^3 - 3x - 5 = g(x)$
129. $g(x) = x^3 - 3x^2 + 2$
131. $k(x) = (x + 1)^3 - 3(x + 1)^2$
133. Discussion and Writing
135. Discussion and Writing
137. [1.2] (a) 38; (b) 38; (c) $5a^2 - 7$; (d) $5a^2 - 7$
138. [1.2] (a) 22; (b) -22; (c) $4a^3 - 5a$; (d) $-4a^3 + 5a$
139. [1.4] $y = -\dfrac{1}{8}x + \dfrac{7}{8}$
140. [1.4] Slope is $\dfrac{2}{9}$; y-intercept is $\left(0, \dfrac{1}{9}\right)$.
141.

143.

145. Start with the graph of $g(x) = [\![x]\!]$. Shift it right $\dfrac{1}{2}$ unit. Domain: all real numbers; range: all integers.
147. Odd   **149.** x-axis, yes; y-axis, no; origin, no
151. x-axis, yes; y-axis, no; origin, no   **153.** 5
155. $(3, 8)$; $(3, 6)$; $\left(\dfrac{3}{2}, 4\right)$ **157.** True

159. $E(-x) = \dfrac{f(-x) + f(-(-x))}{2} = \dfrac{f(-x) + f(x)}{2} = E(x)$

161. (a) $E(x) + O(x) = \dfrac{f(x) + f(-x)}{2} + \dfrac{f(x) - f(-x)}{2} =$
$\dfrac{2f(x)}{2} = f(x)$; **(b)** $f(x) = \dfrac{-22x^2 + \sqrt{x} + \sqrt{-x} - 20}{2} +$
$\dfrac{8x^3 + \sqrt{x} - \sqrt{-x}}{2}$

Exercise Set 2.5

1. 4.5; $y = 4.5x$ **3.** 36; $y = \dfrac{36}{x}$ **5.** 4; $y = 4x$

7. 4; $y = \dfrac{4}{x}$ **9.** $\dfrac{3}{8}$; $y = \dfrac{3}{8}x$ **11.** 0.54; $y = \dfrac{0.54}{x}$

13. 3.5 hr   **15.** 90 g   **17.** About 686 kg   **19.** 40 lb

21. $66\frac{2}{3}$ cm   **23.** 1.92 ft   **25.** $y = \dfrac{0.0015}{x^2}$

27. $y = 15x^2$ **29.** $y = xz$ **31.** $y = \frac{3}{10}xz^2$

33. $y = \dfrac{1}{5} \cdot \dfrac{xz}{wp}$, or $\dfrac{xz}{5wp}$ **35.** 2.5 m   **37.** 36 mph

39. About 108 earned runs   **41.** Discussion and Writing
43. [2.1]

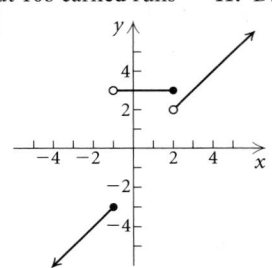

44. [2.4] x-axis, no; y-axis, yes; origin, no
45. [2.4] x-axis, yes; y-axis, no; origin, no
46. [2.4] x-axis, no; y-axis, no; origin, yes

47. \$2.32; \$2.80   **49.** $\dfrac{\pi}{4}$

Review Exercises: Chapter 2

1. True   **2.** False   **3.** True   **4.** True
5. (a) $(-4, -2)$; **(b)** $(2, 5)$; **(c)** $(-2, 2)$
6. (a) $(-1, 0), (2, \infty)$; **(b)** $(0, 2)$; **(c)** $(-\infty, -1)$
7. Increasing: $(0, \infty)$; decreasing: $(-\infty, 0)$; relative minimum: -1 at $x = 0$
8. Increasing: $(-\infty, 0)$; decreasing: $(0, \infty)$; relative maximum: 2 at $x = 0$
9. Increasing: $(2, \infty)$; decreasing: $(-\infty, 2)$; relative minimum: -1 at $x = 2$
10. Increasing: $(-\infty, 0.5)$; decreasing: $(0.5, \infty)$; relative maximum: 6.25 at $x = 0.5$
11. Increasing: $(-\infty, -1.155), (1.155, \infty)$; decreasing: $(-1.155, 1.155)$; relative maximum: 3.079 at $x = -1.155$; relative minimum: -3.079 at $x = 1.155$

12. Increasing: $(-1.155, 1.155)$; decreasing: $(-\infty, -1.155), (1.155, \infty)$; relative maximum: 1.540 at $x = 1.155$; relative minimum: -1.540 at $x = -1.155$
13. $A(l) = l(10 - l)$, or $10l - l^2$ **14.** $A(x) = 2x\sqrt{4 - x^2}$
15. (a) $A(x) = x\left(33 - \dfrac{x}{2}\right)$, or $33x - \dfrac{x^2}{2}$;
(b) $\{x \mid 0 < x < 66\}$;
(c)   **(d)** 33 ft by 16.5 ft

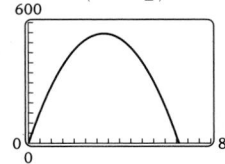

16. (a) $A(x) = x^2 + \dfrac{432}{x}$; **(b)** $(0, \infty)$;
(c) $x = 6$ in., height $= 3$ in.
17.

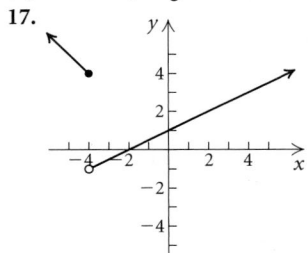

18.

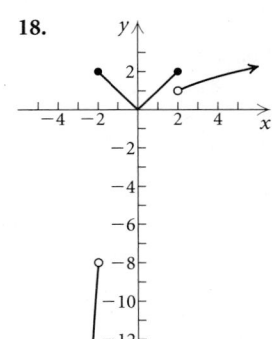

19.

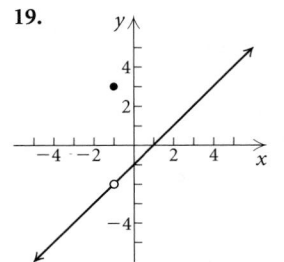

20.

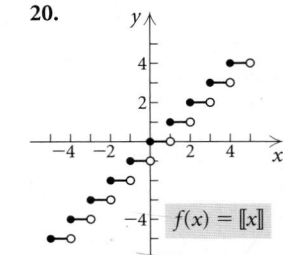

21.

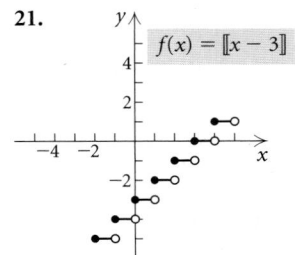

22. $f(-1) = 1$; $f(5) = 2$; $f(-2) = 2$; $f(-3) = -27$
23. $f(-2) = -3$; $f(-1) = 3$; $f(0) = -1$; $f(4) = 3$
24. -33 **25.** 0   **26.** Does not exist

27. (a) Domain of f: $(-\infty, 0) \cup (0, \infty)$; domain of g: $(-\infty, \infty)$; domain of $f + g$, $f - g$, and fg: $(-\infty, 0) \cup (0, \infty)$; domain of f/g: $(-\infty, 0) \cup \left(0, \frac{3}{2}\right) \cup \left(\frac{3}{2}, \infty\right)$

(b) $(f + g)(x) = \frac{4}{x^2} + 3 - 2x; (f - g)(x) = \frac{4}{x^2} - 3 + 2x;$

$(fg)(x) = \frac{12}{x^2} - \frac{8}{x}; (f/g)(x) = \frac{4}{x^2(3 - 2x)}$

28. (a) Domain of f, g, $f + g$, $f - g$, and fg: $(-\infty, \infty)$; domain of f/g: $\left(-\infty, \frac{1}{2}\right) \cup \left(\frac{1}{2}, \infty\right)$;

(b) $(f + g)(x) = 3x^2 + 6x - 1; (f - g)(x) = 3x^2 + 2x + 1;$

$(fg)(x) = 6x^3 + 5x^2 - 4x; (f/g)(x) = \frac{3x^2 + 4x}{2x - 1}$

29. $P(x) = -0.5x^2 + 105x - 6$ **30.** 2 **31.** $-2x - h$

32. $\dfrac{-4}{x(x + h)}$, or $-\dfrac{4}{x(x + h)}$ **33.** 9 **34.** 5

35. 128 **36.** 580 **37.** 7 **38.** -509

39. (a) $(f \circ g)(x) = \dfrac{4}{(3 - 2x)^2}; (g \circ f)(x) = 3 - \dfrac{8}{x^2};$

(b) domain of $f \circ g$: $\left(-\infty, \frac{3}{2}\right) \cup \left(\frac{3}{2}, \infty\right)$; domain of $g \circ f$: $(-\infty, 0) \cup (0, \infty)$

40. (a) $(f \circ g)(x) = 12x^2 - 4x - 1; (g \circ f)(x) = 6x^2 + 8x - 1;$ **(b)** domain of $f \circ g$ and $g \circ f$: $(-\infty, \infty)$

41. $f(x) = \sqrt{x}, g(x) = 5x + 2;$ answers may vary.

42. $f(x) = 4x^2 + 9, g(x) = 5x - 1;$ answers may vary.

43. x-axis, yes; y-axis, yes; origin, yes

44. x-axis, yes; y-axis, yes; origin, yes

45. x-axis, no; y-axis, no; origin, no

46. x-axis, no; y-axis, yes; origin, no

47. x-axis, no; y-axis, no; origin, yes

48. x-axis, no; y-axis, yes; origin, no

49. Even **50.** Even **51.** Odd **52.** Even **53.** Even

54. Neither **55.** Odd **56.** Even **57.** Even

58. Odd **59.** $f(x) = (x + 3)^2$

60. $f(x) = -\sqrt{x - 3} + 4$ **61.** $f(x) = 2|x - 3|$

62.

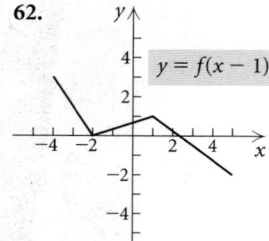

$y = f(x - 1)$

63.

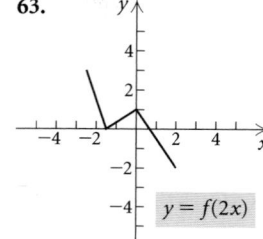

$y = f(2x)$

64.

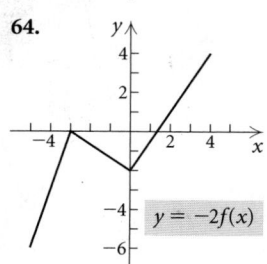

$y = -2f(x)$

65.

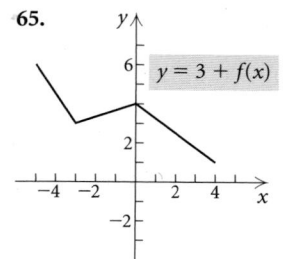

$y = 3 + f(x)$

66. $y = 4x$ **67.** $y = \dfrac{2}{3}x$ **68.** $y = \dfrac{2500}{x}$ **69.** $y = \dfrac{54}{x}$

70. $y = \dfrac{48}{x^2}$ **71.** $y = \dfrac{1}{10} \cdot \dfrac{xz^2}{w}$ **72.** 20 min

73. 75 **74.** 500 watts **75.** A **76.** C **77.** B

78. Discussion and Writing: (a) $4x^3 - 2x + 9$;

(b) $4x^3 + 24x^2 + 46x + 35$; **(c)** $4x^3 - 2x + 42$. **(a)** Adds 2 to each function value; **(b)** adds 2 to each input before finding a function value; **(c)** adds the output for 2 to the output for x.

79. Discussion and Writing: In the graph of $y = f(cx)$, the constant c stretches or shrinks the graph of $y = f(x)$ horizontally. The constant c in $y = cf(x)$ stretches or shrinks the graph of $y = f(x)$ vertically. For $y = f(cx)$, the x-coordinates of $y = f(x)$ are divided by c; for $y = cf(x)$, the y-coordinates of $y = f(x)$ are multiplied by c.

80. Discussion and Writing: (a) To draw the graph of y_2 from the graph of y_1, reflect across the x-axis the portions of the graph for which the y-coordinates are negative. **(b)** To draw the graph of y_2 from the graph of y_1, draw the portion of the graph of y_1 to the right of the y-axis; then draw its reflection across the y-axis. **81.** Let $f(x)$ and $g(x)$ be odd functions. Then by definition, $f(-x) = -f(x)$, or $f(x) = -f(-x)$, and $g(-x) = -g(x)$, or $g(x) = -g(-x)$. Thus, $(f + g)(x) = f(x) + g(x) = -f(-x) + [-g(-x)] = -[f(-x) + g(-x)] = -(f + g)(-x)$ and $f + g$ is odd.

82. Reflect the graph of $y = f(x)$ across the x-axis and then across the y-axis.

Test: Chapter 2

1. [2.1] **(a)** $(-5, -2)$; **(b)** $(2, 5)$; **(c)** $(-2, 2)$

2. [2.1] Increasing: $(-\infty, 0)$; decreasing: $(0, \infty)$; relative maximum: 2 at $x = 0$

3. [2.1] Increasing: $(-\infty, -2.667), (0, \infty)$; decreasing: $(-2.667, 0)$; relative maximum: 9.481 at $x = -2.667$; relative minimum: 0 at $x = 0$

4. [2.1] $A(b) = \frac{1}{2}b(4b - 6)$, or $2b^2 - 3b$

5. [2.1]

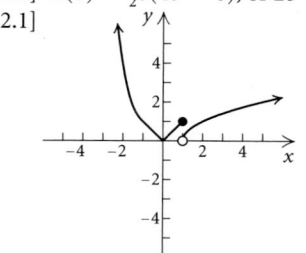

6. [2.1] $f\left(-\frac{7}{8}\right) = \frac{7}{8}; f(5) = 2; f(-4) = 16$ **7.** [2.2] 66

8. [2.2] 6 **9.** [2.2] -1 **10.** [2.2] 0 **11.** [1.2] $(-\infty, \infty)$

12. [1.2], [1.6] $[3, \infty)$ **13.** [2.2] $[3, \infty)$ **14.** [2.2] $[3, \infty)$

15. [2.2] $[3, \infty)$ **16.** [2.2] $(3, \infty)$

17. [2.2] $(f + g)(x) = x^2 + \sqrt{x - 3}$

18. [2.2] $(f - g)(x) = x^2 - \sqrt{x - 3}$

19. [2.2] $(fg)(x) = x^2\sqrt{x - 3}$

20. [2.2] $(f/g)(x) = \dfrac{x^2}{\sqrt{x-3}}$ **21.** [2.2] $\frac{1}{2}$
22. [2.2] $4x + 2h - 1$ **23.** [2.3] 83   **24.** [2.3] 0
25. [2.3] 4
26. [2.3] $(f \circ g)(x) = \sqrt{x^2 - 4}$; $(g \circ f)(x) = x - 4$
27. [2.3] Domain of $(f \circ g)(x) = (-\infty, -2] \cup [2, \infty)$;
domain of $(g \circ f)(x) = [5, \infty)$
28. [2.3] $f(x) = x^4$; $g(x) = 2x - 7$; answers may vary
29. [2.4] x-axis: no; y-axis: yes; origin: no   **30.** [2.4] Odd
31. [2.4] $f(x) = (x - 2)^2 - 1$
32. [2.4] $f(x) = (x + 2)^2 - 3$
33. [2.4]

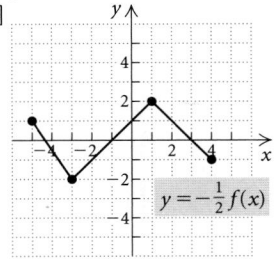

$y = -\frac{1}{2}f(x)$

34. [2.5] $y = \dfrac{30}{x}$ **35.** [2.5] $y = 5x$ **36.** [2.5] $y = \dfrac{50xz^2}{w}$
37. [2.5] 50 ft   **38.** [2.4] C   **39.** [2.4] $(-1, 1)$

CHAPTER 3

Exercise Set 3.1

1. $\sqrt{3}i$ **3.** $5i$ **5.** $-\sqrt{33}i$ **7.** $-9i$ **9.** $7\sqrt{2}i$
11. $2 + 11i$ **13.** $5 - 12i$ **15.** $4 + 8i$ **17.** $-4 - 2i$
19. $5 + 9i$ **21.** $5 + 4i$ **23.** $5 + 7i$ **25.** $11 - 5i$
27. $-1 + 5i$ **29.** $2 - 12i$ **31.** $35 + 14i$
33. $6 + 16i$ **35.** $13 - i$ **37.** $-11 + 16i$
39. $-10 + 11i$ **41.** $-31 - 34i$ **43.** $-14 + 23i$
45. 41   **47.** 13   **49.** 74   **51.** $12 + 16i$
53. $-45 - 28i$ **55.** $-8 - 6i$ **57.** $2i$ **59.** $-7 + 24i$
61. $\frac{15}{146} + \frac{33}{146}i$ **63.** $\frac{10}{13} - \frac{15}{13}i$ **65.** $-\frac{14}{13} + \frac{5}{13}i$
67. $\frac{11}{25} - \frac{27}{25}i$ **69.** $\dfrac{-4\sqrt{3} + 10}{41} + \dfrac{5\sqrt{3} + 8}{41}i$
71. $-\frac{1}{2} + \frac{1}{2}i$ **73.** $-\frac{1}{2} - \frac{13}{2}i$ **75.** $-i$ **77.** $-i$
79. 1   **81.** i **83.** 625   **85.** Discussion and Writing
87. [1.4] $y = -2x + 1$ **88.** [2.2] All real numbers, or
$(-\infty, \infty)$ **89.** [2.2] $\left(-\infty, -\frac{5}{3}\right) \cup \left(-\frac{5}{3}, \infty\right)$
90. [2.2] $x^2 - 3x - 1$ **91.** [2.2] $\frac{8}{11}$
92. [2.2] $2x + h - 3$ **93.** True   **95.** True   **97.** $a^2 + b^2$
99. $x^2 - 6x + 25$

Exercise Set 3.2

1. $\frac{2}{3}, \frac{3}{2}$ **3.** $-2, 10$ **5.** $-1, \frac{2}{3}$ **7.** $-\sqrt{3}, \sqrt{3}$
9. $-\sqrt{7}, \sqrt{7}$ **11.** $-\sqrt{2}i, \sqrt{2}i$ **13.** $-4i, 4i$
15. $0, 3$ **17.** $-\frac{1}{3}, 0, 2$ **19.** $-1, -\frac{1}{7}, 1$
21. (a) $(-4, 0), (2, 0)$; (b) $-4, 2$
23. (a) $(-1, 0), (3, 0)$; (b) $-1, 3$
25. (a) $(-2, 0), (2, 0)$; (b) $-2, 2$ **27.** $-7, 1$
29. $4 \pm \sqrt{7}$ **31.** $-4 \pm 3i$ **33.** $-2, \frac{1}{3}$ **35.** $-3, 5$
37. $-1, \frac{2}{5}$ **39.** $\dfrac{5 \pm \sqrt{7}}{3}$ **41.** $-\dfrac{1}{2} \pm \dfrac{\sqrt{7}}{2}i$
43. $\dfrac{4 \pm \sqrt{31}}{5}$ **45.** $\dfrac{5}{6} \pm \dfrac{\sqrt{23}}{6}i$ **47.** $4 \pm \sqrt{11}$
49. $\dfrac{-1 \pm \sqrt{61}}{6}$ **51.** $\dfrac{5 \pm \sqrt{17}}{4}$ **53.** $-\dfrac{1}{5} \pm \dfrac{3}{5}i$
55. 144; two real   **57.** -7; two imaginary
59. 49; two real   **61.** $2, 6$ **63.** $0.143, 6$
65. $-0.151, 1.651$ **67.** $-0.637, 3.137$ **69.** $-5, -1$
71. $\dfrac{3 \pm \sqrt{21}}{2}$ **73.** $\dfrac{5 \pm \sqrt{21}}{2}$ **75.** $-1 \pm \sqrt{6}$
77. $\dfrac{1}{4} \pm \dfrac{\sqrt{31}}{4}i$ **79.** $\dfrac{1 \pm \sqrt{13}}{6}$ **81.** $\dfrac{1 \pm \sqrt{6}}{5}$
83. $\dfrac{-3 \pm \sqrt{57}}{8}$ **85.** $-1.535, 0.869$ **87.** $-0.347, 1.181$
89. $\pm 1, \pm \sqrt{2}$ **91.** $\pm \sqrt{2}, \pm \sqrt{5}i$ **93.** $\pm 1, \pm \sqrt{5}i$
95. 16   **97.** $-8, 64$ **99.** $1, 16$ **101.** $\frac{5}{2}, 3$
103. $-\frac{3}{2}, -1, \frac{1}{2}, 1$ **105.** About 11.5 sec   **107.** 2010
109. Length: 4 ft; width: 3 ft   **111.** 4 and 9; -9 and -4
113. 2 cm   **115.** Length: 8 ft; width: 6 ft
117. Linear   **119.** Quadratic   **121.** Linear
123. Discussion and Writing   **125.** [1.2] 551,453 associate's
degrees   **126.** [1.2] 660,605 associate's degrees
127. [2.4] x-axis: yes; y-axis: yes; origin: yes
128. [2.4] x-axis: no; y-axis: yes; origin: no
129. [2.4] Odd   **130.** [2.4] Neither   **131.** (a) 2; (b) $\frac{11}{2}$
133. (a) 2; (b) $1 - i$ **135.** 1   **137.** $-\sqrt{7}, -\frac{3}{2}, 0, \frac{1}{3}, \sqrt{7}$
139. $\dfrac{-1 \pm \sqrt{1 + 4\sqrt{2}}}{2}$ **141.** $3 \pm \sqrt{5}$ **143.** 19
145. $-2 \pm \sqrt{2}, \frac{1}{2} \pm \dfrac{\sqrt{7}}{2}i$ **147.** $t = \dfrac{-v_0 \pm \sqrt{v_0^2 - 2ax_0}}{a}$

Visualizing the Graph

1. C   **2.** B   **3.** A   **4.** J   **5.** F   **6.** D   **7.** I
8. G   **9.** H   **10.** E

Exercise Set 3.3

1. (a) $\left(-\frac{1}{2}, -\frac{9}{4}\right)$; (b) $x = -\frac{1}{2}$; (c) minimum: $-\frac{9}{4}$

3. (a) $(4, -4)$; (b) $x = 4$; (c) minimum: -4;
(d)

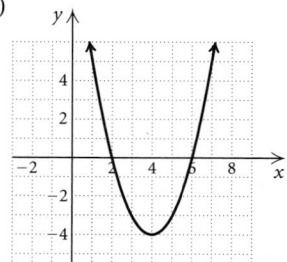

$$f(x) = x^2 - 8x + 12$$

5. (a) $\left(\frac{7}{2}, -\frac{1}{4}\right)$; (b) $x = \frac{7}{2}$; (c) minimum: $-\frac{1}{4}$;
(d)

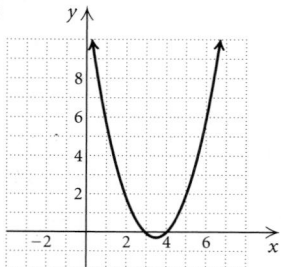

$$f(x) = x^2 - 7x + 12$$

7. (a) $(-2, 1)$; (b) $x = -2$; (c) minimum: 1;
(d)

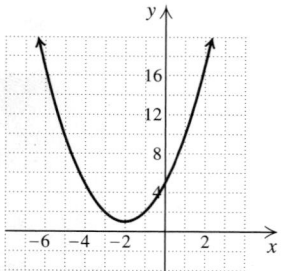

$$f(x) = x^2 + 4x + 5$$

9. (a) $(-4, -2)$; (b) $x = -4$; (c) minimum: -2;
(d)

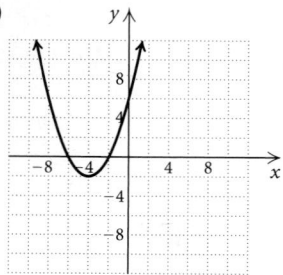

$$g(x) = \frac{x^2}{2} + 4x + 6$$

11. (a) $\left(-\frac{3}{2}, \frac{7}{2}\right)$; (b) $x = -\frac{3}{2}$; (c) minimum: $\frac{7}{2}$;
(d)

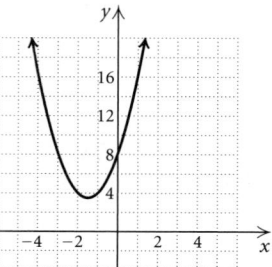

$$g(x) = 2x^2 + 6x + 8$$

13. (a) $(-3, 12)$; (b) $x = -3$; (c) maximum: 12;
(d)

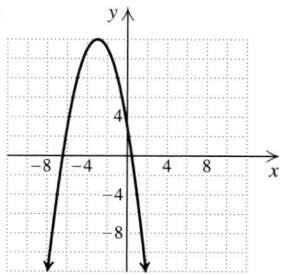

$$f(x) = -x^2 - 6x + 3$$

15. (a) $\left(\frac{1}{2}, \frac{3}{2}\right)$; (b) $x = \frac{1}{2}$; (c) maximum: $\frac{3}{2}$;
(d)

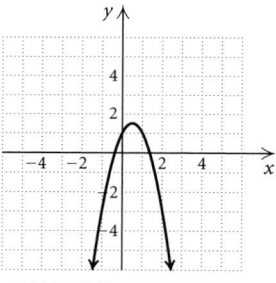

$$g(x) = -2x^2 + 2x + 1$$

17. (f)   **19.** (b)   **21.** (h)   **23.** (c)   **25.** True
27. False   **29.** True   **31.** (a) $(3, -4)$; (b) minimum: -4;
(c) $[-4, \infty)$; (d) increasing: $(3, \infty)$; decreasing: $(-\infty, 3)$
33. (a) $(-1, -18)$; (b) minimum: -18; (c) $[-18, \infty)$;
(d) increasing: $(-1, \infty)$; decreasing: $(-\infty, -1)$
35. (a) $\left(5, \frac{9}{2}\right)$; (b) maximum: $\frac{9}{2}$; (c) $\left(-\infty, \frac{9}{2}\right]$;
(d) increasing: $(-\infty, 5)$; decreasing: $(5, \infty)$
37. (a) $(-1, 2)$; (b) minimum: 2; (c) $[2, \infty)$;
(d) increasing: $(-1, \infty)$; decreasing: $(-\infty, -1)$
39. (a) $\left(-\frac{3}{2}, 18\right)$; (b) maximum: 18; (c) $\left(-\infty, 18\right]$;
(d) increasing: $\left(-\infty, -\frac{3}{2}\right)$; decreasing: $\left(-\frac{3}{2}, \infty\right)$
41. 0.625 sec; 12.25 ft   **43.** 3.75 sec; 305 ft   **45.** 4.5 in.
47. Base: 10 cm; height: 10 cm   **49.** 350 chairs
51. $797; 40 units   **53.** 4800 yd^2 **55.** 350.6 ft

57. Discussion and Writing   **59.** Discussion and Writing
60. [2.2] 3   **61.** [2.2] $4x + 2h - 1$
62. [2.4]

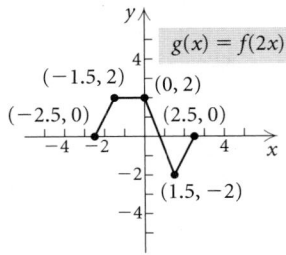

$g(x) = f(2x)$

63. [2.4]   **65.** -236.25

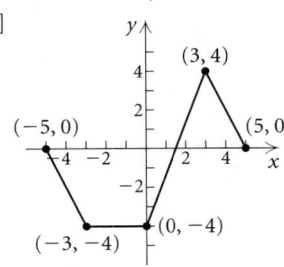

$g(x) = -2f(x)$

67.

$f(x) = (|x| - 5)^2 - 3$

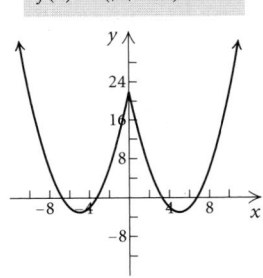

Exercise Set 3.4

1. $\frac{20}{9}$ **3.** 286   **5.** 6   **7.** 6   **9.** 2, 3   **11.** $-1, 6$
13. $\frac{1}{2}, 5$ **15.** 7   **17.** No solution   **19.** $-\frac{69}{14}$ **21.** $-\frac{37}{18}$
23. 2   **25.** No solution
27. $\{x \mid x \text{ is a real number } and\ x \neq 0\ and\ x \neq 6\}$
29. $\frac{5}{3}$ **31.** $\frac{9}{2}$ **33.** 3   **35.** -4 **37.** -5
39. $\pm\sqrt{2}$ **41.** No solution   **43.** 6   **45.** -1 **47.** $\frac{35}{2}$
49. -98 **51.** -6 **53.** 5   **55.** 7   **57.** 2   **59.** $-1, 2$
61. 7   **63.** 7   **65.** No solution   **67.** 1   **69.** 3, 7
71. 5   **73.** -1 **75.** -8 **77.** 81
79. $T_1 = \dfrac{P_1 V_1 T_2}{P_2 V_2}$ **81.** $C = \dfrac{1}{LW^2}$ **83.** $R_2 = \dfrac{RR_1}{R_1 - R}$
85. $P = \dfrac{A}{I^2 + 2I + 1}$, or $\dfrac{A}{(I + 1)^2}$ **87.** $p = \dfrac{Fm}{m - F}$
89. Discussion and Writing   **91.** [1.5] 7.5   **92.** [1.5] 3
93. [1.5] 26.25 million prescriptions   **94.** [1.5] Mall of
America: 96 acres; Disneyland: 85 acres   **95.** $3 \pm 2\sqrt{2}$
97. -1

Exercise Set 3.5

1. $-7, 7$ **3.** 0   **5.** $-\frac{5}{6}, \frac{5}{6}$ **7.** No solution   **9.** $-\frac{1}{3}, \frac{1}{3}$
11. $-3, 3$ **13.** $-3, 5$ **15.** $-8, 4$ **17.** $-1, -\frac{1}{3}$
19. $-24, 44$ **21.** $-2, 4$ **23.** $-13, 7$ **25.** $-\frac{4}{3}, \frac{2}{3}$
27. $-\frac{3}{4}, \frac{9}{4}$ **29.** $-13, 1$ **31.** 0, 1
33. $(-7, 7)$;
35. $[-2, 2]$;
37. $(-\infty, -4.5] \cup [4.5, \infty)$;
39. $(-\infty, -3) \cup (3, \infty)$;
41. $\left(-\frac{1}{3}, \frac{1}{3}\right)$;
43. $(-\infty, -3] \cup [3, \infty)$;
45. $(-17, 1)$;
47. $(-\infty, -17] \cup [1, \infty)$;
49. $\left(-\frac{1}{4}, \frac{3}{4}\right)$;
51. $[-6, 3]$;
53. $(-\infty, 4.9) \cup (5.1, \infty)$;
55. $\left(-\infty, -\frac{1}{2}\right] \cup \left[\frac{7}{2}, \infty\right)$;
57. $\left[-\frac{7}{3}, 1\right]$;
59. $(-\infty, -8) \cup (7, \infty)$;
61. No solution   **63.** Discussion and Writing
65. [1.1] y-intercept   **66.** [1.1] Distance formula
67. [1.2] Relation   **68.** [1.2] Function
69. [1.3] Horizontal lines   **70.** [1.4] Parallel
71. [2.1] Decreasing   **72.** [2.4] Symmetric with respect to
the y-axis   **73.** $\left(-\infty, \frac{1}{2}\right)$ **75.** No solution
77. $\left(-\infty, -\frac{8}{3}\right) \cup (-2, \infty)$

Review Exercises: Chapter 3

1. True   **2.** True   **3.** False   **4.** False   **5.** $-\frac{5}{2}, \frac{1}{3}$
6. $-5, 1$ **7.** $-2, \frac{4}{3}$ **8.** $-\sqrt{3}, \sqrt{3}$ **9.** $-\sqrt{10}i, \sqrt{10}i$
10. 1   **11.** $-5, 3$ **12.** $\dfrac{1 \pm \sqrt{41}}{4}$ **13.** $\dfrac{-1 \pm 2i\sqrt{2}}{3}$
14. $\frac{27}{7}$ **15.** $-\frac{1}{2}, \frac{9}{4}$ **16.** 0, 3   **17.** 5   **18.** 1, 7   **19.** $-8, 1$
20. $(-\infty, -3] \cup [3, \infty)$;

21. $\left(-\frac{14}{3}, 2\right)$;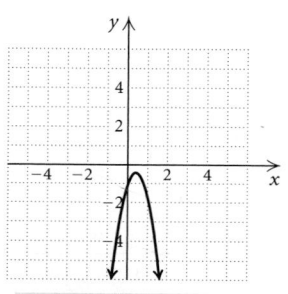

22. $\left(-\frac{2}{3}, 1\right)$;

23. $(-\infty, -6] \cup [-2, \infty)$;

24. $P = \dfrac{MN}{M + N}$ **25.** $-2\sqrt{10}i$ **26.** $-4\sqrt{15}$

27. $-\frac{7}{8}$ **28.** $2 - i$ **29.** $1 - 4i$ **30.** $-18 - 26i$

31. $\frac{11}{10} + \frac{3}{10}i$ **32.** $-i$ **33.** $x^2 - 3x + \frac{9}{4} = 18 + \frac{9}{4}$; $\left(x - \frac{3}{2}\right)^2 = \frac{81}{4}$; $x = \frac{3}{2} \pm \frac{9}{2}$; $-3, 6$ **34.** $x^2 - 4x = 2$; $x^2 - 4x + 4 = 2 + 4$; $(x - 2)^2 = 6$; $x = 2 \pm \sqrt{6}$; $2 - \sqrt{6}, 2 + \sqrt{6}$ **35.** $-4, \frac{2}{3}$ **36.** $1 - 3i, 1 + 3i$

37. $-2, 5$ **38.** 1 **39.** $\pm\sqrt{\dfrac{3 \pm \sqrt{5}}{2}}$ **40.** $-\sqrt{3}, 0, \sqrt{3}$

41. $-2, -\frac{2}{3}, 3$ **42.** $-5, -2, 2$ **43.** (a) $\left(\frac{3}{8}, -\frac{7}{16}\right)$; (b) $x = \frac{3}{8}$; (c) maximum: $-\frac{7}{16}$; (d) $\left(-\infty, -\frac{7}{16}\right]$; (e)

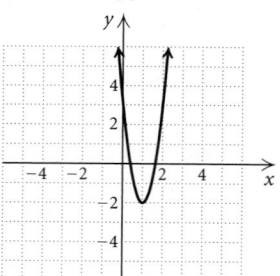

$f(x) = -4x^2 + 3x - 1$

44. (a) $(1, -2)$; (b) $x = 1$; (c) minimum: -2; (d) $[-2, \infty)$; (e)

$f(x) = 5x^2 - 10x + 3$

45. (d)   **46.** (c)   **47.** (b)   **48.** (a)
49. 30 ft, 40 ft   **50.** 6 mph   **51.** 80 km/h
52. $35 - 5\sqrt{33}$ ft, or about 6.3 ft   **53.** 6 ft by 6 ft
54. $\dfrac{15 - \sqrt{115}}{2}$ cm, or about 2.1 cm   **55.** B   **56.** B   **57.** A
58. Discussion and Writing: Write an equation that has only complex-number solutions. Any equation $ax^2 + bx + c = 0$, where the discriminant, $b^2 - 4ac$, is negative, will do.

59. Discussion and Writing: You can conclude that $|a_1| = |a_2|$ since these constants determine how wide the parabolas are. Nothing can be concluded about the h's and the k's.
60. 256   **61.** $4 \pm \sqrt[4]{243}$, or $0.052, 7.948$ **62.** $-7, 9$
63. $-\frac{1}{4}, 2$ **64.** -1 **65.** 9%   **66.** ± 6

Test: Chapter 3

1. [3.2] $\frac{1}{2}, -5$ **2.** [3.2] $-\sqrt{6}, \sqrt{6}$ **3.** [3.2] $-2i, 2i$
4. [3.2] $-1, 3$ **5.** [3.2] $\dfrac{5 \pm \sqrt{13}}{2}$ **6.** [3.2] $\dfrac{3}{4} \pm \dfrac{\sqrt{23}}{4}i$
7. [3.2] 16   **8.** [3.4] $-1, \frac{13}{6}$ **9.** [3.4] 5   **10.** [3.4] 5
11. [3.5] $-11, 3$ **12.** [3.5] $-\frac{1}{2}, 2$
13. [3.5] $[-7, 1]$;
14. [3.5] $(-2, 3)$;
15. [3.5] $(-\infty, -7) \cup (-3, \infty)$;
16. [3.5] $\left(-\infty, -\frac{2}{3}\right] \cup [4, \infty)$;

17. [3.4] $B = \dfrac{AC}{A - C}$ **18.** [3.4] $n = \dfrac{R^2}{3p}$
19. [3.2] $x^2 + 4x = 1$; $x^2 + 4x + 4 = 1 + 4$; $(x + 2)^2 = 5$; $x = -2 \pm \sqrt{5}$; $-2 - \sqrt{5}, -2 + \sqrt{5}$ **20.** [3.4] 3 km/h
21. [3.1] $\sqrt{43}i$ **22.** [3.1] $-5i$
23. [3.1] $3 - 5i$ **24.** [3.1] $10 + 5i$ **25.** [3.1] $\frac{1}{10} - \frac{1}{5}i$
26. [3.1] i **27.** [3.2] $-\frac{1}{4}, 3$
28. [3.2] $\dfrac{1 \pm \sqrt{57}}{4}$ **29.** [3.3] (a) $(1, 9)$; (b) $x = 1$;
(c) maximum: 9; (d) $(-\infty, 9]$;
(e)

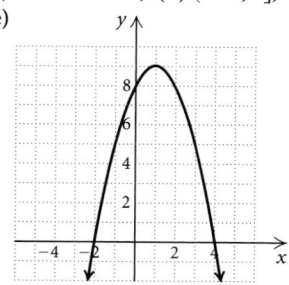

$f(x) = -x^2 + 2x + 8$

30. [3.3] 20 ft by 40 ft   **31.** [3.3] C
32. [3.3], [3.4] $-\frac{4}{9}$

CHAPTER 4

Exercise Set 4.1

1. $\frac{1}{2}x^3$; $\frac{1}{2}$; 3; cubic   **3.** $0.9x$; 0.9; 1; linear
5. $305x^4$; 305; 4; quartic   **7.** x^4; 1; 4; quartic
9. $4x^3$; 4; 3; cubic   **11.** (d)   **13.** (b)   **15.** (c)

17. (a)   **19.** (c)   **21.** (d)   **23.** Yes; no; no
25. No; yes; yes   **27.** -3, multiplicity 2; 1, multiplicity 1
29. 4, multiplicity 3; -6, multiplicity 1
31. ± 3, each has multiplicity 3   **33.** 0, multiplicity 3; 1,
multiplicity 2; -4, multiplicity 1   **35.** 3, multiplicity 2; -4,
multiplicity 3; 0, multiplicity 4   **37.** $\pm\sqrt{3}$, ± 1, each has
multiplicity 1   **39.** -3, -1, 1, each has multiplicity 1
41. ± 2, $\frac{1}{2}$, each has multiplicity 1   **43.** -1.532, -0.347,
1.879   **45.** -1.414, 0, 1.414   **47.** -1, 0, 1
49. -10.153, -1.871, -0.821, -0.303, 0.098, 0.535, 1.219,
3.297   **51.** -1.386; relative maximum: 1.506 at $x = -0.632$,
relative minimum: 0.494 at $x = 0.632$; $(-\infty, \infty)$
53. -1.249, 1.249; relative minimum: -3.8 at $x = 0$, no
relative maxima; $[-3.8, \infty)$ **55.** -1.697, 0, 1.856; relative
maximum: 11.012 at $x = 1.258$, relative minimum: -8.183 at
$x = -1.116$; $(-\infty, \infty)$ **57.** False   **59.** True
61. 19,325,000 cows; 11,460,000 cows   **63.** 1998: 136
deaths; 2001: 49 deaths; 2006: 65 deaths   **65.** $7920
67. $699; $686; $859   **69.** (a) $4\frac{1}{2}\%$; (b) 10%   **71.** (b)
73. (c)   **75.** (a)   **77.** (a) Cubic: $y = 0.000050591548x^3 -$
$0.0056474021x^2 + 0.0177540608x + 14.30443363$; quartic:
$y = -0.0000008087312x^4 + 0.00022028706x^3 -$
$0.0167796743x^2 + 0.2520665814x + 13.57576893$;
(b) quartic: 13.1%; answers may vary   **79.** (a) Cubic:
$y = -1.590909091x^3 + 27.21678322x^2 - 113.3951049x +$
421.3426573; quartic: $y = -0.2966200466x^4 +$
$4.341491841x^3 - 9.860722611x^2 - 39.24009324x +$
399.986014;   (b) cubic: 231,000   **81.** Discussion and Writing
83. [1.1] 5   **84.** [1.1] $6\sqrt{2}$ **85.** [1.1] Center: $(3, -5)$;
radius: 7   **86.** [1.1] Center: $(-4, 3)$; radius: $2\sqrt{2}$
87. [1.6] $\{y \mid y \geq 3\}$, or $[3, \infty)$ **88.** [1.6] $\{x \mid x > \frac{5}{3}\}$, or $\left(\frac{5}{3}, \infty\right)$
89. [3.5] $\{x \mid x \leq -13 \text{ or } x \geq 1\}$, or $(-\infty, -13] \cup [1, \infty)$
90. [3.5] $\{x \mid -\frac{11}{12} \leq x \leq \frac{5}{12}\}$, or $\left[-\frac{11}{12}, \frac{5}{12}\right]$ **91.** 16; x^{16}

Visualizing the Graph

1. H   **2.** D   **3.** J   **4.** B   **5.** A   **6.** C   **7.** I
8. E   **9.** G   **10.** F

Exercise Set 4.2

1. (a) 5; (b) 5; (c) 4   **3.** (a) 10; (b) 10; (c) 9
5. (a) 3; (b) 3; (c) 2   **7.** (d)   **9.** (f)   **11.** (b)

13.
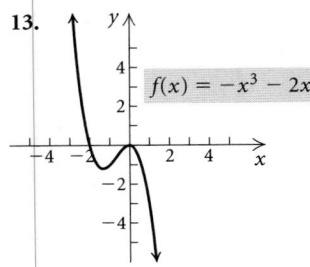
$f(x) = -x^3 - 2x^2$

15.
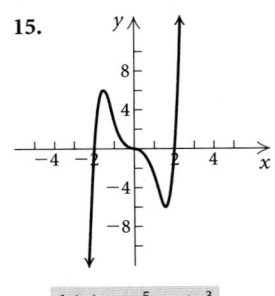
$h(x) = x^5 - 4x^3$

17.
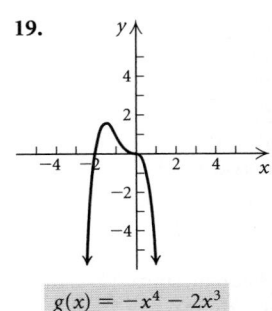
$h(x) = x(x - 4)(x + 1)(x - 2)$

19.
$g(x) = -x^4 - 2x^3$

21.
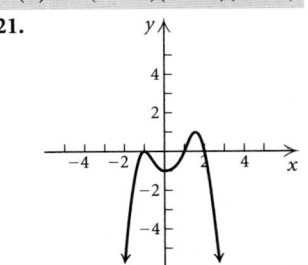
$f(x) = -\frac{1}{2}(x - 2)(x + 1)^2(x - 1)$

23.
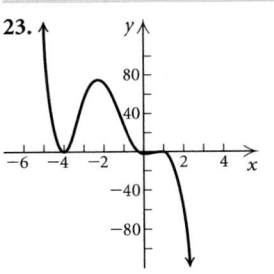
$g(x) = -x(x - 1)^2(x + 4)^2$

25.
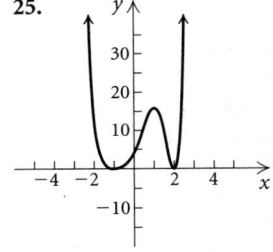
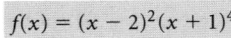
$f(x) = (x - 2)^2(x + 1)^4$

27.
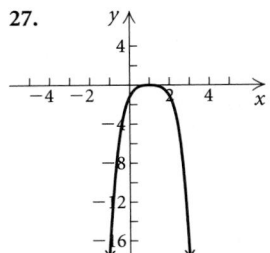
$g(x) = -(x - 1)^4$

29.
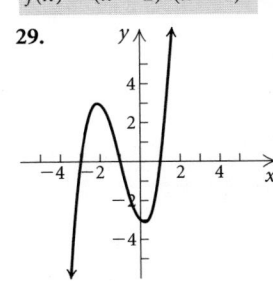
$h(x) = x^3 + 3x^2 - x - 3$

31.
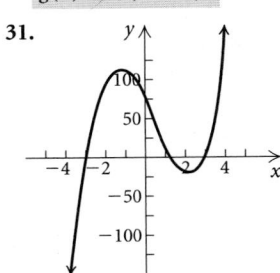
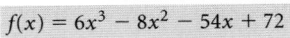
$f(x) = 6x^3 - 8x^2 - 54x + 72$

33. $f(-5) = -18$ and $f(-4) = 7$. By the intermediate value theorem, since $f(-5)$ and $f(-4)$ have opposite signs, then $f(x)$ has a zero between -5 and -4. **35.** $f(-3) = 22$ and $f(-2) = 5$. Both $f(-3)$ and $f(-2)$ are positive. We cannot use the intermediate value theorem to determine if there is a zero between -3 and -2. **37.** $f(2) = 2$ and $f(3) = 57$. Both $f(2)$ and $f(3)$ are positive. We cannot use the intermediate value theorem to determine if there is a zero between 2 and 3.
39. $f(4) = -12$ and $f(5) = 4$. By the intermediate value theorem, since $f(4)$ and $f(5)$ have opposite signs, then $f(x)$ has a zero between 4 and 5. **41.** Discussion and Writing
43. [1.1] (d) **44.** [1.3] (f) **45.** [1.1] (e) **46.** [1.1] (a)
47. [1.1] (b) **48.** [1.3] (c) **49.** [1.5] $\frac{9}{10}$
50. [4.1] $-3, 0, 4$ **51.** [3.2] $-\frac{5}{3}, \frac{11}{2}$ **52.** [1.5] $\frac{196}{25}$

Exercise Set 4.3

1. (a) No; **(b)** yes; **(c)** no **3. (a)** Yes; **(b)** no; **(c)** yes
5. $P(x) = (x + 2)(x^2 - 2x + 4) - 16$
7. $P(x) = (x + 9)(x^2 - 3x + 2) + 0$
9. $P(x) = (x + 2)(x^3 - 2x^2 + 2x - 4) + 11$
11. $Q(x) = 2x^3 + x^2 - 3x + 10, R(x) = -42$
13. $Q(x) = x^2 - 4x + 8, R(x) = -24$
15. $Q(x) = 3x^2 - 4x + 8, R(x) = -18$
17. $Q(x) = x^4 + 3x^3 + 10x^2 + 30x + 89, R(x) = 267$
19. $Q(x) = x^3 + x^2 + x + 1, R(x) = 0$
21. $Q(x) = 2x^3 + x^2 + \frac{7}{2}x + \frac{7}{4}, R(x) = -\frac{1}{8}$
23. $0; -60; 0$ **25.** $10; 80; 998$ **27.** $5,935,988; -772$
29. $0; 0; 65; 1 - 12\sqrt{2}$ **31.** Yes; no **33.** Yes; yes
35. No; yes **37.** No; no
39. $f(x) = (x - 1)(x + 2)(x + 3); 1, -2, -3$
41. $f(x) = (x - 2)(x - 5)(x + 1); 2, 5, -1$
43. $f(x) = (x - 2)(x - 3)(x + 4); 2, 3, -4$
45. $f(x) = (x - 3)^3(x + 2); 3, -2$
47. $f(x) = (x - 1)(x - 2)(x - 3)(x + 5); 1, 2, 3, -5$
49.

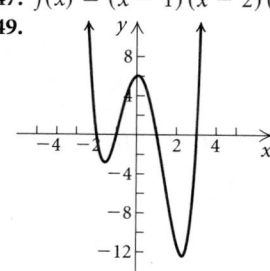

$f(x) = x^4 - x^3 - 7x^2 + x + 6$

51.

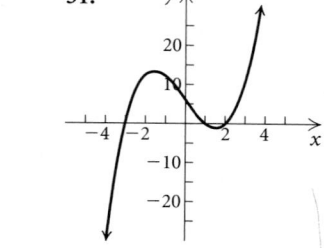

$f(x) = x^3 - 7x + 6$

53.

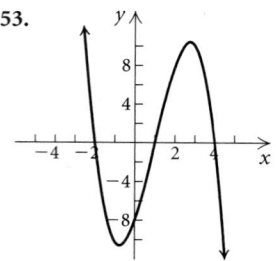

$f(x) = -x^3 + 3x^2 + 6x - 8$

55. Discussion and Writing **57.** [3.2] $\frac{5}{4} \pm \frac{\sqrt{71}}{4}i$
58. [3.2] $-1, \frac{3}{7}$ **59.** [3.2] $-5, 0$ **60.** [1.2] 10
61. [3.2] $-3, -2$ **62.** [1.4] $f(x) = 0.767x + 9.4$; 2005: 13.2 hr; 2010: 17.1 hr **63.** [3.2] $b = 15$ in., $h = 15$ in.
65. (a) $x + 4, x + 3, x - 2, x - 5$;
(b) $P(x) = (x + 4)(x + 3)(x - 2)(x - 5)$; **(c)** yes; two examples are $f(x) = c \cdot P(x)$ for any nonzero constant c; and $g(x) = (x - a)P(x)$; **(d)** no **67.** $\frac{14}{3}$ **69.** $0, -6$
71. Answers may vary. One possibility is $P(x) = x^{15} - x^{14}$.
73. $x^2 + 2ix + (2 - 4i), R -6 - 2i$
75. $x - 3 + i, R\, 6 - 3i$

Exercise Set 4.4

1. $f(x) = x^3 - 6x^2 - x + 30$
3. $f(x) = x^3 + 3x^2 + 4x + 12$
5. $f(x) = x^3 - 3x^2 - 2x + 6$ **7.** $f(x) = x^3 - 6x - 4$
9. $f(x) = x^3 + 2x^2 + 29x + 148$
11. $f(x) = x^3 - \frac{5}{3}x^2 - \frac{2}{3}x$
13. $f(x) = x^5 + 2x^4 - 2x^2 - x$
15. $f(x) = x^4 + 3x^3 + 3x^2 + x$ **17.** $-\sqrt{3}$
19. $i, 2 + \sqrt{5}$ **21.** $-3i$ **23.** $-4 + 3i, 2 + \sqrt{3}$
25. $-\sqrt{5}, 4i$ **27.** $2 + i$ **29.** $-3 - 4i, 4 + \sqrt{5}$
31. $4 + i$ **33.** $f(x) = x^3 - 4x^2 + 6x - 4$
35. $f(x) = x^2 + 16$ **37.** $f(x) = x^3 - 5x^2 + 16x - 80$
39. $f(x) = x^4 - 2x^3 - 3x^2 + 10x - 10$
41. $f(x) = x^4 + 4x^2 - 45$ **43.** $-\sqrt{2}, \sqrt{2}$ **45.** $i, 2, 3$
47. $1 + 2i, 1 - 2i$ **49.** ± 1 **51.** $\pm 1, \pm\frac{1}{2}, \pm 2, \pm 4, \pm 8$
53. $\pm 1, \pm 2, \pm\frac{1}{3}, \pm\frac{1}{5}, \pm\frac{2}{3}, \pm\frac{2}{5}, \pm\frac{1}{15}, \pm\frac{2}{15}$
55. (a) Rational: -3; other: $\pm\sqrt{2}$;
(b) $f(x) = (x + 3)(x + \sqrt{2})(x - \sqrt{2})$
57. (a) Rational: $-2, 1$; other: none;
(b) $f(x) = (x + 2)(x - 1)^2$

59. (a) Rational: -1; other: $3 \pm 2\sqrt{2}i$;
(b) $f(x) = (x + 1)(x - 3 - 2\sqrt{2}i)(x - 3 + 2\sqrt{2}i)$
61. (a) Rational: $-\frac{1}{5}$, 1; other: $\pm 2i$;
(b) $f(x) = 5(x + \frac{1}{5})(x - 1)(x + 2i)(x - 2i)$
63. (a) Rational: $-2, -1$; other: $3 \pm \sqrt{13}$;
(b) $f(x) = (x + 2)(x + 1)(x - 3 - \sqrt{13})(x - 3 + \sqrt{13})$
65. (a) Rational: 2; other: $1 \pm \sqrt{3}$;
(b) $f(x) = (x - 2)(x - 1 - \sqrt{3})(x - 1 + \sqrt{3})$
67. (a) Rational: -2; other: $1 \pm \sqrt{3}i$;
(b) $f(x) = (x + 2)(x - 1 - \sqrt{3}i)(x - 1 + \sqrt{3}i)$
69. (a) Rational: $\frac{1}{2}$; other: $\frac{1 \pm \sqrt{5}}{2}$;
(b) $f(x) = \frac{1}{3}\left(x - \frac{1}{2}\right)\left(x - \frac{1 + \sqrt{5}}{2}\right)\left(x - \frac{1 - \sqrt{5}}{2}\right)$
71. $1, -3$ **73.** No rational zeros   **75.** No rational zeros
77. $-2, 1, 2$ **79.** 3 or 1; 0   **81.** 0; 3 or 1
83. 2 or 0; 2 or 0   **85.** 1; 1   **87.** 1; 0   **89.** 2 or 0; 2 or 0
91. 3 or 1; 1   **93.** 1; 1
95.

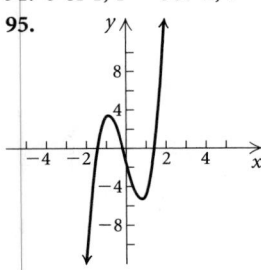

$f(x) = 4x^3 + x^2 - 8x - 2$

97.

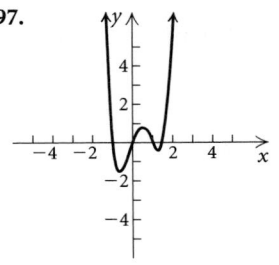

$f(x) = 2x^4 - 3x^3 - 2x^2 + 3x$

99. Discussion and Writing   **101.** [3.3] **(a)** $(4, -6)$;
(b) $x = 4$; **(c)** minimum: -6 at $x = 4$
102. [3.3] **(a)** $(1, -4)$; **(b)** $x = 1$; **(c)** minimum: -4 at $x = 1$
103. [1.5] 10   **104.** [3.2] $-3, 11$
105. [4.1] Cubic; $-x^3$; -1; 3; as $x \to \infty, g(x) \to -\infty$, and as
$x \to -\infty, g(x) \to \infty$ **106.** [3.3] Quadratic; $-x^2$; -1; 2; as
$x \to \infty, f(x) \to -\infty$, and as $x \to -\infty, f(x) \to -\infty$
107. [1.3] Constant; $-\frac{4}{9}$; $-\frac{4}{9}$; zero degree; for all $x, f(x) = -\frac{4}{9}$
108. [1.3] Linear; x; 1; 1; as $x \to \infty, h(x) \to \infty$,
and as $x \to -\infty, h(x) \to -\infty$
109. [4.1] Quartic; x^4; 1; 4; as $x \to \infty, g(x) \to \infty$, and as
$x \to -\infty, g(x) \to \infty$ **110.** [4.1] Cubic; x^3; 1; 3; as $x \to \infty$,
$h(x) \to \infty$, and as $x \to -\infty, h(x) \to -\infty$ **111. (a)** $-1, \frac{1}{2}, 3$;
(b) $0, \frac{3}{2}, 4$; **(c)** $-3, -\frac{3}{2}, 1$; **(d)** $-\frac{1}{2}, \frac{1}{4}, \frac{3}{2}$
113. $-8, -\frac{3}{2}, 4, 7, 15$

Visualizing the Graph

1. A   **2.** C   **3.** D   **4.** H   **5.** G   **6.** F   **7.** B
8. I   **9.** J   **10.** E

Exercise Set 4.5

1. $\{x \mid x \neq 2\}$, or $(-\infty, 2) \cup (2, \infty)$
3. $\{x \mid x \neq 1 \ and \ x \neq 5\}$, or $(-\infty, 1) \cup (1, 5) \cup (5, \infty)$
5. $\{x \mid x \neq -5\}$, or $(-\infty, -5) \cup (-5, \infty)$
7. (d); $x = 2, x = -2, y = 0$ **9.** (e); $x = 2, x = -2, y = 0$
11. (c); $x = 2, x = -2, y = 8x$ **13.** $x = 0$ **15.** $x = 2$
17. $x = 4, x = -6$ **19.** $x = \frac{3}{2}, x = -1$ **21.** $y = \frac{3}{4}$
23. $y = 0$ **25.** No horizontal asymptote
27. $y = x + 1$ **29.** $y = x$ **31.** $y = x - 3$
33. Domain: $(-\infty, 0) \cup (0, \infty)$; no x-intercepts,
no y-intercept;

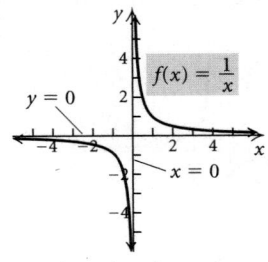

35. Domain: $(-\infty, 0) \cup (0, \infty)$; no x-intercepts,
no y-intercept;

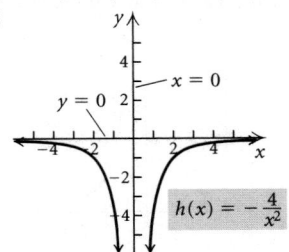

37. Domain: $(-\infty, -1) \cup (-1, \infty)$; x-intercepts: $(1, 0)$ and
$(3, 0)$, y-intercept: $(0, 3)$;

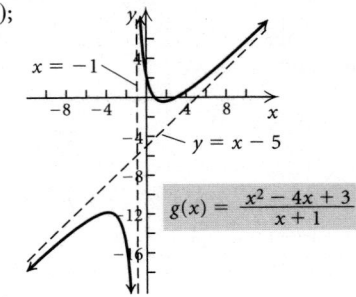

39. Domain: $(-\infty, 5) \cup (5, \infty)$; no x-intercepts,
y-intercept: $\left(0, \frac{2}{5}\right)$;

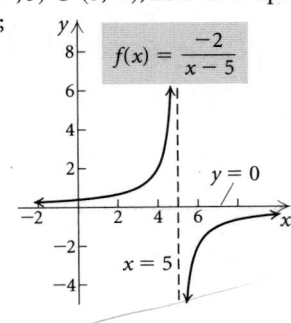

41. Domain: $(-\infty, 0) \cup (0, \infty)$; x-intercept: $\left(-\frac{1}{2}, 0\right)$, no y-intercept;

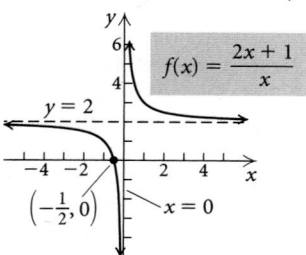

43. Domain: $(-\infty, -3) \cup (-3, 0) \cup (0, \infty)$; no x-intercepts, no y-intercept;

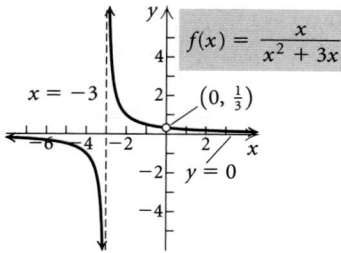

45. Domain: $(-\infty, 2) \cup (2, \infty)$; no x-intercepts, y-intercept: $\left(0, \frac{1}{4}\right)$;

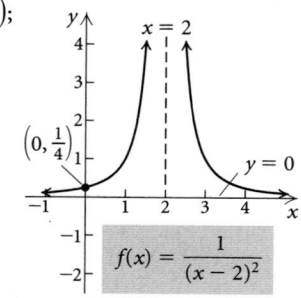

47. Domain: $(-\infty, -3) \cup (-3, -1) \cup (-1, \infty)$; x-intercept: $(1, 0)$, y-intercept: $(0, -1)$;

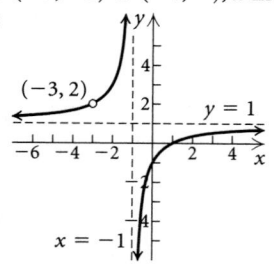

49. Domain: $(-\infty, \infty)$; no x-intercepts, y-intercept: $\left(0, \frac{1}{3}\right)$;

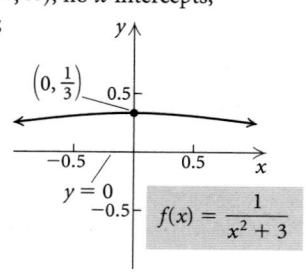

51. Domain: $(-\infty, 2) \cup (2, \infty)$; x-intercept: $(-2, 0)$, y-intercept: $(0, 2)$;

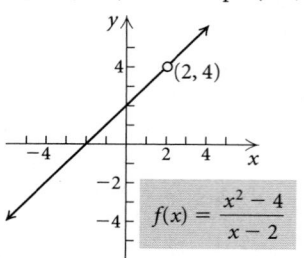

53. Domain: $(-\infty, -2) \cup (-2, \infty)$; x-intercept: $(1, 0)$, y-intercept: $\left(0, -\frac{1}{2}\right)$;

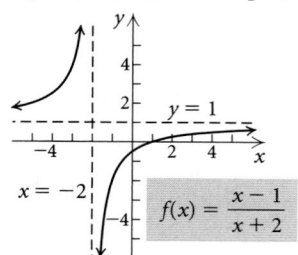

55. Domain: $\left(-\infty, -\frac{1}{2}\right) \cup \left(-\frac{1}{2}, 0\right) \cup (0, 3) \cup (3, \infty)$; x-intercept: $(-3, 0)$, no y-intercept;

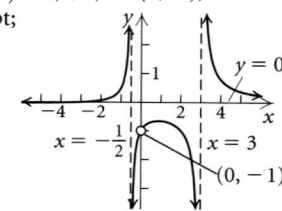

57. Domain: $(-\infty, -1) \cup (-1, \infty)$; x-intercepts: $(-3, 0)$ and $(3, 0)$, y-intercept: $(0, -9)$;

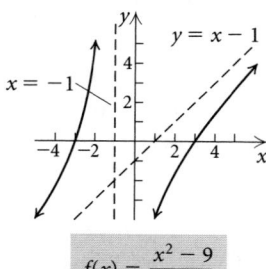

59. Domain: $(-\infty, \infty)$; x-intercepts: $(-2, 0)$ and $(1, 0)$, y-intercept: $(0, -2)$;

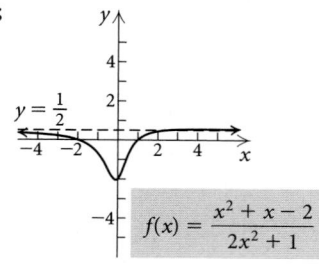

61. Domain: $(-\infty, 1) \cup (1, \infty)$; x-intercept: $\left(-\frac{2}{3}, 0\right)$,
y-intercept: $(0, 2)$;

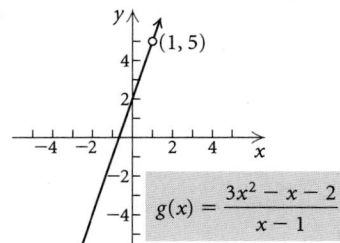

$$g(x) = \frac{3x^2 - x - 2}{x - 1}$$

63. Domain: $(-\infty, -1) \cup (-1, 3) \cup (3, \infty)$; x-intercept: $(1, 0)$,
y-intercept: $\left(0, \frac{1}{3}\right)$;

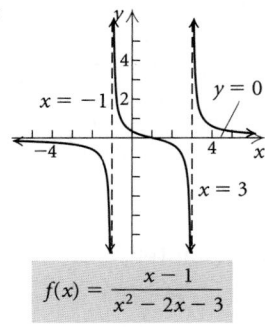

$$f(x) = \frac{x - 1}{x^2 - 2x - 3}$$

65. Domain: $(-\infty, -1) \cup (-1, \infty)$; x-intercept: $(3, 0)$,
y-intercept: $(0, -3)$;

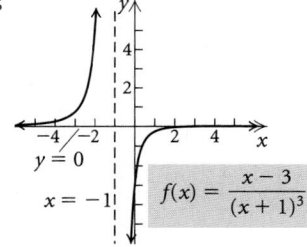

$$f(x) = \frac{x - 3}{(x + 1)^3}$$

67. Domain: $(-\infty, 0) \cup (0, \infty)$; x-intercept: $(-1, 0)$,
no y-intercept;

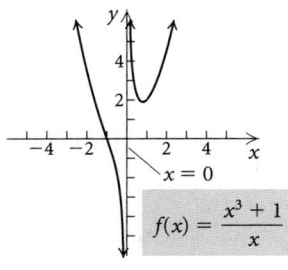

$$f(x) = \frac{x^3 + 1}{x}$$

69. Domain: $(-\infty, -2) \cup (-2, 7) \cup (7, \infty)$;
x-intercepts: $(-5, 0)$, $(0, 0)$, and $(3, 0)$, y-intercept: $(0, 0)$;

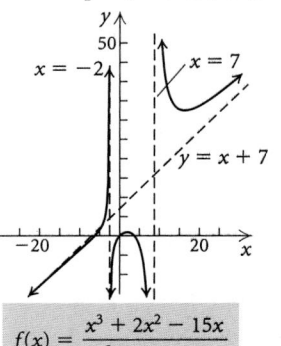

$$f(x) = \frac{x^3 + 2x^2 - 15x}{x^2 - 5x - 14}$$

71. Domain: $(-\infty, \infty)$; x-intercept: $(0, 0)$, y-intercept: $(0, 0)$;

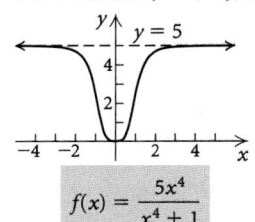

$$f(x) = \frac{5x^4}{x^4 + 1}$$

73. Domain: $(-\infty, -1) \cup (-1, 2) \cup (2, \infty)$; x-intercept: $(0, 0)$,
y-intercept: $(0, 0)$;

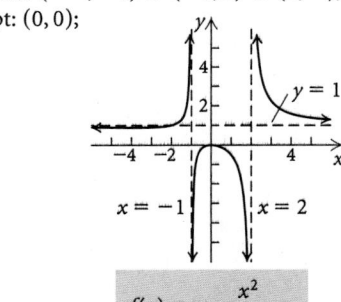

$$f(x) = \frac{x^2}{x^2 - x - 2}$$

75. $f(x) = \dfrac{1}{x^2 - x - 20}$

77. $f(x) = \dfrac{3x^2 + 12x + 12}{2x^2 - 2x - 40}$

79. (a) $N(t)$

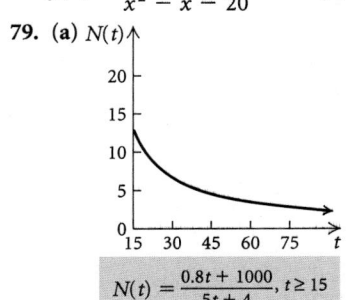

$$N(t) = \frac{0.8t + 1000}{5t + 4}, t \ge 15$$

$N(t) \to 0.16$ as $t \to \infty$;
(b) The medication never completely disappears from the body; a trace amount remains.

81. (a)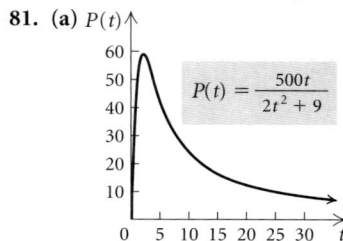

$$P(t) = \frac{500t}{2t^2 + 9}$$

(b) $P(0) = 0$; $P(1) = 45,455$; $P(3) = 55,556$; $P(8) = 29,197$;
(c) $P(t) \to 0$ as $t \to \infty$; **(d)** In time, no one lives in Lordsburg.
(e) 58,926 at $t \approx 2.12$ months

83. $y_1 = \dfrac{x^3 + 4}{x} = x^2 + \dfrac{4}{x}$. As $|x| \to \infty$, $\dfrac{4}{x} \to 0$ and the
value of $y_1 \to x^2$. Thus the parabola $y_2 = x^2$ can be thought of
as a nonlinear asymptote for y_1. The graph confirms this.
85. Discussion and Writing   **86.** [1.2] Domain, range,
domain, range   **87.** [1.3] Slope
88. [1.3] Slope–intercept equation
89. [1.4] Point–slope equation
90. [1.1] x-intercept   **91.** [2.4] $f(-x) = -f(x)$
92. [1.3] Vertical lines   **93.** [1.1] Midpoint formula
94. [1.1] y-intercept   **95.** $y = x^3 + 4$
97.

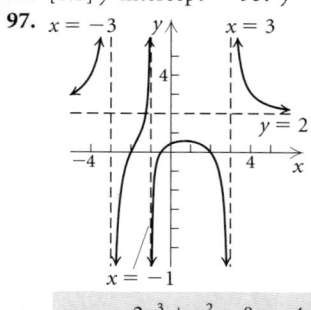

$$f(x) = \frac{2x^3 + x^2 - 8x - 4}{x^3 + x^2 - 9x - 9}$$

Exercise Set 4.6

1. $\{-5, 3\}$ **3.** $[-5, 3]$ **5.** $(-\infty, -5] \cup [3, \infty)$
7. $(-\infty, -4) \cup (2, \infty)$ **9.** $(-\infty, -4) \cup [2, \infty)$
11. $\{0\}$ **13.** $(-5, 0] \cup (1, \infty)$ **15.** $(-\infty, -5) \cup (0, 1)$
17. $(-\infty, -3) \cup (0, 3)$ **19.** $(-3, 0) \cup (3, \infty)$
21. $(-\infty, -5) \cup (-3, 2)$ **23.** $(-2, 0] \cup (2, \infty)$
25. $(-4, 1)$ **27.** $(-\infty, -2) \cup (1, \infty)$
29. $(-\infty, -1] \cup [3, \infty)$ **31.** $(-\infty, -5) \cup (5, \infty)$
33. $(-\infty, -2] \cup [2, \infty)$ **35.** $(-\infty, 3) \cup (3, \infty)$
37. $\varnothing$ **39.** $\left(-\infty, -\frac{5}{4}\right] \cup [0, 3]$ **41.** $[-3, -1] \cup [1, \infty)$
43. $(-\infty, -2) \cup (1, 3)$ **45.** $\left[-\sqrt{2}, -1\right] \cup \left[\sqrt{2}, \infty\right)$
47. $(-\infty, -1] \cup \left[\frac{3}{2}, 2\right]$ **49.** $(-\infty, 5]$
51. $(-\infty, -1.680) \cup (2.154, 5.526)$ **53.** $(-4, \infty)$
55. $\left(-\frac{5}{2}, \infty\right)$ **57.** $(-\infty, 0] \cup [4, \infty)$
59. $\left(-3, -\frac{1}{5}\right] \cup (1, \infty)$
61. $(-\infty, -3) \cup \left[\dfrac{5 - \sqrt{105}}{10}, -\dfrac{1}{3}\right) \cup \left[\dfrac{5 + \sqrt{105}}{10}, \infty\right)$

63. $\left(2, \frac{7}{2}\right]$ **65.** $\left(1 - \sqrt{2}, 0\right) \cup \left(1 + \sqrt{2}, \infty\right)$
67. $(-\infty, -3) \cup (1, 3) \cup \left[\frac{11}{3}, \infty\right)$ **69.** $(-\infty, \infty)$
71. $\left(-3, \dfrac{1 - \sqrt{61}}{6}\right) \cup \left(-\dfrac{1}{2}, 0\right) \cup \left(\dfrac{1 + \sqrt{61}}{6}, \infty\right)$
73. $(-1, 0) \cup \left(\frac{2}{7}, \frac{7}{2}\right)$
75. $\left[-6 - \sqrt{33}, -5\right) \cup \left[-6 + \sqrt{33}, 1\right) \cup (5, \infty)$
77. $(0.408, 2.449)$ **79. (a)** $(10, 200)$; **(b)** $(0, 10) \cup (200, \infty)$
81. $\{n \mid 9 \le n \le 23\}$ **83.** Discussion and Writing
85. [1.1] $(x + 2)^2 + (y - 4)^2 = 9$
86. [1.1] $x^2 + (y + 3)^2 = \frac{49}{16}$ **87.** [3.3] **(a)** $\left(\frac{3}{4}, -\frac{55}{8}\right)$;
(b) maximum: $-\frac{55}{8}$ when $x = \frac{3}{4}$; **(c)** $\left(-\infty, -\frac{55}{8}\right]$
88. [3.3] **(a)** $(5, -23)$; **(b)** minimum: -23 when $x = 5$;
(c) $[-23, \infty)$ **89.** $(-\infty, \infty)$ **91.** $\left[-\sqrt{5}, \sqrt{5}\right]$
93. $\left[-\frac{3}{2}, \frac{3}{2}\right]$ **95.** $\left(-\infty, -\frac{1}{4}\right) \cup \left(\frac{1}{2}, \infty\right)$
97. $(-4, -2) \cup (-1, 1)$
99. $x^2 + x - 12 < 0$; answers may vary
101. $(-\infty, -3) \cup (7, \infty)$

Review Exercises: Chapter 4

1. True   **2.** True   **3.** False   **4.** False   **5.** False
6. (a) $-2.637, 1.137$; **(b)** relative maximum: 7.125 at
$x = -0.75$; **(c)** none; **(d)** domain: all real numbers;
range: $(-\infty, 7.125]$ **7. (a)** $-3, -1.414, 1.414$; **(b)** relative
maximum: 2.303 at $x = -2.291$; **(c)** relative minimum:
-6.303 at $x = 0.291$; **(d)** domain: all real numbers; range: all
real numbers   **8. (a)** $0, 1, 2$; **(b)** relative maximum: 0.202 at
$x = 0.610$; **(c)** relative minima: 0 at $x = 0$, -0.620 at
$x = 1.640$; **(d)** domain: all real numbers; range: $[-0.620, \infty)$
9. $0.45x^4, 0.45, 4$, quartic   **10.** $-25, -25, 0$, constant
11. $-0.5x, -0.5, 1$, linear   **12.** $\frac{1}{3}x^3, \frac{1}{3}, 3$, cubic
13. As $x \to \infty$, $f(x) \to -\infty$, and as $x \to -\infty$, $f(x) \to -\infty$.
14. As $x \to \infty$, $f(x) \to \infty$, and as $x \to -\infty$, $f(x) \to -\infty$.
15. $\frac{2}{3}$, multiplicity 1; -2, multiplicity 3; 5, multiplicity 2
16. $\pm 1, \pm 5$, each has multiplicity 1   **17.** $\pm 3, -4$, each has
multiplicity 1   **18. (a)** 4%; **(b)** 5%   **19. (a)** Linear:
$f(x) = 0.5408695652x - 30.30434783$; quadratic:
$f(x) = 0.0030322581x^2 - 0.5764516129x + 57.53225806$;
cubic: $f(x) = 0.0000247619x^3 - 0.0112857143x^2 +$
$2.002380952x - 82.14285714$; **(b)** the cubic function;
(c) 298, 498
20.

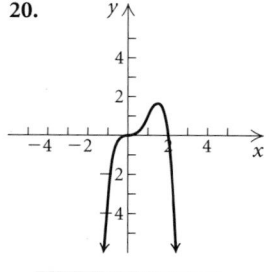

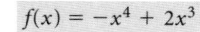

$f(x) = -x^4 + 2x^3$

21.

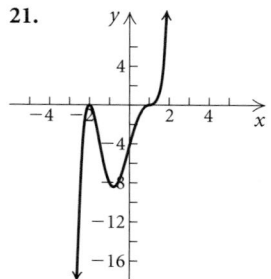

$g(x) = (x - 1)^3(x + 2)^2$

22.

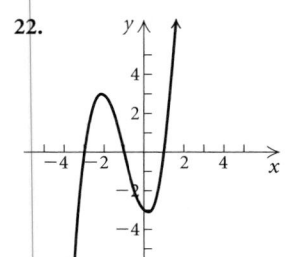

$h(x) = x^3 + 3x^2 - x - 3$

23.

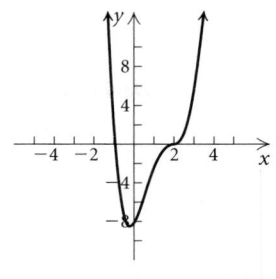

$f(x) = x^4 - 5x^3 + 6x^2 + 4x - 8$

24.

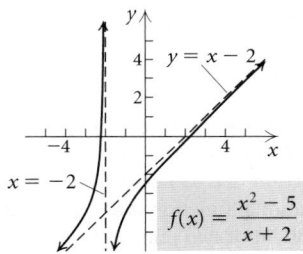

$g(x) = 2x^3 + 7x^2 - 14x + 5$

25. $f(1) = -4$ and $f(2) = 3$. Since $f(1)$ and $f(2)$ have opposite signs, $f(x)$ has a zero between 1 and 2.

26. $f(-1) = -3.5$ and $f(1) = -0.5$. Since $f(-1)$ and $f(1)$ have the same sign, the intermediate value theorem does not allow us to determine whether there is a zero between -1 and 1.

27. $Q(x) = 6x^2 + 16x + 52, R(x) = 155$;
$P(x) = (x - 3)(6x^2 + 16x + 52) + 155$

28. $Q(x) = x^3 - 3x^2 + 3x - 2, R(x) = 7$;
$P(x) = (x + 1)(x^3 - 3x^2 + 3x - 2) + 7$

29. $x^2 + 7x + 22$, R 120   **30.** $x^3 + x^2 + x + 1$, R 0

31. $x^4 - x^3 + x^2 - x - 1$, R 1   **32.** 36   **33.** 0

34. $-141,220$ **35.** Yes, no   **36.** No, yes   **37.** Yes, no

38. No, yes   **39.** $f(x) = (x - 1)^2(x + 4)$; $-4, 1$

40. $f(x) = (x - 2)(x + 3)^2$; $-3, 2$

41. $f(x) = (x - 2)^2(x - 5)(x + 5)$; $-5, 2, 5$

42. $f(x) = (x - 1)(x + 1)(x - \sqrt{2})(x + \sqrt{2})$; $-\sqrt{2}, -1,$
$1, \sqrt{2}$ **43.** $f(x) = x^3 + 3x^2 - 6x - 8$

44. $f(x) = x^3 + x^2 - 4x + 6$

45. $f(x) = x^3 - \frac{5}{2}x^2 + \frac{1}{2}$, or $2x^3 - 5x^2 + 1$

46. $f(x) = x^4 + \frac{29}{2}x^3 + \frac{135}{2}x^2 + \frac{175}{2}x - \frac{125}{2}$, or
$2x^4 + 29x^3 + 135x^2 + 175x - 125$

47. $f(x) = x^5 + 4x^4 - 3x^3 - 18x^2$ **48.** $-\sqrt{5}, -i$

49. $1 - \sqrt{3}, \sqrt{3}$ **50.** $\sqrt{2}$ **51.** $f(x) = x^2 - 11$

52. $f(x) = x^3 - 6x^2 + x - 6$

53. $f(x) = x^4 - 5x^3 + 4x^2 + 2x - 8$

54. $f(x) = x^4 - x^2 - 20$ **55.** $f(x) = x^3 + \frac{8}{3}x^2 - x$

56. $\pm\frac{1}{4}, \pm\frac{1}{2}, \pm\frac{3}{4}, \pm1, \pm\frac{3}{2}, \pm2, \pm3, \pm4, \pm6, \pm12$

57. $\pm\frac{1}{3}, \pm1$ **58.** $\pm1, \pm2, \pm3, \pm4, \pm6, \pm8, \pm12, \pm24$

59. (a) Rational: $0, -2, \frac{1}{3}, 3$; other: none;

(b) $f(x) = 3x(x - \frac{1}{3})(x + 2)^2(x - 3)$

60. (a) Rational: 2; other: $\pm\sqrt{3}$;
(b) $f(x) = (x - 2)(x + \sqrt{3})(x - \sqrt{3})$

61. (a) Rational: $-1, 1$; other: $3 \pm i$;
(b) $f(x) = (x + 1)(x - 1)(x - 3 - i)(x - 3 + i)$

62. (a) Rational: -5; other: $1 \pm \sqrt{2}$;
(b) $f(x) = (x + 5)(x - 1 - \sqrt{2})(x - 1 + \sqrt{2})$

63. (a) Rational: $\frac{2}{3}, 1$; other: none;
(b) $f(x) = 3(x - \frac{2}{3})(x - 1)^2$

64. (a) Rational: 2; other: $1 \pm \sqrt{5}$;
(b) $f(x) = (x - 2)^3(x - 1 + \sqrt{5})(x - 1 - \sqrt{5})$

65. (a) Rational: $-4, 0, 3, 4$; other: none;
(b) $f(x) = x^2(x + 4)^2(x - 3)(x - 4)$

66. (a) Rational: $\frac{5}{2}, 1$; other: none;
(b) $f(x) = 2(x - \frac{5}{2})(x - 1)^4$

67. 3 or 1; 0   **68.** 4 or 2 or 0; 2 or 0

69. 3 or 1; 0

70. Domain: $(-\infty, -2) \cup (-2, \infty)$; x-intercepts: $(-\sqrt{5}, 0)$ and $(\sqrt{5}, 0)$, y-intercept: $(0, -\frac{5}{2})$

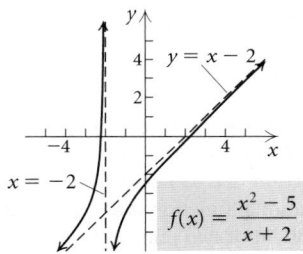

$f(x) = \dfrac{x^2 - 5}{x + 2}$

71. Domain: $(-\infty, 2) \cup (2, \infty)$; x-intercepts: none,
y-intercept: $(0, \frac{5}{4})$

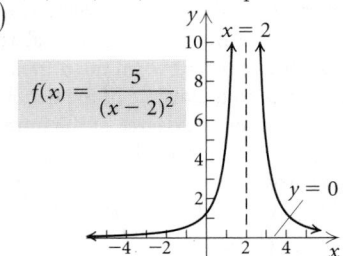

$f(x) = \dfrac{5}{(x - 2)^2}$

72. Domain: $(-\infty, -4) \cup (-4, 5) \cup (5, \infty)$; x-intercepts: $(-3, 0)$ and $(2, 0)$, y-intercept: $(0, \frac{3}{10})$

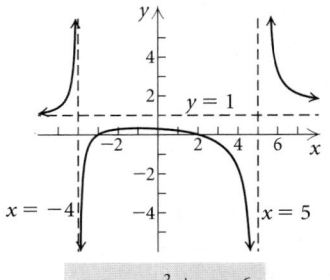

$f(x) = \dfrac{x^2 + x - 6}{x^2 - x - 20}$

73. Domain: $(-\infty, -3) \cup (-3, 5) \cup (5, \infty)$; x-intercept: $(2, 0)$, y-intercept: $\left(0, \frac{2}{15}\right)$

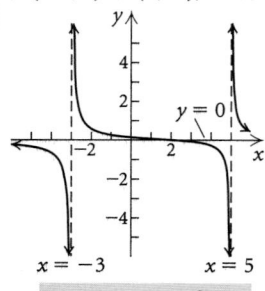

$$f(x) = \frac{x - 2}{x^2 - 2x - 15}$$

74. $f(x) = \dfrac{1}{x^2 - x - 6}$ **75.** $f(x) = \dfrac{4x^2 + 12x}{x^2 - x - 6}$

76. (a) $N(t) \to 0.0875$ as $t \to \infty$; (b) The medication never completely disappears from the body; a trace amount remains.

77. $(-3, 3)$ **78.** $\left(-\infty, -\frac{1}{2}\right) \cup (2, \infty)$

79. $[-4, 1] \cup [2, \infty)$ **80.** $\left(-\infty, -\frac{14}{3}\right) \cup (-3, \infty)$

81. (a) $t = 7$; (b) $(2, 3)$ **82.** $\left[\dfrac{5 - \sqrt{15}}{2}, \dfrac{5 + \sqrt{15}}{2}\right]$

83. A   **84.** C   **85.** B

86. Discussion and Writing: A polynomial function is a function that can be defined by a polynomial expression. A rational function is a function that can be defined as a quotient of two polynomials.

87. Discussion and Writing: Vertical asymptotes occur at any x-values that make the denominator zero. The graph of a rational function does not cross any vertical asymptotes. Horizontal asymptotes occur when the degree of the numerator is less than or equal to the degree of the denominator. Oblique asymptotes occur when the degree of the numerator is 1 greater than the degree of the denominator. Graphs of rational functions may cross horizontal or oblique asymptotes.

88. $\left(-\infty, -1 - \sqrt{6}\right] \cup \left[-1 + \sqrt{6}, \infty\right)$

89. $\left(-\infty, -\frac{1}{2}\right) \cup \left(\frac{1}{2}, \infty\right)$

90. $\{1 + i, 1 - i, i, -i\}$ **91.** $(-\infty, 2)$

92. $(x - 1)\left(x + \dfrac{1}{2} - \dfrac{\sqrt{3}}{2}i\right)\left(x + \dfrac{1}{2} + \dfrac{\sqrt{3}}{2}i\right)$

93. 7   **94.** -4 **95.** $(-\infty, -5] \cup [2, \infty)$

96. $(-\infty, 1.1] \cup [2, \infty)$ **97.** $\left(-1, \frac{3}{7}\right)$

Test: Chapter 4

1. [4.1] $-x^4$, -1, 4; quartic   **2.** [4.1] $-4.7x$, -4.7, 1; linear
3. [4.1] $0, \frac{5}{3}$, each has multiplicity 1; 3, multiplicity 2; -1, multiplicity 3   **4.** [4.1] 3388; 5379; 3514

5. [4.2]

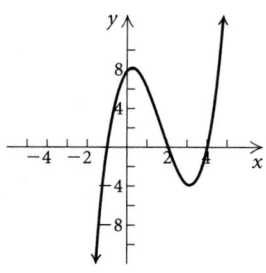

$$f(x) = x^3 - 5x^2 + 2x + 8$$

6. [4.2]

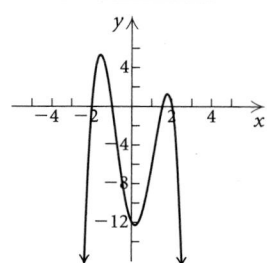

$$f(x) = -2x^4 + x^3 + 11x^2 - 4x - 12$$

7. [4.2] $f(0) = 3$ and $f(2) = -17$. Since $f(0)$ and $f(2)$ have opposite signs, $f(x)$ has a zero between 0 and 2.

8. [4.2] $g(-2) = 5$ and $g(-1) = 1$. Both $g(-2)$ and $g(-1)$ are positive. We cannot use the intermediate value theorem to determine if there is a zero between -2 and -1.

9. [4.3] $Q(x) = x^3 + 4x^2 + 4x + 6$, $R(x) = 1$; $P(x) = (x - 1)(x^3 + 4x^2 + 4x + 6) + 1$

10. [4.3] $3x^2 + 15x + 63$, R 322   **11.** [4.3] -115

12. [4.3] Yes   **13.** [4.4] $f(x) = x^4 - 27x^2 - 54x$

14. [4.4] $-\sqrt{3}, 2 + i$

15. [4.4] $f(x) = x^3 + 10x^2 + 9x + 90$

16. [4.4] $f(x) = x^5 - 2x^4 - x^3 + 6x^2 - 6x$

17. [4.4] $\pm 1, \pm 2, \pm 3, \pm 4, \pm 6, \pm 12, \pm \frac{1}{2}, \pm \frac{3}{2}$

18. [4.4] $\pm \frac{1}{10}, \pm \frac{1}{5}, \pm \frac{1}{2}, \pm 1, \pm \frac{5}{2}, \pm 5$

19. [4.4] (a) Rational: -1; other: $\pm \sqrt{5}$;
(b) $f(x) = (x + 1)\left(x - \sqrt{5}\right)\left(x + \sqrt{5}\right)$

20. [4.4] (a) Rational: $-\frac{1}{2}$, 1, 2, 3; other: none;
(b) $f(x) = 2\left(x + \frac{1}{2}\right)(x - 1)(x - 2)(x - 3)$

21. [4.4] (a) Rational: -4; other: $\pm 2i$;
(b) $f(x) = (x - 2i)(x + 2i)(x + 4)$

22. [4.4] (a) Rational: $\frac{2}{3}$, 1; other: none;
(b) $f(x) = 3\left(x - \frac{2}{3}\right)(x - 1)^3$

23. [4.4] 2 or 0; 2 or 0

24. [4.5] Domain: $(-\infty, 3) \cup (3, \infty)$; x-intercepts: none, y-intercept: $\left(0, \frac{2}{9}\right)$;

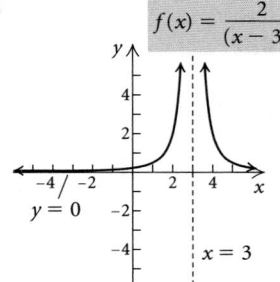

25. [4.5] Domain: $(-\infty, -1) \cup (-1, 4) \cup (4, \infty)$; x-intercept: $(-3, 0)$, y-intercept: $\left(0, -\frac{3}{4}\right)$;

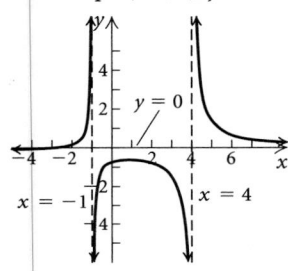

26. [4.5] Answers may vary; $f(x) = \dfrac{x + 4}{x^2 - x - 2}$

27. [4.6] $\left(-\infty, -\frac{1}{2}\right) \cup (3, \infty)$

28. [4.1], [4.6] $(-\infty, 4) \cup \left[\frac{13}{2}, \infty\right)$

29. (a) [4.1] 6 sec; (b) [4.1], [4.6] $(1, 3)$ **30.** [4.2] D

31. [4.1], [4.6] $(-\infty, -4] \cup [3, \infty)$

CHAPTER 5

Exercise Set 5.1

1. $\{(8,7), (8, -2), (-4, 3), (-8, 8)\}$ **3.** $\{(-1, -1), (4, -3)\}$

5. $x = 4y - 5$ **7.** $y^3 x = -5$ **9.** $y = x^2 - 2x$

11.

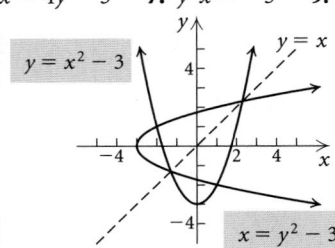

13.

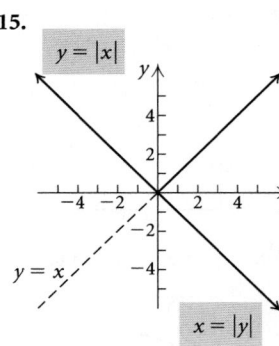

15.

17. Assume $f(a) = f(b)$ for any numbers a and b in the domain of f. Since $f(a) = \frac{1}{3}a - 6$ and $f(b) = \frac{1}{3}b - 6$, we have

$$\frac{1}{3}a - 6 = \frac{1}{3}b - 6$$
$$\frac{1}{3}a = \frac{1}{3}b \qquad \text{Adding 6}$$
$$a = b. \qquad \text{Multiplying by 3}$$

Thus, if $f(a) = f(b)$, then $a = b$ and f is one-to-one.

19. Assume $f(a) = f(b)$ for any numbers a and b in the domain of f. Since $f(a) = a^3 + \frac{1}{2}$ and $f(b) = b^3 + \frac{1}{2}$, we have

$$a^3 + \frac{1}{2} = b^3 + \frac{1}{2}$$
$$a^3 = b^3 \qquad \text{Subtracting } \frac{1}{2}$$
$$a = b. \qquad \text{Taking the cube root}$$

Thus, if $f(a) = f(b)$, then $a = b$ and f is one-to-one.

21. Find two numbers a and b for which $a \neq b$ and $g(a) = g(b)$. Two such numbers are -2 and 2, because $g(-2) = g(2) = -3$. Thus, g is not one-to-one.

23. Find two numbers a and b for which $a \neq b$ and $g(a) = g(b)$. Two such numbers are -1 and 1, because $g(-1) = g(1) = 0$. Thus, g is not one-to-one.

25. Yes   **27.** No   **29.** No   **31.** Yes   **33.** Yes

35. No   **37.** No   **39.** Yes   **41.** No   **43.** No

45. $y_1 = 0.8x + 1.7$,   Domain and range of both f and

$y_2 = \dfrac{x - 1.7}{0.8}$ f^{-1}: all real numbers

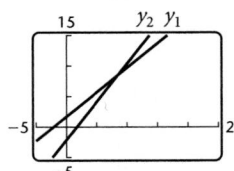

47. $y_1 = \frac{1}{2}x - 4$,   Domain and range of both f

$y_2 = 2x + 8$ and f^{-1}: all real numbers

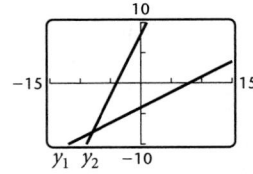

49. $y_1 = \sqrt{x-3}$,
$y_2 = x^2 + 3, x \geq 0$

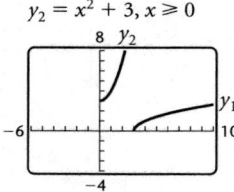

Domain of f: $[3, \infty)$,
range of f: $[0, \infty)$;
domain of f^{-1}: $[0, \infty)$,
range of f^{-1}: $[3, \infty)$

51. $y_1 = x^2 - 4, x \geq 0$; $y_2 = \sqrt{4+x}$

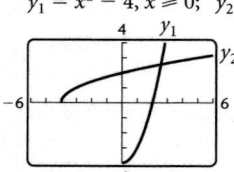

Domain of f: $[0, \infty)$,
range of f: $[-4, \infty)$;
domain of f^{-1}:
$[-4, \infty)$, range of
f^{-1}: $[0, \infty)$

53. $y_1 = (3x-9)^3$, $y_2 = \dfrac{\sqrt[3]{x}+9}{3}$

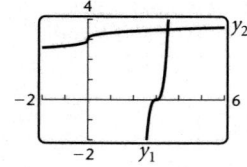

Domain and range of
both f and f^{-1}: all
real numbers

55. (a) One-to-one; (b) $f^{-1}(x) = x - 4$

57. (a) One-to-one; (b) $f^{-1}(x) = \dfrac{x+1}{2}$

59. (a) One-to-one; (b) $f^{-1}(x) = \dfrac{4}{x} - 7$

61. (a) One-to-one; (b) $f^{-1}(x) = \dfrac{3x+4}{x-1}$

63. (a) One-to-one; (b) $f^{-1}(x) = \sqrt[3]{x+1}$

65. (a) Not one-to-one; (b) does not have an inverse that is
a function

67. (a) One-to-one; (b) $f^{-1}(x) = \sqrt{\dfrac{x+2}{5}}$

69. (a) One-to-one; (b) $f^{-1}(x) = x^2 - 1, x \geq 0$

71. $\frac{1}{3}x$ **73.** $-x$ **75.** $x^3 + 5$

77.

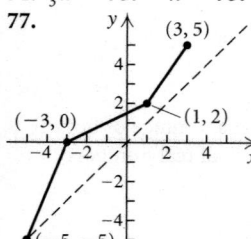

79.

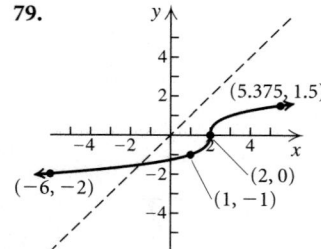

81.

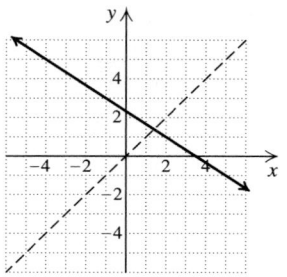

83. $f^{-1}(f(x)) = f^{-1}\left(\frac{7}{8}x\right) = \frac{8}{7} \cdot \frac{7}{8}x = x$;
$f(f^{-1}(x)) = f\left(\frac{8}{7}x\right) = \frac{7}{8} \cdot \frac{8}{7}x = x$

85. $f^{-1}(f(x)) = f^{-1}\left(\dfrac{1-x}{x}\right) = \dfrac{1}{\dfrac{1-x}{x} + 1} =$

$\dfrac{1}{\dfrac{1-x+x}{x}} = \dfrac{1}{\dfrac{1}{x}} = 1 \cdot \dfrac{x}{1} = x$; $f(f^{-1}(x)) = f\left(\dfrac{1}{x+1}\right) =$

$\dfrac{1 - \dfrac{1}{x+1}}{\dfrac{1}{x+1}} = \dfrac{\dfrac{x+1-1}{x+1}}{\dfrac{1}{x+1}} = \dfrac{x}{x+1} \cdot \dfrac{x+1}{1} = x$

87. $f^{-1}(f(x)) = f^{-1}\left(\dfrac{2}{5}x + 1\right) = \dfrac{5\left(\dfrac{2}{5}x+1\right) - 5}{2} =$

$\dfrac{2x+5-5}{2} = \dfrac{2x}{2} = x$; $f(f^{-1}(x)) = f\left(\dfrac{5x-5}{2}\right) =$

$\dfrac{2}{5}\left(\dfrac{5x-5}{2}\right) + 1 = x - 1 + 1 = x$

89. $f^{-1}(x) = \frac{1}{5}x + \frac{3}{5}$; domain of f and f^{-1}: $(-\infty, \infty)$;
range of f and f^{-1}: $(-\infty, \infty)$;

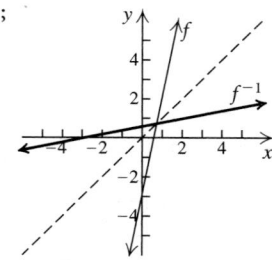

91. $f^{-1}(x) = \dfrac{2}{x}$; domain of f and f^{-1}: $(-\infty, 0) \cup (0, \infty)$;
range of f and f^{-1}: $(-\infty, 0) \cup (0, \infty)$;

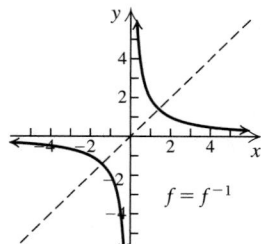

93. $f^{-1}(x) = \sqrt[3]{3x + 6}$; domain of f and f^{-1}: $(-\infty, \infty)$; range of f and f^{-1}: $(-\infty, \infty)$

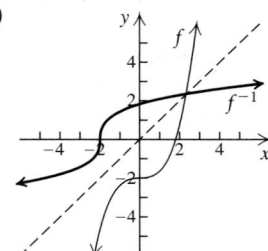

95. $f^{-1}(x) = \dfrac{3x + 1}{x - 1}$; domain of f: $(-\infty, 3) \cup (3, \infty)$;

range of f: $(-\infty, 1) \cup (1, \infty)$;
domain of f^{-1}: $(-\infty, 1) \cup (1, \infty)$;
range of f^{-1}: $(-\infty, 3) \cup (3, \infty)$;

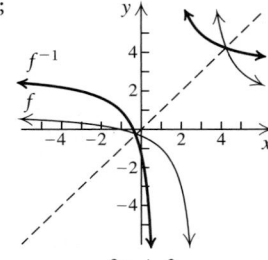

97. 5; a **99. (a)** $3\frac{1}{2}, 6, 6\frac{1}{2}$; **(b)** $s^{-1}(x) = \dfrac{2x + 3}{2}$;

(c) $4\frac{1}{2}, 7, 8\frac{1}{2}$ **101. (a)** 0.5 ft, 11.5 ft, 22.5 ft, 55.5 ft, 72 ft;

(b) $D^{-1}(r) = \dfrac{10r - 5}{11}$; the speed, in miles per hour, that the

car is traveling when the reaction distance is r feet;
(c)
$$y_1 = \frac{11x + 5}{10}, \quad y_2 = \frac{10x - 5}{11}$$

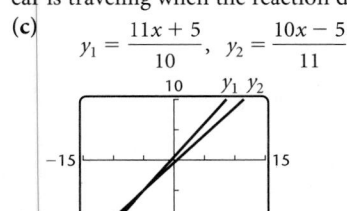

103. Discussion and Writing   **105.** [3.3] (b), (d), (f), (h)
106. [3.3] (a), (c), (e), (g)   **107.** [3.3] (a)   **108.** [3.3] (d)
109. [3.3] (f)   **110.** [3.3] (a), (b), (c), (d)   **111.** Yes
113. No   **115.** $f(x) = x^2 - 3$, for inputs $x \geq 0$;
$f^{-1}(x) = \sqrt{x + 3}$, for inputs $x \geq -3$ **117.** Answers may
vary. $f(x) = 3/x, f(x) = 1 - x, f(x) = x$

Exercise Set 5.2

1. 54.5982   **3.** 0.0856   **5.** (f)   **7.** (e)   **9.** (a)

11.

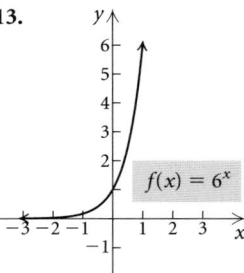

$f(x) = 3^x$

13.

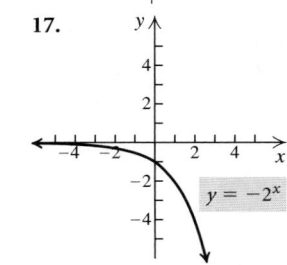

$f(x) = 6^x$

15.

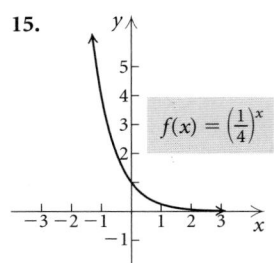

$f(x) = \left(\frac{1}{4}\right)^x$

17.

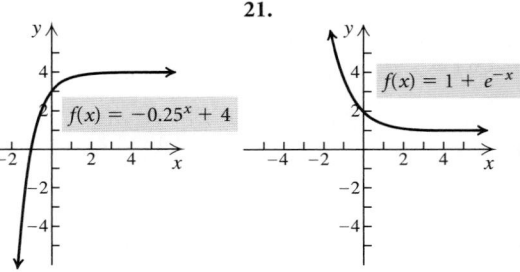

$y = -2^x$

19.

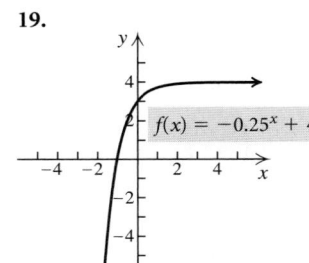

$f(x) = -0.25^x + 4$

21.

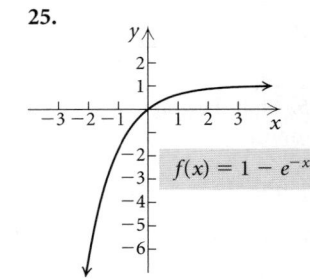

$f(x) = 1 + e^{-x}$

23.

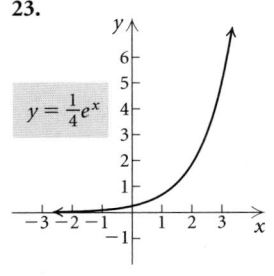

$y = \frac{1}{4}e^x$

25.

$f(x) = 1 - e^{-x}$

27. Shift the graph of $y = 2^x$ left 1 unit.

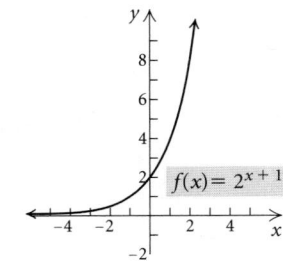

$f(x) = 2^{x+1}$

29. Shift the graph of $y = 2^x$ down 3 units.

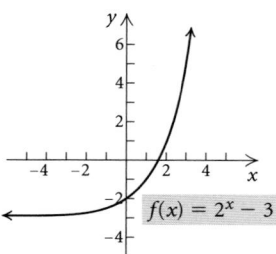

$f(x) = 2^x - 3$

31. Reflect the graph of $y = 3^x$ across the y-axis, then across the x-axis, and then shift it up 4 units.

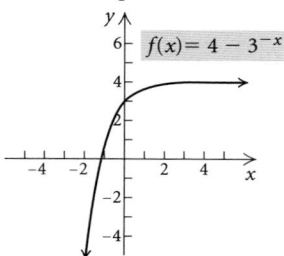

$f(x) = 4 - 3^{-x}$

33. Shift the graph of $y = \left(\frac{3}{2}\right)^x$ right 1 unit.

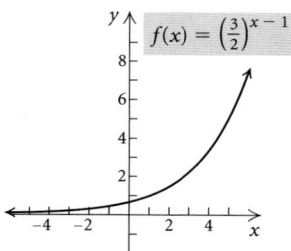

$f(x) = \left(\frac{3}{2}\right)^{x-1}$

35. Shift the graph of $y = 2^x$ left 3 units, and then down 5 units.

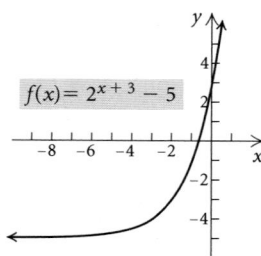

$f(x) = 2^{x+3} - 5$

37. Shrink the graph of $y = e^x$ horizontally.

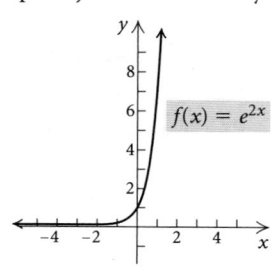

$f(x) = e^{2x}$

39. Shift the graph of $y = e^x$ left 1 unit and reflect it across the y-axis.

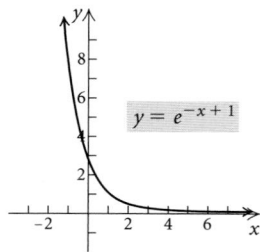

$y = e^{-x+1}$

41. Reflect the graph of $y = e^x$ across the y-axis, then across the x-axis, then shift it up 1 unit, and then stretch it vertically.

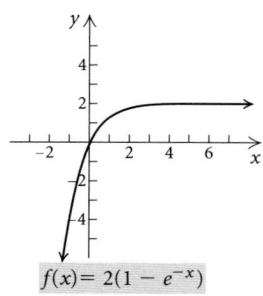

$f(x) = 2(1 - e^{-x})$

43. (a) $A(t) = 82,000(1.01125)^{4t}$
(b) $y = 82,000(1.01125)^{4x}$ **(c)** \$82,000, \$89,677.22, \$102,561.54, \$128,278.90;
(d) about 4.43 yr, or about 4 yr, 5 mo, and 5 days

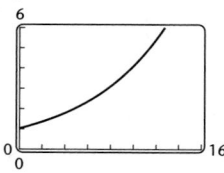

45. \$4930.86 **47.** \$3247.30 **49.** \$153,610.15
51. \$76,305.59 **53.** \$26,086.69
55. \$10.1 billion; \$15.2 billion
57. (a) 5.4 billion gal, 7.1 billion gal;
(b) $y = 1.0283(1.1483)^x$ **(c)** in about 11.4 yr after 1996

59. 3293 GB; 25,004 GB
61. Exports: 2007, \$982.7 billion; 2012, \$2384.8 billion; 2020, \$9851.1 billion; imports: 2007, \$861.3 billion; 2012, \$2029.5 billion; 2020, \$7998.2 billion
63. (a) $y = 595(0.8)^x$ **(b)** \$595; \$476; \$380.80; \$194.97; \$63.89; **(c)** about 4.89 yr, or about 4 yr, 10 mo, and 19 days

65. (a) $y = 29.0626(1.3438)^x$

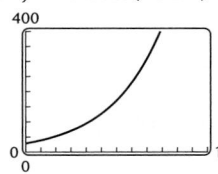

(b) \$39.1 billion; \$171.1 billion; \$558.1 billion; **(c)** about 8.4 yr after 2005

67. (a) $y = 100(1 - e^{-0.04x})$

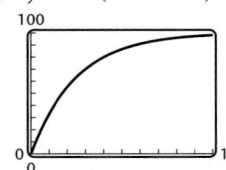

(b) about 63%; **(c)** after 58 days

69. (c)　**71.** (a)　**73.** (l)　**75.** (g)　**77.** (i)
79. (k)　**81.** (m)　**83.** $(1.481, 4.090)$
85. $(-0.402, -1.662), (1.051, 2.722)$　**87.** 4.448
89. $(0, \infty)$　**91.** $2.294, 3.228$　**93.** Discussion and Writing　**95.** Discussion and Writing　**97.** [3.1] $31 - 22i$
98. [3.1] $\frac{1}{2} - \frac{1}{2}i$　**99.** [3.2] $\left(-\frac{1}{2}, 0\right), (7, 0); -\frac{1}{2}, 7$
100. [4.3] $(1, 0); 1$　**101.** [4.1] $(-1, 0), (0, 0), (1, 0); -1, 0, 1$
102. [4.1] $(-4, 0), (0, 0), (3, 0); -4, 0, 3$　**103.** [4.1] $-8, 0, 2$
104. [3.2] $\dfrac{5 \pm \sqrt{97}}{6}$　**105.** $\pi^7; 70^{80}$

107. (a) $y = e^{-x^2}$

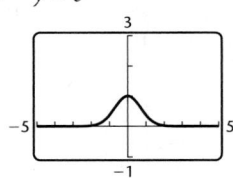

(b) none; **(c)** relative maximum: 1 at $x = 0$

Visualizing the Graph

1. J　**2.** F　**3.** H　**4.** B　**5.** E　**6.** A　**7.** C
8. I　**9.** D　**10.** G

Exercise Set 5.3

1.

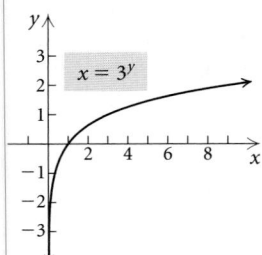

3.

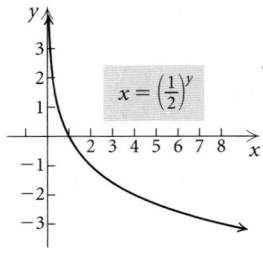

5.

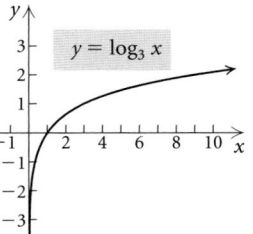

7.

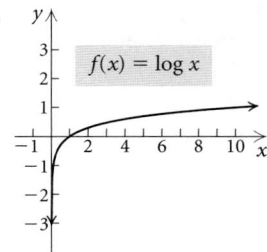

9. 4　**11.** 3　**13.** -3　**15.** -2　**17.** 0　**19.** 1
21. 4　**23.** $\frac{1}{4}$　**25.** -7　**27.** $\frac{1}{2}$　**29.** $\frac{3}{4}$　**31.** 0　**33.** $\frac{1}{2}$
35. $\log_{10} 1000 = 3$, or $\log 1000 = 3$　**37.** $\log_8 2 = \frac{1}{3}$
39. $\log_e t = 3$, or $\ln t = 3$
41. $\log_e 7.3891 = 2$, or $\ln 7.3891 = 2$　**43.** $\log_p 3 = k$
45. $5^1 = 5$　**47.** $10^{-2} = 0.01$
49. $e^{3.4012} = 30$　**51.** $a^{-x} = M$
53. $a^x = T^3$　**55.** 0.4771　**57.** 2.7259　**59.** -0.2441
61. Does not exist　**63.** 0.6931　**65.** 6.6962
67. Does not exist　**69.** 3.3219　**71.** -0.2614
73. 0.7384　**75.** 2.2619　**77.** 0.5880
79. $y_1 = 3^x$, $y_2 = \dfrac{\log x}{\log 3}$

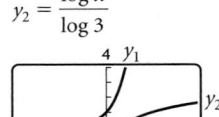

81. $y_1 = \log x$, $y_2 = 10^x$

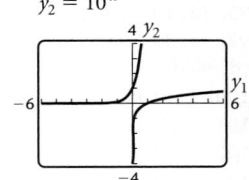

83. Shift the graph of $y = \log_2 x$ left 3 units. Domain: $(-3, \infty)$; vertical asymptote: $x = -3$;

$$y = \frac{\log (x + 3)}{\log 2}$$

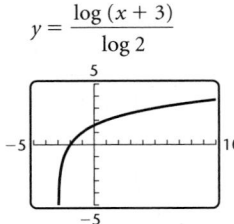

85. Shift the graph of $y = \log_3 x$ down 1 unit. Domain: $(0, \infty)$; vertical asymptote: $x = 0$;

$$y = \frac{\log x}{\log 3} - 1$$

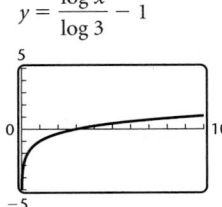

87. Stretch the graph of $y = \ln x$ vertically. Domain: $(0, \infty)$; vertical asymptote: $x = 0$;

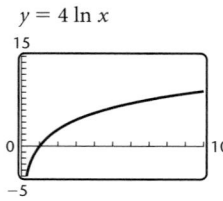

$y = 4 \ln x$

89. Reflect the graph of $y = \ln x$ across the x-axis and shift it up 2 units. Domain: $(0, \infty)$; vertical asymptote: $x = 0$;

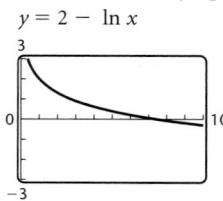

$y = 2 - \ln x$

91. (a) 2.5 ft/sec; **(b)** 2.3 ft/sec; **(c)** 2.1 ft/sec; **(d)** 3.0 ft/sec; **(e)** 2.4 ft/sec; **(f)** 2.2 ft/sec; **(g)** 3.4 ft/sec; **(h)** 1.8 ft/sec
93. (a) 7.85; **(b)** 8.25; **(c)** 9.6; **(d)** 7.85; **(e)** 6.9
95. (a) 10^{-7}; **(b)** 4.0×10^{-6}; **(c)** 6.3×10^{-4}; **(d)** 1.6×10^{-5}
97. (a) 34 decibels; **(b)** 64 decibels; **(c)** 60 decibels; **(d)** 90 decibels   **99.** Discussion and Writing
100. [1.4] $m = \frac{3}{10}$; y-intercept: $\left(0, -\frac{7}{5}\right)$
101. [1.4] $m = 0$; y-intercept: $(0, 6)$
102. [1.4] Slope is not defined; no y-intercept
103. [4.3] -280 **104.** [4.3] -4
105. [4.4] $f(x) = x^3 - 7x$
106. [4.4] $f(x) = x^3 - x^2 + 16x - 16$ **107.** 3
109. $(0, \infty)$ **111.** $(-\infty, 0) \cup (0, \infty)$ **113.** $\left(-\frac{5}{2}, -2\right)$
115. (d)   **117.** (b)
119. (a) $y = x \ln x$

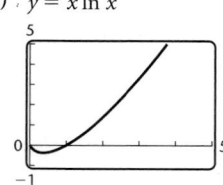

(b) 1; **(c)** relative minimum: -0.368 at $x = 0.368$

121. (a) $y = \dfrac{\ln x}{x^2}$

(b) 1; **(c)** relative maximum: 0.184 at $x = 1.649$

123. $(1.250, 0.891)$

Exercise Set 5.4

1. $\log_3 81 + \log_3 27 = 4 + 3 = 7$
3. $\log_5 5 + \log_5 125 = 1 + 3 = 4$
5. $\log_t 8 + \log_t Y$ **7.** $\ln x + \ln y$ **9.** $3 \log_b t$
11. $8 \log y$ **13.** $-6 \log_c K$ **15.** $\frac{1}{3} \ln 4$
17. $\log_t M - \log_t 8$ **19.** $\log x - \log y$ **21.** $\ln r - \ln s$
23. $\log_a 6 + \log_a x + 5 \log_a y + 4 \log_a z$
25. $2 \log_b p + 5 \log_b q - 4 \log_b m - 9$
27. $\ln 2 - \ln 3 - 3 \ln x - \ln y$
29. $\frac{3}{2} \log r + \frac{1}{2} \log t$ **31.** $3 \log_a x - \frac{5}{2} \log_a p - 4 \log_a q$
33. $2 \log_a m + 3 \log_a n - \frac{3}{4} - \frac{5}{4} \log_a b$ **35.** $\log_a 150$
37. $\log 100 = 2$ **39.** $\log m^3 \sqrt{n}$
41. $\log_a x^{-5/2} y^4$, or $\log_a \dfrac{y^4}{x^{5/2}}$ **43.** $\ln x$ **45.** $\ln (x - 2)$
47. $\log \dfrac{x - 7}{x - 2}$ **49.** $\ln \dfrac{x}{(x^2 - 25)^3}$ **51.** $\ln \dfrac{2^{11/5} x^9}{y^8}$
53. -0.74 **55.** 1.991 **57.** 0.356 **59.** 4.827
61. -1.792 **63.** 0.099 **65.** 3   **67.** $|x - 4|$ **69.** $4x$
71. w **73.** $8t$ **75.** $\frac{1}{2}$ **77.** Discussion and Writing
79. [4.1] Quartic   **80.** [5.2] Exponential
81. [1.4] Linear (constant)   **82.** [5.2] Exponential
83. [4.5] Rational   **84.** [5.3] Logarithmic
85. [4.1] Cubic   **86.** [4.5] Rational   **87.** [1.4] Linear
88. [3.3] Quadratic   **89.** 4   **91.** $\log_a(x^3 - y^3)$
93. $\frac{1}{2} \log_a(x - y) - \frac{1}{2} \log_a(x + y)$ **95.** 7   **97.** True
99. True   **101.** True   **103.** -2 **105.** 3

107. $e^{-xy} = \dfrac{a}{b}$

109. $\log_a \left(\dfrac{x + \sqrt{x^2 - 5}}{5} \cdot \dfrac{x - \sqrt{x^2 - 5}}{x - \sqrt{x^2 - 5}} \right)$

$= \log_a \dfrac{5}{5(x - \sqrt{x^2 - 5})}$

$= -\log_a(x - \sqrt{x^2 - 5})$

Exercise Set 5.5

1. 4   **3.** $\frac{3}{2}$ **5.** 5.044   **7.** $\frac{5}{2}$ **9.** $-3, \frac{1}{2}$ **11.** 0.959
13. 0   **15.** 0   **17.** 6.908   **19.** 84.191   **21.** -1.710
23. 2.844   **25.** $-1.567, 1.567$ **27.** 1.869   **29.** 625
31. 0.0001   **33.** e **35.** $\frac{22}{3}$ **37.** 10   **39.** 4   **41.** $\frac{1}{63}$
43. 2   **45.** 5   **47.** $\frac{21}{8}$ **49.** $-1.518, 0.825$ **51.** $\frac{8}{7}$
53. 0.367   **55.** $-1.911, 4.222$ **57.** 0.621   **59.** -1.532
61. 7.062   **63.** 2.444   **65.** $(4.093, 0.786)$
67. $(7.586, 6.684)$ **69.** Discussion and Writing
70. [3.3] **(a)** $(3, 1)$; **(b)** $x = 3$; **(c)** maximum: 1 when $x = 3$
71. [3.3] **(a)** $(0, -6)$; **(b)** $x = 0$; **(c)** minimum: -6 when $x = 0$ **72.** [3.3] **(a)** $(2, 4)$; **(b)** $x = 2$; **(c)** minimum: 4 when $x = 2$ **73.** [3.3] **(a)** $(-1, -5)$; **(b)** $x = -1$; **(c)** maximum: -5 when $x = -1$ **75.** 0.347   **77.** 10
79. $1, e^4$ or $1, 54.598$ **81.** $\frac{1}{3}, 27$ **83.** $1, e^2$ or $1, 7.389$

85. $0, 0.431$ **87.** $-9, 9$ **89.** e^{-2}, e^2 or $0.135, 7.389$
91. $\frac{7}{4}$ **93.** 5   **95.** $a = \frac{2}{3}b$ **97.** 88

Exercise Set 5.6

1. (a) $P(t) = 6.5e^{0.0114t}$; (b) 6.7 billion, 7.2 billion;
(c) about 18.2 yr after 2006; (d) 60.8 yr
3. (a) 28.9 yr; (b) 1.1% per year; (c) 49.5 yr; (d) 346.6 yr;
(e) 1.8% per year; (f) 1.9% per year; (g) 77.0 yr;
(h) 1.0% per year; (i) 57.8 yr; (j) 26.7 yr
5. In about 637 yr   **7.** (a) $P(t) = 10,000e^{0.054t}$;
(b) \$10,554.85; \$11,140.48; \$13,099.64; \$17,160.07; (c) about
12.8 yr   **9.** About 5135 yr   **11.** (a) 23.1% per minute;
(b) 3.15% per year; (c) 7.2 days; (d) 11 yr; (e) 2.8% per year;
(f) 0.015% per year; (g) 0.003% per year
13. (a) $k \approx 0.0325$; $N(t) = 69,895e^{-0.0325t}$; (b) 12,897 cases;
12,085 cases; (c) about 2037   **15.** (a) $k \approx 0.2119$;
$R(t) = 900e^{0.2119t}$; (b) \$35.9 million; (c) about 3.3 yr;
(d) 48 yr after 1960, or in 2008
17. (a)

$$y = \frac{3500}{1 + 19.9e^{-0.6x}}$$

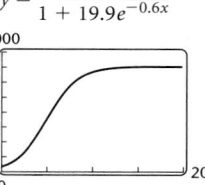

(b) 167; (c) 500; 1758;
3007; 3449; 3495; (d) as
$t \to \infty$, $N(t) \to 3500$; the
number approaches 3500
but never actually reaches it.

19. 46.7°F   **21.** 59.6°F   **23.** (d)   **25.** (a)   **27.** (e)
29. (a) $y = 0.136563665(1.024108508)^x$, where x is the
number of years after 1900; since $r^2 = 0.9709036309$, the
function is a good fit.;
(b)

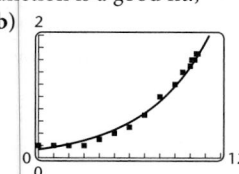

(c) 1.7%; 2.1%; 2.4%

31. (a)

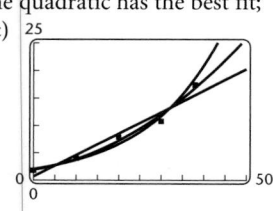

(b) linear: $y = 0.3920710572x + 0.695407279$, $r^2 \approx 0.9465$;
quadratic: $y = 0.0072218083x^2 + 0.1172668587x + 1.973805022$, $r^2 \approx 0.9851$;
exponential: $y = 2.062091236(1.059437743)^x$, $r^2 = 0.9775$;
the quadratic has the best fit;
(c)

(d) linear: 16.8%; quadratic: 18.9%; exponential: 22.0%; the
percentage from the quadratic model seems most realistic.
Answers may vary.
33. (a) $y = 255.0890581(1.277632801)^x$, where x is the
number of years since 1995; (b) 2956 surgeries;
7877 surgeries; (c) in 2011
35. Discussion and Writing   **37.** [1.6] Multiplication
principle for inequalities   **38.** [5.4] Product rule
39. [3.2] Principle of zero products   **40.** [3.2] Principle of
square roots   **41.** [5.4] Power rule
42. [1.5] Multiplication principle for equations
43. \$166.16   **45.** \$14,182.70
47. $t = -\dfrac{L}{R}\left[\ln\left(1 - \dfrac{iR}{V}\right)\right]$ **49.** Linear

Review Exercises: Chapter 5

1. True   **2.** False   **3.** False   **4.** True   **5.** False
6. True   **7.** $\{(-2.7, 1.3), (-3, 8), (3, -5), (-3, 6), (-5, 7)\}$
8. (a) $x = -2y + 3$; (b) $x = 3y^2 + 2y - 1$;
(c) $0.8y^3 - 5.4x^2 = 3y$ **9.** No   **10.** No   **11.** Yes
12. Yes   **13.** (a) Yes; (b) $f^{-1}(x) = \dfrac{-x + 2}{3}$
14. (a) Yes; (b) $f^{-1}(x) = \dfrac{x + 2}{x - 1}$
15. (a) Yes; (b) $f^{-1}(x) = x^2 + 6, x \geq 0$
16. (a) Yes; (b) $f^{-1}(x) = \sqrt[3]{x + 8}$ **17.** (a) No
18. (a) Yes; (b) $f^{-1}(x) = \ln x$
19. $f^{-1}(f(x)) = f^{-1}(6x - 5) = \dfrac{6x - 5 + 5}{6} = \dfrac{6x}{6} = x$;

$f(f^{-1}(x)) = f\left(\dfrac{x + 5}{6}\right) = 6\left(\dfrac{x + 5}{6}\right) - 5 = x + 5 - 5 = x$

20. $f^{-1}(f(x)) = f^{-1}\left(\dfrac{x + 1}{x}\right) = \dfrac{1}{\dfrac{x + 1}{x} - 1} =$

$\dfrac{1}{\dfrac{x + 1 - x}{x}} = \dfrac{1}{\dfrac{1}{x}} = x; f(f^{-1}(x)) = f\left(\dfrac{1}{x - 1}\right) =$

$\dfrac{\dfrac{1}{x - 1} + 1}{\dfrac{1}{x - 1}} = \dfrac{\dfrac{1 + x - 1}{x - 1}}{\dfrac{1}{x - 1}} = \dfrac{x}{x - 1} \cdot \dfrac{x - 1}{1} = x$

21. $f^{-1}(x) = \dfrac{2 - x}{5}$; domain of f and f^{-1}: $(-\infty, \infty)$;
range of f and f^{-1}: $(-\infty, \infty)$;

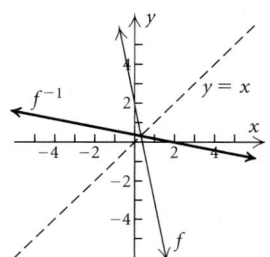

22. $f^{-1}(x) = \dfrac{-2x - 3}{x - 1}$;

domain of f: $(-\infty, -2) \cup (-2, \infty)$;

range of f: $(-\infty, 1) \cup (1, \infty)$;

domain of f^{-1}: $(-\infty, 1) \cup (1, \infty)$;

range of f^{-1}: $(-\infty, -2) \cup (-2, \infty)$

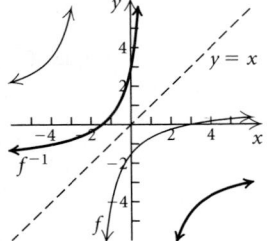

23. 657   **24.** a

25.

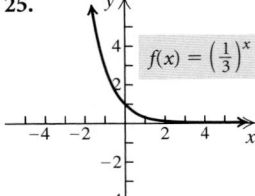

$f(x) = \left(\frac{1}{3}\right)^x$

26.

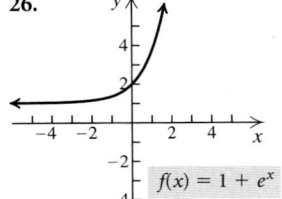

$f(x) = 1 + e^x$

27.

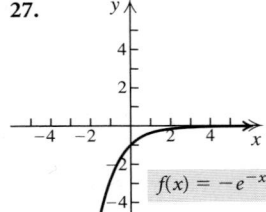

$f(x) = -e^{-x}$

28.

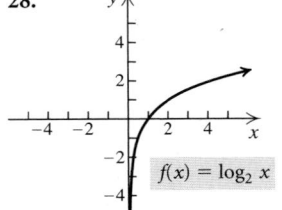

$f(x) = \log_2 x$

29.

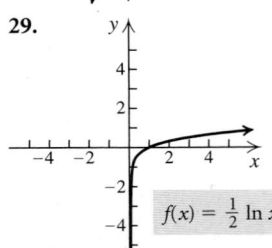

$f(x) = \frac{1}{2}\ln x$

30.

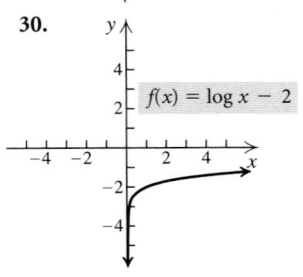

$f(x) = \log x - 2$

31. (c)   **32.** (a)   **33.** (b)   **34.** (f)   **35.** (e)
36. (d)   **37.** 3   **38.** 5   **39.** 1   **40.** 0   **41.** $\frac{1}{4}$
42. $\frac{1}{2}$ **43.** 0   **44.** 1   **45.** $\frac{1}{3}$ **46.** -2 **47.** $4^2 = x$
48. $a^k = Q$ **49.** $\log_4 \frac{1}{64} = -3$ **50.** $\ln 80 = x$, or
$\log_e 80 = x$ **51.** 1.0414   **52.** -0.6308 **53.** 1.0986
54. -3.6119 **55.** Does not exist   **56.** Does not exist
57. 1.9746   **58.** 0.5283   **59.** $\log_b \dfrac{x^3\sqrt{z}}{y^4}$
60. $\ln(x^2 - 4)$ **61.** $\frac{1}{4}\ln w + \frac{1}{2}\ln r$ **62.** $\frac{2}{3}\log M - \frac{1}{3}\log N$
63. 0.477   **64.** 1.699   **65.** -0.699 **66.** 0.233

67. $-5k$ **68.** $-6t$ **69.** 16   **70.** $\frac{1}{5}$ **71.** 4.382   **72.** 2
73. $\frac{1}{2}$ **74.** 5   **75.** 4   **76.** 9   **77.** 1   **78.** 3.912
79. (a) $A(t) = 16{,}000(1.0105)^{4t}$;   (b) $16,000, $20,588.51,
$26,415.77, $33,941.80   **80.** 3837 cases, 8836 cases; 17,222 cases
81. 8.1 yr   **82.** 2.3%   **83.** About 2623 yr   **84.** 5.6
85. 6.3   **86.** 30 decibels   **87.** (a) 2.2 ft/sec;   (b) 8,553,143
88. (a) $k \approx 0.1492$;   (b) $S(t) = 0.035e^{0.1492t}$, where t is the
number of years after 1940;   (c) about $1.459 billion; about
$128.2 billion; about $2534 billion, or $2.534 trillion;   (d) in 2009
89. (a) $P(t) = 3.039e^{0.013t}$, where t is the number of years
since 2005;   (b) 3.201 million, 3.461 million;   (c) in 92 yr;
(d) 53.3 yr   **90.** (a) $y = 2.518123986(1.301660678)^x$;
(b) $y = 2.518123986(1.301660678)^x$

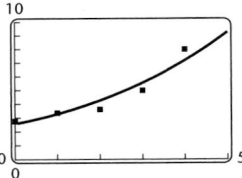

(c) about $27 million
91. No   **92.** (a) $y = 5e^{-x}\ln x$

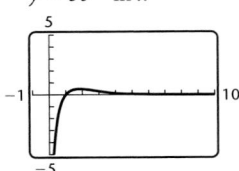

(b) relative maximum: 0.486 at $x = 1.763$; no relative
minimum   **93.** D   **94.** A   **95.** D   **96.** B
97. Discussion and Writing: By the product rule,
$\log_2 x + \log_2 5 = \log_2 5x$, not $\log_2(x + 5)$. Also, substituting
various numbers for x shows that both sides of the inequality
are indeed unequal. You could also graph each side and show
that the graphs do not coincide.
98. Discussion and Writing: The inverse of a function $f(x)$ is
written $f^{-1}(x)$, whereas $[f(x)]^{-1}$ means $\dfrac{1}{f(x)}$.
99. $\frac{1}{64}, 64$ **100.** 1   **101.** 16   **102.** $(1, \infty)$

Test: Chapter 5

1. [5.1] $\{(5, -2), (3, 4), (-1, 0), (-3, -6)\}$
2. [5.1] No   **3.** [5.1] Yes
4. [5.1] (a) Yes;   (b) $f^{-1}(x) = \sqrt[3]{x - 1}$
5. [5.1] (a) Yes;   (b) $f^{-1}(x) = 1 - x$
6. [5.1] (a) Yes;   (b) $f^{-1}(x) = \dfrac{2x}{1 + x}$ **7.** [5.1] (a) No
8. [5.1] $f^{-1}(f(x)) = f^{-1}(-4x + 3) = \dfrac{3 - (-4x + 3)}{4} =$
$\dfrac{4x}{4} = x; f(f^{-1}(x)) = f\left(\dfrac{3 - x}{4}\right) = -4\left(\dfrac{3 - x}{4}\right) + 3 =$
$-3 + x + 3 = x$

9. [5.1] $f^{-1}(x) = \dfrac{4x + 1}{x}$; domain of f: $(-\infty, 4) \cup (4, \infty)$;

range of f: $(-\infty, 0) \cup (0, \infty)$;

domain of f^{-1}: $(-\infty, 0) \cup (0, \infty)$;

range of f^{-1}: $(-\infty, 4) \cup (4, \infty)$;

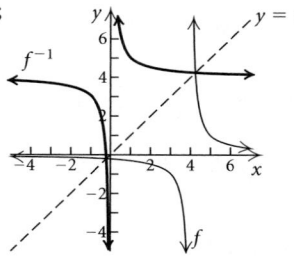

10. [5.2]

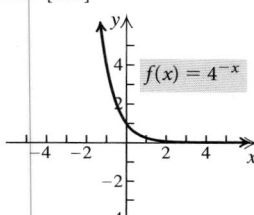

11. [5.3]

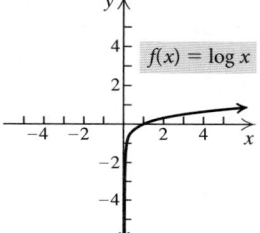

12. [5.2]

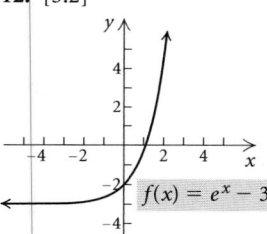

13. [5.3]

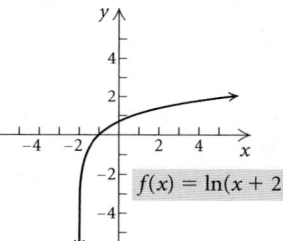

14. [5.3] -5 **15.** [5.3] 1   **16.** [5.3] 0   **17.** [5.3] $\frac{1}{5}$

18. [5.3] $x = e^4$ **19.** [5.3] $x = \log_3 5.4$ **20.** [5.3] 2.7726

21. [5.3] -0.5331 **22.** [5.3] 1.2851   **23.** [5.4] $\log_a \dfrac{x^2 \sqrt{z}}{y}$

24. [5.4] $\frac{2}{5} \ln x + \frac{1}{5} \ln y$ **25.** [5.4] 0.656   **26.** [5.4] $-4t$

27. [5.5] $\frac{1}{2}$ **28.** [5.5] 1   **29.** [5.5] 1   **30.** [5.5] 4.174

31. [5.3] 6.6   **32.** [5.6] 0.0154   **33.** [5.6] (a) 4.5%;

(b) $P(t) = 1000e^{0.045t}$; (c) \$1433.33; (d) 15.4 yr

34. [5.2] C   **35.** [5.5] $\frac{27}{8}$

CHAPTER 6

Exercise Set 6.1

1. $\sin \phi = \frac{15}{17}$, $\cos \phi = \frac{8}{17}$, $\tan \phi = \frac{15}{8}$, $\csc \phi = \frac{17}{15}$, $\sec \phi = \frac{17}{8}$,

$\cot \phi = \frac{8}{15}$

3. $\sin \alpha = \dfrac{\sqrt{3}}{2}$, $\cos \alpha = \dfrac{1}{2}$, $\tan \alpha = \sqrt{3}$, $\csc \alpha = \dfrac{2\sqrt{3}}{3}$,

$\sec \alpha = 2$, $\cot \alpha = \dfrac{\sqrt{3}}{3}$

5. $\sin \phi = \dfrac{7\sqrt{65}}{65}$, $\cos \phi = \dfrac{4\sqrt{65}}{65}$, $\tan \phi = \dfrac{7}{4}$,

$\csc \phi = \dfrac{\sqrt{65}}{7}$, $\sec \phi = \dfrac{\sqrt{65}}{4}$, $\cot \phi = \dfrac{4}{7}$

7. $\csc \alpha = \dfrac{3}{\sqrt{5}}$, or $\dfrac{3\sqrt{5}}{5}$; $\sec \alpha = \dfrac{3}{2}$; $\cot \alpha = \dfrac{2}{\sqrt{5}}$, or $\dfrac{2\sqrt{5}}{5}$

9. $\cos \theta = \frac{7}{25}$, $\tan \theta = \frac{24}{7}$, $\csc \theta = \frac{25}{24}$, $\sec \theta = \frac{25}{7}$, $\cot \theta = \frac{7}{24}$

11. $\sin \phi = \dfrac{2\sqrt{5}}{5}$, $\cos \phi = \dfrac{\sqrt{5}}{5}$, $\csc \phi = \dfrac{\sqrt{5}}{2}$, $\sec \phi = \sqrt{5}$,

$\cot \phi = \dfrac{1}{2}$

13. $\sin \theta = \dfrac{2}{3}$, $\cos \theta = \dfrac{\sqrt{5}}{3}$, $\tan \theta = \dfrac{2\sqrt{5}}{5}$, $\sec \theta = \dfrac{3\sqrt{5}}{5}$,

$\cot \theta = \dfrac{\sqrt{5}}{2}$

15. $\sin \beta = \dfrac{2\sqrt{5}}{5}$, $\tan \beta = 2$, $\csc \beta = \dfrac{\sqrt{5}}{2}$, $\sec \beta = \sqrt{5}$,

$\cot \beta = \dfrac{1}{2}$

17. $\dfrac{\sqrt{2}}{2}$ **19.** 2   **21.** $\dfrac{\sqrt{3}}{3}$ **23.** $\frac{1}{2}$ **25.** 1   **27.** 2

29. 62.4 m   **31.** 9.72°   **33.** 35.01°   **35.** 3.03°

37. 49.65°   **39.** 0.25°   **41.** 5.01°   **43.** 17°36′

45. 83°1′30″   **47.** 11°45′   **49.** 47°49′36″   **51.** 0°54′

53. 39°27′   **55.** 0.6293   **57.** 0.0737   **59.** 1.2765

61. 0.7621   **63.** 0.9336   **65.** 12.4288   **67.** 1.0000

69. 1.7032   **71.** 30.8°   **73.** 12.5°   **75.** 64.4°

77. 46.5°   **79.** 25.2°   **81.** 38.6°   **83.** 45°   **85.** 60°

87. 45°   **89.** 60°   **91.** 30°

93. $\cos 20° = \sin 70° = \dfrac{1}{\sec 20°}$

95. $\tan 52° = \cot 38° = \dfrac{1}{\cot 52°}$

97. $\sin 25° \approx 0.4226$, $\cos 25° \approx 0.9063$, $\tan 25° \approx 0.4663$, $\csc 25° \approx 2.3662$, $\sec 25° \approx 1.1034$, $\cot 25° \approx 2.1445$

99. $\sin 18°49′55″ \approx 0.3228$, $\cos 18°49′55″ \approx 0.9465$, $\tan 18°49′55″ \approx 0.3411$, $\csc 18°49′55″ \approx 3.0979$, $\sec 18°49′55″ \approx 1.0565$, $\cot 18°49′55″ \approx 2.9317$

101. $\sin 8° = q$, $\cos 8° = p$, $\tan 8° = \dfrac{1}{r}$, $\csc 8° = \dfrac{1}{q}$,

$\sec 8° = \dfrac{1}{p}$, $\cot 8° = r$ **103.** Discussion and Writing

104. [5.2]

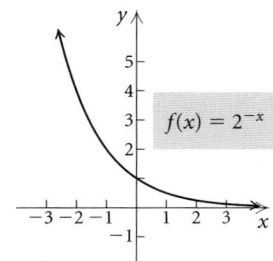

105. [5.2]

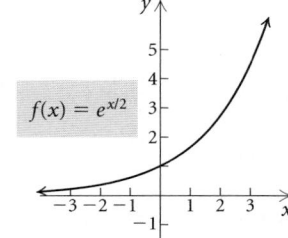

106. [5.3]

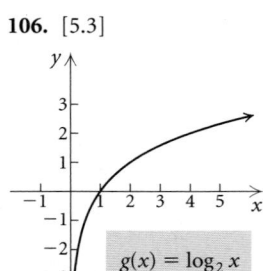

$g(x) = \log_2 x$

107. [5.3]

$h(x) = \ln x$

108. [5.5] 9.21   **109.** [5.5] 4   **110.** [5.5] $\frac{101}{97}$
111. [5.5] 343   **113.** 0.6534
115. Area $= \frac{1}{2}ab$. But $a = c\sin A$, so Area $= \frac{1}{2}bc\sin A$.

Exercise Set 6.2

1. $F = 60°, d = 3, f \approx 5.2$
3. $A = 22.7°, a \approx 52.7, c \approx 136.6$
5. $P = 47°38', n \approx 34.4, p \approx 25.4$
7. $B = 2°17', b \approx 0.39, c \approx 9.74$
9. $A \approx 77.2°, B \approx 12.8°, a \approx 439$
11. $B = 42.42°, a \approx 35.7, b \approx 32.6$
13. $B = 55°, a \approx 28.0, c \approx 48.8$
15. $A \approx 62.4°, B \approx 27.6°, a \approx 3.56$ **17.** Approximately 34°
19. About 62.2 ft   **21.** 61 ft   **23.** 110 ft   **25.** 750 ft
27. About 92.9 cm   **29.** About 599 ft
31. Radius: 9.15 in.; length: 73.20 in.; width: 54.90 in.
33. 17.9 ft   **35.** About 8 km   **37.** About 19.5 mi
39. About 24 km   **41.** Discussion and Writing
42. [1.1] $3\sqrt{10}$, or about 9.487
43. [1.1] $10\sqrt{2}$, or about 14.142
44. [5.3] $\ln t = 4$ **45.** [5.3] $10^{-3} = 0.001$ **47.** 3.3
49. Cut so that $\theta = 79.38°$ **51.** $\theta \approx 27°$

Exercise Set 6.3

1. III   **3.** III   **5.** I   **7.** III   **9.** II   **11.** II
13. $434°, 794°, -286°, -646°$
15. $475.3°, 835.3°, -244.7°, -604.7°$
17. $180°, 540°, -540°, -900°$ **19.** $72.89°, 162.89°$
21. $77°56'46'', 167°56'46''$ **23.** $44.8°, 134.8°$
25. $\sin\beta = \frac{5}{13}, \cos\beta = -\frac{12}{13}, \tan\beta = -\frac{5}{12}, \csc\beta = \frac{13}{5},$
$\sec\beta = -\frac{13}{12}, \cot\beta = -\frac{12}{5}$
27. $\sin\phi = -\frac{2\sqrt{7}}{7}, \cos\phi = -\frac{\sqrt{21}}{7}, \tan\phi = \frac{2\sqrt{3}}{3},$
$\csc\phi = -\frac{\sqrt{7}}{2}, \sec\phi = -\frac{\sqrt{21}}{3}, \cot\phi = \frac{\sqrt{3}}{2}$
29. $\sin\theta = -\frac{2\sqrt{13}}{13}, \cos\theta = \frac{3\sqrt{13}}{13}, \tan\theta = -\frac{2}{3}$
31. $\sin\theta = \frac{5\sqrt{41}}{41}, \cos\theta = \frac{4\sqrt{41}}{41}, \tan\theta = \frac{5}{4}$
33. $\cos\theta = -\frac{2\sqrt{2}}{3}, \tan\theta = \frac{\sqrt{2}}{4}, \csc\theta = -3,$
$\sec\theta = -\frac{3\sqrt{2}}{4}, \cot\theta = 2\sqrt{2}$

35. $\sin\theta = -\frac{\sqrt{5}}{5}, \cos\theta = \frac{2\sqrt{5}}{5}, \tan\theta = -\frac{1}{2},$
$\csc\theta = -\sqrt{5}, \sec\theta = \frac{\sqrt{5}}{2}$
37. $\sin\phi = -\frac{4}{5}, \tan\phi = -\frac{4}{3}, \csc\phi = -\frac{5}{4}, \sec\phi = \frac{5}{3},$
$\cot\phi = -\frac{3}{4}$
39. $30°; -\frac{\sqrt{3}}{2}$ **41.** $45°; 1$ **43.** 0   **45.** $45°; -\frac{\sqrt{2}}{2}$
47. $30°; 2$ **49.** $30°; \sqrt{3}$ **51.** $30°; -\frac{\sqrt{3}}{3}$
53. Not defined   **55.** -1 **57.** $60°; \sqrt{3}$
59. $45°; \frac{\sqrt{2}}{2}$ **61.** $45°; -\sqrt{2}$ **63.** 1   **65.** 0   **67.** 0
69. 0   **71.** Positive: cos, sec; negative: sin, csc, tan, cot
73. Positive: tan, cot; negative: sin, csc, cos, sec
75. Positive: sin, csc; negative: cos, sec, tan, cot
77. Positive: all
79. $\sin 319° = -0.6561, \cos 319° = 0.7547,$
$\tan 319° = -0.8693, \csc 319° \approx -1.5242,$
$\sec 319° \approx 1.3250, \cot 319° \approx -1.1504$
81. $\sin 115° = 0.9063, \cos 115° = -0.4226,$
$\tan 115° = -2.1445, \csc 115° \approx 1.1034,$
$\sec 115° \approx -2.3663, \cot 115° \approx -0.4663$
83. East: about 130 km; south: 75 km
85. About 223 km   **87.** -1.1585 **89.** -1.4910
91. 0.8771   **93.** 0.4352   **95.** 0.9563   **97.** 2.9238
99. $275.4°$ **101.** $200.1°$ **103.** $288.1°$ **105.** $72.6°$
107. Discussion and Writing
109. [4.5]

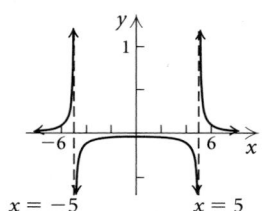

$x = -5$ $x = 5$

$f(x) = \dfrac{1}{x^2 - 25}$

110. [4.2]

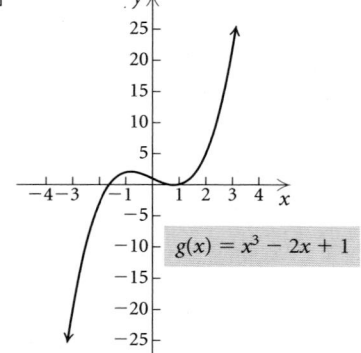

$g(x) = x^3 - 2x + 1$

111. [1.2], [4.5] Domain: $\{x \mid x \neq -2\}$; range: $\{x \mid x \neq 1\}$
112. [1.2], [4.1], [4.5] Domain: $\left\{x \mid x \neq -\frac{3}{2} \text{ and } x \neq 5\right\}$; range: all real numbers

113. [1.5] 12   **114.** [3.2] −2, 3   **115.** [1.5] (12, 0)
116. [3.2] (−2, 0), (3, 0)   **117.** 19.625 in.

Exercise Set 6.4

1.

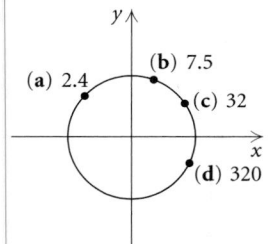

(c) $\dfrac{3\pi}{4}$; (e) $\dfrac{11\pi}{4}$ (a) $\dfrac{\pi}{4}$ (f) $\dfrac{17\pi}{4}$ (d) π (b) $\dfrac{3\pi}{2}$

3.

(b) $\dfrac{2\pi}{3}$ (e) $\dfrac{14\pi}{6}$ (a) $\dfrac{\pi}{6}$ (c) $\dfrac{7\pi}{6}$ (f) $\dfrac{23\pi}{4}$ (d) $\dfrac{10\pi}{6}$

5. $M: \dfrac{2\pi}{3}, -\dfrac{4\pi}{3}; N: \dfrac{3\pi}{2}, -\dfrac{\pi}{2}; P: \dfrac{5\pi}{4}, -\dfrac{3\pi}{4}; Q: \dfrac{11\pi}{6}, -\dfrac{\pi}{6}$

7. (a) 2.4   (b) 7.5   (c) 32   (d) 320

9. $\dfrac{9\pi}{4}, -\dfrac{7\pi}{4}$ **11.** $\dfrac{19\pi}{6}, -\dfrac{5\pi}{6}$ **13.** $\dfrac{4\pi}{3}, -\dfrac{8\pi}{3}$

15. Complement: $\dfrac{\pi}{6}$; supplement: $\dfrac{2\pi}{3}$

17. Complement: $\dfrac{\pi}{8}$; supplement: $\dfrac{5\pi}{8}$

19. Complement: $\dfrac{5\pi}{12}$; supplement: $\dfrac{11\pi}{12}$

21. $\dfrac{5\pi}{12}$ **23.** $\dfrac{10\pi}{9}$ **25.** $-\dfrac{214.6\pi}{180}$, or $-\dfrac{1073\pi}{900}$

27. $-\pi$ **29.** $\dfrac{12.5\pi}{180}$, or $\dfrac{5\pi}{72}$

31. $-\dfrac{17\pi}{9}$ **33.** 4.19   **35.** −1.05   **37.** 2.06

39. 0.02   **41.** 6.02   **43.** 1.66   **45.** −135°
47. 1440°   **49.** 57.30°   **51.** 134.47°   **53.** 225°
55. −5156.62°   **57.** 51.43°

59. $0° = 0$ radians, $30° = \dfrac{\pi}{6}$, $45° = \dfrac{\pi}{4}$, $60° = \dfrac{\pi}{3}$,

$90° = \dfrac{\pi}{2}$, $135° = \dfrac{3\pi}{4}$, $180° = \pi$, $225° = \dfrac{5\pi}{4}$, $270° = \dfrac{3\pi}{2}$,

$315° = \dfrac{7\pi}{4}$, $360° = 2\pi$

61. 2.29   **63.** 5.50 in.   **65.** 1.1; 63°   **67.** 3.2 yd

69. $\dfrac{5\pi}{3}$, or about 5.24   **71.** 3150 $\dfrac{\text{cm}}{\text{min}}$

73. About 12,003 revolutions per hour   **75.** 1047 mph
77. 19,205 revolutions/hr   **79.** About 202
81. Discussion and Writing
83. [5.1] One-to-one   **84.** [6.1] Cosine of θ
85. [5.2] Exponential function
86. [4.5] Horizontal asymptote   **87.** [2.4] Odd function
88. [5.3] Natural   **89.** [5.1] Horizontal line; inverse
90. [5.3] Logarithm   **91.** 111.7 km; 69.8 mi
93. (a) 5°37′30″; (b) 19°41′15″   **95.** 1.676 radians/sec
97. 1.46 nautical miles

Exercise Set 6.5

1. (a) $\left(-\dfrac{3}{4}, -\dfrac{\sqrt{7}}{4}\right)$; (b) $\left(\dfrac{3}{4}, \dfrac{\sqrt{7}}{4}\right)$; (c) $\left(\dfrac{3}{4}, -\dfrac{\sqrt{7}}{4}\right)$

3. (a) $\left(\dfrac{2}{5}, \dfrac{\sqrt{21}}{5}\right)$; (b) $\left(-\dfrac{2}{5}, -\dfrac{\sqrt{21}}{5}\right)$; (c) $\left(-\dfrac{2}{5}, \dfrac{\sqrt{21}}{5}\right)$

5. $\left(\dfrac{\sqrt{2}}{2}, -\dfrac{\sqrt{2}}{2}\right)$ **7.** 0   **9.** $\sqrt{3}$ **11.** 0   **13.** $-\dfrac{\sqrt{3}}{2}$

15. Not defined   **17.** $\dfrac{\sqrt{3}}{2}$ **19.** $-\dfrac{\sqrt{2}}{2}$ **21.** 0   **23.** 0

25. 0.4816   **27.** 1.3065   **29.** −2.1599   **31.** 1
33. −1.1747   **35.** −1   **37.** −0.7071   **39.** 0
41. 0.8391
43. (a)

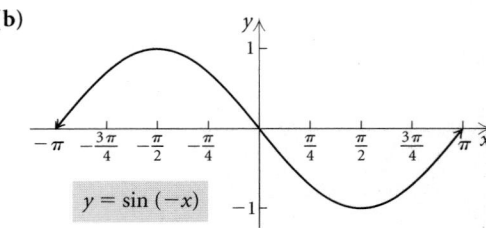

$y = \sin x$

(b)

$y = \sin(-x)$

(c) same as (b); (d) the same
45. (a) See Exercise 43(a);
(b)

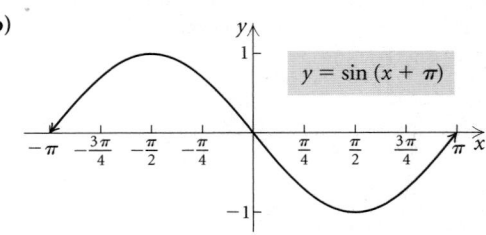

$y = \sin(x + \pi)$

(c) same as (b); (d) the same

47. (a)

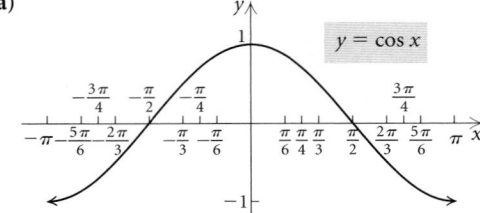

(b)

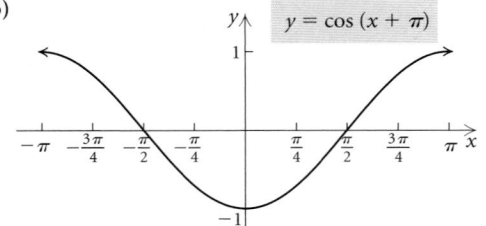

(c) same as (b); **(d)** the same
49. (a)

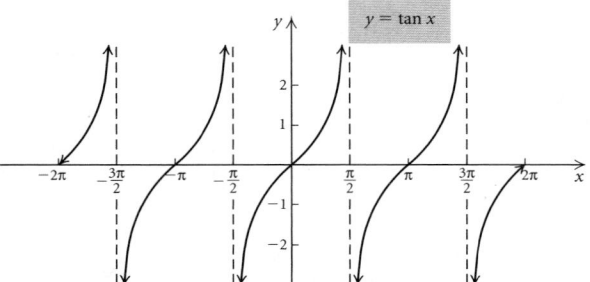

(b)

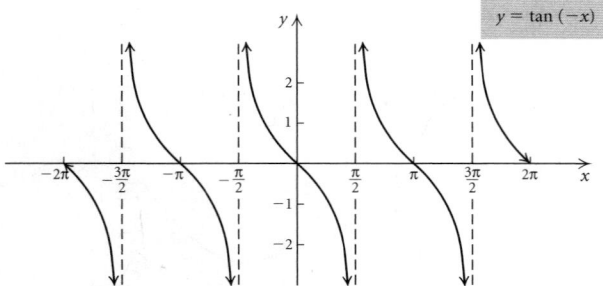

(c) same as (b); **(d)** the same   **51.** Even: cosine, secant; odd:
sine, tangent, cosecant, cotangent   **53.** Positive: I, III;
negative: II, IV   **55.** Positive: I, IV; negative: II, III
57. Discussion and Writing
59. [2.4]

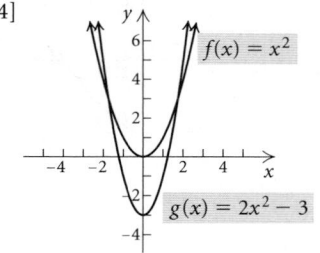

Stretch the graph of f vertically, then shift it down 3 units.

60. [2.4]

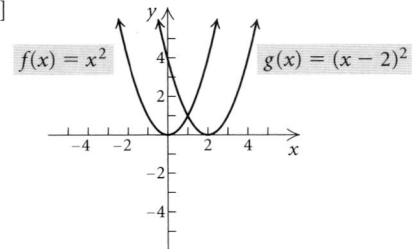

Shift the graph of f right 2 units.
61. [2.4]

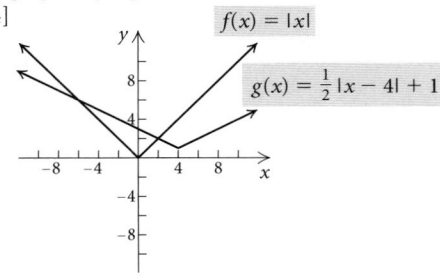

Shift the graph of f to the right 4 units, shrink it vertically,
then shift it up 1 unit.
62. [2.4]

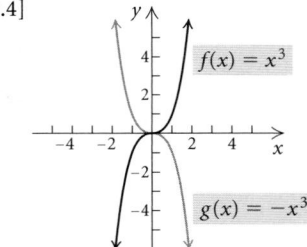

Reflect the graph of f across the x-axis.
63. [2.4] $y = -(x - 2)^3 - 1$ **64.** [2.4] $y = \dfrac{1}{4x} + 3$
65. $\cos x$ **67.** $\sin x$ **69.** $\sin x$ **71.** $-\cos x$
73. $-\sin x$ **75. (a)** $\dfrac{\pi}{2} + 2k\pi, k \in \mathbb{Z}$; **(b)** $\pi + 2k\pi, k \in \mathbb{Z}$;
(c) $k\pi, k \in \mathbb{Z}$ **77.** Domain: $(-\infty, \infty)$; range: $[0, 1]$;
period: π; amplitude: $\dfrac{1}{2}$
79. $\left[-\dfrac{\pi}{2} + 2k\pi, \dfrac{\pi}{2} + 2k\pi\right], k \in \mathbb{Z}$
81. $\left\{x \mid x \neq \dfrac{\pi}{2} + k\pi, k \in \mathbb{Z}\right\}$
83.

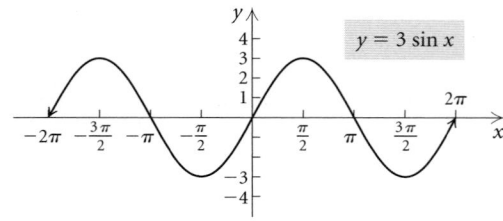

85.

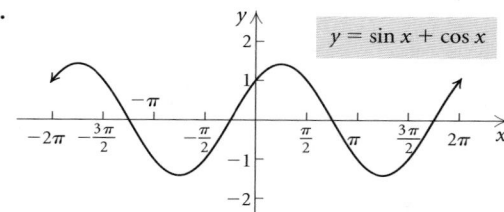

$y = \sin x + \cos x$

87. (a) $\triangle OPA \sim \triangle ODB$;

Thus, $\dfrac{AP}{OA} = \dfrac{BD}{OB}$

$\dfrac{\sin \theta}{\cos \theta} = \dfrac{BD}{1}$

$\tan \theta = BD$

(b) $\triangle OPA \sim \triangle ODB$;

$\dfrac{OD}{OP} = \dfrac{OB}{OA}$

$\dfrac{OD}{1} = \dfrac{1}{\cos \theta}$

$OD = \sec \theta$

(c) $\triangle OAP \sim \triangle ECO$;

$\dfrac{OE}{PO} = \dfrac{CO}{AP}$

$\dfrac{OE}{1} = \dfrac{1}{\sin \theta}$

$OE = \csc \theta$

(d) $\triangle OAP \sim \triangle ECO$

$\dfrac{CE}{AO} = \dfrac{CO}{AP}$

$\dfrac{CE}{\cos \theta} = \dfrac{1}{\sin \theta}$

$CE = \dfrac{\cos \theta}{\sin \theta}$

$CE = \cot \theta$

89. 1

Visualizing the Graph

1. J   **2.** H   **3.** E   **4.** F   **5.** B   **6.** D   **7.** G
8. A   **9.** C   **10.** I

Exercise Set 6.6

1. Amplitude: 1; period: 2π; phase shift: 0

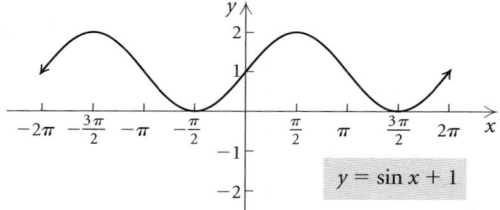

$y = \sin x + 1$

3. Amplitude: 3; period: 2π; phase shift: 0

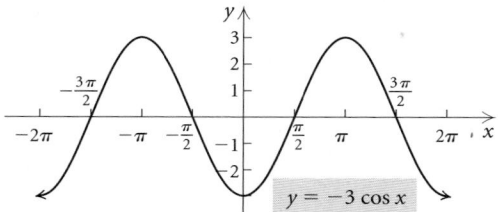

$y = -3 \cos x$

5. Amplitude: $\frac{1}{2}$; period: 2π; phase shift: 0

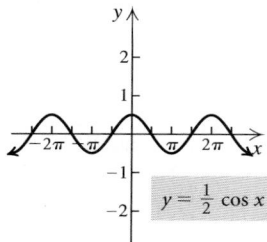

$y = \frac{1}{2} \cos x$

7. Amplitude: 1; period: π; phase shift: 0

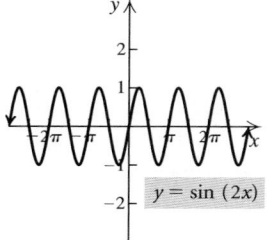

$y = \sin (2x)$

9. Amplitude: 2; period: 4π; phase shift: 0

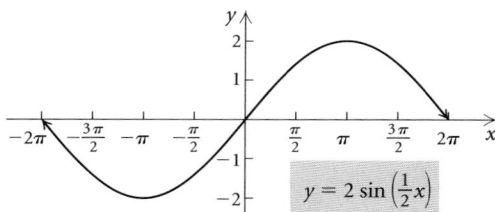

$y = 2 \sin \left(\frac{1}{2}x\right)$

11. Amplitude: $\dfrac{1}{2}$; period: 2π; phase shift: $-\dfrac{\pi}{2}$

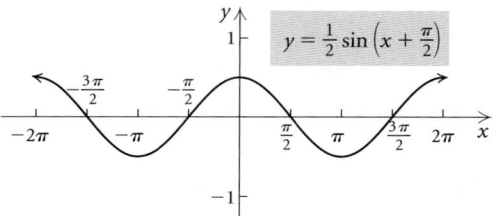

$y = \frac{1}{2} \sin \left(x + \frac{\pi}{2}\right)$

13. Amplitude: 3; period: 2π; phase shift: π

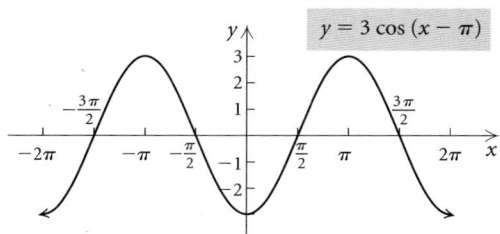

$y = 3 \cos (x - \pi)$

15. Amplitude: $\frac{1}{3}$; period: 2π; phase shift: 0

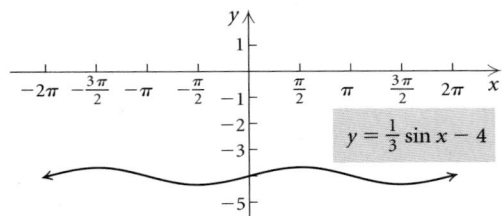

$$y = \frac{1}{3}\sin x - 4$$

17. Amplitude: 1; period: 2π; phase shift: 0

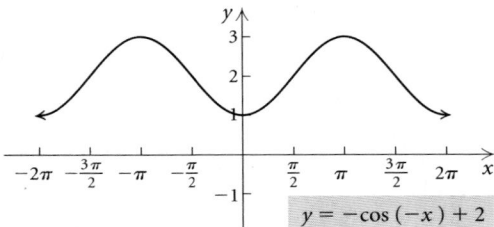

$$y = -\cos(-x) + 2$$

19. Amplitude: 2; period: 4π; phase shift: π

21. Amplitude: $\frac{1}{2}$; period: π; phase shift: $-\dfrac{\pi}{4}$

23. Amplitude: 3; period: 2; phase shift: $\dfrac{3}{\pi}$

25. Amplitude: $\frac{1}{2}$; period: 1; phase shift: 0

27. Amplitude: 1; period: 4π; phase shift: π

29. Amplitude: 1; period: 1; phase shift: 0

31. Amplitude: $\dfrac{1}{4}$; period: 2; phase shift: $\dfrac{4}{\pi}$

33. (b)   **35.** (h)   **37.** (a)   **39.** (f)

41. $y = \frac{1}{2}\cos x + 1$ **43.** $y = \cos\left(x + \dfrac{\pi}{2}\right) - 2$

45.

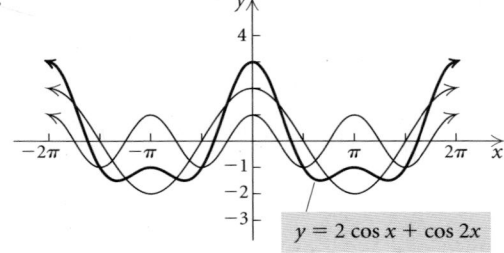

$$y = 2\cos x + \cos 2x$$

47.

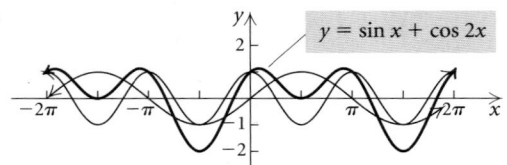

$$y = \sin x + \cos 2x$$

49.

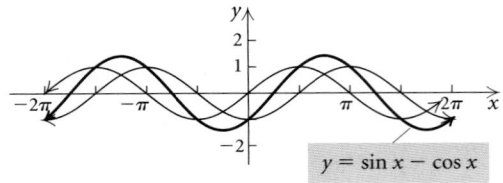

$$y = \sin x - \cos x$$

51.

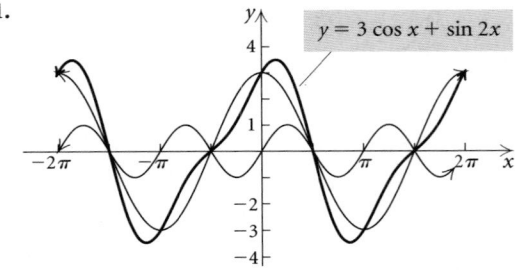

$$y = 3\cos x + \sin 2x$$

53.

$y = x + \sin x$

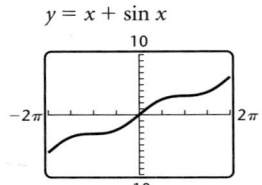

55.

$y = \cos x - x$

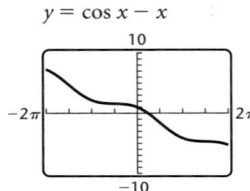

57.

$y = \cos 2x + 2x$

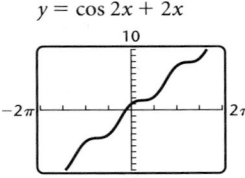

59.

$y = 4\cos 2x - 2\sin x$

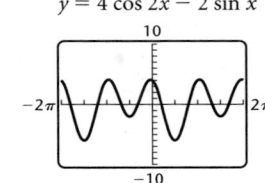

61.

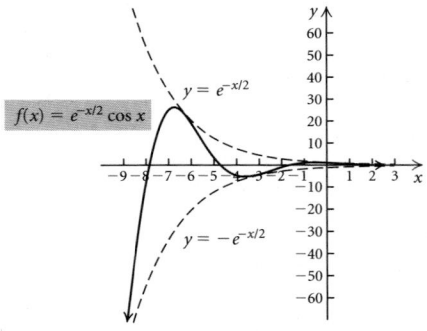

$y = e^{-x/2}$

$f(x) = e^{-x/2}\cos x$

$y = -e^{-x/2}$

63.

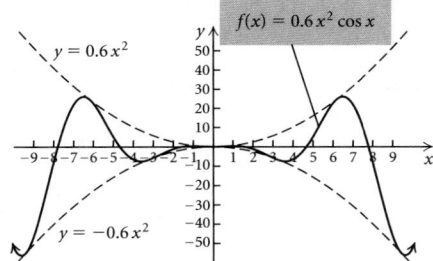

65.

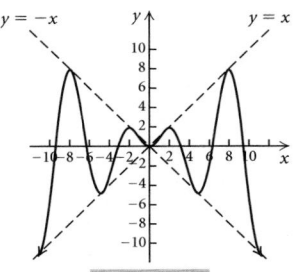

67.

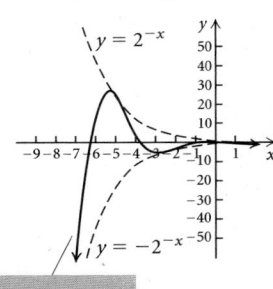

69. Discussion and Writing
71. [4.5] Rational   **72.** [5.3] Logarithmic
73. [4.1] Quartic   **74.** [1.3] Linear
75. [6.6] Trigonometric   **76.** [5.2] Exponential
77. [1.3] Linear   **78.** [6.6] Trigonometric
79. [4.1] Cubic   **80.** [5.2] Exponential
81. Maximum: 8; minimum: 4
83.

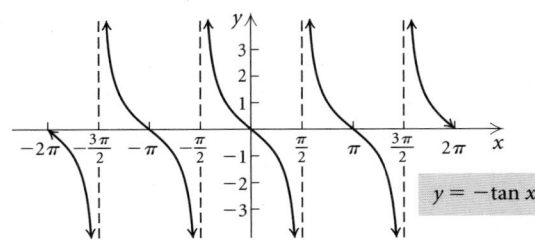

85.

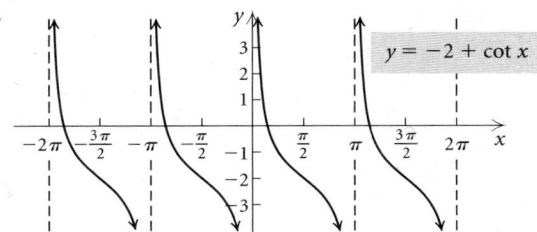

87.

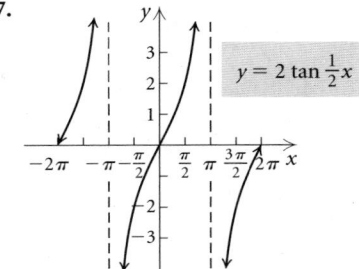

89.

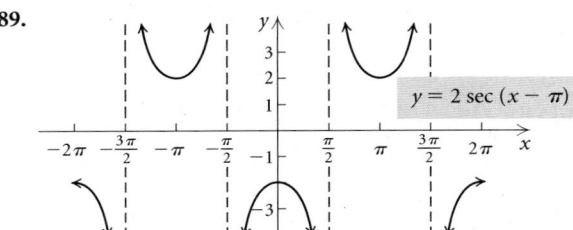

91.

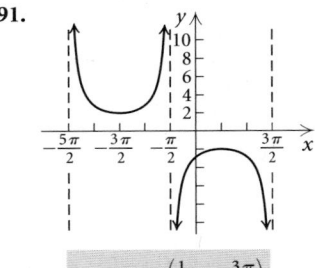

93. $-9.42, -6.28, -3.14, 3.14, 6.28, 9.42$
95. $-3.14, 0, 3.14$
97. (a) $y = 101.6 + 3 \sin\left(\dfrac{\pi}{8}x\right)$;   (b) $104.6°, 98.6°$

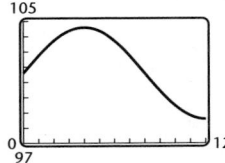

99. Amplitude: 3000; period: 90; phase sh   10
101. 4 in.

Review Exercises: Chapter 6

1. False   2. True   3. True   4. True   5. False

6. False   7. $\sin \theta = \dfrac{3\sqrt{73}}{73}$, $\cos \theta = \dfrac{8\sqrt{73}}{73}$, $\tan \theta = \dfrac{3}{8}$,

$\csc \theta = \dfrac{\sqrt{73}}{3}$, $\sec \theta = \dfrac{\sqrt{73}}{8}$, $\cot \theta = \dfrac{8}{3}$

8. $\cos \beta = \dfrac{3}{10}$, $\tan \beta = \dfrac{\sqrt{91}}{3}$, $\csc \beta = \dfrac{10\sqrt{91}}{91}$,

$\sec \beta = \dfrac{10}{3}$, $\cot \beta = \dfrac{3\sqrt{91}}{91}$ 9. $\dfrac{\sqrt{2}}{2}$ 10. $\dfrac{\sqrt{3}}{3}$

11. $-\dfrac{\sqrt{2}}{2}$ 12. $\dfrac{1}{2}$ 13. Not defined   14. $-\sqrt{3}$

15. $\dfrac{2\sqrt{3}}{3}$ 16. -1 17. $22°16'12''$ 18. $47.56°$

19. 0.4452   20. 1.1315   21. 0.9498   22. -0.9092

23. -1.5282 24. -0.2778 25. $205.3°$

26. $47.2°$ 27. $60°$ 28. $60°$ 29. $45°$ 30. $30°$

31. $\sin 30.9° \approx 0.5135$, $\cos 30.9° \approx 0.8581$,

$\tan 30.9° \approx 0.5985$, $\csc 30.9° \approx 1.9474$, $\sec 30.9° \approx 1.1654$,

$\cot 30.9° \approx 1.6709$ 32. $b \approx 4.5$, $A \approx 58.1°$, $B \approx 31.9°$

33. $A = 38.83°$, $b \approx 37.9$, $c \approx 48.6$ 34. 1748 m

35. 14 ft   36. II   37. I   38. IV   39. $425°, -295°$

40. $\dfrac{\pi}{3}, -\dfrac{5\pi}{3}$ 41. Complement: $76.6°$; supplement: $166.6°$

42. Complement: $\dfrac{\pi}{3}$; supplement: $\dfrac{5\pi}{6}$

43. $\sin \theta = \dfrac{3\sqrt{13}}{13}$, $\cos \theta = \dfrac{-2\sqrt{13}}{13}$, $\tan \theta = -\dfrac{3}{2}$,

$\csc \theta = \dfrac{\sqrt{13}}{3}$, $\sec \theta = -\dfrac{\sqrt{13}}{2}$, $\cot \theta = -\dfrac{2}{3}$

44. $\sin \theta = -\dfrac{2}{3}$, $\cos \theta = -\dfrac{\sqrt{5}}{3}$, $\cot \theta = \dfrac{\sqrt{5}}{2}$,

$\sec \theta = -\dfrac{3\sqrt{5}}{5}$, $\csc \theta = -\dfrac{3}{2}$ 45. About 1743 mi

46.

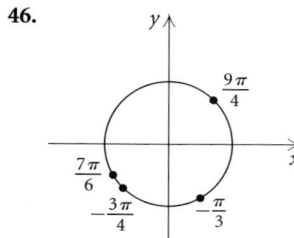

47. $\dfrac{121}{150}\pi$, 2.53   48. $-\dfrac{\pi}{6}$, -0.52 49. $270°$ 50. $171.89°$

51. $-257.83°$ 52. $1980°$ 53. $\dfrac{7\pi}{4}$, or 5.5 cm

54. 2.25, $129°$ 55. About 37.7 ft/min

56. 497,829 radians/hr   57. $\left(\dfrac{3}{5}, \dfrac{4}{5}\right), \left(-\dfrac{3}{5}, -\dfrac{4}{5}\right), \left(-\dfrac{3}{5}, \dfrac{4}{5}\right)$

58. -1 59. 1   60. $-\dfrac{\sqrt{3}}{2}$ 61. $\dfrac{1}{2}$ 62. $\dfrac{\sqrt{3}}{3}$ 63. -1

64. -0.9056 65. 0.9218   66. Not defined   67. 4.3813

68. -6.1685 69. 0.8090

70.

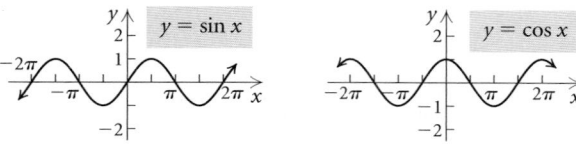

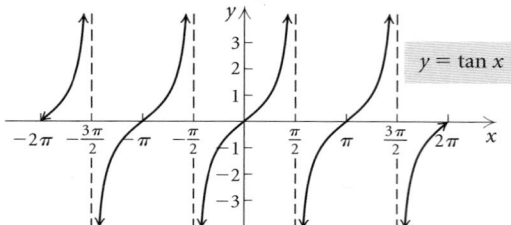

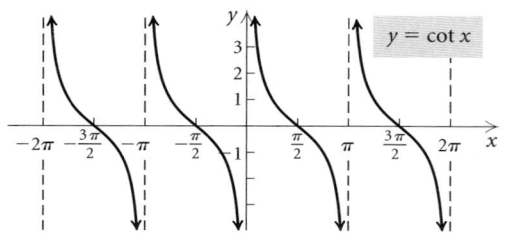

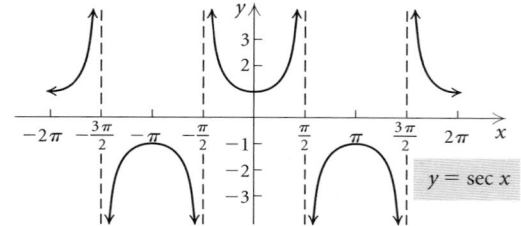

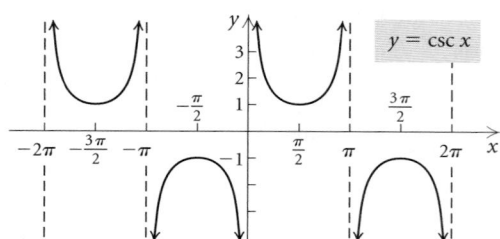

71. Period of sin, cos, sec, csc: 2π; period of tan, cot: π

72.

Function	Domain	Range
Sine	$(-\infty, \infty)$	$[-1, 1]$
Cosine	$(-\infty, \infty)$	$[-1, 1]$
Tangent	$\left\{ x \mid x \neq \dfrac{\pi}{2} + k\pi, k \in \mathbb{Z} \right\}$	$(-\infty, \infty)$

73.

Function	I	II	III	IV
Sine	+	+	−	−
Cosine	+	−	−	+
Tangent	+	−	+	−

74. Amplitude: 1; period: 2π; phase shift: $-\dfrac{\pi}{2}$

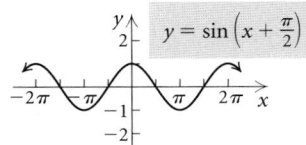

75. Amplitude: $\dfrac{1}{2}$; period: π; phase shift: $\dfrac{\pi}{4}$

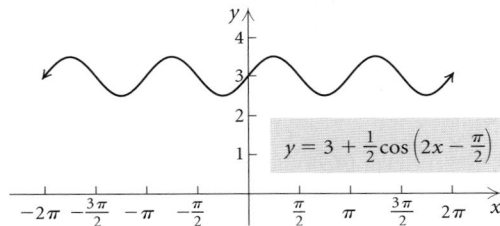

76. (d)   **77.** (a)   **78.** (c)   **79.** (b)
80.   **81.**

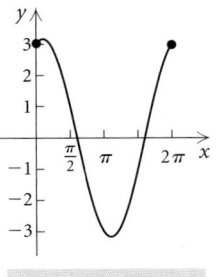

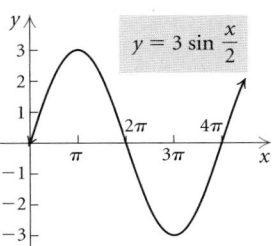

82. C   **83.** B   **84.** B
85. Discussion and Writing: The graph of the cosine function is shaped like a continuous wave, with "high" points at $y = 1$ and "low" points at $y = -1$. The maximum value of the cosine function is 1, and it occurs at all points where $x = 2k\pi$, $k \in \mathbb{Z}$.
86. Discussion and Writing: Both degrees and radians are units of angle measure. A degree is defined to be $\frac{1}{360}$ of one complete positive revolution. Degree notation has been in use since Babylonian times. Radians are defined in terms of intercepted arc length on a circle, with one radian being the measure of the angle for which the arc length equals the radius. There are 2π radians in one complete revolution.

87. Discussion and Writing: No; $\sin x$ is never greater than 1.
88. Discussion and Writing: When x is very large or very small, the amplitude of the function becomes small. The dimensions of the window must be adjusted to be able to see the shape of the graph. Also, when x is 0, the function is undefined, but this may not be obvious from the graph.
89. Domain: $(-\infty, \infty)$; range: $[-3, 3]$; period 4π

90. $y_2 = 2 \sin\left(x + \dfrac{\pi}{2}\right) - 2$

91. The domain consists of the intervals
$$\left(-\dfrac{\pi}{2} + 2k\pi, \dfrac{\pi}{2} + 2k\pi\right), k \in \mathbb{Z}.$$
92. $\cos x = -0.7890$, $\tan x = -0.7787$, $\cot x = -1.2842$, $\sec x = -1.2674$, $\csc x = 1.6276$

Test: Chapter 6

1. [6.1] $\sin \theta = \dfrac{4}{\sqrt{65}}$, or $\dfrac{4\sqrt{65}}{65}$; $\cos \theta = \dfrac{7}{\sqrt{65}}$, or $\dfrac{7\sqrt{65}}{65}$; $\tan \theta = \dfrac{4}{7}$; $\csc \theta = \dfrac{\sqrt{65}}{4}$; $\sec \theta = \dfrac{\sqrt{65}}{7}$; $\cot \theta = \dfrac{7}{4}$

2. [6.3] $\dfrac{\sqrt{3}}{2}$ **3.** [6.3] -1 **4.** [6.4] -1 **5.** [6.4] $-\sqrt{2}$

6. [6.1] $38.47°$ **7.** [6.3] -0.2419 **8.** [6.3] -0.2079
9. [6.4] -5.7588 **10.** [6.4] 0.7827 **11.** [6.1] $30°$
12. [6.1] $\sin 61.6° \approx 0.8796$; $\cos 61.6° \approx 0.4756$; $\tan 61.6° \approx 1.8495$; $\csc 61.6° \approx 1.1369$; $\sec 61.6° \approx 2.1026$; $\cot 61.6° \approx 0.5407$ **13.** [6.2] $B = 54.1°$, $a \approx 32.6$, $c \approx 55.7$

14. [6.3] Answers may vary; $472°$, $-248°$ **15.** [6.4] $\dfrac{\pi}{6}$

16. [6.3] $\cos \theta = \dfrac{5}{\sqrt{41}}$, or $\dfrac{5\sqrt{41}}{41}$; $\tan \theta = -\dfrac{4}{5}$; $\csc \theta = -\dfrac{\sqrt{41}}{4}$; $\sec \theta = \dfrac{\sqrt{41}}{5}$; $\cot \theta = -\dfrac{5}{4}$ **17.** [6.4] $\dfrac{7\pi}{6}$

18. [6.4] $135°$ **19.** [6.4] $\dfrac{16\pi}{3} \approx 16.755$ cm   **20.** [6.6] 1

21. [6.6] 2π **22.** [6.6] $\dfrac{\pi}{2}$ **23.** [6.6] (c)

24. [6.2] About 444 ft   **25.** [6.2] About 272 mi
26. [6.4] $18\pi \approx 56.55$ m/min

27. [6.6]

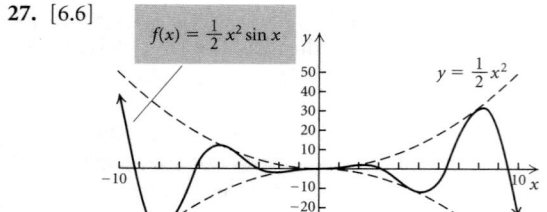

$f(x) = \frac{1}{2}x^2 \sin x$

$y = \frac{1}{2}x^2$

$y = -\frac{1}{2}x^2$

28. [6.6] C

29. [6.5] $\left\{ x \mid -\dfrac{\pi}{2} + 2k\pi < x < \dfrac{\pi}{2} + 2k\pi,\ k \text{ an integer} \right\}$

CHAPTER 7

Exercise Set 7.1

1. $\sin^2 x - \cos^2 x$ **3.** $\sin y + \cos y$ **5.** $1 - 2\sin\phi\cos\phi$

7. $\sin^3 x + \csc^3 x$ **9.** $\cos x (\sin x + \cos x)$

11. $(\sin x + \cos x)(\sin x - \cos x)$

13. $(2\cos x + 3)(\cos x - 1)$

15. $(\sin x + 3)(\sin^2 x - 3\sin x + 9)$ **17.** $\tan x$

19. $\sin x + 1$ **21.** $\dfrac{2\tan t + 1}{3\tan t + 1}$ **23.** 1

25. $\dfrac{5\cot\phi}{\sin\phi + \cos\phi}$ **27.** $\dfrac{1 + 2\sin s + 2\cos s}{\sin^2 s - \cos^2 s}$

29. $\dfrac{5(\sin\theta - 3)}{3}$ **31.** $\sin x \cos x$

33. $\sqrt{\cos\alpha}\,(\sin\alpha - \cos\alpha)$ **35.** $1 - \sin y$

37. $\dfrac{\sqrt{\sin x \cos x}}{\cos x}$ **39.** $\dfrac{\sqrt{2}\cot y}{2}$ **41.** $\dfrac{\cos x}{\sqrt{\sin x \cos x}}$

43. $\dfrac{1 + \sin y}{\cos y}$ **45.** $\cos\theta = \dfrac{\sqrt{a^2 - x^2}}{a}$, $\tan\theta = \dfrac{x}{\sqrt{a^2 - x^2}}$

47. $\sin\theta = \dfrac{\sqrt{x^2 - 9}}{x}$, $\cos\theta = \dfrac{3}{x}$ **49.** $\sin\theta\tan\theta$

51. $\dfrac{\sqrt{6} - \sqrt{2}}{4}$ **53.** $\dfrac{\sqrt{3} + 1}{1 - \sqrt{3}}$, or $-2 - \sqrt{3}$

55. $\dfrac{\sqrt{6} + \sqrt{2}}{4}$ **57.** $\sin 59° \approx 0.8572$

59. $\cos 24° \approx 0.9135$ **61.** $\tan 52° \approx 1.2799$

63. $\tan(\mu + \nu) = \dfrac{\sin(\mu + \nu)}{\cos(\mu + \nu)}$

$= \dfrac{\sin\mu\cos\nu + \cos\mu\sin\nu}{\cos\mu\cos\nu - \sin\mu\sin\nu}$

$= \dfrac{\sin\mu\cos\nu + \cos\mu\sin\nu}{\cos\mu\cos\nu - \sin\mu\sin\nu} \cdot \dfrac{\dfrac{1}{\cos\mu\cos\nu}}{\dfrac{1}{\cos\mu\cos\nu}}$

$= \dfrac{\dfrac{\sin\mu}{\cos\mu} + \dfrac{\sin\nu}{\cos\nu}}{1 - \dfrac{\sin\mu\sin\nu}{\cos\mu\cos\nu}}$

$= \dfrac{\tan\mu + \tan\nu}{1 - \tan\mu\tan\nu}$

65. 0 **67.** $-\dfrac{7}{25}$ **69.** -1.5789 **71.** 0.7071

73. $2\sin\alpha\cos\beta$ **75.** $\cos u$ **77.** Discussion and Writing

79. [1.5] All real numbers   **80.** [1.5] No solution

81. [6.1] 1.9417 **82.** [6.1] 1.6645 **83.** $0°$; the lines are

parallel   **85.** $\dfrac{3\pi}{4}$, or $135°$ **87.** $4.57°$

89. $\dfrac{\cos(x + h) - \cos x}{h}$

$= \dfrac{\cos x \cos h - \sin x \sin h - \cos x}{h}$

$= \dfrac{\cos x \cos h - \cos x}{h} - \dfrac{\sin x \sin h}{h}$

$= \cos x \left(\dfrac{\cos h - 1}{h} \right) - \sin x \left(\dfrac{\sin h}{h} \right)$

91. Let $x = \dfrac{\pi}{5}$. Then $\dfrac{\sin 5x}{x} = \dfrac{\sin\pi}{\pi/5} = 0 \neq \sin 5$.

Answers may vary.

93. Let $\alpha = \dfrac{\pi}{4}$. Then $\cos(2\alpha) = \cos\dfrac{\pi}{2} = 0$, but

$2\cos\alpha = 2\cos\dfrac{\pi}{4} = \sqrt{2}$. Answers may vary.

95. Let $x = \dfrac{\pi}{6}$. Then $\dfrac{\cos 6x}{\cos x} = \dfrac{\cos\pi}{\cos\dfrac{\pi}{6}} = \dfrac{-1}{\sqrt{3}/2} \neq 6$.

Answers may vary.   **97.** $\dfrac{6 - 3\sqrt{3}}{9 + 2\sqrt{3}} \approx 0.0645$

99. $168.7°$ **101.** $\cos 2\theta = \cos^2\theta - \sin^2\theta$, or $1 - 2\sin^2\theta$, or $2\cos^2\theta - 1$

103. $\tan\left(x + \dfrac{\pi}{4}\right) = \dfrac{\tan x + \tan\dfrac{\pi}{4}}{1 - \tan x \tan\dfrac{\pi}{4}} = \dfrac{1 + \tan x}{1 - \tan x}$

105. $\sin(\alpha + \beta) + \sin(\alpha - \beta) = \sin\alpha\cos\beta + \cos\alpha\sin\beta + \sin\alpha\cos\beta - \cos\alpha\sin\beta = 2\sin\alpha\cos\beta$

Exercise Set 7.2

1. (a) $\tan\dfrac{3\pi}{10} \approx 1.3763$, $\csc\dfrac{3\pi}{10} \approx 1.2361$, $\sec\dfrac{3\pi}{10} \approx 1.7013$,

$\cot\dfrac{3\pi}{10} \approx 0.7266$; (b) $\sin\dfrac{\pi}{5} \approx 0.5878$, $\cos\dfrac{\pi}{5} \approx 0.8090$,

$\tan\dfrac{\pi}{5} \approx 0.7266$, $\csc\dfrac{\pi}{5} \approx 1.7013$, $\sec\dfrac{\pi}{5} \approx 1.2361$,

$\cot\dfrac{\pi}{5} \approx 1.3763$

3. (a) $\cos \theta = -\dfrac{2\sqrt{2}}{3}$, $\tan \theta = -\dfrac{\sqrt{2}}{4}$, $\csc \theta = 3$,

$\sec \theta = -\dfrac{3\sqrt{2}}{4}$, $\cot \theta = -2\sqrt{2}$;

(b) $\sin\left(\dfrac{\pi}{2} - \theta\right) = -\dfrac{2\sqrt{2}}{3}$, $\cos\left(\dfrac{\pi}{2} - \theta\right) = \dfrac{1}{3}$,

$\tan\left(\dfrac{\pi}{2} - \theta\right) = -2\sqrt{2}$, $\csc\left(\dfrac{\pi}{2} - \theta\right) = -\dfrac{3\sqrt{2}}{4}$,

$\sec\left(\dfrac{\pi}{2} - \theta\right) = 3$, $\cot\left(\dfrac{\pi}{2} - \theta\right) = -\dfrac{\sqrt{2}}{4}$;

(c) $\sin\left(\theta - \dfrac{\pi}{2}\right) = \dfrac{2\sqrt{2}}{3}$, $\cos\left(\theta - \dfrac{\pi}{2}\right) = \dfrac{1}{3}$,

$\tan\left(\theta - \dfrac{\pi}{2}\right) = 2\sqrt{2}$, $\csc\left(\theta - \dfrac{\pi}{2}\right) = \dfrac{3\sqrt{2}}{4}$,

$\sec\left(\theta - \dfrac{\pi}{2}\right) = 3$, $\cot\left(\theta - \dfrac{\pi}{2}\right) = \dfrac{\sqrt{2}}{4}$

5. $\sec\left(x + \dfrac{\pi}{2}\right) = -\csc x$ **7.** $\tan\left(x - \dfrac{\pi}{2}\right) = -\cot x$

9. $\sin 2\theta = \frac{24}{25}$, $\cos 2\theta = -\frac{7}{25}$, $\tan 2\theta = -\frac{24}{7}$; II

11. $\sin 2\theta = \frac{24}{25}$, $\cos 2\theta = -\frac{7}{25}$, $\tan 2\theta = -\frac{24}{7}$; II

13. $\sin 2\theta = -\frac{120}{169}$, $\cos 2\theta = \frac{119}{169}$, $\tan 2\theta = -\frac{120}{119}$; IV

15. $\cos 4x = 1 - 8\sin^2 x \cos^2 x$, or
$\cos^4 x - 6\sin^2 x \cos^2 x + \sin^4 x$, or $8\cos^4 x - 8\cos^2 x + 1$

17. $\dfrac{\sqrt{2 + \sqrt{3}}}{2}$ **19.** $\dfrac{\sqrt{2 + \sqrt{2}}}{2}$ **21.** $2 + \sqrt{3}$

23. 0.6421 **25.** 0.1735

27. (d); $\dfrac{\cos 2x}{\cos x - \sin x} = \dfrac{\cos^2 x - \sin^2 x}{\cos x - \sin x}$

$= \dfrac{(\cos x + \sin x)(\cos x - \sin x)}{\cos x - \sin x}$

$= \cos x + \sin x$

$= \dfrac{\sin x}{\sin x}(\cos x + \sin x)$

$= \sin x \left(\dfrac{\cos x}{\sin x} + \dfrac{\sin x}{\sin x}\right)$

$= \sin x(\cot x + 1)$

29. (d); $\dfrac{\sin 2x}{2\cos x} = \dfrac{2\sin x \cos x}{2\cos x} = \sin x$ **31.** $\cos x$ **33.** 1

35. $\cos 2x$ **37.** 8 **39.** Discussion and Writing

41. [7.1] $\sin^2 x$ **42.** [7.1] 1 **43.** [7.1] $-\cos^2 x$

44. [7.1] $\csc^2 x$ **45.** [7.1] 1 **46.** [7.1] $\sec^2 x$

47. [7.1] $\cos^2 x$ **48.** [7.1] $\tan^2 x$ **49.** [6.5] (a), (e)

50. [6.5] (b), (c), (f) **51.** [6.5] (d) **52.** [6.5] (e)

53. $\sin 141° \approx 0.6293$, $\cos 141° \approx -0.7772$,
$\tan 141° \approx -0.8097$, $\csc 141° \approx 1.5891$, $\sec 141° \approx -1.2867$,
$\cot 141° \approx -1.2350$ **55.** $-\cos x(1 + \cot x)$ **57.** $\cot^2 y$

59. $\sin \theta = -\frac{15}{17}$, $\cos \theta = -\frac{8}{17}$, $\tan \theta = \frac{15}{8}$

61. (a) 9.80359 m/sec²; **(b)** 9.80180 m/sec²;

(c) $g = 9.78049(1 + 0.005264 \sin^2 \phi + 0.000024 \sin^4 \phi)$

Exercise Set 7.3

1.

$\sec x - \sin x \tan x$	$\cos x$
$\dfrac{1}{\cos x} - \sin x \cdot \dfrac{\sin x}{\cos x}$	
$\dfrac{1 - \sin^2 x}{\cos x}$	
$\dfrac{\cos^2 x}{\cos x}$	
$\cos x$	

3.

$1 - \cos x$	$\dfrac{\sin x}{1 + \cos x}$
$\sin x$	$\dfrac{\sin x}{1 + \cos x} \cdot \dfrac{1 - \cos x}{1 - \cos x}$
	$\dfrac{\sin x(1 - \cos x)}{1 - \cos^2 x}$
	$\dfrac{\sin x(1 - \cos x)}{\sin^2 x}$
	$\dfrac{1 - \cos x}{\sin x}$

5.

$\dfrac{1 + \tan \theta}{1 - \tan \theta} + \dfrac{1 + \cot \theta}{1 - \cot \theta}$	0
$\dfrac{1 + \dfrac{\sin \theta}{\cos \theta}}{1 - \dfrac{\sin \theta}{\cos \theta}} + \dfrac{1 + \dfrac{\cos \theta}{\sin \theta}}{1 - \dfrac{\cos \theta}{\sin \theta}}$	
$\dfrac{\dfrac{\cos \theta + \sin \theta}{\cos \theta}}{\dfrac{\cos \theta - \sin \theta}{\cos \theta}} + \dfrac{\dfrac{\sin \theta + \cos \theta}{\sin \theta}}{\dfrac{\sin \theta - \cos \theta}{\sin \theta}}$	
$\dfrac{\cos \theta + \sin \theta}{\cos \theta} \cdot \dfrac{\cos \theta}{\cos \theta - \sin \theta} +$	
$\dfrac{\sin \theta + \cos \theta}{\sin \theta} \cdot \dfrac{\sin \theta}{\sin \theta - \cos \theta}$	
$\dfrac{\cos \theta + \sin \theta}{\cos \theta - \sin \theta} + \dfrac{\sin \theta + \cos \theta}{\sin \theta - \cos \theta}$	
$\dfrac{\cos \theta + \sin \theta}{\cos \theta - \sin \theta} - \dfrac{\cos \theta + \sin \theta}{\cos \theta - \sin \theta}$	
	0

7.

$\dfrac{\cos^2 \alpha + \cot \alpha}{\cos^2 \alpha - \cot \alpha}$	$\dfrac{\cos^2 \alpha \tan \alpha + 1}{\cos^2 \alpha \tan \alpha - 1}$
$\dfrac{\cos^2 \alpha + \dfrac{\cos \alpha}{\sin \alpha}}{\cos^2 \alpha - \dfrac{\cos \alpha}{\sin \alpha}}$	$\dfrac{\cos^2 \alpha \dfrac{\sin \alpha}{\cos \alpha} + 1}{\cos^2 \alpha \dfrac{\sin \alpha}{\cos \alpha} - 1}$
$\dfrac{\cos \alpha \left(\cos \alpha + \dfrac{1}{\sin \alpha}\right)}{\cos \alpha \left(\cos \alpha - \dfrac{1}{\sin \alpha}\right)}$	$\dfrac{\sin \alpha \cos \alpha + 1}{\sin \alpha \cos \alpha - 1}$
$\dfrac{\cos \alpha + \dfrac{1}{\sin \alpha}}{\cos \alpha - \dfrac{1}{\sin \alpha}}$	
$\dfrac{\dfrac{\sin \alpha \cos \alpha + 1}{\sin \alpha}}{\dfrac{\sin \alpha \cos \alpha - 1}{\sin \alpha}}$	
$\dfrac{\sin \alpha \cos \alpha + 1}{\sin \alpha \cos \alpha - 1}$	

9.

$\dfrac{2 \tan \theta}{1 + \tan^2 \theta}$	$\sin 2\theta$
$\dfrac{2 \tan \theta}{\sec^2 \theta}$	$2 \sin \theta \cos \theta$
$\dfrac{2 \sin \theta}{\cos \theta} \cdot \dfrac{\cos^2 \theta}{1}$	
$2 \sin \theta \cos \theta$	

11.

$1 - \cos 5\theta \cos 3\theta - \sin 5\theta \sin 3\theta$	$2 \sin^2 \theta$
$1 - [\cos 5\theta \cos 3\theta + \sin 5\theta \sin 3\theta]$	$1 - \cos 2\theta$
$1 - \cos (5\theta - 3\theta)$	
$1 - \cos 2\theta$	

13.

$2 \sin \theta \cos^3 \theta + 2 \sin^3 \theta \cos \theta$	$\sin 2\theta$
$2 \sin \theta \cos \theta (\cos^2 \theta + \sin^2 \theta)$	$2 \sin \theta \cos \theta$
$2 \sin \theta \cos \theta$	

15.

$\dfrac{\tan x - \sin x}{2 \tan x}$	$\sin^2 \dfrac{x}{2}$
$\dfrac{1}{2}\left[\dfrac{\dfrac{\sin x}{\cos x} - \sin x}{\dfrac{\sin x}{\cos x}}\right]$	$\dfrac{1 - \cos x}{2}$
$\dfrac{1}{2} \dfrac{\sin x - \sin x \cos x}{\cos x} \cdot \dfrac{\cos x}{\sin x}$	
$\dfrac{1 - \cos x}{2}$	

17.

$\sin (\alpha + \beta) \sin (\alpha - \beta)$	$\sin^2 \alpha - \sin^2 \beta$
$\left(\begin{array}{c}\sin \alpha \cos \beta + \\ \cos \alpha \sin \beta\end{array}\right)\left(\begin{array}{c}\sin \alpha \cos \beta - \\ \cos \alpha \sin \beta\end{array}\right)$	$1 - \cos^2 \alpha -$ $(1 - \cos^2 \beta)$
$\sin^2 \alpha \cos^2 \beta - \cos^2 \alpha \sin^2 \beta$	$\cos^2 \beta - \cos^2 \alpha$
$\cos^2 \beta (1 - \cos^2 \alpha) -$ $\cos^2 \alpha (1 - \cos^2 \beta)$	
$\cos^2 \beta - \cos^2 \alpha \cos^2 \beta -$ $\cos^2 \alpha + \cos^2 \alpha \cos^2 \beta$	
$\cos^2 \beta - \cos^2 \alpha$	

19.

$\tan \theta (\tan \theta + \cot \theta)$	$\sec^2 \theta$
$\tan^2 \theta + \tan \theta \cot \theta$	
$\tan^2 \theta + 1$	
$\sec^2 \theta$	

21.

$\dfrac{1 + \cos^2 x}{\sin^2 x}$	$2 \csc^2 x - 1$
$\dfrac{1}{\sin^2 x} + \dfrac{\cos^2 x}{\sin^2 x}$	
$\csc^2 x + \cot^2 x$	
$\csc^2 x + \csc^2 x - 1$	
$2 \csc^2 x - 1$	

23.

$\dfrac{1 + \sin x}{1 - \sin x} + \dfrac{\sin x - 1}{1 + \sin x}$	$4 \sec x \tan x$
$\dfrac{(1 + \sin x)^2 - (1 - \sin x)^2}{1 - \sin^2 x}$	$4 \cdot \dfrac{1}{\cos x} \cdot \dfrac{\sin x}{\cos x}$
$\dfrac{(1 + 2 \sin x + \sin^2 x) - (1 - 2 \sin x + \sin^2 x)}{\cos^2 x}$	$\dfrac{4 \sin x}{\cos^2 x}$
$\dfrac{4 \sin x}{\cos^2 x}$	

25.

$\cos^2 \alpha \cot^2 \alpha$	$\cot^2 \alpha - \cos^2 \alpha$
$(1 - \sin^2 \alpha) \cot^2 \alpha$	
$\cot^2 \alpha - \sin^2 \alpha \cdot \dfrac{\cos^2 \alpha}{\sin^2 \alpha}$	
$\cot^2 \alpha - \cos^2 \alpha$	

27.

$2 \sin^2 \theta \cos^2 \theta + \cos^4 \theta$	$1 - \sin^4 \theta$
$\cos^2 \theta (2 \sin^2 \theta + \cos^2 \theta)$	$(1 + \sin^2 \theta)(1 - \sin^2 \theta)$
$\cos^2 \theta (\sin^2 \theta + \sin^2 \theta + \cos^2 \theta)$	$(1 + \sin^2 \theta)(\cos^2 \theta)$
$\cos^2 \theta (\sin^2 \theta + 1)$	

29.

$\dfrac{1 + \sin x}{1 - \sin x}$	$(\sec x + \tan x)^2$
$\dfrac{1 + \sin x}{1 - \sin x} \cdot \dfrac{1 + \sin x}{1 + \sin x}$	$\left(\dfrac{1}{\cos x} + \dfrac{\sin x}{\cos x}\right)^2$
$\dfrac{(1 + \sin x)^2}{1 - \sin^2 x}$	$\dfrac{(1 + \sin x)^2}{\cos^2 x}$
$\dfrac{(1 + \sin x)^2}{\cos^2 x}$	

31. Sine sum and difference identities:
$$\sin(x+y) = \sin x \cos y + \cos x \sin y,$$
$$\sin(x-y) = \sin x \cos y - \cos x \sin y.$$
Add the sum and difference identities:
$$\sin(x+y) + \sin(x-y) = 2 \sin x \cos y$$
$$\tfrac{1}{2}[\sin(x+y) + \sin(x-y)] = \sin x \cos y. \quad (3)$$
Subtract the difference identity from the sum identity:
$$\sin(x+y) - \sin(x-y) = 2 \cos x \sin y$$
$$\tfrac{1}{2}[\sin(x+y) - \sin(x-y)] = \cos x \sin y. \quad (4)$$

33. $\sin 3\theta - \sin 5\theta = 2 \cos \dfrac{8\theta}{2} \sin \dfrac{-2\theta}{2} = -2 \cos 4\theta \sin\theta$

35. $\sin 8\theta + \sin 5\theta = 2 \sin \dfrac{13\theta}{2} \cos \dfrac{3\theta}{2}$

37. $\sin 7u \sin 5u = \tfrac{1}{2}(\cos 2u - \cos 12u)$

39. $7 \cos\theta \sin 7\theta = \dfrac{7}{2}[\sin 8\theta - \sin(-6\theta)]$
$$= \dfrac{7}{2}(\sin 8\theta + \sin 6\theta)$$

41. $\cos 55° \sin 25° = \tfrac{1}{2}(\sin 80° - \sin 30°) = \tfrac{1}{2}\sin 80° - \tfrac{1}{4}$

43.

$\sin 4\theta + \sin 6\theta$	$\cot\theta\,(\cos 4\theta - \cos 6\theta)$
$2 \sin \dfrac{10\theta}{2} \cos \dfrac{-2\theta}{2}$	$\dfrac{\cos\theta}{\sin\theta}\left(2 \sin \dfrac{10\theta}{2} \sin \dfrac{2\theta}{2}\right)$
$2 \sin 5\theta \cos(-\theta)$	$\dfrac{\cos\theta}{\sin\theta}(2 \sin 5\theta \sin\theta)$
$2 \sin 5\theta \cos\theta$	$2 \sin 5\theta \cos\theta$

45.

$\cot 4x\,(\sin x + \sin 4x + \sin 7x)$	$\cos x + \cos 4x + \cos 7x$
$\dfrac{\cos 4x}{\sin 4x}\left(\sin 4x + 2 \sin \dfrac{8x}{2} \cos \dfrac{-6x}{2}\right)$	$\cos 4x + 2 \cos \dfrac{8x}{2} \cdot \cos \dfrac{6x}{2}$
$\dfrac{\cos 4x}{\sin 4x}(\sin 4x + 2 \sin 4x \cos 3x)$	$\cos 4x + 2 \cos 4x \cdot \cos 3x$
$\cos 4x\,(1 + 2 \cos 3x)$	$\cos 4x\,(1 + 2 \cos 3x)$

47.

$\cot \dfrac{x+y}{2}$	$\dfrac{\sin y - \sin x}{\cos x - \cos y}$
$\dfrac{\cos \dfrac{x+y}{2}}{\sin \dfrac{x+y}{2}}$	$\dfrac{2 \cos \dfrac{x+y}{2} \sin \dfrac{y-x}{2}}{2 \sin \dfrac{x+y}{2} \sin \dfrac{y-x}{2}}$
	$\dfrac{\cos \dfrac{x+y}{2}}{\sin \dfrac{x+y}{2}}$

49.

$\tan \dfrac{\theta+\phi}{2}(\sin\theta - \sin\phi)$	$\tan \dfrac{\theta-\phi}{2}(\sin\theta + \sin\phi)$
$\dfrac{\sin \dfrac{\theta+\phi}{2}}{\cos \dfrac{\theta+\phi}{2}}\left(2 \cos \dfrac{\theta+\phi}{2} \sin \dfrac{\theta-\phi}{2}\right)$	$\dfrac{\sin \dfrac{\theta-\phi}{2}}{\cos \dfrac{\theta-\phi}{2}}\left(2 \sin \dfrac{\theta+\phi}{2} \cos \dfrac{\theta-\phi}{2}\right)$
$2 \sin \dfrac{\theta+\phi}{2} \cdot \sin \dfrac{\theta-\phi}{2}$	$2 \sin \dfrac{\theta+\phi}{2} \cdot \sin \dfrac{\theta-\phi}{2}$

51. B;

$\dfrac{\cos x + \cot x}{1 + \csc x}$	$\cos x$
$\dfrac{\dfrac{\cos x}{1} + \dfrac{\cos x}{\sin x}}{1 + \dfrac{1}{\sin x}}$	
$\dfrac{\dfrac{\sin x \cos x + \cos x}{\sin x}}{\dfrac{\sin x + 1}{\sin x}} \cdot \dfrac{\sin x}{\sin x + 1}$	
$\dfrac{\cos x\,(\sin x + 1)}{\sin x + 1}$	
$\cos x$	

53. A;

$\sin x \cos x + 1$	$\dfrac{\sin^3 x - \cos^3 x}{\sin x - \cos x}$
	$\dfrac{(\sin x - \cos x)(\sin^2 x + \sin x \cos x + \cos^2 x)}{\sin x - \cos x}$
	$\sin^2 x + \sin x \cos x + \cos^2 x$
	$\sin x \cos x + 1$

55. C;

$\dfrac{1}{\cot x \sin^2 x}$	$\tan x + \cot x$
$\dfrac{1}{\dfrac{\cos x}{\sin x} \cdot \sin^2 x}$	$\dfrac{\sin x}{\cos x} + \dfrac{\cos x}{\sin x}$
$\dfrac{1}{\cos x \sin x}$	$\dfrac{\sin^2 x + \cos^2 x}{\cos x \sin x}$
	$\dfrac{1}{\cos x \sin x}$

57. Discussion and Writing

59. [5.1] **(a)**, **(d)**

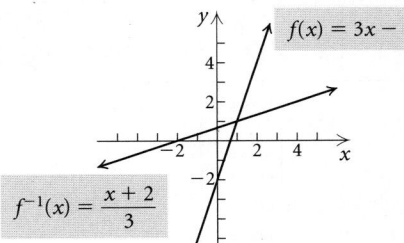

$f(x) = 3x - 2$

$f^{-1}(x) = \dfrac{x+2}{3}$

(b) yes; **(c)** $f^{-1}(x) = \dfrac{x+2}{3}$

60. [5.1] **(a), (d)**

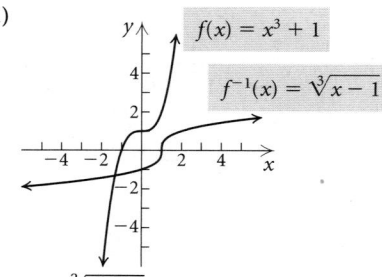

(b) yes; **(c)** $f^{-1}(x) = \sqrt[3]{x-1}$

61. [5.1] **(a), (d)**

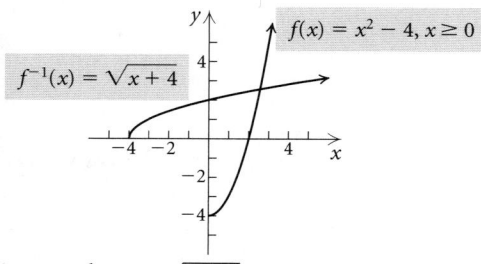

(b) yes; **(c)** $f^{-1}(x) = \sqrt{x+4}$

62. [5.1] **(a), (d)**

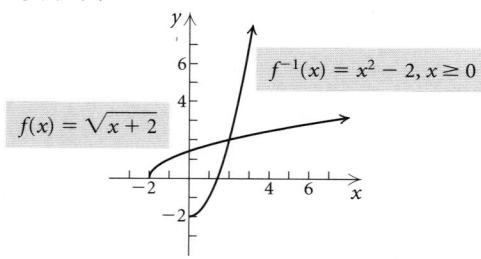

(b) yes; **(c)** $f^{-1}(x) = x^2 - 2, \ x \geq 0$

63. [3.2] $0, \frac{5}{2}$ **64.** [3.2] $-4, \frac{7}{3}$ **65.** [3.2] $\pm 2, \pm 3i$
66. [3.2] $5 \pm 2\sqrt{6}$ **67.** [3.4] 27   **68.** [3.4] 9
69.

$\ln|\tan x|$	$-\ln|\cot x|$
$\ln\left|\dfrac{1}{\cot x}\right|$	
$\ln|1| - \ln|\cot x|$	
$0 - \ln|\cot x|$	
$-\ln|\cot x|$	

71. $\log(\cos x - \sin x) + \log(\cos x + \sin x)$
$\qquad = \log[(\cos x - \sin x)(\cos x + \sin x)]$
$\qquad = \log(\cos^2 x - \sin^2 x) = \log \cos 2x$

73. $\dfrac{1}{\omega C(\tan\theta + \tan\phi)} = \dfrac{1}{\omega C\left(\dfrac{\sin\theta}{\cos\theta} + \dfrac{\sin\phi}{\cos\phi}\right)}$

$\qquad = \dfrac{1}{\omega C\left(\dfrac{\sin\theta\cos\phi + \sin\phi\cos\theta}{\cos\theta\cos\phi}\right)}$

$\qquad = \dfrac{\cos\theta\cos\phi}{\omega C\sin(\theta+\phi)}$

Exercise Set 7.4

1. $-\dfrac{\pi}{3}, -60°$ **3.** $\dfrac{\pi}{4}, 45°$ **5.** $\dfrac{\pi}{4}, 45°$ **7.** $0, 0°$

9. $\dfrac{\pi}{6}, 30°$ **11.** $\dfrac{\pi}{6}, 30°$ **13.** $\dfrac{5\pi}{6}, 150°$

15. $-\dfrac{\pi}{6}, -30°$ **17.** $\dfrac{\pi}{2}, 90°$ **19.** $\dfrac{\pi}{3}, 60°$

21. $0.3520, 20.2°$ **23.** $1.2917, 74.0°$ **25.** $2.9463, 168.8°$
27. $-0.1600, -9.2°$ **29.** $0.8289, 47.5°$
31. $-0.9600, -55.0°$
33. $\sin^{-1}: [-1, 1];\ \cos^{-1}: [-1, 1];\ \tan^{-1}: (-\infty, \infty)$

35. $\theta = \sin^{-1}\left(\dfrac{2000}{d}\right)$ **37.** 0.3 **39.** $\dfrac{\pi}{4}$ **41.** $\dfrac{\pi}{5}$

43. $-\dfrac{\pi}{3}$ **45.** $\dfrac{1}{2}$ **47.** 1 **49.** $\dfrac{\pi}{3}$ **51.** $\dfrac{\sqrt{11}}{33}$

53. $-\dfrac{\pi}{6}$ **55.** $\dfrac{a}{\sqrt{a^2+9}}$ **57.** $\dfrac{\sqrt{q^2-p^2}}{p}$ **59.** $\dfrac{p}{3}$

61. $\dfrac{\sqrt{3}}{2}$ **63.** $-\dfrac{\sqrt{2}}{10}$ **65.** $xy + \sqrt{(1-x^2)(1-y^2)}$

67. 0.9861 **69.** Discussion and Writing
71. Discussion and Writing   **72.** [6.5] Periodic
73. [6.4] Radian measure   **74.** [6.1] Similar
75. [6.2] Angle of depression   **76.** [6.4] Angular speed
77. [6.3] Supplementary   **78.** [6.5] Amplitude
79. [6.1] Acute   **80.** [6.5] Circular
81.

$\sin^{-1}x + \cos^{-1}x$	$\dfrac{\pi}{2}$
$\sin(\sin^{-1}x + \cos^{-1}x)$	$\sin\dfrac{\pi}{2}$
$[\sin(\sin^{-1}x)][\cos(\cos^{-1}x)] +$ $[\cos(\sin^{-1}x)][\sin(\cos^{-1}x)]$ $x\cdot x + \sqrt{1-x^2}\cdot\sqrt{1-x^2}$ $x^2 + 1 - x^2$	1
	1

83.

$\sin^{-1}x$	$\tan^{-1}\dfrac{x}{\sqrt{1-x^2}}$
$\sin(\sin^{-1}x)$	$\sin\left(\tan^{-1}\dfrac{x}{\sqrt{1-x^2}}\right)$
x	x

85.

$\sin^{-1}x$	$\cos^{-1}\sqrt{1-x^2}$
$\sin(\sin^{-1}x)$	$\sin\left(\cos^{-1}\sqrt{1-x^2}\right)$
x	x

87. $\theta = \tan^{-1}\dfrac{y+h}{x} - \tan^{-1}\dfrac{y}{x}; 38.7°$

Visualizing the Graph

1. D   **2.** G   **3.** C   **4.** H   **5.** I   **6.** A   **7.** E
8. J   **9.** F   **10.** B

Exercise Set 7.5

1. $\dfrac{\pi}{6} + 2k\pi, \dfrac{11\pi}{6} + 2k\pi$, or $30° + k \cdot 360°, 330° + k \cdot 360°$

3. $\dfrac{2\pi}{3} + k\pi$, or $120° + k \cdot 180°$

5. $\dfrac{\pi}{6} + 2k\pi, \dfrac{5\pi}{6} + 2k\pi$, or $30° + k \cdot 360°, 150° + k \cdot 360°$

7. $\dfrac{3\pi}{4} + 2k\pi, \dfrac{5\pi}{4} + 2k\pi$, or $135° + k \cdot 360°, 225° + k \cdot 360°$

9. $1.7120, 4.5712$, or $98.09°, 261.91°$ **11.** $\dfrac{4\pi}{3}, \dfrac{5\pi}{3}$, or

$240°, 300°$ **13.** $\dfrac{\pi}{4}, \dfrac{3\pi}{4}, \dfrac{5\pi}{4}, \dfrac{7\pi}{4}$, or $45°, 135°, 225°, 315°$

15. $\dfrac{\pi}{6}, \dfrac{5\pi}{6}, \dfrac{3\pi}{2}$, or $30°, 150°, 270°$ **17.** $\dfrac{\pi}{6}, \dfrac{\pi}{2}, \dfrac{3\pi}{2}, \dfrac{11\pi}{6}$, or

$30°, 90°, 270°, 330°$ **19.** $1.9106, \dfrac{2\pi}{3}, \dfrac{4\pi}{3}, 4.3726$, or

$109.47°, 120°, 240°, 250.53°$ **21.** $0, \dfrac{\pi}{4}, \dfrac{3\pi}{4}, \pi, \dfrac{5\pi}{4}, \dfrac{7\pi}{4}$, or $0°$,

$45°, 135°, 180°, 225°, 315°$ **23.** $2.4402, 3.8430$, or

$139.81°, 220.19°$ **25.** $0.6496, 2.9557, 3.7912, 6.0973$, or

$37.22°, 169.35°, 217.22°, 349.35°$ **27.** $0, \pi, \dfrac{7\pi}{6}, \dfrac{11\pi}{6}$

29. $0, \pi$ **31.** $0, \pi$ **33.** $\dfrac{3\pi}{4}, \dfrac{7\pi}{4}$

35. $\dfrac{2\pi}{3}, \dfrac{4\pi}{3}, \dfrac{3\pi}{2}$ **37.** $\dfrac{\pi}{4}, \dfrac{3\pi}{4}, \dfrac{5\pi}{4}, \dfrac{7\pi}{4}$ **39.** $\dfrac{\pi}{12}, \dfrac{5\pi}{12}$

41. $0.967, 1.853, 4.108, 4.994$ **43.** $\dfrac{2\pi}{3}, \dfrac{4\pi}{3}$

45. $1.114, 2.773$ **47.** 0.515 **49.** $0.422, 1.756$
51. (a) $y = 7 \sin(-2.6180x + 0.5236) + 7$;
(b) $\$10,500, \$13,062$ **53.** Discussion and Writing
55. [6.2] $B = 35°, b \approx 140.7, c \approx 245.4$
56. [6.2] $R \approx 15.5°, T \approx 74.5°, t \approx 13.7$ **57.** [1.5] 36
58. [1.5] 14   **59.** $\dfrac{\pi}{3}, \dfrac{2\pi}{3}, \dfrac{4\pi}{3}, \dfrac{5\pi}{3}$ **61.** $\dfrac{\pi}{3}, \dfrac{4\pi}{3}$
63. 0   **65.** $e^{3\pi/2 + 2k\pi}$, where k (an integer) ≤ -1
67. 1.24 days, 6.76 days   **69.** $16.5°$N   **71.** 1
73. $\dfrac{5}{26}$, or about 0.1923

Review Exercises: Chapter 7

1. True   **2.** True   **3.** True   **4.** False   **5.** False
6. $\csc^2 x$ **7.** 1   **8.** $\tan^2 y - \cot^2 y$
9. $\dfrac{(\cos^2 x + 1)^2}{\cos^2 x}$ **10.** $\csc x (\sec x - \csc x)$
11. $(3\sin y + 5)(\sin y - 4)$
12. $(10 - \cos u)(100 + 10\cos u + \cos^2 u)$ **13.** 1
14. $\frac{1}{2}\sec x$ **15.** $\dfrac{3\tan x}{\sin x - \cos x}$

16. $\dfrac{3\cos y + 3\sin y + 2}{\cos^2 y - \sin^2 y}$ **17.** 1

18. $\frac{1}{4}\cot x$ **19.** $\sin x + \cos x$

20. $\dfrac{\cos x}{1 - \sin x}$ **21.** $\dfrac{\cos x}{\sqrt{\sin x}}$ **22.** $3\sec\theta$

23. $\cos x \cos\dfrac{3\pi}{2} - \sin x \sin\dfrac{3\pi}{2}$

24. $\dfrac{\tan 45° - \tan 30°}{1 + \tan 45° \tan 30°}$

25. $\cos(27° - 16°)$, or $\cos 11°$ **26.** $\dfrac{-\sqrt{6} - \sqrt{2}}{4}$

27. $2 - \sqrt{3}$ **28.** -0.3745 **29.** $-\sin x$
30. $\sin x$ **31.** $-\cos x$

32. (a) $\sin\alpha = -\dfrac{4}{5}, \tan\alpha = \dfrac{4}{3}, \cot\alpha = \dfrac{3}{4}$,

$\sec\alpha = -\dfrac{5}{3}, \csc\alpha = -\dfrac{5}{4}$; (b) $\sin\left(\dfrac{\pi}{2} - \alpha\right) = -\dfrac{3}{5}$,

$\cos\left(\dfrac{\pi}{2} - \alpha\right) = -\dfrac{4}{5}, \tan\left(\dfrac{\pi}{2} - \alpha\right) = \dfrac{3}{4}$,

$\cot\left(\dfrac{\pi}{2} - \alpha\right) = \dfrac{4}{3}, \sec\left(\dfrac{\pi}{2} - \alpha\right) = -\dfrac{5}{4}$,

$\csc\left(\dfrac{\pi}{2} - \alpha\right) = -\dfrac{5}{3}$; (c) $\sin\left(\alpha + \dfrac{\pi}{2}\right) = -\dfrac{3}{5}$,

$\cos\left(\alpha + \dfrac{\pi}{2}\right) = \dfrac{4}{5}, \tan\left(\alpha + \dfrac{\pi}{2}\right) = -\dfrac{3}{4}$,

$\cot\left(\alpha + \dfrac{\pi}{2}\right) = -\dfrac{4}{3}, \sec\left(\alpha + \dfrac{\pi}{2}\right) = \dfrac{5}{4}$,

$\csc\left(\alpha + \dfrac{\pi}{2}\right) = -\dfrac{5}{3}$ **33.** $-\sec x$

34. $\tan 2\theta = \frac{24}{7}, \cos 2\theta = \frac{7}{25}, \sin 2\theta = \frac{24}{25}$; I

35. $\dfrac{\sqrt{2 - \sqrt{2}}}{2}$

36. $\sin 2\beta = 0.4261, \cos\dfrac{\beta}{2} = 0.9940, \cos 4\beta = 0.6369$

37. $\cos x$ **38.** 1   **39.** $\sin 2x$
40. $\tan 2x$
41.

$\dfrac{1 - \sin x}{\cos x}$	$\dfrac{\cos x}{1 + \sin x}$
$\dfrac{1 - \sin x}{\cos x} \cdot \dfrac{\cos x}{\cos x}$	$\dfrac{\cos x}{1 + \sin x} \cdot \dfrac{1 - \sin x}{1 - \sin x}$
$\dfrac{\cos x - \sin x \cos x}{\cos^2 x}$	$\dfrac{\cos x - \sin x \cos x}{1 - \sin^2 x}$
	$\dfrac{\cos x - \sin x \cos x}{\cos^2 x}$

42.

$\dfrac{1 + \cos 2\theta}{\sin 2\theta}$	$\cot \theta$
$\dfrac{1 + 2\cos^2 \theta - 1}{2 \sin \theta \cos \theta}$	$\dfrac{\cos \theta}{\sin \theta}$
$\dfrac{\cos \theta}{\sin \theta}$	

43.

$\dfrac{\tan y + \sin y}{2 \tan y}$	$\cos^2 \dfrac{y}{2}$
$\dfrac{1}{2}\left[\dfrac{\dfrac{\sin y + \sin y \cos y}{\cos y}}{\dfrac{\sin y}{\cos y}}\right]$	$\dfrac{1 + \cos y}{2}$
$\dfrac{1}{2}\left[\dfrac{\sin y(1 + \cos y)}{\cos y} \cdot \dfrac{\cos y}{\sin y}\right]$	
$\dfrac{1 + \cos y}{2}$	

44.

$\dfrac{\sin x - \cos x}{\cos^2 x}$	$\dfrac{\tan^2 x - 1}{\sin x + \cos x}$
	$\dfrac{\dfrac{\sin^2 x}{\cos^2 x} - 1}{\sin x + \cos x}$
	$\dfrac{\sin^2 x - \cos^2 x}{\cos^2 x} \cdot \dfrac{1}{\sin x + \cos x}$
	$\dfrac{\sin x - \cos x}{\cos^2 x}$

45. $3 \cos 2\theta \sin \theta = \frac{3}{2}(\sin 3\theta - \sin \theta)$

46. $\sin \theta - \sin 4\theta = -2 \cos \dfrac{5\theta}{2} \sin \dfrac{3\theta}{2}$

47. B;

$\csc x - \cos x \cot x$	$\sin x$
$\dfrac{1}{\sin x} - \cos x \dfrac{\cos x}{\sin x}$	
$\dfrac{1 - \cos^2 x}{\sin x}$	
$\dfrac{\sin^2 x}{\sin x}$	
$\sin x$	

48. D;

$\dfrac{1}{\sin x \cos x} - \dfrac{\cos x}{\sin x}$	$\dfrac{\sin x \cos x}{1 - \sin^2 x}$
$\dfrac{1}{\sin x \cos x} - \dfrac{\cos^2 x}{\sin x \cos x}$	$\dfrac{\sin x \cos x}{\cos^2 x}$
$\dfrac{1 - \cos^2 x}{\sin x \cos x}$	$\dfrac{\sin x}{\cos x}$
$\dfrac{\sin^2 x}{\sin x \cos x}$	
$\dfrac{\sin x}{\cos x}$	

49. A;

$\dfrac{\cot x - 1}{1 - \tan x}$	$\dfrac{\csc x}{\sec x}$
$\dfrac{\dfrac{\cos x}{\sin x} - \dfrac{\sin x}{\sin x}}{\dfrac{\cos x}{\cos x} - \dfrac{\sin x}{\cos x}}$	$\dfrac{\dfrac{1}{\sin x}}{\dfrac{1}{\cos x}}$
$\dfrac{\cos x - \sin x}{\sin x} \cdot \dfrac{\cos x}{\cos x - \sin x}$	$\dfrac{1}{\sin x} \cdot \dfrac{\cos x}{1}$
$\dfrac{\cos x}{\sin x}$	$\dfrac{\cos x}{\sin x}$

50. C;

$\dfrac{\cos x + 1}{\sin x} + \dfrac{\sin x}{\cos x + 1}$	$\dfrac{2}{\sin x}$
$\dfrac{(\cos x + 1)^2 + \sin^2 x}{\sin x(\cos x + 1)}$	
$\dfrac{\cos^2 x + 2\cos x + 1 + \sin^2 x}{\sin x(\cos x + 1)}$	
$\dfrac{2\cos x + 2}{\sin x(\cos x + 1)}$	
$\dfrac{2(\cos x + 1)}{\sin x(\cos x + 1)}$	
$\dfrac{2}{\sin x}$	

51. $-\dfrac{\pi}{6}, -30°$ **52.** $\dfrac{\pi}{6}, 30°$

53. $\dfrac{\pi}{4}, 45°$ **54.** $0, 0°$ **55.** $1.7920, 102.7°$

56. $0.3976, 22.8°$ **57.** $\frac{1}{2}$ **58.** $\dfrac{\sqrt{3}}{3}$

59. $\dfrac{\pi}{7}$ **60.** $\dfrac{\sqrt{2}}{2}$ **61.** $\dfrac{3}{\sqrt{b^2 + 9}}$ **62.** $-\frac{7}{25}$

63. $\dfrac{3\pi}{4} + 2k\pi, \dfrac{5\pi}{4} + 2k\pi$, or $135° + k \cdot 360°, 225° + k \cdot 360°$

64. $\dfrac{\pi}{3} + k\pi$, or $60° + k \cdot 180°$

65. $\dfrac{\pi}{6}, \dfrac{5\pi}{6}, \dfrac{7\pi}{6}, \dfrac{11\pi}{6}$ **66.** $\dfrac{\pi}{4}, \dfrac{\pi}{2}, \dfrac{3\pi}{4}, \dfrac{5\pi}{4}, \dfrac{3\pi}{2}, \dfrac{7\pi}{4}$

67. $\dfrac{2\pi}{3}, \pi, \dfrac{4\pi}{3}$ **68.** $0, \pi$ **69.** $\dfrac{\pi}{4}, \dfrac{3\pi}{4}, \dfrac{5\pi}{4}, \dfrac{7\pi}{4}$

70. $0, \dfrac{\pi}{2}, \pi, \dfrac{3\pi}{2}$ **71.** $\dfrac{7\pi}{12}, \dfrac{23\pi}{12}$

72. $0.864, 2.972, 4.006, 6.114$

73. 4.917 **74.** No solution in $[0, 2\pi)$

75. B   **76.** A   **77.** C

78. Discussion and Writing:

(a) $2 \cos^2 x - 1 = \cos 2x$
$$= \cos^2 x - \sin^2 x$$
$$= 1 \cdot (\cos^2 x - \sin^2 x)$$
$$= (\cos^2 x + \sin^2 x)(\cos^2 x - \sin^2 x)$$
$$= \cos^4 x - \sin^4 x;$$

(b) $\cos^4 x - \sin^4 x = (\cos^2 x + \sin^2 x)(\cos^2 x - \sin^2 x)$
$$= 1 \cdot (\cos^2 x - \sin^2 x)$$
$$= \cos^2 x - \sin^2 x$$
$$= \cos 2x$$
$$= 2\cos^2 x - 1;$$

(c)

$2\cos^2 x - 1$	$\cos^4 x - \sin^4 x$
$\cos 2x$	$(\cos^2 x + \sin^2 x)(\cos^2 x - \sin^2 x)$
	$1 \cdot (\cos^2 x - \sin^2 x)$
	$\cos^2 x - \sin^2 x$
	$\cos 2x$

Answers may vary. Method 2 may be the more efficient because it involves straightforward factorization and simplification. Method 1(a) requires a "trick" such as multiplying by a particular expression equivalent to 1.

79. Discussion and Writing: The ranges of the inverse trigonometric functions are restricted in order that they might be functions.

80. $108.4°$

81. $\cos(u + v) = \cos u \cos v - \sin u \sin v$
$$= \cos u \cos v - \cos\left(\frac{\pi}{2} - u\right)\cos\left(\frac{\pi}{2} - v\right)$$

82. $\cos^2 x$

83. $\sin\theta = \sqrt{\dfrac{1}{2} + \dfrac{\sqrt{6}}{5}}$; $\cos\theta = \sqrt{\dfrac{1}{2} - \dfrac{\sqrt{6}}{5}}$;
$\tan\theta = \sqrt{\dfrac{5 + 2\sqrt{6}}{5 - 2\sqrt{6}}}$

84. $\ln e^{\sin t} = \log_e e^{\sin t} = \sin t$

85.

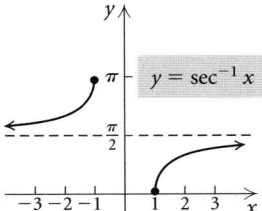

86. Let $x = \dfrac{\sqrt{2}}{2}$. Then $\tan^{-1}\dfrac{\sqrt{2}}{2} \approx 0.6155$ and
$$\dfrac{\sin^{-1}\dfrac{\sqrt{2}}{2}}{\cos^{-1}\dfrac{\sqrt{2}}{2}} = \dfrac{\dfrac{\pi}{4}}{\dfrac{\pi}{4}} = 1.$$ **87.** $\dfrac{\pi}{2}, \dfrac{3\pi}{2}$

Test: Chapter 7

1. [7.1] $2\cos x + 1$ **2.** [7.1] 1 **3.** [7.1] $\dfrac{\cos\theta}{1 + \sin\theta}$

4. [7.1] $2\cos\theta$ **5.** [7.1] $\dfrac{\sqrt{2} + \sqrt{6}}{4}$ **6.** [7.1] $\dfrac{3 - \sqrt{3}}{3 + \sqrt{3}}$

7. [7.1] $\frac{120}{169}$ **8.** [7.2] $\dfrac{\sqrt{5}}{3}$ **9.** [7.2] $\frac{24}{25}$, II

10. [7.2] $\dfrac{\sqrt{2 + \sqrt{3}}}{2}$ **11.** [7.2] 0.9304

12. [7.2] $3\sin 2x$

13. [7.3]

$\csc x - \cos x \cot x$	$\sin x$
$\dfrac{1}{\sin x} - \cos x \cdot \dfrac{\cos x}{\sin x}$	
$\dfrac{1 - \cos^2 x}{\sin x}$	
$\dfrac{\sin^2 x}{\sin x}$	
$\sin x$	

14. [7.3]

$(\sin x + \cos x)^2$	$1 + \sin 2x$
$\sin^2 x + 2\sin x \cos x + \cos^2 x$	
$1 + 2\sin x \cos x$	
$1 + \sin 2x$	

15. [7.3]

$(\csc\beta + \cot\beta)^2$	$\dfrac{1 + \cos\beta}{1 - \cos\beta}$
$\left(\dfrac{1}{\sin\beta} + \dfrac{\cos\beta}{\sin\beta}\right)^2$	$\dfrac{1 + \cos\beta}{1 - \cos\beta} \cdot \dfrac{1 + \cos\beta}{1 + \cos\beta}$
$\left(\dfrac{1 + \cos\beta}{\sin\beta}\right)^2$	$\dfrac{(1 + \cos\beta)^2}{1 - \cos^2\beta}$
$\dfrac{(1 + \cos\beta)^2}{\sin^2\beta}$	$\dfrac{(1 + \cos\beta)^2}{\sin^2\beta}$

16. [7.3]

$\dfrac{1 + \sin\alpha}{1 + \csc\alpha}$	$\tan\alpha$
	$\sec\alpha$
$\dfrac{1 + \sin\alpha}{1 + \dfrac{1}{\sin\alpha}}$	$\dfrac{\sin\alpha}{\cos\alpha}$
	$\dfrac{1}{\cos\alpha}$
$\dfrac{1 + \sin\alpha}{\dfrac{\sin\alpha + 1}{\sin\alpha}}$	$\sin\alpha$
$\sin\alpha$	

17. [7.4] $\cos 8\alpha - \cos\alpha = -2\sin\dfrac{9\alpha}{2}\sin\dfrac{7\alpha}{2}$

18. [7.4] $4\sin\beta\cos 3\beta = 2(\sin 4\beta - \sin 2\beta)$

19. [7.4] $-45°$ **20.** [7.4] $\dfrac{\pi}{3}$ **21.** [7.4] 2.3072

22. [7.4] $\dfrac{\sqrt{3}}{2}$ **23.** [7.4] $\dfrac{5}{\sqrt{x^2 - 25}}$ **24.** [7.4] 0

25. [7.5] $\dfrac{\pi}{6}, \dfrac{5\pi}{6}, \dfrac{7\pi}{6}, \dfrac{11\pi}{6}$ **26.** [7.5] $0, \dfrac{\pi}{4}, \dfrac{3\pi}{4}, \pi$

27. [7.5] $\dfrac{\pi}{2}, \dfrac{11\pi}{6}$ **28.** [7.4] D   **29.** [7.2] $\sqrt{\frac{11}{12}}$

CHAPTER 8

Exercise Set 8.1

1. $A = 121°$, $a \approx 33$, $c \approx 14$ **3.** $B \approx 57.4°$, $C \approx 86.1°$, $c \approx 40$, or $B \approx 122.6°$, $C \approx 20.9°$, $c \approx 14$ **5.** $B \approx 44°24'$, $A \approx 74°26'$, $a \approx 33.3$ **7.** $A = 110.36°$, $a \approx 5$ mi, $b \approx 3$ mi **9.** $B \approx 83.78°$, $A \approx 12.44°$, $a \approx 12.30$ yd **11.** $B \approx 14.7°$, $C \approx 135.0°$, $c \approx 28.04$ cm **13.** No solution **15.** $B = 125.27°$, $b \approx 302$ m, $c \approx 138$ m **17.** 8.2 ft² **19.** 12 yd² **21.** 596.98 ft² **23.** 76.3 m **25.** 787 ft² **27.** About 51 ft **29.** From A: about 35 mi; from B: about 66 mi **31.** About 22 mi **33.** Discussion and Writing **35.** [6.1] 1.348, 77.2° **36.** [6.1] No angle **37.** [6.1] 18.24° **38.** [6.1] 125.06° **39.** [R.1] 5 **40.** [6.3] $\dfrac{\sqrt{3}}{2}$ **41.** [6.3] $\dfrac{\sqrt{2}}{2}$ **42.** [6.3] $-\dfrac{\sqrt{3}}{2}$ **43.** [6.3] $-\dfrac{1}{2}$ **44.** [3.1] 2 **45.** Use the formula for the area of a triangle and the law of sines.

$$K = \frac{1}{2}bc \sin A \quad \text{and} \quad b = \frac{c \sin B}{\sin C},$$

$$\text{so} \quad K = \frac{c^2 \sin A \sin B}{2 \sin C}.$$

$$K = \frac{1}{2}ab \sin C \quad \text{and} \quad b = \frac{a \sin B}{\sin A},$$

$$\text{so} \quad K = \frac{a^2 \sin B \sin C}{2 \sin A}.$$

$$K = \frac{1}{2}bc \sin A \quad \text{and} \quad c = \frac{b \sin C}{\sin B},$$

$$\text{so} \quad K = \frac{b^2 \sin A \sin C}{2 \sin B}.$$

47.

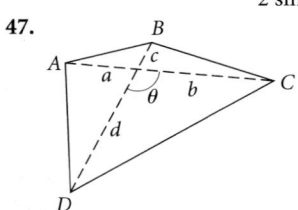

For the quadrilateral $ABCD$, we have

$$\text{Area} = \frac{1}{2}bd \sin \theta + \frac{1}{2}ac \sin \theta$$

$$+ \frac{1}{2}ad(\sin 180° - \theta) + \frac{1}{2}bc \sin (180° - \theta)$$

$$\textit{Note:} \sin \theta = \sin(180° - \theta).$$

$$= \frac{1}{2}(bd + ac + ad + bc) \sin \theta$$

$$= \frac{1}{2}(a + b)(c + d) \sin \theta$$

$$= \frac{1}{2}d_1 d_2 \sin \theta,$$

where $d_1 = a + b$ and $d_2 = c + d$.
49. 44.1 " from wall 1 and 104.3 " from wall 4

Exercise Set 8.2

1. $a \approx 15$, $B \approx 24°$, $C \approx 126°$ **3.** $A \approx 36.18°$, $B \approx 43.53°$, $C \approx 100.29°$ **5.** $b \approx 75$ m, $A \approx 94°51'$, $C \approx 12°29'$ **7.** $A \approx 24.15°$, $B \approx 30.75°$, $C \approx 125.10°$ **9.** No solution **11.** $A \approx 79.93°$, $B \approx 53.55°$, $C \approx 46.52°$ **13.** $c \approx 45.17$ mi, $A \approx 89.3°$, $B \approx 42.0°$ **15.** $a \approx 13.9$ in., $B \approx 36.127°$, $C \approx 90.417°$ **17.** Law of sines; $C = 98°$, $a \approx 96.7$, $c \approx 101.9$ **19.** Law of cosines; $A \approx 73.71°$, $B \approx 51.75°$, $C \approx 54.54°$ **21.** Cannot be solved **23.** Law of cosines; $A \approx 33.71°$, $B \approx 107.08°$, $C \approx 39.21°$ **25.** About 367 ft **27.** About 1.5 mi **29.** $S \approx 112.5°$, $T \approx 27.2°$, $U \approx 40.3°$ **31.** About 912 km **33.** (a) About 16 ft; (b) about 122 ft² **35.** About 4.7 cm **37.** Discussion and Writing **39.** [4.1] Quartic **40.** [1.3] Linear **41.** [6.5] Trigonometric **42.** [5.2] Exponential **43.** [4.5] Rational **44.** [4.1] Cubic **45.** [5.2] Exponential **46.** [5.3] Logarithmic **47.** [6.5] Trigonometric **48.** [3.2] Quadratic **49.** About 9386 ft **51.** $A = \dfrac{1}{2}a^2 \sin \theta$; when $\theta = 90°$

Exercise Set 8.3

1. 5;

3. 1;

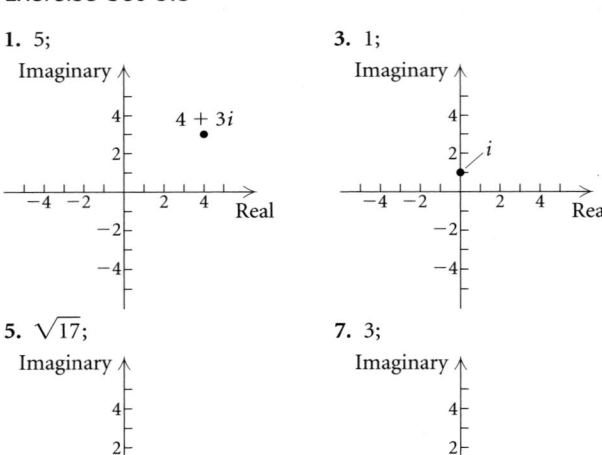

5. $\sqrt{17}$;

7. 3;

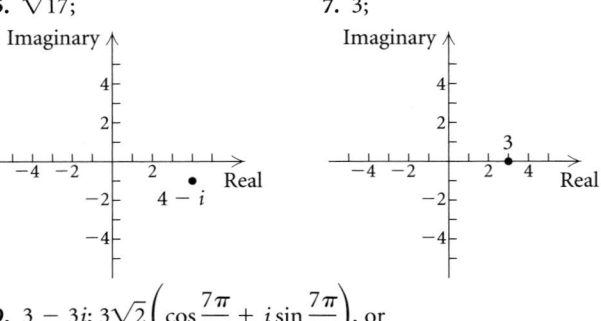

9. $3 - 3i$; $3\sqrt{2}\left(\cos \dfrac{7\pi}{4} + i \sin \dfrac{7\pi}{4}\right)$, or $3\sqrt{2}(\cos 315° + i \sin 315°)$
11. $4i$; $4\left(\cos \dfrac{\pi}{2} + i \sin \dfrac{\pi}{2}\right)$, or $4(\cos 90° + i \sin 90°)$
13. $\sqrt{2}\left(\cos \dfrac{7\pi}{4} + i \sin \dfrac{7\pi}{4}\right)$, or $\sqrt{2}(\cos 315° + i \sin 315°)$

15. $3\left(\cos\dfrac{3\pi}{2} + i\sin\dfrac{3\pi}{2}\right)$, or $3(\cos 270° + i\sin 270°)$

17. $2\left(\cos\dfrac{\pi}{6} + i\sin\dfrac{\pi}{6}\right)$, or $2(\cos 30° + i\sin 30°)$

19. $\dfrac{2}{5}(\cos 0 + i\sin 0)$, or $\dfrac{2}{5}(\cos 0° + i\sin 0°)$

21. $6\left(\cos\dfrac{5\pi}{4} + i\sin\dfrac{5\pi}{4}\right)$, or $6(\cos 225° + i\sin 225°)$

23. $\dfrac{3\sqrt{3}}{2} + \dfrac{3}{2}i$　**25.** $-10i$　**27.** $2 + 2i$　**29.** $2i$

31. $\dfrac{\sqrt{2}}{2} - \dfrac{\sqrt{6}}{2}i$　**33.** $4(\cos 42° + i\sin 42°)$

35. $11.25(\cos 56° + i\sin 56°)$　**37.** 4

39. $-i$　**41.** $6 + 6\sqrt{3}i$　**43.** $-2i$

45. $8(\cos\pi + i\sin\pi)$　**47.** $8\left(\cos\dfrac{3\pi}{2} + i\sin\dfrac{3\pi}{2}\right)$

49. $\dfrac{27}{2} + \dfrac{27\sqrt{3}}{2}i$　**51.** $-4 + 4i$　**53.** -1

55. $-\dfrac{\sqrt{2}}{2} + \dfrac{\sqrt{2}}{2}i, \dfrac{\sqrt{2}}{2} - \dfrac{\sqrt{2}}{2}i$

57. $2(\cos 157.5° + i\sin 157.5°), 2(\cos 337.5° + i\sin 337.5°)$

59. $\dfrac{\sqrt{3}}{2} + \dfrac{1}{2}i, -\dfrac{\sqrt{3}}{2} + \dfrac{1}{2}i, -i$

61. $\sqrt[3]{4}(\cos 110° + i\sin 110°), \sqrt[3]{4}(\cos 230° + i\sin 230°),$ $\sqrt[3]{4}(\cos 350° + i\sin 350°)$

63. $2, 2i, -2, -2i;$

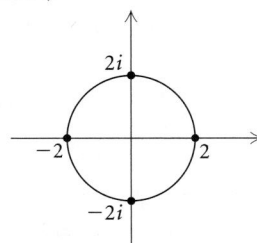

65. $\cos 36° + i\sin 36°,$
$\cos 108° + i\sin 108°, -1,$
$\cos 252° + i\sin 252°,$
$\cos 324° + i\sin 324°;$

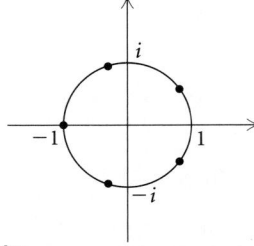

67. $\sqrt[10]{8}, \sqrt[10]{8}(\cos 36° + i\sin 36°), \sqrt[10]{8}(\cos 72° + i\sin 72°),$
$\sqrt[10]{8}(\cos 108° + i\sin 108°), \sqrt[10]{8}(\cos 144° + i\sin 144°), -\sqrt[10]{8},$
$\sqrt[10]{8}(\cos 216° + i\sin 216°), \sqrt[10]{8}(\cos 252° + i\sin 252°),$
$\sqrt[10]{8}(\cos 288° + i\sin 288°), \sqrt[10]{8}(\cos 324° + i\sin 324°)$

69. $\dfrac{\sqrt{3}}{2} + \dfrac{1}{2}i, i, -\dfrac{\sqrt{3}}{2} + \dfrac{1}{2}i, -\dfrac{\sqrt{3}}{2} - \dfrac{1}{2}i, -i, \dfrac{\sqrt{3}}{2} - \dfrac{1}{2}i$

71. $1, -\dfrac{1}{2} + \dfrac{\sqrt{3}}{2}i, -\dfrac{1}{2} - \dfrac{\sqrt{3}}{2}i$

73. $\cos 67.5° + i\sin 67.5°, \cos 157.5° + i\sin 157.5°,$
$\cos 247.5° + i\sin 247.5°, \cos 337.5° + i\sin 337.5°$

75. $\sqrt{3} + i, 2i, -\sqrt{3} + i, -\sqrt{3} - i, -2i, \sqrt{3} - i$

77. Discussion and Writing　**79.** [6.4] $15°$　**80.** [6.4] $540°$

81. [6.4] $\dfrac{11\pi}{6}$　**82.** [6.4] $-\dfrac{5\pi}{4}$　**83.** [R.7] $3\sqrt{5}$

84. [1.1]　　　　　　　　　　　　**85.** [6.5] $\dfrac{\sqrt{3}}{2}$

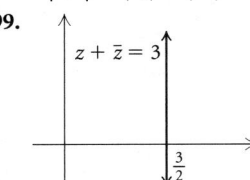

86. [6.5] $\dfrac{\sqrt{3}}{2}$　**87.** [6.5] $\dfrac{\sqrt{2}}{2}$　**88.** [6.5] $\dfrac{1}{2}$

89. $-\dfrac{1 + \sqrt{3}}{2} + \dfrac{1 + \sqrt{3}}{2}i, -\dfrac{1 - \sqrt{3}}{2} + \dfrac{1 - \sqrt{3}}{2}i$

91. $\cos\theta - i\sin\theta$

93. $z = a + bi, |z| = \sqrt{a^2 + b^2}; \bar{z} = a - bi,$
$|\bar{z}| = \sqrt{a^2 + (-b)^2} = \sqrt{a^2 + b^2}, \therefore |z| = |\bar{z}|$

95. $|(a + bi)^2| = |a^2 - b^2 + 2abi| = \sqrt{(a^2 - b^2)^2 + 4a^2b^2}$
$= \sqrt{a^4 + 2a^2b^2 + b^4} = a^2 + b^2,$
$|a + bi|^2 = \left(\sqrt{a^2 + b^2}\right)^2 = a^2 + b^2$

97. $\dfrac{z}{w} = \dfrac{r_1(\cos\theta_1 + i\sin\theta_1)}{r_2(\cos\theta_2 + i\sin\theta_2)}$

$= \dfrac{r_1}{r_2}(\cos(\theta_1 - \theta_2) + i\sin(\theta_1 - \theta_2)),$

$\left|\dfrac{z}{w}\right| = \sqrt{\left[\dfrac{r_1}{r_2}\cos(\theta_1 - \theta_2)\right]^2 + \left[\dfrac{r_1}{r_2}\sin(\theta_1 - \theta_2)\right]^2}$

$= \sqrt{\dfrac{r_1^2}{r_2^2}} = \dfrac{|r_1|}{|r_2|};$

$|z| = \sqrt{(r_1\cos\theta_1)^2 + (r_1\sin\theta_1)^2} = \sqrt{r_1^2} = |r_1|;$
$|w| = \sqrt{(r_2\cos\theta_2)^2 + (r_2\sin\theta_2)^2} = \sqrt{r_2^2} = |r_2|;$

Then $\left|\dfrac{z}{w}\right| = \dfrac{|r_1|}{|r_2|} = \dfrac{|z|}{|w|}.$

99.

$z + \bar{z} = 3$

Visualizing the Graph

1. J　**2.** C　**3.** E　**4.** H　**5.** I　**6.** A　**7.** D
8. G　**9.** B　**10.** F

Exercise Set 8.4

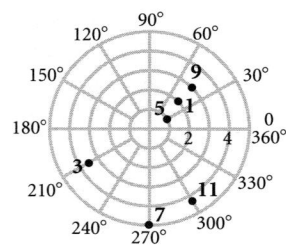

13. *A*: (4, 30°), (4, 390°), (−4, 210°); *B*: (5, 300°), (5, −60°), (−5, 120°); *C*: (2, 150°), (2, 510°), (−2, 330°); *D*: (3, 225°), (3, −135°), (−3, 45°); answers may vary

15. $(3, 270°), \left(3, \dfrac{3\pi}{2}\right)$ **17.** $(6, 300°), \left(6, \dfrac{5\pi}{3}\right)$

19. $(8, 330°), \left(8, \dfrac{11\pi}{6}\right)$ **21.** $(2, 225°), \left(2, \dfrac{5\pi}{4}\right)$

23. $(2, 60°), \left(2, \dfrac{\pi}{3}\right)$ **25.** $(5, 315°), \left(5, \dfrac{7\pi}{4}\right)$

27. (7.616, 66.8°), (7.616, 1.166)
29. (4.643, 132.9°), (4.643, 2.320)

31. $\left(\dfrac{5}{2}, \dfrac{5\sqrt{3}}{2}\right)$ **33.** $\left(-\dfrac{3\sqrt{2}}{2}, -\dfrac{3\sqrt{2}}{2}\right)$

35. $\left(-\dfrac{3}{2}, -\dfrac{3\sqrt{3}}{2}\right)$ **37.** $\left(-1, \sqrt{3}\right)$ **39.** $\left(-\sqrt{3}, -1\right)$

41. $\left(3\sqrt{3}, -3\right)$ **43.** (2.19, −2.05) **45.** (1.30, −3.99)

47. $r(3\cos\theta + 4\sin\theta) = 5$ **49.** $r\cos\theta = 5$ **51.** $r = 6$

53. $r^2\cos^2\theta = 25r\sin\theta$

55. $r^2\sin^2\theta - 5r\cos\theta - 25 = 0$ **57.** $r^2 = 2r\cos\theta$

59. $x^2 + y^2 = 25$ **61.** $y = 2$ **63.** $y^2 = -6x + 9$

65. $x^2 - 9x + y^2 - 7y = 0$ **67.** $x = 5$ **69.** $y = -\sqrt{3}x$

71. 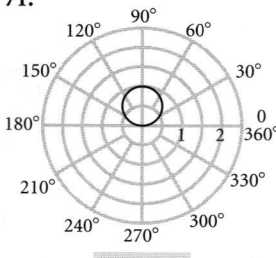 **73.**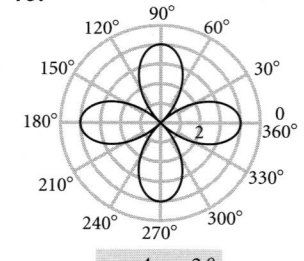

$r = \sin\theta$ $r = 4\cos 2\theta$

75. 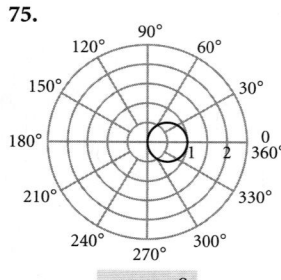 **77.**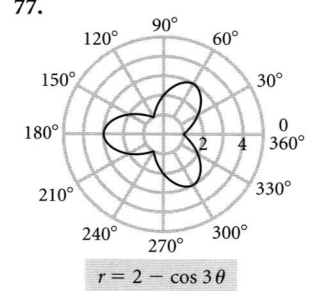

$r = \cos\theta$ $r = 2 - \cos 3\theta$

79. (d) **81.** (g) **83.** (j) **85.** (b) **87.** (e) **89.** (k)
91. 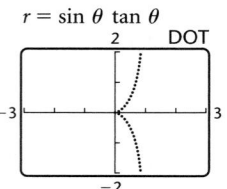 **93.**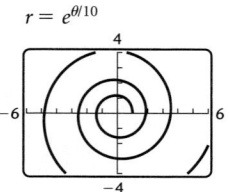

$r = \sin\theta\tan\theta$ $r = e^{\theta/10}$

95. 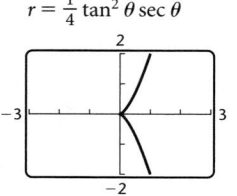 **97.**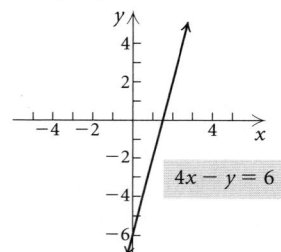

$r = \cos 2\theta\sec\theta$ $r = \frac{1}{4}\tan^2\theta\sec\theta$

99. Discussion and Writing **101.** [1.5] 12 **102.** [1.5] $\frac{1}{5}$
103. [1.3] **104.** [1.3]

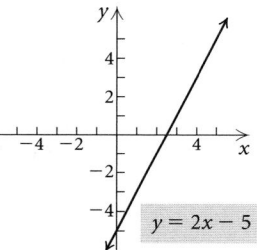 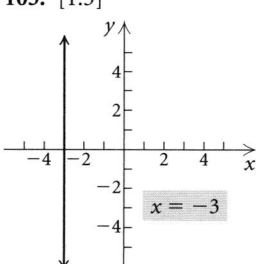

$y = 2x - 5$ $4x - y = 6$

105. [1.3] **106.** [1.3]

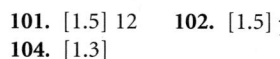 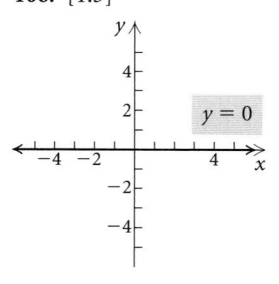

$x = -3$ $y = 0$

107. $y^2 = -4x + 4$

Exercise Set 8.5

1. Yes **3.** No **5.** Yes **7.** No **9.** No **11.** Yes
13. 55 N, 55° **15.** 929 N, 19° **17.** 57.0, 38°
19. 18.4, 37° **21.** 20.9, 58° **23.** 68.3, 18°
25. 11 ft/sec, 63° **27.** 726 lb, 47° **29.** 60°
31. 70.7 east; 70.7 south
33. Horizontal: 215.17 mph forward; vertical: 65.78 mph up
35. Horizontal: 390 lb forward; vertical: 675.5 lb up
37. Northerly: 115 km/h; westerly: 164 km/h
39. Perpendicular: 90.6 lb; parallel: 42.3 lb **41.** 48.1 lb
43. Discussion and Writing **45.** [5.3] Natural
46. [7.2] Half-angle **47.** [6.4] Linear speed
48. [6.1] Cosine **49.** [7.1] Identity

50. [6.1] Cotangent of θ **51.** [6.3] Coterminal
52. [8.1] Sines   **53.** [5.1] Horizontal line; inverse
54. [6.3] Reference angle; acute
55. (a) (4.950, 4.950); **(b)** (0.950, -1.978)

Exercise Set 8.6

1. $\langle -9, 5\rangle$; $\sqrt{106}$ **3.** $\langle -3, 6\rangle$; $3\sqrt{5}$ **5.** $\langle 4, 0\rangle$; 4
7. $\sqrt{37}$ **9.** $\langle 4, -5\rangle$ **11.** $\sqrt{257}$ **13.** $\langle -9, 9\rangle$
15. $\langle 41, -38\rangle$ **17.** $\sqrt{261} - \sqrt{65}$ **19.** $\langle -1, -1\rangle$
21. $\langle -8, 14\rangle$ **23.** 1   **25.** -34
27.   **29.**

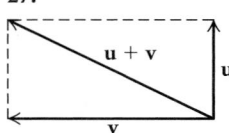   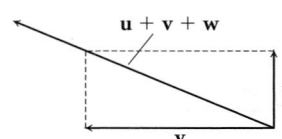

31. (a) $\mathbf{w} = \mathbf{u} + \mathbf{v}$; **(b)** $\mathbf{v} = \mathbf{w} - \mathbf{u}$
33. $\langle -\frac{5}{13}, \frac{12}{13}\rangle$ **35.** $\langle \frac{1}{\sqrt{101}}, -\frac{10}{\sqrt{101}}\rangle$
37. $\langle -\frac{1}{\sqrt{17}}, -\frac{4}{\sqrt{17}}\rangle$ **39.** $\mathbf{w} = -4\mathbf{i} + 6\mathbf{j}$
41. $\mathbf{s} = 2\mathbf{i} + 5\mathbf{j}$ **43.** $-7\mathbf{i} + 5\mathbf{j}$
45. (a) $3\mathbf{i} + 29\mathbf{j}$; **(b)** $\langle 3, 29\rangle$ **47. (a)** $4\mathbf{i} + 16\mathbf{j}$; **(b)** $\langle 4, 16\rangle$
49. $\mathbf{j}$, or $\langle 0, 1\rangle$ **51.** $-\frac{1}{2}\mathbf{i} - \frac{\sqrt{3}}{2}\mathbf{j}$, or $\langle -\frac{1}{2}, -\frac{\sqrt{3}}{2}\rangle$
53. 248°   **55.** 63°   **57.** 50°   **59.** $|\mathbf{u}| = 3$; $\theta = 45°$
61. 1; 120°   **63.** 144.2°   **65.** 14.0°   **67.** 101.3°
69.

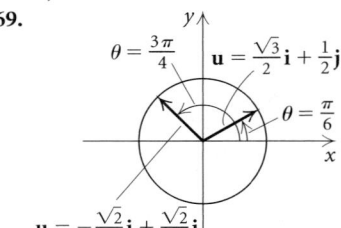

71.

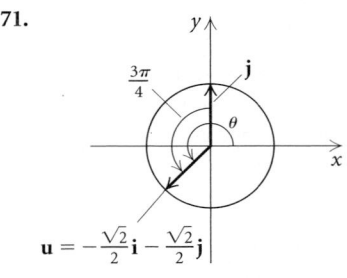

73. $\mathbf{u} = -\frac{\sqrt{10}}{10}\mathbf{i} + \frac{3\sqrt{10}}{10}\mathbf{j}$
75. $\sqrt{13}\left(\frac{2\sqrt{13}}{13}\mathbf{i} - \frac{3\sqrt{13}}{13}\mathbf{j}\right)$

77.

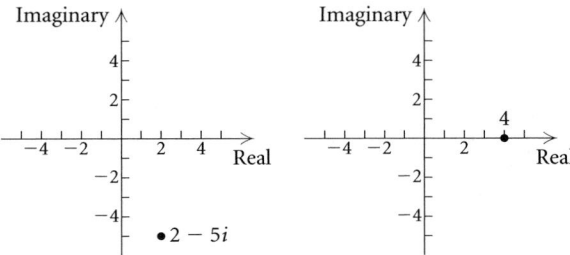

79. 174 nautical mi, S17°E   **81.** 60°
83. 500 lb on left, 866 lb on right
85. Cable: 224-lb tension; boom: 167-lb compression
87. $\mathbf{u} + \mathbf{v} = \langle u_1, u_2\rangle + \langle v_1, v_2\rangle$
$= \langle u_1 + v_1, u_2 + v_2\rangle$
$= \langle v_1 + u_1, v_2 + u_2\rangle$
$= \langle v_1, v_2\rangle + \langle u_1, u_2\rangle$
$= \mathbf{v} + \mathbf{u}$
89. Discussion and Writing
91. [1.3] $-\frac{1}{5}$; $(0, -15)$ **92.** [1.3] 0; $(0, 7)$
93. [4.1] 0, 4   **94.** [3.2] $-\frac{11}{3}, \frac{5}{2}$
95. (a) $\cos\theta = \dfrac{\mathbf{u}\cdot\mathbf{v}}{|\mathbf{u}||\mathbf{v}|} = \dfrac{0}{|\mathbf{u}||\mathbf{v}|}$, $\therefore \cos\theta = 0$ and $\theta = 90°$.
(b) Answers may vary. $\mathbf{u} = \langle 2, -3\rangle$ and $\mathbf{v} = \langle -3, -2\rangle$;
$\mathbf{u}\cdot\mathbf{v} = 2(-3) + (-3)(-2) = 0$
97. $\frac{3}{5}\mathbf{i} - \frac{4}{5}\mathbf{j}, -\frac{3}{5}\mathbf{i} + \frac{4}{5}\mathbf{j}$ **99.** (5, 8)

Review Exercises: Chapter 8

1. True   **2.** False   **3.** False   **4.** False   **5.** False   **6.** True
7. $A \approx 153°, B \approx 18°, C \approx 9°$
8. $A = 118°, a \approx 37$ in., $c \approx 24$ in.
9. $B = 14°50', a \approx 2523$ m, $c \approx 1827$ m
10. No solution   **11.** 33 m^2 **12.** 13.72 ft^2 **13.** 63 ft^2
14. 92°, 33°, 55°   **15.** 419 ft   **16.** About 650 km
17. $\sqrt{29}$;   **18.** 4;

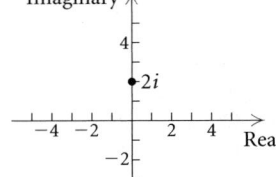

19. 2;   **20.** $\sqrt{10}$;

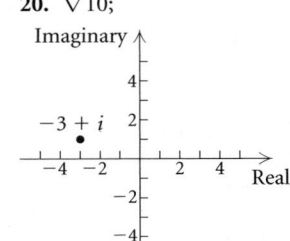

21. $\sqrt{2}\left(\cos \dfrac{\pi}{4} + i \sin \dfrac{\pi}{4}\right)$, or $\sqrt{2}(\cos 45° + i \sin 45°)$

22. $4\left(\cos \dfrac{3\pi}{2} + i \sin \dfrac{3\pi}{2}\right)$, or $4(\cos 270° + i \sin 270°)$

23. $10\left(\cos \dfrac{5\pi}{6} + i \sin \dfrac{5\pi}{6}\right)$, or $10(\cos 150° + i \sin 150°)$

24. $\frac{3}{4}(\cos 0 + i \sin 0)$, or $\frac{3}{4}(\cos 0° + i \sin 0°)$

25. $2 + 2\sqrt{3}i$ **26.** 7   **27.** $-\dfrac{5}{2} + \dfrac{5\sqrt{3}}{2}i$

28. $\sqrt{3} - i$ **29.** $1 + \sqrt{3} + \left(-1 + \sqrt{3}\right)i$

30. $-i$ **31.** $2i$ **32.** $3\sqrt{3} + 3i$

33. $8(\cos 180° + i \sin 180°)$

34. $4(\cos 7\pi + i \sin 7\pi)$ **35.** $-8i$

36. $-\dfrac{1}{2} - \dfrac{\sqrt{3}}{2}i$

37. $\sqrt[4]{2}\left(\cos \dfrac{3\pi}{8} + i \sin \dfrac{3\pi}{8}\right)$,

$\sqrt[4]{2}\left(\cos \dfrac{11\pi}{8} + i \sin \dfrac{11\pi}{8}\right)$

38. $\sqrt[3]{6}(\cos 110° + i \sin 110°)$,
$\sqrt[3]{6}(\cos 230° + i \sin 230°)$, $\sqrt[3]{6}(\cos 350° + i \sin 350°)$

39. $3, 3i, -3, -3i$

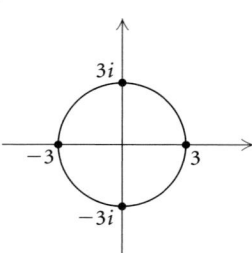

40. $1, \cos 72° + i \sin 72°, \cos 144° + i \sin 144°$,
$\cos 216° + i \sin 216°, \cos 288° + i \sin 288°$

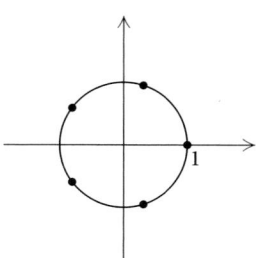

41. $\cos 22.5° + i \sin 22.5°, \cos 112.5° + i \sin 112.5°$,
$\cos 202.5° + i \sin 202.5°, \cos 292.5° + i \sin 292.5°$

42. $\dfrac{1}{2} + \dfrac{\sqrt{3}}{2}i, -1, \dfrac{1}{2} - \dfrac{\sqrt{3}}{2}i$

43. A: $(5, 120°), (5, 480°), (-5, 300°)$; B: $(3, 210°)$,
$(-3, 30°), (-3, 390°)$; C: $(4, 60°), (4, 420°), (-4, 240°)$;
D: $(1, 300°), (1, -60°), (-1, 120°)$; answers may vary

44. $(8, 135°), \left(8, \dfrac{3\pi}{4}\right)$ **45.** $(5, 270°), \left(5, \dfrac{3\pi}{2}\right)$

46. $(5.385, 111.8°), (5.385, 1.951)$
47. $(4.964, 147.8°), (4.964, 2.579)$
48. $\left(\dfrac{3\sqrt{2}}{2}, \dfrac{3\sqrt{2}}{2}\right)$ **49.** $\left(3, 3\sqrt{3}\right)$
50. $(1.93, -0.52)$ **51.** $(-1.86, -1.35)$
52. $r(5 \cos \theta - 2 \sin \theta) = 6$ **53.** $r \sin \theta = 3$
54. $r = 3$ **55.** $r^2 \sin^2 \theta - 4r \cos \theta - 16 = 0$
56. $x^2 + y^2 = 36$ **57.** $x^2 + 2y = 1$
58. $y^2 - 6x = 9$ **59.** $x^2 - 2x + y^2 - 3y = 0$
60. (b)   **61.** (d)   **62.** (a)   **63.** (c)
64. $13.7, 71°$ **65.** $98.7, 15°$
66.   **67.**

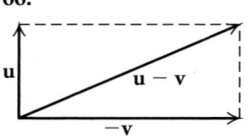 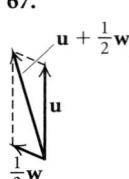

68. 666.7 N, $36°$ **69.** 29 km/h, $149°$
70. 102.4 nautical mi, S43°E   **71.** $\langle -4, 3 \rangle$
72. $\langle 2, -6 \rangle$ **73.** $\sqrt{61}$ **74.** $\langle 10, -21 \rangle$
75. $\langle 14, -64 \rangle$ **76.** $5 + \sqrt{116}$ **77.** 14
78. $\left\langle -\dfrac{3}{\sqrt{10}}, -\dfrac{1}{\sqrt{10}} \right\rangle$ **79.** $-9\mathbf{i} + 4\mathbf{j}$
80. $194.0°$ **81.** $\sqrt{34}; \theta = 211.0°$
82. $111.8°$ **83.** $85.1°$ **84.** $34\mathbf{i} - 55\mathbf{j}$
85. $\mathbf{i} - 12\mathbf{j}$ **86.** $5\sqrt{2}$
87. $3\sqrt{65} + \sqrt{109}$ **88.** $-5\mathbf{i} + 5\mathbf{j}$
89.   **90.**

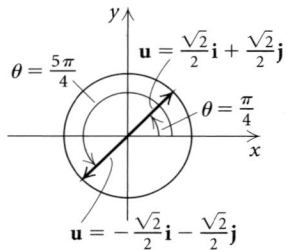 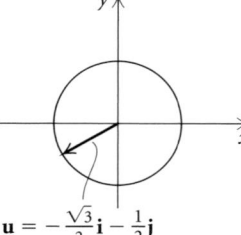

91. $\sqrt{10}\left(\dfrac{3\sqrt{10}}{10}\mathbf{i} - \dfrac{\sqrt{10}}{10}\mathbf{j}\right)$

92. D   **93.** A   **94.** D

95. Discussion and Writing: A nonzero complex number has n different complex nth roots. Thus, 1 has three different complex cube roots, one of which is the real number 1. The other two are complex conjugates. Since the set of reals is a subset of the set of complex numbers, the real cube root of 1 is also a complex root of 1.

96. Discussion and Writing: A triangle has no solution when a sine or cosine value found is less than -1 or greater than 1. A triangle also has no solution if the sum of the angle measures calculated is greater than 180°. A triangle has only one solution if only one possible answer is found, or if one of the possible answers has an angle sum greater than 180°. A triangle has two solutions when two possible answers are found and neither results in an angle sum greater than 180°. **97.** Discussion and Writing: For 150-yd shot, about 13.1 yd; for 300-yd shot, about 26.2 yd **98.** $\frac{36}{13}\mathbf{i} + \frac{15}{13}\mathbf{j}$ **99.** 50.52°, 129.48°

Test: Chapter 8

1. [8.1] $A = 83°$, $b \approx 14.7$ ft, $c \approx 12.4$ ft
2. [8.1] $A \approx 73.9°$, $B \approx 70.1°$, $a \approx 8.2$ m, or $A \approx 34.1°$, $B \approx 109.9°$, $a \approx 4.8$ m
3. [8.2] $A \approx 99.9°$, $B \approx 36.8°$, $C \approx 43.3°$
4. [8.1] About 43.6 cm² **5.** [8.1] About 77 m
6. [8.5] About 930 km
7. [8.3]

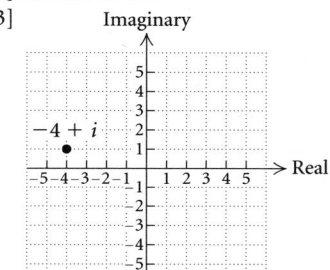

8. [8.3] $\sqrt{13}$

9. [8.3] $3\sqrt{2}(\cos 315° + i \sin 315°)$, or
$$3\sqrt{2}\left(\cos \frac{7\pi}{4} + i \sin \frac{7\pi}{4}\right)$$

10. [8.3] $\frac{1}{4}i$ **11.** [8.3] 16
12. [8.4] $2(\cos 120° + i \sin 120°)$
13. [8.4] $\left(\frac{1}{2}, -\frac{\sqrt{3}}{2}\right)$ **14.** [8.4] $r = \sqrt{10}$
15. [8.4]

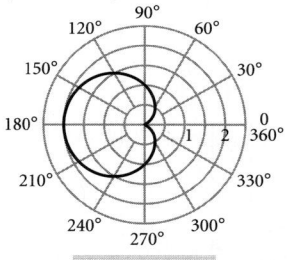

$r = 1 - \cos \theta$

16. [8.5] Magnitude: 11.2; direction: 23.4°
17. [8.6] $-11\mathbf{i} - 17\mathbf{j}$ **18.** [8.6] $-\frac{4}{5}\mathbf{i} + \frac{3}{5}\mathbf{j}$
19. [8.4] A   **20.** [8.1] 28.9°, 151.1°

CHAPTER 9

Visualizing the Graph

1. C   **2.** G   **3.** D   **4.** J   **5.** A   **6.** F   **7.** I
8. B   **9.** H   **10.** E

Exercise Set 9.1

1. (c)   **3.** (f)   **5.** (b)   **7.** $(-1, 3)$ **9.** $(-1, 1)$
11. No solution   **13.** $(-2, 4)$ **15.** Infinitely many solutions; $\left(x, \frac{x-1}{2}\right)$, or $(2y + 1, y)$ **17.** $(5, 4)$
19. $(1, -3)$ **21.** $(2, -2)$ **23.** No solution
25. $\left(\frac{39}{11}, -\frac{1}{11}\right)$ **27.** $(1, -1)$ **29.** $\left(\frac{1}{2}, \frac{3}{4}\right)$
31. Infinitely many solutions; $(x, 3x - 5)$ or $\left(\frac{1}{3}y + \frac{5}{3}, y\right)$
33. $(1, 3)$; consistent, independent
35. $(-4, -2)$; consistent, independent
37. Infinitely many solutions; $(4y + 2, y)$ or $\left(x, \frac{1}{4}x - \frac{1}{2}\right)$; consistent, dependent
39. $(1, 1)$; consistent, independent
41. $(-3, 0)$; consistent, independent
43. $(10, 8)$; consistent, independent
45. True   **47.** False   **49.** True
51. Skiing: 144,400 injuries; snowboarding: 144,000 injuries
53. Second Street: 7972; Main Street: 7712
55. Museums: 24 million visitors; website: 118 million visitors
57. Free rentals: 10; popcorn: 38   **59.** $(15, \$100)$ **61.** 140
63. 6000   **65.** 1.5 servings of spaghetti, 2 servings of lettuce
67. Boat: 20 km/h; stream: 3 km/h
69. 4%: \$6000; 5%: \$9000
71. 6 lb of French roast, 4 lb of Kenyan
73. Plane: 550 mph; wind: 50 mph
75. (a) $r(x) = -0.2020408163x + 113.9112245$; $p(x) = 1.162244898x + 62.08265306$; **(b)** 38 yr after 1995
77. Discussion and Writing
79. [1.5] Hardback: 35 million books; paperback: 75 million books **80.** [1.5] About \$2.8 billion **81.** [3.2] $-2, 6$
82. [3.2] $-1, 5$ **83.** [1.2] 15 **84.** [3.2] 1, 3 **85.** 4 km
87. First train: 36 km/h; second train: 54 km/h
89. $A = \frac{1}{10}$, $B = -\frac{7}{10}$ **91.** City: 294 mi; highway: 153 mi

Exercise Set 9.2

1. $(3, -2, 1)$ **3.** $(-3, 2, 1)$ **5.** $\left(2, \frac{1}{2}, -2\right)$
7. No solution   **9.** Infinitely many solutions; $\left(\frac{11y + 19}{5}, y, \frac{9y + 11}{5}\right)$
11. $\left(\frac{1}{2}, \frac{2}{3}, -\frac{5}{6}\right)$ **13.** $(-1, 4, 3)$ **15.** $(1, -2, 4, -1)$
17. Fish: 149 million; cats: 91 million; dogs: 74 million
19. Brewed coffee: 80 mg; Red Bull: 80 mg; Mountain Dew: 37 mg   **21.** Lettuce: 1 g; asparagus: 3 g; tomato: 8 g
23. Under 10 lb: 60 packages; 10 lb up to 15 lb: 70 packages; 15 lb or more: 20 packages   **25.** $1\frac{1}{4}$ servings of beef, 1 baked potato, $\frac{3}{4}$ serving of strawberries   **27.** 3%: \$1300; 4%: \$900; 6%: \$2800   **29.** Orange juice: \$1.60; bagel: \$2.25; coffee: \$1.50
31. Par-3: 4 holes; par-4: 10 holes; par-5: 4 holes
33. (a) $f(x) = \frac{3}{16}x^2 + \frac{9}{40}x + \frac{42}{5}$; **(b)** \$16.5 billion
35. (a) $f(x) = -0.425x^2 + 5.95x + 179$, or $-\frac{17}{40}x^2 + \frac{119}{20}x + 179$; **(b)** about 148 thousand marriages

37. (a) $f(x) = 0.1822373078x^2 - 11.23256978x + 468.7226133$; **(b)** about 892 morning newspapers
39. Discussion and Writing   **41.** [1.4] Perpendicular
42. [4.1] The leading-term test   **43.** [1.2] A vertical line
44. [5.1] A one-to-one function
45. [4.5] A rational function   **46.** [2.5] Inverse variation
47. [4.5] A vertical asymptote
48. [4.5] A horizontal asymptote
49. $\left(-1, \frac{1}{5}, -\frac{1}{2}\right)$ **51.** $180°$ **53.** $3x + 4y + 2z = 12$
55. $y = -4x^3 + 5x^2 - 3x + 1$
57. Adults: 5; students: 1; children: 94

Exercise Set 9.3

1. 3×2 **3.** 1×4 **5.** 3×3 **7.** $\begin{bmatrix} 2 & -1 & | & 7 \\ 1 & 4 & | & -5 \end{bmatrix}$

9. $\begin{bmatrix} 1 & -2 & 3 & | & 12 \\ 2 & 0 & -4 & | & 8 \\ 0 & 3 & 1 & | & 7 \end{bmatrix}$

11. $3x - 5y = 1$,
   $x + 4y = -2$
13. $2x + y - 4z = 12$,
   $3x \quad + 5z = -1$,
   $x - y + z = 2$
15. $\left(\frac{3}{2}, \frac{5}{2}\right)$ **17.** $\left(-\frac{63}{29}, -\frac{114}{29}\right)$ **19.** $\left(-1, \frac{5}{2}\right)$
21. $(0, 3)$ **23.** No solution   **25.** Infinitely many solutions; $(3y - 2, y)$ **27.** $(-1, 2, -2)$
29. $\left(\frac{3}{2}, -4, 3\right)$ **31.** $(-1, 6, 3)$
33. Infinitely many solutions; $\left(\frac{1}{2}z + \frac{1}{2}, -\frac{1}{2}z - \frac{1}{2}, z\right)$
35. Infinitely many solutions; $(r - 2, -2r + 3, r)$
37. No solution   **39.** $(1, -3, -2, -1)$ **41.** 1:00 A.M.
43. 8%: $8000; 10%: $12,000; 12%: $10,000
45. Discussion and Writing   **47.** [5.2] Exponential
48. [1.3] Linear   **49.** [4.5] Rational   **50.** [4.1] Quartic
51. [5.3] Logarithmic   **52.** [4.1] Cubic   **53.** [1.3] Linear
54. [3.2] Quadratic   **55.** $y = 3x^2 + \frac{5}{2}x - \frac{15}{2}$
57. $\begin{bmatrix} 1 & 5 \\ 0 & 1 \end{bmatrix}, \begin{bmatrix} 1 & 0 \\ 0 & 1 \end{bmatrix}$ **59.** $\left(-\frac{4}{3}, -\frac{1}{3}, 1\right)$
61. Infinitely many solutions; $\left(-\frac{14}{13}z - 1, \frac{3}{13}z - 2, z\right)$
63. $(-3, 3)$

Exercise Set 9.4

1. $x = -3, y = 5$ **3.** $x = -1, y = 1$
5. $\begin{bmatrix} -2 & 7 \\ 6 & 2 \end{bmatrix}$ **7.** $\begin{bmatrix} 1 & 3 \\ 2 & 6 \end{bmatrix}$ **9.** $\begin{bmatrix} 9 & 9 \\ -3 & -3 \end{bmatrix}$
11. $\begin{bmatrix} 11 & 13 \\ 5 & 3 \end{bmatrix}$ **13.** $\begin{bmatrix} -4 & 3 \\ -2 & -4 \end{bmatrix}$ **15.** $\begin{bmatrix} 17 & 9 \\ -2 & 1 \end{bmatrix}$
17. $\begin{bmatrix} 0 & 0 \\ 0 & 0 \end{bmatrix}$ **19.** $\begin{bmatrix} 1 & 2 \\ 4 & 3 \end{bmatrix}$ **21.** $\begin{bmatrix} 1 \\ 40 \end{bmatrix}$

23. $\begin{bmatrix} -10 & 28 \\ 14 & -26 \\ 0 & -6 \end{bmatrix}$ **25.** Not defined

27. $\begin{bmatrix} 3 & 16 & 3 \\ 0 & -32 & 0 \\ -6 & 4 & 5 \end{bmatrix}$

29. (a) $\begin{bmatrix} 300 & 80 & 40 \end{bmatrix}$; **(b)** $\begin{bmatrix} 315 & 84 & 42 \end{bmatrix}$;
(c) $\begin{bmatrix} 615 & 164 & 82 \end{bmatrix}$; the total budget for each area in June and July
31. (a) $C = \begin{bmatrix} 140 & 27 & 3 & 13 & 64 \end{bmatrix}$,
$P = \begin{bmatrix} 180 & 4 & 11 & 24 & 662 \end{bmatrix}$, $B = \begin{bmatrix} 50 & 5 & 1 & 82 & 20 \end{bmatrix}$;
(b) $\begin{bmatrix} 650 & 50 & 28 & 307 & 1448 \end{bmatrix}$, the total nutritional value of a meal of 1 serving of chicken, 1 cup of potato salad, and 3 broccoli spears

33. (a) $\begin{bmatrix} 1.03 & 0.15 & 0.26 & 0.23 & 0.27 \\ 1.10 & 0.14 & 0.24 & 0.21 & 0.25 \\ 1.06 & 0.22 & 0.31 & 0.28 & 0.34 \\ 1.21 & 0.20 & 0.29 & 0.33 & 0.31 \end{bmatrix}$;

(b) $\begin{bmatrix} 65 & 48 & 93 & 57 \end{bmatrix}$;
(c) $\begin{bmatrix} 287.30 & 48.33 & 73.78 & 69.88 & 78.84 \end{bmatrix}$;
(d) the total cost, in dollars, for each item for the day's meals

35. (a) $\begin{bmatrix} 8 & 15 \\ 6 & 10 \\ 4 & 3 \end{bmatrix}$; **(b)** $\begin{bmatrix} 3 & 1.50 & 2 \end{bmatrix}$; **(c)** $\begin{bmatrix} 41 & 66 \end{bmatrix}$;
(d) the total cost, in dollars, of ingredients for each coffee shop
37. (a) $\begin{bmatrix} 6 & 4.50 & 5.20 \end{bmatrix}$; **(b)** $PS = \begin{bmatrix} 95.80 & 150.60 \end{bmatrix}$
39. $\begin{bmatrix} 2 & -3 \\ 1 & 5 \end{bmatrix}\begin{bmatrix} x \\ y \end{bmatrix} = \begin{bmatrix} 7 \\ -6 \end{bmatrix}$

41. $\begin{bmatrix} 1 & 1 & -2 \\ 3 & -1 & 1 \\ 2 & 5 & -3 \end{bmatrix}\begin{bmatrix} x \\ y \\ z \end{bmatrix} = \begin{bmatrix} 6 \\ 7 \\ 8 \end{bmatrix}$

43. $\begin{bmatrix} 3 & -2 & 4 \\ 2 & 1 & -5 \end{bmatrix}\begin{bmatrix} x \\ y \\ z \end{bmatrix} = \begin{bmatrix} 17 \\ 13 \end{bmatrix}$

45. $\begin{bmatrix} -4 & 1 & -1 & 2 \\ 1 & 2 & -1 & -1 \\ -1 & 1 & 4 & -3 \\ 2 & 3 & 5 & -7 \end{bmatrix}\begin{bmatrix} w \\ x \\ y \\ z \end{bmatrix} = \begin{bmatrix} 12 \\ 0 \\ 1 \\ 9 \end{bmatrix}$

47. Discussion and Writing
49. [3.3] **(a)** $\left(\frac{1}{2}, -\frac{25}{4}\right)$; **(b)** $x = \frac{1}{2}$; **(c)** minimum: $-\frac{25}{4}$;
(d)

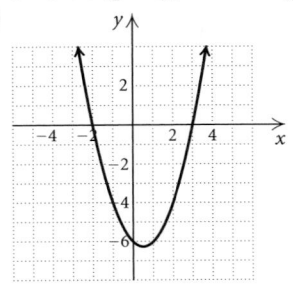

$f(x) = x^2 - x - 6$

50. [3.3] **(a)** $\left(\frac{5}{4}, -\frac{49}{8}\right)$; **(b)** $x = \frac{5}{4}$; **(c)** minimum: $-\frac{49}{8}$;
(d)

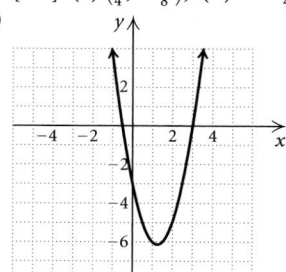

$$f(x) = 2x^2 - 5x - 3$$

51. [3.3] **(a)** $\left(-\frac{3}{2}, \frac{17}{4}\right)$; **(b)** $x = -\frac{3}{2}$; **(c)** maximum: $\frac{17}{4}$;
(d)

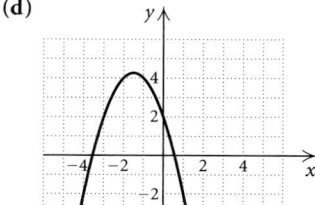

$$f(x) = -x^2 - 3x + 2$$

52. [3.3] **(a)** $\left(\frac{2}{3}, \frac{16}{3}\right)$; **(b)** $x = \frac{2}{3}$; **(c)** maximum: $\frac{16}{3}$;
(d)

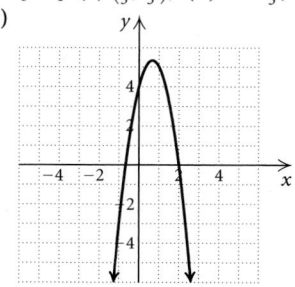

$$f(x) = -3x^2 + 4x + 4$$

53. $(\mathbf{A} + \mathbf{B})(\mathbf{A} - \mathbf{B}) = \begin{bmatrix} -2 & 1 \\ 2 & -1 \end{bmatrix}$; $\mathbf{A}^2 - \mathbf{B}^2 = \begin{bmatrix} 0 & 3 \\ 0 & -3 \end{bmatrix}$

55. $(\mathbf{A} + \mathbf{B})(\mathbf{A} - \mathbf{B}) = \begin{bmatrix} -2 & 1 \\ 2 & -1 \end{bmatrix}$

$$= \mathbf{A}^2 + \mathbf{BA} - \mathbf{AB} - \mathbf{B}^2$$

57. $\mathbf{A} + \mathbf{B} =$

$$\begin{bmatrix} a_{11} + b_{11} & a_{12} + b_{12} & a_{13} + b_{13} & \cdots & a_{1n} + b_{1n} \\ a_{21} + b_{21} & a_{22} + b_{22} & a_{23} + b_{23} & \cdots & a_{2n} + b_{2n} \\ a_{31} + b_{31} & a_{32} + b_{32} & a_{33} + b_{33} & \cdots & a_{3n} + b_{3n} \\ \vdots & \vdots & \vdots & & \vdots \\ a_{m1} + b_{m1} & a_{m2} + b_{m2} & a_{m3} + b_{m3} & \cdots & a_{mn} + b_{mn} \end{bmatrix}$$
$$=$$
$$\begin{bmatrix} b_{11} + a_{11} & b_{12} + a_{12} & b_{13} + a_{13} & \cdots & b_{1n} + a_{1n} \\ b_{21} + a_{21} & b_{22} + a_{22} & b_{23} + a_{23} & \cdots & b_{2n} + a_{2n} \\ b_{31} + a_{31} & b_{32} + a_{32} & b_{33} + a_{33} & \cdots & b_{3n} + a_{3n} \\ \vdots & \vdots & \vdots & & \vdots \\ b_{m1} + a_{m1} & b_{m2} + a_{m2} & b_{m3} + a_{m3} & \cdots & b_{mn} + a_{mn} \end{bmatrix}$$
$$= \mathbf{B} + \mathbf{A}$$

59. $(kl)\mathbf{A} = \begin{bmatrix} (kl)a_{11} & (kl)a_{12} & (kl)a_{13} & \cdots & (kl)a_{1n} \\ (kl)a_{21} & (kl)a_{22} & (kl)a_{23} & \cdots & (kl)a_{2n} \\ (kl)a_{31} & (kl)a_{32} & (kl)a_{33} & \cdots & (kl)a_{3n} \\ \vdots & \vdots & \vdots & & \vdots \\ (kl)a_{m1} & (kl)a_{m2} & (kl)a_{m3} & \cdots & (kl)a_{mn} \end{bmatrix}$

$$= \begin{bmatrix} k(la_{11}) & k(la_{12}) & k(la_{13}) & \cdots & k(la_{1n}) \\ k(la_{21}) & k(la_{22}) & k(la_{23}) & \cdots & k(la_{2n}) \\ k(la_{31}) & k(la_{32}) & k(la_{33}) & \cdots & k(la_{3n}) \\ \vdots & \vdots & \vdots & & \vdots \\ k(la_{m1}) & k(la_{m2}) & k(la_{m3}) & \cdots & k(la_{mn}) \end{bmatrix}$$

$$= k \begin{bmatrix} la_{11} & la_{12} & la_{13} & \cdots & la_{1n} \\ la_{21} & la_{22} & la_{23} & \cdots & la_{2n} \\ la_{31} & la_{32} & la_{33} & \cdots & la_{3n} \\ \vdots & \vdots & \vdots & & \vdots \\ la_{m1} & la_{m2} & la_{m3} & \cdots & la_{mn} \end{bmatrix}$$

$$= k(l\mathbf{A})$$

61. $(k + l)\mathbf{A} =$

$$\begin{bmatrix} (k+l)a_{11} & (k+l)a_{12} & (k+l)a_{13} & \cdots & (k+l)a_{1n} \\ (k+l)a_{21} & (k+l)a_{22} & (k+l)a_{23} & \cdots & (k+l)a_{2n} \\ (k+l)a_{31} & (k+l)a_{32} & (k+l)a_{33} & \cdots & (k+l)a_{3n} \\ \vdots & \vdots & \vdots & & \vdots \\ (k+l)a_{m1} & (k+l)a_{m2} & (k+l)a_{m3} & \cdots & (k+l)a_{mn} \end{bmatrix}$$
$$=$$
$$\begin{bmatrix} ka_{11} + la_{11} & ka_{12} + la_{12} & ka_{13} + la_{13} & \cdots & ka_{1n} + la_{1n} \\ ka_{21} + la_{21} & ka_{22} + la_{22} & ka_{23} + la_{23} & \cdots & ka_{2n} + la_{2n} \\ ka_{31} + la_{31} & ka_{32} + la_{32} & ka_{33} + la_{33} & \cdots & ka_{3n} + la_{3n} \\ \vdots & \vdots & \vdots & & \vdots \\ ka_{m1} + la_{m1} & ka_{m2} + la_{m2} & ka_{m3} + la_{m3} & \cdots & ka_{mn} + la_{mn} \end{bmatrix}$$
$$= k\mathbf{A} + l\mathbf{A}$$

Exercise Set 9.5

1. Yes   **3.** No   **5.** $\begin{bmatrix} -3 & 2 \\ 5 & -3 \end{bmatrix}$ **7.** Does not exist

9. $\begin{bmatrix} \frac{3}{8} & -\frac{1}{4} & \frac{1}{8} \\ -\frac{1}{8} & \frac{3}{4} & -\frac{3}{8} \\ -\frac{1}{4} & \frac{1}{2} & \frac{1}{4} \end{bmatrix}$ **11.** Does not exist

13. $\begin{bmatrix} 0.4 & -0.6 \\ 0.2 & -0.8 \end{bmatrix}$ **15.** $\begin{bmatrix} -1 & -1 & -6 \\ 1 & 0 & 2 \\ 0 & 1 & 3 \end{bmatrix}$

17. $\begin{bmatrix} 1 & 1 & 2 \\ 1 & 1 & 1 \\ 2 & 3 & 4 \end{bmatrix}$ **19.** Does not exist

21. $\begin{bmatrix} 1 & -2 & 3 & 8 \\ 0 & 1 & -3 & 1 \\ 0 & 0 & 1 & -2 \\ 0 & 0 & 0 & -1 \end{bmatrix}$

23. $\begin{bmatrix} 0.25 & 0.25 & 1.25 & -0.25 \\ 0.5 & 1.25 & 1.75 & -1 \\ -0.25 & -0.25 & -0.75 & 0.75 \\ 0.25 & 0.5 & 0.75 & -0.5 \end{bmatrix}$

25. $(-23, 83)$ **27.** $(-1, 5, 1)$ **29.** $(2, -2)$ **31.** $(0, 2)$
33. $(3, -3, -2)$ **35.** $(-1, 0, 1)$ **37.** $(1, -1, 0, 1)$
39. Sausages: 50; hot dogs: 95   **41.** Topsoil: \$239; mulch:
\$179; pea gravel: \$222   **43.** Discussion and Writing
45. [4.3] -48 **46.** [4.3] 194   **47.** [3.2] $\dfrac{-1 \pm \sqrt{57}}{4}$
48. [3.4] $-3, -2$ **49.** [3.4] 4
50. [3.4] 9   **51.** [4.3] $(x + 2)(x - 1)(x - 4)$
52. [4.3] $(x + 5)(x + 1)(x - 1)(x - 3)$
53. $\mathbf{A}^{-1}$ exists if and only if $x \neq 0$. $\mathbf{A}^{-1} = \begin{bmatrix} \frac{1}{x} \end{bmatrix}$

55. $\mathbf{A}^{-1}$ exists if and only if $xyz \neq 0$. $\mathbf{A}^{-1} = \begin{bmatrix} 0 & 0 & \frac{1}{z} \\ 0 & \frac{1}{y} & 0 \\ \frac{1}{x} & 0 & 0 \end{bmatrix}$

Exercise Set 9.6

1. -14 **3.** -2 **5.** -11 **7.** $x^3 - 4x$
9. $M_{11} = 6, M_{32} = -9, M_{22} = -29$
11. $A_{11} = 6, A_{32} = 9, A_{22} = -29$
13. -10 **15.** -10 **17.** -10
19. $M_{41} = -14, M_{33} = 20$
21. $A_{24} = 15, A_{43} = 30$ **23.** 110   **25.** -109
27. $-x^4 + x^2 - 5x$ **29.** $\left(-\frac{25}{2}, -\frac{11}{2}\right)$ **31.** $(3, 1)$

33. $\left(\frac{1}{2}, -\frac{1}{3}\right)$ **35.** $(1, 1)$ **37.** $\left(\frac{3}{2}, \frac{13}{14}, \frac{33}{14}\right)$
39. $(3, -2, 1)$ **41.** $(1, 3, -2)$ **43.** $\left(\frac{1}{2}, \frac{2}{3}, -\frac{5}{6}\right)$
45. Discussion and Writing   **47.** [5.1] $f^{-1}(x) = \dfrac{x - 2}{3}$
48. [5.1] Not one-to-one   **49.** [5.1] Not one-to-one
50. [5.1] $f^{-1}(x) = (x - 1)^3$ **51.** [3.1] $5 - 3i$
52. [3.1] $6 - 2i$ **53.** [3.1] $10 - 10i$ **54.** [3.1] $\frac{9}{25} + \frac{13}{25}i$
55. ± 2 **57.** $\left(-\infty, -\sqrt{3}\right] \cup \left[\sqrt{3}, \infty\right)$ **59.** -34
61. 4   **63.** Answers may vary.   **65.** Answers may vary.

$\begin{vmatrix} L & -W \\ 2 & 2 \end{vmatrix}$ $\begin{vmatrix} a & b \\ -b & a \end{vmatrix}$

67. Answers may vary.
$\begin{vmatrix} 2\pi r & 2\pi r \\ -h & r \end{vmatrix}$

Exercise Set 9.7

1. (f)   **3.** (h)   **5.** (g)   **7.** (b)
9.   **11**

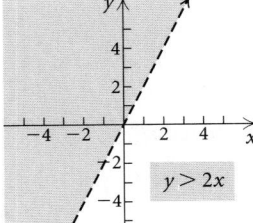
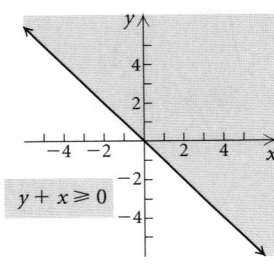

13.   **15.**

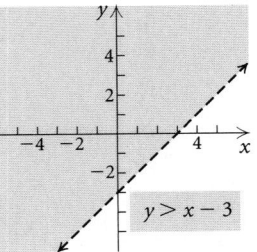
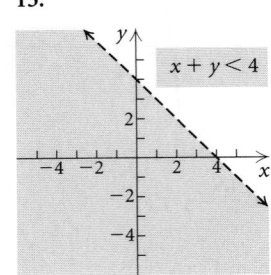

17.   **19.**

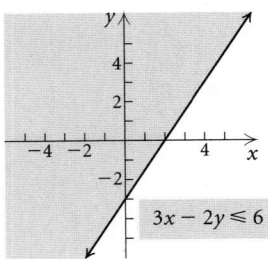
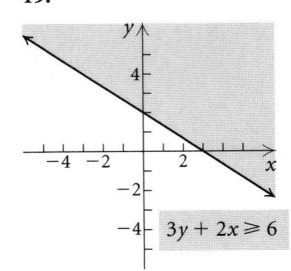

21.

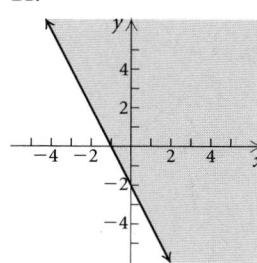

$$3x - 2 \leq 5x + y$$

23.

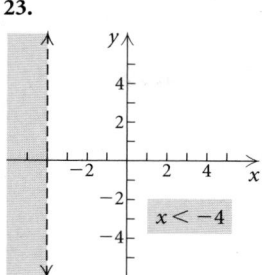

$$x < -4$$

25.

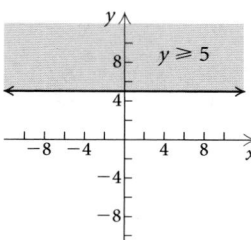

$$y \geq 5$$

27.

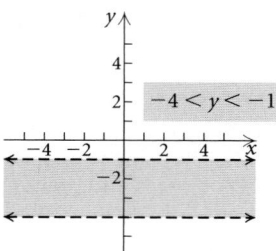

$$-4 < y < -1$$

29.

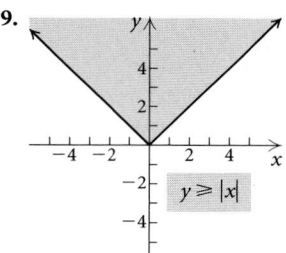

$$y \geq |x|$$

31. (f)   **33.** (a)   **35.** (b)

37. $y \leq -x + 4,$
$y \leq 3x$

39. $x < 2,$
$y > -1$

41. $y \leq -x + 3,$
$y \leq x + 1,$
$x \geq 0,$
$y \geq 0$

43.

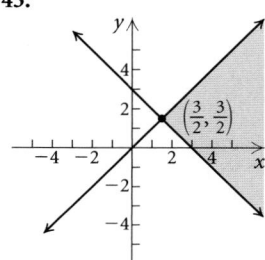

45.

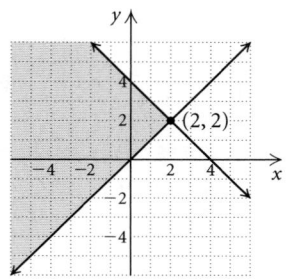

47.

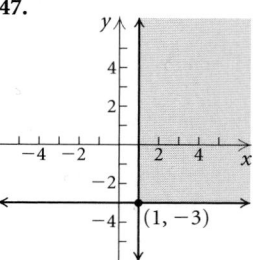

49.

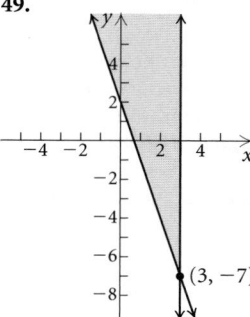

51.

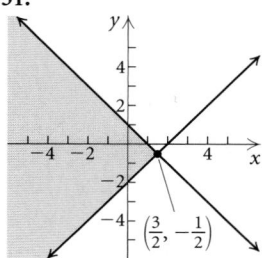

53.

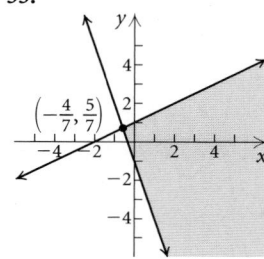

55.

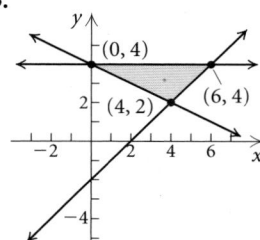

57.

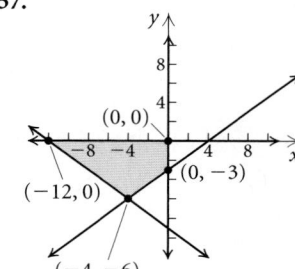

59.

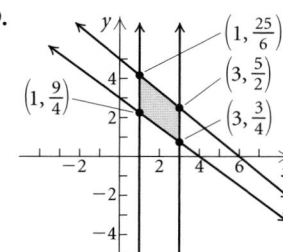

61. Maximum: 179 when $x = 7$ and $y = 0$; minimum: 48 when $x = 0$ and $y = 4$

63. Maximum: 216 when $x = 0$ and $y = 6$; minimum: 0 when $x = 0$ and $y = 0$

65. Maximum income of $18 is achieved when 100 of each type of biscuit are made.

67. Maximum profit of $11,000 is achieved by producing 100 units of lumber and 300 units of plywood.

69. Minimum cost of $36\frac{12}{13}$ is achieved by using $1\frac{11}{13}$ sacks of soybean meal and $1\frac{11}{13}$ sacks of oats.

71. Maximum income of $3110 is achieved when $22,000 is invested in corporate bonds and $18,000 is invested in municipal bonds.

73. Minimum cost of $460 thousand is achieved using 30 P_1's and 10 P_2's.
75. Maximum profit per day of $192 is achieved when 2 knit suits and 4 worsted suits are made.
77. Minimum weekly cost of $19.05 is achieved when 1.5 lb of meat and 3 lb of cheese are used.
79. Maximum total number of 800 is achieved when there are 550 of A and 250 of B.   **81.** Discussion and Writing
83. [1.6] $\{x \mid -7 \le x < 2\}$, or $[-7, 2)$
84. [3.5] $\{x \mid x \le 1 \text{ or } x \ge 5\}$, or $(-\infty, 1] \cup [5, \infty)$
85. [4.6] $\{x \mid -1 \le x \le 3\}$, or $[-1, 3]$
86. [4.6] $\{x \mid -3 < x < -2\}$, or $(-3, -2)$
87.   **89.**

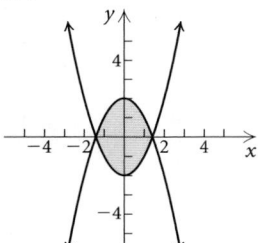   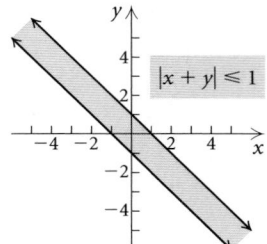

91.

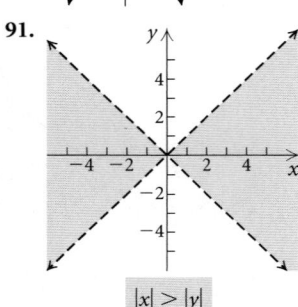

93. Maximum income of $28,500 is achieved by making 30 less expensive assemblies and 30 more expensive assemblies.

Exercise Set 9.8

1. $\dfrac{2}{x-3} - \dfrac{1}{x+2}$ **3.** $\dfrac{5}{2x-1} - \dfrac{4}{3x-1}$

5. $\dfrac{2}{x+2} - \dfrac{3}{x-2} + \dfrac{4}{x+1}$

7. $-\dfrac{3}{(x+2)^2} - \dfrac{1}{x+2} + \dfrac{1}{x-1}$ **9.** $\dfrac{3}{x-1} - \dfrac{4}{2x-1}$

11. $x - 2 + \dfrac{\frac{17}{16}}{x+1} - \dfrac{\frac{11}{4}}{(x+1)^2} - \dfrac{\frac{17}{16}}{x-3}$

13. $\dfrac{3x+5}{x^2+2} - \dfrac{4}{x-1}$ **15.** $\dfrac{3}{2x-1} - \dfrac{2}{x+2} + \dfrac{10}{(x+2)^2}$

17. $3x + 1 + \dfrac{2}{2x-1} + \dfrac{3}{x+1}$

19. $-\dfrac{1}{x-3} + \dfrac{3x}{x^2+2x-5}$

21. $\dfrac{5}{3x+5} - \dfrac{3}{x+1} + \dfrac{4}{(x+1)^2}$ **23.** $\dfrac{8}{4x-5} + \dfrac{3}{3x+2}$

25. $\dfrac{2x-5}{3x^2+1} - \dfrac{2}{x-2}$ **27.** Discussion and Writing
29. Discussion and Writing   **30.** [4.4] 3, $\pm i$
31. [4.4] $-2, \dfrac{1 \pm \sqrt{5}}{2}$ **32.** [4.4] $-2, 3, \pm i$
33. [4.4] $-3, -1 \pm \sqrt{2}$

35. $-\dfrac{\frac{1}{2a^2}x}{x^2+a^2} + \dfrac{\frac{1}{4a^2}}{x-a} + \dfrac{\frac{1}{4a^2}}{x+a}$

37. $-\dfrac{3}{25(\ln x + 2)} + \dfrac{3}{25(\ln x - 3)} + \dfrac{7}{5(\ln x - 3)^2}$

Review Exercises: Chapter 9

1. True   **2.** False   **3.** True   **4.** False   **5.** (a)   **6.** (e)
7. (h)   **8.** (d)   **9.** (b)   **10.** (g)   **11.** (c)   **12.** (f)
13. $(-2, -2)$ **14.** $(-5, 4)$ **15.** No solution
16. Infinitely many solutions; $(-y - 2, y)$, or $(x, -x - 2)$
17. $(3, -1, -2)$ **18.** No solution   **19.** $(-5, 13, 8, 2)$
20. Consistent: 13, 14, 16, 17, 19; the others are inconsistent.
21. Dependent: 16; the others are independent.   **22.** $(1, 2)$
23. $(-3, 4, -2)$ **24.** Infinitely many solutions;
$\left(\dfrac{z}{2}, -\dfrac{z}{2}, z\right)$ **25.** $(-4, 1, -2, 3)$
26. Nickels: 31; dimes: 44   **27.** 3%: $1600; 3.5%: $3400
28. 1 bagel, $\frac{1}{2}$ serving of cream cheese, 2 bananas
29. 75, 69, 82
30. (a) $f(x) = 16x^2 - 38x + 167$;
(b) 197 thousand trademarks

31. $\begin{bmatrix} 0 & -1 & 6 \\ 3 & 1 & -2 \\ -2 & 1 & -2 \end{bmatrix}$ **32.** $\begin{bmatrix} -3 & 3 & 0 \\ -6 & -9 & 6 \\ 6 & 0 & -3 \end{bmatrix}$

33. $\begin{bmatrix} -1 & 1 & 0 \\ -2 & -3 & 2 \\ 2 & 0 & -1 \end{bmatrix}$ **34.** $\begin{bmatrix} -2 & 2 & 6 \\ 1 & -8 & 18 \\ 2 & 1 & -15 \end{bmatrix}$

35. Not defined   **36.** $\begin{bmatrix} 2 & -1 & -6 \\ 1 & 5 & -2 \\ -2 & -1 & 4 \end{bmatrix}$

37. $\begin{bmatrix} -13 & 1 & 6 \\ -3 & -7 & 4 \\ 8 & 3 & -5 \end{bmatrix}$ **38.** $\begin{bmatrix} -2 & -1 & 18 \\ 5 & -3 & -2 \\ -2 & 3 & -8 \end{bmatrix}$

39. (a) $\begin{bmatrix} 0.98 & 0.23 & 0.30 & 0.28 & 0.45 \\ 1.03 & 0.19 & 0.27 & 0.34 & 0.41 \\ 1.01 & 0.21 & 0.35 & 0.31 & 0.39 \\ 0.99 & 0.25 & 0.29 & 0.33 & 0.42 \end{bmatrix}$;

(b) $[32 \quad 19 \quad 43 \quad 38]$;
(c) $[131.98 \quad 29.50 \quad 40.80 \quad 41.29 \quad 54.92]$;
(d) the total cost, in dollars, for each item for the day's meals

40. $\begin{bmatrix} -\frac{1}{2} & 0 \\ \frac{1}{6} & \frac{1}{3} \end{bmatrix}$ **41.** $\begin{bmatrix} 0 & 0 & \frac{1}{4} \\ 0 & -\frac{1}{2} & 0 \\ \frac{1}{3} & 0 & 0 \end{bmatrix}$

42. $\begin{bmatrix} 1 & 0 & 0 & 0 \\ 0 & \frac{1}{9} & \frac{5}{18} & 0 \\ 0 & -\frac{1}{9} & \frac{2}{9} & 0 \\ 0 & 0 & 0 & 1 \end{bmatrix}$

43. $\begin{bmatrix} 3 & -2 & 4 \\ 1 & 5 & -3 \\ 2 & -3 & 7 \end{bmatrix} \begin{bmatrix} x \\ y \\ z \end{bmatrix} = \begin{bmatrix} 13 \\ 7 \\ -8 \end{bmatrix}$ **44.** $(-8, 7)$

45. $(1, -2, 5)$ **46.** $(2, -1, 1, -3)$ **47.** 10   **48.** -18
49. -6 **50.** -1 **51.** $(3, -2)$ **52.** $(-1, 5)$
53. $\left(\frac{3}{2}, \frac{13}{14}, \frac{33}{14}\right)$ **54.** $(2, -1, 3)$

55.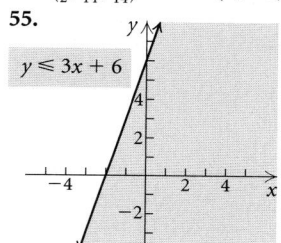
$y \leq 3x + 6$

56.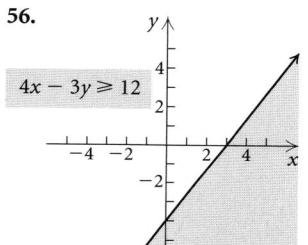
$4x - 3y \geq 12$

57.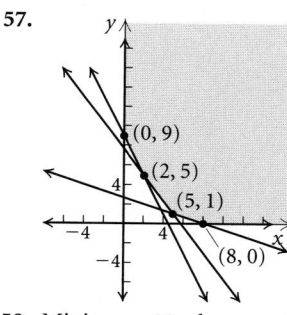
$(0, 9)$
$(2, 5)$
$(5, 1)$
$(8, 0)$

58. Minimum: 52 when $x = 2$ and $y = 4$; maximum: 92 when $x = 2$ and $y = 8$
59. Maximum score of 96 is achieved when 0 group A questions and 8 group B questions are answered.
60. $\dfrac{5}{x+1} - \dfrac{5}{x+2} - \dfrac{5}{(x+2)^2}$ **61.** $\dfrac{2}{2x-3} - \dfrac{5}{x+4}$
62. C   **63.** A   **64.** B
65. Discussion and Writing: During a holiday season, a caterer sold a total of 55 seafood trays and cheese trays. She sold 15 more seafood trays than cheese trays. How many of each were sold?
66. Discussion and Writing: In general, $(\mathbf{AB})^2 \neq \mathbf{A}^2\mathbf{B}^2$. $(\mathbf{AB})^2 = \mathbf{ABAB}$ and $\mathbf{A}^2\mathbf{B}^2 = \mathbf{AABB}$. Since matrix multiplication is not commutative, $\mathbf{BA} \neq \mathbf{AB}$, so $(\mathbf{AB})^2 \neq \mathbf{A}^2\mathbf{B}^2$.
67. 4%: \$10,000; 5%: \$12,000; $5\frac{1}{2}$%: \$18,000
68. $\left(\frac{5}{18}, \frac{1}{7}\right)$ **69.** $\left(1, \frac{1}{2}, \frac{1}{3}\right)$

70.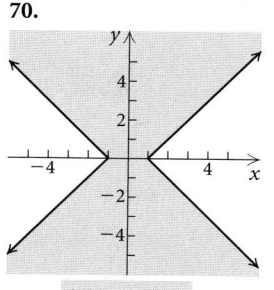
$|x| - |y| \leq 1$

71.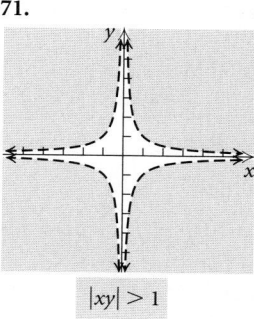
$|xy| > 1$

Test: Chapter 9

1. [9.1] $(-3, 5)$; consistent, independent
2. [9.1] Infinitely many solutions; $(x, 2x - 3)$ or $\left(\dfrac{y+3}{2}, y\right)$; consistent, dependent
3. [9.1] No solution; inconsistent, independent
4. [9.1] $(1, -2)$; consistent, independent
5. [9.2] $(-1, 3, 2)$ **6.** [9.1] Student: 342 tickets; nonstudent: 408 tickets
7. [9.2] Tricia: 120 orders; Maria: 104 orders; Antonio: 128 orders
8. [9.4] $\begin{bmatrix} -2 & -3 \\ -3 & 4 \end{bmatrix}$ **9.** [9.4] Not defined
10. [9.4] $\begin{bmatrix} -7 & -13 \\ 5 & -1 \end{bmatrix}$ **11.** [9.4] Not defined
12. [9.4] $\begin{bmatrix} 2 & -2 & 6 \\ -4 & 10 & 4 \end{bmatrix}$ **13.** [9.5] $\begin{bmatrix} 0 & -1 \\ -\frac{1}{4} & -\frac{3}{4} \end{bmatrix}$
14. [9.4] **(a)** $\begin{bmatrix} 0.95 & 0.40 & 0.39 \\ 1.10 & 0.35 & 0.41 \\ 1.05 & 0.39 & 0.36 \end{bmatrix}$; **(b)** $[26 \quad 18 \quad 23]$;
(c) $[68.65 \quad 25.67 \quad 25.80]$; **(d)** the total cost, in dollars, for each type of menu item served on the given day
15. [9.4] $\begin{bmatrix} 3 & -4 & 2 \\ 2 & 3 & 1 \\ 1 & -5 & -3 \end{bmatrix} \begin{bmatrix} x \\ y \\ z \end{bmatrix} = \begin{bmatrix} -8 \\ 7 \\ 3 \end{bmatrix}$
16. [9.5] $(-2, 1, 1)$
17. [9.6] 61   **18.** [9.6] -33 **19.** [9.6] $\left(-\frac{1}{2}, \frac{3}{4}\right)$
20. [9.7]
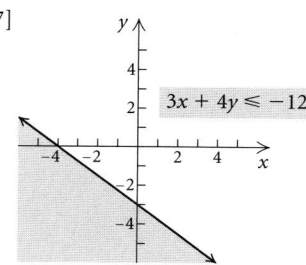
$3x + 4y \leq -12$

21. [9.7] Maximum: 15 when $x = 3$ and $y = 3$; minimum: 2 when $x = 1$ and $y = 0$
22. [9.7] Maximum profit of \$375 occurs when 25 pound cakes and 75 carrot cakes are prepared.
23. [9.8] $-\dfrac{2}{x-1} + \dfrac{5}{x+3}$ **24.** [9.7] D
25. [9.2] $A = 1, B = -3, C = 2$

CHAPTER 10

Exercise Set 10.1

1. (f)   **3.** (b)   **5.** (d)
7. $V: (0, 0)$; $F: (0, 5)$; $D: y = -5$

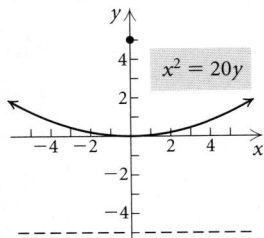

9. $V: (0, 0)$; $F: \left(-\frac{3}{2}, 0\right)$; $D: x = \frac{3}{2}$

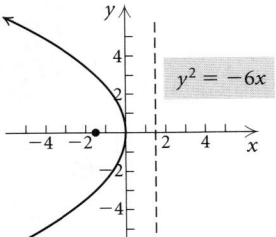

11. $V: (0, 0)$; $F: (0, 1)$; $D: y = -1$

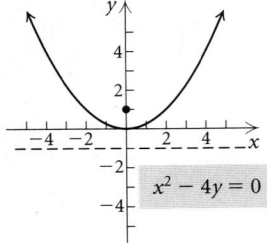

13. $V: (0, 0)$; $F: \left(\frac{1}{8}, 0\right)$; $D: x = -\frac{1}{8}$

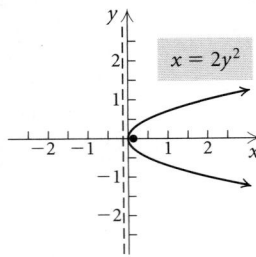

15. $y^2 = 16x$ **17.** $x^2 = -4\pi y$
19. $(y - 2)^2 = 14\left(x + \frac{1}{2}\right)$
21. $V: (-2, 1)$; $F: \left(-2, -\frac{1}{2}\right)$; $D: y = \frac{5}{2}$

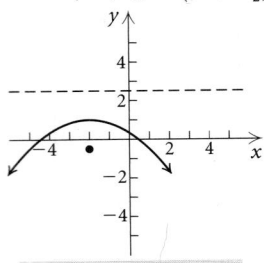

$(x + 2)^2 = -6(y - 1)$

23. $V: (-1, -3)$; $F: \left(-1, -\frac{7}{2}\right)$; $D: y = -\frac{5}{2}$

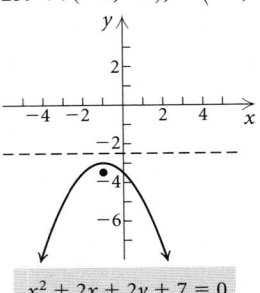

$x^2 + 2x + 2y + 7 = 0$

25. $V: (0, -2)$; $F: \left(0, -1\frac{3}{4}\right)$; $D: y = -2\frac{1}{4}$

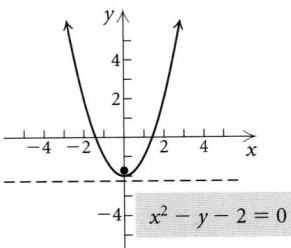

$x^2 - y - 2 = 0$

27. $V: (-2, -1)$; $F: \left(-2, -\frac{3}{4}\right)$; $D: y = -1\frac{1}{4}$

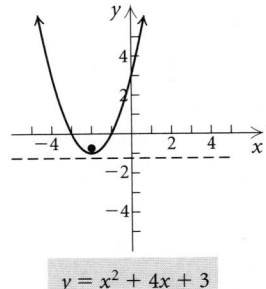

$y = x^2 + 4x + 3$

29. $V: \left(5\frac{3}{4}, \frac{1}{2}\right)$; $F: \left(6, \frac{1}{2}\right)$; $D: x = 5\frac{1}{2}$

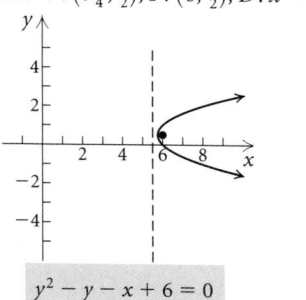

$$y^2 - y - x + 6 = 0$$

31. (a) $y^2 = 16x$; **(b)** $3\frac{33}{64}$ ft   **33.** $\frac{2}{3}$ ft, or 8 in.
35. Discussion and Writing   **37.** [1.1] (h)
38. [1.1], [1.4] (d)   **39.** [1.3] (a), (b), (f), (g)
40. [1.3] (b)   **41.** [1.4] (b)   **42.** [1.1] (f)
43. [1.4] (a) and (g)   **44.** [1.4] (a) and (h); (g) and (h);
(b) and (c)   **45.** $(x + 1)^2 = -4(y - 2)$
47. $V: (0.867, 0.348)$; $F: (0.867, -0.190)$; $D: y = 0.887$
49. 10 ft, 11.6 ft, 16.4 ft, 24.4 ft, 35.6 ft, 50 ft

Exercise Set 10.2

1. (b)   **3.** (d)   **5.** (a)
7. $(7, -2)$; 8   **9.** $(-3, 1)$; 4

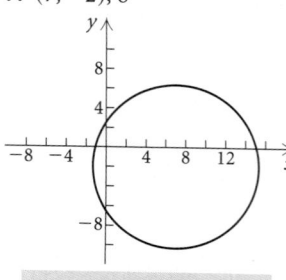

$$x^2 + y^2 - 14x + 4y = 11$$

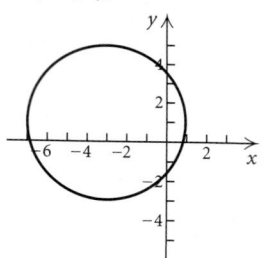

$$x^2 + y^2 + 6x - 2y = 6$$

11. $(-2, 3)$; 5   **13.** $(3, 4)$; 3

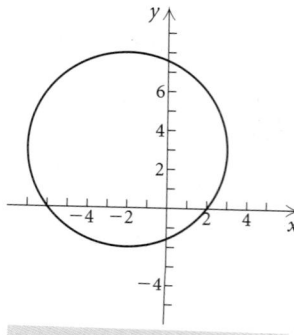

$$x^2 + y^2 + 4x - 6y - 12 = 0$$

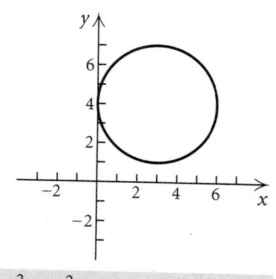

$$x^2 + y^2 - 6x - 8y + 16 = 0$$

15. $(-3, 5)$; $\sqrt{34}$ **17.** $\left(\frac{9}{2}, -2\right)$; $\frac{5\sqrt{5}}{2}$

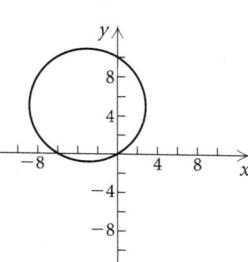

$$x^2 + y^2 + 6x - 10y = 0$$

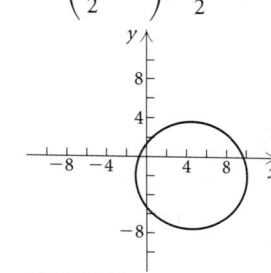

$$x^2 + y^2 - 9x = 7 - 4y$$

19. (c)   **21.** (d)
23. $V: (2, 0), (-2, 0)$;   **25.** $V: (0, 4), (0, -4)$;
   $F: \left(\sqrt{3}, 0\right), \left(-\sqrt{3}, 0\right)$ $F: \left(0, \sqrt{7}\right), \left(0, -\sqrt{7}\right)$

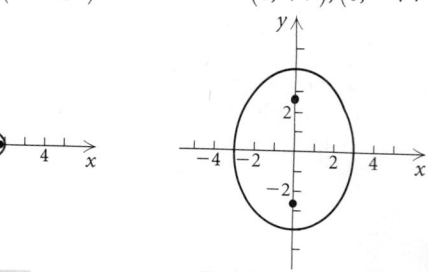

$$\frac{x^2}{4} + \frac{y^2}{1} = 1 \qquad\qquad 16x^2 + 9y^2 = 144$$

27. $V: \left(-\sqrt{3}, 0\right), \left(\sqrt{3}, 0\right)$; $F: (-1, 0), (1, 0)$

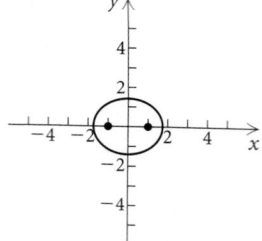

$$2x^2 + 3y^2 = 6$$

29. $V: \left(-\frac{1}{2}, 0\right), \left(\frac{1}{2}, 0\right)$; $F: \left(-\frac{\sqrt{5}}{6}, 0\right), \left(\frac{\sqrt{5}}{6}, 0\right)$

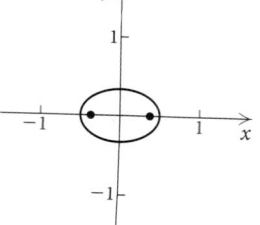

$$4x^2 + 9y^2 = 1$$

31. $\dfrac{x^2}{49} + \dfrac{y^2}{40} = 1$ **33.** $\dfrac{x^2}{25} + \dfrac{y^2}{64} = 1$ **35.** $\dfrac{x^2}{9} + \dfrac{y^2}{5} = 1$

37. $C{:}\,(1,2);\ V{:}\,(4,2),(-2,2);\ F{:}\,\left(1+\sqrt{5},2\right),\left(1-\sqrt{5},2\right)$

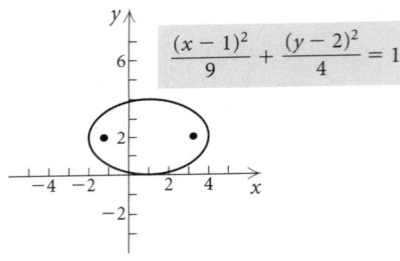

$\dfrac{(x-1)^2}{9} + \dfrac{(y-2)^2}{4} = 1$

39. $C{:}\,(-3,5);\ V{:}\,(-3,11),(-3,-1);\ F{:}\,\left(-3,5+\sqrt{11}\right),$
$\left(-3,5-\sqrt{11}\right)$

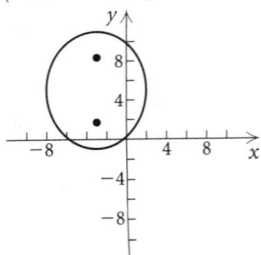

$\dfrac{(x+3)^2}{25} + \dfrac{(y-5)^2}{36} = 1$

41. $C{:}\,(-2,1);\ V{:}\,(-10,1),(6,1);\ F{:}\,(-6,1),(2,1)$

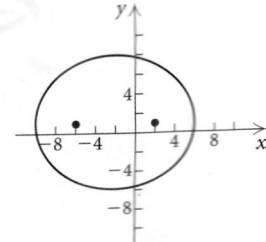

$3(x+2)^2 + 4(y-1)^2 = 192$

43. $C{:}\,(2,-1);\ V{:}\,(-1,-1),(5,-1);\ F{:}\,\left(2+\sqrt{5},-1\right),$
$\left(2-\sqrt{5},-1\right)$

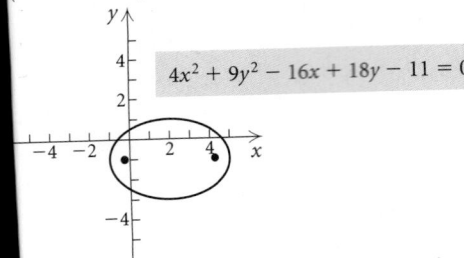

$4x^2 + 9y^2 - 16x + 18y - 11 = 0$

45. $C{:}\,(1,1);\ V{:}\,(1,3),(1,-1);\ F{:}\,\left(1,1+\sqrt{3}\right),\left(1,1-\sqrt{3}\right)$

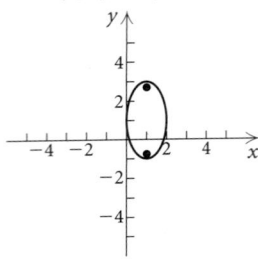

$4x^2 + y^2 - 8x - 2y + 1 = 0$

47. Example 2; $\dfrac{3}{5} < \dfrac{\sqrt{12}}{4}$ **49.** $\dfrac{x^2}{15} + \dfrac{y^2}{16} = 1$

51. $\dfrac{x^2}{2500} + \dfrac{y^2}{144} = 1$ **53.** 2×10^6 mi

55. Discussion and Writing   **57.** [1.1] Midpoint
58. [1.5] Zero   **59.** [1.1] y-intercept
60. [3.2] Two different real-number solutions
61. [4.3] Remainder   **62.** [10.2] Ellipse
63. [10.1] Parabola   **64.** [10.2] Circle
65. $\dfrac{(x-3)^2}{4} + \dfrac{(y-1)^2}{25} = 1$

67. $\dfrac{x^2}{9} + \dfrac{y^2}{484/5} = 1$ **69.** $C{:}\,(2.003,-1.005);$
$V{:}\,(-1.017,-1.005),(5.023,-1.005)$ **71.** About 9.1 ft

Exercise Set 10.3

1. (b)   **3.** (c)   **5.** (a)   **7.** $\dfrac{y^2}{9} - \dfrac{x^2}{16} = 1$

9. $\dfrac{x^2}{4} - \dfrac{y^2}{9} = 1$

11. $C{:}\,(0,0);\ V{:}\,(2,0),(-2,0);\ F{:}\,\left(2\sqrt{2},0\right),\left(-2\sqrt{2},0\right);$
$A{:}\ y=x,\ y=-x$

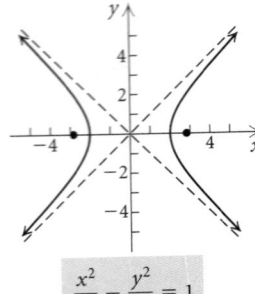

$\dfrac{x^2}{4} - \dfrac{y^2}{4} = 1$

13. $C: (2, -5)$; $V: (-1, -5), (5, -5)$; $F: \left(2 - \sqrt{10}, -5\right)$, $\left(2 + \sqrt{10}, -5\right)$; $A: y = -\dfrac{x}{3} - \dfrac{13}{3}, y = \dfrac{x}{3} - \dfrac{17}{3}$

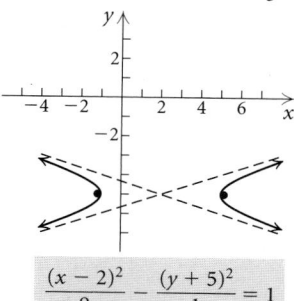

$$\dfrac{(x-2)^2}{9} - \dfrac{(y+5)^2}{1} = 1$$

15. $C: (-1, -3)$; $V: (-1, -1), (-1, -5)$; $F: \left(-1, -3 + 2\sqrt{5}\right), \left(-1, -3 - 2\sqrt{5}\right)$; $A: y = \tfrac{1}{2}x - \tfrac{5}{2}, y = -\tfrac{1}{2}x - \tfrac{7}{2}$

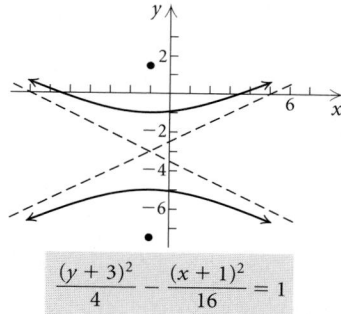

$$\dfrac{(y+3)^2}{4} - \dfrac{(x+1)^2}{16} = 1$$

17. $C: (0,0)$; $V: (-2,0), (2,0)$; $F: \left(-\sqrt{5},0\right), \left(\sqrt{5},0\right)$; $A: y = -\tfrac{1}{2}x, y = \tfrac{1}{2}x$

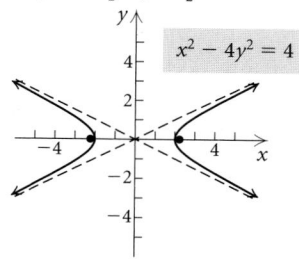

$x^2 - 4y^2 = 4$

19. $C: (0,0)$; $V: (0,-3), (0,3)$; $F: \left(0, -3\sqrt{10}\right), \left(0, 3\sqrt{10}\right)$; $A: y = \tfrac{1}{3}x, y = -\tfrac{1}{3}x$

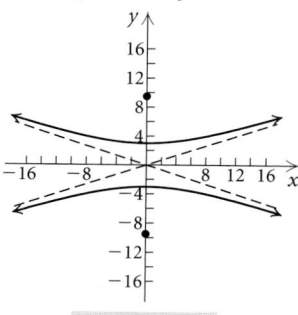

$9y^2 - x^2 = 81$

21. $C: (0,0)$; $V: \left(-\sqrt{2},0\right), \left(\sqrt{2},0\right)$; $F: (-2,0), (2,0)$; $A: y = x, y = -x$

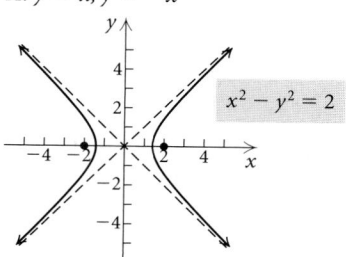

$x^2 - y^2 = 2$

23. $C: (0,0)$; $V: \left(0, -\tfrac{1}{2}\right), \left(0, \tfrac{1}{2}\right)$; $F: \left(0, -\dfrac{\sqrt{2}}{2}\right), \left(0, \dfrac{\sqrt{2}}{2}\right)$; $A: y = x, y = -x$

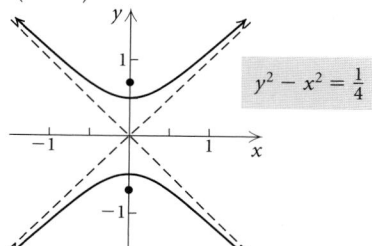

$y^2 - x^2 = \tfrac{1}{4}$

25. $C: (1, -2)$; $V: (0, -2), (2, -2)$; $F: \left(1 - \sqrt{2}, -2\right), \left(1 + \sqrt{2}, -2\right)$; $A: y = -x - 1, y = x - 3$

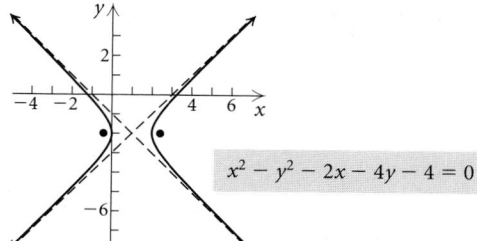

$x^2 - y^2 - 2x - 4y - 4 = 0$

27. $C: \left(\tfrac{1}{3}, 3\right)$; $V: \left(-\tfrac{2}{3}, 3\right), \left(\tfrac{4}{3}, 3\right)$; $F: \left(\tfrac{1}{3} - \sqrt{37}, 3\right), \left(\tfrac{1}{3} + \sqrt{37}, 3\right)$; $A: y = 6x + 1, y = -6x + 5$

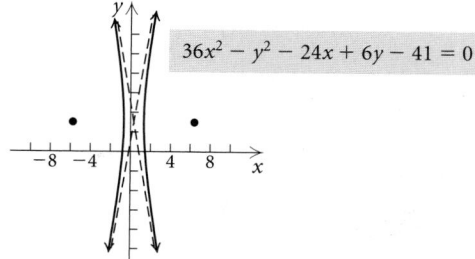

$36x^2 - y^2 - 24x + 6y - 41 = 0$

29. $C: (3, 1)$; $V: (3, 3), (3, -1)$; $F: \left(3, 1 + \sqrt{13}\right)$, $\left(3, 1 - \sqrt{13}\right)$; $A: y = \frac{2}{3}x - 1, y = -\frac{2}{3}x + 3$

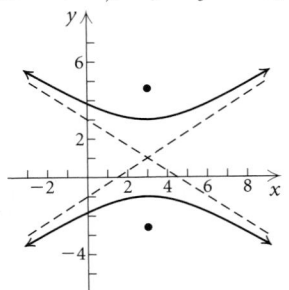

$$9y^2 - 4x^2 - 18y + 24x - 63 = 0$$

31. $C: (1, -2)$; $V: (2, -2), (0, -2)$; $F: \left(1 + \sqrt{2}, -2\right)$, $\left(1 - \sqrt{2}, -2\right)$; $A: y = x - 3, y = -x - 1$

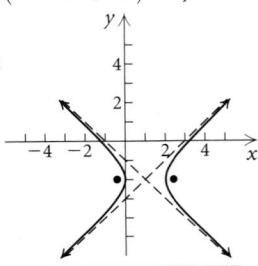

$$x^2 - y^2 - 2x - 4y = 4$$

33. $C: (-3, 4)$; $V: (-3, 10), (-3, -2)$; $F: \left(-3, 4 + 6\sqrt{2}\right)$, $\left(-3, 4 - 6\sqrt{2}\right)$; $A: y = x + 7, y = -x + 1$

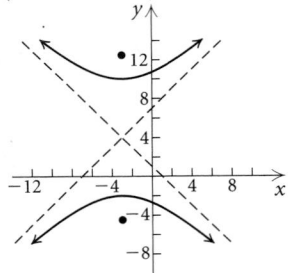

$$y^2 - x^2 - 6x - 8y - 29 = 0$$

35. Example 3; $\dfrac{\sqrt{5}}{1} > \dfrac{5}{4}$ **37.** $\dfrac{x^2}{9} - \dfrac{(y - 7)^2}{16} = 1$

39. $\dfrac{y^2}{25} - \dfrac{x^2}{11} = 1$ **41.** Discussion and Writing

43. [5.1] **(a)** Yes; **(b)** $f^{-1}(x) = \dfrac{x + 3}{2}$

44. [5.1] **(a)** Yes; **(b)** $f^{-1}(x) = \sqrt[3]{x - 2}$

45. [5.1] **(a)** Yes; **(b)** $f^{-1}(x) = \dfrac{5}{x} + 1$, or $\dfrac{5 + x}{x}$

46. [5.1] **(a)** Yes; **(b)** $f^{-1}(x) = x^2 - 4, x \geq 0$

47. [9.1], [9.3], [9.5], [9.6] $(6, -1)$

48. [9.1], [9.3], [9.5], [9.6] $(1, -1)$

49. [9.1], [9.3], [9.5], [9.6] $(2, -1)$

50. [9.1], [9.3], [9.5], [9.6] $(-3, 4)$

51. $\dfrac{(y + 5)^2}{9} - (x - 3)^2 = 1$

53. $C: (-1.460, -0.957)$; $V: (-2.360, -0.957)$, $(-0.560, -0.957)$; $A: y = -1.20x - 2.70, y = 1.20x + 0.79$

55. $\dfrac{x^2}{345.96} - \dfrac{y^2}{22,154.04} = 1$

Visualizing the Graph

1. B   **2.** J   **3.** F   **4.** I   **5.** H   **6.** G   **7.** E
8. D   **9.** C   **10.** A

Exercise Set 10.4

1. (e)   **3.** (c)   **5.** (b)   **7.** $(-4, -3), (3, 4)$
9. $(0, 2), (3, 0)$ **11.** $(-5, 0), (4, 3), (4, -3)$
13. $(3, 0), (-3, 0)$ **15.** $(0, -3), (4, 5)$ **17.** $(-2, 1)$
19. $(3, 4), (-3, -4), (4, 3), (-4, -3)$
21. $\left(\dfrac{6\sqrt{21}}{7}, \dfrac{4i\sqrt{35}}{7}\right), \left(\dfrac{6\sqrt{21}}{7}, -\dfrac{4i\sqrt{35}}{7}\right),$
$\left(-\dfrac{6\sqrt{21}}{7}, \dfrac{4i\sqrt{35}}{7}\right), \left(-\dfrac{6\sqrt{21}}{7}, -\dfrac{4i\sqrt{35}}{7}\right)$
23. $(3, 2), \left(4, \frac{3}{2}\right)$
25. $\left(\dfrac{5 + \sqrt{70}}{3}, \dfrac{-1 + \sqrt{70}}{3}\right), \left(\dfrac{5 - \sqrt{70}}{3}, \dfrac{-1 - \sqrt{70}}{3}\right)$
27. $\left(\sqrt{2}, \sqrt{14}\right), \left(-\sqrt{2}, \sqrt{14}\right), \left(\sqrt{2}, -\sqrt{14}\right), \left(-\sqrt{2}, -\sqrt{14}\right)$
29. $(1, 2), (-1, -2), (2, 1), (-2, -1)$
31. $\left(\dfrac{15 + \sqrt{561}}{8}, \dfrac{11 - 3\sqrt{561}}{8}\right),$
$\left(\dfrac{15 - \sqrt{561}}{8}, \dfrac{11 + 3\sqrt{561}}{8}\right)$
33. $\left(\dfrac{7 - \sqrt{33}}{2}, \dfrac{7 + \sqrt{33}}{2}\right), \left(\dfrac{7 + \sqrt{33}}{2}, \dfrac{7 - \sqrt{33}}{2}\right)$
35. $(3, 2), (-3, -2), (2, 3), (-2, -3)$
37. $\left(\dfrac{5 - 9\sqrt{15}}{20}, \dfrac{-45 + 3\sqrt{15}}{20}\right),$
$\left(\dfrac{5 + 9\sqrt{15}}{20}, \dfrac{-45 - 3\sqrt{15}}{20}\right)$
39. $(3, -5), (-1, 3)$ **41.** $(8, 5), (-5, -8)$ **43.** $(3, 2),$
$(-3, -2)$ **45.** $(2, 1), (-2, -1), (1, 2), (-1, -2)$
47. $\left(4 + \dfrac{3i\sqrt{6}}{2}, -4 + \dfrac{3i\sqrt{6}}{2}\right), \left(4 - \dfrac{3i\sqrt{6}}{2}, -4 - \dfrac{3i\sqrt{6}}{2}\right)$
49. $\left(3, \sqrt{5}\right), \left(-3, -\sqrt{5}\right), \left(\sqrt{5}, 3\right), \left(-\sqrt{5}, -3\right)$
51. $\left(\dfrac{8\sqrt{5}}{5}i, \dfrac{3\sqrt{105}}{5}\right), \left(\dfrac{8\sqrt{5}}{5}i, -\dfrac{3\sqrt{105}}{5}\right),$
$\left(-\dfrac{8\sqrt{5}}{5}i, \dfrac{3\sqrt{105}}{5}\right), \left(-\dfrac{8\sqrt{5}}{5}i, -\dfrac{3\sqrt{105}}{5}\right)$
53. $(2, 1), (-2, -1), \left(-i\sqrt{5}, \dfrac{2i\sqrt{5}}{5}\right), \left(i\sqrt{5}, -\dfrac{2i\sqrt{5}}{5}\right)$

55. True   **57.** True   **59.** 6 cm by 8 cm
61. Length: $\sqrt{3}$ m; width: 1 m   **63.** 30 yd by 75 yd
65. 4 in. by 5 in.   **67.** 16 ft, 24 ft   **69.** (b)
71. (d)   **73.** (a)
75.

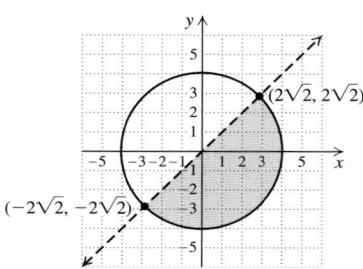

77.   **79.**

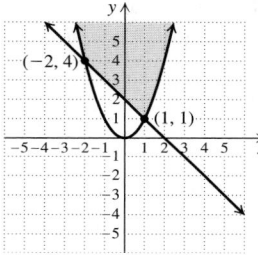   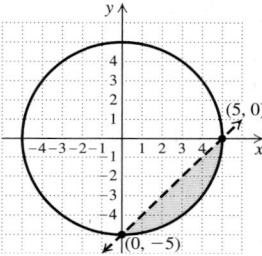

81.   **83.**

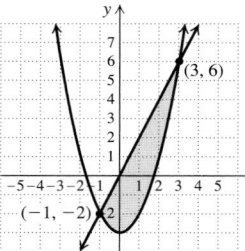   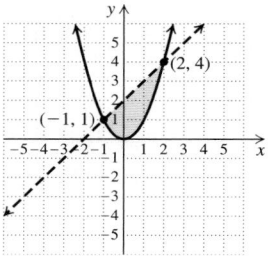

85. Discussion and Writing   **87.** [5.5] 2   **88.** [5.5] 2.048
89. [5.5] 81   **90.** [5.5] 5   **91.** $(x-2)^2 + (y-3)^2 = 1$
93. $\dfrac{x^2}{4} + y^2 = 1$ **95.** $(x-1)^2 + (y-2)^2 = 25$

97. There is no number x such that $\dfrac{x^2}{a^2} - \dfrac{\left(\frac{b}{a}x\right)^2}{b^2} = 1$,

because the left side simplifies to $\dfrac{x^2}{a^2} - \dfrac{x^2}{a^2}$, which is 0.
99. $\left(\frac{1}{2}, \frac{1}{4}\right), \left(\frac{1}{2}, -\frac{1}{4}\right), \left(-\frac{1}{2}, \frac{1}{4}\right), \left(-\frac{1}{2}, -\frac{1}{4}\right)$
101. Factor: $x^3 + y^3 = (x+y)(x^2 - xy + y^2)$. We know
that $x + y = 1$, so $(x+y)^2 = x^2 + 2xy + y^2 = 1$, or
$x^2 + y^2 = 1 - 2xy$. We also know that $xy = 1$, so
$x^2 + y^2 = 1 - 2 \cdot 1 = -1$. Then
$x^3 + y^3 = 1 \cdot (-1 - 1) = -2$.   **103.** $(2, 4), (4, 2)$
105. $(3, -2), (-3, 2), (2, -3), (-2, 3)$
107. $\left(\dfrac{2 \log 3 + 3 \log 5}{3(\log 3 \cdot \log 5)}, \dfrac{4 \log 3 - 3 \log 5}{3(\log 3 \cdot \log 5)}\right)$
109. $(1.564, 2.448), (0.138, 0.019)$

111. $(0.871, 1.388)$ **113.** $(1.146, 3.146), (-1.841, 0.159)$
115. $(0.965, 4402.33), (-0.965, -4402.33)$
117. $(2.112, -0.109), (-13.041, -13.337)$
119. $(400, 1.431), (-400, 1.431), (400, -1.431),$
$(-400, -1.431)$

Exercise Set 10.5

1. $(0, -2)$ **3.** $\left(1, \sqrt{3}\right)$ **5.** $\left(\sqrt{2}, 0\right)$ **7.** $\left(\sqrt{3}, 1\right)$
9. Ellipse or circle   **11.** Hyperbola   **13.** Parabola
15. Hyperbola   **17.** Ellipse or circle
19.

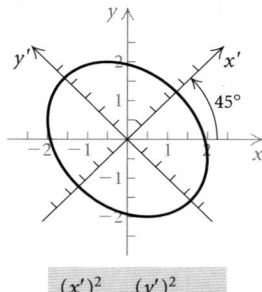

$$\frac{(x')^2}{3} + \frac{(y')^2}{5} = 1$$

21.

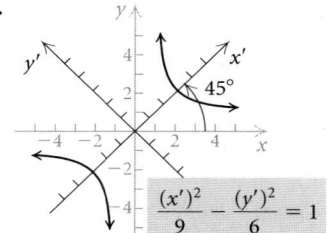

$$\frac{(x')^2}{9} - \frac{(y')^2}{6} = 1$$

23.

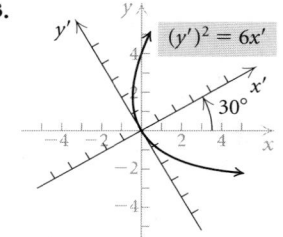

$$(y')^2 = 6x'$$

25.

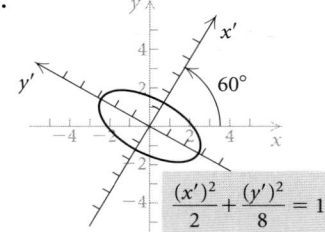

$$\frac{(x')^2}{2} + \frac{(y')^2}{8} = 1$$

27.

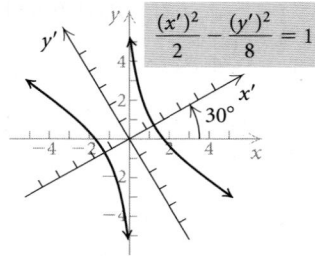

$$\frac{(x')^2}{2} - \frac{(y')^2}{8} = 1$$

29.

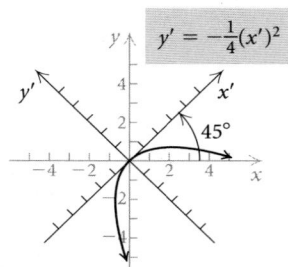

$$y' = -\frac{1}{4}(x')^2$$

31.

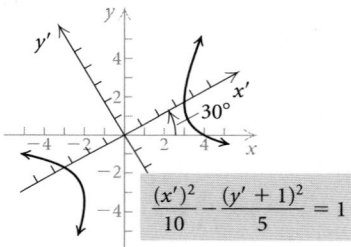

$$\frac{(x')^2}{10} - \frac{(y' + 1)^2}{5} = 1$$

33.

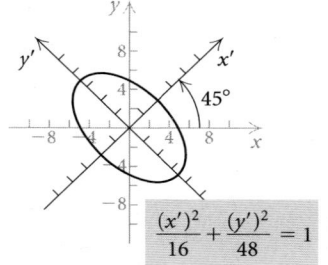

$$\frac{(x')^2}{16} + \frac{(y')^2}{48} = 1$$

35.

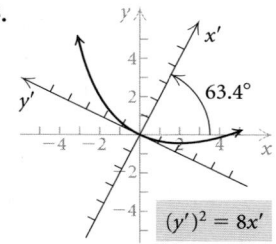

$$(y')^2 = 8x'$$

37.

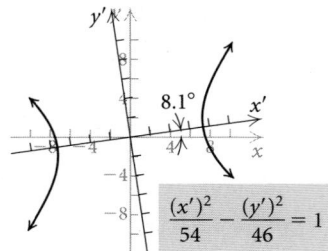

$$\frac{(x')^2}{54} - \frac{(y')^2}{46} = 1$$

39. Discussion and Writing   **41.** [6.4] $\dfrac{2\pi}{3}$

42. [6.4] $-\dfrac{7\pi}{4}$ **43.** [6.4] $60°$ **44.** [6.4] $135°$

45. $x = x' \cos \theta - y' \sin \theta, \ y = x' \sin \theta + y' \cos \theta$

47. $A' + C' = A \cos^2 \theta + B \sin \theta \cos \theta + C \sin^2 \theta$
$\qquad\qquad + A \sin^2 \theta - B \sin \theta \cos \theta + C \cos^2 \theta$
$\qquad = A(\sin^2 \theta + \cos^2 \theta) + C(\sin^2 \theta + \cos^2 \theta)$
$\qquad = A + C$

Exercise Set 10.6

1. (b)   **3.** (a)   **5.** (d)   **7.** (a) Parabola; (b) vertical, 1 unit to the right of the pole; (c) $\left(\frac{1}{2}, 0\right)$;
(d) $\qquad r = \dfrac{1}{1 + \cos \theta}$

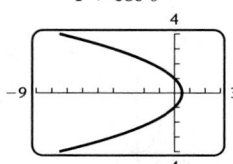

9. (a) Hyperbola; (b) horizontal, $\frac{3}{2}$ units below the pole;
(c) $\left(-3, \dfrac{\pi}{2}\right), \left(1, \dfrac{3\pi}{2}\right)$;
(d) $\qquad r = \dfrac{15}{5 - 10 \sin \theta}$

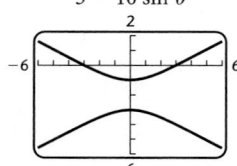

11. (a) Ellipse; (b) vertical, $\frac{8}{3}$ units to the left of the pole;
(c) $\left(\frac{8}{3}, 0\right), \left(\frac{8}{9}, \pi\right)$;
(d) $\qquad r = \dfrac{8}{6 - 3 \cos \theta}$

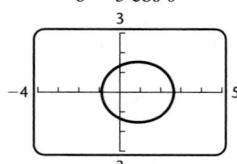

13. (a) Hyperbola; **(b)** horizontal, $\frac{4}{3}$ units above the pole;

(c) $\left(\frac{4}{5}, \frac{\pi}{2}\right), \left(-4, \frac{3\pi}{2}\right)$;

(d) $r = \dfrac{20}{10 + 15 \sin \theta}$

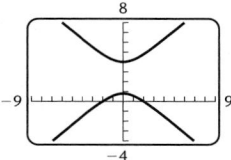

15. (a) Ellipse; **(b)** vertical, 3 units to the right of the pole;
(c) $(1, 0), (3, \pi)$;

(d) $r = \dfrac{9}{6 + 3 \cos \theta}$

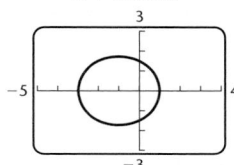

17. (a) Parabola; **(b)** horizontal, $\frac{3}{2}$ units below the pole;

(c) $\left(\frac{3}{4}, \frac{3\pi}{2}\right)$;

(d) $r = \dfrac{3}{2 - 2 \sin \theta}$

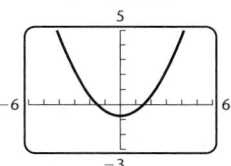

19. (a) Ellipse; **(b)** vertical, 4 units to the left of the pole;
(c) $(4, 0), \left(\frac{4}{3}, \pi\right)$;

(d) $r = \dfrac{4}{2 - \cos \theta}$

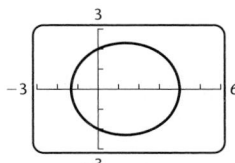

21. (a) Hyperbola; **(b)** horizontal, $\frac{7}{10}$ units above the pole;

(c) $\left(\frac{7}{12}, \frac{\pi}{2}\right), \left(-\frac{7}{8}, \frac{3\pi}{2}\right)$;

(d) $r = \dfrac{7}{2 + 10 \sin \theta}$

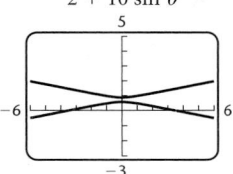

23. $y^2 + 2x - 1 = 0$ **25.** $x^2 - 3y^2 - 12y - 9 = 0$

27. $27x^2 + 36y^2 - 48x - 64 = 0$

29. $4x^2 - 5y^2 + 24y - 16 = 0$

31. $3x^2 + 4y^2 + 6x - 9 = 0$ **33.** $4x^2 - 12y - 9 = 0$

35. $3x^2 + 4y^2 - 8x - 16 = 0$

37. $4x^2 - 96y^2 + 140y - 49 = 0$ **39.** $r = \dfrac{6}{1 + 2 \sin \theta}$

41. $r = \dfrac{4}{1 + \cos \theta}$ **43.** $r = \dfrac{2}{2 - \cos \theta}$

45. $r = \dfrac{15}{4 + 3 \sin \theta}$ **47.** $r = \dfrac{8}{1 - 4 \sin \theta}$

49. Discussion and Writing

51. [1.2] $f(t) = (t - 3)^2 + 4$, or $t^2 - 6t + 13$

52. [1.2] $f(2t) = (2t - 3)^2 + 4$, or $4t^2 - 12t + 13$

53. [1.2] $f(t - 1) = (t - 4)^2 + 4$, or $t^2 - 8t + 20$

54. [1.2] $f(t + 2) = (t - 1)^2 + 4$, or $t^2 - 2t + 5$

55. $r = \dfrac{1.5 \times 10^8}{1 + \sin \theta}$

Exercise Set 10.7

1. $x = \frac{1}{2}t, \ y = 6t - 7; \ -1 \leqslant t \leqslant 6$

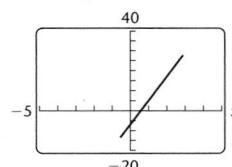

$y = 12x - 7, -\frac{1}{2} \leqslant x \leqslant 3$

3. $x = 4t^2, \ y = 2t; \ -1 \leqslant t \leqslant 1$

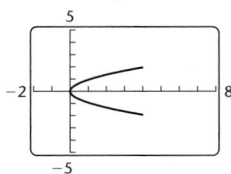

$x = y^2, -2 \leqslant y \leqslant 2$

5. $x = t^2, \ y = \sqrt{t}; \ 0 \leqslant t \leqslant 4$

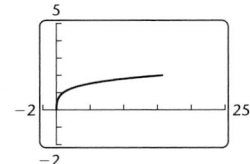

$x = y^4, 0 \leqslant y \leqslant 2$

7.
$$x = t + 3, \quad y = \frac{1}{t+3}; \quad -2 \le t \le 2$$

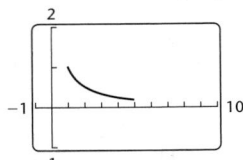

$$y = \frac{1}{x}, \quad 1 \le x \le 5$$

9.
$$x = 2t - 1, \quad y = t^2; \quad -3 \le t \le 3$$

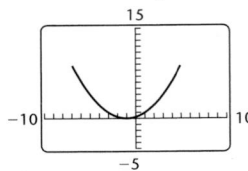

$$y = \frac{1}{4}(x+1)^2, \quad -7 \le x \le 5$$

11.
$$x = e^{-t}, \quad y = e^t; \quad -\infty < t < \infty$$

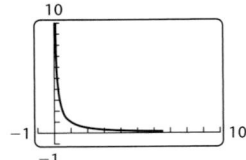

$$y = \frac{1}{x}, \quad x > 0$$

13.
$$x = 3\cos t, \quad y = 3\sin t; \quad 0 \le t \le 2\pi$$

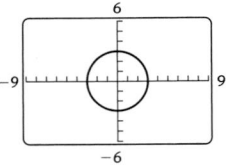

$$x^2 + y^2 = 9, \quad -3 \le x \le 3$$

15.
$$x = \cos t, \quad y = 2\sin t; \quad 0 \le t \le 2\pi$$

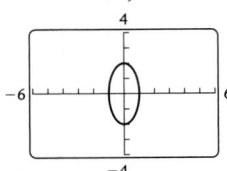

$$x^2 + \frac{y^2}{4} = 1, \quad -1 \le x \le 1$$

17.
$$x = \sec t, \quad y = \cos t; \quad -\frac{\pi}{2} < t < \frac{\pi}{2}$$

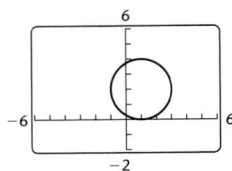

$$y = \frac{1}{x}, \quad x \ge 1$$

19.
$$x = 1 + 2\cos t, \quad y = 2 + 2\sin t; \quad 0 \le t \le 2\pi$$

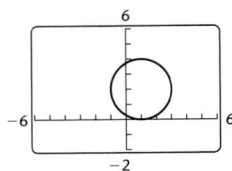

$$(x-1)^2 + (y-2)^2 = 4, \quad -1 \le x \le 3$$

21. 0.7071 **23.** −0.2588 **25.** 0.7265 **27.** 0.5460

29. Answers may vary. $x = t, y = 4t - 3; x = \frac{t}{4} + 3,$

$y = t + 9$

31. Answers may vary. $x = t, y = (t-2)^2 - 6t; x = t + 2,$
$y = t^2 - 6t - 12$

33. (a) $x = 40\sqrt{3}\,t, y = 7 + 40t - 16t^2;$
(b) $x = 40\sqrt{3}\,t, \quad y = 7 + 40t - 16t^2$ (c) 31 ft, 23 ft;

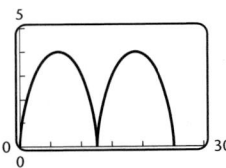

(d) about 2.7 sec; (e) about 187.1 ft; (f) 32 ft

35.
$$x = 2(t - \sin t), \quad y = 2(1 - \cos t); \quad 0 \le t \le 4\pi$$

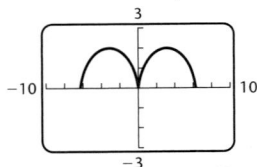

37.
$$x = t - \sin t, \quad y = 1 - \cos t; \quad -2\pi \le t \le 2\pi$$

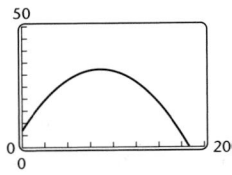

39. Discussion and Writing
41. [1.1]

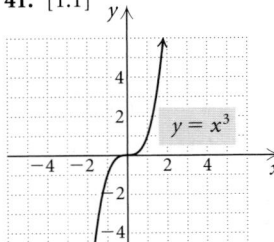

42. [1.1]

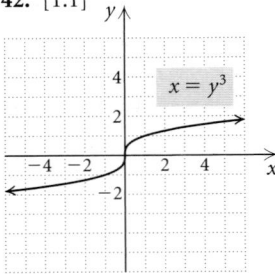

43. [1.2]

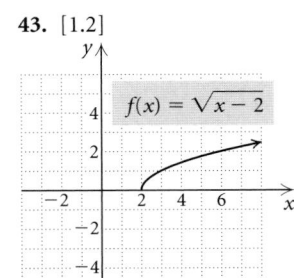

$f(x) = \sqrt{x-2}$

44. [4.5]

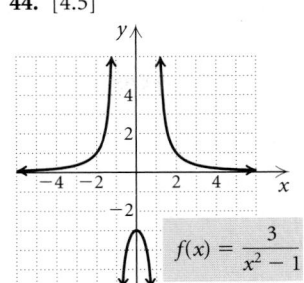

$f(x) = \dfrac{3}{x^2 - 1}$

45. $x = 3\cos t, y = -3\sin t$

Review Exercises: Chapter 10

1. True   **2.** False   **3.** True   **4.** False   **5.** False
6. (d)   **7.** (a)   **8.** (e)   **9.** (g)   **10.** (b)   **11.** (f)
12. (h)   **13.** (c)   **14.** $x^2 = -6y$ **15.** $F: (-3, 0)$;
$V: (0,0); D: x = 3$ **16.** $V: (-5, 8); F: \left(-5, \frac{15}{2}\right); D: y = \frac{17}{2}$
17. $C: (2, -1); V: (-3, -1), (7, -1); F: (-1, -1), (5, -1)$;

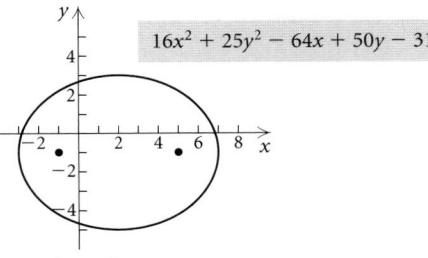

$16x^2 + 25y^2 - 64x + 50y - 311 = 0$

18. $\dfrac{x^2}{9} + \dfrac{y^2}{16} = 1$

19. $C: \left(-2, \frac{1}{4}\right); V: \left(0, \frac{1}{4}\right), \left(-4, \frac{1}{4}\right)$;

$F: \left(-2 + \sqrt{6}, \frac{1}{4}\right), \left(-2 - \sqrt{6}, \frac{1}{4}\right)$;

$A: y - \frac{1}{4} = \frac{\sqrt{2}}{2}(x + 2), y - \frac{1}{4} = -\frac{\sqrt{2}}{2}(x + 2)$

20. 0.167 ft   **21.** $\left(-8\sqrt{2}, 8\right), \left(8\sqrt{2}, 8\right)$

22. $\left(3, \dfrac{\sqrt{29}}{2}\right), \left(-3, \dfrac{\sqrt{29}}{2}\right), \left(3, -\dfrac{\sqrt{29}}{2}\right), \left(-3, -\dfrac{\sqrt{29}}{2}\right)$

23. $(7, 4)$ **24.** $(2, 2), \left(\frac{32}{9}, -\frac{10}{9}\right)$ **25.** $(0, -3), (2, 1)$
26. $(4, 3), (4, -3), (-4, 3), (-4, -3)$
27. $\left(-\sqrt{3}, 0\right), \left(\sqrt{3}, 0\right), (-2, 1), (2, 1)$ **28.** $\left(-\frac{3}{5}, \frac{21}{5}\right), (3, -3)$
29. $(6, 8), (6, -8), (-6, 8), (-6, -8)$
30. $(2, 2), (-2, -2), \left(2\sqrt{2}, \sqrt{2}\right), \left(-2\sqrt{2}, -\sqrt{2}\right)$ **31.** 7, 4
32. 7 m by 12 m   **33.** 4, 8   **34.** 32 cm, 20 cm
35. 11 ft, 3 ft

36.

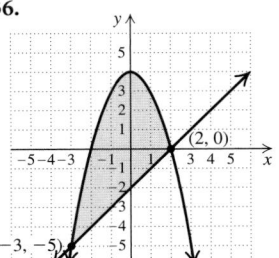

37.

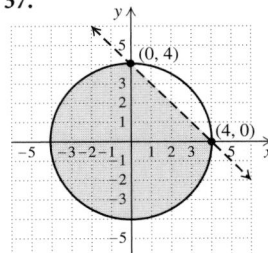

38.

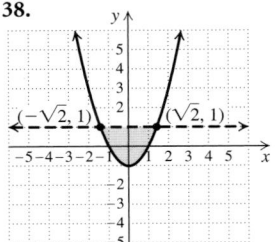

39.

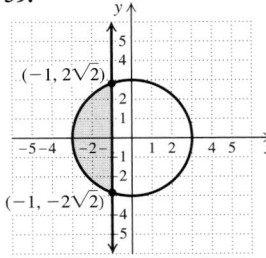

40.

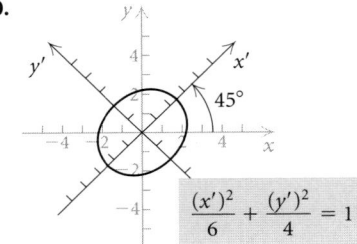

$\dfrac{(x')^2}{6} + \dfrac{(y')^2}{4} = 1$

41.

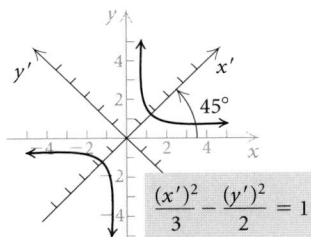

$\dfrac{(x')^2}{3} - \dfrac{(y')^2}{2} = 1$

42.

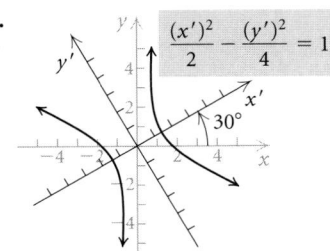

$\dfrac{(x')^2}{2} - \dfrac{(y')^2}{4} = 1$

43.

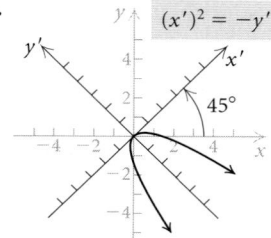

$(x')^2 = -y'$

45°

44.

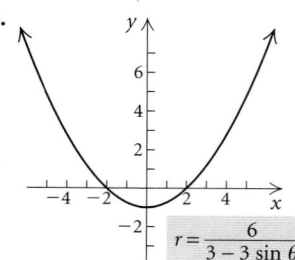

$r = \dfrac{6}{3 - 3\sin\theta}$

Horizontal directrix 2 units below the pole; vertex: $\left(1, \dfrac{3\pi}{2}\right)$

45.

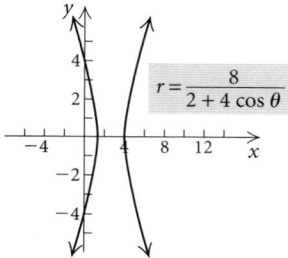

$r = \dfrac{8}{2 + 4\cos\theta}$

Vertical directrix 2 units to the right of the pole; vertices: $\left(\dfrac{4}{3}, 0\right), (-4, \pi)$

46.

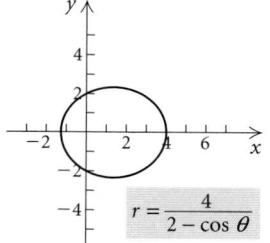

$r = \dfrac{4}{2 - \cos\theta}$

Vertical directrix 4 units to the left of the pole; vertices: $(4, 0), \left(\dfrac{4}{3}, \pi\right)$

47.

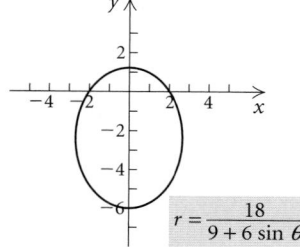

$r = \dfrac{18}{9 + 6\sin\theta}$

Horizontal directrix 3 units above the pole;

vertices: $\left(\dfrac{6}{5}, \dfrac{\pi}{2}\right), \left(6, \dfrac{3\pi}{2}\right)$

48. $x^2 - 4y - 4 = 0$ **49.** $3x^2 - y^2 - 16x + 16 = 0$

50. $3x^2 + 4y^2 - 8x - 16 = 0$

51. $9x^2 + 5y^2 + 24y - 36 = 0$

52. $r = \dfrac{1}{1 + \frac{1}{2}\cos\theta}$, or $r = \dfrac{2}{2 + \cos\theta}$ **53.** $r = \dfrac{18}{1 - 3\sin\theta}$

54. $r = \dfrac{4}{1 - \cos\theta}$ **55.** $r = \dfrac{6}{1 + 2\sin\theta}$

56.

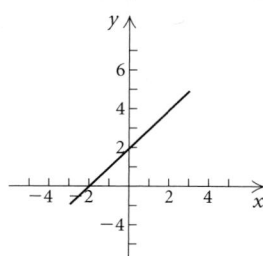

$x = t,\ y = 2 + t;\ -3 \le t \le 3$

$y = 2 + x, -3 \le x \le 3$

57.

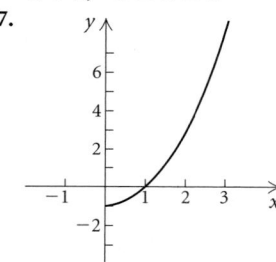

$x = \sqrt{t},\ y = t - 1;\ 0 \le t \le 9$

$y = x^2 - 1, 0 \le x \le 3$

58.

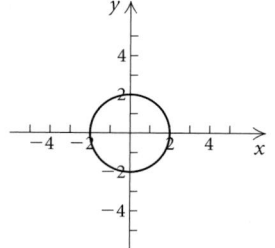

$x = 2\cos t,\ y = 2\sin t;\ 0 \le t \le 2\pi$

$x^2 + y^2 = 4$

59.

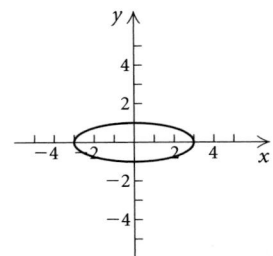

$$x = 3 \sin t, \quad y = \cos t; \quad 0 \leqslant t \leqslant 2\pi$$

$$\frac{x^2}{9} + y^2 = 1$$

60. Answers may vary. $x = t, y = 2t - 3; x = t + 1,$
$y = 2t - 1$
61. Answers may vary. $x = t, y = t^2 + 4; x = t - 2,$
$y = t^2 - 4t + 8$
62. (a) $x = 75\sqrt{2}t, y = 75\sqrt{2}t - 16t^2;$ (b) 174.2 ft, 60.4 ft;
(c) about 6.6 sec; (d) about 700.0 ft; (e) about 175.8 ft
63. B   **64.** D   **65.** C
66. Discussion and Writing: Although we can always visualize the real-number solutions, we cannot visualize the imaginary-number solutions.
67. Discussion and Writing: The equation of a circle can be written as

$$\frac{(x - h)^2}{a^2} + \frac{(y - k)^2}{b^2} = 1,$$

where $a = b = r$, the radius of the circle. In an ellipse, $a > b$, so a circle is not a special type of ellipse.

68. $\dfrac{8}{7}, \dfrac{7}{2}$ **69.** $(x - 2)^2 + (y - 1)^2 = 100$

70. $x^2 + \dfrac{y^2}{9} = 1$ **71.** $\dfrac{x^2}{778.41} - \dfrac{y^2}{39,221.59} = 1$

Test: Chapter 10

1. [10.3] (c)   **2.** [10.1] (b)   **3.** [10.2] (a)   **4.** [10.2] (d)
5. [10.1] $V: (0,0); F: (0,3); D: y = -3$

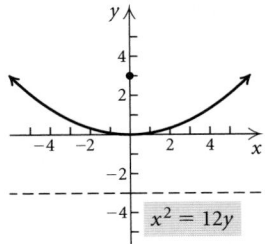

$x^2 = 12y$

6. [10.1] $V: (-1,-1); F: (1,-1); D: x = -3$

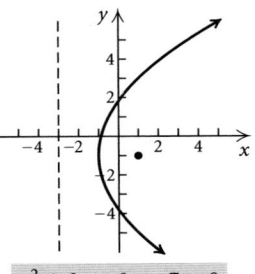

$y^2 + 2y - 8x - 7 = 0$

7. [10.1] $x^2 = 8y$
8. [10.2] Center: $(-1, 3)$; radius: 5

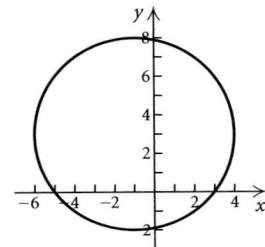

$x^2 + y^2 + 2x - 6y - 15 = 0$

9. [10.2] $C: (0,0); V: (-4,0), (4,0); F: \left(-\sqrt{7},0\right), \left(\sqrt{7},0\right)$

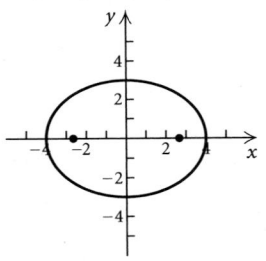

$9x^2 + 16y^2 = 144$

10. [10.2] $C: (-1,2); V: (-1,-1), (-1,5);$
$F: \left(-1, 2 - \sqrt{5}\right), \left(-1, 2 + \sqrt{5}\right)$

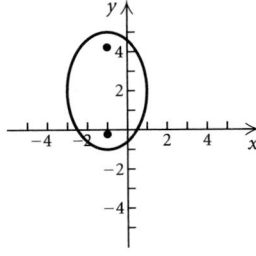

$\dfrac{(x+1)^2}{4} + \dfrac{(y-2)^2}{9} = 1$

11. [10.2] $\dfrac{x^2}{4} + \dfrac{y^2}{25} = 1$
12. [10.3] $C: (0,0); V: (-1,0), (1,0); F: \left(-\sqrt{5},0\right), \left(\sqrt{5},0\right);$

$A: y = -2x, y = 2x$

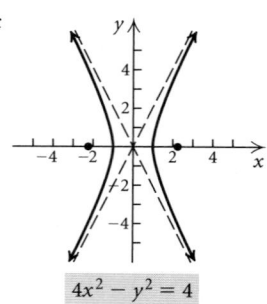

$4x^2 - y^2 = 4$

13. [10.3] $C: (-1, 2)$; $V: (-1, 0), (-1, 4)$; $F: \left(-1, 2 - \sqrt{13}\right)$, $\left(-1, 2 + \sqrt{13}\right)$; $A: y = -\frac{2}{3}x + \frac{4}{3}, y = \frac{2}{3}x + \frac{8}{3}$

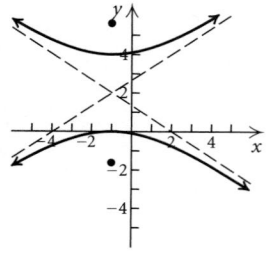

$\dfrac{(y - 2)^2}{4} - \dfrac{(x + 1)^2}{9} = 1$

14. [10.3] $y = \dfrac{\sqrt{2}}{2}x, y = -\dfrac{\sqrt{2}}{2}x$ **15.** [10.1] $\dfrac{27}{8}$ in.

16. [10.4] $(1, 2), (1, -2), (-1, 2), (-1, -2)$
17. [10.4] $(3, -2), (-2, 3)$ **18.** [10.4] $(2, 3), (3, 2)$
19. [10.4] 5 ft by 4 ft   **20.** [10.4] 60 ft by 45 ft
21. [10.4]

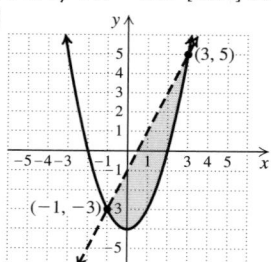

22. [10.5] After using the rotation of axes formulas with $\theta = 45°$, we have $\dfrac{(x')^2}{9} + (y')^2 = 1$.

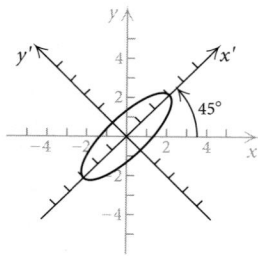

23. [10.6]

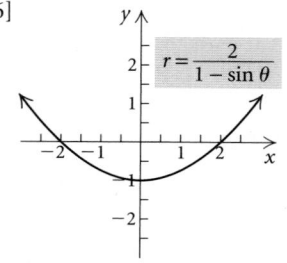

Horizontal directrix 2 units below the pole; vertex: $\left(1, \dfrac{3\pi}{2}\right)$

24. [10.6] $r = \dfrac{6}{1 + 2\cos\theta}$

25. [10.7]

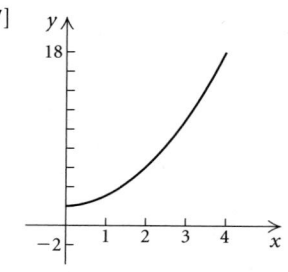

$x = \sqrt{t}, \ y = t + 2; \ 0 \le t \le 16$

26. [10.7] $x^2 + y^2 = 9, -3 \le x \le 3$
27. [10.7] Answers may vary. $x = t, y = t - 5; x = t + 5,$
$y = t$
28. [10.7] **(a)** $x = 125\sqrt{3}t, y = 10 + 125t - 16t^2$;
(b) 119 ft, 241 ft; **(c)** about 7.9 sec; **(d)** about 1710.4 ft;
(e) about 254.1 ft   **29.** [10.1] A
30. [10.2] $(x - 3)^2 + (y + 1)^2 = 8$

CHAPTER 11

Exercise Set 11.1

1. 3, 7, 11, 15; 39; 59   **3.** $2, \frac{3}{2}, \frac{4}{3}, \frac{5}{4}; \frac{10}{9}; \frac{15}{14}$
5. $0, \frac{3}{5}, \frac{4}{5}, \frac{15}{17}; \frac{99}{101}; \frac{112}{113}$ **7.** $-1, 4, -9, 16; 100; -225$
9. 7, 3, 7, 3; 3; 7   **11.** 34   **13.** 225   **15.** $-33,880$
17. 67

19.

n	u_n
1	2
2	2.25
3	2.3704
4	2.4414
5	2.4883
6	2.5216
7	2.5465
8	2.5658
9	2.5812
10	2.5937

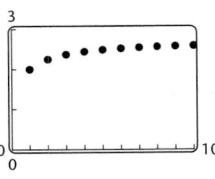

21.

n	u_n
1	2
2	1.5538
3	1.4988
4	1.4914
5	1.4904
6	1.4902
7	1.4902
8	1.4902
9	1.4902
10	1.4902

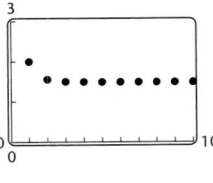

23. $2n$ **25.** $(-1)^n \cdot 2 \cdot 3^{n-1}$ **27.** $\dfrac{n+1}{n+2}$

29. $n(n+1)$ **31.** $\log 10^{n-1}$, or $n-1$ **33.** 6; 28

35. 20; 30   **37.** $\frac{1}{2} + \frac{1}{4} + \frac{1}{6} + \frac{1}{8} + \frac{1}{10} = \frac{137}{120}$

39. $1 + 2 + 4 + 8 + 16 + 32 + 64 = 127$

41. $\ln 7 + \ln 8 + \ln 9 + \ln 10 = \ln(7 \cdot 8 \cdot 9 \cdot 10) =$ $\ln 5040 \approx 8.5252$

43. $\frac{1}{2} + \frac{2}{3} + \frac{3}{4} + \frac{4}{5} + \frac{5}{6} + \frac{6}{7} + \frac{7}{8} + \frac{8}{9} = \frac{15,551}{2520}$

45. $-1 + 1 - 1 + 1 - 1 = -1$

47. $3 - 6 + 9 - 12 + 15 - 18 + 21 - 24 = -12$

49. $2 + 1 + \frac{2}{5} + \frac{1}{5} + \frac{2}{17} + \frac{1}{13} + \frac{2}{37} = \frac{157,351}{40,885}$

51. $3 + 2 + 3 + 6 + 11 + 18 = 43$

53. $\frac{1}{2} + \frac{2}{3} + \frac{4}{5} + \frac{8}{9} + \frac{16}{17} + \frac{32}{33} + \frac{64}{65} + \frac{128}{129} + \frac{256}{257} + \frac{512}{513} +$ $\frac{1024}{1025} \approx 9.736$ **55.** $\displaystyle\sum_{k=1}^{\infty} 5k$ **57.** $\displaystyle\sum_{k=1}^{6} (-1)^{k+1} 2^k$

59. $\displaystyle\sum_{k=1}^{6} (-1)^k \frac{k}{k+1}$ **61.** $\displaystyle\sum_{k=2}^{n} (-1)^k k^2$ **63.** $\displaystyle\sum_{k=1}^{\infty} \frac{1}{k(k+1)}$

65. $4, 1\frac{1}{4}, 1\frac{4}{5}, 1\frac{5}{9}$ **67.** $6561, -81, 9i, -3\sqrt{i}$ **69.** $2, 3, 5, 8$

71. **(a)** 1062, 1127.84, 1197.77, 1272.03, 1350.90, 1434.65, 1523.60, 1618.07, 1718.39, 1824.93; **(b)** \$3330.35

73. 1, 2, 4, 8, 16, 32, 64, 128, 256, 512, 1024, 2048, 4096, 8192, 16,384, 32,768, 65,536   **75.** 1, 1, 2, 3, 5, 8, 13

77. **(a)** $a_n = -3.257142857n^2 + 13.51428571n +$ 174.4714286; **(b)** 138.3 thousand patents; 109.5 thousand patents; 74.1 thousand patents   **79.** Discussion and Writing

80. [9.1], [9.3], [9.5], [9.6] $(-1, -3)$

81. [9.1], [9.3], [9.5], [9.6] 127 takeout meals; 81 meals eaten in restaurants   **82.** [10.2] $(3, -2)$; 4

83. [10.2] $\left(-\dfrac{5}{2}, 4\right); \dfrac{\sqrt{97}}{2}$ **85.** $i, -1, -i, 1, i; i$

87. $\ln(1 \cdot 2 \cdot 3 \cdot \cdots \cdot n)$

Exercise Set 11.2

1. $a_1 = 3, d = 5$ **3.** $a_1 = 9, d = -4$ **5.** $a_1 = \frac{3}{2}, d = \frac{3}{4}$

7. $a_1 = \$316, d = -\$3$ **9.** $a_{12} = 46$ **11.** $a_{14} = -\frac{17}{3}$

13. $a_{10} = \$7941.62$ **15.** 33rd   **17.** 46th   **19.** $a_1 = 5$

21. $n = 39$ **23.** $a_1 = \frac{1}{3}; d = \frac{1}{2}; \frac{1}{3}, \frac{5}{6}, \frac{4}{3}, \frac{11}{6}, \frac{7}{3}$ **25.** 670

27. 160,400   **29.** 735   **31.** 990   **33.** 1760   **35.** $\frac{65}{2}$

37. $-\frac{6026}{13}$ **39.** 1260 poles   **41.** 4960¢, or \$49.60

43. Yes; 32; 1600 ft   **45.** 3 plants; 171 plants   **47.** Yes; 3

49. Discussion and Writing

50. [9.1], [9.3], [9.5], [9.6] $(2, 5)$

51. [9.2], [9.3], [9.5], [9.6] $(2, -1, 3)$

52. [10.2] $(-4, 0), (4, 0); \left(-\sqrt{7}, 0\right), \left(\sqrt{7}, 0\right)$

53. [10.2] $\dfrac{x^2}{4} + \dfrac{y^2}{25} = 1$ **55.** n^2

57. $a_1 = 60 - 5p - 5q; d = 5p + 2q - 20$ **59.** 6, 8, 10

61. $5\frac{4}{5}, 7\frac{3}{5}, 9\frac{2}{5}, 11\frac{1}{5}$

63. Insert 16 arithmetic means between 1 and 50 with $d = \frac{49}{17}$.

65.
$$m = p + d$$
$$\underline{m = q - d}$$
$$2m = p + q \qquad \textbf{Adding}$$
$$m = \frac{p + q}{2}$$

Visualizing the Graph

1. J   **2.** A   **3.** C   **4.** G   **5.** F   **6.** H   **7.** E

8. D   **9.** B   **10.** I

Exercise Set 11.3

1. 2   **3.** -1 **5.** -2 **7.** 0.1   **9.** $\dfrac{a}{2}$ **11.** 128

13. 162   **15.** $7(5)^{40}$ **17.** 3^{n-1} **19.** $(-1)^{n-1}$

21. $\dfrac{1}{x^n}$ **23.** 762   **25.** $\frac{4921}{18}$ **27.** True   **29.** True

31. True   **33.** 8   **35.** 125   **37.** Does not exist   **39.** $\frac{2}{3}$

41. $29\frac{38,569}{59,049}$ **43.** 2   **45.** Does not exist   **47.** \$4545.$\overline{45}$

49. $\frac{160}{9}$ **51.** $\frac{13}{99}$ **53.** 9   **55.** $\frac{34,091}{9990}$ **57.** \$2,684,354.55

59. **(a)** About 297 ft; **(b)** 300 ft   **61.** \$23,841.50

63. \$523,619.17   **65.** \$86,666,666,667   **67.** Discussion and Writing   **69.** [2.3] $(f \circ g)(x) = 16x^2 + 40x + 25$; $(g \circ f)(x) = 4x^2 + 5$ **70.** [2.3] $(f \circ g)(x) = x^2 + x + 2$; $(g \circ f)(x) = x^2 - x + 3$ **71.** [5.5] 2.209   **72.** [5.5] $\frac{1}{16}$

73. $\left(4 - \sqrt{6}\right)/\left(\sqrt{3} - \sqrt{2}\right) = 2\sqrt{3} + \sqrt{2}$,
$\left(6\sqrt{3} - 2\sqrt{2}\right)/\left(4 - \sqrt{6}\right) = 2\sqrt{3} + \sqrt{2}$; there exists a common ratio, $2\sqrt{3} + \sqrt{2}$; thus the sequence is geometric.
75. (a) $\frac{13}{3}; \frac{22}{3}; \frac{34}{3}, \frac{46}{3}, \frac{58}{3}$; (b) $-\frac{11}{3}; -\frac{2}{3}, \frac{10}{3}, -\frac{50}{3}, \frac{250}{3}$ or 5; 8, 12,
18, 27   **77.** $S_n = \dfrac{x^2(1 - (-x)^n)}{x + 1}$
79. $\dfrac{a_{n+1}}{a_n} = r$, so $\ln \dfrac{a_{n+1}}{a_n} = \ln r$. But $\ln \dfrac{a_{n+1}}{a_n} = \ln a_{n+1} - \ln a_n = \ln r$. Thus, $\ln a_1, \ln a_2, \ldots$ is an arithmetic sequence with common difference $\ln r$.   **81.** 512 cm^2

Exercise Set 11.4

1. $1^2 < 1^3$, false; $2^2 < 2^3$, true; $3^2 < 3^3$, true; $4^2 < 4^3$, true; $5^2 < 5^3$, true

3. A polygon of 3 sides has $\dfrac{3(3 - 3)}{2}$ diagonals. True; A polygon of 4 sides has $\dfrac{4(4 - 3)}{2}$ diagonals. True; A polygon of 5 sides has $\dfrac{5(5 - 3)}{2}$ diagonals. True; A polygon of 6 sides has $\dfrac{6(6 - 3)}{2}$ diagonals. True; A polygon of 7 sides has $\dfrac{7(7 - 3)}{2}$ diagonals. True.

5. S_n: $2 + 4 + 6 + \cdots + 2n = n(n + 1)$
S_1: $2 = 1(1 + 1)$
S_k: $2 + 4 + 6 + \cdots + 2k = k(k + 1)$
S_{k+1}: $2 + 4 + 6 + \cdots + 2k + 2(k + 1)$
$= (k + 1)(k + 2)$
(1) *Basis step*: S_1 true by substitution.
(2) *Induction step*: Assume S_k. Deduce S_{k+1}.
Starting with the left side of S_{k+1}, we have
$$2 + 4 + 6 + \cdots + 2k + 2(k + 1)$$
$$= k(k + 1) + 2(k + 1) \quad \textbf{By } S_k$$
$$= (k + 1)(k + 2). \quad \textbf{Distributive law}$$

7. S_n: $1 + 5 + 9 + \cdots + (4n - 3) = n(2n - 1)$
S_1: $1 = 1(2 \cdot 1 - 1)$
S_k: $1 + 5 + 9 + \cdots + (4k - 3) = k(2k - 1)$
S_{k+1}: $1 + 5 + 9 + \cdots + (4k - 3) + [4(k + 1) - 3]$
$$= (k + 1)[2(k + 1) - 1]$$
$$= (k + 1)(2k + 1)$$
(1) *Basis step*: S_1 true by substitution.
(2) *Induction step*: Assume S_k. Deduce S_{k+1}.
Starting with the left side of S_{k+1}, we have
$$1 + 5 + 9 + \cdots + (4k - 3) + [4(k + 1) - 3]$$
$$= k(2k - 1) + [4(k + 1) - 3] \quad \textbf{By } S_k$$
$$= 2k^2 - k + 4k + 4 - 3$$
$$= 2k^2 + 3k + 1$$
$$= (k + 1)(2k + 1).$$

9. S_n: $2 + 4 + 8 + \cdots + 2^n = 2(2^n - 1)$
S_1: $2 = 2(2 - 1)$
S_k: $2 + 4 + 8 + \cdots + 2^k = 2(2^k - 1)$
S_{k+1}: $2 + 4 + 8 + \cdots + 2^k + 2^{k+1} = 2(2^{k+1} - 1)$
(1) *Basis step*: S_1 is true by substitution.
(2) *Induction step*: Assume S_k. Deduce S_{k+1}.
Starting with the left side of S_{k+1}, we have
$$\underbrace{2 + 4 + 8 + \cdots + 2^k}_{} + 2^{k+1}$$
$$= \quad 2(2^k - 1) \quad + 2^{k+1} \quad \textbf{By } S_k$$
$$= 2^{k+1} - 2 + 2^{k+1}$$
$$= 2 \cdot 2^{k+1} - 2$$
$$= 2(2^{k+1} - 1).$$

11. S_n: $n < n + 1$
S_1: $1 < 1 + 1$
S_k: $k < k + 1$
S_{k+1}: $k + 1 < (k + 1) + 1$
(1) *Basis step*: Since $1 < 1 + 1$, S_1 is true.
(2) *Induction step*: Assume S_k. Deduce S_{k+1}. Now
$$k < k + 1 \quad \textbf{By } S_k$$
$$k + 1 < k + 1 + 1 \quad \textbf{Adding 1}$$
$$k + 1 < k + 2. \quad \textbf{Simplifying}$$

13. S_n: $2n \le 2^n$
S_1: $2 \cdot 1 \le 2^1$
S_k: $2k \le 2^k$
S_{k+1}: $2(k + 1) \le 2^{k+1}$
(1) *Basis step*: Since $2 = 2$, S_1 is true.
(2) *Induction step*: Let k be any natural number. Assume S_k. Deduce S_{k+1}.
$$2k \le 2^k \quad \textbf{By } S_k$$
$$2 \cdot 2k \le 2 \cdot 2^k \quad \textbf{Multiplying by 2}$$
$$4k \le 2^{k+1}$$
Since $1 \le k$, $k + 1 \le k + k$, or $k + 1 \le 2k$.
Then $2(k + 1) \le 4k$.   **Multiplying by 2**
Thus, $2(k + 1) \le 4k \le 2^{k+1}$, so $2(k + 1) \le 2^{k+1}$.

15.
$$S_n: \quad \frac{1}{1 \cdot 2 \cdot 3} + \frac{1}{2 \cdot 3 \cdot 4} + \frac{1}{3 \cdot 4 \cdot 5} + \cdots$$
$$+ \frac{1}{n(n + 1)(n + 2)} = \frac{n(n + 3)}{4(n + 1)(n + 2)}$$
$$S_1: \quad \frac{1}{1 \cdot 2 \cdot 3} = \frac{1(1 + 3)}{4(1 + 1)(1 + 2)}$$
$$S_k: \quad \frac{1}{1 \cdot 2 \cdot 3} + \frac{1}{2 \cdot 3 \cdot 4} + \cdots + \frac{1}{k(k + 1)(k + 2)}$$
$$= \frac{k(k + 3)}{4(k + 1)(k + 2)}$$
$$S_{k+1}: \quad \frac{1}{1 \cdot 2 \cdot 3} + \frac{1}{2 \cdot 3 \cdot 4} + \cdots + \frac{1}{k(k + 1)(k + 2)}$$
$$+ \frac{1}{(k + 1)(k + 2)(k + 3)}$$
$$= \frac{(k + 1)(k + 1 + 3)}{4(k + 1 + 1)(k + 1 + 2)} = \frac{(k + 1)(k + 4)}{4(k + 2)(k + 3)}$$

(1) *Basis step*: Since $\dfrac{1}{1 \cdot 2 \cdot 3} = \dfrac{1}{6}$ and

$\dfrac{1(1+3)}{4(1+1)(1+2)} = \dfrac{1 \cdot 4}{4 \cdot 2 \cdot 3} = \dfrac{1}{6}$, S_1 is true.

(2) *Induction step*: Assume S_k. Deduce S_{k+1}.

Add $\dfrac{1}{(k+1)(k+2)(k+3)}$ on both sides of S_k and simplify the right side. Only the right side is shown here.

$\dfrac{k(k+3)}{4(k+1)(k+2)} + \dfrac{1}{(k+1)(k+2)(k+3)}$

$= \dfrac{k(k+3)(k+3) + 4}{4(k+1)(k+2)(k+3)}$

$= \dfrac{k^3 + 6k^2 + 9k + 4}{4(k+1)(k+2)(k+3)}$

$= \dfrac{(k+1)^2(k+4)}{4(k+1)(k+2)(k+3)}$

$= \dfrac{(k+1)(k+4)}{4(k+2)(k+3)}$

17.

S_n: $\quad 1 + 2 + 3 + \cdots + n = \dfrac{n(n+1)}{2}$

S_1: $\quad 1 = \dfrac{1(1+1)}{2}$

S_k: $\quad 1 + 2 + 3 + \cdots + k = \dfrac{k(k+1)}{2}$

S_{k+1}: $\quad 1 + 2 + 3 + \cdots + k + (k+1) = \dfrac{(k+1)(k+2)}{2}$

(1) *Basis step*: S_1 true by substitution.

(2) *Induction step*: Assume S_k. Deduce S_{k+1}.
Starting with the left side of S_{k+1}, we have

$\underbrace{1 + 2 + 3 + \cdots + k} + (k+1)$

$= \dfrac{k(k+1)}{2} + (k+1)$ **By S_k**

$= \dfrac{k(k+1) + 2(k+1)}{2}$ **Adding**

$= \dfrac{(k+1)(k+2)}{2}.$ **Distributive law**

19. S_n: $\quad 1^3 + 2^3 + 3^3 + \cdots + n^3 = \dfrac{n^2(n+1)^2}{4}$

S_1: $\quad 1^3 = \dfrac{1^2(1+1)^2}{4}$

S_k: $\quad 1^3 + 2^3 + 3^3 + \cdots + k^3 = \dfrac{k^2(k+1)^2}{4}$

S_{k+1}: $\quad 1^3 + 2^3 + 3^3 + \cdots + k^3 + (k+1)^3$
$\qquad = \dfrac{(k+1)^2[(k+1)+1]^2}{4}$

(1) *Basis step*: S_1: $1^3 = \dfrac{1^2(1+1)^2}{4} = 1$. True.

(2) *Induction step*: Assume S_k. Deduce S_{k+1}.

$1^3 + 2^3 + \cdots + k^3 = \dfrac{k^2(k+1)^2}{4}$ $\quad S_k$

$1^3 + 2^3 + \cdots + k^3 + (k+1)^3 = \dfrac{k^2(k+1)^2}{4} + (k+1)^3$

Adding $(k+1)^3$

$= \dfrac{k^2(k+1)^2 + 4(k+1)^3}{4}$

$= \dfrac{(k+1)^2}{4}[k^2 + 4(k+1)]$

$= \dfrac{(k+1)^2}{4}(k^2 + 4k + 4)$

$= \dfrac{(k+1)^2(k+2)^2}{4}$

21. S_n: $\quad 1^5 + 2^5 + 3^5 + \cdots + n^5$
$\qquad = \dfrac{n^2(n+1)^2(2n^2 + 2n - 1)}{12}$

S_1: $\quad 1^5 = \dfrac{1^2(1+1)^2(2 \cdot 1^2 + 2 \cdot 1 - 1)}{12}$

S_k: $\quad 1^5 + 2^5 + 3^5 + \cdots + k^5$
$\qquad = \dfrac{k^2(k+1)^2(2k^2 + 2k - 1)}{12}$

S_{k+1}: $\quad 1^5 + 2^5 + 3^5 + \cdots + k^5 + (k+1)^5$
$\qquad = \dfrac{(k+1)^2[(k+1)+1]^2[2(k+1)^2 + 2(k+1) - 1]}{12}$

(1) *Basis step*: S_1: $1^5 = \dfrac{1^2(1+1)^2(2 \cdot 1^2 + 2 \cdot 1 - 1)}{12}$. True.

(2) *Induction step*: Assume S_k:

$1^5 + 2^5 + \cdots + k^5 = \dfrac{k^2(k+1)^2(2k^2 + 2k - 1)}{12}.$

Then $1^5 + 2^5 + \cdots + k^5 + (k+1)^5$

$= \dfrac{k^2(k+1)^2(2k^2 + 2k - 1)}{12} + (k+1)^5$

$= \dfrac{k^2(k+1)^2(2k^2 + 2k - 1) + 12(k+1)^5}{12}$

$= \dfrac{(k+1)^2(2k^4 + 14k^3 + 35k^2 + 36k + 12)}{12}$

$= \dfrac{(k+1)^2(k+2)^2(2k^2 + 6k + 3)}{12}$

$= \dfrac{(k+1)^2(k+1+1)^2(2(k+1)^2 + 2(k+1) - 1)}{12}.$

23. S_n: $\quad 2 + 6 + 12 + \cdots + n(n+1) = \dfrac{n(n+1)(n+2)}{3}$

S_1: $\quad 1(1+1) = \dfrac{1(1+1)(1+2)}{3}$

S_k: $\quad 2 + 6 + 12 + \cdots + k(k+1) = \dfrac{k(k+1)(k+2)}{3}$

S_{k+1}:

$2 + 6 + 12 + \cdots + k(k+1) + (k+1)[(k+1)+1]$
$= \dfrac{(k+1)[(k+1)+1][(k+1)+2]}{3}$

(1) *Basis step*: S_1: $\quad 1(1+1) = \dfrac{1(1+1)(1+2)}{3}$. True.

(2) *Induction step:* Assume S_k:

$$2 + 6 + 12 + \cdots + k(k+1) = \frac{k(k+1)(k+2)}{3}.$$

Then $2 + 6 + 12 + \cdots + k(k+1) + (k+1)(k+1+1)$

$$= \frac{k(k+1)(k+2)}{3} + (k+1)(k+2)$$

$$= \frac{k(k+1)(k+2) + 3(k+1)(k+2)}{3}$$

$$= \frac{(k+1)(k+2)(k+3)}{3}$$

$$= \frac{(k+1)(k+1+1)(k+1+2)}{3}.$$

25. $S_n:$ $a_1 + (a_1 + d) + (a_1 + 2d) + \cdots +$
$$[a_1 + (n-1)d] = \frac{n}{2}[2a_1 + (n-1)d]$$

$S_1:$ $a_1 = \frac{1}{2}[2a_1 + (1-1)d]$

$S_k:$ $a_1 + (a_1 + d) + (a_1 + 2d) + \cdots +$
$$[a_1 + (k-1)d] = \frac{k}{2}[2a_1 + (k-1)d]$$

$S_{k+1}:$ $a_1 + (a_1 + d) + (a_1 + 2d) + \cdots +$
$$[a_1 + (k-1)d] + [a_1 + ((k+1)-1)d]$$
$$= \frac{k+1}{2}[2a_1 + ((k+1)-1)d]$$

(1) *Basis step:* Since $\frac{1}{2}[2a_1 + (1-1)d] = \frac{1}{2} \cdot 2a_1 = a_1$, S_1 is true.

(2) *Induction step:* Assume S_k. Deduce S_{k+1}. Starting with the left side of S_{k+1}, we have

$\underbrace{a_1 + (a_1 + d) + \cdots + [a_1 + (k-1)d]} + [a_1 + kd]$

$$= \frac{k}{2}[2a_1 + (k-1)d] \qquad\qquad + [a_1 + kd]$$
$$\textbf{By } S_k$$

$$= \frac{k[2a_1 + (k-1)d]}{2} + \frac{2[a_1 + kd]}{2}$$

$$= \frac{2ka_1 + k(k-1)d + 2a_1 + 2kd}{2}$$

$$= \frac{2a_1(k+1) + k(k-1)d + 2kd}{2}$$

$$= \frac{2a_1(k+1) + (k-1+2)kd}{2}$$

$$= \frac{2a_1(k+1) + (k+1)kd}{2}$$

$$= \frac{k+1}{2}[2a_1 + kd].$$

27. Discussion and Writing
28. [9.1], [9.3], [9.5], [9.6] $(5, 3)$
29. [9.2], [9.3], [9.5], [9.6] $(2, -3, 4)$
30. [9.1], [9.3], [9.5], [9.6] Hardback: 50; paperback: 30
31. [9.2], [9.3], [9.5], [9.6] 1.5%: \$800; 2%: \$1600; 3%: \$2000
33. $S_n:$ $x + y$ is a factor of $x^{2n} - y^{2n}$.
$S_1:$ $x + y$ is a factor of $x^2 - y^2$.
$S_k:$ $x + y$ is a factor of $x^{2k} - y^{2k}$.
$S_{k+1}:$ $x + y$ is a factor of $x^{2(k+1)} - y^{2(k+1)}$.

(1) *Basis step:* $S_1:$ $x + y$ is a factor of $x^2 - y^2$. True.
$\qquad\qquad$ $S_2:$ $x + y$ is a factor of $x^4 - y^4$. True.
(2) *Induction step:* Assume $S_{k-1}:$ $x + y$ is a factor of $x^{2(k-1)} - y^{2(k-1)}$. Then $x^{2(k-1)} - y^{2(k-1)} = (x+y)Q(x)$ for some polynomial $Q(x)$.
Assume $S_k:$ $x + y$ is a factor of $x^{2k} - y^{2k}$. Then $x^{2k} - y^{2k} = (x+y)P(x)$ for some polynomial $P(x)$.
$x^{2(k+1)} - y^{2(k+1)}$
$\quad = (x^{2k} - y^{2k})(x^2 + y^2) - (x^{2(k-1)} - y^{2(k-1)})(x^2 y^2)$
$\quad = (x+y)P(x)(x^2 + y^2) - (x+y)Q(x)(x^2 y^2)$
$\quad = (x+y)[P(x)(x^2 + y^2) - Q(x)(x^2 y^2)]$
so $x + y$ is a factor of $x^{2(k+1)} - y^{2(k+1)}$.

35. $S_2:$ $\log_a (b_1 b_2) = \log_a b_1 + \log_a b_2$
$S_k:$ $\log_a (b_1 b_2 \cdots b_k) = \log_a b_1 + \log_a b_2 + \cdots + \log_a b_k$
$S_{k+1}:$ $\log_a (b_1 b_2 \cdots b_{k+1}) = \log_a b_1 + \log_a b_2 + \cdots + \log_a b_{k+1}$

(1) *Basis step:* S_2 is true by the properties of logarithms.
(2) *Induction step:* Let k be a natural number $k \geq 2$. Assume S_k. Deduce S_{k+1}.

$\log_a (b_1 b_2 \cdots b_{k+1})$ $\quad$ **Left side of S_{k+1}**
$\quad = \log_a (b_1 b_2 \cdots b_k) + \log_a b_{k+1}$ $\quad$ **By S_2**
$\quad = \log_a b_1 + \log_a b_2 + \cdots + \log_a b_k + \log_a b_{k+1}$

37. $S_2:$ $\overline{z_1 + z_2} = \bar{z}_1 + \bar{z}_2$:
$\overline{(a+bi) + (c+di)} = \overline{(a+c) + (b+d)i}$
$\qquad\qquad = (a+c) - (b+d)i$
$\overline{(a+bi)} + \overline{(c+di)} = a - bi + c - di$
$\qquad\qquad = (a+c) - (b+d)i.$
$S_k:$ $\overline{z_1 + z_2 + \cdots + z_k} = \bar{z}_1 + \bar{z}_2 + \cdots + \bar{z}_k.$
$\overline{(z_1 + z_2 + \cdots + z_k) + z_{k+1}}$
$\quad = \overline{(z_1 + z_2 + \cdots + z_k)} + \overline{z_{k+1}}$ $\quad$ **By S_2**
$\quad = \bar{z}_1 + \bar{z}_2 + \cdots + \bar{z}_k + \bar{z}_{k+1}$ $\quad$ **By S_k**

39. $S_1:$ i is either i or -1 or $-i$ or 1.
$S_k:$ i^k is either i or -1 or $-i$ or 1.
$i^{k+1} = i^k \cdot i$ is then $i \cdot i = -1$ or $-1 \cdot i = -i$ or $-i \cdot i = 1$ or $1 \cdot i = i$.

41. $S_1:$ 3 is a factor of $1^3 + 2 \cdot 1$.
$S_k:$ 3 is a factor of $k^3 + 2k$, i.e., $k^3 + 2k = 3 \cdot m$.
$S_{k+1}:$ 3 is a factor of $(k+1)^3 + 2(k+1)$.
Consider

$(k+1)^3 + 2(k+1) = k^3 + 3k^2 + 5k + 3$
$\qquad\qquad = (k^3 + 2k) + 3k^2 + 3k + 3$
$\qquad\qquad = 3m + 3(k^2 + k + 1).$

A multiple of 3

Exercise Set 11.5

1. 720 $\quad$ **3.** 604,800 $\quad$ **5.** 120 $\quad$ **7.** 1 $\quad$ **9.** 3024 $\quad$ **11.** 120
13. 120 $\quad$ **15.** 1 $\quad$ **17.** 6,497,400 $\quad$ **19.** $n(n-1)(n-2)$
21. n $\quad$ **23.** $6! = 720$ $\quad$ **25.** $9! = 362,880$
27. $_9P_4 = 3024$ $\quad$ **29.** $_5P_5 = 120$; $5^5 = 3125$

31. $_5P_5 \cdot _4P_4 = 2880$ **33.** $8 \cdot 10^6 = 8{,}000{,}000$; 8 million

35. $\dfrac{9!}{2!\,3!\,4!} = 1260$ **37.** (a) $_6P_5 = 720$; (b) $6^5 = 7776$;

(c) $1 \cdot _5P_4 = 120$; (d) $1 \cdot 1 \cdot _4P_3 = 24$

39. (a) 10^5, or 100,000; (b) 100,000

41. (a) $10^9 = 1{,}000{,}000{,}000$; (b) yes

43. Discussion and Writing   **44.** $[1.5]\ \frac{9}{4}$, or 2.25

45. $[3.2]\ -3, 2$ **46.** $[3.2]\ \dfrac{3 \pm \sqrt{17}}{4}$ **47.** $[4.4]\ -2, 1, 5$

49. 8   **51.** 11   **53.** $n - 1$

Exercise Set 11.6

1. 78   **3.** 78   **5.** 7   **7.** 10   **9.** 1   **11.** 15   **13.** 128

15. 270,725   **17.** 13,037,895   **19.** n **21.** 1

23. $_{23}C_4 = 8855$ **25.** $_{13}C_{10} = 286$

27. $\dbinom{10}{7} \cdot \dbinom{5}{3} = 1200$ **29.** $\dbinom{52}{5} = 2{,}598{,}960$

31. (a) $_{31}P_2 = 930$; (b) $31^2 = 961$; (c) $_{31}C_2 = 465$

33. Discussion and Writing   **34.** $[1.5]\ -\frac{17}{2}$

35. $[3.2]\ -1, \frac{3}{2}$ **36.** $[3.2]\ \dfrac{-5 \pm \sqrt{21}}{2}$

37. $[4.4]\ -4, -2, 3$ **39.** $\dbinom{13}{5} = 1287$ **41.** $\dbinom{n}{2}; 2\dbinom{n}{2}$

43. 4   **45.** 7

47. Line segments:

$$_nC_2 = \frac{n!}{2!\,(n-2)!} = \frac{n(n-1)(n-2)!}{2 \cdot 1 \cdot (n-2)!} = \frac{n(n-1)}{2}$$

Diagonals: The n line segments that form the sides of the n-gon are not diagonals. Thus the number of diagonals is

$$_nC_2 - n = \frac{n(n-1)}{2} - n$$
$$= \frac{n^2 - n - 2n}{2} = \frac{n^2 - 3n}{2}$$
$$= \frac{n(n-3)}{2},\ n \geq 4.$$

Let D_n be the number of diagonals of an n-gon. Prove the result above for diagonals using mathematical induction.

S_n: $D_n = \dfrac{n(n-3)}{2}$, for $n = 4, 5, 6, \ldots$

S_4: $D_4 = \dfrac{4 \cdot 1}{2}$

S_k: $D_k = \dfrac{k(k-3)}{2}$

S_{k+1}: $D_{k+1} = \dfrac{(k+1)(k-2)}{2}$

(1) *Basis step*: S_4 is true (a quadrilateral has 2 diagonals).
(2) *Induction step*: Assume S_k. Note that when an additional vertex V_{k+1} is added to the k-gon, we gain k segments, 2 of which are sides of the $(k+1)$-gon, and a former side $\overline{V_1 V_k}$

becomes a diagonal. Thus the additional number of diagonals is $k - 2 + 1$, or $k - 1$. Then the new total of diagonals is $D_k + (k - 1)$, or

$$\begin{aligned} D_{k+1} &= D_k + (k-1) \\ &= \frac{k(k-3)}{2} + (k-1) \qquad \textbf{By } S_k \\ &= \frac{(k+1)(k-2)}{2} \end{aligned}$$

Exercise Set 11.7

1. $x^4 + 20x^3 + 150x^2 + 500x + 625$

3. $x^5 - 15x^4 + 90x^3 - 270x^2 + 405x - 243$

5. $x^5 - 5x^4y + 10x^3y^2 - 10x^2y^3 + 5xy^4 - y^5$

7. $15{,}625x^6 + 75{,}000x^5y + 150{,}000x^4y^2 +$ $160{,}000x^3y^3 + 96{,}000x^2y^4 + 30{,}720xy^5 + 4096y^6$

9. $128t^7 + 448t^5 + 672t^3 + 560t + 280t^{-1} + 84t^{-3} +$ $14t^{-5} + t^{-7}$ **11.** $x^{10} - 5x^8 + 10x^6 - 10x^4 + 5x^2 - 1$

13. $125 + 150\sqrt{5}\,t + 375t^2 + 100\sqrt{5}\,t^3 + 75t^4 +$ $6\sqrt{5}\,t^5 + t^6$

15. $a^9 - 18a^7 + 144a^5 - 672a^3 + 2016a - 4032a^{-1} +$ $5376a^{-3} - 4608a^{-5} + 2304a^{-7} - 512a^{-9}$ **17.** $140\sqrt{2}$

19. $x^{-8} + 4x^{-4} + 6 + 4x^4 + x^8$ **21.** $21a^5b^2$

23. $-252x^5y^5$ **25.** $-745{,}472a^3$ **27.** $1120x^{12}y^2$

29. $-1{,}959{,}552u^5v^{10}$ **31.** 2^7, or 128

33. 2^{24}, or 16,777,216   **35.** 20   **37.** $-12 + 316i$

39. $-7 - 4\sqrt{2}i$ **41.** $\displaystyle\sum_{k=0}^{n} \binom{n}{k}(-1)^k a^{n-k}b^k$

43. $\displaystyle\sum_{k=1}^{n} \binom{n}{k} x^{n-k}h^{k-1}$ **45.** Discussion and Writing

46. $[2.2]\ x^2 + 2x - 2$ **47.** $[2.2]\ 2x^3 - 3x^2 + 2x - 3$

48. $[2.3]\ 4x^2 - 12x + 10$ **49.** $[2.3]\ 2x^2 - 1$

51. $3, 9, 6 \pm 3i$ **53.** $-4320x^6y^{9/2}$

55. $-\dfrac{35}{x^{1/6}}$ **57.** 2^{100} **59.** $[\log_a (xt)]^{23}$

61. (1) *Basis step*: Since $a + b = (a + b)^1$, S_1 is true.
(2) *Induction step*: Let S_k be the statement of the binomial theorem with n replaced by k. Multiply both sides of S_k by $(a + b)$ to obtain

$$\begin{aligned} & (a + b)^{k+1} \\ &= \left[a^k + \cdots + \binom{k}{r-1}a^{k-(r-1)}b^{r-1} \right. \\ &\qquad + \left. \binom{k}{r}a^{k-r}b^r + \cdots + b^k \right](a + b) \\ &= a^{k+1} + \cdots + \left[\binom{k}{r-1} + \binom{k}{r} \right]a^{(k+1)-r}b^r \\ &\qquad + \cdots + b^{k+1} \\ &= a^{k+1} + \cdots + \binom{k+1}{r}a^{(k+1)-r}b^r + \cdots + b^{k+1}. \end{aligned}$$

This proves S_{k+1}, assuming S_k. Hence S_n is true for $n = 1, 2, 3, \ldots$.

Exercise Set 11.8

1. (a) 0.18, 0.24, 0.23, 0.23, 0.12; **(b)** Opinions may vary, but it seems that people tend not to pick the first or last numbers.
3. 11,700 pieces   **5. (a)** T, S, R, N, L; **(b)** E; **(c)** yes
7. (a) $\frac{2}{7}$; **(b)** $\frac{5}{7}$; **(c)** 0; **(d)** 1   **9.** $\frac{350}{31,977}$ **11.** $\frac{1}{108,290}$
13. $\frac{33}{66,640}$ **15. (a)** HHH, HHT, HTH, HTT, THH, THT, TTH, TTT; **(b)** $\frac{3}{8}$; **(c)** $\frac{7}{8}$; **(d)** $\frac{7}{8}$; **(e)** $\frac{3}{8}$ **17.** $\frac{9}{19}$ **19.** $\frac{1}{38}$
21. $\frac{18}{19}$ **23.** $\frac{9}{19}$ **25.** Answers will vary.
27. Discussion and Writing   **28.** [1.5] Zero
29. [5.1] One-to-one   **30.** [1.2] Function; domain; range; domain; range   **31.** [1.5] Zero   **32.** [11.6] Combination
33. [2.5] Inverse variation   **34.** [4.3] Factor
35. [11.3] Geometric sequence
37. (a) 36; **(b)** $\dfrac{36}{_{52}C_5} \approx 1.39 \times 10^{-5}$
39. (a) $(13 \cdot {}_4C_3) \cdot (12 \cdot {}_4C_2) = 3744$; **(b)** $\dfrac{3744}{_{52}C_5} \approx 0.00144$
41. (a) $4 \cdot \dbinom{13}{5} - 4 - 36 = 5108$; **(b)** 0.00197
43. (a) $\dbinom{10}{1}\dbinom{4}{1}\dbinom{4}{1}\dbinom{4}{1}\dbinom{4}{1}\dbinom{4}{1} - 4 - 36 = 10,200$; **(b)** 0.00392

Review Exercises: Chapter 11

1. True   **2.** False   **3.** True   **4.** False   **5.** $-\frac{1}{2}, \frac{4}{17}, -\frac{9}{82}$, $\frac{16}{257}; -\frac{121}{14,642}; -\frac{529}{279,842}$ **6.** $(-1)^{n+1}(n^2 + 1)$
7. $\frac{3}{2} - \frac{9}{8} + \frac{27}{26} - \frac{81}{80} = \frac{417}{1040}$

8.

n	u_n
1	0.3
2	2.5
3	13.5
4	68.5
5	343.5
6	1718.5
7	8593.5
8	42968.5
9	214843.5
10	1074218.5

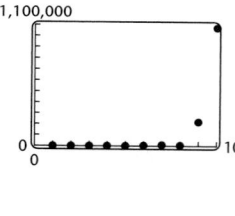

9. $\displaystyle\sum_{k=1}^{7} (k^2 - 1)$ **10.** $\frac{15}{4}$
11. $a + 4b$ **12.** 531   **13.** 20,100   **14.** 11   **15.** -4
16. $n = 6, S_n = -126$ **17.** $a_1 = 8, a_5 = \frac{1}{2}$
18. Does not exist   **19.** $\frac{3}{11}$ **20.** $\frac{3}{8}$ **21.** $\frac{241}{99}$
22. $5\frac{4}{5}, 6\frac{3}{5}, 7\frac{2}{5}, 8\frac{1}{5}$ **23.** 167.3 ft   **24.** $45,993.04
25. (a) $7.38; **(b)** $1365.10   **26.** $88,888,888,889

27. S_n: $1 + 4 + 7 + \cdots + (3n - 2) = \dfrac{n(3n - 1)}{2}$

S_1: $1 = \dfrac{1(3 - 1)}{2}$

S_k: $1 + 4 + 7 + \cdots + (3k - 2) = \dfrac{k(3k - 1)}{2}$

S_{k+1}: $1 + 4 + 7 + \cdots + (3k - 2) + [3(k + 1) - 2]$
$= 1 + 4 + 7 + \cdots + (3k - 2) + (3k + 1)$
$= \dfrac{(k + 1)(3k + 2)}{2}$

(1) *Basis step:* $\dfrac{1(3 - 1)}{2} = \dfrac{2}{2} = 1$ is true.
(2) *Induction step:* Assume S_k. Add $(3k + 1)$ on both sides.

$1 + 4 + 7 + \cdots + (3k - 2) + (3k + 1)$

$= \dfrac{k(3k - 1)}{2} + (3k + 1)$

$= \dfrac{k(3k - 1)}{2} + \dfrac{2(3k + 1)}{2}$

$= \dfrac{3k^2 - k + 6k + 2}{2}$

$= \dfrac{3k^2 + 5k + 2}{2}$

$= \dfrac{(k + 1)(3k + 2)}{2}$

28. S_n: $1 + 3 + 3^2 + \cdots + 3^{n-1} = \dfrac{3^n - 1}{2}$

S_1: $1 = \dfrac{3^1 - 1}{2}$

S_k: $1 + 3 + 3^2 + \cdots + 3^{k-1} = \dfrac{3^k - 1}{2}$

S_{k+1}: $1 + 3 + 3^2 + \cdots + 3^{(k+1)-1} = \dfrac{3^{k+1} - 1}{2}$

(1) *Basis step:* $\dfrac{3^1 - 1}{2} = \dfrac{2}{2} = 1$ is true.
(2) *Induction step:* Assume S_k. Add 3^k on both sides.

$1 + 3 + \cdots + 3^{k-1} + 3^k$

$= \dfrac{3^k - 1}{2} + 3^k = \dfrac{3^k - 1}{2} + 3^k \cdot \dfrac{2}{2}$

$= \dfrac{3 \cdot 3^k - 1}{2} = \dfrac{3^{k+1} - 1}{2}$

29.

$$S_n: \quad \left(1 - \frac{1}{2}\right)\left(1 - \frac{1}{3}\right) \cdots \left(1 - \frac{1}{n}\right) = \frac{1}{n}$$

$$S_2: \quad \left(1 - \frac{1}{2}\right) = \frac{1}{2}$$

$$S_k: \quad \left(1 - \frac{1}{2}\right)\left(1 - \frac{1}{3}\right) \cdots \left(1 - \frac{1}{k}\right) = \frac{1}{k}$$

$$S_{k+1}: \quad \left(1 - \frac{1}{2}\right)\left(1 - \frac{1}{3}\right) \cdots \left(1 - \frac{1}{k}\right)\left(1 - \frac{1}{k+1}\right)$$
$$= \frac{1}{k+1}.$$

(1) *Basis step*: S_2 is true by substitution.

(2) *Induction step*: Assume S_k. Deduce S_{k+1}. Starting with the left side of S_{k+1}, we have

$$\left(1 - \frac{1}{2}\right)\left(1 - \frac{1}{3}\right) \cdots \left(1 - \frac{1}{k}\right)\left(1 - \frac{1}{k+1}\right)$$

$$= \frac{1}{k} \cdot \left(1 - \frac{1}{k+1}\right) \qquad \text{By } S_k$$

$$= \frac{1}{k} \cdot \left(\frac{k+1-1}{k+1}\right)$$

$$= \frac{1}{k} \cdot \frac{k}{k+1}$$

$$= \frac{1}{k+1}. \qquad \textbf{Simplifying}$$

30. $6! = 720$ **31.** $9 \cdot 8 \cdot 7 \cdot 6 = 3024$

32. $\binom{15}{8} = 6435$ **33.** $24 \cdot 23 \cdot 22 = 12,144$

34. $\dfrac{9!}{1! \, 4! \, 2! \, 2!} = 3780$ **35.** $3 \cdot 4 \cdot 3 = 36$

36. **(a)** $_6P_5 = 720$; **(b)** $6^5 = 7776$; **(c)** $_5P_4 = 120$;
(d) $_3P_2 = 6$ **37.** 2^8, or 256

38. $m^7 + 7m^6 n + 21m^5 n^2 + 35m^4 n^3 + 35m^3 n^4 + 21m^2 n^5 + 7mn^6 + n^7$

39. $x^5 - 5\sqrt{2}\,x^4 + 20x^3 - 20\sqrt{2}\,x^2 + 20x - 4\sqrt{2}$

40. $x^8 - 12x^6 y + 54x^4 y^2 - 108x^2 y^3 + 81y^4$

41. $a^8 + 8a^6 + 28a^4 + 56a^2 + 70 + 56a^{-2} + 28a^{-4} + 8a^{-6} + a^{-8}$ **42.** $-6624 + 16,280i$

43. $220a^9 x^3$ **44.** $-\binom{18}{11} 128 a^7 b^{11}$ **45.** $\frac{1}{12}; 0$ **46.** $\frac{1}{4}$

47. $\frac{6}{5525}$ **48.** $\frac{86}{206} \approx 0.42, \frac{97}{206} \approx 0.47, \frac{23}{206} \approx 0.11$

49. **(a)** $a_n = 0.175n + 10.0165$; **(b)** 11.417 million self-employed workers   **50.** B   **51.** A   **52.** D

53. Discussion and Writing: A list of 9 candidates for an office is to be narrowed down to 4 candidates. In how many ways can this be done?

54. Discussion and Writing: Someone who has managed several sequences of hiring has a considerable income from the sales of the people in the lower levels. However, with a finite population, it will not be long before the salespersons in the lowest level have no one to hire and no one to sell to.

55. S_1 fails for both (a) and (b).

56. $\dfrac{a_{k+1}}{a_k} = r_1, \dfrac{b_{k+1}}{b_k} = r_2$, so $\dfrac{a_{k+1} b_{k+1}}{a_k b_k} = r_1 r_2$, a constant.

57. **(a)** No (unless a_n is all positive or all negative); **(b)** yes; **(c)** yes; **(d)** no (unless a_n is constant); **(e)** no (unless a_n is constant); **(f)** no (unless a_n is constant)

58. $-2, 0, 2, 4$ **59.** $\frac{1}{2}, -\frac{1}{6}, \frac{1}{18}$ **60.** $\left(\log \dfrac{x}{y}\right)^{10}$ **61.** 18

62. 36   **63.** -9

Test: Chapter 11

1. [11.1] -43 **2.** [11.1] $\frac{2}{3}, \frac{3}{4}, \frac{4}{5}, \frac{5}{6}, \frac{6}{7}$
3. [11.1] $2 + 5 + 10 + 17 = 34$

4. [11.1]

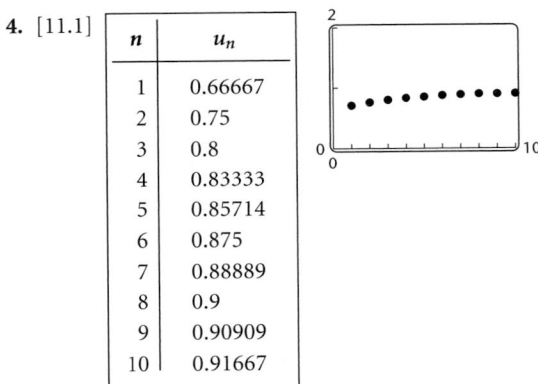

n	u_n
1	0.66667
2	0.75
3	0.8
4	0.83333
5	0.85714
6	0.875
7	0.88889
8	0.9
9	0.90909
10	0.91667

5. [11.1] $\displaystyle\sum_{k=1}^{6} 4k$

6. [11.1] $\displaystyle\sum_{k=1}^{\infty} 2^k$ **7.** [11.1] $3, 2\frac{1}{3}, 2\frac{3}{7}, 2\frac{7}{17}$ **8.** [11.2] 44

9. [11.2] 38   **10.** [11.2] -420 **11.** [11.2] 675
12. [11.3] $\frac{5}{512}$ **13.** [11.3] 1000   **14.** [11.3] 510
15. [11.3] 27   **16.** [11.3] $\frac{56}{99}$
17. [11.1] \$10,000, \$8000, \$6400, \$5120, \$4096, \$3276.80
18. [11.2] \$12.50   **19.** [11.3] \$74,399.77
20. [11.4]

$$S_n: \quad 2 + 5 + 8 + \cdots + (3n - 1) = \frac{n(3n + 1)}{2}$$

$$S_1: \quad 2 = \frac{1(3 \cdot 1 + 1)}{2}$$

$$S_k: \quad 2 + 5 + 8 + \cdots + (3k - 1) = \frac{k(3k + 1)}{2}$$

$$S_{k+1}: \quad 2 + 5 + 8 + \cdots + (3k - 1) + [3(k + 1) - 1]$$
$$= \frac{(k + 1)[3(k + 1) + 1]}{2}$$

(1) *Basis step*: $\dfrac{1(3 \cdot 1 + 1)}{2} = \dfrac{1 \cdot 4}{2} = 2$, so S_1 is true.

(2) *Induction step*:

$$2 + 5 + 8 + \cdots + (3k - 1) + [3(k + 1) - 1]$$
$$= \frac{k(3k + 1)}{2} + [3k + 3 - 1] \qquad \text{By } S_k$$
$$= \frac{3k^2}{2} + \frac{k}{2} + 3k + 2$$
$$= \frac{3k^2}{2} + \frac{7k}{2} + 2$$
$$= \frac{3k^2 + 7k + 4}{2}$$
$$= \frac{(k + 1)(3k + 4)}{2}$$
$$= \frac{(k + 1)[3(k + 1) + 1]}{2}$$

21. [11.5] 3,603,600   **22.** [11.6] 352,716

23. [11.6] $\dfrac{n(n - 1)(n - 2)(n - 3)}{24}$ **24.** [11.5] $_6P_4 = 360$

25. [11.5] **(a)** $6^4 = 1296$; **(b)** $_5P_3 = 60$

26. [11.6] $_{28}C_4 = 20,475$ **27.** [11.6] $_{12}C_8 \cdot {}_8C_4 = 34,650$

28. [11.7] $x^5 + 5x^4 + 10x^3 + 10x^2 + 5x + 1$

29. [11.7] $35x^3y^4$ **30.** [11.7] $2^9 = 512$ **31.** [11.8] $\frac{4}{7}$

32. [11.8] $\frac{48}{1001}$ **33.** [11.1] B   **34.** [11.5] 15

APPENDIX

1. Acute   **2.** Obtuse   **3.** Straight   **4.** Acute
5. Obtuse   **6.** Right   **7.** 79°   **8.** 7°   **9.** 23°
10. 85°   **11.** 32°   **12.** 58°   **13.** 61°   **14.** 36°
15. 177°   **16.** 126°   **17.** 41°   **18.** 167°   **19.** 95°
20. 51°   **21.** 78°   **22.** 135°   **23.** Congruent
24. Congruent
25. $m\angle 2 = 67°$, $m\angle 3 = 33°$, $m\angle 4 = 80°$, $m\angle 6 = 33°$
26. $m\angle 1 = 56°$, $m\angle 3 = 82°$, $m\angle 5 = 42°$, $m\angle 6 = 82°$
27. (a) $\angle 1$ and $\angle 3$, $\angle 2$ and $\angle 4$, $\angle 8$ and $\angle 6$, $\angle 7$ and $\angle 5$;
(b) $\angle 2$, $\angle 3$, $\angle 6$, $\angle 7$;   **(c)** $\angle 2$ and $\angle 6$, $\angle 3$ and $\angle 7$
28. (a) $\angle 1$ and $\angle 5$, $\angle 2$ and $\angle 6$, $\angle 3$ and $\angle 7$, $\angle 4$ and $\angle 8$;
(b) $\angle 3$, $\angle 4$, $\angle 5$, $\angle 6$;   **(c)** $\angle 3$ and $\angle 5$, $\angle 4$ and $\angle 6$
29. $m\angle 6 = m\angle 2 = m\angle 8 = 125°$;
$m\angle 5 = m\angle 3 = m\angle 7 = m\angle 1 = 55°$
30. $m\angle 2 = m\angle 6 = m\angle 4 = 34°$;
$m\angle 3 = m\angle 5 = m\angle 7 = m\angle 1 = 146°$

31. $\angle ABE \cong \angle DCE$, 95°; $\angle BAE \cong \angle CDE$; $\angle AEB \cong \angle DEC$; $\angle BED \cong \angle AEC$
32. $\angle ABC \cong \angle DCB$, 133°; $\angle ABE \cong \angle DCF$, 47°
33. $\angle ABC \cong \angle C$; $\angle A \cong \angle CDA$; $\angle AEB \cong \angle CED$; $\angle ABC \cong \angle BED$
34. $\angle TPQ \cong \angle TRS$, $\angle TQP \cong \angle TSR$
35. $\angle R \leftrightarrow \angle A$, $\angle S \leftrightarrow \angle B$, $\angle T \leftrightarrow \angle C$; $\overline{RS} \leftrightarrow \overline{AB}$, $\overline{RT} \leftrightarrow \overline{AC}$, $\overline{ST} \leftrightarrow \overline{BC}$
36. $\angle D \leftrightarrow \angle C$, $\angle E \leftrightarrow \angle N$, $\angle F \leftrightarrow \angle B$; $\overline{DE} \leftrightarrow \overline{CN}$, $\overline{EF} \leftrightarrow \overline{NB}$, $\overline{FD} \leftrightarrow \overline{BC}$
37. $\angle C \leftrightarrow \angle W$, $\angle B \leftrightarrow \angle J$, $\angle S \leftrightarrow \angle Z$; $\overline{CB} \leftrightarrow \overline{WJ}$, $\overline{CS} \leftrightarrow \overline{WZ}$, $\overline{BS} \leftrightarrow \overline{JZ}$
38. $\angle B \leftrightarrow \angle S$, $\angle V \leftrightarrow \angle T$, $\angle W \leftrightarrow \angle A$; $\overline{BV} \leftrightarrow \overline{ST}$, $\overline{VW} \leftrightarrow \overline{TA}$, $\overline{WB} \leftrightarrow \overline{AS}$

39. $\angle A \leftrightarrow \angle R$, $\angle B \leftrightarrow \angle S$, $\angle C \leftrightarrow \angle T$; $\dfrac{AB}{RS} = \dfrac{AC}{RT} = \dfrac{BC}{ST}$

40. $\angle P \leftrightarrow \angle S$, $\angle Q \leftrightarrow \angle T$, $\angle R \leftrightarrow \angle V$; $\dfrac{PQ}{ST} = \dfrac{PR}{SV} = \dfrac{QR}{TV}$

41. $\angle M \leftrightarrow \angle C$, $\angle E \leftrightarrow \angle L$, $\angle S \leftrightarrow \angle F$; $\dfrac{ME}{CL} = \dfrac{MS}{CF} = \dfrac{ES}{LF}$

42. $\angle S \leftrightarrow \angle W$, $\angle M \leftrightarrow \angle L$, $\angle H \leftrightarrow \angle K$; $\dfrac{SM}{WL} = \dfrac{SH}{WK} = \dfrac{MH}{LK}$

43. $\dfrac{PS}{ND} = \dfrac{SQ}{DM} = \dfrac{PQ}{NM}$ **44.** $\dfrac{GH}{UT} = \dfrac{GK}{UV} = \dfrac{HK}{TV}$

45. $\dfrac{TA}{GF} = \dfrac{TW}{GC} = \dfrac{AW}{FC}$ **46.** $\dfrac{JK}{ON} = \dfrac{JL}{OM} = \dfrac{KL}{NM}$

47. $QR = 10$, $PR = 8$ **48.** $AM = 24$, $GT = 25$
49. 36 ft   **50.** 33 ft   **51.** 100 ft   **52.** 72 ft   **53.** 17
54. $\sqrt{34} \approx 5.831$ **55.** $\sqrt{32} \approx 5.657$ **56.** $\sqrt{98} \approx 9.899$
57. 12   **58.** 5   **59.** 4   **60.** $\sqrt{31} \approx 5.568$ **61.** 26
62. 13   **63.** 12   **64.** 24   **65.** 2   **66.** 1
67. $\sqrt{2} \approx 1.414$ **68.** $\sqrt{8} \approx 2.828$ **69.** 5
70. $\sqrt{50} \approx 7.071$ **71.** 3   **72.** $\sqrt{5} \approx 2.236$
73. $\sqrt{211,200,000}$ ft $\approx 14,533$ ft
74. $\sqrt{1850}$ yd ≈ 43.012 yd   **75.** 240 ft   **76.** 25 ft
77. $\sqrt{18}$ cm ≈ 4.243 cm   **78.** $\sqrt{75}$ m ≈ 8.660 m
79. $\sqrt{208}$ ft ≈ 14.422 ft   **80.** $\sqrt{26,900}$ yd ≈ 164.012 yd

Index

Abscissa, 63
Absolute value, 6, 55
   of complex numbers, 664, 712
   equations with, 284, 289
   inequalities with, 286, 289
   and roots, 47, 55
Absolute value function, 203
Absorption, coefficient of, 463
ac-method of factoring, 27
Acceleration due to gravity, 597, 632
Acute angle, 472, 500, 971
Addition
   associative property, 5, 55
   commutative property, 5, 55
   of complex numbers, 238
   of exponents, 10, 55
   of functions, 182, 228
   of logarithms, 426, 432, 464
   of matrices, 757, 759, 805
   of ordinates, 557
   of polynomials, 19, 20
   of radical expressions, 48
   of rational expressions, 39, 40
   of vectors, 690, 700, 701, 713
Addition principle
   for equations, 33, 55, 129, 156
   for inequalities, 150, 156
Addition properties
   for matrices, 759, 805
   for real numbers, 5, 55

Additive identity
   for matrices, 758, 759
   for real numbers, 5, 55
   for vectors, 701
Additive inverse
   for matrices, 758, 759
   for real numbers, 5, 55
Aerial navigation, 513
Algebra, fundamental theorem of, 332, 372
Algebra of functions, 182, 228
   and domains, 184
Alternate interior angles, 977
Alternating sequence, 891
Ambiguous case, 645
Amount of an annuity, 916
Amplitude, 538, 549, 555, 567
Angles, 498
   acute, 472, 500, 971
   alternate interior, 977
   between lines, 587
   between vectors, 706, 713
   central, 521
   complementary, 480, 501, 973
   congruent, 975
   corresponding, 977
   coterminal, 499
   decimal degree form, 478
   of depression, 488
   direction, 703
   D°M′S″ form, 478

   drift, 692
   of elevation, 488
   initial side, 498
   interior, 977
   negative rotation, 498
   obtuse, 500, 971
   positive rotation, 498
   quadrantal, 506
   radian measure, 518
   reference, 507, 508
   right, 472, 500, 971
   standard position, 498
   straight, 500, 971
   supplementary, 501, 974
   terminal side, 498
   vertex, 498
   vertical, 976
Angular speed, 522
Annuity, 916
Applications, *see Index of Applications*
Arc length, 521
Area, territorial, 95
Area of a triangle, 649, 712
Argument of $a + bi$, 665
Arithmetic mean, 909
Arithmetic progressions, *see* Arithmetic sequences
Arithmetic sequences, 900
   common difference, 900, 965
   nth term, 901
   sums, 904, 965

Arithmetic series, 903
ASK mode, 132
Associative properties
    for matrices, 759, 805
    for real numbers, 5, 55
Astronomical unit, 16
Asymptotes
    of exponential functions, 397
    horizontal, 347, 349, 351, 373
    of a hyperbola, 834, 837
    of logarithmic functions, 411
    oblique, 350, 351, 373
    slant, 350
    vertical, 345, 351, 373
Atmospheric pressure, 462
AU (astronomical unit), 16
Augmented matrix, 748, 770
AUTO mode, 67
Average, 909
Average rate of change, 103, 185
Axes, 62. *See also* Axis.
    of an ellipse, 824
    of a hyperbola, 833, 834
    rotation of, 855–860, 882
Axis
    conjugate, 834
    imaginary, 664
    major, 824
    minor, 824
    polar, 676
    real, 664
    of symmetry, 262, 289, 814
    transverse, 833
    x-, 62
    y-, 62

Back-substitution, 722
Bar graph, 62
Base
    change of, 415, 464
    converting base b to base e, 455
    of an exponential function, 394, 395
    in exponential notation, 9
    of a logarithmic function, 410
Base e logarithm, 414
Base–exponent property, 435, 464
Base-10 logarithm, 413
Basis step, 923
Bearing, 489, 511
Beer–Lambert law, 463
Bel, 425
Bevel, 659
Binomial, *see* Binomials

Binomial coefficient notation, 940, 949
Binomial expansion
    using factorial notation, 949, 965
    $(k + 1)$st term, 950, 966
    using Pascal's triangle, 948
Binomial theorem, 948, 949, 965
Binomials, 19
    product of, 21, 55
Boiling point and elevation, 95
Book value, 909
Boyle's law, 226
Break-even point, 733

Calculator. *See also* Graphing calculator.
    logarithms on, 413
    trigonometric functions on, 479, 507, 509, 534
Canceling, 37
Carbon dating, 451, 453
Cardioid, 681
Carry out, 133, 156
Cartesian coordinate system, 62, 63
Catenary, 820
Center
    of circle, 71, 822
    of ellipse, 824
    of hyperbola, 833
Central angle, 521
Change-of-base formula, 415, 464
Checking the solution, 33, 34, 132, 133, 152, 156, 277, 281, 619
Circle APP, 823
Circle graph, 62
Circles, 71, 814, 822
    center, 71, 822
    equation, standard, 72, 156, 822
    graph, 72, 823
    radius, 71, 822
    unit, 79, 514
Circular arrangements, 937
Circular functions, 531, 567
Cissoid, 687
Clearing fractions, 276
Closed interval, 4
Coefficient, 18
    of absorption, 463
    binomial, 940, 949
    of determination, 306
    of linear correlation, 123
    leading, 18, 296
    of a polynomial, 18, 332
    of a polynomial function, 333, 334

Coefficient matrix, 748
Cofactor, 777
Cofunctions, 481, 566, 589, 590, 633
Collecting like (or similar) terms, 19
Column matrix, 763
Columns of a matrix, 748
Combination, linear, of unit vectors, 702
Combinations, 938–943, 965
Combinatorics, 928. *See also*
    Combinations; Permutations.
Combined variation, 222, 228
Combining like (or similar) terms, 19
Common denominator, 39, 40
Common difference, 900, 965
Common factors, 23
Common logarithms, 413
Common ratio, 918, 965
Commutative properties
    for matrices, 759, 805
    for real numbers, 5, 55
Complementary angles, 480, 501, 973
Completing the square, 247, 263, 817
Complex conjugates, 240, 289
    as zeros, 333
Complex numbers, 236, 237, 289
    absolute value, 664, 712
    addition, 238
    argument, 665
    conjugates of, 240, 289
    division of, 241, 669, 712
    graphical representation, 664
    imaginary, 237
    imaginary part, 237
    multiplication of, 239, 668, 712
    polar notation, 665
    powers of, 670, 712
    powers of i, 239, 240
    pure imaginary, 237
    real part, 237
    roots of, 671, 712
    standard notation, 665
    subtraction, 238
    trigonometric notation, 665, 712
Complex rational expressions, 41
Component form of a vector, 697
Components of a vector, 692
    scalar, 697
Composite function, 189, 191
Composition of functions, 191
    and inverse functions, 388, 612, 613, 634
    trigonometric, 612, 613, 634
Compound inequality, 151

Compound interest, 14, 17, 55, 398, 447, 464
Congruent angles, 975
Conic sections, 814, 865
    circles, 814, 822
    classifying equations, 839
    degenerate, 861
    directrix, 865
    discriminant, 861, 882
    eccentricity, 831, 841, 865, 866
    ellipses, 814, 824, 867, 882
    focus, 865
    hyperbolas, 814, 833, 867, 882
    parabolas, 814, 867, 882
    polar equations, 867, 882
    and rotation of axes, 855–860, 882
Conics APP, 816, 823, 828, 836
Conjugate axis of hyperbola, 834
Conjugates
    of complex numbers, 240, 289
        as zeros, 333
    of radical expressions, 50
Conjunction, 151
CONNECTED mode, 174, 343
Connecting the Concepts, 30, 90, 141, 143, 256, 262, 301, 328, 397, 411, 540, 543, 611, 659, 839, 943
Consistent system of equations, 721
Constant function, 97, 166
Constant of proportionality, 219, 221
Constant term, 18
Constant, variation, 219, 221
Constraints, 790
Continuous compounding of interest, 447, 464
Continuous functions, 298
Converting base b to base e, 455
Converting between degree and radian measure, 519
Converting between D°M′S″ and decimal degree notation, 478
Converting between exponential and logarithmic equations, 412
Converting between parametric and rectangular equations, 873–876
Converting between polar and rectangular coordinates, 678, 679
Converting between polar and rectangular equations, 679, 680, 870
Converting between trigonometric and standard notation, 666, 667
Cooling, Newton's law of, 460

Coordinate system
    Cartesian, 62, 63
    polar, 676
Coordinates, 63
Correlation, coefficient of, 123
Correspondence, 80. See also Functions; Relation.
Corresponding angles, 977
Cosecant function, 473, 502, 531, 543, 566, 567
    graph, 542
Cosine function, 472, 473, 502, 531
    amplitude, 538, 540
    domain, 537, 540
    graph, 536, 540
    period, 538, 540
    range, 537, 540
Cosines, law of, 655, 712
Costs, 114
Cotangent function, 473, 502, 531, 542, 543, 566, 567
Coterminal angles, 499
Counting principle, fundamental, 929, 965
Cramer's rule, 779–782, 806
Critical value, 367
Cube root, 45
Cube root function, 203
Cubes, factoring sums and differences, 29, 55
Cubic function, 296, 297
Cubic regression, 306
Cubing function, 203
Curve fitting, 120. See also Mathematical models; Regression.
Curve, plane, 872, 873
Cycloid, 878

Damped oscillation, 559
Dead Sea scrolls, 453
Decay, exponential, 451, 464
Decay rate, 451, 452
Decibel, 425
Decimal degree form, 478
Decimal notation, 2, 3
Decomposing a function, 195
Decomposition, partial fractions, 798, 806
Decreasing function, 166
Degenerate conic section, 861
Degree
    of a polynomial, 18, 19
    of a polynomial function, 296
    of a term, 19

Degree measure, 478, 499
    converting to radians, 519
    decimal form, 478
    D°M′S″ form, 478
DeMoivre, Abraham, 670
DeMoivre's theorem, 670, 712
Demand, 728
Denominator
    least common, 39, 40
    rationalizing, 49, 50
Dependent equations, 722
Dependent variable, 66
Depreciation, straight-line, 113
Depression, angle of, 488
Descartes, René, 63
Descartes' rule of signs, 338, 372
Descending order, 18
Determinants, 776–782, 805, 806
Determination, coefficient of, 306
DIAGNOSTIC, 123, 306
Diagonals of a polygon, 371
Difference. See also Subtraction.
    common, 900, 965
    of cubes, factoring, 29, 55
    of functions, 182, 228
    of logarithms, 428, 432, 464
    square of, 21
    of squares, factoring, 28
Difference identities, 580–584, 633
Difference quotient, 185
Dimensions of a matrix, 748
Direct variation, 219, 228
Direction angle, 703
Direction of a vector, 689
Directly proportional, 219
Directrix
    of a conic section in polar form, 865
    of a parabola, 814
Discontinuous function, 298
Discriminant, 251, 252, 861, 882
Disjunction, 152
Displacement, 688
Distance. See also Distance formula.
    of a fall, 253
    on the number line, 6
    projectile motion, 268
Distance formula. See also Distance.
    for motion, 135, 156
    between points, 69, 156
Distributive property
    for matrices, 759, 805
    for real numbers, 5, 55
Dividend, 324

Division
    of complex numbers, 241, 669, 712
    of exponential expressions, 10, 55
    of functions, 182, 228
    of polynomials, 323–327, 372
    of radical expressions, 47, 55
    of rational expressions, 38, 39
    synthetic, 325, 326
Divisor, 324
D°M′S″ form, 478
Domain
    and the algebra of functions, 184
    of a function, 81, 87, 88
        exponential, 397
        logarithmic, 411
        polynomial, 298
        rational, 343
        trigonometric, 537, 540, 543, 611
    of a rational expression, 36
    of a relation, 82
    restricting, 388
DOT mode, 174, 343
Dot product, 706, 713
Double angle identities, 590–592, 633
Doubling time, 447
    and growth rate, 449, 464
DRAW menu, 73, 786, 787
Drift angle, 692

e, 400, 464
Earthquake magnitude, 421
Eccentricity
    of a conic in polar form, 865
    of an ellipse, 831, 866
    of a hyperbola, 841, 866
Economic multiplier effect, 917
Element
    in a function's domain, 81
    of a set, 3
Elevation
    angle of, 488
    and boiling point, 95
Eliminating the parameter, 873
Eliminating the xy-term, 859, 882
Elimination
    Gauss–Jordan, 751
    Gaussian, 737
Elimination method, 722
Ellipse, 814, 824, 881, 882
    applications, 829
    center, 824
    eccentricity, 831, 866
    equations, 825, 827, 861, 867, 881, 882
    foci, 824, 825, 827

graphs, 826
major axis, 824
minor axis, 824
vertices, 824, 825, 827
x-, y-intercepts, 824, 825
Ellipse APP, 828
Empty set, 286
Endpoints, 3
Entries of a matrix, 748
Equality of matrices, 756
Equally likely outcomes, 957
Equation, 32, 129
    with absolute value, 284, 289
    of circles, 72, 156, 822, 861, 867, 881, 882
    of ellipses, 825, 827, 861, 867, 881, 882
    equivalent, 32, 129
    exponential, 435
    graphs of, 64, 65. See also Graphs.
    of hyperbolas, 834, 837, 861, 867, 882
    of inverse relations, 380
    linear, 32, 106, 472, 736
    logarithmic, 439
    logistic, 450
    matrix, 763
    in one variable, 129
    of parabolas, 815, 817, 861, 867, 881, 882
    parametric, 873
    point–slope, 116, 119, 156
    polar, 679
        of conics, 867, 882
    Pythagorean, 984
    quadratic, 32, 244, 289
    quadratic in form, 252
    radical, 279
    rational, 276
    rectangular, 679
    reducible to quadratic, 252
    regression, 122, 306, 454, 460, 462, 742
    related, 361, 785
    roots of, 244. See also Solutions of equations.
    slope–intercept, 106, 115, 119, 156
    solutions of, 32, 63, 129, 143, 256, 301, 720, 736
    solving, 32, 33, 55, 129, 156, 244, 247, 249, 252, 276, 279, 284, 289, 435, 439
    solving systems of, 720, 721, 722, 737, 749, 751, 772, 779–782, 805, 806, 843
    systems of, 720, 843

translating to, 133, 156
    trigonometric, 619
Equation-editor screen, 66
Equilateral triangle, 53
Equilibrium, forces in, 707
Equilibrium point, 728
Equilibrium price, 728
Equivalent equations, 32, 129
    systems, 724
Equivalent expressions, 38
Equivalent inequalities, 150
Equivalent vectors, 689, 698
Euler, Leonhard, 400
Evaluating a determinant, 776–779, 805
Evaluating a function, 84
Even functions, 202, 228
Even multiplicity of zeros, 303
Even roots, 47, 55
Event, 957
Experimental probability, 954, 955, 966
Exploring with Technology, 100, 105, 261, 396, 535, 537, 539, 575, 673, 681, 902, 912
Exponent(s), 9
    integer, 9
    and logarithms, 412, 464
    properties of, 10, 55
    rational, 50, 56
    of zero, 9
Exponential decay model, 451, 464
Exponential equations, 435
Exponential function, 394, 464
    base, 394, 395
    base e, 400
    converting base b to base e, 455
    domain, 397
    graphs, 395, 401
    inverses, 408. See also Logarithmic functions.
    properties, 397
    range, 397
    y-intercept, 397
Exponential growth model, 446
    and doubling time, 447, 449, 464
    growth rate, 446
Exponential notation, 9
Exponential regression, 454
Expressions
    complex rational, 41
    equivalent, 38
    radical, 46
    rational, 36
    with rational exponents, 50, 56

Factor, common, 23
Factor theorem, 327, 372
Factorial notation, 931, 965
   and binomial expansion, 949
Factoring
   ac-method, 27
   common factors, terms with, 23
   completely, 29
   differences of cubes, 29, 55
   differences of squares, 28
   FOIL method, 26
   by grouping, 23
   grouping method for trinomials, 27
   polynomial functions, 327, 332, 372
   polynomials, 23–30, 55
   squares of binomials, 29
   strategy, 30
   sums of cubes, 29, 55
   trinomials, 24–28
Factors of polynomials, 23, 327, 372
Familiarize, 133, 156
Feasible solutions, 791
Fibonacci sequence, 898
Finite sequence, 890
Finite series, 893
First coordinate, 63
Fitting a curve to data, 122
Fixed costs, 114
Foci, *see* Focus
Focus
   of a conic in polar form, 865
   of an ellipse, 824, 825, 827
   of a hyperbola, 833, 834, 837
   of a parabola, 814
FOIL, 21
   and factoring, 26
Force, 688
Forces in equilibrium, 707
Formulas, 141
   change-of-base, 415, 464
   compound interest, 14, 17, 55
   diagonals of a polygon, 371
   distance (between points), 69, 156
   games in a sports league, 310
   Heron's, 663
   for inverses of functions, 385
   midpoint, 70, 156
   mortgage payment, 58
   motion, 135, 156
   quadratic, 249, 289
   recursion, 895
   rotation of axes, 857, 882
   savings plan, 17

simple interest, 136, 156
solving for a given variable, 141
FRAC feature, 51
Fractional equations, *see* Rational
   equations
Fractional exponents, 50, 56
Fractional expressions, 36. *See also*
   Rational expressions.
Fractions
   clearing, 276
   partial, 798, 806
Free fall, 253
Functions, 80, 81
   absolute value, 203
   algebra of, 182, 228
   amplitude, 538
   circular, 531, 567
   cofunctions, 481, 566
   composite, 189, 191
   composition of, 191, 612, 613, 634
   constant, 97, 166
   continuous, 298
   as correspondences, 81
   cube root, 203
   cubic, 296, 297
   cubing, 203
   decomposing, 195
   decreasing, 166
   defined piecewise, 171
   difference of, 182, 228
   difference quotient, 185
   discontinuous, 298
   domain, 81, 87, 88
   element, 81
   evaluating, 84, 85
   even, 202, 228
   exponential, 394, 464
   graphs, 85
   greatest integer, 173
   horizontal-line test, 384
   identity, 97, 203
   increasing, 166
   input, 83
   inverse trigonometric, 608, 611, 634
   inverses of, 383
   library of, 203. *See also* inside back
      cover.
   linear, 97
   logarithmic, 408, 410, 464
   logistic, 450
   nonlinear, 97
   nonpolynomial, 298
   notation for, 84
   objective, 790

odd, 202, 228
one-to-one, 382, 383, 464
output, 83
periodic, 538
phase shift, 554
piecewise, 171
polynomial, 296, 372
product of, 182, 228
quadratic, 244, 289, 297
quartic, 296
quotient of, 182, 228
range, 81, 88
rational, 342, 373
reciprocal, 203
   trigonometric, 473, 566
reflections, 199, 206, 207, 212, 228
relative maxima and minima, 167, 168
sequence, 890
square root, 203
squaring, 203
stretching and shrinking, 208, 212, 228
sum of, 182, 228
transformations, 203–212, 228,
   547–557, 567
translations, 203–205, 212, 228, 547,
   552, 555, 567
trigonometric, 472, 473, 502, 509, 531,
   540, 543, 566
values, 84, 85
vertical-line test, 86
zeros of, 139, 143, 156, 244, 256, 301,
   304, 333, 334, 335, 338, 372
Fundamental counting principle, 929, 965
Fundamental theorem of algebra, 332, 372

Games in a sports league, 310
Gauss, Karl Friedrich, 737, 751
Gauss–Jordan elimination, 751
Gaussian elimination, 737, 749
General term of a sequence, 891, 892
Geometric progressions, *see* Geometric
   sequences
Geometric sequences, 910
   common ratio, 910, 965
   infinite, 890
   *n*th term, 911, 965
   sums, 912, 915, 965
Geometric series, 912
   infinite, 913, 915
Grad, 529
Grade, 102
Graph styles, 785
Graphing calculator
   and absolute value, 6, 665

Graphing calculator (*continued*)
  ANGLE menu, 478
  and circles, 73, 823
  coefficient of correlation, 123
  coefficient of determination, 306
  and combinations, 941
  and complex numbers, 238, 239, 241,
      665, 666, 667, 668, 670, 671, 673
  and composition of functions, 193
  CONNECTED mode, 174, 343
  converting to/from degrees/radians, 519
  converting to/from D°M′S″ and
      decimal degree form, 478
  and cube roots, 46
  cubic regression, 306
  DEGREE mode, 478
  and determinants, 779, 781
  DIAGNOSTIC, 123, 306
  DMS feature, 478
  D°M′S″ notation, 478
  DOT mode, 174, 343
  DRAW menu, 73, 786, 787
  and ellipses, 826
  equation-editor screen, 66
  and exponential functions, 400
  exponential regression, 454
  and factorial notation, 931
  FRAC feature, 51
  and function values, 84
  graph styles, 577, 785
  graphing equations, 66
  graphing functions, 85
      exponential, 397
      logarithmic, 417
  and greatest integer function, 174
  and hyperbolas, 836, 838
  and identities, 577
  and inequalities, 152, 785–787, 849
  INEQUALZ APP, 786, 789
  INTERSECT feature, 130
  and inverse trigonometric functions,
      480, 610
  and inverses of functions, 386, 480, 610
  linear regression, 122
  logarithmic regression, 460
  and logarithms, 413
  logistic regression, 462
  and matrices, 750, 752, 757–759, 761,
      762, 771
  MAXIMUM feature, 168
  MINIMUM feature, 168
  and *n*th roots, 46
  and nonlinear systems of inequalities,
      849

and order of operations, 13
  and parabolas, 816, 819
  and parametric equations, 535, 537, 673
  and permutations, 930, 934
  and polar equations, 681
  and polynomial inequalities, 362–367
  and probability, 959
  quadratic regression, 742
  quartic regression, 306
  RADIAN mode, 519, 534
  random-number generator, 962
  and range, 267
  regression, 122, 306, 454, 631, 742
  and relative maxima and minima, 168
  row-equivalent operations, 750
  scatterplot, 123
  and scientific notation, 12
  and sequences, 891, 892, 894, 895,
      905, 913
  SEQUENTIAL mode, 577
  Shade, 786, 787
  sine regression, 631
  and solving equations, 130, 131, 140
  and square roots, 46
  squaring the window, 73
  standard viewing window, 66
  and systems of equations, 720, 750,
      752, 773
  and systems of inequalities, 789
  TABLE feature, 67, 132
  TEST menu, 152
  and trigonometric function values,
      479, 507, 509, 534
  VALUE feature, 84
  viewing window, 66
  *x*th root feature, 46
  and zeros of functions, 140, 304
  ZERO feature, 140
  ZOOM menu, 67, 173
Graphs, 64
  asymptotes, *see* Asymptotes
  bar, 62
  circle, 62
  of circles, 72, 73, 823
  of complex numbers, 664
  of conics, polar equations, 867–869
  cosecant function, 542
  cosine function, 536
  cotangent function, 542
  ellipses, 826
  equations, 64
  exponential functions, 395, 401
  functions, 85
  hole in, 173

horizontal lines, 98, 101
hyperbolas, 836, 838
inequalities, 151, 152, 784–788, 806
intercepts, 64, 106. *See also x*-intercept;
    y-intercept.
intervals, 4
inverse function, 386
    trigonometric, 609
inverse relations, 380
inverses of exponential functions, 408
leading-term test, 299, 300, 372
line, 62
linear functions and equations, 64, 65,
    107, 121
linear inequalities, 784–788, 806
logarithmic functions, 408, 410,
    417, 464
nonlinear, 67, 120
of ordered pairs, 63
parabolas, 261, 263, 289, 816
parametric equations, 872–876
piecewise-defined functions, 171
of points, 63
of polar equations, 680
of polynomial functions, 313–318
quadratic functions, 261, 263, 289
rational functions, 351
reflection, 199, 206, 207, 228
rotation of axes, 855–860, 882
scatterplot, 120, 123
secant function, 542
of sequences, 892
shrinking, 208, 212, 228
sine function, 536, 540
solving equations using, 141
stretching, 208, 212, 228
of sums, 557
symmetry, 198–201, 228
systems of equations, 720
systems of inequalities, 788, 849
tangent function, 541
transformations, 203–212, 228,
    547–557, 567
translation, 203–205, 212, 228, 547,
    552, 555, 567
turning point, 313
vertical lines, 98, 101
x-intercept, 64, 140, 143, 256, 301, 313,
    411, 825
y-intercept, 64, 105, 106, 397, 824, 825
Gravity, acceleration due to, 597, 632
Greater than (>), 3
Greater than or equal to (≥), 3
Greatest integer function, 173

Grouping, factoring by, 23, 27
Growth, limited, 430, 464
Growth model, exponential, 446
Growth rate, exponential, 446
   and doubling time, 449, 464

Half-angle identities, 593, 633
Half-life, 452, 453, 464
Half-open interval, 4
Half-plane, 785
Handshakes, number of, 371
Hanoi, tower of, 928
Height estimates, 108, 113
Heron's formula, 663
Hole in a graph, 173
Hooke's law, 225
Horizontal asymptotes, 347, 349, 351, 373
Horizontal component of a vector, 693
   scalar, 697
Horizontal line, 98, 119, 156
   slope, 101
Horizontal stretching and shrinking, 208, 212, 228, 550, 551, 555, 567
Horizontal translation, 204, 205, 212, 228, 552, 555, 567
Horizontal-line test, 384
Hyperbola, 814, 833
   applications, 840
   asymptotes, 834, 837
   center, 833
   conjugate axis, 834
   eccentricity, 841, 866
   equations, 834, 837, 861, 882
   foci, 833, 834, 837
   graphs, 836, 838
   transverse axis, 833
   vertices, 833, 834, 837
Hyperbola APP, 836
Hypotenuse, 48, 472, 984

i, 236
   powers of, 239, 240
Identities
   additive
      matrix, 758, 759
      real numbers, 5, 55
   cofunction, 481, 566, 589, 590, 633
   multiplicative
      matrix, 768
      real numbers, 5, 55
   Pythagorean, 574–576, 633

trigonometric, 481, 566, 574–576, 580–584, 589–594, 602, 603, 633, 634
   proving, 598, 599
Identity, 574. *See also* Identities.
Identity function, 97, 203
Identity matrix, 758
Imaginary axis, 664
Imaginary numbers, 237
Imaginary part, 237
Inconsistent system of equations, 721
Increasing function, 166
Independent equations, 721
Independent variable, 66
Index
   of a radical, 46
   of summation, 894
Induction, mathematical, 923, 965
Induction step, 923
Inequalities, 150
   with absolute value, 286, 289
   compound, 151
   equivalent, 150
   graphing, 151, 152, 785–788, 806
   linear, 150, 784
   polynomial, 360
   quadratic, 360
   rational, 366
   solutions, 150
   solving, 150, 156, 286, 289, 364, 368
   systems of, 788, 849
INEQUALZ APP, 786, 789
Infinite sequence, 890
Infinite series, 893, 913
Initial point of a vector, 689
Initial side of an angle, 498
Input, 83
Integers, 2
   as exponents, 9
Intercepts, 64, 105, 106. *See also* x-intercept; y-intercept.
Interest
   compound, 14, 17, 55, 398
      continuously, 447, 464
   simple, 136, 156
Interior angles, 977
Intermediate value theorem, 318, 319, 372
INTERSECT feature, 130
Intersect method, 130
Interval notation, 3, 4
Inverse of a function, *see* Inverses
Inverse relation, 380
Inverse variation, 221, 228

Inversely proportional, 221
Inverses
   additive
      of matrices, 758
      of real numbers, 5, 55
   function, 383
      and composition, 388
      exponential, 408
      formula for, 385
      and reflection across $y = x$, 386
      trigonometric, 608, 611, 634
   multiplicative
      of matrices, 769
      of real numbers, 5, 55
   relation, 380
Invertible matrix, 772
Irrational numbers, 2, 3
Irrational zeros, 334
Irreducible quadratic factor, 798
Isosceles triangle, 476

Joint variation, 222, 228
Jordan, Wilhelm, 751

$(k + 1)$st term, binomial expansion, 950, 965
Knife bevel, 659

Latitude, 597
Law of cosines, 655, 712
Law of sines, 642, 712
Leading coefficient, 18, 296
Leading 1, 751
Leading term, 296
Leading-term test, 299, 300, 372
Least common denominator (LCD), 39, 40
Legs of a right triangle, 48, 984
Length of a vector, 689, 698, 713
Less than (<), 3
Less than or equal to (≤), 3
Libby, Willard E., 451
Library of functions, 203. *See also* inside back cover.
Light-year, 12
Like terms, 19
Limit, infinite geometric series, 915
Limited population growth, 430, 464
Limiting value, 450
Line graph, 62
Line of symmetry, *see* Axis of symmetry
Linear combination of unit vectors, 702
Linear correlation, coefficient of, 123

Linear equations, 106
  in one variable, 32, 129
  solving, 33, 55
  in three variables, 736
  in two variables, 720
Linear functions, 97
  zero of, 139
Linear graph, 120
Linear inequalities, 784
  systems of, 788
Linear programming, 790, 836
Linear regression, 122
Linear speed, 522, 523, 567
Lines
  angle between, 587
  graphs of, 64, 66, 106
  horizontal, 98, 119, 156
  parallel, 117, 119, 156
  perpendicular, 118, 119, 156
  point–slope equation, 116, 119, 156
  regression, 123
  secant, 185
  slope of, 99, 100, 119, 156
  slope–intercept equation, 106, 115,
    119, 156
  vertical, 98, 119, 156
Lithotripter, 829
ln, 414
log, 413
Logarithm, 410, 412, 464. *See also*
    Logarithmic functions.
  base *e*, 414
  base–exponent property, 435, 464
  base-10, 413
  a base to a logarithmic power, 431
  of a base to a power, 431
  on a calculator, 413
  change-of-base, 415, 464
  common, 413
  and exponents, 412, 464
  natural, 414
  power rule, 427, 432, 464
  product rule, 426, 432, 464
  properties, 426–432, 464
  quotient rule, 428, 432, 464
Logarithm function, *see* Logarithmic
    functions.
Logarithmic equality, principle of,
    436, 464
Logarithmic equation, 439
Logarithmic functions, 410, 464. *See also*
    Logarithm.
  base, 410
  domain, 411

graphs, 408, 410, 417, 464
  properties, 411
  range, 411
  x-intercept, 411
Logarithmic regression, 460
Logarithmic spiral, 687
Logistic function, 450
Logistic regression, 462
LORAN, 840
Loudness of sound, 425

Magnitude of an earthquake, 421
Magnitude of a vector, 689, 698, 713
Main diagonal, 748
Major axis of an ellipse, 824, 825
Mathematical induction, 923, 965
Mathematical models, 97, 119, 305, 454,
    631, 741, 848. *See also* Regression.
Matrices, 748
  addition of, 757, 805
  additive identity, 758
  additive inverses of, 758
  augmented, 748, 770
  coefficient, 748
  cofactor, 777
  column, 763
  columns of, 748
  determinant of, 776–782, 805, 806
  dimensions, 748
  entries, 748
  equal, 756
  and Gauss–Jordan elimination, 751
  and Gaussian elimination, 749
  identity, multiplicative, 768
  inverses of, 769
  invertible, 772
  main diagonal, 748
  minor, 776
  multiplicative inverse, 769
  multiplying, 760, 805
    by a scalar, 759
  nonsingular, 772
  opposite of, 758
  order, 748
  reduced row-echelon form, 751, 805
  row, 763
  row-echelon form, 750, 751, 805
  row-equivalent, 748
  row-equivalent operations, 749, 805
  rows of, 748
  and scalar multiplication, 759, 805
  singular, 772
  square, 748
  subtraction, 757, 805

and systems of equations, 749,
    751, 772
  zero, 758
Matrix, *see* Matrices
Matrix equations, 763
Maximum
  linear programming, 791
  quadratic function, 262, 289
  relative, 167, 168, 313
MAXIMUM feature, 168
Mean, arithmetic, 909
Midpoint formula, 70, 156
Mil, 529
Minimum
  linear programming, 791
  quadratic function, 262, 289
  relative, 167, 168, 313
MINIMUM feature, 168
Minor, 776
Minor axis of an ellipse, 824
Models, mathematical, 97, 119, 305,
    454, 631, 741, 848. *See also*
    Regression.
Monomial, 19
Mortgage payment formula, 58
Motion, projectile, 268, 877
Motion formula, 135, 156
Multiple, scalar, 699
Multiplication
  associative property, 5, 55
  commutative property, 5, 55
  complex numbers, 239, 668, 712
  of exponential expressions, 10, 55
  of exponents, 10, 55
  of functions, 182, 228
  of matrices, 760, 805
    and scalars, 759
  of ordinates, 559
  of polynomials, 20, 21, 55
  of radical expressions, 47, 55
  of rational expressions, 38, 39
  scalar
    for matrices, 759
    for vectors, 699, 701, 713
Multiplication principle
  for equations, 33, 55, 129,
    156
  for inequalities, 150, 156
Multiplication properties
  for matrices, 763, 805
  for real numbers, 5, 55
Multiplicative identity
  matrices, 768
  real numbers, 5, 55

Multiplicative inverse
  matrices, 769
  real numbers, 5, 55
Multiplicity of zeros, 302, 303

nth partial sum, 893
nth root, 45
nth roots of unity, 672
nth term
  arithmetic sequence, 901, 965
  geometric sequence, 911, 965
Nanometer, 16
Nappes of a cone, 814
Natural logarithms, 414
Natural numbers, 2
Nautical mile, 597
Negative exponents, 9
Negative rotation, 498
Newton's law of cooling, 459
Nondistinguishable objects,
    permutations of, 935, 965
Nonlinear function, 97
Nonlinear graph, 67, 120
Nonlinear systems of equations, 843
Nonlinear systems of inequalities, 849
Nonnegative root, 46
Nonpolynomial functions, 298
Nonreal zeros, 333
Nonsingular matrix, 772
Notation
  binomial coefficient, 940
  combination, 939
  decimal, 2, 3
  exponential, 9
  factorial, 931, 965
  function, 84
  interval, 3, 4
  inverse function, 382
  for matrices, 756
  polar, 665
  radical, 46
  scientific, 11
  set, 4, 37
  sigma, 894
  standard, for complex number, 665
  summation, 894
  trigonometric, for complex numbers,
    665, 712
Number line, 3
Numbers
  complex, 236, 237, 289
  imaginary, 237
  integers, 2
  irrational, 2, 3

natural, 2
pure imaginary, 237
rational, 2
real, 2, 3
whole, 2
Numerator, rationalizing, 49, 50

Objective function, 790
Oblique asymptotes, 350, 351, 373
Oblique triangle, 640
Obtuse angle, 500, 971
Odd functions, 202, 228
Odd multiplicity of zeros, 303
Odd roots, 47, 55
One-to-one function, 382, 464
  and inverses, 383
Open interval, 3
Operations, order of, 13
Opposite of a matrix, 758
Order
  descending, 18
  of a matrix, 748
  of operations, 13
  of real numbers, 3
Ordered pair, 63
Ordered triple, 736
Ordinate, 63
Ordinates
  addition of, 557
  multiplication of, 559
Origin, 62
  symmetry with respect to, 199, 200,
    228
Orthogonal vectors, 711
Oscillation, damped, 559
Oscilloscope, 557
Outcomes, 957
  equally likely, 957
Output, 83

Pair, ordered, 63
Parabola, 261, 566
  applications, 267, 819, 820
  axis of symmetry, 262, 289, 814
  directrix, 814
  equations, 815, 817
  focus, 814
  graphs, 261, 263, 289, 816
  vertex, 262, 266, 289, 814
Parabola APP, 819
Parallel lines, 117, 119, 156
  properties, 978
Parallelogram law, 691
Parameter, 873

Parametric equations, 873
  converting to a rectangular equation,
    873–876
Parsec, 16
Partial fractions, 798, 806
Partial sum, 893
Pascal's triangle, 946
  and binomial theorem, 948
Peanut, 687
Perimeter, 138
Periodic function, 538, 551, 555, 567
Permutation, 928, 929
  of n objects, k at a time, 932, 933, 965
    allowing repetition, 934
  of n objects, total number, 930, 965
  of nondistinguishable objects, 935, 965
Perpendicular lines, 118, 119, 156
pH, 424
Phase shift, 554
Pi (π), 3
Piecewise-defined function, 171
Pixels, 174
Plane curve, 872, 873
Plotting a point, 63, 676
Point
  coordinates of, 63
  plotting, 63, 676
Point–slope equation, 116, 119, 156
Poker hands, 945, 963
Polar axis, 676
Polar coordinates, 676
  converting to/from rectangular
    coordinates, 678, 679
Polar equation
  of a conic, 867, 882
  converting to a rectangular equation,
    679, 680, 870
  graphing, 680
Polar notation, 665
Pole, 676
Polygon, number of diagonals, 371
Polynomial function, 296, 372. See also
    Polynomials.
  coefficients, 333, 334
  degree, 296
  domain, 298
  factoring, 327, 332, 372
  graph of, 313–318
  with integer coefficients, 334
  leading coefficient, 296
  leading term, 296
  leading-term test, 299, 300, 372
  as models, 305
  with rational coefficients, 334

Polynomial function (*continued*)
  with real coefficients, 333
  values, 325
  zeros of, 301, 304, 313, 333
Polynomial. *See also* Polynomial functions.
  addition, 19, 20
  binomial, 19
  coefficients, 18, 332
  constant term, 18
  degree of, 18, 19
  degree of a term, 19
  descending order, 18
  division of, 323–327, 372
  factoring, 23–30, 55
  factors of, 327, 372
  leading coefficient, 18
  monomial, 19
  multiplication of, 20, 21, 55
  in one variable, 18
  prime, 30
  in several variables, 19
  subtraction of, 19, 20
  term of, 18
  trinomial, 19
Polynomial inequalities, 360
  solving, 364
Population growth, 446, 464
  doubling time, 447, 464
  limited, 430, 464
Position vector, 697
Positive rotation, 498
Power rule
  for exponents, 10, 55
  for logarithms, 427, 432, 464
Powers
  of complex numbers, 670, 712
  of i, 239, 240
  logarithms of, 427, 432, 464
  principle of, 279
Present value, 463
Pressure
  atmospheric, 462
  at sea depth, 113
Price, equilibrium, 728
Prime polynomial, 30
Principal root, 46
Principle(s)
  addition, 33, 55, 129, 150, 156
  fundamental counting, 929, 965
  of mathematical induction, 923, 965
  multiplication, 33, 55, 129, 150, 156
  P (experimental probability), 955, 965
  P (theoretical probability), 958, 965

of powers, 279, 289
of square roots, 33, 55, 244
of zero products, 33, 55, 244
Probability, 954
  experimental, 954, 955, 966
  properties, 959, 966
  theoretical, 954, 958, 966
Problem-solving steps, 133, 156
Product. *See also* Multiplication.
  dot, 706, 713
  of functions, 182, 228
  logarithm of, 426, 432, 464
  of matrices, 760
  raised to a power, 10, 55
  scalar
    for matrices, 759
    for vectors, 699, 701, 713
  of a sum and a difference, 21, 55
  of vectors, 706, 713
Product-to-sum identities, 602, 634
Product rule
  for exponents, 10, 55
  for logarithms, 426, 432, 464
Profit, total, 274
Programming, linear, 790, 806
Progressions, *see* Sequences
Projectile motion, 268, 877
Properties
  of addition, 5, 55
  base–exponent, 435, 464
  distributive, 5, 55
  of exponential functions, 397
  of exponents, 10, 55
  of logarithmic equality, 436, 464
  of logarithmic functions, 411
  of logarithms, 426–432, 464
  of matrix addition, 759, 805
  of matrix multiplication, 763, 805
  of multiplication, 5, 55
  of parallel lines, 978
  probability, 959, 966
  of radicals, 47, 55
  of real numbers, 5, 55
  of scalar multiplication
    for matrices, 759, 805
    for vectors, 701
  of vector addition, 701
  vertical angle, 976
Proportional, *see* Variation
Proportions and similar triangles, 982
Proving trigonometric identities, 598, 599
Pure imaginary numbers, 237
Pythagoras, 985

Pythagorean equation, 475, 984
Pythagorean identities, 574–576, 633
Pythagorean theorem, 49, 56, 984

Quadrantal angle, 506
Quadrants, 62
Quadratic equations, 32, 244, 289. *See also* Quadratic function.
  discriminant, 251, 252
  solving, 33, 34, 55, 244, 247, 249, 289
  standard form, 244
Quadratic form, equation in, 252
Quadratic formula, 249, 289
Quadratic function, 244, 289, 297. *See also* Quadratic equations.
  graph, 261, 263, 289. *See also* Parabola.
  maximum value, 262, 289
  minimum value, 262, 289
  zeros of, 244
Quadratic inequalities, 360
Quadratic regression, 741, 742
Quartic function, 296
Quartic regression, 306
Quotient
  difference, 185
  of functions, 182, 228
  logarithm of, 428, 432, 464
  of polynomials, 324
  raised to a power, 10, 55
Quotient rule
  for exponents, 10, 55
  for logarithms, 428, 432, 464

Radian measure, 518, 522
  converting to degrees, 519
Radical, 46. *See also* Radical expressions.
Radical equation, 279
Radical expressions, 46. *See also* Roots.
  addition, 48
  conjugate, 50
  converting to exponential notation, 50, 56
  division, 47, 55
  multiplication, 47, 55
  properties of, 47, 55
  rationalizing denominators or numerators, 49, 50
  simplifying, 46, 47, 55
  subtraction, 48
Radicand, 46
Radius, 71, 822
Raising a power to a power, 10, 55

Raising a product or quotient to a power, 10, 55
Ramp for the disabled, 103
Random-number generator, 962
Range
    of a function, 81, 88
        exponential, 397
        logarithmic, 411
        trigonometric, 537, 611
    of a relation, 82
Rate of change, average, 99, 103
Ratio, common, 910, 965
Rational equations, 276
Rational exponents, 50, 56
Rational expression, 36
    addition, 39, 40
    complex, 41
    decomposing, 798, 806
    division, 38, 39
    domain, 36
    least common denominator, 39, 40
    multiplication, 38, 39
    simplifying, 37, 41
    subtraction, 39, 40
Rational function, 342, 372
    asymptotes, 347, 349, 350, 351, 373
    domain, 343
    graphs of, 351
Rational inequalities, 366
    solving, 368
Rational numbers, 2
Rational zeros theorem, 335, 372
Rationalizing denominators or
        numerators, 49, 50
Reaction distance, 113
Reaction time, 113
Real axis, 664
Real numbers, 2, 3
    order, 3
    properties, 5, 55
Real part, 237
Reciprocal, in division, 38
Reciprocal function, 203
    trigonometric, 473
Rectangular coordinates, converting to/
        from polar coordinates, 678, 679
Rectangular equation
    converting to/from parametric
        equations, 873–876
    converting to/from polar equation,
        679, 680
Recursion formula, 895
Recursively defined sequence, 895

Reduced row-echelon form, 751, 805
Reducible to quadratic equation, 252
Reference angle, 507, 508
Reference triangle, 501
Reflection, 199, 206, 207, 228
    and inverse relations and functions, 386
    on unit circle, 531
Regression
    cubic, 306
    exponential, 454
    linear, 122
    logarithmic, 460
    logistic, 462
    quadratic, 741, 742
    quartic, 306
    sine, 631
Regression line, 123
Related equation, 361, 785
Relation, 82
    domain, 82
    inverse, 380
    range, 82
Relative maxima and minima, 167,
        168, 313
Remainder, 324
Remainder theorem, 325, 372
Repeating decimals, 2
    changing to fraction notation, 915
Representing a vector, 692
Resolving a vector into its components,
        692
Restricting a domain, 388
Resultant, 690
Richter scale, 421
Right angle, 472, 500, 971
Right triangle, 48, 472, 984
    solving, 485
Rise, 99
Roots. *See also* Radical expressions.
    of complex numbers, 671, 712
    cube, 45
    even, 47, 55
    nth, 45
    nonnegative, 46
    odd, 47, 55
    principal, 46
    square, 45
Roots of equations, 244
Rose-shaped curve, 681, 682
Rotation of a ray, 498
Rotation of axes, 855–860, 882
Row matrix, 763
Row-echelon form, 750, 751, 805

Row-equivalent matrices, 748
Row-equivalent operations, 749, 805
Rows of a matrix, 748
Run, 99

Salvage value, 909
Sample space, 957
Scalar, 697, 699, 759
Scalar components of a vector, 697
Scalar multiplication
    for matrices, 759
    for vectors, 699, 701, 713
Scalar product
    for matrices, 759
    for vectors, 699, 701, 713
Scatterplot, 120, 123
Scientific notation, 11
Secant function, 473, 502, 531, 543,
        566, 567
    graph, 542
Secant line, 185
Second coordinate, 63
Semicubical parabola, 687
Semiperimeter, 663
Sequences, 890
    alternating, 891
    arithmetic, 900
    Fibonacci, 898
    finite, 890
    general term, 891, 892
    geometric, 910
    graph, 892
    infinite, 890
    partial sum, 893
    recursive definition, 895
    of statements, 922
    terms, 891
Series
    arithmetic, 903
    finite, 893
    geometric, 912
    infinite, 893
    partial sum, 893
Set-builder notation, 4, 37
Sets
    element of, 3
    empty, 286
    notation, 4, 37
    solution, 32, 150, 785
    subsets, 3
    union of, 152
Shade above or shade below, 786, 787

Shift of a function, *see* Translation of function
Shrinkings, 208, 212, 228, 550, 555, 567
Sigma notation, 894
Signs of trigonometric functions, 503, 567
Similar figures, 980
Similar terms, 19
Similar triangles, 474, 981
Simple-interest formula, 136, 156
Simplifying radical expressions, 46, 47, 55
Simplifying rational expressions, 37
    complex, 41
Simplifying trigonometric expressions, 577
Sine function, 472, 473, 502, 531, 566, 567
    amplitude, 538, 540
    domain, 537, 540
    graph, 536, 540
    period, 538, 540
    range, 537, 540
Sine regression, 631
Sines, law of, 642, 712
Singular matrix, 772
Slant asymptote, 350
Slope, 99, 100, 119, 150
    applications, 102–105
    of a horizontal line, 101
    as rate of change, 99, 103
    of a vertical line, 101
Slope–intercept equation, 106, 115, 119, 156
Solution set, 32, 150, 785
Solutions
    of equations, 32, 63, 129, 143, 256, 301
    feasible, 791
    of inequalities, 150, 785
    of systems of equations, 720, 736
Solving equations, 32, 33, 55, 129, 156, 244, 247, 249, 252, 276, 279, 284, 289, 435, 440, 619. *See also* Solving systems of equations.
Solving formulas, 141
Solving inequalities, 150, 156, 286, 289, 364, 368
    systems of, 849
Solving systems of equations
    using Cramer's rule, 779–782, 806
    elimination method, 722
    Gauss–Jordan elimination, 751
    Gaussian elimination, 737, 749
    graphically, 720
    using the inverse of a matrix, 772, 805
    nonlinear, 793
    substitution method, 721

Solving triangles, 485, 640
    oblique, 640, 654
    right, 485
Sound
    loudness, 425
    speed in air, 90
Speed
    angular, 522
    linear, 522, 523, 567
    of sound in air, 90
Spiral
    of Archimedes, 687
    logarithmic, 687
Sports league, number of games, 310
Square matrix, 748
    determinant, 776–782, 805, 806
Square root, 45
Square root function, 203
Square roots, principle of, 33, 55, 244
Squares of binomials, 21, 55
Squares, differences of, 28
Squaring function, 203
Squaring the window, 73
Standard form
    for equation of a circle, 72, 156, 822
    for equation of an ellipse, 825, 827
    for equation of a hyperbola, 834, 837
    for equation of a parabola, 815, 817
    for quadratic equation, 244
Standard notation, complex numbers, 665
Standard position
    for an angle, 498
    for a vector, 697
Standard viewing window, 66
State the answer, 133, 156
Statements, sequences of, 922
Stopping distance, 113
Straight angle, 500, 971
Straight-line depreciation, 113
Stretchings, 208, 212, 228, 550, 555, 567
Strophoid, 687
Study tips, 47, 100, 136, 168, 184, 199, 237, 248, 281, 318, 326, 339, 410, 448, 491, 525, 530, 577, 599, 607, 627, 643, 660, 665, 728, 763, 802, 817, 835, 857, 879, 891, 901, 932, 954
Subsets, 3, 939
    total number, 951, 966
Substitution method, 721
Subtraction
    of complex numbers, 238
    of exponents, 10, 55
    of functions, 182, 228

of logarithms, 428, 432, 464
of matrices, 757
of polynomials, 19, 20
of radical expressions, 48
of rational expressions, 39, 40
of vectors, 700, 713
Sum and difference identities, 580–584, 633
Sum and difference, product of, 21, 55
Sum-to-product identities, 603, 634
Summation, index of, 894
Summation notation, 894
Sums
    of arithmetic sequences, 904, 965
    of cubes, factoring, 29, 55
    of functions, 182, 228
    of geometric sequences, 912, 915, 965
    graphs of, 557
    of logarithms, 426, 432, 464
    partial, 893
    squares of, 21
    of vectors, 690, 700, 701, 713
Supplementary angles, 501, 974
Supply and demand, 728
Symmetry, 198
    axis of, 262, 289
    with respect to the origin, 199, 200, 228
    with respect to the *x*-axis, 199, 200, 228
    with respect to the *y*-axis, 199, 200, 228
Synthetic division, 325, 326
Systems of equations, 720
    consistent, 721
    in decomposing rational expressions, 801
    dependent equations, 721
    equivalent, 724
    inconsistent, 721
    independent equations, 721
    in matrix form, 763, 764
    nonlinear, 843
    and partial fractions, 801, 802
    solutions, 720, 736
    solving, *see* Solving systems of equations
    in three variables, 736
    in two variables, 720
Systems of inequalities, 788
    nonlinear, 849

TABLE feature, 67, 132
Tangent function, 473, 502, 531, 541, 543, 566, 567
Term, leading, 296
Terminal point of a vector, 689
Terminal side of an angle, 498

Terminating decimal, 2
Terms of a polynomial, 18
    constant, 18
    degree of, 19
    like (similar), 19
Terms of a sequence, 891, 901
Territorial area, 95
TEST menu, 152
Tests for symmetry, 200, 228
Theorem
    binomial, 948, 949, 965
    DeMoivre's, 670, 712
    factor, 327, 372
    fundamental, of algebra, 332, 372
    intermediate value, 318, 319, 372
    Pythagorean, 49, 56, 984
    rational zeros, 335, 372
    remainder, 325, 372
Theoretical probability, 954, 958, 966
Threshold weight, 310
Time of a free fall, 253
Total cost, profit, revenue, 274
Total number of subsets, 951, 966
Tower of Hanoi problem, 928
Trade-in value, 909
Transformations of functions, 203–212,
    228, 547–557, 567
Translate to an equation, 133, 156
Translation of a function, 203–205, 212,
    228, 547, 552, 555, 567
Transversal, 976
Transverse axis of a hyperbola, 823
Tree diagram, 929
Triangles
    area, 648, 712
    equilateral, 53
    isosceles, 476
    oblique, 640
    Pascal's, 946
    reference, 501
    right, 48, 472, 984
    semiperimeter, 663
    similar, 474, 981
    solving, 485, 640, 654
Trigonometric equations, 619
Trigonometric expressions, simplifying,
    577
Trigonometric functions
    of acute angles, 472, 473, 478, 566
    of any angle, 502, 509, 567
    circular, 531, 567
    cofunctions, 481, 566, 589, 590, 633
    composition, 612, 613, 634
    domain, 537, 540, 543, 611

graphs, 536, 540–543
    inverse, 608, 611, 634
    of quadrantal angles, 506
    range, 537, 611
    reciprocal, 473, 566
    signs of, 503, 567
    value of 30°, 60°, 45°, 475–477, 566
Trigonometric identities, 481, 574–576,
    580–584, 589–594, 602, 603,
    633, 634
    proving, 598, 599
Trigonometric notation for complex
    numbers, 665, 712
Trigonometric ratios, see Trigonometric
    functions
Trinomial, 19
    factoring, 24–28
Triple, ordered, 736
Turning point, 313
Twisted sister, 687

Undefined slope, 101
Union of sets, 152
Unit circle, 79, 514
    reflections on, 531
Unit vector, 702
Unity, nth roots of, 672

VALUE feature, 84
Values
    critical, 367
    function, 84, 85
Variable, 66
Variable costs, 114
Variation
    combined, 222, 228
    direct, 219, 228
    inverse, 221, 228
    joint, 222, 228
Variation constant, 219, 221
Variations in sign, 338, 372
Vectors, 688, 689
    addition of, 690, 700, 701, 713
    additive identity, 701
    angle between, 706, 713
    component form, 697
    components, 692
        scalar, 697
    direction, 689
    direction angle, 703
    dot product, 706, 713
    equivalent, 689, 698
    initial point, 689
    length, 689, 698, 713

as linear combinations of unit vectors,
    702
    magnitude, 689, 698, 713
    orthogonal, 711
    position, 697
    representing, 692
    resolving into components, 692
    resultant, 690
    scalar multiplication, 699, 701, 713
    standard position, 697
    subtraction, 700, 713
    sum of, 690
    terminal point, 689
    unit, 702
    zero, 701
Velocity, 688
Vertex
    of an angle, 498
    of an ellipse, 824, 825, 827
    of a hyperbola, 833, 834, 837
    of a parabola, 262, 266, 289, 814
Vertical angles, 976
Vertical asymptotes, 345, 351, 373
Vertical component of a vector, 693
    scalar, 697
Vertical line, 98, 119, 156
    slope, 101
Vertical stretching and shrinking, 202,
    212, 228, 548–550, 555, 567
Vertical translation, 204, 212, 228, 547,
    555, 567
Vertical-line test, 86
Vertices, see Vertex
Viewing window, 66
    squaring, 73
    standard, 66
Visualizing the Graph, 74, 109, 215, 271,
    320, 356, 422, 561, 629, 685, 730,
    850, 918

Walking speed, 419
Weight
    minimum ideal, 113
    threshold, 310
Whispering gallery, 829
Whole numbers, 2
Windmill power, 310
Window, 66
    squaring, 73
    standard, 66

x-axis, 62
    symmetry with respect to, 199,
        200, 228

x-coordinate, 63
x-intercept, 64, 140, 143, 256, 301
    ellipse, 825
    logarithmic function, 411
    polynomial function, 313
xy-term, eliminating, 859, 882
*x*th root, 46

y-axis, 62
    symmetry with respect to, 199, 200, 228
y-coordinate, 63

y-intercept, 64, 105, 106
    ellipse, 824, 825
    exponential function, 397

Zero, exponent, 9
ZERO feature, 140
Zero of a function, *see* Zeros of functions
Zero matrix, 758
Zero method, 140
Zero products, principle of, 33, 55, 244
Zero slope, 101

Zero vector, 701
Zeros of functions, 139, 143, 156, 244, 256, 301, 304, 333, 334, 335, 338, 372
    multiplicity, 302, 303
ZOOM menu, 67, 173

Index of Applications

ASTRONOMY

Distance to Pluto, 16
Distance between points on the earth, 530
Distance to a star, 12
Earth's orbit, 16, 832
Hyperbolic mirror, 841
Light-years, 16
Linear speed of the earth, 528
Linear speed of an earth satellite, 523
Lunar crater, 650
Parabolic orbit, 872
Satellite location, 565, 632
Weight of an astronaut, 226
Weight on Mars, 225

AUTOMOTIVE

1957 Studebaker Golden Hawk, 458
Auto insurance, 310
Car distance, 169
Gas mileage, 736
License plates, 937
Maximizing mileage, 795
Road grade, 128

BIOLOGY/LIFE SCIENCES

Animal speed, 91
Bacteria growth, 898
Bald eagle population, 144
Eagle's flight, 696
Escherichia coli growth, 405
Mass of a proton, 15
Maximizing animal support in a forest, 797
Newton's law of cooling, 459
Rabbit population growth, 455, 898
Reaction distance, 392
Reaction time, 113
Temperature and depth of the earth, 159
Territorial area of an animal, 95
Water weight, 146

BUSINESS

Advertising, 176, 406, 424, 462, 755, 921
Allocation of resources, 797
Average cost, 359
Bestsellers, 149
Break-even point, 733
Business travel, 11
Chain business deals, 968
Coffee mixtures, 734
Coffee sales, 379, 404, 745
Consumer electronics, 732
Corn-based ethanol, 404
Cost, 775
Cost of business on the Internet, 154
Cost of political conventions, 461
Dairy profit, 762
DVD videos produced and shipped, 456
e-commerce, 735, 744, 927
Fixed costs, 114
Food service management, 766, 809, 811
Honey production, 61, 111
iPod sales, 394
Junk mail, 961
Mail-order business, 733, 744
Maximizing profit, 274, 791, 795, 796, 812
Minimizing cost, 274, 795
Morning newspapers, 746
Online sales, 406
Packaging and price, 149
Patents issued, 899
Periodic sales, 564
Power line costs, minimizing, 181
Pricing, 163
Processing telephone orders, 811
Produce, 765
Production, 760, 766, 767, 908
Profit, 766, 767
Quality control, 956
Recording studio, 639, 653, 656
Sales, 125, 144, 460, 631, 734, 735, 744, 775
Sales promotions, 733
Salvage value, 406, 898, 909, 970
Snack mixtures, 725
Ticket sales, 811
Total cost, 114, 158, 162
Total cost, revenue, and profit, 188, 229
Total profit, 370
Trade-in value, 909
Trademarks, 808
Travel bookings online, 104
U.S. mail, 162
Wendy's restaurant chain, 952
Where the textbook dollar goes, 144

CHEMISTRY

Antifreeze mixtures, 736
Boiling point and elevation, 95

Boiling point of water, 190
Carbon dating, 453, 457, 463, 466
Freezing point of water, 190
Half-life, 452
Hydronium ion concentration, 424, 466
Mass of a neutron, 12
Nuclear disintegration, 16
pH of substances, 424
Radioactive decay, 457

CONSTRUCTION

Arch of a circle in carpentry, 79
Beam deflection, 331
Box construction, 179
Bridge expansion, 53
Carpentry, 832, 852
Chesapeake Bay Bridge-Tunnel, 16
Construction of picnic pavilions, 497
Corral design, 178
Dog pen, 229
Fencing, 852, 886
Floor plans, 967
Framing a house, 222
Golden Gate Bridge, 494
Ladder height, 991
Ladder safety, 480
Landscaping, 852, 886
Norman window, 274
Office dimensions, 853
Play space, 177
Pole stacking, 906, 907
Rafters for a house or building, 487, 495
Ramps for the disabled, 103
Road grade, 102
Sidewalk width, 291
Storage area, 170
Surveying, 49, 990
Swimming pool, 79
Theater seating, 908

CONSUMER

Baskin-Robbins ice cream, 944
Book club purchases, 735
Cab fare, 138, 145
Cable subscribers, 121, 122
Cheese consumption, 392
Concert ticket prices, 733
Cordless telephones, 991
Dining out, 133
First-class postage, 382

Mother's Day spending, 743
Moving costs, 154, 163
Museum admission prices, 732
Oil consumption, 144
Parking costs, 181
Price increases, 744
Price of school supplies, 775
Red meat and poultry consumption, 734
Snack food cost, 744
Spring break vacation, 75
Stamp purchase, 755
Uninsured, 147
Veterinary expenses, 134
Volume and cost, 227

ECONOMICS

Debt held by foreigners, 310
Decreasing value of the dollar, 95
Economic multiplier, 917, 921, 967
Filing tax returns electronically, 127
Individual income taxes, 15
Manhattan Island, value of, 456
Social Security, 126, 467, 937
Straight-line depreciation, 113, 909
Supply and demand, 462, 728, 733
World trade with China, 394, 405

EDUCATION

Associate's degrees conferred, 260
Bachelor's degrees earned by women, 405
College applications, 461
Faculty at two-year colleges, 160
Grading on a curve, 936
Preprimary school enrollment, 746
Student loans, 136, 146
Study time versus grades, 127
Test options, 944, 970
Test scores, 231, 808, 809

ENGINEERING

Beam weight, 225
Bridge arch, 833
Bridge expansion, 53
Bridge supports, 831
Canyon depth, 663
Data storage, 144, 405
Digital hubs, 154
Distance across a river or lake, 482, 717, 989
F-106 blueprint, 983

Field microphone, 821
Gears, 652
Guy wire, 57, 59, 573, 616, 991
Headlight mirror, 821
Height of a cliff, 990
Height of a tree, 989
Pumping rate, 225
Pumping time, 231
Satellite dish, 821, 886
Spotlight, 821, 884
Suspension bridge, 822
Windmill power, 310

ENVIRONMENT

Cloud height, 489
Daylight hours, 631
Earthquake magnitude, 421, 424, 466, 469
Erosion, 147
Fire direction, 651
Forest fire, distance to, 490
Lightning detection, 495
Poachers, 661
Recycling rechargeable batteries, 226
River current, 293
River speed, 528
Solid waste, 15
Temperature conversion, 160
Tree height, 494
Volcanic activity, 147
Water from melting snow, 220
Weather balloon height, 494
Wind, 695, 715
World's ten largest earthquakes, 91

FINANCE

Amount of an annuity, 916, 920, 967, 970
Borrowing, 145, 755
Budget, 765
Checking account plans, 155
College fund, 17
Compound interest, 14, 17, 398, 403, 404, 447, 456, 469, 890, 898
Credit card debt, 112, 144
Growth of a stock, 407
Interest compounded annually, 310, 374, 898
Interest compounded continuously, 447, 456, 469
Interest compounded n times per year, 14
Interest in a college trust fund, 404
Interest on a CD, 404

Internet banking, 62
Investment, 145, 146, 154, 155, 160, 407, 466, 734, 740, 744, 775, 808, 810, 853, 908, 927
Loan repayment, 920
Mortgage payments, 58
Present value, 463
Retirement account, 17
Saving for college, 466
Savings account interest, 292
Total savings, 908

GENERAL INTEREST

Aerial photography, 493
Band formation, 908
Big sites, 284
Blouse sizes, 197
Boarding stable, 651
Book arrangements, 967
Card drawing, 958, 959, 961, 967
Circular arrangements, 937
Code symbols, 967
Coin arrangements, 936
Coin tossing, 962
Coins, 808
Committee selection, 943, 959
Decline in land-line phones, 112
Dining habits, 899
Dog years, 312
Doubling the thickness of paper, 920
Drawing coins, 961, 970
Flag displays, 967
Forgetting, 424, 460
Fraternity officer selection, 944
Fraternity–sorority names, 967
Garden plantings, 908
Garlic supply, 404
Hands on a clock, 530
Handshakes, number of, 371
Height of a flagpole, 983, 989
Jolts of caffeine, 743
Knife bevel, 659
Letter arrangements, 967
Linguistics, 961
Loading ramp, 494
Low-carbohydrate gardening, 743
Marbles, 959, 961, 970
Mason Dots, 960
Milk cows, 309
Money combinations, 953
Norman Rockwell painting, 458
Numerical relationship, 854, 884

Organic pet food, 155
Phone numbers, 936
Prize choices, 967
Program planning, 936
Random best-selling novels, 962
Random-number generator, 962
Scuba club officers, 970
Select a number, 960
Sheep and lambs, 126
Spring cleaning, 743
Street names, 732
Surveillance cameras, 454
The Ellipse, 832
Ticket line, 736
Time of return, 755
Tomb in the Valley of the Kings, 457
Total gift, 967
Tower of Hanoi problem, 928
Typing speed, 406
Visiting the Smithsonian, 733
Women's shoe sizes, 392
Zip codes, 937

GEOMETRY

Angle of elevation, 494, 495
Angle of inclination, 616
Angle measure, 145, 146
Antenna length, 496
Area of a backyard, 650
Area of a circle, 227
Area of an isosceles triangle, 663
Area of a parallelogram, 652, 713
Area of a quadrilateral, 653
Area of a triangular garden, 649
Banner design, 852
Building, height of, 495
Carpet area, 177
Diagonal of a soccer field, 991
Diagonal of a square, 991
Diagonals of a polygon, 371
Dimensions of a box, 291, 854
Dimensions of a piece of land, 848
Dimensions of a rectangle, 884
Dimensions of a rug, 259
Dimensions of a sail, 662
Distance from a lighthouse, 495
Easel display, 494
Enclosing an area, 493
Flower garden, 713
Garden area, 176
Garden dimensions, 146
Gas tank volume, 177

Graphic design, 852
Guy wire, 616
Height of a mural, 568, 618
Inscribed cylinder, volume, 181
Inscribed pentagon, 494
Inscribed rectangle, area, 178, 229
Inscribed rhombus, 177
Isosceles trapezoid, 662
Kite, height of, 495, 572
Legs of a right triangle, 291
Length of a pole, 651
Lines and triangles from points, 944
Maximizing area, 267, 274, 293, 330
Maximizing volume, 273, 291
Minimizing area, 275, 359
Molding plastics, 178
Parking lot dimensions, 163
Perimeter, 884
Petting zoo dimensions, 259
Picture frame dimensions, 259, 852
Pipe diameter, 496
Poster dimensions, 146
Radius of a circle, 884
Radius of the earth, 497
Sandbox, 713
Seed test plots, 813, 853
Sides of an isosceles triangle, 714
Sign dimensions, 852
Surface area, minimizing, 229
Tablecloth area, 229
Test plot dimensions, 146
Triangular pennant, 232
Triangular scarf, 176
Volume of a box, 178
Volume of a tree, 223

HEALTH/MEDICINE

Calcium content of foods, 147
Cases of tuberculosis, 457
Child's age and recommended daily amount of fiber, 80
Cholesterol level and risk of heart attack, 374
Fat intake, 225
Female smokers, 745
Heart rate, maximum, 127
Height estimates, 108
HIV cases, 111
Ibuprofen in the bloodstream, 305
Medical dosage, 359, 376
Minimizing nutrition cost, 796
Minimum ideal weight, 113

Nutrition, 145, 734, 744, 765, 808
Obesity-related surgeries, 461
Prescription drug sales, 126, 735, 898
Spread of an epidemic, 459
Temperature during an illness, 176, 355, 370, 564, 632
Threshold weight, 310
Triplet births, 125
Twin births, 125
Whooping cough, 466
Winter sports injuries, 732

LABOR

Auto mechanic's earnings, 219
Commissions, 145, 734
Daily doubling salary, 916, 920
Hourly wages, 145, 905, 970
Income plans, 153, 155
Maximizing income, 795, 796
Minimizing salary cost, 797
Minimum wage, 158
Salary, 898
Self-employed workers, 235, 259, 968
Unemployed, 311
Work rate, 112

PHYSICS

Acceleration due to gravity, 597, 632
Alternating current, 606
Angular speed, 528, 529, 530
Aperture, relative, 225
Atmospheric pressure, 462
Beer–Lambert law, 463
Bouncing golf ball, 967
Bouncing ping-pong ball, 920
Boyle's law, 226
Damped oscillations, 565
Depth of a well, 269
Drag, atmospheric, 226
Electric current, power of, 231
Electrical theory, 606
Electricity, 463
Force, 691, 694, 695, 696, 710, 715
Free fall, time of, 253, 258, 259
Height of a ball, 273
Height of a cliff, 275
Height of an elevator shaft, 275
Height of a projectile, 273
Height of a rocket, 268, 273, 376, 378
Height of a thrown object, 370
Hooke's law, 225

Intensity of light, 226
Linear speed, 528, 569, 572
Loudness of sound, 425, 467
Mechanics, 606
Nanowires, 16
Nuclear cooling tower, 842
Parachutist free fall, 908
Pitch, musical, 225
Pressure at sea depth, 113
Projectile motion, 309, 877, 880, 885, 887
Ripple spread, 197
Rotating beacon, 565
Shipping crate, 693, 696
Sound, speed in air, 90
Sound of an airplane, 497
Stopping distance, 113, 226, 233
Suspended block, 707
Tension, 710
Water wave, 565

SOCIAL SCIENCES

Anthropology estimates, 113
Arson damage to churches, 467
Decline in teen smoking, 112
Election poll, 968
Fiancé visas, 406
Flags of nations, 934
International adoptions, 160
Memorial flag case, 493
Production unit, 961
Senate committees, 944
Sleep-starved Americans, 284
Small group interaction, 908
Sociological survey, 955
Time of murder, 459
Walking speed, 419, 424, 467

SPORTS/ENTERTAINMENT

500 Festival mini-marathon, 62
Baseball, 695
Baseball bunt, 662
Baseball card, value of, 458
Batting orders, 934
Bridge hands, 944
Bungee jumping, 920
Circus guy wire, 587
Circus high wire act, 661
Climbing wall, 587
Country music stations, 741
Dart throwing, 957, 962
Diagonal of a soccer field, 991

Dimensions of a softball diamond, 986
Distance between bases, 482
Double-elimination tournaments, 937
Drive-in movie sites, 159
Earned-run average, 226
Ferris wheel seats, 514
Flush, 945, 964
Four of a kind, 963
Full house, 945, 963
Games in a sports league, 310, 944, 945
Golf, 745
Golf: distance versus accuracy, 716
Golf distance finder, 177
Gondola aerial lift, 471, 488
Hiking at the Grand Canyon, 486
Hot-air balloons, 477, 694, 695
In-line skater, 662
International NBA players, 394
Michigan lottery, 942, 962
NASCAR attendance, 162
NCAA advertisement for a basketball tournament, 75
NCAA men's Final Four basketball competition, 889
NFL stadium elevation, 147
Nielsen ratings, 145
Poker hands, 944, 962, 963
Quilt design, 495
Rock concert, 650
Rolling dice, 957, 958, 960, 967
Rope course, 587, 991
Roulette, 962
Royal flush, 963
Running rate, 112
Running vs. walking, 149
Sand Dunes National Park, 494
Setting a fishing reel line counter, 493
Single-elimination tournaments, 937
Slow-pitch softball, 662
Soccer fields, 138, 146
Sports fans, 62
Straight, 964
Straight flush, 963
Survival trip, 662
Swimming pool, 663
Television viewers, 145
Theater attendance, 747
Theme park attendance, 126
Three of a kind, 963
Tour of Flanders, 528
Two pairs, 964
U.S. Cellular Field, 490
"Wheel of Fortune," 961

STATISTICS/DEMOGRAPHICS

Allowance, average weekly amount, 225
Children as a percentage of the population, 406
Favorite pets, 719, 743
Fibonacci sequence and population growth, 898
Foreign-born population, 310
Hours online, 330
House of Representatives, 225
Limited population growth, 450
Marriages, number of, 745
Number of farms, 457
Percentage of Americans 85 and older, 460
Personal space in Hong Kong, 1, 16
Population growth, 359, 370, 376, 446, 450, 455, 456, 466, 467, 469, 920
Prison population, 147
Television ratings, 956
Tornado fatalities, 295, 309
U.S. farm acreage, 311
Young people killed by gunfire in the United States, 377

TRANSPORTATION

Aerial navigation, 511, 513
Airplane speed, 135
Airplane speed and direction, 692, 704
Airplane, distance to airport, 53, 177
Airplanes, 662, 695, 709, 714, 715
Airport distance, 990
Airport security violations, 745
Angle of depression, 616
Angle of revolution, 525, 527
Angular velocity, 569
Bicycling speed, 254
Boating, 694, 709, 727
Bus chartering, 392
Bus travel, 221
Capstan, angular speed, 524
Declining number of railroad miles in the United States, 305
Distance between towns, 495
Distance traveled, 146
Flying into a headwind or with a tailwind, 146, 160
Helmet use, 104
Hot-air balloon, distance from rangefinder, 165, 176
International travel, 309
John Deere tractor, 528
Lighthouse, 651
Lobster boat, 496
Location, 569, 572
Location of airplanes, 717
Lost luggage, 16
Luggage wagon, 695
Mackinac Island, 652
Minimizing transportation cost, 796
Motion, 291, 734, 735, 736
Nautical mile, 597, 632
Navigation, 842, 885
Rate of travel, 225
Reconnaissance airplane, 651, 663
Rescue mission, 644
Richmond International Raceway, 111
Safety line to raft, 493
Ships, 662, 694, 709, 715
Train speeds, 146
Transcontinental railroad, 747
Traveling up- or downstream, 146
U.S. transportation systems, 155
Valve cap on a bicycle, 514
Visitor to National Parks, 145
Wheelbarrow, 695

Geometry

Plane Geometry

Rectangle
Area: $A = lw$
Perimeter: $P = 2l + 2w$

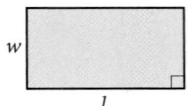

Square
Area: $A = s^2$
Perimeter: $P = 4s$

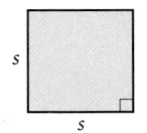

Triangle
Area: $A = \frac{1}{2}bh$

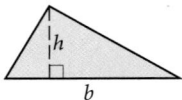

Sum of Angle Measures
$A + B + C = 180°$

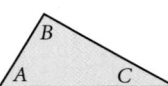

Right Triangle
Pythagorean theorem
(equation):
$a^2 + b^2 = c^2$

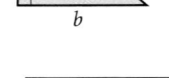

Parallelogram
Area: $A = bh$

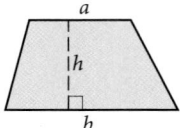

Trapezoid
Area: $A = \frac{1}{2}h(a + b)$

Circle
Area: $A = \pi r^2$
Circumference:
$C = \pi d = 2\pi r$

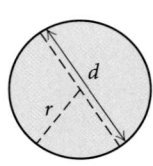

Solid Geometry

Rectangular Solid
Volume: $V = lwh$

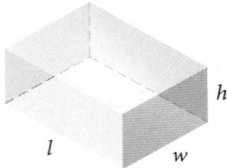

Cube
Volume: $V = s^3$

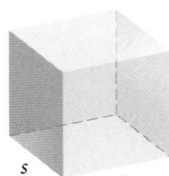

Right Circular Cylinder
Volume: $V = \pi r^2 h$
Lateral surface area:
$L = 2\pi rh$
Total surface area:
$S = 2\pi rh + 2\pi r^2$

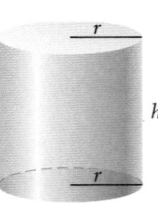

Right Circular Cone
Volume: $V = \frac{1}{3}\pi r^2 h$
Lateral surface area:
$L = \pi rs$
Total surface area:
$S = \pi r^2 + \pi rs$
Slant height:
$s = \sqrt{r^2 + h^2}$

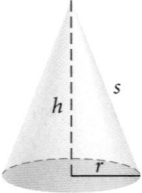

Sphere
Volume: $V = \frac{4}{3}\pi r^3$
Surface area: $S = 4\pi r^2$

Algebra

Properties of Real Numbers

Commutative:	$a + b = b + a; \quad ab = ba$
Associative:	$a + (b + c) = (a + b) + c;$
	$a(bc) = (ab)c$
Additive Identity:	$a + 0 = 0 + a = a$
Additive Inverse:	$-a + a = a + (-a) = 0$
Multiplicative Identity:	$a \cdot 1 = 1 \cdot a = a$
Multiplicative Inverse:	$a \cdot \dfrac{1}{a} = 1, \ a \neq 0$
Distributive:	$a(b + c) = ab + ac$

Exponents and Radicals

$$a^m \cdot a^n = a^{m+n} \qquad \frac{a^m}{a^n} = a^{m-n}$$

$$(a^m)^n = a^{mn} \qquad (ab)^m = a^m b^m$$

$$\left(\frac{a}{b}\right)^m = \frac{a^m}{b^m} \qquad a^{-n} = \frac{1}{a^n}$$

If n is even, $\sqrt[n]{a^n} = |a|$.

If n is odd, $\sqrt[n]{a^n} = a$.

$$\sqrt[n]{a} \cdot \sqrt[n]{b} = \sqrt[n]{ab}, \ a, b \geq 0$$

$$\sqrt[n]{\frac{a}{b}} = \frac{\sqrt[n]{a}}{\sqrt[n]{b}}$$

$$\sqrt[n]{a^m} = \left(\sqrt[n]{a}\right)^m = a^{m/n}$$

Special-Product Formulas

$$(a + b)(a - b) = a^2 - b^2$$

$$(a + b)^2 = a^2 + 2ab + b^2$$

$$(a - b)^2 = a^2 - 2ab + b^2$$

$$(a + b)^3 = a^3 + 3a^2b + 3ab^2 + b^3$$

$$(a - b)^3 = a^3 - 3a^2b + 3ab^2 - b^3$$

$$(a + b)^n = \sum_{k=0}^{n} \binom{n}{k} a^{n-k} b^k, \quad \text{where}$$

$$\binom{n}{k} = \frac{n!}{k!\,(n-k)!}$$

$$= \frac{n(n-1)(n-2)\cdots[n-(k-1)]}{k!}$$

Factoring Formulas

$$a^2 - b^2 = (a + b)(a - b)$$

$$a^2 + 2ab + b^2 = (a + b)^2$$

$$a^2 - 2ab + b^2 = (a - b)^2$$

$$a^3 + b^3 = (a + b)(a^2 - ab + b^2)$$

$$a^3 - b^3 = (a - b)(a^2 + ab + b^2)$$

Interval Notation

$$(a, b) = \{x \mid a < x < b\}$$

$$[a, b] = \{x \mid a \leq x \leq b\}$$

$$(a, b] = \{x \mid a < x \leq b\}$$

$$[a, b) = \{x \mid a \leq x < b\}$$

$$(-\infty, a) = \{x \mid x < a\}$$

$$(a, \infty) = \{x \mid x > a\}$$

$$(-\infty, a] = \{x \mid x \leq a\}$$

$$[a, \infty) = \{x \mid x \geq a\}$$

Absolute Value

$$|a| \geq 0$$

For $a > 0$,

$$|X| = a \rightarrow X = -a \quad \text{or} \quad X = a,$$

$$|X| < a \rightarrow -a < X < a,$$

$$|X| > a \rightarrow X < -a \quad \text{or} \quad X > a.$$

Equation-Solving Principles

$$a = b \rightarrow a + c = b + c$$

$$a = b \rightarrow ac = bc$$

$$a = b \rightarrow a^n = b^n$$

$$ab = 0 \leftrightarrow a = 0 \quad \text{or} \quad b = 0$$

$$x^2 = k \rightarrow x = \sqrt{k} \quad \text{or} \quad x = -\sqrt{k}$$

Inequality-Solving Principles

$$a < b \rightarrow a + c < b + c$$

$$a < b \text{ and } c > 0 \rightarrow ac < bc$$

$$a < b \text{ and } c < 0 \rightarrow ac > bc$$

(Algebra continued)

Algebra (continued)

The Distance Formula

The distance from (x_1, y_1) to (x_2, y_2) is given by

$$d = \sqrt{(x_2 - x_1)^2 + (y_2 - y_1)^2}.$$

The Midpoint Formula

The midpoint of the line segment from (x_1, y_1) to (x_2, y_2) is given by

$$\left(\frac{x_1 + x_2}{2}, \frac{y_1 + y_2}{2} \right).$$

Formulas Involving Lines

The slope of the line containing points (x_1, y_1) and (x_2, y_2) is given by

$$m = \frac{y_2 - y_1}{x_2 - x_1}.$$

Slope–intercept equation: $\quad y = f(x) = mx + b$

Horizontal line: $\quad\quad\quad y = b \quad \text{or} \quad f(x) = b$

Vertical line: $\quad\quad\quad\quad x = a$

Point–slope equation: $\quad\quad y - y_1 = m(x - x_1)$

The Quadratic Formula

The solutions of $ax^2 + bx + c = 0, a \neq 0$, are given by

$$x = \frac{-b \pm \sqrt{b^2 - 4ac}}{2a}.$$

Compound Interest Formulas

Compounded n times per year: $\quad A = P\left(1 + \frac{i}{n}\right)^{nt}$

Compounded continuously: $\quad P(t) = P_0 e^{kt}$

Properties of Exponential and Logarithmic Functions

$\log_a x = y \leftrightarrow x = a^y \qquad\qquad a^x = a^y \leftrightarrow x = y$

$\log_a MN = \log_a M + \log_a N \qquad \log_a M^p = p \log_a M$

$\log_a \dfrac{M}{N} = \log_a M - \log_a N$

$\log_b M = \dfrac{\log_a M}{\log_a b}$

$\log_a a = 1 \qquad\qquad\qquad\qquad \log_a 1 = 0$

$\log_a a^x = x \qquad\qquad\qquad\qquad a^{\log_a x} = x$

Conic Sections

Circle: $\qquad (x - h)^2 + (y - k)^2 = r^2$

Ellipse: $\qquad \dfrac{(x - h)^2}{a^2} + \dfrac{(y - k)^2}{b^2} = 1,$

$\qquad\qquad \dfrac{(x - h)^2}{b^2} + \dfrac{(y - k)^2}{a^2} = 1$

Parabola: $\qquad (x - h)^2 = 4p(y - k),$

$\qquad\qquad (y - k)^2 = 4p(x - h)$

Hyperbola: $\qquad \dfrac{(x - h)^2}{a^2} - \dfrac{(y - k)^2}{b^2} = 1,$

$\qquad\qquad \dfrac{(y - k)^2}{a^2} - \dfrac{(x - h)^2}{b^2} = 1$

Arithmetic Sequences and Series

$a_1, a_1 + d, a_1 + 2d, a_1 + 3d, \ldots$

$a_{n+1} = a_n + d \qquad\qquad a_n = a_1 + (n - 1)d$

$S_n = \dfrac{n}{2}(a_1 + a_n)$

Geometric Sequences and Series

$a_1, a_1 r, a_1 r^2, a_1 r^3, \ldots$

$a_{n+1} = a_n r \qquad\qquad a_n = a_1 r^{n-1}$

$S_n = \dfrac{a_1(1 - r^n)}{1 - r} \qquad S_\infty = \dfrac{a_1}{1 - r}, \ |r| < 1$

Trigonometry

Trigonometric Functions

Acute Angles

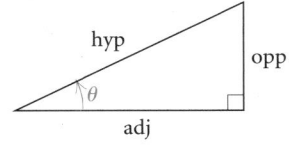

$$\sin \theta = \frac{\text{opp}}{\text{hyp}}, \quad \csc \theta = \frac{\text{hyp}}{\text{opp}},$$

$$\cos \theta = \frac{\text{adj}}{\text{hyp}}, \quad \sec \theta = \frac{\text{hyp}}{\text{adj}},$$

$$\tan \theta = \frac{\text{opp}}{\text{adj}}, \quad \cot \theta = \frac{\text{adj}}{\text{opp}}$$

Any Angle

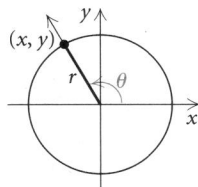

$$\sin \theta = \frac{y}{r}, \quad \csc \theta = \frac{r}{y},$$

$$\cos \theta = \frac{x}{r}, \quad \sec \theta = \frac{r}{x},$$

$$\tan \theta = \frac{y}{x}, \quad \cot \theta = \frac{x}{y}$$

Real Numbers

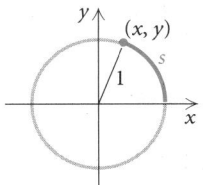

$$\sin s = y, \quad \csc s = \frac{1}{y},$$

$$\cos s = x, \quad \sec s = \frac{1}{x},$$

$$\tan s = \frac{y}{x}, \quad \cot s = \frac{x}{y}$$

Basic Trigonometric Identities

$$\sin (-x) = -\sin x,$$
$$\cos (-x) = \cos x,$$
$$\tan (-x) = -\tan x,$$

$$\tan x = \frac{\sin x}{\cos x},$$

$$\cot x = \frac{\cos x}{\sin x},$$

$$\csc x = \frac{1}{\sin x},$$

$$\sec x = \frac{1}{\cos x},$$

$$\cot x = \frac{1}{\tan x}$$

Pythagorean Identities

$$\sin^2 x + \cos^2 x = 1,$$
$$1 + \cot^2 x = \csc^2 x,$$
$$1 + \tan^2 x = \sec^2 x$$

Identities Involving $\pi/2$

$$\sin (\pi/2 - x) = \cos x,$$
$$\cos (\pi/2 - x) = \sin x, \quad \sin (x \pm \pi/2) = \pm\cos x,$$
$$\tan (\pi/2 - x) = \cot x, \quad \cos (x \pm \pi/2) = \mp\sin x$$

Sum and Difference Identities

$$\sin (u \pm v) = \sin u \cos v \pm \cos u \sin v,$$
$$\cos (u \pm v) = \cos u \cos v \mp \sin u \sin v,$$
$$\tan (u \pm v) = \frac{\tan u \pm \tan v}{1 \mp \tan u \tan v}$$

Double-Angle Identities

$$\sin 2x = 2 \sin x \cos x,$$
$$\cos 2x = \cos^2 x - \sin^2 x$$
$$= 1 - 2 \sin^2 x$$
$$= 2 \cos^2 x - 1,$$

$$\tan 2x = \frac{2 \tan x}{1 - \tan^2 x}$$

Half-Angle Identities

$$\sin \frac{x}{2} = \pm\sqrt{\frac{1 - \cos x}{2}}, \quad \cos \frac{x}{2} = \pm\sqrt{\frac{1 + \cos x}{2}},$$

$$\tan \frac{x}{2} = \pm\sqrt{\frac{1 - \cos x}{1 + \cos x}} = \frac{\sin x}{1 + \cos x} = \frac{1 - \cos x}{\sin x}$$

(continued)

Trigonometry *(continued)*

The Law of Sines

In any $\triangle ABC$,

$$\frac{a}{\sin A} = \frac{b}{\sin B} = \frac{c}{\sin C}.$$

The Law of Cosines

In any $\triangle ABC$,

$$a^2 = b^2 + c^2 - 2bc \cos A,$$
$$b^2 = a^2 + c^2 - 2ac \cos B,$$
$$c^2 = a^2 + b^2 - 2ab \cos C.$$

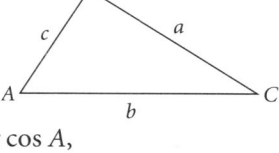

Trigonometric Function Values of Special Angles

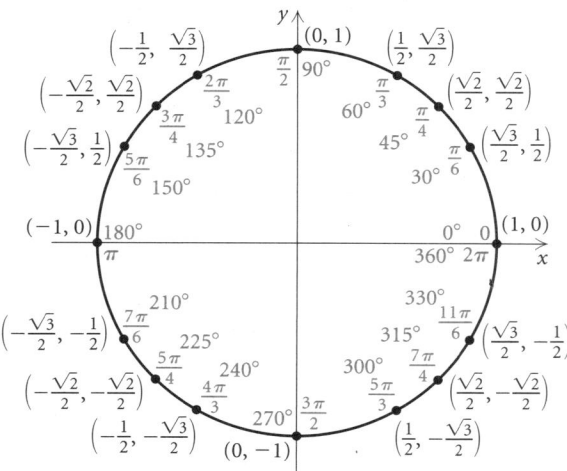

Graphs of Trigonometric Functions

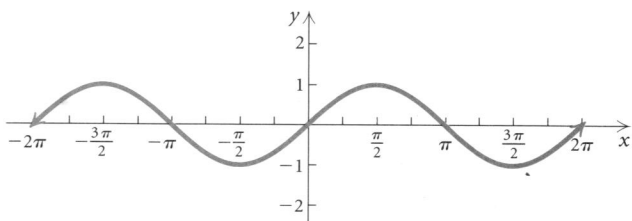

The sine function: $f(x) = \sin x$

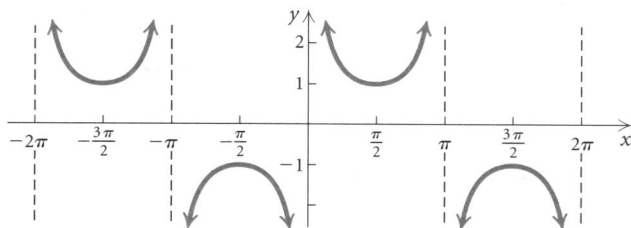

The cosecant function: $f(x) = \csc x$

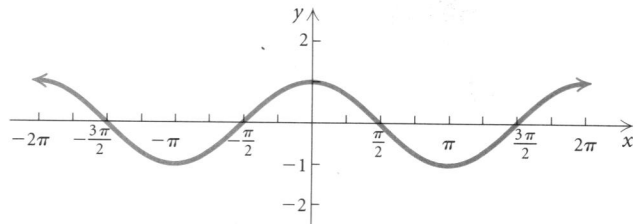

The cosine function: $f(x) = \cos x$

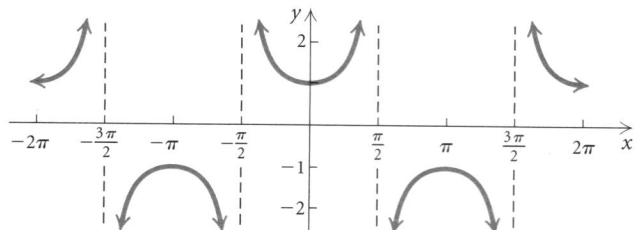

The secant function: $f(x) = \sec x$

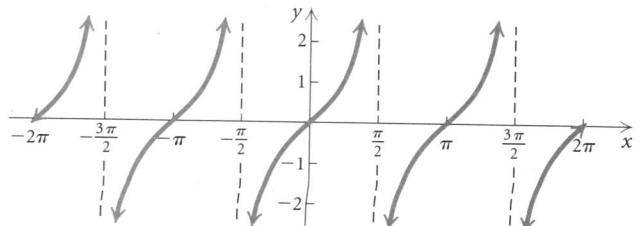

The tangent function: $f(x) = \tan x$

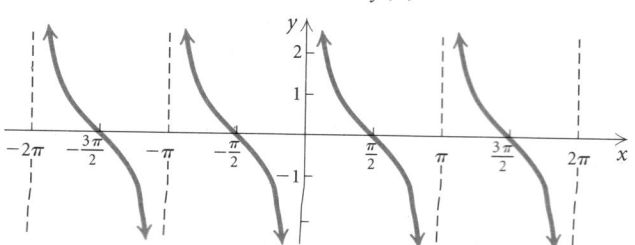

The cotangent function: $f(x) = \cot x$